GEOTECHNICAL HAZARDS

PROCEEDINGS OF THE XITH DANUBE – EUROPEAN CONFERENCE ON SOIL MECHANICS AND GEOTECHNICAL ENGINEERING/POREČ/CROATIA/25-29 MAY 1998

GEOTECHNICAL HAZARDS

Edited by

Božica Marić
Conex, Co., Zagreb, Croatia

Zvonimir Lisac
Civil Engineering Institute of Croatia, Zagreb, Croatia

Antun Szavits-Nossan
Faculty of Civil Engineering, University of Zagreb, Croatia

Taylor & Francis
Taylor & Francis Group

LONDON AND NEW YORK

The texts of the various papers in this volume were set individually by typists under the supervision of each of the authors concerned.

Authorization to photocopy items for internal or personal use, or the internal or personal use of specific clients, is granted by Taylor & Francis, provided that the base fee of US$ 1.50 per copy, plus US$ 0.10 per page is paid directly to Copyright Clearance Center, 222 Rosewood Drive, Danvers, MA 01923, USA. For those organizations that have been granted a photocopy license by CCC, a separate system of payment has been arranged. The fee code for users of the Transactional Reporting Service is: 90 5410 957 2/98 US$ 1.50 + US$ 0.10.

Published by Taylor & Francis
2 Park Square, Milton Park, Abingdon, Oxon, OX14 4RN
270 Madison Ave, New York NY 10016

Transferred to Digital Printing 2007

ISBN 90 5410 957 2
© 1998 Taylor & Francis

Publisher's Note
The publisher has gone to great lengths to ensure the quality of this
reprint but points out that some imperfections in the original
may be apparent

Geotechnical Hazards, Marić, Lisac & Szavits-Nossan (eds) © 1998 Taylor & Francis, ISBN 90 5410 957 2

Table of contents

Earthquake geotechnical engineering

Environmental geotechnics

Foundations and soil improvement

Geotechnical parameters

Landslides and slope stability

Geotechnical Hazards, Marić, Lisac & Szavits-Nossan (eds) © 1998 Taylor & Francis, ISBN 90 5410 957 2

Foreword

Danube European Conferences for Soil Mechanics and Geotechnical Engineering (DECSMGE) have already become a tradition. The last Conference held in 1995 in Mamaia, Romania, attracted not only local geotechnical engineers and researchers but also a number of specialists from other countries. The wider appeal of the Conference, proving the ever increasing acceptance of the geotechnical profession, set new standards for future hosts of Danube European Conferences. Members of the Croatian Society for Soil Mechanics and Geotechnical Engineering were well aware of these standards when representatives of the Danubian countries in Mamaia accepted their invitation to host the XIth Conference in 1998 in Poreč, Croatia.

The Croatian Society set Geotechnical hazard as the Conference theme, acknowledging the exceptional diversity of local ground conditions and environmental factors which have played a decisive role in designing engineering structures in Danubian countries through centuries. Landslides, difficult foundation soils, earthquake induced instabilities of saturated sands, foundation soil improvement, and various other foundation problems and solutions were reported in this region much earlier than the emergence of modern soil mechanics. Many well-preserved historic monuments are the evident proof of ingenious local builders. The disappearance of others by causes of nature proves that hazard was and still is a way of life in this region.

It has to be emphasised that the Croatian Society for Soil Mechanics and Geotechnical Engineering was presented by the honour to organise the Danube European Conference. In order to understand such an honour it should be remembered what it was like in Croatia in 1995, and particularly what it was like in Vukovar upon Danube in 1991. It also has to be emphasised that this present was further extended upon the Croatian Society by the more than generous support provided by the leading authorities of the International Society for Soil Mechanics and Geotechnical Engineering: the past president Prof. Michele Jamiolkowski, and the past vice president for Europe Prof. William F. Van Impe; the actual president Prof. Kenji Ishihara, and the actual vice president for Europe Prof. Heinz Brandl. Their help and advice are deeply acknowledged. Many thanks to all our special guests who delivered Keynote lectures at the Conference and made it a real feast for the geotechnical community.

The Conference Topics were selected by the Organising Committee according to the contributed abstracts which reflected a great interest for geotechnical hazard among researchers and practising engineers all over the world. Numerous authors from Danubian countries and many other countries in Europe, Asia, and America sent their papers to the Organising Committee. A selection of these papers was made by the Paper Review Board making every possible effort to accept for publication fresh research results and interesting case histories. The Organising Committee greatly acknowledges the co-operation of authors in holding as closely as possible to high paper presentation standards set out by the publisher.

It is our hope, perhaps too ambitious, that papers included in this volume will help make a step forward in the rationalisation and reduction of geotechnical hazard. The Organising Committee of the XIth DECSMGE wishes to thank all those who devoted themselves to the publication of these Proceedings. Special thanks go to authors of papers whose professional efforts make the essence of this Conference.

Editors

Invited lectures

Geotechnical Hazards, Marić, Lisac & Szavits-Nossan (eds) © 1998 Taylor & Francis, ISBN 90 5410 957 2

Foundation strengthening and soil improvement for scour-dangered river bridges

H. Brandl

Technical University Vienna, Austria

ABSTRACT: The scouring of bridges leading over water represents a geotechnical as well as a hydraulic problem world-wide. This is demonstrated here on hand of relevant case histories: on the river Danube and on its mountainous tributaries, Drau and Inn. The first case represents a prevention measure. The second one describes the failure of a major bridge on Europe's busiest trans-Alpine highway which caused a traffic jam of about 70 km length, directly affecting three countries and involving about 10 million people. The scouring excavated the riverbed to a depth of 14 m, whereby causing a river pier settlement of up to 1.35 m and heavy tilting. The collapse of the bridge (3 decks of 450 to 480 m length) could be avoided by applying several geotechnical emergency measures. This paper reports on the reasons for the failure, the emergency activities, and the final rehabilitation. The unexpected ductile behaviour of the three bridge decks resting on the scoured pier changed the international standards for prestressed reinforced concrete structures.

1 INTRODUCTION

During the last 30 years about 1000 bridges (out of approximately 600.000) have collapsed in the USA, 60 % of those collapses were due to scouring of the foundation. In the Danube-European region and the Alpine areas of Europe scouring has become an increasing problem because of the gradual self-deepening of the mobile riverbeds. This goes back to former excessive regulatory training of rivers and torrents which now exhibit a significantly shorter flow way (and reduced bed load sediment). The increased hydraulic gradients represent a high erosion potential. Riverbed deepening may also be man-made during the construction of hydropower plants. In that case an excavation of the downstream riverbed increases the gross height and energy output of the facility. Scouring is also a danger to bridges in coastal zones where the tides may cause significant seabed erosion to great depths. Fig. 1 shows a bridge with 0.6 m horizontal and 0.4 m vertical movement due to a 12 m deep scour.

Scouring and riverbed or seabed excavation requires the protection of river bridges by foundation strengthening and/or soil improvement. Such measures need a close co-operation among geotechnical, hydraulic, and structural engineers,

whereby the main responsibility is that of the geotechnical engineer.

2 STRENGTHENING AND UNDERPINNING OF RIVER PIER FOUNDATIONS

2.1 Bridges crossing the river Danube

The construction of the hydraulic power plant in Vienna caused a backwater level which is up to 8.3 m higher than the former level of the river Danube. This required the raising of several bridges within a range of 1.7 to 4.7 m. Fig. 2 shows the cross section of one of the river piers of a 23-field railway bridge (constructed in 1870) which consists of two parallel superstructures with a total length of 850 m each. These bridge decks had been already lifted by 1.70 m several decades before. In connection with the new hydropower plant the piers and foundations had to be strengthened in order to provide sufficient safety factors: Due to the lifted superstructures, the forces, especially moments changed significantly, and more-over, an increasing traffic volume had to be taken into consideration. Foundation strengthening should comprise not only an improvement of the lateral capacity (especially in the longitudinal direction of

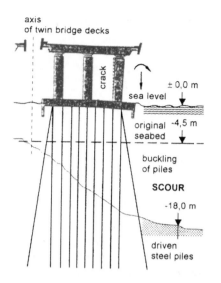

Fig. 1 Failure of a motorway-bridge due to a 12 m deep scouring of the seabed.

the bridge) but also of the stability against ground failure whereby 6 m deep scours had to be taken into account. This required an entire ringing and under-pinning of the old foundation with a box-shaped load-bearing member.

The subsoil consists of quaternary sandy gravel with boulders (4 to 7 m deep) underlain by tertiary sediments (mainly silty clay). Around the piers, the river bed is protected with rockfill against scouring which locally reach a depth of about 5 m.

For underpinning old river piers box-like enclosures consisting of piles or jet grouting columns or diaphragm walls have proved very effective. Due to the shipping lanes and the high flow velocity of the river Danube (locally up to 4 – 6 m/s) it was not possible to fill a temporary island in this case. Furthermore, piling from a ship was too risky: Secant piles could not be installed precisely enough to obtain sufficient overlapping along the entire pile length, which is an essential prerequisite for a fully composite behaviour of box-shaped foundations.

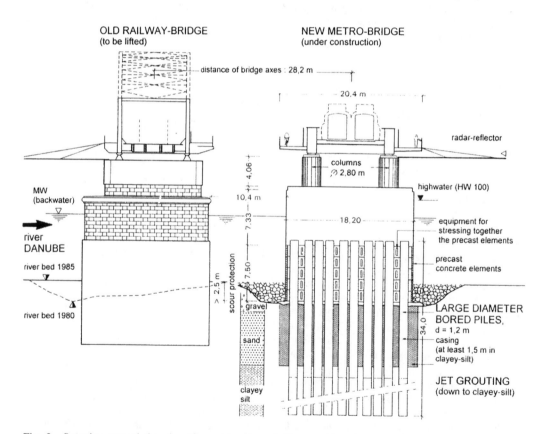

Fig. 2 Scouring around the pier of an old railway bridge crossing the river Danube (upstream) requires strengthening of the deteriorated foundation. Simultaneous construction of a new Metro-bridge downstream.

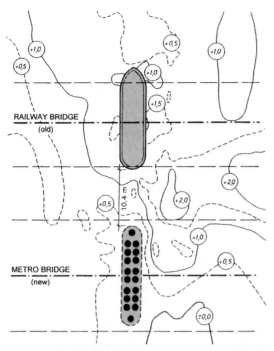

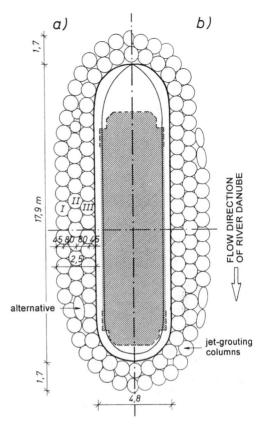

Fig. 3 Riverbed monitoring around the close bridge piers; to Fig. 2. Depth of scours inspite of local refilling (isolines in metres).

Fig. 4 Foundation strengthening of the river piers of a railway bridge crossing the river Danube by a box-shaped enclosure with jet grouting columns. Ground plan of the alternatives a) and b): construction sequence I to III.

Large boulders and fossil trees near the interface between quaternary and tertiary sediments were also an obstacle to piling. Another constraint was the simultaneous construction of a new bridge at the distance of only 10.4 m. Fig. 2 illustrates this difficult situation and the tendency of scouring around the pier of the old railway bridge. The close position of the bridges increased local riverbed deepening and required permanent refilling and monitoring (Fig. 3).

The solution was a strengthening and underpinning by installing overlapping jet grouting columns in a box-shaped form around the existing river pier, whereby two alternatives were considered (Fig. 4). Jet grouting did not require an island but could be performed from a platform hanging on the superstructure (trussed steel girders) - Fig. 5. In case of high-water the platform could be pulled up. With jet grouting a statically clearer mode of load transfer between old and new foundation could be achieved than in case of contiguous bored piles. In order to minimise the loss of suspension, grouting started along the outer ring (columns I in Fig. 5). Moreover, this made a higher jetting pressure along the inner

ring possible, hence achieving an intensive bond between old foundation and jet grouting columns. With increasing composite behaviour the maximum edge pressures of the foundation could be reduced from $\sigma_{max} = 1.1$ to 0.65 MN/m².

The main problem was how to grout through the rock fill without any overburden in the river bed and how to avoid a washing out of the fresh grout material. The soil required a grouting pressure of 50 to 60 MPa.

The first design proposed underwater concrete to fill the voids within the rockfill and to place a 0.5 m thick concrete cover on it. This would have been very difficult and would have needed a local river back up. According to shipping requirements, the concrete cover could be considered only an auxiliary

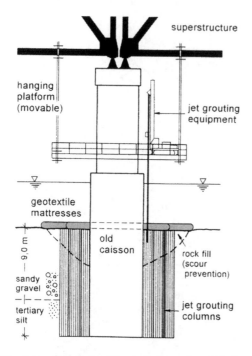

hanging
platform
(movable)

superstructure

jet grouting
equipment

geotextile
mattresses

old
caisson

rock fill
(scour
prevention)

9.0 m

sandy
gravel

tertiary
silt

jet grouting
columns

Fig. 5 Cross section to Fig. 4. Construction of the jet grouting columns from a moveable platform hanging on the superstructure.

measure and was to be removed after bridge lifting. Moreover the scour prevention should remain a flexible rock fill and not be changed into a rigid block.

In order to provide a flexible and removable cover, "geotextile mattresses" were placed on the river bed around the pier (Fig. 6). These elements consisted of geocomposite bags (2.0 x 2.0 x 0.5 m) filled with a special underwater concrete (type K5 with 350 kg/m^3 cement). The geotextile mattresses were manufactured on a special barge near the river pier and placed by a crane which was mounted on another ship close to the pier (Fig. 6a). Accurate placing was controlled by laser from a third ship and by sonic logging as well as by hand (with rods). Several fixations and winches provided a precise handling.

The special barge carried several boxes which served as formwork for manufacturing the geotextile mattresses. These contained a geosynthetic net (mesh width 45 mm, edge rope 16 mm) to carry the elements as indicated in Figs. 5, 6a). After placing the mattresses, the top of the net was cut off. The flexible geocomposite and fresh concrete adapted themselves very closely to the rough river bed. The filled mattresses overlap each other by 0.3 m, thus

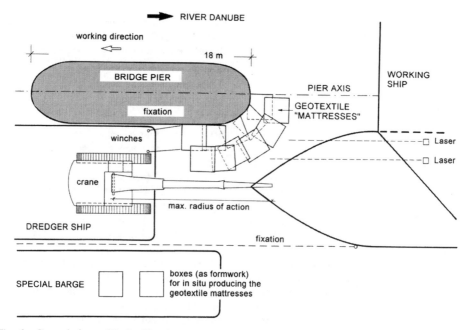

Fig. 6   Ground plan to Fig.5 with scheme of producing and placing the geotextile mattresses around the river pier.

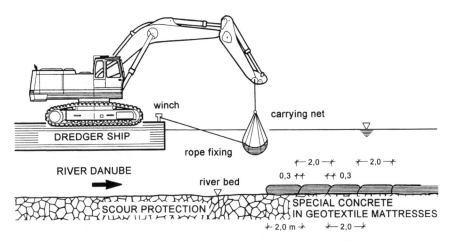

Fig. 6a   Placing of the geotextile mattresses from a dredger ship.

forming a continuous surface.

While the net served only for carrying purposes, the geocomposite served as a container. It consisted of a 60µ membrane (100 % polyethylene) and a staple fibre nonwoven geotextile (100 % polypropylene). The total thickness of the composite was about 3.5 mm with a weight of appr. 550 g/m². The strip tensile strength was more than 11 kN/m in both directions, with an elongation at break of at least 80 %. The sheets were welded with hot-air or by flaming.

The geotextile mattresses proved very suitable, and jet grouting could be performed without any problem. During grouting the movements of the pier were continuously monitored by rotating laser (with acoustic signals). In case of values of more than 0.1 mm the grouting procedure was instantly modified.

The quality of the jet grouting columns was checked by boring and sampling. After 90 days, the unconfined compressive strength varied between $q_u = 6.2$ and 15.5 MN/m² (sample slenderness 2:1) and the modulus between E = 400 to 1000 MN/m². Both parameters increased from the outer to the inner columns and exhibited their maximum close to the interface between old and new foundation, thus providing an intensive land effect with the old caisson.

Bridge lifting (11000 tons) was performed within five days.

2.2 Bridges crossing the river Drau

Along the Austrian part of the river Drau a continuous series of hydropower plants were constructed during the past decades. In several cases

the riverbed was excavated downstream the new facilities in order to increase their gross height, hence their efficiency. The excavation depth was up to 6 m, which changed the hydraulic characteristics of the river significantly and required an underpinning of several bridge piers.

Underpinning was generally performed with box-shaped pile foundations around the old footings. The bored piles could be installed from temporary islands in an overlapping or contiguous form.

As experience shows, box-shaped deep foundations have proved very successful for such cases. Predominantly they consists of bored piles or diaphragm walls, but jet-grouted walls or sheet-pile walls are in use too. The bearing capacity, earthquake- and scouring resistance, and the resisting moment towards lateral forces (from embankments on soft soils or from creeping slopes) of such box-shaped foundations are significantly higher than for conventional pile groups of the same pile number, because the system acts as a compound body: The core of enclosed soil cannot move laterally and takes part in bearing the external loads. The foundation system may be compared with a filled pot, turned upside down.

In order to provide a sufficient bond effect, the bored piles along the circumference of the box foundation must be installed in a secant (overlapping) or at least contiguous form so that shear stresses can be taken over along their connecting line. For evaluating the bearing capacity of box-foundations, two methods have proved successful in designing practice which should be used simultaneously (Brandl, 1985, 1987, 1988):

• Checking the bearing capacity of the single pile (or diaphragm panel, get grouting column) ⇒

7

safety factor F_1

- Checking the bearing capacity and settlement of the box-foundation as a unit (monolith theory) $\Rightarrow$ safety factor F_2

Evaluating the bearing capacity of single elements provides only fictitious limit values because the compound effect between concrete elements and enclosed soil core is neglected. Thus, maximum loads are calculated for piles, diaphragm wall panels or jet grouting columns. But actually, single elements cannot fail because of the composite behaviour of the structural system and the rigid (reinforced) connection of the deep foundation with the capping reinforced concrete structure. Therefore, very low safety factors are sufficient for this theoretical model: Usually $F_1 \geq 1.15$; for short construction stages or catastrophic conditions even values of $F_1 \geq 1.05$ have been allowed. Contrary to the monolith-theory, the skin friction may be taken into consideration along the outside and inside circumference of the foundation-box, but not between the single elements.

The other limit value hypothesis is called „monolith-theory". A composite interaction between deep foundation elements and the enclosed soil is assumed. This compound body may be considered approximately a large-area „shallow foundation" reaching relatively deeply into the subsoil. The cross sectional area comprises the full circumference of the foundation if secant piles, diaphragm walls or jet grouting columns are installed. In case of contiguous piles, the theoretical, calculatory area should be reduced by at least one pile diameter. Only skin friction along the outside face of the quasi-monolithic foundation box may be taken into account. This limit hypothesis provides the minimum loads for the concrete elements (piles etc.). A full composite effect

will occur only theoretically but hardly in practice. Therefore, relatively high safety factors must be demanded for the idealised model: usually about $F_2 \geq 3.0$ if conventional methods for evaluating the base failure of shallow foundations are used.

According to the above theory and practical experience, box-shaped new foundations consisting of overlapping piles (or diaphragm walls or overlapping jet grouting columns) have been installed since about 20 years to increase the stability of old river bridges. In case of contiguous or intermittent piling, the spaces between the piles (and old footing) were grouted as indicated in Fig. 7. Fig. 8 shows the underpinning of an old railway bridge founded on caissons, which was required because of an excavation of the riverbed of at least 3.5 m and an increased traffic load.

3 SCOUR REPAIR AND BRIDGE REHABILITATION

3.1 General; bridge system

In July 1990 a major bridge on Europe's busiest trans-Alpine motorway failed due to an aggressive river scour which developed unnoticed after a breakdown in checking procedures during a relatively small flood event. The disaster occurred in Kufstein close to Austria's border to Germany and near to Italy. (Fig. 9). It caused the greatest traffic jam that Europe ever experienced involving about 10 million people (at the weekend of peak holiday season).

The 450 to 480 m long structure of a triple bridge deck consists of three hollow box girders carrying the A2-highway and a federal main road (Fig. 10). It

Fig. 7 Strengthening and underpinning of an old river bridge pier with bored piles in a box-shaped pattern. Grouting of the space between piles and old foundations to achieve a composite interaction.

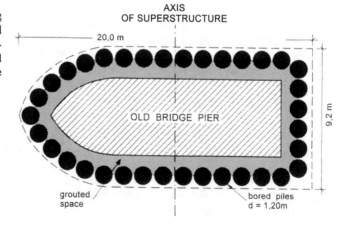

8

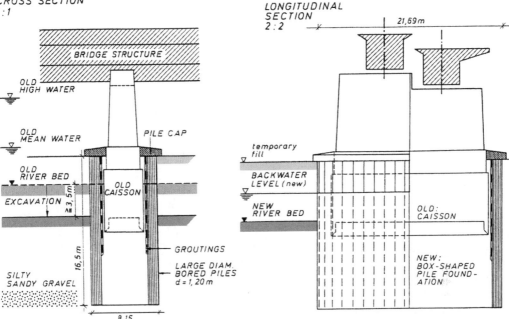

CROSS SECTION
1:1

BRIDGE STRUCTURE

OLD HIGH WATER

OLD MEAN WATER

PILE CAP

OLD RIVER BED

EXCAVATION ≈ 3,5 m

OLD CAISSON

SILTY SANDY GRAVEL

16,5 m

GROUTINGS

LARGE DIAM. BORED PILES
d = 1,20 m

8,15

LONGITUDINAL SECTION
2:2

21,69 m

temporary fill

BACKWATER LEVEL (new)

NEW RIVER BED

OLD: CAISSON

NEW: BOX-SHAPED PILE FOUND-ATION

Fig. 8 Underpinning of an old bridge pier in the river Drau. Excavation of the river bed for a hydropower plant required a new box-foundation with contiguous bored piles and grouting.

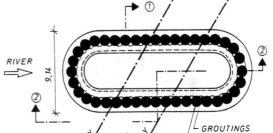

GROUNDPLAN

RIVER

9,14

RAILWAY

RAILWAY

GROUTINGS
BORED PILES

crosses the river Inn, another main road and the international railway line. The bridge decks had been constructed by incremental launching in 1968, and the 52 m long main pier in the river rested on a flat foundation ringed by sheet pile walls reaching 7 m beneath the riverbed.

The hollow box girders were constructed as normal steel-reinforced concrete members with a relatively high percentage of non-prestressed reinforcement and a separate longitudinal prestressing inside the webs using concentrated prestressing tendons.

The foundation of the damaged river pier is illustrated in Figs. 11, 16, 17. It is a 52 m long strip footing (52 x 8.5 x 2 m) with a mean base pressure of σ_m = 450 kN/m². The maximum edge pressure was σ_{max} = 650 kN/m² (before the failure). Sheet pile walling was considered only as temporary measure during pit excavation but had no significant permanent effect. A possible riverbed scouring of 3 m depth was taken into account in the foundation design.

The shaft of the river pier is not reinforced except the minimum steel rods. It was clad with granite blocks which were set in concrete during the construction.

3.2 Ground properties

The ground consists of heterogeneous (inter-)layers as common in Alpine valleys: Young river sediments of sand to silt are underlain by sandy gravel with boulders, also including laminated uniform sands and silt, or lenses of fines (clayey). Fig. 12 shows the grain size distribution of the alluvial deposits along

9

Fig. 9 Location of the Kufstein bridge as a major bridge on Europe's busiest trans-Alpine highway.

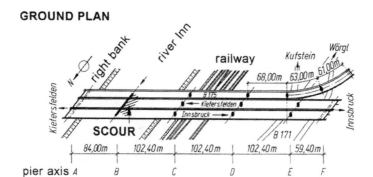

Fig 10 Highway Bridges Kufstein crossing the river Inn: Ground-plan, longitudinal view and cross section B171 and B175 are federal main roads.

GROUND PLAN

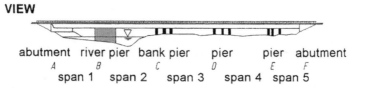

VIEW

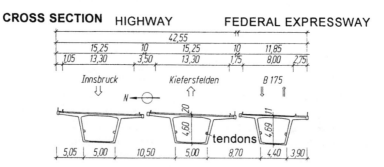

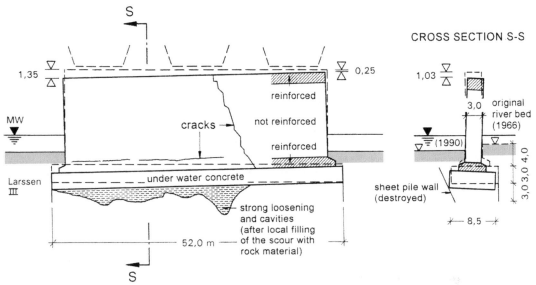

CROSS SECTION S-S

Fig. 11 Foundation of the river pier and position after excessively differential settlement and rotation in the longitudinal and transversal direction. Indicated is also the torn off and destroyed sheet pile wall on the scoured pier side.

the bridge. The package varies from loose to medium, and finally to dense in greater depth. The hydraulic conductivity ranges between $k = 5 \cdot 10^{-3}$ m/s in sandy gravels to $k = 10^{-8} - 10^{-10}$ m/s in fine

Fig. 12 Grain size distribution of the alluvial deposits along the Inn-Bridge Kufstein. Ranges of the main soils and scatter of single values.

deposits, whereby a clear unisotropic behaviour prevails: $k_h = 5$ to $30\ k_v$. The soil properties exhibited a wide scatter both in the vertical and horizontal direction.

The old design was based on the results of exploratory borings which had not shown any critical uniform sand/silt deposits in the location of the central river pier (2 borings were drilled there). After the bridge failure, 11 core borings and numerous soundings were performed for detailed ground investigation and scour localisation. About 100 soil samples were taken to optimise the rehabilitation measures involving soil improvement and foundation strengthening/underpinning.

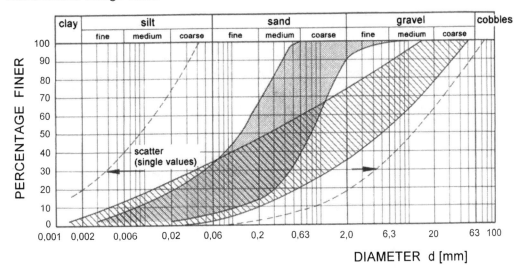

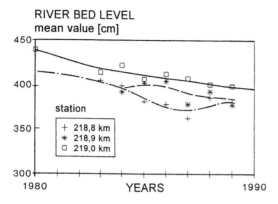

RIVER BED LEVEL
mean value [cm]

station
+ 218,8 km
* 218,9 km
□ 219,0 km

YEARS

Fig. 13 Gradual riverbed erosion of the river Inn in Kufstein between 1980 and 1990.

3.3 *Mode and causes of the failure*

The damage was discovered on July 11, 1990 at 10 p.m. by a car driver who recognised a step in the pavement (near the bridge abutment A – see Fig. 15). Shortly afterwards, all traffic routes were closed on and beneath the bridge. A crisis committee met that same night and decided on immediate emergency measures to stop the excessive bridge movements which were obviously caused by scouring of the river pier (B in Figs. 10,15).

The scour was plugged with about 29000 tons of rock material and dredged stone, and additional 11000 tons were filled to stabilise the riverbed. At that time the river pier had already settled about 0.2 m on the upstream side and nearly 1.3 m at the downstream side. Simultaneously, the pier exhibited a strong tilting in the longitudinal direction of the bridge, and the head of the pier hat moved 1.1 m from his original position.

The failure occurred due to the superposition of several reasons:

- A long-term tendency of riverbed erosion (Fig. 13) which gradually weekend the gravel covering fine sediments.
- Lenses of fine, uniform sands on the left side of the river pier which were extremely sensitive to erosion and scouring.
- Increased flow velocity and scour aggressivity of the river during a flood.
- Constraints of the riverbed due to the installation of groynes and previous construction measures. The latter comprised access fillings to temporary bridge supports which were necessary for bridge maintenance works in 1989. These fillings were not yet completely removed.

Fig. 14 Groundplan of the scour after emergency refilling near the river pier. Also indicated (broken lines) are the hydraulic constraints due to new groynes and previous fills for a temporary bridge support.

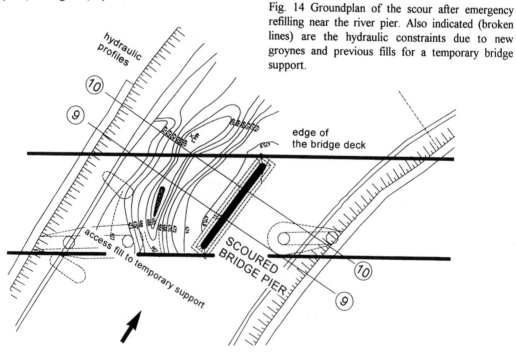

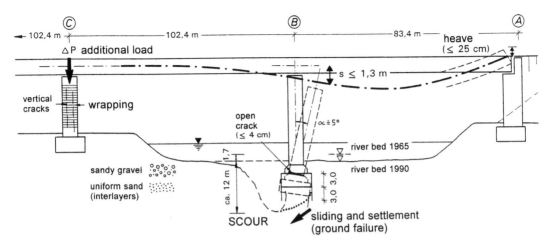

Fig. 15 Scheme of the bridge failure (not to scale); downstream view. Indicated is also the overloading (ΔP) of the pier row C.

The interaction of these factors caused a deep scour along the left side of the river pier (locally up to 15 m depth) – Fig 10,14. The transversal movement of the pier was similar to that of a base failure (Figs. 15, 16) which proved that the scour had occurred rather quickly. Otherwise it would have rotated in the other direction according to the gradual loss of lateral earth pressure. The rotation of the pier in its longitudinal direction caused a differential settlement up to 1.1 m (Figs. 11,17). Fig. 18 gives a detailed view of the old and new position of the river pier, also comprising statically relevant data. The reconstruction of the stages of ground erosion / pier movements is given in Fig. 19. The severe settlement of the river pier caused a heave of the bridge abutment (A) which fortunately was recorded in time by a car driver (a civil engineer). Simultaneously the pier row C was overloaded (ΔP in Fig. 15) and exhibited intensive vertical cracking.

The differential movements caused several cracks in the river pier, moreover a part of the strip foundation was sheared off (downstream – Fig. 18). The bridge bearings were completely demolished and the bridge decks exhibited numerous severe cracks. Accordingly, the entire structure was in an unstable limit equilibrium, close to collapse. Fig. 21 shows the critical phase which illustrates clearly, that the bridge would have collapsed if the quick emergency refilling of the deep scour hollow had not been performed. A collapse would have caused a sudden rise of the river level up to 5 m, hence a disaster in the upstream parts of the city of Kufstein.

In order to quantify the failure reasons (especially for legal aspects, as the matter was brought to court), hydraulic model tests were conducted. Some results are illustrated in Fig. 22. They disclosed that the deep scour occurred due to a local breakthrough in the riverbed gravel and subsequent aggressive erosion of sandy-silty lenses that finally led to an overall ground removal.

Fig. 16 The leaning bridge pier B before partial re-tilting. Upstream view.

13

Fig. 17 View of the bridge pier with large settlements towards downstream and stabilising measures: Wrapping of the toe-zone with steel loops, tying back of the head-zone with prestressed anchors (strands) across the river, support of the bridge decks with steel plates and hydraulic jacks.

On the temporary island some grouting points for ground improvement (underpinning) are visible; in the background is the pontoon-bridge between river bank and temporary island.

3.4 *Stabilising and rehabilitation measures*

3.4.1 Decision making

Shortly after scouring, the bridge structure exhibited such a low stability that an additional settlement of the river pier of only few centimetres would have caused collapse. Accordingly, the predominating opinion of the structural engineers was, that the bridge cannot be saved but must be demolished. This would have caused not only severe international traffic constraints and high construction costs but also enormous compensation payments due to the closure of a main European traffic route. The next question was whether one could allow the workers to be exposed to the danger of working on and underneath the bridge. Thus, the bridge saving and security measures were preliminarily a challenge to geotechnical engineering – in connection with structural engineering.

The time for making the decision to demolish or save the bridge was limited to one day being under high political and medial pressure. There was no time for specific geotechnical investigations: experience and engineering intuition had to solve the problem.

3.4.2 Overview of measures

In addition to the emergency measures immediately during scouring, numerous other activities were required from geotechnical, structural and hydraulic point of view (e.g. Figs. 23, 24).

- Access fill round the scoured pier.
- Heavy rockfill as further scour prevention.
- Temporary support of the bridge girders with steel plates and hydraulic jacks.
- Wrapping and prestressing of the cracked river bank pier (C in Fig. 15) with greased strands.
- Installation of auxiliary support structures (trussed steel columns) aside the pier row C and under the bridge spans (Fig. 24).
- Lateral support of the slipped bridge decks with jacks.
- Cooling of the bridge deck with sprinkling water during hot summer days to minimise temperature constraints which could have caused collapse (Fig. 25).
- Wrapping and prestressing of the cracked river pier (B) with flat steel tendons (13 MN) – see Fig. 17. These elements remained for permanent strengthening (see Fig. 27).
- Tying back of the river pier with steel tendons and prestressed anchors to prevent further rotation and make re-tilting easier.

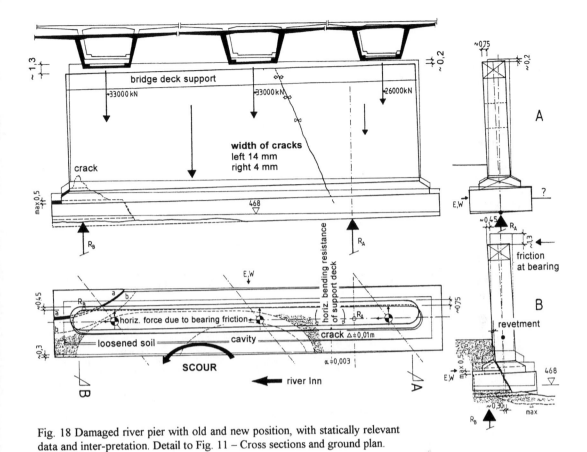

Fig. 18 Damaged river pier with old and new position, with statically relevant data and inter-pretation. Detail to Fig. 11 – Cross sections and ground plan.

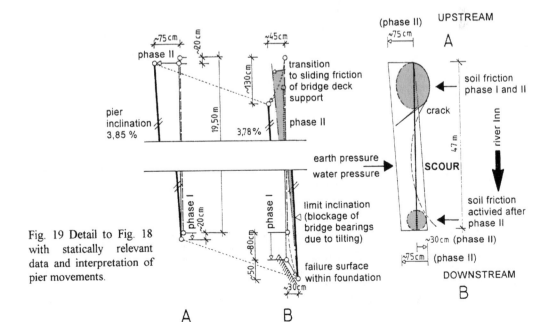

Fig. 19 Detail to Fig. 18 with statically relevant data and interpretation of pier movements.

Fig. 20 Partial view of the crack pattern within the hollow box girder of the downstream bridge deck. Crack spacing was between 0,3 to 1,5 m.

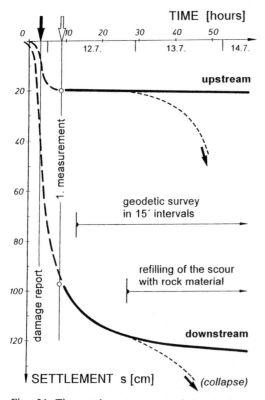

Fig. 21 Time-settlement curves of the river pier during the first phase of the failure. Measuring points on the upstream and downstream superstructure.
The quick refilling of the scour saved the bridge from collapse.

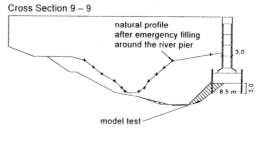

Cross Section 9 – 9

natural profile
after emergency filling
around the river pier

model test

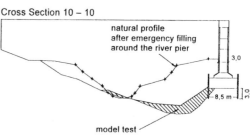

Cross Section 10 – 10

natural profile
after emergency filling
around the river pier

model test

Fig 22 Results of hydraulic model tests in comparison with the in-situ riverbed shape after scouring and refilling.

- Tying back of the downstream bridge deck on the left bridge abutment (Fig. 26).
- Grouting of caves and voids in the loosened and rearranged subsoil beneath the river pier.
- Ground improvement and partial re-tilting of the bridge pier with soil fracturing.

16

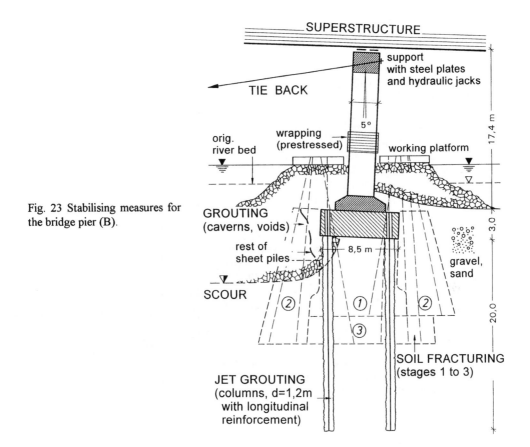

Fig. 23 Stabilising measures for the bridge pier (B).

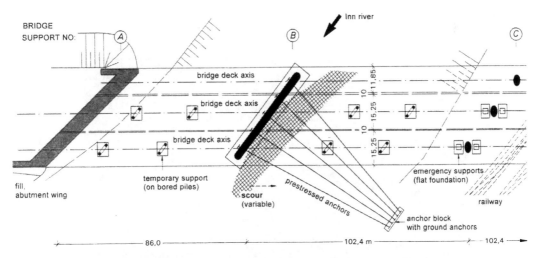

Fig. 24 Groundplan of the bridge between the axes A and C including the auxiliary supports of the superstructures during the exchange of the prestressed tendons.
Temporary columns (trussed steel structures) at the pier row C and tying back of the river pier (B) with prestressed anchors also indicated.

17

- Ground stabilisation, drilling of relief borings to locally extract soil in order to partially re-tilt the river pier.
- Underpinning of the river pier with jet grouting columns (with longitudinal reinforcement) ringing the old foundation and reaching 20 m below the original foundation base.
- Installation of a transverse barrier fill (ground sill) in the river downstream the bridge, to reduce the high flow velocity ($v = 10$ m/s) and the local turbulences of the river.

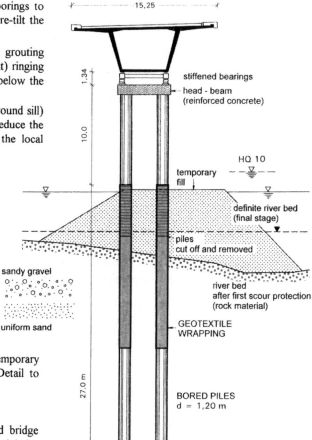

stiffened bearings
head - beam
(reinforced concrete)

HQ 10

temporary fill

definite river bed
(final stage)

piles
cut off and removed

sandy gravel

uniform sand

river bed
after first scour protection
(rock material)

GEOTEXTILE
WRAPPING

BORED PILES
d = 1,20 m

Fig. 24a Cross section through the temporary support of the downstream bridge deck. Detail to Fig. 24.

Fig. 25 Sprinkling of the heavily deformed bridge decks during hot summer days to minimise temperature constraints in the statically sensitive continuous girder.

18

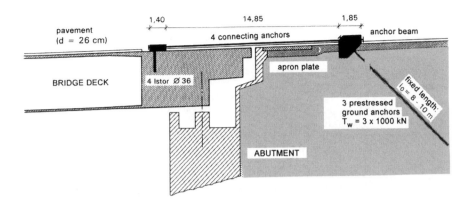

Fig. 26 Tying back of the downstream bridge deck on the left bridge abutment (axis F in Fig. 10).

- Removal of the 35 cm thick pavement of the bridge and installation a thin asphaltic layer. This reduced the dead loads by a magnitude which was higher than the maximum traffic load.
- Gradual lifting of the bridge decks, grouting of the cracks with a width > 0.2 mm and installation of an additional centric prestressing.
- Uncovering the toe of the pier shaft and the upper part of the old foundation (after the jet grouting columns exhibited sufficient strength).
- Strengthening of the lower pier zone and grouting of the interface between old foundation and the subsoil improved by soil fracturing.
- Wrapping of the leaning, cracked pier shaft with reinforced concrete, thus achieving vertical faces again (Fig. 27).
- Casting reinforced concrete supports on top of the settled river pier and installation of new bearings.
- Installation of temporary bridge deck supports (trussed steel columns on bored piles) in the river to make an exchange of the deteriorated prestressing members within the hollow box girders possible.
- Permanent stabilisation of the riverbed with selected quarried rock and dredged stone and gravel.

The auxiliary supports of the cracked pier row C represented an emergency measure. In contrary, the auxiliary supports beneath the river spans of the bridge were installed after securing, strengthening and lifting of the structure. They were required for the exchange of the concentrated prestressing tendons inside the webs of the mean and downstream bridge girder. This measure had been envisaged due to inspection results already before the bridge failure

The supports were founded on twin piles bored from access islands. In order to avoid bulging of the piles in the scoured ground, and to make an easy removal possible, the pile reinforcement was wrapped with a highly flexible non woven geotextile (Fig. 24a).

3.4.3 Monitoring

According to the low stability of the structure and the sensitive subsoil beneath the scoured river pier, intensive monitoring was required. The results were used for continuous safety evaluation and the optimisation of the interacting geotechnical and structural measures:

- Permanent survey of the river pier with precision levelling. In the first month within 15'-intervals.
- Equipping of the cracked zones of the downstream bridge with inductive distance sensors with electronic data transfer to a computerised monitoring centre. Additionally, wire extensometers, distance sensors and temperature sensors were installed and operating within 4 days.
- Survey of the river pier with 6 electronic tube level monitors, 4 inclinometers and 2 fissurometers.
- Control of lateral ground deformation around the river pier with additional 5 inclinometers. These results were especially valuable during soil fracturing and jet grouting.
- Continuous levelling of the riverbed, because the flow line and secondary scouring varied permanently according to several access measures.
- Measurement of groundwater flow velocity to assess the risk of an inner soil erosion due to the local differences of the water level in the river (local backwater zones etc.)
- Control of grouting schemes and parameters, and the interaction between fill grouting, soil fractu-

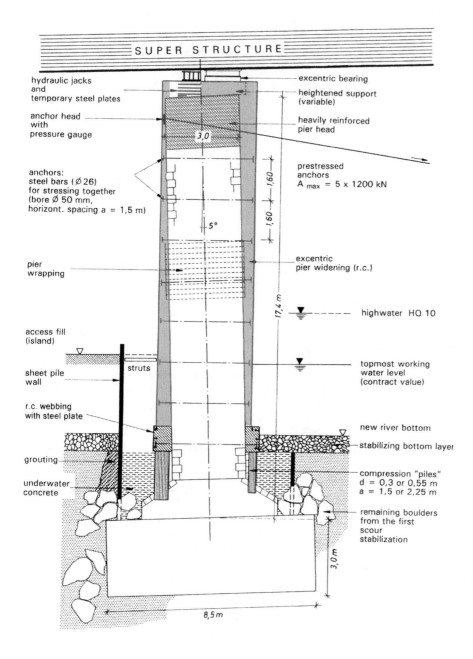

SUPER STRUCTURE

hydraulic jacks
and
temporary steel plates

anchor head
with
pressure gauge

anchors:
steel bars (∅ 26)
for stressing together
(bore ∅ 50 mm,
horizont. spacing a = 1,5 m)

pier
wrapping

access fill
(island)

sheet pile
wall

r.c. webbing
with steel plate

grouting

underwater
concrete

excentric bearing

heightened support
(variable)

heavily reinforced
pier head

prestressed
anchors
A max = 5 x 1200 kN

excentric
pier widening (r.c.)

highwater HQ 10

topmost working
water level
(contract value)

new river bottom

stabilizing bottom layer

compression "piles"
d = 0,3 or 0,55 m
a = 1,5 or 2,25 m

remaining boulders
from the first
scour
stabilization

struts

3,0

5°

1,60

1,60

17,4 m

3,0 m

8,5 m

Fig. 27 Emergency stabilisation (wrapping and tying back) and permanent strengthening of the leaning river pier (B). Vertical reshaping of the lateral faces by an excentric reinforced concrete ringing of the shaft. Upstream view.
Underpinning of the foundation not drawn.

ring and jet grouting, and the movements of soil and river pier.
• Exploratory and control borings and soundings; investigation of soil and grout samples etc.

Because of the extremely low safety of the structure, remote reading was performed in 5 sec.-

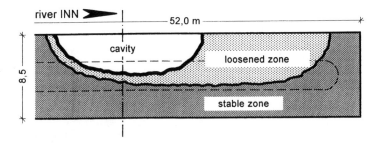

Fig. 28 Unstable subgrade of the river pier during the first week after scouring. Caves inspite of refilling the scour. The foundation was supported safely only along the dark area.

intervals in the first phase, then in 10 sec.-intervals, 24 hrs a day. The data were registered and fed into a two-step alarm system: The lower critical value led to a pre-warning and to a meeting of the crisis-staff. The higher value would have stopped immediately the railway and road traffic, which was – to a limited extent – possible already some weeks after the disaster. The monitoring centre was occupied day and night, and remote reading was checked daily by manual measurements.

3.4.4 Ground improvement and foundation strengthening/underpinning

These measures started with fill grouting and soil fracturing followed by jet grouting from an access island around the damaged river pier. The interval of settlement measurement was 15 minutes in the first weeks, and then gradually increased to 6 hours, until the jet grouting columns exhibited sufficient strength.

3.4.4.1 *Fill grouting*

Scouring had caused large caves and voids beneath the existing footing. During the subsidence more than one third of the foundation had lost its subgrade (Fig. 28). The local refilling with rock in turbulent water provided only a very loose density with large voids and even caves. Moreover, the original soil was heavily loosened and rearranged, and the washing out of fine gravel, sand and silt went on.

Consequently, a rapid grouting of the voids and caves, resp. cavities was the most urgent measure to avoid a complete bridge collapse. But because of the high flow velocity of the water (up to 10 m/s) and the local turbulences it could not be performed in a conventional way. The grout material would have

been washed out immediately. Therefore "geotextile expander bags" were fixed on toe of the grouting hose or sleeve pipes as indicated in Fig. 29. The following procedure proved to be most appropriate.

- Underground investigation by drilling boreholes (with casing).
- Installation of tightly folded "geotextile expander bags" by means of an auxiliary rod or stiff grouting hose.
- Pushing the geotextile expander bag (or bags) in the open void or cave.
- Grouting of a rather plastic suspension (rich in cement and limestone filler) with low pumping rate to about 80 % of the theoretical volume of the geotextile bags.
- Adding of an accelerating agent (water glass) along a separate hose or pipe.
- Entire filling of the geotextile bags in steps and finally increasing the grouting pressure up to 200 kPa, thus expanding the bags significantly.

The grout pipes were tubes à manchettes with 20 cm spaces of openings. Thus, grouting could be adapted very precisely to local subsoil conditions.

The geotextile expander bags consisted of a needle punched nonwoven made from polypropylene endlessfibres. It showed the following characteristics:

- thickness d = 3.0 mm
- weight g = 350 g/m²

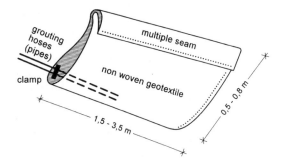

Fig. 29 Scheme of the geotextile expander bags for grouting caves in the underground.

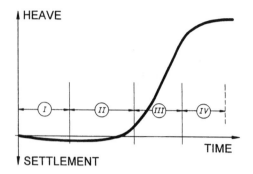

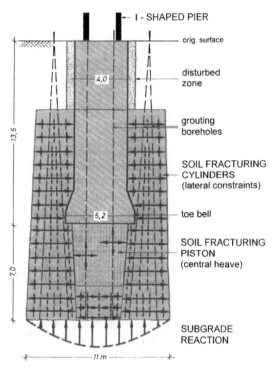

Fig. 30 The soil fracturing method - schematical. Phases of movements, I to IV.

- tear strength ≥ 21 kN/m at elongations of ε ≥ 50/80 %
- CBR-puncture resistance ≥ 3.5 kN
- hydraulic properties: opening size $D_w = 0.08$ mm vertical k-value = 4×10^{-3} m/s

According to Fig. 29 the nonwoven geotextile was at first needled to form a bag at the end of the grouting pipe or hose. Then it was intensively folded and finally fixed to the grouting pipe with a clamp.

Due to their flexibility, the expanding geotextile bags filled the voids during grouting. Moreover, these bags formed a primary "grain" structure (skeleton) within the caves, thus making a subsequent grouting of the whole underground possible.

3.4.4.2 Soil fracturing

Soil fracturing was performed rather simultaneously with the grouting of voids and caves. The deformation measurements during soil fracturing showed results which are characteristic of this technique, as schematically sketched in Fig. 30: Small settlements in the first stage of drilling and grouting (phase I). Setting in of first heaves and gradual heave acceleration during phase II. Commonly, phase I + II takes the most time. Phase III leads to the greatest heave within a relatively short period and needs an increasing fracturing pressure. The residual deformations and the duration of the post grouting phase IV depend on the long-term behaviour of the soil; in most cases it is shorter than phase II, III, unless the ground beneath the improved soil has a considerable tendency towards creeping.

Fig. 31 Scheme for re-lifting a bridge pier by means of soil fracturing. Case history from a previous bridge failure.
Model of the stress state in the subsoil caused by repeated grouting.

The aim of soil fracturing was not only a soil improvement but also a partial re-tilting of the leaning bridge pier. From other site experiences (Brandl, 1991) the following scheme of grouting sequences was chosen (Fig. 31):

- Lateral soil improvement, whereby mainly horizontal grout-lamellas develop. A gradually increasing horizontal stress and constraints make the formation of vertical lamellas possible.
- Improvement of the subsoil in a greater depth beneath the foundation base.
- Grouting of the remaining soil core within the improved soil ring. This acts like a piston which heaves the foundation. During this phase the coefficient of lateral earth pressure is about K = 1.

Soil fracturing was performed very cautiously in steps with fracturing pressures between 35 to 120 bar, and subsequent grouting pressures between 20 and 100 bar. Sometimes an intermediate grouting

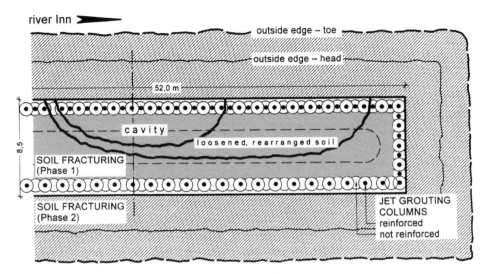

river Inn ➤

outside edge – toe

outside edge – head

52,0 m

8,5

cavity

loosened, rearranged soil

SOIL FRACTURING
(Phase 1)

SOIL FRACTURING
(Phase 2)

JET GROUTING
COLUMNS
reinforced
not reinforced

Fig. 32 Partial ground plan of the river pier foundation with soil improvement and underpinning measures. Scoured area roughly indicated.

was necessary according to the results of monitoring. In total 240 grouting boring (up to 21 m depth) were drilled and 1550 m³ of grout material injected; this represents about 12 % of the improved soil volume which coincides well with the hitherto experience. The cumulative grout quantity versus time is shown in Fig. 39. Core borings (∅ 120 mm) served for quality control and disclosed a quasi-monolithic block (Fig. 32).

Fill grouting and soil fracturing began two weeks after the aggressive scouring. Fig. 33, 34 illustrate how sensitive the subsoil was and that even the daily temperature fluctuations caused movement fluctuations of the river pier head. The influence of bore-hole drilling and grouting, and of a secondary scour due to a new flood, is also clearly visible.

The entire soil improvement and foundation underpinning had to be finished till March 1991, because the access island had to be removed before the expected spring flood. Consequently, the ground engineering works were performed in day and night shifts. Fig. 35 shows the settlement behaviour in this critical stage. It illustrates that a small re-tilting of the leaning pier could be achieved which was sufficient to bring back the resultant force into the statically relevant core of the foundation. The pier head could be removed by 11 cm towards his original position. The final re-levelling of the bridge support ought to be achieved with jacks on the pier head. Moreover, a further re-tilting with soil fracturing, relief borings and tying back of the river would have lasted too long.

3.4.4.3 Jet grouting

Fill grouting and soil fracturing represented only a soil improvement, whereby progressive inner erosion could not be excluded in the long-term. Thus, for strengthening and underpinning of the weakened old foundation 136 jet grouting columns were installed, 100 of them with a centric longitudinal reinforcement, ∅ 50 mm (Fig. 32). The overlapping columns (∅ 1.2 m) ring the old foundation completely and reach 20 m below its base. Due to the prevailing ground conditions and the required quality, the triple-technique of jet grouting was used.

Quality control was performed during grouting and by core drillings after 28 and 90 days. The results of the unconfined compressive tests on "undistribed" samples showed a wide scatter which is characteristic of jet grouting under such conditions. It is influenced by several factors:

- Wide local scatter of grain size distribution, density and permeability of the subsoil.
- Withdrawal rate and pressures during grouting.
- Sort and quantity of the cement (resp. W/C-factor).
- Portion of soil included in the soil-grouting-mixture.
- Loosening of the improved soil during drilling and sampling.
- Age and mode of curing of samples.
- Enclosures of grout material, boulders etc. in the sample (scale-effect).

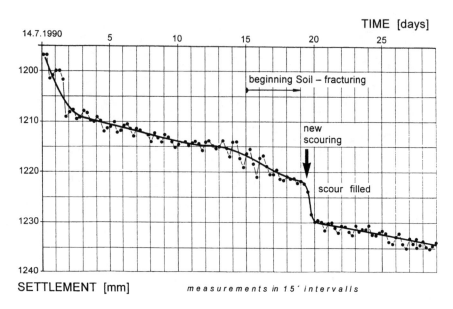

Fig. 33 Settlement of the river pier (B) beneath the downstream superstructure during the critical initial phase of stabilisation. Readings in 15sec.-intervalls. Influence of temperature (July/August) and of a small new scour.

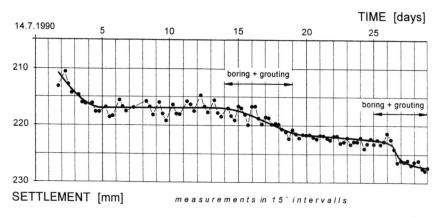

Fig. 34 Settlement of the river pier (B) beneath the upstream superstructure during the critical initial phase of stabilisation. Readings in 15sec.-intervalls. Influence of temperature (July/August) and of the grouting sequence during the initial phase of soil fracturing (according to phase I in Fig. 30).

The above factors have much more influence on the relatively small-scale samples ($\varnothing$ 120 mm) than on the in-situ jet grouting body. This is illustrated in Figs. 36 to 38.

The unconfined compressive tests were performed on samples with a slenderness of h : d = 2 : 1. 28 days after jet grouting values between q_u = 6 and 11 MN/m² were observed. From experience it is known that drilling and sampling already 28 days after jet grouting yields lower limit values due to unavoidable

loosening of the fresh columns. The strain at failure usually varied between ε = 1.0 to 2.0 %, whereby curing until 4 month was of no relevant influence.

Fig. 36 shows some stress-strain curves after long-term curing, illustrating the wide scatter and the rather linear behaviour of the improved material under service load. From such straight lines, which exist already after 28 days of hardening, an appropriate elasticity modulus can be obtained for design.

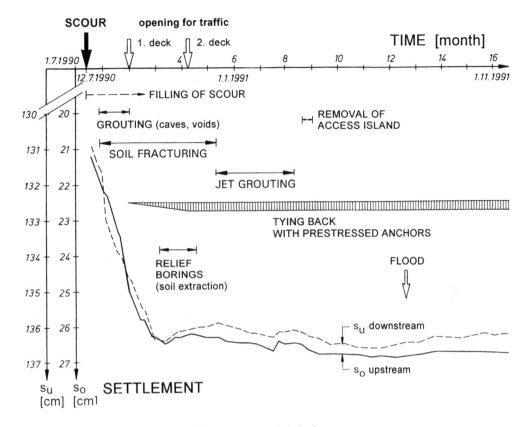

Fig. 35 Settlement of the upstream and downstream part of the river pier (No. B in Figs. 10, 15) during the stabilising measures. Measuring points not the same as in Figs. 33, 34.
Dates of the re-opening of the traffic; the 3rd superstructure was opened in 1992.

Similar to concrete or rock, jet grouted soil shows a fair correlation between the modulus E and the compressive strength q_u, whereby the dimensions have to be especially considered (e.g. Fig. 37):

$$E = k \cdot \sqrt{q_u}$$

The correlation coefficient was found to be $r = 0.90$ which is quite sufficient. The empirical factor, k, increases, with time and changes with the dimension of E and q_u.

The dry density of the jet grouting soil samples exhibited also a similarly wide scatter. The reasons were the same as mentioned above, but additionally a strongly varying density (up to $\rho_s = 2,83$ t/m³) and suspensions rich in cement played a role.

3.5 Anchorages

The leaning river pier was tied back with 5 strands leading to a reinforced concrete block at the right river bank, and this block was fixed to the ground by 5 prestressed grout anchors. (Figs. 17, 23, 24). This measure increased the safety factor against overturning and made a certain re-tilting possible. Loading was performed incrementally up to $T_{max} = 6000$ kN (Fig. 39) and required continuous monitoring and adapting to the structure and soil performance. Thus, the eccentric resultant force in the pier could be moved back by 0.7 m towards the pier axis. Re-tilting and seasonal temperature fluctuations caused a repeated temporary decreasing of the anchor forces which had to be re-adjusted then according to the monitoring results. Final unloading of the prestressed strands was performed again in steps to avoid stress constrains in the bridge bearings.

The downstream bridge girder was in danger to slip from the bearings. Therefore, it was tied back with strands to an anchored reinforced concrete beam behind the apron plate of the left bridge abutment (Fig. 26).

25

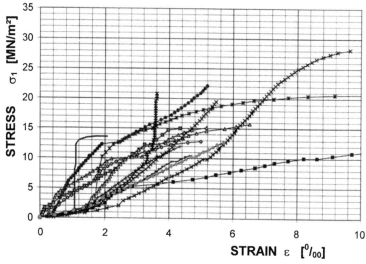

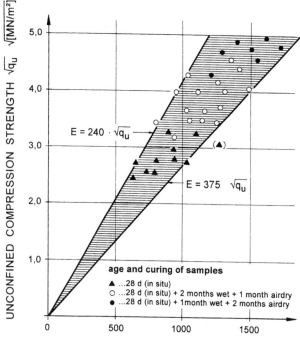

Fig. 36 Stress-strain curves of core samples from jet grouting columns beneath the river pier. Unconfined compression tests; ratio of sample height to diameter about 2:1. Samples taken 28 days after jet grouting, then on year of curing (moist).

Fig. 37 Correlation between the modulus, E, and the unconfined compressive strength, q_u, of core samples from jet grouting columns beneath the river pier. Samples taken 28 days after jet grouting. Influence of duration and way of sample curing.

3.6 *Bearing capacity of the strengthened river pier foundation*

Fill-grouting, soil-fracturing and reinforced jet-grouting created a foundation system which can be considered a composite body reaching about 25 m below riverbed. A subgrade reaction model of this complex system is illustrated in Fig. 40. Due to the overlapping of the jet grouting columns a box-shaped structure is formed. For calculation, theoretical idealisation are unavoidable. Experience and monitoring have disclosed that the quasi-monolith theory and the single-element failure theory comprise every possibility of interaction, hence leading to maximum and minimum values as basis for design, safety assessment and engineering judgement.

The safety factors can be calculated as described in chapter 2.2.

3.7 Re-opening of the traffic routes

Ten days after the aggressive scouring (11.7.1990) the failing bridge and unstable riverbed were secured so far that the traffic beneath the bridge could be reopened (i.e. international railway line and federal motorway). The upstream bridge deck was opened on 31.8.1990 (at first only with traffic restrictions), the second bridge deck on 8.11.1990. Thus, within four months after the failure, the entire international traffic could pass again the bridge though several remedial and strengthening measures were still in progress. The third bridge deck (downstream) which was damaged most, was re-levelled in October 1990. It then underwent a general rehabilitation (replacing of the concentrated tendons in the prestressed girder) which had been planed already before scouring occurred. This work was finished in June 1992. The safety factors of the triple grider bridge are now higher than ever before.

4 PRACTICAL CONCLUSIONS

Monitoring of riverbeds and old bridges over water has disclosed that the safety against scouring is clearly below modern standards in many cases. This is underlined by numerous bridge failures world-wide in the past decades. Practical conclusions derived from comprehensive river bridge inspection, scour repair and bridge rehabilitation, resp. foundation strengthening are:

- Inspection of riverbeds (and seabeds) near bridge piers should be intensified. Government regulations and stringent rules resp. should call for inspections after flood events as well as once yearly.

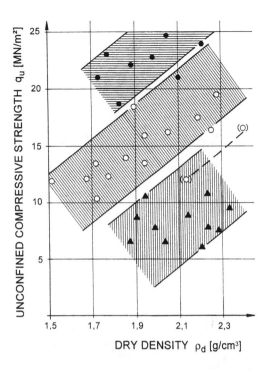

Fig. 38 Correlation between dry density, ρ_d, and the unconfined compressive strength, q_u, of core samples from jet grouting columns beneath the river pier. Sample taken 28 days after jet grouting. Influence of duration and way of sample curing. Symbols similar to Fig. 37.

- Bridge inspectors should be trained to evaluate scour, and geotechnical / hydraulic engineers should be required to evaluate bridge inspection reports where scour is identified.

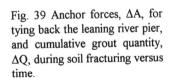

Fig. 39 Anchor forces, ΔA, for tying back the leaning river pier, and cumulative grout quantity, ΔQ, during soil fracturing versus time.

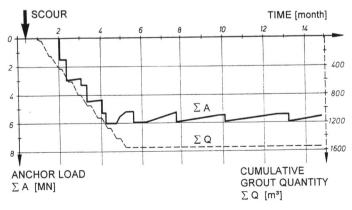

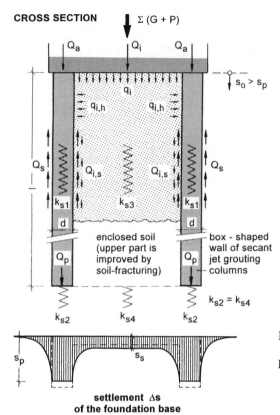

CROSS SECTION $\Sigma (G + P)$

Q_a Q_i Q_a

$s_o > s_p$

q_i

$q_{i,h}$ $q_{i,h}$

Q_s $Q_{i,s}$ $Q_{i,s}$ Q_s

k_{s1} k_{s3} k_{s1}

d d

enclosed soil
(upper part is
improved by
soil-fracturing)

box - shaped
wall of secant
jet grouting
columns

Q_p Q_p

$k_{s2} = k_{s4}$

k_{s2} k_{s4} k_{s2}

s_p s_s

**settlement Δs
of the foundation base**

Fig. 40 Subgrade reaction model of the strengthened
foundation of the river pier showing soil responses to
superimposed loads. Box-shaped new foundation
consisting of overlapping jet grouting columns and
soil improvement (soil fracturing) in the upper part.
Hatching on the diagram (below) indicates the
difference between actual settlement and idealised
model in case of load increase.
For calculation of safety factors see chapter 2.2.

- The highway and railway authorities should
develop a "crucial element checklist" for each
bridge, to identify such structural or geotechnical elements which, when deteriorated
would independently cause a sudden unexpected
collapse of a section of the bridge. This list
should then become part of the bridge inspection
report. On discovery of deterioration of an
element (including foundation, possibility of
scouring etc.) on the crucial list the local /
federal authorities should be required to
immediately close the bridge or repair it.

- A box-shaped ringing of old foundations with
bored piles, jet grouting columns or diaphragm
walls has proved very suitable as failure
prevention in unstable riverbeds. Scouring then
has not the aggressive effect like on conventional
pile groups. Moreover, inner erosion in the
encapsulated subsoil beneath the old foundation
is excluded.

- The outstanding behaviour of the scoured river
bridge in Kufstein disclosed that prestressed
reinforced concrete structures are much more
ductile than hitherto assumed. The fact, that the
continuous girder superstructure did not collapse
inspite of 1.35 m pier settlement had a significant
influence on Eurocode 2. It gave rise to the new
view that prestressed concrete is not to be
considered concrete under pressure, but rather
as steel-reinforced concrete with additionally
activated reinforcement.

REFERENCES

Brandl, H. 1985. Die Reichsbrücke über die Donau
in Wien. Heft 3 der Mitteilungen für Grundbau,
Bodenmechanik und Felsbau, Techn. Universität
Wien.

Brandl, H. 1987. Deep box-foundations with piles
and diaphragm walls in weak soils. Proceedings of
the 9th Asisan Geotechnical Conference, Bankok:
6-15 to 6-28.

Brandl, H. 1988. Die Übertragung konzentrierter
hoher Lasten im weichen Untergrund – Transfer
of concentrated high loads into weak soil. 2nd
Baltic Conference on Soil Mech. and Found. Eng.
Proceedings, Tallin/Estonia.

Brandl, H. 1991. Stabilisation of excessively settling
bridge piers. Proceedings of the 10th European
Conference on Soil Mechanics and Foundation
Engineering, Firenze. A.A. Balkema, Rotterdam.

Brandl, H. 1996. Bodenmechanik und Spezialtiefbau
bei der Wiederherstellung der Innbrücke Kufstein.
Österr. Betonverein, Heft 26, Wien.

Brandl, H., Pauser A., Seeber, G. 1996. Innbrücke
Kufstein der Inntal-Autobahn A12. Unpublished
expertise (Geotechnics – Structural Engineering –
Hydraulics), Vienna.

Wittke, W. 1991. Damage incurred at the bridges
over the Inn River at Kufstein (Austria), and their
repair. Structural Engineering 1/91: 28 - 34

Geotechnical Hazards, Marić, Lisac & Szavits-Nossan (eds) © 1998 Taylor & Francis, ISBN 90 5410 957 2

The practice of unsaturated soil mechanics

D.G. Fredlund
Unsaturated Soils Group, Department of Civil Engineering, University of Saskatchewan, Sask., Canada

ABSTRACT: The primary emphasis of classical soil mechanics has been on the behavior of saturated soils. Two primary weaknesses associated with classical soil mechanics have revolved around a lack of understanding of the behavior of unsaturated soils and the lack of a method to quantify the moisture flux boundary condition imposed by climatic conditions. Engineering practice reveals that many of the problems encountered involves unsaturated soils from above the water table.

The basic elements and relationships for unsaturated soil mechanics theories are defined and illustrated in this paper. Emphasis is placed on the classical areas of seepage, shear strength and volume change. A brief summary is presented on procedures that have been used for measuring the unsaturated soil property functions. The use of the soil-water characteristic curve for estimating the unsaturated soil property functions is described. The basic principles underlying the estimation of long-term, equilibrium soil suction profiles is also presented.

INTRODUCTION

Soil mechanics is generally considered to be one of the younger areas in the field of civil engineering. Its youthfulness is primarily the result of difficulties encountered in defining the state of stress required for describing the mechanical behavior of a soil. The state of stress for a saturated soil was defined by Terzaghi (1936) and the discipline of soil mechanics took on the character of being a science.

Research papers presented at the First International Conference on Soil Mechanics and Foundation Engineering in 1936, showed that there was a keen interest in the behavior of soils both below and above the water table (i.e., saturated and unsaturated soils). Subsequent conferences showed that much of the research was gravitating towards understanding saturated soil behavior. However, much of the infrastructure required to support urban and societal development involved unsaturated soil from above the groundwater table. Recent concerns regarding proper environmental stewardship has also given impetus to the need for understanding flow and contaminant transport phenomena in the zone above the groundwater table. The pore-water pressures were negative in this region and proved to be difficult to measure both in the laboratory and in the field. A clear description has been slow to emerge for the stress state required to represent all mechanical behavior for unsaturated soils.

The objective of this paper is to present a state-of-the-art, conceptual framework for the analysis of geotechnical problems involving soils with negative pore-water pressures (i.e., unsaturated soils), regardless of the soil type involved. An attempt is made to parallel the classical areas which have been delineated for saturated soil mechanics. The scope is limited to setting an introductory framework for unsaturated soil mechanics.

TOPICS COMMON TO CLASSICAL SOIL MECHANICS

Classical soil mechanics grew rapidly in the 1930's with the synthesis of a series of applied mechanics problems which had been formulated over the past 100 years or so (Skempton, 1985). The book, "Theoretical Soil Mechanics", (Terzaghi, 1943), contained a synthesis of what was considered to be the main types of soil mechanics problems.

Retaining wall design problems and an understanding of the active earth pressure state proved to be an important subject. Equally important was the design of anchored bulkheads and the passive earth pressure state. This brought in the concepts of plastic equilibrium which were also applied to spread footing and pile bearing capacity design. These formulations were for dry soils but were subsequently extended to the case of submergence with water. The initial emphasis was on the mechanics of the problems and not so much on the selection of appropriate soil parameters.

Many problems in soil mechanics required that there be a means of calculating the total normal and shear stresses at all points throughout a soil continuum. Stress distribution theories for a one phase solid (Boussinesq, 1885) was used for computing the stresses in a particulate continuum. This allowed for the computation of the total stress state under various loading conditions.

Seepage problems were subdivided into confined and unconfined type problems (Casagrande, 1937). Graphical flow net solutions were proposed for relatively simple geometric and soil conditions. The difficulties associated with quantifying the climatic flux boundary condition, and characterizing the permeability of the soil above the phreatic line, resulted in empirical graphical solutions for simple unconfined flow cases without a flux boundary at ground surface.

The theory of consolidation provided a mathematical solution to problems related to the amount and rate of settlement of soft, saturated, compressible clays (Taylor, 1948). This provided an analytical solution for settlement problems and also produced a lucid visualization of the effective stress principle "at-work". The most important contribution of the theory of consolidation may have been the mathematical and conceptual tool it provided for teaching students to "think" the way a soil "behaves".

The concept of limiting equilibrium (i.e., the plastic state occurring along a relatively thin zone), found its primary application in the analysis of the stability of slopes. Analytical procedures were proposed for both total and effective stress cases (Terzaghi and Peck, 1948). However, it was the effective stress formulation which has enjoyed increasing acceptance. The effective stress, slope stability analysis and the settlement analysis have most effectively illustrated the use of the effective stress state for saturated soil mechanics problems.

Other soil mechanics problems that were addressed involved the design of supports for excavations, tunnels and shafts. Another area of study was related to vibration problems. Vibration problem analyses, as well as many of the other problems analyzed in the 1930's, assumed that the soil was dry. In general, soil mechanics formulations appeared to address conditions where the soil was either dry (i.e., no pore pressures required), or where the soil was saturated (i.e., the effective stress concept applied). Little attention was given to formulations where the soil was unsaturated and the effect of negative pore-water pressures needed to be considered.

THE MAIN SOIL PARAMETERS ARISING OUT OF CLASSICAL SOIL MECHANICS

Plastic equilibrium analyses and limiting equilibrium analyses assume that all or a part of the soil mass is in a state of plastic equilibrium. The soil parameters required for an analysis are the effective cohesion, c', and the effective angle of internal friction, ϕ'. These soil parameters apply when the soil is dry as well as when the soil is saturated. The shear strength parameters are required for bearing capacity, lateral earth pressure and slope stability type problems. If the soil is saturated, the pore-water pressures must also be predicted. Difficulties in predicting the pore-water pressures for many situations (e.g., bearing capacity problems), has led to the use of so-called total stress analyses with simulated shear strength parameters. Total stress soil parameters are obtained by attempting to match, in the laboratory, the water content (and degree of saturation) and stress state anticipated in the prototype.

Classical stress distribution theories required a knowledge of the elastic parameters of the soil. For a linear, elastic soil subjected to small strains, the parameters are Young's modulus, E, and Poisson's ratio, μ. Estimates of these soil parameters were generally sufficient for the prediction of stresses in a continuum. The prediction of elastic deformations proved to be much more difficult and it took several decades before testing techniques (mainly insitu procedures) proved to yield acceptable results. These parameters should correspond to the effective stress state of a saturated soil for analysis purposes; however, simulation total stress, elastic parameters have once again, often been used due to difficulties in predicting the pore-water pressures.

The settlement of soft saturated clays was historically analyzed as a one-dimensional problem where the required elastic parameters were obtained from an oedometer test. The primary soil parameter was the nonlinear coefficient of compressibility, a_v, which was usually linearized on a semi-logarithm plot to give the compressive index, C_c. This procedure has proven to be quite useful in engineering practice. At the same time, there seems to have been some difficulty in quantifying appropriate elastic soil parameters for settlement and deformation analyses involving two and three-dimensional, numerical models. Volume change and shear strain problems have proven to be difficult from an analytical standpoint for saturated soils.

The coefficient of permeability, k, is the soil parameter required when performing a seepage analysis. The coefficient of permeability is known to range over many orders of magnitude but its use in engineering practice has generally taken the form of a constant value for a specific strata of saturated soils. It was seepage type problems involving saturated/unsaturated soil systems that brought an awareness of the need for soil property functions when analyzing unsaturated soils problems. The need for soil property functions for other unsaturated soils problems has subsequently become evident.

THE DEVELOPMENT OF SUPPORT INDUSTRIES FOR SOIL MECHANICS

Numerous support areas or industries have arisen with time to assist in implementing soil mechanics theories. For example, subsurface exploration has become a part of solving geotechnical engineering problems. As a result, drilling and sampling equipment has been designed and manufactured. The equipment has largely been produced with the need to address saturated soils problems located at a considerable depth. Unsaturated soils are, by nature, usually quite near to the ground surface. Equipment and sampling procedures have often not taken into account the special needs associated with obtaining undisturbed samples in the unsaturated soil zone. As an example, the frequency and depth of samples in the unsaturated zone is often not well designed into the exploration program.

Laboratory testing equipment has also been largely designed giving attention to saturated soils. Testing procedures commonly require that the soil specimen be immersed in water at the start of the test. For unsaturated soils, the design of the equipment and the test procedures used for testing saturated soils, are not appropriate for obtaining satisfactory data for engineering analyses.

Field instrumentation, particularly piezometers, have been designed and manufactured for the measurement of positive pore-water pressures. Highly negative pore-water pressures have proven to be difficult to measure and numerous attempts have met with limited success (and many failures). There is need for more research and attention to be given to the measurement of negative pore-water pressures because of its important role in implementing unsaturated soil mechanics theories. Other areas of field instrumentation such as the measurement of vertical and horizontal movement, and the measurement of total stresses, are developed and apply equally for saturated and unsaturated soils.

Numerical modelling of soil continua is a support industry that has grown in parallel with the computer industry. The need for computers has become increasingly evident with attempts to address the unsaturated soil portion of the profile along with the saturated soil portion. Unsaturated soils formulation are, in general, nonlinear and require considerable computing capability in order to perform simulations. Probably the most significant area of change has been in the seepage area where the use of saturated/unsaturated seepage modelling has become routine engineering practice. It appears that the use of knowledge-based computer systems may play an increasingly important role in allowing saturated/unsaturated numerical modelling to be performed for all the classic areas of soil mechanics (e.g., slope stability, soil collapse, soil heave and others).

WEAKNESSES WHICH HAVE BECOME APPARENT IN SOIL MECHANICS

Two primary areas of weakness have become apparent in classical soil mechanics. First, the theories and formulations for saturated soil behavior did not accurately represent the unsaturated soil region where the pore-water pressures are negative (i.e., the zone above the water table is called the vadose zone). Second, the interaction between the atmosphere above ground surface and the soil near to the ground surface, produced a moisture flux boundary condition, had not been adequately addressed. These two areas of weakness have meant that the geotechnical engineer did not have adequate theoretical formulations for problems in the vadose zone of the soil profile where most of the infrastructure for society was located.

The ground surface flux boundary condition plays a dominant role in many unsaturated soil problems. For example, the amount of shrinking or swelling around a slab-on-grade is largely controlled by the surrounding environment or the atmospheric conditions at the site. Quantifying the ground surface moisture flux has experienced difficulties, primarily in predicting the actual evaporative flux which will occur from the ground surface under vegetated and non-vegetated conditions. Significant strides have now been made in quantifying actual evaporative fluxes at ground surface (Wilson, 1990).

Another weakness associated with the practice of unsaturated soil mechanics has been difficulty related to the laboratory testing of unsaturated soils. Laboratory testing for unsaturated soils has proven to be costly and demanding from an experimental standpoint.

PROGRESS TOWARDS SOLVING UNSATURATED SOIL MECHANICS PROBLEMS

Research in the 1970's (Fredlund and Morgenstern, 1977), showed that the stress state must be defined differently in the unsaturated soil zone than it had been defined for the saturated zone. Constitutive relations for unsaturated soils have now been proposed and tested for uniqueness for the classic problem areas of soil mechanics. With the constitutive relations has come the realization that many of the unsaturated soil properties can no longer be defined as constants. Rather, the soil properties become nonlinear functions of the stress state. These functions also produce nonlinearity in the theoretical formulations but computers now have little difficulty in providing numerical solutions. Also, there now appears to be excellent potential for the use of unsaturated soil property functions in engineering practice. The unsaturated soil property functions are indirectly estimated from soil-water characteristic curves and saturated soil properties.

The rapid development of microcomputers has meant that the geotechnical engineer needs to view the role of numerical modeling differently than in the past. The model for a relatively complex problem can now be set up in little time and the engineer is able to answer a range of "What if?" possibilities. Volume change and deformation type problems still remain a challenge in terms of both their formulation and the assessment of unsaturated soil property functions.

The difficulties associated with measuring highly negative pore-water pressures have meant that it was not possible to practice geotechnical engineering for unsaturated soils in a manner similar to that used for saturated soils problems. The ability to measure pore-water pressures means that the stress state can be determined and the engineer has a powerful "tool" for decision-making.

WHY DID UNSATURATED SOIL MECHANICS EMERGE SO SLOWLY?

There has been a tendency to organize conferences around various problematic soil types that did not tend to fit the mold of classical soil mechanics. These soils are often regional in nature and labeled as problematic soils. Examples are the international conferences on residual soils, expansive soils and collapsing soils. The *International Conference on Engineering Problems of Regional Soils* held in Beijing, 1988, concentrated on several problematic soils. With the exception of soft coastal clays, the other problematic soils dealt with at the Beijing Conference, had one thing in common; namely, negative pore-water pressures. In other words, the negative pore-water pressures render the soils unsaturated and attempts to apply saturated soil theories give rise to difficulties.

Compacted soils have also been referred to as problematic soils and once again, the pore-water pressures upon compaction are negative. Rather than organize independent conferences on each of the problematic type soils, it would appear to be better to attempt to extend classical soil mechanics formulations to embrace the behavior of unsaturated soils. It was not until 1995 that the *First International Conference on Unsaturated Soils* was held in Paris, France. This conference provided a synthesis of the theory and showed that there was, in general, agreement on the form of the basic theories associated with unsaturated soils. This appeared to be largely due to the common use of two independent stress state variables to describe all mechanical behaviour of an unsaturated soil.

TYPICAL EXAMPLES INVOLVING UNSATURATED SOIL MECHANICS

In many routine engineering problems, a saturated/unsaturated soil mechanics approach along with a consideration of the surface flux boundary conditions has been found to produce superior simulations of geotechnical engineering problems. A few examples involving unsaturated problems which are encountered in engineering practice are itemized in this section, under the general soil mechanics categories of i.) seepage, ii.) shear strength and iii.) volume change. The list is by no means complete.

Seepage Problems

1. Contamination of the vadose zone and the groundwater from chemical spills which occur near ground surface.
2. Quantification of the actual ground surface flux boundary condition which is a function of the suction in the soil.
3. Design of compacted clay covers to control the amount of water flow through waste management facilities.
4. Design of silt and sand (and sandwiched) covers to limit infiltration and maximize precipitation storage, to underlying wastes.
5. Analysis to predict the amount of mounding that may occur to the water table below a waste facility.
6. Design aspects associated with compacted earth fills.

7. Prediction of the pore-water pressures in a compacted, earthfill such as a dam.

8. Prediction of the pore-water pressures in a compacted earthfill when going through the transition from the end of construction to steady state conditions.

9. Prediction of the pore-water pressures associated with heavy rainfalls on the surface of a dam.

10. Modelling of the movement of surface water (e.g., precipitation) into an expansive soil.

11. Long-term predictions for the "closure" or decommissioning design of a mining operation, taking into account the climatic conditions.

Shear Strength Problems

1. Modelling of natural slopes to ascertain the precipitation conditions under which a slope is likely to fail.

2. Assessment of the stability of loosely compacted, unsaturated fills as a result of an increased degree of saturation.

3. Assessment of the stability of cuts or trenches for pipelines, often cut into unsaturated soil.

4. Assessment of the stability of temporary excavations around construction which are subjected to varying climatic conditions.

5. Design for remedial measures to ensure the stability of cuts and excavations.

6. Designs for the lateral earth pressure against retaining structures where a cohesive soil may have been placed behind the wall.

7. Prediction of lateral earth pressures produced by wetting, expansive soils placed adjacent to a relatively rigid wall.

6. Prediction of the bearing capacity of shallow footings founded above the water table.

7. Design of roads, railways and airfields as part of the infrastructure.

Volume Change Problems

1. Prediction of the volume changes which take place in the soil below shallow footings as a result of the precipitation.

2. Design procedures to mitigate against the shrinkage of high volume change soils under drying conditions.

3. Prediction of the amount of collapse that may occur in a meta-stable-structured soils such as loess or poorly compacted silts and sand.

4. Prediction of the depth of cracking as it relates to problems such as earth pressures and slope stability.

5. Prediction of volume change in compacted fills such as dams and embankments as a result of changes in matric suction.

6. Determination of conditions under which cracking will occur in covers over waste containment sites.

THE TERMINOLOGY
"SATURATED/UNSATURATED SOIL
MECHANICS"

The term, "saturated/unsaturated soil mechanics", is used in the sense that the theories and formulations apply to the high majority of soil conditions encountered in engineering practice. Saturated-/unsaturated soil mechanics applies to soils near the ground surface as well as those at greater depths. It applies to soils above the water table as well as those below the water table. It applies to unsaturated soils and saturated soils alike using the same general theories and formulations. It can be shown that theories and formulations which embrace the unsaturated portion of a soil profile have saturated soil behavior as a special case (i.e., there is one unified theory for soil mechanics). In other words, there is a smooth transition between the unsaturated and the saturated cases, and the entire soil profile can be treated as one continuum where saturated soil behavior becomes a special case of unsaturated soil behavior.

IMPORTANCE OF A PROPER STRESS STATE DESIGNATION

The stress state of a saturated soil is completely defined by effective stress variables (Terzaghi, 1936),

$$[\sigma'] = \begin{bmatrix} (\sigma_x - u_a) & \tau_{yx} & \tau_{zx} \\ \tau_{xy} & (\sigma_y - u_a) & \tau_{zy} \\ \tau_{xz} & \tau_{yz} & (\sigma_z - u_a) \end{bmatrix} \qquad [1]$$

where

σ_x, σ_y, σ_z = total stresses in x, y, z directions, respectively
u_w = pore-water pressure.

Using a similar stress state approach, the stress state for an unsaturated soil can be defined using two sets of independent stress state variables (Fredlund and Morgenstern, 1977). There are three sets of possible stress state variables, of which only two are independent. It was possible to form independent stress tensors from the proposed stress state variables. The stress tensors associated with the three sets of independent stress state variables are:

$$
\begin{bmatrix}
(\sigma_x - u_a) & \tau_{yx} & \tau_{zx} \\
\tau_{xy} & (\sigma_y - u_a) & \tau_{zy} \\
\tau_{xz} & \tau_{yz} & (\sigma_z - u_a)
\end{bmatrix}
\qquad [2a]
$$

$$
\begin{bmatrix}
(\sigma_x - u_w) & \tau_{yx} & \tau_{zx} \\
\tau_{xy} & (\sigma_y - u_w) & \tau_{zy} \\
\tau_{xz} & \tau_{yz} & (\sigma_z - u_w)
\end{bmatrix}
\qquad [2b]
$$

$$
\begin{bmatrix}
(u_a - u_w) & 0 & 0 \\
0 & (u_u - u_w) & \tau_{zx} \\
0 & 0 & (u_a - u_w)
\end{bmatrix}
\qquad [2c]
$$

In the special case of a saturated soil, the air pressure, u_a, is equal to the water pressure, u_w, and the three stress tensors reduce to one single stress tensor (i.e., Eq. [1]).

The stress state variables most often used in the formulations of unsaturated soils problems are as follows:

$$
\begin{bmatrix}
(\sigma_x - u_a) & \tau_{yx} & \tau_{zx} \\
\tau_{xy} & (\sigma_y - u_a) & \tau_{zy} \\
\tau_{xz} & \tau_{yz} & (\sigma_z - u_a)
\end{bmatrix}
\qquad [3a]
$$

$$
\begin{bmatrix}
(u_a - u_w) & 0 & 0 \\
0 & (u_u - u_w) & \tau_{zx} \\
0 & 0 & (u_a - u_w)
\end{bmatrix}
\qquad [3b]
$$

The stress state at a point in an unsaturated soil element, in terms of the above two sets of stress state variables, are shown in Fig. 1 (Fredlund and Morgenstern, 1977). Figure 2 makes use of a simplified visualization aid to illustrate the smooth transition in the stress state when going from the saturated to the unsaturated portions of the soil profile.

EXTENSION OF THE THEORY OF SHEAR STRENGTH TO UNSATURATED SOILS

The Mohr-Coulomb shear strength equation for a saturated soil can be written in terms of effective shear strength parameters.

$$
\tau = c' + (\sigma - u_a)\tan\phi'
\qquad [4]
$$

The shear strength equation for an unsaturated (or saturated/unsaturated) soil can be formulated as a linear combination of two independent stress state variables, (i.e., $(\sigma_n - u_a)$ and $(u_a - u_w)$):

$$
\tau = c' + (\sigma - u_a)\tan\phi' - (u_a - u_w)\tan\phi^b
\qquad [5]
$$

Three-dimensional representations of Eq. 5 are presented in Fig. 3. It has been found that the shear strength relationship involving suction can be either linear (Fredlund et al, 1978). or nonlinear (Vanapalli et al., 1996). In general, it is possible to linearize the nonlinear shear strength versus suction relationship over a selected range of soil suctions. However, it is also possible to represent nonlinear forms of the saturated/unsaturated shear strength envelope in a mathematical form. The nonlinear form can be estimated through the use of the soil-water characteristic curve and will be presented later in this paper.

EXTENSION OF THE THEORY OF SEEPAGE TO UNSATURATED SOILS

Darcy's flow law which states that the velocity of flow is proportional to the hydraulic head gradient, is applicable to both saturated and unsaturated soil media (Childs and Collis-George, 1950). For a saturated soil, Darcy's flow equation can be written as follows for the case where the cartesian coordinate are the same as the direction of the major and minor coefficients of permeability.

$$
v_x = k_x \frac{dh}{dx}; \, v_y = k_y \frac{dh}{dy}
\qquad [6]
$$

In a saturated soil, the coefficient of permeability (or hydraulic conductivity) is a constant. The constant coefficient of permeability in a saturated soil is related to the essentially constant value of the water content of the saturated soil.

For an unsaturated soil, Darcy's flow equation must have the coefficient of permeability as a function of the negative pore-water pressure (or matric suction).

$$
v_x = k_x(u_w)\frac{dh}{dx}; \, v_y = k_y(u_w)\frac{dh}{dy}
\qquad [7]
$$

The water content, to a large degree, controls the cross-sectional area of the soil available for water flow. The water content of an unsaturated soil decreases as a soil desaturates. The dependence of the coefficient of permeability on the water content is based on the assumption that water can only flow through the wetted portion of the soil. An integration along the soil-water characteristic curve provides a measure of the quantity of water in the soil.

Gardner (1958) proposed a form for the permeability function for an unsaturated soil. The equation required two soil properties for its definition; one related to the air entry value of the soil and the other related to the rate at which the soil desaturates. Other permeability functions have subsequently been suggested which allow greater flexibility in defining the coefficient of permeability of an unsaturated soil over a wider range of soil suctions (Fredlund et al., 1994). Engineers can now select the type permeability function to be used in a numerical solution for an unsaturated flow problem in a manner similar to the way different methods of slope stability analyses can be selected.

The steady state, partial differential equation which must be solved for the flow of water through a saturated soil is obtained by satisfying conservation of mass for a representative elemental volume. When coefficients of permeability in the x- and y- directions are equal, the LaPlacian partial differential equation (i.e., Eq. 8) is obtained which can readily be solved using the flow net technique. Numerical methods are more commonly used and can easily handle cases of greater heterogeneity.

$$k_x \frac{\partial^2 h}{\partial x^2} + k_y \frac{\partial^2 h}{\partial y^2} = 0 \qquad [8]$$

The steady state, partial differential flow equation for an unsaturated soil is also formulated by satisfying conservation of mass for a representative elemental volume. The partial differential equation is nonlinear because the flow law is nonlinear. The

$$k_x(u_w) \frac{\partial^2 h}{\partial x^2} + k_y(u_w) \frac{\partial^2 h}{\partial y^2} + \frac{\partial k_x(u_w)}{\partial x} \frac{\partial h}{\partial x} +$$

$$\frac{\partial k_y(u_w)}{\partial x} \frac{\partial h}{\partial x} = 0 \qquad [9]$$

pore-water pressure changes from point to point in the soil and these changes are reflected in the partial differential flow equation. Therefore, an unsaturated soil is similar to a heterogeneous soil with a constantly changing coefficient of permeability.

Numerical methods such as the finite element technique, have become a necessary tool for solving saturated/unsaturated flow problems. There has been wide acceptance of saturated/unsaturated modeling in engineering practice since the 1980's and today there are several software packages available which

provide great flexibility in solving seepage problems.

EXTENSION OF THE THEORY OF VOLUME CHANGE TO UNSATURATED SOILS

The volume change constitutive behavior of an unsaturated soil can be written in several possible forms (Fredlund, 1979). The classical soil mechanics form for the one-dimensional, K_o loading of a saturated soil involved the relationship between void ratio and effective stress.

$$de = a_v d(\sigma_y - u_w) \qquad [10]$$

Similar classical soil mechanics forms for the volume change constitutive relations for an unsaturated soil involve the use of void ratio, e, water content, w and/or degree of saturation, S. The need for two independent constitutive relations for an unsaturated soil can be demonstrated through the differentiation of the basic volume-mass relationship.

$$\int_{e_o}^{e_f} S \, de + \int_{S_o}^{S_f} e \, dS = G_s \int_{w_o}^{w_f} dw \qquad [11]$$

The independent stress state variables can be used to formulate the constitutive relations for an unsaturated soil element. The volume change constitutive equation for isotropic loading, $(\sigma_c - u_a)$, written in terms of void ratio, e, is,

$$de = a_t d(\sigma_c - u_a) + a_m d(u_a - u_w) \qquad [12]$$

where:

$a_t = \dfrac{\partial e}{\partial(\sigma - u_a)}$ coefficient of compressibility with respect to a change in $(\sigma - u_a)$

and

$a_m = \dfrac{\partial e}{\partial(u_a - u_w)}$ coefficient of compressibility with respect to a change in $(u_a - u_w)$.

The void ratio and water content constitutive surfaces corresponding to the above equations can be represented graphically as three-dimensional surfaces where the water content, w, or void ratio, e, are plotted against the ordinates, $(\sigma - u_a)$ and $(u_a - u_w)$ (Fredlund and Morgenstern, 1976). The void ratio constitutive surfaces for the loading and

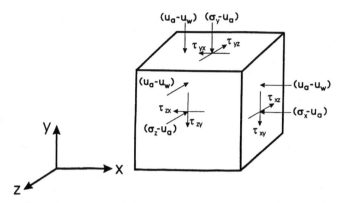

Figure 1. Definition of the state of stress for an unsaturated soil

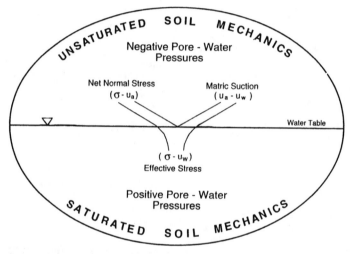

Figure 2. A visualization aid illustrating the stress state for the world of saturated/unsaturated soil mechanics

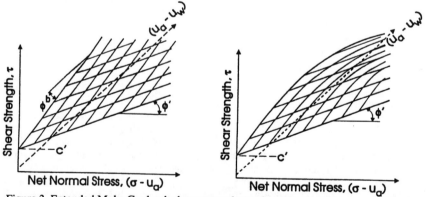

Figure 3. Extended Mohr-Coulomb shear strength envelope showing the strength parameters for a saturated/unsaturated soil; (a) linear strength envelope, (b) nonlinear strength envelope.

unloading of a stable-structured soil are shown in Fig. 4.

The void ratio change can be independent of the water content change for an unsaturated soil element. For a complete volume-mass characterization, a second constitutive relationship is required. The water content constitutive relationship is generally used and can be written as follows for the case of isotropic loading,

$$dw = b_t \, d(\sigma_c - u_a) + b_m \, d(u_a - u_w) \qquad [13]$$

where :

$b_t = \dfrac{\partial w}{\partial(\sigma - u_a)}$ coefficient of water content change with respect to a change in $(\sigma - u_a)$

and

$b_m = \dfrac{\partial w}{\partial(u_a - u_w)}$ coefficient of water content change with respect to a change in $(u_a - u_w)$

Similar sets of constitutive equations can also be written for non-isotropic loading conditions. When the net confining stress on the soil is equal to zero, the resulting equation is a relationship between water content and soil suction, commonly referred to as the soil-water characteristic curve.

$$dw = b_m \, d(u_a - u_w) \qquad [14]$$

The equation is often written in terms of volumetric water content, θ_w.

$$d\theta_w = m_2^w (u_a - u_w) \qquad [15]$$

The nature of the soil-water characteristic curve is shown in Fig. 5. The primary features of the soil-water characteristic curve are the air entry value and the residual water content conditions. The soil exhibits hysteresis upon wetting and drying and the soil suction corresponding to zero water content is approximately 1,000,000 kPa.

A similar approach to the formulation of the constitutive relations can be applied within the critical state framework as proposed by Wheeler (1991), which use three stress parameters, p, q, r, along with two additional state parameters (i.e., water content, w, and specific volume, v). The specific volume, v, and void ratio, e, are related as follows:

$$v = 1 + e \qquad [16]$$

Figure 6 makes use of a visualization aid to summarize the basic constitutive relations for saturated and unsaturated soils for the three classic areas of soil mechanics.

THE MEASUREMENT OF UNSATURATED SOIL PROPERTIES

The measurement of saturated soil parameters has developed to the point where testing has become routine in commercial laboratories. The laboratory measurement of unsaturated soil parameters has been developed at universities and research laboratories but commercial testing has, in general, been found to be too costly for most engineering projects. Unsaturated soil properties also become nonlinear soil property functions which require more testing for their definition. While the ability to measure the unsaturated soil properties associated with constitutive relations is important, the laboratory testing of unsaturated soils may never enjoy the same level of routine as has been the case for saturated soils. The possibilities surrounding the indirect estimation or prediction of unsaturated soil property functions will be discussed later in this paper.

The implementation of unsaturated soil mechanics also requires the ability to control and verify behavior in the field. This means that it is important to be able to monitor changes in the stress state in the field. The task of monitoring soil suction insitu has proven to be an ongoing challenge for geotechnical engineers. Some of the significant steps forward in measuring soil suction are mentioned below.

MEASUREMENT OF SOIL SUCTION

The direct measurement of matric suction values (insitu and in the field) from zero up to several atmospheres provides the greatest challenge for the geotechnical engineer. The recent developments of Ridley (1993) and Guan and Fredlund (1997) have given new hope that the direct measurement of matric suction may become a part of engineering practice. The key to producing a sensor for the direct measurement of high suctions has been the conditioning of the water used in the sensor and the care in manufacturing the sensor. Figure 7 shows a cross-section of the sensor developed by Guan and Fredlund (1997).

The indirect measurement of soil suction is also of great value in implementing unsaturated soil mechanics. The use of thermal conductivity sensors (Fredlund, 1992), has shown encouraging success but the technique requires further research. The research is required on all aspects of the sensor

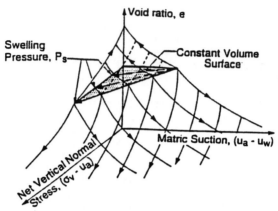

Figure 4. Void ratio constitutive surfaces for loading and unloading of a stable-structured soil.

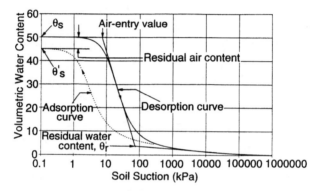

Figure 5. Definition of variables associated with the soil-water characteristic curve.

design ranging from the manufacturing of the ceramic to the performance of the electronic components.

The measurement of total suction is also of good potential in cases where the soil suctions are relatively large. Both the thermocouple psychrometers and the filter paper technique have been used with similar degrees of success in practice The techniques are simple to use but produce considerable scatter in the data. More research, particularly on the use of the filter paper technique is justified. The calibration curve for the filter paper is in reality a soil-water characteristic curve for filter paper as shown in Fig. 8.

The osmotic suction does not appear to be commonly measured but can be successfully determined using the squeezing technique (Krahn and Fredlund, 1972).

LABORATORY MEASUREMENT OF SHEAR STRENGTH VERSUS SUCTION

The laboratory measurement of the shear strength of a soil versus soil suction can be performed using either a triaxial apparatus or a direct shear apparatus. In each case, the equipment must be modified to allow for the independent measurement (or control) of the pore-air and the pore-water pressures (Fredlund and Rahardjo, 1993). The pore-air pressure can be applied through a coarse or extremely low air entry value ceramic. The pore-water pressures can be measured or controlled below a high air entry disk which is usually sealed with epoxy to the lower base platen.

The direct shear apparatus has proven to be superior for testing unsaturated soils because the reduced length of drainage path provides for more rapid equalization of the pore-water pressures. Even so, testing in the direct shear apparatus will often

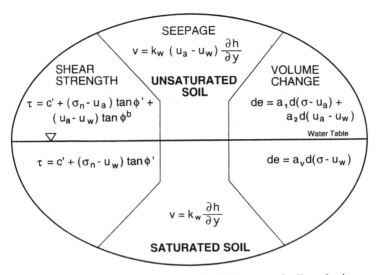

Figure 6. Constitutive relations for the classic areas of soil mechanics

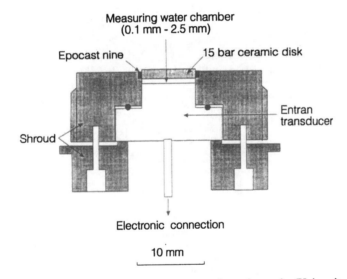

Figure 7. Direct measurement, high range matric suction probe (University of Saskatchewan design).

require two days for equalization under the applied stress state and then one day to apply the shear stress to failure. Since the shear strength relationship is nonlinear, several specimens or multi-stage testing is required. Triaxial testing can require considerably more time. The end result is that unsaturated soil testing is time consuming and the testing procedures require that great care be exercised.

LABORATORY MEASUREMENT OF COEFFICIENT OF PERMEABILITY VERSUS SUCTION

Steady state and unsteady state testing procedures have been successfully used to measure the coefficient of permeability of a soil under varying

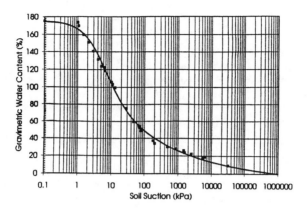

Figure 8. The soil-water characteristic curve for Whatman No. 40
filter paper used as a calibration curve to measure soil suction
(modified from Hamblin, 1981).

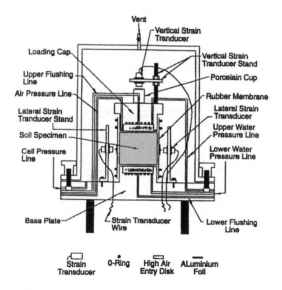

Figure 9. Triaxial permeameter for determining the coefficient of
permeability of unsaturated soils (Huang, 1994).

applied soil suctions (and therefore varying degrees
of saturation) (Fredlund and Rahardjo, 1993).
Unsteady state testing procedures require the
application of a transient flow analysis in the
interpretation of the data.

The steady state testing procedure is relatively
easy to perform provided the applied suctions do not
exceed one atmosphere of suction. Each specimen
may require one or more weeks for testing at
different applied suctions. Figure 9 shows the

apparatus used by Huang (1994) for the steady state
permeability testing. The limits on the range for
which the coefficient of permeability can be
measured are controlled by the permeability of the
high air entry disk (i.e., the highest coefficient of
permeability) and the accuracy of the smallest flow
measurement. Extending the apparatus beyond 100
kPa of suction requires the use of an axis-translation
technique.

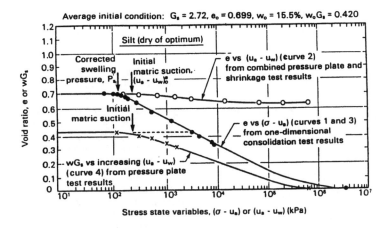

Figure 10. Volume change relationships for an unsaturated, compacted silt (Ho et al., 1992).

A more complete summary of steady state and transient flow, laboratory tests which can be performed to quantify the unsaturated coefficient of permeability of a soil can be found in the reference by Benson and Gribb (1997).

LABORATORY MEASUREMENT OF VOLUME CHANGE PROPERTIES AND THE SOIL-WATER CHARACTERISTIC CURVE

Three independent laboratory tests can be performed to define the extreme limits on the overall volume change and the water content constitutive relationships. The first laboratory test defines the basic relationship between void ratio and net total stress, $(\sigma - u_a)$. This relationship can be defined under zero suction conditions using a conventional one-dimensional oedometer test or a triaxial test (isotropic test). The test results are referred to as the compression curve for the soil. This relationship is equivalent to the water content versus net total stress curve.

The second laboratory test involves the use of the pressure plate apparatus to define the relationship between gravimetric or volumetric water content and matric suction, $(u_a - u_w)$. The test is most commonly performed with the net total stress equal to zero although testing can also be performed under various applied total stresses. The test results are referred to as the soil-water characteristic curve, SWCC, for the soil. At high suctions, (i.e., up to 1,000,000 kPa), vacuum desiccators can be used to complete the water content versus total suction relationship. Reference will later be made to the soil-water characteristic curve because it becomes the basis for

indirectly estimating the unsaturated soil property functions.

The third laboratory test is the shrinkage test for defining the ratio of the slope of the void ratio and the water content change of the soil in response to a change in soil suction. Data from the above three tests can be used to define the extremes or limits on the void ratio and water content constitutive surfaces. Figure 10 shows typical data for a compacted silt (Ho et al, 1992).

Little research has been done to define the shape of the constitutive surface between the extremes or limits, mentioned above. The defining of the remainder of the constitutive surfaces is generally assumed to have a particular character but this is a subject requiring further research. Research is required on soils with both a stable-structure and a meta-stable structure.

INDIRECT ESTIMATION OF UNSATURATED SOIL PROPERTY FUNCTIONS

The reason that unsaturated soils respond differently to changes in pore-water pressure than to changes in total stress, is primarily due to the different cross-sectional area over which each of the stresses act. The degree of saturation of the soil provided the first attempt at accounting for the difference between the effect of the two stress state variables (Bishop, 1959). More recently, the soil-water characteristic curve has been used for estimating the change in soil properties which must be associated with the use of each of the stress state variables (Fredlund, 1996; Fredlund, 1997). Several mathematical equations have been proposed for the soil-water characteristic curve (Fredlund and Xing, 1994). The equation proposed by Fredlund and Xing (1994) simulates the soil-water characteristic curve over the entire range of soil suctions up to 1,000,000 kPa. The equation is as follows:

$$\theta = C(u_a - u_w) \frac{\theta_s}{\left\{ \ln \left[e + ((u_a - u_w)/a)^n \right] \right\}^m}$$ [17

and

$$C(u_a - u_w) = 1 - \frac{\ln\{1 + (u_a - u_w)/(u_a - u_w)_r\}}{\ln\{1 + 1000000/(u_a - u_w)_r\}}$$

The change in shear strength as a soil desaturates is related to the amount of water over the cross-sectional area of the failure plane. There must first be an assumption made regarding the relationship between the area representation of the amount of water and the volume representation of the amount of water in the soil. The assumption required is called Green's Theorem (Fung, 1977), which applies for a "Representative Elemental Volume", REV. It is then possible to use the soil-water characteristic curve as a means of estimating the shear strength function of an unsaturated soil.

The change in coefficient of permeability as a soil desaturates is related to the amount of water in the soil. The function defining the change in coefficient of permeability will not be directly related to water content due to tortuosity changes and other effects. However, the soil-water characteristic curve can provide the basis for estimating the coefficient of permeability function of an unsaturated soil.

The change in volume of a soil as it desaturates can also be related to the amount of water in the soil. Therefore, the soil-water characteristic curve can once again be used to estimate the volume change function of an unsaturated soil with respect to a change in soil suction. The volume change constitutive relations are quite complex and have not, as yet, been fully formulated.

The large amount of data available on soil-water characteristic curves should prove to be of value in assisting practicing engineers in the prediction of unsaturated soil property functions. The data needs to be put into a knowledge-based software system for easy use in engineering practice. There also appears to be good promise for the estimation of the soil-water characteristic curve from grain size distribution curves (Fredlund, 1997).

ESTIMATION OF THE SHEAR STRENGTH FUNCTION

The relationship between soil suction and the shear strength of an unsaturated soil is referred to as the shear strength function. This relationship can be predicted with sufficient accuracy for many engineering problems through a knowledge of the saturated shear strength parameters and the soil-water characteristic curve.

The analysis is started with the understanding that a change in pore-water pressure produces the same change in shear strength as a change in total stress (under drained conditions), as long as the soil is saturated. The ϕ^b angle corresponding to suctions below the air entry value of the soil, will be equal to ϕ'. As the soil desaturates, the ϕ^b angle will decrease because water no longer covers the entire void space on the failure plane. The decrease in shear strength is related to the volumetric water content of the soil; however, the relationship need not be directly proportional to volumetric water content.

Two procedures have been suggested for predicting the shear strength function (Vanapalli et al, 1996). One procedures involve assuming a residual water content and the other procedure involves an assuming an additional soil property, during the integration process along the soil-water characteristic curve.

The first procedure involving the use of an estimated residual suction, performs integration along the soil-water characteristic curve from a suction of zero up to the residual soil suction. The second procedure involves the use of the entire soil-water characteristic curve but applies a p power to the dimensionless volumetric water content during integration. The following shear strength expression as a function of matric suction and the effective angle of internal friction was proposed by Fredlund et al (1995).

$$\tau(u_a - u_w) = c' + (\sigma - u_a)\tan\phi'$$
$$+ \tan\phi' \int_0^{u_a - u_w} [\Theta(u_a - u_w)]^p \, d(u_a - u_w)$$ [18]

The use of the p factor appears to give a better fit between the predictions and experimental data than does the method using an estimate of the residual suction. The p factor appears to range from 1 to 5 and appears to increase with the plasticity of the soil. A p factor of 1 can often be used for the lower suction range.

Figure 11 shows a comparison between the measured and predicted shear strength function for a glacial till (Vanapalli et al, 1996). The shear strengths were measured using a modified direct

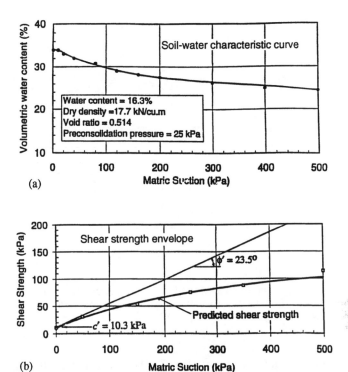

Figure 11. Prediction of the unsaturated shear strength function from the soil-water characteristic curve; (a) soil-water characteristic curve, (b) shear strength envelope (Vanapalli et al., 1996).

shear box. The predictions were made using a *p* factor of 2.2 during integration.

ESTIMATION OF THE COEFFICIENT OF PERMEABILITY FUNCTION

The relationship between soil suction and the coefficient of permeability of an unsaturated soil is referred to as the permeability function (Huang et al, 1994). This relationship can be predicted with sufficient accuracy for many engineering problems through a knowledge of the saturated coefficient of permeability and the soil-water characteristic curve.

The analysis is started with the understanding that the saturated coefficient of permeability of the soil applies for suctions below the air entry value of the soil. The remainder of the permeability function is predicted by integrating along the soil-water characteristic curve. The numerical integration is performed by dividing the soil-water characteristic curve into several water content increments. The integration is performed over the range from the saturated volumetric water content to the residual

water content of the soil. It is more convenient to perform the integration on the logarithmic scale of soil suction. It is also possible to improve the flexibility (and accuracy) of the predictions by applying a *q* factor to the dimensionless water content in much the same way as the *p* factor was applied in the shear strength prediction. A *q* factor of 1 can generally be used for the lower suction range.

Figure 12 shows a comparison between the measured and predicted coefficient of permeability function for a silt (Huang, 1994). The coefficients of permeability were measured in a specially designed permeameter. The predictions were made based on an estimated residual water content.

ESTIMATION OF THE WATER STORAGE FUNCTION

A modulus defining the amount of water stored in the soil is required when performing an unsteady state or transient seepage analysis on saturated/unsaturated soils. For a saturated soil, the

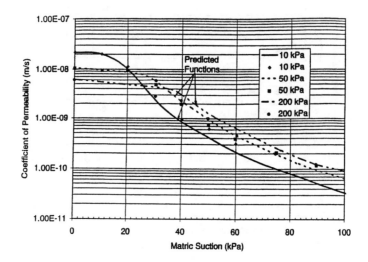

Figure 12. Comparison of a measured coefficients of permeability and permeability functions predicted from the soil-water characteristic curves, for a silt preconsolidated at 10, 50 and 200 kPa, respectively (Huang, 1994).

water storage property is referred to as the coefficient of volume change, m_v (i.e., the relationship between effective stress and volume change). When a suction is applied to the soil, the voids can desaturate. For unsteady state seepage analysis in an unsaturated soil, it is the slope of the soil-water characteristic curve which defines the amount of water stored in the soil.

The relationship between soil suction and the water storage modulus of an unsaturated soil, m_2^w, is referred to as the water storage function. This relationship can be predicted with sufficient accuracy for many engineering problems through a knowledge of the saturated coefficient of volume change and the soil-water characteristic curve. The water storage modulus, m_2^w, can be computed as the slope of the soil-water characteristic curve.

ESTIMATION OF THE VOLUME CHANGE FUNCTION

The relationship between soil suction and the volume change of an unsaturated soil is referred to as the volume change function. To-date, little research has been done on an analytical prediction for the volume change function. However, it is anticipated that this relationship could also be predicted with sufficient accuracy for many engineering problems through the use of the compression curve for the saturated soil and the soil-water characteristic curve.

In addition to predicting the extremes or limiting conditions of the constitutive surfaces, it is necessary to be able to estimate the entire volume change and water content change constitutive surfaces. This may prove to be a fairly difficult task.

USE OF UNSATURATED SOIL PROPERTY DATABASES

To-date, little use has been made of soil property databases in classical soil mechanics. Soil tests have been performed for a specific purpose and the use of the laboratory test results for approximating the soil properties at other sites has been limited. Alternatively, it has been shown that the soil property functions only need to be approximated for many geotechnical engineering analyses involving unsaturated soils (Fredlund, 1997). This reality gives rise to the potential use of data bases of soils information on unsaturated soils. Of particular importance is the data related to the measurement of the soil-water characteristic curves.

The pressure plate tests (and pressure membrane tests) used for the measurement of the soil-water characteristic curve (SWCC), originated in the Soil Sciences and Agricultural disciplines. Over the years, thousands of SWCC tests have been performed in many countries of the world. Also, specially designed pressure plate cells have been designed and built which are more appropriate for

44

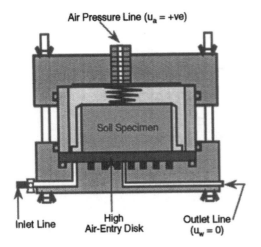

Figure 13 Pressure plate apparatus for testing a single specimen (University of Saskatchewan design)

$$P_p(d) = \cfrac{1}{\ln\left[\exp(1) + \left(\dfrac{g_a}{d}\right)^{g_n}\right]^{g_m}}$$

$$\times\left[1 - \left\{\cfrac{\ln\left(1 + \dfrac{d_r}{d}\right)}{\ln\left(1 + \dfrac{d_r}{d_m}\right)}\right\}^7\right] \qquad [19]$$

where:

$P_p(d)$ = percent passing a particular grain size, d,

g_a = fitting parameter related to the initial break in the grain size curve,

g_n = fitting parameter related to the steepest slope of the grain size curve,

g_m = fitting parameter related to the curvature of the grain size curve,

d = particle diameter (mm),

d_r = residual particle diameter (mm), and

d_m = minimum particle diameter (mm).

use in geotechnical engineering. An example of such a pressure plate cell is shown in Figure 13.

Some of this information has already been assembled into a data base format (Fredlund et al, 1997). As a result, it is possible to generate algorithms which convert the data bases into powerful knowledge-based computer systems. Queries of the data base can be performed to retrieve all the data on soils which have similar grain size distributions, densities, and other properties, as desired. In most cases it is the classification properties of the soil which is most useful in locating relevant information in the data base.

VALUE OF THE GRAIN SIZE DISTRIBUTION CURVE

The grain size distribution curve of a soil can be used in unsaturated soils engineering in a couple of ways. First, the curve can be used to identify similar soils in the data base. Second, the grain size distribution curve has been used as the basis for the direct estimation of the soil-water characteristic curve (Fredlund et al, 1997). For either of the above two purposes, it is useful to first perform a best-fit equation through the grain size distribution data. This can be done using a modification of the Fredlund and Xing (1994) equation used to fit the soil-water characteristic curve. The equation takes the following form (Fredlund et al 1997).

The grain size curve can then be analyzed as an incremental series of particle sizes from the smallest to the largest in order to build an overall soil-water characteristic curve. Once the entire grain size distribution curve is incrementally analyzed, the individual soil-water characteristic curves are superimposed to give the soil-water characteristic curve for the entire soil. Figure 14a shows a grain size distribution curve which has been best-fit using the modified Fredlund and Xing equation. Figure 14b shows the soil-water characteristic curve which has been predicted from the grain size distribution curve along with an experimentally measured set of data. The predictions of the soil-water characteristic curves from the grain size distribution curves have been found to be quite accurate for sands and reasonably accurate for silty soils. Clays, tills and loams are more difficult to predict but the results to-date appear to be promising. Using the predictive algorithm within a knowledge-based data base system allows the algorithm to be better trained with time.

EQUILIBRIUM STATES FOR THE SOIL SUCTION PROFILE

The prediction of the long term stress state boundary conditions has also proven to be difficult for unsaturated soils problems. The ground surface is a dynamic boundary which is controlled largely by the environmental or weather conditions. While the ground surface is truly a moisture flux boundary condition, it is also useful to understand the meaning

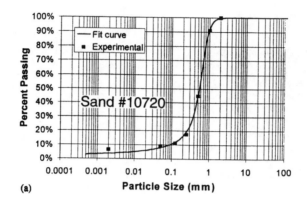

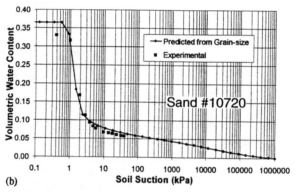

Figure 14. Estimation of the soil-water characteristic curve, a.) grain-size distribution fit for Sand #10720, b.) comparison between experimental and predicted curves for sand #10720.

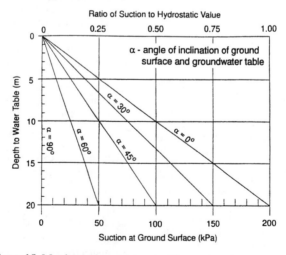

Figure 15. Matric suction at ground surface versus the depth to the water table for situations where the ground surface flux is zero.

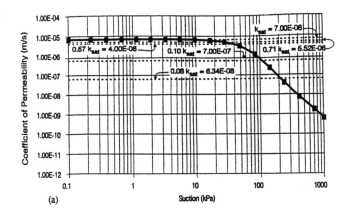

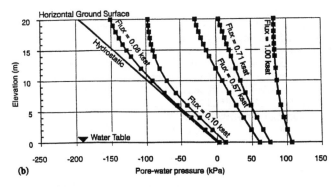

Figure 16. Effect of steady state infiltration on the matric suction at ground surface and the depth of the water table; (a) coefficient of permeability function, (b) change in pore-water pressures for various flux conditions.

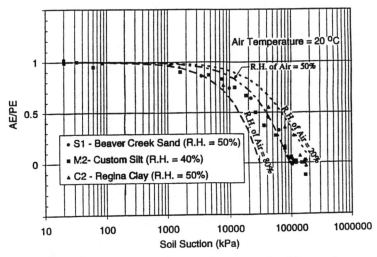

Figure 17. The relationship between soil suction and the ratio of the actual evaporation to the potential evaporation from the ground surface.

47

of the long-term, equilibrium conditions. Croney et al, 1958, made reference to the importance of the location of the groundwater table, while at the same time noting that the suction conditions near to the surface were related to climatic conditions (e.g., Thornwaithe Index). If the groundwater table is within the upper 30 meters, the hydrostatic conditions generally provide a useful reference state for analytical purposes.

Figure 15 shows the matric suction stress state which would exist at the ground surface under hydrostatic conditions (i.e., zero ground surface flux). Also shown are the ground surface matric suction conditions if the ground surface were sloping and the ground water table were parallel to the ground surface. Measured soil suctions should be analyzed with reference to hydrostatic conditions in order to assess whether there is a long-term, net condition of infiltration or evapotranspiration.

A condition of steady state infiltration does not necessarily mean that the suctions in the soil will be significantly reduced. The suctions in the upper portion of the profile will only be significantly reduced when the infiltration rate over a significant period of time approaches the saturated coefficient of permeability of the soil (Fig. 16). Rainfall and infiltration rates should be analyzed relative to the permeability function for the soil.

QUANTIFICATION OF THE GROUND SURFACE MOISTURE FLUX

The primary hindrance to the evaluation of ground surface moisture fluxes is related to the prediction of the actual evaporation (and evapotranspiration) from the ground surface. Recent studies using a coupled heat and mass transport theory for the soil-atmosphere zone have shown that it is possible to estimate the actual evapotranspiration from the ground surface (Wilson, 1990).

Figure 17 shows the relationship between soil suction and the ratio of the actual evaporation to the potential evaporation from a soil surface. Experimental results have shown that the relationship is the same for all types of soil (e.g., sand, silt or clay). This finding is of great significance for the analysis of ground surface, moisture flux boundary conditions and has been successfully applied in engineering design (Wilson et al, 1994).

IMPLEMENTATION AND THE NEED FOR AN OBSERVATIONAL APPROACH

The role of numerical modeling appears to be of ever increasing importance as engineering practice moves towards simultaneously modeling the saturated and unsaturated soil portions of the profile. Numerical models can be quickly entered into the computer and a number of possible scenarios can be reviewed. This provides for a new approach to many geotechnical problems which would previously have to be handled in an empirical manner.

Saturated/unsaturated numerical modeling requires that the unsaturated soil property functions be generated in a reasonably economical manner. Even though estimation procedures are used for determining the unsaturated soil property functions, the end result would appear to provide a distinct advantage over empirical procedures and rules-of - thumb.

It is important that the observational approach be maintained, whenever possible, in engineering practice associated with unsaturated soils. Engineers need to observe field behavior and publish case histories on the performance of engineered structures where saturated/unsaturated soil modeling has been involved. Field instrumentation, including the measurement of soil suction, will assist in providing confidence in the estimation techniques used in design.

REFERENCES

Benson, C. and Gribb, M. (1997), "Measuring unsaturated hydraulic conductivity in the laboratory and the field", Geotechnical Special Publication No. 68, ASCE Geo-Institute, eds by Sandra L. Houston and Delwyn G. Fredlund, Logan, Utah, July 15-19, pp. 113-168.

Bishop, A. W. (1959), "The principle of effective stress", Teknisk Ukeblad, Vol. 106, No. 39, pp. 859-863.

Boussinesq, J. (1885), "Application des potentiels a l'etude de l'equilibre et du mouvement des solides elastiques", Paris, Gauthier-Villard.

Casagrande, A. (1935), "Seepage through dams", Journal of New England Water Works Association, Vol. 51, No. 2, pp. 131 - 172.

Childs, E. C. and Collis-George, N. (1950), "The permeability of porous materials", Proc. of the Royal Society, Vol. 201A, pp. 392-405.

Croney, D., Coleman, J. D. and Black, W. P. M. (1958), "Movement and distribution of water in soil in relation to highway design and performance", Water and its Conduction in Soils, Highway Research Board, Special Report, Washington, D. C., No. 40, pp. 226-252.

Fredlund, D.G. (1979), "Appropriate concepts and technology for unsaturated soils", Second

Canadian Geotechnical Colloquium: Can. Geot. Jour., Vol. 16, No. 1, pp. 121-139.

Fredlund, D. G. (1992), "Background, theory and research related to the use of thermal conductivity matric suction measurements", Advances in Measurement of Soil Physical Properties : Bringing Theory into Practice, SSSA Special Publication No. 30, Soil Science of America, Madison, WI, pp. 249-262.

Fredlund, D. G. (1995), "Prediction of unsaturated soil functions using the soil-water characteristic curve, Proc. of the Bength B. Broms Symposium on Geotechnical Engineering, Singapore, 13-15 December, pp. 113-133.

Fredlund, D. G. (1997), "From theory to the practice of unsaturated soil mechanics", Proc. of the Third Brazilian Symposium on Unsaturated Soils, NONSAT'97, Rio de Janeiro, Brazil, April.

Fredlund, D. G. (1997), "An introduction to unsaturated soil mechanics", Geotechnical Special Publication No. 68, ASCE Geo-Institute, eds by Sandra L. Houston and Delwyn G. Fredlund, Logan, Utah, July 15-19, pp. 1-37.

Fredlund, D. G. and Morgenstern, N. R. (1976), "Constitutive relations for volume change in unsaturated soils", Can. Geot. Jour., Vol. 13, No. 3, pp. 261-276.

Fredlund, D.G. and Morgenstern, N.R. (1977) "Stress state variables for unsaturated soils", ASCE Jour. of the Geot. Engineering Division, DT5, Vol. 103, pp. 447-466.

Fredlund, D.G. and Rahardjo, H. (1993) "Soil mechanics for unsaturated soils", John Wiley & Sons, New York, 560pp.

Fredlund, D. G. , Morgenstern, N. R. and Widger, A. (1978), "The shear strength of unsaturated soils", Can. Geot. Jour., Vol. 15, No. 3 , pp. 313-321.

Fredlund D. G. and Xing, A. (1994). "Equations for the soil-water characteristic curve". Can. Geot. Jour., Vol. 31, No. pp. 521-532.

Fredlund, D. G., Xing, A., Fredlund, D. G., and Barbour, S. L. (1995), "The relationship of the unsaturated shear strength to the soil-water characteristic curve", Can. Geot. Jour., Vol. 33, pp. 440-448.

Fredlund, D.G., Xing, A. and Huang, S. (1994), "Predicting the permeability function for unsaturated soils using the soil-water characteristic curve", Can. Geot. Jour., Vol. 31, No. 4, pp. 533-546.

Fredlund, M. D., Fredlund, D. G. and Wilson, G. W. (1997), "Prediction of the soil-water characteristic curve from grain-size distribution and volume-mass properties", Proc. of the Third Brazilian Symposium on Unsaturated Soils, NONSAT '97, Rio de Janeiro, Brazil, April 22-25, Vol. 1, pp. 13-24.

Fredlund, M. D., Fredlund, D. G. and Wilson, G. W. (1997) "Estimation of unsaturated soil properties using a knowledge-based system", Proc. of the Fourth Congress of Computing in Civil Engineering, ASCE, Philadelphia, PA., June 16-18, pp. 501-510.

Fung, Y. C. (1977), "A first course in continuum mechanics", 2nd edition, Englewood Cliffs, N.J., Prentice-Hall, 340 pp.

Gardner, W. R. (1958), "Some steady state solutions of the moisture flow equation with application to evaporation from a water table", Soil Science, Vol. 85, pp. 228-232.

Guan Yun, Fredlund, D. G. (1997), "Direct measurement of high soil suction", Proc. of the 3rd Brazilian Symposium on Unsaturated Soils, NONSAT '97, Rio de Janeiro, Brazil, April 22-25, Vol. 1, pp.

Hamblin, A. P. (1981), "Filter paper method for routine measurement of field water potential", Jour. of Hydrology, Vol. 53, No. 3/4, pp. 355-360.

Ho, D. Y. F., Fredlund, D. G. and Rahardjo, H. (1992), "Volume change indices during loading and unloading of an unsaturated soil", Can. Geot. Jour., Vol. 29, No. 2, pp. 195-207.

Huang, Shangyan (1994), "Evaluation and laboratory measurement of the coefficient of permeability in deformable, unsaturated soils", Ph.D. dissertation, University of Saskatchewan, Saskatoon, Saskatchewan, Canada, 272 pp.

Huang, Shangyan, Fredlund, D. G. and Barbour, S. L. (1994), "A history of the coefficient of permeability function", Proc. of the Sino-Canadian Symposium on Unsaturated/Expansive Soils, Wuhan, China, June 7-8, pp. 57-80.

Krahn, J. and Fredlund, D. G. (1972), "On total, matric and osmotic suction", Jour. of Soil Science, Vol. 114, No. 5, pp. 339-348.

Ridley, A. M. (1993), "The measurement of soil moisture suction", Ph.D. dissertation, Imperial College, London, U.K., 218p.

Skempton, A. W. (1985), "A history of soil properties, 1717 - 1927", Golden Jubilee Volume, Proc. of the Eleventh Int. Conf. Soil Mech. And Foundation Engineering, San Francisco, pp. 95-121.

Taylor, D. W. (1948), "Fundamentals of soil mechanics", John Wiley & Sons, New York, 700 pp.

Terzaghi, K. (1936), "The shear resistance of saturated soils", Proc. of the First Int. Conf. Soil Mechanics and Foundation Engineering, Cambridge, MA, Vol. 1, pp. 54-56.

Terzaghi, K. (1943), "Theoretical soil mechanics", John Wiley & Sons, New York, 510 pp.

Terzaghi, K. and Peck, R. B. (1948), "Soil mechanics in engineering practice", John Wiley & Sons, New York.

Vanapalli, S., Fredlund, D. G., Pufahl, D. E., and Clifton, A. W. (1996), "Model for the prediction of shear strength with respect to soil suction", Can. Geot. Jour., Vol. 33, pp. 379-392.

Wheeler, S.J. (1991) "Alternative framework for unsaturated soil behavior", Geotechnique, Vol. 41, No. 2. pp. 257-261.

Wilson, G. W. (1997) "Surface flux boundary modelling for unsaturated soils" Geotechnical Special Publication No. 68, ASCE Geo-Institute, eds by Sandra L. Houston and Delwyn G. Fredlund, Logan, Utah, July 15-19, pp. 38-67.

Wilson, G.W. (1990) "Soil evaporative fluxes for geotechnical engineering problems", Ph.D. dissertation, University of Saskatchewan, Saskatoon, Canada.

Wilson, G. W., Fredlund, D. G. and Barbour, S. L. (1994), "Coupled soil-atmosphere modeling for evaporation", Can. Geot. Jour., Vol. 31, No. 2, pp. 151-161.

DEFINITION OF SYMBOLS

a - parameter which is approximately equal to the air entry value

a_v - coefficient of compressibility

a_m - coefficient of compressibility with respect to a change in matric suction

a_s - coefficient of compressibility with respect to a change in net normal stress

b_m - coefficient of water content change with respect to a change in matric suction

b_s - coefficient of water content change with respect to a change in net normal stress

c' - effective cohesion intercept

e - void ratio

e - 2.718.....

f - reference to the final state

G_s - specific gravity of soil solids

h - total hydraulic head

k_{sat} - saturated coefficient of permeability of the soil

$k_w(u_a - u_w)$ - coefficient of permeability with respect to the water phase, as a function of matric suction, $(u_a - u_w)$.

k_x - coefficient of permeability in the x-direction

k_y - coefficient of permeability in the y-direction

m - soil parameter related to the residual water content

m_v - coefficient of volume change for a saturated soil

m_2^w - coefficient of water volume change with respect to a change in soil suction

n - soil parameter related to the rate of desaturation

o - reference to the original state

S - degree of saturation

u_a - pore-air pressure

u_w - pore-water pressure

$u_a - u_w$ - matric suction

v_w - flow rate of water

v_x - velocity of water flow in the x-direction

v_y - velocity of water flow in the y-direction

w - water content

x - horizontal coordinate direction

y - vertical coordinate direction

ϕ' - effective angle of internal friction

ϕ^b - rate of increase in shear strength with respect to a change in soil suction

θ_w - volumetric water content

π - osmotic suction

σ - total stress

σ_c - isotropic confining stress

σ_n - total normal stress on the failure plane

$\sigma - u_a$ - net normal stress

$\sigma - u_w$ - effective stress

τ - shear strength

θ_w - volumetric water content

θ_s - volumetric water content at saturation

$\theta(u_a - u_w)$ - volumetric water content at any suction

$\Theta(u_a - u_w)$ - θ_w/θ_s (dimensionless volumetric water content)

$\partial h_w / \partial y$ - hydraulic head gradient in the y-direction.

Geotechnical Hazards, Marić, Lisac & Szavits-Nossan (eds) © 1998 Taylor & Francis, ISBN 90 5410 957 2

Soil-pile interaction in liquefied deposits undergoing lateral spreading

K. Ishihara
Department of Civil Engineering, Science University of Tokyo, Japan

M. Cubrinovski
Kiso-Jiban Consultants Co., Ltd, Japan

ABSTRACT: For the design of foundation piles subjected to lateral spreading of the ground, it is necessary to specify the lateral force acting on the pile body from the surrounding soil deposit which is moving as a result of liquefaction. In view of many design codes incorporating spring-supported beam system to model the soil-pile interaction, it may be reckoned reasonable to investigate the effects of the lateral flow in terms of how much the spring constant should be degraded as a result of softening of the soil due to liquefaction. In an attempt to clarify this aspect, back-analyses were made for cases of damage to foundations for which field data such as soil conditions and damage features are made available. As a result, it was found that the degree of degradation in the spring constants is dependent on the relative movement of the piles and surrounding soils. Analysis results showed that, for example, if the pile displacement is about 50 % of the surrounding soil, the stiffness in the spring constant should be reduced to 1/1000 ~ 1/100 of the value which is generally assumed in the design practice considering non-liquefied conditions.

1 INTRODUCTION

Design of piles considering the effects of seismic motions is generally performed through the use of a soil-pile interaction model in which a vertically-placed beam is supported by a series of spring elements. The beam represents the performance of the pile, and soil properties are represented by the spring constants. The effects of horizontal seismic motions on piles are allowed for by incorporating a horizontal force at the top of the pile, which is equivalent to the inertia force from superstructures, but for this type of analysis the spring constants are determined for conditions of no softening of the soils due to liquefaction. When liquefaction is of concern the stiffness of the soils is reduced and these effects need to be considered. The Japanese Highway Code stipulate, for example, that for the majority of cases the spring constants be reduced by a factor of 1/6 - 2/3 for making the analyses. However, if it comes to the effects of lateral spreading of once liquefied soils, there has been no requirement stipulated in the code for the design of pile foundations. Concerns have been kindled on these effects since the Kobe Earthquake in 1995 because of the extensive occurrence of damage to foundation piles apparently due to the lateral spreading. When piles are subjected to the lateral flow of once liquefied soils, lateral forces would be applied directly to the pile body throughout the depth of liquefaction. In assessing this force in the design, there would be two types of approach. The first method consists of assessing directly the lateral force on the pile body either based on empiricism or by means of the concept of viscous flow (Chaudhuri et al., 1995). This may be called "Force-based approach". In either way, it would be difficult to introduce a parameter which is indicative of the degree of destructiveness of the ground failure. Thus the specification of the lateral force would have to be made irrespective of whether the ground displacement is destructively large or small. In the second method, the lateral displacement of the ground is specified through the depth of the deposit where the lateral spreading is induced. This prescribed displacement is applied to the spring system inducing lateral forces acting on the pile body. This procedure may be called "Displacement-based approach". One of the advantages of this method is that it is possible to specify the amount of ground displacement which is indicative of the degree of destructiveness or severity of the lateral spreading. However, the choice of the spring constants has a profound influence on the magnitude of the lateral force induced, and as such difficulty is encountered in evaluating correctly this value. It is expected that the spring constant in laterally spreading soils is much smaller than that in the case of the back and forth movement of soils as stipulated in the Japanese Code of Highway Design as mentioned above. While the

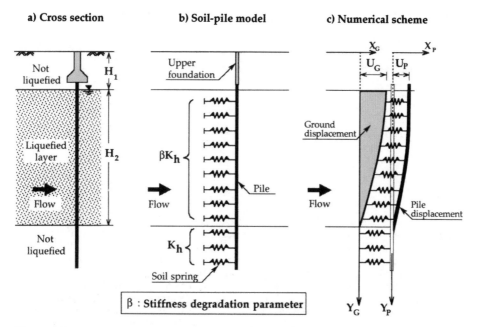

a) Cross section   **b) Soil-pile model**   **c) Numerical scheme**

Figure 1. Soil-pile model and numerical computation scheme

code basically stipulates 1/6 to 2/3 reduction in the coefficient of subgrade reaction, the reduction is anticipated to be much more drastic if the effects of lateral spreading are allowed for. Thus, it becomes necessary to know the order of magnitude by which the conventionally used coefficient of subgrade reaction should be degraded to account for the interaction phenomenon taking place in the course of the lateral spreading. Some preliminary calculation was performed in a previous paper (Ishihara, 1997), but in needs of more data additional case studies were undertaken recently in the above context and the results of such investigation are presented below.

2 SUBGRADE REACTION MODEL

In this interaction model, the soil is assumed to be a Winkler-type material in which the lateral force, F, is proportional to the relative displacement between the pile and the soil,

$$F = \beta k \ (U_G - U_P) \qquad (1)$$

where U_G and U_P are lateral displacements of the ground and pile, respectively, and k is the coefficient of subgrade reaction. If the soil is brought to a state of liquefaction thereby developing lateral spreading, the stiffness of the soil would be reduced drastically leading to a reduction of the k-value. The degree of this stiffness reduction is expressed by β in Eq. (1)

which will be referred to as the "Stiffness degradation parameter". This may be alternatively interpreted as representing highly nonlinear characteristics of the soil in a state of lateral spreading. The soil-pile interaction model adopted in this study is displayed schematically in Fig. 1. Main features of the model are outlined as follows.

(1) The stiffness of the springs is assumed to decrease by a factor β upon liquefaction and lateral spreading of the soil through the depth of liquefaction, H_2. The spring constant in the underlying non-liquefied zone is assumed not to be degraded.

(2) Generally, there is an unliquefied layer to a certain depth H_1 near the surface. This depth may be roughly defined to be equal to the depth to the ground water table. The movement of this soil mass in the unliquefied layer may be modeled generally in two ways. One is to assume the surface layer to move in unison with the underlying liquefied stratum. In this case, it may be assumed that an earth pressure is applied to the upper portion of the pile in the same direction as the flow of the underlying liquefied soil layer, as illustrated in Fig. 2a. In this type of surface soil movement, the displacement of the pile head is considered always smaller than the overall movement of the surrounding soil. The other conceivable mode of movement would be to assume the surface soil to displace by an amount smaller than that occurring in the underlying deposit, as illustrated in Fig. 2b. In this case, the pile head is considered to move by an

52

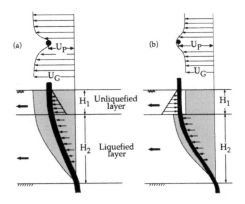

Figure 2. Modes of surface soil displacement and direction of force application

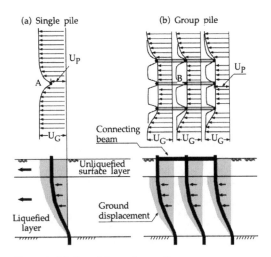

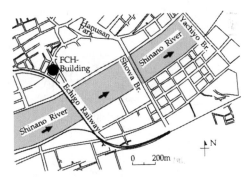

Figure 3. Modes of the surface soil movement

amount equal to or greater than the surrounding soil in the upper layer and an earth pressure is postulated to act in the direction opposite to the soil movement. When the pile head moves by an amount equal to the surrounding soil, the earth pressure becomes equal to zero. It is to be noticed that, whichever the mode of displacement is, the movement of the soil in the underlying liquefied deposit should be assumed to be greater than the displacement of the pile. Otherwise, the direction of the force to the pile body is reversibly oriented and the basic assumption of the liquefied soil pushing the pile would become invalid.

It appears that the first type of the surface soil movement might be encountered in the case such as a single pile where there is no object in its vicinity as illustrated in Fig. 3a. In the case of group piles, free movement of the surface soil within the pile group might be restricted due to constraints from the surrounding piles, or by the presence of horizontal connecting beams as illustrated schematically in Fig. 3b. Thus, the second type of movement of the surface soil illustrated in Fig. 2b will become valid. It is to be noticed that the mode of displacement of the surface layer would be significantly influenced by various factors including the stiffness of the piles and characteristics of the foundation raft.

It will be assumed in the following back-analysis that the constraints as above will not exert any influence on the overall movement and flow in the underlying liquefied soil deposit. Thus, no matter whichever the mode of displacement of the surface layer is, the interaction of the pile in the underlying layer is not influenced.

(3) The ground displacement is specified and given to the springs in the liquefied portion of the soil deposit, and displacement and bending moment of the pile are calculated, whereby the pile is assumed to deform in an elasto-plastic manner.

Figure 4. Location of the building in Niigata investigated by excavation

3 CASE STUDIES

3.1 *Family Court House Building*

In the area of extensive liquefaction at the time of the Niigata Earthquake in 1964, lateral flow of liquefied sands took place exerting deleterious effects on foundations of structures. Among many of those injured by the lateral spreading, the reinforced concrete piles (RC-piles) supporting a three-story reinforced concrete building were those investigated in details by excavating the surrounding soil deposits to a depth of about 10 m (Yoshida and Hamada, 1991). The location of the building is shown in Fig. 4. The movement of the ground as identified by virtue of the air-photo interpretation is also reported in the paper as shown in Fig. 5, where it may be seen roughly that the movement of the ground consequent upon liquefaction was about 1.0 m in the north-east

53

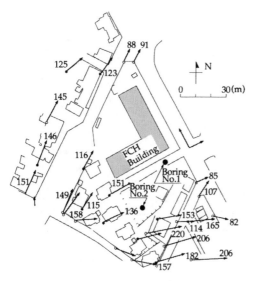

Figure 5. Permanent ground displacements (cm) near the FCH Building (Yoshida and Hamada, 1991)

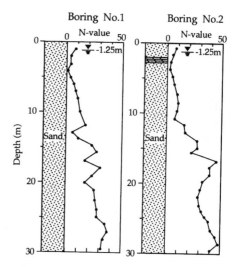

Figure 7. Soil boring data in the vicinity of FCH Building (Yoshida and Hamada, 1991)

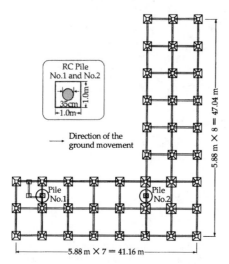

Figure 6. Foundation plan of the FCH Building (Yoshida and Hamada, 1991)

direction at the place where the Family Court House was located. The layout of the piled foundations are displayed in Fig. 6 where the exact locations of two investigated footings (No. 1 and No. 2) are indicated. These footings 1m x 1m in size were supported by single reinforced concrete piles 35 cm in diameter as indicated in the inset of Fig. 6. The investigations were conducted in 1989, 25 years after the earthquake. Such an opportunity presented itself

when the building was completely demolished. Sheet piles were driven to enclose the foundation pile and while being dewatered, the ground was excavated to expose the complete body of the pile as it had been buried.

Soil profiles were investigated in 1987 by means of borings at two sites near the building as indicated in Fig. 5. Results of the soil investigation are shown in Fig. 7 where it can be seen that deposits of clean sand prevail to a depth of 30 m with SPT N-values less than 10 to a depth of about 12 m.

The outcome of field measurements on the lateral deformation of the two piles made in the excavated pits is demonstrated in Fig. 8. According to the design chart, both piles were designed to have a length of 14 m, but perhaps because of a stiff sand layer encountered, the pile driving for the Pile No. 1 must have been stopped at a depth of 10-12 m. This situation is conjectured from the absence of any injury to the Pile No. 1 around the bottom of the excavated pit (Yoshida and Hamada, 1991). The Pile No. 2 appeared to have been driven to a deeper stratum with the design length of 14 m as evidenced by many cracks that had developed near the bottom of the pit. This is indicative of a large bending moment having exerted during the lateral flow of the liquefied sand. The Pile No. 2 was completely chopped off about 2 m below the pile head, as indicated by the discontinuity in the measured pile deformation shown in Fig. 8.

When the excavation was conducted in 1989, the pit was dug first to a depth of 2.0 m while breaking and removing the footing slab. According to the design chart, the pile top was located at an elevation 1.65 m below the ground surface. The elevation of

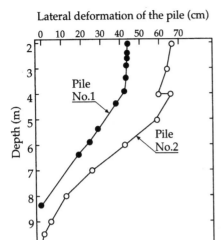

Lateral deformation of the pile (cm)

Figure 8. Lateral deformation of Piles No. 1 and
No. 2 in FCH Building (Yoshida and Hamada, 1991)

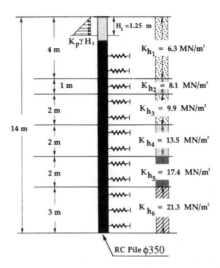

Figure 9. Spring constants for the RC-Pile No. 2 in
FCH Building, Niigata

the ground water table could be somewhat variable,
but based on the design chart and soil investigations it
may be assumed to have been about 1.0 - 1.5 m deep
as accordingly indicated in Fig. 7. The feature in
which the pile head was connected with the footing
slab is not known unfortunately. When the piles
suffered injury due to the lateral flow at the time of the
1964 Niigata Earthquake, the soil mass above the
ground water table may be assumed to have moved
together with the liquefied soil deposit underlying the
ground water table. This appears to have been the
case, however, in the free field conditions where there
was no constraint near the piles. In the foundation
system of the FCH Building, a number of footing
slabs were connected to each other by underground
horizontal beams about 1.0 m in height and 0.8 m in
width. Existence of such beams is considered to have
exercised some constraint on the movement of the
surface soil deposit in the vicinity of the single piles
being studied. Thus, it may well be assumed that the
deformation pattern as illustrated in Fig. 2b has
occurred at the time of the lateral spreading in 1964
Niigata Earthquake. If this assumption is valid the
lateral force must be acting in the direction opposite to
that of the flow and its magnitude may be assumed
to be equal to the passive earth pressure $K_p \gamma H_1^2 / 2$. It
is also envisaged that the near surface deposit might
have been ravaged by violent motions of the ground
due to liquefaction involving cracking and spurting of
mud water. Thus, the earth pressure acting on the pile
would have been much smaller than that mobilized in
the passive earth pressure conditions. As an extreme
case, there would have been no lateral force acting on
the pile-top. This could be the case where the pile and

surrounding surface-soil move together without any
relative displacement. With the considerations as
above, it will be of interest to see the effects of the
lateral force by assuming two cases, that is, zero earth
pressure and passive earth pressure $K_p \gamma H_1^2 / 2$.

Though the connection between the footing and
pile top is not known, it may be acceptable to assume
for simplicity that the pile-footing system is modeled
by a single continuous beam throughout the depth as
illustrated in Fig. 1. The single pile is postulated to be
supported by a series of springs and subjected to the
zero or passive earth pressure to a depth of
$H_1 = 1.25$ m as illustrated in Fig. 9. Regarding the
lateral force below the depth H_1, it is assumed to have
been applied through the springs which are deformed
in consistence with an amount of displacement
specified at each depth of the liquefied layer.

In the analysis of the Piles No. 1 and No. 2 in the
building of the Family Court House, the spring
constants were evaluated by using an empirical
formulae for the coefficient of subgrade reaction k
stipulated in the Japanese Code of Bridge Design.

The spring constant K_h was determined by
multiplying the coefficient of subgrade reaction k by
an effective width of the pile

$$K_h = k\,d \qquad (2)$$

The value of K_h evaluated in this way with reference
to the SPT N-value of Fig. 7 is shown in Fig. 9 for
the representative soil profile at the site of Family
Court House.

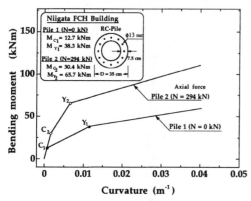

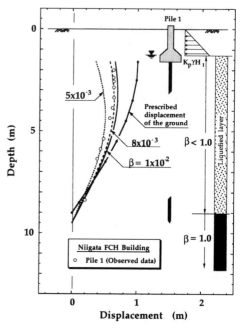

Figure 10. Bending moment versus curvature relation for the RC-piles in the FCH Building

In making the analysis, it is necessary to specify the relation between the bending moment and curvature of the beam representing the flexural characteristics of the reinforced concrete pile. This relation was established based on the cross section and deformation characteristics of the piles as shown in Fig. 10. The analysis of the piles undergoing the lateral displacement were performed by a series of computations as described below.

(1) First, the lateral displacement is prescribed throughout the depth. In the case of the Family Court House site, the sand deposits are postulated to have developed liquefaction to a depth of 9.0 m and the entire soil mass above it to have moved with a cosine distribution of lateral displacement with its maximum value of 1.0 m at a depth of 1.25 m.

(2) The displacement of the ground as postulated above was applied to the equivalent linear spring model and analysis was made for the spring-supported beam to obtain the lateral displacement of the pile. In conducting the analysis, the pile head was assumed to be free to move horizontally whereas the lower end of the pile was assumed to be free only to rotate for the Pile No. 1 at the depth of 9.5 m and fixed for the Pile No. 2 at the depth of 14.0 m. In the analysis, the spring constant for the supposedly liquefied layer was reduced in a wide range taking the stiffness degradation parameter of $\beta = 5 \times 10^{-4} - 1 \times 10^{-2}$ as defined by Eq (1).

(3) As briefly mentioned above, the passive earth pressure was applied in the upper portion of the pile to a depth of $H_1 = 1.25$ m, where the angle of internal friction of $\phi = 30°$ and $\gamma_t = 17.6$ kN/m³ were assumed to be the case. In the remaining portion of the pile to the depth of 9.0 m, the displacement with a cosine distribution was applied to the springs. Analysis was made as well for the case of no earth pressure in the upper portion of the piles with the same conditions otherwise.

Figure 11. Lateral displacements of Pile No. 1 computed for different β-values: Analysis with passive earth pressure (Niigata FCH Building)

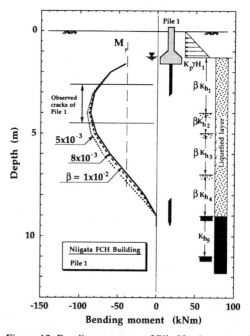

Figure 12. Bending moments of Pile No. 1 computed for different β-values: Analysis with passive earth pressure (Niigata FCH Building)

56

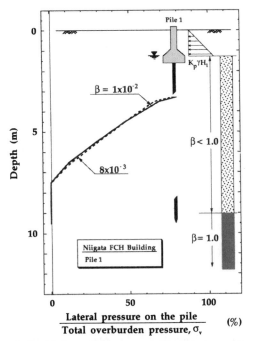

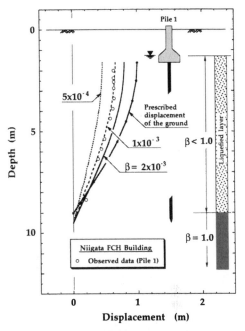

Figure 13. Lateral pressure on Pile No. 1 computed for different β-values: Analysis with passive earth pressure (Niigata FCH Building)

Figure 14. Lateral displacements of Pile No. 1 computed for different β-values: Analysis without earth pressure (Niigata FCH Building)

(4) From the layout of the piled foundations shown in Fig. 6, the Pile No. 1 and No. 2 may be considered as single piles and as such the analysis was conducted for a pile with a width of d = 35 cm.

The outcome of the analysis for the Pile No. 1 is presented in Fig. 11 through Fig. 16 and the results for the Pile No. 2 are shown in Figs. 17 through 22.

As shown in Figs. 11 and 14, the lateral displacement for the Pile No. 1 is assumed to be zero at the depth of 9.5 m. Assuming this to be reasonable, Fig. 11 shows that the case of analysis with passive earth pressure using the degradation parameter $\beta = 8\times10^{-3}$ can provide the best fit to the displacements observed in the in-situ investigation. It is to be noticed that the displacement of the pile at the top of the liquefied layer or at the bottom of the unliquefied surface layer was shown to be 0.6 m which is smaller than the value of 1.0 m of the ground displacement prescribed in the analysis at this elevation. The degradation parameter $\beta = 8\times10^{-3}$ is plotted in Fig. 31 versus the relative displacement $U_G - U_P$ normalized to the thickness of the liquefied layer, H_2.

In the case of the analysis with no earth pressure in the surface layer, the results of calculations are presented in Fig. 14. It may be seen that the best fit between the observed and calculated behaviour is achieved with $\beta = 1\times10^{-3}$ where the lateral pile-top

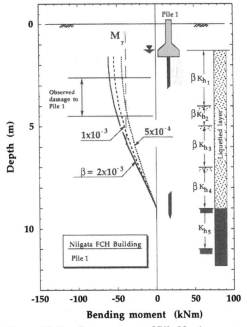

Figure 15. Bending moments of Pile No. 1 computed for different β-values: Analysis without earth pressure (Niigata FCH Building)

57

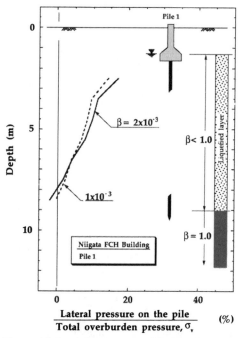

Figure 16. Lateral pressure on Pile No. 1 computed for different β-values: Analysis without earth pressure (Niigata FCH Building)

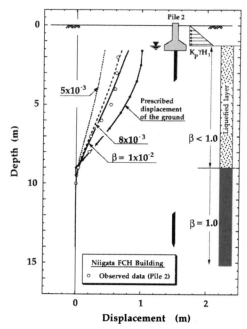

Figure 17. Lateral displacements of Pile No. 2 computed for different β-values: Analysis with passive earth pressure (Niigata FCH Building)

displacement is obtained also as 0.6 m in the analysis. The value of the normalized relative displacement $(U_G - U_P)/H_2$ is plotted in Fig. 31 versus the value of β.

Thus, if the liquefied layer is assumed to have induced lateral spreading overall in the area of FCH Building without any constraint from surrounding objects, a relative displacement of 1.0 - 0.6 = 0.4 m is considered to have occurred at the top of the liquefied layer. Then, the stiffness degradation parameter is reckoned to have taken a value between $\beta = 1 \times 10^{-3}$- 8×10^{-3} which depends generally upon the magnitude of the lateral pressure acting on the upper portion of the pile in the unliquefied surface layer.

The bending moment induced in the pile body is shown in Fig. 12 for the case of passive earth pressure application. It may be seen that at depths of 1.5 m to 7.5 m the induced bending moment far exceeds the yield value M_y at which larger cracks are considered to develop in the concrete. Observation of the in-situ investigation revealed that predominant cracks existed at the depths of 2.5 m to 4.5 m as accordingly indicated in Fig. 12. In the case of no earth pressure application, the range of depth at which the calculated bending moment exceeds the yield value coincides reasonably well with the depth of observed cracks as indicated in Fig. 15, but the value of the calculated moment for this case is smaller as com-

pared to the moment computed by considering the passive earth pressure. It is to be noticed that the bending moments are the least affected by the degradation parameter β, but they are significantly influenced by the condition whether or not there is a lateral pressure from the surface layer. It would be of interest to see how much lateral force is applied to the pile body in the lower layer from the surrounding soil deposit when it is undergoing lateral spread. For this reason, the lateral pressure per unit area of the pile body was computed and normalized to the total overburden pressure σ_v at each depth. The results of such analysis are demonstrated in Figs. 13 and 16 where it may be seen that, while the influence of the stiffness degradation parameter β is very small, effects of the lateral pressure in the unliquefied surface layer is significant on the lateral force. In the case of no earth pressure application, the lateral pressure on the pile is found to be somewhere between 5 and 20 % of the total overburden pressure whereas that computed by considering the passive earth pressure is predominantly in the range between 10 % and 70 % of σ_v.

The outcome of the analysis for the Pile No. 2 at FCH Building foundation is demonstrated in Figs. 17 through 22. With respect to the pile deformation, the best degree of coincidence is obtained, as shown in Figs. 17 and 20, when a value of $\beta = 8 \times 10^{-3}$ and

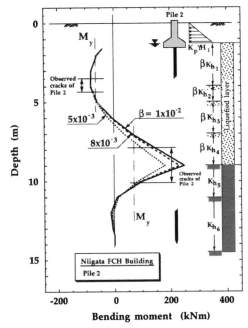

Figure 18. Bending moments of Pile No. 2 computed for different β-values: Analysis with passive earth pressure (Niigata FCH Building)

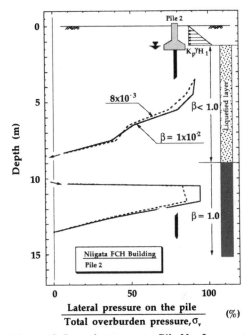

Figure 19. Lateral pressure on Pile No. 2 computed for different β-values: Analysis with passive earth pressure (Niigata FCH Building)

3.3×10^{-3} is assumed for the degradation parameter. Regarding the bending moment and lateral stress, the value of β does not generate much difference, but the magnitude of the earth pressure in the unliquefied surface layer is a dominant factor. In both cases being considered, the computed bending moment is in excess of the yielding moment and far above the crack-inducing moment within the depth range in which cracks were observed at the time of the in-situ investigation. Therefore, the bending moment would not be regarded as a quantity by which validity of the assumption is checked regarding the influence of the earth pressure in the surface layer.

For the Pile No. 2, Figs. 19 and 22 indicate that the lateral pressure on the pile in the liquefied deposit is in the range of 10 to 80 % and 5 to 50 % of the total overburden pressure for the cases with and without earth pressure application, respectively.

3.2 Building in Fukaehama

The damage to foundation piles of a 3-story building at the time of the 1995 Kobe Earthquake is reported by Tokimatsu et al. (1997). The building was situated 6 m inland from the quay wall on a reclaimed fill in Higashi-Nada, Kobe. It was constructed early in 1980's and supported by prestressed concrete piles (PC-piles) 40 cm in diameter and about 20 m long.

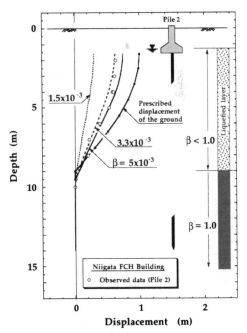

Figure 20. Lateral displacements of Pile No. 2 computed for different β-values: Analysis without earth pressure (Niigata FCH Building)

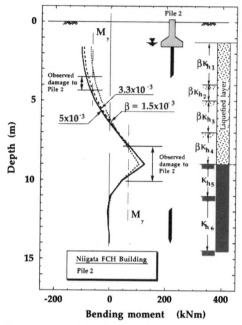

Figure 21. Bending moments of Pile No. 2 computed for different β-values: Analysis without earth pressure (Niigata FCH Building)

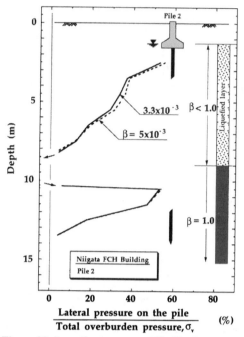

Figure 22. Lateral pressure on Pile No. 2 computed for different β-values: Analysis without earth pressure (Niigata FCH Building)

The building sustained a tilt of 3 degrees due to the lateral spreading of the soils. The piles beneath the foundation were exposed and detailed investigations were carried out in-situ to see features of injury on precast reinforced concrete piles. The investigations included visual observation of cracks around inside walls of the hollow cylindrical piles by means of a television camera and measurements of tilt through the depth by using a slope indicator lowered into the inside hole of the piles.

The plan view of the building foundation is shown in Fig. 23 with precise locations of the piles investigated. The soil profiles obtained at locations of boreholes No. 1 and No. 2 are presented in Fig. 24 where it may be seen that the reclaimed fills of the Masado soil exist to a depth of about 8.5 m. Liquefaction and lateral spread seem to have occurred in this reclaimed deposit. The results of the ground survey reported by Tokimatsu et al. (1997) are shown in Fig. 25 in terms of the lateral displacement of the ground surface plotted versus the distance from the waterfront. It may be seen that the ground displacement at the place of Pile S-7, 6 m inland, was about 1.2 m towards the sea. From an independent survey involving measurements of the inclination of the building structure and displacement of the roof, the displacement of the head of the Pile S-7 was estimated to be about 80 cm.

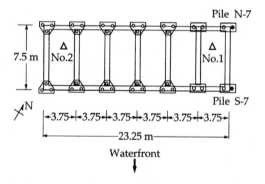

Figure 23. Arrangements of the foundations in plan view of the building in Higashi-Nada, Kobe (Tokimatsu et al., 1997)

The scheme of the back analysis for the performance of Pile S-7 is indicated in Fig. 26. The pile 20 m long is assumed to be rigidly-connected to the overlying footing slab and embedded to stiff strata extending from 9 m to 20 m which did not develop liquefaction. The soil above the ground water table may be assumed to have moved together with the underlying liquefied soil, but because of the constraint

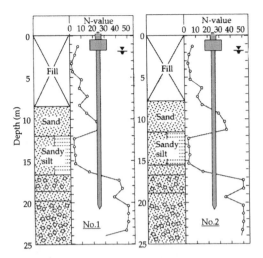

Figure 24. Soil profile beneath the foundations of the building in Higashi-Nada, Kobe (Tokimatsu et al., 1990)

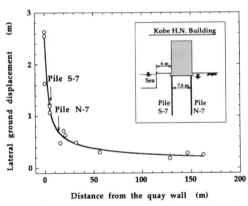

Figure 25. Lateral displacement of the ground surface at the site of the building in Higashi-Nada, Kobe (Tokimatsu et al., 1990)

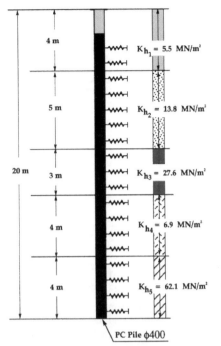

Kobe H.N. Building

Figure 26. Spring constants for the PC-pile of the building in Higashi-Nada, Kobe

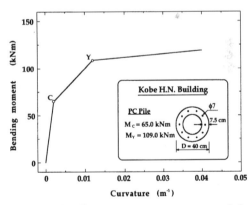

Figure 27. Bending moment versus curvature relation for the PC-pile of the building in Higashi-Nada, Kobe

produced by the horizontal underground beams between the front and back rows of the footing slabs, the soil in the unliquefied surface layer appears not to have been free to move with the underlying liquefied soil stratum. Thus, it appears highly likely that the pile in the unliquefied layer has moved nearly in unison with the surrounding soil stratum. With this assumption, the earth pressure in the surface layer may be neglected. If the pile is assumed to have moved ahead of the surrounding soil, the condition as illustrated in Fig. 2b might have been produced, thereby inducing the passive earth pressure acting against the movement of the pile. The computation for

this case is not presented in this paper.

The model of the beam-spring system used in the analysis is illustrated in Fig. 26 where K_h-values estimated by the empirical formula in the design code are indicated. The nonlinear stiffness characteristics of the PC-pile are shown in Fig. 27 in terms of bending moment versus curvature relation.

61

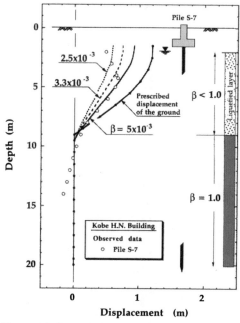

Figure 28. Lateral displacements of the Pile S-7 computed for different β-values: Analysis without earth pressure (building in Higashi-Nada, Kobe)

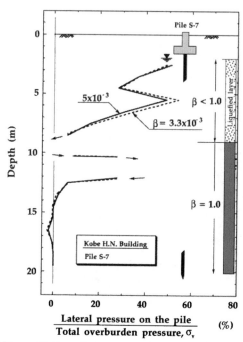

Figure 30. Lateral pressure on the Pile S-7 computed for different β-values: Analysis without earth pressure (building in Higashi-Nada, Kobe)

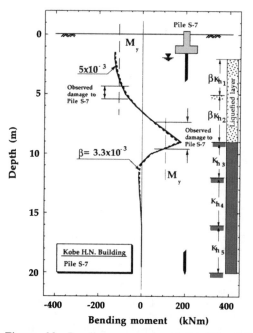

Figure 29. Bending moments of the Pile S-7 computed for different β-values: Analysis without earth pressure (building in Higashi-Nada, Kobe)

Displacement distributions computed in analyses of Pile S-7 for the case of zero earth pressure in the upper layer are shown in Fig. 28. Superposed in this diagram is the displacement of the pile measured by means of the inclinometer. Comparison of the observed displacement with the computed one appears to indicate that the stiffness degradation parameter should take a value of $\beta = 5 \times 10^{-3}$ in order for the displacement of the pile head to become equal to the observed value of 0.8 m. The value of $\beta = 5 \times 10^{-3}$ obtained above is shown in Fig. 31 for comparison sake plotted versus the normalized relative displacement $(U_G - U_P)/H_2$. Distribution of the computed bending moment is shown in Fig. 29 versus depth where it may be seen that the moment in excess of the yield value M_y occurs at two depth ranges which are roughly coincident with the depths of crack development observed by virtue of the video camera. Figure 30 shows the lateral pressure acting on the pile which is normalized to the total overburden pressure σ_v. It may be seen that the lateral pressure on the pile is less than 50 % of the total overburden pressure.

3.3 *Foundation of spherical tank*

At the time of the Kobe Earthquake in 1995, spherical tanks in a man-made island were damaged by extensive

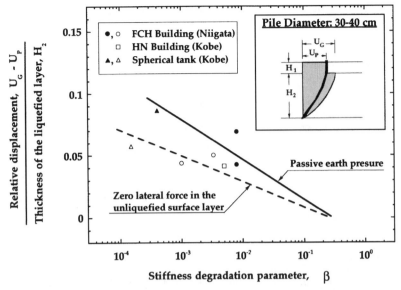

Figure 31. Normalized relative displacement between the ground and pile versus the stiffness degradation parameter β : Summary plot for the three case histories

liquefaction and consequent lateral flow of the ground. Details of the damage feature and accounts of pile behaviour were introduced elsewhere (Ishihara, 1997) with reference to the back analysis made for the damaged piles. The back analysis was conducted in the same fashion as described above in order to assess the range of the stiffness reduction factor for piles subjected to lateral flow of the liquefied deposit. In that analysis scheme, the earth pressure on piles in the upper unliquefied layer was assumed to be a passive earth pressure. It was possible to perform the back analysis by assuming zero earth pressure on the piles in the surface layer. The results of the analysis for the two extreme cases as above are presented in Fig. 31 in terms of the relation between the normalized relative displacement and the stiffness degradation parameter.

4 DISCUSSIONS

In the routine practice, the design of piles is often made through the use of a Winkler-type model in which soil-pile interaction is represented by the coefficient of subgrade reaction k. When soil deposits are identified to be softened as a result of liquefaction, this coefficient is reduced drastically. For the level ground conditions the Japanese Code of Highway Bridge foundation recommends that the effect of liquefaction should be taken into account by reducing the value of k by a factor ranging between 1/6 and 2/3 for design of embedded foundations. If this norm

is to be extended to include sloping grounds undergoing lateral spreading, it would be logical for all practical purposes to seek how much reduction in the k-value would farther be necessary to allow for the effect of lateral spreading on the design of piles embedded in such ground. While many laboratory model tests have been performed using 1-g shaking table to evaluate effects of level-ground liquefaction on the k-value, little has been known on the effects of the lateral spreading on the reduction of the k-value.

As it is generally difficult to conduct model tests, it would be worthwhile trying to investigate these effects based on back-analysis of known behaviour of piles in the field during earthquakes. Thus, the above data of back-analyses were arranged so as to find out governing factors influencing the stiffness degradation parameter. In an effort in this vein, it was discovered that the stiffness degradation parameter, β, is closely related with the amount of pile deflection relative to the displacement of the ground. It was further shown that the deflection of the pile does depend on the thickness of the liquefied stratum H_2. Thus, the normalized relative displacement $(U_G - U_P)/H_2$ was chosen as a parameter influencing the value of β. In this context, plots were made of the back-analyzed data as shown in Fig. 31. It may be seen in this diagram that the value of β tends to decrease with increasing normalized relative displacement. Note that the above relation is dependent on the flexural characteristics of the pile being considered. Thus, the relation shown in Fig. 31 does hold valid only for the case of reinforced concrete piles with a diameter of 30-40 cm. If the pile of greater stiffness is considered such as cast-in-place concrete pile with a diameter of 1.0-1.5 m, the relation would be different.

As seen in Fig. 31, the correlation between the normalized relative displacement and the stiffness degradation parameter tends to change depending upon the lateral force being applied in the upper portion of the pile in the non-liquefied surface soil deposit. It is apparent that, if the passive earth pressure is applied in the direction opposite to the flow, the value of β giving an equal normalized displacement would become larger as compared to the case of no lateral force.

In all the cases studied above, the piles being considered were located within the space enclosed by neighboring piles or close to the horizontal connecting beams. Thus, strong constraints appears to have been present for the upper portion of the piles under consideration. This indicates that the surface unliquefied soil layer must have moved together with the piles in question thereby producing no earth pressure against the upper portion of the piles. With this fact in mind, the relation for the case of no earth pressure in Fig. 31 is regarded as reflecting actual situations developed in the field during the earthquakes. It is to be noted that the piles in the liquefied layer must have been displaced behind the soil deposit, producing a fairly large relative displacement. Then attention should be drawn solely to the soil-pile interaction in the liquefied deposit. It is apparent that the greater the reduction in the stiffness of the liquefied soil, the larger the relative displacement would be between the soil and piles.

5 CONCLUSIONS

As a result of the back analyses of three case histories of piles injuries during the past earthquakes, it was found that the spring constants representing the soil-pile interaction are by a factor of 1×10^{-3}-1×10^{-2} smaller in the laterally spreading soils as compared to the normal soil condition where there is no liquefaction. It was also pointed out that the degree of stiffness degradation in the laterally flowing soil deposits is related to the displacement of the pile relative to the surrounding soil, with the stiffness degradation becoming more pronounced with increasing relative displacement between the pile and the ground. The lateral force acting on the pile during the lateral spreading was found to be less than 50-60 % of the total overburden pressure.

ACKNOWLEDGMENTS

In making the analyses, additional pieces of information provided by Dr. N. Yoshida, Sato Kogyo Co. Ltd, and Dr. S. Fujii, Taisei Corporation, were of great help. The authors wish to express deep thanks to these individuals.

REFERENCES

Chaudhuri, D., Toprak, S. and O'Rourke, T.D. 1995. Pile response to lateral spread: A benchmark case. *Lifeline Earthquake Engineering; Proceedings 4th U.S. Conference, San Francisco, California* :1-8.

Ishihara, K. 1997. Geotechnical aspects of the 1995 Kobe Earthquake. *Proceedings 14th International Conference on Soil Mechanics and Foundation Engineering, Terzaghi Oration, Hamburg.*

Tokimatsu, K., Oh-oka, H., Shamoto, Y., Nakazawa, A., & Asaka, Y. 1997. Failure and deformation modes of piles caused by liquefaction-induced lateral spreading in 1995 Hyogoken-Nambu Earthquake. *Geotechnical Engineering in Recovery from Urban Earthquake Disaster, KIG FORUM '97 Kobe, Japan, Kansai Branch of the Japanese Geotechnical Society* : 239-248.

Yoshida, N. & Hamada, M. 1991. Damage to foundation pile and deformation pattern of ground due to liquefaction-induced permanent ground deformations. *Proceedings 3rd Japan-U.S. Workshop on Earthquake Resistant Design of Lifeline Facilities and Countermeasures for Soil Liquefaction, National Center for Earthquake Engineering Research, NCEER-91-0001:* 147-156.

Geotechnical Hazards, Marić, Lisac & Szavits-Nossan (eds) © 1998 Taylor & Francis, ISBN 90 5410 957 2

Design parameters of granular soils from in situ tests

Michele Jamiolkowski, Diego C.F. Lo Presti & Francesco Froio
Politecnico di Torino, Italy

ABSTRACT: The capability of in situ testing techniques to assess the mechanical properties of granular soils is here reviewed after a brief qualitative picture of the mechanical soil behaviour. Particular attention is paid to recent innovations and the capabilities of in situ testing methods to assess the relative density, stiffness and secant angle of shear resistance of coarse grained soils. In particular, the advantages and limitations of penetration, dilatometer, pressuremeter and seismic tests are critically reviewed.

1. INTRODUCTION

Despite the progress made in retrieving undisturbed sand and gravel samples in Japan (Yoshimi et al. 1978, 1984, Hatanaka et al. 1985, Konno et al. 1991, Goto et al. 1992, Adachi and Tokimatsu 1994) and North America (Singh et al. 1982, Konrad 1990), the greatest majority of geotechnical projects around the world are based on design parameters that have been inferred from a large variety of in situ tests. Considering that this situation, except of large projects, is destined to continue in the future, the present paper is devoted to a brief review of the in situ techniques which allow the mechanical characterisation of coarse grained soil deposits. Bearing in mind the complexity and width of this subject the authors have decided to cover a limited number of selected topics and to pose the following restrictions:

- only tests performed in completely drained conditions are considered

- the interpretation of in situ tests is based on simple models of soil behaviour, as, with very few exceptions, it is not yet possible by now to infer the parameters of more complex soil models from in situ tests.

- although the authors are aware that there are other parameters of prominent design interest, the present work is limited to the assessment of stiffness [E, G, M] and angle of shear resistance for sands and gravels from in situ tests.

The following kinds of in situ tests have, in particular, been considered: Standard penetration Test (SPT), Cone Penetration Test (CPT), Marchetti Dilatometer Test (DMT), Research Dilatometer Test (RDMT), Pressuremeter Tests and Seismic Tests. As far as recent or quite recent innovations concerning the above mentioned tests are concerned, it is worthwhile mentioning the following aspects:

- Recognition of the importance of the energy delivered to the rods during execution of SPT (Schmertmann and Palacios 1979, Kovacs and Salomone 1982).

- Development, calibration and use of dynamic penetration tests with samplers larger than that employed for SPT (Yoshida 1988, Yoshida et al. 1988, Hatanaka et al. 1988 and 1989, Konno et al. 1991, Goto et al. 1992, Suzuki et al. 1992 and 1993, Crova et al. 1992, Harder and Seed 1986). This kind of test, named Large Penetration Test (LPT), is especially suited for coarse grained gravely soils.

- Incorporation of porous stone into the standard electric CPT tip which allows continuous measurement of the pore pressure during penetration at preselected elevations (Baligh et al. 1981, Campanella and Robertson 1981, Muromachi 1981, Tumay et al. 1981, De Ruiter 1981, Smits 1982).

- Development of Marchetti's flat dilatometer, followed by implementation of the device with a sensor that allows one to monitor the process of expansion of the dilatometer blade in a continuous manner (Marchetti 1980, Motan and Khan 1988, Fretti et al. 1993).

- Creation of a number of multipurpose CPT and CPTU probes which, in addition to the cone resistance q_c, penetration pore pressure u and local shaft friction f_s, are able to assess additional parameters of the penetrated soil thus making the test interpretation easier or/and more comprehensive. Some of these multipurpose tools, such as the Seismic Cone (SCPT) and the Cone Pressuremeter (CPM) will later be discussed in more detail.

The following two categories of interpretation methods can be distinguished, while considering different in situ techniques for granular soils:

Direct Methods

The response of geotechnical-engineered-construction of interest is directly correlated to the in situ test results as, for example, the ultimate bearing capacity of a single pile and penetration resistance (Bustamante and Gianneselli 1982, Reese and O'Neill 1988) or the susceptibility of sand deposits to liquefaction and penetration resistance (Seed et al. 1983, Robertson et al. 1985, Seed et al. 1985).

Indirect Methods

The parameters which describe the mechanical behaviour of a soil element are inferred from in situ test results. These parameters are used to solve the relevant boundary value problem through the computational models that are available in geotechnical engineering practice.

In virtue of what has so far been stated, the present paper only deals with indirect methods. These methods can be subdivided into the following three basic categories:

- In situ tests, in which all soil elements follow a similar effective stress path (ESP), i.e. seismic tests and, though with some limitations, ideally inserted Self-Boring Pressuremeter Tests (SBPT). In these cases, the elemental soil properties can be inferred by adopting appropriate constitutive relationships.

- In situ tests during which the soil elements follow different ESPs. For this kind of test it is possible to infer the parameters that reflect the overall response of the tested volume of the soil, with the assumption of oversimplified constitutive relationships. The Plate Load Test (PLT) is a typical example of this category of in situ methods, which provides a highly non-linear operational stiffness of the tested soil.

- All tests that involve push-in devices, such as static and dynamic penetration tests, induce a very large strain in the soil surrounding the probe during the insertion process. In this case, only empirical correlations between the ground properties and test results can be established.

From this, it would appear that, with a few exceptions, the interpretation of in situ tests is always quite complex and unavoidably involves the use of simplified procedures. Within this context it is worthwhile to recall the basic requirements

suggested by Wroth (1988) in order to establish reliable correlations between ground properties and in situ test results:

- any reliable relationship should be start from a comprehensive understanding of the stress-strain-time behaviour of natural soils.

- it should be based on the physical appreciation of why the considered properties can be related to the in situ test results.

- it should be set against a background of a theory, however idealised.

- it should be expressed in terms of dimensionless variables, taking advantage of continuum mechanics scaling laws.

2. MECHANICAL BEHAVIOUR - A BRIEF SUMMARY

On the basis of what has been stated, it is possible to obtain a more rational and reliable interpretation of in situ test results if these results are analysed within a general framework of mechanical soil behaviour. This paragraph is devoted to describing the most important features of the mechanical behaviour of granular soils in the light of the most recent progress in this field.

A better understanding of the mechanical behaviour of soils was mostly related, in recent years, to the improvement of the quality and reliability of laboratory testing, which has benefited from the following:

- Extensive use of microcomputer based systems which allow one to perform feed-back controlled experiments.

- Development of new techniques to locally measure strains of up to $5 \cdot 10^{-6}$ with a high degree of confidence.

- A better accuracy in the measurement of the vertical load imposed on the sample, thanks to the load cells located inside the cell.

- Development of 'new generation' apparatuses that allow one to investigate the mechanical behaviour of soils over a wide range of imposed stress states.

- Development of new techniques, such as bender elements (BE) to measure the small strain stiffness during laboratory tests.

The need for a more precise and accurate measurement of mechanical soil properties has, in turn, originated from the recognition, since the early eighties, of the fact that displacements of well designed geotechnical-engineered-constructions were generally quite small and overpredicted when using soil parameters that were inferred from conventional laboratory tests (Simpson et al. 1979, Burland 1989, Tatsuoka, et al 1995a, Tatsuoka et al. 1997).

Mechanical soil behaviour is hereafter discussed, referring to the scheme proposed by Jardine (1985, 1992, 1995) which is shown in Fig. 1.

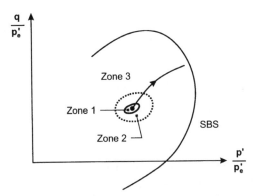

Figure 1 Stress-strain behaviour of soils simplified picture Jardine(1985, 1992); Jardine et al (1991)

This scheme, which basically follows the ideas of Mroz (1967) and Hashiguchi (1985), considers two additional kinematic sub-yield surfaces inside the State Boundary Surface (SBS). The SBS is the large scale yield envelope of the Cam Clay Model family. Multiple yield surface models give a more realistic picture of soil behaviour. A similar scheme has been worked out by Stallebrass (1990) and Stallebrass and Taylor (1997).

The essential features of this model are shown in the normalised stress plane in Fig. 1 where, by referring to an axisymmetric stress state such as that encountered in the Triaxial (TX) and Torsional Shear (TS) tests, one obtains:

$$p' = (\sigma'_a + 2 \cdot \sigma'_r)/3$$

$$q = \sigma'_a - \sigma'_r$$

p'_e = equivalent pressure on the isotropic virgin compression line (VCL) (Hvorslev 1937)

σ'_a, σ'_r = axial and radial effective stresses, respectively

The normalised stress plane p'/p'_e vs. q/p'_e can be divided into three distinct zones in order to match the behaviour of large variety of soils and soft rocks, which has mainly been observed in monotonic and cyclic TX and TS tests.

Zone 1

Within this zone, for all practical purposes, the soil exhibits a linear elastic stress-strain response. The Young's modulus E_o and shear modulus G_o within this zone, can be regarded as the initial stiffness of the relevant stress-strain curves of a given soil. Both these moduli, if properly normalised with respect to the void ratio and effective stresses, result to be equal or similar in magnitude regardless of the type of loading (monotonic or cyclic), applied shear strain level, number of loading cycles N, the strain rate and the stress/strain history (Fig. 2).

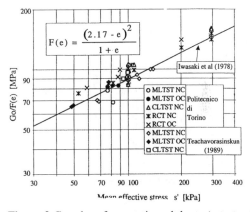

Figure 2 G_0 values from static and dynamic tests at isotropic stress states on Toyoura sand (Jamiolkowski et al. 1994) (TX: monotonic triaxial tests, ML and CLTST: monotonic and cyclic torsional shear tests, RCT: torsional RC tests and BE: Bender Element tests measuring $(V_s)_{vh}$)

In reality, a measurable influence of the strain rate on E_o and/or G_o has been observed, even in the case of granular soils (Tatsuoka et al. 1997). However, this influence is quite small and, in practice, it is possible to consider that the initial stiffness only depends on the current soil state assessed via the void ratio e, effective consolidation stresses and the material fabric originated by the depositional and post depositional processes. The following empirical formulae can be used to express the initial stiffness:

$$(G_o)_{ij} = C_{ij} \cdot F(e) \cdot \sigma'^n_i \cdot \sigma'^n_j \cdot p^{1-2n}_a \quad (1)$$

$$(E_o)_{ii} = C_{ii} \cdot F(e) \cdot \sigma'^{2n}_i \cdot p^{1-2n}_a \quad (2)$$

where: C_{ij} and C_{ii} are dimensionless material constants that incorporate the effect of the structure, F(e) is the void ratio function which accounts for the soil density and σ'_i and σ'_j are the normal effective stresses acting in the i and j directions.

It is obvious from eqs. (1) and (2) that while the shear modulus depends on two stress components, Young's modulus depends only on the relevant stress component. This kind of pressure dependence of the moduli at small strains is well documented in literature (Roesler 1979, Stokoe et al. 1985, 1991, Lo Presti and O'Neill 1991, Lo Presti et al. 1995, Tatsuoka et al. 1995b, Tatsuoka and Kohata 1995, Tatsuoka et al. 1997). Figure 3 shows some experimental data concerning carbonate Kenya sand, which evidence the dependence of Young's modulus on one stress component.

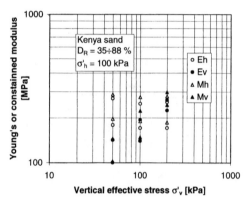

Figure 3a Influence of vertical effective stress on Young's and constrained moduli resulting from seismic tests

As far as the material fabric is concerned, it is worthwhile to point out that, under constant consolidation stresses, both G_o and E_o increase with

time when subject to drained creep. By referring to eqs. (1) and (2), this means that constants C_{ij} or C_{ii} increase with ageing. The increase of the small strain stiffness has been observed in the laboratory by many researchers for both reconstituted and "undisturbed" samples. In the case of the so-called "undisturbed samples" this phenomenon also reflects a partial recovery of the sampling disturbance. The stiffness-increase is much more pronounced in fine grained soils. In particular, the stiffness-raise rate with time increases with an increasing plasticity index (PI) for these soils. The increase of G_o with time after the end of primary consolidation (EOP) is often quantified by means of the following empirical formula (Anderson and Stokoe 1978):

$$G_o(t) = G_o(t = t_p) \cdot \left[1 + N_G \cdot \log \frac{t}{t_p} \right] \quad (3)$$

where
$G_o(t) =$ small strain shear modulus at $t > t_p$
$G_o(t = t_p) =$ small strain shear modulus at $t = t_p$
$t =$ any generic time larger than t_p
$t_p =$ time to complete the primary consolidation
$N_G = \dfrac{G_o(t) - G_o(t = t_p)}{G_o(t = t_p)} \cdot \dfrac{1}{\log(t / t_p)}$ normalised rate
of increment of the small strain shear modulus per log cycle of time

The N_G values range from between 0.05 and 0.20 in clays and increase with the PI or the secondary compression index (Kokusho 1987, Mesri 1987).The values of N_G are smaller in coarse grained soils and fall in the range of 0.01 to 0.03 except for special

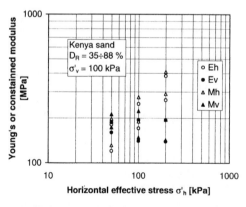

Figure 3b Influence of horizontal effective stress on Young's and constrained moduli resulting from seismic tests

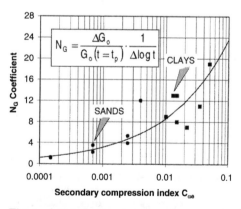

Figure 4 Relationship between the parameter N_G and the secondary compression index $C_{\alpha e}$; $t_p =$ time at EOP (Lo Presti et al. 1996)
soils such as glauconite and carbonate sands where it

can attain much higher values, comparable to those of clays. The dependence of N_G on the secondary compression index sounds more rational (Fig. 4).

The structure that originated during depositional and post-depositional processes is, in general, anisotropic. Both G_o and E_o result to be dependent on the direction of the applied effective stress path, in the case of the volume element tests, and on the directions of propagation and polarisation in the case of the seismic body waves (Lee and Stokoe 1986, Stokoe et al. 1995, Jamiolkowski et al. 1995a, Tatsuoka and Kohata 1995, Bellotti et al. 1996, Fioravante et al. 1998a). This directional dependency of stiffness reflects the elastic transverse-isotropy of many natural soils. The small strain anisotropy reflects two basic types of phenomena:

- Inherent or fabric anisotropy which a given soil exhibits when subject to an isotropic state of stress.
- Stress induced anisotropy which reflects the directional variation of stiffness as a function of the applied anisotropic stress system.

The relatively limited experimental data collected so far for granular soils are contradictory:

- According to Tatsuoka et al. (1997) the ratio of E_h / E_v in Zone 1 falls within the range of 0.45 to 1.0. The lower limit of the ratio holds for gravels while the higher limit is for medium to fine sands. The above results have been inferred from triaxial tests by performing small unload-reload cycles of vertical and horizontal stresses and by making an assumption for the Poisson ratio in the horizontal plane (ν_{hh}).

- On the other hand, the performance of seismic tests in Calibration Chambers on reconstituted specimens of sands and fine gravels has provided values of the G_{hh} / G_{vh} and E_h / E_v ratios of about 1.2 (Stokoe et al. 1985, Stokoe et al. 1991, Lo Presti and O'Neill 1991, Bellotti et al. 1996, Fioravante et al. 1998a, 1998b). In this case shear moduli were directly determined from the propagation velocities of shear waves. Young's moduli were inferred from the propagation velocities of rod-compression-waves, i.e. longitudinal resonant column tests (Stokoe et al. 1991) or from those of plane-compression-waves. In this case it was necessary to make some assumptions concerning the shape of the wave front and the interpretative model (Bellotti et al. 1996).

(One should notice that $G_{hh} = G_o$ on the horizontal plane, $G_{vh} = G_o$ on the vertical plane, $E_h = E_o$ in the horizontal direction, $E_v = E_o$ in the vertical direction)

From the above information it is not possible to draw any definitive conclusion. Only in the case of sands is it possible to state that the inherent anisotropy is quite negligible.

- The stress induced anisotropy in soils plays an important role in the directional variation of small strain stiffness. The impact of the anisotropic stress state on the initial stiffness can be clearly understood by considering that the anisotropic shear and Young's moduli do not depend on the same stress components, as stated by eqs. (1) and (2). Therefore, the stiffness ratios, as well as the ν_{vh} Poisson ratio, depend on the applied effective stress ratio $K = \sigma_a' / \sigma_r'$ (Lo Presti 1995, Bellotti et al 1996, Tatsuoka and Kohata 1995, Pallara 1995, Tatsuoka et al. 1997). Figure 5 shows the stiffness ratios as inferred from seismic tests performed both in situ and in the laboratory on cohesionless soils (Jamiolkowski et al. 1995b).

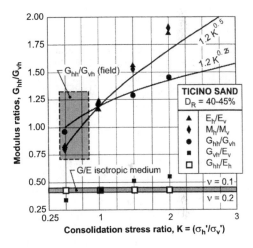

Figure 5 Small strain stiffness anisotropy: field versus laboratory data (Adapted from Bellotti et al, 1996)

The small strain viscous damping (D_o) in Zone 1, when determined via reliable laboratory experiments (Tatsuoka et al. 1995b, Stokoe et al. 1995), generally results to be less than 1 %. D_o, depends on the same factors already mentioned concerning small strain stiffness, although their quantitative impact on D_o is not necessarily the same as that observed in the case of G_o and E_o. In particular, it is worth mentioning the high sensitivity of D_o to the loading frequency f and hence to the strain rate, as recently pointed out (Shibuya et al. 1995, Tatsuoka et al. 1995b, Stokoe et al. 1995, Lo Presti et al. 1997a, 1997b).

The border between Zones 1 and 2, defined in terms of strain, is usually called linear threshold strain. According to the type of laboratory test in question this term can be applied either to the shear strain ($\varepsilon_{sr}^e, \gamma_t^e$) or to the axial strain (ε_{at}^e). Hereafter the symbol ε_t^e will be used to indicate the 'linear threshold strain' without specifying what kind of test it has been inferred from.

Generally the values of ε_t^e are inferred from TX, TS and RC tests. From the large set of experimental data available at present the following can be concluded, as far as the magnitude of ε_t^e is concerned:

- In uncemented and unaged granular soils the values of ε_t^e usually fall within the narrow range of $7 \cdot 10^{-6}$ to $2 \cdot 10^{-5}$.

- Ageing, cementation and other diagenetic processes tend to increase the value of ε_t^e to up to one order of magnitude (Tatsuoka and Kohata 1995, Tatsuoka et al. 1997).

- Moreover, the linear threshold strain ε_t^e increases with strain rate (Isenhower and Stokoe 1981, Tatsuoka and Shibuya 1992, Lo Presti et al. 1996,1997b) This is a consequence of the fact that, even in the so-called linear domain, the soil behaviour is rate dependent, especially for fine grained soils.

Zone 2

When soil is strained beyond ε_t^e the ESP penetrates into Zone 2 (Fig. 1). In this zone the stress-strain response starts to be non linear. Consequently the secant stiffness G and E depend not only on the current state of the soil but also on the imposed shear stress or strain level. Moreover G and E are influenced by many other factors such as strain rate, ageing, OCR, recent stress history or direction of perturbing ESP, etc. The terms "recent stress history" and "perturbing stress path" were introduced and used by Richardson (1988), then by Stallebrass (1990), Smith (1992) and Jardine (1995). These terms simply indicate respectively the re-consolidation and shearing stress-paths That are applied to the specimen in the laboratory.

The decay of E and G, with an increasing strain level, generally does not exceed 20-30 % of their initial value. Moreover, in cyclic tests, the stiffness is only moderately affected by the number of loading cycles and after few cycles the stress-strain response becomes stable. This indicates that the plastic strains inside Zone 2 are negligible and therefore the soil behaviour can be modelled as non-linear elastic or quasi-elastic.

Very little is known about the stiffness anisotropy in Zone 2. As a first approximation, considering the relatively small impact of the plastic phenomena on the observed stress-strain response, which implies only a secondary modification of the soil fabric, it can be postulated that the anisotropy of G and E preserves the features of that of Zone 1.

That which has been previously mentioned for the deformation moduli, applies, as a first approximation, to D which appears however, to be much more affected by strain rate than E and G.

The limit between Zones 2 and 3 can again be defined in terms of the strain level, which is called the volumetric threshold strain ε_t^v (Dobry et al. 1982). The concept behind the ε_t^v, refers to the onset of important permanent volumetric strains (ε_v^p) and either positive or negative residual excess pore pressure (Δu) in drained and undrained tests respectively. Certainly, when ε_t^v is exceed during cyclic tests, N starts to influence the degradation of stiffness in a definitive manner and the stress-strain response never becomes stable, i.e. Δu and ε_v^p continue to accumulate in undrained and drained tests respectively with an increase of the number of loading cycles (see Dobry et al. 1982, Chung et al. 1984, Lo Presti 1989, Vucetic 1994).

The values of ε_t^v are at least one order of magnitude higher than those of ε_t^e (Vucetic 1994, Stokoe et al. 1995, Dobry et al. 1982, Chung et al. 1984, Lo Presti 1989). Moreover, the values of ε_t^v are influenced by the following factors and phenomena: creep, moderate cyclic loading at strains larger than ε_t^v, overconsolidation ratio or prestressing, direction of the perturbing stress path and strain rate of the perturbing stress path.

An increase of ε_t^v by one order of magnitude was observed by Dobry et al. (1982) and Chung et al. (1984) for prestrained sands. A similar increase of ε_t^v was observed for overconsolidated (OCR=6) or prestrained ($\gamma_p = 0.05\%$) Ticino sand specimens (Lo Presti 1989). Creep deformations and cyclic prestraining can lead to a similar and noticeable increase of ε_t^v in sands (Tatsuoka et al. 1997). However, in some cases, the application of large cyclic prestraining can produce a destructuration of the granular materials. This has been observed in the case of crushable sands and in the case of gravels (Tatsuoka et al. 1995b, 1997).

Moreover ε_t^v can increase with an increase of the relative angle between the direction of the recent ESP and that of the perturbing ESP. Atkinson et al. (1990) and Jardine et al. (1991) have noticed this increase in the case of reconstituted and intact clays.

Data given by Lo Presti et al. (1995) on reconstituted Toyoura sand specimens that were sheared in triaxial compression after different consolidation stress-paths would suggest that similar considerations hold in the case of sands. The increase of ε_t^v with strain rate of the perturbing stress-path is quite well documented in literature for fine and granular soils (Lo Presti et al. 1996, 1997a, Tatsuoka et al. 1997).

Zone 3

When the deformation process engages Y_2, the soil starts to yield and the plastic deformation becomes important. As the ESP proceeds towards the State Boundary Surface (SBS) which coincides with Y_3 (Fig. 1), the ratio of the plastic shear strain to the total shear strain increases approaching values close to one at Y_3 (Fig. 6).

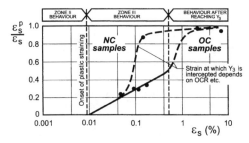

Figure 6 Relationship between permanent and total shear strain in Magnus till (Jardine, 1992)

In Zone 3 the stress-strain response of soils becomes highly non-linear. G, E and D depend to a great extent on the shear stress and strain level. Factors such as strain rate, creep and OCR greatly influence the magnitudes of these parameters.

Factors such as the recent stress history and the direction of perturbing ESP continue to influence the stress-strain response within Zone 3. However, their relevance decreases with an increase in distance from Y_2 (Richardson 1988, Stallebrass 1990, Jardine 1995).

Moreover, the stress-strain response to cyclic loading is no longer stable and a continuous degradation of the mechanical properties of soil is observed. In the case of undrained cyclic loading of granular soils, this leads to a continuous accumulation of the pore pressure and eventually to liquefaction.

The above qualitative scheme of soil behaviour is characterised by the kinematic nature of the Y_1 and Y_2 surfaces which are always dragged with current stress point. On the contrary the SBS is relatively immobile so that any sharp change of the ESP from the Y_3 inwards leaves its position unchanged except

in soils with highly developed fabric in which the collapse of the structure can determine its contraction.

One of the most relevant aspects of the behaviour of coarse grained soils at failure is their curvilinear strength envelope (Berezantzev 1964, 1970, De Beer 1965, Vesic and Clough 1968) which reflects the close link between the shear strength at peak and the soil dilatancy (Rowe 1962, Bolton 1986). Several researchers have proposed failure criterions that are capable to taking the curvilinear nature of the strength envelope of sands and gravels into account (Jaroshenko 1964, Baligh 1976, Maksimovic 1989). The criterion proposed by Baligh (1976) is formulated as follows:

$$\tau_{ff} = \sigma_{ff}' \left[\tan \varphi_o' + \tan \alpha \left(\frac{1}{2.3} - \text{Log} \frac{\sigma_{ff}'}{p_a} \right) \right] \quad (4)$$

where: τ_{ff} =shear stress on the failure plane at failure; σ_{ff}' =effective normal stress on the failure plane at failure; φ_o' =secant angle of shear resistance at $\sigma_{ff}' = 2.72 \cdot p_a$; p_a =atmospheric pressure expressed with the same unit system adopted for shear and normal stresses; α =angle which describes the curvature of the failure envelope.

Hereafter, the secant angle of shear resistance at peak $\varphi_p' = a \tan(\tau_{ff} / \sigma_{ff}')$ will always be considered.

Table 1a reports the values of the parameters of eq. (4) for various sands which have been used in Italy in Calibration Chambers.

Table 1a Parameters of curvilinear strength envelope of test sands

Sand	D_R [%]	φ_o' [°]	α
Ticino	25	33.1	0.15
	45	37.6	5.7
	65	40.9	11.7
	85	43.4	9.8
Quiou	65	41.7	7.4
	75	42.5	15.6
	95	45.3	21.9
Kenya	35	40.2	1.05
	85	48.7	15.2
Toyoura	35	33.0	1.7
	55	37.7	6.7

As already mentioned, the curvilinear nature of the failure envelope of sands and gravels is explainable within Rowe's (1962) stress-dilatancy theory framework. A simplified and very useful formulation of this theory was proposed by Bolton (1986) who approximated the difference between φ_p'

and the angle of shear resistance at the critical state φ'_{cv}, (also called at constant volume), by means of the following formulae:

(axisymmetric) $\quad \varphi'_p - \varphi'_{cv} = 3 \cdot DI$ (5)

(plane strain) $\quad \varphi'_p - \varphi'_{cv} = 5 \cdot DI$ (6)

with $\quad\quad DI = D_R (Q - \ln p'_f)$ (7)

where: DI = Relative dilatancy index; D_R = relative density; p'_f = mean effective stress at failure (kPa); Q = factor which depends on the mineralogical composition and crushability of the grains.

Table 1b Values of φ'_{cv} and Q obtained from triaxial tests for different sands

Sand	φ'_{cv} [°]	Q
Ticino	33.1	10.8
Quiou	40.7	7.5
Kenya	40.2	9.5
Toyoura	33.0	9.8

Table 2 shows the values of Q which were suggested by Bolton (1986) for different mineralogical compositions of the grains. Table 1b reports the values of φ'_{cv} and Q which were inferred from triaxial test results performed on different sands by writers.

Table 2 Values for the Q parameter suggested by Bolton (1986)

Q	Grain mineral
10	Quartz and feldspar
8	Limestone
7	Anthracite
5.5	Chalk

What stated above gives a simplified picture of the shear strength of granular materials. In particular, the following relevant questions were intentionally disregarded because the semiempirical and simplified approaches available to assess φ'_p from in situ test results do not allow one to take them into account:

- the dependence of the peak angle of shear resistance on the intermediate principal effective stress σ'_2 (Ko and Scott 1968, Lade and Duncan 1973, Yamada and Ishihara 1979, Ochiai and Lade 1983, Lam and Tatsuoka 1988).
- the anisotropic nature of peak shear strength in sands and gravels (Arthur and Menzies 1972, Oda 1972, Oda et al. 1978, Lam and Tatsuoka 1988)

The assessment of φ'_p, which is the design parameter of interest, from in situ test results is implicitly based on the following assumptions:

- φ'_p is controlled by both D_R and σ'_{ff};

- soil is cohesionless (c'=0), therefore the failure criterion of eq. (4) holds;
- the lower limit of φ'_p is φ'_{cv}.

3. EVALUATION OF RELATIVE DENSITY

Since the classical work by Terzaghi and Peck (1948) the looseness and denseness of granular soils have been linked to shear strength, stiffness, liquefaction susceptibility and other important engineering characteristics by means of the relative density (D_R) parameter (Holtz 1972). This parameter has been correlated to the results of Standard Penetration Tests (SPT) (Gibbs and Holtz 1957) and subsequently to static Cone Penetration Tests (CPT) (Schmertmann 1976). Despite the many problems encountered when assessing D_R (Tavenas et al. 1972), this parameter continues to preserve an important role in everyday engineering practice. In view of this, a brief summary of the state of the art in assessing the relative density by means of in situ tests is here presented.

The already mentioned original work by Gibbs and Holtz (1957) was further extended and validated by Bieganousky and Marcuson (1976, 1977). Skempton (1986) presented a very comprehensive review of the current practice of SPT. In this work, it is pointed out that, in order to correctly evaluate D_R from the blow-count (N_{SPT}), the following aspects should be considered:

- the correlations by Gibbs and Holtz (1957) and Bieganousky and Marcuson (1976, 1977) were obtained from the results of large scale laboratory tests carried out on freshly deposited uniform siliceous sands. These correlations are not capable of taking the influence of ageing, grading and oversonsolidation on N_{SPT} into account.
- N_{SPT} values should be normalised with respect to the effective overburden stress (N_1) and the rated energy delivered to the SPT rods (N_{60}). Therefore, the operational correlations between SPT blow-count and D_R can be expressed by means of the following empirical relationships:

$$(N_1)_{60} = A \cdot D_R^2$$ (8)

where: $(N_1)_{60}$ = SPT blow-count for an effective overburden stress of 1 bar and an energy ratio equal to 60%.

$$(N_1)_{60} = N_{SPT} \cdot C_N \cdot \frac{ER}{60}$$ (9)

where: N_{SPT} = measured blow-count, C_N = coefficient which normalises the measured blow-count to the effective overburden stress of 1 bar, ER = E^x / E_r = rod energy ratio ($E^x = \frac{1}{2} \cdot m \cdot v^2$ =

Table 3 Typical values of ER for SPT procedures adopted in different countries (adapted from Decourt 1989)

AVERAGE SPT ROD ENERGY RATIO			$N_{60} = N_{SPT} \dfrac{ER}{60} = C N_{SPT}$	
	Hammer		**ER**	**C**
Country	**Type**	**Release**	**(%)**	**(-)**
Argentina	Donut	Rope-cathead	45	0.75
Brazil	Pinweight	Manual	72	1.20
China	Donut	Free-fall	60	1.00
Colombia	Donut	Rope-cathead	50	0.83
Italy	Donut	Free-fall	65	1.08
Japan	Donut	Free-fall	78	1.30
	Donut	Rope-cathead	68	1.13
Paraguay	Pinweight	Manual	72	1.20
U.K.	Donut	Free-fall	60	1.00
	Donut	Rope-cathead	50	0.83
U.S.A.	Donut	Rope-cathead	45	0.75
	Safety	Rope-cathead	60	1.00
	Safety	Free-fall	85	1.40
Venezuela	Donut	Rope-cathead	43	0.72

theoretical free-fall energy, E_r = energy effectively delivered to the rod stem)

The values of C_N can be evaluated on the basis of the criteria formulated by Seed et al. (1983) and Liao and Whitmann (1986). According to Skempton (1986) the $(N_1)_{60}$ values can be evaluated as follows:

- $C_N = 2/(1+\sigma'_{vo})$ for fine sands of medium density
- $C_N = 3/(2+\sigma'_{vo})$ for dense coarse sands
- typical values of ER in different countries, where different SPT driving are used, are reported in Table3

Skempton (1986) recommended the following values of parameter A for eq. (8) for NC deposits: A=35 for freshly deposited sands; A=40 for recent man-made fills; A=55 for natural aged sands. Skempton (1986) also indicated in which way parameter A should be modified in the case of OC deposits.

What is here stated holds for sands. In the case of gravely deposits, there is a lack of clearly defined procedures for evaluating the relative density from SPT results. Over the last decade, many researchers have used the results of LPT to find correlations with SPT results or to develop independent correlations against the relative density (see as an example Crova et al. 1992).

Schmertmann (1976) proposed a correlation between the cone resistance (q_c) the relative density and the vertical effective stress, using the results of CPT's performed on sands in the Calibration Chamber (CC) of the University of Florida. The analytical expression of such a correlation, also used by other researchers (Harman 1976, Tumay 1976) is below reported:

$$q_c = C_o \cdot \left(\sigma'_{vo}\right)^{C_1} \cdot \exp(C_2 \cdot D_R) \qquad (10)$$

$$D_R = \frac{1}{C_2} \ln\left[\frac{q_c \cdot \left(\sigma'_{vo}\right)^{C_1}}{C_o}\right] \qquad (11)$$

Since the pioneering work by Schmertmann, many CCs have been put into operation in North America, Europe and Japan generating a large data-base of CPTs performed in different sands and providing a deeper insight into the merits and limitations of this kind of large-scale laboratory tests. The key points that have emerged from these experiments can be summarised as follows:

- due to the finite dimensions of the CC, the measured cone resistance is affected by an error in comparison to that obtainable in the case of an infinite sand deposit with the same relative density. This phenomenon, named chamber size effect (Parkin and Lunne 1982, Ghionna 1984, Zohrabi 1993) leads to an underestimate or overestimate of the field q_c whose magnitude depends on the crushability and compressibility of the test sand, the ratio of the CC specimen diameter (D_c) to that of the cone (d_c), D_R, confining stresses applied to the CC specimen and boundary conditions imposed on the CC specimen during the cone penetration.

- the correlation of q_c vs. D_R and σ'_{vo} holds only for NC sands. A correlation for NC and OC deposits should refer to the effective mean in situ stress σ'_{mo} instead of σ'_{vo}.

- thanks to the works by (Dusseault and Morgenstern 1979 and Barton and Palmer 1989) which have investigated the effect of geological time on porosity, fabric and mechanical properties of coarse grained soil deposits, it is obvious that the empirical correlations based on the results of tests performed on laboratory reconstituted specimens, are applicable only in the case of young, unaged NC soils.

Figures 7 and 8 show the updated correlations which were obtained by Garizio (1997) using a data-base of more than 400 CC tests performed on pluvially deposited uncrushable to moderately crushable siliceous sands. Figure 7 shows the correlation $D_R = f(q_c, \sigma'_{vo})$ and Figure 8 shows that of $D_R = f(q_c, \sigma'_{mo})$ which is valid for NC and OC sands respectively. For the first time these correlations have been obtained after correction of the measured q_c for the chamber size effect, according to the suggestions given by Tanizawa (1992) and Salgado (1993).

Recently, the possibility of inferring in situ D_R using other tests has been explored. Among the many tests, the in-situ measurement of the propagation velocity of body waves by means of seismic tests and the in situ measurement of the electrical resistivity are probably those of major interest. Their merits are however beyond the scope of the present work.

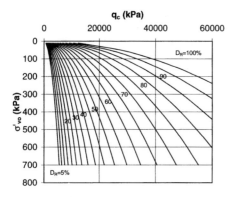

Figure7 Empirical correlations $q_c=f(\sigma'_{vo}, D_R)$ for NC siliceous test sands

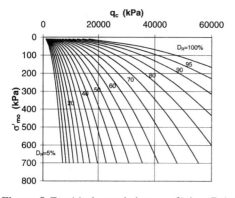

Figure 8 Empirical correlations $q_c=f(\sigma'_{mo}, D_R)$ for NC and OC siliceous test sands

4. STRENGTH OF COARSE GRAINED MATERIALS FROM IN SITU TESTS

The evaluation of the shear strength of sands and gravels in every day practice is usually based on the results of SPTs and CPTs. Recently, Felice (1997) also proposed an empirical correlation between the lateral stress index K_D from DMTs and the peak angle of shearing resistance $\varphi'_p(TC)$ from triaxial compression tests. This correlation is based on the experimental results obtained in the laboratory on reconstituted specimens of three different siliceous sands.

The available procedures for the assessment of φ'_p from SBPTs (Hughes et al. 1977, Wroth 1982, Robertson and Hughes 1986) are more rational and elaborate. Basically, the pressuremeter test is regarded as the expansion of an infinite cylindrical cavity. Consequently, the obtainable angle of shear resistance corresponds to that mobilised in plane strain conditions.

Additional assumptions, which are usually adopted for the interpretation of any kind of in situ tests, are: the cohesion is equal to zero; tests occur under fully undrained conditions; it is always necessary to refer the φ'_p, inferred from in situ test, to the level of σ'_{ff} which is relevant for the design problem under consideration.

As far as SPT is concerned, the empirical correlations $N_{SPT} = f(\varphi'_p, \sigma'_{vo})$ proposed by Peck et al. (1953) and De Mello (1971) are the most frequently used in practice. It is generally accepted that these correlations provide a conservative estimate of $\varphi'_p(TC)$. It is not known to which level of σ'_{ff} the obtained values of $\varphi'_p(TC)$ should be referred.

In the nineties, Goto et al. (1992), Suzuki et al. (1993), Hatanaka and Uchida (1996) proposed new correlations between N_{SPT} and φ'_p, which were based on laboratory tests performed on undisturbed samples of sands and gravels retrieved from the ground by means of the freezing method. The most recent correlation between N_1 and φ'_p for sands is reported by Hatanaka and Uchida (1996):

$$\varphi'_p(TC) = \sqrt{20 \cdot N_1} + 17 \pm 3 \qquad (12)$$

where $\varphi'_p(TC)$ is in degrees and N_1 has been normalised according to Liao and Withman (1986).

This relationship has been obtained using the Japanese trip hammer which delivers 78 % of the nominal energy (ER=78) to the rod stem. The database, which was used to develop eq. (12), refers to soils with the following physical characteristics: $0.15 \leq D_{50} \leq 0.75$ [mm]; $1.6 \leq U_c \leq 22.3$; $34 \leq D_R \leq 81\%$: Moreover, this database is characterised by the following intervals of vertical stress, normalised blow-count and angle of shear resistance: $39.2 \leq \sigma'_{vo} \leq 137$ [kPa]; $4 \leq N_1 \leq 28$; $28 \leq \varphi'_p(TC) \leq 40°$.

Hatanaka and Uchida (1996) also report a tentative correlation between the Japanese LPT blow-count N_{LI} and $\varphi'_p(TC)$ for gravely soils:

$$\varphi'_p(TC) = \sqrt{20 \cdot N_{LI}} + 20 \qquad (13)$$

This formula holds for the LPT developed in Japan (Yoshida 1988). For this kind of LPT, the following relationship, which allows one to compare the N_{LI} to N_1, holds:

$$N_1 = 1.5 \cdot N_{LI} \qquad (14)$$

The prediction of φ'_p from CPT can be made by means of one of the following approaches:

- inverse use of the formulae aimed at predicting the bearing capacity of deep foundations. A simple approach consists in the use of the classical bearing capacity theory of a rigid plastic body (Durgunoglu and Mitchell 1973, Janbu and Sennesset 1974). The evaluation of φ'_p, by means of this theory, requires the knowledge of q_c, σ'_{vo} and u_o (hydrostatic pressure). This approach neglects the effects of material compressibility and, for this reason, underpredicts φ'_p. The bearing capacity theory of Durgunoglu and Mitchell (1973) takes the influence of the initial in situ horizontal stress into account and therefore requires the knowledge of such a parameter. An example of this kind of correlation $q_c = f(\varphi'_p, \sigma'_{vo}, K_o)$ is shown in Figure 9.

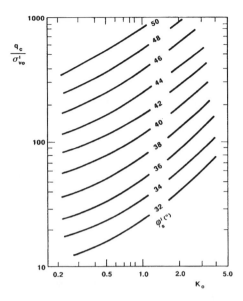

Figure 9 Peak secant friction angle $\varphi'_p(TC)$ for ticino sand from CPT

- use of theories of the expanding cylindrical and spherical cavities in elastic-perfectly plastic soil (Vesic 1972, Baligh 1975, Keaveny 1985, Salgado 1993). This approach requires the knowledge of q_c, σ'_{vo}, K_o and u_o, as in the previous method. Moreover, it is necessary to know the shear modulus G and the plastic volumetric strain ε_v^p that allow computation of the rigidity index:

$$I_R = \frac{G}{\sigma' \cdot \tan \varphi_p} \qquad (15)$$

and, in the case of contractant behaviour $\left(\varepsilon_v^p > 0 \right)$ the reduced rigidity index:

$$I_{RR} = \frac{I_R}{1 + \varepsilon_v^p \cdot I_R} \qquad (16)$$

where: $\sigma' = \sigma'_{ho}$ (cylindrical cavity) or σ'_{mo} (spherical cavity)

The fact that the term $\tan \varphi_p$ appears in the right side of eq. (15) involves the use of an iterative procedure when assessing the angle of shear resistance. Mitchell and Keaveny (1986) have shown that the cavity expansion approach (Vesic 1972, 1975, 1977) provides an estimate of φ_p for nine different siliceous sands with an error of less than 5 % as far as the values obtained from triaxial compression tests are concerned. Mitchell and Keaveny (1986) assumed that $G = G_{50}$, i.e. the shear modulus to be introduced in eq. (15) is that corresponding to a mobilisation factor of 50 %. As already mentioned, this approach requires the knowledge of a greater number of parameters than other approaches. The parameters which concern soil deformability and compressibility are rarely available in every day practice.

- the assessment of φ'_p from CPT is also possible using the stress dilatancy theory (Bolton 1986). The applicability of this approach was shown by Jamiolkowski and Robertson (1988) and Jamiolkowski (1989). This method requires the knowledge of q_c, σ'_{vo}, K_o and φ'_{cv} and assumes that D_R in situ can be assessed in a reliable manner from SPT, CPT or DMT. The evaluation of φ'_p also involves an iterative procedure in this case as summarised in Figure 10.

The use of this approach in the case of siliceous sands $(Q \cong 10)$ provide a good estimate of φ'_p:

$$\frac{\varphi'_p (TC)(laboratory)}{\varphi'_p (TC)(CPT)} = 1.03 \pm 0.04$$

Figure 11 shows the values of $\varphi'_p(TC)$ (i.e. φ'_o in eqs. 4 and 17) for Ticino and Hokksund sands evaluated at $\sigma'_{ff} = 267[kPa]$. $\varphi'_{cv} = 34°$ was assumed for both sands.

It is worthwhile to remember that $\varphi'_p(TC)$ depends on σ'_{ff} and that it is difficult to define and to determine the σ'_{ff} value for a CPT. However, it is suggested to evaluate σ'_{ff} as a fraction of q_c $(\sigma'_{ff} \cong \beta \cdot q_c)$. The suggested values for β are reported in Table 4.

The value of $\varphi'_p(TC)$ obtained from a CPT is related to the in situ density and σ'_{ff} developed during the test. A different level of σ'_{ff} is usually encountered when solving a given boundary value problem. It is therefore necessary to correct the obtained value of $\varphi'_p(TC)$ to take its stress dependence into account. This correction can be made by means of eq. (4) (Baligh 1976) which can be written in the following way:

$$\tan \varphi'_p = \tan \varphi'_o + \tan \alpha \left(\frac{1}{2.3} - Log \frac{\sigma'_{ff}}{P_a} \right) \quad (17)$$

where: φ'_o = secant peak angle of shear resistance at $\sigma'_{ff} = 267[kPa]$

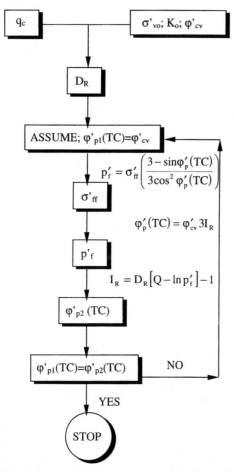

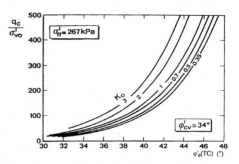

Figure 10 $\varphi'_p(TC)=f(q_c)$ of sand using Bolton's stress-dilatancy theory

Figure 11 Angle of shering resistance $\varphi'_p(TC)$ obtained using Bolton (1986) stress-dilatancy theory

Table 4 Average effective stress at failure on failure plane around penetrating cone (Delladonna 1988)

D_R [%]	Delladonna (1988)	Fleming et al. (1992) (*)
15	0.155	0.090 - 0.151
35	0.129	0.075 - 0.129
65	0.066	0.074 - 0.042
85	0.061	0.061 - 0.035
100	0.046	0.046 -0.026

(*) Upper limit for $\sigma'_{vo} = 100$ [kPa], lower limit for $\sigma'_{vo} = 300$ [kPa]

An appropriate account of the non linearity of the failure envelope of coarse granular materials is responsible for the fact that all the above mentioned methods unavoidably involve a trial and error approache when assessing φ'_p.

A semi-empirical method for determining the plane strain peak angle of shearing resistance $\varphi'_p(PS)$ and the dilatancy angle ν from SBPT was proposed by Hughes et al. (1977). This method, that combines the sand behaviour observed during drained simple shear tests and Rowe's (1962) stress-dilatancy theory, suggests that $\varphi'_p(PS)$ and υ can be determined by means of the following expressions:

$$\sin\varphi'_p(PS) = \frac{S}{1+(S-1)\cdot\sin\varphi'_{cv}} \quad (18)$$

$$\sin\upsilon = S + (S-1)\cdot\sin\varphi'_{cv} \quad (19)$$

where: S = slope of the $\log p' - \log\varepsilon$ curve that is obtained from a pressuremeter test when a well developed plastic zone exists around the probe; p' = current cavity effective stress; ε = current cavity strain.

This method is based on a simplified soil model (elasto-plastic-dilatant sand behaviour) which assumes that $\varphi'_p(PS)$ is constant at failure and that a linear relationship exists between volumetric ε_v and shear strains $\gamma = 2\cdot\varepsilon_\theta$ during the expansion of the probe (Fig. 12).

The application of this method involves some difficulties, among which, the following should be mentioned:

- in the case of an ideal probe installation, the initial point of the expansion curve should be given by $\varepsilon = 0$ and $p' = \sigma'_{ho}$, i.e. the in situ effective horizontal stress prior to the probe insertion. As a matter of fact, the experimentally determined expansion curve is affected by the disturbance caused by the probe insertion. Consequently, the slope S of the $(\log p' - \log\varepsilon)$ curve and the $\varphi'_p(PS)$ and υ values obtained from a pressuremeter test are

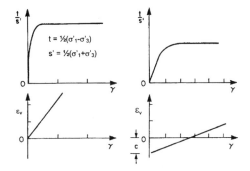

Figure 12 Simplified soil model assumed by Hughes et al.(1977) (Robertson and Hughes, 1986)

influenced by disturbance, due to the probe insertion (Fahey and Randolph 1984). In order to mitigate the influence of disturbance, it has been suggested by Fahey and Randolph (1984) to determine the S slope of the corrected $(\log p' - \log\varepsilon_c)$ curve, where $\varepsilon_c = \varepsilon - \Delta\varepsilon$, ε = current measured cavity strain and $\Delta\varepsilon$ = cavity strain at $p' = \sigma'_{ho}$. This procedure involves a value of the coefficient of earth pressure at rest K_o being assumed. As an alternative, Mair and Wood (1987) and Clark (1992) suggest choosing a value of $\Delta\varepsilon$ so as to maximise the part of the expansion phase which exhibits a linear relationship between $\log\varepsilon_c$ and $\log p'$. The chosen $\Delta\varepsilon$ should obviously fall within an interval of values with a physical meaning.

- the maximum cavity strain obtainable with currently available SBP probes is $\varepsilon \cong 12$ to 15 %. In the case of loose or medium dense sands this strain level is not sufficient to approach the condition where the plastic zone around expanding probe is sufficiently developed in order to obtain a linear relationship between $\log p'$ and $\log\varepsilon$. As a consequence, the determined S slope is too low with an underestimate of $\varphi'_p(PS)$ and υ. Robertson (1982) and Robertson and Hughes (1986) proposed a correction for $\varphi'_p(PS)$ and υ obtained from SBPT with a maximum cavity strain of about 10 % using the Hughes et al (1977) method. This correction is based on the results of laboratory tests performed on Ottawa sand and is incorporated in the Robertson and Hughes (1986) charts that show the $\varphi'_p(PS)$ and ν which one should obtain from a test in which S were inferred at sufficiently large cavity strains (i.e. $\varepsilon \gg 10$ %). The Figures 13 and 14 report the Robertson and Hughes (1986) correction charts in which ascissa corresponds to the S values determined from SBPTs performed at maximum $\varepsilon \gg 10$ to 15 %.

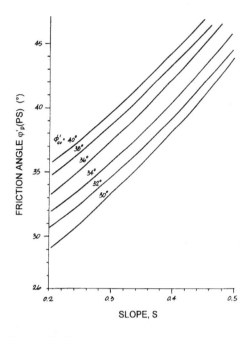

Figure 13 Correlation between S and φ'_p(PS) for SBP tests data (Robertson and Hughes 1986)

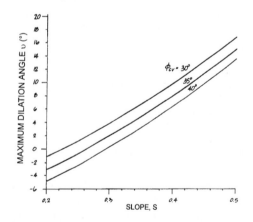

Figure 14 Correlation between S and υ for SBP tests data (Robertson and Hughes 1986)

- the pressuremeter probes have a finite length to diameter ratio (L/D). This leads to a displacement field which, at large expansion strain, exhibits an important departure from that based on the theory of cylindrical cavity with an infinite length This factor is important when determining the ultimate cavity

stress p_u and shear strength. The existing studies indicate that as the L/D ratio decreases, one obtains an increasing overestimate of p_u and φ'_p(PS) (Laier 1973, Ajalloeian and Yu 1996). In particular, the φ'_p(PS) is overestimated by 10 to 15 % with the Hughes et al (1977) method, in the case of L/D=6.

- the knowledge of φ'_{cv} is required to determine φ'_p(PS) and ν from SBPT. The φ'_{cv} can be determined from drained triaxial compression tests performed on loose samples. An approximate estimate of φ'_{cv} can be done with help of data reported in Table 5.

Table 5 Typical values of φ'_{cv} (Robertson 1982, Hanna and Youssef (1987)

Soil type	φ'_{cv} [°]
well graded siliceous sand and gravel	34 - 37
uniform coarse siliceous sand	33 - 35
uniform medium siliceous sand	33 - 35
uniform fine siliceous sand	30 - 33
uniform fine to medium carbonate sand	38 - 42

the upper and lower limits of the φ'_{cv} values refer to angular and tounded particles respectively.

A more elaborate approach for SBPT result interpretation has been proposed by Manassero (1989). This approach can be applied to all sands, it yields the entire stress-strain curve and removes some restrictive assumptions of the Hughes et al. (1977) method. In particular, it is not necessary to assume a constant dilatancy during expansion and a constant angle of shearing resistance at failure which permits one to model strain softening after peak.

The above mentioned methods to derive the strength parameter from SBPT should also be applicable to Cone Pressuremeter tests, at least in principle. However, data shown by Withers et al. (1989) and Ghionna et al. (1995) indicate that additional research is needed with this respect. At present, the evaluation of φ'_p(PS) from Cone Pressuremeter tests relies mostly on semiempirical correlations (Yu et al. 1994, Ghionna et al 1995)

5. STIFFNESS

5.1 *Seismic Tests*
Seismic tests are conventionally classified into borehole and surface methods. These methods enable one to determine the velocity of body waves

[compressional (P) and/or shear (S)] and surface waves [Rayleigh] respectively which induce very small strain levels into the soil, i.e. $\varepsilon_{ij} < 0.001$ % (Woods 1978). It is therefore possible, on the basis of the measured wave velocities, to obtain the small strain deformation characteristics according to the well known relationships:

$$G_o = \rho V_s^2 \qquad (20)$$
$$M_o = \rho V_p^2 \qquad (21)$$
$$V_R / V_s \cong (0.862 + 1.14\nu) / (1 + \nu) \qquad (22)$$
$$\nu = (V_p^2 - 2V_s^2) / 2(V_p^2 - V_s^2) \qquad (23)$$

where: G_o, M_o = small strain shear and constrained moduli respectively; ρ = mass density; V_s, V_p, V_R = velocities of shear, compressional and Rayleigh waves respectively; ν = Poisson ratio

The above relationships are based on the hypotheses of elasticity and isotropy.

The P wave velocity measured below the water table is more or less coincident with that of sound propagation in water ($V_{water} \cong$ 1500-1600 m/s, see Mitchell et al. 1994 as an example) and does not represent a propagation velocity in the soil skeleton which is usually less than V_{water}. Under these conditions the assessment of both constrained modulus and Poisson Ratio is no longer possible using the above listed equations (Woods and Stokoe 1985).

Recently researchers have paid much more attention to the possibility of inferring the small strain damping ratio, D_o from seismic tests and various examples can be found in literature, as mentioned later on.

Among the various borehole seismic methods, the well known Cross-Hole (CH) and Down Hole (DH) tests should be mentioned as being capable of providing a shear modulus profile. More recently, the Seismic Cone Penetrometer Tests (SCPT) both with a downhole configuration (Campanella and Robertson 1984) and a crosshole configuration (Baldi et al 1988) are useful for such a determination. The Spectral-Analysis-of-Surface-Waves (SASW) (Nazarian and Stokoe 1983, 1984, Nazarian et al. 1983) technique is the most important innovation in the field of seismic surface methods. Stokoe et al (1994) provided a very useful review of the method for both terrestrial and offshore applications of SASW.

Current practice and recent innovations of borehole methods for seismic exploration are covered by many comprehensive works (Auld 1977,

Stokoe and Hoar 1978, Woods 1978, Woods and Stokoe 1985, Woods 1991 and 1994, Jamiolkowski et al. 1995b). In the following, the fundamental aspects of borehole and surface seismic testing are briefly summarised:

a) a good coupling between boreholes and the surrounding soil is the key point to obtain useful measurements.

b) a check of the borehole verticality by means of inclinometer measurements is also highly recommended in order to accurately determine the length of wave travel path.

c) the repeatability of generated waveforms, the possibility of polarity inversion and generation of either compression or shear specialised body waves are desired features for any energy source.

d) the use of a pair of geophones, as receivers, located at a fixed distance (Patel 1981) can greatly increase the accuracy and the resolution of the test allowing one to use the true interval method for data interpretation.

e) dedicated portable waveform analysers or computer based data acquisition systems, in addition to conventional electronic equipment (oscilloscopes and seismographs) have greatly increased the capability of seismic tests especially as far as data interpretation is concerned (Hall 1985). Thanks to these enhancements, travel time measurement can easily be accomplished using more accurate methods in comparison to the so called "By eye" method. In particular the cross-correlation algorithm (Roesler 1978) can be used to determine travel time for seismic tests performed with a pair of receivers located at a known distance d.

It is worthwhile to notice that the SCPT satisfies the above requirements.

Generally, the shear wave velocities inferred from various seismic tests are in good agreement (see the example in Fig. 15 concerning the Po River sands). However the results from CH tests are generally considered as the most reliable. In Figure 16 the ratio of V_s measured using SASW, SCPT and another borehole method to that obtained from CH tests is plotted against depth. The data of this Figure concern granular deposits in USA seismic areas. It is possible to see that a relatively good agreement exists. Very low values of V_s have been obtained near the surface from SASW tests. The authors do not know the experimental work here discussed in detail but it is supposed that a large and systematic underestimate of V_s from SASW method near the surface could be due to one of the following reasons:

- use of geophones that are not capable of operating in the high frequency range

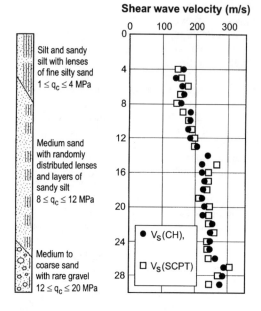

Figure 15 Shear wave velocity in Po river sand

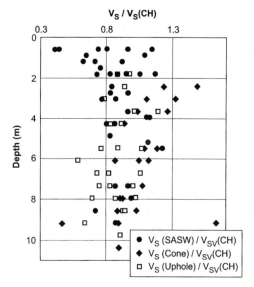

Figure 16 Seismic tests: comparision of field data obained by different methods (Data from Mitchell et al, 1994)

- insufficient discretization of the shallow strata where the velocity can strongly change with depth due to mechanical heterogeneity or other phenomena.

It is worthwhile to point out that the stiffness inferred from seismic tests represents the soil stiffness in Zone 1. As already discussed, this stiffness depend only on the soil state and therefore this stiffness represents an initial reference value for both static and cyclic loading problems. Moreover, if one considers that the shear wave velocity reflects the soil state in terms of density and applied pressure it can be considered as a sort of state parameter. Its assessment, especially in granular soils where laboratory tests on undistured samples cannot easily be performed, is of great importance.

5.2 *Penetration Tests*

In the following, the possibility of inferring soil stiffness from SPT, CPT, DMT and RDMT is discussed.

In spite of the fact that penetration resistance represents the soil's response to very large deformations, it is possible to establish good correlations between small strain moduli and these parameters. The reason why it is possible to correlate small strain moduli to the penetration resistance is that both are mainly dependent on the soil state.

Many empirical correlations between the penetration resistance from SPT (Ohta and Goto 1978, Imai and Tonouchi 1982) or CPT (Sycora and Stokoe 1983, Robertson and Campanella 1983, Rix 1984, Baldi et al 1986, 1989a, 1989b, Bellotti et al. 1986, Lo Presti and Lai 1989, Rix and Stokoe 1991) and the small strain shear modulus G_o have been established using different databases. Among the many available correlations it is worthwhile to mention the following:

a) Otha and Goto (1978), adapted by Seed et al. (1986)

$$V_s = 69 \cdot N_{60}^{0.17} \cdot Z^{0.2} \cdot F_A \cdot F_G \qquad (24)$$

where: V_s =shear wave velocity (m/s), N_{60} =number of blow/feet from SPT with an Energy Ratio of 60%, Z=depth (m), F_G =geological factor (clays=1.000, sands=1.086), F_A =age factor (Holocene=1.000, Pleistocene=1.303)

b) Rix and Stokoe (1991)

$$G_o = 1634 \cdot q_c^{0.250} \cdot \sigma_{vo}^{'0.375} \qquad (25)$$

where: G_o, q_c and σ'_{vo} are expressed in [kPa]. This correlation is based on field data and CC tests.

c) Jamiolkowski et al. (1988) have shown that the ratio G_o / q_c mainly depends on relative density and is only moderately influenced by overburden stress. Figure 17, which is based on field and CC data, can be used to infer the small strain shear modulus from CPT.

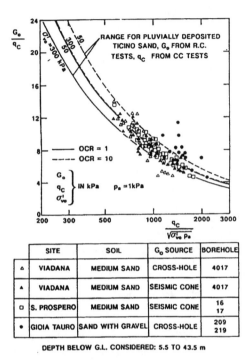

Figure 17 q_c versus G_0 correlation for uncemented predominantly quartz sand

The success in obtaining good correlations between G_0 and the penetration test results, should encourage one to establish such empirical correlations for a specific site. This approach makes it possible to consider the spatial variability of soil properties in a very cost effective way. However, due to their purely empirical nature, these correlations, when applied to sites which are different than those considered in the original database, can provide just an approximate estimate of G_0 which in many cases can be quite poor with a possible error of ± 100 %.

Conventional dilatometer tests (Marchetti 1980, 1997) can provide the dilatometer modulus E_D according to the following formula:

$$E_D = 34.7 \cdot (p_1 - p_o) \quad (26)$$

where: p_o is the conventional lift off pressure and p_1 the pressure corresponding to a displacement of 1.1

mm of the central part of the steel membrane. These pressures are both corrected to take the membrane stiffness into account.

Research dilatometer tests (Campanella et al. 1985, Jamiolkowski et al. 1988, Motan and Khan 1988, Fretti et al. 1992, Bellotti et al. 1989, 1997) can provide a complete expansion curve similar in shape to that observed during pressuremeter tests and allow the performance of unload-reload loops during membrane expansion. In this case, besides the dilatometer modulus E_D, it is possible to determine an unload-reload modulus according to the following formula:

$$E_D^{UR} = 38.2 \cdot \frac{(p_u - p_r)}{\Delta \rho} \quad (27)$$

where: p_u and p_r are the pressures which correspond to the apexes of the loop, while $\Delta \rho$ is the displacement observed during the loop (see Fig. 18).

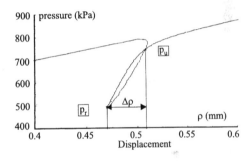

Figure 18 Unload-reload loop from RDMT

In the following the experimental results obtained from DMTs performed in a CC on reconstituted specimens of two silica sands (Ticino and Toyoura sands) are discussed. The dilatometer modulus was correlated, for both Ticino and Toyoura sands, with the small strain shear modulus G_0 which was inferred from RC and torsional shear tests (Lo Presti 1989, Pallara 1995). In particular, Fig. 19 shows the ratio G_0 / E_D vs. K_D for Ticino sand. A similar trend was observed for Toyoura sand. The parameters of the fitting equations are reported on the Figure for both Ticino and Toyoura sands. The K_D parameter is computed as $K_D = (p_o - u_o) / \sigma_{vo}$ and is mainly influenced by the relative density. It is possible to observe that the G_0 / E_D ratio is not influenced by the OCR. This means that not only the small strain stiffness, but also the dilatometer modulus is rather insensitive to OCR (Fretti et al. 1992). Moreover, the trend of G_0 / E_D vs. K_D is very similar to that of

G_o / q_c vs. D_R (Jamiolkowski et al. 1988) which confirms that the dilatometer modulus can be regarded as a very large strain parameter, like the point resistance. The good correlation between G_o and E_D once again suggests the possibility of establishing such a correlation for a specific site for the already discussed reasons.

To better understand the engineering significance of the dilatometer moduli E_D and E_D^{UR} measured during dilatometer tests, they were compared with those that result from triaxial, torsional shear and resonant column tests. In particular, an attempt was

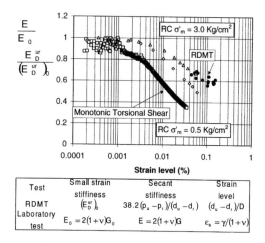

Test	Small strain stiffness	Secant stiffness	Strain level
RDMT	$(E_D^{ur})_o$	$38.2 (p_u - p_r)/(d_u - d_r)$	$(d_u - d_r)/D$
Laboratory test	$E_o = 2(1+\nu)G_o$	$E = 2(1+\nu)G$	$\varepsilon_a = \gamma/(1+\nu)$

Figure 21 E_D^{ur} vs. laboratory test stiffnes for Toyoura sand

$$(E_D^{ur})_o = \frac{2(1+\nu)}{(1-\nu^2)} \cdot 900 \cdot \frac{(2.17-e)^2}{1+e} \cdot \sigma_m^{0.4}, \quad \sigma_m = \frac{1}{3}\left[\sigma_v + (p_u + p_r)\right], \quad \nu = 0.2$$

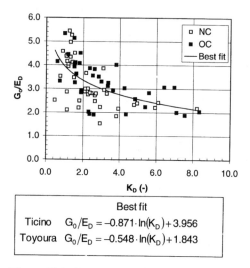

Best fit	
Ticino	$G_0/E_D = -0.871 \cdot \ln(K_D) + 3.956$
Toyoura	$G_0/E_D = -0.548 \cdot \ln(K_D) + 1.843$

Figure 19 G_0/E_D vs. K_D for Ticino sand

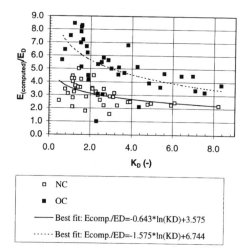

☐ NC

■ OC

—— Best fit: Ecomp./ED=-0.643*ln(KD)+3.575

······ Best fit: Ecomp./ED=-1.575*ln(KD)+6.744

Figure 20 E (computed)/E_D vs. K_D for Ticino sand

made to correlate the dilatometer modulus E_D to a large strain secant Young modulus and, in particular, to the Young modulus corresponding to an axial strain ε_a of 0.1 % in a TX test. It is believed that this value of strain corresponds to the average vertical strain involved in the settlement of well designed shallow foundations on sands. This Young's modulus was computed by means of the following equation (Fahey and Carter 1993):

$$E = E_o \cdot \left[1 - f \cdot (q/q_{max})^g\right] \qquad (28)$$

The experimental coefficients f and g as well as the expression to compute the small strain Young's modulus (E_o) and the maximum deviator stress (q_{max}) have been summarised by Pallara (1996). These parameters were obtained from a huge database of TX test results. Figure 20 shows the ratio E(computed)/E_D vs. K_D. It is possible to observe two different trends for NC and OC tests. This difference again confirms that E_D is a measure of the sand response after it has been subjected to large straining along a complicated path during penetration of the dilatometer blade, i.e. E_D represents an elasto-plastic soil response at large strains which is not sensitive to the stress history of sand since the penetration of the dilatometer blade almost completely obliterates the influence of the stress-history on the conventional dilatometer stiffness.

The unload-reload moduli E_D^{ur} are about one order of magnitude greater than the conventional modulus E_D and can be compared to the stiffness obtained in the laboratory from cyclic tests. The E_D^{ur} values were compared to the stiffness obtained from RC tests and monotonic loading torsional shear tests. Fig. 21 shows this comparison for Toyoura sand. The unload reload moduli from RDMT are consistent with those obtained in the laboratory from cyclic tests if the strain level and pressure are taken into account. On the other hand the RDMT stiffness is much greater than that inferred from monotonic loading tests. The above considerations apply to reconstituted sands but nothing is known about the aged sands.

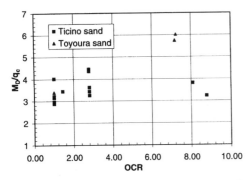

Figure 22 M_D/q_c ratio for reconstituted silica sand

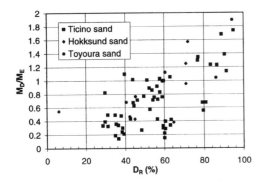

Figure 23 M_D/M_E ratio for reconstituted silica sand

The DMT is also used to infer the constrained modulus M according to well defined procedures (Marchetti 1980, 1997). Figure 22 shows the M_D/q_c ratio vs. OCR for very dense reconstituted silica sands (M_D = dilatometer constrained modulus). Figure 23 shows the M_D/M_E ratio vs. D_R for reconstituted NC silica sands (M_E = constrained modulus from unloading stage).

5.3 Self Boring Pressuremeter Tests

Different pressuremeter devices are used to in situ assess the stiffness of coarse grained, mostly sandy soils:

- Menard Pressuremeter (MP) is generally inserted into a pre-bored hole. This device was developed in the early sixties in France and is widely used in the case of direct interpretation methods (Baguelin et al. 1978);

- Self Boring Pressuremeter (SBP) was developed in the early seventies in UK and France (Baguelin and Jezequel 1973 and Wroth and Hughes 1973). This device, thanks to the fact that the probe insertion occurs with a very limited soil disturbance, allows, at least in principle, the assessment of the in situ total horizontal stress σ_{ho} and of the soil stiffness.

- Cone Pressuremeter (CPM) is a standard (1000 mm²) or larger (1500 mm²) cone with a pressuremeter module incorporated behind the tip (Withers et al. 1986, 1989, Schnaid 1990, Yu 1990, Ghionna et al. 1995). CPM tests provide the results which can be obtained from a typical CPT and those obtainable from pressuremeter tests.

A pressuremeter test consists of the expansion of a cylindrical cavity which has a finite length L and diameter D. During the test, the applied cavity pressure (p) and the corresponding circumferential strain at the cavity wall ε are measured. The test yields an expansion curve of the type shown in Figure 24 which allows, at least in principle, the direct determination of the following shear moduli:

- G_o from the initial slope of the expansion curve

- G_{ur} from small unload-reload cycles which can be performed during the expansion curve or even during the contraction phase of the test.

These moduli can be directly determined from the expansion curve because in both cases it is possible to neglect soil non-linearity as a first approximation. However, the initial shape of the expansion curve is sensitive to soil disturbance and to the compliance of the measuring system even in the case of SBP tests. Consequently, the direct assessment of G_o from the expansion curve is not possible in practice, while the determination of a pseudo-elastic stiffness from small unload-reload cycles, as suggested by Wroth (1982), seems quite suitable and reliable if appropriate assumptions are made (Robertson and Hughes 1986, Bellotti et al. 1989, Byrne et al. 1990, Fahey 1991, Fahey and Carter 1993, Ghionna et al. 1994).

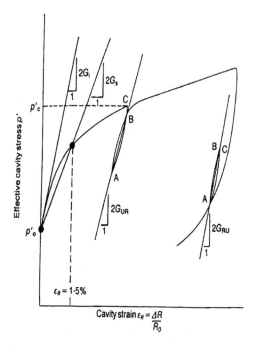

Figure 24 Schematic of shear moduli from SBP tests

- relative depth of the loop

$$R = \frac{p'_C - p'_A}{2 \cdot \sin\varphi'_{PS} \cdot p'_C /(1 + \sin\varphi'_{PS})} \qquad \text{(Fahey \quad 1991)}$$

(φ'_{PS} = plane strain angle of shear resistance which can be inferred from the expansion curve as suggested by Hughes et al. 1977)

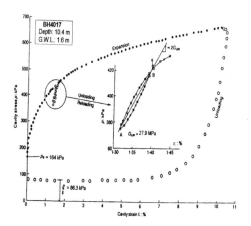

Figure 25 Example of SBPT in Po river sand

By referring to the SBP expansion curve obtained at the Viadana site (Bruzzi et al 1986 and Bellotti et al. 1989) in Holocene Po river sand (Figure 25), the G_{ur} value of the loop can be computed by means of the following expression:

$$G_{ur} = \frac{1}{2} \frac{(p_B - u_o) - (p_A - u_o)}{\varepsilon_B - \varepsilon_A} \qquad (29)$$

where: p_A, p_B = total cavity stress applied at points A and B respectively; ε_A, ε_B = corresponding circumferential strains at the cavity wall; u_o = hydrostatic pore pressure.

If one considers the pressure-dependency and non-linearity of soil stiffness, recalled in section 2, it is evident that any rational interpretation of the ur-loop should attempt to link the G_{ur} value to the following factors:
- average plane strain effective stress $s' = \frac{1}{2}(\sigma'_1 + \sigma'_3) = \frac{1}{2}(\sigma'_r + \sigma'_\theta)$ existing at the start of the loop which is a function of p'_C (σ'_r = radial effective stress; σ'_θ = circumferential effective stress, p'_C = effective cavity stress at the start of unloading)

The R relates the current depth of the unload loop ($p'_C - p'_A$) to the amount of stress unload at which the reverse plasticity can occur (see Wroth 1982). In practice, R plays the same role as the shear stress ratio τ / τ_{max} when describing the non-linearity of soil stiffness measured in laboratory tests (τ / τ_{max} = mobilisation factor, i.e. the ratio of the current shear stress to that at failure).

Reliable values of G_{ur} can be obtained if tests are performed according to the following suggestions:
- after the expansion pressure p_C is attained, the test should be paused for a rest period from 10' to 15' at constant cavity pressure, before starting unloading, in order to anticipate creep deformation. This is aimed at obtaining a regular loop with a clearly defined cross-over point B which, otherwise, could be obscured by the on going creep deformation.
- the depth of the ur-loop should not exceed $0.5 \cdot R$, as recommended by Fahey (1991). In this case, the condition that the stress-state on the horizontal plane, at point A, is isotropic ($\sigma'_r = \sigma'_\theta$) is satisfied. This precaution largely reduces the amount of hysteresis that the soil elements experience during the test.

84

Due to the compliance and limited accuracy of the system which measures the cavity displacements, the lower limit of the measurable $\varepsilon_B - \varepsilon_A$ amplitude results to be from one to two orders of magnitude higher than the linear threshold strain.

As far as the use of G_{ur} is concerned in practice, the two following approaches can be envisaged:

- G_{ur} values are used to back extrapolate G_o (Bellotti et al. 1989, Byrne et al. 1990, Fahey 1991, Fahey and Carter 1993, Ghionna et al. 1994). In order to infer G_o from measured G_{ur}, two fundamental steps are required. First, a $G_o(SBP)$ is inferred from the measured G_{ur} by means of one of the commonly used quasi-linear relationships. The so obtained $G_o(SBP)$ is then corrected to take the pressure dependence of G_o into account. In fact, when the ur-loop is performed the average plane strain effective stress around the cavity is much higher than that existing at the beginning of the expansion curve. This correction is carried out by means of the following equation (Bellotti et al. 1989)

$$G_o(SBP)_c = G_o(SBP) \cdot \left(\frac{s_o'}{s_{AV}'} \right)^n \qquad (30)$$

where: $G_o(SBP)_c = G_o(SBP)$ corrected for the stress level; $s_{AV}' = $ average plane strain effective stress around the cavity (point C); $s_o' = \sigma_{ho}' = $ average plane strain effective stress before expansion; n = stress exponent which ranges from 0.4 and 0.5 for poorly graded sands and increases with an increase of the coefficient of uniformity up to 0.6.

An example of application of the approach proposed by Ghionna et al. (1994) to the results of Camkometer tests performed on Ticino sand in a CC and in the Holocene Po river sand deposit is shown in Fig. 22. The $G_o(SBP)_c$ values inferred from the CC tests have been compared to those obtained in the laboratory from Resonant Column tests on the same sand. The $G_o(SBP)_c$ values inferred from in situ tests are compared to those measured at the same site by means of CH tests.

The comparison proposed by Bellotti et al. (1989) also takes into account the problems of stiffness anisotropy and of different numbers of cycles experienced by the soil element in different types of tests.

The assessment of G_o and τ_{max} in situ from pressuremeter tests, allows one to define the parameters of conventional quasi-elastic relationships.

- the second approach compares the measured G_{ur} values to the $G/G_o - \gamma$ decay curves that result from laboratory tests. It is also mandatory in this case to link the measured G_{ur} to the average values of shear strain (γ_{AV}) and mean plane effective stress experienced by the soil around the expanding cavity. A simplified approach of this kind can be found in the work by Bellotti et al. (1989) who assume that soil behaves as an elastic-perfectly-plastic medium.

What has been here mentioned refers to the experience gained from SBP tests. Considering low sensitivity of G_{ur} to the disturbance caused by probe installation, at least in principle, other types of pressuremeter tests offer the possibility of assessing a meaningful value of shear modulus from an ur-loop in unaged uncemented coarse grained soils (i.e. Ghionna et al. 1995). However, in aged deposits the insertion of other than self-boring probes causes an unavoidable destructuration of the soil fabric and consequent understimate of the stiffness.

6. MATERIAL DAMPING

Various methods are currently used to determine the damping ratio D_o from seismic tests, among these the following are worth mentioning:

1) The spectral ratio method (Mok 1987, Fuhriman 1993) is based on the following assumptions which hold only in the far field:
- the amplitude of the body waves decreases in proportion to r^{-1}, where r is the distance from the input source, because of the so called geometric damping
- body wave attenuation, due to material damping, is proportional to the frequency
- the soil-receiver transfer function can be considered to be identical for both receivers

Based on the above assumptions, the damping ratio can be computed by means of the following equation:

$$D(f) = \frac{\ln[A_1(f) \cdot r1 / A_2(f) \cdot r2]}{\Phi(f)} \qquad (31)$$

where: r1 and r2 are the distances of a pair of receivers from the source, $A_1(f)$ and $A_2(f)$ are the amplitude spectra at the two receivers and $\Phi(f)$ is the phase of the wave velocity or phase difference between the two receivers.

2) The spectral slope method, originally developed for downhole measurements (Redpath et

al. 1982, Redpath and Lee 1986), differs from the previous one in that it assumes that material damping is frequency independent and that it is not necessary to define the law for geometric damping.

The attenuation constant, defined as the ratio of attenuation to frequency $k = \alpha / f$ represents the spectral slope, i.e. the slope of the spectral ratio against frequency:

$$k = \frac{-\Delta\{\ln[A_1(f)/A_2(f)]\}}{\Delta f(r2 - r1)} \qquad (32)$$

therefore material damping can be computed by means of the following expression:

$$D(f) = \frac{-\Delta\{\ln[A_1(f)/A_2(f)]\}}{\Delta f \cdot 2\pi \cdot \Delta t(f)} \qquad (33)$$

Campanella and Stewart (1990) studied the applicability of the above described methods to downhole SCPT's. They found that the spectral slope method provides more realistic values of material damping.

However, none of the available methods seem to be sufficiently consolidated and reliable enough for every day use.

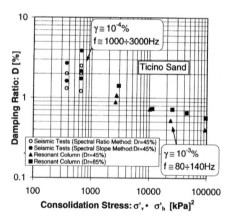

Figure 26 Damping ratio from laboratory and in situ tests on reconstitute sand

Examples of damping computation, with spectral ratio and spectral slope methods, are given in Figure 26 (Puci and Lo Presti 1998). In this Figure the damping ratio values obtained for dry, reconstituted, Ticino sand from RC tests and from seismic tests performed on large size specimens are compared. It is possible to see that the small strain damping ratio from seismic tests is systematically greater than that obtained from RC tests. Different reasons can be offered to explain the above discrepancy: 1) the use, in the case of seismic tests, of frequencies that are much higher than those adopted in RC tests, 2) the

assumption of a geometric law for radiation damping which is probably not realistic in the case of heterogeneous and anisotropic media like soils (Rix et al 1997).

The back analysis of in situ soil model tests (Konno et al. 1991) and the use of the logarithm decrement method for the analysis of the transient response of free field earthquake records (Huerta et al. 1994) have been used to obtain in situ D vs. γ curves.

Typical D vs. γ curves from laboratory tests on reconstitutted granular soils are shown in Figure 27.

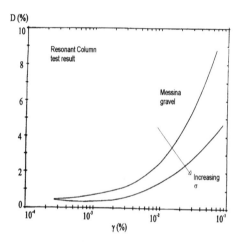

Figure 27a Damping ratio vs. shear strain for Messina gravel

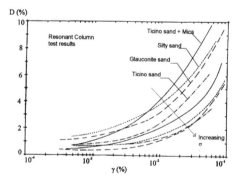

Figure 27b Damping ratio vs. shear strain for various sand

7. FINAL REMARKS

The present capability of in situ testing techniques to assess the relative density, stiffness and secant peak angle of shearing resistance of granular soils have been reviewed in this paper. In particular, the advantages and limitations of penetration, dilatometer, pressuremeter and seismic tests have been critically reviewed. Hereafter, lights and shadows issuing from this review are briefly summarised:

- the assessment of the relative density of unaged, uncemented sandy soils using CPT DMT and SPT is quite well established. The major uncertainties arise as far as structured deposit or gravely soils are concerned.

- similar considerations hold as far as the assessment of the secant peak angle of shearing resistance using penetration tests is concerned.

- the assessment of φ'_p using pressuremeter tests follows a more rational approach which requires the knowledge of a set of input parameters not always available in everyday practice.

- the enhancement and development of seismic tests (CH, DH, SCPT, SASW) to assess G_o and its use, as a reference value, in design practice is one of the most important goal obtained over the last decade. It is also possible to obtain good correlations between G_o and the penetration test results. The determination of such empirical correlations for a specific site makes it possible to consider the spatial variability of soil properties in a very cost effective way.

- the assessment of soil stiffness using SBPT and in particular interpreting ur-loops requires the knowledge of average values of shear strain (γ_{AV}) and mean plane effective stress experienced by the soil around the expanding cavity. In fact, it is mandatory to link the measured G_{ur} to these values. However, it is not known to which extent the expansion of the probe can cause destructuration of the soil fabric in the case of aged and cemented soil deposits. This point deserve further research.

- the assessment of damping ratio from in-situ tests is still restricted to research purpose.

8. REFERENCES

Adachi K. & K. Tokimatsu 1994. Development in Sampling of Cohensioless Soils in Japan. *Proc. XIII ICSMFE, New Delhi, Inde.*

Ajalloeian R & H.S. Yu. 1996. Chamber Studies of the Effects of Pressuremeter Geometry on test Results in Sand. *RR. 14109, Dept. of Civil Engineering and Surveying, The University of Newcastle, New South Wales, Australia.*

Anderson, D.G. & K.H. Stokoe 1978. Shear Modulus: a Time Dependent Soil Property. *Dynamic Geotechnical Testing ASTM STP 654*: 66-90.

Arthur J.R.F. & B.K. Menzies 1972. Inherent Anisotropy in Sand. *Géotechnique.* 22(1):115-128.

Atkinson J.H., Richardson D. & S.E. Stallebrass 1990. Effect of Stress History on the Stiffness of Overconsolidated Soil. *Géotechnique.* 40(4):531-540.

Auld B. 1977. Cross-Hole and Down-Hole V_S by Mechanical Impulse. *Journal GED, ASCE.* 103 (GT12): 1381-1398.

Baguelin F. & J.F. Jezequel 1973. Le Pressiometre Autoforeur. *Annales de l'ISTBTP.* 97:135-159.

Baguelin F., Jezequel J.F. & D.H. Shields 1978. The Pressuremeter and Foundation Engineering. *Trans. Tech. Publication.*

Baldi, G., Bellotti, R., Ghionna, V., Jamiolkowski, M.& E. Pasqualini 1986. Interpretation of CPTs and CPTUs - Part II: Drained Penetration in Sands. *Proc. of 4th International Geotechnical Seminar on Field Instrumentation and In Situ Measurements, Singapore.*

Baldi G., Bruzzi, D., Superbo, S., Battaglio, M. & M. Jamiolkowski 1988. Seismic Cone in Po River Sand. *Proc. ISOPT-1, Orlando, Fla.* 2: 643-650.

Baldi, G., Bellotti, R., Ghionna, V.N., Jamiolkowski, M. & D.C.F. Lo Presti 1989a Modulus of Sands from CPT's and DMT's. *Proc. XII ICSMFE, Rio de Janeiro.* 1: 165-170.

Baldi, G., Jamiolkowski, M., Lo Presti, D.C.F., Manfredini, G. and G.J. Rix 1989b. Italian Experience in Assessing Shear Wave Velocity from CPT and SPT. *Earthquake Geotechnical Engineering, Proc. of Discussion Session on Influence of Local Conditions on Seismic Response, XII ICSMFE, Rio de Janeiro, Brasil*: 157-168

Baligh M.M. 1975. Theory of Deep Site Static Cone Penetration Resistance *RR. R-75-56, Cambridge, Mass.*

Baligh M.M. 1976.Cavity Expansion in Sands with Curved Envelopes. *Journal of GED, ASCE*: 102 (GT11): 1131-1146.

Baligh, M.M., Azzouz, A.S., Wissa, A.Z.E., Martin, R.T. & M.J. Morrison 1981. The Piezocone Penetrometer. *Proc. of Symposium on Cone Penetration Testing and Experience, ASCE National Convention, St Louis, Missouri.*

Barton M.E. & S.N. Palmer 1989. The Relative Density of Geologically Aged British Fine and

Fine-medium Samds. *Quaterly Journal of Engineering Geology.* 1: 49-58.

Bellotti, R., Ghionna, V.N., Jamiolkowski, M., Lancellotta, R. & G. Manfredini 1986. Deformation Characteristics of Cohesionless Soils from in Situ Tests. *Use of In Situ Tests in Geotechnical Engineering, Geotechnical Special Pubblication, No. 6, S.P. Clemence, Ed., ASCE*: 47-73.

Bellotti, R., Ghionna, V.N., Jamiolkowski, M., Robertson, P.K. & R.W. Peterson 1989 Interpretation of Moduli from Self-Boring Tests in Sand *Géotechnique.* 39(2): 269-292.

Bellotti R., Jamiolkowski M., Lo Presti D.C.F. & D.A. O'Neill 1996. Anisotropy of Small Strain Stiffness in Ticino Sand. *Geotechnique.* 46(1): 115-131.

Bellotti R., Benoit J., Fretti C. & M. Jamiolkowski 1997. Stiffness of Toyoura Sand from Dilatometer Tests. *Journal of GGE, ASCE*: 123(9): 836-846.

Berezantzev W.G. 1964. Calculation of Foundations Basis *Edit. Construction Literature, Leningrad* (in Russian)

Berezantzev W.G. 1970. Calculation of the Construction Basis *Edit. Construction Literature, Leningrad* (in Russian)

Bieganousky W.A. & W.F. Marcuson 1976 Laboratory Standard Penetration Tests on Reid Bedford Model and Ottawa Sands *Research Report S-76-2, No. 1, Waterways Experiment Station, Vicksburg.*

Bieganousky W.A. & W.F. Marcuson 1977 Laboratory Standard Penetration Tests on Platte River Sand and Standard Concrete Sand *Research Report S-76-2, No. 2, Waterways Experiment Station, Vicksburg.*

Bolton M.D. 1986. The Strength and Dilatancy of Sands" *Geotechnique.*36(1): 65-7

Bruzzi D., Ghionna V.N., Jamiolkowski M., Lancellotta R. & G. Manfredini 1986. Self-Boring Pressuremeter in Po River sand. *Proc. 2nd Int. Symposium on Pressuremeter and Its Marine Applications. ASTM STP 950*: 57-74

Burland, J.B. 1989.Small Is Beautiful-The Stiffness of Soil at Small Strains *Ninth Laurits Bjerrum Memorial Lecture, Canadian Geotechnical Journal.* 26(4): 499-516.

Bustamante M. & L. Gianneselli 1982. Pile Bearing Capacity Prediction by means of Static Penetrometer CPT. *Proc. ESOPT II, Amsterdam.* 2: 493-500, Balkema.

Byrne P.M., Salgado F.M. & Howie J.A. 1990. Relationship between the Unload Shear Modulus from Pressuremeter Tests and the Maximum

Shear Moduli of Sand. *Proc. 3rd Int. Symposium Pressuremeter, Oxford*: 231-242.

Campanella, R.G. & P.K. Robertson 1981. Applied Cone Research. *Dept. of Civil Engineering, University of British Columbia, Vancouver, Soil Mechanics Series*, N. 46.

Campanella, R.G. & P.K. Robertson 1984- A Seismic Cone Penetrometer to Measure Engineering Properties of Soil. *Society Of Exploration Geophisics, Atlanta, Ga.*

Campanella R.G., Robertson P.K. Gillespie D & J. Greig 1985. Recent Development in In-Situ Testing of Soils. *Proc. of XI ICSMFE, S. Francisco.* 2: 849-54. Balkema, Rotterdam

Campanella, R.G. & W.P. Stewart 1990. Seismic Cone Analysis Using Digital Signal Processing for Dynamic Site Characterization. *43rd Canadian Geotechnical Engineering Conference, Quebec City.*

Chung, R.M., Yokel, F.Y. & H. Wechsler 1984. Pore Pressure Buildup in Resonant Column Tests. *Journal of GED, ASCE.* 110(2): 247-261.

Crova, R. Jamiolkowski M., Lancellotta R. & D.C.F. Lo Presti 1992. Geotechnical Characterization of Gravelly Soils at Messina Site *Selected Topics, Proceedings of the Wroth Memorial Symposium, Thomas Telford, London*: 199-218.

De Beer E.E. 1965 Influence of the mean normal stress on the Shear Strength of Sand *Proc. VI ICSMFE, Montreal.*

Decourt L. 1989. General Report Discussion Session 2. SPT, CPT, Pressuremeter Testing and Recent Developments in In-Situ Testing- Part 2: The SPT State of the Art Report. *Proc. XII ICSMFE, Rio de Janeiro, 1989, Balkema* 4: 2405-2416

Delladonna F. 1988. Interpretation of Penetration Tests Performed in Calibration Chamber *M. Sc. Thesis, Department of Structural Engineering, Politecnico di Torino* (in Italian).

De Mello V.F.B 1971 The Standard Penetration Test *SOA Report, Proc. IV Panamerican Conf. on SMFE, San Juan, Puerto Rico.*

De Ruiter J. 1981. Current Penetrometer Practice. *Proc. Symposium on Cone Penetration Testing and Experience, ASCE National Convention, St Louis, Missouri.*

Dobry, R., Ladd, R.S., Yokel, F.Y., Chung, R.M. & D. Powell 1982. Prediction of Pore Water Pressure Build-up and Liquefaction of Sands During Earthquake by the Cyclic Strain Method. *National Bureau of Standards Building, Science Series 138, Washington, D.C.*

Durgunoglu N.T. & J.K. Mitchell 1973. Static Penetration Resistance of Soils. *R.R. prepared for NASA, University of California, Berkley*

Dussealt M.B. & Morgenstern N.R. 1979. Locked Sands. *Quaterly Journal of Engineering Geology.* 2: 117-131

Fahey, M. & M.F. Randolph 1984. Effect of Disturbance on Parameters Derived from Self-Boring Pressuremeter Tests in Sand *Géotechnique.* 1: 81-97.

Fahey, M. 1991. Measuring Shear Modulus in Sand with the Self-Boring Pressumeter *Proc. X ECSMFE, Florence, Italy, Balkema, Rotterdam.* 1: 73-76.

Fahey, M. & J.P. Carter 1993. A Finite Element Study of the Pressuremeter Test in Sand Using a Non-Linear Elastic Plastic Model. *Canadian Geotechnical Journal.* 30: 348-362.

Felice A. 1997. Determinazione dei Parametri Geotecnici e in Particolare di Ko da Prove Dilatometriche. *M.Sc. Thesis, Department of Structural Engineering, Politecnico di Torino.*

Fioravante V., Fretti C., Froio F., Jamiolkowski M., Lo Presti D.C.F. & S. Pedroni 1998a Anisotropy of elastic stiffness in Kenya Sand. *Submitted to Géotechnique.*

Fioravante V., Jamiolkowski M., Lo Presti D.C.F., Manfredini G. & S. Pedroni 1998b. Assessment of coefficient of earth pressure at rest from shear wave velocity measurements *Géotechnique, in print.*

Fretti, C., Lo Presti, D.C.F. & R. Salgado 1993 The Research Dilatometer: in Situ and Calibration Chamber Test Results. *Rivista Italiana di Geotecnica.* XXVI(4): 237-243.

Fuhriman, M.D. 1993. Cross-Hole Seismic Tests at two Northern California Sites Affected by the 1989 Loma Prieta Earthquake. *M. Sc. Thesis, The University of Texas at Austin.*

Garizio G.M. 1997. Determinazione dei Parametri Geotecnici e in Particolare di Ko da Prove Penetrometriche *M.Sc. Thesis, Department of Structural Engineering, Politecnico di Torino.*

Ghionna V.N. 1984. Influence of Chamber Size and Boundary Conditions on the Measured Cone Resistance. *Proc. Seminar on Cone Penetration Testing in the Laboratory, Dept. of Civil Engineering, University of Southampton.*

Ghionna, V., Karim, M. & S. Pedroni 1994 Interpretation of Unload-Reload Modulus from Pressuremeter Tests in Sand *Proc. XIII ICSMFE New Delhi, India*: 115-120.

Ghionna V.N., Jamiolkowski M., Pedroni S. & S. Piccoli 1995. Cone Pressuremeter Tests in Po River Sand. *Proc. of the Conference on The Pressuremeter and its New Avenue, Sherbrooke, Canada*: 471-480.

Gibbs H.J. & W.G. Holtz 1957. Research on Determining the Density of Sands by Spoon Penetration testing. *Proc. IV ICSMFE.* 1: 35-39, London

Goto, S., Suzuki, Y., Nishio, S. & Oh Oka, H., 1992 Mechanical Properties of Undisturbed Tone-River Gravel obtained by In-Situ Freezing Method. *Soils and Foundations.* 32(3): 15-25.

Hashiguchi K. 1985. Two and Three-Surface Models of Plasticity. *Proc. V Int. Conference on Numerical Methods in Geomechanics.*

Hall, J.R. 1985. Limits on Dynamic Measurements and Instrumentation. *Richart Commemorative Lectures, ASCE*: 108-119.

Hanna A.M. & H. Youssef 1987. Evaluation of Dilatancy Theories for Granular Materials. *Proc. IS on Prediction and Performance in Geotechnical Engineering*: 227-236. Calgary

Harder, L.F. & H.B. Seed 1986. Determination of Penetration Resistance for Coarse-Grained Soils Using the Becker Hammer Drill. *Report No. UCB/EERC-86/06, University of California, Berkeley, Ca.*

Harman E.D. 1976 A Statistical Study of Static Cone Bearing Capacity. *M. Sc. Thesis, University of Florida, Gainsville, Fla.*

Hatanaka M., Sugimoto M. & Suzuki Y. 1985. Liquefaction Resistance of Two Alluvial Volcanic Soils Sampled by In Situ Freezing. *Soils and Foundations.* 25(3): 49-63.

Hatanaka, M., Suzuki, Y., Kawasaki, T. & M. Endo 1988. Cyclic Undrained Shear Properties of High Quality Undisturbed Tokyo Gravel. *Soil and Foundations.* 28(4): 57-68.

Hatanaka M. & Uchida A. 1996. Empirical Correlation between Penetration Resistance and Internal Friction Angle of Sandy Soils. *Soils and Foundations.* 36(4): 1-10.

Holtz W.G. 1972. The Density Approach: Uses, Testing Requirements, Reliability and Shortcomings *ASTM STP 523*: 5-17.

Huerta, C.I., Acosta, J., Roesset, J.M. & K.H. Stokoe 1994. In Situ Determination of Soil Damping from Earthquake Records. *Proc. of ERCAD Berlin.* Balkema. 1: 227-234.

Hvorslev, M.J. 1937. On the Strength Properties of Remoulded Cohesive Soils. *Thesis, Danmarks Naturvidenskabelige Samfund, Ingenior videnskabelige Skrifter, Copenhagen.*

Hughes Y.M.O, Wroth C.P. & Windle D. 1977 Pressuremeter Tests in Sands *Géotechnique.* 27(4): 455-477.

Imai, T. & Tonouchi, K. 1982 Correlations of N-Values with S-Wave Velocity. *Proc. ESOPT II, Amsterdam.* 2: 67-72.

Isenhower, W.M. & Stokoe, K.H. 1981. Strain Rate Dependent Shear Modulus of San Francisco Bay Mud. *International Conference on Recent*

Advances in Geotechnical Earthquake Engineering and Soil Dynamics, St. Louis, Missouri.

Jamiolkowski, M., Ghionna, V.N., Lancellotta, R. & Pasqualini, E. 1988 New Correlations of Penetration Test in Design Practice. *Proc. of ISOPT-I, Orlando, Fla.*

Jamiolkowski M. & Robertson P. 1988 Closing address: future trends for penetration testing. *Geotechnology Conference: Penetration Testing in the UK, Birmingham*: 321-342, Thomas Telford

Jamiolkowski M. 1989. Shear Strength of Cohesionless Soils from CPT *De Mello Volume*: 191-204.

Jamiolkowski, M., Lancellotta, R. & Lo Presti, D.C.F., 1995a Remarks on the Stiffness at Small Strains of Six Italian Clays *Keynote Lecture 3, IS Hokkaido:* 2:. 817-836, Balkema

Jamiolkowski M., Lo Presti D.C.F. & Pallara O. 1995b Role of in Situ Testing in Geotechnical Earthquake Engineering *SOA7, Third ICRAGEE&SD, S. Louis 1995*: III: 1523-1546.

Janbu N. & Sennesset K. 1974 Effective Stress Interpretation of In Situ Static Penetration Tests *Proc. ESOPT I, Stockolm.* 2.2: 181-193.

Jardine, R.J. 1985. Investigations of Pile-Soil Behaviour with Special Reference to the Foundations of Offshore Structures. *Ph.D. Thesis, University of London, London.*

Jardine, R.J., St John, H.D., Hight, D.W. & Potts, D.M. 1991. Some Pratical Applications of a Non-Linear Ground Model. *Proc. of X ECSMFE, Florence.* 1: 223-228. Balkema

Jardine, R.J. 1992. Some Observations on the Kinematic Nature of Soil Stiffness. *Soils and Foundations.* 32(2): 111-124

Jardine, R.J. 1995. One Perspective of the Pre-Failure Deformation Characteristics of Some Geomaterial. *Keynote Lecture 5, IS Hokkaido.* 2. 855-886. Balkema

Keaveny J.M. 1985. In Situ Determination of Drained and Undrained Soil Strength Using the Penetration Test. *Ph. D. Thesis, University of California, Berkeley*

Jaroshenko W.A. 1964. Interpretation of Static Cone Penetration Test in Sandy Soils. *Fundamento Proekt. No. 1964* (in Russian)

Ko H. Y. and Scott R.F. 1968. Deformation of Sand at Failure *Journal of SMFE, ASCE.* 94(SM4): 883-898.

Kokusho, T. 1987. In-situ Dynamic Soil Properties and Their Evaluation. *Proc. 8th Asian Regional Conference on SMFE.* 2: 215-340.

Konno, T., Suzuki, Y., Tateishi, A., Ishihara, K., Akino, K. & Iizuka, S. 1991. Gravelly Soil Properties by Field and Laboratory Tests. *Proc. of 3rd International Conference on Case Histories in Geotechnical Engineering, S. Louis.* Paper 3.12.

Konrad J-M. 1990. Sampling of Saturated and Unsaturated Sands by Freezing *Geotechnical Testing Journal.* 13(2): 88-96.

Kovacs, W.D. & Salomone, L.A. 1982 SPT Hammer Energy Measurements. *Journal of GED, ASCE.* No. GT4.

Lade P. V. & Duncan J.M. 1973. Cubical Triaxial Tests on Cohesionless Soil. *Journal of SMFE, ASCE.* 99(SM10): 793-813.

Laier J. 1973. Effect of Pressuremeter Probe Length to Diameter Ratio and Borehole Disturbance on Pressuremeter Test Results in Dry Sand. *Ph. D. Thesis, University of Florida, Gainsville, Fla.*

Lam W-K. & Tatsuoka F. 1988 Effects of Initial Anisotropic Fabric and σ_2 on Strength and Deformation Characteristics of Sand. *Soils and Foundations.* 28(1): 89-106.

Lee, S.H. & Stokoe, K.H. II 1986. Investigation of Low Amplitude Shear Wave Velocity in Anisotropic Material. *Geotechnical Engineering Report GR 86-6, University of Texas at Austin.*

Liao S.C. & Withmann R.V. 1986. Overburden Correction Factors for SPT in Sand. *Journal of GED, ASCE.* 112(3): 373-377.

Lo Presti D.C.F. 1989. Proprietà Dinamiche dei Terreni. *Proc. XIV Conferenza Geotecnica di Torino, Dipartimento di Ingegneria Strutturale, Poltecnico di Torino.*

Lo Presti, D.C.F. & Lai, C,. 1989. Shear Wave Velocity in Soils from Penetration Tests. *R.R. Dip. di In. Strutturale, Politecnico di Torino.* (21).

Lo Presti, D.C.F. & O'Neill, D.A. 1991. Laboratory Investigation of Small Strain Modulus Anisotropy in Sand. *Proc. of ISOCCT1, Postdam, NY.*

Lo Presti, D.C.F. 1995. Measurement of Shear Deformation of Geomaterials from Laboratory Tests. *General Report Session 1a, IS Hokkaido.* 2: 1067-1088. Balkema

Lo Presti D.C.F., Jamiolkowski M., Pallara O., Pisciotta V. & Ture S. 1995. Stress Dependence of Sand Stiffness. *3rd International Conference on Recent Advances in Geotechnical Earthquake Engineering and Soil Dynamic.* 1: 71-76.

Lo Presti D.C.F., Jamiolkowski M., Pallara O. & Cavallaro A.. 1996. Rate and Creep Effect on the Stiffness of Soils. *GSP No. 61, ASCE*: 166-180

Lo Presti D.C.F., Jamiolkowski M., Pallara O., Cavallaro A. & Pedroni S. 1997a. Shear Modulus and Damping of Soils *Géotechnique Symposium in Print.* 43(3): 603-617.

Lo Presti D.C.F., Pallara O. & Cavallaro A. 1997b. Damping Ratio of Soils from Laboratory and In-Situ Tests *Proc. XIV ICSMFE, Seismic Behaviour of Ground and Geotechnical Structures,* Balkema, Rotterdam: 391-400.

Mair R.J & Wood D.M. 1987. Pressuremeter Testing Methods and Interpretation. *CIRIA-Report, Edit Butterworths, London.*

Maksimovic 1989. Non linear Failure Envelope for Coarse Grained Soils. *Proc. XII ICSMFE.* 1: 731-734.

Manassero M. 1989. Stress-Strain Relationship from Drained Self-Boring Pressuremeter Tests in Sands *Géotechnique.* 39(2): 293-307.

Marchetti, S. 1980. In Situ Tests by Flat Dilatometer *Journal of GED, ASCE.* 106(GT3): 299-321.

Marchetti S. 1997. The Flat Dilatometer: Design Applications. *Proc. III Int. Geotechnical Engineering Conference, Cairo.* 421-428.

Mesri, G. 1987 Fourth Law of Soil Mechanics. *Proc. IS on Geotechnical Engineering of Soft Soils, Mexico City, Mexico.*

Mitchell J.K. & Keaveny J.M. 1986. Determining Sand Strength by Cone Penetrometer. *In Situ 86 Proc. Spec. Conf. GED ASCE, Virginia Tech., Blacksburg,* 1-86.

Mitchell, J.K., Lodge, A.L., Coutinho, R.Q., Kayen, R.E., Seed, R.B., Nishio, S. & Stokoe, K.H. II, 1994. In Situ Test Results from four Loma Prieta Earthquake Liquefaction Sites: SPT, CPT, DMT and Shear Wave Velocity. *EERC Report No. UCB/EERC - 94/04.*

Mok, Y.J. 1987. Analytical and Experimental Studies of Borehole Seismic Methods. *Ph.D. Thesis, Univ. of Texas at Austin, Austin, TX.*

Motan, S.E. & Khan, A.Q. 1988 "n Situ Shear Modulus of Sands by a Flat Plate Penetrometer: a Laboratory Study. *Geotechnical Testing Journal.* 11(4):. 257-262.

Mroz Z. 1967 On the Description of Anisotropic Work Hardening. J*our. of Mech. Phys. Solids.* (15): 163-175.

Muromachi, T. 1981. Cone Penetration Testing in Japan. *Proc. of Symposium on Cone Penetration Testing and Experience, ASCE National Convention, St Louis, Missouri.*

Nazarian S. & Stokoe K.H. 1983. Use of the Spectral Analysis of Surface Waves for Determination of Moduli and Thickness of Pavement Systems *Transortation Research Recod, No. 954, Washington, D.C.*

Nazarian, S. and Stokoe, K.H. 1984 In Situ Shear Wave Velocities from Spectral Analysis of Surface Waves *Proc. of the 8th World Conference on Earthquake Engineering,* San Francisco, CA. 3: 31-38.

Nazarian, S. and Stokoe, K.H. & Hudson, W.R., 1983 Use of the Spectral Analysis of Surface Waves Method for Determination of Moduli and Thickness of Pavement Systems *Research Recod, No. 930, Transortation Research Board*: 38-45.

Ochiai H. & Lade P.V. 1983. Three-Dimensional Behaviour of Sand with Anisotropic Fabric *Journal GED., ASCE.* 109(GT10): 1313-1328.

Oda M. 1972. The Mechanism of Fabric Changes During Compressional Deformation of Sand. *Soils and Foundations.* 12(2): 1-18.

Oda M., Koishikawa I. & Higuchi T. 1978. Experimental Study on Anisotropic Shear Strength of Sand by Plane Strain Test. *Soils and Foundations.* 18(1): 25-38.

Ohta, Y. & Goto, N. 1978. Empirical Shear Wave Velocity Equations in Terms of Characteristic Soil Indexes. *Earthquake Engineering and Structural Dynamics.* 6.

Pallara O. 1995. Comportamento sforzi-deformazioni di due sabbie soggette a sollecitazioni monotone e cicliche. *Ph. D. Thesis, Department of Structural Engineering, Politecnico di Torino.*

Pallara O. 1996. Verifica delle leggi elastiche quasi-lineari nelle sabbie mediante prove di laboratorio. *Rapporto di Ricerca, Aprile 1996, Department of Structural Engineering, Politecnico di Torino.*

Parkin A.K. & Lunne T. 1982. Boundary Effects in the Laboratory Calibration of a Cone Penetrometer for Sand. *Proc. of ESOPT II, Amsterdam*: 761-768.

Patel, N.S. 1981. Generation and Attenuation of Seismic Waves In Downhole Testing. *M. Sc. Thesis GT81-1, Department of Civil Engineering, University of Texas at Austin.*

Puci I & Lo Presti D.C.F. 1998. Damping Measurement of Reconstituted Granular Soils in Calibration Chamber by means of Seismic Tests. *Submitted to the 11 European Conference on Earthquake Engineering.*

Redpath, B.B., Edwards, R.B., Hale, R.J. & Kintzer, F.C. 1982 Development of Field Techniques to Measure Damping Values for Near Surface Rocks and Soils. *Prepared for NSF Grant No. PFR-7900192.*

Redpath B.B. & Lee R.C. 1986. In-Situ Measurements of Shear-Wave Attenuation at a Strong-Motion Recording Site. *Prepared for USGS Contract No. 14-08-001-21823.*

Reese L.C. & M.W. O'Neill 1988. Drilled Shafts: Construction Procedures and Design Methods. *FHWA-HI-88-42. Publication ADSC TL-4.*

Richardson, D. 1988. Investigations of Threshold Effects In Soil Deformation. *Ph. D. Thesis, The City University, London.*

Rix, G.J. 1984. Correlation of Elastic Moduli and Cone Penetration Resistance. *M. Sc. Thesis, The University of Texas at Austin.*

Rix, G.J. & Stokoe, K.H. II 1991. Correlation of Initial Tangent Modulus and Cone Penetration Resistance. *Proc. I ISOCCT, Postdam, New York.*

Rix G.J., Lai C.G. & Spang W. A. 1997. In Situ Measurement of Damping Ratio using Surface Waves. *Submitted to Journal of GGE, ASCE.*

Robertson P.K. 1982 In Situ Testing of Soil with Emphasis on Its Application to Liquefaction Assessment. *Ph. D. Thesis, The University of British Columbia, Vancouver.*

Robertson P.K. & Campanella R.G. 1983. Interpretation of Cone Penetration Tests - Part I: Sands. *Canadian Geotechnical Journal.* 20(4): 718-733.

Robertson P.K. & Campanella R.G. 1985. Liquefaction Potential of Sands Using the Cone Penetration Test. *Journal GED, ASCE.* 111(3): 383-403.

Robertson P.K. & Hughes J.M.O. 1986 Determination of Properties of Sand from Self-Boring Pressuremeter Tests. *Proc. II IS on the Pressuremeter and Its Marine Applications, ASTM STP 950*: 283-302.

Roesler, S.K. 1978. Correlation Methods in Soil Dynamic. *Proc. of Dynamic Method in Soil and Rock Dynamics, Karlrushe,* Balkema, Rotterdam. 1:. 309-334.

Roesler, S.K. 1979. Anisotropic Shear Modulus Due to Stress- Anisotropy *Journal of GED, ASCE.* 105(GT7): 871-880.

Rowe P.W. 1962 The Stress-Dilatancy Rotation for Static Equilibrium of an Assembly of Particles in Contact *Proc. of the Royal Society.* 269A.

Salgado R. 1993 Analysis of Penetration Resistance in Sands. *Ph. D. Thesis, University of California, Berkeley*

Schmertmann J.H. 1976. Updated qc-Dr Correlation. *Unpublished Report, University of Florida, Gainsville, Fla.*

Schmertmann, J.H. & Palacios, A. 1979. Energy Dynamics of SPT. *Journal of GED, ASCE.* 105(GT8): 909-926.

Schnaid F. 1990. A Study of the Cone Pressuremeter in Sand. *Ph. D. Thesis, Oxford University.*

Seed H.B., Idriss I.M. & Arango I. 1983. Evaluation of Liquefaction Potential Using Field Performance Data. *Journal of GED, ASCE.* 109(3): 458-482.

Seed H.B., Tokimatsu K., Harder L.F. & Chung R. 1985. Influence of SPT Procedures in Soil Liquefaction Resistance Evaluations. *Journal GED, ASCE.* 111(12): 1425-45.

Seed H.B., Wong. R.T., Idriss I.M. & Tokimatsu K. 1986. Moduli and Damping Factors for Dynamic Analysis of Cohesionless Soils. *Journal GED, ASCE.* 112(11): 1016-1032.

Shibuya S., Mitachi T., Fukuda F. & Degoshi T. 1995. Strain Rate Effect on Shear Modulus and Damping of Normally Consolidated Clay. *Geotechnical Testing Journal.* 18(3): 365-375.

Simpson B., O'Riordan N.J. & Croft D.D. 1979. A Computer Model for the Analysis of Ground Movements in London Clay. *Geotechnique.* 29(2): 149-175.

Singh S., Seed H.B. & Chan C.K. 1982. Undisturbed Sampling of Saturated Sands by Freezing. *Journal GED, ASCE.* 108(GT2): 247-264.

Skempton, A.W. 1986. Standard Penetration Tests Procedures & the Effects in Sands of Overburden Pressure, Relative Density, Particle Size, Ageing and Overconsolidation. *Géotechnique.* 36(3): 425-447.

Smith, P.R. 1992. The Behaviour of Natural High Compressibility Clay with Special Reference to Construction on Soft Ground. *Ph.D. Thesis, Imperial College, London.*

Smits, F.P. 1982. Penetration Pore Pressure Measured with Piezometer Cones. *Proc. of II ESOPT, Amsterdam.* Balkema

Stallebrass, S.E. 1990 Modelling the Effect of Recent Stress History on the Deformation of the Overconsolidated Soils. *Ph.D. Thesis, The City University of London.*

Stallebrass S.E. & Taylor R.N. 1997. The Development and Evaulation of a Constitutive Model for the Prediction of Ground Movements in Overconsolidated Clay. *Géotechnique.* 47(2): 235-254.

Stokoe K.H. II & Hoar, R.J. 1978. Variable Affecting In Situ Seismic Measurements. *Proc. of the Conference on Earthquake Engineering and Soil Dynamics, ASCE,* Pasadena, CA.2: 919-939.

Stokoe, K.H. II, Lee, S.H.H. and Knox, D.P. 1985. Shear Moduli Measurement Under True Triaxial Stresses. *Proc. of Advances in the Art of Testing Soils Under Cyclic Loading Conditions, ASCE Convention, Detroit*

Stokoe, K.H. II, Lee, J.N.K. & Lee, S.H.H. 1991. Characterization of Soil in Calibration Chambers with Seismic Waves. *Proc. ISOCCT1, Postdam, NY*

Stokoe, K.H. II, Wright, S.G., Bay, J.A. & Roësset, J.M. 1994. Characterization of Geotechnical Sites by Sasw Method. Proc. XIII ICSMFE New Delhi, India TC # 10: 15-25.

Stokoe, K.H. II, Hwang S.K., Lee, J.N.K. & Andrus R.D. 1995. Effects of Various Parameters on the Stiffness and Damping of Soils at Small to

Medium Strains. *Keynote Lecture 2, IS Hokkaido.* 2: 785-816. Balkema

Suzuki, Y., Hatanaka, M., Konno, T., Ishihara, K. & Akino, K. 1992. Engineering Properties of Undisturbed Gravel Sample. *Proc. 10th World Conference on Earthquake Engineering, Madrid.*

Suzuki, Y., Goto, S., Hatanaka, M. & Tokimatsu, K. 1993. Correlation Between Strengths and Penetration Resistances for Gravelly Soils. *Soils and Foundations.* 33(1): 92-101.

Sykora, D.W. & Stokoe, K.H. II 1983. Correlations of in Situ Measurements in Sands with Shear Wave Velocity. *Geotechnical Engineering Report GR83-33, University of Texas at Austin, Austin, Texas.*

Tanizawa F. 1992 Correlations Between Cone Resistance and Mechanical Properties of Uniform Clean Sand. *Internal report ENEL CRIS, Milan.*

Tatsuoka, F. & Shibuya, S. 1992. Deformation Characteristics of Soil and Rocks from Field and Laboratory Tests. *Keynote Lecture, IX Asian Conference on SMFE, Bangkok.* 2: 101-190.

Tatsuoka F. & Kohata Y. 1995 Stiffness of Hard Soils and Soft Rocks in Engineering Applications. *Keynote Lecture 8, IS Hokkaido 1994.* 2: 947-1066. Balkema

Tatsuoka F., Kohata Y., Ochi K. & Tsubouchi T., 1995a. Stiffness of Soft Rocks in Tokyo Metropolitan Area from Laboratory Tests to Full-Scale Behaviour. *International Workshop on Rock Foundation of Large Scaled Structures, Tokyo 30th Sept. 1995.*

Tatsuoka F., Lo Presti D.C.F. & Kohata Y. 1995. Deformation Characteristics of Soils and Soft Rocks Under Monotonic and Cyclic Loads and Their Relations. *3rd International Conference on Recent Advances in Geotechnical Earthquake Engineering and Soil Dynamic, State of the Art 1.* 2: 851-879.

Tatsuoka F., Jardine R.J., Lo Presti D.C.F., Di Benedetto H. & Kodaka T. 1997. Characterising the Pre-Failure Deformation Properties of Geomaterials. *Theme Lecture, Plenary Session 1, XIV ICSMFE, Hamburg, in print.* Balkema

Tavenas F.A, Ladd R.S. & La Rochelle P. 1972. Accuracy of Relative Density Measurements: Results of a Comparative Test Program. *ASTM STP 523*: 18-60.

Terzaghi K. & Peck R.B. 1948. *Soil Mechanics in Engineering Practice.* Edit. Wiley, New York.

Tumay M. 1976. Cone Bearing vs. Relative Density Correlation in Cohesionless Soils. *Dozen Dissertation. Bogazici Univ.* Istanbul.

Tumay, M.T., Boggers, R.L. & Acar, Y. 1981. Subsurface Investigations with Piezocone Penetrometer. *Proc. of Symposium on Cone Penetration Testing and Experience, ASCE National Convention, St Louis, Missouri.*

Vesic A.S. & Clough G.W. 1968 Behaviour of Granular Material under High Stresses. *Journal SMFD, ASCE,* 94(SM3): 661-688.

Vesic A.S. 1972. Expansion of Cavities in Infinite Soil Mass. *Journal SMFD, ASCE.* 98(SM3): 265-290.

Vesic A.S. 1975. Principles of Pile Foundation Design. *Soil Mechanics Series No. 38, Duke University, Durham, NC.*

Vesic A.S. 1977. Design of Pile Foundations *T.R.B. Nat. Coop Highway Research Program. Synthesis of Highway Practice 42.*

Vucetic, M. 1994. Cyclic Threshold Shear Strains in Soils *Journal of GED, ASCE.* 120(12): 2208-2228.

Withers, N.J., Schaap, L.H.J., Kolk, K.J. & Dalton, J.C.P. 1986. The Development of the Full Displacement Pressuremeter. *Proc. of 2nd IS on The Pressuremeter and Its Marine Applications, University of Texas, ASTM STP 950*: 38-56

Withers N.J., Howie J.A., Hughes J.M.O. & Robertson P.K. 1989. Performance and Analysis of Cone Pressuremeters Tests in Sand. *Géotechnique.* 39(3): 433-54.

Woods R.D. 1978 Measurement of dynamic soil properties. *Earthquake Engineering and Soil Dynamics, Pasadena CA.* 1: 91-179.

Woods, R.D. and Stokoe, K.H. II 1985. Shallow Seismic Exploration in Soil Dynamics. *Richart Commemorative Lectures, ASCE.* 120-156.

Woods, R.D. 1991. Field and Laboratory Determination of Soil Properties at Low and High Strains. *Proc. of 2nd International Conference on Recent Advances in Geotechnical Earthquake Engineering and Soil Dynamics, St. Louis, Missouri, SOA1.*

Woods, R.D. 1994. Borehole Methods in Shallow Seismic Exploration. *Proc. of XIII ICSMFE New Delhi, India TC # 10*: 91-100.

Wroth C.P. & Hughes J.H.O. 1973. An Instrument for the In-Situ Measurement of the Properties of Soft Clays. *Proc. VIII ICSMFE.* 1.2: 487-494.

Wroth C.P. 1982. British Experience with the Self-Boring Pressuremeter *Proc. I Symposium on the Pressuremeter and Its Marine Applications, Paris*: 143-164.

Wroth, C.P. 1988. Penetration Testing - A More Rigorous Approach to Interpretation. *Proc. I ISOPT, Orlando, Florida.*

Yoshida, Y. 1988. A Proposal on Application of Penetration Tests on Gravelly Soils. *Abiko Research Laboratory, Rep. No. U 87080* (in Japanese).

Yoshida, Y., Kokusho, T. & Motonori, I. 1988. Empirical Formulas of STP Blow Counts for Gravelly Soils. *Proc. I ISOPT, Orlando, Florida*: 381-387.

Yoshimi Y., Hatanaka M. & Oh-oka H. 1978. Undisturbed Testing of Saturated Sands by Freezing. *Soils and Foundations*. 18(3): 59-73.

Yoshimi Y., Tokimatsu K., Kaneko O. & Makihara Y. 1984. Undrained Cyclic Shear Strength of a Dense Niigata Sand. *Soils and Foundations*. 24(4): 131-145.

Yu H.S. 1990. Cavity Expansion Theory and Its Application to the Analysis of Pressuremeter. *Ph. D. Thesis. Oxford University.*

Yu H.S., Schnaid F. and Collins I.F. 1994. Analysis of Cone Pressuremeter Tests in Sands. *R.R. 105.09, Dept. of Civil Engineering and Surveying, The University of Newcastle, New South Wales, Australia.*

Yamada Y. and Ishihara K. 1979. Anisotropic Deformation Characteristics of Sand Under Three-Dimensional Stress-Conditions. *Soils and Foundations*. 19(2): 79-94.

Zohrabi M. 1993. Calibration of Penetrometers and Interpretation of Pressuremeter in Sand. *Ph. Dissertation, Dept. of Civil Engineering, University of Southampton, UK.*

Geotechnical Hazards, Marić, Lisac & Szavits-Nossan (eds) © 1998 Taylor & Francis, ISBN 90 5410 957 2

Slope movements – Geotechnical characterization, risk assessment and mitigation

Serge Leroueil
Department of Civil Engineering, Université Laval, Sainte-Foy, Que., Canada

Jacques Locat
Department of Geology and Geological Engineering, Université Laval, Sainte-Foy, Que., Canada

ABSTRACT:
In this paper, the Authors apply a geotechnical characterization of slope movements to risk assessment, mitigation, and selection of warning systems for such slope movements. Because the geotechnical characterization rationally describes the involved phenomena, it appears to be a very powerful framework for that purpose.

1 INTRODUCTION

Starting with the idea of developing an expert system for slope engineering (Faure *et al.* 1988), it was found that knowledge in this domain was rather scattered and quite often related to local geological conditions, including types of materials, topography, seismic activity, etc. It thus appeared very difficult, if not impossible, to feed an expert system in these conditions. Several remarks were then made:

1) The links between geology and landslide activity were well-established, and detailed geomorphological classifications of slope movements had been proposed (Varnes 1978, Hutchinson 1988). These classifications relate, in particular, the type of movement to the type of soil involved.

2) Even if slope movements result from the mechanical behaviour of soils and rocks, this aspect had received moderate consideration, and only few classifications referred to it (Sassa 1985, Hutchinson 1988, Meunier 1995). Moreover, when geotechnical aspects were considered, the focus was placed on stability and strength parameters whereas problems often concern movements appearing before or after failure.

3) It appeared that the mechanical behaviour of geomaterials cannot be adequately described by considering only 3 classes, as done in Varnes' classification.

3) Similarly to what is done when a laboratory shear test is performed, it appeared extremely useful to consider 4 different possible stages in slope movements: pre-failure stage, failure, post-failure stage, and reactivation (Fig. 1).

From these remarks, Leroueil *et al.* (1996) proposed a geotechnical characterization of slope movements which, essentially, extends geomorphological classifications to incorporate aspects of soil and rock mechanics. This geotechnical characterization is summarized in Section 2 of the present paper.

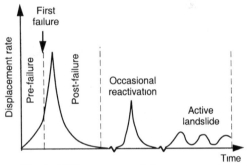

Fig. 1 - Different stages of slope movements.

Concurrently, Vaunat *et al.* (1992), proposed a general methodology for evaluating the risk associated with slope movements. To do so, they followed the general approach proposed by Varnes *et al.* (1984), but classified the information concerning the slope into three classes, as suggested by Champetier de Ribes (1987): predisposition factors, triggering or aggravating factors, and revealing factors. These classes were used in the geotechnical characterization previously mentioned. As shown in Section 3 of the paper, they are extremely useful for assessing the risk. Finally, in Section 4 of the paper, it is shown how the geotechnical characterization can be used for the selection of appropriate warning devices and mitigation methods.

Due to limited time for preparing this paper, and also because the described work is in development, only the framework concerning the geotechnical characterization and its application to risk assessment, mitigation and selection of warning systems is presented here. It is also worth noting that the authors are not experts in risk assessment and that this paper should be considered as humble thoughts on this difficult problem.

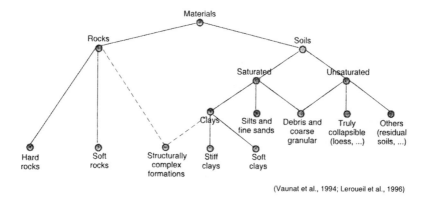

(Vaunat et al., 1994; Leroueil et al., 1996)

Fig. 2 - Material types considered in the geotechnical characterization.

2 GEOTECHNICAL CHARACTERIZATION OF SLOPE MOVEMENTS

Varnes (1978) and Hutchinson (1988) proposed classifications of slope movements which consider the type of movement and the type of material, and could thus be presented in a two-dimensional matrix. The types of movements are essentially those defined by Cruden *et al.* (1994) and Cruden & Varnes (1996), *i.e.* falls, topples, slides, lateral spreads, flows, and combinations of these basic types. For the materials, Varnes (1978) considered only 3 classes, *i.e.* bedrock, debris and earth. It is obviously not sufficient to adequately describe the mechanical behaviour of soils and rocks in general, and Vaunat *et al.* (1994) and Leroueil *et al.* (1996) suggested the 9 classes indicated in Fig. 2.

To this two-dimensional matrix, Vaunat *et al.* (1994) and Leroueil *et al.* (1996) suggested the addition of a third axis to take into account the four stages of movements:
- The pre-failure stage, when the soil mass is essentially overconsolidated, intact, and continuous. This stage is mostly controlled by progressive failure phenomena and creep.
- The onset of failure characterized by the formation of a continuous shear surface through the entire soil or rock mass. As failure is influenced by strain rate effects, strain softening, progressive failure and discontinuities, the relevant parameters are often difficult to define.
- The post-failure stage which includes movement of the soil or rock mass involved in the landslide, from just after failure until it essentially stops. The behaviour of the sliding material during this stage mostly depends on the redistribution of the potential energy available at failure into friction energy, disaggregating or remolding energy, and kinetic energy. The behaviour during this stage depends not only on the mechanical characteristics of the involved material, but also on its physical properties and the geometrical characteristics of the slope.

- The reactivation stage, when the soil or rock mass slides along one or several pre-existing shear surfaces. This reactivation can be occasional or continuous with seasonal variations of the rate of movement. The behaviour is then essentially controlled by the residual friction angle of the soil.

Leroueil *et al.* (1996) give details for these four stages and the associated behaviours. In particular, they point out that the involved mechanical phenomena, controlling laws and parameters are very different from one stage to another. For example, if slope movements associated to pre-failure and reactivation stages are, to a large extent, related to creep, the controlling laws are completely different. During the pre-failure stage, there is creep of the entire soil mass which is controlled by stress level, accumulated strain, and time or strain rate; during the reactivation stage, creep is localized along the pre-existing failure surface and is controlled only by the shear stress level.

The 3-D matrix (type of material; type of movement; stage of movement) forms the characterization matrix which is schematized in Fig. 3. Obviously, not all the elements of the matrix are representative of real situations. However, for each relevant element of the characterization matrix, a characterization sheet is proposed, which contains the following elements:
- the controlling laws and parameters;
- the predisposition factors;
- the triggering or aggravating factors;
- the revealing factors; and
- the consequences of the movement.

Figure 4 shows such a characterization sheet for slides in stiff clay or clay shale during the reactivation stage.

This geotechnical characterization has been applied to soft clays (Vaunat 1998), structurally complex and stiff jointed clayey soils from southern Italy (D'Elia *et al.* 1998), and will be examined for residual soils, in a workshop to be held in Rio (June 1998).

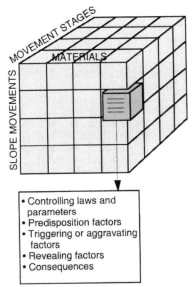

- Controlling laws and parameters
- Predisposition factors
- Triggering or aggravating factors
- Revealing factors
- Consequences

(Vaunat et al., 1994; Leroueil et al., 1996)

Fig. 3 - Schematic slope movement characterization.

3 RISK ASSOCIATED TO SLOPE MOVEMENTS

3.1 Introduction

The geotechnical characterization constitutes a general framework for analyzing slope movements, and the geotechnical characterization sheets provide guidelines for specific types of movement at a given stage and for a given material. Real cases can be more complex, often involving several materials and several types of movement. However, this approach, which examines slope behaviour from the mechanical point of view, appears to be useful for evaluating the risk associated to slope movements.

As in any risk analysis, the first step consists of defining all possible dangers, *i.e.* in our case, all possible slope movements which can affect a given area. Then, each danger has to be characterized. To do so, characterization sheets such as that shown in Fig. 4 are recommended because they rationally organize the information and force the engineer to answer a series of questions essential to the understanding of the situation:

- what is the type of movement and its geometrical characteristics?
- what is the stage of movement. This information helps define the controlling laws?
- what are the materials involved. This information helps define the geotechnical parameters?
- what are the predisposition factors?
- what are the triggering or aggravating factors?
- what are the revealing factors?
- and what are the consequences of the movement, information which is obviously essential in a risk analysis?

Movement: slide	Stage: reactivation

Material: stiff clay and clay shale

Controlling laws and parameters

- Residual shear strength: $\tau_{fr} = \sigma'_n \tan \phi'_r$.
- Rate of displacement or sliding: $\upsilon = A \left(\dfrac{\tau}{\sigma'_n \tan \phi'_r} \right)^n$.

Predisposition factors

- Pre-existing shear surface(s).
- Soil particules which can be reoriented.

Triggering or aggravating factors

- Increase in pore pressure in the vicinity of shear surfaces.
- Increase of shear stresses by:
  - erosion at the toe of the slope or, loading at the top of the slope.
- Seismic loading

Revealing factors

- Localized displacements on vertical profiles.
- Geometry and movements evidencing sliding of essentially rigid blocks.

Movement consequences

- Rate of displacement: generally very slow (less than 1 m/year).
- Remedial works can be considered only in some cases.

(from Leroueil et al., 1996)

Fig. 4 - Slope movement characterization sheet for reactivated slides in stiff clays and clay shales.

When associated to slopes, risk has often been considered in terms of failure. This indeed constitutes an important proportion of the encountered problems, particularly when risk is examined in terms of loss of lives. This, however, does not satisfy all the geotechnical engineering needs. In particular, for cases of linear structures and infrastructures, such as bridges, roads, railway tracks, pipelines, etc. and in zones of precarious stability, the questions are related to movements and rates of movement rather than to failure as such. Also, failure itself does not indicate the possibility of retrogression or the run-out distance which, quite often, may or may not put a given element at risk. The geotechnical characterization which considers different stages of slope movements provides a general and rational framework for examining all these aspects.

3.2 Total risk associated to slope movements

Varnes *et al.* (1984) defined the total risk R_T as the set of damages resulting from the occurrence of a phenomenon. It can be described by the following relationship (se also Wu *et al.* 1996):

$$R_T = \Sigma \; H \; R_i \; V_i \qquad (1)$$

in which:
- H is the hazard or the phenomenon occurrence probability within a given area and a given time period;
- R_i (for i = 1 to n) are the elements at risk, potentially damaged by the phenomenon;
- V_i is the vulnerability of each element represented by a damage degree comprised between 0 (no loss) and 1 (total loss).

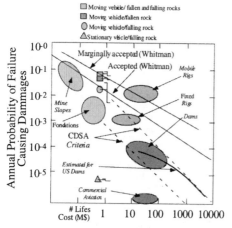

(Modified after Bunce et al. 1997)

Fig. 5 - Risk for selected engineering projects

3.3 *Acceptable risk*

Whitman (1984) presented the annual probability of failure for different engineering projects as a function of the consequences (Fig.5), and suggested "accepted" and "marginally accepted" levels. To apply such an approach to slope failures, it must first recognized that such levels cannot be defined by geotechnical engineers only. Because they have or may have social, economical, environmental, political and legal implications, they depend on the will and the means of the society. Also, as discussed in particular by Fell (1994) and Finlay & Fell (1997), they may depend on human behaviour and perception. Several of these aspects can be mentioned: a) much higher risks, up to three orders of magnitude, are more accepted for voluntary risk than for involuntary risk; b) the risk acceptance is larger for natural dangers than for man-made structures, and thus, in particular, for slopes on which remediation works have been performed; c) there is a difference between a risk which is tolerated and one which is accepted; d) the acceptable risk to property loss only is different from the acceptable risk to loss of life. Consequently, no general level of acceptance can be established.

These aspects are being discussed in Canada, mostly in relation to the loss of life, and several risk acceptance criteria have been proposed (Morgan 1992, 1997, Cave 1992, Hungr *et al.* 1993, Evans 1997). Fig. 6 shows some of them.

When the evaluated total cost associated to an event is very high for a given society, both the risk, as calculated with Eq.1, and the cost have to be examined concurrently. Even if for such a case the probability of occurrence is extremely small, it is essential to think about what could be done if it happens the following year. Lessons on this aspect can probably be learned from dam engineering in which the design is verified for a "probable maximum flood".

3.4 *Elements at risk and their vulnerability*

The French risk exposure maps (PER) program put the elements at risk into 4 categories: individuals, properties and goods, activities, and social functions (Asté 1991). These elements can suffer consequences directly from the phenomenon or from induced phenomena. For example, destruction of a house by a landslide is a direct consequence of the event whereas flooding of a town due to the damming of a river by a landslide is an indirect consequence. Obviously, all these elements have to be considered in a risk analysis.

In the context of the geotechnical characterization, the elements at risk as well as their vulnerability are obtained from "movement consequences" (Fig.7). In particular, the rate of movement, which is highly variable from case to case, is of major importance when the element "loss of life" is considered; indeed, when the rate of movement is relatively low, people have time to escape from the involved area. The rate of movement as well as the accumulated displacement would both be important factors when the element at risk is a linear structure, for example a pipeline.

3.5 *Hazard*

Hazard is the probability that a phenomenon, which can be a first slope failure or an active landslide reaching a given rate, occurs. When the failure stage is concerned, the hazard is directly related to the probability of the triggering factor to reach a critical value leading to failure. For the pre-failure and reactivation stages, the hazard associated to a rate of movement is related to the probability that the aggravating factor reaches a given value leading to this rate (Fig. 8).

For the post-failure stage, the hazard associated to a movement with given characteristics is mostly related to the materials involved and the predisposition factors (Fig. 8) and is thus more difficult to define. An example of link between the behaviour of a soil mass during the post-failure stage and initial conditions is given by Lee *et al.* (1988) who studied rainfall-induced landslides in the San Francisco Bay .

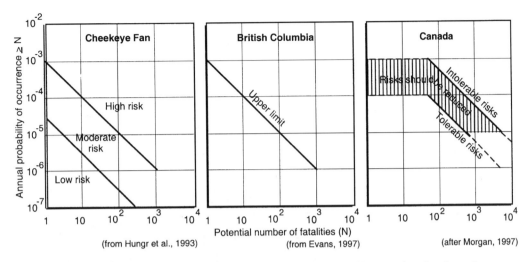

Fig. 6 - Proposed guidelines for assessing risks to life from naturally occurring slope hazards.

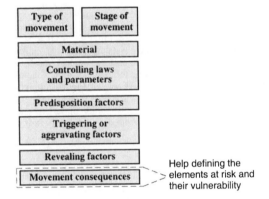

Fig. 7 - Geotechnical characterization and the elements at risk and their vulnerability.

These authors found an association between initial stress conditions above the steady-state line (contracting soil or positive state parameter [Been & Jefferies 1985]) and debris flow occurrence, whereas initial states near or below the steady-state line (dilating materials or negative state parameter) corresponded to the occurrence of slow-moving nondisintegrated failures. The position of the initial stress conditions compared to the steady-state line is thus a predisposition factor which strongly influences the post-failure stage.

When considering a landslide, two stages are involved in the geotechnical characterization: the failure stage followed by the post-failure stage. As a consequence, the hazard has to be subdivided into 2 parts: the hazard associated to the possibility of having a failure (H_f), and the hazard associated to the possibility that the post-failure stage has specific characteristics (H_{post-f}).

$$H = H_f \times H_{post-f} \qquad (2)$$

This is illustrated in Fig. 9b. The probability that a landslide reaches the house shown on the figure is the hazard associated to a first failure multiplied by the hazard associated to the possibility that the runout distance can be larger than L. When elements, such as vehicles, are moving or are temporarily in the area at risk, the probability analysis has to be modified accordingly. An example is given by Bunce et al. (1997) who assessed the hazard from rock fall on a highway from British Columbia, Canada. They considered three possibilities: a falling rock hitting a moving or a stationary vehicle, and a moving vehicle hitting a fallen rock.

Several authors described the probability function H_{post-r} as the probability of spatial impact, i.e. the probability that the landslide impacts a given element.

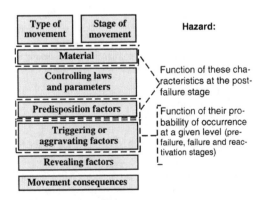

Fig. 8 - Geotechnical characterization and hazard.

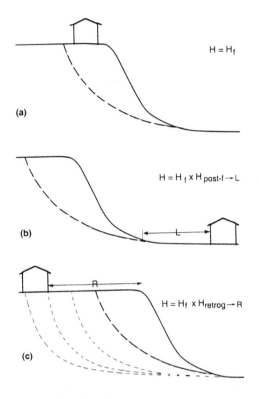

(a)

$H = H_f$

(b)

$H = H_f \times H_{post\text{-}f} \rightarrow L$

L

(c)

R

$H = H_f \times H_{retrog} \rightarrow R$

Fig. 9 - Hazard associated to (a) simple failure, (b) failure with a run-out distance of L, and (c) retrogressive failure.

A similar situation can be found in the case of retrogressive landslides. To have a retrogressive landslide, there must be a first failure, and then, conditions such that the backscarp becomes unstable, thus triggering other landslides. The hazard associated to a landslide retrogressing to a distance R (Fig. 9c) is thus the hazard associated to a first failure (H_f) multiplied by the hazard associated to the possibility of having retrogression to a distance R ($H_{retrog.}$):

$$H = H_f \times H_{retrog} \qquad (3)$$

A major difficulty in risk analysis is uncertainty. Morgenstern (1995) distinguishes three sources of uncertainty:
- parameter uncertainty;
- model uncertainty;
- human uncertainty.

If it is considered that there is no human uncertainty and that the model (the calculation model and the method for determining the parameters) is fully representative of the considered problem, the uncertainty would only be on the parameters. This uncertainty can be subdivided into two parts:
- a) The uncertainty that depends on the spatial variation of the parameters characterizing the material

and the predisposition factors. This uncertainty also depends on the extent and the quality of the investigation.
- b) The uncertainty due to the temporal variation of the triggering or aggravating factors. It is essentially because of these variations that there can be failure or a change in the rate of movement. When the triggering of a landslide or the possibility to reach a given rate of movement depends on several factors (high water level and erosion for example), the probability that it occurs is the sum of the probabilities that these factors reach a given level, plus the probability that these different factors can combine to produce the same event.

To define the hazard associated to parameter uncertainty, a probabilistic model can be applied to a deterministic solution. In the case of failure, the model can be based on limit equilibrium, and failure would be supposed to occur at a calculated factor of safety of 1.0. The probability of failure then becomes the probability that F becomes smaller than 1.0.

Examination of the influence of parameter uncertainty, advocated in particular by Lacasse & Nadim (1994), is a necessary step in a reliability analysis. The advantage of considering this aspect can be illustrated by the case presented by Lacasse & Nadim (1994). As shown in Fig. 10, a factor of safety of 1.79 obtained with a high level of uncertainty on input parameters may correspond to a probability of failure significantly higher than a factor of safety of 1.40 established with a low uncertainty on input parameters. The factor of safety is thus not sufficient for quantifying a probability of failure and has to be qualified, for example, by a reliability index.

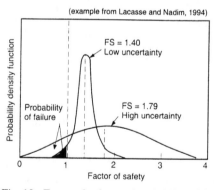

Fig. 10 - Factor of safety and probability of failure

As mentioned by Morgenstern (1995), the model can also be a source of uncertainty. This is probably true when stability is considered because the involved processes are generally complex (influence of structure and anisotropy, effects of strain rate, progressive failure, influence of geological anomalies, etc.) and not fully represented by existing methods. It would also be difficult to consider probabilistic methods for pre-failure and post-failure stages which are even less understood than failure. The stage for which known methods would be the most appropriate is the reactivation stage. In such a case: there is sliding of essentially rigid blocks over an essentially rigid base; the strength parameters are the residual ones; and the conditions are thus close to the assumptions made in limit equilibrium stability models. Moreover, such a case can often be backanalysed in order to ascertain the strength parameters.

Morgenstern (1995) also mention human uncertainty. It is obviously there, but generally difficult to take into account.

In many cases, the link between hazard and controlling factors is not easy to define. For example, when rain is the triggering or aggravating factor, it has to be realized that its influence on slope movements depends on infiltrations and how infiltrations result in pore pressures. Also, at the post-failure stage, there are generally several influencing factors which often have a quite indirect relation with the movement. So, in many cases, it is difficult to quantitatively establish the hazard.

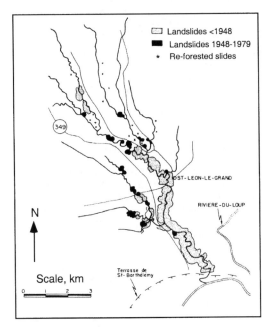

Figure 11. Landslides along the Chacoura River watershed between 1948 and 1979 (after Locat *et al.* 1984).

In such conditions, the hazard has to be defined semi-quantitatively or qualitatively. For that purpose, Hungr (1997) suggests a useful scale (Table 1) in which the class limits have been selected so as to possess a certain physical meaning. An annual probability of more than 1/20 means that the hazard may be imminent, an annual probability of 1/100 signifies a situation where an event should be

expected within the lifetime of a person or a typical structure, etc.

Table 1 - Suggested semi-quantitative probability scale for landslide hazard magnitude or intensity.

Term	Range of frequency (1/year)
Very high probability	> 1/20
High	1/100 - 1/20
Medium	1/500 - 1/100
Low	1/2500 - 1/500
Very low	< 1/2500

(from Hungr, 1997)

Engineers generally have several sources of information which can help in defining hazard semi-quantitavely or qualitatively. Several examples are given hereafter:
- *Historic information.* An example illustrating the benefits which can be gained from historic information is provided by Locat *et al.* (1984). A close examination of air photographs taken between 1948 and 1979 along the Chacoura River Valley was performed, and Fig. 11 presents the observed landslide distribution. It was shown that erosion is the major triggering factor in this area, landslides generally occurring in the outer part of river meanders and in the upstream part of the valley, where the river is still deepening and the longitudinal stream profile is relatively steep. In particular, it was observed that one out of five erosion scars observed in 1948 evolved into a landslide during the following 30 years. As shown in Fig. 12, the study also gave information on the types of landslides occurring along the Chacoura River and their size.

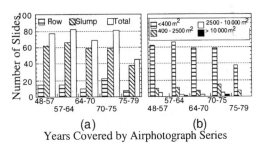

Fig. 12 - Number of slides per period of airphotograph coverage (a), and surface classes distribution (b) between 1948 and 1979 (Locat & Leroueil 1997).

- *Geomorphological information.* For example, Lebuis *et al.* (1983) compiled geometrical characteristics of slopes from the St-Ambroise area, Québec, in a slope height vs slope angle diagram. As shown in Fig. 13, they found that slopes presenting signs of instability are on the right hand side of the curve coresponding to failure (F = 1) for c' = 7 kPa and $\Phi' = 31°$. For slopes from this area which would be on the right hand side of the curve, the probability

101

of failure is very high. On the other hand, it can be thought that the farther on the left of the line a slope is, the lower the probability of failure is.

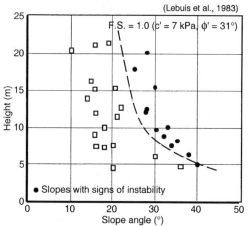

Fig. 13 - Height and angle of slopes in clay, Saint-Ambroise area, Québec.

- *Climatic-landslide activity relationships.* In particular, several relationships have been established for specific geologic and climatic contexts between rainfall characteristics and the occurrence and the severity of landslides. Fig.14 shows one example presented by Kim *et al.* (1992) for landslides in South Korea. The number of events is given as a function of the hourly rainfall and the rainfall cumulated during the previous two-day period. Kim *et al.* (1992) also mention geomorphological and geological factors which could also be used in specifying the hazard.

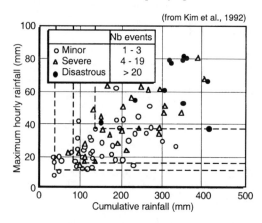

Fig. 14 - Relationship between landslide events and rainfall in South Korea.

- *Geotechnical information.* For example, Tavenas *et al.* (1983) and Leroueil *et al.* (1996) enumerate the predisposition factors leading to retrogression in soft clays, which can thus be used to estimate H_{retrog} in Eq. 3 and Fig. 9c. They are:

a) *The ability of the clay to be remoulded.* This depends on the height of the slope (H), which governs the available potential energy at failure, and on the mechanical and physical characteristics of the clay. Leroueil *et al.* (1996) indicate that full remoulding of the clay is obtained for stability numbers, $N_c = \gamma H/c_u$, at least equal to 4 for a plasticity index, I_p, of about 10, and to 7 or 8 when I_p reaches values of about 40.

b) *The ability of the clay to flow out of the crater when remoulded.* This is directly related to the consistency of the remoulded material, *i.e.* to the remoulded shear strength or the liquidity index, or the yield strength (Locat 1997) of the clay. In fact, it can flow if the water content is equal or higher than the liquid limit of the clay. Also, as shown in Fig. 15, the retrogression distance has a tendency to increase when the remoulded shear strength of the clay decreases.

c) *A topography which permits the evacuation of the liquefied debris.*

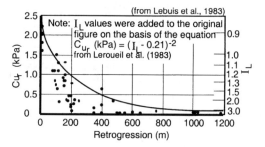

Fig. 15 - Distance of retrogression as a function of the consistency of the remoulded material.

As previously indicated, uncertainty depends, among other factors, on the quality of the geological and geotechnical investigation performed. As indicated by Sällfors *et al.* (1996), the new Swedish rules for slope stability analysis take this aspect into account by requiring minimum factors of safety which depend on the risk involved, and on the degree of sophistication of the investigation. These required minimum values are given in a table by Sällfors *et al.* (1996), but the principle can be schematically described, as shown in Fig. 16. In such an approach, for a given level of risk, the required minimum factor of safety which could be 1.62 for a rough investigation (illustration on Fig. 16) could decrease to 1.45 if a detailed investigation is performed.

In conclusion, evaluating hazard is the most difficult task when dealing with risk assessment related to slope movements. However, as indicated by Lacasse & Nadim (1994), it is not a reason to omit defining it or establishing its significance.

3.6 Total risk assessment

Having defined the elements at risk, their vulnerability, and the hazard, the total risk associated to a given slope movement can be evaluated by using Eq. 1. Fig. 17 (steps 1 to 6) shows the corresponding risk management procedure.

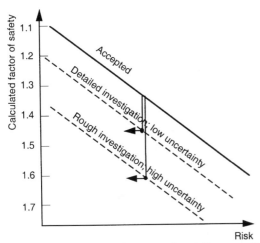

Fig. 16 - Consideration on the influence of uncertainty on a minimum required factor of safety (schematic).

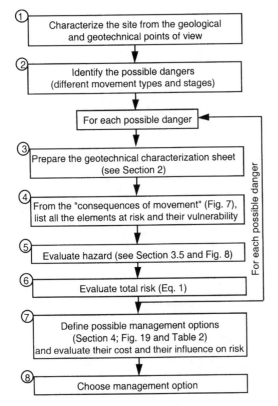

Fig. 17 - Risk management procedure.

4 WARNING DEVICES AND MITIGATION

When dealing with slopes of precarious stability, there are several options. The first one consists in improving the investigation to specify the risk; and, if this latter is considered too high, three options remain: a) do nothing to the slope ; b) do nothing, but put a warning system in order to insure or improve the safety of people; c) improve the safety of the slope to a satisfactory level in order to reduce the risk. Indications for the selection of warning devices and mitigation techniques are briefly examined in the following section, in the context of the geotechnical characterization.

For the solutions considered to improve a given situation, reductions in hazard and risk have to be examined as a function of the cost in order to select the most appropriate one on a cost/benefit basis. Figure 18, modified after Einstein (1997), illustrates this aspect. It is worth mentioning here that, as indicated by Morgenstern (1995), even if the absolute risk and hazard are not well-known, the hazard and risk improvements are more precise since a number of the uncertainties are the same for all the considered solutions. It is also worth noting that warning systems do not modify the hazard but contribute to reducing the consequences of the movement and thus the risk.

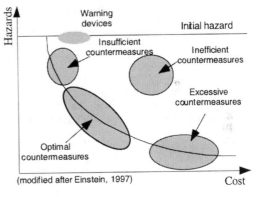

(modified after Einstein, 1997)

Fig. 18 - Hazard reduction-cost of intervention analysis

4.1 Warning systems.

The geotechnical characterization of the movement as well as the stage of the movement can be used for defining the appropriate type of warning system (Fig. 19). At a pre-failure stage, the warning system can be applied either to revealing factors or to aggravating or triggering factors. The revealing factors could be the opening of fissures or the movement of given points in the slope; in such cases, the warning criterion can be a magnitude or a rate of movement, which are then small. When associated to triggering or aggravating factors, there is a need to firstly define the relation between the magnitude of the controlling factors and

103

Table 2 - Approaches for mitigating slope movements and their consequences.

	Eliminates the problem or reduces the consequences	Decreases driving forces	Increases resisting forces
Intervention on:			
• **Material, controlling laws and parameters**			
• Removal of unstable material	●		
• Total or partial substitution of the sliding mass with material with better mechanical and drainage characteristics			●
• Electro-osmosis			●
• Soil treatment with lime, cement or other additives			●
• Thermal treatment			●
• Etc.			
• **Predisposition factors**			
• Nailing, piling, anchoring, bolting			●
• Earthworks for decreasing driving forces		●	
• Buttress or counterweight fills		●	
• Etc.			
• **Trigerring or aggraving factors**			
• Surface drainage			●
• Subsurface drainage (trenches, subhorizontal drains, drainage wells, etc.)			●
• Protection against erosion		●	
• Etc.			
• **Movement consequences**			
• Protection against falling or sliding materials (catch nets and walls, sheds, tunnels, etc.)	●		
• Dykes for containing mudflows, debris flows, etc.	●		
• Etc.			

the stability or the rate of movement of the slope. The warning criterion can be a given level of the water table, a given stage of erosion, a minimum suction in a loess deposit, etc.

At the failure stage, the warning system can only be linked to revealing factors, generally a sudden acceleration of movements or the disappearance of a target.

At the post-failure stage, the warning system has to be associated to the expected consequences of the movement. Its choice will, in particular, depend on the rate of movement and runout distance.

Finally, for active or reactivated slides, the warning system can be applied to the triggering or aggravating factors, the revealing factors, or some of the possible consequences of the movement.

4.2 *Mitigation.*

Mitigation requires a clear understanding of the processes that cause the slope movements, and the geotechnical characterization provides it. There are many possibilities to mitigate the associated risk, and they will not be described here. However, the different approaches can be classified in association with the classes of factors provided by the geotechnical characterization (Table 2). Their effect,

which can be to eliminate the problem or reduce the consequences, decrease the driving forces or increase the resisting forces are also indicated in Table 2.

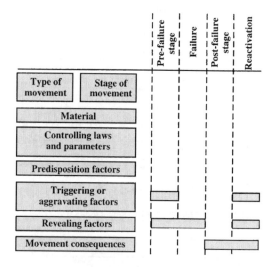

Fig. 19 - Possible avenues for warning systems, depending on the stage of movement.

5 CONCLUSION

The assessment of risk associated to slope movements and the selection of mitigation approaches are extremely difficult. However, it is necessary and an important research effort is needed. In this paper, the geotechnical characterization of slope movements proposed by Vaunat et al. (1994) and Leroueil et al. (1996) is considered. Because it rationally describes a slope movement in a given material, with its controlling law and parameters, predisposition factors, triggering and aggravating factors, revealing factors and movement consequences, the geotechnical characterization appears to be a useful tool:

- The elements at risk and their vulnerability are obtained from "movement consequences"; except for the post-failure stage, the hazard depends on the probability that triggering or aggravating factors reach a level leading to failure or a given rate of movement.
- The warning systems can be applied to the triggering or aggravating factors, the revealing factors or some of the movement consequences, depending on the stage of the movement considered.
- Mitigation requires a good understanding of the process that causes the slope movement and the geotechnical characterization provides it. The possible approaches are classified in relation with this characterization method.

6 ACKNOWLEDGEMENTS

The Authors wish to thank the Fonds pour la formation des chercheurs et l'aide à la recherche (F.C.A.R.), Québec, and the National Science and Engineering Researh Council of Canada for supporting this research. They would also like to thank J. Vaunat, R.-M. Faure, and L. Picarelli for fruitful discussions.

7 REFERENCES

Asté, J.-P. 1991. Personal communication.
Been, K. & Jefferies, M.G. 1985. A state parameter for sands. Géotechnique, Vol.35:99-112.
Bunce, C.M., Cruden, D.M. & Morgenstern, N.R. 1997. Assessment of the hazard from rock fall on a highway. Canadian Geotechnical J., Vol.34(3):344-356.
Cave, P.W. 1992. Natural hazards risk assessment and land use planning in British Columbia. Geotechnique and Natural Hazards, Canadian Geotechnical Society, Vancouver, 1-11.
Champetier de Ribes, G. 1987. La cartographie des mouvements de terrains: des ZERMOS aux PER. Bulletin de liaison des laboratoires des Ponts et Chaussées, Vol.150-151:9-19.
Cruden, D.M. & Varnes, D.J. 1996. Landslides types and processes. In Landslides: Investigation and Mitigation, Transportation Research Board, Washington, Special Report No. 247, pp.36-75,.
Cruden D.M., Krautner, E., Beltram, L., Lefevre, G., Ter-Stepanian, G.I. & Zhang, Z.Y. 1994.

Describing landslides in several languages: the Multilingual Landslide Glossary. Proc.7th Int. Congress of the Int. Association of Engineering Geology, Lisbon, Vol.3:1325-1333.
D'Elia, B., Picarelli, L., Leroueil, S. & Vaunat, J. 1998. Geotechnical characterization of slope movements in structurally complex clay soils and stiff jointed clays. Rivista Italiana di Geotecnica. In print.
Einstein, H.H. 1997. Landslide risk - Systematic approaches to assessment and management. Proc. Int. Workshop on Landslide Risk Assessment, Honolulu, 25-50.
Evans, S.G. 1997. Fatal landslides and landslide risk in Canada. Proc. Int. Workshop on Landslide Risk Assessment, Honolulu, 185-196.
Faure, R.M., Leroueil, S., Rajot, J.-P., La Rochelle, P., Sève, G. & Tavenas, F. 1988. Xpent, système expert en stabilité des pentes. Proc. 5th Int. Symp. on Landslides, Lausanne, Vol.1:625-629.
Fell, R. 1994. Landslide risk assessment and acceptable risk. Canadian Geotechnical J., 31(2):261-272.
Finlay P.J. & Fell, R. 1997. Landslides: risk perception and acceptance. Canadian Geotechnical J., Vol.34(2):169-188.
Hungr, O. 1997. Some methods of landslide hazard intensity mapping. Proc. Int. Workshop on Landslide Risk Assessment, Honolulu, 215-226.
Hungr, O., Sobkowicz, J. & Morgan, G.C. 1993. How to economize on natural hazards. Geotechnical News, 11(1):54-57.
Hutchinson, J.N. 1988. Morphology and geotechnical parameters of landslides in relation to geology and hydrogeology. Proc.5th Int. Symp. on Landslides, Lausanne, Vol.1:3-35.
Kim, S.K., Hong, W.P. & Kim, Y.M. 1992. Prediction of rainfall-triggered landslides in Korea. Proc. 6th Int. Symp. on Landslides, Christchurch, Vol.2:989-994.
Lacasse, S. & Nadim, F. 1994. Reliability issues and future challenges in geotechnical engineering for offshore structures. Plenum paper, Proc. 7th Int. Conf. on the Behaviour of Offshore Structures, BOSS' 94, MIT, Cambridge, USA, 9-38.
Lebuis, J., Robert, J.-M. & Rissman, P. 1983. Regional mapping of landslide hazard in Québec. Proc. Int. Symp. on Slopes on Soft Clays, Linköping, Swedish Geotechnical Institute Report No. 17:205-262.
Lee, H.J., Ellen, S.D. & Kayen, R.E. 1988. Predicting transformation of shallow landslides into high-speed debris flows. Proc. 5th Int. Symp. on Landslides, Lausanne, Vol.1:713-718.
Leroueil, S., Tavenas, F. & Le Bihan, J.-P. 1983. Propriétés caractéristiques des argiles de l'est du Canada. Canadian Geotechnical J., Vol.20(4):681-705.
Leroueil, S., Vaunat, J., Picarelli, L., Locat, J., Lee, H. & Faure, R. 1996. Geotechnical characterization of slope movements. Proc. 7th Int. Symp. on Landslides, Trondheim, 53-74.
Locat, J. 1997. Normalized rheological behaviour of fine muds and ther flow properties in a

pseudoplastic regime. *ASCE 1st International Conference on Debris-flow Hazard Mitigation: Mechanics, Prediction and Assessment*. San Francisco, pp.: 260-269.

Locat, J., Demers, D., Lebuis, J. & Rissmann, P. 1984. Prédiction des glissements de terrain: application aux argiles sensibles, rivière Chacoura, Québec, Canada. *Proc. 4th Int. Symp. on Landslides*, Toronto, Vol.2:549-555.

Locat, J. & Leroueil, S. 1997. Landslide stages and risk assesment issues in sensitive clays and other soft sediments. *Proc. Int. Workshop on Landslide Risk Assessment*, Honolulu, 261-270.

Meunier, M. 1995. Classification of stream flows. *Proc. Pierre Beghin Int. Workshop on Rapid Gravitational Mass Movements*, Grenoble, pp.231-236.

Morgan, G.C. 1992. Quantification of risks from slope hazards. Geological Hazards' 91 Workshop, B.C. Geological Survey Branch Open File 1992-15, pp.57-70.

Morgan, G.C. 1997. A regulatory perspective on slope hazards and associated risks to life. *Proc. Int. Workshop on Landslide Risk Assessment*, Honolulu, 285-295.

Morgenstern, N.R. 1995. Managing risk in geotechnical engineering. *Proc. 10th Pan American Conf. on Soil Mechanics and Foundation Engineering*, Guadalajara, Vol.4. In press.

Sällfors, G., Larsson, R. & Ottosson, E. 1996. New Swedish national rules for slope stability analysis. *Proc. 7th Int. Symp. on Landslides*, Trondheim, Vol.1:377-380.

Sassa, K. 1985. The geotechnical classification of landslides. *Proc. 4th Int. Conf. and Field Workshop on Landslides*, Tokyo, Vol.1:31-40.

Tavenas, F., Flon, P., Leroueil, S. & Lebuis, J. 1983. Remolding energy and risk of slide retrogression in sensitive clays. *Proc. Symp. on Slopes on Soft Clays*, Linköping, SGI Report No 17:423-454.

Varnes, D.J. 1978. Slope movement. Types and processes. Transportation Research Board Report 176 "Landslides. Analysis and Control", pp.11-33.

Varnes, D.J. & the IAEG Commission on Landslides and Other Mass Movements on Slopes. 1984. Landslide Hazard Zonation - A review of the Principles and Practice. UNESCO, Paris.

Vaunat, J. 1998. Contribution à l'élaboration d'un système expert de prévision des risques liés à un mouvement de terrain. Ph.D. Thesis in preparation, Laval University, Québec.

Vaunat, J., Leroueil, S. & Tavenas, F. 1992. Hazard and risk analysis of slope stability. *Proc. 1st Canadian Symp. on Geotechnique and Natural Hazards*, Vancouver, Vol.1:397-404.

Vaunat, J., Leroueil, S. & Faure, R. 1994. Slope movements: a geotechnical perspective. *Proc. 7th Int. Congress of the Int. Association of Engineering Geology*, Lisbon, 1637-1646.

Whitman, R.V. 1984. Evaluating calculated risk in geotechnical engineering. *J. of Geotechnical Engineering Division*, ASCE, 110:145-188.

Wu, T.H., Tang, W.H. & Einstein, H.H. 1996. Landslide hazard and risk assessment. Chapter 6 of "Landslide. Investigation and mitigation", Transportation Research Board, Special Report 247:106-118.

Geotechnical Hazards, Marić, Lisac & Szavits-Nossan (eds) © 1998 Taylor & Francis, ISBN 90 5410 957 2

Recent advances in geotechnical earthquake engineering

Shamsher Prakash
University of Missouri-Rolla, Mo., USA

ABSTRACT: Systematic studies on soil dynamics and geotechnical earthquake engineering started in the US after the Niigata and Anchorage earthquakes of 1964. The amount of literature generated in 3-decades around the world has helped in earthquake hazard mitigation. However several problems still defy clear understanding. Three such problems, e.g. liquefaction, seismic displacements of rigid retaining walls, and dynamic soil-pile-structure interaction have been briefly described.

INTRODUCTION

Systematic studies on soil dynamics and geotechnical earthquake engineering started in the US after the Niigata and the Anchorage earthquakes of 1964. The amount of literature generated in 3-decades around the world has helped in earthquake hazard mitigation. However several problems still defy clear understanding. Important developments take place in almost all facets of geotechnical earthquake engineering everyday e.g.: ground motion characterization, soil properties, liquefaction, dynamic soil structure interaction, retaining structures, earth dams and abutments and bridges and others. There have been several important recent events e.g. Mexico City earthquake 1985, Loma Prieta earthquake 1989, Northridge earthquake 1994 and Kobe earthquake 1995, which resulted in renewed studies on the behavior of geotechnical structures, focussed attention on several new issues and accelerated pace of studies on several topics which had already been investigated enough.

It is impossible to address such a large array of topics and events in this presentation. I have, therefore, selected only 3-topics for discussion ie liquefaction, retaining walls and soil-pile interactions.

LIQUEFACTION OF SOILS

Several important state of the art reports have been prepared by learned authors on the subject (Finn 1981, 1991, Dobry 1995). NRC Workshop (1985) had reviewed the progress on the subject from the very beginning of liquefaction studies (Seed and Lee 1966) till that date. I will address a few specific issues.

a) Liquefaction resistance of sands in the Laboratory versus Field

b) Liquefaction of silts and clays.

Liquefaction of sands has been investigated extensively for about 30 years since the first systematic laboratory studies by Seed and Lee (1966). It has been determined that the factors which control liquefaction resistance of sands are: 1) Grain size, 2) Grain size distribution, 3) Relative density, 4) Initial effective confining pressure, 5) Dynamic stress level, 6) Number of pulses of dynamic stress level, 7) Initial K_c conditions, 8) Overconsolidation ratio (OCR), 9) Soil fabric, 10) Aging, 11) Previous strain History, and 12) Cementation

Among these factors, 1 to 9 can be studied in laboratory environment (Finn, 1981; Sandoval 1989). The last three factors, which include aging, previous strain history, and cementation and to a degree, soil fabric and K_c, cannot be studied satisfactorily in laboratory tests. For studying the influence of these factors on the liquefaction potential of sands, a novel approach based on correlation of SPT data and the stress causing liquefaction from field tests, holds promise (Guo 1998).

There are two methods at present (1998) for evaluating liquefaction resistance of sands: 1) analysis based upon laboratory tests on reconstituted samples; and 2) analysis based upon data from *in-situ* tests, e.g., Standard-Penetration Test (SPT) or Cone-Penetration Test (CPT). Preliminary studies show that these two approaches do not give similar results. The differences between the results of the two approaches are attributed to the fact that the specimens tested in the laboratory do not reflect the influences of the last three factors mentioned earlier.

These approaches to obtain the shear stress causing liquefaction will be briefly described below.

SHEAR STRESS CAUSING LIQUEFACTION FROM LABORATORY DATA

In the "simplified procedure" proposed by Seed and Idriss (1971), the shear stress causing liquefaction for a deposit is evaluated by examining the correlation between 1)grain size, 2) relative density, and 3) number of cycles associated with a ground motion. A factor, c_r, is adopted to adjust the lab test result under an isotropic stress condition to liquefaction under an anisotropic stress condition as in the field. Figure 1 shows the laboratory test results of a number of triaxial tests with different grain sizes of the samples at relative density of about 50%. The stresses required to cause liquefaction for sands at other relative densities are estimated based on the assumption that for relative densities up to about 80%, the shear stress to cause initial liquefaction is approximately proportional to the relative density and is estimated from Equation 1, (Seed and Idriss, 1971):

$$(\frac{\tau_{liq}}{\sigma'_o})_{lD_r} = (\frac{\sigma_{dc}}{2\overline{\sigma}_a})_{L50} \, c_r \, \frac{D_r}{50} \qquad (1)$$

and the shear stress causing liquefaction is:

$$\tau_{liq} = (\frac{\tau_{liq}}{\sigma'_0})_{lD_r} \, \sigma'_0 \qquad (2)$$

where
$(\tau/\sigma'_0)_{lDr}$ = stress ratio causing liquefaction under field conditions;
$(\sigma_{dc}/2\overline{\sigma}_a)_{L50}$ = stress ratio causing liquefaction in triaxial test, σ_{dc} is the cyclic deviator stress and $\overline{\sigma}_a$ is the initial ambient pressure under which the sample was consolidated;
σ'_0 = effective overburden pressure;

c_r = correction factor for stress condition from triaxial to simple shear;
D_r = relative density (%);
τ_{liq} = shear stress causing liquefaction.

In-situ relative density can be estimated based on *in-situ* tests such as SPT tests. Marcuson (1978), Skempton (1986), Youshida (1988) and, Kulhaway and Mayne (1990) proposed empirical relationships of the following types:

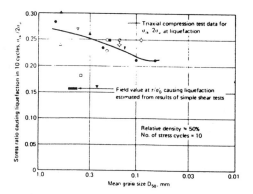

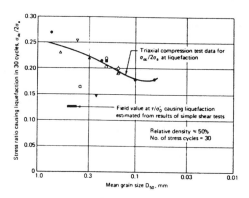

Figure 1. Stress condition causing liquefaction of sands in 10 cycles and 30 cycles (after Seed and Idriss, 1971)

Skempton (1986):

$$D_r = \sqrt{\frac{N'_{70}}{32 + 0.288 \, \sigma'_0}} \qquad (3)$$

where:
N'_{70} = normalized SPT blow count corrected to 70 percent energy;
σ'_0 = effective over burden pressure (in kPa).

Youshida (1988):

$$D_r = 25 \, \sigma'_0{}^{(-0.12)} (N_1)_{60}{}^{(0.46)} \qquad (4)$$

where:
$(N_1)_{60}$ = SPT blow count normalized for 60 percent energy and overburden pressure of 100 kPa;
σ'_0 = effective over burden pressure (in kPa).

These will be used to describe relative density, D_r, in Eqs. 1 and 2.

SHEAR STRESS CAUSING LIQUEFACTION FROM SPT

The liquefaction assessment chart developed by Seed *et al.* (1985) is often used in practice, Figure 2. The independent variable, $(N_1)_{60}$, in this figure is the SPT blow count corrected to a vertical effective overburden pressure of 100 kPa and an energy level of 60 percent of the free-fall energy of the hammer. The curves drawn in the chart represent the boundary lines between liquefiable and nonliquefiable level sandy sites with various percentages of fines in the sands for an earthquake of magnitude 7.5. The effects of the earthquake are characterized by the cyclic stress ratio (CSR) defined as the ratio of shear stress over the effective overburden pressure.

For a site if the point plots on or above the curve, the site is susceptible to liquefaction. Figure 2 is designed for earthquake magnitude M = 7.5. The stress causing liquefaction can be computed by timing the cyclic stress ratio (CSR) obtained in Figure 2 with effective overburden pressure.

LIQUEFACTION OF SAND-SILT MIXTURES

In the past three decades, the phenomenon of liquefaction of sand had been the subject of intensive research. As a consequence, a significant database exists on liquefaction of sands. However, the problem of liquefaction of sands mixed with fines content is equally important.

Seed *et al* (1985) chart Figure 2 shows the boundary line between liquefiable and nonliquefiable level sandy sites with various percentages of fines in the sands for an earthquake of magnitude of 7.5. Based on this information it was concluded that (NRC 1985):

(1) It is clear that, for soils with the same $(N_1)_{60}$, ignoring the presence of fines can be *conservative*, and that the fine content should be noted in evaluating the liquefaction susceptibility of a sand deposit.
(2) However, it still is not possible to evaluate the likelihood of liquefaction of a silty sand with the same confidence as for clean sand.

Figure 3 shows a plot of CSR vs. fines content for sands for $(N_1)_{60}$ of 5, 10 and 20. CSR increases with increase in fines content at constant values of $(N_1)_{60}$. This type of plot leads to an erroneous belief that

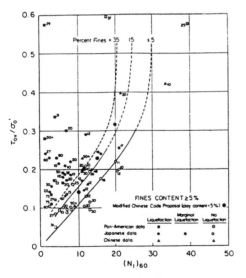

Figure 2. Relationships between stress ratio causing liquefaction and $(N_1)_{60}$ values for sands for magnitude 7.5 earthquakes. (After Seed *et al.*, 1985)

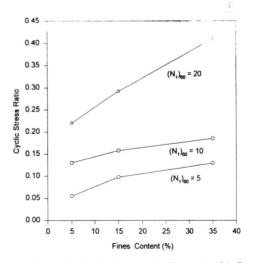

Figure 3. Cyclic stress ratio for sands with fines content (after Seed et al 1985)

increasing fines content in a sand results in increased liquefaction resistance.

Troncoso (1990) compared the cyclic strength of tailing sands with different silt contents ranging from 0 to 30% at a constant void ratio of 0.85, Figure 4. It was found that the cyclic strength *decreased* with the increase of fine content. Figure 5 shows the influence of fine content on cyclic stress ratio for liquefaction at different cycles (N_l). The increase of fine content reduced the liquefaction resistance (CSR). This conclusion is in apparent contradiction with that of Seed *et al* (1985). Note, however, that the difference between the two results is from the different criteria used in their studies. In Troncoso's (1990) study, the cyclic shear strength of tailing sands was examined at a constant void ratio, while in the study of Seed *et al.*, (1985) cyclic stress ratio was investigated at the same SPT values.

Prakash and Sandoval (1992) examined the data of Ishihara and Koseki (1989) on sixteen reconstituted samples of Toyoura sand with a different percentage of kaolinite, Kanto loam and Tailing. Only three samples with plasticity index of 2-4% are of interest. Ishihara and Koseki (1989) stated that, in their investigation, void ratio is not considered as an independent parameter and it might be desirable to investigate the effects of fines content on the cyclic strength under identical values of void ratio. It was concluded that Plasticity Index is an extremely important variable controlling CSR for liquefaction.

nonplastic silt, two things may happen:

1) the mechanism of pore pressure generation changes, leading to higher pore pressures; and
2) plasticity imparts some cohesive character to this mixture and therefore increased resistance to liquefaction.

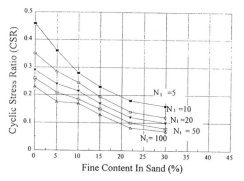

Figure 5 CSR versus fines content of tailings dam

It is the interplay of these factors which will determine whether the liquefaction resistance of silt-clay mixtures increases or decreases compared to that of the pure silts.

Typical test data on liquefaction of silts and silt-clay mixtures has been critically evaluated and some definite trends on their liquefaction behavior are determined (Guo, 1998).

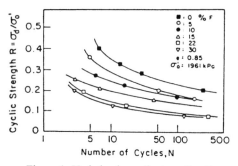

Figure 4. Variation in cyclic strength with number of cycles (Troncoso 1990)

LIQUEFACTION OF SILTS AND SILT CLAY MIXTURES

In nonplastic silts, the mechanism of pore pressure generation is about the same as that for sands. If a small percentage of highly plastic material is added to

Undisturbed Samples

El Hosri *et al* (1984) performed tests on six silt samples in their natural undisturbed state. The soil specimens came from two sites at depths from 20 to 40 m. The soils tested were mainly sandy silt and silty clay (ML-CL or ML-MH). The samples had been tested at different void ratio (e_o). We are not able to isolate the effect of only one variable, *i.e.,* e_0 or PI. It was, therefore, decided to normalize the CSR for all samples for a common void ratio of 0.644. It was assumed as a first approximation that CSR for initial liquefaction is directly proportional to void ratio (Guo, 1998). Table 1 has been determined by normalizing the CSR for $e_0 = 0.644$.

In Figure 6, normalized CSR is plotted against PI. It is seen that the cyclic stress ratio of undisturbed samples decreases first with increasing plasticity index up to PI of about 5. The lowest cyclic stress ratio is presented by sample B with a plasticity index of 5. If the plasticity index of the undisturbed silt sample equals to 15, its cyclic stress ratio is even higher than that for sample A, which is a sand-silt mixture. Therefore, it may be concluded that increase of plasticity index decreases the liquefaction resistance of soil in the low range of plasticity index.

Table 1 Comparison of Normalized Test Results (Guo 1998) (CSR normalized to initial void ratio $e_o = 0.644$)

Sample No.	PI	e_o	Normalized e_o	Normalized CSR
A	-	0.644	0.644	0.295
B	5	0.478	0.644	0.238
F	6.5	0.600	0.644	0.312
C	8	0.548	0.644	0.225
D	9	0.654	0.644	0.310
E	15	0.914	0.644	0.463

Source: El Hosri *et al* 1984

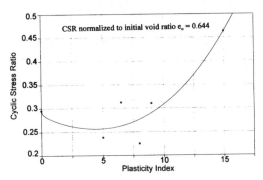

Figure 6 Normalized CSR vs PI of undisturbed samples (Data of El Hosri et al 1984)

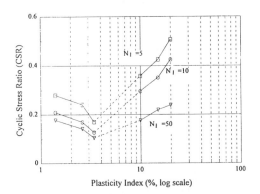

Figure 7 CSR vs PI for reconstituted silt-clay mixtures (Data of Sandoval 1989 and Puri 1984)

Reconstituted Samples

Sandoval(1989), Prakash and Sandoval (1992) performed tests on silts (with PI = 1.7%, 2.6, 3.4%), and Puri 1984 (with PI 10-20%) to determine the relationship between cyclic stress ratio and number of cycles to cause initial liquefaction (u = $\bar{\sigma}_o$). Test results of the two studies above are shown in Figure 7. The plasticity index range for a lowest value of CSR is between PI values of 4 and 10. The two studies, on both undisturbed and reconstituted samples suggest that the hypothesis suggested previously is justified.

Also, on undisturbed samples, structure, aging and cementation have not been investigated. Therefore, there is a need of further systematic study of liquefaction behavior of these soils (Guo 1998).

CONCLUSION

It may be concluded that 1) for liquefaction resistance of sands the effects of structure, aging and other factors need more studies, 2) The importance of structure in fine grained soils such as silts needs to be recognized when evaluating the pore pressure generation and strain developments, 3) There is no definite criterion for evaluating the liquefaction potential of silty soils, and there is confusion on the influences of clay content, plasticity index as well as void ratio. For proper understanding the seismic behavior of silt and silt clay mixtures, those factors need further study.

RETAINING STRUCTURES

We have summarized critical research on computation of seismic displacements of rigid retaining walls in a companion paper to these proceedings (Wu and Prakash, 1998).

I want to discuss now primarily provisions of Eurocode-8 Ch.7 (1994) and their influence on the design of retaining walls.

EUROCODE-8 ch7

Comprehensive provisions for aseismic design of rigid retaining walls are included in codes of India, Japan (IAEE, 1992) and Eurocode-8(1994). However, only Eurocode-8 proposed the concept of limiting displacements smaller than 300*α (mm) for free gravity walls, where α is the maximum ground acceleration ratio. Also non-linear soil properties for base soil and backfill in analysis of sliding and rocking displacements are recommended. The seismic constant of horizontal acceleration for design is α_h ($= \alpha/r$) which is constant along the wall height and r is 2 for free gravity walls.

a. The general expression for computing the lateral dynamic force is

$$P_{ae} = \tfrac{1}{2}\ \gamma^*\ (1 \mp \alpha_v)\ k_{ae}H^2 + P_{ws} + P_{wd} \qquad (5)$$

Where k_{ae} is dynamic earth pressure coefficient (Mononobe-Okabe method).

$$k_{ae} = \frac{\cos^2(\phi - \psi - \alpha)}{\cos\psi\cos^2\alpha\cos(\delta + \alpha + \psi)\left[1 + \sqrt{\dfrac{\sin(\phi + \delta)\sin(\phi - i - \psi)}{\cos(\alpha - i)\cos(\delta + \alpha + \psi)}}\right]^2} \qquad (6)$$

φ is the friction angle of backfill, δ is the friction angle of the wall-backfill interface and ψ is defined as

$$\psi = \tan^{-1}\left[\alpha_h \div (1 \pm \alpha_v)\right] \qquad (7)$$

and varies with different field conditions. P_{ws} is static water force and act at the 1/3 of the height above the base, P_{wd} is hydrodynamic force and lies at 0.4 of this saturated layer from the base, α_h and α_v are horizontal and vertical ground acceleration coefficients and γ^* is appropriate unit weight of soil according to the field conditions as explained further.

b. For dry backfill
$P_{ws} = P_{wd} = 0$ in equation 5.

c. For dynamically impervious backfill (below the water table):

$$\gamma^* = \gamma_{sat} - \gamma_w \qquad (8)$$

$$\psi = \tan^{-1}\left[\gamma_{sat} \div (\gamma_{sat} - \gamma_w)\right] \times \left[\alpha_h \div (1 \pm \alpha_v)\right] \qquad (9)$$
$$P_{wd} = 0$$

d. For dynamically (highly) pervious backfill (below the water table):

$$\gamma^* = \gamma_{sat} - \gamma_w \qquad (10)$$

$$\psi = \tan^{-1}\left[\gamma_d \div (\gamma_{sat} - \gamma_w)\right] \times \left[\alpha_h \div (1 \pm \alpha_v)\right] \qquad (11)$$

$$P_{wd} = 7/12 * \alpha_h * \gamma_w * H' \qquad (12)$$
where H' is the height of saturated layer

ANALYSIS

A typical wall (Fig. 8) 6m high with the following base and backfill properties was designed for FOS against sliding ($\geq$ 1.5), overturning ($\geq$ 1.5), bearing capacity failure ($\geq$ 2.5) and Eccentricity ($\leq$ B/6) under static condition.

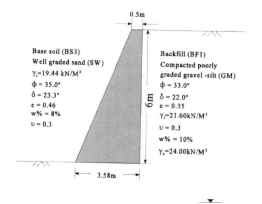

Figure 8. Dimension of wall and soil properties used

This wall was analyzed for displacements for the following field conditions 1-7. (Fig. 9). Field conditions 1-4 are specified in the Eurocode and conditions 5-7 were selected in addition.

The displacements were computed for the following cases

Cases 1

The base width has been designed as for field condition 1 and displacements computed for El-Centro earthquake and different field conditions. The computed displacements are plotted in Figure 10 and listed in Table 2. According to Eurocode, the permissible displacement is 8.7cm ie (300 * 0.29).

In field condition 3 and 4 of Eurocode, the retaining wall exceeds permissible displacements. Field condition 7 is very much a practical case. Here also, displacements exceed the permissible displacement.

Cases 2

Dynamic displacements were computed for different wall dimensions which were determined by different static field conditions. Computed displacements are plotted in Figure 11 and listed in Table 3.

	Field Conditions	Used Parameters
	Condition 1 moist BF* moist BS*	$\gamma^* = \gamma_t$ $\psi = \tan^{-1}\left(\dfrac{\alpha_h}{1 \pm \alpha_v}\right)$ $P_{wd} = 0$
	Condition 2 moist BF saturated BS	$\gamma^* = \gamma_t$ $\psi = \tan^{-1}\left(\dfrac{\alpha_h}{1 \pm \alpha_v}\right)$ $P_{wd} = 0$
	Condition 3 submerged with impervious BF	$\gamma^* = \gamma_{sat} - \gamma_w$ $\psi = \dfrac{\gamma_{sat}}{\gamma_{sat} - \gamma_w} \tan^{-1}\left(\dfrac{\alpha_h}{1 \pm \alpha_v}\right)$ $P_{wd} = 7/12 * \alpha_h * \gamma_w * H'$
	Condition 4 submerged with pervious BF	$\gamma^* = \gamma_{sat} - \gamma_w$ $\psi = \dfrac{\gamma_d}{\gamma_{sat} - \gamma_w} \tan^{-1}\left(\dfrac{\alpha_h}{1 \pm \alpha_v}\right)$ $P_{wd} = 2*7/12 * \alpha_h * \gamma_w *H'$
	Condition 5 perched with impervious BF	$\gamma^* = \gamma_{sat} - \gamma_w$ $\psi = \dfrac{\gamma_{sat}}{\gamma_{sat} - \gamma_w} \tan^{-1}\left(\dfrac{\alpha_h}{1 \pm \alpha_v}\right)$ $P_{wd} = 0$
	Condition 6 perched with pervious BF	$\gamma^* = \gamma_{sat} - \gamma_w$ $\psi = \dfrac{\gamma_d}{\gamma_{sat} - \gamma_w} \tan^{-1}\left(\dfrac{\alpha_h}{1 \pm \alpha_v}\right)$ $P_{wd} = 7/12 * \alpha_h * \gamma_w * H'$
	Condition 7 saturated BF with sloping drain	$\gamma^* = \gamma_{sat} - \gamma_w$ $\psi = \dfrac{\gamma_{sat}}{\gamma_{sat} - \gamma_w} \tan^{-1}\left(\dfrac{\alpha_h}{1 \pm \alpha_v}\right)$ $P_{wd} = 0$

*BF = backfill   BS = base soil

Figure 9. Field conditions and corresponding parameters for dynamic displacements.

Table 2. Computed displacements for Case 1 after 10 sec of El-Centro earthquake.

Field Conditions	Wall Displacements (cm)
1	6.26
2	6.51
3	10.58
4	13.30
5	7.40
6	9.76
7	11.46

Table 3. Computed displacements with different field conditions and corresponding wall dimensions.

Field Conditions	Wall Dimension (m)	Wall Displacements (cm)
1	3.58	6.26
2	3.58	6.51
3	2.63	14.35
4	2.63	17.72
5	10.38	2.86
6	10.38	3.39
7	3.69	11.08

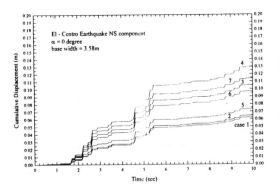

Figure 10. Computed displacemens for case 1.

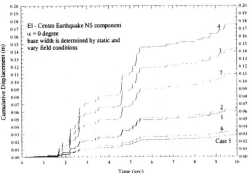

Figure 11. Computed displacemens for case 2

113

CONCLUSION

We have developed a realistic method to compute displacements of rigid retaining walls for different field conditions and considering soil non-linearity and accounting for sliding and rocking motions of the walls.

A design procedure for permissible displacements may then be perfected (Wu 1998).

DYNAMIC   SOIL-PILE-STRUCTURE INTERACTIONS

The behavior of structures under static or dynamic loads, founded on soils, is different from that of similar structures founded on rock. Problems involving dynamic loads differ from the corresponding static problems in two important aspects: the time varying nature of the excitation; and the role played by inertia.

The stress-strain behavior of soils is nonlinear even at very low strains. One of the commonly used procedures in practice to incorporate the effect of soil nonlinearity consists of performing a linear analysis at each time step by using the soil properties (modulus and damping), consistent with the strains resulting from the previous time step (Roesset and Tassoulas, 1982).

A parametric analysis was performed to study the effect of nonlinear behavior of soil on natural frequency response and other parameters of structures (Kumar 1996).

Simplified model to perform DSSI

The response of the system, excited by a modified base motion having acceleration of $\ddot{x}_g$, is governed by Equation (1)

$$[M]\{\ddot{x}\} + [C]\{\dot{x}\} + [K]\{x\} = -[M_F]\{I\}\ddot{x}_g \quad (6)$$

where [M], [C], and [K] are mass, damping, and stiffness matrices of the system, respectively; $\{x\}$ is the column vector of displacements relative to the position of the structure after free-field displacement; $[M_F]$ is the matrix to calculate force at each degree of freedom; a dot superscript denotes differentiation with respect to time, t; and $\{I\}$ is a column vector of ones. It should be noted that $[M_F]$ is differentiated from [M] is Equation (13) (Kumar 1996).

Parameters used

Structure: Three structures (2-, 4-, and 6-story) are considered. The structures are assumed to consist of: story height = 3.5 m; size of each column in a story = 300 x 300 mm (2 columns in each story); mass of each story = 19.9 Mg/m^3; and mass moment-of-inertia of each story about its own axis = 38.2 kN-m-sec^2. Stiffnesses of the structures are calculated using the procedures given by Chopra (1995).

Foundation: The structures are assumed to be supported on pile; each column has four HP 12X72 steel H-piles at a center-to-center spacing of 0.9 m. Length of each pile is 8 m. Pile cap is 1.6 m X 1.6 m X 1.0 m. Pile caps have actual and effective depths of embedment of 0.8 m and 0.6 m, respectively. Stiffnesses and damping of foundations are calculated as recommended by Gazetas (1991).

Soil: Soil is assumed as linear elastic and nonlinear. A wide range of soil shear modulii, from 15,000 to 170,000 kPa is used to perform the parametric study.

Base Motion: Two types of base input motions, sinusoidal and recorded earthquake time history (El Centro earthquake of 1940, N-S component), are used. Both the recorded time history and sinusoidal base motions are scaled to a maximum acceleration of 0.1 g at rock. The frequency of sinusoidal base motion assumed is 0.92 Hz. Free-field motion is used as the base input motion.

EFFECT OF DSSI AND SOIL NONLINEARITY ON *FUNDAMENTAL* NATURAL FREQUENCIES

Linear soil

The *fundamental* natural frequency response of the 2-, 4-, and 6-story structures with respect to the low strain soil shear modulus (G_{max}) is shown in Figure 12. The *fundamental* natural frequency of the 2-story structure, fixed at the base, is 2.13 Hz, and for flexible based structure, it varies from 1.50 Hz at G_{max} = 15,000 kPa to 1.98 Hz at G_{max} = 170,000 kPa. For the 4-story structure the fixed base fundamental frequency is 1.20 Hz and that for flexible based structure the variation is from 0.75 Hz at G_{max} = 15,000 kPa to 1.06 Hz G_{max} = 1,70,000 kPa. The *fundamental* natural frequency of the 6-story structure, fixed at the base, is 0.83 Hz, whereas the for flexible base it varies from 0.47 Hz at G_{max} = 15,000 kPa to 0.71 Hz at G_{max} = 170,000 kPa.

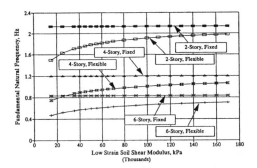

Figure 12. Variation of *fundamental* natural frequencies with low strain soil shear modulus (Linear soil)

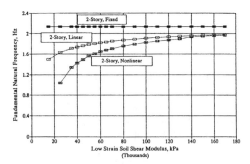

Figure 13. Effect of soil nonlinearity on *fundamental* natural frequency response of 2-story structure, sinusoidal base motion

The *fundamental* natural frequency of the system for the flexible based structure is lower than the corresponding value for the fixed base structure. Also, the *fundamental* natural frequency of the flexible based structures increases nonlinearly with the increase in soil shear modulus and approaches the fundamental frequency of a fixed base structure.

Nonlinear soil

It is understood that the natural frequencies of a system, consisting of soil and structures, depend only on the properties of the soil and structure and not on the base input motion. However, this is true only when soil is assumed to behave as linear elastic. When nonlinear behavior of soil is considered, the stiffness of soil changes with time, depending on the level of strains in the soil. Also, at any particular time, the strains in soil, and thus the foundation stiffness, will not be same for any two different base input motions.

Figure 13 shows the effect of soil nonlinearity on the fundamental natural frequency of the 2-story structure, Similar response curves for the 4- and 6-story structures are presented else where (Kumar 1996, Kumar and Prakash 1997). These results show that, for all structures, soil nonlinearity resulted in lower fundamental natural frequencies of the system for all values of the low strain soil shear modulus compared to the frequencies obtained assuming soil as linear elastic. This is because of the fact that the effect of soil nonlinearity is to reduce the soil modulus which means reduction in the foundation stiffness and, therefore, lower *fundamental* natural frequency response.

The above results depend upon the realistic prediction of response of pile, as explained further.

COMPARISON OF PREDICTED RESPONSE WITH OBSERVED RESPONSE OF SINGLE PILES AND PILE GROUPS

Several lateral dynamic load tests on full-sized single piles with pile cap only (no super structure) were performed to check if the predicted response tallied with the measured response, (Gle, 1981; Woods, 1984). Also, Novak and El-Sharnouby (1984) performed tests on a group of model piles to compare predictions with performance. The predicted response did not tally with the measured response in either case.

Tests of full-size single piles

Fifty-five steady-state lateral vibration tests were performed on 11 pipe piles 14 in. in outside diameter with wall thickness of 0.188 in. to 0.375 in. (0.47 cm to 0.94 cm) at three sites in southeast Michigan (Woods, 1984). The end-bearing piles were 50 to 160 ft (15 to 48 m) long.

Figure 14 shows response curves for the pile GP 13-7, 157 ft (47.1 m) long in soft clay. The pile was excited in steady-state oscillation by attaching an eccentric weight vibrator (Lazan oscillatory) to the head of the pile. It was observed that the frequency of maximum response decreased as the force level increased, indicating non-linear response. A PILAY computer program was used by Woods (1984) to determine stiffness and damping of the pile (Novak and Aboul-Ella 1977). However, PILAY assumed that the soil surrounding the pile in a given layer is the same at all distances from the pile.

A dynamic response curve with this solution is shown in Figure 15 along with the field data. The correlation between predicted and measured response is very poor. In all tests, computed response based on stiffness and damping from PILAY and measured response showed that the amplitudes of motion were greater than predicted and the frequency of maximum response was lower than predicted.

In an attempt to match the measured response with the computed response the following two approaches were adopted (Woods 1984).

1. For predicting the response, only a fraction of the rocking and translation stiffness computed by PILAY was used. It was found that even with a wide variation in rocking stiffness, the observed amplitudes in the frequency range just above the horizontal translation peak was still higher than the predicted amplitude. The observed increase is more likely due to change in soil parameters caused by pile driving.

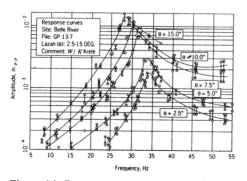

Figure 14 Response curves; a decrease in resonant frequency with increasing amplitudes (Gle, 1981).

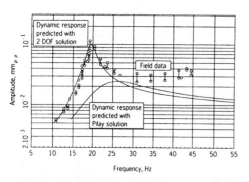

Figure 15 Typical response curves predicted by PILAY superimposed on measured pile response (Woods, 1984).

2. Because of the poor correlation achieved in the initial attempt, a second correlation with the analytical procedure -PILAY 2- was attempted. PILAY 2 permits an inclusion of a *"softened"* or *"weakened"* zone surrounding the pile, simulating the disturbance to the soil caused by pile installation.

A good match of the measured and predicted response could be obtained by a considerably reduced soil moduls in the softened zone (one-tenth to two-tenths of the original value) and the extent of the softened zone (one-half to one times the pile radius). A loss of contact of the soil with pile for a short length close to the ground surface also improved the predicted response.

El-Sharnouby and Novak (1984) performed dynamic tests on a 102 steel pipe piles group and analyzed the data by several methods. The predicted response curves were no where close to the measured ones.

In the case of pile groups, the pile-group interaction factors also appear. Therefore, the match of predicted and measured response are farther apart.

The question arises that if realistic predictions of response of piles without the super structure is not possible, how can the realistic response of super structures supported on piles be predicted?

CONCLUSION

Predicted solutions of dynamic soil-pile-structure interactions do not match the measured ones. Further research is needed to define modification to the soil properties ie stiffness and damping for a better match.

ACKNOWLEDGMENT

The draft was commented upon by Tony Guo, and Alex Wu and typed with painstaking effort by Charlena Ousley. Figures had also been prepared by Guo and Wu. All help is acknowledged.

REFERENCES

Chopra, A.K. *Dynamics of structures-theory and applications to earthquake engineering.* Prentice Hall, Englewood Cliffs, New Jersey, 1985.

Dobry, R. (1995) "Liquefaction and Deformation of Soils and Foundations Under Seismic Conditions" State of the Art, *Proc. Third International Conf. On*

Recent Advances in Geotechnical Earthquake Engineering and Soil Dynamics Ed. Shamsher Prakash Vol. III, pp. 1465-1490, University of Missouri-Rolla, Rolla, Missouri.

El Hosri, M. S., Biarez, H., Hicher, P. Y., (1984), "Liquefaction characteristics of silty clay," Proc. Eighth World Conference on Earthquake Engineering, Prentice-Hall, Inc., Englewood Cliffs, New Jersey, 1984, Vol. III, pp 277-284

EUROCODE 8 (EUROPEAN PRESTANDARD 1994) "Design Provisions for Earthquake Resistance of Structures- Part 5: Foundations, Retaining Structures and Geotechnical Aspects", The Commission of the European Communities.

Finn, W. D. Liam (1981), "Liquefaction potential: Developments since 1976," *International Conference on Recent Advances in Geotechnical Earthquake Engineering and Soil Dynamics*, Ed. Shamsher Prakash University of Missouri-Rolla, Rolla, Missouri, April 26-May 3, 1981, Vol. II, pp 665-681

Finn, W.D.L. (1991), "Assessment of Liquefaction Potential and Post-Liquefaction Behavior of Earth Structures 1981-1991" *Proc. Second International Conf. On Recent Advances in Geotechnical Earthquake Engineering and soil Dynamics*, Ed. Shamsher Prakash Vol. III, pp. 1833-1850, University of Missouri-Rolla, Rolla, Missouri.

Gle, D.R., (1981), "The Dynamic Lateral Response of Deep Foundations", Ph.D. Dissertation, The University of Michigan, Ann Arbor.

Gazetas, G. "Foundation vibrations" *Foundation Engineering Handbook*, H.Y. Fang, ed., Van Nostrand Reinhold, 553-593, 1991.

Guo, T. (1998), "Liquefaction of sands and silts," Ph.D. Thesis, University of Missouri-Rolla, Missouri, 1998. (Under progress)

IAEE (1992) "*Earthquake Resistant Regulations A World List-1992*" International Association for Earthquake Engineering.

Ishihara, K. and J. Koseki, (1989) "Cyclic shear strength of fines-containing sands," Earthquake and Geotechnical Engineering, Japanese Soc. SM&FE, pp 101-106

Kulhaway, F. H. and P. W. Mayne (1990), "Manual on estimating soil properties for foundation design," *Report No. EL-6800*, Electric Power Research Institute, Palo Alto, CA

Kumar, S. "Dynamic Response of Low-Rise Buildings Subjected to Ground Motion Considering Nonlinear Soil Properties and Frequency-Dependent Foundation Parameters". Ph.D. Dissertation, University of Missouri-Rolla, Rolla, Mo. 1996.

Kumar, S. And S. Prakash, "Natural Frequency Response of Structures Considering Soil-Structure Interaction", *Special Volume, Earthquake Geotechnical Engineering*, XIV International Conference of Soil Mechanics and Foundation Engineering, Hamburg, 1997, pp -

Marcuson, W. F., III (1978), "Determination of *in-situ* density of sands," *Dynamic Geotechnical Testing*, ASTM STP 654, American Society for Testing and Material, 1978, pp 318 -340

National Research Council, (1985), "Liquefaction of soils during earthquakes," Report No. CETS-EE-001, Com. Earthq. Eng., National Academy Press, Washington, D.C.

Novak, M. And Aboul-Ella, E., (1977), "PILAY -- A Computer Program for Calculation of Stiffness and Damping of Piles in Layered Media", Report No. SACDA 77-30, University of Western Ontario, London, Ontario, Canada,.

Novak and El-Sharnouby, B., (1984), "Evaluation of dynamic Experiments on Pile Group", J. Geotech. Eng. Div., ASCE, Vol. 110, No. 6, pp 738-756.

Prakash S. and V.K. Puri, (1982) "Liquefaction of loessial soils," Proc. Third International Conference on Seismic Microzonation, Seattle, Wash. June 28-July 1

Prakash, S. and J.A. Sandoval, (1992) "Liquefaction of low plasticity silts," Journal of Soil Dynamics and Earthquake Engineering, Vol. 71 No. 7 pp. 373-397.

Puri, V.K., (1984) "Liquefaction behavior and dynamic properties of loessial (silty) soils," Ph.D. Thesis, University of Missouri-Rolla, Mo., 1984.

Roesset, J.M. and J.L. Tassoulas, "Nonlinear soil-Structure Interaction: an Overview", *Earthquake*

Ground Motion and its Effects on Structures, ASME, AMD, 53, 59-76, 1982.

Sandoval, J. A.. (1989), "Liquefaction and settlement characteristics of silt soils," Ph.D. Dissertation, University of Missouri-Rolla

Seed, H. B. and I. M. Idriss (1971), "Simplified procedure fro evaluating soil liquefaction potential," *Journal of Soil Mechanics*, Foundation Division, ASCE, Vol. 97, No. SM 9, pp. 1249-1273

Seed, H. B., and K. L. Lee (1966), "Liquefaction of saturated sands during cyclic loading," *Journal of Soil Mechanics*, Foundation Division, ASCE, Vol. 92, No. SM6, PP 105-134

Seed, H. B., T. K. Harder, and R. M. Chung (1985), "Influence of SPT Procedures in soil liquefaction resistance evaluations," *Journal of Geotechnical Engineering*, Vol. 111, No. 12, pp. 1425-1445

Skempton, A. W. (1986), "Standard penetration test procedures...," *Geotechnique*, Vol. 36, No. 3, pp 425-447

Troncoso, J. H. (1990) "Failure risks of abandoned tailings dams," Proceedings, International Symposium on Safety and Rehabilitation of Tailings Dams, ICOLD, Sydney, Australia, May, PP.82-89.

Woods, R.D., (1984), "Lateral Interaction between Soil and Pile", Proceedings International Symposium on Dynamic soil Structure Interaction, Minneapolis, MN, pp 47-54.

Wu, Y. (1998), "Displacement Based Analysis and Design of Rigid Retaining Walls During Earthquake", Ph.D. Dissertation, Univ. Of Missouri-Rolla, USA (Under preparation).

Wu, Y. and Prakash, S (1998), "On Comparative Seismic Displacements of Rigid Retaining Walls" XI Danube-European Conference On Soil Mechanics and and Foundation Engineering, Poreè, 25-29 May.

Yoshida, I. et al., (1988), "Empirical formulas of SPT blow counts for gravelly soils," 1st ISOPT vol. 1, pp 381-387

Geotechnical Hazards, Marić, Lisac & Szavits-Nossan (eds) © 1998 Taylor & Francis, ISBN 90 5410 957 2

The failure of the Port of Nice: An example of static liquefaction of sand

F. Schlosser
Ecole Nationale des Ponts et Chaussées, Paris, France

ABSTRACT : The large and sudden failure of the fill of the dike of the Port of Nice in 1979 has led to extensive investigations in order to determine the cause of the failure. It has been found that the sliding was due to a static liquefaction of loose sand seams trapped in a silt layer. Liquefaction was initiated by a rapid drawdown of the sea level caused by the arrival of an off-shore tidal wave. Use of piezocone in soil investigation and of typical static liquefaction laboratory tests are required for the prediction of the corresponding liquefaction potential.

1. INTRODUCTION

Among the different cases of failure due to liquefaction of loose sand, a particular one is related to seams of loose sand trapped in a thick layer of a much less permeable soil, like silt or clay.

Such a stratigraphy can often be found in deltaic deposits of coastal areas and it can induce serious risks of liquefaction and as a consequence, it can result in failure of natural or artificial marine slopes. Generation of excess pore water pressure in such seams of sand cannot dissipate because of the presence of impervious soil confining the sand. Moreover, these seams are generally horizontal or sub-horizontal in the slope direction, which constitute a very critical factor for the stability. It must be noted that it is difficult to detect such seams in classical soil investigations.

The failure of the new Port of Nice (France), is a typical example of this case. It occured in 1979, eight months after the construction of the main embankment and resulted in a quasi-instantaneous slide of about 10 millions m³ of material.

Nice on the south-east coast of France (Fig. 1). The slide involved about 2 to 3 million m³ of fill and also about 7 million m³ of the underlying clayey silt and silty sandy alluvial deposits on which the fill had been placed. The time required for sliding of the approximately 10 million m³ of soil was about 4 minutes.

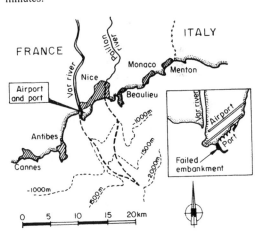

Figure 1. Project location of the new port of Nice

2. DESCRIPTION OF THE FAILURE

On the beginning of the afternoon of October 16, 1979 a major slide occured in the fill which had been placed to construct the dike of the new port at

The slide debris from the port moved out to sea, first down the sloping face of the delta deposit on which it was constructed and then along an off-shore canyon and finally along the sea floor, rupturing two sets of sea-floor communications cables located at distances of about 90 and 120 kms off-shore from Nice. Both sets of cables were moved about 15 kms from their original positions. The velocity of flow on the almost flat sea bed between the cables was about 7 km/h.

A typical soil profile through the delta in the location where the slide movements started is shown in Figure 3. The main fill was kept at least 150 m from the crest of the outer slope of the delta deposits

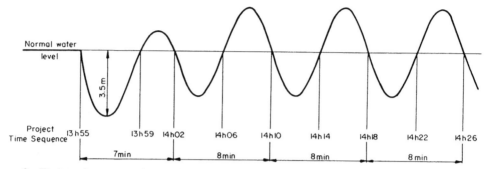

Figure 2. Timing of sequence of water level fluctuations at the port based on observations of witnesses

An off-shore soil investigation, in the area where the slide mass came to rest, show that an undersea flow slide of several hundred million m³ of soil was probably involved in causing the cable breaks. Since the port slide involved only about 10 million m³, a very large volume of supplementary slide debris would have to have been generated from some submarine location off-shore to account for this very large volume of sediment which finally came to rest about 150 km off-shore.

At about the same time as the sliding of the port fill occurred, a tidal wave was observed along a section of about 120 km of the coast of south-eastern France. This tidal wave had an amplitude of about 3 m near Nice and Antibes (about 12 km from Nice) and decreased in amplitude to about 0.25 m at Ile du Levant (about 90 km west of Nice). Its period at all locations along the coast was 8 mn and the records of the sea level movements clearly showed that its first manifestation was a lowering of the sea level of 3.5m at the port followed by a long sequence of rising and lowering fluctuations (Fig. 2). The flooding associated with this wave caused the loss of several lives and did considerable damage to local communities and harbors.

Only the outer portion of the fill dike, which had been placed to construct a port facility, was involved in the slide. The fill, much of which was deposited through or in water, was constructed on a deltaic deposit of stratified clayey silt and silty sand, the outer boundary of which sloped off-shore at about 15° to the horizontal.

in order that it would not have any significant effect on the stability of the outer slopes. The delta is cut by two submarine canyons, which extend to a distance of about 15 km off-shore where they merge together (Figure 1). These are the underwater extensions of the Var and Paillon rivers on land.

It had long been recognized that pervious layers of sandy gravel and gravelly sand existed in the delta deposits, which connected with water levels on land and created an artesian pressure condition ($\Delta u_o = 50$ kPa) at depths between 40 m and 50 m in the delta deposits (Fig. 3). This condition was taken into account during the planning of the fills and in the evaluation of pre-construction and post-construction slope stability.

There were a number of unusual circumstances associated with the slide. They included the following:

1. The port fill and the underlying supporting soils were generally similar to those involved in the construction of an extension to the adjacent airport where they had proved to be adequately stable over a period of several years.

2. The fill and underlying soils were indicated to be stable on the basis of geotechnical studies made prior to and during construction. While the factor of safety of the outer parts of the delta slopes was somewhat less than that normally required for permanent construction (F ≈ 1.3 to 1.4), it

was consistent with that generally accepted as adequate for the period of construction.

3. The soil fill proved to be adequately stable over a period of about 8 months prior to the slide, during which time very little fill was placed in the port area. Since virtually no fill had been placed for a period of about eight months, piezometers showed, as anticipated, a reduction in the pore water pressures induced by the new fill in the delta deposits and a corresponding

8. Sensitive seismographs at Antibes (about 10 km from Nice) showed a marked change in both the frequency and amplitude of background seismic waves starting just before the port slide occurred, but they showed no earthquake activity on the day of the slide. In fact the most recent earthquake was a magnitude 2 event occurring about 20 kms from Nice on Oct. 1, about 16 days before the slide.

The early interpretation of these events made by a special commission of the French authorities in 1980

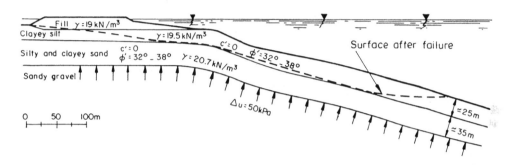

Figure 3. Slope configuration before failure and after failure.

improvement in the stability of the fill. This improvement would be expected to continue as the pore pressures dissipated with time.

4. There was no prior indication, from surface observations or from the piezometer readings, of instability of the port fill.

5. The slide was preceded by several days of very heavy rainfall (about 250 mm in 4 days) causing the Var and Paillon rivers to be running full and with greater than usual capacity. The rainfall increased the artesian pressure in the delta deposits by about 1 m of head, as evidenced by piezometer readings in the delta deposits underlying the adjacent airport area.

6. Although there was no warning of an impending slide, the slide in the port fill occurred rapidly in a period of about 4 minutes.

7. The effects of the tidal wave were recorded on maregraphs installed along the coast-line at Nice, Villefranche and Ile du Levant.

led to the conclusion, that the slide in the port fill had triggered a massive under-water landslide, which in turn had caused the tidal wave. However this hypothesis left unanswered the question of what caused the occurence of the slide in the port fill in the first place.

Since the tidal wave and the failure of the port fill resulted in the loss of lives and property, a comprehensive investigation was initiated by the French authorities. The studies and legal implications lasted more than 10 years. Terrasol was requested in 1981 by the local authorities in Nice and then by the Ministry of Equipement (Public Works) to contribute with Prof. H.B. Seed to the determination of the cause of this major coastal slide.

3. INVESTIGATIONS AFTER FAILURE

3.1 Investigations about the tidal wave

The early interpretation leading to the conclusion that the slide of the port fill triggered the tidal wave, was due to the fact that no seismic event had been recorded and that it has been considered that such a large tidal wave could not occur without an earthquake activity.

However a careful review of the testimony of witnesses to the tidal wave activity gave strong evidence that tidal wave effects, involving a lowering of the sea, occured 2 minutes before the start of the port slide.

Therefore, in order to further explore the source of the tidal movements observed along the coast, studies on a reduced scale model and numerical analyses were performed in order to estimate the amplitudes of tidal movements caused by a slide of the port fill. The results gave values of the amplitudes which were only about 1/10 of the actual amplitudes observed along the coast and provided further evidence to indicate that the tidal wave was not caused by the port slide, but must have some other source, probably a major off-shore slide.

In order to examine this possibility a study of the bathymetry in the vicinity of Nice was performed. The results of bathymetry surveys in 1973 and in 1979, just after the slide, indicated that the main area of loss of material occured about 15 km off-shore from Nice in the canyons. The estimated volume of material missing was between 100 million and 300 million m^3. There was thus good evidence that a large submarine slide occured in the canyon, close to the submarine confluent of the Var and Paillon rivers, where there is a change in the submarine slope angle. It is therefore reasonable to believe that this slide has generated a significant tidal wave and that the port failure occured in association with this wave.

3.2 Soil and failure investigations

Before construction and following the failure numerous borings and in situ tests were performed. As shown in figure 1, the delta deposits in the sliding area consisted of :

1) a layer of soft clayey silt and silty clay, about 25m thick;
2) a layer of fine silty sand and clayey sand, about 35 m thick;
3) a substratum of sandy gravel.

The sliding surface which was observed after failure, was relatively flat and located at the base of the silt layer with a curvature directed downwards. These very particular features, associated with the rapidity

of the sliding (4 mn) and the fact that all the slided materials have passed through a kind of bottleneck, suggested a liquefaction type failure. A soil liquefaction phenomenon could also explain why the port has failed with a static global safety factor of 1.35.

It is why a careful soil investigation was performed in order to determine what soils in the delta deposits could fail suddenly due to apparently minor changes in loading conditions. The clayey silt and silty clay were eliminated because they did not indicate characteristics of the type exhibited by quick-clays (Atterberg limits and water content). A serie of piezocone penetration tests performed in the area of the fill adjacent to the point where sliding stopped and also off-shore in areas close to the landslide zone(Fig. 4), in order to investigate the silty sand layer, showed two important results :

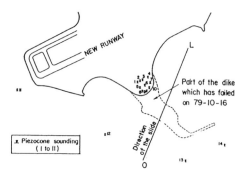

Figure 4. Soil investigation with piezocone

1) at the base of the clayey silt and silty clay layer was located a succession of loose sand seams, apparently continuous and 15 to 50 cm thick, which had not been detected in the initial investigation before construction;

2) the use of various correlations (Schmertmann, 1977; Seed, 1982) between the tip resistance and the relative density showed that on-shore the relative density of the sand in the seams varies between 20 and 45 %, and off-shore it is comprised between 15 an 25 %, which is characteristic of a potentially very liquefiable sand.

Figure 5 presents typical test results obtained with the piezocone. Silt is characterized by a small but regular tip resistance and positive excess pore pressures. When the cone tip reaches a seam of loose sand, a light peak of tip resistance as well as a pore pressure increase often followed by a decrease can be observed (Fig. 5a). Recording of the dissipation of the excess pore water pressures during penetration stops allow to determine whether the tip of the piezocone is in the silt or in the loose sand. In the sand a quasi-complete dissipaption is reached within 2 to 3 minutes, whereas in the silt, 20 to 30 minutes are necessary (Fig. 5b).

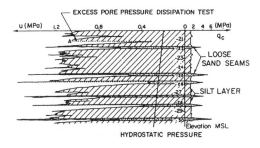

a) *Measurements of the tip resistance and of the pore pressures*

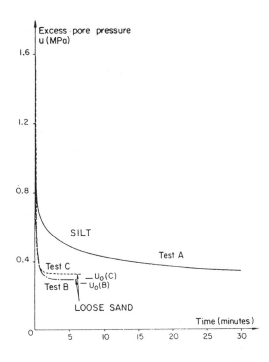

b) *Excess pore pressure dissipation tests*

Figure 5. Results of piezocone tests

This stratigraphy with seams of loose sand trapped in a silt layer is very sensitive to liquefaction since any generation of excess pore water pressure in the sand of the seams cannot be dissipated because of the pressure of impervious soil confining the sand. Morever, in the present case the sand seams were inclined in the slope direction, which constituted a very critical factor for the stability.

However, as no earthquake activity had been recorded in the vicinity of the port site around the time of the failure, only static liquefaction could be considered.

4. FAILURE OF THE PORT BY STATIC LIQUEFACTION

Dynamic liquefaction of loose sands submitted to vibrations has been the subject of numerous studies and papers. The evaluation of the corresponding dynamic liquefaction potential can be done either with laboratory tests (dynamic triaxial test, torsion test) or with in-situ tests, mainly SPT and CPT.

On the contrary, static liquefaction, which consists in a loss of shear strength of a loose sand submitted to a rapid variation of total stresses in undrained conditions, has been much less studied. G. Castro (1969) first and then H.B. Seed (1983) have described and studied this phenomenon with the triaxial apparatus.

Figure 6 shows the difference in behaviour in a triaxial test between the Nice loose sand in drained conditions and the same sand in undrained conditions, both having the same initial anisotropic ($\sigma'_{10} = K_c \sigma'_{30}$) state of effective stresses. On the vertical axis of the stress-strain curve is plotted the ratio τ/σ'_0 of the shear stress τ over the initial normal stress σ'_0 both of these stresses beeing exerted on the potential failure plane in the sample. It can be noted that in undrained conditions a very small increase of the ratio τ/σ'_0 or of the axial deformation ε_1 is sufficient to cause a drop in the sand resistance. This drop, corresponding to a static liquefaction, is the more important as the relative density of the sand is lower.

Considering that static liquefaction is initiated for the maximum of the shear stress τ or of the stress ratio τ/σ'_0 and taking this value as the failure criterion, Seed (1983) has determined the relationship existing between the stress ratio at failure (τ_f/σ'_0) and the one at the initial stage (τ_0/σ'_0).

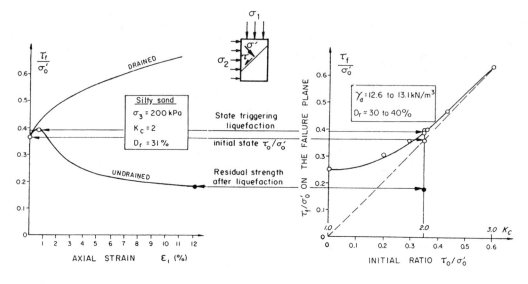

Figure 6. Relationship between initial and at failure stress ratio on failure plane in undrained triaxial tests on Nice loose sand.

Figure 6 shows the corresponding curve obtained for the loose sand of Nice. It is clear from this diagram that under undrained conditions a small increase of 2 to 5 % of the initial stress ratio ranging from 0.4 to 0.5 is sufficient for causing failure of the sample by static liquefaction.

Figure 7 shows the evolution of the excess pore water pressure generated during a static liquefaction test performed on the Loire River sand (Canou, 1985) and the corresponding effective stress path

followed in the $(p = (\sigma'_1 + \sigma'_3)/2 ; q = (\sigma'_1 - \sigma'_3)/2)$ plane. It can be noted that in this case $(D_r \approx 20\%,$ constant rate of deformation in the test, initial isotropic state of stress) a complete liquefaction of the sample is obtained. The pore water pressure increases continuously until complete liquefaction.

In order to evaluate the potential for slide movements due to static liquefaction in a sand seam near the base of the clayey silt layer as a result of sea level lowering it is necessary to determine :

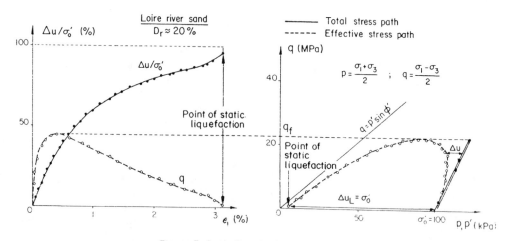

Figure 7. Static liquefaction triaxial test

124

1) *The initial stress conditions at points along a potential failure surface*; these results obtained from a FEM analysis are shown in Figure 8 at points along the surface after failure.

2) *The increase in the shear stress* τ *at points along the potential failure surface resulting from a sudden lowering of the sea water level.* This is a standard drawdown analysis in soil mechanics and the results of such a study can be made either by limit equilibrium methods or by finite element method. The analysis was made by finite element method which provides information on distribution of stress changes rather than a gross average for a potential failure surface. The finite element analyses showed that for the point on the failure surface having the highest ratio of initial shear stress/effective normal stress ($\tau/\sigma'_0 = 0{,}51$), he increase in shear stress due to a 3 m lowering of the sea level was about 1.5 to 2%.

3) *The stress increases which must be applied to each soil element along the potential failure surface in order to bring it to a condition of incipient liquefaction.* This data obtained by laboratory tests on reasonably representative samples, is presented in Fig. 6. Comparison of these required stress increases with those induced by lowering of the sea level, as discussed in (2) above, indicates whether liquefaction is likely to be triggered or not at any given point.

Applying these principles to the soil conditions in the slide area at Nice, it may be noted from Fig. 8 that for the most highly stressed elements along the potential failure surface, the initial shear stress/normal effective stress ratio is 0.51. For this initial stress ratio (τ_0/σ'_0), the data in Fig. 6b indicates that failure and liquefaction of the element will occur for a stress increase of 2 to 3%. The results of the drawdown analysis show that a lowering of the sea level by 3 m would increase the shear stress by about 1.5 % to 2 %. Thus within the limits of geotechnical engineering accuracy, and especially in view of the facts that 1) the in-situ condition of the sand is likely to be looser than that of the samples used in the laboratory testing program ($D_r \approx 20$ % compared to 30 %), and 2) the initial in-situ stress ratios are likely to be somewhat higher than those shown in Fig. 8 because all construction pore pressures were not dissipated which reduces the values of σ'_0, these conditions are representative of a potential failure condition in a zone of limited extent along the potential failure surface.

Once liquefaction had been induced in this small zone, however, it might be expected that the accompanying loss of shear resistance will result in a transfer of shear stress to the immediately adjacent zone which will then lose shear resistance and, by the repeated development of this type of progressive failure, liquefaction will develop along all the failure line shown in Fig. 2.

CONCLUSION

The failure of the Port of Nice has been the first case in France of a sea-coastal slide due to static liquefaction of loose sand seams trapped in a silt layer. As these seams of loose sand had not been detected in the soil investigations carried out before the construction, the effect of a seismic event had only been taken into account by performing classical pseudo-static stability analyses of the slopes. However as liquefaction of loose sand is easier under dynamic loading than under static loading (Seed, 1983), an earthquake in the vicinity of Nice would have also produced the same rapid failure of the dike and may be more.

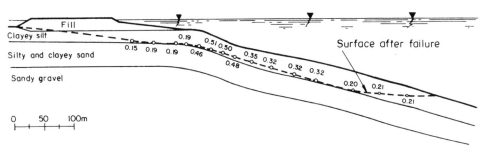

Figure 8. Computed values of τ/σ' *at points along the slip surface (artesian pressure* $\Delta u_a = 60$ *kPa).*

The rapid drawdown of the sea level along the coast has been the triggering mechanism of the static liquefaction of the loose sand seams. It has been generated by a major submarine landslide of 200 to 300 million of m^3 located 15 to 20 km offshore, the origin of which has not been completely established.

Some years later, in 1982 and 1984, rapid failures occured in the submarine sandy slopes of the port of Dunkirk, after the construction of a large quay (West Ore Quay). Soil investigations performed after the second failure with the piezocone showed the existence of a silt layer containing loose sand seams. Static liquefaction of the loose sand has been identified as the phenomenon responsible for the slides.

Recommendations have been made in 1989 by the French Port Authorities. It is indicated that silt layers in deltaic deposits must be considered as areas of potential instability for submarine slopes because they often contain thin layers of loose sand that can be statically or dynamically liquefied and result in serious failures, even if the angle of the initial slope is small. Careful investigations of these silt layers have to be performed with piezocone penetration tests and excess pore pressure dissipation tests.

REFERENCES

Canou J., (1985) - "Study of static liquefaction of a loose sand with the triaxial apparatus", Internal report CERMES. Ecole Nationale des Ponts et Chaussées. Paris.

Canou J. (1989) - "Contribution à l'étude et à l'Evaluation des Propriétés de liquéfaction d'un sable". Thèse de doctorat de l'ENPC. Juin 1989.

Castro G. (1969) - "Liquefaction of sands". Harvard Soil Mechanics Series N° 81. Cambridge, Mass.

Schlosser F., Corté J.F., Dormieux L. (1987) - "Dynamic effects of groundwater". State of the Art Report. Proc. of the 9th ECSMFE. Dublin.

Schlosser F., (1985) - "Liquefaction of loose sand seams in marine slopes". Session 7B. Seismic stability of marine slopes. 11th ICSMFE. San Francisco.

Seed H.B. (1983) - "Stability of Port Fills and Coastal Deposits". Proc. of 7th Asian Reg. Conf. on SMFE. Technion Institute of Technology. Haïfa. Israel. Aug. 14-19, 1983.

Geotechnical Hazards, Marić, Lisac & Szavits-Nossan (eds) © 1998 Taylor & Francis, ISBN 90 5410 957 2

Environmental geotechnics – ITC5-reports and future goals

W.F. Van Impe
Laboratory of Soil Mechanics, Ghent University, Belgium

ABSTRACT : The effect of the ISSMGE to support the large interest of a major group of geotechnical researchers for discussion topics with environmental impact, resulted already in three ICEG conferences and many specialized Seminars. The work of ITC5 has always been the major triggering effect to organise those scientific events. In this paper the accomplishments of the ITC5 group of experts will be briefly summarized.

1 INTRODUCTION

The work of the ITC-5 within the ISSMGE, covering broadly the many environmental aspects in our Geotechnical profession, has been mainly undertaken by a group of core members and colleagues, listed in the first draft report (XIVth Int. Conf. on Soil Mechanics and Foundation Engineering, 6-12 September 1997), cfr. Table 1. The outcome of their work has been briefly compiled here in this paper, adapted and extended with personal comments and suggestions on the drafts of the subcommittees SC2 - Dr. Shackelford, SC3 - Dr Wojnarowiez, SC6 - Prof. Kamon, SC7 - Dr L. de Mello and the report (SC4) on containment and waste disposal - Dr M. Manassero et al. (Osaka, November 1996).

Table 1
TC 5 working program

TC5-SC2 C.D. Shackelford	Contaminant migration-test methods, modelling and monitoring
TC5-SC3 M. Wojnarowicz	Waste stability - Waste classification and characterisation - Stability and bearing capacity of solid waste land-fills
TC5-SC4 M. Manassero	Controlled landfills - Design of lining systems - Capping systems and sealing barriers - Regulatory aspects and quality assurance - Leachate collection and leak detection systems
TC5-SC5 M. Loxham	Contaminated land reclamation (design, construction and management)
TC5-SC6 M. Kamon	Assessment of geo-environmental hazards from dredging materials
TC5-SC7 L. De Mello	Assessment of geo-environmental hazards from non-traditional geotechnical construction materials
TC5-SC8 R.G. CLARK	Long term behaviour under extreme loading

2 GEOTECHNICAL PROPERTIES OF MSW RELATED TO LANDFILLING

The Geotechnics of Waste Landfills is fully governed by the geotechnical properties of waste with particular reference to MSW on the one hand, the stress-strain, seepage behaviour and pollutant transport mechanism of the lining boundaries on the other hand. Both major problems are being discussed extensively in our report of Osaka - 1996, 2nd ICEG. Some of this work was devoted to the mechanics of the MSW material.

The design and operation of sanitary landfills involve a variety of geotechnical problems, for which reliable knowledge of the geotechnical properties of mainly domestic solid waste (unit weight, compressibility, etc...) is required in order to perform even basic engineering design.

The quantification of these properties is very difficult because :

1. Municipal solid waste is inherently heterogeneous and variable among different geographic locations (see table 2) ;
2. It is difficult to obtain samples of relevant size to be representative of in-situ conditions;
3. There are no generally accepted sampling and testing procedures for waste materials ;
4. The properties of the waste materials change more drastically with time
5. The level of training and education of the personnel on site should be high enough in order to deal with all necessary basic interpretation and understanding of the measurements.

Important physical characteristics of domestic waste include moisture content, particle size distribution, organic content and unit weight. However, one has to bear in mind even more specific characteristics of waste materials.

Indeed, waste fill generally consists of many different types of constituents, and these constituents are usually degradable and interacting. Detailed discussion on some of those physical characteristics are given in our key-note paper (2nd ICEG, Osaka, November 1996).

Waste type commonly are divided into two groups:
1. Soil-like,
2. other wastes

In case 1, soil mechanics' procedures for laboratory tests and design procedures are commonly used. For the case of solid municipal waste, it is however difficult to use soil mechanics methods. One possible approach is to try and identify the waste by running a gradation curve (Jessberger, 1994) for the various portions, ending up with gradation curves somewhat similar to soil. Fig.1 shows the results of such grain size analysis on different type of municipal wastes. It is interesting to note the tendency of the percentage of fine grained material to increase with ageing of the waste. This tendency can be explained by the various degradation processes the waste is usually undergoing in time.

Proper assessment of the hydraulic characteristics of the waste itself is an important design element because of the potential impacts related to uncontrolled migration of leachate and the stability problem. Table 3 provides data of the hydraulic conductivity for wastes; at first sight it seems that

Table 2
MSW-components as weight percentage
for different cities (adapted Bouazza et al.,1996)

MSW	1	2	3	43	5	6	7	8	9	10
metals	1	1	3	3	5	2	2.5	4	1	3
paper cardboard	25	5	12	3	22	10	31	19	2	16
plastics	-	1	5	-	-	3	9.5	7	3	20
leather wood rubber	7	1	-	7	3	6	4	4	1	-
textiles	3	-	-	10	-	3	5	-	-	-
putrescible materials	44	45	74	15	20	61	28	59	71	58
glass	1	1	4	10	6	1	9	2	1	2
others	19	46	2	22	46	14	11	5	21	1

1. Bangkok (Thailand), 2. Pekin (China), 3. Nairobi (Kenya),   4. Hong-Kong, 5. New-York (USA), 6. Istanbul (Turkey), 7. Geneva (Switzerland), 8. Athens (Greece), 9. Cochabamba (Bolivia), 10. Wollongong(Australia)

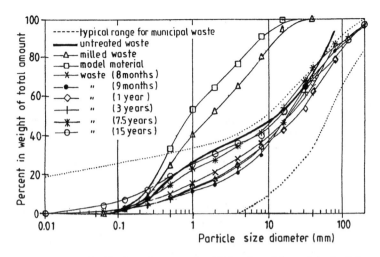

Fig. 1. Grain size distribution of untreated milled and model municipal solid waste
(from Jessberger, 1994a)

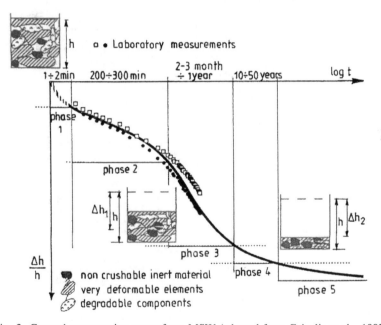

Fig. 2. General compression curve from MSW (adapted from Grisolia et al., 1992)

the measured values could be associated with "clean fine sand". One has however to keep in mind that these values are influenced by the degree of compaction, waste ageing etc... One therefore needs to assess the hydraulic conductivity on a case-by-case basis. However, it seems that a value of 10^{-5} m/s can reasonably be suggested as a first approximation.

The determination of the mechanical properties of municipal solid waste is an even harder task due to the several influencing factors mentioned earlier.

For fine-grained, soil-like waste, mechanical properties - time dependant compressibility, shrinkage and swelling behaviour, shear strength - are determined using conventional geotechnical test methods, like compression or shear apparatus. For mixed and coarse-grained soil-like waste some modifications of the testing equipment may be necessary, e.g. triaxial compression tests or shear tests on very large scale samples. For other wastes, it is necessary to undertake in situ tests on trial areas. Tests using very large scale shear or triaxial equipment may also be useful.

The mechanisms governing domestic waste stress-strain behaviour are many and complex, even more so than for a soil due to the extreme heterogeneity of the waste, their own "particle"

deformability, the degradable nature of the material and the large voids present in the initial refuse fill. The main mechanisms involved in refuse settlement are discussed by some authors, (Sowers, (1973), Huitric (1981) and Gilbert & Murphy (1987), Van Impe & Bouazza (1996) (fig. 2).

Our general proposal can be summarized as (fig. 2) :

1. Physical compression due to mechanical distortion, bending, crushing and reorientation.

Table 3

Summary of determination of hydraulic conductivity of domestic waste

Source	Unit weight (kN/m^3)	Hydraulic conductivity (m/s)	Method
Fungaroli et al. (1979)*	1.1-4 (milled waste)	10^{-5} to 2×10^{-4}	Lysimetres determination
Koriates et al. (1983)*	8.6	5.1×10^{-5} to 3.15×10^{-5}	Laboratory tests
Oweis & Khera (1986)*	6.45	10^{-5}	Estimation from field data
Oweis et al. (1990)*	6.45 9.4-14 6.3-9.4	10^{-5} 1.5×10^{-6} 1.1×10^{-5}	Pumping test Falling head field tests Test pit
Landva & Clark (1990)*	10.1-14.4	1×10^{-5} to 4×10^{-4}	Test pit
Gabr & Valero (1995)	-	10^{-7} to 10^{-5}	Laboratory tests
Blengino et al (1996)	9-11	3.10^{-7} to 3.10^{-6}	Deep boreholes (30-40m). Falling head field tests
Manassero (1990)	8-10	$1.5.10^{-5}$ to $2.6.10^{-4}$	Pumping tests (15 + 20 in depth)
Beaven & Powrie (1995)	5-13	10^{-7} to 10^{-4}	Laboratory tests under contining pressure from 0 to 600 kPa
Brandl (1990)	11-14 (roller comp.) 13-16 (roller+DC)	2.10^{-5} to 7.10^{-6} 5.10^{-6} to 3.10^{-7}	Falling head field tests Test pit
Brandl (1994)	9-12 (pretreated)	2.10^{-5} to 1.10^{-6}	Laboratory tests
Brandl (1994)	9-12 (pretreated)	5.10^{-4} to 3.10^{-5}	Laboratory tests
Brandl (1994)	13-17 (very compacted)	2.10^{-6} to 3.10^{-8}	Laboratory tests

2. Ravelling settlement due to migration of small particles into voids among large particles.
3. Viscous behaviour and consolidation phenomena involving both solid skeleton and single particles or components,
4. Decomposition settlement due to the biodegradation of the organic components
5. Collapse of components due to physico-chemical changes such as corrosion oxidation ; degradation of inorganic components.

This proposal for subdividing the MSW load settlement curve also closely matches the indications (adapted from a proposal by Grisolia et al. 1992).

The factors affecting the magnitude of settlement (under own weight as well as under overloads) are many and are influenced by each other (Edil et al., 1990).

The term consolidation in the above suggested steps of MSW load-settlement curves, refers to settlement resulting from the dewatering of the freshly deposited saturated materials. Shrinkage is the process by which organic solids and moisture are gradually decomposed and converted to carbon dioxide and methane, resulting in a corresponding decrease in the volume of the fill. Compaction is defined as the re-orientation of solids into a more dense configuration due to the gradual loss of rigidity in solids from the creep of solids under overburden or from solid decomposition. It's suggested also that such waste solids may initially "bridge" across voids, but lateron collapse, which indeed may be judged to be potentially the most significant feature of settlement of sanitary fills. The addition of inert waste results in an increase of the initial waste density and may reduce significantly the (bio)technical interactions, and consequently the overall long term settlements.

Under its own weight, refuse settlement typically ranges from 5% to 30% of the original thickness with most of the settlement occurring in the first year or two. In all settlement sequences, the composition of the waste as far as leachate generation capabilities, the organic content relevant to biodegration, etc... are of outmost importance.

Settlement predictions for solid waste landfills are complicated due to this random nature and decompositional characteristics of the waste, short-term and long-term environmental conditions, and operational methods. Additionally unpredictable differential settlement can occur over relatively short distances from deterioration or collapse of containers, appliances, and similar materials. Settlement predictions can be further complicated for landfills constructed on compressible foundations that may exhibit complex settlement characteristics. The understanding of settlement characteristics of waste is a prerequisit to an acceptable design of landfills, since the differential settlements can cause a number of major closed landfill problems such as : breakage of gas or leachate drainage pipes, surface profile changes with possible depressions and percolating water masses, etc...

Although landfill settlement is frequently evaluated using the theory of one-dimensional consolidation, this approach is complicated since :
- the use of compression and recompression ratios are dependent on initial values of e or H and these properties are often not reliably known
- the (e) vs. log(p) or log (t) relationship are often not linear and therefore, compression coefficients C_c and C_α vary considerably with the initial stresses within the landfill while these stresses change with time (Fassett et al. 1994) ;
- the amount of primary settlement depends on the effective stress, which in turn is a function of refuse unit weight and the level of leachate within the landfill (both of which may be poorly known and may change with time).

Some data from laboratory tests suggest that the primary consolidation, over a relative short time lapse, is by far the most important factor (over more than 70 %) in the overall MSW settlement to be expected finally. This justifies the application of dynamic compaction methods during and after landfill completion, so as to accelerate and amplify the primary consolidation under own weight.

Published records of laboratory measurements for waste settlement are very scarce, only some were reported recently. Landva & Clark (1990) compressed old wastes (age unknown) from different sites in a 0.5 m diameter consolidometer. The results are given in fig. 3 and show the high compressibility of the waste. Values of CR in the range of 0.2 to 0.5, depending on the stress level, were reported. Whereas $C_{\alpha\varepsilon}$ was found to be in the range of 0.2 % to 3 % per log cycle of time; and it appeared also that $C_{\alpha\varepsilon}$ increased with increasing organic content.

A proposal for the $C_{\alpha\varepsilon}$ values estimation in case of MSW, is given in fig. 4.

Compression behaviour of mixed or municipal solid waste can be assessed by the use of various field tests, each with their own specific application boundaries, such as plate load tests (fig. 5) for surface layers, pressuremeter tests in depth, more advanced spectral analysis of surface waves (Van Impe et al 1993) and overall direct settlement measurements.

At this stage, a geotechnical parameter describing

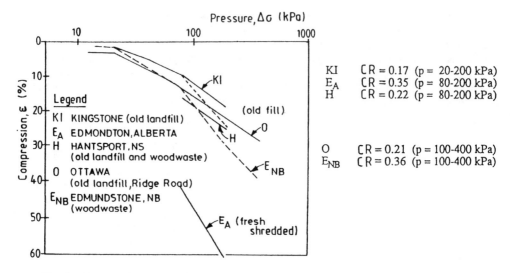

Fig. 3a. Compressive strain versus pressure for various fills (from Landva & Clark, 1990)

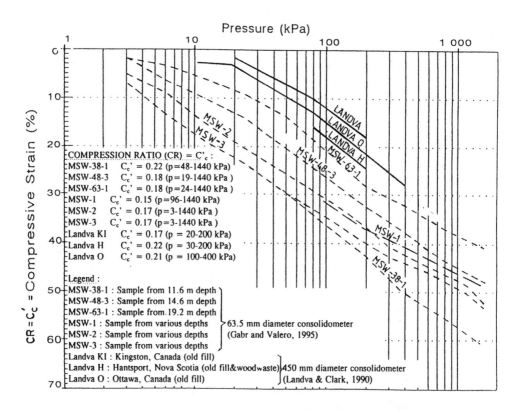

Fig. 3b. Compressibility characteristics as measured from conventional and unconventional tests

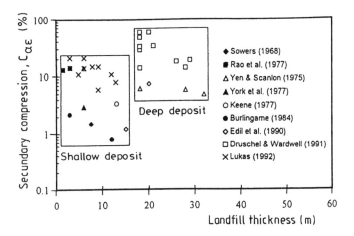

Fig. 4. Variation of $C_{\alpha\epsilon}$ with landfill thickness

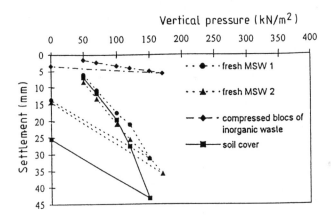

Fig. 5. Plate load tests at the Appels Landfill, Dendermonde, Belgium
(W.F. Van Impe et al., 1994)

the compressive behaviour of municipal waste, such as the stiffness modulus, is still not enough used. It can be determined either by the use of laboratory tests, back analysis of landfill settlements or better from relevant field testing at corresponding strain levels. Nowadays, the small strain overall stiffness moduli for municipal and other waste disposal sites, in many cases can reliably be estimated with non-destructive methods such as Spectral Analysis of Surface Waves (SASW). In the case of laboratory tests, Jessberger & Kockel (1993) stressed that the stiffness modulus is derived from the virgin compression line and therefore only yields in-formation can be obtained on settlements of waste that has never been subjected before to loads higher than those applied in the confined compression tests.

Fig. 6 shows the load dependent range (at various levels of strain) for the stiffness modulus of MSW reported by several authors. The variation of the stiffness modulus is delimited by an upper bound and a lower bound depending on several factors such as waste composition, state of compaction, soil cover, plate diameter (in some cases), type of tests and ageing. It should be pointed out that, according to the experiences recorded by Haegeman & Van Impe (1995) and Haegeman (1995) the SASW values can be assumed to show some 30 % decay when translated in plate load test results.

Not many data are available on the topic of dynamic properties of solid waste. Some reference here is made to Kavazanjian et al. 1995, Husmand et al. 1990 and recently Matasovic & Vucetic, 1993.

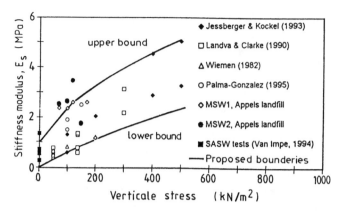

Fig. 6. Variation of stiffness modulus with vertical stress

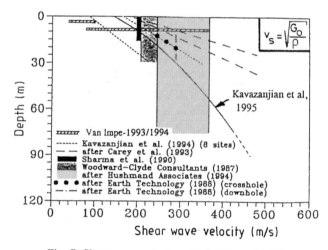

Fig. 7. Shear wave velocity of MSW from SASW

Most of them are related to landfill material consisting mainly of old residential, commercial and industrial solid wastes, during strong motion earthquakes. Some relevant information is gathered in figures 7. More recent data are available concerning the dynamic compaction response of MSW (Van Impe & Bouazza 1995), as shown in fig. 8a,b.

Shear resistance is a geotechnical parameter of primary concern in describing the properties of domestic solid wastes. Equivalent to soil mechanics, shear angle or angle of internal friction, ϕ, and "intercept cohesion", c, are used in design calculation. Four general approaches are used to estimate the shear strength of domestic solid wastes ;
1. laboratory testing in TX-conditions

2. back calculation from field tests and operational records
3. in-situ testing ;
4. direct shear tests of large dimensions

Laboratory testing has been performed on reconstituted samples from landfills, samples in which various substitution were made, and on samples collected from drive samplers. Large triaxial compression cells or large direct shear apparatus can be used. As specified earlier, interpretation of the tests on waste by means of concepts derived from soil behaviour can be useful, at least at the present state of knowledge. On this basis, the concepts of shear angle, and "cohesion" intercept are commonly used. On the other hand, wastes are usually non saturated. Therefore, interpretation of test results in terms of

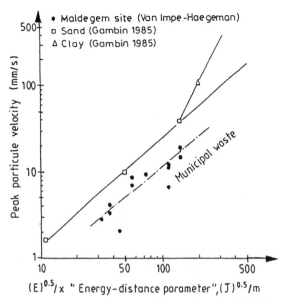

Fig. 8a. Peak particle velocity versus drop energy scaled-distance for the Maldegem
waste site and other type pf soils (after Van Impe & Bouazza, 1995)

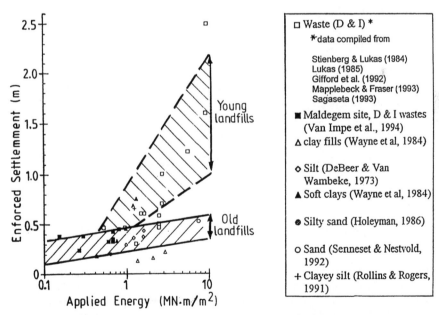

Fig. 8b. Enforced settlement versus applied energy for waste and other type of soils
(after Van Impe & Bouazza, 1995)

135

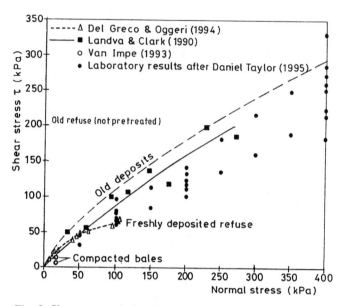

Fig. 9. Shear-stress relationship for MSW (from direct shear tests)

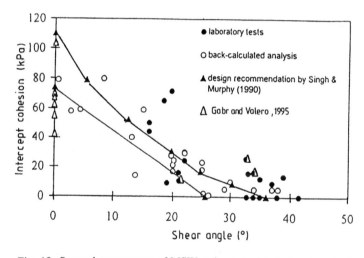

Fig. 10. Strength parameters of MSW estimated with different methods

undrained situation with no volume change ($\phi = 0$) may be a very unrealistic approach and an analysis in terms of equivalent c "intercept" - ϕ can be more adequate (Sanchez-Alciturri et al. 1993a).

On the basis of TX results on MSW, it is apparent that the ultimate state boundary surface at failure cannot be clearly defined because of reachable strain limits of this kind of test equipments. Nevertheless,.it is obvious that "failure" parameters ϕ and c according to the Mohr-Coulomb criteria are

for operational problems, (stability analysis...) more reliably defined on the basis of allowable strain (p-q-e-envelopes varying with shear strain level).

The variation of the shear stress (τ) with normal stress (σ) from direct shear tests, as reported by several authors (at conventional strain levels of about 10 to 15 %) is shown in fig. 9. It is interesting to note that some aspects are similar to the behaviour of conventional material such as soil. Indeed in the case of compacted waste bales, higher values

Table 4

Shear characteristics of MSW

Reference	Shear strength parameters		Data type	Comments
	c'(kPa)	φ' (°)		
Jessberger(1994)	7	38	not stated	reporting Gay et al.(1978) (MSW)
Jessberger(1994)	7	42	simple shear	reporting Gay et al.(1981) 9 month old MSW
Jessberger(1994)	28	26.5	simple shear	reporting Gay et al.(1981) fresh MSW
Fasett et al.(1994)	10	32	suggested values	reporting Jessberger & Kockel (1991)
Fasett et al.(1994)	10	23	suggested values	suggested by authors
Kolsch (1995)	15	15	suggested values	suggested by authors
Cowland et al.(1993)	10	25	bach analysis	deep trench cut in waste. Suggested values by authors
Del Greco&Oggeri (1993)	15.7	21	direct shear	tests on baled waste. Lower density bales
Del Greco&Oggeri (1993)	23.5	22	direct shear	tests on baled waste. Higher density bales
Landva&Clark(1986)	19	42	direct shear	old refuse
Landva&Clark(1986)	16	38	direct shear	old refuse
Landva&Clark(1986)	16	33	direct shear	old refuse + 1year
Landva&Clark(1986)	23	24	direct shear	fresh, shredded refuse
Landva&Clark(1986)	10	33.6	direct shear	woodwaste/refuse mixture
Landva&Clark(1990)	22→19	24→39	direct shear	σ ≈ 480 kPa
Golder Associates (1993)	0	41	direct shear	project specific testing
Kavazanjian et al(1995)	-	25	back analysis	overburden ≈ 180 kPa
Kavazanjian et al(1995)	-	28	back analysis	overburden ≈ 110 kPa
Kavazanjian et al(1995)	-	30	back analysis	overburden ≈ 45 kPa
Kavazanjian et al(1995)	-	34	back analysis	overburden ≈ 60 kPa
Richardson&Reynolds (1991)	10	18→43	in site ; large direct shear	14 kPa < σ < 38 kPa
Van Impe et al (1996)	20	0	back analysis	overburden ≤ 20kPa
Van Impe et al (1996)	0	38	back analysis	20 kPa < overburden≤ 60kPa
Van Impe et al (1996)	20	30	back analysis	overburden > 60 kPa

of friction angle are attained at low normal stress levels. Whereas the interlocking is revealed under higher vertical stresses. Overall the shear angles have a low value due mainly to the presence of large amounts of plastic materials in the tested bales. In the case of old refuse, higher friction angles and intercept cohesion are obtained due to the mixed matrix of the material (soil - waste) and also due to the range of stress level. A curved linear failure envelop can be fitted through the data to account for the stress level.

Various authors have presented values of shear strength of waste, obtained from Laboratory, in situ tests, or from back analysis, table 4.

It has become acceptable to present the reported values in terms of Mohr-Coulomb strength parameters, intercept cohesion (c) and shear angle (φ) in a figure such as fig. 10.

The very wide scatter in the results observed in fig. 10 makes it difficult to draw any structive conclusion. It should be borne in mind that compilation of such data is always difficult due to the lack of

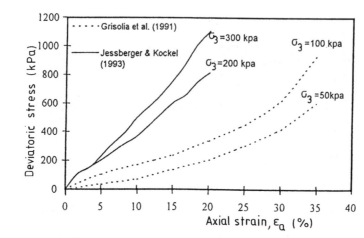

Fig. 11. Typical stress-strain relationship for MSW

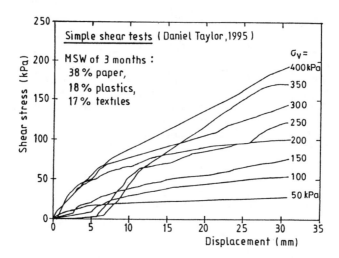

Fig. 12. Typical shear stress vs. Displacement Relationship

information and details (especially strain levels) about the case study or test.

Typical shear test and TX stress strain curves are given in fig. 11 and 12. From this figure, it is clear that failure conditions have not been reached and that there is an increase in mobilised shear strength with displacement. From Taylor's tests, the assessment of the mobilised friction angle at three normal stresses increased from arount 7° at a dis- placement of 5 mm, to between 27 and 30° at a displacement of 30 mm.

The same principle of implementing the strain versus strength mobilisation in the design is obviously equally valid for the liners. Typically,

mineral liners such as remoulded clay and bentonite enhanced soils would only need to strain around 3 % to reach their peak shear strength. Geosynthetic materials reach their peak interface shear strengths at small displacements, (for example a smooth HDPE geomembrane/nonwoven geotextile reaches peak values at around 1 to 3 mm and a textured HDPE geomembrane/nonwoven geotextile reaches peak at strains up to 10 mm ; Jones (1997)). There is therefore clearly also a serious strain compatibility issue landfill designers need to address, since waste, mineral liner and geosynthetics, are mobilising their peak strengths at different strain levels.

A curved linear failure envelope can be fitted

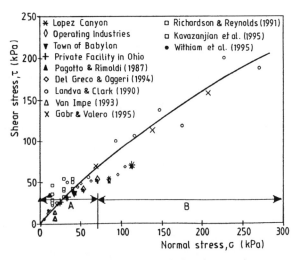

Fig. 13. Domestic solid waste strength data from various sources

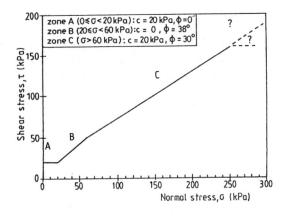

Fig. 14. Domestic solid waste strength data design recommendation

through many of the obtained data, a bilinear envelope can be assumed for the sake of simplicity. Two distinct zones can be distinguished : 1) zone A corresponding to low stress levels where the ϕ values are higher and c values very low. 2) zone B corresponding to higher stress levels ; fig. 13.

Finally, the most common assumptions are either to go out from purely cohesive ($\varphi = 0$) or from purely frictional (c = 0) behaviour. For domestic waste, there is no firm basis for such assumptions. As a consequence, for field tests it is better to not assume c or ϕ being zero.

Our proposal would be the suggestion in fig. 14 (and table 4) which should give approximate starting design values of c and ϕ according to three distinct zones:

- zone A corresponding to very low stresses (0 kPa $\leq \sigma_v < 20$ kPa) where the domestic solid waste behaviour can be only cohesif. In this case, c ≈ 20 kPa.
- zone B corresponding to low to moderate stresses (20 kPa $\leq \sigma_v < 60$ kPa). In this case, c = 0 kPa and $\varphi \approx 38°$.
- zone C corresponding to higher stresses ($\sigma_v \geq 60$ kPa). In this case, c $\geq$ 20 kPa and $\varphi \approx 30°$

3 WASTE CONTAINMENT SYSTEMS

It's worthwhile to firstly repeat that miscible contaminant transport in porous media in general, is

139

Table 5

Chemical and biological processes affecting miscible contaminant transport (modified after National Research Council 1990)

Process	Definition	Significance
Sorption	Partitioning of contaminant between pore water and porous medium	Adsorption reduces rate of contaminant migration and makes contaminant removal difficult
Radioactive Decay	Irreversible decline in the activity of a radionuclide	Important attenuation mechanism when half-life for decay is $\leq$ residence time ; results in byproducts
Dissolution Precipitation	Reactions resulting in release of contaminants from solids or removal of contaminants as solids	Dissolution is significant at the source or at the migration front ; precipitation is an important attenuation mechanism, particularly in high pH system (pH $>$ 7)
Acid/Base	Reactions involving a transfer of protons (H^+)	Controls other reactions (e.g., dissolution/ precipitation)
Complexation	Combining of anions and cations into a complex form	Affects speciation that can affect sorption, solubility etc.
Hydrolysis Substitution	Reaction of a halogenated organic compound with water or a componention of water (hydrolysis) or with another anion (substitution)	Typically makes an organic compound more susceptible to biodegradation and more soluble
Oxidation Reduction (Redox)	Reactions involving a transfer of electrons	Important attenuation mechanism in terms of controlling precipitation of metals
Biodegradation	Reactions controlled by micro-organisms	Important attenuation mechanism for organic compounds ; may result in undesirable byproducts

controlled by a variety of physical, chemical and biological processes, (table 5).

In the past, hydraulic barriers, simply consisting of a layer of compacted clay or of a single geomembrane, were used for waste containment. In many cases, no specific measures were adopted to control pollutant migration for landfills underlain by low permeability natural subsoils.

Recently, several institutions for environmental protection, have defined guidelines for the construction of containment systems suggesting design procedures that take into account pollutant transport (USCFR, 1992 ; ETC8, 1993 ; CIRIA, 1995). These activities have greatly contributed to establish gene-

ral principles for construction and quality control of engineered barriers used as pollutant containment systems.

One could focus on the three basic components of a system for waste containment : (1) bottom liners, (2) sidewall liners, and (3) covers.

Three scenarios are important with respect to waste containment by engineered barriers (Shackelford, 1989, 1993 ; Manassero & Shackelford, 1994a) : (1) pure diffusion, (2) diffusion with positive advection, and (3) diffusion with negative advection. Each of these three scenarios is illustrated with respect to vertical and horizontal barriers in fig. 15. The pure diffusion case may result when a com-

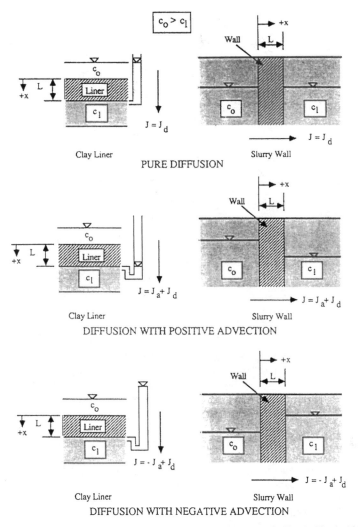

$c_o > c_l$

Wall

Clay Liner

Slurry Wall

PURE DIFFUSION

$J = J_d$

Clay Liner

Slurry Wall

DIFFUSION WITH POSITIVE ADVECTION

$J = J_a + J_d$

Clay Liner

Slurry Wall

DIFFUSION WITH NEGATIVE ADVECTION

$J = -J_a + J_d$

Fig. 15. Waste containment scenarios (after Manassero & Shackelford, 1994a)

pacted clay barrier is placed below the water table or when a slurry wall is placed around an existing contaminated area for remediation. The most common scenario for new landfills is the case of diffusion with positive advection, where the compacted clay liner is placed above the water table. Examples for the case of diffusion with negative advection are when a clay liner is placed below the water table or over an artesian aquifer and when the groundwater table, within a contaminated area, contained by a slurry wall, is lowered (e.g. by pumping) to induce inward advective flux (Gray & Weber, 1984 ; Manassero, 1991 ; Manassero & Pasqualini, 1992). In principle the diffusion with negative advection is the

most effective containment system but it is not always used due to practical problems involved with the continuous use of the dewatering systems.

There are both closed-form and numerical solutions for the advective-dispersive-reactive equation (ADRE) available for describing the containment system behaviour referring to the aforementioned scenarios (van Genuchten & Alves, 1982 ; Manassero & Shackelford, 1995, 1996 ; Rabideau et al., 1996).

Numerical or semi-analytical solutions can take into account nowadays layered barriers and other types of boundary conditions allowing a more accurate description of the actual field conditions.

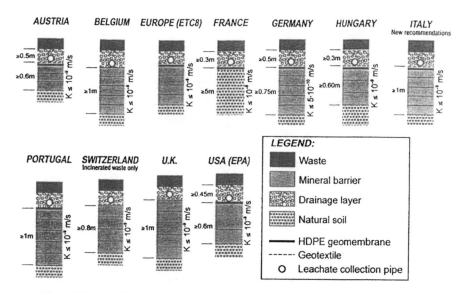

Fig. 16. Bottom lining system for municipal waste landfills from different regulations and recommendations

Among these, the solution of Rowe & Booker (1985), for example can simulate the following boundary conditions :
- initial finite contaminant mass at the entrance boundary, with the possibility of taking into account decay, withdrawal by the leachate collection system, and time history by step variations;
- removal of contaminant at the exit boundary by a flushing stream resulting in constant concentration $c_1 < c_0$ at steady state conditions where c_1 is a function of the stream velocity.

Since there is a wide range of possibilities offered by both analytical and numerical contaminant transport models, it is important to evaluate the reliability of input parameters.

Looking at the design of bottom liners, an important topic pertains to the present state of regulations adopted by industrialized countries. Bouazza & Van Impe (1995) collected a comprehensive list of different regulations and recommendations. In spite of the incentive toward the unification of these norms due to the progress of many international research programs, different approaches are still apparent countrywide. Looking at the bottom liner for municipal waste in fig. 16, the trend toward the composite sealing barriers is evident. Nevertheless, for example, the French regulation still considers the possibility of using a single geomembrane even if there exists a natural underlying formation of

unrestrictive permeability values, provided that the formation thickness exceeds 5 m. Similarly, the minimum requirements of Belgium and the UK provide for a single compacted clay layer as well as in Switzerland where only stabilized ash from incineration of waste is landfilled. As far as the vulnerability of natural subsoil, it is interesting to observe that the USA trend is to define the sealing lining systems independently from the natural subsoil conditions, whereas some European countries take into account to some extent the natural subsoil features for the definition of the artificial lining system.

Aside from regulations and guidelines, the design of modern landfills should be based on the fundamental principles listed below :
- the mineral barrier is the basic component of the sealing system for long-term performance ($t > 50$ years) ;
- the requirements and characteristics of the mineral sealing layer in order of importance are : (1) low hydraulic conductivity (HC) at the field scale, (2) long-term compatibility with the chemicals to be contained, (3) high sorption capacity, and (4) low diffusion coefficient ;
- the composite lining systems can give important advantages both in the short and long-term due to: (1) reduction of HC due to the attenuation of local defects of both geomembrane and compacted clay as shown in table 6 (Giroud & Bona-

parte, 1989 ; Giroud et al., 1992 ; Daniel, 1993), (2) enhancement of flow within the drainage layers toward the collection pipes (i.e. minimization of pounding leachate on the liner) and, (3) the geomembrane on the top of the clayey barrier can delay direct contact between clay and leachate long enough for consolidation of the clay portion of the composite system due to the establishment of high effective stresses when the waste is landfilled. In this way it is possible to reduce or avoid compatibility problems (Rowe et al., 1995) ;

- the drainage efficiency has a very important effect in reducing the height of ponding leachate on the barrier and, consequently, also the advective migration ;
- construction details play a fundamental role in the final efficiency of the lining system in term of full-scale HC.

The advantages of composite liners in terms of preventing advective transport are apparent (tables 6 and 7) especially for poor quality mineral barrier (k > 10^{-9} m/s) as observed by Daniel (1991). The fol-

Tabel 6

Ratem of leachate migration through various types of liners

(after Giroud et al., 1994)

	Rate of leachate in liters per hectare per day (Iphd)
Mineral liner	**100.000 Iphd** for soil liner having k ≈ 10^{-7} m/s
	10.000 Iphd for soil liner having k ≈ 10^{-8} m/s
	1000 Iphd for soil liner having k ≈ 10^{-9} m/s
Geomembrane liner	**100 Iphd** for a geomembrane installed with strict construction quality assurance and containing a typical amount of defects
Simple composite liner	**10 Iphd** for a composite liner constructed with a soil having k ≈ 10^{-8} to 10^{-7} m/s
	0.1 Iphd for a composite liner constructed with a soil having k ≈ 10^{-9} m/s
Double composite liner	≈ **0 Iphd** for a double composite liner where each of the two liners is constructed with a soil having k ≈ 10^{-9} m/s

Table 7

Results of permeation tests through simple and composite lining

(after Jessberger, 1994b)

	Barrier	HDPE 1 mm	HDPE 1 mm gravel 160mm	HDPE 1 mm clay 85 mm	HDPE 1 mm clay 160 mm	HDPE > 2 mm clay 0.6-1m
Pollutant						
Water sulubilit	permeant	permeation rate (g/m^2 d)				
good	ACETON	0.9	2.7	1.9	0.68	
	THF	1.2	2.2	2.7	1.23	
	MET	0.7	1.2	0.8	0.75	
poor	i-OCTAN	0.6	0.05	0.001	0.0001	
	TOLUOL	2.4	0.06	0.015	0.0004	→ 0 ?
	XYLOL	2.0	0.08	0.001	0.0001	
	Trichloroethyl	3.3	0.1	0.001	0.0001	
	Tetrachloroethyl	2.9	0.04	0.001	0.0001	
	Chlorobenzol	2.1	0.02	0.0002	0.0001	
	Sum	16.1	6.4	5.4	2.7	

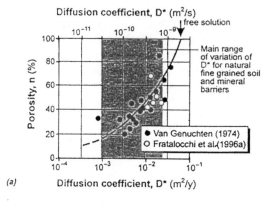

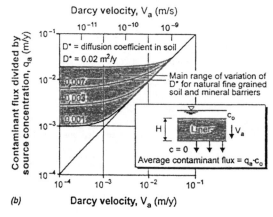

Fig. 17. (a) Range of diffusion coefficient for soils and mineral barriers, (b) relative importance of diffusive and advective transport through mineral barriers (after van Genuchten, 1974 ; Jessberger, 1995)

lowing general observations can made :
- given the common ranges for the key parameters for contaminant transport modelling, the reduction of advective pollutant migration due to the presence of the geomembrane in composite liners is significant whenever the mineral component has $k > 10^{-10}$ m/s. This results in an important decrease of both contaminant flux and concentration into the underlying aquifer ;
- diffusion coefficient of the geomembrane is, in general, orders of magnitude lower than that of the mineral layers. However, because the geomembrane is generally very thin, the reduction of the diffusive flux is limited, in particular for some organic compounds ;
- for HC lower than 10^{-10} m/s, the improvement given by the geomembrane, especially for organic compounds, becomes more negligible because of the predominance of diffusive transport. Indeed, the diffusive transport can be of the same order of

magnitude both in the mineral layer and in the geomembrane.

The most important advantages in using geomembranes for reducing the contaminant flux through lining systems can be fully exploited in the case of average to poor quality mineral components. Moreover, it must be considered that the duration of these advantages is related to the active life of the geomembrane before degradation estimated at about 50 to 100 years, at best, on the basis of the present knowledge.

Construction details play a fundamental role in the global efficiency of the landfill liners. Some of these details are shown in fig. 18, i.e. the position of the leachate collection shaft and of the shaft penetration of the liners. In particular, the most favourable position for the leachate removing shaft is illustrated in fig. 18b with leachate collection sumps and risers emplaced along the side slopes.

Due to the need to reduce the superficial area of

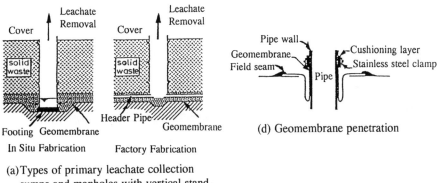

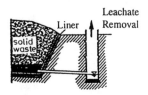

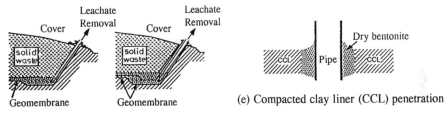

(a) Types of primary leachate collection sumps and manholes with vertical stand-pipe going through the waste and cover

(b) Types of primary (left) and secondary (hight) leachate collection sumps and pipe risers going up the side slopes

(c) Primary leachate collection system with standpipe outside the landfill and connected with sub-horizontal pipeline

(d) Geomembrane penetration

(e) Compacted clay liner (CCL) penetration

(f) Geosynthetic clay liner (GCL) penetration

Fig. 18. Various schemes for leachate removal systems and for pipe penetrations through different types of barrier (adapted after Daniel & Koerner, 1995)

impact of new landfills, the inclination of side slopes is generally increased to improve the ratio between the volumetric capacity and the print of the landfill. Different kinds of sidewall lining systems that are able to achieve the same safety level as the bottom composite liners and, at the same time, to allow the construction on slope angle up to around 60°, have been recently proposed.

In fig. 19, three different alternatives for steep slope side liners are shown. In the first case (fig. 19a) the mineral layer has been constructed with natural clay in horizontal lifts achieving the final slope profile by means of a finishing excavation. The natural clay can be mixed on site before compaction with 5 % to 10 % by weight of cement in order to achieve the strength that assures the stability of the slope (Manassero & Pasqualini, 1993b). The clay-cement mixture on the side slopes provides an increase in the interface strength between the mineral and the geomembrane (fig. 20). Preliminary results of direct shear tests show a significant increase in contact strength (Pasqualini & Stella, 1994). In some cases the clay-cement mixture can produce an additional positive effect such as a decreased of the

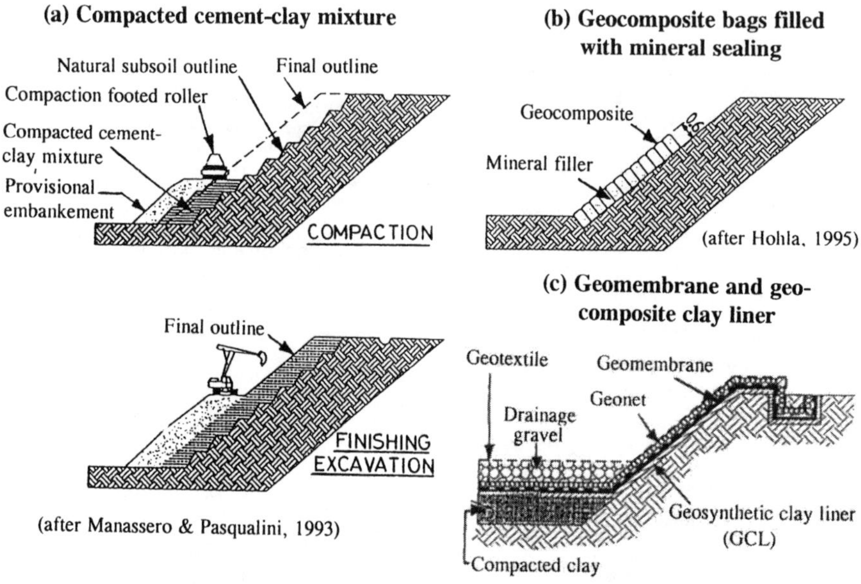

(a) Compacted cement-clay mixture

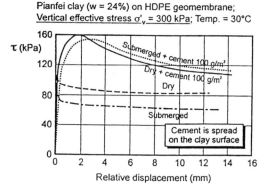

Natural subsoil outline   Final outline

Compaction footed roller

Compacted cement-
clay mixture

Provisional
embankement

COMPACTION

**(b) Geocomposite bags filled
with mineral sealing**

Geocomposite

Mineral filler

(after Hohla, 1995)

**(c) Geomembrane and geo-
composite clay liner**

Geotextile   Geomembrane

Geonet

Drainage
gravel

Geosynthetic clay liner
(GCL)

Compacted clay

Final outline

FINISHING
EXCAVATION

(after Manassero & Pasqualini, 1993)

Fig. 19. Alternative liners for steep side slopes

Pianfei clay (w = 24%) on HDPE geomembrane;
Vertical effective stress σ'_v = 300 kPa; Temp. = 30°C

τ (kPa)

Submerged + cement 100 g/m³

Dry + cement 100 g/m³

Dry

Submerged

Cement is spread
on the clay surface

Relative displacement (mm)

Fig. 20. Direct shear tests results on compacted clay-geomembrane interfaces
(after Pasqualini & Stella, 1994)

HC, as in the example shown in fig. 21. The second alternative type of liner for steep sides (fig. 19b) consists of composite geotextile-geomembrane bags filled with plastic concrete or cement-bentonite (CB) self-hardening slurries. The third alternative type of steep slope liners (fig. 19c) comprises geosynthetic clay liners (GCL) consisting of geosynthetics in combination with dry bentonite.

The main purposes of the cover system can be listed as follows (Daniel & Koerner, 1993) : (1) to raise the ground surface and provide appropriate slopes for promoting runoff and controlled drainage

of surface water, (2) to separate buried waste from vegetation, animals and humans, (3) to minimize infiltration of water into the waste, and (4) to control the release of gas out of the waste. Not all the aforementioned functions must be simul tane-ously satisfied for all landfills since the specific requirements are dependent from many factors such as : (1) location of the landfill, (2) climate conditions, (3) phase of landfill activity, (4) general strategy of the landfill management, and (5) type of waste.

Most cover systems are composed of multiple

7th day
18th day
23rd day
STANDARD PROCTOR COMPACTED SAMPLES
▲ Clayey silt (Clay content 25% PI=21% w=31%)
● Clayey silt with 5% by weight of blast furnace cement immediately permeated after compaction with hydraulic gradient i=50
■ Clayey silt with 5% by weight of blast furnace cement. Time between compaction and permeation 100 days
Hydraulic conductivity k (m/s)
10^{-9}
10^{-10}
10^{-11}
0.0
0.05
0.10
0.15
0.20
Pore volume of flow PVF (-)

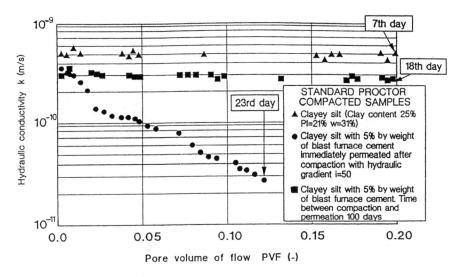

Fig. 21. Influence of blast furnace cement and curing conditions on hydraulic conductivity of a compacted soil (after Manassero & Pasqualini, 1993a)

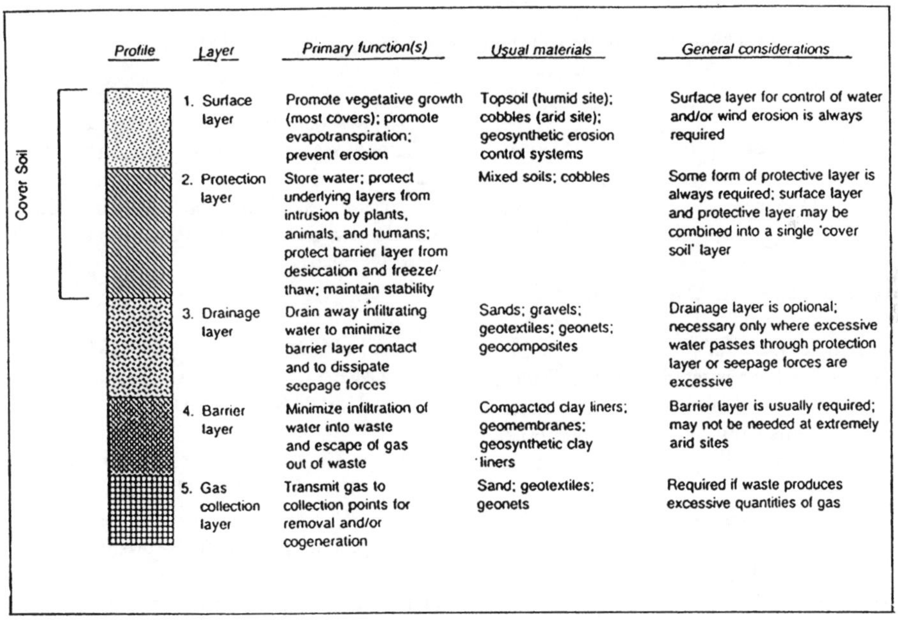

Profile
Layer
Primary function(s)
Usual materials
General considerations
Cover Soil
1. Surface layer
Promote vegetative growth (most covers); promote evapotranspiration; prevent erosion
Topsoil (humid site); cobbles (arid site); geosynthetic erosion control systems
Surface layer for control of water and/or wind erosion is always required
2. Protection layer
Store water; protect underlying layers from intrusion by plants, animals, and humans; protect barrier layer from desiccation and freeze/thaw; maintain stability
Mixed soils; cobbles
Some form of protective layer is always required; surface layer and protective layer may be combined into a single 'cover soil' layer
3. Drainage layer
Drain away infiltrating water to minimize barrier layer contact and to dissipate seepage forces
Sands; gravels; geotextiles; geonets; geocomposites
Drainage layer is optional; necessary only where excessive water passes through protection layer or seepage forces are excessive
4. Barrier layer
Minimize infiltration of water into waste and escape of gas out of waste
Compacted clay liners; geomembranes; geosynthetic clay liners
Barrier layer is usually required; may not be needed at extremely arid sites
5. Gas collection layer
Transmit gas to collection points for removal and/or cogeneration
Sand; geotextiles; geonets
Required if waste produces excessive quantities of gas

Fig. 22. Five possible components in a cover system (after Daniel & Koerner, 1993)

components. As shown in fig. 22, the components of cover system can be grouped into five categories (Daniel & Koerner, 1993). In the same figure, the specific functions of these covering components, the different materials that can be used, and some considerations such as the possibility to avoid a layer in some specific conditions, are detailed.

The different types of suggested cover systems by some country regulations and recommendations are shown in fig. 23. Based on these recommendations, a compacted clay barrier at least 0.45 m thick must be provided in order to minimize water infiltration

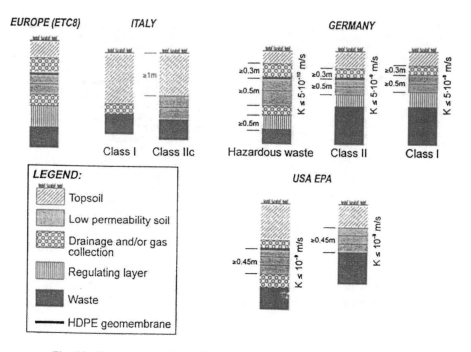

Fig. 23. Cover systems from different regulations and recommendations

and gas migration.

Focusing the attention on the sealing layer of the capping system, the following main aspects should be considered : (1) comparison between compacted clay liners (CCL) and geosynthetic clay liner (GCL), and (2) capillary barriers (cfr. Van Impe, W.F., Manassero, M., Bouazza, A., Osaka 1996).

The report of the ITC-5 subcommittee SC4 finally also discusses the suitability of mineral sealing compounds must be demonstrated in each individual case by suitability tests. These tests must be undertaken only by qualified institutes with appropriate equipment, and most of all with sufficient experience, to look for :
- composition and characteristics of the individual compounds of the mix
- characteristics of the fresh sealing compound
- workability and solidification behavior of the sealing compound
- strength and stress deformation behavior of the hardened sealing compound
- permeability of the hardened sealing compound
- unit mass and water content
- physico-chemical properties related to specific requirements

Examples of mixtures for different types of walls are given in the ITC-5 report. For one-phase walls

ready mixed construction materials are widely used in Europe, reaching very good results, without requiring time for swelling of bentonite. For two-phase walls sealing compounds were developed with high density and high resistance against pollutant attack.

4 ASSESSMENT OF GEO-ENVIRONMENTAL HAZARDS FROM DREDGED AND CONSTRUCTION SITE SLUDGES

Within the United States, about 500 million m^3 of sludge is dredged annually (US Amy Corps of Engineers : COE, 1987). Throughout the Great Lakes, about 5 million m^3 of sludge is dredged annually to maintain navigation in channels and harbors for commercial, military, and recreational users, and as parts of environmental remediation projects (EPA 1990). About one half of the total sludge dredged in the Great Lakes is sufficiently contaminated to be problematic sludge and require placement in a confined disposal facility (CDF). The contaminated to be problematic sludge and require placement in a confined disposal facility (CDF). The sludge requires special consideration during dredging and disposal operations because of the potentially adverse impact on water quality and local organisms.

In the 1970s, protocols for the control of disposal of dredged material were set up, two of which are the London (Dumping) Convention and the Oslo and Paris Convention. They were set up primarily to regulate the disposal of noxious substances into the oceans, but they included the regulation of dredged sediment as well. This inclusion was to be expected, given that the annual volume of dredged material disposed at sea greatly exceeds any other material.

In the north-east Atlantic/North Sea regions approximately 150.000.000 wet tonnes of dredged material were disposed of in 1990 compared with 10.000.000 tonnes of sewage sludge and less than 2.000.000 tonnes of chemical waste. This draws the attention of those who wish to regulate dumping to dredged material. Recognising that dredged material consists mainly of natural sediment and only a small proportion of the total volume dredged annually is contaminated, a number of organisations, including CEDA, PIANC and IAPH, launched a campaign to change the image s that gradually dredged "spoil" became known as dredged "material". This term is now embedded in the conventions and has been a contributory factor in getting dredged material treated as a special case.

EC Legislation

Several Western European countries have developed their own disposal policies or guidelines, but certain EC Directives govern the disposal and/or use of dredged material in EC countries.

Classification of dredged material in the EC region

An EC Directive regarding waste (91/692/EEC) specifies that the amount of waste produced should be minimised, beneficial use options for material should be sought, failing that, disposal of the waste may be undertaken.

One beneficial use option is the spreading of material on agricultural land. An EC guideline concerning the use of dredged material in agriculture (86/278/EEC) is binding for all EC members and states :

Limit values A1

Limits for the concentration of heavy metals in soilsl (mg/kg dry weight) with pH 6 or 7. These standards which were designed for soils are applied to dredged material.

Limit values A2

Limits for the concentration of heavy metals in sludge for spreading on agricultural land

These limit values are covered in the description of the Conventions, Codes and Conditions; Land Disposal

In addition, an EC Directive on Environmental

Impact Assessments (EIA) can mean that certain dredging and disposal options require EIAs or environmental reviews to be undertaken.

Belgium
Classification guidelines for the storage of dredged material

Based on the 86/278/EC Directive, OVAM (the Public Company for Waste in Flanders) has prepared a guideline for the classification of dredged material in view of storage. This classification comprises five groups with increasing concentration of heavy metals.

In the above mentioned classification, no clear distinction is made between the terms "soil", whether or not in its natural condition, and "sludge" either before dredging or in the storage site. In Belgium, dredged material from navigable waterways is regarded as a special case of waste and is therefore subject to the Waste Management Decree of 2 july 1981.

Germany
In the absence of sediment quality criteria from the Dredged Material Guidelines of the Oslo Commission for the assessment of contaminant concentrations in dredged material, quality criteria for trace metals and organic contaminants were established by the German Federal Ministry of Transport for management purposes within the Federal Water and Navigation Administration. The German quality criteria represent management values and are neither ecotoxiclogical quality criteria nor quality targets, and they are only applicable to dredged material from German federal waterways.

Italy
The legislation governing the disposal of dredged material in Italy involves a number of decrees which have been issued over the last 20 years.

Statutes and regulations
The main piece of legislation in Italy for the management of dredged material disposal is the DECRETO 24 gennaio 1996. This legislation replaces the considerable number of previous decrees.

Procedure, standards and permits
The sea disposal of dredged material which can be classified as toxic/noxious is not allowed under Italian legislation. The legislation outlines the contaminant groups which require assessment. Sea disposal is allowed for dredged material which is not

toxic/noxious, where toxic/noxious is defined under a different piece of legislation (D.P.R. 915/82,27 luglio 1984). However, a permit is required but there is a requirement to prove that the other disposal routes, such as to land are not available and better in terms of environmental risks.

The authorisation of a permit is ministered by the Minister of the Environment (Ministero Dell' Ambiente). Other organisations involved include the Sea Authority (Compartimento Maritima) and the USL.

There is a standard procedure to be followed in issuing a permit which is clearly outlined in the legislation. Dredged material from the entrance to ports and channels which need to be cleared immediately may undertake a different procedure which is faster.

The Netherlands
The general environmental quality standards for the Netherlands were on the basis of a risk approach. Following long-term ecotoxicological research it was realised that for a number of substances the existing standards could result in harmful effects on the reproduction and grouwth of organisms. Further, the standards for water and sediment were not attumed to each other. Therefore in 1994 a new set of quality criteria was set up for water and sediment, based on an estimate of the effects on the quatic ecosystem. The quality criteria are described in the "Evaluation Note on Water" of March 1994. The new set comprises five standard values : target value, limit value, reference value, intervention value and signal value.

Statutes and regulations
The London Convention and the Dredged Material Guidelines are implement in the Dutch Seawater Pollution Act. The general principle of this Act is the prohibition of the disposal of "waste material". This implies that each disposal at sea of lightly contaminated dredged material is subject to licensing under the Seawater Pollution Act.

Many classification system for dredged material has recently been revised. Five quality levels exist which are outlined in the Evaluation Note (table 8). These quality levels are as follows :

Target value : indicates the level below which risks to the environment are considered to be negligible, at the present state of knowledge.
Limit value : Concentration at which the water sediment is considered as relatively clean. The limit value is the objective for the year 2000.
Reference value : Is a reference level indicating

Table 8
Target value, limit value, reference value, intervention value and signal value for water sediments (Evaluation Note on Water, March 1994). Water sediments

Parameter	Unit	Target value	Limit value	Reference value	intervention value	Signal value
arsenic	mg/kg ds	29	55	55	55	150
cadmium	mg/kg ds	0.8	2	7.5	12	30
chromium	mg/kg ds	100	380	380	380	1000
copper	mg/kg ds	35	35	90	190	400
mercury	mg/kg ds	0.3	0.5	1.6	10	15
lead	mg/kg ds	85	530	530	530	1000
nickel	mg/kg ds	35	35	45	210	200
zinc	mg/kg ds	140	480	720	720	2500
PAH Total 10 PAK	mg/kg ds	1	1	10	40	-
PCB-28	µg/kg ds	1.0	4	30	-	-
PCB-52	µg/kg ds	1.0	4	30	-	-
PCB-101	µg/kg ds	4.0	4	30	-	-
PCB-118	µg/kg ds	4.0	4	30	-	-
PCB-138	µg/kg ds	4.0	4	30	-	-
PCB-153	µg/kg ds	4.0	4	30	-	-
PCB-180	µg/kg ds	4.0	4	30	-	-
Total 6 PCB	µg/kg ds	20/0	-	-	-	-
Total 7 PCB	µg/kg ds	-	-	200	1000	-

whetherdredging spoil is still dit for discharge in surface water, under certain conditions, or should be treated otherwise. It indicates the maximum allowable level above which the risks for the environment are unacceptable.

Intervention value : An indicative value, indicating that remediation may be urgent, owing to increased risks to public health and the environment.

Signal value : only for heavy metals. Concentration level of heavy metals above which the need for cleaning up should be investigated.

Norway

Dredging activities are of minor importance in Norway, annually representing a few hundred thousand tonnes.

Statutes and regulations

Dredging activities are currently regulated through the general terms of the Pollution Control Act (1981), (section 7) regarding the duty to prevent pollution, and the need for a permit when performing activities which may give rise to pollution (section 11). Disposal of dredged material is regulated through. Regulations on Dumping and Incineration of Substances and Materials at Sea (1980).

Spain

The Spanish framework for the management of dredged material is at present in the form of recommendations intended to have legally binding powers in the near future. The management framework mainly implements the Oslo and London Conventions, but in addition it includes characterisation criteria for the dredged sediment, and sediment classification categories associated with different management techniques.

Procedure, standards and permits

The Centre for Studies and Experimentation on Public Works (CEDEX) in collaboration with the Spanish Institute of Oceanography has prepared a document titled. "Recommendations for the management of dredged material from Spanish ports and harbours" (RMDM). They aim to control the management of dredged material in Spanish waters. The recommendations have no legal powers at present but it is the intention to be given such a status in th near future. These recommendations have been approved by a number of authorities which manage more than 95 % of Spain's dredged material.

The United Kingdom

A case-by-case system is adopted throughout the UK and there are no legislative or guideline standards. England, Scotland and Northern Ireland have separate but equivalent bodies that implement marine disposal of dredged material.

Statutes and regulations

Dredging disposal to sea is governed by part II of the Food and Environmental Protection Act 1985 (FEPA). Currently FEPA legislation relates to the disposal of material only and not to the dredging operation. The Act provides the framework which controls all deposit in the sea and provides for a high degree of environmental protection consistent with national policy and international obligations arising from the conventions to which the United Kingdom is a signatory.

Operational systems for dredging are designed in such a way that during dredging no spillage occurs. Dredging technology nowadays is capable of greatly reducing turbidity and resuspension of bottom sediment ; however, special equipment has to be deployed and advanced operational methods must be used (Herbich, 1995). It gradually became obvious that the dredging would be the most efficient and cost-effective way of cleanup contaminated sediment. The appropriate dredging methods can be considered such as mechanical, hydraulic, and pneumatic dredging for removal of contaminated sediment.

Table 9 summarizes the dredging equipment capabilities in removal of contaminated sediment Properties of dredged sludges vary greatly due to the differences in respective origin, sedimented area, dredging and excavation methods employed, additive materials etc. One of the main common characteristics of all sludges is the high water content. In particular, sedimented sludges have extremely high water contents such as 200-250 %. The unit densities of sludges are shown in fig. 24. The sludges also may have high organic contents. Figure 25 shows for example the relationship between Ig-loss and COD (Chemical Oxygen Demand) in Japan.

Some discharged sludges with high water content can be treated by dehydration resulting in volume decrease. Many types of inorganic or organic flocculants have been developed and utilized in many dehydration plants.

As modern remediation strategies there is accelerated drainage and consolidation (fig. 26) capping or natural burials for the improved sludge, with increased strength and compressibility of the material. There are a number of in situ tests that

Table 9

Type	Production	Depth Limitation m(ft)	Resuspension of Sediment	Comments
Mechanical				
Open clamshell wateright	low	9.1-12.2 (30-40)	high	
Watertight clamshell bucket	low	9.1-12.2 (30-40)	low	Experiments conducted in St. Johns River
Cable-arm bucket	low	9.1-12.2 (30-40°		Experiments conducted at Hamilton Harbor, Ontario, Canada
Mechanical-Hydraulic				
"Mud Cat"	moderate	4.6-7.6 (15-25)	low to moderate	New development
"Mud Cat ENV"	moderate		low	Experiments conducted in Sydney, Nova Scotia, Canada
Remotely controlled "Mud Cat"	low	4.6 (15)	low to moderate	New development
"Cleanup" system	moderate	21.3 (70)	low to moderate	Extensively used in Japan
Cutterhead	moderate to high	12.2 (40)	low	Pilot study in New Bedford, Massachusetts
Hydraulic-Suction				
"Refresher"	moderate to high	18.2-35.0 (60-115)	low	Extensively used in Japan
"Matchbox"	moderate to high	25.9 (85)	low to moderate	Experiments conducted at Calumet Harbor
"Wide Sweeper"	moderate	30.5 (100)	low	Used in Japan
Pneumatic				
"Pneuma"	low to moderate	60.9 (200)	low	Evaluated by USAE Waterways Experiment station
"Oozer"	moderate to high	18.0 (59)	low	Used extensively in Japan
Mechanical-Hydraulic-Pneumatic				
Screw impeller	low to moderate	6.1 (20)	low	Used in Japan (high-density)
Airtight bucketwheel	low to moderate	4.6 (15)	low	Used in Japan (high density)

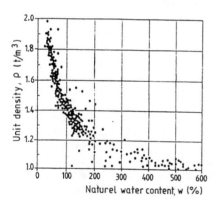

Fig. 24. Natural properties of sludges

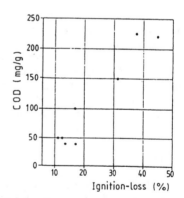

Fig. 25. Chemical properties of sludges

have been developed for soft aquatic sludge such as SASW devices to measure the increased material stiffness.

When waste sludge, is simply disposed in water depots, there are certain risks of dispersion of contaminants to the environment :

(1) pore water can be pushed out by consolidation of the deposits, settlements caused by the weight of the overlying sediments, and

(2) infiltration of groundwater by differences in concentration of contaminants between pore water and surrounding groundwater.

Transport by diffuse ways can be hampered by :

(1)   sandwich liners,

(2)   layers with active carbon,

(3)   clay layers

(4) layers with precipitated materials, like calcium, hydroxide, and

(5) combinations of the above possibilities.

The regulatory requirements for the disposal of dredged material are determined by both the type and level of the contaminants associated with the dredged material. The confined disposal facility (CDF) design criteria based on contaminated level and pathway is shown in fig. 27.

Innovative methods for separation and de-

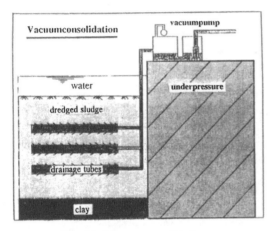

Fig. 26. Vacuumconsolidation

watering of the fine grained parts of sludge have been developed in Europe. Technologies as shown in fig. 28 are used successfully for and annual sludge throughput of about 2 million m^3. The separation of the coarse and fine fractions of the sludge is done by hydrocyclones and fluidized bed classifiers. The contaminated fines are dewatered until a soil mechanical consistency index higher than 0.5 is reached. This material has an undrained shear strength high enough for deposition in hill shaped landfills with total heights of 30 to 40 m above ground level and slopes with an inclination of about 1:6. Contaminated water from the dewatering process is treated in a separate multiphase waste water plant consisting of tricking filters, rotating biological contactors etc. for nitrification. As indicated before typical accelerated consolidation procedures for dredged sludge have been worked out in Belgium and in the Netherlands. This technique seems to be very promising for future storage increase of this material (fig. 26).

Moreover, mixtures of excavated soil and water can also be discharged from many kinds of excavation works. Generally, it is free from contaminated substances, but contains lots of fines difficult to dehydrate because it includes dispersants such as bentonite or polymers to be used for regulating viscosity. In solidifying sludges, a large amount of

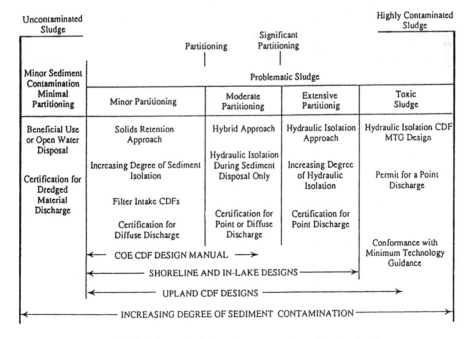

Fig. 27. CDF design criteria based on contaminant level and pathway

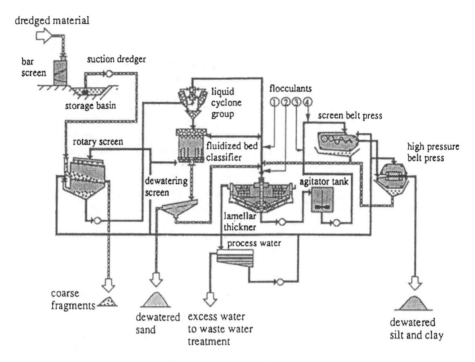

Fig. 28. Operating principle of a plant for mechanical treatment of dredged material

stabilizer (e.g., cement) is needed to harden the waste sludge to attain proper strength.

Years ago the authorities realized that the waste sludge was contaminated and if no measures were taken to prevent inflow of pollutants from direct sources that for ages the waste sludge would be contaminated. The main goal is to reach a situation that all waste sludges are not contaminated and can be freely reused. In the past most pollution came from industrial sources.

Waste sludges are subdivided in 5 contamination levels :

Class O :

Measured values are lower than background value and signal value. Waste sludge can be used on land, according to the rules for reuse of light conta-minated soil, under water use is allowed if the quality of the subsoil does not decrease.

Class 1 :

Measured values are between background value and signal value. Waste sludge can be used on land, according to the rules for reuse of light conta-minated soil, under water use is allowed if the quality of the subsoil does not decrease.

Class 2 :

Measured values are between signal value and intervention level. Dredging materials can be used on land close to 20 m from the canal/ditch/river. Reuse on land should be in accordance with the rules for reuse of light contaminated soil, under water use is allowed if the quality of the subsoil does not decrease.

Class 3 :

Measured values are between above intervention level. The use of dredging material under water should be limited as much as possible, storage only under Isolation, Control and Monitoring (ICM) Control conditions.

Class 4 :

Measured value are above intervent levels. Reuse is not allowed, storage only using ICM rules.

Background or target values should be corrected for the amount of clay and/or organic constituents of the sludge.

Soil type correction formula are proposed in the corresponding ITC-5 subcommittee SC-6 report.

5 ASSESSMENT OF GEO-ENVIRONMENTAL HAZARDS FROM NON-TRADITIONAL GEO TECHNICAL CONSTRUCTION MATERIALS

About the characterization and evaluation of geotechnical and geohydrological problems related to :

- combustion residues,
- MSW and sludge ashes
- paper industry sludge
- phosphogypsum
- used tires

as well as the major geotechnical problem related to mine tailings as construction materials in disposal areas and tailings dams, several research groups start to work right now.

Non-Traditional Construction Materials include all residue generated in industrial processes, energy production utilities, mining and beneficiation activities, being non-profitable end products which were routinely disposed. A major effort to reuse/recycle these materials helps the preservation of natural resources, and minimizes the need of disposal areas.

The potential environmental impacts of such uses cannot be forgotten, and the proper analysis of the potential geo-environmental hazards generated has to be included in the feasibility and design studies, using as reference the existing standards. It is believed that further development in standards for environmental assessment and quality assurance systems are required (Hartlén, et al. 1996).

Particular use of many different by-products (cfr. ITC-5 - SC 7) are, at this time, being studied in different parts of the world. It is worth mentioning that broad international experience has been gathered in the use of distinct industrial residue in road construction as substitute for natural aggregates.

6 CONCLUSIONS

The composition of waste contained within a landfill is influenced by the affluence of the catchment area, the time of year, age, etc.... It is this variability of the composition of the waste that makes settlement prediction very difficult. For this reason the waste mass should be divided into filling periods, with an accurate composition of the waste assigned to several distinct periods. Waste composition alters with age as biodegradation takes place and the organic content of the waste degrades into inert waste. This reaction occurs concurrently with other chemical reactions altering the composition of the waste. These changes in composition should equally be taken into consideration during settlement calculations.

The ITC-5 activities in the framework of the ISSMGE efforts towards the environmental geotechnics are elaborated. Mainly the progress in modelling of migration of pollutants, waste classification and the above discussed mechanical characterisation, design of engineered landfills and the assessment of geo-environmental hazards are remarkable.

It looks important to the future terms of reference of this ITC-5 committee to at least stress much more on the MSW waste mechanics and also on the remediation philosophy related to geo-environmental problems of dredged and non-traditional con- struction materials.

REFERENCES

Blower, T.; Cowland, J.W. & Tang, K.Y. (1996), The stability of Hong Kong Landfill Slopes, Engineering Geology of Waste Disposal, S.P. Bentley Ed., Engineering Geology Special Publication N° 11, The Geotechnical Society.

Del Greco, O. & Oggeri, C. (1993), Geotechnical parameters of sanitary wastes, Proc. Sardinia '93, 4th Int. Landfill Symposium, Cagliari, Italy, pp. 1421-1431.

Grisolia, M., Napoleoni, Q., Tancredi, G. (1992), Considerazioni sulla compressibilità dei rifiuti solidi urbani, Proc. 1st Italian-Brazilian symposium on sanitary and Environmental engineering, Rio de Janeiro.

Grisolia, M., Napoleoni, Q., Sirini, P., Tancredi, G. (1991),Geotechnical behaviour of sanitary landfill based on laboratory and in situ tests, Proc. 15th conferences of geotechnics of Torino, Società Ingegneri e Architetti in Torino (in Italian).

Haegeman, W.F. (1995), Non destructive evaluation of a road construction by the SASW method, Proc. Danube European conference on soil mechanics and foundation engineering, Mamaia, Romania, vol. 1.

Haegeman, W. Van Impe, W.F. (1995), Development on the SASW method, Proc. 11th ECSMFE, Copenhagen, Denmark, Danish Geotechnical Society, Bulletin 11, vol. 2.

Fassett, J.B., Leonards, G.A. & Repetto, P.C. (1994), Geotechnical properties of municipal solid wastes and their use in landfill design, Proc.

Waste Tech. '94, National solid wastes management association, Charleston, South Carolina.

Jones, D.R.V. (1997), The design of landfill lining systems using geosynthetics, the Nottingham trent University, PhD thesis in preparation

Jessberger, H.L. (1994), Geotechnical aspects of landfill design and construction, part 2, Material parameters and test methods, Proc. Instn. Civil engineering geotechnical engineering, 107, Apr., pp. 105-113

Jessberger, H.L. & Kockel, R. (1993), Determination and assessment of the mechanical properties of waste materials, proc. Sardinia '93, 4th Int. Landfill symposium, Cagliari, Italy.

Kolsch, F. (1995), Material values for some mechanical properties of domestic waste, Proc. Sardinia '95, 5th Int. Landfill symposium, Cagliari, Italy

Landva, A. O. & Clark, J.I. (1986), Geotechnical testing of wastefill, Proc. 39th Canadian geotechnical conference, Ottawa, Ontario, pp. 371-385.

Report of the ISSMFE Technical Committee TC5 on Environmental Geotechnics, SC2 - Contaminant migration test methods, modelling and monitoring C.D. Shackelford

Report of the ISSMFE Technical Committee TC5 on Environmental Geotechnics, SC3 - Waste stability M. Wojnarowicz

Report of the ISSMFE Technical Committee TC5 on Environmental Geotechnics, SC4 - Controlled landfills M. Manassero

Report of the ISSMFE Technical Committee TC5 on Environmental Geotechnics, SC6 - Assessment of geo-environmental hazards from dredging materials M. Kamon

Report of the ISSMFE Technical Committee TC5 on Environmental Geotechnics, SC7 - Assessment of geo-environmental hazards from non-traditional geotechnical construction materials L. De Mello

Jones, F. ; Taylor, D. ; Dixon, N. (1997), Shear strength of waste and its use in landfill stability analysis, Geoenvironmental engineeing, Thomas Telford, London, 1997.

Taylor, D.P. (1995), Strength testing of domestic waste by plate bearing and shear box tests, MSc dissertation, University of Durham School of Engineering.

Van Impe, W.F. (1993), Stability analysis of the Appels landfill, International report, Soil mechanics laboratory, Ghent University (unpublished) (in Flemish).

Van Impe, W.F. (1994), Municipal and industrial waste improvement by heavy tamping, Proc. Meeting of geotechnical engineering Geotechnics in the Design and construction of Controlled waste landfills, Milazzo (ME), Associazione Poligeotecnici Riuniti.

Van Impe, W.F., Bouazza, A. (1995), Compactage dynamique des dechets menagers, Journée Louis Menard (in French)

Van Impe, W.F., Bouazza, A. (1996), Densification of waste fills by dynamic compaction, Canadian geotechnical journal, vol. 33, Nr. 6, pp 879-887.

Van Impe, W.F., Bouazza, A., Haegeman, W., (1996), Quality control of dynamic compaction in waste fills, Proc. 2nd ICEG, Osaka, ISSMFE, A.A. Balkema, Rotterdam.

Van Impe, W.F. ; Manassero, M. ; Bouazza, A. (1996), Waste disposal and containment, Second International Congres on Environmental Geotechnics, 5-8 November 1996, Osaka-Japan, Vol. Stniate of the arts reports pp. 193-242.

Geotechnical Hazards, Marić, Lisac & Szavits-Nossan (eds) © 1998 Taylor & Francis, ISBN 90 5410 957 2

A consistent macroscopic mathematical model for soil consolidation problems

S. Arnod, M. Battaglio, N. Bellomo, D. Costanzo, S. Foti, R. Lancellotta & L. Preziosi
Politecnico di Torino, Italy

ABSTRACT: This paper is aimed at presenting a nonlinear model for consolidation problems. The reference configuration is assumed to be the configuration of the soil skeleton and, consistently, balance equations for the fluid are also given in a Lagrangian formulation. Both the problems of partition of the stress tensor between the two phases and the use of the Darcy's law as a momentum balance equation are discussed. Finally the one-dimensional case is validated throughout the data of a case history.

1 INTRODUCTION

Since the publication of the pioneering work by Terzaghi (1923), there has been a growing interest in consolidation theory. This interest arises from both theoretical requirements, linked to the mechanics of porous media, as well as from requirements linked to engineering applications, related for example to the prediction of settlement rate, changes of soil properties with the evolution of state of the porous media, oil production, diffusion of pollutants, transient phenomena occurring in earthquakes and wave propagation. It is also relevant to observe that similar problems are of interest in biomechanics, where the porous medium is composed by the bone structure with the circulating blood. Therefore, in its essence, the problem is relevant in all scientific areas involved into the analysis of "the coupling between the evolution of a porous deformable medium and the motion of the fluid in the connected porosity".

This rather open perspective, also put into evidence at the recent Euromech Colloquium held in Essen (Germany) in June 1997, explains the large amount of research work devoted to this topic, started by the extension of the governing equations for static problems (Biot 1941) to the dynamic ones (Biot 1956, 1962), and requires some introductory notes in order to built a framework of the relevant contributions.

1.1 Microscopic approach and averaging theories.

The mechanics of a porous medium can be described by following two different approaches.
The "microscopic approach" takes into account all the phases constituting the porous medium, and each of them is treated in its own domain as a single body by accounting for the interaction mechanisms throughout the internal boundaries. Then the obtained results can be related to the macroscopic domain by using "averaging strategies". Examples of these averaging strategies are referred by Slattery (1967), Whitaker (1966) and Bear & Bachmat (1991).

1.2 Macroscopic approach

The "macroscopic approach" starts from an a-priori assumption of homogenized phases and is based on the axioms of the theory of mixtures together with the concept of volume fractions. This second procedure is here preferred for the following two reasons: when dealing with natural soils, there is no information about the internal boundaries of pore structures, so that the averaging procedure applied to the microstructure is a rather virtual one; secondly, the porous response measured in experiments is already in itself the response of an implicit macroscopic assumption, so that it appears more consistent to link the experimental results to the macroscopic description.

Fundamentals of the theory of mixtures date back to the works of Truesdell (1957), Truesdell & Toupin (1960), Bowen (1976, 1980, 1982). The paper of de Boer (1996) is particularly relevant for its review character.

It is significant to note that in this theory the kinematics of the fluid phase is described by using a classic Eulerian approach, whereas the kinematics of the solid structure is described by introducing a Lagrangian description. In order to overcome shortcomings deriving from this mixed description, more recently Coussy (1989, 1995), Bourgeois & Dormieux (1996) and Wilmanski (1996) have suggested to introduce a unified Lagrangian description, by assuming the soil skeleton as a material reference volume, and by referring the fluid motion to the soil skeleton.

The short introduction above justifies why to the present work has been given the title "a consistent macroscopic mathematical model...". The adjective "consistent" arises from the fact that the partition of the total stress tensor between the soil skeleton and the fluid phase is not introduced a-priori, but is derived by considering the porous medium as an open thermodynamic system. In this respect it is also shown in which circumstances the stress partition introduced by Terzaghi (1923, 1936) can be applied to a porous medium.

Finally, the suggested one-dimensional nonlinear finite deformations model is validated throughout the results of a case history.

2 LAGRANGIAN FORMULATION OF BALANCE EQUATIONS

The porous medium is assumed to be a two phases system, composed by a matrix with an incompressible and inviscid fluid. The two phases are chemically inert, so that mass and energy exchanges between the phases of chemical origin are not considered. In addition the evolution process is considered to be isotherm.

The representative elementary volume of the porous medium is defined according to Bear (1972): no matter where we place it within a porous medium domain, it will always contain a persistent solid phase and a void space.

The structure composition of the porous medium is described throughout the introduction of the volume fractions

$$n = \frac{e}{1+e} \quad \text{and} \quad (1-n) = \frac{1}{1+e} \tag{1}$$

corresponding to the fluid phase and to the solid matrix respectively. The porous medium is then described as two superimposed continuous bodies, each of them occupying the same space region at the same time.

The conservation of mass of the solid component is identically satisfied in the Lagrangian description, i.e.

$$\frac{d}{dt}\left[J\rho_s(1-n)\right] = 0 \tag{2}$$

being ρ_s the mass density of the solid matrix and

$$J = \det \mathbf{F} = \frac{1+e}{1+e_0} \tag{3}$$

where e_0 is the void ratio in the reference configuration.

The deformation gradient tensor $\mathbf{F}$ is the tensor which maps at any time a neighbourhood of a point $\mathbf{X}$ in the reference configuration onto a neighbourhood of a point $\mathbf{x}$ in the current configuration, i.e.

$$d\mathbf{x} = \mathbf{F} \cdot d\mathbf{X} \tag{4}$$

Accordingly, the Lagrangian finite deformation tensor is given by:

$$\mathbf{E} = \tfrac{1}{2}(\mathbf{F}^T\mathbf{F} - \mathbf{1}) \tag{5}$$

In order to give the Lagrangian formulation of mass balance law for the fluid, reference is made to the following considerations.

Let dV_0 be the initial volume of the porous medium element, that in the current configuration is mapped into $dV = JdV_0$. The volume of the fluid phase in the reference configuration is given by n_0dV_0, and in the current one by $ndV = nJdV_0$.

The balance of mass for the fluid phase over a material control volume V_m fixed on the solid matrix and bounded by the surface A_m writes:

$$\frac{d}{dt}\int_{Vm}\left(\rho_w \cdot \frac{e}{1+e}\right)dV + \int_{Am}\left(\rho_w \cdot \frac{e}{1+e}\right)(\mathbf{v}^w - \mathbf{v}) \cdot \mathbf{n}da = 0 \tag{6}$$

being $\mathbf{v}^w$ and $\mathbf{v}$ the velocity of the liquid constituent and of the solids.

Recalling that the initial oriented area $d\mathbf{A} = \mathbf{N}dA$ is mapped into $d\mathbf{a} = \mathbf{n}da$ in the current configuration, i.e.

$$J(\mathbf{F}^{-1})^T \cdot \mathbf{N}dA = \mathbf{n}da \tag{7}$$

by transforming the left hand side of (6) in an integral over the reference volume and by using the Gauss theorem in order to transform the surface integral in an integral over the reference volume, one gets:

$$\frac{d}{dt}\left(\frac{e}{1+e_0}\right) + Div\left[\frac{e}{1+e_0}\mathbf{F}^{-1}(\mathbf{v}^w - \mathbf{v})\right] = 0 \tag{8}$$

where the operator Div outlines that the operation is performed in terms of Lagrangian coordinates.

Sometimes it is convenient to introduce, according to Biot (1977) (see also Coussy, 1995), as "m" the change of fluid mass referred to the initial volume, so that the quantity $(m \cdot dV_0)$ indicates the difference of fluid mass passing from the initial to the current configuration

$$m = J\rho_w n - \rho_{w0} n_0 \qquad (9)$$

and to define the Eulerian vector

$$\mathbf{w} = \rho_w n(\mathbf{v}^w - \mathbf{v}) = \rho_w n \mathbf{v}^r \qquad (10)$$

representing the mass flux relative to the skeleton. With this variables, equation (8) writes:

$$\frac{dm}{dt} + Div\left[J\left(\mathbf{F}^{-1}\right) \cdot \mathbf{w} \right] = 0 \qquad (11)$$

Consider now the momentum balance for the porous medium and let $\rho, \mathbf{a}, \mathbf{a}^w$ be the mass density of the porous medium as a whole, the acceleration of the solid skeleton and the acceleration of the fluid. The Cauchy's first law of motion is expressed in the form

$$\int_V \left[\rho\mathbf{a} + n\rho_w(\mathbf{a}^w - \mathbf{a}) \right] dV = \int_V \rho\mathbf{b}dV + \int_S \mathbf{t}dS \qquad (12)$$

where $\mathbf{b}$ is the body force for unit mass and $\mathbf{t} = \mathbf{n} \cdot \mathbf{T}$ is the surface traction per unit area, being $\mathbf{T}$ the Cauchy stress tensor.

The application of Gauss theorem to the surface integral yields

$$\nabla \cdot \mathbf{T} + \rho(\mathbf{b} - \mathbf{a}) - n\rho_w(\mathbf{a}^w - \mathbf{a}) = 0 \qquad (13)$$

which is the local formulation of the momentum balance.

In order to obtain the Lagrangian formulation we need to introduce the second Piola-Kirchhoff stress tensor

$$\tilde{\mathbf{T}} = J\left(\mathbf{F}^{-1} \cdot \mathbf{T} \cdot (\mathbf{F}^{-1})^T\right) \qquad (14)$$

and by using the transport formula

$$\mathbf{F} \cdot \tilde{\mathbf{T}} \cdot \mathbf{N}dA = \mathbf{T} \cdot \mathbf{n}da \qquad (15)$$

we obtain

$$Div\left(\mathbf{F} \cdot \tilde{\mathbf{T}}\right) + (\rho_0 + m)(\mathbf{b} - \mathbf{a}) - (n_0\rho_{w0} + m)(\mathbf{a}^w - \mathbf{a}) = 0 \qquad (16)$$

When considering the momentum balance of each phase taken separately, the internal interaction with the other phase must compare explicitly into the balance equation, so that for the fluid phase one should write

$$\nabla \cdot \mathbf{T}^w + \rho_w(\mathbf{b}^w - \mathbf{a}^w) + \mathbf{S}_{int} = 0 \qquad (17)$$

where partial stress tensor for the fluid phase as well

as the above mentioned interaction term have been introduced.

There are two main difficulties in using such an equation: the first one is concerned with the fact that the introduction of the partial stress tensor for the fluid phase requires an a-priori assumption about the partition of the total stress tensor between the two phases. Secondly, the interaction force cannot specified at a macroscopic level. For these reasons it is customary to substitute the balance equation of the fluid phase by Darcy's law

$$\frac{\mathbf{w}}{\rho_w} = \frac{\mathbf{K}}{\rho_w g}\left(-\nabla p + \rho_w \mathbf{g}\right) \qquad (18)$$

In equation (18) the hydraulic conductivity tensor $\mathbf{K}$ is a positive definite tensor, linked to the current configuration of the solid skeleton.

The permeability tensor $\mathbf{K}$ has been proved by Neuman (1977) to be a symmetric tensor.

If the principal directions of anisotropy coincide with the assumed coordinate axes, then

$$k_{ij} = 0 \quad \text{for } i \neq j \qquad (19)$$

i.e. the permeability tensor reduces to a diagonal tensor.

This could be the case in the reference frame, supposed that the soil mass is not a heterogeneous anisotropic system in which the principal directions of anisotropy vary from one formation to another.

In practice it is quite often assumed that the porous medium is isotropic in its reference configuration, so that

$$\tilde{\mathbf{K}} = k\mathbf{I} \qquad (20)$$

However, having defined $\tilde{\kappa}$ in the current configuration, it follows that:

$$\tilde{\mathbf{K}} = J\mathbf{F}^{-1} \cdot \mathbf{K} \cdot (\mathbf{F}^{-1})^T \qquad (21)$$

which gives in general the dependence of the permeability tensor on the deformation gradient. In addition also the components of $\tilde{\kappa}$ may depend on $\mathbf{F}$, e.g. on J.

Note that in the current configuration $\mathbf{K}$ is still a diagonal tensor if the deformation preserves the principal axes directions. In this case $\mathbf{K}$ depends on $\mathbf{F}$ through its determinant and in Soil Mechanics it is usual to express this dependence by using empirical relations between k and the void ratio.

There are however some basic aspect to be discussed in order to accept the Darcy's law as an expression equivalent to a momentum balance equation (see Bear 1972, Neuman 1977).

First, equation (18) is founded on empirical evidence based on steady flow of Newtonian fluids at small Reynolds numbers and it cannot be derived

159

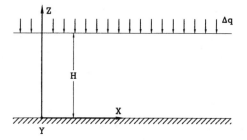

Figure 1. One dimensional consolidation problem.

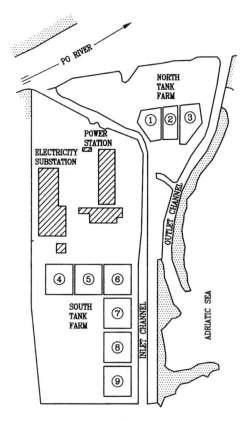

Figure 2. Location of the Porto Tolle thermal power plant.

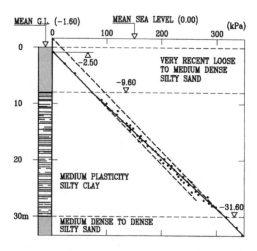

Figure 3. Soil profile and ground water level.

Neuman (1977), which showed that the Darcy's law can be derived from the Navier-Stokes equations by using a formal averaging procedure

$$\langle v_i^f \rangle = -\frac{k_{ij}\rho^w}{\mu}\frac{\partial\langle\phi\rangle}{\partial x_i} - \frac{k_{ij}}{\mu n V}\int_{\Gamma} p n_i d\Gamma \qquad (22)$$

In equation (22), μ and Γ are the dynamic fluid viscosity and fluid-solid interface, and $\langle\phi\rangle$ is the pore-volume average of the potential (energy per unit mass)

$$\phi = -x_i g_i + \frac{p}{\rho^w} \qquad (23)$$

In order to reduce equation (23) to the Darcy's law it is necessary that the surface integral vanishes. A sufficient condition in this respect is that each component of the surface normal vector n_i be symmetrically distributed about a zero average value. This is the case for all porous media having a point, an axis or a plane of symmetry, so that most of natural soils fall in this category.

3. THE POROUS MEDIUM AS A THERMO-DYNAMIC OPEN SYSTEM

Once the momentum balance equation for the fluid phase has been introduced, the problem still remains about the partition of the total stress tensor between the porous medium components. In Soil Mechanics it is usual to decompose the total stress into the effective stress, assigned to the soil skeleton, and the pore water pressure, according to the Terzaghi's definition of effective stress (Terzaghi 1936).

as a momentum balance in terms of macroscopic quantities. As such, it is open to debate the question whether it can be used for nonsteady conditions, or for compressible fluid or for media with nonuniform porosity.

To answer these questions it is particularly relevant the theoretical contribution made by

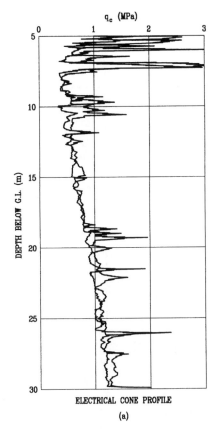

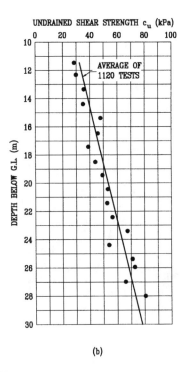

Figure 4. a) Electrical CPT profile, b) Vane Test profile.

This partition is heuristic in its nature, and cannot be regarded as a principle, because of lack of generality when applied to porous media other than soils.

Note that depending on the nature of the problem we are dealing with, the fundamental question to be answered is the following: "what is the combination of total stress tensor and the pore pressure which produces deformations of the porous medium?".

In this respect, a more general approach consider the porous medium as an open thermodynamic system, i.e. the soil skeleton represent a control volume and the flux of matter through it is taken into account (Biot 1977, Coussy 1995).

If for the purpose of the present note we restrict our analysis to a thermoporoelastic system, it is possible to assume that the free energy

$$\Psi = J\psi \tag{24}$$

is a function of the deformation tensor $\mathbf{E}$ and "m".

Then, the two state equations for the porous medium are the following (Coussy 1989,1995)

$$\tilde{\mathbf{T}} = \frac{\partial \Psi}{\partial \mathbf{E}} \qquad g_m = \frac{\partial \Psi}{\partial m} \tag{25}$$

being g_m the free enthalpy of the unit fluid mass. Adding the fluid state equation

$$\frac{1}{\rho_w} = \frac{dg_m}{dp} \tag{26}$$

yields

$$d\tilde{\mathbf{T}} = \mathbf{C} : d\mathbf{E} - M\mathbf{B}\frac{dm}{\rho_w}$$
$$dp = M\left(-\mathbf{B} : d\mathbf{E} + \frac{dm}{\rho_w}\right) \tag{27}$$

where

161

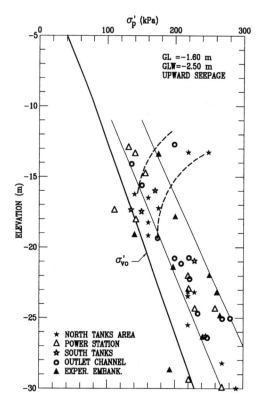

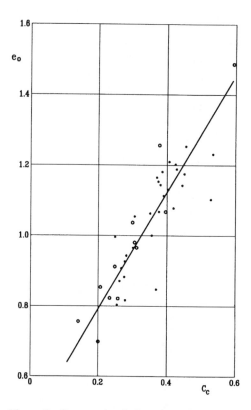

Figure 5. Effective overburden stress and preconsolidation stress.

Figure 6. Compression Index vs. void ratio (Bilotta & Viggiani 1975).

$$\mathbf{C} = \frac{\partial^2 \Psi}{\partial \mathbf{E}^2} \qquad M = \rho_w^2 \frac{\partial^2 \Psi}{\partial m^2}$$

$$\frac{-M\mathbf{B}}{\rho_w} = \frac{\partial^2 \Psi}{\partial \mathbf{E} \partial m} \qquad (28)$$

Equations (27) give the response of the porous medium in terms of finite deformations. In particular, as discussed in details by Coussy (1995), by setting $dm/dt = 0$, its behaviour is fully undrained and the tensor $\mathbf{C}$ assumes in this case the meaning of the tensor of elastic undrained tangent modulii.

An alternative formulation of equations (27) is the following

$$d\widetilde{\mathbf{T}} = \mathbf{C}_0 : d\mathbf{E} - \mathbf{B}\, dp$$
$$\mathbf{C}_0 = \mathbf{C} - M\, \mathbf{B} \otimes \mathbf{B} \qquad (29)$$

and in this case it can be observed that, by setting $dp/dt = 0$, the behaviour of the porous medium will

be fully drained, and the tensor $\mathbf{C}_0$ has the meaning of tensor of the elastic tangent drained modulii.

Note that in this way the soil response has been characterized without introducing any a-priori assumption about the partition of the total stress tensor.

It can be proved (see Bourgeois & Dormieux 1996) that the following transport formula holds

$$\mathbf{T} + p\mathbf{I} = \frac{1}{J}\mathbf{F} \cdot \left(\widetilde{\mathbf{T}} + \mathbf{B}\,p\right) \cdot \mathbf{F}^T \qquad (30)$$

which finally proves that the effective stress tensor in the current configuration, as defined by Terzaghi, is linked to the tensor $(\widetilde{\mathbf{T}} + \mathbf{B}p)$, relative to the reference configuration, and confirms the possibility of using the Terzaghi partition in all circumstances where $\mathbf{B} = \mathbf{1}$.

In order to give a physical meaning to the tensor $\mathbf{B}$ it is necessary to explore the compatibility between the parameters of the macroscopic model and those of the constituent phases. If for sake of simplicity it is assumed that the porous medium is a

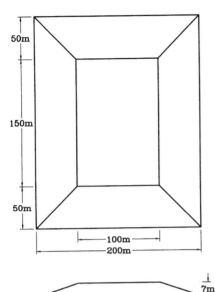

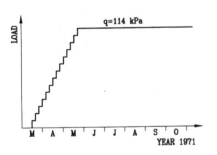

Figure 7. Trial embankment and loading sequence.

linear and isotropic medium, than

$$\mathbf{B} = b\mathbf{1} \tag{31}$$

and the scalar quantity "b" (Coussy 1995, Lade & De Boer 1997), called Biot's coefficient, has the expression

$$b = 1 - \frac{K}{K_s} \tag{32}$$

being K and K_s the bulk modulus of the soil skeleton as a whole and the bulk modulus of the matrix. With such a meaning, if the solids grains are assumed to be incompressible, the Terzaghi partition expression holds.

4. THE ONE-DIMENSIONAL CONSOLIDATION MODEL

The one-dimensional model is particularly relevant in geotechnical applications, so that it is of interest to show how a general model can be derived without any restrictive assumption from the above equations.

Consider in Figure 1 a soil stratum of infinite extent in the plane OXY and transversely isotropic about the OZ axis. The upper surface is supposed to be a free draining boundary, so that

$$p(t, Z = H) = 0 \tag{33}$$

The lower boundary is supposed to be impermeable, and the related boundary condition is given by

$$\mathbf{w}(t, Z = 0) \cdot \mathbf{e}_z = 0 \tag{34}$$

or, by using the Darcy's law, can also be expressed in terms of excess pore pressure

$$\frac{\partial u}{\partial Z}(t, Z = 0) = 0 \tag{35}$$

At the instant $t = 0$ the upper boundary is being loaded with a uniform load

$$\mathbf{t}(t \geq 0, Z = H) = -\Delta q\, \mathbf{e}_z \tag{36}$$

By accounting for the one-dimensional nature of the problem, the deformation gradient tensor reduces to

$$\mathbf{F} = \mathbf{1} + \frac{\partial \xi}{\partial Z}\, \mathbf{e}_z \otimes \mathbf{e}_z \tag{37}$$

and the vertical component of the Piola-Kirchoff tensor is linked to that of the Cauchy tensor according to

$$\tilde{T}_{zz} = \frac{T_{zz}}{1 + \dfrac{\partial \xi}{\partial Z}} \tag{38}$$

The Lagrangian mass balance equation for the solid phase reduces to

$$\frac{\partial z}{\partial Z} = \frac{1 + e}{1 + e_0} \tag{39}$$

being e_0 the void ratio in the reference configuration. The Lagrangian mass balance formulation (8) reduces to

$$\frac{\partial}{\partial t}\left(\rho_w \frac{e}{1 + e} \frac{\partial z}{\partial Z}\right) + \frac{\partial}{\partial Z}\left[\rho_w \frac{e}{1 + e}(v^w - v)\right] = 0 \tag{40}$$

By substituting the (39) into (40) one gets

163

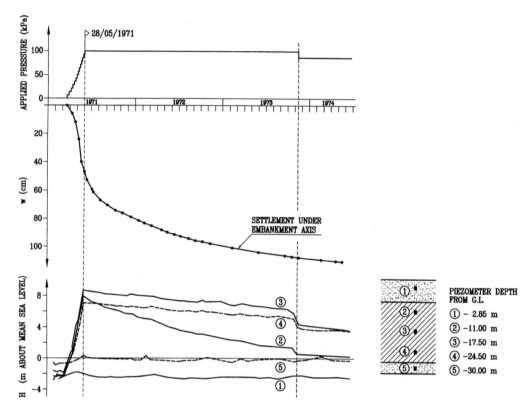

Figure 8. Settlement and pore pressure under the trial embankment (Bilotta & Viggiani 1975).

Table 1. Index properties within the silty clay stratum.

Depth below G.L. (m)	γ_n (kN / m³)	w (%)	w_L (%)	PI (%)	CF (%)
10 ÷ 13	18.5 ± 0.7	36.2 ± 6.2	49.5 ± 8.2	29.0 ± 7.9	26.3 ± 5.4
13 ÷ 16	18.4 ± 0.7	37.2 ± 6.5	53.8 ± 7.9	31.8 ± 7.7	33.0 ± 8.6
16 ÷ 19	18.4 ± 0.6	37.4 ± 4.7	53.8 ± 7.0	30.9 ± 6.7	37.6 ± 6.8
19 ÷ 22	18.5 ± 0.6	35.3 ± 4.4	51.3 ± 4.9	29.9 ± 5.3	37.7 ± 4.3
22 ÷ 25	18.8 ± 0.5	34.3 ± 4.1	50.3 ± 6.0	28.5 ± 5.7	33.6 ± 3.0
25 ÷ 28	18.4 ± 0.5	38.0 ± 4.4	54.9 ± 7.1	32.6 ± 6.2	35.0 ± 8.2

$$\frac{\partial e}{\partial t} + (1 + e_o)\frac{\partial}{\partial Z}\left[\frac{e}{1+e}(v^w - v)\right] = 0 \qquad (41)$$

The balance of linear momentum equation (16), when neglecting the inertial terms, gives

$$Div\left(\mathbf{F} \cdot \tilde{\mathbf{T}}\right) + (\rho_o + m)\mathbf{b} = 0 \qquad (42)$$

and, by using the definition of "m", one gets

$$\frac{\partial T_{zz}}{\partial Z} + g\frac{e\rho_w + \rho_s}{1 + e_o} = 0 \qquad (43)$$

Finally the Lagrangian formulation of Darcy's law is given by

164

Table 2. Undrained strength at Porto Tolle site.

Type of Test	c_u / σ'_{1c}	Legend
TX - CKoU - CL	0.31 ± 0.03	triaxial compression loading
TX - CKoU - EL	0.19 ± 0.04	triaxial extension loading
DSS - CKoU	0.26 ± 0.02	direct simple shear
PSD - CKoU - CL	0.39 ± 0.03	plane strain device
FVT	$c_u / \sigma'_{v0} = 0.29$ $ST = 2 \div 3$	GEONOR field vane test

σ'_{1c} : major principal consolidation stress in lab tests $\approx$ effective overburden stress for in situ tests
σ'_{v0} : vertical effective overburden stress
c_u : undrained strength
ST: sensitivity from FVT

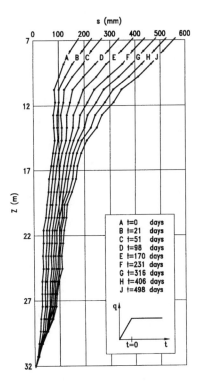

Figure 9. Settlement at different depth.

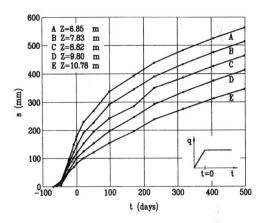

Figure 10. Settlement evolution with time.

as well as the definition of effective stress with sign convention as used in Soil Mechanics ($T'_{zz} = T_{zz} - p$), one obtains the Lagrangian equation of one-dimensional consolidation

$$\frac{\partial e}{\partial t} + (1+e_0)\frac{\partial}{\partial Z}\left[\frac{k}{g\rho_w}\left(\frac{\rho_s - \rho_w}{1+e}g + \frac{\partial T'_{zz}}{\partial Z}\frac{1+e_0}{1+e}\right)\right] = 0$$

(45)

The above equation has been derived without any assumption regarding the constitutive relationships, as well as the self-weight of the consolidating stratum or the smallness of deformation, so that it represents a general formulation. It has been proved by Lancellotta & Preziosi (1997) that Terzaghi's (1923) theory as well as the nonlinear formulation

$$\frac{e}{1+e}(v^w - v) = -\frac{k}{g\rho_w}\left[\left(1+\frac{\partial\xi}{\partial Z}\right)^{-1}\frac{\partial p}{\partial Z} + \rho_w g\right]$$

(44)

By inserting equation (44) into (40), and using (43)

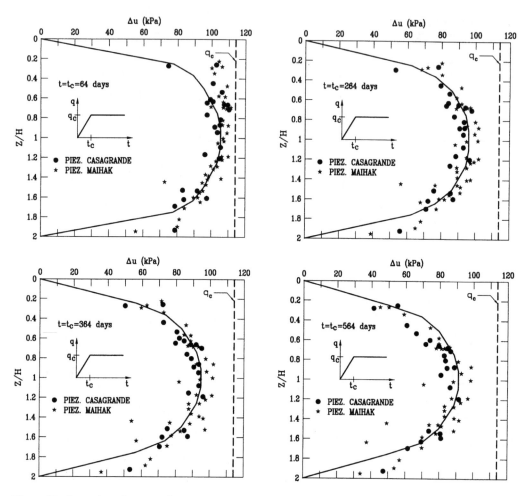

Figure 11. Examples of measured excess pore pressure isochrones.

given by Davis & Raymond (1965) or the more general formulation given by Gibson et al. (1967), as well as other models suggested in literature (see Lancellotta 1995), can be obtained as particular cases of it.

5. OBSERVED BEHAVIOUR OF A TRIAL EMBANKMENT

The construction of the 2.4 GW ENEL thermal power plant of Porto Tolle required an intensive programme of site investigation, installation of vertical drains, preloading embankments for the storage tanks and controlled water tests. This intensive effort was linked to the location of the power plant on recent sediments of the Po river delta. The expected large settlements of nine floating-roof storage tanks also suggested an extended monitoring programme, so that a huge amount of data was collected related to both soil properties as well as to the behaviour of the structures.

Part of this data was already presented by Appendino et al. (1979), Battaglio et al. (1981), Berardi & Lancellotta (1997), Bilotta & Viggiani (1975), Croce et al. (1973), Ghionna et al. (1981), Hegg et al. (1983).

Hegg et al. (1983) and Berardi & Lancellotta (1997) have in particular focused the attention on the performance of the storage tanks during the water testing programme and the observed consolidation rates. Bilotta & Viggiani (1975) have referred on the

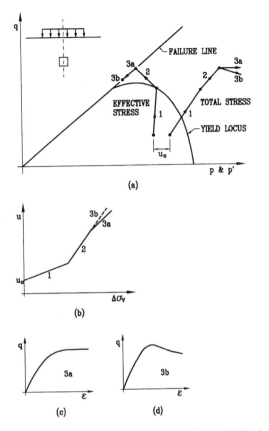

(a)

(b)

(c)   (d)

Figure 12. Idealised soil behaviour (Parry & Wroth 1976)

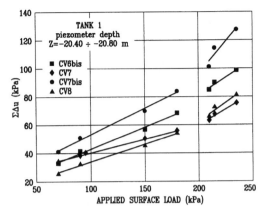

Figure 13. Pore pressure response during the loading stage (Berardi & Lancellotta 1997).

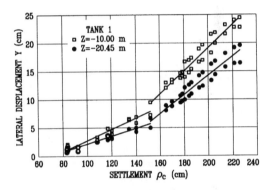

Figure 14. Trend of lateral displacements vs. the vertical settlement (Berardi & Lancellotta 1997).

behaviour of a trial embankment, whose data are re-analysed in the present context.

5.1 Soil profile

The whole area covers a very flat delta plain (Fig. 2) with original elevation ranging between -1 and -2 m.s.l.. This grade was raised to elevation 0.00 ÷ 0.50 m by placing a sand fill during 1974-75.

Soil profile (Fig. 3) consists of a stratum of very recent loose to medium silty sand with embedded thin layers and lenses of soft cohesive and organic sediments.

Below this stratum a 20 ÷ 22 m thick deposit of light brown soft silty-clay is encountered, with frequent seams and lenses of fine silty-sand of few centimetres. In its lower part, the presence of organic matter in the silty-clay matrix in quite frequent and during the borings a number of small pockets of organic gas were encountered. The soft silty-clay deposit terminates abruptly at the depth of 28 ÷ 30 m in a medium dense sand.

The ground water level in the upper sand stratum ranges between -1 to -1.5 m.s.l.; in the lower sand stratum exceeds the level found in the upper one by 1.5 ÷ 2 m. As a consequence the clay is subjected to a slight upward seepage, the hydraulic gradient being of about 0.12.

Examples of electrical CPT and Vane Tests profiles are shown in Figure 4 and average values of the index properties, as obtained from more than 300 samples of silty-clay, are summarised in Table1.

5.2 Stress history

The local geology indicates the presence of holocene deposits up to 150 ÷ 200 m thick. Since the soft clay deposits do not exceed 6000 ÷ 10,000 years in age, this thickness corresponds to a very high sedimentation rate. In particular, the upper silty sand

167

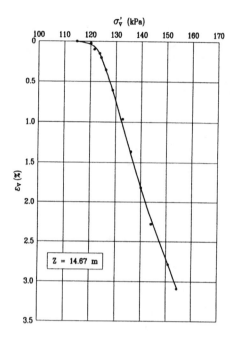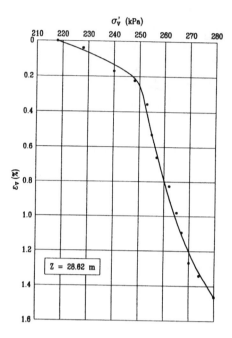

Figure 15. In situ compression curves as deduced from in situ measurements.

strata were deposited within the past 200 years. This geological information indicates that the soft cohesive deposit is only slightly overconsolidated, due to ground water oscillation, as it could be partially inferred from the trend with depth of the yield stress σ'_p compared to that of the effective overburden stress σ'_{v0} in Figure 5.

The value of the coefficient of earth pressure at rest, measured by means of in situ total stress cells, ranges between 0.48 and 0.56 and this value is in agreement with measurements made with instrumented oedometer tests.

5.3 Compressibility

The compression index, as deduced from oedometer tests (Bilotta & Viggiani 1975), ranges between 0.35 and 0.55, its value being related to the in situ void ratio (Fig. 6).

5.4 Shear strength

Finally, as far as the shear strength parameters are concerned, Table 2 summarises the undrained strength values as deduced from different laboratory and in situ tests.

5.5 Monitoring programme and observed behaviour

As already mentioned, because of the very large expected settlements, a huge effort was spent to monitor the behaviour of a test embankment 200×250 m wide, with a height of 7 m, resulting in a uniform pressure on the soil of 114 kPa (Fig.7). This embankment was built in less than 3 months.

The monitoring programme included: surface settlement plate sensors; 8 boreholes instrumented with deep settlement gauge and 110 piezometers (62 electrical and 48 Casagrande type).

Figure 8 shows an example of settlement evolution with time as well as examples of pore pressure decay with time, as measured at various depths.

Figure 9 reports settlement distribution with depth. Each line represents an isochrone in terms of settlements and the settlement evolution with time, as shown in Figure 10, enables one to obtain the evolution of the vertical strain component.

Figure 11 finally reports examples of pore pressure isochrones and it is relevant to note that the soft silty clay horizon behaves as a whole homogeneous stratum, because the sand lenses are not interconnected.

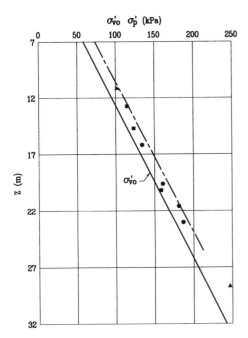

Figure 16. Yield stress as deduced from field data.

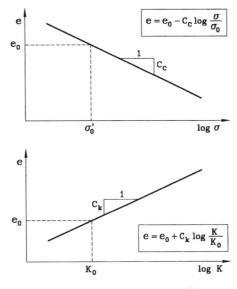

Figure 17. Constitutive laws.

6. IDEALISED SOIL BEHAVIOUR AND DATA INTERPRETATION

6.1 Occurrence of yielding

Although the soft cohesive deposit under consideration has a rather recent origin, it can be expected that its behaviour exhibits the features of a lightly overconsolidated clay, due to changes in ground water level or delayed secondary compression.

For this reason even a very simplified model must have some prerequisite in order to capture the essential features of the soil response. Among these prerequisites there is the need to predict the occurrence of local yielding. As suggested by Parry & Wroth (1976), when a lightly overconsolidated soil is reloaded (Fig. 12) it moves into a space whose projection into the p - q plane is bounded by a yield surface. It can be assumed that soil response is elastic until the loading path reaches the failure or the yielding condition. If it reaches the yield surface, the soil continues to deform plastically until it reaches failure. In this framework it is relevant to recall that excess pore pressure developed during undrained tests are governed by the volumetric characteristics of the soil, so that a yield condition would be indicated by a sharp increase in pore pressure.

Figure 13 reports experimental data from one of the tanks under consideration. It must be observed that in order to avoid the influence of pore pressure dissipation, $\Sigma \Delta u$ in this figure represents the summation of the increments of excess pore pressure developed during each single loading stage.

A similar behaviour has been observed by other authors (Holtz & Holm 1979, Tavenas et al. 1979, Leroueil et al. 1990) with respect to the trend of lateral displacements as compared to the vertical settlements. Within the yield locus, soil exhibits low compressibility and lateral displacements are of small amount if compared with settlements. But as soon as a local yield condition is reached, there is a sharp increase of the above relation between the maximum lateral displacement and the vertical settlement under the centre, as illustrated for the present case in Figure 14.

In addition to all these data, Figure 15 reports two examples of in situ compressibility curves, reconstructed by using vertical strain values from Figure 10, associated with vertical effective stress deduced from measured pore pressure decay with time and Figure 16 shows the trend of the yield stress with depth as deduced from these curves.

In conclusion, despite the geological evidences related to the recent origin of this deposit, all the available data related to the observed soil response suggest the existence of a mechanical overconsolidation phenomenon and requires to account, even in a very simplified way, for the occurrence of yielding in order to predict the behaviour of the structures (Berardi & Lancellotta 1997).

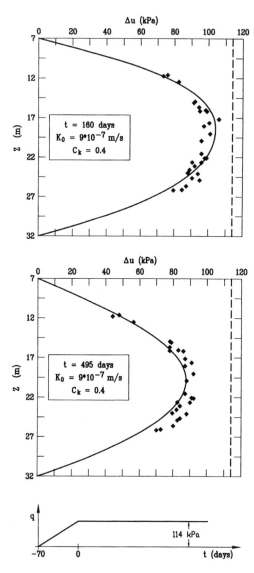

Figure 18. Computed versus measured pore pressure decay with time.

In the present analysis the compressibility parameters have been introduced according to the values deduced from laboratory tests (Fig. 6), whereas the hydraulic conductivity has been back-figured from field performance.

Examples of this kind of back analysis are presented in Figure 18, where a comparison of the calculated excess pore pressure isochrone in the clay layer (continuos line) and the measured values at the same time is shown. The values of k_0 (initial value of hydraulic conductivity, corresponding to a value of e_0 of the void ratio) and C_k used for the analysis are also shown.

7. FINAL REMARKS

The increased interest in the behaviour of deformable porous material requires a generalised mathematical formulation of the coupling between the evolution of the porous medium deformations and the motion of the fluid in the connected porosity.

The first part of this work presented a general model which accounts for finite strain and material nonlinearity, consistent with a macroscopic description, which does not require an a-priori partition of the total stress between the soil skeleton and the pore fluid.

The one-dimensional Lagrangian formulation of this model has been applied to the case history of Porto Tolle, in order to show its capability in capturing the most relevant aspects of the problem as presented by the influence of self-weight and full nonlinearity.

ACKNOWLEDGMENTS

The authors would like to thank Mr. R. Maniscalco for his helping in preparing drawings and discussing experimental data.

REFERENCES

Appendino, M. & M. Jamiolkowski & R. Lancellotta 1979. Pore pressure of NC clay around driven displacements piles. *Proc. Conf. on Recent Developments in the Design and Construction of Piles.* 169-175, ICE, London.

Battaglio, M. & M. Jamiolkowski & R. Lancellotta & R. Maniscalco 1981. Piezometer probe test in cohesive deposits. *Proc. Symp. on Cone Penetration Testing and Experience, ASCE National Convention, St. Louis, Missouri.*

Bear, J. 1972. *Dynamics of fluids in porous media.* Elsevier.

Bear, J. & Y. Bachmat 1991. *Introduction to modelling of transport phenomena in porous*

6.2 Consolidation phenomenon

The above discussed case history has been used as an example of application of the mathematical model presented in Section 4. In this respect we note that the use of equation (45) requires specification of constitutive laws, i.e. changes of void ratio with effective stress, as well as changes of hydraulic conductivity with void ratio, as schematically shown in Figure 17.

media. Kluwer A. Publ.

Berardi, R. & R. Lancellotta 1997. Yielding from Field Performance and Its Influence on Differential Settlements of Oil Tanks. (Submitted for publication).

Bilotta, E. & C. Viggiani 1975. Un'indagine sperimentale in vera grandezza sul comportamento di un banco d'argille normalmente consolidate. *XII Convegno Nazionale di Geotecnica, AGI, Cosenza, Italy*.

Biot, M.A. 1941. General theory of three dimensional consolidation. *J. Appl. Phys.*, 12, 155-164.

Biot, M.A. 1956. Theory of propagation of elastic waves in a fluid saturated porous solid. *J. Acoustic Soc. of America*, 28, 168-191.

Biot, M.A. 1962. Mechanics of deformation and acoustic propagation in porous media. *J. Appl. Phys.*, 33, 1482-1498.

Biot, M.A. 1977. Variational Lagrangian thermodynamics of nonisothermal finite strain mechanics of porous solids and thermomolecular diffusion. *Int. J. Solids Structures*, 13, 579-597.

Bourgeois, E. & L. Dormieux 1996. Consolidation of a nonlinear poroelastic layer in finite deformations. *Eur. J. Mech., A/Solids*, 15, 4, 575-598.

Bowen, R.M. 1976. *Theory of mixtures - Continuum physics*. III/1, 317, Ed. A.C. Eringen, Academic Press.

Bowen, R.M. 1980. Incompressible porous media models by use of the theory of mixtures. *Int. J. Eng. Sc.*, 18, 1129-1148.

Bowen, R.M. 1982. Compressible porous media models by use of the theory of mixtures. *Int. J. Eng. Sc.*, 20, 6, 697-735.

Coussy, O. 1989. Thermodynamics of saturated porous solids in finite deformations. *Eur. J. Mech., A/Solids*, 8, 1-14.

Coussy, O. 1995. *Mechanics of porous continua*. Wiley.

Croce, A. & G. Calabresi & C. Viggiani 1973. In situ investigations on Pore Pressure in Soft Clays. *Proc. 8th ICSMFE, Moscow*.

Davis, E.H. & G.P. Raymond 1965. A nonlinear theory of consolidation. *Geotechnique*, 15, 161-173.

De Boer, R. 1996. Highlights in the historical development of the porous media theory: toward a consistent macroscopic theory. *Applied Mechanics Review*, 49, 201.

Ghionna, V. & M. Jamiolkowski & R. Lancellotta & L. Tordella & C.C. Ladd 1981. Performance of Self-Boring Pressuremeter Test in Cohesive Deposits. *MIT Report, Cambridge, Mass.*

Gibson, R.E. & G.L. England & M.J.L. Hussey 1967. The theory of one-dimensional consolidation of saturated clays. *Geotechnique*, 17, 261-273.

Hegg, U. & M. Jamiolkowski & R. Lancellotta & E. Parvis 1983. Behaviour of Oil Tanks on Soft Cohesive Ground Improved by Vertical Drains. *Proc. 8th ECSMFE, Helsinki*, 2, 627-632.

Holtz, R.D. & G. Holm 1979. Test embankment on an organic silty clay. *Proc. 7th ECSMFE, Brighton*, 3, 79-86.

Lade, P.V. & R. de Boer 1997. The concept of effective stress for soil, concrete and rock. *Geotechnique*, 47, 61-78.

Lancellotta, R. 1995. *Geotechnical Engineering*. Balkema

Lancellotta, R. & L. Preziosi 1997. A general nonlinear mathematical model for soil consolidation problems. *Int. J. Engng. Sc.*, 35, 10/11, 1045-1063.

Leroueil, S. & J.P. Magnan & F. Tavenas 1990. *Embankments on soft clays*. Ellis Horwood.

Neuman, S.P. 1977. Theoretical derivation of Darcy's law. *Acta Mechanica*, 25, 153-170.

Parry, R.H.G. & C.P. Wroth 1976. Pore pressure in soft ground under surface loading: theoretical considerations. *U.S. Army Eng. Waterways Experiment Station, Contract Report* S-76-3.

Slattery, J.C. 1967. Flow of viscoelastic fluids through porous media. *Am. Inst. Chem. Eng. J.*, 13, 1066.

Tavenas, F. & C. Mieussens & F. Burges 1979. Lateral displacements in clay foundations under embankments. *Canadian Geotech. Journal* 16, 532-550.

Terzaghi, K. 1923. Die Berechnung der Durchlassigkeitsziffer des Tones aus dem Verlauf der Hydrodynamishen Spannungserscheinungen. *Sitznugshrichte Akad. Wissen. Wien of Mathem. Naturw. Kl.*, 132, 125-138.

Terzaghi, K. 1936. The shearing resistance of saturated soils and the angle between the planes of shear. *1st ICSMFE, Cambridge, Mass.*, 1, 54-56.

Truesdell, C. 1957. Sulle basi della termodinamica. *Rend. Lincei*, 22, 33-38, 158-166.

Truesdell, C. & R.A. Toupin 1960. *The classical field theories*. Handbuch der Physik, Band III/1, 226-902, Springer-Verlag.

Whitaker, S. 1966. The equations of motion in porous media. *Chem. Eng. Sc.*, 21, 291.

Wilmanski, K. 1996. Porous media at finite strains: the new model with balance equation for porosity. *Arch. Mech.*, 48, 4, 591-628.

Earth structures

Geotechnical Hazards, Marić, Lisac & Szavits-Nossan (eds) © 1998 Taylor & Francis, ISBN 90 5410 957 2

Post-failure analysis of earth retaining wall

L. Areias, W. Haegeman & W. F. Van Impe
Laboratory of Soil Mechanics, Ghent University, Belgium

P. Van Calster
Geologica N.V., Bertem, Belgium

ABSTRACT: Results of a post-failure stability analysis are presented for an earth retaining wall that failed while a shallow trench was being excavated at the toe to install underground pipes. The results were obtained using the classical Coulomb's trial-wedge method and a commercially available PLAXIS finite element program. It is shown that either of these methods could have prevented this accident, when used to verify the stability of the wall before the start of excavation.

1 INTRODUCTION

A rigid concrete earth retaining wall located in Genk, Belgium failed while a shallow 1.8-m-deep trench was being excavated along its toe. A cross-section view of the wall is shown in Figure 1.

A field investigation conducted after failure reported lateral movements of approximately 12 cm at the base of the wall. A significant rotation at the top of the wall was also noted. The failure caused irreparable damage to the earth retaining structure and the railway located on top.

A post-failure investigation was performed using Coulomb's earth pressure (trial-wedge) method to document the cause of the accident. Coulomb's earth pressure theory is commonly used to evaluate earth pressures behind retaining walls, and is known to provide satisfactory results (Pradel 1994, Lambe & Whitman 1979).

A second stability analysis was performed using the commercially available finite element program PLAXIS (version 6.1), developed by Plaxis B. V. of the Netherlands, for comparison with the classical Coulomb's method.

2 COULOMB'S METHOD

2.1 *Equilibrium forces*

The forces in equilibrium with the wall and the surrounding soil are presented in Figure 2. These forces take into consideration the presence of a surcharge load from the railway and the contribution of friction from the concrete wall. The corresponding load diagram for the wall system is shown schematically in Figure 3. The following notation is used in this paper to define the equilibrium state of the soil and wall system :

a = centroidal distance from point of rotation, m.
e = effective width of wall, m.
B = bearing surface width, m.
b_1 = distance to resultant active thrust (P_a), m.
b_2 = distance to resultant passive thrust (P_p), m.
b_3 = height of wall key, m.
d = centroidal distance to resultant normal force (N_R), m.
F_1–F_4 = resultant of N_1–N_4 and T_1–T_4, kN/m.
H = height of wall, m.
h = height of passive soil wedge, m.
N_R = resultant of normal stresses from wall system (including surcharge), kN/m.
N_1–N_4 = resultant of normal stresses on assumed soil/soil or soil/wall planes, kN/m.
P_a = active thrust, kN/m.
P_{av} = vertical component of active thrust, kN/m.
P_{ah} = horizontal component of active thrust, kN/m.
P_p = passive thrust, kN/m.
P_{pv} = vertical component of passive thrust, kN/m.
P_{ph} = horizontal component of passive thrust, kN/m.
q_s = equivalent surcharge stress from railway, kN/m^2.
R_o = toe of wall and point of rotation.
T_1–T_4 = frictional resistance between soil/soil and soil/wall, kN/m.
T_R = resultant of frictional resistance between soil and wall, kN/m.
W_1–W_7 = weight of soil and/or concrete elements, kN/m.
W_R = resultant weight of wall, kN/m.

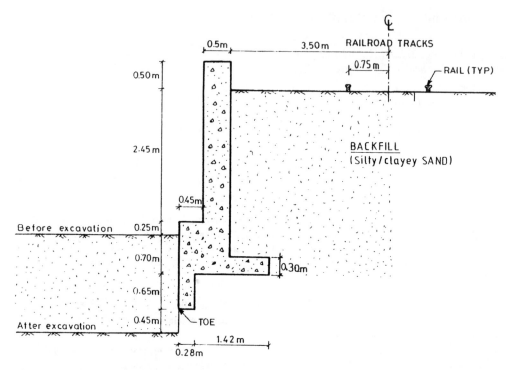

Figure 1. Cross-section view of earth retaining wall.

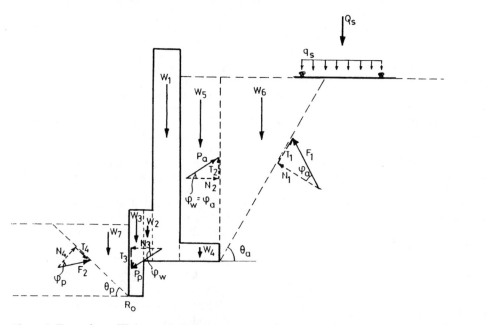

Figure 2. Forces in equilibrium with wall and surrounding soil.

2.2 Parameters used in analysis

Backfill:

Material: silty/clayey SAND.

φ_a = friction angle in active wedge = 30°. Estimated from cone penetration tests.

φ_p = friction angle in passive wedge = 27.5°. Estimated from cone penetration tests.

γ_a = unit weight of active wedge = 17 kN/m³ dry (20.3 kN/m³ saturated). Measured.

γ_p = unit weight of passive wedge = 17 kN/m³. Measured.

θ_a = angle of inclination of assumed active failure surface.

θ_p = angle of inclination of assumed passive failure surface.

Concrete:

γ_c = unit weight f concrete = 24 kN/m³. Typical for normal-weight reinforced concrete mixtures (Mehta 1986).

φ_w = friction angle between backfill and concrete = $(2/3)\varphi_a$. From Sanglerat et al.(1985).

Surcharge:

Q_s = railway surcharge load = 600 kN. Calculated from static weight of car plus load.

q_s = equivalent stress exerted on active soil wedge due to surcharge load = 38 kN/m².

2.3 Active thrust

The free body diagram and force polygon for the active soil wedge are shown in Figures 4 & 5, respectively. Applying the law of sines, the active thrust (P_a) can be expressed as :

$$P_a = \frac{\left(W_6 + Q_s\right)\sin\left(\theta_a - \varphi_a\right)}{\sin\left(90 + \varphi_w + \varphi_a - \theta_a\right)} \tag{1}$$

Since movement in the active wedge will likely take place entirely in the backfill material, we can set $\varphi_a = \varphi_w$ in (1) and obtain:

$$P_a = \frac{\left(W_6 + Q_s\right)\sin\left(\theta_a - \varphi_a\right)}{\sin\left(90 + 2\varphi_a - \theta_a\right)} \tag{2}$$

The weights W_6 and Q_s can further be expressed as a function of the inclination angle (θ_a) of the active wedge by the following equations:

$$W_6 = \frac{1}{2}\gamma H^2 \cot\theta_a \; ; \text{ and} \tag{3}$$

$$Q_s = q_s H \cot\theta_a \tag{4}$$

Substituting for W_6 and Q_s in (2) gives :

$$\frac{P_a}{\frac{1}{2}\gamma_a H^2 + q_s H} = \frac{\cot\theta_a \; \sin\left(\theta_a - \varphi_a\right)}{\sin\left(90 + 2\varphi_a - \theta_a\right)} \tag{5}$$

The horizontal stress resulting from the surcharge load Q_s is assumed to be distributed totally and uniformly over the active soil wedge, and its resultant force located halfway along the depth H of the wall (Fig. 3).

The equilibrium equation (5) is expressed as a function of the variable θ_a, which defines the boundary of the trial wedge through the soil. Since Coulomb's earth pressure method is based on the assumption that full shear strength of the soil is mobilised behind the wall, we need to find the largest trial wedge that will give the largest force against the wall. The largest force against the wall was calculated using equation (5) by evaluating P_a for various values of θ_a, as shown in Table 1.

Table 1. Variation of P_a with θ_a for active wedge and $\varphi_a = 30°$.

θ_a	$\cot\theta_a$	$\sin$ $(\theta_a-\varphi_a)$	$\sin(90+$ $2\varphi_a-\theta_a)$	$\dfrac{P_a}{\frac{1}{2}\gamma_a H^2 + q_a H}$
50.0	0.839	0.342	0.985	0.291
52.5	0.767	0.383	0.991	0.296
55.0	0.700	0.423	0.996	0.297
57.5	0.637	0.462	0.999	0.295
60.0	0.577	0.500	1.000	0.288

The largest wedge is one for which θ_a is approximately 55.0°, as shown in Table 1. The active thrust is therefore:

$$P_a = 0.297\left(\frac{1}{2}\gamma_a H^2 + q_s H\right) \tag{6}$$

The resultant thrust P_a from the active soil wedge is assumed to act at the lower third point of the depth H of the wall (Fig. 3).

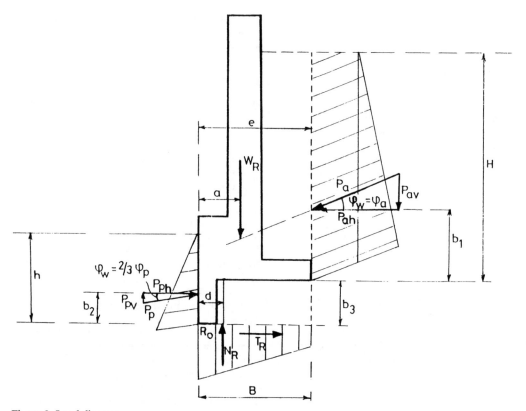

Figure 3. Load diagram.

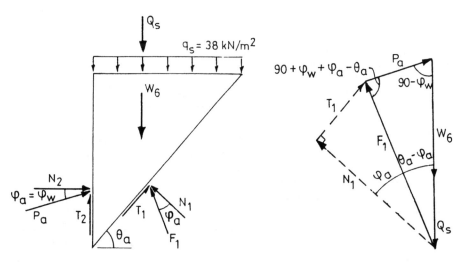

Figure 4. Free body - active wedge.   Figure 5. Force polygon - active wedge.

2.4 Passive thrust

The passive soil wedge free body diagram and force polygon are shown in Figures 6 & 7, respectively. Applying the law of sines, the passive thrust (P_p) can be expressed by:

$$P_p = \frac{W_7 \sin\left(\theta_p + \varphi_p\right)}{\sin\left(90 - \varphi_p - \theta_p - \varphi_w\right)} \qquad (7)$$

The weight of the passive block (W_7) is given by :

$$W_7 = \frac{1}{2} \gamma_p h^2 \cot \theta_p \qquad (8)$$

Substituting for W_7 in (7), and rearranging, we obtain :

$$\frac{P_p}{\frac{1}{2} \gamma_p h^2} = \frac{\cot \theta_p \sin\left(\theta_p + \varphi_p\right)}{\sin\left(90 - \varphi_p - \theta_p - \varphi_w\right)} \qquad (9)$$

The equilibrium equation (9) is expressed as a function of the variable θ_p, which defines the boundary of the passive trial wedge through the soil.

The free body that leads to the smallest value of Pp gives the maximum possible passive thrust that the soil can take in equilibrium. This value is calculated iteratively, in Table 2, for various values of θ_p.

Table 2. Maximum thrust for equilibrium of passive trial wedge

θ_p	$\cot \theta_p$	$\sin$ $(\theta_p\text{-}\varphi_p)$	$\sin(90\text{-}$ $\varphi_p\text{-}\theta_p\text{-}\varphi_w)$	$\frac{P_p}{\frac{1}{2}\gamma_p h^2}$
16.0	3.487	0.688	0.473	5.08
18.0	3.078	0.713	0.442	4.97
20.0	2.747	0.737	0.410	4.94
22.0	2.475	0.760	0.378	4.98
24.0	2.246	0.783	0.345	5.09

The maximum possible thrust P_p is, therefore, one that corresponds to approximately $\theta_p = 20.0°$, as shown in Table 2. The value of P_p is thus:

$$P_p = 4.94 \left(\frac{1}{2} \gamma_p h^2\right) \qquad (10)$$

The resultant thrust P_p from the passive soil wedge is assumed to act at the lower third point of the depth h of the wall (Fig. 3).

2.5 Other parameters

In addition to the active and passive thrusts presented above, other parameters, needed to evaluate the stability of the wall, were also calculated. These parameters are represented graphically in Figure 3. A summary of calculated values are summarised in Table 3.

Table 3. Summary of calculated values

Parameter	Value	Units
a	0.93	m
b_1	1.43	m
b_2	0.45	m
b_3	0.65	m
B	1.70	m
d	0.51	m
e	1.70	m
h	1.35	m
H	3.40	m
$N_R^{(1)}$	151	kN/m
$N_R^{(2)}$	139	kN/m
P_a	73	kN/m
P_p	76	kN/m
P_{ah}	63	kN/m
P_{av}	36	kN/m
P_{ph}	73	kN/m
P_{pv}	24	kN/m
$T_R^{(1)}$	55	kN/m
$T_R^{(2)}$	50	kN/m
W_R	114	kN/m

Notes :
[1] Ignores passive resistance.
[2] Includes passive resistance.

2.6 Factor of safety against sliding

The factor of safety against sliding (F_s) is given by:

$$F_s = \frac{T_R}{T_L} \qquad (11)$$

where:

$$T_R = N_R \tan (\varphi_w) \qquad (12)$$

$$\varphi_w = \frac{2}{3} (\varphi_a) \qquad (13)$$

$$T_L = P_{ah} - \frac{1}{2} \left(P_{ph}\right) \; ; \text{and} \qquad (14)$$

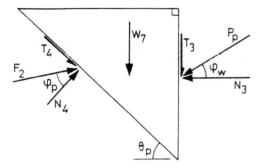

Figure 6. Free body - passive wedge.

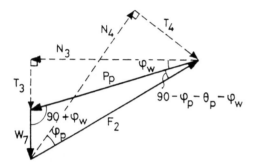

Figure 7. Force polygon - passive wedge.

N_R = the total normal load from the wall, including the surcharge weight due to the railroad. This load is corrected to take into account the effects of wall friction and passive soil wedge.

The frictional force T_L represents a net pushing force on the wall from the active soil wedge. A reduction factor of 2 is applied to the resistance force (P_{ph}) from the passive soil wedge before calculating T_L in equation (14), as suggested by Lambe & Whitman (1979).

Before excavation, the factor of safety against sliding is thus (Table 3):

$$F_s = \frac{50}{26} = 1.9 \qquad (15)$$

This indicates that the wall is stable before excavation.

After excavation, the term P_{ph} in (14) becomes zero, and the factor of safety against sliding is, therefore:

$$F_s = \frac{55}{63} = 0.9 \qquad (16)$$

The calculated $F_s < 1$, which indicates that the wall is unstable after excavation. This is consistent with observations made during construction of the trench, which reported lateral movements of approximately 12 cm at the base of the wall and irreparable damage to the structure.

2.7 Overturning

In addition to lateral movements, rotation was also observed at the wall after failure. Overturning moments were calculated assuming that the excavation did not proceed below the toe of the wall, which corresponds to the removal of the passive soil wedge. The results showed that the wall was stable against overturning, with a ratio of 3.7 obtained for the moment of weight over the overturning moment.

In reality, the trench was excavated 0.45 m below the toe of the wall, as shown in Figure 1. This over-excavation is believed to have caused soil displacement at the base of the wall, thereby shifting the point of rotation (R_o) back from the toe, making it possible for the wall to rotate.

It is concluded, therefore, that the observed rotation was caused by over-excavation of the toe, as described. The mechanism of over-excavation was not investigated further in this study.

3 PLAXIS FINITE ELEMENT METHOD

For comparison with the Coulomb method, a post-failure analysis was performed using the finite element program PLAXIS (version 6.1), developed by Plaxis B.V. of the Netherlands. The program allows introducing interface elements to evaluate soil-structure interaction.

It is also possible to simulate construction and excavation by adding and removing elements, respectively. This allows a realistic assessment of stresses and displacements caused by an excavation, and is used in this study to simulate construction of the trench that led to the failure of the wall presented in this paper.

The mesh configuration with material sets used for the analysis is presented in Figure 8. The following material parameters of Poisson's ratio (ν), Young's modulus (E), and shear modulus (G) used in the analysis are presented in Table 4. All other parameters are the same as those used in the Coulomb analysis.

A staged excavation at the toe of the wall was simulated using PLAXIS to evaluate the effects of the excavation on the stability of the wall. The results of this simulation are presented in Figure 9.

The results show that excessive displacements are reached at a factor of staged construction of only 0.33, approximately. A factor of staged construction of 1 is obtained in this simulation when full excavation can be performed without excessive displacements (failure).

These results are consistent with those obtained using the Coulomb method, in that both results confirmed failure of the wall during excavation at the toe.

Table 4. Material parameters for analysis in PLAXIS.

Material set	ν	$E^{(1)}$ (MN/m²)	$G^{(2)}$ (kN/m²)
1	0.3	-	5000
2	0.25	30,000	-
3	0.3	-	5000

Notes :
[1] Typical for normal weight concrete, (Mehta 1986).
[2] From cone penetration tests.

4 CONCLUSIONS

Post-failure analyses, performed using the classical Coulomb earth pressure method (trial wedge) and PLAXIS (version 6.1) finite element program, confirmed the failure of an earth retaining wall caused by the excavation of a shallow trench along its toe. The results show that either one of the methods used in the post-failure analysis could have predicted this failure before the start of construction, and thus have prevented the accident.

REFERENCES

Lambe, T.W. & R.V. Whitman 1979. *Soil mechanics, SI version.* John Wiley & Sons.
Mehta, P.K. 1986. *Concrete: structure, properties and materials.* Prentice-Hall: International Series in Civil Engineering and Engineering Mechanics.

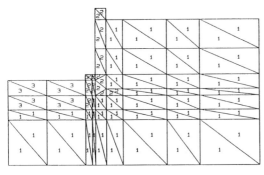

Figure 8. Mesh configuration - PLAXIS.

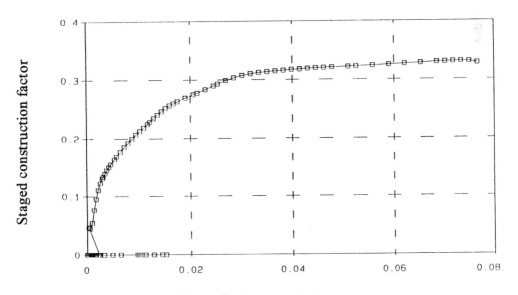

Total displacement (m)

Figure 9. Simulation of staged excavation - PLAXIS.

Pradel, D. 1994. Active pressure distribution in cohesive soils. *Proceedings of the XIII ICSMFE, New Delhi, India*: 795-798.

Sanglerat, G., G. Omivari & B. Cambou 1985. Practical problems in soil mechanics and foundation engineering.

Geotechnical Hazards, Marić, Lisac & Szavits-Nossan (eds) © 1998 Taylor & Francis, ISBN 90 5410 957 2

Settlement computation – A simplified method

D.K. Baidya
Department of Civil Engineering, I.I.T. Kharagpur, India

ABSTRACT: One dimensional consolidation theory is generally used to calculate the consolidation settlement. Estimation of average pressure increase, Δp due to foundation load over the compressible layer is one of the important steps in settlement calculation and several approximate methods are developed for this purpose. This paper presents a simplified method to estimate the average pressure increase, Δp over the compressible layer due to foundation load. Pressure increase, Δp is obtained by dividing the area of the influence diagram for vertical stress under the compressible layer by the thickness of the layer. Area of the influence diagram under the compressible layer is obtained by direct integration of the influence factor over the depth and is expressed in the form of equation. Improvements of present method over the existing methods are shown with an illustrative example.

1 INTRODUCTION

1.1 *Causes of settlement*

Most of the structures which engineers build rest upon soil, and the imposed load from the structure causes the soil to subside. Two distinct kinds of action within the foundation soil may cause the subsidence or settlement of structure resting on soil. When the shearing stresses developed due to imposed load exceed the shearing strength of the material, the soil fails by sliding downward and laterally, and the structure settles and perhaps tips out of vertical alignment. In the other case, a structure settles by virtue of the compressive stress and the accompanying strain which are developed in the soil as a result of the load imposed upon it. But settlement or subsidence is not necessarily an adverse characteristic of a structure, provided it is uniform throughout and does not reach excessive proportion. If the settlements are unequal, that is one corner or one end goes down more than the rest of the structure serious consequences may result.

1.2 *Effect of settlement*

In the building structure, differential settlements cause plaster to crack badly, door and window frames to bend, brick and masonry work to loosen, and floors to crack and fault. As a result there will be a general increase in maintenance costs and rapid deterioration of the value of the building. In an extreme case, unequal settlements may impair the structural integrity of the

frame work and cause the building to be condemned. If the structure is a tall smokestack, monument or church spire, unequal settlements of the foundation may cause the structure to lean in an unsightly manner. In an extreme case, it may lean far enough to become dangerously unstable. Hence, limiting the settlement/differential settlement is one of the basic requirements to be considered in successful design of foundation.

1.3 *Existing methods*

Settlement due to the shear failure of material is out of the scope of this paper. Other settlements i.e., settlement due to compressive stress and the accompanying strain which are developed in the soil as a result of imposed load upon it, mainly consists of two parts viz., (1) immediate or elastic settlement and (2) consolidation settlement. In sandy soil consolidation settlement is absent and immediate or elastic settlement is the main concern whereas consolidation settlement is the main concern for clayey soil since elastic settlement is negligible compared to the consolidation settlement in clay. Elastic settlement of different shapes and type of foundations can be obtained from Terzaghi (1943), Christian & Carrier (1978), Janbu et al (1956), Steinbrenner (1934). For the consolidation settlement, one dimensional consolidation theory proposed by Terzaghi (1943) is effectively used to estimate the probable magnitude and time rate of settlement of structure resting on soil. Related equations popularly used to estimate the consolidation settlement are given below

$$\Delta H = H m_v \Delta p \qquad (1)$$

$$\Delta H = \frac{C_c}{1+e_0} H \log_{10}(1 + \frac{\Delta p}{\sigma v}) \qquad (2)$$

where,

  ΔH = total consolidation settlement
  H = thickness of compressible layer
  m_v = coefficient of volume compressibility
  C_c = compression index
  e_0 = initial void ratio
  σ_v = effective overburden pressure at the middle of the compressible layer
  Δp = increase of pressure on the compressible layer due to imposed load on the soil.

Estimation of pressure increase on compressible layer, Δp due to imposed load is one of the important steps in calculating consolidation settlement by both the methods. Pressure increase at a particular point below the footing can be estimated effectively by Boussinesq equation. In calculating settlement using Equation 1 or 2, pressure increase over the compressible layer is necessary. Hence, several approximate methods are developed to estimate this. Different approximate methods which are currently used are summarised below:
(1) Average of the stresses calculated at the top and bottom of the compressible strata assuming certain slope (generally 2:1) of spreading stress over depth.
(2) Average of the stresses calculated at the top and bottom of the compressible strata using Boussinesq stress distribution.
(3) Stress intensity at the center of the compressible strata calculated either by Boussinesq distribution or assuming some slope of spreading stress over depth.
(4) Dividing the area of influence diagram under the compressible layer by the thickness of the compressible layer.

1.4 Scope of the paper

In methods (1) to (3), the thinner the layer the better is the accuracy. Hence, for a very thick compressible strata it is necessary to divide the whole layer into a number of thin layers and apply equation (1) or (2) separately to calculate settlement of the sublayer. Method (4) is proposed by Bowels (1988) where numerical integration is suggested to find out the area under the influence diagram. This method can be satisfactory if very small interval is chosen to calculate the area. But, for thick compressible layer this

procedure becomes tedious. To overcome these difficulties it is aimed at in this paper to develop an accurate method for the estimation of Δp by direct integration of influence diagram.

2 ANALYSIS

Vertical stress intensity, σ_v at any radial distance, r and at any depth, z due to a point load applied on the surface of the homogeneous elastic isotropic material is as given by (Timoshenko & Goodier 1951)

$$\sigma_v = \frac{3Pz^3}{2\pi(r^2+z^2)^{5/2}} \qquad (3)$$

But each foundation has some specific area and vertical stress intensity at any depth below the foundation can be obtained integrating the equation over the area. Vertical stress intensity along the axis of the circular loaded area of radius r_0 and load intensity, q can be written from Equation 3 as

$$(\sigma_v)_{r-0} = \int_0^{r_0} \frac{3qz^3 r\, dr}{(r^2+z^2)^{5/2}} \qquad (4)$$

Contact stress intensity, q below the foundation may not be constant throughout the area and it depends on type of soil below the foundation and rigidity of the foundation (Taylor 1948). Hence in the analysis different possible contact stress distributions are assumed.

2.1 Contact stress distribution

(1) Rigid base type contact stress distribution:

$$q_R = \frac{Q}{2\pi r_0 \sqrt{r_0^2 - r^2}} \qquad 0 \le r \le r_0 \qquad (5)$$

$$or \quad (q_R)_{r-0} = q_{RO} = \frac{Q}{2\pi r_0^2} = \frac{q_{av}}{2} \qquad (5a)$$

(2) Uniform contact stress distribution:

$$q_U = q_{UO} = \frac{Q}{\pi r_0^2} = q_{av} \qquad at\ any\ r \qquad (6)$$

184

(3) Parabolic contact stress distribution:

$$q_P = \frac{2Q(r_0^2 - r^2)}{\pi r_0^4} \qquad 0 \le r \le r_0 \qquad (7)$$

$$or \quad (q_P)_{r=0} = q_{PO} = \frac{2Q}{\pi r_0^2} = 2q_{av} \qquad (7a)$$

2.2 Influence factor for vertical stress

The above three different contact stress distributions are shown in Figure 1. Substituting the expressions for q from Equations 5-7 in Equation 4 and integrating over the limit, vertical stress intensity, σ_z at any depth, z and along the axis of the circular area can be written as,

$$\sigma_z = q_0 I_z \qquad (8)$$

where $q_0 = q_{RO}$, q_{UO} and q_{PO} respectively for rigid base, uniform and parabolic contact stress distributions

$$I_z = I_{zR} = \frac{3(z/r_0)^2 + 1}{[1 + (z/r_0)^2]^2} \qquad (8a)$$

for rigid base

$$I_z = I_{zU} = 1 - \frac{(z/r_0)^3}{[1 + (z/r_0)^2]^{3/2}} \qquad (8b)$$

for uniform

$$and \quad I_z = I_{zP} = 1 - 2(z/r_0)^2 + \frac{2(z/r_0)^3}{\sqrt{1 + (z/r_0)^2}} \qquad (8c)$$

for parabolic contact stress distribution.

2.3 Settlement

Figure 2 shows the typical variation of I_z with depth and a compressible layer within the influence zone. Consolidation settlement for an infinitesimal layer dz at depth z within the compressible layer as shown in

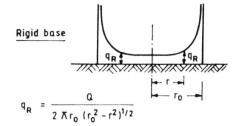

Rigid base

$$q_R = \frac{Q}{2\pi r_0 (r_0^2 - r^2)^{1/2}}$$

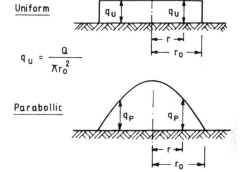

Uniform

$$q_u = \frac{Q}{\pi r_0^2}$$

Parabollic

$$q_P = \frac{2Q(r_0^2 - r^2)}{\pi r_0^4}$$

$q_{R,U,P}$ = Stress at a distance, r, from the centre

Q = Total load on the footing

Figure 1. Various contact stress distributions used in the analysis

Figure 2, can be written from Equation 1 as given below

$$d(\Delta H) = m_v q_0 I_z dz \qquad (9)$$

where I_z is the influence factor for vertical stress intensity at any depth, z as given in Equation 8a-8c.

Substituting the expressions for I_z from Equation 8a-8c for different contact stress distributions in Equation 9, total consolidation settlement of the compressible layer as shown in Figure 2, can be expressed as,

$$\Delta H = q_0 m_v r_0 F_z \qquad (10)$$

where

$$F_z = [2\tan^{-1}(h) - \frac{h}{1 + (h)^2}]_{h_1}^{h_2} \qquad (10a)$$

for rigid base

185

$$F_s = [h - \frac{2+h^2}{[1+h^2]^{1/2}}]_{h_1}^{h_2} \qquad (10b)$$

for uniform

$$F_s = [h - \frac{3}{2}h^3 + 2 [\frac{1+h^2}{3} - \sqrt{1+h^2}]]_{h_1}^{h_2} \qquad (10c)$$

for parabolic contact stress distribution.

where h = nondimensional depth = z/r_0 and $q_0 = q_{R0}, q_{U0}$ and q_{P0} respectively for rigid, uniform and parabolic contact stress distribution as explained earlier. Comparing Equation 10 with Equation 2, Δp for the layer can also be expressed as given below:

$$\Delta p = \frac{F_s r_0 q_0}{H} \qquad (11)$$

Using the expression for Δp obtained from Equation 11 in Equation 2, consolidation settlement of any compressible layer can also be obtained. The following illustrative example shows the methods of calculation and comparisons with different existing methods.

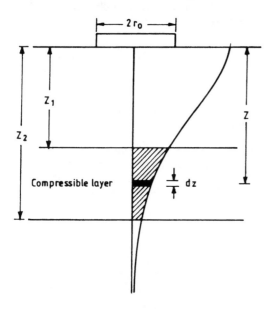

Figure 2. Typical influence diagram for vertical stress and a compressible layer within the influence zone.

3 ILLUSTRATIVE EXAMPLE

A footing of radius 1.5 m carrying a load of 600 kN resting on a 2.5 m thick sand layer as shown in Figure 3. Other related soil data of the compressible layer and the sand layer are given in the Figure. Consolidation settlement for the compressible layer as shown in the Figure are computed by different methods as follow:

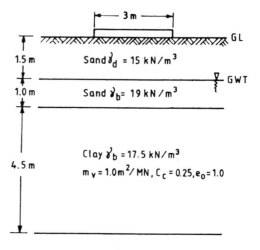

Figure 3. Schematic diagram of soil profile under the footing of example problem.

Present method: 1
$h_1 = z_1/r_0 = 1.67, h_2 = z_2/r_0 = 4.67$
from Equation 10 b

$$F_s = [z/r_0 - \frac{h^2 + 2}{\sqrt{1 + h^2}}]_{1.67}^{4.67} = 0.475$$

Using Equation 10

$$settlement = m_v q_0 F_s r_0 = \frac{1.0 \times 600 \times 0.475 \times 1.5}{\pi \times 1.5^2} = 60.5 mm$$

Method: 2
Effective overburden pressure σ_v at the middle of the compressible layer = 54.375 kN/m²
Using Equation 11

$$\Delta p = \frac{F_s r_0 q_0}{H} = 13.44 \ kN/m^2$$

settlement from equation (2)

$$\Delta H = \frac{C_c}{1+e_0} H \log_{10}(1+\frac{\Delta p}{\sigma v})$$

$$= \frac{0.25}{1+1} 4.5 \log_{10}(1+\frac{13.44}{48.375}) = 60.0 \, mm$$

Conventional method
Method 1
Obtaining Δp from Equation 8 & 8b at the middle of layer and using it in Equation 1, settlement of the clay layer is obtained as 50.66 mm. Similarly, obtaining Δp from the same equation at top and bottom of the layer and using average of it in Equation 1, settlement is obtained as 82.79 mm. Since the layer is 4.5 m thick, it is also divided into number of sub-layers (2, 4, 8) and settlements obtained by both the methods mentioned above are presented in Table 1 for comparison.
Method 2
Obtaining Δp in similar way and using it in Equation 2, settlements obtained by different methods are presented in Table 2 for comparison. Since uniform contact pressure distribution is usually used in most of the conventional methods, all computations are done based on uniform contact stress distribution.

Table 1 Comparison of settlements between existing methods and present method using Equation 1

No. of sublayer	Existing method		Present
	Using Δp from Equation 8 & 8b at the middle of the layer	Using ave of Δp at top and bottom of the layer obtained from Equation 8 & 8b	
1	50.66	82.79	
2	57.45	66.73	60.5
4	59.86	62.09	
8	60.48	60.98	

4 DISCUSSION ON RESULTS

Consolidation settlement of any compressible layer can be obtained accurately by a single step using Equations 10-10c for different contact stress distributions. Values of F_z for infinitely homogeneous compressible layer can be obtained from Equations 10a-10c using limits of h from o to ∞. The values of F_z for infinitely

Table 2 Comparison of settlements between existing methods and present method using Equation 2

No. of sublayer	Existing method		Present
	Using Δp from Equation 8 & 8b at the middle of the layer	Using ave of Δp at top and bottom of the layer obtained from Equation 8 & 8b	
1	51.1	78.97	
2	60.6	68.97	60.0
4	63.74	65.75	
8	64.5	64.96	

homogeneous compressible layer can be obtained from Equations 10a-10c using limits of h from o to ∞. The values of F_z for infinitely homogeneous compressible layer are 3.14, 2 and 1.33 for rigid base, uniform and parabolic contact stress distribution respectively and corresponding consolidation settlements are 1.57 $m_v\, q_{av}\, r_0$, 2 $m_v\, q_{av}\, r_0$, and 2.66 $m_v\, q_{av}\, r_0$ respectively. This brings out the order of variation due to type of contact stress distribution. A numerical example is presented to compare settlements by present method with the conventional methods. ΔH obtained in single step by present method and by conventional methods assuming single layer and also dividing the whole layer into number of sublayers (2, 4, 8) are presented in Tables 1-2. Settlements obtained based on coefficient of volume compressibility (Equation 1) are presented in Table 1 and settlements obtained based on compression index (Equation 2) are presented in Table 2. Settlements are calculated by conventional method obtaining pressure increase, Δp over the compressible layer by two important methods among many existing methods. It can be seen from Tables 1-2 that there are significant differences between the settlements obtained by conventional method assuming a single layer and present method by a single step. It can be further observed that results obtained by conventional methods using large number of sublayers (i.e., 8) approach the results obtained by present method in a single step. Results are presented here for circular footing. For other footing shapes, like square and rectangle, results can be obtained without losing much of accuracy using equivalent circle method, where radius of equivalent circle $r_0 = $ Sqrt (A/π), and A = contact area of the footing.

5 CONCLUSIONS

Increase of pressure, Δp in the compressible layer due to footing load at the surface is very important in calculating consolidation settlement by any method. A simplified method in the form of equation to estimate Δp is presented in this paper. Results are presented for different contact stress distributions which can be used to meet various engineering judgements depending on the soil type and rigidity of the footing. The method of analysis is also illustrated with a numerical example. It is observed that present method estimates the consolidation settlement accurately in a single step with very little calculation. Analysis is presented for circular foundation but results can be obtained for other noncircular footings using equivalent circular method.

6 REFERENCES

Boussinesq, J (1885), " Application des potentials a' l'etude de l'equilibre et de mouvement des solids elastiques. Gauthier - villars, Paris.

Bowels, J.E. (1988), "Foundation analysis and design", McGraw-Hill Book Co., New York.

Christian J.T. & Carrier, W.D. (1978), "Janbu, Bjerrum and kjaernsli's chart Reinterpreted, Canadian Geotechnical Journal, vol, 15

Janbu et al (1956) veiledning veldlqsning afundamentering soppgarer, Pub no. 6, Norawegian Geotechnical Institute.

Steinbrenner W.(1934) Taflen zur setzungsberechnung, Die strasse, vol 1, 121-124

Taylor, D. W (1948) fundamentals of Soil Mechanics, John Wiley and sons, New York.

Terzaghi K. (1943), "Theoretical soil mechanics" John wiley, New York.

Timoshenko, S.P. & Goodier, J.N. (1970) "Theory of Elasticity" McGraw Hill International Edition

Geotechnical Hazards, Marić, Lisac & Szavits-Nossan (eds) © 1998 Taylor & Francis, ISBN 90 5410 957 2

Zur Mobilisierung des passiven Erddrucks in trockenem Sand

U. Bartl & D. Franke
Institut für Geotechnik, TU Dresden, Germany

ABSTRACT: Kenntnisse zur Mobilisierung des passiven Erddrucks stellen eine wesentliche Grundlage zur Verbesserung bzw. Interpretation von Verformungsberechnungen bei Stützkonstruktionen dar. Ein entsprechendes Mittel zur Gewinnung entsprechender Informationen sind Modellversuche. Im nachfolgenden werden Ergebnisse eigener Untersuchungen mittels Zentrifugen- und 1g- Modellversuchen zur genannten Problematik für den Fall der sogenannten Kopfpunktdrehung auszugsweise vorgestellt.

1 EINLEITUNG

Mit den klassischen, analytischen Methoden der Erddrucktheorie kann nur der Grenzwert des Erdwiderstands, der passive Erddruck berechnet werden. Aussagen zum Mobilisierungsverhalten des Erdwiderstandes bis hin zum passiven Erddruck sind damit nicht möglich. Die Kenntnis hierzu ist jedoch für Verformungsberechnungen bzw. zur Berechnung des Erddrucks für bestimmte Verformungsbedingungen notwendig. Bei der Berechnung von Stützkonstruktionen wird bis heute meist ein verminderter passiver Erddruck als stützender Erddruck angesetzt. Diese Vorgehensweise ist unbefriedigend, da die tatsächlichen Wandbewegungen und Erdwiderstände unbestimmt bleiben. Erddruckansätze nach diesen Verfahren können damit entweder zu unwirtschaftlichen oder schlimmer, zu gebrauchsuntauglichen Stützkonstruktionen führen. Auch beim Einsatz von numerischen Berechnungsverfahren stellen Messungen eine wesentliche Grundlage zur Kalibrierung bzw. Interpretation der Berechnungsergebnisse dar.

In der aktuellen deutschen "Erddruck-Norm" (DIN V 4085-100 1996) ist z.B. das Bild 1 aufgeführt. Darin sind z.B. zwei Mobilisierungsfunktionen für die Erdwiderstandskraft qualitativ dargestellt. Im Falle eines nichtbindigen Bodens wird hier zwischen der Mobilisierung bei einer mitteldichten bis dichten Lagerung (Kurve 1 im Bild 1) und einer lockeren Lagerung (Kurve 2 im Bild 1) unterschieden. Zum

quantitativen Mobilisierungsverhalten beschränken sich jedoch die Angaben auf Anhaltswerte zur erforderlichen Wandbewegung für die Mobilisierung von 50% des passiven Erddrucks.

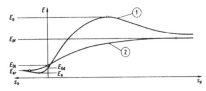

Bild 1. Prinzipdarstellung der Erddruck - Bewegungsabhängigkeit nach DIN V 4085-100 (1996)

In der Literatur sind weitere Angaben zu finden (WEISSENBACH 1975, EAU 1996, ÖNORM B 4434 1993 u.a.). Hierbei handelt es sich in der Regel um empirische Durchschnittswerte, die aus Versuchen verschiedener Autoren gewonnen wurden. Bei diesen "Erddruckmessungen" lagen schon aus Gründen der verschiedenen Versuchsanordnungen die unterschiedlichsten Randbedingungen vor, so daß eine direkte Vergleichbarkeit der Ergebnisse nur bedingt gegeben ist. Zu erkennen ist daraus aber, daß die Mobilisierung des Erdwiderstandes als eine Funktion in Abhängigkeit von verschiedenen Einflußgrößen angegeben werden muß. Um dies zu untersuchen, sind Versuchsreihen mit gezielter Parametervariation notwendig. Einzelversuche an wirkli-

189

chen Bauwerken sind hier wenig geeignet, da sie allenfalls Messungen im Sinne von Stichproben darstellen. Für systematische Meßreihen kommen vor allem Modellversuche in Frage.

Am Institut für Geotechnik der TU Dresden wurde deshalb in der ersten Untersuchungsstufe, mittels 1g-Modellversuchen, der Einfluß einzelner Größen, wie Wandbewegungsart, Lagerungsdichte des Bodens, geometrische Randbedingungen und Beschaffenheit der Wandoberfläche das Mobilisierungsverhalten des Erdwiderstandes auf eine starre, ebene Wand in trockenem Sand untersucht.

die Wandhöhe (mit den Anteilen e_{pn} und e_{pt}) erfaßt werden. Über Ergebnisse wurde bereits auszugsweise von Bartl (1995) berichtet.

Grundlage für die spätere Nutzung der Versuchsergebnisse für die Berechnung von Stützwänden im natürlichen Maßstab sind Kenntnisse über den Einfluß von Maßstabseffekten, speziell des Spannungsniveaus. Zur Untersuchung dieses Einflusses konnten mit der freundlichen Unterstützung von Herrn Prof. Pregl Zentrifugenversuche an der Universität für Bodenkultur Wien durchgeführt werden. Das hierbei eingesetzte Modell (Bild 2 und Tabelle 1)

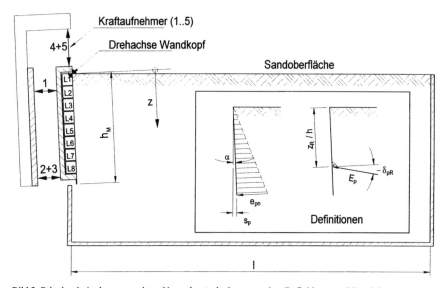

Bild 2. Prinzipschnitt der verwendeten Versuchsstände & verwendete Definitionen und Bezeichnungen

2 VERSUCHSBEDINGUNGEN

Für die Versuche, die auch von der Deutschen Forschungsgemeinschaft (DFG) im Rahmen einer Sachbeihilfe unterstützt wurden, stand der an unserem Institut dafür errichtete Versuchsstand zur Verfügung (Bild 2 und Tabelle 1). In diesen Versuchsstand können gegenwärtig Modelle mit Wandhöhen bis zu 60 cm eingebaut werden. Möglich ist dabei sowohl die Ausführung von Wandbewegungen mit Parallelverschiebung, mit Drehung um eine festgelegte Achse, als auch mit kombinierten (überlagerten) Bewegungsabläufen. Bei den Versuchen kann neben dem auf die Wand wirkenden resultierenden Erddruck (Erddruckkraft mit Lage und Angriffsrichtung) auch die Erdwiderstandsverteilung über

erlaubte die Ausführung von Versuchen mit einer Wandhöhe bis 12cm und bis zu einer Beschleunigung von 35g. Somit war es hierbei möglich eine charakteristische Spannung im Modell zu erzeugen, die einer charakteristischen Spannung an einem bis zu 420cm ($35 \cdot 12$cm) hohem Prototypen entspricht.

Tabelle 1. Auf die jeweilige Wandhöhe h_M bezogene Abmessungen d. Versuchseinrichtungen

	b_{VW} / h_M [1]	b_{MW} / h_M [1]	l / h_M [1]
1g-Versuche, $h_{M,\,1g} = 564$ mm	1.77	0.40	3.19
Z-Versuche, $h_{M,\,Z} = 120$ mm	2.08	1.25	3.83

Als Versuchsmaterial kam jeweils der selbe trockene, eng gestufte, fein- bis mittelkörnige Quarzsand (Dresdner Hellersand mit: $d_{50} = 0,4$ mm; $U \approx 3$; $\rho_S = 2,65$ g/cm³; min $\rho_d = 1,501$ g/cm³; max $\rho_d = 1,813$ g/cm³) zum Einsatz.

Die bei den Versuchen gemessene Erddruckkraft E'_p wird in normierter Form dargestellt. So wird der dimensionsfreie Betrag des Normalanteils der mobilisierten passiven Erddruckkkraft mit

$$K'_{pnR} = \frac{2E'_{pn}}{n_g \cdot \gamma \cdot h^2} \qquad (1)$$

ermittelt.

n_g...   Maßstabsfaktor der Erdbeschleunigung (Maß für das Spannungsniveau)
γ...   Wichte des Bodens unter Erdbeschleunigung
h...   Wandhöhe des Modells ($h = h_M$)

Bei allen Versuchen wurde weiterhin der mobilisierte Erddruckneigungswinkel (δ'_{pR}) und die Angriffshöhe der Erddruckkraft (z_R) gemessen.

$$\tan \delta'_{pR} = \frac{E'_{pt}}{E'_{pn}} \qquad (2)$$

3 ERGEBNISSE ZENTRIFUGENVERSUCHE

Beispielhaft werden hier zwei Versuchsreihen mit jeweils dichter Lagerung des Modellsandes und zwei unterschiedlichen Wandoberflächen vorgestellt. Bei diesen Oberflächen handelt es sich zum einen um poliertes Hartaluminium (Alu) und zum anderen um eine mit Schleifpapier der Körnung 220 beklebte Versuchswand (S220).

Tabelle 2.   Ausgangslagerungsdichte (Mittelwerte) je Versuchsreihe

	Anzahl d. Vers.	D [1]	e [1]	$I_D = D_r$ [1]
Alu, dicht	10	0.83	0.506	0.85
S220, dicht	8	0.80	0.513	0.83

Im Bild 3 sind die Ergebnisse zur Mobilisierung des Normalanteils der passiven Erddruckkraft K'_{pnR} über die Wandbewegung s_p/h für die Versuchsreihe mit Aluminiumoberfläche der Wand und dichter Lagerung des trockenen Sandes (Alu, dicht) dargestellt. Über die entsprechenden Versuchsergebnisse der Reihe mit rauherer Wandoberfläche (S220, dicht) wurden bereits von Bartl & Franke (1997) berichtet.

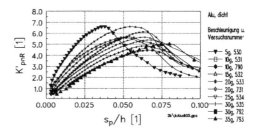

Bild 3. Normalanteil der Erdwiderstandskraft über s_p/h

3.1 Passiver Grenzzustand

Als normierte passive Erddruckkraft (Grenzzustand) wird hier das jeweils auftretende Maximum der Erdwiderstandskraft (Peak- Wert) nach Gleichung (3) definiert.

$$K_{pnR} = \max K'_{pnR} \qquad (3)$$

Im Bild 4 sind die Versuchsergebnisse der normierten passiven Erddruckkraft - Normalkomponente K_{pnR} - über den Maßstab der Beschleunigung n_g dargestellt. Die aus diesen Ergebnissen ermittelten Regressionskurven auf der Grundlage einer Potenzfunktion nach Gleichung (4) zeigen eine Abnahme des Betrages der passiven Erddruckkraft mit anwachsendem Spannungsniveau.

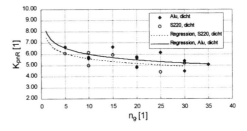

Bild 4. Normalanteil der passiven Erddruckkraft über n_g

$$K_{pnR} = K_{pnR0} \cdot n_g^{\beta_K} \qquad (4)$$

Die aus der Regression ermittelten Koeffizienten für die Gleichung (4) sind in der Tabelle 3 angegeben.

Tabelle 3. Parameter der Regressionsfunktionen in Bild 4

	K_{pnR0} [1]	β_K [1]
Alu, dicht	8.08	-0.13
S220, dicht	7.42	-0.12

Die normierte Grenzverschiebung s_{pG}/h ist im Bild 5 über das Spannungsniveau n_g aufgetragen. Wie aus der Abbildung hervorgeht, ist mit zunehmendem Spannungsniveau eine deutliche Zunahme der passiven Grenzverschiebung zu verzeichnen.

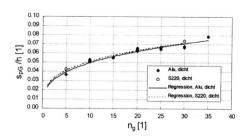

Bild 5. Normierte passive Grenzverschiebung über n_g

Der Zusammenhang zwischen dem Spannungsniveau und der Grenzverschiebung läßt sich für die Ergebnisse der beiden Versuchsreihen gut mit einer Potenzfunktion nach Gleichung (5) erfassen.

$$s_{pG}/h = \left(s_{pG}/h\right)_0 \cdot n_g^{\beta_s} \qquad (5)$$

Die sich aus der entsprechenden Regression ergebenden Parameter sind in der Tabelle 4 zusammengestellt.

Tabelle 4. Parameter der Regressionsfunktionen in Bild 5

	$(s_{pG}/h)_0$ [1]	β_s [1]
Alu, dicht	0.023	0.31
S220, dicht	0.025	0.33

3.2 Mobilisierungsverhalten

Zur Beschreibung des Last-Verformungsverhaltens (Mobilisierungsverhalten) kommt in der bodenmechanischen Literatur häufig eine Potenzfunktion zur Anwendung. So wird z.B. zur Beschreibung der Mobilisierung des Betrages der passiven Erddruck-

kraft in Abhängigkeit vom Wandbewegungsbetrag von Ziegler (1987) ein Ansatz nach Gleichung (6) benutzt, wobei er für den hier betrachteten Fall einer "passiven Kopfpunktdrehung" die Koeffizienten zu $a = 0.93$ und $b = 212$ vorschlägt. Diese Angaben basieren auf der Grundlage von Triaxialversuchen und FEM- Berechnungen.

$$\frac{2 \cdot \Delta E}{\gamma \cdot h^2} = b \cdot \left(\frac{s}{h}\right)^a \qquad (6)$$

Mit:

$$\Delta E = E'_p - E_0 \qquad (7)$$

Im Bild 6 sind die vorliegenden Versuchsergebnisse der Reihe Alu, dicht - Bild 3 - in Anlehnung an die obige Formulierung als Differenzwert $\Delta K'_{pnR}$ nach Gleichung (8) in doppelt-logarithmischer Form über den Betrag der Wandbewegung dargestellt. Eine Potenzfunktion stellt sich darin als Gerade dar.

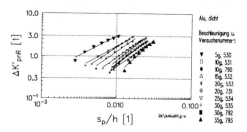

Bild 6. Mobilisierung der Erdwiderstandskraft über s_p/h

$$\Delta K'_{pnR} = \frac{2 \cdot \Delta E'_{pn}}{\gamma \cdot h^2} = \left(K'_{pnR} - K_0\right) \qquad (8)$$

Zu jedem Versuch sind außerdem die sich aus der Regression nach Gleichung (9) ergebenden Mobilisierungsfunktionen eingetragen.

$$\Delta K'_{pnR} = b \cdot \left(\frac{s_p}{h}\right)^a \qquad (9)$$

Die Regression wurde hierbei jeweils für einen eingeschränkten Wertebereich ("Gebrauchszustand") von

$$\frac{\left(K_{pnR} - K_0\right)}{10} \leq \Delta K'_{pnR} \leq \frac{\left(K_{pnR} - K_0\right)}{2}$$

durchgeführt.

Im Bild 7 sind die entsprechenden Koeffizienten aus den Regressionen nach Gleichung (9) für beide Versuchsreihen in Abhängigkeit vom Spannungsniveau n_g aufgeführt.

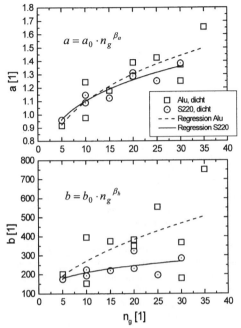

Bild 7. Koeffizienten nach Gl. (9) für Funktionen in Bild 5

Zur Darstellung des Zusammenhangs zwischen dem Spannungsniveau und den Koeffizienten den Mobilisierungsfunktion nach Gleichung (9) werden im Bild 7 ebenfalls Potenzfunktionen verwendet. Die sich danach ergebenden Parameter für diese Ansätze sind in der Tabelle 5 zusammengestellt.

Tabelle 5. Parameter der Regressionsfunktionen in Bild 7

	a_0 [1]	β_a [1]	b_0 [1]	β_b [1]
Alu, dicht	0.62	0.25	87	0.50
S220, dicht	0.71	0.19	123	0.23

Im Bild 8 sind die Versuchsergebnisse (jeder 10. Meßwert) zur Mobilisierung von K'_{pnR} der Reihe Alu, dicht in einer auf die entsprechenden Grenzwerte bezogenen Form nach den Gleichungen (10) und (11) aufgeführt.

$$mob\ \Delta K'_{pnR} = \frac{K'_{pnR} - K_0}{K_{pnR} - K_0} \qquad (10)$$

$$mob\ \Delta s_p = \frac{s_p - s_A}{s_{pG} - s_A} \qquad (11)$$

s_A... "Vorbewegung" im Versuch vor dem Start des Wandantriebs (Bartl & Franke 1997), i.d.R.: $s_A = 0$

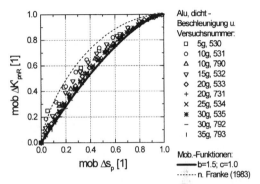

Bild 8. Bezogene Mobilisierung des Betrages der Erdwiderstandskraft (Normalanteil)

Ein signifikanter Einfluß des Spannungsniveaus auf die bezogene Mobilisierung des Betrages der Erdwiderstandskraft ist hierbei nicht zu erkennen. Das entstandene "Werteband" wird vor allem auf eine "normale" Versuchsstreuung - z.B. Streuung der Ausgangslagerungsdichten u.a. - zurückgeführt.

In das Bild 8 sind auch zwei Mobilisierungsfunktionen nach Gleichung (12) eingetragen. Zum einen der Vorschlag von Franke (1983) mit den Exponenten $b = 2.0$ und $c = 0.7$ und zum anderen eine Mobilisierungsfunktion mit $b = 1.5$ und $c = 1.0$. Dies soll verdeutlichen, daß auch ein solcher Ansatz nach entsprechender Koeffizientenanpassung durchaus zur Beschreibung des Mobilisierungsverhaltens dienen könnte.

$$mob\ \Delta K'_{pnR} = \left(1 - \left(1 - mob\ \Delta s_p\right)^b\right)^c \qquad (12)$$

4 ERGEBNISSE 1g-VERSUCHE

Versuchsergebnisse einer Meßreihe mit Aluminiumwandoberfläche (Alu - Tabelle 6) und drei verschiedenen Lagerungsdichten sind im Bild 9 dargestellt. Hierbei handelt es sich um die Parameter der mobilisierten Erddruckkraft über den Betrag der Wandbewegung in jeweils dimensionsfreier Darstellungsform.

Tabelle 6. Versuche der Reihe Alu

Versuch	$D_r = I_D$ [1]	e [1]	D [1]	ρ [g/cm³]
EPK033	0.30	0.673	0.26	1.584
EPK035	0.51	0.611	0.46	1.645
EPK032	0.68	0.558	0.64	1.701

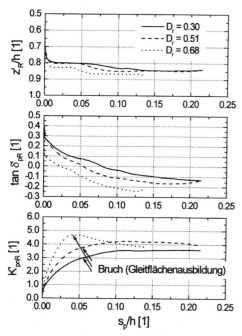

Bild 9. Mobilisierung der passiven Erddruckkraft

Bei dem Versuch mit einer Ausgangslagerungsdichte von $D_r = 0.68$ kommt es während der Mobilisierung der Erdwiderstandskraft zur Ausbildung eines deutlichen Peak- Wertes ($K_{pnR} = \max K'_{pnR}$ im Bild 9). Erst nach dem Überschreiten dieses Maximums wurde in den eingebauten Sandmarkierungsschichten und an der Sandoberfläche die Ausbildung einer Gleit- bzw. Bruchfläche sichtbar.

Bei den beiden Versuchen mit geringerer Lagerungsdichte ($D_r = 0.30$ und 0.51) bildet sich jeweils die Gleitfläche bereits vor dem Erreichen eines Kraft- Maximums durchgehend aus. Aus diesem Grund wird im nachfolgenden als passiver Grenzzustand der Zustand bei vollständiger Ausbildung des Bruches (Gleitflächenausbildung) festgelegt.

4.1 Passiver Grenzzustand

In der Tabelle 7 sind die normierten Parameter zur Resultierenden im passiven Grenzzustand (passive Erddruckkraft) zusammengestellt.

Tabelle 7. Parameter der Erddruckkraft im Grenzzustand

Versuch D_r [1]	s_{pG}/h [1]	K_{pnR} [1]	z_R/h [1]	$\tan \delta_{pR}$ [1]
0.30	0.057	3.02	0.80	0.06
0.51	0.056	3.77	0.81	-0.01
0.68	0.041	4.68	0.83	-0.06

Die sich im Grenzzustand einstellende Angriffshöhe der Erddruckkraft von $z_R = 0.80...0.83h$ weist bereits auf eine deutliche Abweichung der sich einstellenden Erddruckverteilung von der "klassischen" Erddruckverteilung mit $z_R = 0.67h$ hin. Es zeigt sich gleichzeitig, daß sich die Angriffshöhe der Resultierenden mit zunehmender Lagerungsdichte weiter zum Fuß der Wand verlagert.

Die gemessene Erddruckverteilung über die Wandhöhe (Bild 10) bestätigt dies. Sie zeigt eine überproportionale Zunahme des passiven Erddrucks mit der Tiefe z. Zur Erfassung dieser Erddruckverteilungen erscheint eine Potenzfunktion nach Gleichung (13) brauchbar.

$$e_{pn} = e_{pnF} \cdot (z/h)^n \tag{13}$$

e_{pnF}.. passiver Erddruck (Normalkomponente) am Fuß der Wand

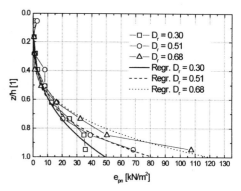

Bild 10. Erddruckverteilung im Grenzzustand

Für die im Bild 10 neben den Meßwerten eingetragenen Verteilungsfunktionen wurden dabei ganzzahlige Exponenten n gewählt (Tabelle 8), die sich an der Angriffshöhe der Resultierenden z_R/h orientieren.

Tabelle 8. Gewählte Exponenten für Gl. (13)

Versuch: D_r [1]	n [1]
0.30	3
0.51	4
0.68	4

Von Fang et al. (1994) wurden ähnliche Ergebnisse zur Erddruckverteilung festgestellt. Allerdings ist dort nur ein Versuch mit Kopfpunktdrehung und lockerer Lagerung dokumentiert.

Die durch die Glasscheiben im Sand registrierten Bruchbilder (Bild 11) zeigen für alle drei Versuche die Ausbildung einer gekrümmten Gleitfläche. Ein signifikanter Einfluß der Lagerungsdichte auf die Gleitflächenform kann jedoch dabei nicht festgestellt werden.

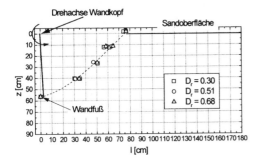

Bild 11. Bruchbilder an den Seitenwänden des Kastens

4.2 Mobilisierungsverhalten

Im Bild 12 sind die Versuchsergebnisse zur Mobilisierung des Normalanteils der Erdwiderstandskraft in der auf die entsprechenden Grenzwerte bezogenen Form nach den Gleichungen (10) und (11) aufgeführt. Ebenfalls eingetragen sind die zwei Mobilisierungsansätze nach Gleichung (12), die bereits im Bild 8 verwendet wurden.

An Hand der Darstellung ist kein besonderer Einfluß der Lagerungsdichte auf die bezogene Mobilisierung des Normalanteils der Erdwiderstandskraft festzustellen.

5 ZUSAMMENFASSUNG

5.1 Zentrifugenversuche

Ein Einfluß der zwei unterschiedlichen Wandoberflächen auf die Größe, die Richtung und den Angriffspunkt der passiven Erddruckkraft ist nicht festzustellen.

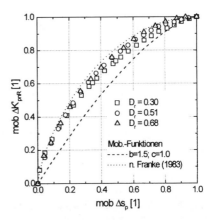

Bild 12. Auf Grenzwerte bezogene Mobilisierung des Betrages der Erdwiderstandskraft

Eine deutliche Abhängigkeit vom Maßstab der Beschleunigung, und somit dem Spannungsniveau, zeigen
- der Betrag des Normalanteils der passiven Erddruckkraft (E_{pn} bzw. K_{pnR}) - *Er nimmt mit zunehmender Beschleunigung (n_g) ab.* - und
- der Betrag der Grenzverschiebung (s_G) - *Er nimmt mit zunehmender Beschleunigung (n_g) zu.*

Eine geringe bis keine Abhängigkeit vom Spannungsniveau zeigen (Bartl & Franke 1997)
- die Angriffsrichtung der passiven Erddruckkraft (tan δ_{pR}),
- die Angriffshöhe der passiven Erddruckkraft (z_R) und
- die Bruchkörperlänge (l_B).

Ein Einfluß des Spannungsniveaus auf die bezogene ("prozentuale") Mobilisierung der Erddruckkraft ist nicht erkennbar (Bild 8).

5.2 1g-Versuche

Bei den Versuchen mit lockerer bis mitteldichter Lagerung bildet sich bereits vor dem Erreichen einer maximalen Erdwiderstandskraft im Sand eine durchgehende Gleitfläche aus. Der dazu erforderliche Wandbewegungsbetrag ist bei diesen zwei Versuchen nahezu gleich groß.

Die registrierte Gleitflächenform und -geometrie weist keine besondere Abhängigkeit von der Lagerungsdichte auf.

Die gemessene Erddruckverteilung im Grenzzustand (Normalanteil e_{pn}) läßt sich bei allen Versuchen mittels einer Potenzfunktion gut annähern. Mit zunehmender Lagerungsdichte kommt es zu einer leicht zunehmenden Verlagerung des Erddruckes zum Wandfuß hin (Exponent n bzw. z_R wird größer).

Ein besonderer Einfluß der untersuchten Lagerungsdichten auf die bezogene ("prozentuale") Mobilisierung der Erddruckkraft (Bild 12) ist nicht feststellbar.

6 LITERATUR

Bartl, U. (1995): Untersuchungen zum Erdwiderstand auf ebene Wände am Beispiel von 1g-Modellversuchen mit Kopfpunktdrehung. In: Festschrift zum 60. Geburtstag von Prof. Dr.-Ing. habil. Dietrich Franke. Mitteilungen Institut für Geotechnik, TU Dresden, Heft 3, S. 201-216

Bartl, U., Franke, D. (1997): Ergebnisse von Modellversuchen zur Untersuchung der Abhängigkeit des stützenden Erddrucks von Wandbewegungen - am Beispiel von Zentrifugenversuchen mit Kopfpunktdrehung. In: Franke, D. (Hrsg.): Ohde-Kolloquium 1997. Mitteilungen Institut für Geotechnik, TU Dresden, Heft 4, S. 1-27

DIN V 4085-100 (1996): Baugrund; Berechnung des Erddrucks - Teil 100: Berechnung nach dem Konzept mit Teilsicherheitsbeiwerten.

EAU 1996 (1997): Empfehlungen des Arbeitsausschusses "Ufereinfassungen" Häfen und Wasserstraßen: EAU 1996. Hrsg: Arbeitsausschuß "Ufereinfassungen" der Hafenbautechnischen Ges. e.V. und der Dt. Ges. für Geotechnik e.V., 9. Aufl., Berlin, Ernst & Sohn

Fang, Y.-S.; Chen, T.-J.; Wu, B.-F. (1994): Passive earth pressures with various wall movements. Journal of Geotechnical Engineering, 120. Jg., H. 8, S. 1307-1323.

Franke, D. (1983): Beiträge zur praktischen Erddruckberechnung. Habilitationsschrift, TU Dresden

Ziegler, M. (1987): Berechnung des verschiebungsabhängigen Erddrucks in Sand. Veröffentlichungen des Inst. f. Bodenmechanik und Felsmechanik der Universität Karlsruhe, H. 101

ÖNORM B 4434 (1993): Erd- und Grundbau - Erddruckberechnung.

Weißenbach; A. (1975): Baugruben - Teil 2. Ernst & Sohn, Berlin, München u. Düsseldorf

Geotechnical Hazards, Marić, Lisac & Szavits-Nossan (eds) © 1998 Taylor & Francis, ISBN 90 5410 957 2

Analysis of causes of catastrophic failures of reinforced earth embankments

O.Y. Eschenko & K.Sh. Shadunts
Department of Civil Engineering, Kuban State Agrarian University, Russia

ABSTRACT: The reasons of the catastrophic failures of the embankments and the bulkheads made of the reinforced soil with the rigid face panels and the reinforcement in the form of bands or strips are analysed in the article. The results of numerical calculations in the reinforcement under different conditions of failure are given. The calculation model is built on the basis of "domino" principle, i.e. a gradual failure of the construction as a whole is considered in it when the gradual failure of the separate reinforcing bands takes place. The main reasons for the catastrophic failures are found on the grounds of the analytical investigations and the recommendations are given to prevent such accidents.

1 INTRODUCTION

The use of the principles of the bionics is one of the possible directions of improvement of the structures made of reinforced soil. For example, the use of the principle of the predetermined form of the stability loss allows to increase safety and "survivability" of the construction, and it is very important for such objects as the piers or dams. This principle demands that when the external loads exceeding the calculated ones take place, the construction will be destructed in the only, predetermined way. Thus, it is necessary to analyse possible causes of failure of reinforced embankments in order to use them.

A large number of investigations is devoted to this question (Eschenko 1991), and the modern principles of computation of the constructions made of the reinforced soil are based on the study and the exposure of possible strains which take place when the subgrade of such constructions is destructed. From the point of view of the soil mechanics the variety of forms of stability loss can be divided into two types:

1. The soil suffers considerable plastic strains which damp relatively slowly in the course of time. As a rule, no rupture of the reinforcing elements takes place. This type of failure is characterized by the fact that smaller main stress along the shift surface after failure has the same value as before it:

$$\sigma_3^P = K\sigma_3 \qquad (1)$$

where K is proportionality factor, $K<1$,

and the larger one remains almost without changes:

$$\sigma_1^P \approx \sigma_1 \qquad (2)$$

2. The soil strains are of continuous character, they take place during relatively small period of time (as a rule, from several seconds to several minutes) and are accompanied by rupture of continuity of the medium. Both main stresses along the shift surface tend to zero:

$$\sigma_1^P, \sigma_3^P \to 0 \qquad (3)$$

Such type of failure is called the catastrophic one.

Both failure types can take place due to the same reasons (Sondermann 1983). But it remains unclear why in some cases, for example, the mechanical damages of the reinforcement lead to the catastrophic failures and in other cases not.

2 ANALYSIS OF CAUSES

In order to solve a task being set we have carried out the analysis of the experimental data on locality and the model ones published during the last twenty years in two directions:

1. failure type dependence on the kind of faceplate;

2. failure type dependence on the kind of reinforcement.

The analysis results are given in the generalized form in Table 1.

These data show that in all cases there is a connection between the reinforcement strength, the type of facing and the character of failure. The construc-

do not lead to the stabilization of the strains. Thus, the cause of a catastrophic failure should be searched not only in the type of reinforcement, but also in the conditions of its interaction with faceplate.

Table 1. Character of failures of different constructions of reinforced soil embankments

Type of rein-forcement	Continuity of covering	Reinforcement rupture strains	Character of em-bankment failure	Type of faceplate	Mechanism of failure
Steel bands Plastic bands Flexible rods Nets	Interrupted	Less then maximal soil tension strains	Catastrophic fail-ure with rein-forcement rup-ture	Rigid (reinforced concrete plates, steel, brickwork, etc.)	As a rule, it begins from the lowest, maximal loaded lay-ers of reinforcement and develops up-wards
Nets	Solid				
Geotextiles Geomembranes		More then maximal soil tension strains	Sufficient strains, usually without reinforcement rupture	Flexible (reverse bend of rein-forcement, steel, aluminium)	Is deformed as a me-dium without conti-nuity rupture

tions with rigid facing and the reinforcement in the form of the separate bands made of the high - tensile material have the greatest tendency to the cata-strophic failures. As a rule, such reinforcement has a small value of the relative strain by rupture ε'.

If it is compared with maximal values of tension strain ε_3^s in typical soils of the reinforced embank-ments, one can see that, as a rule, the following con-dition is fulfilled under the catastrophic failure:

$$\varepsilon' \leq \max(\varepsilon_3^s) \qquad (4)$$

Vice versa, a destruction according to the first type is characteristic for the materials which can suf-fer a large extension without rupture under load (for example, polypropylene).

But further analysis has shown that high strength of reinforcement is not the only and even not the main cause of the catastrophic failures. It has been noted in many papers that any failure of the rein-forced embankment is accompanied by significant displacements of the face surface of the construc-tion. It is known that such displacements (for any bulkheads) are accompanied by a sharp drop of pres-sure of the soil on the limiting surface. That's why if the failure is connected with the increase of lateral pressure of the soil (for example, due to water satu-ration), the failure process should be stabilized (at least for the short period of time), and it takes place only for the reinforcement which is strained consid-erably. As a rule, in case of the embankments rein-forced by separate strips of high - tensile rein-forcement, the horizontal displacements of faceplate

2.1 Calculation methods

In order to find the faceplate rigidity influence and the character of its connection with reinforcement on the stress-strain state of the construction, a series of calculations has been carried out in accordance with "ARMDAM" program. The construction had a height of 16.06m and was reinforced with the nets of TENSAR SR-2 type. Coarse sand with an angle of internal friction of 31°, unit adhesion of 1 kPa and unit weight of 18.2 kN/m^3 was considered as a soil. The soil was modelled by an ideal flexible plastic medium with Poisson's ratio 0.3. In two series of calculations only the extreme cases with faceplate were considered: it was assumed either ideally flexi-ble (reverse bend of reinforcement) or very rigid (reinforced concrete). Two variants of fixing of rein-forcement were envisaged for the embankments with reinforced concrete faceplate:

1. stationary fixing to faceplate;
2. reinforcement had a possibility to be displaced independently from faceplate in the vertical direc-tion and together with it in the horizontal direction.

The displacements and the stresses were regis-tered in all the elements of the soil and reinforce-ment during the computation. A simplified strained diagram of the construction in the special scale of the displacements, a faceplate settlements diagram (point "A") and a graph of the factors of safety (FS) averaged according to the horizontal layers in the soil elements were displayed for the convenience of the analysis of the results being obtained. The typi-cal examples of the strained state of the construc-tions are shown in Figure 1.

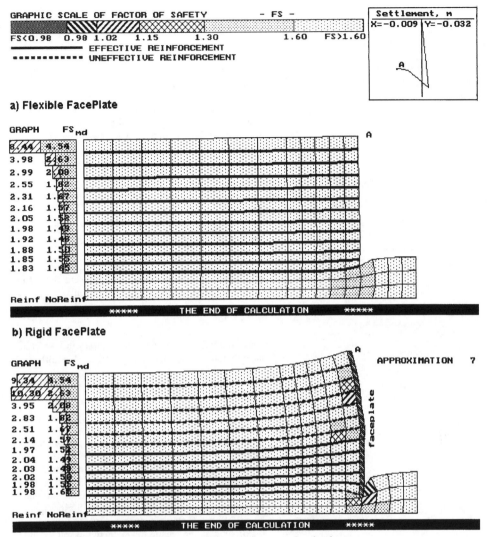

Figure 1. Faceplate rigidity influence on stress - strain state of embankment

2.2 Calculation results

It is seen from the figures that at least two negative factors appear additionally if there is no possibility of free displacement of reinforcement along the faceplate surface in the vertical direction (it takes place in the constructions being considered).

Firstly, due to the difference of module of deformation of the soil and faceplate during the erection of the construction and consolidation of the soil it looks like as reinforcement hangs on faceplate with its one end. It leads to a rise of the additional tensile forces in reinforcement (as compared with the case when the reinforcement has a possibility to be dis-

placed freely along faceplate in the vertical direction, the increase has reached 22%).

Secondly, rigid faceplate forms a local zone of large plastic stresses on the contact with the foundation (Fig. 1b) which makes it necessary to reinforce the foundation in this place or to construct a footing. If the reinforcement has an opportunity to be displaced independently and vertically, rigid faceplate promotes, on the contrary, the increase of stability of the construction due to the creation of "the compression" effect.

At last, thirdly, when the reinforcement is fixed immovable, a considerable redistribution of stresses in the reinforcement and the factors of safety in the soil takes place as compared with other cases being considered. When the figures are compared, it can be seen that in case of flexible faceplate all the reinforcing interlayers participate effectively in the perception of the load, and the averaged factor of safety in the horizontal layers of the soil is increased upwards monotonously.

a) after failure

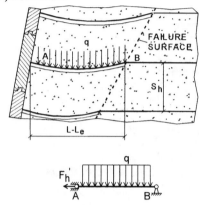

b) interrupted reinforcement

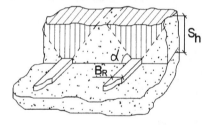

Figure 2. Calculation scheme

In case of immovable fixing of the reinforcement on faceplate only the reinforcement layers arranged in the lower part of the embankment take place in the acceptance of tensile forces effectively. The reinforcing interlayers arranged in the upper part of the construction accept only very small forces. It leads to the change of the stress - strain state of the soil which is seen in graph FS_{md}. In this case the graph of the change of the factors of safety of the soil according to the height of the embankment has a multyextremal character and according to the absolute value it has larger values than in other cases being considered, as a rule.

Thus, the availability of rigid faceplate in the construction in the combination with the stationary fixing of the reinforcement leads to a considerable change of the operating conditions of the reinforcing

elements. In case of a rupture of one of the reinforcement layers (as a rule, one of the lowest layers) and the formation of a wedge of the failure, these changes are of qualitative character: the reinforcing elements begin to work as a flexible band fixed with its one end to the rigid faceplate and with its other end to the stationary part of the construction, and loaded by the weight of the soil layer situated above (Fig. 2a). By such type of loading the maximal tensile force in the reinforcement is determined as:

$$H_R = \sqrt[3]{\frac{q^2 E_R F_R}{24}} \qquad (5)$$

where q = linear load on the calculated row of reinforcement, kN; E_R = modulus of strain of the reinforcement, kPa; F_R = cross section of the reinforcement, m^2.

In this formula q value depends on the type of reinforcement. Thus, for solid reinforcement the load can be determined according to the formula:

$$q = \gamma S_h (L - L_e) \qquad (6)$$

For the interrupted one it is necessary to take into account the spacial character of loading (Fig. 2b):

$$q = \gamma S_h [S_R ctg(\alpha) + B_R](L - L_e) \qquad (7)$$

where S_h = the distance between the layers of reinforcement, m; B_R = band width of reinforcement, m; L = total length of reinforcement, m; L_e = length of reinforcement outside failure surface, m.

The test calculations show that at the moment of failure the band reinforcement (Fig. 2a) takes the load which is considerably larger than under normal operation. This difference is very large for the upper layers of reinforcement where the load zone is maximal: here the tensile force due to the soil weight influence can be ten times as much than the calculated value of the force. It will lead to a quick rupture of reinforcement and the load redistribution to other reinforcing bands, etc. As a result, the reinforcing elements are destructed according to "domino" principle, but downwards now: a rupture of one band of reinforcement leads to the consecutive rupture of the rest.

3 CONCLUSIONS

The constructive peculiarities of the reinforced soil structures with rigid faceplate and stationary fixing of reinforcement lead to a change in the embankment operation type in the process of failure.

The principal measures for the prevention of catastrophic failures of the reinforced soil constructions originate from the expressions (4)-(7).
They consist of:

1. the use of solid reinforcement of the type of geonets, geotextiles and geomembranes;

2. the provision of the possibility of free vertical displacements of the reinforcement along the faceplate;

3. the reduction of the distance between the layers of reinforcement to $0.2 \div 0.25$ m, in the upper part of the construction especially;

4. the replacement of small quantity of high - tensile reinforcement by greater quantity of low - tensile reinforcement or by their combination;

5. the use of soils with a large value of the angle of internal friction;

6. the selection of strength of reinforcement taking into consideration its acceptance of the loads determined by the expression (5).

REFERENCES

Eschenko O.Y. 1991. *Reinforced soils and basements.* Thesis for the candidate degree, Krasnodar: KSAU.

Sondermann W. 1983. *Sprengungen und Verformungen bei bewehrter Erde.* - Braunschweig

Geotechnical Hazards, Marić, Lisac & Szavits-Nossan (eds) © 1998 Taylor & Francis, ISBN 90 5410 957 2

Compactibility of soils by means of a pulsator

M.J.Glinicka
Technical University of Bialystok, Poland

ABSTRACT: In this paper laboratory tests for the compactibility of fine-grained soils by using a pulsator are described. The tests made according to the pulsation method gave good results as compared with the standard methods. This laboratory method for determining of compactibility parameters is found to be more effective and less time and labour-consuming than standard method. Many factors that influenced the compaction of soils by pulsation method were analysed. Mechanical properties of the tested soil compacted by the pulsation method and the standard method were considered. The paper also presents test results of in situ soil compaction using pulsatory plate compactors. The purpose of tests was to determine compaction conditions by using a pulsator in the field.

1 INTRODUCTION

In the construction of highway embankments, earth dams, and many other engineering structures, loose soils must be compacted to increase their unit weights. Efficient compaction makes it possible to substantially improve the bearing capacity and stability of a fill, increase the impermeability and, in most cases, practically eliminate settlement. Soil compaction is the process whereby soil particles are mechanically constrained to pack closely together through a reduction in the air voids. Compaction is measured quantitatively in terms of the dry density of the soil. The increase in the dry density of soil produced by compaction depends on the type of soil, on its water content, and on the amount and manner of application of the compacting energy.

Commonly used tests are the tests performed according to the compaction method corresponding either to Standard or Modified Proctor Methods. In some countries compaction methods based on vibration are also used (Forsblad 1981). In Poland Standard Proctor test is generally applied (Polish Standard PN-88/B-04481).

The purpose of the reported investigation was an elaboration of a laboratory method for determining compactibility parameters using a pulsator as a method which is more effective than the standard one used up to now. The tests results of in situ soil compaction using pulsatory plate compactors are also presented. The investigations included fine-grained soils, which in most cases were commonly used for embankments.

The range of the investigations included:
-laboratory tests to determine the optimum water content (w_{opt}) and corresponding maximum dry density (ρ_{dmax}) of fine-grained soils by means of a pulsator and standard methods,
- additional laboratory tests to determine the influence of various factors on soil compactibility,
- model tests concerning the effectiveness of pulsatory compactors for soil compaction.

2 LABORATORY TESTS

For laboratory tests a pulsator "Wiep-4" was used. The author of the device was Eyman (Eyman 1984). A pulsator, which was a kind of a pulse press, originally developed for concrete mix compaction, was modified for soil compaction tests.

To increase the compaction force the vibrator was modified so as to produce asymmetrical vibrations. It was achieved as superposition of differently oriented harmonic vibrations. The following conditions are necessary:
- amplitudes of all the vibrations have to be the same,
- rotation velocities of each pair of eccentric weights have to be in proportion like the sequence of natural numbers.
In this way, after a full rotation of the slowest pair of eccentric weights there is concentration of all centrifugal forces giving the desired impulse. However, between impulses the centrifugal forces eliminate each other.

The pulsator "Wiep-4" mechanism used in tests consists of a set of four pairs of eccentric weights. Impulse vertical forces are produced due to an appropriately chosen velocity and phases of eccentric weight revolutions. The relation between the impulse forces produced by pulsator and time is shown in Figure 1. The frequencies used in the tests were 30 Hz and 35 Hz which corresponded to the maximum attained impulse forces of 8000 N and 11000 N, respectively.

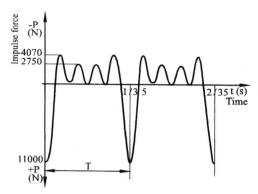

Figure 1. The relation between the impulse forces produced by a pulsator and time (frequency of 35 Hz).

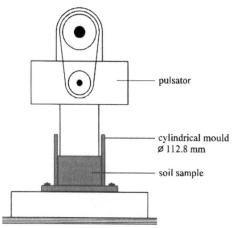

Figure 2. The test set-up for laboratory compaction of soils by means of a pulsator.

For laboratory tests a pulsator was equipped with cylindrical rammers working in standard moulds. The cylinders of the diameters 112.8 mm and 152.4 mm were applied. The tests, in most cases, were carried out on soil samples placed in a mould with diameter of 112.8 mm. In Figure 2 the test set-up for laboratory compaction of soils by means of a pulsator is shown.

By means of this method the soil was compacted at various water contents in order to determine compactibility parameters ρ_{dmax} and w_{opt}. Soil samples of varying number of layers (from 1 to 5) were compacted for different periods of time. Laboratory tests according to Standard and Modified Proctor procedures were also performed. Results of the pulsator compaction experiments were compared with the results of Standard and Modified Proctor Methods.

The tests were made on fine-grained soils such as: clay, clayey sand, sandy clay, silty clay, fine sand, and medium-grained sand. The compaction curves for different types of soils obtained by means of a pulsator are shown in Figure 3 and those obtained with Standard Proctor Method in Figure 4. In the case of the tested soils, the results were obtained by compacting the soil placed in the mould in one layer for the period of 60 s using frequency of 30 Hz. The shape of compaction curves determined by means of pulsator tests was similar to these obtained by standard methods. The results showed that water content has a great influence on the effect of compaction achieved by a given soil. Besides water content, another important factor, that affects compaction is soil type. In the case of cohesive soils the tests showed that compaction of soil samples in one layer using a pulsator for the period of 60 s and frequency of 30 Hz gave good results compared with these obtained with Standard Proctor Method. However, in the case of the tested sands, good results (as compared with Standard Proctor Method) were obtained by compacting soil in one layer for period of 30 s using a frequency of 30 Hz (Glinicka 1990).

On the basis of the results of pulsator tests and the criterion of short time limit it was determined that in order to obtain the same compaction effect as in Standard Proctor Method it is necessary to continue pulsation for 60 s for cohesive soils and 30 s for non-cohesive ones. The soil for testing should be placed in the mould in one layer. The pulsation method has been adopted to a standard mould. It was possible to achieve a uniform compaction of soil in single layer throughout the whole sample. The performed tests are thought to be correct, as far as reproductibility of results is concerned.

It was found that the frequencies applied in the method had no significant influence on the values of compactibility parameters.

The compaction energy corresponding to pulsation during 60 s was estimated to 2.8 x 10^5 Nm/m^3. In the case of Standard Proctor Method the energy was 5.9 x 10^5 Nm/m^3.

The time of vibration is an important factor in the compaction process both for cohesive and non-cohesive soils. This relationship was analysed by many researchers (Brand 1973, Dobry & Whitman 1973).

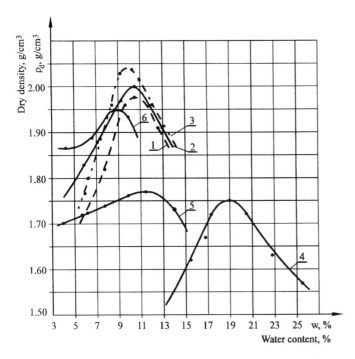

Figure 3. Compaction curves for different types of soils obtained with pulsation method (one layer, compaction time 1 min, frequency 30Hz): 1 - clay, 2 - clayey sand, 3 - sandy clay, 4 - silty clay, 5 - fine sand, 6 - medium - grained sand.

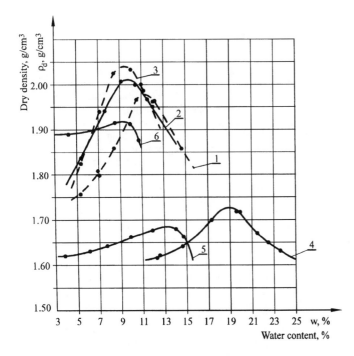

Figure 4. Compaction curves for different types of soils obtained with Standard Proctor Method: 1 - clay, 2 - clayey sand, 3 - sandy clay, 4 - silty clay, 5 - fine sand, 6 - medium - grained sand.

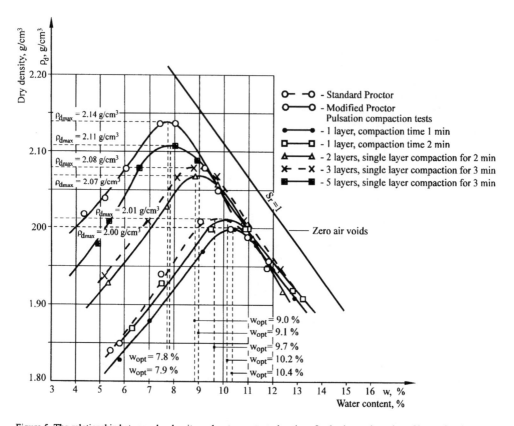

Figure 5. The relationship between dry density and water content, duration of pulsation and number of layers for clayey sand.

The results of tests showed that compactibility of soils by means of pulsator depend on the time of pulsation and on the number of layers to be compacted. In order to obtain better results of compaction by means of a pulsator the number of layers and the time of pulsation were increased so as to get comparable results with Modified Proctor Method. The influence of the compaction time and a number of layers on the maximum dry density value for clayey sand, is shown in Figure 5. As far as the Modified Proctor Method is concerned, insignificant differences in the maximum dry density values and in the optimum water content values were noticed. In this case the compaction was made in five layers using a pulsator; each layer was compacted for 3 min (total compaction time 15 min) and the frequency was 30 Hz. Similar results were also obtained for clay (Glinicka 1996). A good compaction effect by means of a pulsator as compared with Modified Proctor Method was obtained for clay in three layers. Total compaction time was 12 min. Further increase of the number of layers and time of compaction had no influence on the maximum dry density.

It should be noted that a wide range of compaction periods was investigated. Comparing the test results with the results obtained using Modified Proctor Method the following conclusions can be drawn.

In the case of non-cohesive soils compaction during 60 s in one layer using frequency 30 Hz gives the results in good agreement with Modified Proctor Method. In the case of cohesive soils the maximum dry density values obtained in that way were approximately 6.5% to 10.5% lower than Modified Proctor Method results (Glinicka 1992).

Similar discrepancy of the results was also obtained using vibration method developed by Dynapac (as compared to Modified AASHO procedure) (Forsblad 1981).

Supplementary tests included the determination of mechanical properties of the compacted soil. Triaxial tests for clay samples compacted by pulsation and Standard Proctor Method were performed. It was found that for a pulsation compacted soil the cohesion was higher but internal friction angle slightly lower than for a soil compacted with Standard Proctor Method. However, the shear strength of soil was higher when the pulsation was used. The results are shown in Figure 6.

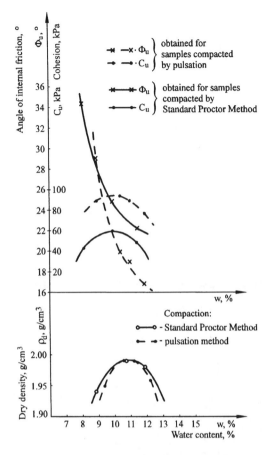

Figure 6. The internal friction angle and cohesion of clay as a function of water content and dry density.

The influence of those different compaction methods on the CBR of clay was also investigated. Penetration tests in the CBR apparatus was performed. It should be noted that the cylinder used (1 dm³) was not exactly the same as it should be in this method. Obtained CBR values are therefore denoted CBR* and are used to compare the relative performance of the soil compacted by means of two different methods. It should be observed that the penetration resistance recorded for the soil compacted by means of a pulsator was higher than for the soil compacted using standard method provided that water content was close to optimal (Figure 7). When increasing the water content above w_{opt} the difference in penetration resistance was diminishing.

3 MODEL TESTS ON EXPERIMENTAL PLOTS

Most of the compaction in the field is done with rollers. Vibratory rollers are used mostly for the densification of granular soils. The vibrating plates can be used for effective compaction of granular soils over a limited area (Braja 1994). In addition to soil type and water content, other factors must be considered to achieve the desired unit weight of compaction in the field. These factors include the thickness of layers and the number of passes of the vibratory compactors and the type of compacting equipment.

The model tests consisted of experimental compaction tests on a number of fields. The purpose of the tests was to prove the possibility of the application of the pulsator for soil compaction. In situ tests were performed on experimental fields in layers of various thickness and revealed compaction effects for variable numbers of machine passes and for various surfaces of compaction plate. Soil compaction state was determined for trial fields consisting of sandy clay, clayey sand and medium-grained sand. To make the investigations possible a pulsator was modified to become a pulsatory plate compactor by attaching a bottom plate.

The fields used in the experiments were filled with sandy clay. The following thickness of sandy clay were tested: 0.14 m, 0.19 m, 0.30 m. The effect of compaction was analysed after the following number of passes 2, 4 and 6. The compaction was carried out using the "Wiep-4" pulsatory compactor. Its mass was 54 kg, the size of the plate was 0.22 x 35 m. The frequency of 30 Hz was used which corresponded to maximum impulse force of 8000 N. The required degree of compaction was achieved after 2 passes of the pulsator for layer of 0.14 m and 4 passes for the layer of 0.19 m. For the layer of 0.30 m to achieve the required degree of compaction it was necessary to perform 8 passes (Glinicka 1997). The results of the experiments are shown in Figure 8.

Additional experiments were conducted on other fields, for example, at the Metro building site in Warsaw. The experiments concerned clayey sand and sandy clay as well as medium-grained sand. Interesting results were obtained for medium-grained sand of three different thicknesses: 0.30 m, 0.50 m and 0.70 m. Two different models of pulsators were applied:
-"Wiep-3" - mass of 40 kg, plate size 0.22 x 0.35 m, maximum impulse force of 8000 N, frequency of 30 Hz,
-"Wiep-4" - mass of 60 kg, plate size 0.27 x 0.42 m, maximum impulse force of 11000 N, frequency of 35 Hz.

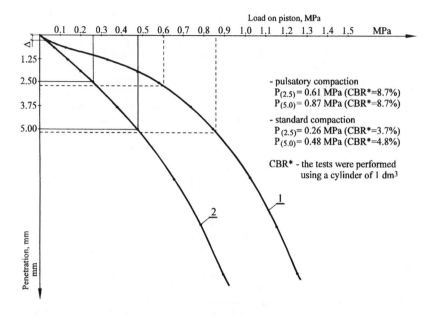

Figure 7. Penetration curves: 1 - for clay compacted by means of a pulsator (one layer, compaction time 1 min, frequency 30 Hz) w = 10.9 %, 2 - for clay compacted with standard method, w = 11.1 %.

Moreover the pulsator "Wiep-4" equipped with plate of 0.50 x 0.50 m was used especially for the fields made in layers of 0.50 m and 0.70 m in thickness. Figure 9 shows a variation of dry density with depth for given number of pulsatory compactors passes. In the case of medium-grained sand plots the results were the following:
- with "Wiep-3" for the layer of 0.30 m and 0.50 m the required degree of compaction was reached after 2 passes,
- "with Wiep-4" the required degree was achieved for 0.30 m after 2 passes.
It was impossible to get a proper compaction of 0.50 m and 0.70 m layers using a compactor with a large plate. In order to achieve a required degree of compaction the mass of the pulsator should be increased or the maximum impulse force should be higher.

In site compaction by means of a pulsator depends on impulse force and the mass of the pulsator, the dimension of compaction plate, as well as the thickness of the layer to be compacted and the number of passes performed. The efficiency of the pulsator depends on the value of impulse force per unit area of the compactor plate. The pulsator can be used in the field to compact both cohesive and non-cohesive soils.

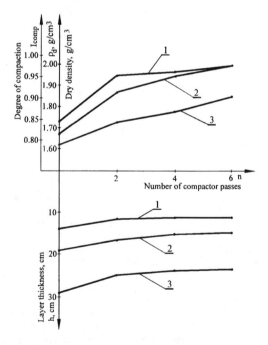

Figure 8. Compaction test results for sandy clay obtained at three experimental plots: 1 - plot no 1, 2 - plot no 2, 3 - plot no 3.

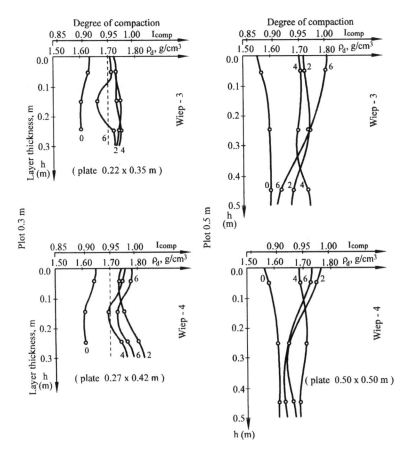

Figure 9. The results of pulsatory compaction of medium - grained sand at experimental plots.

4 CONCLUSIONS

Performed laboratory and field tests confirmed that for testing soil compaction and determination of geotechnical paramaters of compactibility the pulsator can be applied.

A laboratory compaction method by means of pulsator gives good results in comparison to commonly used methods. It can be succesfully applied to a number of fine-grained soils. The proposed method for laboratory determination of compactibility parameters is more effective and less time and labour-consuming than other generally applied methods.

Pulsatory plate compactors can be applied for field compaction of soil. Further tests are necessary to determine an optimal set of compaction parameters for specific application.

REFERENCES

Braja, M. Das 1994. *Principles of Geotechnical Engineering*. Chap.4: 88-128. Boston: PWS Publishing Company.

Brand, E.W. 1973. Some Observations on the Control of Density by Vibration. Evaluation of Relative Density and Its Role in Geotechnical Projects Involving Cohesionless Soils. ASTM SPT 523.

Dobry, R. & R. V. Whitman 1973. Compaction of Sand on a Vertically Vibrating Table. Evaluation of Relative Density and Its Role in Geotechnical Projects Involving Cohesionless Soils. ASTM SPT 523.

Eyman, K. 1984. Compaction of sand concretes. *XXX Scientific Conference KILiW PAN-KN PZITB*, (in Polish) Krynica/Poland, 37-42.

Forsblad, L. 1981. *Vibratory Soil and Rock Fill Compaction*. Sweden: Dynapac.

Glinicka, M.J. 1990. New Laboratory Tests for the Compactibility of Fine-Grained Soils by Means of a Pulsator. *IX National Conference on Soil Mechanics and Foundation Engineering*, (in Polish) Krakow/Poland, Vol.1: 189-194.

Glinicka, M.J. 1992. Compactibility of Soils by Means of a Pulsator and Evaluation of the Test Results. *Symposium, National Experience in Subgrade Improvements*, (in Polish) Gdansk/Poland, 137-142.

Glinicka, M.J. 1996. Investigation of the Influence of the Pulsation Method on the Compactibility of Fine-Grained Soils. *XLI Scientific Conference KILiW PAN-KN PZITB*, (in Polish) Krynica/Poland, Vol.7: 29-36.

Glinicka, M.J. 1997. Investigation of the Influence of the Pulsation Method on the Soil Compaction in Situ. *XI National Conference on Soil Mechanics and Foundation Engineering. Geotechnics in Civil Engineering*, (in Polish) Gdansk/Poland, Vol.2: 77-82.

Polish Standard PN-88/B-04481 Building soils. Laboratory tests. (in Polish) Poland.

Geotechnical Hazards, Marić, Lisac & Szavits-Nossan (eds) © 1998 Taylor & Francis, ISBN 90 5410 957 2

Moderne Verfahren zur Ertüchtigung von Erdbauwerken bestehender Eisenbahnstrecken

C.Göbel
Hochschule für Technik und Wirtschaft Dresden, Germany

Zusammenfassung: In Ostdeutschland müssen gegenwärtig viele bestehende Eisenbahnstrecken auf deutlich höhere Fahrgeschwindigkeiten ausgebaut werden. Die Erdbauwerke dieser Eisenbahnstrecken sind vor über 100 Jahren hergestellt worden. Sie entsprechen in der Regel nicht den heute gültigen Anforderungen an Dichte, Tragfähigkeit und Standsicherheit und müssen bei ständiger Aufrechterhaltung des Eisenbahnbetriebes durch bautechnische Maßnahmen „ertüchtigt" werden. Die damit verbundenen geotechnischen Probleme und Risiken und einige moderne Verfahren zur Ertüchtigung dieser Erdbauwerke werden an zwei praktischen Beispielen dargestellt.

1 EINFÜHRUNG

Im Rahmen der Schienenverkehrsprojekte Deutsche Einheit müssen vor allem in Ostdeutschland bestehende Eisenbahnstrecken auf deutlich höhere Fahrgeschwindigkeiten ausgebaut werden. Die Erdbauwerke dieser Eisenbahnstrecken (Dämme, Einschnitte, Anschnitte) sind in der Regel in der zweiten Hälfte des 19. Jahrhunderts unter den damaligen äußerst bescheidenen technischen Möglichkeiten hergestellt worden. Sie wurden ohne wesentliche Verdichtung geschüttet und weisen oft nur eine lockere Lagerung auf.

Während der Oberbau den im Laufe der Zeit gestiegenen Belastungen und Fahrgeschwindigkeiten ständig angepaßt wurde, sind die Erdbauwerke nicht oder nur wenig verstärkt worden. Sie entsprechen nicht mehr den heute gültigen Anforderungen an Dichte, Tragfähigkeit und Standsicherheit und müssen so „ertüchtigt" werden, daß das heutige Sicherheitsniveau erreicht wird. Verbunden damit ist die Lösung einer Vielzahl von geotechnischen und geometrischen Problemen bei ständiger Aufrechterhaltung des Eisenbahnbetriebes.

Bei der Lösung insbesondere der geotechnischen Probleme gerät der geotechnische Gutachter in einen gewissen Zwiespalt. Einerseits erfüllt ihn Anerkennung für die ingenieurtechnischen Leistungen der Erbauer dieser Eisenbahnstrecke, denn schließlich sind die Erdbauwerke ohne wesentliche Schäden noch vorhanden, andererseits ist er sich aber der geotechnischen Risiken und Gefahren voll bewußt. Da es eine pauschale Lösung insbesondere für die geotechnischen Probleme nicht gibt, sind die Fragen von Risiko, Sicherheit und Wirtschaftlichkeit für jedes Erdbauwerk gesondert zu stellen, zu beantworten und vor allem zu verantworten. Das bedeutet, daß der geotechnische Gutachter für jedes einzelne Erdbauwerk aus dem Spannungsfeld von Risiko, Sicherheit und Wirtschaftlichkeit heraus eine bautechnische Lösung finden muß, die dem heutigen Sicherheitsniveau entspricht.

Die wesentlichsten geotechnischen und geometrischen Probleme, die bei der Instandhaltung und dem Ausbau bestehender Eisenbahnstrecken auf den heute gültigen Standard auftreten können, sind nachfolgend zusammengefaßt worden:

- Inhomogenitäten der geschütteten Böden mit ständig wechselnden bodenmechanischen Eigenschaften und Verwendung ungeeigneten Bodens;
- unzureichende Dichte der geschütteten Erdbauwerke;
- unzureichende Tragfähigkeit im Bereich der Dammkrone;
- übersteilte Böschungen an den Erdbauwerken mit unzureichender Standsicherheit ;
- unzureichende Breite des vorhandenen Planums;
- unzureichende Abmessungen des gesamten Bahnkörpers (Nichteinhaltung des Regelprofils).

Für die Lösung der geometrischen Probleme ist in der Regel eine Planumsverbreiterung erforderlich, in deren Folge Dammverbreiterungen bzw. Einschnittserweiterungen notwendig werden. Diese geometrischen Veränderungen zur Herstellung des Regelprofils für Ausbaustrecken können mit erheblichen Eingriffen in die bestehenden Dämme und Einschnitte verbunden sein und haben geotechnische Risiken und Gefahren zur Folge, die sicher beherrscht werden müssen. Die Lösung dieser Problematik hat in den letzten Jahren zu einem deutlichen Innovationsschub geführt. In dessen Folge wurden viele neue und rationelle Bauverfahren entwickelt, erpobt und erfolgreich angewendet. Einige dieser neuen Verfahren sollen nachfolgend am Beispiel der Sanierung eines Eisenbahndammes und am Beispiel der Erweiterung eines Einschnittes erläutert werden.

2 SANIERUNG EINES EISENBAHNDAMMES

Ein bis zu 13 m hoher Eisenbahndamm einer bestehenden Eisenbahnstrecke ist vor etwa 120 Jahren entsprechend den damaligen Möglichkeiten geschüttet worden. Die Dammschüttmassen, die überwiegend aus Verwitterungsböden des Muschelkalks bestehen, wurden ohne wesentliche Verdichtung in den Damm eingebaut. Im Ergebnis der geotechnischen Untersuchung des Dammes wurden die Konsistenz der Dammschüttmassen als weich bis steif eingeschätzt und die Rechenwerte der Bodenkennwerte ermittelt. Mit diesen Rechenwerten und den Verkehrslasten für Ausbaustrecken sind Standsicherheitsberechnungen nach DIN 4084 durchgeführt worden, die für die größten Dammhöhen Sicherheitsbeiwerte zwischen 1,05 und 1,10 ergaben. Diese Werte liegen damit erheblich unter dem nach DIN 4084 für den Lastfall 1 und die Verwendung eines Lamellenverfahrens erforderlichen Sicherheitsbeiwert von 1,40. Es war zu entscheiden, ob das Risiko einer solch geringen Standsicherheit zu verantworten ist oder ob standsicherheitserhöhende bautechnische Maßnahmen erforderlich sind. Dabei war zu berücksichtigen, daß der Eisenbahnbetrieb eine hohe Sicherheitsrelevanz besitzt und künftig höhere dynamische Verkehrslasten in den Damm eingetragen werden. Aus diesen Gründen wurde entschieden, bautechnische Maßnahmen zur Erhöhung der Standsicherheit auf das erforderliche Niveau vorzu-

sehen. Da die vorhandene Planumsbreite für das Regelprofil einer zweigleisigen Ausbaustrecke nicht ausreicht und zusätzliches Gelände nicht in Anspruch genommen werden durfte, muß die zu wählende Sanierungsmaßnahme gleichzeitig folgende Aufgaben erfüllen:

- Erhöhung der Standsicherheit der Dammböschungen auf das erforderliche Niveau
- Erhöhung der Tragfähigkeit des Dammplanums auf das erforderliche Niveau
- Verbreiterung des vorhandenen Dammplanums auf das für Ausbaustrecken erforderliche Regelprofil ohne Inanspruchnahme zusätzlichen Geländes

Nach vergleichenden Betrachtungen verschiedener Sanierungsverfahren wurden das Fräs-Misch-Injektionsverfahren (FMI-Verfahren) und das Hydro-Zementations-Verfahren (HZ-Verfahren) für die Sanierung ausgewählt, weil mit diesen Verfahren die obengenannten Bedingungen am besten erfüllt werden konnten. Diese Verfahren sind in mehreren Veröffentlichungen (Feuerbach 1996, Bause 1997) ausführlich beschrieben. Beiden Verfahren ist gemeinsam, daß eine Verbesserung des anstehenden Bodens mit Zement an Ort und Stelle erfolgt, wobei der Boden durch spezielle Geräte mit Zement vermischt wird.

Die unzureichende Standsicherheit der Dammböschungen wurde durch Stützkörper aus Erdbeton, die in der Böschung nach dem HZ-Verfahren hergestellt worden sind, erhöht. Dabei wurden mit speziellen Schreitbaggern 2 m breite Schlitze bis unter die potentielle Gleitfläche ausgehoben, wobei der Boden seitlich zwischengelagert wurde. Über eine Schlauchleitung wird die Zementsuspension in den Schlitz eingebracht und unter Zugabe des zwischengelagerten Bodens mit der Baggerschaufel vermischt (mixed in place). Auf diese Weise entstehen 2 m breite Erdbeton-Stützkörper in der Böschung. Die Sicherung der Dammböschungen durch die Erdbeton-Stützkörper wurde zuerst ausgeführt, um eine ausreichende Sicherheit für die Durchführung des FMI-Verfahrens auf der Dammkrone zu gewährleisten.

Beim FMI-Verfahren wird der Erdbetonkörper ohne Bodenaushub hergestellt, wobei eine geländegängige Fräs-Misch-Maschine mit einem bis zu 9 m langen Fräsbaum den anstehenden Boden auffräst und gleichzeitig mit Zementsuspesion vermischt. Mit diesem Verfahren wurde das Planum des bestehenden

Eisenbahndammes stabilisiert und verbreitert (Bild 1).

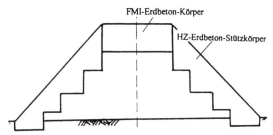

Bild 1. Anordnung der FMI-Erdbeton-Körper und der HZ-Erdbeton-Körper im Dammbereich

Die Abstände der Erdbeton-Stützkörper sowie die Dicke des FMI-Körpers wurden unter der Voraussetzung berechnet, daß sowohl die Standsicherheit der Dammböschungen als auch die globale Standsicherheit des gesamten Dammes erfüllt sein müssen. Während der Dammsanierung traten keine verfahrensbedingten Probleme auf. Dieser Damm wird voll begrünt, so daß die Stabilisierungsmaßnahmen später nicht mehr wahrgenommen werden.

3 AUFWEITUNG EINES EINSCHNITTES

Zur Herstellung des Regelprofils für Ausbaustrecken war in einem über 100 Jahre alten Einschnitt eine Planumsverbreiterung erforderlich, die nur über eine Aufweitung des bestehenden Einschnitts erreicht werden konnte.

Die hydrologischen und geotechnischen Verhältnisse des betrachteten Einschnitts sind sehr kompliziert. In den oberen Bereichen der Einschnittsböschung wurden geringmächtige Schichten von Sanden und Kiesen festgestellt, während im unteren Bereich Schluffe, Braunkohlentone und auch Braunkohle mit sehr geringer Wasserdurchlässigkeit anstanden. Die unteren Schichten wirkten als Wasserstauer und führten dazu, daß das Sickerwasser gesammelt wurde und etwa in Höhe des Böschungsfußes aus der Böschung austrat.

Da keine weiteren Grundstücksflächen in Anspruch genommen werden durften und die begrünte natürliche Einschnittsböschung weitestgehend erhalten werden sollte, bestand die bisherige Standardlösung darin, unverankerte, im Baugrund eingespannte Stahlspundwände am Böschungsfuß zu schlagen und den Böschungsbereich vor der Spundwand abzugraben (Bild 2). Diese Lösung ist ohne Risiko, da die Standsicherheit auch im Bauzustand gewährleistet ist. Sie ist aber sehr kostenintensiv, ästhetisch unbefriedigend und wenig innovativ. Weitere Nachteile bestehen im Korrosionsproblem und darin, daß aufwendige Einrichtungen zur Abführung des Sickerwassers durch die Spundwand hindurch vorzusehen sind. Deshalb wurde nach Alternativlösungen gesucht.

Gewählt und ausgeführt wurde ein Fußanschnitt der Böschung im Bauzustand und eine endgültige Stützung des Fußanschnitts durch eine Stützbauwerk aus Gabionen (Bild 2). Dabei ist sowohl während des Bauzustandes als auch im Endzustand eine ausreichende Standsicherheit der Böschung zu gewährleisten. Für den Nachweis der Standsicherheit im Endzustand mit Stützbauwerk müssen die Forderungen nach DIN 4084 erfüllt werden.

Der kritische Zustand ist jedoch der Bauzustand, bei dem die durch den Fußanschnitt „geschwächte" Böschung nur über einen begrenzten Zeitraum ohne Stützung verbleibt und deshalb als temporäre Böschung bezeichnet werden kann. Für eine solche temporäre Böschung ist es gerechtfertigt, die Kohäsion c' der anstehenden Böden in vertretbarer Größe in Ansatz zu bringen, wofür es zwei verschiedene Möglichkeiten gibt.

Die erste Möglichkeit besteht darin, Standsicherheitsberechnungen für den Bauzustand nach DIN 4084 durchzuführen, wofür die in Tabelle 1 für temporäre Böschungen angegebenen Kohäsionswerte angesetzt werden dürfen (Bilz 1994).

Bei der zweiten Möglichkeit wird die Einschnittsböschung im Bauzustand als Baugrubenböschung nach DIN 4124 aufgefaßt, und es werden die dort angegebenen Böschungsneigungen ohne Stand-

Tabelle 1. Köhäsionswerte c' für temporäre Böschungen.

Bodenart nach DIN 18196	γ [kN/m^3]	φ' [°]	c' [kN/m^2]
TA	20	15	25
TM	20	17,5	20
UA	18,5	20	10
ST, SU, GT, GU	20	30	10
UM	18,5	25	7
Feinsand, mitteldicht	18,5	30	10*
Feinsand, locker	17	25	8*
SW, SI, SE mitteldicht	18,5	35	5*
SW, SI, SE locker	17	32	4*
GW, GI, GE mitteldicht	18,5	37	2*
GW, GI, GE locker	18,5	32	1*

* Kapillarkohäsion

213

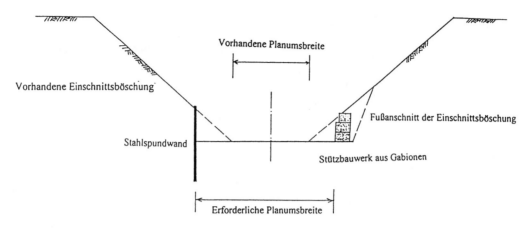

Bild 2. Aufweitung eines Einschnittes

sicherheitsnachweis zugelassen, wenn die übrigen Randbedingungen nach DIN 4124 erfüllt sind:

- nichtbindiger oder weicher bindiger Boden
  $$\beta = 45^0$$
- steifer oder halbfester bindiger Boden
  $$\beta = 60^0$$

Die nach DIN 4124 zulässigen Böschungsneigungen beruhen auf dem Ansatz ähnlicher Kohäsionswerte wie in Tabelle 1.

Unter Abwägung der überschauhbaren Risiken und Gefahren wurde entschieden, die im Bauzustand im Fußbereich angeschnittene Einschnittsböschung als Baugrubenböschung aufzufassen und im Bereich der Braunkohlentone und der Braunkohlen einen Böschungswinkel von 60^0 und im Bereich der übrigen Böden ein Böschungswinkel von 45^0 vorgesehen. Diese Entscheidung kann dadurch begründet werden, daß die Böschung im Bauzustand sowohl zeitlich als auch räumlich nur begrenzt geschwächt wird und mit der gewählten Lösung keine Gefährdung des Eisenbahnbetriebes und der öffentlichen Ordnung verbunden war.

Durch die bewußte Übernahme eines „kalkulierten Risikos" konnte somit auf die Anordnung einer Stahlspundwand verzichtet und eine Lösung ausgeführt werden, die wirtschaftliche und ästhetische Vorteile aufwies. Während der gesamten Bauarbeiten traten keine verfahrensbedingten Störungen auf.

LITERATUR

Bause, M. 1997. Anwendung des Fräs-Misch-Injektionsverfahrens (FMI) bei der Sanierung von Eisenbahndämmen. EI - Der Eisenbahn-Ingenieur 48(1997) Heft 7.

Bilz, P. 1994. Mögliche Abmessungen ungesicherter Bereiche an Grabenwänden und aufgelösten Stützwänden. Geotechnik 17(1994).

Feuerbach, J. 1996. Bodenverbesserung mit dem Hydro - Zementations- und mit dem Fräs-Misch-Injektionsverfahren. Beton-Informationen 36(1996).

DIN 4084. Baugrund; Böschungs- und Geländebruchberechnungen, Entwurf Juni 1990.

DIN 4124. Baugruben und Gräben; Böschungen, Arbeitsraumbreiten, Verbau, August 1981.

Geotechnical Hazards, Marić, Lisac & Szavits-Nossan (eds) © 1998 Taylor & Francis, ISBN 90 5410 957 2

Geotechnical microzonation in the Arctic related to climate warming

Branko Ladanyi
Ecole Polytechnique, Montréal, Canada

ABSTRACT: The present paper attempts to quantify the sensitivity of frozen ground strength to temperature increase that may result from general climate warming, and proposes a simple strength-sensitivity index that could be used as a basis for microzonation mapping of the zones of potentially stable and unstable permafrost, with the purpose of estimating the potential damage to the existing structures, and drafting the design guidelines for future projects.

1 INTRODUCTION

Geotechnical microzonation in the Arctic related to climate change deals primarily with the effects such change may have on the occurrence and state of permafrost. The term "permafrost" describes a thermal condition in which any material in the earth crust, such as soil and rock, stays below 0 °C for two or more years. Permafrost underlies 20 per cent of the world's land area, being widespread in North America, Eurasia and Antarctica. In the northern hemisphere, this includes about half of Russia and Canada, and 80 per cent of Alaska. It is also found on Greenland, northern Scandinavia, Mongolia and northern China. It occurs also at high elevations, such as Tibet Plateau and in mountainous regions of the world. The approximate distribution of permafrost in the Northern Hemisphere is shown in Figure 1, while Figure 2 shows a typical ground temperature envelopes in the ground in a permafrost region. On a smaller scale, its distribution and thickness are controlled by microclimatic, hydrologic, geologic, topographic and biologic factors. The surface of permafrost is generally overlain by a seasonally frozen active layer, a vegetative mat, a seasonal snow cover, and a complex boundary layer in the air. Consequently, changes in meteorological conditions, such a air temperature and snow cover, result in the changes in the surface temperature that are complex and difficult to predict.

The physical and mechanical properties of permafrost are generally temperature dependent, and at temperatures close to the melting point, they depend strongly on temperature. Most of the civil engineering concerns related to the climate warming can be classified into those related to an increase in permafrost temperature, those related to increases in the active layer thickness (annual thaw depth), and those related to the degradation of permafrost. Engineering concerns related to a general warming of the permafrost result primarily from the decrease of its mechanical strength in its frozen and eventually thawed state. Continued climatic warming will eventually cause much of the discontinuous permafrost to thaw, resulting in increasing rates of thaw settlement of structures in thaw-sensitive regions of the Arctic.

While there is still a considerable disagreement among the climatologists about the most probable rate of future climate warming, it is nevertheless not too early to start thinking about the kind of effects a certain amount of warming would have on the behavior and stability of natural human-made structures in the North. In order to quantify the possible impacts, there has been a trend in recent years to establish the sensitivity of the physical, biological, social and economic systems to climate and changes in climate. In the area of permafrost science and engineering, there have been some proposals for estimating the sensitivity of predictions on: permafrost thermal regime, active layer thickness, and permafrost strength reduction.

The present paper attempts to quantify the sensitivity of frozen ground strength to temperature increase, and proposes a simple strength-sensitivity

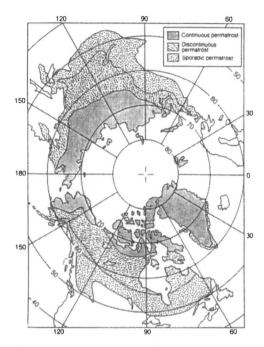

Figure 1. Distribution of permafrost in the Northern Hemisphere.

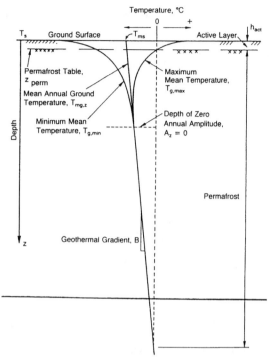

Figure 2. Schematic extreme ground temperature curves in a permafrost region.

index, that could be used as a basis for mapping the zones of potentially stable and unstable permafrost, with the purpose of estimating the potential damage to the existing structures, and drafting the design guidelines for future projects.

2 CLIMATE WARMING TRENDS IN THE ARCTIC

Alaska

In Alaska, global air temperatures have increased significantly over the last century, with warming in the higher latitudes being greater than at lower latitudes. According to Esch and Osterkamp (1990), air temperatures in Alaska have generally followed the global trend with some variations. Most Alaskan stations showed a cooling from about 1940 to about 1976, and then an abrupt warming, beginning in 1976 (Fig. 3). Current climate models predict a 3° to 12°C rise in mean annual air temperatures for northern latitudes by the middle of the next century, assuming an effective doubling of greenhouse gas concentrations in the atmosphere by that time (Esch & Osterkamp, 1990; Esch, 1993).

As the response of permafrost to air temperature variation depends on thermal processes at the ground surface and within the active layer, the permafrost temperature variations due to climatic changes are generally difficult to predict. For Alaska, the analyses of deep permafrost temperature profiles from the North Slope indicate a 20th century warming trend of 2° to 4°C at the surface of permafrost over much of the area. Esch and Osterkamp (1990) suggest that engineers designing for these latitudes, which include most of Alaska, may be justified in assuming in the design a temperature warming of about 1° to 2 °C per decade, until more refined forecasts become available.

Canada

Although the climate warming appears to slightly vary going from the west to the east along the Canadian arctic coast, with the eastern portion showing even some temporary cooling, there are clear signs that a warming trend exists along the Mackenzie Valley (Burgess & Riseborough, 1990; Etkin, 1990). According to Nixon (1990) and Nixon et al.,(1990), based on data supplied by David Etkin of the Canadian Climate Centre, Environment Canada, the mean annual temperatures for Inuvik, Norman Wells and Fort Simpson show that a warming of about 1° C has

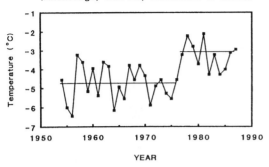

Alaskan Annual Temperatures
(Anchorage, Barrow, Fairbanks & Nome)

Figure 3. Mean annual air temperatures, averaged for four Alaskan locations (After Esch and Osterkamp, 1990).

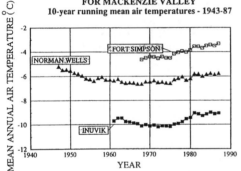

LONG TERM MEAN ANNUAL TEMPERATURE VARIATIONS FOR MACKENZIE VALLEY
10-year running mean air temperatures - 1943-87

Figure 4. Long-term mean annual temperature variations for Mackenzie valley (After Nixon, 1990).

occurred over the last 15 years. On the average, as seen in Figure 4, from about 1975, the rate of air temperature increase was about 0.1° C/year for Fort Simpson and Inuvik, and about 0.06° C/year for Norman Wells. However, the same records indicate that a cooling trend over the two decades preceding 1975 also took place. Although the current warming trend may be explained as just another fluctuation in a series of cyclic variations, a continuous increase in CO_2 and other reflective gases in the atmosphere may help to continue further warming without substantial cooling cycles. If the current warming trend of 0.05° to 0.10° C continues for the next 25 years or more, then structures currently founded on warming permafrost may experience some negative effects

(Nixon, 1990; Nixon et al., 1990).

Russia

Permafrost distribution in Russia and its climate sensitivity is shown in Figure 5, which will be discussed later in the paper.

3 EFFECT OF CLIMATE WARMING ON ENGINEERING STRUCTURES IN PERMAFROST REGIONS

3.1 *Ground temperature variation*

It is understood that a steady increase of air temperature will result in two related effects on the ground temperature:

(a) An increase in the mean annual temperature at the ground surface, which will slowly propagate to the depth and, depending on latitude, produce either a thinning or a complete disappearance of the permafrost layer, and

(b) An increase in the annual amplitude of seasonal ground temperature variation, damped with depth, and affected by the related changes in precipitation (snow cover), groundwater hydrology and vegetation.

The potential effects of warming of the mean annual temperature on permafrost will be very different for continuous and discontinuous permafrost zones. In the continuous zones, the effect on the permafrost would be to warm it and possibly to change the depth of the active layer. Thawing at the base of permafrost would start several centuries later and would proceed at a rate of about a centimeter or more per year. (Esch & Osterkamp, 1990; Osterkamp & Lachenbruch, 1990). In the discontinuous zone, the effects of a few degrees warming in the mean annual temperature of permafrost would be extremely serious. Since most of this permafrost is within a few degrees of thawing, it should eventually disappear. Although many centuries will be required for the permafrost to disappear entirely, increases in the active layer and thawing of the warmest permafrost from the top could begin almost immediately (Esch & Osterkamp, 1990; Osterkamp & Lachenbruch, 1990).

In general, the effects of the mean annual temperature increase on permafrost are difficult to predict, because they are affected by the microclimate, and the soil type, ice content and salinity (Smith & Riseborough, 1983, 1985; Riseborough, 1990).

217

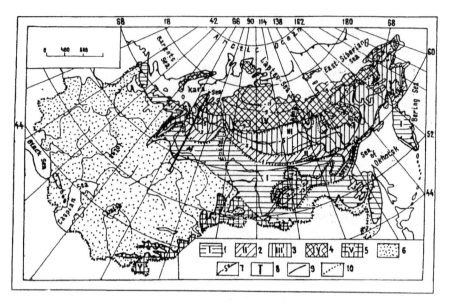

Figure 5. Map of geothermal zones in Russia, according to Vyalov et al.(1993), based on the mean annual ground temperature, T_z:
Zone I: $T_z = 0$ to -1 °C; Zone II: $T_z = -1$ to -3 °C; Zone III: $T_z = -3$ to -7 °C; Zone IV: $T_z =$ below -7 °C. Zone 5: permafrost in mountains; Zone 6: regions with positive temperature.

3.2 *Engineering concerns*

The physical and mechanical properties of frozen soils are generally temperature dependent, and this effect is most pronounced at temperatures within one or two degrees C of thawing. Most of the permafrost engineering concerns related to warming of permafrost have been summarized by Esch and Osterkamp (1990), as follows:

1. Warming of permafrost body at depth.
    a. Increase in creep rate of existing piles and footings.
    b. Increased creep of embankment foundations.
    c. Eventual loss of adfreeze bond support for piling.
2. Increases in annual thaw. (active layer)
    a. Thaw settlements during seasonal thawing.
    b. Increased frost-heave forces on pilings.
    c. Increased total and differential frost heaving during winter.
3. Development of residual thaw zones (taliks).
    a. Decrease of effective length of piling in permafrost.
    b. Progressive landslide movements.
    c. Progressive surface settlements.

4 EFFECTS OF WARMING ON FROZEN GROUND BEHAVIOUR

Contrary to frozen rocks and dense gravels, whose strength depends mainly on mineral bonds and internal friction, it is known that the major portion of the mechanical strength of fine-grained frozen soils is due to ice bonding. An increase in temperature increases the unfrozen water content in such soils and decreases the ice bonding (cohesion) of soil particles. In other words, at a steady temperature increase, the frozen soil will steadily weaken, and, eventually, on complete thawing loose all its additional strength due to ice cementation.

4.1 *Frozen soil strength sensitivity*

Similarly to most other properties of frozen soil, its strength dependes not only on the temperature, but also on its density, ice content and salinity. In addition, it is also affected by the degree of confinement and the applied strain rate. For a general strength comparison, the uniaxial compression strength is the most useful measure. Its strain rate and temperature dependence can be expressed by :

$$q_f = q_{fo} \, f(\theta) \qquad (4.1)$$

with

$$q_{fo} = \sigma_{co} \, f(\dot{\epsilon}) \qquad (4.2)$$

Here, $f(\theta)$ expresses the effect of temperature on strength and creep. Its most often used empirical form is

$$f(\theta) = (1 + \theta/\theta_o)^w \qquad (4.3)$$

while $f(\dot{\epsilon})$ expresses the effect of strain rate and time on strength.

$$f(\dot{\epsilon}) = (\dot{\epsilon}_1/\dot{\epsilon}_c)^{1/n} \qquad (4.4)$$

In the above equations:

θ = $-T$ = frost temperature, $^\circ$C
θ_o = 1 $^\circ$C
$\dot{\epsilon}_1$ = uniaxial compression strain rate, s^{-1}
$\dot{\epsilon}_c$ = reference strain rate, s^{-1}
σ_{co} = reference stress (MPa) at $\dot{\epsilon}_c$ and when T tends to 0 $^\circ$C
q_f = uniaxial compression strength, MPa
n = creep exponent for stress
$w \le 1$ = temperature exponent

Equation (4.1), combined with Eqs.(4.2) and (4.4), gives the usual temperature- and strain-rate dependence of strength. For a given soil, the effect of temperature alone (at the same strain rate) can be expressed by combining Eqs.(4.1) and (4.3):

$$q_f = q_{fo} \, (1 + \theta/\theta_o)^w \qquad (4.5)$$

Figure 6 shows schematically a typical strength variation of frozen soil with temperature, as expressed by Eq.(4.5), and Figure 7 is an example of such variation measured in uniaxial compression tests on a frozen fine sand (Sayles, 1968).

Alternatively, if one wants to evaluate the effect of temperature on creep or creep rate, the following equations are valid.

For creep:

$$\epsilon_1 = \left(\frac{q}{\sigma_{co} f(\theta)} \right)^n \left(\frac{\dot{\epsilon}_c t}{b} \right)^b \qquad (4.6)$$

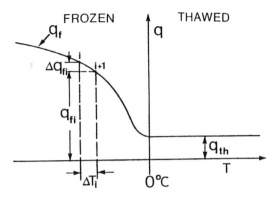

Figure 6. Schematic frozen soil strength variation with temperature change.

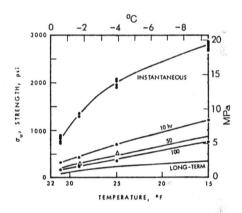

Figure 7. Temperature-time dependence of uniaxial compression strength for a frozen fine sand (After Sayles, 1968).

For creep rate:

$$\dot{\epsilon}_1 = \dot{\epsilon}_c \left(\frac{q}{\sigma_{co} f(\theta)} \right)^n \left(\frac{\dot{\epsilon}_c t}{b} \right)^{b-1} \qquad (4.7)$$

where t = time (s), b = creep exponent of time, q = the applied uniaxial compression stress, and ϵ_1 = the resulting creep strain. For all the rest equal, Eqs. (4.6) and (4.7) show that the effect of temperature change on creep and creep rate follows a law of the form $[\sigma_{co} f(\theta)]^{-n}$. So, for example, for two temperatures, θ_1, θ_2, the corresponding creep and creep rate ratios will be

219

$$\frac{\epsilon_{101}}{\epsilon_{102}} = \frac{\dot{\epsilon}_{101}}{\dot{\epsilon}_{102}} = \left[\frac{f(\theta_1)}{f(\theta_2)}\right]^{-n} = \left(\frac{\theta_1 + \theta_o}{\theta_2 + \theta_o}\right)^{-wn} \quad (4.8)$$

For example, for $w = 1$, $n = 3$, a temperature increase by 1 °C, from $\theta_2 = 2°$ C to $\theta_1 = 1°$ C, will give a 3.375 times higher creep strains and creep strain rates.

As seen in Figure 6, when temperature increases by $\Delta T_i = -\Delta\theta_i$, the strength will drop by Δq_{fi}. The strength drop depends on the strength variation gradient $(\Delta q/\Delta\theta)_i$ and on the temperature change $\Delta\theta = -\Delta T_i$. For $f(\theta)$ given by Eq. (4.3), the gradient of strength change at $q_f = q_{fi}$ is equal to

$$(dq_f/d\theta)_i = q_{fo}[df(\theta)/d\theta] = q_{fo}(w/\theta_o)(1 + \theta/\theta_o)^{w-1} \quad (4.9)$$

which can also be written as

$$(dq_f/d\theta)_i = q_{fo}w[f(\theta)/(\theta + \theta_o)] \quad (4.10)$$

It is proposed that the strength sensitivity index, S_T, for frozen soil be defined by the ratio

$$S_T = \frac{\Delta q_{fi}}{q_{fi}} \quad (4.11)$$

where

q_{fi} = strength at temperature θ_i, and
Δq_{fi} = its variation due to temperature change.
As from Eq.(4.1):

$$q_{fi} = q_{fo}f(\theta_i) \quad (4.12)$$

and from Eq.(4.10):

$$\Delta q_{fi} = (dq_f/d\theta)_i \Delta\theta_i = q_{fo} w[f(\theta_i)/(\theta_i + \theta_o)] \quad (4.13)$$

Equation (4.11) yields the strength sensitivity index

$$S_T = w\frac{\Delta\theta_i}{\theta_i + \theta_o} \quad (4.14)$$

The index S_T is seen to be a function of the initial permafrost temperature, θ_i, the temperature change, $\Delta\theta_i$, and the frozen soil sensitivity to temperature change, expressed by w.

For example, an ice-rich silt ($w = 0.60$) at -7 °C will, for a temperature increase of 1 °C, undergo a strength loss of $S_T = 0.60[1/(7 + 1)] = 7.5$ %, while the same soil at temperature of -3 °C, would loose $S_T = 0.60[1/(3 + 1)] = 15$ % of its original strength.

(Note: As shown in Ladanyi (1997), there are several other possibilities for expressing the effect of temperature on frozen soil strength, proposed by different authors.)

The above defined strength sensitivity index, S_T, may be a useful measure for evaluating the strength loss of frozen soils in permafrost regions where climatic warming is not expected to result in a complete permafrost thawing. The index requires a certain knowledge on temperature sensitivity of strength and creep of some typical arctic soils. Although some such information already exists (See e.g. in Andersland and Ladanyi (1994) or Ladanyi (1997)), much more should be done in this direction by further laboratory and field testing of permafrost soils. Eventually, combining information on permafrost occurrence, soil types and their characteristics with the climate warming trends, it may be possible to construct climate warming effect microzonation maps, that would show not only the trends to active layer increase and permafrost thawing, but also the reduction of permafrost strength due to warming.

Permafrost sensitivity maps of this kind may become a useful basis for predicting the effect of climate warming on the existing constructed facilities in the Arctic, and for establishing guidelines for the design of the new ones. A proposal in this direction was made by Vyalov et al. (1993), Figure 5, who propose to establish a permafrost sensitivity zonation in the Arctic, based on the average annual ground temperature of permafrost, T_z. Four zones were proposed: Zone I (very unstable) with T_z from 0 to -1 °C, Zone II (mildly unstable) with T_z from -1 to -3 °C, Zone III (stable) with T_z from -3 to -7° C, and Zone IV (very stable) with T_z below -7 °C.

As for the effect of warming climate on particular construction elements, such as foundations, roads and dams, eventual design guidelines should first establish if and how much the present normal safety margins cover the effects of climate warming during the lifetime of the structure, and how and when the climate warming will become a decisive factor in the design. However, at this moment, lacking such guidelines, and considering the observed warming trends, a warming rate of about 0.1 °C/year seems a reasonable assumption in the design of future constructed facilities in the Arctic.

5 CONCLUSIONS

1. Climate warming should be taken into account in engineering design where appropriate and where the effect would represent an important component of the geothermal design.

2. Climate warming could be introduced into a code or manual of practice, if necessary.

3. In engineering projects, climate warming may

be a factor, if its effects go beyond those anticipated in an existing conservative approach.

    4. Sensitivity maps for the loss of strength of permafrost due to climate warming can be established by using a simple strength sensitivity index, as the one proposed in this paper.

6 REFERENCES

Andersland, O.B. & Ladanyi, B. 1994. *An introduction to frozen ground engineering,* Chapman & Hall, New York, 352p.

Burgess, M.M. & Riseborough, D.W.1990. Observations on the thermal response of discontinuous permafrost terrain to development and climatic change - An 800 km transect along the Norman Wells pipeline. *5th Canad. Permafrost Conf.,* Collection Nordicana, No.54, Laval University, Québec, pp. 291-297.

Esch, D.C. 1993. Impacts of northern climate changes on Arctic engineering Practice. In: *"Impacts of Climatic Change on Resource Management in the North",* 4th Canada-US Symp. & Biennal AES/DIAND Meeting on Northern Climate, pp. 185-192.

Esch, D.C. and Osterkamp, T.E. 1990. Cold regions engineering: Climatic warming concerns for Alaska. *J. of Cold Regions Engineering, ASCE,* Vol.4, No.1, pp. 6-14.

Etkin, D. 1990. Greenhouse warming: Consequences for arctic climate. *J. of Cold Regions Engineering, ASCE,* Vol.4, pp. 54-67.

Ladanyi, B. 1972. An engineering theory of creep of frozen soils. *Canad. Geotech. J.,* Vol.9, pp. 63-80.

Ladanyi, B. 1997. Mechanical properties data base for ground freezing applications. *Int. Symp. on Ground Freezing and Frost Action in Soils,* Luleå (S. Knutsson, ed.), Balkema, Rotterdam, pp. 43-52.

Nixon, J.F. (Derick) 1990. Effect of climate warming on pile creep in permafrost. *J. of Cold Regions Engineering, ASCE,* Vol.4, No.1, pp. 67-73.

Nixon, J.F. (Derick), Sortland, K.A. & James, D.A. 1990. Geotechnical aspects of northern gas pipeline design. *5th Canad. Permafrost Conf.,* Collection Nordicana No.54, Laval University, Québec, pp. 299-307.

Osterkamp, T.E. & Lachenbruch, A.H. 1990. Thermal regime of permafrost in Alaska and predicted global warming. *J. of Cold Regions Engineering, ASCE,* Vol.4, No.1, pp. 38-42.

Riseborough, D.W. 1990. Soil latent heat as a filter of the climate signal in permafrost. *5th Canad. Permafrost Conf.,* Collection Nordicana No.54, Laval University, Québec, pp. 199-205.

Sayles, F.H. 1968. *Creep of frozen sands.* U.S. Army C.R.R.E.L., Hanover, NH, Tech. Report 190.

Smith, M.W. 1990. Potential responses of permafrost to climatic change. *J. of Cold Regions Engineering, ASCE,* Vol.4, No.1, pp. 29-37.

Smith, M.W. & Riseborough, D.W. 1983. Permafrost sensitivity to climate change. *4th Int. Permafrost Conf.,* Fairbanks, AK, Nat. Academy Press, Washington, DC, pp. 1178-1183.

Smith, M.W. & Riseborough, D.W. 1985. The sensitivity of thermal predictions to assumptions in soil properties. *4th Int. Symp. on Ground Freezing,* Sapporo, Japan, Balkema, Rotterdam, Vol.1, pp. 17-23.

Vyalov, S.S., Gerasimov, A.S., Zolotar, A.J. & Fotiev, S.M. (1993). "Ensuring structural stability and durability in permafrost ground areas at global warming of the Earth's climate." *6th Int. Permafrost Conf.,* Beijing, China, South China University of Technology Press, pp. 955-960.

Geotechnical Hazards, Marić, Lisac & Szavits-Nossan (eds) © 1998 Taylor & Francis, ISBN 90 5410 957 2

Analysis and performance of an embankment on organic subsoil

Z. Lechowicz, J. Bakowski & S. Rabarijoely
Department of Geotechnics, Warsaw Agricultural University, Poland

ABSTRACT: Paper presents an analysis and performance of the staged construction of main dam embankment of Nielisz reservoir. The results of laboratory and in situ tests performed for 2-staged embankment with preloading on soft subsoil, consists of mud and organic mud layers are presented. Field vane tests, cone penetration tests and dilatometer tests were carried out to obtain the design profiles of mechanical parameters for soft soils. Laboratory tests were used to compare with the results of in situ tests. The complementary treatment of field and laboratory test results made it possible to estimate the actual state of the complex subsoil conditions and safety factor for staged construction. Prediction of the increase in undrained shear strength of organic soils was carried out. Consolidation prediction was performed based on large strain analysis taking into account non-linear characteristics and secondary compression of organic soils. Organic soils strengthening as well as predicted settlements and excess pore pressure were approved by the field measurements performed during construction.

1 INTRODUCTION

Embankments are sometimes have to be located on organic subsoil. Using construction by stages the increase in shear strength due to consolidation can be utilised (Hartlen & Wolski 1996). Use of preloading technology made it possible to reduce the postconstruction settlements. In case of staged construction the main problems which should be solved in design stage are evaluation of a safe loading rate and prediction of consolidation performance).

Due to high compressibility and very low virgin shear strength of organic soils it is necessary to determine very carefully the initial geotechnical conditions as well as their change during embankment construction). The level of geotechnical hazard increases when embankment is constructed on the soft subsoil with organic soil layers with significant spatial variability in thickness and mechanical properties. Properly accounting for spatial variability of organic soils properties reduced substantially the uncertainties associated with the design and performance of staged construction.

2 DESCRIPTION OF THE TEST SITE

Nielisz test site is located in south-eastern Poland in the Wieprz river valley where in period 1994 - 1996 extensive testing programme including laboratory and in situ tests was carried out (Lechowicz & Rabarijoely 1997).

Due to appearance of soft soils the main dam embankment of Nielisz reservoir was constructed by stages utilising the increase in shear strength due to consolidation. Calculations performed in design stage indicated that the use of preloading technology enabled to reduce the postconstruction settlements and to limit the time of upstream construction. Necessity to finish the dam construction in three years forced very close cooperation between designer and geotechnical engineers. Because spatial variability in thickness and mechanical properties of soft soils high differences in settlements caused by embankment loading were expected. Due to significant differences in settlements more flexible sealing and protection of upstream slope as well as drainage system were selected (Fig. 1). Finally main dam embankment was constructed in 2-stages with preloading fills (Fig. 2).

At the Nielisz site soft subsoil consists of mineral and organic sediments. The original thickness of soft soils at the site varied from 1.0 to 5.0 m. The upper about 0.5 - 1.0 meter mainly consists of the sandy silt or silt. Further down, there is the layer of mud or organic mud with thickness of 1.0 - 4.0 m mostly divided into two layers by the silt layer. Below the sand layer is found. The results of the index properties of organic soils are given in Table 1.

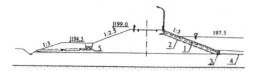

Figure 1. Main dam embankment of Nielisz reservoir: 1 - riprap on the geotextile, 2 - sealing by the geomembrane on the geotextile covered by sand layer, 3 - stone toe, 4 - sealing by geomembrane covered by sand layer, 5 - stone toe drain surrounded by geotextile.

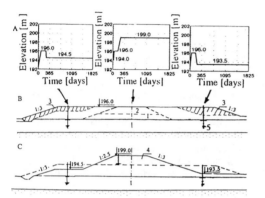

Figure 2. Dam embankment constructed in 2-stages with preloading: A - construction schedule, B - first stage with preloading fills, C - second stage with removal of preloading fills and increase of embankment crest; 1 - existing embankment, 2 - removal of existing embankment to elevation 194.0 m, 3 - preloading fill, 4 - increase of embankment crest up to final elevation 199.0 m, 5 - settlement gauge.

Table 1. Index properties of organic soils

Properties	Organic mud	mud
Water content w_n [%]	120 - 200	65 - 120
Unit density ρ [t m^{-3}]	1.2 - 1.3	1.3 - 1.5
Specific density ρ_s [t m^{-3}]	2.1 - 2.3	2.3 - 2.5
Liquid limit w_L [%]	130 - 220	70 - 130
Organic content [%]	21 - 35	8 - 20

At the site slightly organic soils appear as mud with an organic content between 8% and 20%. Organic soils classified as organic mud appear with the organic content between 21% and 35% with higher values in the upper layer and lower values in the lower layer.

Outside of the existing embankment under the downstream berm and the upstream slope the soft soils are overconsolidated with an overconsolidation ratio, OCR, decreasing from 3 to 2 with depth. At the end of first stage the effective vertical stress was

higher than the initial preconsolidation pressure. During second stage in soft subsoil under the embankment crest the effective vertical stress exceed the initial preconsolidation pressure several times.

In order to obtain the realistic picture of subsoil condition at the Nielisz site the results of laboratory and in situ tests were analysed to evaluate the spatial variability of soft soils. Special attention was paid to properly accounting for the thickness of soft subsoil, soil layer sequence, stress history and mechanical soil parameters. From analysis the typical cases of soil layer sequence were selected. The most compressible organic subsoil consists of organic mud divided into two layers by the silt layer. As an example the organic subsoil in cross-section hm 4+50 is presented below.

The profiles of material index I_D, lateral stress index K_D, dilatometer modulus E_D and pore pressure index U_D from dilatometer test with the average values and ± one standard deviation calculated for selected layers for virgin soft soils in cross-section hm 4+50 are shown in Figure 3. Figure 4 shows the profiles of cone resistance q_c, sleeve friction f_s and friction ratio R_f from cone penetration test in the same location.

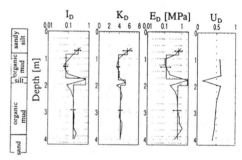

Figure 3. Index parameters I_D, K_D, E_D and U_D profiles from dilatometer test in virgin soft subsoil in cross-section hm 4+50.

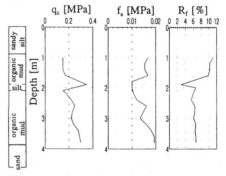

Figure 4. Profiles q_c, f_s and R_f from cone penetration test in virgin soft subsoil in cross-section hm 4+50.

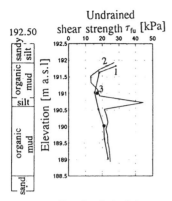

Figure 5. Profiles of undrained shear strength in virgin soft subsoil in cross-section hm 4+50: 1 - field vane shear test, 2 - dilatometer test, 3 - CK_0U triaxial tests.

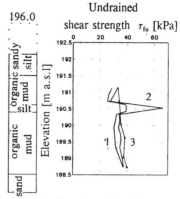

Figure 6. Profiles of undrained shear strength in cross-section hm 4+50 at the end of first stage: 1 - field vane shear test, 2 - dilatometer test, 3 - predicted.

On the basis of in situ test results undrained shear strength profiles were determined. Correction factors estimated according to the SGI recommendations (Larsson et al. 1984, Wolski et al. 1988) were used to evaluate undrained shear strength from field vane shear tests. To obtain the undrained shear strength from dilatometer test the equation proposed by Roque et al. (1988) with bearing capacity factor N_c estimated from the equation proposed by Lechowicz and Rabarijoely (1997) was used. A comparison between undrained shear strength obtained from CK_0U triaxial tests and in situ tests for organic soils in cross-section hm 4+50 is shown in Figure 5. The evaluated undrained shear strengths of organic soils are low and amount to 15 - 20 kPa in the upper organic mud layer and 20 - 25 kPa in the lower organic mud layer.

3 STRENGTHENING OF ORGANIC SOILS

For the design of the schedule of embankment construction the prediction of increase in shear strength due to consolidation was carried out. Increase in shear strength was predicted using method developed by Lechowicz (1994) based on the non-linear relationship between undrained shear strength and effective stress both in the overconsolidated and normally consolidated states.

An example of predicted profile of undrained shear strength in cross-section hm 4+50 for consolidated subsoil beneath the preloading fill from the upstream side at the end of first stage is shown in Figure 6. A considerable increase in shear strength both in the upper and lower organic mud layers is evident at the end of first stage.

A comparison of the predicted undrained shear strength with values obtained from in situ tests shows quite good agreement.

4 CONSOLIDATION ANALYSIS

The prediction of settlements and excess pore pressures was carried out using numerical method developed by Szymanski (1994) based on the large-strain consolidation analysis taking into account non-linear characteristics and secondary compression of organic soils. Predicted values were compared with field measurements of settlement gauges and pore pressures measured by BAT system.

An example of predicted settlements of organic subsoil in cross-section hm 4+50 under the downstream berm are shown in Figure 7. The calculations of settlements were performed for the embankment with the elevation of 194.5 m (final design berm) and 196.0 m (preloading fill). Calculation results indicate, that final settlement caused by design berm equals 0.19 m.

Under the preloading fill the total settlement expected for the design berm was predicted after 5 months of consolidation. Field measurements show that before removal of preloading fill the settlement equals 0.26 m. A comparison between predicted and measured rates of settlement indicates a good agreement.

Field measurements show that after removal of preloading fill only small amount of swelling was observed. After swelling the observed settlements were much higher than expected values caused by design berm.

An example of predicted excess pore pressures in organic subsoil in the same location for the embankment with the elevation of 194.5 m and 196.0 m is shown in Figure 8. A comparison of predicted values indicates that under preloading fill the excess pore pressures are higher after half a year. Higher values of excess pore pressure after removal of

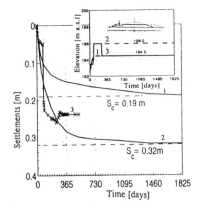

Figure 7. Comparison of predicted and measured settlements of subsoil in cross-section hm 4+50 under downstream berm: 1 - predicted, for elevation 194.5 m, 2 - predicted for elevation 196.0 m, 3 - measured.

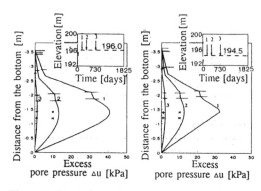

Figure 8. Comparison of excess pore pressure in cross-section hm 4+50 under downstream berm: 1 - after half a year, 2 - after one year, 3 - after two years; x, + - values measured by BAT system.

preloading fill quite fast dissipate which is confirmed by pore pressure measured by BAT system (Fig. 8).

Results of settlement prediction for soft subsoils under design downstream berm in other cross-sections indicate that expected settlements varied from 0.05 m to 0.30 m. Field measurements show that expected settlements were passed after half a year of consolidation. Before removal of preloading fill at the end of the stage measured settlements varied from 0.15 m to 0.45 m.

5 CONCLUSIONS

Main dam embankment of Nielisz reservoir is an example of staged construction with preloading in difficult geotechnical conditions. Short time period for dam construction with significant variability in the subsoil conditions caused necessity to perform more detailed geotechnical investigation and analysis. The complementary treatment of field and laboratory test results made it possible to estimate the actual state of the complex geotechnical conditions. Good agreement has been obtained between the predicted values of settlement and undrained shear strength and those measured at the test site. The analysis of the results indicated that the use of preloading technology in case of 2-staged construction of Nielisz dam enabled to speed up the performance of sealing and protection of upstream slope as well as drainage systems.

REFERENCES

Hartlen, J. & W. Wolski 1996. *Embankments on organic soils.* Elsevier, Amsterdam.

Larsson, R., U. Bergdahl & L. Eriksson 1984. Evaluation of shear strength in cohesive soils with special reference to Swedish practice and experience. *Swedish Geotechnical Institute,* Linköping, Information 3: 1-32.

Lechowicz, Z. & S. Rabarijoely 1997. Use of dilatometer test in evaluation of organic subsoil strengthening. *Conf. on Recent Advances in Soft Soil Engineering.* Malaysia. Kuching: 185-196

Lechowicz, Z. 1994. An evaluation of the increase in shear strength of organic soils. *Proc. Advances in Understanding and Modelling the Mechanical Behaviour of Peat.* Delft: 167-179.

Roque, R., N. Janbu & K. Senneset 1988. Basic interpretation procedures of flat dilatometer tests. *Proc. Int. Sym. on Penetration Testing ISOPT-1,* Orlando, 1,: 577-587.

Szymanski, A. 1994. The use of constitutive soil models in consolidation analysis of organic subsoil under embankment. *Proc. Advances in Understanding and Modelling the Mechanical Behaviour of Peat.* Delft: 231-240.

Wolski W., A. Szymanski, J. Mirecki, Z. Lechowicz, R. Larsson, J. Hartlen, K. Garbulewski & U. Bergdahl 1988. Two stage-constructed embankments on organic soils. *Swedish Geotechnical Institute, Linköping, Report* 32.

Geotechnical Hazards, Marić, Lisac & Szavits-Nossan (eds) © 1998 Taylor & Francis, ISBN 90 5410 957 2

Time settlement monitoring of earthfill piers

S. Manea
Technical University of Civil Engineering Bucharest, Romania

R. Ciortan
Design Institute for Roads, Air and Water Transports (IPTANA SA), Bucharest, Romania

ABSTRACT: The earthfill for the pier designated "I S" and "II S" and intended to contain handling in the new Constantza South harbour has been constructed between 1986-1994 from local clayey materials and by using various techniques. For modelling the specific behaviour of these fills, some computer calculations were processed, by taking into account various hypotheses and geotechnical parameters obtained from special laboratory tests. Also, in situ measurements have been performed over a period of about two and a half years. A set of in-situ tests was performed on "I S" pier consisting of platform loading. The paper presents a comparison between measured and calculated settlements, as well as possible explanations concerning some specific effects as remarked when such harbour earth structures are constructed (zones where high porosity of earthfill were recorded as a result of construction techniques, driving out of earth fragments through submerged surrounding dikes and so on).

1 GENERAL

A rather cheap solution for acquiring new territories from the sea is to resort to embankments made of local materials, even if their mechanical proprieties are not so suitable.

The I S and II S piers in Constantza harbour were realised in the period 1986-1994 (Figure 1) by filling with earthen materials, mostly resulted from excavations performed for the Danube – Black Sea canal. The earthfills were surrounded by gravity quay walls (Figure 2) and laid by various procedures over different periods – from several months to a couple of years – so they are appreciably non – homogeneous in terms of geotechnical properties and behaviour. Thus the size of deformations as well as their evolution in time were variable, both during construction phase and later on.

The completion of piers, their putting into operation and the variety of their use in the framework of harbour activities imposed the choice of the most suitable solutions of structural systems and founding methods, taking into account the predicted settlements and their charges in the course of time.

2 GEOTECHNICAL FEATURES OF THE SITE AND OF PIER STRUCTURES

Geotechnical studies were performed by a team of the Design Institute IPTANA S.A. and in the Soil Mechanics Laboratories of the Technical University of Civil Engineering, Bucharest, aiming to determine the nature and properties of soil deposits resulted from filling works in the body of I S and II S harbour piers.

By previous investigations it had been ascertained that the that the original sea bottom on the site was located at 14 to 18 m depth, whereas the base limestone rock was found at 20 to 30 m below sea level. Consequently a 5 to 20 m thick sedimentary deposit was present, consisting of alternating non-cohesive materials (fine sand, silt and clay) which the actual fills were laid upon.

Other research borings and laboratory test revealed the non-homogenous character of fills in pier body, formed by a cohesive earth mass. On the top of these fills some layers of various thickness (0.6 to 3.5 m) were placed, consisting of coarse fragments such as stones and blocks at I S pier, or hard, well-compacted clay, at II S pier.

Non-homogenous local material fills are 14 to 15 m thick at I S and 18 to 19 m at II S pier.

All geotechnical laboratory tests on the above-mentioned fills were generally performed according to Romanian norms. Mechanical parameters were evaluated by incremental compression tests both in oedometers and consolidometers and by shear tests in reversible direct shear or triaxial compression apparatuses.

The laboratory test results have been processed by an original methodology (Andrei & Manea 1980)

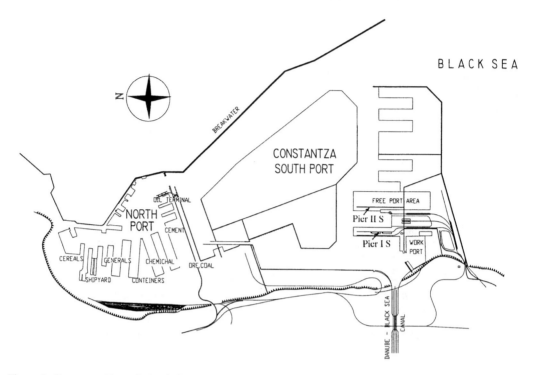

Figure 1. Constantza Port. General view.

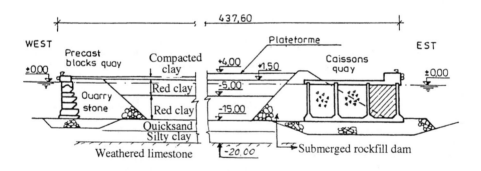

Figure 2. Cross section through piers I S and II S.

based on "prints" and "state diagrams", enabling the systematisation of geotechnical information and the classification of on-site materials (fills and natural ground) by types of characteristic soils, in order to predict their subsequent behaviour.

On this line the following geotechnical features of the pier fills were determined:
– the type of clay materials which mainly constitute these fills: fat clays, reacting with sea water, among them prevailing some expansive red clays resulted, as mentioned, from Danube-Black Sea canal excavations.
– the under-consolidated state of clayey materials, as regarding the time elapsed from the lay-down and the present overburden (OCR decreases until values as low as 0.4). This aspect is more conspicuous for the I S pier, where the quick construction rate and

228

the covering of fill by the quay platform resulted in the slowing down of the drainage process. On the contrary, for the II S pier, the longer duration of construction phase and some lithological circumstances allowed a higher consolidation degree of fills.

– owing to underconsolidation an increased compressibility of generally soft and plastic-consistent clayey materials was observed. The values of oedometric moduli for various pressure increments are appreciably different on each pier, so the value of $M_{200-300}$ lies between 2 and 5 MPa for I S pier, whereas for II S pier it reaches 4 to 7 MPa.

– a long-time consolidation process due to the clayey character of fills was noted. It is expressed by consolidation coefficient $c_v = n.10^{-4}$ to 10^{-5} cm^2/s and permeability coefficient $k = n.10^{-8}$ to 10^{-9} cm/s.

As concerning the original soil materials in the foundation of piers the following conclusions are to be noted:

– clayey silts, silts and muds encountered beneath fills down to 15-20 m depth below sea level are generally in a soft-plastic state, behaving as highly or extremely compressible sediments.

– clay and silty clay layers representing the transition to limestone base rock are in a stiff state, obviously displaying better mechanical properties. For instance, Figure 3 presents some clayey fill and in-situ ground materials prints.

3 SOME ASPECTS RELATED TO TIME BEHAVIOUR OF EARTHFILLS AS RESULTED FROM THEORETICAL ANALYSIS AND IN SITU MEASUREMENTS

Consolidation settlement under soil overburden is typical for all kinds of earthfills. The development of this process is influenced by soil type and physical state, construction methods, the amount of loads and their variation during construction and operation. The prediction of the behaviour of studied pier structures has been established by calculations and in-situ measurements on these piers.

A model and a computer programme has been elaborated even in 1994 for II S pier by taking into account some particular characteristics of this structure, especially form constructional and operational viewpoint. Based on previous studies, a prediction of settlement evolution has been carried out, considering both pier's own weight and external loads (Manea, Ciortan 1995, 1996). It resulted than about 50% of overall settlement occurred during construction period (1986-1994); as for the remainder a settlement rate of 3-4 cm/year has been presumed, with a damping trend.

Between 1994 and 1996 a set of measurements on the fills in the pier II S were performed, on topographic marks placed in the north-east part of the pier, over an area of about 70,000 m^2 (Figure 4).

The measurements revealed some specific effects of this kind of marine structures. Thus when earthen materials are laid as fills in closed spaces mostly below water level, an amount of low-consistency mass is accumulated on various thickness, in the final construction stages; in this case a soft, loose sandy-

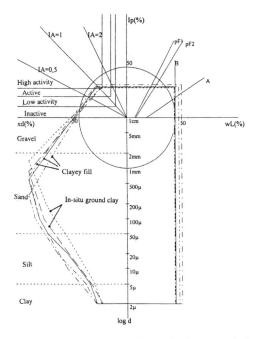

Figure 3. Prints for clayey fill and in-situ ground clay.

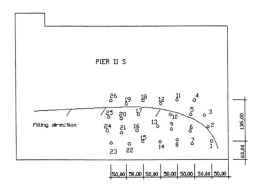

1-26 Topographic marks

Figure 4. Measurements location.

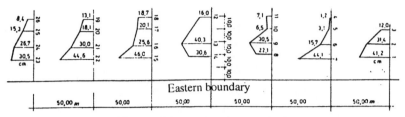

Figure 5. Measured settlements (cm)

clayey silt where $I_c \approx 0.1$ and $n \approx 40\%$ has been locally found on layers as thick as 3 to 4 meters.

On the other side, by wave permanent action on water level oscillations in the harbour basin, an exfiltration flow occurs through the submerged rockfill dam surrounding the working area, which may lead to the drawing out of particles if a suitable filter is not provided (Figure 5). Consequently on the peripheral zone of the pier larger deformations than in the central zone may appear.

Settlements measured over a period of two years in the eastern side of the fill, on a 350 m long and 50 m wide area, were as large as 25 to 45 cm, namely up to five times greater than those observed on the centre line of the pier.

Because the above-mentioned phenomena are practically not felt in the middle part, located at 100 to 150 m from the limit of the precinct, the measured settlements are not larger than 5 to 8 cm; they arose from consolidation process under the own weight of the fill. These values are in accordance with the values calculated.

During the subsequent 15 months (April 1996 to July 1997) measurements on boundary marks revealed a diminution of settlement rate, in accordance with the analytical prediction.

These aspects, correlated with future monitoring, are to be observed when the coming container terminal will be designed and constructed.

As for I S pier, now in current operation for depositing various goods, the need arose in 1997 to install a new terminal for cereals, consisting of 10 silos 30 m diameter and 30 m high, also imposing some special settlement restrictions. It is to note that in the course of two or three years of exploitation, under a current overburden pressure of 100 kPa no damage or settlement of pier platform was recorded.

Based on the behaviour of II S pier and taking into account by considering the geotechnical characteristics of the filling material a set of in-situ tests was found to be useful on the location of I S pier consisting of platform loading and settlement measurements on the site of designed structures.

These tests were performed by incrementally surcharging the pier platform (Figure 6) with steel plates on a circular area ($D = 23$ m, $A = 415$ m^2) exerting contact pressures up to 100 kPa.

Deformations were measured along two radial directions until 30 m apart from the centre of loaded area. Each load increment attained 15 kPa and was maintained constant over two days.

Pier platform "I S"

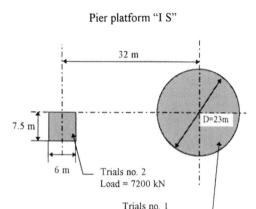

Figure 6. Experimental trials position on the platform

The final results of these full-scale experiments are shown in Figure 7. Settlements reached 114 mm in the centre zone, decreasing to 3-7 mm at a distance of 20 m from loaded area; they agreed with calculated values.

In the same time a similar load (150kPa) was applied on a smaller area ($A = 45$ m^2), the maximum measured settlements were about 7 mm, which emphasises the importance of the size of loaded area and consequently of the mass influenced by loading.

The ratio between the diameter of loaded area and the thickness of highly deformable soil layer was 1.4 in the first case and 0.4 in the second one.

The lack of settlements during operation, but also the occurrence of sudden deformations during load tests may be explained by the peculiar features of this embankment, which did not achieve but a small part of consolidation settlements, because the material was underconsolidated and had small drainage facilities.

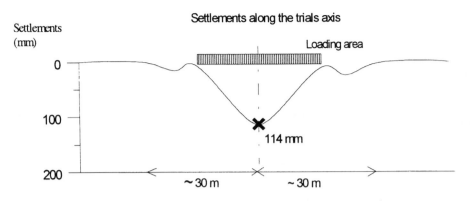

Settlements along the trials axis

Curves of equal measured settlements (mm)

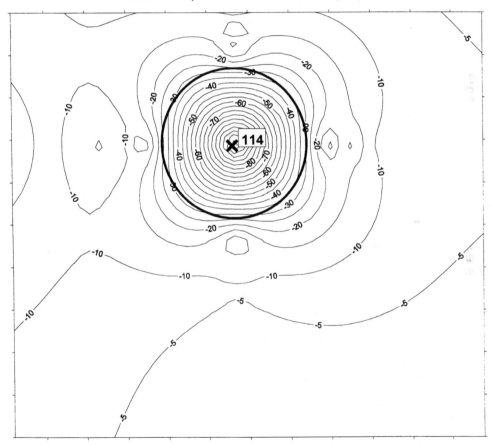

Figure 7. Settlement distribution within studied area

4 CONCLUSIONS

The geotechnical investigation of pier fills revealed the time behaviour of the material; it was also found that geotechnical parameters in the present time are different from those existing in the past, when earthfills were realised. Nevertheless the time-settlement curves based on analysis enabled with sufficient accuracy the prediction of settlement evolution during the first years of pier operation.

In the course of a period of 10 years of pier construction a considerable part (almost a half) of final settlement may occur, due to the own weight of the fill, if the body of the pier may be drained (as in the case of II S pier). If this is not possible (the I S pier) the material remains unconsolidated because undrained and the settlement process is prolonged until local conditions are modified, as a consequence of the clayey nature of the soil.

Loading tests obviously confirmed the importance of the size of loaded area and of the manner in which the load is applied as well.

Construction procedures under seaside conditions may lead to the formation of zones where the earthfill is loose or soft, thus needing certain improvement measures. On the zone close to the surrounding quay walls an inverted filter system is needed, in order to avoid the drawing out of fine particles from inside the pier, as a result of wave action and water level oscillations.

The monitoring of earthfill behaviour by in-situ measurements and full-scale tests allowed a realistic knowledge of on-site situation, ascertaining and completing geotechnical information and calculations carried out on this base. The processing of all data allows the most appropriate construction and founding solution to be found on his type of sediments and fills.

REFERENCES

Andrei, S. & Manea, S. 1980. Prediction on the expansive clay behaviour. *Proc. 4th Int. Conf. on Soils - Characterisation and Treatment of Expansive Soils*. Engineering Design, Denver, Colorado.

Andrei, S. & Manea, S. 1987. Forecast of moisture and volume changes in unsaturated soils. *Proc. of 9th ECSMFE*, Dublin.

Ciortan, R. et al. 1982. Soil parameters improvement for harbour depots by intensive dynamic consolidation. *Intensive dynamic consolidation of weak soils for transportation foundations*, in Romanian, IPTANA, Bucharest.

Costet, J. & Sanglerat, G. 1981. *Cours pratique de mécanique des sols*. Dunod. Paris.

Giroud, J.P. 1973. *Tables pour le calcul des fondations*. Dunod. Paris.

Manea, S. & Ciortan, R. 1995. Prediction of settlement evolution for a pier embankment in Constantza South Port. *Proc. of 10th Danube-European Conf. on Soil Mechanics and Foundation Engineering*, vol. 2, pp 355-364, Mamaia, Romania.

Manea, S. & Ciortan, R. 1996. Model and experimental study of earthfill behaviour within the "II S" mole at Constantza harbour. *Proc. of 8th National Conf. for Soil Mechanics and Foundation Engineering*, Iasi, Romania.

Geotechnical Hazards, Marić, Lisac & Szavits-Nossan (eds) © 1998 Taylor & Francis, ISBN 90 5410 957 2

Long-term filtration performance of nonwoven geotextiles

G. Mannsbart
Polyfelt GesmbH, Linz, Austria

Barry R. Christopher
Christopher Consultants, USA

ABSTRACT: This paper evaluates the long-term performance of needlepunched nonwoven geotextiles used as filters in coastal and river bank applications. Filtration and mechanical properties of nonwoven geotextiles exhumed from five separate projects in Europe and the Far East sites are reviewed. Hydraulic conditions varied from wave action to stream flow and the filtered soils ranged from sands to marine clays. Performance was based on the condition of the site and the hydraulic and mechanical condition of the exhumed geotextiles as well as comparison of the geotextiles used with respect to current filter design criteria.

1 INTRODUCTION

Needlepunched nonwoven geotextiles have been widely used as filter material in hydraulic construction works for more than 20 years. In order to investigate the long-term filter performance of needlepunched nonwoven geotextiles, a study has been carried out by Delft Hydraulics Laboratories (The Netherlands) where excavated samples from 5 locations in Europe and the Far East were tested with respect to their hydraulic and mechanical performance.

The objectives of the investigation were to evaluate the geotextile filtration performance after six to fourteen years in service and to qualify existing filtration design criteria adopted for extreme soil conditions. This was achieved by the execution of a series of hydraulic index (permeability and opening size) and performance tests on the exhumed geotextiles. The mechanical performance of the filter is mainly focussed on the installation. On the opposite the long term retention and hydraulic performance was evaluated.

2 PROJECTS EVALUATED

Three locations in Europe and two in Malaysia were selected on the basis of their locations and representation of a variety of subsoil and water flow conditions:

(a) River Danube at Greifenstein (Austria)
(b) Elbe - Lübeck canal near Mölln (Germany)
(c) Coastal protection at Lacanau (France)
(d) Coastal protection at Sungai Buntu (Malaysia)
(e) Coastal protection at Pantai Murni (Malaysia)

2.1 Greifenstein (Austria)

- Type of construction: river bank protection, 150 m downstream of a weir and shipping lock (see Figure 1).
- Year of construction: 1981-82
- Geotextile type: polypropylene continuos filament needlepunched nonwoven, opening size $O_{90,w}$ = 0.09 mm, permeability at 2 kPa = 4×10^{-3} m/s, mean elongation at break > 60%, puncture resistance > 3200 N.
- Soil: clay and silt 10 %, sand 59 %, gravel and stones 31 % (see Figure 1).
- Revetment: riprap, 50 - 200 kg stones, 1 m thickness, slope 2:3.
- High turbulence and wave action due to operation of the turbines of the power station; the excavation location was under water for 6 months of the year; water level variation is several meters.

Observations:
(a) Behaviour during and after installation:
The sample was taken near the average water level. The sample showed some damage had occurred to

the geotextile, most likely due to high installation stresses. During construction stones were reportedly dropped onto the geotextile from heights of up to three meters (or more) and hammered down with the excavator bucket to get a smooth surface. The riprap was placed directly on top of the geotextile, without a cushion layer of finder stones or sand, against normal construction recommendations.

The sample which was taken had a size of 5 m². It showed 2 holes which were directly beneath (i.e.

2.2 Elbe-Lübeck-Canal, Mölln (Germany)

- Construction type: navigation canal (Figure 2).
- Year of construction: 1978.

Geotextile type: special grade composite: polypropylene continuos filament nonwoven plus polypropylene staple fibres. The continuous filament nonwoven filter layer had the following properties: Opening size $O_{90,w}$ = 0.09 mm,

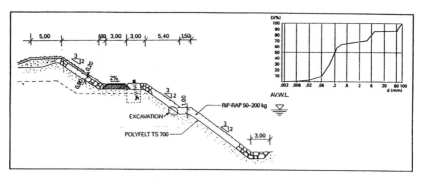

Figure 1. Cross Section and Soil Gradation Curve (Greifenstein)

not in between) the riprap stones. This puncturing had occurred due to unusal high installation stress.

(b) Functional behaviour of filter:
- The revealed soil surface still looked flat, no sign of erosion was observed. Since the construction of the embankment in 1981, no problems have arisen. For test results see Tables 1 to 4 in the Test Results section following.

Permeability at 2 kPa = 4 x 10^{-3} m/s, mean elongation at break > 60%, Puncture resistance > 3850 N. The additional nonwoven was to be used as a protective layer for the filter layer
- Soil: clay and silt 1,5 %, sand 62 %, gravel 36,5 % (see Figure 2).
- Revetment: Originally gabion mattresses;
Wave action: ship action, water level variation limited (0.5 m).

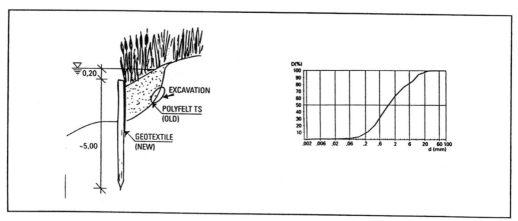

Figure 2. Cross Section and Soil Gradation Curve (Mölln)

Observations:

(a) Behaviour during and after installation:
The sample showed no damage due to installation stress.
The excavation took place just along the borderline of water and canal bank. The geotextile was covered with soil and a layer of riprap about 20 cm thick. In 1987 these gabions failed due to corrosion. Therefore, a wooden sheet pile with geotextile was installed. Behind the sheet piling the original geotextile was covered with soil (see Figure 2).

(b) Functional behaviour of filter:
The geotextile sample had no tears or other damage caused by construction or its use in the past 14 years. During these past 14 years, no problems have been reported. For test results see Tables 1 to 4 in the Test Results section following.

Observations:

a) Behaviour during and after installation:
The geotextile showed large deformations without any tear, illustration the good adaptability of nonwoven geotextiles to the shape of the revetment stones. The deformations measured on the geotextile were more than 50%.

b) Functional behaviour of filter:
The structure is in service since 1984; no complaints have been reported. For test results see Tables 1 to 4 in the Test Results section following.

2.4 Sungai Buntu (Malaysia)

• Type of construction: coastal protection (see Figure 10).
• Year of construction: 1986.

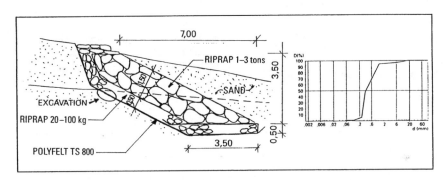

Figure 3. Cross Section and Soil Gradation Curve (Lacanau)

2.3 Lacanau (France)

• Construction type: coastal protection (Figure 3).
• Year of construction: 1984.
• Geotextile type: polypropylene continuous filament needlepunched nonwoven, opening size $O_{90,w}$ = 0.09 mm, permeability at 2 kPa = 4 x 10^{-3} m/s, mean elongation at break > 60%, puncture resistance > 4000 N
• Soil: 95 % sand (see Figure 3).
• Revetment: riprap, 50 cm thick with 20 - 100 kg stones + 150 cm thick with 1 to 3 tons stoned; slope 2:1; everything partly covered with sand; aim of the revetment is to protect the dunes against wave action.
• Wave action: full ocean stress from the Atlantic Ocean; during low tide, the construction is well above the sea level.

• Geotextile: polypropylene continuous filament needlepunched nonwoven, effective opening size $O_{90,w}$ = 0.09 mm, permeability at 2 kPa = 4 x 10^{-3} m/s, mean elongat. at break > 60%, punct. resist. > 2300 N
• Soil: clay and silt 63 %, sand 31 %, gravel 6 % (see Figure 4).
• Revetment: riprap, 5 - 20 kg stones, covered by 100 - 300 kg stones, slope 1:3 (see Figure 4).
• Wave action: very severe tidal action.

Observations:

a) Behaviour during and after installation:
The sample was taken from near the average water level. No damage was visible on the excavated geotextile sample.
b) Functional behaviour of filter:
No problems have been reported during the 8 years of installation. For test results see Tables 1 to 4.

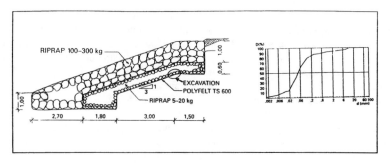

Figure 4. Cross Section and Soil Gradation Curve (Sungai Buntu)

2.5 Pantai Murni (Malaysia)

- Type of construction: coastal protection (see Figure 5).
- Year of construction: 1986.
- Geotextile: polypropylene continuous filament needlepunched nonwoven, op. size $O_{90,w} = 0.09$ mm, permeability at 2 kPa = 4×10^{-3} m/s, mean elongation at break > 60%, puncture resistance > 3200 N.
Soil: clay and silt 14 %, sand 67 %, gravel 19 % (see Figure 5).

A thick clay deposit was found above the primary armor, accumulated over the years by the incoming tide. The revetment profile had changed, but the filter conformed well to the deformed revetment system.

(b) Functional behaviour of filter
No problems have been reported during the 8 years of installation. The samples were taken from 2 different locations. For test results see Tables 1 to 4 in the Test Results section following.

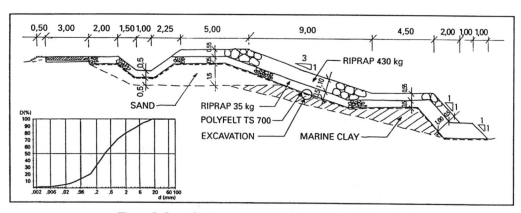

Figure 5. Cross Section and Soil Gradation Curve (Pantai Murni)

- Revetment: 50 cm thick (35 kg stones) + 110 cm (430 kg stones); slope 1:3 (see Figure 5).
- Wave action: very severe tide action.

Observations:
(a) Behaviour during and after installation:
The samples showed no damage due to installation stress.

3 TEST RESULTS

Functional properties
The permeability and the Effective Opening Size (090,w) tested according to Franzius Institute and Delft Hydraulics were determined for both the exhumed and the cleaned geotextiles samples from all locations.

236

The average results of each location are given in Tables 1 and 2. The results show that there is no significant difference in the average value of the opening size of the exhumed or cleaned geotextiles as compared to their original values. Sufficient pore sizes are still available to ensure high water flow and permeability.

Hydraulic Properties
Investigations on the permeability of the geotextiles

Long term functional properties
The subsoils from three sites were also obtained and carefully handled to preserve their original properties.
The subsoil and the geotextile were placed into a filtration test cylinder and normal pressure similar to that on the site was applied. These tests allowed the determination of both the soil permeability and also, more importantly, the influence of the presence of the geotextile in the permeability of the adjacent soil.

Table 1. Comparative opening size data for original (new), exhumed and cleaned geotextiles.

location geotextile	Opening Size O90,w			Percentage with regard to the original value		Ratio of exhumed to cleaned
	orig.[1] (mm)	exhumed[2] (mm)	cleaned[2] (mm)	exhumed (%)	cleaned (%)	(%)
Greifenstein, TS 700	0.09	0.09	0.101	100	112	89
Mölln, MPV 8061	-	0.084	0.080	-	-	105
Lacanau, TS 800	0.08	0.095	0.101	119	126	94
Sungai Buntu, TS 600	0.11	0.140	0.152	127	138	92
Pantai Murni, TS 700	0.09	0.120	0.111	133	123	108

[1] tested by the Franzius Institute   [2] tested by Delft Hydraulics

Table 2. Comparative permeability data for original, exhumed and cleaned geotextiles.

location/geotextile	Water permeability (k) $\Delta h= 0.10$ m			Percentage with regard to the original value		Ratio of exhumed to cleaned
	orig.[1] (l/m^2/s)	exhumed[2] (l/m^2/s)	cleaned[2] (l(m^2/s)	exhumed (%)	cleaned (%)	(%)
Greifenstein, TS 700	190	76	112	40	59	68
Möll, MPV 8061	47	23	44	49	94	52
Lacanau, TS 800	120	41	62	34	52	66
Sungai Buntu, TS 600	250	79	210	32	84	38
Pantai Murni, TS 700	190	52	125	27	68	42

[1] tested by the Franzius Institute   [2] tested by Delft Hydraulics

(see Table 2) support the above results by the fact that a reduction factor of only 1.5 to 3.2 was obtained within the period where the geotextiles were in service. This seems to be very small in comparison with a safety factor of 100 ($k_{geotextile}$ vs. k_{soil}) used in the design. The weight of the samples was measured to assess the quantity of soil particles inside the filter layer. (Table 4) The remaining porosity is very high, which explains the small reduction of permeability.

The lowest migration of fine particles is obtained for the sites of Lacanau and Mölln because they are both coarse or uniform sand. On the other hand it can be stated that for fine soils (Greifenstein, Sungai, Pantai) the clogging decreases with the thickness of the filter layer.

The results showed that the k-value of the zone „geotextile plus 5 cm adjacent soil" was up to 18 times higher than the permeability of the soil.

Table 3. Filter performance test results

Location	Greifenstein	Mölln	Lacanau
Year	1981	1978	1984
Geotextile	280 g/m2 nonwoven	860 g/m2 nonwoven	400 g/m2 nonwoven
Duration	86 hours	12 days	8 hours
k_s (soil)	5 x 10^{-6} m/s	5 x 10^{-6} m/s	3.5x10^{-6} m/s
k_r(gtx+soil)	90 x 10^{-6} m/s	50x10^{-6} m/s	3.5x10^{-6} m/s
k_r/k_s	18	10	1

At the site of Lacanau this higher permeability was not observed because the soil to be filtered was a uniform sand. In this case the soil skeleton is formed by the smallest particle and no "secondary" mineral filter will build up.

Installation related properties
Mechanical tests (weight as well as strip, CBR-

(d) The needlepunched nonwoven geotextiles which were installed in various hydraulic structures for 6 to 14 years all performed well. The opening to size did not change, the permeabilities are still although they are reduced slightly, and the mechanical properties changed within the expected and acceptable range.While on one of the projects the geotextile was visisbly damaged, the ability of the

Table 4. Weight, thickness and strength test results of original, exhumed and cleaned geotextile samples.

location/geotextile	Weight			Pores filled[3]		Strip Tensile Strength (N/5 cm)			CBR Puncture Strength (N)		
	orig.[1] (g/m²)	exhum. (g/m²)	cleaned[2] (g/m²)	(%)	orig.[1] (mm)	orig.	exhum.	+/-	orig.	exhumed	+/-
Greifenstein, TS 700	280	1560	299	18	2.6	780	676	-13%	3000	2340	-22%
Mölln, MPV 8061	865	1305	905	4	5.8	-	-	-	-	-	-
Lacanau, TS 800	400	715	400	3	3.3	1080	1000	-7%	3400	3485	+3%
Sungai B., TS 600	200	1242	228	11	2.0	560	535	-4%	2150	1970	-8%
Pantai M., TS 700	280	2523	313	17	2.6	780	713	-9%	-	-	-

[1] tested by the manufacturer   [2] tested by Delft Hydraulics   [3] estimated
Note: original samples tested by the Manufacturer and exhumed samples tested by Delft Hydraulics.

puncture and grab strength - where the size of the sample allowed the performance of such tests) were also carried out on the exhumed samples. The results are summarized on Tables 4 and 5. They indicate that the reduction in tensile strength properties on the order of 5 to 15% has occurred. It is felt by the authors that this small amount of strength loss may have occurred during installation. Such losses have been reported in the literature.

4 SUMMARY AND CONCLUSIONS

(a) The results of the opening size tests showed no significant difference between as exhumed and cleaned geotextiles.
(b) The permeability / permittivity tests (Table 2) showed a reduction in the permeability of the as-exhumed geotextiles as compared to the clean geotextiles ranging from a factor of 1.5 to 3.2 but their permeability still well exceeded the permeability of the soils tested (Table 3). This is due to low filtration thicknesses of the filters. If more resistance to puncturing is needed the additional thickness should not influence the filter performance.
(c) The performance tests showed an increasing permeability in the direction of the geotextile, indicating the creation of a natural filter at the geotextile-soil interface.

geotextile to conform with the riprap and the underlying soil apparently prevented soil migration problems at these isolated locations. The elongation of the geotextile is an important property to prevent failure due to high installation stresses.
(e) A need of good contact between the geotextile and the soil is needed to build up a "secondary" filter layer and to avoid local clogging due to migration of fines
(f) In general it can be stated that the weight of the products used are not very high for coastal protection (300-400 g/m²) This probably explains the good filter behaviour but also some problems due to installation stress. The elongation at break plays an gimportant role regarding this „survivability" of the filter geotextile.

5 LITERATURE

De Groot and Verheij (1993), Long-term filter performance of Polyfelt geotextiles, Delft Hydraulics Laboratory, Report No Q 1539.
Lafleur, Eichenauer and Werner (1996); geotextile Filter Retention criteria for well graded cohesionless soils, Geofilters 96 conference, Montreal .
Giroud (1996), Granular Filters and Geotextile Filters, Geofilters 96 conference, Montreal.

Geotechnical Hazards, Marić, Lisac & Szavits-Nossan (eds) © 1998 Taylor & Francis, ISBN 90 5410 957 2

Indeterminacy and risks in the earth-mechanical prognoses

D. N. Milewski
Federation of Scientific and Technical Unions, Sofia, Bulgaria

ABSTRACT: The construction practice is full with examples of indeterminacy leading to unexpected consequences, disappointment or even considerable injuries.
It is very risky to rely on the geotechnical model of engineering-geology space, based on variable soil as well as on the safety factor widely used in the engineering practice. The safety factor is not sufficient to prevent risk. It could only decrease the risk or lead to reinsuring. The risk requires additional consideration and evaluation of the adequacy between the real space and its model. This calls for continuous control from the beginning of the geotechnical research to the end of construction work. In the cases when the first class engineering equipment is concerned the control should continue even during the operation.
Methodical instructions for decreasing the indeterminacy and for choosing the proper decision are given in the paper.
Examples are given for the case when untraditional approach should be applied in order to estimate the risk and to choose the final decision.

1. INTRODUCTION

The final product of the solution of soil mechanical problems is the prognosis about the conduct of construction soil, in result of the natural changes in the composition and its stress - strain state under the expected loading during operation. The purpose is to obtain the quantity evaluation of possible undesirable reflects in the time and the selection of protection. Similar to the traditional medical doctors practice is to be used, following the sequence: case history, diagnosis, treatment. For soil structures, it is necessary to apply the systematic approach: purposeful information about the study of the engineering - geological area's and of the subject of protection in the frame of their interaction - geotechnical modelling, grounded with a prognosis concerning the advisability.

Regardless if the said activities would be fulfilled by one person or by different specialists, the rule is that the good doctor treats not the decease but the ill person and that the success depends on the solution of the dilemma if we could rely on the present information related with the engineering - geological elements** and with the geotechnical model made with them.

2. INSTABILITY AND INDETERMINACY OF EARTH CHARACTERISTICS:

Using the generalisations of G.K.Bondarik, we could note the following:
The process of forming the properties of rocks and earth is not fully determining and is complicated by a casual component.

---

* *Engineering and geological area- area of the earth crust studied for the purposes of the engineering solutions and prognoses.*

** *Engineering and geological elements- geological corpses forming the engineering and geological area, having a proved individual characteristics.*

The indices about the qualities of the lithological variations, if studied through their co-ordinates - p = f (x,y,z,t), form casual areas, which, due to their complicated configuration, could be described mathematically only with the help of distribution of probabilities on a wide scheme of realisations.

It is impossible during the experiments to repeat exactly the individual realisation of the area of the characteristics, regardless of the exactitude in measuring the detailed structure of the area.

There are basic and general regimes of variation for the geological corpses related to the features of composition, status and qualities of the lithological variations in space and time, which reflect the fundamental laws for their distribution. Using the morphological and the geostratigraphical analysis, the said particularity permits to reveal the genetic determination of geological space with nonhomogeneous structure.

It is well known as well that the methods of observation and studies, together with the experience of the lab assistant and the specialist, are as well sources of casualty, having subjective character and it is accumulated on the natural objective casualty.

The totality of variations, coming out from the listed factors, form the so called "indefiniteness of the situation". When the casualties result from theoretical or methodological positions, the indefiniteness is fundamental. Intermediate group - mainly characteristics of soil, for which soil mechanics couldn't find out the way to include its mathematical apparatus (mineral composition of soil, quantity and chemical composition of absorbed ions, relying only on the theory of elasticity or only on the law of Coulomb, etc.)

The practice of construction is abundant of examples about indefiniteness, leading to surprises, to disappointments or even to great damages. The relying on the geotechnical model of engineering and geological space formed of different soils, as well as on the factor of safety, widely used in the engineering practice, is full of risk. The factor of safety is not sufficient for risk protection. It is able only to decrease the risk or leads to a reinsuring.

For the cases of casual indefiniteness, statistic methods of probability are applicable for evaluation of variation of soil characteristics, for stratification of engineering - geological space, of engineering - geological elements and for introducing of guaranteed calculated soil mechanical features with standardised reliability. The specialised literature is rich of methods for statistic analysis, but in that case also the risk is inevitable. It is smaller in smaller sites and in the case of multiple constructions on the same geological formation.

The less reliable evaluation of safety with the biggest risk for the engineering solutions about the nature and the size of the measures is when the sites are: earthfill dams, built with different sorts of soils or heterogeneous soil without having selected the soils by zones in the dam; revival of ancient landslide; hydraulic filled tailings dams etc.

3. EXAMPLE

The "Shishmanov val" dam is a typical example. The dam is built in 1963. Investigations for evaluation its stability have been made in 1972 and 1980. Every time when using circular slip surface model, we achieve a static factor of safety less than one - inadequacy between the real state and the geotechnical model. The program for investigation of the dam corresponds to the following requirements:

- The methods for investigation of the characteristics of the soil material and the state of the dam have to allow going into details;
- Not to use average parameters but experimentally obtained characteristics used for assessment and decision;
- The mathematical model for assessment of the reliability of the dam have not to use acceptance of the surface and uniform shear strength, etc, thus leading to conventionality.

In order to fulfil these requirements for "Shishmanov val" dam, we made following parallel investigations:

on disturbed and undisturbed soil samples from the body of the dam and its base;
with cone penetration (CPT);
with VAN-TEST method;
with pressiometer.

Research of souls in according with Bulgarian governement standarts (BGS).

Such "monitoring" diagrams show and prove the following:
-diversity of the properties of the soils as earthfill material and as already built earthfill dam;

deviations of the water content from W_{opt} and the density from the standard maximum density (Fig. 1);

-the dam body contains soils ranging from dispersive (Class C in Fig. 2) to swelling soils(Class A-c and A-d$_{1,2}$);

- a permeability reaching $k_f = 5 \times 10^{-4}$ cm/s and a specific yield up to $\mu = 0.08$ is characteristic for the first, and for the second $k_f < 1 \times 10^{-5}$ cm/s and $\mu < 0$;

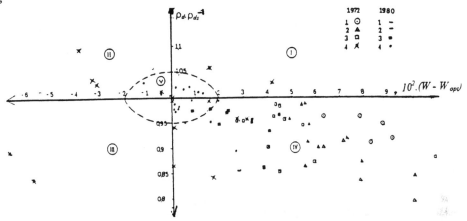

Fig. 1. *Diagram of deviation from optimum state W - W $_{opt}$ and $\rho_d \cdot \rho_{ds}^{-1}$ with the example as in fig. 1; I quadrant - overmoisted, overcompacted soils; II undermoisted - overcompacted soils; III undermoisted - undercompacted soils; IV overmoisted - undercompacted soils; V centre for optimum state of compacted soils*

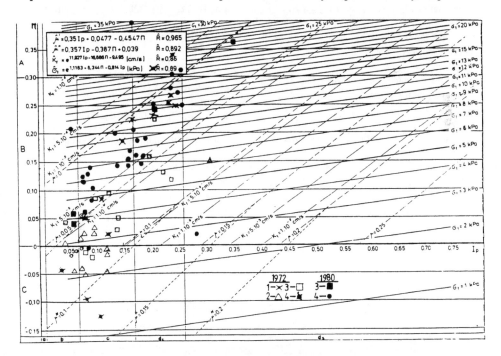

Fig 2. Nomogram for determining the specific yield for clayey soils μ' and for fine clayey sands μ''; permeability k_f (cm/s); tensile strength of practically saturated soils - σ_t , (kPa); $\Pi = (\rho_s W_L \cdot \rho_w^{-1} - e) \cdot (1+e)^{-1}$; ($\rho_s$ -specific density, ρ_w- water density).

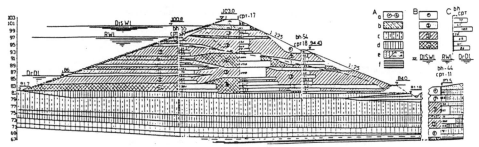

Fig. 3. Cross section of " Shishmanov val" dam. A. Lithological varieties: a (1 to 4) earthfill body; b- silty clay; c- loess clay ; d-loess; e- silty sand and pliocene small size clayey sand; f- pliocene compact clay. B. Zoning of dam after penetration resistance; q - CPT.

-the content of silty and fine sands ($Ip < 0.07$ and $\Pi < 0.05$) is found, which can hold water in the pores more than W_L even under standard maximum compaction, and their tensile strength varies from 1 to 6 kPa;

-soils for which a durable pore water pressure is characteristic are found; fortunately their share is limited, mainly beneath the phreatic surface.

Altogether 3826 readings of the penetration resistance test were carried out in the dam, and correspondingly so many readings of shear strength.

From the parallel investigation data objective equations of regression with low and very low correlation relation are obtained.

Figures 1 and 2 which allow application of claster and trend analysis for characterization of the soil are made.

Cross section profiles with different total shear strength are obtained (Fig. 3).

With the exception of the method of finite elements, all other known methods are modifications of assessing the safety of the dam after the limit state. The methods considered as elementary and used to obtain the factor of safety, meet to a higher extent the requirements:

1. Maslov - (1977) : $\eta = \tau_i / \tan \beta \sum\limits_{1}^{i} \gamma_i h_i$

2. Bromhead* (1986):

   $\eta = (\tau_i / \gamma_i h_i) \sec\beta \, \mathrm{cosec}\, \beta$

3. Gibson & Morgenstern (1962)

$\eta = N_{(\beta)} (\tau_i / \sum\limits_{1}^{i} \gamma_i h_i)[1 - (H\gamma_w / \sum\limits_{1}^{i}\gamma_i h_i)]$

where τ_i is the total shear strength of the separate layers with thickness h_i and volume weight γ_i; γ_w - volume weight of the water;

β - angle of dame slope;

$H = \sum\limits_{1}^{i} h_i$; N- tabulated coefficient depending on β.

A profile of the dam defined by isolines - geometrical place of points with equal values of static factor of safety is made (fig.4).

Analysis for assessment of the erosion resistance of the dam soil is made. As the result of the analysis we show that the dam has zones with $\eta \leq 1$ but they are closed as in capsule by soil with higher stability.

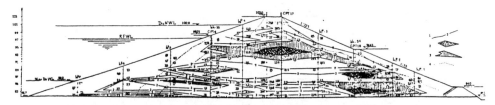

Fig.4. Cross section with zoning according to the different degree of static equilibrium. 1. Locus of points of equal values of the equilibrium factor; 2. Zones of $\eta < 1$; 3. Zones of $1 \leq \eta \leq 1.5$; 4. Outlines of the zones at factor of safety $\eta = 1$.

The weak zones enclosed in capsules in the dam are proved and indicated. At present the dam slopes are under no danger of sliding.

A threat for the dam is the water seeping through the soil foundation and outflanking the dam, and it will remain until the outflow of mineral particles stops.

We prove that the future operation of the "Shishmanov val" dam have to be treated against hydraulic disruption. All characteristics are with measured reliability.

The above exposed, as well as the position of the phreatic surface, permits without complicated calculations, to see that the geotechnical model also has indefiniteness but leads to more clear valuations and solutions.

4. CONCLUSION

Every geotechnical model has indefiniteness leading sometimes to undesirable results. The probability statistic methods give the possibility to evaluate the bigger or smaller realization of the soil mechanic prognoses, but concerning the indefiniteness of the situation. The fundamental indefiniteness which arise from the non utilization of the characterizing feachers, look for a solution in future. It is necessary to increase and standardize the methods of geotechnical studies.

An information with modifications of the reliability is a necessary but insufficient condition for the reliability of geotechnical models and prognoses, but the unreliable information determines the unreliability of the models as well as of the prognoses. Even a complicated mathematical apparatus could not help for elimination of this defect.

The geofunds organized in view to the operational multiple utilization of the data base have proved their utility.

5. REFERENCES

Bondarik G.K., 1968- "Theoretical bases and methods of engineering and geological studies of the rocks", in "Engineering geology & state planning", Moscow, "Nauka" - in Russian

Brilluen L, 1968 - "Scientifically indefiniteness and information", Moscow, Ed. "Mir"- in Russian

Bromhead E.N., 1986 . The Stability of slopes, p.131

Istomina V.S. 1956. Seepage stability of cohesive soils. in "Problems of seepage calculation of geotechnical structures", p.140, Moscow (in Russian)

Gaurilkevitch Y., 1970-"Perspectives of changes of the informative analysis of mechanics of soils", Ed."Lodz" (in Russian)

Gibson K.E., Morgensterm N.K.,1962. A note on the shear stability in normally consolidated clays. Geotechnique 12, No.3

Maslov N.N. 1977. Soil mechanics in practice, Moscow (in Russian)

Milewski D. 1977. Reliability of investigation information and prognoses of engineering, geological and hydrogeological studies, review "Hydrology & Meteorology ", year XXVI, No.3, Sofia (in Bulgarian).

Milewski D. 1993 ."Characterization of the status and valuation of the security of earthfill dams, built of heterogeneous soils without zoning". Fourth Bulgarian national congress on dam construction, Sofia(inBulgarian).

Geotechnical Hazards, Marić, Lisac & Szavits-Nossan (eds) © 1998 Taylor & Francis, ISBN 90 5410 957 2

Flood risk of protected floodplain basins

L. Nagy
National Water Authority, Budapest, Hungary

ABSTRACT: This paper is going to introduce briefly the determination of separate flood plain basins, the selection and determination of characteristic flood stages inducing typical economic impacts, the principles of taking the safety factor or the probability of structural failure of the defences into consideration in flood risk mapping.

1. INTRODUCTION

Flood risk map is a cartographical representation of flood and flood damage characteristics of different probability. They are basic tools in flood prone areas for land use planning, for priority setting in the field of investments for the establishment or improvement of flood security, and they are also essential for insurance planning and for increasing the public awareness of risk.

Important characteristics of floods influencing possible damages are the expected water level (or the expected depth of flooding), the frequency or return period of different water levels, flow velocity conditions, flood duration. All these characteristics can be represented in a flood risk map (Marco 1992).

Flood risk maps are usually made for unprotected floodplains of river or creek valleys. In such cases the surface of the water flowing in the river bed can be computed as a variable unsteady flow in an open channel. Different water surface corresponding to discharges of different probability are to be determined, and the horizontal projection of the respective water levels to the terrain will indicate the limits of flood of different probability. Characteristic depth of flooding are easy to derive from detailed topographic maps or digital terrain models. Such flood risk maps are usually applied for land zoning or for the planning of structural flood alleviation schemes.

In Hungary, where 97 % of the flood plains are already protected, we believe that the risk of damages can also be related to the stability or safety of the flood defence structures, dikes, confinement dikes. The length of the Hungarian flood dikes is more than 4200 km, so the flood risk is firstly depend of the stability of the dikes.

2. DETERMINATION OF THE EXTENSION OF THE SEPARATE FLOOD PLAIN BASINS

A separate flood plain basin is understood as a particular part of the flood plain, which is bounded on the one side by natural higher terrain extending along the edge of the flood plains, on the other side - in case of protected flood plains - by the flood embankment, and which is inundated in the event of a breach in the defences without the inundation spreading into the adjacent flood plain basins.

The boundary and size of the flood plain basins was of interest in the early periods of flood protection already, to allocate the costs of flood defence to the various parties interested. In the past, at the contemporary level and possibilities of engineering, the extension of flood prone areas was determined simply by horizontal projection of the peak stages recorded so far, to the terrain.

The progress made in surveying and computation technology, further in hydrologic and hydraulic research have introduced the possibility of modelling of the actual physical phenomenon: the process of flow and storage on the terrain of the water pouring in through a breach in the embankment at different initial levels having different probability. Thus extension of possible inundation in the protected flood plain with different probabilities may be determined. (Balo 1979)

To determine the size and the contours of the separate flood plain basins, one or several breach points in the defences were assumed, depending on the size of the particular basin, which result in the largest inundation area there. The flood hydrographs of 1 % and 1 %o probability were then transformed to the breach point. As a result of the above work the map scaled 1:50 000 and the records of the endangered areas for each flood plain basin likely to be inundated in case of a levee failure.

Flow in through the breach was computed using the general weir formula. The shape of the breach was assumed to be a hexatic parabola. The length of the breach was computed on the base of geotechnical and hydraulic considerations taking into account practical experience too. The breach was assumed to reach its final size within two hours so the peak discharge of water entering the basin was assumed to be reached after the second hour (Toth 1997).

Basically two version of inundation process was distinguished. In case of a basin with a relatively small capacity equalization of the level of inundation with the level of the river is expected over the altitude of the terrain on the protected side. In case of large basins the volume of water flowed in through the breach is not sufficient to fill the basin until the falling river level reaches the altitude of the terrain at the breach. Storage will be realised only in the deepest parts of the basin at a level corresponding to the volume of the water flowed in.

The surface of the water flowing in the valley to the lowest "reservoir" was computed starting from the storage level of the lowest "reservoir" upwards section by section, as a variable unsteady flow in an open channel, solving the differential equation describing the process of inundation by the method of finite differences, by successive approximation.

3. DETERMINATION THE CONVENTIONAL SAFETY FACTOR

The floods during the past 50 years have caused 92 embankment failures, of which 59 were due to overtopping (52 during the 1956 ice-jam flood on the Danube), 13 to hydraulic soil failure, 10 to saturation and 2 to leakage along structures, while no cause could be identified positively in the case of 8 (Nagy 1993). In the protected flood plain basins the occurrence of the various loss types can be related to the flood stages affecting the stability or safety of the flood defences (Zorkoczy et al. 1987).

Improvements over the past 150 years involved but rarely any change in the original trace of the embankments. Explorations of the subsoil and soil mechanical tests have been introduced as late as 35-40 years ago, which revealed but recently that the original trace passes over areas with adverse soil conditions, where the soil profile contains
- the meander crossings with its different soil layers,
- layers of organic soil or peat,
- dispersive soils,
- loose, poorly graded fine sands in the vicinity of the surface, etc.

The programme for investigation 4200 km flood dikes has been compiled in the eighties for exploring the subsoil of flood embankments and for identifying the potential sections of piping failure. The basic considerations underlying the method are as follows:
-the subsoil under long embankments of moderate height must be investigated,
-the soil profile must be explored continuously (virtually by metres),
-the subsoil consists generally of a cohesive cover over layers becoming increasingly coarser with depth.

In order to carry out the investigation on the stability of the dikes, must be divided into characteristic sections, within which the following should be presumed more or less constant (Galli 1976):
- the high of the crest,
- the stratification of foundation soil and the quality of the layers,
- material of the existing dike as well as that of the reinforcement or new defences.

- typical cross-section of the existing dikes,
- phenomena, observed along the dikes during floods.

The sectioning conforming to the characteristics of the foundation soil has a special importance and needs special care. In the course of investigation the safety of the embankments against piping failure is determined by successive approximation involving several disciplines, like geophysics, hydraulics, soil mechanics, surveying.

To determining the longitudinal profile of the long dikes and the individual sections, one of the best methods is the permanent horizontal geoelectric probing with 1,0 meter electrode distance. The application of this method makes the exploration of continuos stratification possible. This method reduces the cost of exploration, while the application of more expensive methods may be required less often, only for identification of the layers at easily determinable points.

Controlling the safety factor of the embankment divided into characteristic sections must be accomplished section by section, according to standard methods specified in appropriate guidelines and standards (MSZ-10 429, MSZ-10 110, MI-10 268, MI-10 422).

The conventional safety factor

$$n = R/Q$$

where R is the resistance (or strength), and Q is the action effect (load). Used and transforming the equations determining the safety factor of the defences at actual water stages, the flood levels corresponding to previously selected safety factors can be determined.

So we have the opportunity of defining the flood hydrograph peaking at the level corresponding the loading capacity of the defence structure. Since the most vulnerable cross sections of the defences are also known, the flood hydrographs representing the loading capacity are to be transformed to these possible breachpoints. Repeating the computations carried out earlier in order to define the extension of the floodplain of 1 % probability of inundation, the extension of the flood plain section threatened by the stage corresponding to the loading capacity of the defences will be determined (Toth 1994).

4. THE PROBABILITY OF FAILURE AT FLOOD DIKES

Advanced dimensioning methods consider both the impacts inducing (Q) or hindering (R) the breach to be independent and probabilistic variables. It is obvious that from the viewpoint of stability all the combinations of load and resistance are disadvantageous where $R < Q$, represented in the figure with the barred territory. The size of this territory is equal with the failure probability and therefore is appropriate for characterising the magnitude of risk of the given section.

The relation between load and resistance may be expressed by the safety margin:

$$SM = R - Q,$$

which is also a probabilistic variable.

The failure probability expresses the probability of the opportunity of load exceeding resistance

$$p_f = P(Q > R)$$

or

$$p_f = P(SM \leq 0)$$

The failure probability can be determined either from the available soil physical data, applying probabilistic design methods for the whole calculation system or from the traditionally calculated safety factors using semi-deterministic approach.

For flood dikes the value of failure probability is generally must be

$$p_f < 0,01,$$

which means that among all possible combinations of load and resistance values only 1 % would lead to breach, in other words in 1 % of possible cases will be $Q > R$. The acceptable probability of failure can estimate between

$$0,0005 < p_f < 0,005$$

It would only be proper to ask why would be practical the use of failure probability instead of the safety factor that we got accustomed to in practice? The answer is:

- we can characterize the system of defence structures;
- answer can be given to the reliability of our results, that is the uncertainties can be handled;
- evaluation of risk will be possible.

The failure probability of a dike with conventional geotechnical methods we can calculate for a given water stage. Repeating the calculation for more water stages gives the failure probability as a function of the height. Figure 1 represents the results of the calculated values of the failure probability in the possible range of water stages, also the probability of occurrence of water stages in case of a given profile of an embankment (Nagy 1996). Since the failure probability and the occurrence of water stages are independent, the probability of their joint occurrence can be calculated as the product of the multiplication of their probability, that is $R(x) * Q(x)$.

Investigating the $R(x)*Q(x)$ function, the risk of failure of a profile of an embankment can be characterised by
- the maximum value of $R(x)*Q(x)$ function, or
- the area under (beside) the $R(x)*Q(x)$ function.
These considerations are interesting enough for further investigation.

5. CONCLUSIONS

How safe a dike? The answer is provided by a probabilistic risk assessment, the benefits of which were described along with a standard for tolerable risk. It was stressed that in the absence of analytical techniques, the difficulty of assigning probabilities can be addressed by the use of engineering judgement by experienced engineers expert in the area in question, familiar with the dike with all investigations and previous studies at their disposal. It was proposed that a risk could become a systematic and comprehensive framework for the application of engineering judgement.

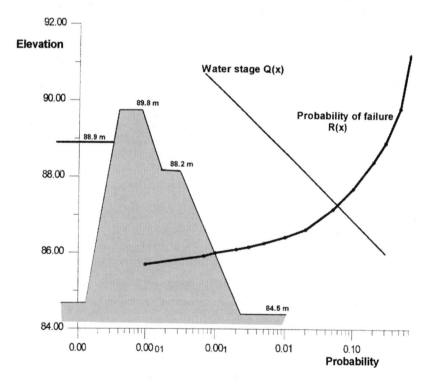

Failure probability of a dike section
Figure 1.

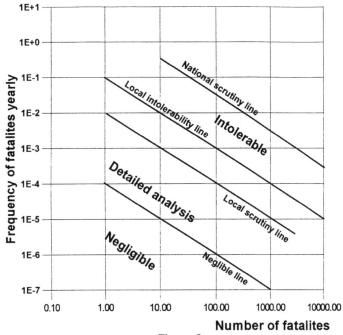

Figure 2.

Risk is the product of failure probability and consequences of the failure. The application of failure probability in the evaluation of existing and also in design of new flood defence structures gives us the possibility of adaptation these problems to the risk standards. A standard for tolerable risk is needed in conjunction with a risk analysis to evaluate dam safety, its purpose being to permit decisions on dike safety remedial work to be based directly on risk in a consistent and quantifiable manner. The risk criteria developed for major hazards of transport studies in UK as an example shown in Figure 2 On this way consideration of risk management is now became an integral of the judicial, legislative and regulatory processes in Hungary.

6. REFERENCES

Balo, Z. 1979. *Determination of flood plain basins in Hungary*. VITUKI Bulletins No. 23., Budapest

Galli, L. 1976. *Stability analysis of flood defence earthworks*. Manual, OVH, Budapest.

Marco, J. B. 1992. *Flood risk mapping*. NATO ASI on "Coping with Floods", Erice, November 3-15, 1992., Kluwer Academic Publishers, 1994. pp. 353-374.

Nagy, L. 1993. *Investigation of dikes with insufficient safety factor*. R&D Report for the Ministry of Transport, Communication and Water Management, Budapest. Manuscript.

Nagy, L. 1996. *Risk of failure of flood defence earthworks*. R&D Report for the Ministry of Transport, Communication and Water Management, Budapest. Manuscript.

Toth, S. 1994. *Flood Risk Mapping and Analysis with Special Regards to the Vulnerability of Protected Floodplain Basins*. Proceedings of the NATO ASI on Defence from Floods and Floodplain Management, Budapest, April 26 - May 7, 1994, Kluwer Academic Publishers 1995. pp. 429-442.

Toth, S. 1997. *Flood risk mapping of protected floodplain basins*. MAFF Conference, Keel,

Zorkoczy et al. 1987. *Flood defence*. Manual. VIZDOK, Budapest. pp. 247-248.

MSZ-10 429 Load and safety factors of flood embankments

MSZ-10 110 Hydraulic earth structures. Density requirements

MI-10 268 Exploration of the subsoil and borrow material of flood embankments.

MI-10 422 Dimensioning flood embankments

Geotechnical Hazards, Marić, Lisac & Szavits-Nossan (eds) © 1998 Taylor & Francis, ISBN 90 5410 957 2

An investigation of hazard in reinforced embankments

Ali Porbaha
Technical Research Institute, TOA Corporation, Yokohama, Japan

Masaki Kobayashi
Port and Harbour Research Institute, Yokosuka, Japan

ABSTRACT: Hazard identification is the primary stage of any geotechnical design in which all potential failure mechanisms must be identified and checked. This study presents the failure mechanisms of unreinforced and geotextile reinforced embankments of 63.4^o(1H:2V) on firm foundations that failed under self-weight using a geotechnical centrifuge. A numerical technique was applied to understand different aspects of mechanism through development of maximum shear strain contours, plastic yield zones, displacement vectors and deformed meshes. It was found that the developed failure mechanism was closely associated with the reinforcement length. For the case of short reinforcement (L/H=0.5), the slip surfaces obtained from physical tests developed further back the embankment face, leading to external failure. However, gradual increase of reinforcement length pushed the slip surface inside the reinforced zone toward the face of the embankment, leading to internal failure.

1 INTRODUCTION

Failures of geological and soil materials can be divided into a number of classes. In a broad sense, there are three components of failure: *mechanism*, *properties* and *analysis*, as discussed by Scott (1987). Before any subsequent steps can be taken, the designers must have a clear picture of the *mechanics* involved in a failure. The initial emphasis of failure investigations is usually directed to the elucidation of the mechanism. Examples are the failure of dams, or the collapse of a structure in an earthquake. The sequence of events is important. Secondly lies the determination of both the qualitative and the quantitative nature of the material *properties* which rendered the failure possible. Was the soil easily erodible? Was it susceptible to creep? Was the stress-strain relation unstable? Then, subsequently, what were the peak and residual shearing strengths, modulus values etc.? When these two components have been classified, *analyses* can be attempted, incorporating mechanism and properties.

In parallel to this discussion, the objective of this investigation is to present the first class of hazard, i.e. the failure *mechanism*, associated with reinforced embankments with slope angle of 63.4^0 (1H:2V)

through experimental and numerical techniques. The effect of varying reinforcement lengths on failure mechanism were investigated, and discussed in detail. Several researchers have applied small scale tests to model geotechnical problems (see, for example, Bolton et al., 1978; Ovesen, 1984). The literature contains a large number of cases in which finite element analysis was applied for predicting the performance of field or laboratory tests (see, for example, Rowe and Soderman, 1985; Almeida et al., 1986; Wu et al.,1992). This paper aims to synthesize these two physical and numerical procedures to investigate the behaviors of unreinforced and geotextile reinforced embankments that failed under self-weight using geotechnical centrifuge.

2 HAZARD IDENTIFICATION

A hazard is something with the potential to cause harm. The hazard identification stage is similar to the initial geotechnical design stage, in which all the potential failure mechanisms must be identified and design checks completed. These represent the limit states which must be checked for serviceability and ultimate conditions. The limit state can develop in either a gradual ductile manner or a rapid brittle way.

Table 1. Ductile and Brittle failure states (Nicholson, 1994)

Feature	Ductile	Brittle
Nature of failure	Gradual development of movements	Abrupt with progressive failure
Governing limit state	Serviceability	Ultimate
Prediction ability	Numerical models provide reasonable deformation predictions Many case histories for extrapolation	Modelling complex because of rupture and stain softening Few case histories to back-analyse
Monitoring systems	Simple monitoring can be used Time to implement contingency plans	Simple monitoring system cannot detect pre-failure movement Progressive failure so rapid that contingency plans cannot be implemented
Influence on the observational method	Good Gradual deformation enables people to be evacuated Damage to structures can be avoided or controlled to acceptable levels while still operating close to most probable conditions Good opportunities for savings	Bad Work in small stages to minimize risk to people and structures Need to work conservatively with comfortable factors of safety In soil/structure applications the structural failure should be ductile even if the local soil mode is brittle Poor opportunities for savings

This latter condition is similar to the progressive failure mechanism referred to by Peck (1969). The differences between ductile and brittle hazards or limit states are summarized in Table 1 (Nicholson, 1994). The observational method works best where ductile failure mechanisms occur as these provide the opportunity for monitoring the development of failure and carrying out contingency plans where necessary. For brittle limit states the observational method can only limit or localize failure so that the risk/severity is minimized. Ductile mechanisms do not normally cause injuries because the movements develop gradually so that there is sufficient time for people to be evacuated. Brittle failure mechanisms can occur quickly so that there is insufficient time to evacuate people. Such failures can result in injuries and financial loss.

One major advantage of geosynthetic reinforced soil structures is that they are inherently flexible, compared to conventional rigid retaining systems. Therefore, the hazard caused by these structures are expected to fall within the ductile-type of failure.

3 CENTRIFUGE MODEL TEST

The advantages of using the centrifuge to achieve self-weight and stress path similarity have been discussed by Schofield (1980). In this study Hydrite Kaolin was used to construct the backfill, the retained fill, and the foundation, of model retaining walls (liquid limit is 49% and plastic limit is 33%).

Shear strengths of kaolin were obtained from direct shear tests on specimens taken from the model after failure occurred in the centrifuge, and exposed to the normal stress equal to the maximum experienced by the specimen during the test. The cohesion ranged between 16.3 kN/m^2 and 23.8 kN/m^2 with friction angle ranging between 18.4^0 and 21.7^0.

The geotextile used in this study is a non-woven interfacing fabric. The maximum tensile strength of the geosynthetic simulant, using ASTM wide-width test (D4595) and zero-span test, was measured to be 0.053 kN/m and 0.117 kN/m, respectively.

Model embankments were built on firm compacted clay foundations with a dry unit weight of 13.5 kN/m^3. The first layer of reinforcement was placed on the exposed portion of the foundation, a layer of soil placed, and the geotextile folded back 32 mm into the soil to provide a flexible facing for the model against a temporary support. A compressive stress was then applied increasing slowly to produce a lift of backfill and retained fill with dry unit weight of 12.3 kN/m^3. This process was repeated for successive layers, each of which had finished thicknesses of 19 mm, until the model reached the desired height of 152 mm.

After the centrifuge test, the coordinates of failure surfaces were recorded using a profilometer

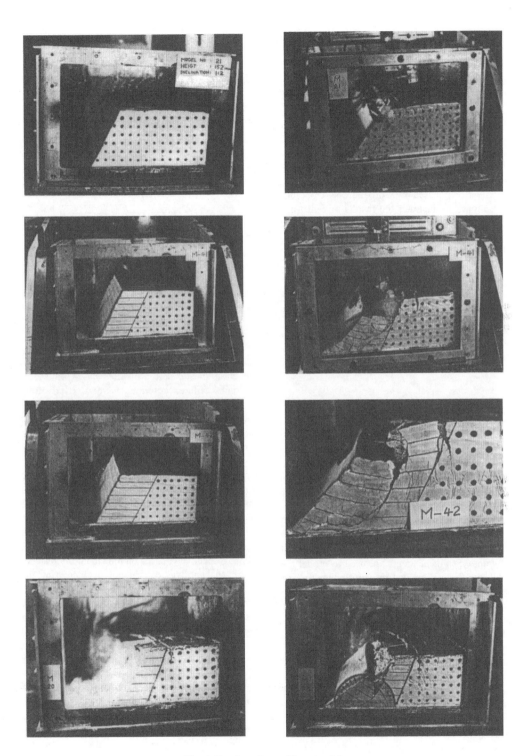

Fig. 1 Results of centrifuge model tests

253

measuring the vertical profile at 10 mm horizontal intervals through various model cross-sections.

Table 2 presents the model geometry and the material properties of the 63.4° (1H:2V) embankments constructed on firm foundations. Reinforcement length varied from no reinforcement, to a maximum reinforcement length of 114 mm, or 0.75 times model height. All models were constructed with eight layers of uniform reinforcements. The behaviors of individual models in terms of crack development and foundation rigidity were discussed by Porbaha and Goodings (1996). Figure 1 illustrates the results of centrifuge tests.

Table 2. Geometry and material properties

Model No.	Length (mm)	L/H ratio	Cohesion (kN/m²)	Friction angle (deg.)
M-21	0	0	19.5	18.4
M-41	76	0.50	17.3	20.7
M-42	100	0.66	24.6	18.3
M-20	114	0.75	22.4	20.1

L/H= Length of reinforcement per model height

4 FEM ANALYSIS

The two dimensional finite element mesh used for the analysis is shown in Figure 2. The geometry of the mesh is selected to ensure proper modeling of the wall identical to the physical model. The finite element mesh consists of 336 nodes and 141 elements to simulate various components of the reinforced embankment. Kobayashi (1984) developed the FEM program for the analysis of geotechnical problems.

Different types of elements are used to model the backfill, the foundation, and the reinforcement. The simulation comprises of eight-noded quadrilateral isoparametric solid elements to model the soil in the backfill and the foundation, and three-noded bar elements for the reinforcement. The six-noded joint elements is commonly used to simulate the soil-interface interaction and to apply the properties based on the results obtained from laboratory pullout tests. However, the implementation of such a process seems obscure in this study. The reason is that the backfill material is a partially saturated soil and the hydraulic characteristics of nonwoven geotextile allows drainage in both sides of the reinforcement. Accordingly, the moisture content of the soil which tends to be peak value at the center of the soil layer

reaches a minimum value at the interface in the vicinity of the reinforcement, and thus creating a complex soil profile that is difficult to model in ordinary pullout tests. For this reason to resolve this issue the interface between soil and geotextile is considered to be the same as the soil model, despite some deficiencies with the complex real situations.

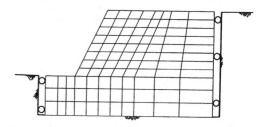

Fig. 2 Finite element mesh and boundary conditions

The constitutive models commonly used to study the behavior of reinforced retaining structures at failure are based on Mohr-Coulomb yield criterion. The elasto-plastic soil properties based on Mohr-Coulomb failure criterion were derived from laboratory tests for each model slope, as listed in Table 2. Poisson ratio of 0.30 was adopted for the analysis. The reinforcement was modeled as beam element with axial stiffness of 0.1 kN/m and no flexural rigidity (due to the extensibility of the reinforcement). The tensile strength of the geotextile from zero-span laboratory test was 0.117 kN/m that was input to the program. The fixities at the boundaries allow deformations in vertical direction (roller hinge in y-direction). Unit weights of the backfill and of the foundation soil are 17.3 and 18.2 kN/m³, respectively, based on the laboratory results.

Each analysis was carried out by increasing the gravitational acceleration at a very small load increment to reach the point where no convergence was acquired. Then, the analyses continued with steps back by oscillating back and forth around the smaller increments to ensure convergence with an accuracy of 0.0001g. A large number of iterations was applied based on the fictitious viscoplasticity algorithm (Zienkiewicz et al., 1975; Kobayashi, 1984 and 1988; Zienkiewicz and Taylor, 1990) to calculate the load and consequently the collapse height of the slopes. During the analysis the vertical deformation of the crest is measured, as the gravitational acceleration increases (see Figure 3).

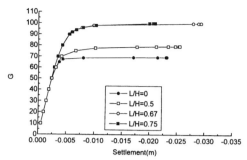

Fig. 3 Gravitational acceleration versus crest settlement

5 DISUSSION OF RESULTS

The centrifugal accelerations and prototype equivalent heights, resulted from physical and numerical modeling of different model embankments, are presented in Tables 3 and 4. The characterization of tension crack is not possible during the analysis due to material discontinuity which results from cracking. However, it is possible to estimate the centrifugal acceleration at the time when tension crack occurs from the load-settlement curves for each model. The criterion is to determine the gravitational acceleration at which the initial tangent in the elastic region intersects the tangent to the curve at a small deformation resulting from development of tension crack. The failure, however, is clearly defined when large deformation occurs at constant stress levels, as shown in Figure 3.

Table 3. Comparison of centrifugal accelerations from physical and numerical approaches

Model No.	$(N_c)_{EXP}$	$(N_f)_{EXP}$	$(N_c)_{FEM}$	$(N_f)_{FEM}$
M-21	65	67	68	69.0
M-41	73	77	70	79.3
M-42	85	100	90	100.1
M-20	92	102	90	100.4

$(N_c)_{EXP}$=gravitational acceleration at tension crack obtained from centrifuge tests
$(N_f)_{EXP}$=gravitational acceleration at failure obtained from centrifuge tests
$(N_c)_{FEM}$=estimated gravitational acceleration at tension crack obtained from FEM analyses
$(N_c)_{FEM}$=gravitational acceleration at failure obtained from FEM analyses

5.1 Prototype Equivalent Height

In terms of gravitational acceleration at tension crack

the prediction by this approximate technique is within 4 to 6% of the experimental values. In terms of gravitational acceleration at failure, which represents the collapse prototype equivalent height, the maximum difference is 1.6 g that accounts for the prototype equivalent height of 0.24 m, which indeed, is practically not significant. These numerical results increase the credibility of FEM to predict the centrifugal accelerations and prototype equivalent heights resulted from centrifuge tests.

Table 4. Comparison of prototype equivalent heights from physical and numerical approaches

Model No.	$(H_c)_{EXP}$	$(H_f)_{EXP}$	$(H_c)_{FEM}$	$(H_f)_{FEM}$
M-21	9.9	10.2	10.3	10.5
M-41	11.1	11.7	10.6	12.0
M-42	12.9	15.2	13.7	15.2
M-20	14.0	15.5	13.7	15.3

$(H_c)_{EXP}$= prototype equivalent height at tension crack obtained from centrifuge tests
$(H_f)_{EXP}$= prototype equivalent height at failure obtained from centrifuge tests
$(H_c)_{FEM}$= estimated prototype equivalent height at tension crack obtained from FEM analyses
$(H_c)_{FEM}$= prototype equivalent height at failure obtained from FEM analyses

5.2 Failure Surfaces

Figures 4 and 5 show the contours of maximum shear strains, displacement vectors, yield zones with plastic deformations, and deformed meshes after failure for both unreinforced and reinforced embankments studied here. The traces of slip surfaces obtained from centrifuge model tests are also plotted in those Figures.

In the case of unreinforced model (M-21) the slip surface coincides with the peaks of shear strain contours at the bottom of the embankment and diverges at the crest. For the case of models with short reinforcements (M-41, M-42), the slip surfaces from physical tests are behind the contours of maximum shear strains. This is not surprising when the failure characteristics of these models are taken into consideration, as shown in Figure 1. Based on the post-test observations the failure surfaces of these models occurred either far beyond the back of the reinforcement, as in model M-41, or just occurred behind the reinforced zone, as in the case of M-42. Unlike model M-20 in which the failure was entirely internal, in both these cases the slip surfaces occurred outside the reinforced zone. One source of discrepancy with the numerical prediction may also

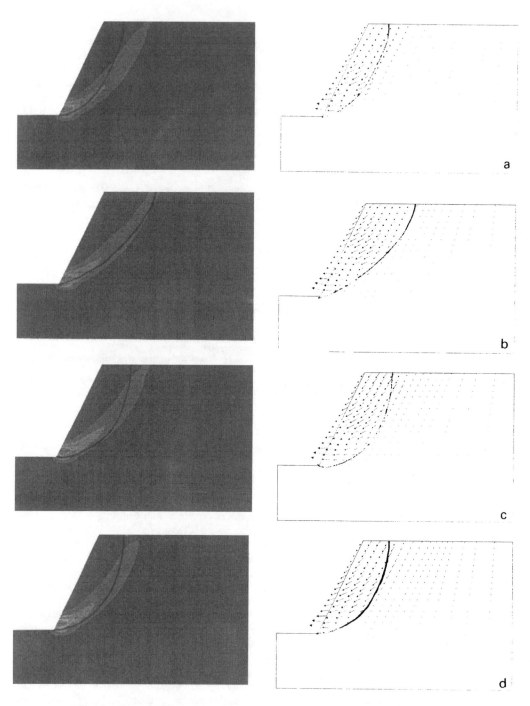

Fig. 4   Contours of maximum shear strain (left) and displacement vectors
(right) for model embankments: (a) M-21, (b) M-41, (c) M-42, and (d) M-20

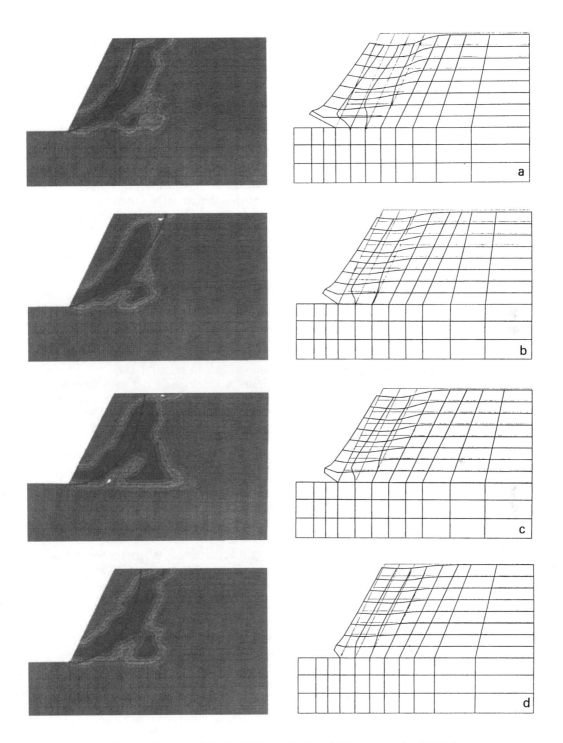

Fig. 5   Contours of plastic yield zone (left) and deformed meshes (right) for
model embankments: (a) M-21, (b) M-41, (c) M-42, and (d) M-20

be attributed to the initial state of the stresses in the model which was not modeled in this study due to the complexities in stress path during model construction and the partially saturated soil of the backfill.

The meshes for both unreinforced and reinforced cases, as shown in Figure 5, represent the deformation of the embankments. Increase of reinforcement length reduced the deformation at the toe and at the crest. The deformed mesh at the toe of the models will be improved as a result of selecting a finer mesh size.

The slip surfaces from physical modeling are located within these yield zones, as shown in Figure 5, even though the identification of clear-cut slip surfaces at the crest of the embankments are not as explicit as the body of the models. In the case of unreinforced model, however, the slip surface from physical test is closer to the face of the embankment.

CONCLUSIONS

Geotechnical hazard caused by inadequate reinforcement in reinforced embankments of 63.4^0 was investigated in this study. It was emphasized that the failures of geological and soil materials can be divided into three classes; mechanism, properties and analysis. This study investigated the failure mechanism of reinforced embankments on firm foundations. The results of numerical simulation were compared with physical models and discussed in terms of tension crack development, gravitational acceleration, prototype equivalent height, and failure surface.

It was found that the developed failure mechanism was closely associated with the reinforcement length. For the case of short reinforcement (L/H=0.5), the slip surfaces obtained from physical tests developed further back the embankment face, leading to external failure. However, gradual increase of reinforcement length pushed the slip surface inside the reinforced zone toward the face of the embankment, leading to internal failure.

One major advantage of geosynthetic reinforced soil structures is that they are inherently flexible, compared to conventional rigid retaining systems. Therefore, the hazard caused by these structures are expected to fall within the ductile-type of failure.

REFERENCES

Almeida, M.S.S., Britto, A.M., and Parry, R.H.G. 1986. Numerical modeling of a centrifuged embankment on soft clay, *Canadian Geotechnical Journal*, 23, 103-114.

Bolton, M. D., Choudhury, S. P., and Pang, R.P.L. 1978. Reinforced earth walls: A centrifuge model study, Symposium on earth reinforcement, ASCE Annual convention, Pittsburgh, 252-281.

Kobayashi, M. 1984. Stability analysis of geotechnical structures by finite elements, *Report of Port and Harbour Research Institute*, Vol. 23, No.1: 83-110.

Kobayashi, M. 1988. Stability analysis of geotechnical structures by adaptive finite element procedure, *Report of Port and Harbour Research Institute*, Vol. 27, No.2, June: 3-22.

Nicholson, D. P. 1994. The observational method in geotechnical engineering, *Geotechnique,* Vol. 44, No.4, 613-618.

Ovesen, N.K. 1984. Centrifuge tests of embankments reinforced with geotextiles on soft clay, Proc. of Int'l Symposium on Geotechnical Centrifuge Model Testing, Tokyo, 14-21.

Peck, R.B. 1969. Advantages and limitations of observational method in applied soil mechanics, *Geotechnique,* 19, No.2, 171-187.

Porbaha, A. and Goodings, D. J. 1996. Centrifuge modeling of geotextile reinforced steep clay slopes, Canadian Geotechnical Journal, 33, 696-704.

Rowe, R.K., and Soderman, K.L. 1985. An approximate method for estimating the stability of geotextile reinforced embankments, *Canadian Geotechnical Journal*, 22, No.3: 392-398.

Schofield, A. N. 1980. Cambridge geotechnical centrifuge operations, *Geotechnique,* 30 (3), 227-268.

Scott, R.F. 1987. Failure, Geotechnique 37, No.4, 423-466.

Wu, T.H.J., Christopher, B. Siel, D., Nelson, N.S., Chou, H., and Helwany, B. 1992. The effectiveness of geosynthetic reinforced embankments constructed over weak foundation, *Geotextiles and Geomembranes*, 11: 133-150.

Zienkiewicz, O.C., and Humpheson, C., and Lewis, R.W. 1975. Associated and non-associated viscoplasticity and plasticity in soil mechanics, Geotechnique, Vol. 25: 671-689.

Zienkiewicz, O.C., and Taylor, R.L. 1990. The finite element method, McGraw Hill, fourth edition, Vol.2: 256-260.

Geotechnical Hazards, Marić, Lisac & Szavits-Nossan (eds) © 1998 Taylor & Francis, ISBN 90 5410 957 2

Die Harmonisierung des Regelwerkes für Geotechnik in Europa: Entwicklung und Bedeutung für die Praxis

W. Sadgorski

Bayer. Landesamt für Wasserwirtschaft, München, Germany

ABSTRACT: Since 1981 various groups of experts of european countries are preparing an european norm for geotechnics. A so called pre-norm ENV 1997-1 Geotechnical design - Part 1: General rules was accepted 1994 and recomended to the member countries for a trial application. Among the new principles introduced by this pre-norm the concept of partial factors is the most important one. Even if on this point a general agreement was reached, the way of application of partial factors created several problems. To obtain a decision basis , a large amount of trial calculation was carried out, a coincidence with the results following traditional methods being a criterion for the reliability of the new method. According to the variability of natural conditions and often empirical charakter of present practice in geotechnics, the future EN 1997-1 will be a "umbrella code", which may be supplemented by more detailed national regulations.

1. EINFÜHRUNG - ZIEL UND ERGEBNISSE DER EUROPÄISCHEN BAUNORMUNG

In ihrer Bestrebung, die Märkte ihrer Mitgliedsländer auch für Bauleistungen zu öffnen, hat die Kommission der Europäischen Gemeinschaften (KEG) Mitte der siebziger Jahre festgestellt, daß dazu eine Harmonisierung der Bauvorschriften der einzelnen Länder erforderlich ist. Deshalb wurde beschlossen, gemeinsame Regeln - EUROCODES genannt - für den Entwurf, die Bemessung und die Ausführung von Bauwerken aufzustellen. Das entsprechende Programm der Kommission sah zuletzt 10 Eurocodes für die verschiedenen Bauwerksarten vor.

Der erste Schritt zur Bearbeitung der EUROCODES war die Anpassung einer evtl. existierenden Richtlinie, aufgestellt von einem Fachverband (z.B. vom Europäischen Betonverein CEB) oder die Aufstellung einer solchen (falls keine vorhanden) als sog. "model code". Nach Billigung der großen Linie eines "model code" in den Ländern fand eine Überarbeitung in Gruppen von 6 - 7 Personen ("project teams" genannt) statt. Der so entstandene Entwurf wurde, nunmehr durch das mit der weiteren Bearbeitung beauftragte Europäische

Normungsinstitut CEN, an die Mitgliedsländer zur Stellungnahme und Zustimmung übermittelt und nach Berücksichtigung der Einwände als Europäische Vornorm ENV in den drei CEN-Sprachen veröffentlicht.

Die Mitgliedsländer sollten dann die verabschiedete ENV als Alternative zu ihrer eigenen Normung zur probeweisen Anwendung zulassen. Die erste solche Vornorm war die ENV 1992-1 (EUROCODE 2) für Stahlbeton, mehrere weitere Vornormen folgten. Jede Vornorm wurde in den einzelnen Ländern unter Begleitung eines Nationalen Anwendungsdokumentes (NAD) adaptiert, das die Verbindung zur bestehenden Nationalen Normung herstellt und einige Zahlenwerte festschreibt, die in der ENV in Klammern ("boxed values") angegeben sind. In erster Linie handelt es sich um die Zahlenwerte der Teilsicherheitsbeiwerte.

2. BEARBEITUNG DER CEN - NORM FÜR GEOTECHNIK

Nach vorangehenden Kontakten zwischen der KEG und der ISSMFE entstand 1981 eine Arbeitsgruppe der Nationalen Gesellschaften für Geotechnik der damals 10 Länder der EG unter Vorsitz von Prof. N. Krebs-Ovesen aus Dänemark, welche die Bearbeitung eines EUROCODE (EC) 7 für Geotechnik aufnahm. Später kamen Vertreter aus Portugal und Spanien hinzu. Die Tätigkeit konzentrierte sich auf den 1. Teil der geplanten Norm : "Geotechnical Design - general rules" ("Entwurf, Berechnung und Bemessung in der Geotechnik") und begann mit einer Sichtung der vorliegenden nationalen Normen der Teilnehmerländer.

Es stellte sich ein höchst uneinheitliches Bild heraus. Während einige Länder über ein weitentwickeltes Normungssystem auf diesem Gebiet verfügten, z.B. gab es in Deutschland (einschl. der Normen für Labor und Feldversuche) bereits ca. 50 Normen mit insgesamt 800 Seiten, lagen in mehreren Ländern gar keine Normen vor. Man bediente sich in solchen Ländern teilweise der Normen anderer Länder oder faßte die Lehrunterlagen der angesehensten Hochschulen des Landes als eine Art Normen auf. Das letztere war besonders in kleineren Ländern der Fall. Auch Umfang, Detailierungs- und Verbindlichkeitsgrad unterschieden sich stark voneinander. Bild 1 zeigt den Umfang der 1990 bestehenden Normen für Geotechnik der CEN-Mitgliedsländer ohne Wertung des Inhaltes.

Auch die Aussagekraft und die rechtliche Bedeutung der Normenwerke der teilnehmenden Länder unterschieden sich erheblich, was nicht zuletzt auf prinzipiell unterschiedliche Auffassungen darüber zurückzuführen ist, was denn eine Norm darstelle und zu leisten habe. Es stellten sich zwei Grundtendenzen heraus, die man als Legalitäts- bzw. Legitimitätsprinzip bezeichnen kann. Die erste Richtung, z.B. in Deutschland vorherrschend erlaubt grundsätzlich (fast) jedermann, einen Standsicherheitsnachweis vorzunehmen, wenn die relevante Norm exakt befolgt wird. Die zweite Richtung, typisch z.B. für Großbritannien, erlaubt den durch gesonderte Prüfungen qualifizierten Personen ("chartered engineers"), nach eigenem Ermessen Standsicherheitsnachweise durchzuführen.

Dementsprechend sind die Normen für ingenieurmäßige Bearbeitung in Deutschland auch auf dem Gebiet der Geotechnik ausführlich und verbindlich abgefaßt. Sie sollen dazu beitragen, nicht nur ein Mindestmaß an Sicherheit zu garantieren, sondern auch die Streuung von durch verschiedene Ingenieure erhaltenen Ergebnissen zur einer bestimmten Aufgabe zu minimieren.

In Großbritannien dagegen stellen die "Codes of Practice" lediglich sehr umfangreiche Checklisten dar, welche dem Bearbeiter eine Richtschnur für die vorzunehmenden Schritte im Entwurf und in der Bemessung anbieten. In anderen englischsprachigen Ländern besteht sogar die Meinung, die wichtigste Aufgabe einer Norm sei, dem Ingenieur keine Einschränkungen aufzuerlegen, damit er "eine gute Leistung erbringen könne"!

Innerhalb von sechs Jahren (bis 1987) wurde dennoch der "model code" fertiggestellt und den beteiligten Nationalen Gesellschaften übermittelt, wobei das Werk noch nicht als reif für die direkte Anwendung in der Praxis erachtet wurde. Ein weiterer Schritt in Richtung Praxisreife wurde von einem "project team" aus sieben Mitgliedern gemacht, das 1991 den Entwurf der Vornorm ENV 1997-1 vorlegte. Nach einer kritischen Durchsicht in den nunmehr 18 Mitgliedsländern des CEN und anschließender Diskussion im zuständigen Unterausschuß CEN/TC 250/SC 7 wurde 1993 die Vornorm verabschiedet. Die Mitgliedsländer wurden aufgefordert, die probeweise Anwendung der Vornorm in der Praxis zu ermöglichen; dieser Aufforderung kamen sie in unterschiedlichem Maße nach.

Im Sommer 1997, nach zweieinhalbjähriger Probezeit wurde eine Umfrage durchgeführt, bei der sich die Mehrheit der Mitgliedsländer für eine Überarbeitung der Vornorm in eine endgültige Europäische Norm EN 1997-1 aussprach. Für die Überarbeitung gingen zahlreiche Vorschläge ein; eine Arbeitsgruppe mit Beteiligung aller interessierten Mitgliedsländer unter Vorsitz von Herrn Prof. U. Smoltczyk, Stuttgart, nimmt seit März 1997 die Überarbeitung vor.

3. DAS DEUTSCHE NORMENWERK FÜR GEOTECHNIK

Eine Normung auf dem Gebiet der Geotechnik besteht in Deutschland seit 1934, dem Jahr der Veröffentlichung der ersten DIN 1054 "Zulässige Belastung des Baugrundes" durch den Deutschen Baugrundausschuß. Derselbe gab 1935 die "Richtlinien für bautechnische Bodenuntersuchungen" heraus (1997 erschien ein Faksimiledruck der 2. Auflage dieser Richtlinie von 1937). Derzeit besteht das Normenwerk des DIN für Geotechnik aus etwa 50 Einzelnormen, die sich in Gruppen einordnen lassen:

− Grundlegende Normen, 3 Stck.,
− Normen für Baugrunduntersuchung u.ä., 5 Stck.

- Normen für bodenmechanische Labor- und Feldversuche, ca. 20 Stck.
- Normen für Gründungselemente, 8 Stck.,
- Normen für Gründungsverfahren, 3 Stck., und
- Berechnungsnormen, 5 Stck.

Zu den meisten Normen (bis auf die Versuchsnormen) werden auch Erläuterungen und wenn nötig, auch Zahlenbeispiele herausgegeben.

Die DIN-Normen für Geotechnik sind auch in zwei Sammelbänden zusammengefaßt.

Außer den DIN-Normen bestehen zahlreiche anspruchsvolle Empfehlungen der Arbeitskreise der Deutschen Gesellschaft für Geotechnik DGGT (z.B. EAB, für Tunnel- und Felsbau u.a.), der Hafenbautechnischen Gesellschaft (EAU) u.a.

4. WICHTIGSTE NEUERUNGEN IN DER ENV 1997-1

Im Vergleich zur traditionellen geotechnischen Normung, z.B. in Österreich, Deutschland, Frankreich u.a. weist die ENV 1997-1 einige erhebliche Neuerungen auf:

1. Einführung der Geotechnischen Kategorien
2. Einführung "charakteristischer" Werte der Bodenkenngrößen
3. Aufforderung, alle denkbaren Bruchmechanismen zu untersuchen
4. Klar definierte Grenzzustände
5. Legalisierung der Beobachtungsmethode
6. Teilsicherheitsbeiwerte anstelle der bisherigen globalen Sicherheitsbeiwerte

Durch die drei eingeführten Geotechnischen Kategorien (§ 2.1) soll gewährleistet werden, daß Umfang und Qualität der geotechnischen Untersuchungen, Berechnungen und Überwachungsmaßnahmen der Komplexität des vorliegenden geotechnischen Problems angepaßt werden. Bei Einstufung eines konkreten Falles in eine der drei Kategorien sind: Art und Größe des Bauwerkes bzw. seiner Teile, die Baugrund- und Grundwasserverhältnisse, die Umgebung (Nachbarbebauung, Verkehr usw.) sowie die regionale Erdbebentätigkeit und weitere Umwelteinflüsse zu berücksichtigen. Das System der Geotechnischen Kategorien wurde bereits 1990 in die entsprechende DIN 4020 übernommen und hat sich sehr gut bewährt.

Bei geotechnischen Standsicherheitsnachweisen wird für die Baugrundkenngrößen von sog. "charakteristischen Werten" (§ 2.4.3) ausgegangen, die für den konkreten Fall als vorsichtige Schätz-werte ("conservative best estimate") aufgrund der Ergebnisse der Baugrunduntersuchung festzulegen sind. Dabei sind insbesondere das für die Erfassung eines Bruchmechanismus aktivierte Bodenvolumen und die innerhalb dieses Volumens möglichen Streuungen zu beachten. Die sich daraus ergebenden Konsequenzen wurden von von Soos (1990) ausführlich behandelt, danach ist bei den meisten bodenmechanischen Problemen die Aussagewahrscheinlichkeit des geschätzten Mittelwertes maßgebend. Normalerweise werden sich jedoch die anzunehmenden charakteristischen Werte kaum von den bisher verwendeten Rechenwerten (in Deutschland "cal-Werte" genannt), unterscheiden.

Die bestehenden, meist sukzessiv entstandenen Normenwerke europäischer Länder behandeln einzelne Versagensarten im Detail und lassen häufig rechnerisch nicht genau erfaßbare Bruchmechanismen unerwähnt. Die ENV 1997-1 listet dagegen in den Spezialabschnitten 6 bis 9 die für jede Art geotechnischer Bauwerke grundsätzlich möglichen und zu untersuchenden Versagensmechanismen auf. Da jedoch die einzelnen Formulierungen zum Teil Kompromisse darstellen, könnten beim gegenwärtigen Text mitunter Irritationen entstehen, die bei der Überarbeitung in eine EN beseitigt werden sollen.

Die Definition der zu untersuchenden Grenzzustände für geotechnische Bauwerke wurden im "model code" grundsätzlich definiert als:

- Grenzzustand der Tragfähigkeit (ULS) 1B, bei dem durch große Verformungen (jedoch ohne Bruchmechanismus) im Untergrund sich ein Bruchmechanismus, z.B. statisches Versagen, in einem Überbau (Bauwerk, Straße usw.) einstellt,
- Grenzzustand der Tragfähigkeit (ULS) 1C, bei dem im Untergrund bzw. im Erdbauwerk ein Bruchmechanismus entsteht, und
- Grenzzustand der Gebrauchstauglichkeit (SLS), bei dem durch große Verformungen im Untergrund exzessive Verformungen im Überbau entstehen, welche den Gebrauch des Bauwerkes stark einschränken.

Bild 2 veranschaulicht anhand einer stark verallgemeinerten Beanspruchungs-Verformungslinie die Bereiche bzw. Entstehungsbedingungen für diese Grenzzustände. Im Zuge der weiteren Bearbeitung der Vornorm fanden verschiedene Umformulierungen statt, später wurden die Begriffe "Fälle" ("cases") "B" und "C" in der grundlegenden Europäischen Norm ENV 1991-1 (EC 1) weitestgehend als Ersatz für die oben formulierten Grenzzustände eingeführt. Dabei sind einige Mißverständnisse aufgetreten, die jedoch bei der Überarbeitung der ENV 1997-1 in eine EN beseitigt werden sollen. Wichtig ist dabei die Feststellung, daß für einen bestimmten Versagens- (Bruch-) Mechanismus nur der Nachweis eines Grenzzustandes durchzuführen sein wird.

Die bereits von Karl von Terzaghi und Arthur Casagrande in die geotechnische Praxis eingeführte Beobachtungsmethode (observational method) wird in § 2.7 der Vornorm definiert und es werden die erforderlichen Vorkehrungen bei Anwendung dieser Methode aufgeführt. Es lassen sich dabei erhebliche wirtschaftliche Vorteile bei Beibehaltung des tatsächlichen Sicherheitsniveaus erreichen.

5. ANWENDUNG VON TEILSICHERHEITS-BEIWERTEN IN DER GEOTECHNIK

Im konstruktiven Ingenieurbau wurde die Bemessung mit Teilsicherheitsbeiwerten bereits seit 1952 in der damaligen Sowjetunion und im darauffolgenden Jahrzehnt in mehreren weiteren osteuropäischen Ländern eingeführt. Für geotechnische Probleme begründete Brinch Hansen 1956 die Anwendung von Teilsicherheitsbeiwerten und veranlaßte gleichzeitig ihre Übernahme in die dänische Norm für Gründungen DN 415. Beweise für die Anwendbarkeit des Verfahrens wurden in mehreren Arbeiten, insbesondere von Meyerhof (s .z.B. Meyerhof 1993) erbracht.

In Deutschland hat ein Arbeitsausschuß für die Grundlagen der Bauwerkssicherheit mit geotechnischer Beteiligung 1981 die Einführung der Bemessung mit Teilsicherheiten dringend empfohlen (GruSiBau, DIN 1981). Ursprünglich bestand die Absicht, nicht nur probabilistische Überlegungen, sondern auch probabilistische Verfahren bei der Festlegung des anzuwendenden Systems von Teilsicherheitsbeiwerten anzuwenden. Für einfache Fälle von Gründungen wurden hierzu Anfang der 80-er Jahre in Schweden (Olsson & Stille 1984), Deutschland, England u.a. umfangreiche Untersuchungen durchgeführt, die jedoch zahlreiche Schwierigkeiten offenbarten. Insbesondere die naheliegende Bestrebung, bei der Umstellung der Sicherheitsnachweise weder Wirtschaftlichkeitsnoch Sicherheitseinbußen zu erleiden zeigte, daß "exaktere" probabilistische Verfahren nicht in die Praxis des Alltages eindringen können.

Als Kompromiß bot sich zunächst der seit vier Jahrzehnten in Dänemark praktizierte Weg der konstanten Teilsicherheitsbeiwerte für Einwirkungen und Bodenwiderstände mit einer geringen Anzahl von Werten an. Die dabei am schwersten zu beantwortende Frage war, ob Einwirkungen aus dem Untergrund, insbesondere Erddruck als charakteristische Größen errechnet und unmittelbar vor der Bemessung eines Konstruktionselementes, z.B. einer Spundwand, mit Teilsicherheiten für Einwirkungen belegt werden oder die charakteristischen Bodenkenngrößen tan φ' und c' noch vor der Erddruckermittlung mit Teilsicherheiten zu Bemessungsgrößen umgerechnet werden. Der erste Weg entspricht der Praxis des konstruktiven Ingenieurbaues, der zweite dagegen der bisherigen dänischen Praxis.

Die Reduzierung der Scherfestigkeit eines Bodens durch Division durch einen konstanten Wert unabhängig vom Mittelwert der Scherfestigkeit (zur Vereinfachung sei von tan φ' ausgegangen) führt zu Abschlägen in tan φ', welche proportional dem Mittelwert sind. Mit diesen Abschlägen sollen die Unsicherheiten bei der Ermittlung und Anwendung des betreffenden Bodenkennwertes abgedeckt werden.

Bereits Ohde (1952) hat jedoch darauf hingewiesen, daß nicht etwa die Variationskoeffizienten von tan φ', sondern die Standardabweichungen σ für alle Bodenarten etwa gleich sind. Auswertungen der Ergebnisse von rd. 270 Versuchen in den Bildern 3 und 4 bestätigen die These von Ohde, bei der als angemessener Sicherheitsabstand 2σ eingesetzt wird. Die Anwendung eines konstanten Teilsicherheitsbeiwertes $\gamma_m = 1,25$ unabhängig von der Bodenart würde zu einer starken Überdimensionierung für "gute" Böden (Sande und Kiese) und zu einer nicht vertretbaren Unterdimensionierung bei "schwachen" Böden wie erstkonsolidierten Tonen führen. Solange in der Bemessungsgleichung nur oder überwiegend Einwirkungen aus dem Untergrund auftreten, wie z.B. bei der Sicherheit gegen Böschungsbruch, ist der beschriebene Umstand noch hinnehmbar. Gegenüberstellungen von Einwirkungen infolge z.B. Erddruck und anderen Einwirkungen und Widerständen, etwa bei der Bemessung eines Stützbauwerkes können zu unvertretbaren Über- oder Unterdimensionierungen führen.

Gleichzeitig wurden in mehreren Ländern in größerem Stil Vergleichsberechnungen aufgenommen, durch welche das "neue" Verfahren und seine Elemente an das "alte" kalibriert werden sollte.

Wichtige Ergebnisse der Vergleichsberechnungen sind in DIN (1996a), ENPC (1996) und Sänger/Voigt (1995) veröffentlicht worden. Sie alle führten vor, daß insbesondere für die Bemessung der tragenden Elemente von Stützbauwerken nur der "Fall B" zu weitgehend gleichen Ergebnissen führt wie die bisherige Praxis (s. Bild 5). Diese Tatsache wird bei der Überarbeitung der ENV 1997-1 gebührend berücksichtigt.

6. AUSWIRKUNGEN AUF DIE NATIONALE NORMUNG

Angesichts der sehr unterschiedlichen Situation der nationalen Normung sowie der Bemessungspraxis der Mitgliedsländer des CEN auf geotechnischem Gebiet (über die Vielfalt der Ergebnisse bei der Bemessung einer Spundwand s. z.B. Sadgorski (1995)) war eine vollständige Angleichung der geotechnischen Bemessung in allen CEN-Mitgliedsländern von vorne herein kaum möglich. Hinzu kamen zwei weitere Gesichtspunkte:

a) Die Vorstellungen, was eine Norm für Geotechnik zu leisten hätte, sind von Land zu Land immer noch recht unterschiedlich, und

b) Die Sicherheit eines geotechnischen Bauwerkes hängt keineswegs nur oder nicht einmal vorwiegend von den Zahlenwerten von Teilsicherheiten, sondern auch von Umfang und Qualität der Baugrunduntersuchung sowie der Ableitung der Bodenkennwerte unter Berücksichtigung der vorhandenen und zu erwartenden Spannungszustände (Spannungswege), von der zutreffenden Erfassung verschiedener Randbedingungen (z.B. durch die Nachbarbauwerke), von der Qualität und Intensität der Baukontrolle u.a. ab.

Aus den aufgeführten Gründen fiel die ENV 1997-1 wesentlich allgemeiner aus, als z.B. die Vornormen für Stahlbeton- und Stahlbau ENV 1992-1 und ENV 1993-1 (s. z.B. Sadgorski 1995). Es zeigte sich damit, daß zumindest die "erste Generation" der künftigen Europäischen Norm EN 1997-1 nur eine "Rahmennorm" ("umbrella code") sein kann, die in einzelnen Ländern durch weiterführende Normen ergänzt werden kann. Diese Regelung wurde vom Hauptausschuß für die EUROCODES - dem TC 250 - im September 1996 gebilligt. Außerdem werden nicht nur die Teilsicherheitsbeiwerte als Klammerwerte ("boxed values") bleiben, es werden vielmehr weitere Elemente wie Sicherheitsklassen, Lastfälle und Modellfaktoren für die

einzelnen Länder unterschiedlich ausfallen müssen, damit alle Teilnehmerländer der etwa im Jahre 2000 zu verabschiedenden Norm zustimmen können. Es besteht aber unter den beteiligten Mitgliedern des SC 7 die feste Absicht, die Zahlenwerte in so knappen Grenzen wie möglich streuen zu lassen.

Bereits während der Bearbeitung des "model code" für EC 7 begann in den teilnehmenden Ländern eine freiwillige und sukzessive Anpassung an seine Grundsätze.

So wurden in Deutschland die durch das neue Teilsicherheitskonzept betroffenen Normen gründlich überarbeitet und als Vornormen mit dem Zusatz -100 neu herausgegeben. Es sind die grundlegende Norm für Geotechnik DIN V 1054-100 Sicherheitsnachweise im Erd- und Grundbau sowie sechs Berechnungsnormen: DIN V 4017 Grundbruchberechnungen, DIN V 4019 Setzungsberechnungen, DIN V 4084 Böschungs- und Geländebruchberechnungen, DIN V 4085 Erddruckberechungen sowie DIN 4126 Schlitzwandberechnungen. Diese Vornormen sind, zusammen mit der deutschen Fassung der ENV 1997-1, in einem gesonderten Heft publiziert worden. Gemäß der in Deutschland üblichen Normungpraxis, zu jeder Norm so weit erforderlich eine Erläuterung zu verfassen, wurde zur ENV 1997-1 sowie zu den erwähnten zugehörigen Vornormen ein Kommentar herausgegeben (Sadgorski & Smoltczyk 1996).

In Frankreich wurde das frühere System mit getrennten Regeln für öffentliche und private Bauvorhaben aufgegeben und es befinden sich mehrere weiterführende Normen des AFNOR nach dem Konzept mit Teilsicherheiten in zügiger Bearbeitung.

In den Niederlanden entstand in wenigen Jahren ein kompaktes Normenwerk für Geotechnik, bestehend aus drei NEN - Normen des NNI. In einigen anderen Ländern dürften entsprechend der landesüblichen Praxis entweder nur ganz knappe oder gar keine Zusatzdokumente zur künftigen EN 1997-1 erforderlich sein.

7. FOLGERUNGEN

Die Harmonisierung der Normen für Geotechnik der Mitgliedsländer des CEN und die Bearbeitung einer gemeinsamen Norm für Geotechnik gestalteten sich wesentlich schwieriger, als es auf den Gebieten des konstruktiven Ingenieurbaues der Fall war. Dabei kam immer wieder zum Vorschein, daß die Geotechnik in allen ihren Zweigen - Bodenmechanik, Erdstatik, Grundbau usw. sehr stark durch die örtlichen, vornehmlich geologischen Gegebenheiten und sich daran orientierenden Untersuchungs- sowie empirischen und halbempirischen Bemessungsmethoden geprägt ist. In mehreren Ländern waren direkte Bemessungsmethoden entstanden, die aus den Ergebnissen z.B. von Pressiometerversuchen oder Drucksondierungen zu den Abmessungen von Flach- und Tiefgründungen führen. Die darauf basierende Erfahrung galt es unbedingt zu erhalten.

Die Einführung des Teilsicherheitskonzeptes führte zu einer größeren Vielfalt der bevorzugten Wege und der zu berücksichtigenden Faktoren als ursprünglich angenommen wurde.

Insbesondere entstanden bei der für das Teilsicherheitskonzept erforderlichen Trennung von Einwirkungen und Widerständen erhebliche Schwierigkeiten und Widersprüche. Sie sind umso größer, je mehr eine Koppelung zwischen einem Überbau und seiner Gründung zu berücksichtigen ist.

Eine Lösung der geschilderten Probleme wurde durch die Anerkennung der Notwendigkeit gefunden, die gemeinsame europäische Norm EN 1997-1 als Rahmennorm ("umbrella code") zu betrachten die eventuell durch weiterführende nationale Normen ergänzt werden darf. Diese dürfen freilich den Forderungen ("principles") der EN nicht widersprechen, sondern nur Festlegungen von Anwendungsregeln und Zahlenwerte enthalten. Von dieser Möglichkeit machen bereits zahlreiche Länder Gebrauch.

Mit der Veröffentlichung des durch die zur Überarbeitung der Vornorm ENV 1997-1 eingesetzte Arbeitsgruppe ausgearbeiteten Entwurfes als prEN ist für 1999 zu rechnen.

Zur weiteren Harmonisierung der geotechnischen Praxis wurden Entwürfe für zwei weitere Normteile: ENV 1997-2 Geotechnische Bemessung unterstützt durch Laborversuche und ENV 1997-3 Geotechnische Bemessung unterstützt durch Feldversuche, aufgestellt. Dadurch wird eine Harmonisierung der Versuchspraxis angestrebt, wobei die eigentlichen Versuchsnormen der Mitgliedsländer ihre Gültigkeit behalten werden. Diese Normteile werden voraussichtlich demnächst als Vornormen verabschiedet werden.

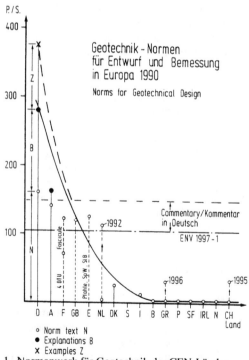

1. Normenwerk für Geotechnik der CEN-Länder

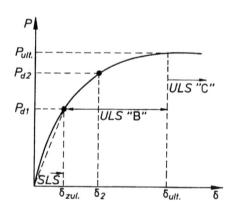

2. Verallgemeinertes Beanspruchungs-Verformungs-Diagramm mit Bereichen der Grenzzustände;
Generalised Load- (Stress-) Deformation with ranges of the limit states in ground. P = action, stress, stress ratio etc. in ground; δ = real or relative displacement, distorsion, strain etc. in ground; SLS - defined by deformation in (super) structure and ground; ULS 1B - defined by failure in structure and deformation in ground; ULS 1C - defined by failure in ground.

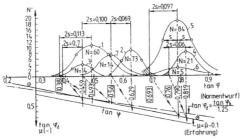

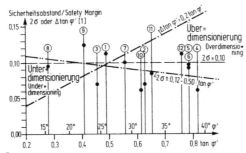

3. Verteilung versuchsmäßig bestimmter Scher-festigkeiten tan φ' für verschiedene Boden-arten; 1, 2 und 3 nach Ohde (1952) im Rah-menschergerät; 4, 5 und 6 nach Melzer (1963) aus Sondierungen; 7 für Abraummaterial (Verfasser).

- Distribution of shear strength tan φ' of different soil types in test series,

4. Vergleich der Sicherheitsabschläge mit γ_m und 2σ; 1 bis 7 s. Bild 3; 8 bis 11 nach Schultze (diverse)
   Comparsion of safety margins with γ_m and 2σ.

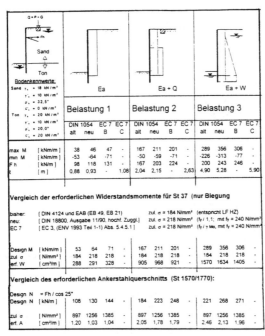

	Belastung 1				Belastung 2				Belastung 3			
	Ea				Ea + Q				Ea + W			
	DIN 1054 alt	EC 7 neu	EC 7 B	C	DIN 1054 alt	EC 7 neu	EC 7 B	C	DIN 1054 alt	EC 7 neu	EC 7 B	C
max M [kNm/m]	38	46	47	-	167	211	201	-	289	356	306	-
min M [kNm/m]	-53	-64	-71	-	-50	-59	-71	-	-226	-313	-77	-
F h [kN/m]	98	118	131	-	167	203	224	-	200	243	246	-
t [m]	0,88	0,93	-	1,08	2,04	2,15	-	2,63	4,90	5,28	-	5,90

Vergleich der erforderlichen Widerstandsmomente für St 37 (nur Biegung)

		zul. σ = 184 N/mm² (entspricht LF HZ)
bisher:	[DIN 4124 und EAB (EB 49, EB 21)	
neu:	[DIN 18800, Ausgabe 11/90, höchst. Zuggl.]	zul. σ = 218 N/mm² (fy / 1,1; mit fy = 240 N/mm²
EC 7	[EC 3, (ENV 1993 Teil 1-1) Abs. 5.4.5.1]	zul. σ = 218 N/mm² (fy / γ Mo, mit fy = 240 N/mm²

Design M [kNm/m]	53	64	71	-	167	211	201	-	289	356	306	-
zul σ [N/mm²]	184	218	218	-	184	218	218	-	184	218	218	-
erf. W [cm³/m]	288	291	328	-	905	968	921	-	1570	1634	1405	-

Vergleich des erforderlichen Ankerstahlquerschnitts (St 1570/1770):

Design N = Fh / cos 25°												
Design N [kN/m]	108	130	144	-	184	223	248	-	221	268	271	-
zul σ [N/mm²]	897	1256	1385	-	897	1256	1385	-	897	1256	1385	-
erf. A [cm²/m]	1,20	1,03	1,04	-	2,05	1,78	1,79	-	2,46	2,13	1,96	-

5. Ergebnisse der Vergleichsberechnungen von Sänger (1995) für gegebene Bodenverhältnisse;
   Results of comparative calculations by Sänger (1995) for given ground conditions.

8. LITERATUR

Breitschaft, G. & J. Hanisch 1978: Neues Sicherheitskonzept im Bauwesen aufgrund wahrscheinlichkeitstheoretischer Überlegungen - Folgerungen für den Grundbau unter Einbeziehung der Probenahme und der Versuchstechnik. Vorträge der Baugrundtagung 1978 in Berlin. Essen: Eigenverlag der Deutschen Gesellschaft für Erd- und Grundbau.

DIN Deutsches Institut für Normung 1981: Grundlagen zur Festlegung von Sicherheitsanforderungen für bauliche Anlagen. Berlin/Köln: Beuth Verlag.

DIN Deutsches Institut für Normung 1996: Vornorm DIN V ENV 1997-1 Entwurf, Berechnung und Bemessung in der Geotechnik, Teil 1: Allgemeine Regeln.

DIN Deutsches Institut für Normung 1996a: Vornorm DIN V 1054-100 Sicherheitsnachweise im Erd- und Grundbau, Teil 100: Berechnung mit dem Konzept mit Teilsicherheitsbeiwerten.

DIN Deutsches Institut für Normung 1997: Eurocode 7 Entwurf, Berechnung und Bemessung in der Geotechnik, Beispiele. Berlin: Beuth Verlag

ENPC 1996: EUROCODE 7 "Calcul Géotechnique": presentations et perspectives. Journées d'etude de l'ecole des ponts. Paris: Ponts formation edition.

Jappelli, R. 1996: Eurocodice 7: progettazione geotecnica - Scopi, principi e compatibilitá con le norme italiane. Rivista Italiana di Geotecnica, Anno XXX, n.2/3, p. 5-35.

Mayerhof, G.G. 1993: Development of geotechnical limit state design. Keynote address to Internat. Symposium on limit state design in geotechnical engineering in Kopenhagen, Vol. 1/3 S. 1-12.

Ohde, O. 1952: Die Berechnung der Standsicherheit von Böschungen und Staudämmen. Vorträge der Baugrundtagung 1952 in Essen, Bautechnik Archiv, Heft 6. Berlin: W. Ernst & Sohn

Olsson, L. & H. Stille 1984: Partialkoefficientmetoden i geotekniken. Rapport R52. Stockholm: Byggforsningsradet.

Sadgorski, W. 1995: Die Vornorm ENV 1997-1 - grober allgemeiner Rahmen oder Zeitbombe ? GEOTECHNIK 18(3), S. 131-137.

Sadgorski, W. 1995: European Final Draft for a Geotechnical Engineering Norm ENV 1997-1 and its Meaning for Tunneling. Trends in Design & Construction of Mechanical Tunneling.

Proc. of Internat. Lecture Series, Hagenberg, 14./15.12.1995. Rotterdam: Balkema.

Sadgorski, W. & U. Smoltczyk 1996: Sicherheitsnachweise im Erd- und Grundbau, Kommentar zu DIN V ENV 1997-1 und den zugehörigen DIN - Normen. Berlin: Beuth.

Sänger, Ch. & Th. Voigt 1995: Berechnungsbeispiele von verankerten Spundwänden nach DIN 1054 alt/ DIN V 1054-100/ ENV 1997-1. Vortragsmanuskript.

Schultze, E. 1971: Contribution to Probability in Soil Mechanics. Proceedings of the First International Congress on Application of Statistics and Probability to Soil and Structural Engineering in Hong Kong.

Smoltczyk, U. 1994: Ein neues Zeitalter beginnt: Das GeoCen. GEOTECHNIK 13(4), S. 212-216.

Geotechnical Hazards, Marić, Lisac & Szavits-Nossan (eds) © 1998 Taylor & Francis, ISBN 90 5410 957 2

Geotechnische Untersuchung eines Schadensfalles an einer Uferböschung

H. Schulz
Institut für Bodenmechanik und Grundbau, Universität der Bundeswehr, München, Germany

H.-J. Köhler
Bundesanstalt für Wasserbau, Karlsruhe, Germany

ABSTRACT: Im Zuge des Ausbaus der oberen Tideems zu einer Wasserstraße für das 1000 t-Schiff sind im Bereich zwischen Papenburg und Herbrum als Teilstück des Dortmund-Ems-Kanals ab 1982 beide Ufer als durchlässige Böschungssicherungen hergestellt worden. Dabei wurden als Filter Geotextilien verwendet. Nach nur 10 Jahren zeigten sich etwa auf Höhe des Tideniedrigwassers Aufwölbungen der Geotextilen Filter, so daß es zur Verlagerung von Steinen und zum Bloßliegen der Geotextilien auf Flächen von 1-2 m^2 Größe und zu zunehmenden Böschungsverformungen kam. Die Ursachen der Schäden werden mit Hilfe eines geotechnischen Standsicherheitskonzeptes untersucht.

BAUWERK UND SCHADEN

Der Schadensbereich liegt an einem Abschnitt der oberen Tideems mit 50 m mittlerer Wasserspiegelbreite und 5 m mittlerer Wassertiefe bei mittlerem Tideniedrigwasser (MTnw), Bild 1. Die Böschungsneigung beträgt 1:3. Der anstehende Untergrund besteht überwiegend aus holozänen Fein- und Mittelsanden. Seine Festigkeit kann durch einen charakteristischen Wert des Reibungswinkels von 32,5° gekennzeichnet werden. Bild 1 zeigt die Situation in einem repräsentativen Querschnitt nach Alberts, 1995.

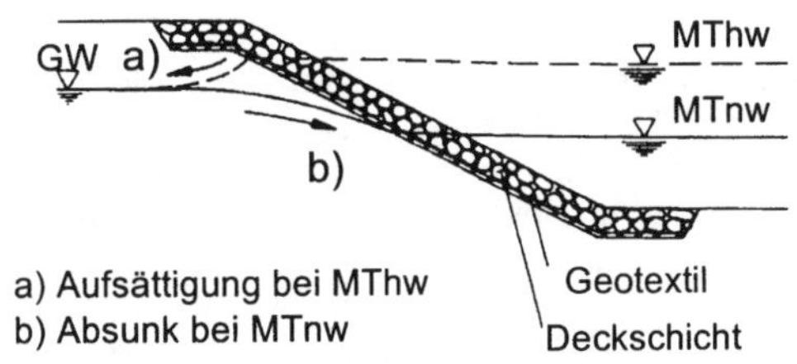

a) Aufsättigung bei MThw
b) Absunk bei MTnw
Geotextil
Deckschicht

Bild 1: Querschnitt der Uferböschung im Schadensbereich nach Alberts, 1995

Die Böschungssicherung besteht aus einer Deckschicht aus Wasserbausteinen Klasse III nach TLW (siehe Literaturverzeichnis) mit einer Steinlänge von 15 - 45 cm bei einer mittleren Dicke von 45 cm. Der Übergang zum anstehenden Untergrund erfolgt durch geotextile Filter unterschiedlicher Hersteller. Die Geotextilien werden vor Baubeginn entsprechend RPG (siehe Literaturverzeichnis) auf Filterfestigkeit gegenüber dem anstehenden Boden untersucht.

Die hydraulischen Verhältnisse lassen sich wie folgt zusammenfassen:

Mittlerer Grundwasserstand: +0,85 mNN
Mittleres Tideniedrigwasser: - 0,95 mNN

Maximale schiffserzeugte hydraulische Belastungen am Böschungsfuß sind bei mittleren Tideniedrigwasser mit einer Wasserspiegelabsenkung von z_a = 0,6 m anzusetzen (Alberts, 1995). Die Dauer der Absenkung des Wasserspiegels aus Schiffsvorbeifahrten beträgt etwa t_a = 3 - 5 Sekunden, danach bleibt der Wasserspiegel während der Schiffspassage etwa auf diesem Niveau.

Für diese Belastungen und die damit verbundenen Wellenhöhen und Rückströmgeschwindigkeiten sind Wasserbausteine der

Klasse III nach TLW im allgemeinen lagestabil gegenüber den angegebenen Einwirkungen.

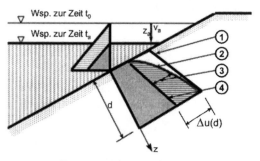

Legende: ① Porenwasserdruck zur Zeit t=0

② Porenwasserdruck zur Zeit t=t$_a$

③ Porenwasserüberdruck Δu zur Zeit t=t$_a$

④ Porenwasserdruck zur Zeit t=t$_\infty$

Bild 2: Porenwasserüberdruck unter einer Uferböschung infolge einer schnellen Wasserspiegelabsenkung

URSACHE DER SCHÄDEN

Bei der Suche nach der Ursache der Schäden ist zu überprüfen, ob die geostatisch erforderliche Dicke der Böschungssicherung gegen Abgleiten infolge Schiffsabsunk und Tideveränderungen ausreicht. Soweit die Gleitsicherheit für diese Einwirkungen gegeben ist, sind weitere Schadensursachen zu untersuchen.

In der Regel ist der Porenwasserüberdruck, der bei der schnellen Absenkung des Wasserspiegels infolge Tide-, Wind- und/oder Schiffswellen entsteht, als auslösend für Schäden anzusehen (Bild 2). Ursache für den Porenwasserüberdruck wiederum ist die Kompressibilität des Porenwassers. Bei Druckschwankungen im Porenwasser infolge von schnellen Wasserspiegeländerungen im Gewässer treten Volumenänderungen des in Form feiner Bläschen eingeschlossenen Luftvolumens auf, welches mehr als 10 % des Porenvolumens ausmachen kann. Die Veränderung des Luftvolumens ist aber zunächst nicht möglich, weil sich dazu das Wasservolumen verändern muß, d.h. es muß Wasser ab- oder auch zuströmen können. Der dazu erforderliche Gradient wird in den unterhalb des abgesenkten Wasserspiegels liegenden Zonen durch den in Richtung zum freien Wasser sich einstellenden Druckunterschied aufgebaut. Dieser Gradient reduziert die zur Verfügung stehende Normalspannung in böschungsparallelen Gleitfugen und damit die widerstehenden Scherkräfte des Bodens in dieser Gleitfuge.

Wenn der Scherfestigkeitsabnahme durch den Porenwasserüberdruck nicht Rechnung getragen wird, können infolge der Absenkung bodenmechanische Grenzzustände eintreten. Daher ist die Kenntnis des bei schnellen Wasserspiegelabsenkungen entstehenden Porenwasserüberdrucks der Schlüssel zur standsicheren Ausbildung von Deckwerken. Hierzu wurden, unabhängig vom hier beschriebenen Schadensfall, über viele Jahre hinweg durch die Bundesanstalt für Wasserbau (BAW) Messungen des Porenwasserüberdrucks in situ, in 1:1-Modellen und in Laborversuchen durchgeführt /Köhler, 1989/ und mit theoretischen Untersuchungen verglichen, Sleath, 1970. Mit dem Ansatz einer Exponentialfunktion und einen, die wesentlichen Eigenschaften des Korngerüstes und der Porenflüssigkeit beinhaltenden Parameter, dem Porenwasserdruckparameter b, erhält man den Porenwasserüberdruck Δu als Funktion der Tiefe z normal zur Böschungsfallinie:

$$\Delta u(z,t) = z_a \gamma_w (1 - \exp(-b \cdot z)) \qquad (1)$$

Für die praktische Anwendung in einem allgemeinen Bemessungsansatz zur Standsicherheit durchlässiger Böschungssicherungen wurden die theoretischen Zusammenhänge und die Ergebnisse der Messungen für den b-Wert in einem einfachen Diagramm zusammengefaßt (Bild 3).

Die in Bild 3 dargestellte Abhängigkeit des b-Wertes von der Durchlässigkeit k beinhaltet aufgrund der Korrelation mit Meßwerten den theoretisch nicht exakt erfaßbaren Einfluß des Luftanteils im Porenwasser sowie der Kompressibilität des Korngerüstes unter den sehr niedrigen Spannungen, die unter Deckwerken vorliegen, für den Fall einer Wasserspiegelabsenkung infolge Schiffahrt. Die Absunkdauer beträgt hierbei im allgemeinen 3 - 5 Sekunden.

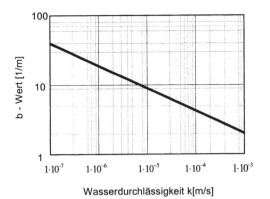

Bild 3: b-Wert als Funktion des Durchlässigkeitsbeiwertes k

Der theoretische Zusammenhang des Porenwasserdruckbeiwertes b mit der Zusammendrückbarkeit von Boden und Porenwasser und mit dem Durchlässigkeitsbeiwert k ist durch folgenden Ausdruck gegeben (Schulz, 1986, Schulz und Köhler, 1989):

$$b = \sqrt{\frac{\gamma_w (n \cdot \beta_w + \frac{1}{E_s})}{k \cdot t_a}} \qquad (2)$$

STANDSICHERHEITSBETRACHTUNGEN

Bei der allgemeinen Betrachtung der Standsicherheit von Böschungssicherungen an Wasserstraßen unter dem Einfluß von Wellen, schnellem Wasserspiegelabsunk und Strömungen sind folgende Grenzzustände zu untersuchen:

- Lagestabilität der Elemente der Deckschicht
- Filterstabilität
- Sicherheit des Deckwerks gegen
  - Abgleiten
  --Abheben bei Fußstützung, wenn diese Gleiten verhindert.

Die Lagestabilität der Elemente der Deckschicht, bei durchlässigen Böschungssicherungen i.d.R. Wasserbausteine, wird nach hydraulischen Kriterien gewährleistet, die im allgemeinen entsprechend der Hudson-Formel aufgebaut sind (Hudson,1959,1975).

Die Filterstabilität wird durch bekannte Filterregeln bzw. auf der Basis örtlicher Erfahrung sicher gestellt. Die bekannten Ansätze beruhen in der Regel auf geometrischen Kriterien und beinhalten daher eine gewisse Sicherheit. Ansätze, bei denen der hydraulische Gradient berücksichtigt wird, sind bisher für Deckwerksfragen noch nicht allgemein akzeptiert. Allerdings werden bei der Anwendung von geotextilen Filtern an den Wasserstraßen der Bundesrepublik Deutschland Systemtests mit Boden im Kontakt zum Geotextil vorgenommen (RPG).

In den Grenzzuständen Gleiten bzw. Abheben wird ausschließlich untersucht, wie schwer ein Deckwerk sein muß, um die destabilisierende Wirkung der Porenwasserüberdrücke bei schnellen Wasserspiegelabsenkungen infolge aller Arten von Wellen, Tide und ablaufendem Hochwasser durch zusätzliche Auflast zu kompensieren. Im Grundsatz entsprechen diese Nachweise den bekannten geotechnischen Standsicherheitsnachweisen für den GZ1C (ENV 1997, Teil 1 bzw. DIN V 4084-100).

Die bei derartigen Nachweisen nach der kinematischen Methode erforderliche Variation der Gleitlinientiefe kann hier entfallen, da die Tiefe für die kritische böschungsparallele Gleitlinie aus dem Ausdruck

$$d_{krit} = \frac{1}{b} \, (ln(\tan\varphi_{B'} \cdot \gamma_W \cdot z_a \cdot b)$$
$$- ln\,(\cos\beta \cdot \gamma_{B'} \cdot (\tan\varphi_{B'} - \tan\beta))) \qquad (3)$$

ermittelt werden kann. Mit d_{krit} kann das erforderliche wirksame Gewicht g' der Deckschicht über das Gleichgewicht des Krafteckes in Bild 4 berechnet werden. Es gilt folgender Ausdruck:

$$g' = \gamma_D' d_D = \frac{\Delta u \cdot \tan\varphi_{B}' - c_{B}'}{\cos\beta \cdot \tan\varphi_{B}' - \sin\beta} - (\gamma_F' d_F + \gamma_B' d_{krit}) \qquad (4)$$

Im vorliegenden Fall treten Schiffswellen mit Absunkwerten z_a = 0,6 m auf. Für diese Absunkwerte ergibt sich bei einem Reibungswinkel des Untergrundes von φ' = 32,5° und einem b-Wert von b = 5,5 (für k = 5 · 10^{-5} m/s aus Bild 3) eine erforderliche Dicke d_D der Deckschicht von

d_D = 1,1 m.

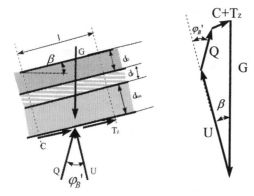

Bild 4: links System, rechts Krafteck zur Beschreibung der Standsicherheit eines Deckwerkes.

Dabei ist zu bemerken, daß bei einem geotextilen Filter für das Flächengewicht $g_F = 0$ gilt. Die Wichte der Deckschicht ergibt sich bei einem Porenanteil von $n = 0,45$ und einer Steinrohdichte $\rho_s = 2,3$ t/m^3 (Mindestwert nach TLG (siehe Literaturverzeichnis)) zu:

$$\gamma' = (1-n)(\gamma_s-\gamma_w) = 7,15 \text{ kN/m}^3.$$

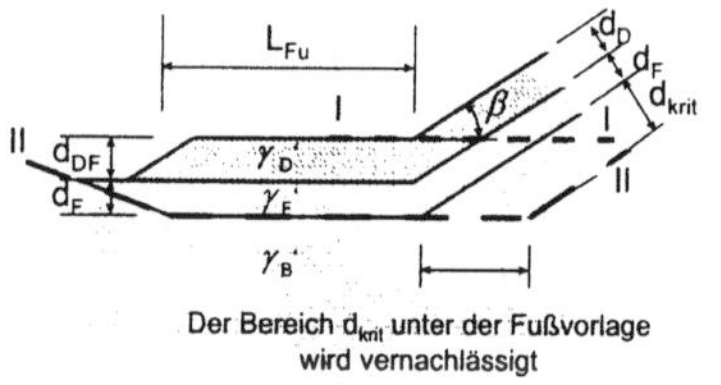

Der Bereich d_{krit} unter der Fußvorlage wird vernachlässigt

Bild 5: Fußvorlage eines Deckwerkes und zu untersuchende Bruchmechanismen

Da die Deckschichtdicke in der oben berechneten Stärke nicht aufgebracht worden ist, muß überprüft werden, ob eine Fußstützung des Deckwerks in der Lage ist, die fehlende Stabilität gegen Abgleiten sicher zu stellen. An Wasserstraßen und schiffbaren Flüssen ist dies häufig möglich, wobei sich drei unterschiedliche Formen bewährt haben, wenn die Randbedingungen, die deren Anwendung zugrunde liegen, beachtet werden. Die Bilder 5 und 6 zeigen die 2 Arten der Fußstützung „Fußvorlage" und „Fußverlängerung" (MAR (siehe Literaturverzeichnis)) und die zu untersuchenden

Bruchmechanismen: Gleiten in der Deckschichten selbst (Mechanismus I) und Gleiten im Boden um die Fußstützung herum (Mechanismus II).

Bild 7 zeigt die Fußstützung durch eine Fußspundwand nach MAR. Hierbei wird die zur Gleitsicherheit fehlende Stützkraft F, der Spundwand zugewiesen, wobei gleichzeitig ein

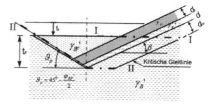

Bild 6: Fußverlängerung eines Deckwerkes und zu untersuchende Bruchmechanismen

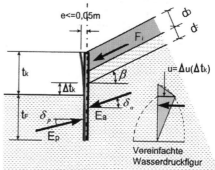

Bild 7: Stützung eines Deckwerkes durch eine Fußspundwand

Kolk vor der Spundwand entstehen darf. Die erforderliche Stützkraft greift als Einwirkung am Kopf der Wand an, im Boden muß sowohl vor als auch hinter der Wand der Porenwasserüberdruck Δu infolge des schnellen Wasserspiegelabsunks nach Glg (1) berücksichtigt werden. Für den Erdwiderstand kann ein geschlossener Ausdruck abgeleitet werden.

Für diese Mechanismen wurden Algorithmen entwickelt, die eine deutliche Reduzierung der Deckschichtdicke gegenüber dem Fall der reinen Reibungshaftung ermöglichen. Die Formeln sind im MBB (siehe Literaturverzeichnis) wiedergegeben, einschließlich ihrer Herleitung im Anhang des MBB.

Im behandelten Schadensfall spielt die Fußstützung eine Rolle. Unter Ansatz einer 2 m langen Fußvorlage als Fußstützung, wie sie an Wasserstraßen der Bundesrepublik nach MAR bei erosionsunempfindlichen Untergrund als eine der Regelbauweisen für die Fußausbildung bei Wasserstraßen mit 4m Wassertiefe angewandt wird, ergibt sich eine Reduzierung der erforderlichen Deckschichtdicke d_D auf

$$d_D = 0,9 \text{ m}.$$

Diese Reduzierung ist darauf zurückzuführen, daß die Schubkräfte der Deckschicht in den Fußbereich geleitet werden. Bei der Überprüfung hat sich der Bruchmechanismus II-II nach Bild 4 durch die Deckschicht als maßgebend erwiesen. Der Scherwiderstand der Deckschicht wurde mit $\varphi'_D = 55°$ angesetzt. Bei einer ausgeführten Deckschichtdicke von 0,45 m ist also ein Schaden aus ungenügender Auflast erklärbar.

Analoge Untersuchungen zum Einfluß der Absenkung des Außenwasserspiegels gegenüber dem Grundwasserspiegel aus dem Ablaufen der Tide ergaben, daß aus Gründen der geotechnischen Stabilität kein dickeres Deckwerk erforderlich ist. Im Fall der Tide ist wegen der verlängerten Absunkdauer der b-Wert im Verhältnis der Wurzeln der Absunkzeiten von

Schiffsabsunk zu Tideabsunk $\sqrt{\dfrac{t_a}{t_{Tide}}}$

zu verringern.
Dadurch entstehen im Untergrund bei Tideabsunk kleinere Porenwasserüberdrücke.

Als eine weitere Schadensursache muß darüber hinaus die Möglichkeit des Dichtsetzens des geotextilen Filters und der Steinschüttung durch Schlickfall untersucht werden. Tatsächlich hat sich nach den Angaben bei Alberts, 1995, der Schlickfall ab 1990, also etwa 2 Jahre vor dem Auftreten der Schäden, erheblich erhöht. Für diese Möglichkeit spricht, daß erst ca. 10 Jahre nach Beginn der Ausbaumaßnahmen Schäden aufgetreten sind.

Die im Wasser enthaltenen Schwebstoffe werden bei ihrem Absetzvorgang von dem geotextilen Filter zurückgehalten. Bei Niedrigwasser entsteht kein ausreichender Gradient aus dem Untergrund heraus, um den Filter wieder frei zu spülen. Auch die bei Schiffsvorbeifahrten aus dem Untergrund heraus auftretenden Gradienten

$$i = \frac{1}{\gamma_w} \frac{d}{dz} \cdot \Delta u(z,t), \qquad (5)$$

die Werte bis i = 4 erreichen, sind offensichtlich zu klein, um ein Freispülen des geotextilen Filters und der Steinschüttung von Schwebstoffen zu erreichen. Durchlässigkeitsversuche an entnommenen Proben der geotextilen Filter ergaben Durchlässigkeitswerte der bodenbesetzten Proben von $k = 8 \cdot 10^{-6}$ m/s, die deutlich unter der Durchlässigkeit des anstehenden Bodens liegen.

Um unter diesen Bedingungen eine standsichere Böschungssicherung zu erhalten, sollte diese wenigstens an jeder Stelle den Wasserüberdruck aus Schiffsvorbeifahrten aufnehmen können. Dazu ist, wenn keine Fußstützung angesetzt wird, ein wirksames Flächengewicht (Sicherheit gegen Abgleiten) entsprechend der Gleichung (4) erforderlich, welches sich durch die Verringerung der Wasserdurchlässigkeit des geotextilen Filters im vorliegenden Fall mit dem Porenwasserdruckparameter b = 10 zu g' = 8,6 kN/m² errechnet. Die Dicke der Deckschicht mit den bisher hierzu genannten Daten ergäbe sich damit zu 1,2 m. Dieser Wert ist bei weitem höher als die ausgeführte Stärke von 45 cm. Daher konnten sich Filter- und Deckschicht unter Ausnutzung der Zugfestigkeit des Geotextils verformen und Bodenumlagerungen traten ein.

Bei einem sich dicht setzenden Geotextil wird sich auch infolge der ablaufenden Tide ein Überdruck einstellen. Da gleichzeitig aber wegen der nicht vollständigen Abdichtung des Gewässerbettes eine Entwässerung stattfindet, läßt sich der unter einer dichten Böschungssicherung entstehende Wasserüberdruck nicht einfach berechnen. Leider liegen auch keine Meßwerte vor.

Unterstellt man, daß auch im Fall einer dichten Deckschicht eine Fußstützung mobilisiert wird, die alle Längskräfte aufnehmen kann, dann muß jedoch mindestens ein wirksames Flächengewicht gegen Abheben von

$$g' = \frac{z_a \cdot \gamma_w}{\cos \beta} = 6,3 kN/m^2 \qquad (6)$$

bzw. eine erforderliche Dicke von:

$$d_D = \frac{g'}{\gamma_D'} = 0,89 m \qquad (7)$$

vorhanden sein, um eine Destabilisierung des Bodens unter der Deckschicht und damit langsam zunehmende Deformationen der Böschungssicherung, die auch als hydrodynamische Bodendeformation bezeichnet werden (MBB), zu verhindern. Ein Kornfilter hätte zusätzliches Gewicht auf den Untergrund gebracht, weil in ihm keine Porenwasserüberdrücke mehr wirken. Dies hätte die Standsicherheit verbessert. Möglicherweise hätte er einem Dichtsetzen besser entgegen gewirkt als ein Geotextil, weil wegen seiner fehlenden Zugfestigkeit örtliche Aufbrüche den Porenwasserüberdruck abbauen können und dem Dichtsetzen entgegengewirkt wird.

ZUSAMMENFASSUNG

Anhand eines Schadensfalles an der Oberen Tideems konnte gezeigt werden, daß aufgrund geotechnischer Standsicherheitskriterien die Ursachen des Schadens darin erkannt werden können, daß das Deckwerksgewicht für die hydraulischen Einwirkungen aus der Schiffahrt zu gering war. Die zusätzliche Belastung aus Schlickfall hat ein Dichtsetzen des Geotextils bewirkt und damit die Schäden und Verformungen an der Uferböschung verstärkt.

Die infolge des schnellen Absunks des Wasserspiegels von 60 cm aus der Vorbeifahrt von Schiffen sich einstellenden Wasserdrücke unter der Böschungssicherung erfordern bei Aufnahme von Längskräften im Fußbereich eine ca. 90 cm dicke Schutzschicht aus Wasserbausteinen. Tatsächlich waren 45 cm eingebaut.

LITERATUR

Alberts, D., 1995, Deckwerksschäden an der Oberen Tideems. HANSA-Schiffahrt-Schifbau-Hafen, 132. Jahrgang - 1995 - Nr. 6

DIN V 4084-100, Baugrund, Böschungs- und Geländebruchberechnungen, Teil 100: Berechnung nach dem Konzept mit Teilsicherheitsbeiwerten

ENV 1997-1, Eurocode 7: Entwurf, Berechnung und Bemessung in der Geotechnik - Teil 1: Allgemeine Regeln

Köhler, H.-J., 1989, Messung von Porenwasserüberdrücken im Untergrund; Mitteilungsblatt der Bundesanstalt für Wasserbau, Karlsruhe Nr. 66

Hudson, R.Y., 1959, Laboratory investigations of rubble-mound breakwaters, Journal of Waterways and Harbors, ASCE Sept. 1959

Hudson, R.Y.,1975, Reliability of rubble-mound breakwater stability models (Schlußbericht); Miscellaneous Paper H-75-5

MAG, Merkblatt Anwendung von Geotextilen Filtern an Wasserstraßen, Ausgabe 1993; Bundesanstalt für Wasserbau, Karlsruhe

MAR, Merkblatt Anwendung von Regelbauweisen für Böschungs- und Sohlensicherungen an Wasserstraßen, Bundesanstalt für Wasserbau, Karlsruhe

MBB Merkblatt Bemessung von Böschungs- und Sohlensicherungen, Bundesanstalt für Wasserbau, Karlsruhe (in Vorbereitung)

RPG, Richtlinien für die Prüfung von geotextilen Filtern im Verkehrswasserbau, BAW Karlsruhe

Schulz, H. ,1986, Kompressibilität und Porenwasserüberdruck - Bedeutung für Gewässersohlen. Mitteilungsblatt der Bundesanstalt für Wasserbau Nr. 76

Schulz, H. und Köhler, H.-J., 1989, The developments in Geotechnics concerning the dimensioning of revetments for inland navigable waterways, Pianc-Bulletin, Nr. 64

Sleath, J.F.A., 1970, Wave-Induced Pressures in Beds of Sand. Journal of the Hydraulic Division ASCE, Febr. 1970

TLG, Technische Lieferbedingungen für geotextile Filter, zu beziehen über die Drucksachenstelle bei der WSD Mitte, Am Waterlooplatz 9, 30169 Hannover

TLW, Technische Lieferbedingungen für Wasserbausteine, zu beziehen über die Drucksachenstelle bei der WSD Mitte, Am Waterlooplatz 9, 30169 Hannover

SYMBOLE:

$\Delta u(z,t)$ - Porenwasserüberdruck [kN/m^2] als Funktion der Tiefe z unter der Böschungsoberfläche zum Zeitpunkt t = t_a des maximalen Absunks z_a

z_a - maximaler Absunk des Wasserspiegels [m]

γ_w - Wichte des Wassers [kN/m^3]

b - Porenwasserdruckparameter [1/m] nach Bild 3 für eine Absunkdauer von t_a = 3-5 Sekunden

z - Tiefe, normal zur Böschungsoberfläche [m]

k - Durchlässigkeit [m/s]

E_s - Steifemodul des Bodens [kN/m^2]

β_w - Kompressionsmodul des lufthaltigen Wassers [m²/kN]:

$$\beta_w = 5 \cdot 10^{-7} + \frac{1-S}{(p_{atm} + p_{hydst})}$$

n - Porenanteil des Bodens

S - Sättigungsgrad [-]

p_{atm} - Atmosphärendruck [kN/m²]

p_{hydst} - hydrostatischer Druck in der betrachteten Tiefe [kN/m²]

t_a - Dauer des Absunks [s]

t_{Tide} - Absunkdauer der Tide zwischen Tidehöhe des Grundwasserstandes und maßgebendem Tideniedrigwasser

d_{krit} - kritische Tiefe [m] der böschungswinkel [°]

$\gamma_B{}'$ - Wichte des Bodens unter Auftrieb [kN/m³]

$\varphi_B{}'$ - effektiver Reibungswinkel des Bodens [°]

$c_B{}'$ - effektive Kohäsion des Bodens [kN/m²]

$b(t)$ - Porenwasserdruckparameter [1/m] nach Bild 3 für eine Absunkdauer von t_a = 3 - 5 Sekunden

g' - Flächengewicht in [kN/m²]

d_D - Dicke der Deckschicht [m]

$\gamma_D{}'$ - Wichte der Deckschicht unter Auftrieb [kN/m³]

$\gamma_F{}'$ - Wichte des Filters unter Auftrieb [kN/m³], bei geotextilen Filtern $\gamma_F{}'$ = 0

d_F - Dicke des Filters [m], bei geotextilen Filtern d_F = 0.005

i - hydraulischer Gradient [-]

d_{krit} - kritische Tiefe [m] nach Glg. (7-4)

d_{krithG} - kritische Tiefe in der Sohle eines Gewässers (β = 0) bei Porenwasserüberdrücken infolge schneller Wasserspiegelabsenkungen [m]

γ_S - Wichte der Wasserbausteine [kN/m³]

Geotechnical Hazards, Marić, Lisac & Szavits-Nossan (eds) © 1998 Taylor & Francis, ISBN 90 5410 957 2

Piping processes and stability of hydrotechnical constructions

K. Sh. Shadunts & V. V. Podtelkov
Kuban State Agrarian University, Krasnodar, Russia

ABSTRACT: This paper is devoted to analysis of the causes for piping processes and using some original construction to stop the activation of the escape of the dust-like particles from the wharf wall soil basement and covering plates of the dips. Authors suggested use the filtering concrete for the drainage constructions and gave theoretical material to calculate drainage parameters.

1 INTRODUCTION

The change of the operation mode of a water development, a lock on a water storage in connection with the reduction of the navigation on the Kuban river has led to a unilateral flow of underground water and a considerable activation of the escape of the dustlike and fine particles from covering of the dips of the earth structures downstream. The deformations of the wharf wall took place, one of its sections was lowered by 35 cm and deviated from the vertical line by 40 cm (Fig. 1). The voids under the plates of covering of the dips have reached 100-120 cm in depth (Fig. 2). The situation becomes complicated by worsening of the operation of intercepting drainage which is choked up with the clay particles and iron salts during a long period of operation of the water storage.

Figure 1. Deformed wharf wall downstream

Figure 2. Holes in the slope covering plates for specifying of the piping escape volumes

2 DRAINAGE CONSTRUCTIONS

It is possible to stop further deformations by restoration of the drainage structures which alter the character of the depression and prevent piping. But it is difficult to construct the new drainage structures, because it is necessary to dig deep trenches. The absence of special effective excavating machines, the necessity of shuttering, laying of the drainpipes and the use of multilayer drainage filling lead to the fact that the construction becomes very expensive.

The experience of the use of the filtering concretes is taken as a basis of the drainage constructions suggested by K. Sh. Shadunts (Patent of Russia, priority dated 11 March, 1997).

The use of the sheet piles as the guards which prevent water from entering the foundation pits is known. We have elaborated the construction of reinforced concrete sheet piles acting as a drainage. A section of the sheet piles determined by the calculation which takes into account the natural underground water level and the necessary water lowering is filled with filtering concrete during manufacture and a drainage hole is made. During the submersion of the sheet piles the coinciding holes form a pipe which is given the necessary slope. The final part is formed as a head of the structure. The drainage placed along the slope is arranged in the following way (Fig. 3). For the cutting off drainage which intercepts the underground flow the sheet piles are manufactured in the same way and have an external wall which faces the bottom of the slope and is permeable to water; this wall is made of dense concrete which fences a filtering part.

2.1 Grain - Size Distribution of Aggregates of Porous Concrete

Gravel or granite crushed stone can be used as aggregates for porous concretes. Crushed brick, crushed expanded clay aggregate and slag can be used as light - weight aggregates. Strength and decrepitude period of the structure where porous concrete will be used depend on the aggregate. The investigation have been carried out with porous concrete with aggregate made of granite crushed stone.

Initially the experiments were carried out with porous concrete made of homogeneous crushed stone and their purpose was to select porous concrete distribution for the contact with hydraulic sand of Kiev hydroelectric station. Further investigations were carried out with porous concrete on the aggregates of different size and homogeneity beginning from small crushed stone with a size of the fractions of 0.1-10 mm to crushed stone with a size of the fractions of 7-80 mm.

Puzzolan and slag portland cements are used as a binder for porous concrete which are not less than M 300 as well as portland cements M 400-500. It is necessary to use the corresponding additives when the filters operate under the conditions of aggressive water.

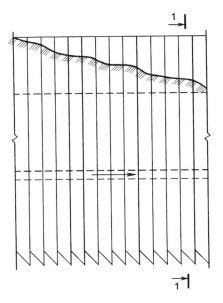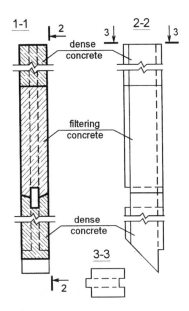

Figure 3. Drainage structure made of sheet piles with a filtering section

Cement-to-crushed stone ratio should be 1:6 or 1:7 according to weight when water-to-cement ratio is 0.32-0.42. Smaller ratios should be accepted for large sized aggregates, because they have smaller specific surface of the particles. When porous concrete is manufactured, quantity of water plays a significant role On the one hand, the amount of water should be enough to make cement slurry movable; on the other hand, there should be no drain off of cement slurry and clogging of the pores.

It is better to select water - to - cement ratio with the help of the trial batches. During the process of manufacture of porous concrete it is necessary to watch aggregate humidity, and if the changes of more than 0.7-1% take place, it is necessary to change water - to - cement ratio.

2.2 Manufacture of Porous Concrete

As a rule, porous concrete is consolidated by vibration though consolidation by means of rodding is also possible.

As the experiments have shown the consolidation depends on fluctuation amplitude and vibration duration insignificantly. Practically when vibration has frequency nearly 6000 per minute, the complete consolidation is achieved during 0.5-1.2 minutes. When hardening of porous concrete takes place, one should observe the same mode of temperature and humidity as for usual concrete.

Any crushed stone can be used as an aggregate for porous concrete. If heterogeneity of crushed stone of even large size is changed by means of adding of granite waste or sand, one can get porous concrete with small pores which can be used for the contact with sand or cohesive soils.

Volumetric weight of porous concrete can be changed from 1.65 to 2.2 tons per m^3 and depends mainly on heterogeneity of the aggregate. If heterogeneity is increased, volumetric weight is increased.

Opened (or effective) porosity which determines filtration and piping properties of porous concrete is reduced with the increase of heterogeneity.

Strength of porous concrete is 4.0-4.5 MPa, and it is even 10.0-10.5 MPa for concrete with small pores. No reduction in strength of the porous concrete is found during the increased filtration of tap water during the year. Vice versa, its insignificant increase was observed. It is confirmed by the operation of the vertical porous drainage during many years at Zimlyansky and Volgogradsky water developments.

2.3 Filtering parameters of porous concrete constructions

Filtration capacity of porous concrete is changed within great limits - from 30 to 5000 m per day and depends both on large size and heterogeneity of the aggregate.

Depending on grain-size distribution of the aggregate the size of the pores plays a decisive role for the selection of the size of the components which are in contact with the soil as well as for the determination of the possibility of mud injection of the pores by the soil.

277

A number of experiments has been carried out in order to determine the intensity of clogging of the pores by salts which are in water. These experiments have shown that sedimentation of various oxides in the pores takes place not in porous concrete, but in usual filter material (crushed stone, gravel) and in all other materials, including bitumens and plastic. Sediments are formed in the pores due to availability of oxygen in the pores.

Many investigators are busy with the problems concerning the protection of the filters from clogging. At present there are no safe recommendations concerning the elimination of this phenomenon.

Inspection pits of the drainage system made of the sheet piles with a filtering part can be arranged sinking the cylindrical shells at a certain distance.

If the sections made of filtering concrete are used in the shells, they will provide water absorption in addition to the function of sediment and water purification.

The drainage structures of the type being suggested have one more advantage over trench drainage: besides drainage, they retain soil. Having sufficient strength the sheet piles prevent the formation of the collectors in the soil which promote piping of the clay particles and increase slope stability.

3 CALCULATION OF DRAINAGE

The calculation of the influence zone of drainage structure arranged in clay soils with a jet flow of water along the interlayers can give only an approximate result due to variability of filtration factors and values of the gradients of head in the soil along the construction.

Horizontal drainage with a raw of sheet piles as a basement is an effective anti-landslide and anti-piping construction. Such constructions provide a complete interception of water-bearing horizon under complicated water development conditions during unilateral influx of ground water when confining layer depth comprises 6 or 7 m.

Water influx along an inclined seam to the cut off (arranged transverse to the slope) drainage construction per linear unit of its length can be determined according to the formula

$$q = 2 K_f d (180° - \alpha) / 180°, \qquad (1)$$

where K_f - filtration factor of water-bearing soil, d - width of a slot of the drainage structure, α - angle of incidence of confining layer surface.

Value d being necessary for passing of water flowing to drainage is

$$d = (H I / 2) [180° / (180° - \alpha)] , \qquad (2)$$

where I - hydraulic gradient; H - water-bearing layer power.

The total inflow of water to the drainage structure is

$$Q = q l , \qquad (3)$$

where l - length of the drainage structure.

When the drainage structures are arranged along the direction of the landslide displacement, the distance between the drains ($2 a$)can be determined with the help of the formula:

$$h_{max} = 0.773 (q / k) lg\{1 + [(2 a) / (q / K_f)]\} , \qquad (4)$$

where h_{max} - excess of the maximal mark of the depression over the drain bottom mark; q - bilateral influx of water per 1 m of drain,

$$q = [(H - h_0) / 2] K_f I_a , \qquad (5)$$

where H - power of water - bearing layer over the drain; h_o - height of water layer in the drain when the mode is installed; I_a - average grade of the depression curve.

4 CONCLUSIONS

1. The piping processes are the cause of the development of the deformation of many water development works. Their activation is often caused by the drawbacks in the operation of the drainage structures intercepting the soil flow.

2. The new water lowering systems under the conditions of the operating water development can be arranged by sinking of the special sheet piles with a section of filtering concrete in which a water conductive hole is made.

REFERENCES

Guide on design and arrangement of engineering structures being put deeper. 1986. Moscow: Stroyizdat, (NIISK of Gosstroy of the USSR).

Mits I.S. 1968. *Properties of porous concrete as a material for drainages.* Collection "Voprosy geotekhniki", No. 13. p. 64-71. Kiev.

Reference book on design of engineering preparation of the area where the buildings are erected on. 1983. Kiev: "Budivelnik" 192 p.

Geotechnical Hazards, Marić, Lisac & Szavits-Nossan (eds) © 1998 Taylor & Francis, ISBN 90 5410 957 2

Confining degree of reinforced soil structures

R.A. Sofronie
UNESCO Chair in Integrated Rural Development, Bucharest, Romania

V. Feodorov
IRIDEX Group, Bucharest, Romania

ABSTRACT: The paper presents, with the aid of a simplified model of analysis, the value of confining force. Permanent, variable and accidental actions in two loading cases are considered. The degree of confining is defined by the ratio between confining force and structure self-weight. It depends by the two geometrical characteristics of soil structures: aspect ratio and slope inclination. The result of analysis shows the advantage of confining the reinforced soil structures especially when seismic actions occur. However the suggested method has its own structural limitations.

1 INTRODUCTION

Earth reinforcement is often used for retaining walls and steep embankments. For stability reasons such soil structures are usually designed with a high width/height ratio. Values of $R = 0.6$ and higher are typical for these massive structures (Jones 1996). They support well the static loads but under dynamic actions high inertial forces are developing.

Reports on recent Kobe earthquake show that geogrid-reinforced retaining walls exhibited a rather satisfactory seismic stability. Such high stability was explained by the monolithic behavior of reinforced zones of the soil provided with an aspect ratio much higher that those of gravity type walls (Faccioli et al. 1996). However, it was not proved yet that the assumptions adopted for the analysis of external stability are equally valid for static and dynamic actions. For example, according to Eurocode 8 recommendations, the point of application of the dynamic resultant force should be assumed to act at the midheight of the wall but some recent pseudostatic analysis found that point at $0.6H$. Shaking table tests of scale models of gravity retaining walls are expected to bring new useful data to check and improve the analysis methods of external stability (Crewe et al.1998).

The philosophy of reinforced soils is based on the interaction between grid reinforcement and granular fill. Grid interlocks with soil to develop resistance to pullout. Stresses from the soil, taken in both friction and bearing on transverse members of the grid, are transferred through the junctions to the longitudinal carrying members. This mechanism of stress transfer was proved for static loads by shear box tests. The

pullout failure is one of the two conditions of internal stability. The phenomenon mainly depends on the gravity pressure of the soil on grid reinforcement. Preloading and even prestressing of the reinforced soil (Tatsuoka et al. 1996) can enhance this vertical pressure. The idea of inducing additional compressive stresses in soil to improve its behavior was proposed since 1973 for Pisa Tower (Leonhardt 1996).

With respect to gravity the reinforced soil walls are self-retaining structures. In order to reduce and control their response, especially under seismic actions, the confinement of these structures was recently suggested (Sofronie & Feodorov 1995, 1996). Confinement is also related to gravity but means much more than retaining. The scope of confining is to close soil structures and induce in them, with the aid of gravity, a three axial state of compression. Technologically, the procedure can become rather complicated when naturally curved surfaces are considered. In a crude simplified approach, referring only to the existing retaining walls and steep embankments, the procedure of confining consists in applying an external force with two components: one horizontal, near the top of soil structure, and another one vertical or inclined, parallel to structure facing. Since the confining force is oppositely inclined about the resultant of disturbing forces its magnitude can be chosen to center the soil structure on its base. Mathematically, that means to cancel the eccentricity of resultant of all forces acting on structure base (Feodorov 1997). The paper presents a simplified model of analysis for the confining force. The degree of confining is then defined by the ratio between confining force and structure self-weight. Its variation with aspect ratio and slope inclination for different cases of loading is further discussed (Gulvanessian & Holicky 1996).

2 CONFINING FORCE

Since the purpose of the paper is demonstrative a simplified model of analysis was adopted. Accordingly, only six forces are considered in analysis: self-weight W, the horizontal component of active force E_{agh} and confining force S, as permanent actions, surcharge p and the horizontal component of active force from surcharge E_{aph}, as variable actions and seismic force Q, as an accidental action (Figure1).

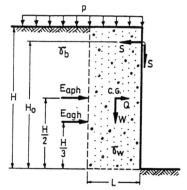

Figure 1. Forces acting on an analyzing model of a retaining vertical wall.

For a given aspect ratio $R = L/H$ only the value of confining force is unknown. When assessing the external stability of a soil structure bearing and tilt failure should be firstly checked. Assuming, in accordance with BS 8006 and DIN 4017, a Meyerhoff distribution the factored bearing pressure acting on the base of the wall is

$$q_r = \frac{R_V}{L - 2e} \tag{1}$$

where

$$R_V = W + S + pL \tag{2}$$

and e is the eccentricity of resultant load R_V about the centerline of the base. Since

$$e = \frac{S\left(H_0 - \dfrac{L}{2}\right) - \left(\dfrac{E_{agh}}{3} + \dfrac{E_{aph} + Q}{2}\right)H}{R_V} \tag{3}$$

by canceling the eccentricity e one finds the expression of confining force

$$S = \frac{\left(\dfrac{E_{agh}}{3} + \dfrac{E_{aph} + Q}{2}\right)H}{H_0 - \dfrac{L}{2}} \tag{4}$$

where always $H_0 > L/2$. Therefore the force of confining has a uniquely determined value for the combination of actions considered above. For other compatible combinations of actions but the same value of S, as defined by (4), the eccentricity no longer remains zero. It will be accepted only when resultant R_V lies in the middle third of the base, i.e. $e < L/6$.

3 ROLE OF ASPECT RATIO

The procedure of confining involves a technological effort. In order to assess this effort the confining force S is compared with the self-weight of soil structure W. The ratio S/W was termed degree of confining. By limiting the expression (4) only to permanent actions one finds

$$\frac{S}{W} = \frac{K_{ab}}{6} \frac{\gamma_b}{\gamma_w} \frac{1}{\dfrac{L}{H}\left(0.9 - 0.5\dfrac{L}{H}\right)} \tag{5}$$

where K_{ab} is coefficient of active earth pressure for the backfill, γ_b unit weight of unreinforced soil, γ_w unit weight of reinforced fill, and 0.9 is a value for the ratio H_0/H. The graphic of function (5) shows that for usual aspect ratios the degree of confining takes reduced values, between 15% and 18%, what proves the procedure is economic and worth to be applied (Figure 2). The most important advantage of confining is the possibility to reduce the aspect ratio of reinforced soil structures from 0.6 to 0.4 what means a remarkable economy of 33%.

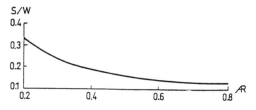

Figure 2. Confining degree versus aspect ratio for permanent actions.

By including in (5) the effect of variable actions one finds that expression of S/W explicitly depends on the surcharge p and overall height of reinforced soil structure H, namely

$$\frac{S}{W} = \frac{\dfrac{K_{ab}}{6}\dfrac{\gamma_b}{\gamma_w}\left(1 + \dfrac{3p}{\gamma_b H}\right)}{\dfrac{L}{H}\left(0.9 - 0.5\dfrac{L}{H}\right)} \quad (6)$$

In this case the degree of confining increases for $R = 0.4$ from 21% to 26% as overall height reduces from $12\ m$ to $4\ m$ (Figure 3). The result is explained by the high non-linearity of expression (6) with respect to the overall height H.

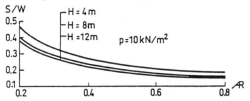

Figure 3. Confining degree versus aspect ratio for permanent and variable actions.

By adding to (6) the effect of accidental actions one finds

$$\frac{S}{W} = \frac{\dfrac{K_{ab}}{6}\dfrac{\gamma_b}{\gamma_w}\left(1 + \dfrac{3p}{\gamma_b H} + \dfrac{3k\gamma_w}{K_{ab}\gamma_b}\dfrac{L}{H}\right)}{\dfrac{L}{H}\left(0.9 - 0.5\dfrac{L}{H}\right)}. \quad (7)$$

In this case the degree of confining drastically increases from simple to almost double. However, for a typical aspect ratio $R = 0.4$ and a seismic intensity $k=0.15$ it remains under 30% what is still acceptable for practical purposes (Figure 4).

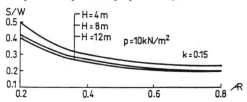

Figure 4. Confining degree versus aspect ratio for permanent, variable and accidental actions.

In all the three cases discussed above for massive soil structures with an aspect ratio higher than 0.6 the procedure of confining does not present a practical interest. On the other side the so-called slender soil structures with an aspect ratio lower than 0.4, although the degree of confining increases, they present a special interest for steep embankments and slip repairs. However, the aspect

ratio can not be reduced much under 0.4 due to stability reasons of reinforced soil.

4 ROLE OF SLOPE INCLINATION

The inclination of slope is denoted by α and can take positive or negative values. The forces acting on this simplified model of analysis are the same like those used above for a retaining vertical wall (Figure 5).

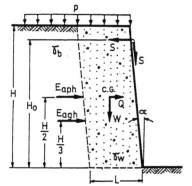

Figure 5. Forces acting on an analyzing model of a confined steep slope.

For permanent actions one finds the degree of confining

$$\frac{S}{W} = \frac{\dfrac{E_{agh}}{3W} - \dfrac{\tan\alpha}{2}}{0.9 - 0.5\dfrac{L}{H}\cos\alpha} \quad (8)$$

limited to the positive inclination

$$\alpha_{lim} = \tan^{-1}\frac{2E_{agh}}{3W} \quad (9)$$

as shown in Figure 6.

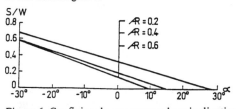

Figure 6. Confining degree versus slope inclination for permanent actions.

Due to confining negative slopes are also allowed. They are not limited although the degree of confining increases. Naturally, slender structures need higher confining degrees than those massive. The financial effort is worth as long as the volume of work reduces up to three times when the aspect ratio decreases from *0.6 to 0.2*. For practical purposes two or more different, always slopes following a convex polygon, can be provided.

For permanent, variable and accidental actions the expression of confining degree becomes

$$\frac{S}{W} = \frac{\dfrac{E_{agh}}{3W} + \dfrac{E_{aph} + Q}{2W} - \dfrac{tan\alpha}{2}}{0.9 - 0.5\dfrac{L}{H}cos\alpha} \qquad (10)$$

being also limited to the positive inclination

$$\alpha_{lim} = tan^{-1}\frac{\dfrac{2}{3}E_{agh} + E_{aph} + Q}{W}. \qquad (11)$$

As shown in Figure 7 the non-permanent actions increase the confining degree limiting in this way the range of inclinations of practical interest to small values like $\pm 5^0$.

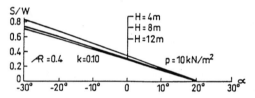

Figure 7. Confining degree versus slope inclination for permanent, variable and accidental actions.

In the cases presented above the state of confining was induced by a single force S applied at the height H_o. The procedure is simple and easy to be carried out. Similarly, two or more confining forces can be applied, all equal between them or with variable intensities along the height. In this way the so called active reinforcing is achieved. The most suitable method is a continuing confinement when the facing of reinforced soil structure is accordingly shaped.

The value of confining force is not at all arbitrary. In all cases for a given or adopted aspect ratio the state of confining remains unique as was defined by the equilibrium equation. One can not exist in the same time two or more confining states. When the aspect ratio is progressively increased the confining degree assumes a minimum value. Under this limit value any confinement disappears. For usual values $(S/W)_{min} = 0.15$ and does not present any practical interest. A special situation occurs when the confining force yields, i.e. $S{\to}0$. It is easy to find from equation (3) that even only for permanent

actions and an aspect ratio of *0.4* the eccentricity becomes twice of $L/6$. That means bearing and tilt failure. In the case of a slope with $\alpha=5^0$ the eccentricity is *32.3%* smaller but remains over $L/6$.

5 CONCLUSIONS

1. The procedure of confining applies to reinforced soils rather than to unreinforced ones. Grid reinforcement allows a more uniform distribution and transfer of stresses. Confined soil structures are centred on their bases and therefore they essentially differ by reinforced soil structures.

2. The confining force is uniquely determined for a certain state of loading from an equilibrium equation. The role of this force is to maintain the confined structures almost centred on their bases for the rest of all loading states.

3. The confining degree is a decisional tool. It allows assessing the opportunity of confining the reinforced soil structures.

4. The reduction of aspect ratio under *0.6* brings both safety and economy. The confining degree helps to choose the most appropriate aspect ratio and then to check both external and internal stability.

5. The serviceability of the confined soil structures of under any state of loading can be permanently controlled. That means the confining degree can be used to assess their reliability.

REFERENCES

Crewe, A.J. et al. 1998. Shaking table tests of scale models of gravity retaining walls. In: *Proceedings of the Sixth SECED International Conference, Oxford 26-27 March 1998.*
Faccioli, E. & R. Paolucci 1996. *Seismic Behaviour and Design of Foundations and Retaining Structures.* ECOEST/PREC8, Report No. 2
Feodorov, V. 1997. *Confined Structures of Reinforced Soil.* Bucharest: Doctoral Thesis.
Gulvanessian, H. & M. Holicky 1996. *Designers' handbook to Eurocode 1.* Part 1: Basis of design. London: Thomas Telford.
Jones, C.J.F.P. 1996. *Earth reinforcement and soil Structures.* London : Thomas Telford.
Leonhardt, F. 1997. The Committee to Save the Tower of Pisa: A Personal Report. *SEI 3/97: 201.*
Sofronie, R. A. & V. Feodorov 1995. *Method of active reinforcing the soil structures.* Bucharest: Patent Office RO 110 964.
Sofronie, R. A. & V. Feodorov 1996. *Method of confining the reinforced soil structures.* Bucharest: Patent Office RO 112 040.
Tatsuoka, F. et al. 1996. Preloaded and Prestressed Reinforced Soil. *Soil and Found. Journal.* (Prep.)

Geotechnical Hazards, Marić, Lisac & Szavits-Nossan (eds) © 1998 Taylor & Francis, ISBN 90 5410 957 2

Rapid quality control method of compaction of non-cohesive soil embankment

M.J.Sulewska

Faculty of Building and Environmental Engineering, Bialystok Technical University, Poland

ABSTRACT: Required compaction of earth bulk provides sufficient load capacity, stability and durability of the embankment. The lightweight dynamic deflectometer can be applied in quick control of compaction in embankment made of non-cohesive soil. The laboratory tests were carried out in order to find out dependencies of modulus E_D on geotechnical parameters.

1 INTRODUCTION

Construction of various earth engineering structures of high load bearing capacity, stability and durability requires regular compaction of each built-in layer of soil. Control tests of achieved degree of compaction is conducted in required progress of earth works.

The following methods of quality of compaction control tests are most often used in Poland:
- direct - based on measurements of relative compaction I_s (BN-77/8931-12),
- indirect - method of dynamic sounding using cone penetrometer (PN-74/B-04452) or rigid plate load tests (BN-64/8931- 02).

The rigid plate load tests method and direct method applied in compaction control of soil built in embankment are time-consuming, especially when constructed earthworks are of considerable volume or they are spread on a large area. On the contrary, results of dynamic sounding with cone penetrometers can be interpreted from certain depth known as critical depth.

Indirect method applying the lightweight dynamic deflectometer (TP BF-StB Teil B 8.3) can be considered as one of quick ways of current control of surface layer of embankment soils. The lightweight dynamic deflectometer is a device enabling generation of short duration force impulses being then transmitted on surface of tested medium. The dynamic modulus of soil deformation E_D is obtained as a result of test.

The aim of investigations presented in the work is evaluation of formula for dynamic modulus of deformation E_D for non-cohesive fill soil in relation to geotechnical parameters defining soil compaction i.e. relative compaction I_s (the percent relation between the dry soil density obtained from the field equipment and the maximum value corresponding to the laboratory compaction test), volumetric density of soil skeleton ρ_d (PN-88/B-04481) and modulus of deformation under the first E_{po} and the second E_p plate trial loading of soil (BN-64/8931-02).

2 THE LIGHTWEIGHT DYNAMIC DEFLECTOMETER

The lightweight dynamic deflectometer type ZFG 01 (TP BF-StB Teil B 8.3) was used for soil testing and its diagram is presented in the Fig. 1.

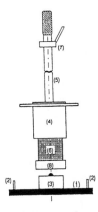

Fig.1. Diagram of the lightweight dynamic deflectometer, where (1) = trust plate; (2) = holders; (3) = detector controlling settlement; (4) = falling weight; (5) = guide bar; (6) = shock absorber; (7) = gripping device of falling weight; (8) = safety lock of stability.

Test with the lightweight dynamic deflectometer consists of loading the subsoil with circular steel plate (1) of radius $r = 0,15\ m$ and surge of falling weight (4) mass of $10\ kg$ along guide bar (5) on shock absorber (6) from the height of $h = 0,83\ m$.

On the theoretical basis of considered problem (Weingart 1981; Kudla et al. 1991) it is assumed that loading of soil with the lightweight dynamic deflectometer can be treated as a problem of short duration quasistatical pressure of the plate on the elastic half-space.

3 LABORATORY TESTS

Tests were performed on laboratory setup (Sulewska 1993) on the modeled substratum made of medium sand (sieve-analysis curve presents Fig. 2). Investigated layer of soil thickness of $a = 0,3\ m$ or $0,5\ m$ was spread on the subbase layer made of medium sand thickness of $a_l = 0,3 \div 0,4\ m$ and $I_s \geq 1,0$. Soil moisture ranged from 3,5 to 6,7 percent. Volumetric density near the loading plate was tested using sand volumeter (BN-77/8931-12), moisture was measured applying drying method, maximum volumetric density of soil skeleton ($\rho_{ds} = 1,800$ g/cm^3) and optimum moisture content ($w_{opt} = 10,0$ %) were obtained from Proctor I method (PN-88/B-04481). Modulus of soil deformation tests were performed with steel loading plate diameter of $D = 0,30\ m$, statically loaded in one test point (BN-64/8931-02) and dynamically loaded in two another test points (TP BF-StB Teil B 8.3).

Value of dynamical modulus of deformation for tested soil were computed from:

$$E_D = 1,5\frac{r\sigma_D}{u_D} = \frac{22,5}{u_D} \qquad [\text{MPa}] \qquad (1)$$

where σ_D = amplitude of dynamic stress under loading plate; $\sigma_D = 0,1\,\text{MPa}$; u_D = average settlement of loading plate as result of three test impacts, [mm].

Value of statical modulae of deformation under the first loading (E_{po}) and the second one (E_p) were calculated from:

$$E_{po}; E_p = D\frac{\Delta\sigma}{\Delta u} = \frac{22,5}{\Delta u} \qquad [\text{MPa}] \qquad (2)$$

where $\Delta\sigma = \sigma_2 - \sigma_1$ = range of stresses for computed modulae value $E_{po}; E_p$,[MPa];
$\Delta\sigma = 0,125 - 0,050 = 0,075 MPa$;

$\Delta u = u_2 - u_1$ = difference of loading plate settlement under σ_2 and σ_1 correspondingly, [mm].

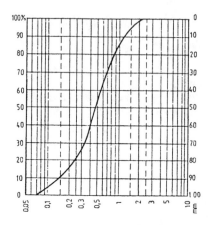

Fig. 2. Medium sand sieve - analysis curve

30 sets of variables I_s, ρ_d, E_D, E_{po}, E_p, a were obtained as results of modeled subbase. Statistical parameters of variables are presented in Table 1.

Verification of hypothesis concerning equality of mean values in the groups applying method of ONE-WAY analysis of variance based on Last Significant Differences was performed in order to determine whether the thickness of tested soil layer affects values of statical and dynamical modulus of deformation. They were taken for significance level $\alpha = 0,798 \div 0,952$ and values of F-Ratio $F = 0,069 \div 0,004$ with $1\ on\ 28$ degrees of freedom. Multiple range tests were also conducted.

Satisfaction of the following assumptions (Podgórski 1995)
- normality of characteristics distribution in the groups applying Kolmogorov - Smirnov's test for significance level $\alpha = 0,05$,
- variance homogeneity check in groups using Cochran's C test and Bartlett's test for significance level $\alpha = 0,05$
were verified for legalisation of conclusions resulting from ONE-WAY analysis of variance. Results of the above analysis are presented in Table 2.

The above analysis allows to formulate the conclusion that the thickness of tested layer of soil in the range from 0,3 m to 0,5 m does not affect test results of statical modulus of deformation E_{po} ; E_p and dynamical one E_D.

Regression between variables in the set of all measurements was then determined. Table 3 presents matrix of linear correlation coefficients for tested variables. Analysis of correlation matrix shows significant dependencies between dynamic modulus of deformation and statical modulus of deformation as well as relationship between degree of compaction and volumetric density of soil skeleton. Then best fitted regression models between two variables were chosen after simple regression analysis. Adequate model was chosen checking significance of ratio of sum of squares of lack-of-fit model to sum of squares of pure error from repetitions due to the measurements being conducted with repetitions.

Selected relationships $E_D = f(I_s)$, $E_D = f(\rho_d)$, E_{po}; $E_p = f(E_D)$ are presented in Fig. 3 ÷ 5.

Table 1. Statistical parameters of variables I_s, ρ_d, E_D, E_{po}, E_p dependent on the thickness of tested soil layer

a [m]	Geotechnical parameter	Number of observations	Minimum value	Maximum value	Average value	Standard deviation	Coefficient of variation [%]	Skewness ratio	Kurtosis
0,3	I_s [-]	7	0,907	0,993	0,962	0,030	3,09	-1,26	0,96
	ρ_d [g/cm³]	7	1,632	1,787	1,732	0,054	3,10	-1,28	1,01
	E_D [MPa]	14	16,11	36,17	28,63	6,22	21,72	-0,85	-0,15
	E_{po} [MPa]	7	11,25	39,47	28,12	9,48	33,71	-0,74	0,64
	E_p [MPa]	7	38,14	112,50	91,50	25,59	27,97	-1,91	3,75
0,5	I_s [-]	8	0,922	0,988	0,963	0,021	2,23	-1,01	0,77
	ρ_d [g/cm³]	8	1,660	1,788	1,733	0,038	2,22	-1,02	0,75
	E_D [MPa]	16	10,96	38,79	29,32	8,03	27,39	-1,10	0,35
	E_{po} [MPa]	8	13,24	36,89	27,93	7,80	27,93	-1,07	0,53
	E_p [MPa]	8	44,12	112,50	92,17	22,96	24,91	-1,61	2,20

Table 2. Analysis of assumptions legalising analysis of variance

Geotechnical parameter	Kolmogorov - Smirnov's test P*		Tests		Assumptions	
	$a=0,3\ m$	$a=0,5\ m$	Cochran's	Bartlett's	normal distribution	variance homogeneity check
I_s	0,133	0,250	0,242	0,239	not rejected	not rejected
ρ_d	0,125	0,222	0,228	0,226	not rejected	not rejected
E_D	0,594	0,279	0,349	0,353	not rejected	not rejected
E_{po}	0,904	0,349	0,488	0,487	not rejected	not rejected
E_p	0,122	0,167	0,706	0,706	not rejected	not rejected

P* = probability value determining bottom level where assumption is rejected.

Table 3. Matrix of linear correlation coefficients for tested variables

Geot. param.	I_s	ρ_d	E_D	E_{po}	E_p	a
I_s	1,00					
ρ_d	0,999	1,000				
E_D	0,920	0,919	1,000			
E_{po}	0,951	0,951	0,925	1,000		
E_p	0,954	0,954	0,911	0,902	1,000	
a	0,016	0,012	0,050	-0,012	0,015	1,000

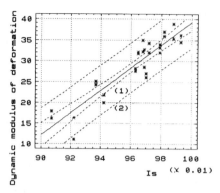

Fig. 3. Dependence $E_D = f(I_s)$ for medium sand:
$E_D = 270,51 \, I_s - 231,34$; $r = 0,920$, where $r =$ correlation ratio, (1),(2) = range of confidence limits and prediction limits computed for probability 95%.

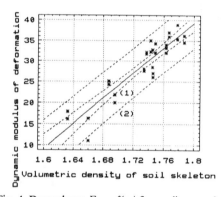

Fig. 4. Dependence $E_D = f(\rho_d)$ for medium sand:
$E_D = 150,03 \, \rho_d - 230,91$; $r = 0,919$.

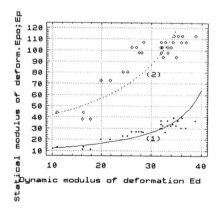

Fig. 5. Dependence E_{po} ; $E_p = f(E_D)$ for medium sand: (1) $E_{po} = 1/(0,106 - 0,002 \, E_D)$; $r = -0,896$
(2) $E_p = 1/(0,030 - 0,001 \, E_D)$; $r = -0,852$.

4 CONCLUSIONS

Results of tests and their statistical analysis presented in the paper enable to formulate the following conclusions:
- positive correlation exists between dynamic modulus of soil deformation and geotechnical parameters defining compaction of non-cohesive soil i.e., degree of compaction and volumetric density of soil skeleton,
- clear positive correlations exist between dynamic and statical modulus of soil deformation,
- the thickness of subsoil layer measured with lightweight dynamic deflectometer can vary from 0,3 m to 0,5 m.

Calibration in a manner presented in this work should precede using of the lightweight dynamic deflectometer in testing of the compaction. Calibration shall be conducted on the trial field of soil deposit which is foreseen for embankment construction and required correlation between dynamic modulus of deformation and other geotechnical parameters are to be elaborated. Lightweight dynamic deflectometer can be applied in current control of quality and uniformity of compaction as a kind of alternative or additional test to statical test with loading plate.

REFERENCES

BN-77/8931-12 *Determination of soil degree of compaction.* (in Polish).
BN-64/8931-02 *Motor roads. Determination of modulus of deformation for flexible surfaces and subsoil usin plate loading test.* (in Polish) .
Kudla, W. & R. Floss & Ch.Trautmann 1991: *Dynamischer Plattendruckversuch - Schnellprüfverfahren für die Qualitätssicherung von ungebundenen Schichten.* Straße und Autobahn Nr 2.
PN-74/B-04452 *Building soils. Site investigation.* (in Polish).
PN-88/B-04481 *Building soils. Laboratory tests.* (in Polish).
Podgórski, J. 1995: *Statistic with computer. Statgraphics version 5&6.* MIKOM Warsaw.
Sulewska, M.J. 1993: *Modulus of deformation for non-cohesive soil determined applying dynamic method.* Ph.D. Research. Bialystok Technical University Faculty of Building and Environmental Engineering.
Technische Prüvorschrift für Boden und Fels im Straßenbau TP BF- StB Teil B 8.3. 1992: *Dynamischer Plattendruckversuch mit Hilfe des Leichten Fallgewichtsgerätes.*
Weingart, W. 1981: *Probleme der dynamischen Tragfähigkeitsprüfung mit Fallgeräten.* Die Strasse Nr 11.

Geotechnical Hazards, Marić, Lisac & Szavits-Nossan (eds) © 1998 Taylor & Francis, ISBN 90 5410 957 2

The factors determining the deformations of organic subsoil under embankment

A. Szymański & W. Sas
Department of Geotechnics, Warsaw Agricultural University, Poland

ABSTRACT: The results of field and laboratory investigations on two test sites located on organic soils in northwestern Poland are presented. Comprehensive investigations comprising observations at four test embankments and laboratory investigations were carried out to study the behaviour of consolidation process in organic subsoil. It was shown that large deformations, both vertically and horizontally occur during and after construction period. Due to this foundations of embankments on soft organic soils involves subsoil improvement prior to construction. One of the improvement methods is preloading with the application of vertical drains to accelerate the consolidation process. The observations of organic subsoil behaviour show that vertical strip drains may effectively improve the bearing capacity of amorphous peat and gyttja for stage-constructed embankments. Moreover, it was shown that there is a significant difference in pore pressure distribution under embankments with and without vertical drains. As a result of the vertical drain installation under part of slopes, a significant decrease in the horizontal displacements took place.

1 INTRODUCTION

Construction of embankments on soft organic soils gives rise to special problems. The most obvious ones are large deformations that may occur during and after the construction period, both vertically and horizontally. The settlements often appear quickly but may also continue for very long time periods due to creep. The low strength often causes stability problems, and consequently the load sometimes has to be placed in stages or, alternatively, the soil must be improved through prior treatment.

The selection of suitable construction schedule involves estimation of the final settlement as well as prediction of the subsoil deformation course and pore pressure dissipation. Data on the current state of elevation of each subsoil layer and the effective stress distribution make it possible to determine the shear strength increase and to evaluate stability.

The purpose of this paper is to introduce up to date knowledge of how to correctly evaluate the soil deformation and consolidation course in organic subsoil under embankments.

The paper consists of two parts:

1. The first part is dedicated to the general behaviour of organic soil under load as well as analysis of subsoil deformation and consolidation under embankments;

2. The second part is devoted to the analysis of the influence of the vertical drains on the consolidation performance.

Organic soils, which according to definition contain a varying proportion of organic matter, include peat (remains of dead vegetation in various stages of decomposition), gyttja (plant and animal remains deposited in lakes), and organic silts and clays. The above mentioned types of soils differ greatly with respect to engineering properties. Organic clays and silts behave rather similarly to inorganic cohesive soils, whereas peats, particularly those that are fibrous with a small degree of humification, have extraordinary properties. Having an exceptionally high water content, which can even exceed 2000%, the fibrous peat is extremely compressible and the dominant process during its deformation under load is secondary compression.

The negative opinion of some practising engineers concerning the application of vertical drains in organic soils is mainly due to the high secondary compression, large deformation, and relatively high permeability of this type of soil. It is also suspected that an organic environment may have a harmful effect on drain material.

Several years of observations carried out on test embankments built on amorphous peat and calcareous gyttja have provided some information concerning the performance of vertical prefabricated drains in organic subsoil. They show, to some extent, the influence of vertical drains on the consolidation process in organic soils and elucidate the behaviour of strip drains in organic subsoils. Investigations were carried out by the Department of Geotechnics at the Warsaw Agricultural University in the years 1976-1981 (Furstenberg et al. 1983) and were continued in cooperation with the Swedish Geotechnical Institute from 1982 to 1988 (Wolski et al. 1988), and next were performed in the years 1989 to 1997 (Szymański 1997).

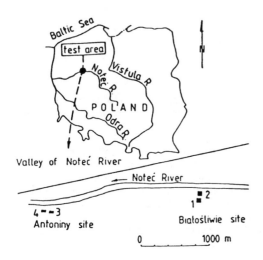

Figure 1. Location of the test site.

2 DESCRIPTION OF THE TEST AREA

The two test sites were located in northwestern Poland in the Noteć river valley, the first near the village of Białośliwie, the second near the village of Antoniny. The distance between the two sites is approximately 2 km (Fig. 1). The river valley is about 10 km wide, and the area is relatively flat, seasonally flooded, and covered with grass vegetation. The upper soft soils in the area consist of a layer of amorphous peat on top of layer of fine-grained calcareous soil, namely gyttja. Gyttja is organic soil that originates from the remains of plants and animals rich in fats and proteins, in contrast with peat which is formed from the remains rich in carbohydrates. These soft organic soils were underlain by dense sand. A more detailed description concerning the Antoniny site can be found in Wolski et al. (1988).

At the Białośliwie site, the 4-m-thick organic subsoil consists of 2-m-thick peat layer and a 2-m-thick gyttja layer. A schematic soil profile is shown in Fig. 2a. The static groundwater table is located 0.5 m below the ground surface.

At the Antoniny site the organic subsoil consists of a 3.1-m-thick peat layer and a 4.7-m-thick gyttja layer. A large number of boreholes, field-vane tests, and standard penetration tests (SPTs) have been performed. Soil properties of the organic stratum are shown in Fig. 2b. The groundwater table (GWT) is present in the peat layer at a depth of 0.5-0.8 m below the surface. The GWT in the gyttja and sand layer is 0.6-1.6 m higher than that of the upper peat stratum because of artesian pressure in the sand layer.

From the results of constant rate of strain (CRS) oedometer tests carried out with different rates of strain it is clear that the peat and gyttja are preconsolidated, with overconsolidation ratios (OCR) of 3 to 4 and 1 to 2, respectively.

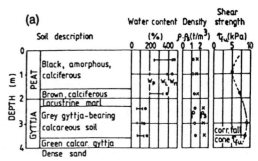

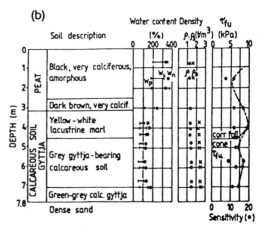

Figure 2. Organic subsoil properties at (a) the Białośliwie site and (b) the Antoniny site. w_L, liquid limit; w_n, natural water content; w_p, plastic limit.

3 FIELD INVESTIGATIONS

At the Białośliwie site two test embankments were constructed in one stage to a height of 2.0 m (Fig. 3a). Underneath the crest of test embankment 2, vertical drains, Geodrains with paper filters, were installed in a square pattern with a 1.0 m spacing. For monitoring the subsoil behaviour, surface and

288

deep settlement gauges and piezometers were installed.

At the Antoniny site two embankments were built in stages. To reach the final height of 4.0 m (Fig. 3b) the embankment construction had to be divided into three stages. Geodrains in a square pattern with a 1.2 m spacing were installed under embankment 4. The subsoil behaviour was monitored by means of piezometers, various types of settlement gauges, and inclinometers that allowed measurements of vertical and horizontal displacements and pore pressures.

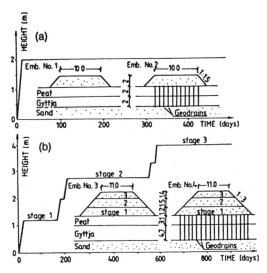

decrease of horizontal displacement was associated with an increase of undrained shear strength (Lechowicz et al. 1987; Koda et al. 1993).

To explain the background of this effect, which can be termed confining, some observation data will be analysed. The initial pore pressure conditions are

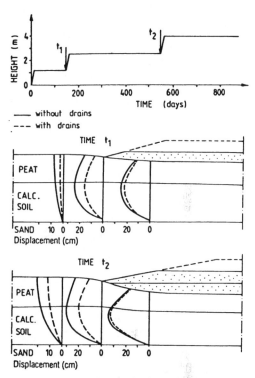

Figure 3. Construction schedule for the test embankments. (a) The Białośliwie site, 1978-1981. (b) The Antoniny site, 1983-1978.

Figure 4. Vertical and horizontal displacements under embankment slopes.

The measured deformation and pore pressure response, at both experimental sites provided the basis for the estimation of the influence of geodrains on organic subsoil deformation.

Observation of vertical displacements in the subsoil was performed by means of settlement gauges of 4 types: hose, plate, screw and magnetic. The horizontal displacements in organic subsoil have been calculated from inclinometer readings. The magnitude of subsoil deformation at the end of each construction stage is presented in Fig. 4.

During observations of the behaviour of embankments at the Antoniny site, an additional effect of vertical drains installed in highly compressible soil was detected. When the subsoil of the embankment with vertical drains was compared with that without drains it was found that the performance of vertical drains changed the pattern of horizontal displacements, i.e., they were smaller and their distribution was more confined. This

characterised by a small artesian pressure in the sand layer, underlying the compressible organic formation. After loading, the dissipation of the excess pore pressure was more rapid in the part of the subsoil with Geodrains. Distributions of pore pressure in the subsoil under embankments with and without Geodrains are shown in Fig. 5. The pore pressures, measured almost 1 year after loading with the second stage of the embankment, are significantly lower within the part with Geodrains than outside. In other words, in the subsoil of the embankment with Geodrains the pore pressure was lower under the embankment than in its perimeter, i.e., the confining zone. The same pattern of pore pressure distribution was obtained during observation after the first and third stages of loading.

Because of the hydraulic gradient evolving between the parts of the embankments subsoil without and with Geodrains, seepage towards the

Geodrains is likely to develop in a confining zone (see arrows marked in Fig. 5). Thus the created seepage pressure, being of the opposite direction to the horizontal stresses entailed by loading, had to diminish the horizontal displacements. Those measured under the toe of the embankment slope with Geodrains were 30-50% lower than those without Geodrains (Fig. 4). Moreover, under the embankment with Geodrains the horizontal displacements were confined to a narrower zone than under then embankment without Geodrains.

It is important to note that horizontal displacement under the centre of embankments are similar in both cases, with and without Geodrains. This might prove that the effect of mechanical reinforcement by vertical drains can be disregarded.

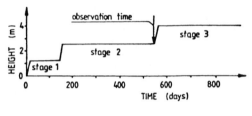

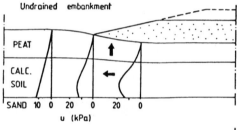

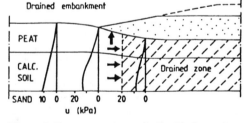

Figure 5. Excess pore-pressure distribution under embankment slope.

4 PREDICTION OF SUBSOIL DEFORMATION

The analysis of field equipment readings indicates that the deformation development in organic soil under embankments demonstrates large vertical and horizontal displacements rarely encountered in mineral soils. Therefore, the calculation of settlement development and excess pore pressure dissipation can be done by means of consolidation theory taking into account additional factors determining the deformation process under embankment e.g.:

1. Large strain effect using the reduced or convective coordinate systems;

2. One-dimensional pore water flow during the calculation of subsoil settlement under centre of embankment crest.

3. Axisymetrical state of pore water flow while the vertical drains in the subsoil are applied.

4. Two-dimensional state of strain and pore water flow during the calculation of subsoil deformation under the embankment slopes.

5. Nonlinear stress-strain relationship for soil skeleton expressed in terms of void ratio.

6. Nonlinear void ratio - permability relationship describing the limitation of pore water flow in consolidation process.

7. Confining effect on subsoil deformation under embankment slopes which is caused by vertical drains.

In the case of deep subsoil and relatively small base of embankment, where large horizontal displacements may appear, the deformation prediction can be carried out by a two-dimensional consolidation analysis with application of the convective coordinate system. The numerical computations have been performed using the following differential equation (Szymanski 1994):

$$\frac{\partial}{\partial \xi_v}\left[\frac{k_v}{\gamma_w}\frac{\partial u}{\partial \xi_v}\right]+\frac{\partial}{\partial \xi_h}\left[\frac{k_h}{\gamma_w}\frac{\partial u}{\partial \xi_h}\right]+\frac{e}{1+e}\frac{de}{d\sigma}\frac{Du}{Dt}\bigg|_x =0 \qquad (1)$$

where e = void ratio; γ_w = unit weight of water; u = excess pore pressure; σ = effective stress; k_v and k_h = permability coefficient in vertical and horizontal directions; ξ_v and ξ_h = convective coordinates; t = time.

Application of numerical approach in the computation of settlement course in organic subsoil, taking into account primary and secondary compression is proposed. In this procedure the governing equation will take the following form:

$$\frac{\partial}{\partial \xi}\left[\frac{k_v}{\gamma_w}\frac{\partial u}{\partial \xi}\right]+\frac{\partial}{\partial \rho}\left[\frac{k_h}{\gamma_w}\frac{u}{\rho}+\frac{k_h}{\gamma_w}\frac{\partial u}{\partial \rho}\right]=$$
$$= -\frac{1}{1+e}\frac{de_p}{d\sigma'}\frac{Du}{Dt}\bigg|_x -\frac{1}{1+e}\frac{de_s}{dt} \qquad (2)$$

where ρ = radial coordinate; k_v = coefficient of vertical permeability; k_h = coefficient of horizontal permeability; de_p = the change of void ratio during the primary consolidation:

$$\frac{d\,e_p}{d\sigma'} = C_r \log \frac{\sigma'_p}{\sigma'_{vo}} + C_c \log \frac{\sigma'_f}{\sigma'_p} \tag{3}$$

where C_r = recompression index; slope of e-$log\sigma'_v$, curve in the recompression range; C_c = compression index, slope of e-$log\sigma'_v$ curve; σ'_{vo} = in situ effective stress; σ'_p = preconsolidation pressure; σ'_f = final effective vertical stress. de_s- the change of void ratio during the secondary compression:

$$\frac{de_s}{dt} = e_0 - C_a \log \frac{t}{t_0} \tag{4}$$

where de_s= the change of void ratio during the secondary compression; e_o = initial void ratio; C_a = coefficient of secondary compression; slope of e-$logt$ curve after the end of primary consolidation for different effective stress; t = calculation time; t_o = previous step of calculation time.

The dependence of consolidation prediction on the boundary conditions limited by the drain discharge capacity decrease (Factor A) and nonlinear soil permeability characteristics (Factor B) has been analysed on the basis of field and laboratory investigations and numerical calculations using convective coordinate systems (Fig. 6). Factor A in numerical computations is taken into account by means of variable boundary condition whereas soil permeability (Factor B) by nonlinear variability of coefficient k. Use of differential equation for estimation of the consolidation of the loaded subsoil requires application of numerical solution because of nonlinear variability of soil parameters. The numerical solution of the equation can be done by the method of finite differences in which the structure disturbance effect around the drain is described by characteristics: e-k'_h - for the disturbed zone and e-k_h – for undisturbed zone, while the flow resistance in drain is defined by the modification of boundary condition:

$$\frac{k_w}{\gamma_w} \frac{\partial}{\partial \xi} \left[\frac{\partial u}{\partial \xi} \right] = \frac{2}{r_w} \left[\frac{\partial u}{\partial \rho} \right]_{\rho=r_w} \tag{5}$$

The consolidation equation can be written in shape of:

$$\Omega = \frac{\partial u}{\partial t} = L\,u \tag{6}$$

$$\Omega = -\frac{1}{1+e} \frac{de}{d\sigma'} \quad \text{coefficient} \tag{7}$$

$$Lu = (L_\xi + L_\rho)u \quad \text{- operator} \tag{8}$$

$$L_\xi u = \frac{\partial}{\partial \xi} \left(\alpha \frac{\partial u}{\partial \xi} \right) \qquad \alpha = \frac{k_v}{\gamma_w} \tag{9}$$

$$L_{\rho}u = \frac{\partial}{\partial \xi} \left(\beta \frac{u}{\partial \xi} \right) \qquad \text{or} \tag{10}$$

$$L_{\rho}u = \frac{\partial}{\partial \rho} \left(\beta \frac{u}{\rho} + \beta \frac{\partial u}{\partial \rho} \right) \qquad \beta = \frac{k_h}{\gamma_w} \tag{11}$$

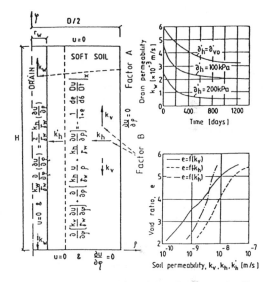

Fig. 6. The factors determining the pore water flow in consolidation of embankment subsoil.

While assuming the determination:

$$u^n = u \ (t = n\,\Delta t)$$

where Δt = step of diskretization after the t time and correspondingly; Λ^n = differential shape of the L operator; α and β coefficients are calculated at the moment when $t = n\Delta t$.

The consolidation equation can be written after Samarski (1977) in the shape of the system:

$$\left(E\Omega^n - \frac{\Delta t}{2} \Lambda^n_\xi \right) \vartheta = L^n u^n \tag{13}$$

$$\left(E\Omega^n - \frac{\Delta t}{2} \Lambda^n_\rho \right) u_t = \vartheta \Omega^n \tag{14}$$

$$u^{n+1} = u^n + \Delta t\, u_t \tag{15}$$

where E = unit operator; v and u_t = auxilary variables.

Differential operators $\wedge_\xi$ and $\wedge_\rho$ have been assumed in the shape:

$$\wedge_\rho u = \frac{1}{\Delta\rho}\left[\beta_{i,j-05}\left(\frac{u_{i,j-05}}{\rho_{i,j-05}} + \frac{u_{i,j-1}-u_{i,j}}{\Delta\rho}\right) - \beta_{i,j-05}\left(\frac{u_{i,j-05}}{\rho_{i,j-05}} + \frac{u_{i,j}-u_{i,j-1}}{\Delta\rho}\right)\right] \tag{16}$$

$$\wedge_\xi u = \frac{\alpha_{i+0.5j}\dfrac{u_{i+1,j}-u_{i,j}}{\Delta\xi_{i,j}} - \alpha_{i-0.5,j}\dfrac{u_{i,j}-u_{i-1,j}}{\Delta\xi_{i-1,j}}}{0.5\left(\Delta\xi_{i-1,j}+\Delta\xi_{i,j}\right)} \tag{17}$$

where $u_{ij} = u(\xi_{ij},\rho_j)$; $\rho_j = j\,\Delta\rho$; $\Delta\rho$ = discretization step along the radius (constant); $\Delta\xi_{ij}$ = discretization step along the vertical axis (variable) in the j^n-the profile.

The discretization step $\Delta\xi_k$ variable in time can be calculated from the relationship:

$$\Delta\xi_{i,j}^{n+1} = \frac{1+e_{i,j}^{n+1}}{1+e_{i,j}^{n}}\Delta\xi_{i,j}^{n} \tag{18}$$

where $\Delta\xi_{ij}^n$ = discretization step at the moment n; e_{ij}^n = void ratio in the point i,j at the moment n.

The current value of the void ratio is calculated from the material characteristics e-σ'.
The effective stress is equal to:

$$\sigma'_{i,j} = \sigma'_{v0} + \left(\sigma_{i,j}-u_{i,j}\right) \tag{19}$$

where σ'_{vo} = "in situ" effective stress; σ_{ij} = total stress increment.

In organic subsoil of the embankment considerable secondary deformations depending on time occur. Reckoning this in the numerical estimation of changes of the void ratio e-determined from the characteristics e-σ' - the change of e resuming from creep of the skeleton is taken into consideration. Hence at the points where the excess pore water pressure u_{ij} equals to zero, the change of void ratio e_{ij} can be forecasted by the relationship:

$$e_{i,j}^{n+1} = e_{i,j}^{n} - C_\alpha \log\frac{t^{n+1}}{t^n} \tag{20}$$

where C_α = coefficient of secondary compression.

5 CALCULATION RESULTS

The prediction of soil displacements and pore pressure dissipation in subsoil under the test embankment was based on the theoretical consolidation analysis considering two–dimensional state of strain and axissimetrical pore water flow. The calculations of subsoil consolidation were carried out using the compression characteristics describing the variability of C_c and C_r (Fig. 7) as well as C_α (Fig. 8) and relationship between vertical and horizontal strains (Fig. 9).

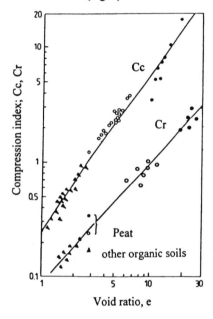

Figure 7. Relationship between compression index C_c, C_r and void ratio.

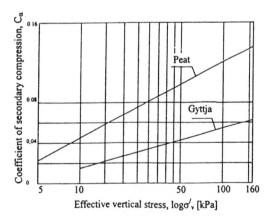

Figure 8. Relationship between coefficient of secondary compression C_α and σ_v'.

A comparison of the computation results with the field observations has shown quite good coincidence in pore water pressure and displacements when the presented in this paper factors determining the deformation process in organic subsoil have been taking into consideration in numerical calculations (Fig. 10 and Fig. 11).

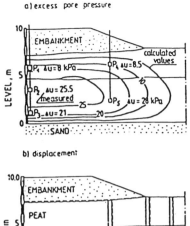

a) excess pore pressure

b) displacement

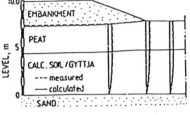

Figure 11. Measured and calculated results at the II stage construction (two-dimensional analysis).

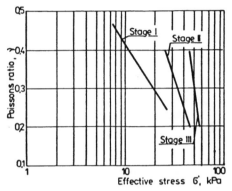

Figure 9. Relationship between horizontal (ε_h) and vertical (ε_v) unit displacements observed under test embankment slopes.

6 CONCLUSIONS

Observations of the consolidation process in organic soils demonstrate large values and nonlinear character of deformation. Therefore, the prediction of consolidation performance in organic subsoil should be carried out by methods taking into account the variation of soil parameters and large strains analysis. One of the methods to obtain the correct estimation of consolidation course is the application of convective coordinate system in governing equation.

Field tests showed that vertical strip drains, despite earlier views, may effectively improve the bearing capacity of decomposed (amorphous) peat and gyttja for stage constructed embankments. Embankment subsoil composed of peat and gyttja layers, underlain by sand, settles more than twice as fast when drains are installed. Acceleration of the consolidation process is much greater in the later phase of consolidation when the permeability of organic soils is essentially diminished.

In organic soils that demonstrate large horizontal displacements, an additional effect can be observed. This is the confining effect that consists in diminishing the horizontal displacement in the subsoil adjacent to the zone with drains. This in turn, produces a slight increase of shear strength, thus improving conditions for staged construction.

The use of consolidation theory for the prediction of soil displacements under embankments requires

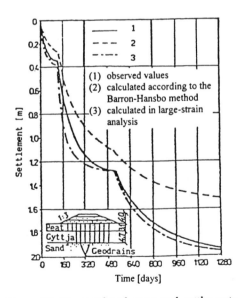

Figure 10. Measured and computed settlements of embankment subsoil with the vertical drains at the Antoniny site.

to take into consideration additional factors determining consolidation process in organic subsoil e.g. large vertical and horizontal movements, limitation effects under embankment slopes, decrease in compression and permeability parameters.

REFERENCES

Furstenberg, A., Z. Lechowicz, A. Szymański & W. Wolski 1983. Effectiveness of vertical drains in organic soils. *Proc. of the 8th European Conference on Soil Mechanics and Foundation Engineering,* Helsinki, vol. 2, pp. 611-616.

Koda, E., A. Szymański & W. Wolski 1993. Field and laboratory experience with the use of strip drains in organic soils. *Canadian Geotechnical Journal, vol.30, pp. 308-318.*

Larsson, R. 1986. Consolidation of soils. *Report No 29. Swedish Geotechnical Institute.*

Lechowicz, Z., A. Szymański & W. Wolski 1987. Effects of groundwater on embankment subsoil deformation. *Proceedings 9th European Conference on Soil Mechanics and Foundation Engineering,* Dublin, vol. 1, pp. 451-454.

Samarski, A. 1977. *Teorja raznostnych schiem.* Nauka, Moscow.

Szymański, A. 1994. The use of constitutive soil models in consolidation analysis of organic subsoil under embankment. *Proceeding of the Workshop on Advances in Understanding and modelling the Mechanical behaviour of Peat.,* Balkema, Rotterdam: 231-240.

Szymanski, A. 1997. Numerical analysis of consolidation performance in layered soft subsoil. *Conf. On Recent Advances in Soft Soil Engineering.* Malaysia. Kuching: 230-241.

Wolski, W., A. Szymański, J. Mirecki, Z. Lechowicz, R. Larsson, J. Hartlen, K. Garbulewski & U. Bergdahl 1988. Two staged-constructed embankments on organic soils. *Swedish Geotechnical Institute, Report 32,* Linkoping.

Geotechnical Hazards, Marić, Lisac & Szavits-Nossan (eds) © 1998 Taylor & Francis, ISBN 90 5410 957 2

Geotechnical hazards and educational aspects

I. Vanicek
Czech Technical University Prague, Czech Republic

ABSTRACT: In the paper two basic topics connected with geotechnical hazards are not only discussed but it is also shown how they are treated at the CTU Prague. The first one is connected with earth and rockfill dams. To show the major sources of risk of the dam failure the individual questions as a deformability, cracks and internal erosion are discussed with the help of event tree - logical diagram. The similar approach is used for the definition of the risk of the thousands small pond earth dams, which were constructed some centuries ago. Educational principles in environmental geotechnics are also shown for problems connected with old burdens - (old contaminated sites) - what all steps have to be made during the decision making process.

1 RISK OF DAM FAILURE

1.1 New earth and rockfill dams

By observing many embankment dams and by statistical analysis it was proved that the classical slope stability failure is the very exceptional case. In most cases different problems and failures started due to the differential settlement, which can be the main source of tensile cracks. Water passing through these cracks can create the internal erosion. If the design of the filter is not appropriate for this process, as well as the design of drains, internal erosion can lead to the failure.

During the educational process at the CTU the emphasis is given not only to the numerical modelling but the main effort is concentrated on the logical scheme (diagram), which can be used during the decision making process. One version - Vanicek (1988) - of such logical diagram is shown in Fig. 1 and represents a certain modification of the Whitman's version (1984). From this diagram the students can see the mutual linkage of the individual factors influencing the risk of failure and what all aspects have to be taken into consideration.

The answer to the first question about the crack development probability covers all reasons which can have some influence on the crack development in the core of the dam. In the wider sense this question can cover not only the cracks in the dam core but also the joints along the abutment, seepage channel along the outlet or in subsoil.

The numerical modelling of the stress - strain behaviour of the dam body is very useful because from it the probability of the crack development can be judged. But the answer to the question about this probability can be found also with the help of the logical diagram of the second order - Vanicek (1985).

Cracks are influenced by:

- geometrical conditions:

• dam valley	- regular
	- irregular
• valley abutment	- flat
	- steep
• longitudinal axis of dam	- arching upstream
	- downright
• core	- wide, slightly inclined
	- thin, vertical

- material conditions:

• core	- flexible, plastic
	- brittle
• interaction of individual parts	- small
	- big
• embankment compressibility	- low
	- high
• settlement of subsoil (total, differential)	- low
	- high

- other conditions:

• speed of reservoir filling	- small
	- high
• additional interference into core	- no
	- yes

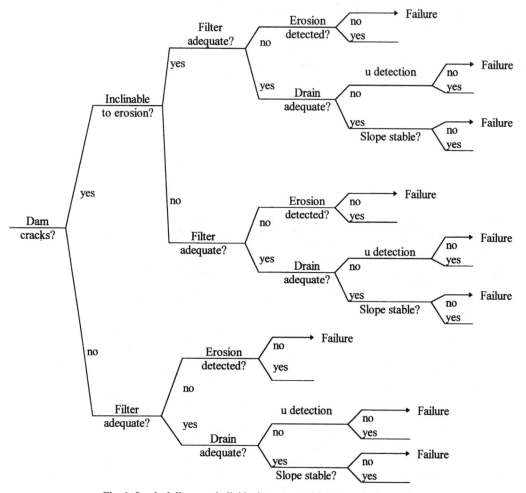

Fig. 1: Logical diagram, individual steps on which depends dam safety

The great advantage of this logical diagram of the second order is that each aspect can be not only discussed but its sensitivity can be checked by numerical modelling. Each group of students can prepare a small study dealing with the influence of individual aspect.

Also the other answers to the questions presented in the logical diagram of the first order can be obtained with the help of additional questions. For example the sensitivity to the internal erosion will be dependent on:

- basic soil is   - flocculate   - resistant
                                - inclinable to erosion
             - dispersive
- swelling potential is   - high
                              - small

Similarly the filter adequacy depends on the following factors:

- basic soil - liability to segregation   - small
                                           - big
- filter properties - filter   - grain
                                   - geotextile
            - filtration criteria   - for perfect filter
                                   - for classical filter
                                   - based on aggregates
                                   - not met any
- other conditions: - hydraulic gradient - low
                                        - high
                - construction supervision   - careful
                                          - none

Similarly it is possible to judge the quality of drains, the quality of monitoring system for pore pressure u, internal erosion etc. Only after the question about slope stability is also important.

The explanation of the whole problem with the help of this logistic scheme can be very useful for the students, because on the one side the complexity is shown and on the other one the mutual connection of the partial factors with which the overall risk of failure is connected. This procedure is not in contradiction with numerical methods because it helps to give to the numerical methods another dimension, to show the possibilities for alternative calculations for changed initial assumptions.

1.2 Old small dams

In the Czech Republic there is still roughly 30 000 old small dams of ponds which were constructed in the Middle Ages. Significance of such ponds in countryside is very important especially from the environmental point of view. But the financial profit, first of all from fishing industry (carp), is very low.

We know for high dams, according to the report of ICOLD Paris 1974, that for earth and rockfill dams the percentage of accidents is 4.77%, and of failures is 1.48%. Certainly for old embankments this percentage is decreasing but for level 0.2 - 0.5% still we have roughly between 60 and 150 problems each year.

After political changes in our country and after the restitution process the responsibility was returned to old owners or to new owners. But the quality of ponds or embankments is not so good as was 50, 60 years ago. So the Ministry of Agriculture started a new project the main aim of which is to obtain overall information about existing situation and after that to use this information in decision making process. The results can be summarised in the following.

- General information about level of protection against floods, what is the quality of water, how much sludge is in the reservoir and what is its quality. A certain pressure on new owners can be expected to improve this situation but with a certain financial support from the state.
- Information about impact of the failure on surroundings - height of dam, volume of reservoir, vicinity of estate below the reservoir - results can be used for the definition of duties of the new owner.
- Information about new factors which have the influence on the stability - results can be used by new owners to decrease the risk of failure.

For the last point the following factors were recognised as most important:

- Quality of the old outlet (mostly from wood) and especially its connection with a new discharge object (today mostly from concrete) - the investigation with the help of TV camera can be very useful,
- What sort of transportation is going over dam crest - heavy transportation can be reason for additional settlement, cracks, lost of connection etc.

What sort of vegetation was and is growing on the embankment - during last few decades the pollution had a negative impact on old trees. After dying or cutting down of the trees this fact has a negative impact on safety (e.g. old root system can be a preferential path for underground water).

After finishing this project the final database will be opened and will help to decrease the risk of failure. But it was shown again that small additional aspects can change the overall look on the solved problem.

2 OLD BURDENS - OLD CONTAMINATED SITES

During last 50 years in the Czech Republic many places were strongly polluted (by army, chemical industry etc.) and now it is necessary to start with remediation process.

A certain advantage is the fact that the worst old burdens are solved not only with the money of new owners of buildings and lands but also from the Fund of National Property, which obtained some money during a process of privatisation. The typical example of successful remediation of the old chemical waste dump were described by Vanicek et al. (1997).

Because there exist many similar old burdens and the sum of money is limited, it is necessary to determine the priorities. It is necessary first of all to solve the problems which are most acute.

During the lectures we are putting the emphasise on the following questions - Yong (1996), which can help to determine this priority:

Q1: How serious is the contamination? i.e. what is the extent (spatial) and degree of contamination ?

Q2: Is the contamination a threat to public health and the environment?

Q3: How can the „threat" be managed? i.e. Does one need to „contain", remediate ? Or can the „threat" be ignored?

Yong also emphasises the multi-disciplinary approach to the solved problem and therefore a great attention is given to the equanimity of individual professions involved in this problem.

When we are trying to find the answer to the above mentioned questions and after that to solve the problem, the following phases must be taken into account:

- phase of investigation - effective way how to obtain the information about contamination origin, range, type, interaction with environment etc.,
- phase of definition of the potential hazard - at the moment of investigation and as a function of time,
- phase of determination - main aim is to define whether the remediation is really needed, or the restriction for the contaminated area is adequate for this case. For the case of remediation the decision making process has to distinguish between containment and decontamination,

Phase	Main aim	Methods
phase of investigation	input information	■ methods of investigation ■ sampling of soil and water
phase of definition of the potential hazard	information about contamination degree in space and time	■ acceptable limits for different type, exploitation ■ numerical modelling of the contaminant transport
phase of determination	selection between individual methods	■ financial tools (information about prices)
phase of the selection of the most appropriate methods of remediation	selection of the most appropriate method	■ methods of containment (vertical, horizontal barriers) ■ methods of decontamination
phase of realisation	assure the quality	■ methods of supervision
phase of post realisation	monitoring	■ methods of monitoring ■ observational method of design

Fig. 2: Phases of remediation process

- phase of the selection of the most appropriate methods of remediation - for both basic cases the selection of most appropriate technologies,
- phase of realisation - the main aim is to assure the best quality - supervision - to guarantee required quality,
- phase of post realisation - classical monitoring to assure the quality or monitoring as part of design process - together with observational method can improve the efficiency of the whole process of remediation.

For the students it is necessary to show not only this vertical hierarchy, but also horizontal hierarchy in which different methods and technologies are clarified in more details - (Fig. 2).

3 CONCLUSION

During the educational process, especially when dealing with typical problems of environmental geotechnics, the logical diagram or different tables according which the risk analysis can be evaluate more easily, proved to be a very useful tool. Therefore two examples were presented here. The students can see very clearly the mutual connection between individual aspects, which have a great influence on geotechnical hazards. Only after it the sensitivity of the individual aspects can be checked with the help of numerical methods.

4 LITERATURE

Vanicek, I. 1985. Cracks in clay core and their behaviour (in Czech). Doctor Science thesis. Prague: CTU, 326 p.

Vanicek, I. 1988. *Development and behaviour in clay core of earth and rockfill dams* (in Czech). Prague: SNTL, 167 p.

Vanicek, I., Bohac, J., Ricica, J. & Zaruba, J. 1997. Remedy of a chemical waste damp with foundation of a new damp on its surface. In: Proc. 14th ICSMFE, Hamburg. Vol. 3: 1857-1860.

Yong, R.N. 1996. Multi-disciplinarity of environmental geotechnics. In: Proc. 2nd International Congress on Environmental Geotechnics. Preprint of special lectures: 29-47. Osaka: Japanese Geotechnical Society.

Whitman, R.V. 1984. Evaluating calculated risk in geotechnical engineering. Journal of Geotechnical Eng. Div.: ASCE, No. 2, pp. 143-188.

Geotechnical Hazards, Marić, Lisac & Szavits-Nossan (eds) © 1998 Taylor & Francis, ISBN 90 5410 957 2

Oozy with peat foundation soils and influences on behaviour of the Flam Dam

D.Veliciu & R.Stånescu
Institute for Land Reclamation, Bucharest, Romania

E. Marchidanu
Technical University of Civil Engineering, Bucharest, Romania

ABSTRACT: Paper presents a case history from Danube Delta describing relevant field investigations, results of laboratory tests on soil samples, monitoring and quality control. The Flam Valley Dam has been built up for retention of tailings deposits supplied by the Tulcea Alumina Plant. Constituted from compacted loess, the dam is 30 m high, 500 m long and it will be arisen to be 40 m high in the near future. The dam is founded on high compressible oozy soils with interbedded peat. Geotechnical characteristics of soil foundation are discussed in correlation with observation of the dam behaviour vs. time.

1. INTRODUCTION

The Flam Valley Dam has been originally built up to form a tailings dam in order to deposit the residual sludge resulting from the dressing of bauxite at the aluminium plant in Tulcea (Danube Delta).

During the first stage of construction, an embankment of 20 m high was erected of compacted loess.

Due to the tailings dam clogging in a second stage the dam has been elevated 10 m, increasing its storage volume from 2.8 to 5.1 millions cubic meters. At the present the tailings dam is almost completely clogged and consequently it implies a new stage to heighten the dam with 10 m more. This third stage will be final as the Flams Valley morphology does not allow to exceed the +45 m level.

For a total height of 40 m the volume of sludge behind the dam will reach approximately 11.5 millions cubic meters. In Figure 1 successive stages in building up the dam and the increase of storage capacity for the residual slam are shown.

2. GEOLOGICAL CONDITIONS OF SITE

In the site of the dam, the Flam's Valley exhibits a trapezoidal section approximately 100 m wide at its base (level +5 m) and 420 m wide at the level of +45 m.

From a geological point of view the setting of the dam is characterised by shallow deposits 30 -40 m thick overlaying the bedrock formed by Cretaceous formations in sandy-calcareous facies. In the flood plain zone the shallow deposits of Holocene age are constituted by high compressible soft soils composed of clays and silts, and oozy with peat. Below Holocene deposits are Pleistocene deposits with predominately clayish facies.

The valley banks are covered by loess with thickness of 20-25 m. A geological cross-section through the valley is shown in Figure 2.

3. SITE INVESTIGATIONS OF FOUNDATION SOIL.

The dam location area was properly prospected according to each construction stage. The most extensive works being performed both in the field and in the laboratory, during the 3-rd stage.

Geotechnical boreholes have been drilled as deep as 20-30 m. Due to the presence of high compressible soft soil horizons, with remarkable thickness, quasi-state cone penetration tests (QCPT) were performed with good results.

In situ compressibility tests were carried out in a borehole at three depth levels corresponding to the three different soil types.

Laboratory geotechnical analyses were performed on the samples collected from boreholes. The

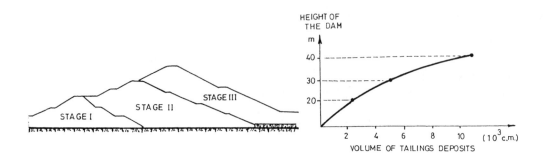

HEIGHT OF
THE DAM

Figure 1. Stages in the construction of the dam and variation graph of the tailings deposits

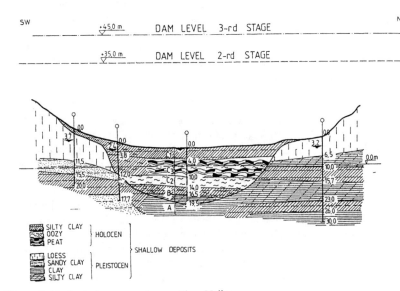

Figure 2. Geological cross section on Flam Valley

processing and interpretation of the observational data enabled to achieve a lithological profile as it is presented in Figure 3.

The water table level was encountered in boreholes at various depths from 0.0 to 2.0 m.

In order to select a consolidation method for the saturated soft soils which constitute the foundation soil, a laboratory model was studied having in view a vertical drain network.

4. GEOTECHNICAL CHARACTERISTICS

A large number of samples (about 200) from all soil types were analysed in laboratory. The main geotechnical characteristics of the analysed soils are presented in Table 1.

The soil compressibility was determined both in situ and in laboratory by oedometer tests.

In Figure 4 are shown typical oedometer curves corresponding to the soils from the lithologic section.

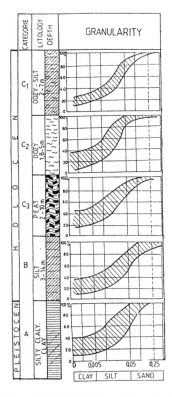

Figure 3. Lithologic profile on the foundation soil

Table 1. Geotechnical characteristics

Soil type	m %	P.I. %	I_L	γ kN/m³	N %	om %
oozy silt C1	19-35 25	11-29 19	0.28-0.90 0.55	18-21 20	36-49 43	0.2-3 1.6
oozy C2	22-128 51	10-91 14	0.19-0.48 0.40	13-19 17	42-80 56	1-8 3.5
peat C3	30-471 29			11-17 13	55-93 79	3-46 15
silt B	19-25 23	10-38 24	0.40-1.00 0.68	19-21 20	33-41 38	
clay A	17-28 22	12-38 24	0.40-1.00 0.72	33-41 38	33-41 38	

Keys: 19-35 is interval of variation and 25 is average; letters (C1, C2 etc.) correspond to formations from geological cross-section; m is moisture content; P.I. is plasticity index; I_L is liquidity index; γ is unit weight; n is porosity; omc is organic material content

The values of E-modulus and oedometer modulus determined by usual oedometer tests and long time oedometer tests (10-30 days) are shown in Table 2.

Shear strength characteristics were determined on saturated samples (unconsolidated-undrained UU and consolidated-undrained CU) by direct simple shear and by compression triaxial tests.

During in situ tests the shear strength of the soils has been investigated with the vane test method. The graphs from Figure 5 show the variation range of the shear strength ($\Phi = 0°$) resulted from both the triaxial and the vane test. The graphs indicate a fair correlation among the results obtained with the two methods mentioned above. Variation ranges of "Φ" and "c" parameters are plotted and presented in Table 3 referring to the different shearing test types.

5. PREDICTED SETTLEMENT OF THE FOUNDATION SOIL

The vertical settlement of the foundation soil has been investigated on seven characteristic transverse cross-section, using the EDEF software. In Figure 6 is shown an example of calculated section and the settlement values variation corresponding to the lithology of the foundation soil and the loading induced by the dam into the respective section.

6. STABILITY ANALYSIS OF DAM-FOUNDATION SOIL

In order to evaluate the stability of the dam-foundation soil under the load for the maximum rising of the dam, four cross-sections were selected to calculate the stability factor. Stability of the slope was evaluated by applying the Fellenius, Bishop and Janbu methods using the STABTZ and G-SLOPE software. The computation was accomplished by taking into account different hypotheses. Values for shear strength of the soils from the foundation soil are indicated in Table 4 for these hypotheses. The adopted geometry of the dam used for modelling does not consider a stage-constructed-fill at the toe slope (hypothesis 1), and takes into account a stage-constructed-fill of different size at the toe of the slope (hypothesis 2). Computation has been performed successively for the static loading and the seismic effect (seismic coefficient of $K_s = 0.08$ corresponding to the

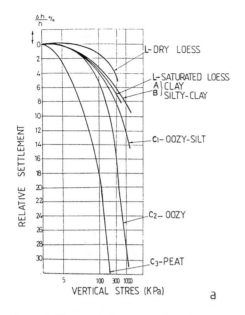
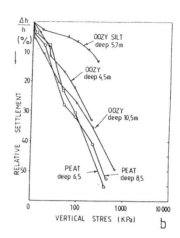

Figure 4. Characteristic compression curves a. Ordinary oedometric tests. b. Consolidations tests

Table 2. Results of in situ and laboratory tests for E and oedometer moduli.

Soil type	In situ E kPa	Oedometer modulus M										
		compression test					consolidation test					
		load increment (kPa) -					load increment (kPa) -					
		0-100	100-200	200-300	300-400	400-500	0-10	10-25	25-50	50-100	100-200	200-400
C1	2800	2400	6600	7700	11000	10000	180	125 0	182 0	223 0	558 0	6680
C2	1200	1800	3100	4300	4000	2200	47-124	170-210	480-220	560-700	810-830	1060
C3	1400	550	1100	1100	2000	1400	70-80	240-280	110-150	440-720	580-760	500-980
A+B		3500	8300	10000	14300	12500						

Keys: C1...B are as in Table 1.

intensity of 7 on the MKS seismic scale). The calculated values of the stability factors are summarised in Table 5

7. ADOPTED SOLUTION TO INCREASE SLOPE STABILITY

An expected improvement of the dam slope stability will be achieved applying the following solutions:
- construction of a vertical drainage system at the 15-20 m depth by filling holes (diameter 425 mm) with two types of material (sand in the external cylinder in order to form a crown shape filter and gravel with high permeability in the coaxial internal cylinder
- construction of a compacted loess stage fill (15 m thick and 25 m wide) at the toe of the slope laying on a gravel blanket 1 m thick with high permeability. Drainage system is connected to the gravel blanket (Fig. 7).
Designing of the drainage columns network was sustained by numerous laboratory tests and

302

mathematical modelling, having as starting point two options: (1) square pattern and (2) triangular pattern. The calculated required time for a consolidation degree of U = 85-90% was obtained considering (a) no drainage columns; (b) with drainage columns located in a network spaced at 2 m, 4 m as well as 6 m. The graphs in Figure 8 illustrate the results of the calculations.

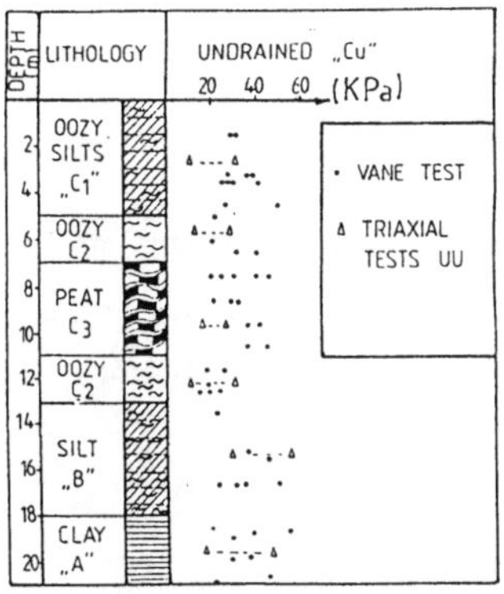

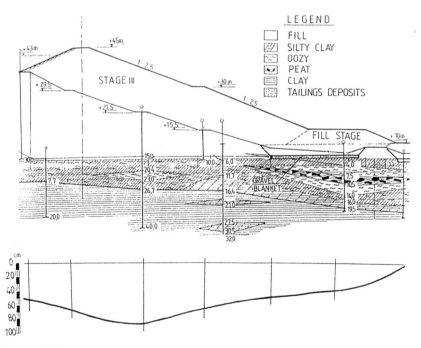

Figure 5. Variation range of shear strenght resulted from triaxial and vane tests

Table 3 Results of shear tests

Soil type	Simple shear test (UU)		Triaxial (UU)			
			Total stress		Effective stress	
	Φ_u	c_u kPa	Φ_u	c_u kPa	Φ'	c' kPa
C1	12-17	4-30	1-17	9-27	5-34	3-48
C2	13-23	5-24	0-4	11-24	25-28	5-18
C3			0-2	17-31		
B	17-31	14-46	2-4	34-57	2-30	1-39
A	6-28	10-70	4-21	18-47	15-29	18-28

Keys: UU is unconsolidated-undrained sample; Φ is angle of internal friction and c is unit cohesion.

Table 4. Shear strength characteristics

Soil type	Total stress		Effective stress		γ	r_u	
	Φ_u	c_u kPa	Φ_u'	c_u' kPa	kN/m³	f.sl	d.fl
d.fl	14	10	22	10	20		0.00
C1	5	15	10	25	20	0.25	0.10
C2	2	10	8	10	17		0.25
C3	1	10	10	25	13		0.10
B	7	15	12	30	20	0.50	0.25
A	7	20	12	35	20		0.50

Keys: r_u is pore pressure ratio; d.fl is dam fill; f.sl is foundation soil; C1... A, Φ, c and γ as in Table 3.

Figure 6. Soil foundation settlement graph calculated on a characteristic transverse cross section of the dam.

Table 5. Stability factor values for down stream slope (no drenage column)

Section	Fellenius		Bishop		Janbu	
	static	dynamic	static	dynamic	static	dynamic
3-3'	0.907	0.747	1.011 a 1.071 b 1.100 c	0.827	0.918 a 0.971 b 1.021 c	0.749
4-4'	1.052	0.817	1.066	0.887	0.970	0.808
5-5'	1.462	1.100			1.219	
8-8'	1.026	0.825			0.837	

Keys: stage-constructed-fill of a)10 m high / 20 m wide; b)15 m high / 20 m wide; c)15 m high / 25 m wide;

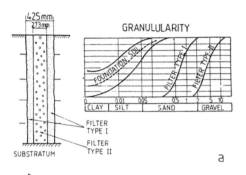

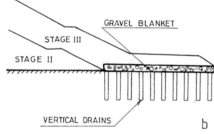

Figure 7. Vertical drainage network
a. Vertical drainage column filled with sand and gravel
b. Drainage blanket and vertical drains.

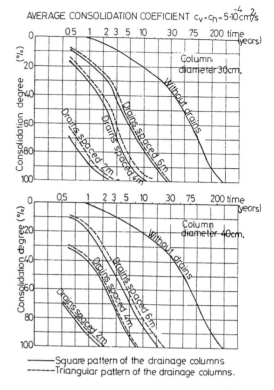

AVERAGE CONSOLIDATION COEFICIENT $c_v = c_h = 5 \cdot 10^{-4} cm^2/s$

———Square pattern of the drainage columns
- - - - Triangular pattern of the drainage columns.

Figure 8. Graphs of variation for the consolidation degree vs. time.

8. DISCUSSION

From the models adopted resulted that a time interval of approximately 150 years will be necessary for the soft high compressible soils to reach a consolidation degree U = 85-90% without a drainage system. Adopting a drainage blanket connected to a drainage system with elements spaced at 1.6 m and a stage-constructed-fill 15-20 m thick and 25 m wide, the required time interval for the same consolidation degree will be reduced at 2-3 years.

A more rapid consolidation of the foundation soil using the proposed drainage system will bring many advantages as: (1) an increase to the double values of the shear strength parameter until the end of consolidation time period; (2) time period for the ultimate settlement is reduced by a factor of ten;

(3) expecting time between loading stages of the foundation soil is reduced accordingly; (4) stability of the dam-foundation soil for all situations including seismic safety is assumed.

Geotechnical Hazards, Marić, Lisac & Szavits-Nossan (eds) © 1998 Taylor & Francis, ISBN 90 5410 957 2

Failures and failure modes of dams and embankments caused by geotechnical problems

A.Vogel
Data Station for Dam Failures Vienna, Austria

ABSTRACT: Over 100 large dams or embankments have failed by geotechnical problems, worldwide. Case histories and reports of some of them will be presented. For precalculation of dam break waves and disaster mitigation it is important to know failure modes of dams and embankments due to geotechnical problems. Also of interest is the contribution of the location of geotechnical hazards in relation to the height of the risk in the case of a failure.

1 INTRODUCTION

According to latest informations of DSDF-VIENNA, 712 dams have failed worldwide. 188 of them affected embankments, the failures of them caused by geotechnical problems.

The mean geotechnical causes, which led to failures of embankments or embankment dams are:

A) Structural parameters:
  - Marginal bondings between embankments and appurtenant works.
  - Marginal slope protections.
  - Inadequate designs.
B) Soil parameters:
  - Of the dam body itself.
  - Of the foundation.
C) External unforeseen influences:
  - Landslides or land subsidences.
  - Natural or induced earthquakes.

Failure scenarios have a direct connection to this mean geotechnical causes. Failures caused by sliding of complete dam bodies or upstream slips or downstream slips have often their origin in marginal structural parameters.

Seepages or piping, which can happen through the foundation or through the dam body or along the outlet have the origin in marginal soil parameters or bad compaction of dam material near the outlet.

Different settlements of the foundation or different movements in the dam body led to cracking, followed sometimes by hydraulic fracturing and last but not least by seepage or piping.

Earthquakes are events, which mostly reach dimensions of natural catastrophes. In this century, in average, every year over 10.000 victims were registered, caused by seismic events. The earthquake design of dams is a particularly important problem in view of the catastrophic consequences of a dam failure. It is obviously, that seismic failures or damages of dams are partly neglect in the official dam literature.

During Tangshan earthquake in China at July 28,1976 more than 100 small dams failed. Ojika earthquake in Japan at May 1,1939 destroyed 12 homogeneous earth dams. In Niigata earthquake in Japan at June 16,1964 eight earth dams failed completely. Only these three events led to the destruction of more than 120 earth dams.

For the precalculation of dam break waves and a disaster mitigation it is important to know the failure modes of dams and embankments due to geotechnical problems, from which result the most highly downstream hazards. It is necessary to investigate possible connections between the failure cause, the time-table of failure process and the velocity and the extent of the destruction of the dam body.

2 SIGNIFICANT FAILURES OF EMBANK-MENTS OR EMBANKMENT DAMS CAUSED BY GEOTECHNICAL PROBLEMS

Table 1. List of failed embankments or embankments dams caused by geotechnical problems

NAME	COUNTRY	TYPE	FAIL. YEAR	CONS. YEAR	HEIGHT m	LENGTH m	RES.VOL. hm3	CAUSE
AHRAURA	INDIA	TE	1953	1953	25,9	---	61,0	CF
ALAMO ARROYO 2	USA	TE	1960	1960	21,0	---	6,6	IE
ALEXANDER	USA	TE	1930	------	38,1	189,6	2,8	SF
AMOROS	CHILE	TE	1984	1979	42,5	220,0	60,0	IE
ANACONDA	USA	TE	1938	1898	22,0	---	0,3	FF
APISHAPA	USA	TE	1923	1920	35,0	178,6	23,0	IE
ASHTI	INDIA	ER	1883	------	17,4	---	---	SF
BALDWIN HILL	USA	TE	1963	1951	80,0	198,0	11,0	CR
BARAHONA	CHILE	TD	1928	------	63,0	---	---	SE
BATTLE RIVER	CANADA	TE	1956	------	14,0	548,0	15,0	IE
BEAR GULCH	USA	TE	1914	1896	19,2	222,5	---	SF
BELLA VISTA	CHILE	TD	1965	------	20,0	---	0,1	SE
BELLE FOURCHE	USA	TE	1933	1909	35,0	1981,2	---	SF
BILBERRY	GR.BRITAIN	TE	1852	1845	20,0	310,0	0,3	IE
BLACK MOSS	GR.BRITAIN	TE	1810	1810	---	---	---	IE
BLACK ROCK	USA	TE	1909	1907	21,3	208,0	18,0	IE
BLACKBROOK	GR.BRITAIN	TE	1799	1795	28,0	160,0	0,2	CF
BOLAN	PAKISTAN	TE	1976	1960	22,9	533,0	90,0	IE
BON ACCORD	SOUTH AFR.	TE	1937	1925	18,0	---	31,0	IE
BON ESPERANCA	BRAZIL	TE	1977	1976	17,0	450,0	37,0	FF
BRUSH HOLLOW	USA	TE	1928	1910	30,5	266,7	5,2	SF
BUFFALO CREEK	USA	TE	1972	1970	13,0	152,0	0,5	SF
CALAVERAS	USA	TE	1918	1914	73,2	384,1	201,0	SF
CARSINGTON	GR.BRITAIN	TE	1984	1984	32,0	1250,0	---	SF
CAULK LAKE	USA	TE	1973	1950	20,0	134,0	0,7	IE
CENTER CREEK	USA	TE	1973	1869	18,9	182,9	0,4	IE
CERRO NEGRO 3	CHILE	TD	1965	------	20,0	---	0,1	SE
CERTEJ SACARIMB	ROMANIA	TD	1971	------	27,0	---	---	SF
CHINGFORD	GR.BRITAIN	TE	1937	------	10,4	---	---	SF
CLENDENING	USA	TE	1934	1934	21,3	---	---	IE
COLD SPRINGS	USA	TE	1912	------	15,2	---	---	IE
COLEMAN	USA	TE	1954	------	-----	---	---	SE
CORPUS CHRISTI	USA	TE	1930	------	31,7	1243,6	78,9	IE
COSTILIA	USA	TE	1924	1920	36,9	182,9	19,4	IE
COSTILIA	USA	TE	1941	1920	36,9	182,9	19,4	SF
CUBA RESERVOIR	USA	TE	1868	1851	16,0	---	0,5	FF
CWM CARNE	GR.BRITAIN	TE	1875	1792	12,0	---	0,1	IE
D.M.A.D.	USA	TE	1983	1960	10,4	823,0	12,6	FF
DAGGS DAM	USA	TE	1973	1919	19,2	---	4,9	IE
DALEDYKE	GR.BRITAIN	TE	1864	1863	28,9	382,0	3,1	CR
DAVIS	USA	TE	1914	1914	12,2	---	58,0	IE
DECKERS	USA	TE	1983	1953	5,5	365,8	0,1	CF
DEEP BROOK	CANADA	TE	1949	1949	14,0	---	---	CR
DESABIA FOREBAY	USA	TE	1932	1903	16,2	315,5	0,4	IE

EAST LEE	USA	TE	1968	1965	7,6	144,8	0,5	IE
EILDON	AUSTRALIA	TE	1929	1927	40.0	213,4	3390,0	CR
EKLUTNA	USA	TE	1964	1929	7,9	169,2	---	SE
EL COBRE NEW	CHILE	TD	1965	------	15,0	---	0,5	SE
EL COBRE OLD	CHILE	TD	1965	------	35,0	---	1,9	SE
EL SALTO	BOLIVIA	TE	1976	------	15,0	30,0	0,5	IE
ELANDSDRIFT	SOUTH AFR.	TE	1975	1974	28,0	600,0	3,3	CF
EMA	BRAZIL	TE	1940	1932	18,5	370,0	10,0	IE
EMERY	USA	TE	1966	1850	16,0	130,0	0,5	IE
EMPIRE	USA	TE	1909	1906	12,0	---	47,0	IE
ENGLISH WATER S.	USA	TE	1965	1965	15,8	148,0	0,6	IE
FLAGSTAFF	AUSTRALIA	ER	1963	1963	15,7	---	---	IE
FONTENELLE	USA	TE	1965	1964	47,9	1524,0	440,0	FF
FORSYTHE	USA	TE	1921	1920	19,8	---	---	IE
FORT PECK	USA	TE	1938	1938	76,0	---	24000,0	SF
FRED BURR	USA	TE	1948	1947	16,0	99,0	0,6	IE
FRUIT GROWERS	USA	TE	1937	1898	11,0	277,4	41,2	SF
GARZA	USA	TE	1926	1926	37,0	3352,8	239,0	SF
GRAHAM LAKE	USA	TE	1923	1922	34,1	335,3	200,0	FF
GREENLICK	USA	TE	1904	1901	19,0	259,0	0,6	IE
GROS VENTRE	USA	TE	1927	1925	54,9	---	80,2	IE
HANS STRIJDOM	SOUTH AFR.	ER	1980	1977	18,5	400,0	---	CF
HATCHTOWN	USA	TE	1914	1908	18,9	237,7	16,0	IE
HAUSER LAKE	USA	STD	1908	1907	21,3	192,0	66,0	FF
HEBRON	USA	TE	1914	1913	17,2	1127,8	---	IE
HELL HOLE	USA	ER	1964	1964	30,0	---	2,6	IE
HIERO VIEJO	CHILE	TD	1965	------	5,0	---	---	SE
HINDS LAKE	CANADA	TE	1982	1980	12,0	5180,0	7500,0	IE
HOLMES CREEK	USA	TE	1924	1903	19,8	---	---	SF
HORSE CREEK	USA	TE	1914	1912	16,9	5150,0	20,9	IE
HOSOROGI	JAPAN	TE	1948	------	8,5	505,5	---	SE
HYOKIRI	COREA	TE	1961	1940	15,6	109,5	0,2	CF
IBRA	GERMANY	TE	1977	1977	10,0	---	---	CF
JENNINGS CREEK 3	USA	ER	1963	1962	21,0	92,0	0,4	FF
JENNINGS CREEK 16	USA	ER	1964	1960	16,8	113,4	0,3	IE
JULESBURG	USA	TE	1910	1891	16,8	---	34,7	IE
JUMBO	USA	TE	1910	1905	18,3	1219,2	30,0	IE
KAILA	INDIA	TE	1959	1955	26,3	213,4	14,0	SF
KANTALAI	INDIA	TE	1986	600	26,8	2524,0	135,0	IE
KEDAR NALA	INDIA	TE	1964	1964	20,0	---	17,0	IE
KRONE	SOUTH AFR.	TE	1969	1958	7,0	---	114,0	SE
KÜDDOW	GERMANY	TE	1930	1929	6,0	300,0	3,0	IE
LA LAGUNA	MEXICO	TE	1969	1912	16,0	675,0	4,3	IE
LA PATAGUA	CHILE	TD	1965	------	15,0	---	0,1	SE
LA REGADERA	COLOMBIA	TE	1937	------	36,9	---	---	FF
LAC LAURENT	FRANCE	TE	1219	1191	20,0	---	---	IE
LAFAYETTE	USA	TE	1928	1928	36,6	---	---	FF
LAKE CAWNDILLA	AUSTRALIA	TE	1962	1961	12,5	65,8	730,0	IE
LAKE FRANCES 1	USA	TE	1899	1899	15,2	302,4	0,9	IE
LAKE FRANCES 2	USA	TE	1935	1901	24,0	396,0	2,3	IE

LAKE LITCHFIELD	USA	TE	1975	1975	19,0	133,0	956,0	IE
LAKE TOXAWAY	USA	TE	1916	1902	18,9	117,4	10,0	FF
LAWN LAKE	USA	TE	1982	1902	7,3	115,5	0,9	IE
LEADER MIDDLE 15	USA	TE	1969	1963	7,0	---	---	IE
LEEUW GAMKA	SOUTH AFR.	TE	1928	1920	15,0	548,0	10,0	IE
LITTLE DEER CREEK	USA	TE	1963	1962	25,9	109,7	1,8	IE
LITTLEFIELD	USA	ER	1929	1929	37,0	91,0	---	IE
LIVERINGA	AUSTRALIA	TE	1957	1957	6,4	---	---	IE
LLIU LLIU	CHILE	TE	1985	1934	20,0	550,0	3,0	SE
LLUEST WEN	GREAT BRIT.	TE	1969	1896	19,5	222,5	1,0	IE
LO OVALLE	CHILE	TE	1985	1932	12,5	1520,0	13,0	SE
LONG TOM	USA	TE	1916	1906	15,2	---	---	IE
LOS MAQUIS 3	CHILE	TD	1965	------	15,0	---	0,1	SE
LOWER HOWELL	USA	TE	1906	1877	11,0	---	---	SE
LOWER KHAJURI	INDIA	TE	1949	1949	16,0	---	43,0	FF
LOWER VAN NORMAN	USA	ER	1971	1921	43,0	664,5	25,0	SE
LYMAN	USA	TE	1915	1913	19,8	256,0	43,0	IE
LYNDE BROOK	USA	TE	1876	1871	19,8	87,5	2,9	IE
MAFETENG	LESOTHO	TE	1988	1988	23,0	500,0	---	CF
MAGIC VALLEY	USA	TE	1911	1909	39,6	884,0	236,8	IE
MANIVALI	INDIA	TE	1976	1975	18,4	---	4,8	IE
MARSHALL CREEK	USA	TE	1937	------	25,6	472,4	---	FF
MARSHALL LAKE	USA	TE	1909	1890	26,2	661,4	12,6	IE
MASTERSON	USA	TE	1951	1950	18,3	---	---	IE
MENA	CHILE	TE	1888	1885	17,0	200,0	0,1	IE
MILBURN	USA	TE	1893	1893	---	---	0,2	FF
MILL CREEK	USA	TE	1957	1899	20,0	83,8	0,3	IE
MILL RIVER	USA	TE	1874	------	12,9	183,0	---	FF
MINATARE	USA	TE	1920	1915	19,2	1127,8	---	SF
MOCHIKOSHI 1	JAPAN	TD	1978	------	30,5	152,4	0,5	SE
MOHEGAN PARK	USA	TE	1963	1853	6,1	---	0,2	IE
MOLTENO	SOUTH AFR.	TE	1882	1881	15,0	800,0	0,2	CF
NANAKSAGAR	INDIA	TE	1967	1962	16,5	19300,0	---	IE
NARRAGUINNEP	USA	TE	1951	1910	29,6	1280,2	23,0	SF
NASH BOG POND	USA	TE	1969	------	10,1	274,3	6,2	CR
NECAXA	MEXICO	TE	1909	1909	59,0	297,2	---	SF
NIZHNE SVIRSKAYA	USSR	TE	1935	1934	28,0	1860,0	1190,0	SE
NORTH DIKE	USA	TE	1907	1905	25,0	---	---	SF
ODIEL	SPAIN	ER	1970	1970	35,0	154,0	3,3	CF
OWEN RESERVOIR	USA	TE	1914	1914	17,1	59,7	---	IE
PALAKMATI	INDIA	TE	1953	1942	14,6	---	---	SF
PAMPULHA	BRAZIL	TE	1954	1941	18,0	350,0	18,0	IE
PLEASANT VALLEY	USA	TE	1928	1928	19,2	---	76,5	IE
PORTLAND	USA	TE	1893	1889	13,7	---	0,6	IE
PROSPECT	AUSTRALIA	TE	1888	1888	26,0	2204,0	50,0	CR
PROSPECT	AUSTRALIA	TE	1980	1910	13,7	1842,5	7,5	IE
QUAIL CREEK	USA	TE	1988	1984	24,0	610,0	50,0	IE
RAMAYANA	CHILE	TD	1965	------	5,0	---	0,2	SE
RESERVOIR 4	USA	TE	1912	------	15,2	304,8	---	SF

ROGERS	USA	TE	1954	------	---	---	---	SE
ROPPATJERN	NORWAY	TE	1976	------	8,0	---	3,2	IE
RUAHIHI	NEW ZEALA.	ER	1981	1981	32,0	67000,0	31,0	IE
SAINT LUCIEN	ALGERIA	TE	1862	1861	27,1	---	2,0	IE
SALUDA	USA	TE	1930	1930	63,4	2389,0	---	IE
SAN LUIS	USA	TE	1981	1967	116,5	5630,0	3083,0	SF
SANTA HELENA	BRAZIL	ER	1985	1979	17,0	400,0	24,0	FF
SANTO AMARO	BRAZIL	TE	1907	1905	19,2	1585,0	---	SF
SCHAEFFER	USA	TE	1921	1911	30,0	335,0	---	CR
SHEEP CREEK	USA	TE	1970	1969	18,0	330,0	1,4	IE
SHEFFIELD	USA	TE	1925	1917	8,0	215,0	---	SE
SHELTON 1	USA	TE	1903	1881	---	---	0,2	IE
SINKER CREEK	USA	TE	1943	1910	21,5	335,0	3,3	IE
SMARTT SINDICATE	SOUTH AFR.	TE	1961	1912	28,0	2802,0	98,0	IE
SNAKE RAVINE	USA	TE	1898	1893	19,5	89,6	---	CF
STANDLEY LAKE	USA	TE	1914	1911	34,4	2133,6	---	SF
STANDLEY LAKE	USA	TE	1916	1911	34,4	2133,6	---	SF
STOCKTON CREEK	USA	ER	1950	1949	29,0	100,0	0,5	IE
SWANSEA	USA	TE	1879	1867	24,4	---	---	IE
TABLE ROCK COVE	USA	TE	1928	1927	42,7	228,6	34,5	IE
TAPPAN	USA	TE	1934	1934	15,9	---	---	FF
TERRACE	USA	TE	1957	1912	51,2	182,9	22,2	SF
TETON	USA	TE	1976	1975	93,0	945,0	355,0	IE
TORESON	USA	TE	1953	1898	15,0	96,0	1,4	IE
TORSIDE	GREAT BRIT.	TE	1855	1854	31,0	270,0	6,7	FF
TOWANDA	USA	TE	1939	1882	---	---	9,5	IE
TUPELO BAYOU	USA	TE	1973	1973	15,2	442,0	4,0	IE
UPPER STAVA	ITALY	TD	1985	------	22,0	300,0	---	SF
UTICA	USA	TE	1902	1873	21,3	---	0,7	SF
VIRGIN RIVER	USA	ER	1929	1929	19,8	---	---	CF
WACO	USA	TE	1961	1961	42,7	5468,4	895,5	SF
WADI GHATARA	LYBIA	TE	1977	1972	39,0	217,0	5,5	IE
WALTER BOULDIN	USA	TE	1975	1967	50,0	2268,0	---	SF
WARMWITHENS	GREAT BRIT.	TE	1970	1860	10,0	---	---	IE
WASSY	FRANCE	TE	1883	------	16,0	---	---	SF
WEISSE DESSE	CSSR	TE	1916	1915	18,0	244,0	0,4	IE
WHITEWATER BROOK	USA	TE	1972	1943	18,6	137,0	0,5	IE
XONXA	SOUTH AFR.	TE	1974	1972	24,0	300,0	---	CF
ZOEKNOG	SOUTH AFR.	TE	1993	1992	40,0	685,0	9,2	IE
ZUNI	USA	TE	1909	1907	21,3	219,5	19,5	IE

TE	Earth dam	IE	Internal Erosion	SE	Seismic Failure
ER	Rockfill dam	SF	Sliding Failure	FF	Foundation Failure
STD	Steel dam	CR	Cracking Failure		
TD	Tailing dam	CF	Construction Failure		

3 FAILURES CAUSED BY STRUCTURAL PARAMETERS

It is obviously, that homogeneous earth dams are vulnerable to all kinds of erosion phenomena and reach quickly critical behaviours in the case of loss of slope stability. Slope stability failures during construction led often to critical situations, when first filling of reservoirs is in process at the same time. Although the knowledge of soil behaviour has increased in the last 50 years, failures during construction happen also nowadays.

Carsington Dam failed 1984 due to a slip of the upstream side with a volume of about 500.000 cubic metres.

First filling of a reservoir is always a critical time for the behaviour of a dam. Inadequate design or marginal slope protection multiply with the storage the hazards to public safety.

In 1977 during its test period, a 10 m high homogeneous earth dam failed in a recreation park in Germany. Based on water-bearing sandstone containing open seams, the dam was constructed with an extended thin membrane on the upstream slope. A second membrane tightened the valley 30 m into the reservoir was fitted into the valley loam. A small concrete beam placed on fill material at the upstream toe of the dam connected these two membranes. A bottom drainage drawn forth to the dam toe should discharge the water, flow from the subsoil as well as the water, which penetrated into the dam body.

Caused by erodible fill material, which consisted of sandstone of varying quality and the poor connection of the dam membrane to the discharge tunnel and the concrete beam at the upstream fill toe, seepages started and increased. One day before the failure after 6 month of full storage, seepages increased considerably near the outlet and later washed out a part of 8 m length of the dam body.

4 FAILURES CAUSED BY SOIL PARAMETERS

Seepage effects through embankments or foundations of embankment dams are directly influenced by the properties of the soil materials of the dam body or the foundation. In the analysis of causes of dam incidents seepage effects were often the final results of cracking, internal erosion and foundation failures.

4.1 Cracking

In many cases dams have failed due to seepage within a very short time after the first reservoir filling. If we take false construction out of consideration, it is obviously impossible for the leakage in such a short time to pass through a well compacted soil structure or through a core of a dam having low permeability. Dangerous leakage in such cases can be the result of a concentrated weakness, only, such as a crack. Cracks in embankments are effects, which have their origin in:
- Differential settlements of the foundation and the abutments.
- Shrinkage cracks in soil horizons of the dam body caused by working stops during construction. If later the minor principal stress acting within the embankment is lower than the later hydrostatic water pressure, hydraulic fracturing is possible.
- Bad compaction of soil beneath outlets.
- Burrowing animals.
- Anisotropy of adjacent soil layers, caused by construction practise and different water contents.

If cracking is the cause of an embankment failure, penetrating water from the reservoir led to an aprubt change of the stress acting on the plane of the crack, resulting in erosion of soil and increasing of crack width, as long as water pressure is high enough for a progressive event.

In September 1965 the Fontenelle Dam in USA failed nearly caused by cracks in the foundation of the right abutment, which consisted of sandstone, shale and siltstone. A single-line grout curtain at the side of this abutment seemed enough prevention against seepage. Soon after the start of impounding, first seepage was observed, which reached the maximum level of 2000 litres per second. After a wet area could be observed on the downstream slope, it needed 36 hours to erode roughly 8000 cubic metres of the embankment. After reducing the water level and dumping heavy rocks in the hole of the eroded cavern, the leakage decreased to safe rates and the construction could be rescued.

Cracking of the insufficiently compacted dam body was the cause of the failure of Dale Dyke Dam in England 1864. Most of the embankment contained a mixture of shale and rock, which were excavated from the reservoir floor. Layer thickness up to 1,8 m did not allow proper compaction. As usual in England, this embankment dam construction had a central core of wet puddle clay with a top width of 1.2 m and a base width of 4.9 m. The fill on either side of this core was porous and permeable. A longitudinal crack along the downstream slope, 3 m down from the top with an opening width of 13 mm was observed some hours before its sudden breach.

Although the fills of embankment dams are well compacted today, zones of bad compacted soil beneath outlets are the cause of failures of dams also nowadays.

Zoeknog Dam in South Africa was constructed 1991/1992 and impounding started in December 1992. Caused by heavy rainfall in the catchment area, the reservoir level increased from 12 m to 18 m in the first three weeks of January 1993, which represented 22 % of the reservoir capacity. A 9.65 m diameter shaft of a morning glory spillway was located in the homogeneous earth dam. Along the left side of the outlet piping occured and the dam was washed out completely from the crest to river bed on a length of 70 m.

Mafeteng Dam in Lesotho, also a homogeneous earth dam failed in 1988 under similary circumstances, as seepage near the concrete wall of the spillway led to the erosion of the dam.

4.2 Internal Erosion

Internal erosion is often a process, which happens so gradually, that the removal of material in the embankment or in the core is not visible at a field examination of the dam body. Therefore dangerous situations of embankments, which are primary caused by internal erosion could be detected and remedied before they become serious, if contol mechanisms for supervision of seepage are installed in the dam construction.

Failures by internal erosion are processes, which concerned often constructions, which stood in operation for many years. Center Creek Dam in USA with 104 years, CWM Carne Reservoir in England with 83 years, Kantalai Dam in India with 1400 years, Lluest Wen Dam in England with 73 years, Mohegan Park Dam in USA with 110 years or Warmwithens Dam in England with 110 years in operation are good examples for this fact.

Failures by internal erosion immediately after first impoundings are often the results of cracking or hydraulic fracturing.

In homogeneous dams made of dispersive clays a special form of internal erosion can happen, called piping. The failures started with an initial leakage, caused by cracking or hydraulic fracturing, which gradually eroded a tunnel. In cases of low earth dams in Australia with low storage heights, the water discharged more or less gradually, leaving a tunnel in the dam body after the end of the event. In other cases the erosion of the tunnel proceeded to a point, where the roof collapsed, forming then the final breach.

In dams of dispersive clay the construction soil inclined to shrink by desiccation and penetrating water in shrinkage cracks led to a suspension of the soil particles together with other colloidal and dissolved matters. After their easily erosion the soil structure loose their resistance against shear stresses. An Australian study showed, that the velocity of this process is also partly dependent from the chemistry of pore water and from the stored water.

Piping failures along the outlet or piping as a result of anisotropy of adjacent soil layers also happen in modern high dams.

An earth dam with a central impervious core of moraine failed 1976 in Norway. To take care of the normal seepage through the construction, a 7 m wide and 0.5 m thick drainage blanket was laid over the whole length downstream of the dam, connected with a toe drain 2 m deep with two outlets. A bottom outlet was built through the dam. After an extremely dry summer and autumn in 1975, the reservoir was only partly filled and the dam body stayed uncovered with snow during the following winter 1975/76 with extremely deep temperatures but low snow fall. The poor compacted material surrounding the culvert got frozen to a depth of 2 m into the dam body. At the beginning of May 1976 the reservoir was quickly filled, caused by high temperatures and sudden melting of snow. Due to the absence of water pressure through the first operation period, the frozen material along the culvert was not consolidated and the surplus of water led to a loosing of the shear resistance. The sudden high water pressure of full storage pushed out slowly the wet soil parts along the outlet, until a free water passage was made trough the embankment.

5 FAILURES OF TAILING DAMS

As in table 1 shown also tailing dams failed by geotechnical causes. The Buffalo Creek failure with 125 victims and the Lower Stava catastrophe, which caused more than 220 deaths show, that tailing dam failures reached alarming proportions in the past. According to the embankment design options, whether tailing dams are constructed with the upstream or downstream method, whether the body itself is made of tailings or compacted natural soils, these constructions if a failure occure, include public hazards, which are absolutely equivalent to those impounding water.

The risk of failing by earthquakes is however equal for both types of dams. When comparing the failures of dams during Fukui earthquake 1948, Nevada earthquake 1954 and Alaska earthquake 1964 with those of tailing dams in La Ligua earthquake 1965 and Izu-Oshima-Kinkai earthquake 1978, interesting differences can be recognized. While tailing dams failed immediately or during the earthquakes, standard type dams in Fukui, Nevada and Alaska failed within a period of 24 hours. La Ligua earthquake showed also, that tailing dams, which were not actively used, did not fail. The effect of liquefaction is obviously connected with the quantity of saturation, which is necessary for the result of an outflow in the case of a failure.

The mean geotechnical factor leading to a potential instability of an embankment is the ammount of loose saturated cohesionsless soils in a dam body itself or in the foundation, which may also liquify during an earthquake.

Good examples for this fact are the severe incidents of La Palma Dam and La Marquesa Dam, which were damaged extensively by the Central Chile earthquake 1985. Both were built of silty clayey sands, using light equipment and having therefore a relatively low degree of compaction.

La Marquesa Dam with a height of 10 m and a length of 220 m was suffered by longitudinal cracks with 0.8 m maximum width and up to 2 m depths. Field observations showed horizontal displacements in the part of greatest damage, which reached about 11 m at the toe of the upstream slope and 6.5 m at the toe of the downstream slope.

The middle third of La Palma Dam, which was 10 m high and 140 m long, breached into blocks between longitudinal cracks. A major crack 80 m long with an maximum width of 1.2 m developed along the crest, its upstream side had settled more than 80 cm relative to its downstream side. A second major crack 60 m long developed 2 m below the crest in the upstream slope, with a surface width of about 80 cm and a drop of 1.5 m between the downstream and the upstream sides.

The slope failures at La Marquesa and La Palma dams happened due to loose sand layers near the base of both embankments. Total failures of both embankments did not occure, because the water levels in the dams were relatively low at the time of the earthquake.

Similary displacements and ruptures in blocks were observed at the failed embankment of Belci Dam in Romania, which failed by overtopping in 1991. A longitudinal trench on the dam crest was the reason for its unfavourable failure szenario.

6 ASPECTS TO PUBLIC HAZARDS

Most of the failures have been caused by internal erosion of the dam body or piping through the foundation. In modern dams as a result of adequately designed filters and good foundation monitoring this type of failure should not occure. But old dams often not posess such safety systems and therefore the assessment of safety of old constructions against this type of hazard is not possible. In every case, where it was not possible to lower the reservoir level in a very short time, internal erosion led to the destruction of the dam over ist full height. Only in low earth dams with low storage heights, failures by piping ended sometimes in the erosion of a tunnel. Failures by internal erosion always led to the loss of the embankment and result in catastrophic outflows. Warmwithens Dam demonstrated 1970, that the time between detecting a major leak and the complete destruction of the dam possibly reached only 3.5 hours. The complete failure of the 93 m high Teton Dam happened in 4 hours.

Investigations of foundation failures and failures by cracking show, that the destruction velocities of dam bodies are much higher than those by internal erosion. Mill River Dam was eroded within 1 hour, releasing three quarter of its storage in 20 minutes.

In our era of total unhuman economy it is usual to discuss what risk of failure is acceptable. Economic consequences of failures can be calculated easily but the costs of human life will never be determined in any satisfactory way.

7. REFERENCES

Charles, J.A. 1993. Embankment dams and their foundations: safety evaluation for static loading. *Proceedings of the International Workshop on Dam Safety Evaluation.* Vol. 4, 47-75, Grindel - wald.

De Alba P.A. et al. 1988. Analyses of dam failures in 1985 Chilean earthquake. *Journal of Geotechnical Engineering,* Vol. 114, 1414-1434

ICOLD Bulletin 99. 1995. Dam Failures. Statistical Analysis.

Jansen , R.B. 1980. Dams and Public Safety. *U.S. Department of the Interior.* Denver.

Loukola, E. et al. 1993. Embankment Dams and their Foundations: evaluation of erosion. *Proceedings of the International Workshop on Dam Safety Evaluation.* Vol 4, 17-46, Grindelwald.

Okusa, S. et al. 1980. Liquefaction of Mine Tailings in the 1978 Izu-Oshima-Kinkai Earthquake, Central Japan. *Proceedings of the 7th World Conference on Earthquake Engineering.* Istanbul.

Rallings, R.A. 1966. An Investigation into the Causes of Failure of Farm Dams in the Brigalow Belt of Central Queensland. *Water Research Foundation of Australia.* Bulletin No. 10.

Seed, H.B. & Idriss, I.M. 1967. Analysis of Soil Liquifaction: Niigata Erathquake. *Journal of the Soil Mechanics and Foundations Division* Vol. 93, 83-108.

Seed, H.B. 1980. Lessons from the performance of earth dams during earthquakes. *Proceedings of the Conference on Dams and Earthquake.* 97 - 104. Thomas Telford Ltd., London.

Sherard, J.L. et al. 1972a. Hydraulic Fracturing in Low Dams of Dispersive Clay. *Proceedings of the Speciality Conference on Performance of Earth and Earth-Supported Structures.* Vol. 1, 653-689, New York.

Sherard, J.L. et al. 1972b. Piping in Earth Dams of Dispersive Clay. *Proceedings of the Speciality Conference on Performance of Earth and Earth-Supported Structures.* Vol. 1, 589-626, New York.

Vick, S.G. et al. 1985. Risk Analysis for Seismic Design of Tailings Dams. *Journal of Geotechnical Engineering.* Vol. 111, 916-933.

Vogel, A. 1985. Bibliography of the History of Dam Failures. *Data Station for Dam Failures DSDF-VIENNA.*

Vogel, A. 1987a. Safety Problems of Tailing Dams *Hydraulic Structures and Rock Mechanics.* Vol. 23, 117-122, Bucharest.

Vogel, A. 1987b. Failures of Earth Dams Due to Piping Along The Outlet. *Hydraulic Structures and Rock Mechanics.* Vol. 23, 123-134, Bucharest.

Vogel A. 1993. Research for Risk Evaluation of Dam Failures by Defining a Break-Dimension-Time Factor. *Proceedings of the International Workshop on Dam Safety Evaluation.* Vol. 1, 21-31. Grindelwald.

Vogel, A. 1994. Erdbebenschäden an Talsperren. State of Art Report. *Data Station for Dam Failures DSDF-VIENNA.*

Earthquake geotechnical engineering

Geotechnical Hazards, Marić, Lisac & Szavits-Nossan (eds) © 1998 Taylor & Francis, ISBN 90 5410 957 2

Uniform risk in site-specific seismic hazard analysis

Atilla M. Ansal & Recep İyisan

İstanbul Technical University, Faculty of Civil Engineering, Maslak, Turkey

ABSTRACT: A site-specific seismic hazard analysis may be considered as composed of four consecutive stages that can be assumed independent and thus evaluated separately. The first stage is the estimation of the design earthquake magnitude based on seismological and geological data in the region. The second stage is the estimation of the source distance of the design earthquake. The third stage is the estimation of the design earthquake characteristics at the bedrock based on attenuation relationships. The fourth stage is the estimation of design earthquake characteristics on the ground surface based on the local geotechnical site conditions. Each of these stages involves various degrees of uncertainties therefore probabilistic approaches need to be adopted to determine the exceedence probabilities in these four stages to evaluate the overall uncertainty. A case study conducted for a site in Bursa located in western Turkey will be presented for determining the design earthquake characteristics with respect to a constant exceedence probability in these four stages.

1. INTRODUCTION

In assessing the earthquake recurrence for a site, tectonic formations that can generate earthquakes and the seismic history in the region need to be evaluated in a probabilistic manner (Allen 1995). In areas with active seismicity and complex tectonic formations, it may be realistic to assume a single tectonic areal source with a fixed radius around the investigated site to determine the earthquake recurrence relationship, to calculate exceedence probabilities with respect to earthquake magnitudes and thus define the design earthquake magnitude corresponding to a selected exceedence probability based only on seismic data compiled for the region.

In the second stage the source distance of the design earthquake need to be estimated in accordance with the geological and tectonic formations in the region. However, no matter how this selection is made, it will contain some amount of uncertainty. This uncertainty can be evaluated by adopting a probabilistic approach in the determination of the source distance.

The third stage requires the use of a suitable attenuation relationship. There are large number of attenuation relationships developed based on different data sets obtained in different parts of the World (Ambraseys 1995; Campbell & Bozorgnia 1993; Iai et al. 1993). Recently obtained large number of strong motion records made it possible to account for the differences in the source mechanisms, as well as for the site conditions. However, even with this new set of attenuation relationships (Campbell 1993), it is essential to take into account the variability in terms of exceedence probabilities.

The fourth stage involves the estimation of earthquake characteristics on the ground surface at the selected site. One option is to perform site response analysis using a suitable earthquake acceleration time history as the input motion at the base rock for local soil profiles. However, observations have shown that due to regional tectonic formations and different earthquake source mechanisms, each earthquake may have unique properties. Even earthquakes occurring in the same fault zone could have important differences. This aspect of source characteristics introduces a major uncertainty in the selection of the input motion for site response analysis and need to be evaluated in a probabilistic manner.

An attempt was made to evaluate the uncertainties involved in these four stages of site-specific seismic hazard analysis to establish a procedure for the selection of earthquake magnitude, source distance, attenuation, and site effects by keeping the exceedence levels constant in all four stages. A case study for a site in Bursa located in western Turkey will be presented to demonstrate the applicability of the proposed approach in determining the earthquake design spectrum and peak ground acceleration.

2. EARTHQUAKE MAGNITUDE

The earthquake occurrence for the investigated region can be determined based on Gutenberg-Richter relationship expressed as;

$$\text{Log } N = a - b M \qquad (1)$$

where **M** is the earthquake magnitude, **N** is the number of earthquakes larger than magnitude **M** and **a** and **b** are the coefficients representing the seismicity of the investigated region.

Earthquake records for the historical era (approximately between 10 AD and 1900) for the region were compiled based on available earthquake catalogues (Erdik, et al. 1985; Ergin, Güçlü & Uz 1967; Sipahioğlu 1984). Since these records are in terms of intensities, the relation developed for Turkey has been adopted to convert the intensities to magnitudes. The analysis to evaluate the seismic hazard was performed in terms of these calculated magnitudes. Due to the nature of the earthquake records for this era, it appears more reliable to base the study on medium strong and strong earthquakes. Therefore only earthquakes with intensity $(I \geq VI)$ have been used.

The earthquake records for the instrumental era (1900-1996) were compiled based on earthquake

catalogues, reports, papers and data bases (Alsan, Tezuçan, & Bath 1975; Ayhan 1988; Ayhan, et al. 1989; BUKORC 1987a & 1987b; Erdik, et al. 1985; BUKOERI 1996; Ergin, Güçlü & Uz 1967 & 1971; Gençoğlu, İnan & Güler 1990; Güçlü, Altınbaş & Eyidoğan 1986; Sipahioğlu 1984).

One shortcoming with the use of instrumental records is the limited time interval for the compiled data that may not represent the tectonic regime. Thus an assessment based only on instrumental data may not be very realistic. On the other hand historical data compiled for much longer time period, may not be very accurate with respect to epicentres, dates and intensities. Therefore instead of an analysis based only on instrumental or on historical data, it would be more reliable to evaluate the earthquake records from historical and instrumental era together to estimate the seismic hazard. A weighted averaging procedure was adopted with weights as 40% and 60 % for historical and instrumental data, respectively. The **a** and **b** coefficients of Gutenberg-Richter frequency magnitude relationship were determined as the average of the values for historical and instrumental era obtained by regression analysis. These average values were used to calculate the average variation of return periods and exceedence probabilities.

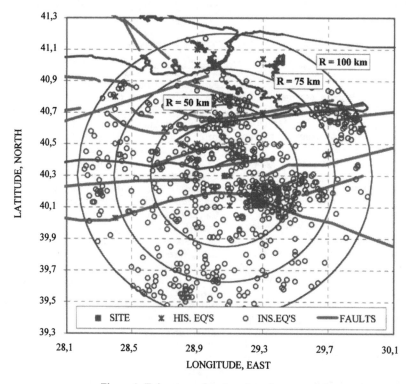

Figure 1. Epicenters of past earthquakes around Bursa

The other factor that may govern the earthquake probabilities is the size or the radius of the selected areal source. A parametric study was conducted with respect to the radius of the areal source to assess the effect of the selected size of the source zone on seismic hazard and earthquake magnitude probabilities.

All epicentres for the historical (I≥V) and instrumental earthquakes (M≥4), as well all for earthquakes with M≥3 after 1984 (BUKOERI, 1996) and estimated fault locations (Barka, 1991, 1992) are marked to show the level of seismic activity with respect to source radius of 100, 75, and 50 km in Figure 1.

Regression analyses to determine the Gutenberg-Richter frequency magnitude relationship were performed for historical (I≥VI) and instrumental (M≥4.0) era separately for source zone radius of 100, 75 and 50 km using the data given in Table 1.

Table 1. Earthquakes of the historical and instrumental era for R=100, 75 and 50 km

Historical era	Number of Occurrence		
Earthquake Intensity	R=100 km	75 km	50 km
X	4	3	1
IX	12	4	2
VIII	40	6	2
VII	46	2	1
VI	58	21	16
Total number	162	36	22
Instrumental era			
Magnitude interval	R=100 km	75 km	50 km
6.4<M	1	-	-
6.2<M≤6.4	1	1	-
6<M≤6.2	2	1	-
5.6<M≤5.8	5	3	3
5.4<M≤5.6	2	-	-
5.2<M≤5.4	7	4	1
5<M≤5.2	7	4	2
4.8<M≤5	9	4	2
4.6<M≤4.8	8	5	3
4.4<M≤4.6	10	7	5
4.2<M≤4.4	21	17	12
4.≤M≤4.2	10	8	7
Total number	83	54	35

Adopting Poisson's distribution, the variation of the exceedence probabilities and return periods were calculated for weighted average of historical and instrumental data with respect to earthquake magnitudes for source radius of R = 100, 75 and 50 km as shown in Figure 2. No normalisation with respect to area size was made since only one areal source was used. The design earthquake magnitude

was determined as M = 7.1 corresponding to arbitrarily selected exceedence probability of 10% and return period of 1000 years for source radius of 50 km considering that most damaging earthquakes are near-field shallow earthquakes.

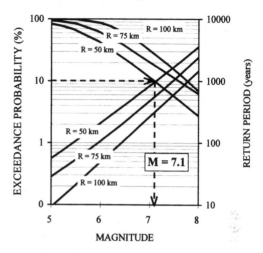

Figure 2. Variation of return period and exceedence probability with respect to earthquake magnitude

3. SOURCE DISTANCE

The studies concerning the geology and tectonic characteristics of the region (Barka 1991, 1992) have shown that the active faults in the region have a very complex structure. It would not be unrealistic to assume that the probable epicentre and the related fault for a strong earthquake can be anywhere within 50 km away from the site in the selected tectonic areal source zone.

Since most of the attenuation relationships adopt source distance as the closest distance to the surface projection of the fault rupture, it is important to select a realistic source distance in accordance with the geological and tectonic features. However, if this selection is made deterministically, it will not be possible to assess the uncertainty and related risk levels. To have overall constant risk level equal to the exceedence level selected for the design earthquake magnitude, it is necessary to adopt a probabilistic approach in the determination of the source distance.

Due to the tectonic features of the region most of the earthquakes were shallow earthquakes with average focal depth of 10 km. Therefore it is assumed that the epicentral distances of all instrumentally recorded earthquakes with magnitude M≥3, which were medium to small magnitude crustal earthquakes, could be taken as possible source distances (Ambraseys, Simpson and Bommer 1996). The statistical distribution of these epicentral distances less than 50 km was evaluated using different probability density functions. There were not very significant differences among these different density functions. However, since Beta probability density function was observed to give the best fit with respect to observed data, the epicentral distance corresponding to risk level (probability of being less) of 10% was determined as 17 km using the Beta probability density function as shown in Figure 3.

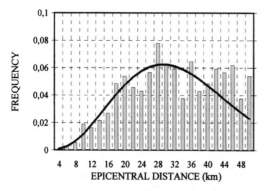

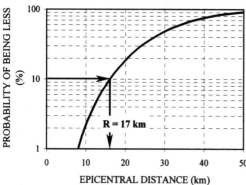

Figure 3. Probability of being less with respect to epicentral distance

An attempt was made to modify the attenuation relationship proposed for Europe by Ambraseys (1995) by performing a new regression analysis using only recorded accelerations from Turkish earthquakes. In this analysis the correlation coefficient was calculated as r = 0.785 with standard deviation of 42% of the mean.

In the proposed probabilistic approach, the bedrock peak accelerations need to be determined with respect to exceedence probabilities. It is assumed that the scatter in the attenuation relationship can be modelled by normal probability density function with standard deviation as 42% of the mean as calculated from the regression analysis.

In this situation, peak ground acceleration on the bedrock, corresponding to the exceedence probability of 10%, was calculated as Ap=450 gal as shown in Figure 4.

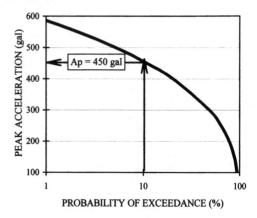

Figure 4. Exceedence probability for peak ground acceleration

4. GEOTECHNICAL SITE CONDITIONS

Insitu seismic tests, namely down-hole, cross-hole and Suspension PS logging measurements were conducted to determine the shear and pressure wave velocity profiles at 4 locations at the selected site.

Down-hole seismic wave measurements were conducted at four boreholes by generating P and S wave pulses on the ground surface while determining the first arrival times by a three component geophone located in the borehole. Due to this testing technique and method of interpretation, it is only possible to determine the average seismic wave velocities for the soil layers encountered at the soil profile. It is not possible to detect thin soft or hard layers since the wave velocities measured are the average wave velocities for all the soil layers between the ground surface and the geophone elevation. In all four soil profiles the shear wave velocities measured ranged between approximately 200 m/sec to 450 m/sec as in the case of previous findings. The shear wave velocities in two boreholes were fairly uniform with shear wave velocities slightly higher between the depths of 5 to 20 meters that may be due to the overconsolidation effect of the fluctuating ground water table. In other two boreholes the shear wave velocity profile show a uniform increase.

S and P wave velocities of the soil layers encountered at the site were also measured by cross-hole seismic wave velocity measurements. The tests were conducted at three locations between two boreholes drilled approximately 5 m apart. In this method one of the boreholes was used to generate shear wave pulse by a inhole shear hammer while the other hole was used for determining the first arrival times by three orthogonally placed geophones. In the tests conducted at the site, the wave velocity measurements were conducted approximately at every two meters during going down and coming up stages in the boreholes. In all three soil profiles the shear wave velocities measured ranged between 250 m/sec near the ground surface and 500 m/sec around 30m depth indicating fairly uniform soil stratification composed of dense and very stiff soil layers.

Suspension PS logging measurements were conducted in 4 boreholes that were properly cased and filled with water, at approximately every 2 meters. There were not very significant differences among the shear wave velocity profiles obtained at four locations. In addition shear wave velocities measured approximately at the top 30 m of all soil profiles have shown a fairly uniform distribution ranging between 250 m/sec to 500 m/sec as shown in Figure 5. These findings support the presence of stiff to very stiff silty clay layers, medium dense to dense silty sandy layers and weathered partly cemented sandstone-siltstones as observed during the borings.

The site response analysis program SHAKE developed by Schnabel, Seed and Lysmer (1971) was used to determine the earthquake characteristics on the ground surface for the free field conditions. For this purpose four shear wave velocity profiles obtained from Suspension PS logging measurements as shown in Figure 5, as well as the three shear wave velocity profiles obtained from cross-hole tests were modelled as composed of 2m layers extending down to 50 m depth. Site response analyses were conducted for these seven shear wave velocity profiles.

5. SITE RESPONSE ANALYSIS

The fourth reason for the variability in the design earthquake parameters is due to the earthquake characteristics (McGuire 1995) and coupled effects of earthquake and site conditions (Ansal and Siyahi 1995). It is possible to demonstrate this source and site coupling based on recorded accelerations at the same station during different earthquakes.

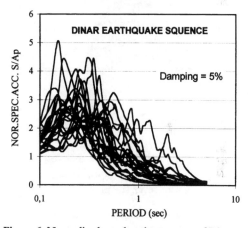

Figure 6. Normalised acceleration spectra of Dinar earthquakes

As shown in Figure 6, there are significant differences among the normalised elastic acceleration response spectra for different earthquakes recorded at Dinar station during the earthquake sequence that caused extensive damage in October 1995. All of these recorded accelerations were from shallow near-field medium strong earthquakes with epicentres ranging from 2 to 19 km and magnitudes ranging between M_L=4.1-5.9 (GDDA-ERD 1997). The recorded peak accelerations were in the range of 69-330 gals representing medium strong to strong ground motion and some inelastic behaviour of soil layers. The differences in predominant soil periods and amplification ratios are different for each

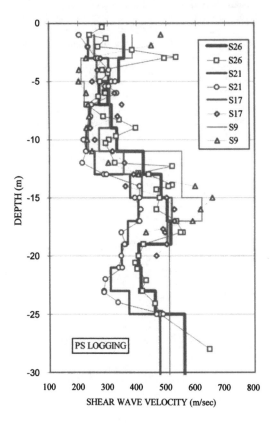

Figure 5. Variation of shear wave velocity with depth for four soil profiles

earthquake relatively independent of magnitude, epicentral distance and peak acceleration values.

Therefore to account for the variability and to assess the characteristics of the design earthquake on the ground surface, it would be suitable to carry out a statistical study (Ansal & Lav 1995). For this purpose a parametric study was conducted using 20 near field acceleration records obtained during previous earthquakes in western Turkey under similar tectonic formations at various strong motion stations in Turkey (GDDA-ERD 1997). All of these records were scaled to peak acceleration of 450 gal corresponding to the selected design earthquake of magnitude M=7.1 with possible epicentre 17 km away. Site response analyses were performed using all 20 acceleration time records as the input motion at the bedrock.

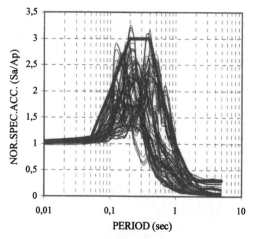

Figure 7. Calculated normalised acceleration spectra for seven soil profiles

The elastic acceleration response spectra normalised with respect to calculated peak ground accelerations for the seven shear wave velocity profiles were determined on the ground surface for all 20 acceleration records. The normalised elastic acceleration spectra calculated using shear wave velocity profiles obtained by cross-hole and PS Logging were similar. In addition the average spectral amplification were also similar for different soil profiles. Even though there were differences among the calculated spectra, these differences can be neglected for all practical purposes. The results also indicate that it is possible to adopt a conservative design spectrum as the outer envelope of calculated normalised elastic acceleration response spectra using 20 scaled earthquakes for 7 shear wave velocity profiles as shown in Figure 7.

For engineering applications peak ground acceleration and normalised elastic acceleration response spectrum can be regarded as two parameters representing the earthquake characteristics on the ground surface. One possible approach is to assume that these parameters are independent and it is possible to carry out a probability analysis for each of them separately. Based on this assumption, the variations in the effects of input earthquake motion, in terms of peak ground acceleration and normalised elastic acceleration response spectrum can be modelled statistically by adopting a normal probability density function.

The peak horizontal accelerations calculated at the ground surface using 20 different records for 7 shear wave velocity profiles were observed to vary within a maximum value of 0.68g and minimum value of 0.32g. It is assumed that the selected acceleration records for this study represent the characteristics of possible earthquakes that may take place in the near vicinity of the site since the records used were obtained in similar tectonic regions. In this way the effects of differences that may be encountered in the characteristics of a probable earthquake can be taken into account with respect to the safety level required for the design of structures.

To be consistent in the final evaluation of the design earthquake characteristics, the probability level selected for the peak ground acceleration should match the probability level adopted in the seismicity study when selecting the magnitude of the design earthquake. Assuming normal probability distribution and using the calculated 140 peak ground accelerations as the data set with mean value of 0.483 g and standard deviation of $\sigma = 0.067$, the peak ground acceleration corresponding to the exceedence probability of 10%, was calculated as $a_p = 0.57$ g as shown in Figure 8.

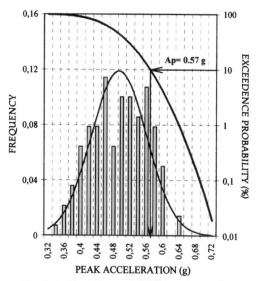

Figure 8. Exceedence probability of peak ground acceleration

Similar statistical approach was also applied for determining design normalised elastic acceleration response spectrum. The spectrum corresponding to 10% exceedence probability was calculated for all periods independently by assuming normal distribution and using mean and standard deviations determined for each period value of the 140 calculated normalised acceleration response spectra as shown in Figure 9.

Based on this study conducted to determine the required design earthquake characteristics for the structures to be constructed on the selected site, the variation of the normalised acceleration response spectrum or the Spectrum Coefficient, S, calculated with respect to structural periods is shown in comparison to the suggested design spectrum in the 1997 Turkish Earthquake Code as well as in comparison to the envelop spectrum in Figure 9.

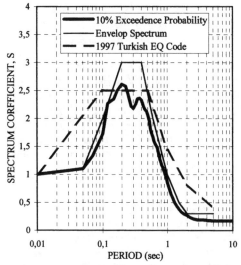

Figure 9. Design spectrum from 1997 Turkish Earthquake Code, constant exceedence probability and envelope spectrum obtained in this study

CONCLUSIONS

A site-specific seismic hazard analysis was considered as composed of statistically independent four consecutive stages that can be evaluated separately. The basic purpose in the suggested procedure was to keep the exceedence probabilities or the risk levels constant throughout the four evaluation stages of the site-specific seismic hazard analysis. This is necessary in order to control and assess the overall exceedence (risk) levels for the design earthquake characteristics taken as the peak ground acceleration and normalised spectral acceleration and to evaluate the related cost of the construction.

In the first stage assuming a single areal source zone with fixed radius around the site, the earthquake magnitude exceedence probabilities were calculated based on seismic data compiled for the region using conventional Gutenberg-Richter magnitude frequency relationship.

In the second stage assuming that source distance for the region can be represented by the epicentral distance, epicentres of all instrumentally recorded earthquakes could be possible epicentres in future. The probabilities for the epicentral distance was calculated using a Beta probability density function obtained by modelling the statistical distribution of the epicentral distances of past earthquakes with magnitude M≥3.

In the third stage the exceedence probabilities of the peak acceleration at the bedrock were determined assuming that the scatter in the attenuation relationship proposed by Ambraseys (1995) can be modelled by a normal probability density function.

The geotechnical site conditions were investigated based on in-situ suspension PS logging, cross-hole and down-hole seismic wave velocity measurements. A parametric study was conducted using 20 Turkish acceleration records, scaled to peak acceleration amplitude obtained in the previous stage, as the input motion at the bedrock for the 7 in-situ measured shear wave velocity profiles.

In the fourth and last stage, earthquake characteristics on the ground surface taken as the peak horizontal ground acceleration and normalised elastic acceleration spectrum were determined with respect to exceedence probabilities.

Assuming that the acceleration records obtained in Turkey represent the range of possible earthquakes that may take place in the future, the variation of peak ground acceleration calculated from site response analyses were modelled statistically by a normal probability density function and the exceedence probabilities were determined.

A site-specific design spectrum was determined in a similar manner by statistically modelling the calculated normalised acceleration response spectra value at all periods by a normal probability density function. The spectrum corresponding to the selected exceedence probability was calculated for all periods independently.

REFERENCES

Allen,C.R. 1995. Earthquake Hazard Assessment: Has Our Approach Been Modified in the Light of Recent Earthquakes. *Earthquake Spectra* 11(3):357-366.

Alsan,E., L.Tezuçan,. & M.Bath 1975. *An Earthquake Catalogue for Turkey for the Interval*

1913-1970. Kandilli Observatory, Istanbul, Turkey and Seismological Institute, Uppsala, Sweden.

Ansal,A.M. & B.G.Siyahi 1995. Effects of Coupling Between Source and Site Characteristics During Earthquakes. *Proc. of 5th SECED Conference on European Seismic Design Practice:* 83-92. Rotterdam, Balkema.

Ansal,A.M. & M.A.Lav 1991. Effect of Variability of Input Motion Characteristics on Ground Response Spectra. *Proc. of 4th Int. Conf. on Seismic Zonation:* 131-138. Stanford, Cal.

Ambraseys,N.N., K.A.Simpson, & J.J.Bommer 1996. Prediction of Horizontal Response Spectra in Europe. *Earthquake Engineering and Structural Dynamics.* 25:371-400.

Ambraseys,N.N. 1995. The Prediction of Earthquake Peak Ground Acceleration in Europe. *Earthquake Engineering and Structural Dynamics.* 24:467-490.

Ayhan,E. 1988. Strong Earthquakes ($M_s \geq 5.5$) in Turkey Between 1981-1988 and Their Results. *Bulletin of Earthquake Research.* 15(61):5-121 (in Turkish).

Ayhan,E. et al. 1989. The Active Fault Zones and Earthquake Activity in Western Anatolia between 1976-1986. *Bulletin of Earthquake Research.* 16(64):5-192 (in Turkish).

Barka,A.A. 1992. The North Anatolian Fault Zone. *Annales Tectonica.* VI:164-195.

Barka,A.A. 1991. Tectonic Formations Controlling the Seismicity of Istanbul and a Microzonation Study for Istanbul. *Istanbul and Earthquakes Symposium:* 35-56. Chamber of Civil Engineers. Istanbul (in Turkish).

Borcherdt,R.D. 1994. Estimates of Site Dependent Response Spectra for Design (Methodology and Justification). *Earthquake Spectra.* 10(4):617-654.

BUKORC, Boğaziçi University Kandilli Observatory Research Centre 1987a. Earthquake Activity in western Turkey during 1984-1985-1986. *Bulletin of Earthquake Research.*14(58):5-111(in Turkish).

BUKORC, Boğaziçi University Kandilli Observatory Research Centre 1987b. Earthquake Activity in western Turkey during 1983. *Bulletin of Earthquake Research.* 14(57):93-127 (in Turkish).

BUKOERI, Boğaziçi University Kandilli Observatory and Earthquake Research Institute 1996. Seismic Data for Marmara Region. ftp://boun.edu.tr/kandilli/pub/

Campbell,K.W. 1993. Comparison of Contemporary Strong-Motion Attenuation Relationships. *Proc. of Int. Workshop on Strong Motion Data*: (2)49-70.

Campbell,K.W. & Y.Bozorgnia 1993. Near-Source Attenuation of Peak Horizontal Acceleration from Worldwide Acceleograms Recorded from 1957 to 1993. *Proc. of Int. Workshop on Strong Motion Data*: (2)71-81.

Erdik,M., V.Doyuran, P.Gülkan & N.Akkaş 1985. *Evaluation of Earthquake Risk for Turkey with a Statistical Approach.* METU Earthquake Research Centre. Ankara

Ergin,K., U.Güçlü, & Z.Uz 1967. *Earthquake Catalogue of Turkey and Its Vicinity(11 - 1964).* I.T.U. Mining Faculty, Earth Physics Institute, Publication No: 24.

Ergin,K., U.Güçlü, & Z.Uz 1971. *Earthquake Catalogue of Turkey and Its Vicinity (1965-1970).* I.T.U. Mining Faculty, Earth Physics Institute, Publication No: 28.

GDDA-ERD, General Directoriat of Disaster Affairs, Earthquake Research Department 1996. *Data Bank of Strong Motion Records for Turkey.* Ankara, Turkey

Gençoğlu,S., E.İnan, & H.Güler 1990. *Earthquake Hazard in Turkey.* Chamber of Geophysical Engineering Publication. Ankara.

Güçlü,U., G.Altınbaş & H.Eyidoğan 1986. *Earthquake Catalogue of Turkey and Its Surroundings (1971-1975).* I.T.U. Earth Sciences Research Centre, Seismology and Seismotectonic Section, Publication No: 30.

Iai,S., Y.Matsunaga, T.Morita, H.Sakurai, E.Kurata & K.Mukai 1993. Attenuation of Peak Ground Accelerations in Japan. *Proc. of Int. Workshop on Strong Motion Data*: (2)3-22.

Marcellini,A. 1995. Probabilistic Hazard Evaluation in Terms of Response Spectra. *Proc. of 3rd Turkish National Earthquake Engineering Conference:* 407-420. Istanbul, Turkey.

McGuire,K.R. 1995. Probabilistic Seismic Hazard Analysis and Design Earthquakes: Closing the Loop. *BSSA.* 85(5):1275-1284.

Schnabel,P.B. J.Lysmer, & H.B.Seed 1972. *Shake-A Computer Program for Earthquake Analysis of Horizontally Layered Sites.* EERC Report No.72-12. Uni.of Cal., Berkeley, Cal.

Sipahioğlu,S. 1984. An Investigation of Earthquake Activity for the North Anatolian Fault Zone and Its Vicinity. *Bulletin of Earthquake Research Institute.* 11(45):1-108 (in Turkish).

Geotechnical Hazards, Marić, Lisac & Szavits-Nossan (eds) © 1998 Taylor & Francis, ISBN 90 5410 957 2

Effective stress analysis of seismic vertical array sites at Kobe

M.Cubrinovski
Kiso-Jiban Consultants Co., Ltd, Japan

K.Ishihara & M.Hatano
Department of Civil Engineering, Science University of Tokyo, Japan

ABSTRACT: This paper presents acceleration records recovered at three vertical array sites that manifested liquefaction during the 1995 Kobe Earthquake and discusses the characteristics of the ground responses and liquefaction by using an effective stress method of analysis. A detailed account on extensive liquefaction of reclaimed deposits of gravelly sands is given. The analyses performed demonstrate the potential of the effective stress method of analysis for quantitative evaluation of ground responses influenced by excess pore pressures.

1 INTRODUCTION

Since the Niigata Earthquake of 1964 liquefaction has been recognized as a major geotechnical hazard related to earthquakes. A recent reminder of the destructive effects of liquefaction was the Hyogoken-Nanbu Earthquake of January 17, 1995, that severely hit the area of Kobe, Japan. A significant characteristic of this earthquake of magnitude M=7.2 (JMA) is that its causative faults run through the central part of the city of Kobe thus exposing a great number of engineering structures to an unprecedented level of shaking intensity. The main shock of the quake caused widespread liquefaction over large areas of reclaimed lands in the coastal zone of Kobe. The exceptionally extensive and massive liquefaction inflicted numerous cases of destruction including slumping of revetment lines, excessive ground deformation and failure of foundations. A peculiar feature of the liquefied fill material is that it is a sandy soil containing a fairly large portion of gravel, commonly over 30 % and often as much as 60 % by weight. In view of the above facts, there is no doubt that the Kobe Earthquake represents an earthquake event and liquefaction case history of unparalleled proportion.

Many strong motion records of the main shock of the quake were obtained at sites that manifested liquefaction. Among these records particularly valuable are those recovered with several seismic downhole arrays since they include acceleration records at different depths of the soil profiles. This paper presents acceleration records obtained at three such sites in the Kobe area, and discusses the characteristics of the ground response and liquefaction by using an effective stress method of analysis.

2 SEISMIC VERTICAL ARRAYS

Figure 1 indicates the locations of the three seismic vertical array sites. The Port Island site (PI-site) and the Higashi-Kobe site (HK-site) are located at the man-made islands, Port Island and Fukaehama, respectively. These sites are in proximity to the most heavily damaged zone of the city of Kobe, which in this range stretches roughly along the Hanshin and Tokaido lines. The Takasago site (TKS-site), on the other hand, is located about 25 km toward northwest, on the opposite side of the fault rupture zone.

2.1 Port Island site

The reclaimed land of Port Island, with an area of 436 ha, was constructed in the period between 1966 and 1980. The soil used for land filling is a decomposed weathered granite, locally known as Masado. As indicated in Figure 2, the down-hole array site is located at the northwest corner of the island. The array contains strong motion accelerometers at the ground surface and at depths of 16, 32 and 83 m. Characteristics of the soil profile including SPT N-values and shear wave velocities are shown in Figure 3 where it may be seen that the fill deposit exists down to 18 m depth with SPT blow counts ranging between 5 to 10. Beneath the reclaimed fills there is a silty clay layer which is identified as the seabed deposit before the reclamation. The ground water level is at about 3 m depth.

Laboratory tests on intact samples of Masado soils provided a wealth of information about the cyclic strength or liquefaction resistance of the fill material. Shown in Figure 4 are data compiled by Ishihara et al.

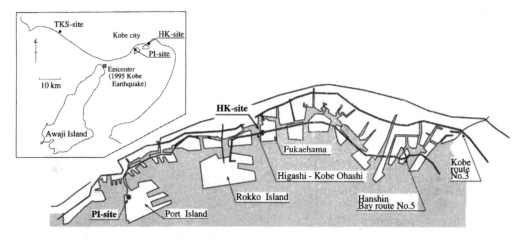

Figure 1. Locations of three seismic vertical arrays in the area of Kobe

(1996) from two series of cyclic triaxial tests on undisturbed samples of Masado soils. These samples have been recovered by conventional sampling techniques from Port Island, prior to the earthquake. Superimposed in Figure 4 are results from cyclic triaxial tests on undisturbed samples of Masado soil recovered by means of ground freezing technique, reported by Hatanaka et al. (1997). These samples have been recovered from a site in close proximity to the down-hole array site, after the Kobe Earthquake. Having in mind that the fill material is well-graded and contains a significant fraction of gravel, it appears that both the SPT blow count number and liquefaction resistance of the Masado soils are relatively low.

The low liquefaction resistance of the reclaimed soil was dramatically manifested during the Kobe Earthquake. The violent shaking caused widespread liquefaction resulting in an average settlement of 30-40 cm and a 15-20 cm thick layer of sand and mud littered on the ground surface over a large portion of Port Island. The extensiveness of the liquefaction is illustrated in Figure 2 where the area covered by sand boils is shown. The caisson-type quay walls moved towards the sea as much as 3 to 4 m which was accompanied by an equally large amount of lateral spreading of the backfill soil. Based on data from comprehensive ground surveying measurements along a number of cross sections in directions perpendicular to the revetment line, Ishihara et al. (1997) established a relation between the movement of the ground due to lateral spreading and the distance from the waterfront. A summary plot of the lateral ground displacement versus distance from the revetment line is shown for the north or south facing quay walls in Figure 5. According to these data the site of the down-hole array is beyond the zone affected by lateral spreading.

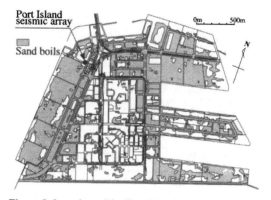

Figure 2. Location of the Port Island seismic array

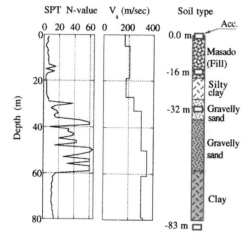

Figure 3. Soil profile of the Port Island array site

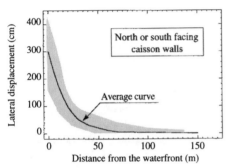

Figure 4. Cyclic strength of undisturbed samples of Masado soils recovered from Port Island

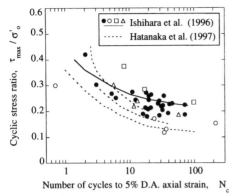

Figure 5. Horizontal ground displacement due to lateral spreading (Ishihara et al., 1997)

The N-S acceleration time histories recorded at the Port Island array site are presented in Figure 6. Noteworthy is that the top two accelerograms have been recorded in the reclaimed fill layer that liquefied during the earthquake. Signs of liquefaction are apparent in the decrease of the acceleration amplitudes, loss of high frequency response and elongation of the predominant period of the motion at the ground surface. The ground motion at this site exhibited very pronounced directionality as illustrated by the trajectory of the horizontal accelerations at the ground surface shown in Figure 7. The seismic motion is oriented in the northwest-southeast (NW-SE) direction or perpendicular to the direction of the fault line.

2.2 Higashi-Kobe site

The seismic vertical array site at Fukaehama resembles in many aspects the site of Port Island. Fukaehama is also a man-made island which has been reclaimed using the same type of fill materials (Masado soils) approximately in the same period as Port Island. As shown in Figure 8, the reclaimed fill

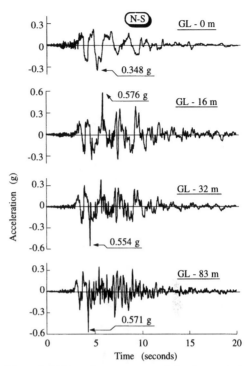

Figure 6. N-S accelerograms recorded at the vertical array site of Port Island (Toki, 1995)

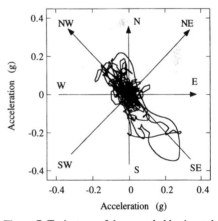

Figure 7. Trajectory of the recorded horizontal accelerations at the ground surface (Port Island site)

is about 15 m thick and overlies the original seabed layer of silty clay. The ground water level is at about 3 m depth. The seismic array consists of two sets of accelerometers located at the ground surface and in the gravelly sand layer at depth of 33 m, respectively. Except for the larger SPT N-values in the top 3-4 m, the characteristics of the Higashi-Kobe site are very

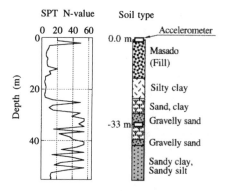

Figure 8. Soil profile of the Higashi-Kobe array site

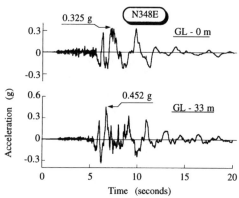

Figure 9. Recorded N348E accelerograms at the down-hole array site of Higashi-Kobe

similar to those of the Port Island site. Recorded accelerograms at the down-hole array site of Higashi-Kobe are displayed in Figure 9. These data are in the N348E direction which is a direction close to that of the maximum shaking intensity.

2.3 *Takasago site*

Unlike the two seismic arrays described above, the Takasago down-hole array is installed in a natural soil deposit. Figure 10 presents the profile of this site with the locations of the three sets of accelerometers at the ground surface, and at depths of 25 and 100 m, respectively. The profile includes various soils which are highly stratified down to a depth of 100 m. With exception of a couple of thin layers, the soil below 16 m depth is characterized by relatively high SPT N-values and shear wave velocities in excess of 300 m/s. The liquefiable layer of interest is the silty sand layer between 3.5 and 8.5 m. The ground water level is at about 2.5 m.

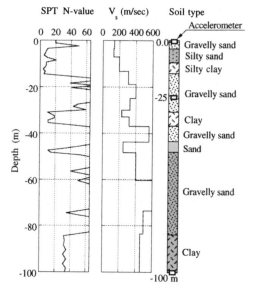

Figure 10. Soil profile of the Takasago array site

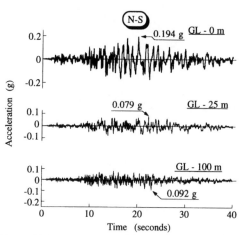

Figure 11. Recorded N-S accelerograms at the vertical array site of Takasago

Sato et al. (1996) reported that sand boils have been observed at the Takasago array site after the earthquake. No detailed account on the manifestation of the liquefaction is given, however. It appears from the recorded accelerograms with this seismic array, shown in Figure 11, that the liquefaction at this site, if any, had very different features from that of the Port Island and Fukaehama. Unlike the response at these two sites, the motion at the Takasago site is significantly amplified at the ground surface. In addition, the intensive part of the shaking has longer duration and the motion has less pronounced directionality. These features of the ground response at the Takasago site are in accordance with the fact that this site is more distant from the source zone and presumably subject to a lesser extent and intensity of liquefaction, if any.

3 ANALYSIS AND DISCUSSION

Presented acceleration records themselves provide considerable information about the ground responses at the three array sites. In order to fully understand the extent and effects of the liquefaction, however, it is of utmost importance to examine other essential characteristics of the ground responses such as development and distribution of excess pore pressures and shear strains throughout the depth of the soil deposits. For that purpose, series of effective stress analyses of the array sites were carried out.

3.1 *Numerical procedures*

A fully coupled effective stress method of analysis of saturated soil is used in which, besides the soil nonlinearity, the influence of the excess pore pressures on soil behaviour is considered. The employed elastoplastic deformation law of soil is described in Cubrinovski and Ishihara (1998). To facilitate the material modelling, in the case of non-liquefiable soils the effects of the excess pore pressures were disregarded and such soils were modeled either as nonlinear elastoplastic materials or linear materials with an equivalent stiffness. The analyses were performed using the finite element code DIANA-J.

Analyses in different horizontal directions (e.g., N-S, E-W, etc.) were carried out using a single horizontal component of a recorded motion. Acceleration time histories recorded at lower levels of the down-hole arrays were applied as base input motions to soil-column models composed of finite elements. The side boundaries of such soil-column model were specified to shear identical displacements thus enforcing a simple shear mode of deformation. Migration and dissipation of excess pore pressures were allowed both during and after the shaking.

3.2 *Port Island analysis*

Results from an analysis of a shallow ground model comprising of the reclaimed Masado layer and the original seabed layer of silty clay are discussed in the following. The northwest-southeast (NW-SE) component of the recorded accelerations at 32 m depth was used as a base excitation in the analysis. It is to be recalled that the NW-SE direction corresponds to the direction of maximum shaking intensity at this site (Figure 7).

Due to insufficient laboratory tests data for the Masado soils, the stress-strain and state parameters of the employed elastoplastic soil model were approximated with those of Toyoura sand with a relative density of $D_r = 60$ % whereas the dilatancy parameters were determined by simulating the liquefaction resistance curve shown with the solid line in Figure 4. The silty clay was modeled as a linear material with degraded shear modulus to 30 % from the initial modulus. This degree of stiffness reduction was estimated from the recorded accelerations based on the observed time lag in the wave propagation between the monitoring points at 16 and 32 m depth.

Comparison of the computed and recorded accelerograms at the ground surface and 16 m depth are shown in Figure 12 where a very good agreement between the measured and the analytical data may be seen. These results prove the analysis relevant for more detailed evaluation of the ground response.

A time history of computed excess pore pressures at 10.5 m depth of the Masado layer is displayed in Figure 13. This plot shows a typical pattern in the pore pressure build-up for the part of the fill deposit bellow 5 m depth. It may be seen in this figure that the initiation of strong acceleration pulses was accompanied with a sharp rise in the excess pore pressure which reached the initial vertical effective stress (σ'_v) after only one and a half to two cycles of

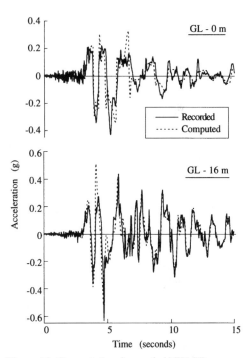

Figure 12. Computed and recorded NW-SE accelerograms at the ground surface and at 16 m depth (Port Island site)

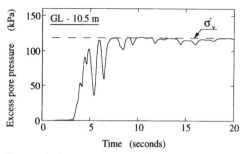

Figure 13. Computed excess pore pressures at 10.5 m depth (Port Island analysis)

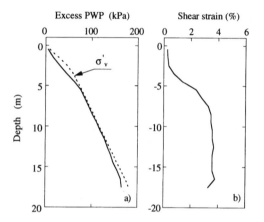

Figure 14. Distribution of the maximum excess pore pressures and shear strains along the depth of the Masado layer (Port Island analysis)

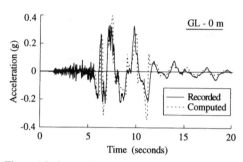

Figure 15. Comparison of computed and recorded accelerations at the ground surface (Higashi-Kobe site)

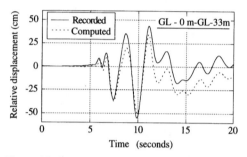

Figure 16. Comparison of computed and 'recorded' relative displacements between the ground surface and 33 m depth (Higashi-Kobe site)

strong shaking. Signs of cyclic mobility as indicated by the temporary drops in the excess pore pressures are also apparent in Figure 13. Distribution of the maximum excess pore pressures and shear strains in the fill deposit are presented in Figure 14. These results reveal that the reclaimed Masado layer completely liquefied at depths greater than 5 m and experienced maximum shear strains on the order of 3-4 %.

3.3 *Higashi-Kobe analysis*

The ground response computed in the analyses of Higashi-Kobe array site was very similar to that of the Port Island site, and therefore, the results for the HK-site are not presented in details. The results shown herein are obtained from an analysis performed by using the recorded N348E accelerogram at 33 m depth as a base input motion. Except for the input motion and small differences in the thicknesses of the layers, other features of the Higashi-Kobe analysis are identical to those of the Port Island analysis.

Figure 15 shows a comparison between the computed and the recorded accelerations at the ground surface, whereas Figure 16 shows comparison of relative displacements which is the difference in displacements between the ground surface and 33 m depth at each instant of time. Here the 'recorded' relative displacements actually were computed by a double integration of the recorded acceleration time histories. In both figures a reasonable degree of agreement between the computed and recorded data may be seen.

3.4 *Takasago analysis*

Laboratory test data on soils from the Takasago array site were not available for the analysis, and therefore, precise material modelling was not possible. All soils were model as nonlinear elastoplastic materials, and the effects of the excess pore pressures were considered only for the potentially liquefiable layer of silty sand at 3.5 to 8.5 m depth. The stress-strain curves of the soils were determined by simulating conventional G-γ relationships (degradation of the secant shear modulus with increasing shear strain) for sand, clay or gravel, depending on the prevalent soil fraction of each layer. The initial shear moduli were determined

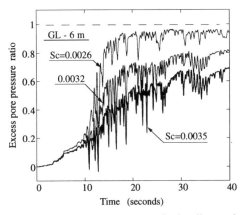

Figure 17. Excess pore pressures in the silty sand computed in analyses with different cyclic strength (different S_c -values)

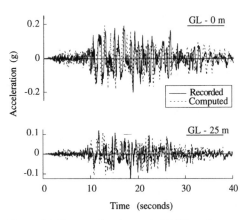

Figure 18. Computed and recorded E-W accelerograms at the ground surface and at 25 m depth (Takasago site)

using the shear wave velocities shown in Figure 10.

Since the cyclic strength curve of the liquefiable silty sand layer was not known, a parametric study was carried out to investigate the effects of the excess pore pressures on the ground response. In a series of analyses, a dilatancy parameter (S_c) of the employed constitutive soil model, which controls the pore pressure build-up and hence the liquefaction resistance, was varied so as to induce different levels of excess pore pressures in the silty sand layer. An example of the calculation is presented in Figure 17 where excess pore pressures at 6 m depth computed in three independent analyses with different S_c -values are shown. In addition to the analyses cases shown in Figure 17 a total stress analysis was conducted in which the excess pore pressures were not allowed.

It was found from the parametric study that the best simulation of the recorded accelerations is achieved for S_c = 0.0035 or when the maximum excess pore pressures in the silty sand layer are on the order of 60-75 % from the initial vertical effective stress (σ'_v). In the case of complete liquefaction and especially when no pore pressures were allowed (total stress analysis) notably lower level of agreement between the computed and recorded data was achieved. These results suggests that although relatively high excess pore pressures were induced in the sandy silt layer, complete or at least extensive liquefaction did not take place. In support of such indication is also the fact that the computed and 'recorded' relative displacements between the ground surface and 25 m depth at this site were about 4-5 cm or approximately ten times smaller than the corresponding relative displacements at the Port Island and Higashi-Kobe array sites.

Comparisons of computed and recorded data at the ground surface and 25 m depth are shown in

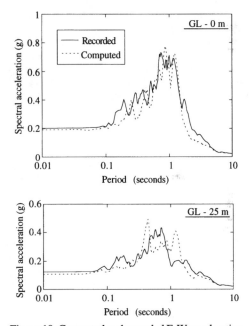

Figure 19. Computed and recorded E-W acceleration response spectra at the ground surface and at 25 m depth (Takasago site)

Figures 18 and 19 where E-W acceleration time histories and corresponding response spectra are shown respectively. The computed data shown in these figures is from the analysis in which the excess pore pressure for S_c = 0.0035 shown in Figure 17 was computed. The smaller level of agreement between the computed and recorded data obtained in this analysis is considered to be mostly due to the lack of experimental data for the soils at this site.

4 CONCLUSIONS

1. Most of the undensified fill deposits of Masado soils liquefied during the 1995 Kobe Earthquake. The liquefaction was accompanied with maximum shear strains of 3-4 % and relative displacements within the fill deposit of about 40-50 cm.

2. The ground response at the Takasago array site was affected by the high excess pore pressures that developed in the silty sand layer, however, complete and extensive liquefaction did not occur.

3. The applied effective stress analyses provided accurate simulations of the recorded ground motions and valuable additional information for the ground responses at the three array sites.

ACKNOWLEDGMENTS

The ground motion records at the Port Island and Takasago arrays were obtained by the Kobe-city and Kansai Electric Power Company respectively, and were provided through the Committee on Earthquake Observation and Research in the Kansai Area. The ground motion records at the Higashi-Kobe array were obtained and provided by the Civil Engineering Research Center of the Ministry of Construction.

REFERENCES

Cubrinovski, M. & K. Ishihara 1998. State concept and modified elastoplasticity for sand modelling. *Soils and Foundations* (to be published).

Hatanaka, M., Uchida, A. & J. Ohara 1997. Liquefaction characteristics of a gravelly fill liquefied during the 1995 Hyogo-Ken Nanbu Earthquake. *Soils and Foundations* 37(3):107-115.

Ishihara, K., Yasuda, S. & H. Nagase 1996. Soil characteristics and ground damage. *Soils and Foundations*, Special Issue January 1996: 109-118.

Ishihara, K., Yoshida, K. & M. Kato 1997. Characteristics of lateral spreading in liquefied deposits during the 1995 Hanshin-Awaji Earthquake. *Journal of Earthquake Engineering* 1(1): 23-55.

Sato, K., Kokusho, T., Matsumoto, M. & E. Yamada 1996. Nonlinear seismic response and soil property during strong motion. *Soils and Foundations*, Special Issue January 1996: 41-52.

Toki, K. (1995). Report of Committee on Earthquake Observation and Research in the Kansai Area. (in Japanese).

Geotechnical Hazards, Marić, Lisac & Szavits-Nossan (eds) © 1998 Taylor & Francis, ISBN 90 5410 957 2

Design of tunnel lining constructed in grouted soil for seismic effects

N. N. Fotieva & Y. I. Klimov
Department of Materials Mechanics, Tula State University, Russia

N. S. Bulychev
Department of Underground Construction, Tula State University, Russia

ABSRACT: The paper describes the method of designing tunnel linings of an arbitrary cross-section shape upon Earthquake seismic effects taking into account artificially strengthened soil zone around the opening. The method allows the most unfavourable stress state of the lining at any possible combinations and directions of long longitudinal and shear seismic waves propagating in the plane of tunnel cross-section to be determined.

1 INTRODUCTION

In tunnels being constructed in complicated hydro-geological conditions special techniques such as grouting (cementation) the surrounding soil mass are applied. Strengthening by grouting allows anisotropy of soils around a tunnel to be reduced and their deformation modulus to be raised. In this case a soil zone around the opening the mechanical properties of which are different from those of the rest massif is being created. This circumstance exerts a substantial influence on the lining stress state.

At present for designing concrete or Ferro-concrete tunnel linings of an arbitrary cross-section shape upon the actions of the soil's own weight and external water pressure taking into account the influence of grouted zone there are methods by Fotieva & Sammal (1993) and Fotieva, Bulychev & Sammal (1993) based on the analytical solutions of the corresponding plane contact problems for a double-layer non-circular ring in a linearly deformable medium. In the paper presented the method of designing tunnel linings upon the Earthquake seismic effects is described.

2. DESIGN METHOD

The design method developed is based on a general approach proposed in the book by Fotieva (1980) according to which the design consists in the determination of the most unfavourable lining stress state at any possible combinations and directions of long longitudinal (compression-tension) and shear seismic waves spreading in the plane of a tunnel cross-section.

With that aim two plane elasticity theory quasi-static contact problems are analysed, the design schemes of which are shown in Figure 1 a, b. Here the S_1 infinite medium simulating the soil mass is characterised by the E_1 deformation modulus and the v_1 Poisson ratio. The S_2 external ring layer of the Δ_1 thickness the material of which has the E_2 deformation modulus and the v_2 Poisson ratio simulates the soil zone strengthened by grouting. The S_3 internal ring layer of the Δ_2 thickness with the E_3 and v_3 characteristics simulates the tunnel lining.

The ring layers and the medium undergo deformation together, that is conditions of continuity of stresses and displacements vectors are fulfilled on the L_i ($i = 1, 2$) contact lines. The L_3 internal outline is free from loads.

In the first problem (Figure 1, a) the S_1 medium is subjected upon infinity to a double-axis compression with non-equal components directed under an arbitrary α angle to the co-ordinates axis simulating the action of long arbitrary directed longitudinal wave in the compression phase. The stresses on the infinity are expressed by formulae:

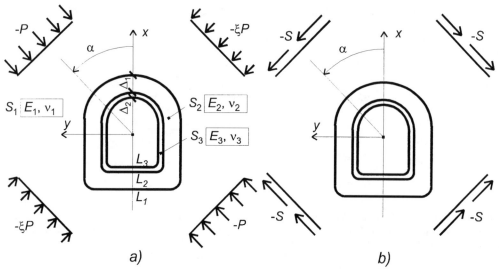

Figure 1. The design scheme of the tunnel lining undergoing the action of longitudinal (a) and shear (b) waves.

$$P = \frac{1}{2\pi} A k_1 \gamma c_1 T_0, \quad \xi = \frac{\nu_1}{1 - \nu_1} \quad (1)$$

where A = the coefficient corresponding to tne Earthquake intensity, k_1 = the coefficient taking admissible damages into account, γ = the soil unit weight, c_1 = the speed of longitudinal waves spreading, T_0 = the prevailing period of the soil particles oscillations.

In the second problem (Figure 1, b) the S_1 medium is subjected to a pure shear upon infinity simulating the action of a long arbitrary directed shear wave. Here

$$S = \frac{1}{2\pi} A k_1 \gamma c_2 T_0, \quad (2)$$

where c_2 = the speed of shear waves spreading.

From the solution of the first problem (Fig. 1, a) the $\sigma^{(P)}$ stresses (here the σ symbol signifies all components of stress tensor), appearing in the lining due to the action of a long longitudinal wave falling at an α arbitrary angle are determined; from the solution of the second problem (Fig. 1, b) the $\sigma^{(S)}$ stresses called forth by a shear wave are obtained.

Further, the sum and the difference of general expressions for $\sigma_\theta^{(P)}$ and $\sigma_\theta^{(S)}$ normal tangential stresses characterising the lining stress state

caused by mutual actions of longitudinal and shear waves passing simultaneously (the worst case) are investigated in every point of the L_3 internal outline on the extremuma relatively the α angle of the waves falling. With this aim the following equations are solved

$$\frac{\partial}{\partial \alpha} [\sigma_\theta^{(P)} \pm \sigma_\theta^{(S)}] = 0 \quad (3)$$

and for every point such a combination of waves and such an angle of their falling are determined at which normal tangential stresses in the point considered are maximal by their absolute. It allows the envelope diagram of normal tangential stresses on the L_3 internal outline to by obtained analytically.

The stresses upon the external L_2 outline, the N longitudinal forces and the M bending moments in every lining normal section are determined at such a combination and such a direction of waves at which the σ_θ normal tangential stress in that section has a maximal absolute value.

The stresses and forces obtained by that way are assumed to have the signs "plus" and "minus" and summed up with stresses and forces appearing due to other acting loads in their most unfavourable combinations. After that a sections strength test upon compression and tension is made.

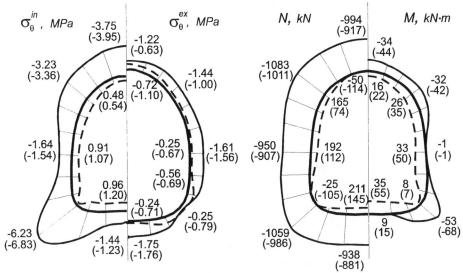

Figure 2. The results of the tunnel lining design.

If the lining is not anchored to the massif and is designed with an allowance of fissure forming we assume that the tensile normal loads are not transferred upon the lining. In this case the action of the longitudinal waves in the tension phase is not to be taken into account and the design is made on the base of two different envelope diagrams of normal tangential stresses, obtained using the maximal absolute values of the compressive (negative) stresses and tensile (positive) ones, called forth by mutual actions of shear waves and longitudinal waves in the compression phase.

The analytical solutions of the elasticity theory plane contact problems (Fig. 1, a, b) have been obtained by Klimov (1991) with the application of the complex variable analytic functions theory using the apparatus of conform mapping and complex series   (Muskhelishvili, 1966).

The computer program for designing tunnel linings constructed with the application of soil grouting undergoing Earthquake's effects has been elaborated.

3. EXAMPLES OF THE DESIGN

Results of designing the railway tunnel are given in Figure 2. The shape and sizes of the lining cross-section are shown in Figure 3.

Calculations were made at the following input data:

$\Delta_1 = 4\ m,\ \ \Delta_2 = 0.4\ m,\ \ E_1 = 1000\ \ MPa,\ \ \nu_1 = 0.3,$

$E_2 = 1800\ MPa,\ \ \nu_2 = 0.3,\ \ E_3 = 23000\ MPa,$

$\nu_3 = 0.2,\ \ \gamma = 23\ kN\,/\,m^3,\ \ A = 0.4,\ \ k_1 = 0.25,$

$T_0 = 0.5\ s.$

The lining is designed with the allowance of fissures.

The distributions of maximal compressive and tensile σ_θ^{in} normal tangential stresses on the internal outline of the lining cross-section, corresponding them σ_θ^{ex} normal tangential stresses on the external outline, the N longitudinal forces and the M bending moments are shown in Figure 2 by solid and dotted lines correspondingly.

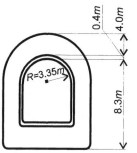

Figure 3. The lining cross-section.

For the comparison values of the same stresses and forces obtained in the case when the grouted soil zone is absent are given in brackets.

In conclusion we can mark that the approach described based on evaluating the most unfavourable lining stress state allows to increase the safety factor of tunnels being constructed in seismic regions.

REFERENCES

Fotieva, N. N. & A. S. Sammal 1993. Design of linings of large cross - section tunnels undergoing construction in rock. *Safety and Environmental Issues in Rock Engineering. Int. Symp. EUROCK'93*: 545 - 549. Lisbon, Portugal.

Fotieva, N.N., Bulychev, N.S., A.S.Sammal 1993. Design of tunnel linings constructed in weak water bearing rocks improved by cement grouting. *Int. Sympozium on Geotechnical Engineering of Hard Soil - Soft Rocks:* 1439-1444. Athene, Greece.

Fotieva, N.N., 1980. *Design of Underground Structures Support in Seismic Active Regions.* Moscow, Nedra.

Klimov, Y.I., 1991. Design of tunnel linings upon the Earthquake effects taking into account the cementation of rocks. Underground Structures Mechanics, Tula.

Muskhelishvili, N.I., 1966. *Some Basic Problems of Mathematical Elasticity Theory.* Moscow, Nauka.

Geotechnical Hazards, Marić, Lisac & Szavits-Nossan (eds) © 1998 Taylor & Francis, ISBN 90 5410 957 2

Liquefaction hazard evaluation by Swedish weight sounding test

J. Kuwano – *Department of Civil Engineering, Tokyo Institute of Technology, Japan*

K. Ogawa – *Fudo Corporation, Tokyo, Japan (Formerly: Science University of Tokyo), Japan*

T. Kimura – *Kurita Corporation, Tokyo, Japan (Formerly: Science University of Tokyo), Japan*

H. Aoki – *Kajima Corporation, Tokyo, Japan (Formerly: Science University of Tokyo), Japan*

ABSTRACT: The paper describes the liquefaction hazard evaluation by the Swedish weight sounding test which seems to be suitable for housing sites because this penetration test method has the advantages of portability, easy operation and enough penetration ability. The Swedish weight sounding tests were carried out in the model grounds as well as in the fields where liquefaction damages were observed during the past earthquakes. A chart was prepared to discriminate between occurrence and non-occurrence of ground rupturing due to liquefaction. Results of the evaluation showed reasonably good agreement with the damage of the site.

1 INTRODUCTION

In seismic active regions such as Japan, strong earthquakes have occurred repeatedly causing heavy damages. Liquefaction is one of the hazards which causes damage of wooden houses. Methods of liquefaction potential evaluation widely used so far incorporate standard penetration and/or cone penetration test results at levels of up to 20 to 30 m below the ground surface (e.g. Seed & Idriss 1971, Iwasaki et al. 1978, Tokimatsu & Yoshimi 1983). Although they have been used successfully for the liquefaction potential evaluation into the deep foundation ground of big structures such as bridges and oil tanks, they are not necessarily suitable for housing sites. This is because in the case of housing the main concern is the shallow subsurface layers and several points are needed to be investigated for better evaluation of the damage of each house. The paper describes the liquefaction hazard evaluation by the Swedish weight sounding test which seems to be suitable for housing sites because this penetration test method has the advantages of portability, easy operation and enough penetration ability.

2 SITE INVESTIGATION

Liquefaction hazards of a residential quarter in Wakami-cho of Akita Prefecture, which suffered from extensive liquefaction damage by the 1983 Nihonkai-chubu earthquake, were evaluated by the simplified method as explained in the following. Wakami-cho is the town located along the west shore of Lake Hachirogata in the north west of Japan main island. Liquefaction evaluation was also carried out in Azuma-mura, the village of Ibaraki Prefecture, which was damaged by the 1987 Chibaken-toho-oki earthquake. Azuma-mura is located on the left bank of the lower course of the Tone River.

2.1 Microtopography of investigated sites

It is known that the sites of certain microtopography such as former river channel is vulnerable to liquefaction (Wakamatsu 1991). Hence the geomorphological classification map was used for the seismic microzonation (Kotoda et al. 1988). The standard of liquefaction vulnerability based on microtopography was proposed by National Land Development Technology Center (NLDTC) of Japan in 1991 as summarized in Table 1.

Table 1. Liquefaction vulnerability classification based on microtopography (NLDTC 1991)

Liquefaction vulnerability	Microtopographical unit
high	former river channel or pond, reclaimed land, margin of natural levee, lowland between sand dunes, gentle slope at the margin of sand dune, riverbed with fine sand, fill with high water table
medium	delta, natural levee, back swamp, fan with gentle slope, delta, valley plain, reclaimed land by drainage
low	sand dune, fan, gravelly riverbed

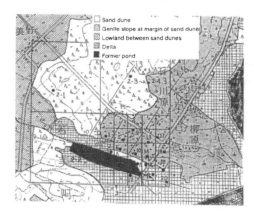

Figure 1. Microtopography of Wakami-cho

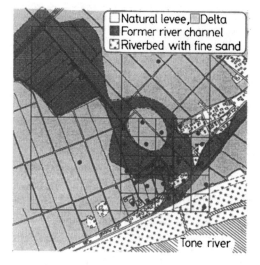

Figure 2. Microtopography of Azuma-mura

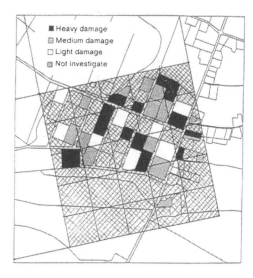

Figure 3. Damage of houses caused by liquefaction (Wakami-cho)

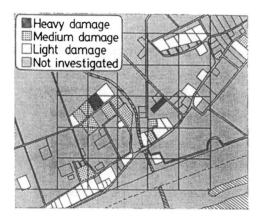

Figure 4. Damage of houses caused by liquefaction (Azuma-mura)

The investigated area in Wakami-cho consists of sand dune, former pond etc. as shown in Figure 1. According to the standard of NLDTC, the liquefaction potential is high in former pond, gentle slope at margin of sand dune and lowland between sand dunes. It is medium in delta and low in sand dune.

The Azuma-mura investigation area is covered with natural levee, former river channel and so on as shown in Figure 2. Liquefaction potential would be high in the former river channel and riverbed with fine sand. It could be medium in natural levee and delta.

2.2 Damage of houses caused by liquefaction

Damage of houses in the areas were investigated through interviews with the residents in terms of settlement, tilting and structural damage of building which were graded into 0 to 3. The extent of damage of the houses, which were mostly small wooden houses of two stories at most, was assessed based on the information thus obtained. Remedial works seemed to reflect the extent of the damage in general. The damage of houses caused by liquefaction is summarized in Figures 3 and 4.

2.3 In-situ test

The Swedish weight sounding penetrometer (Broms & Bergdahl 1982) consists of a screw point, rods and

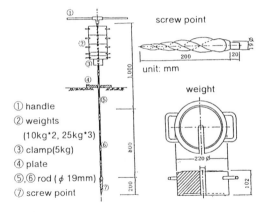

① handle
② weights
   (10kg*2, 25kg*3)
③ clamp(5kg)
④ plate
⑤,⑥ rod (φ 19mm)
⑦ screw point

Figure 5. Swedish weight sounding test equipment

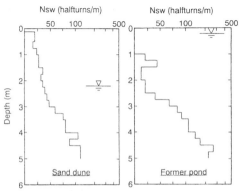

Figure 6. Examples of the penetration test results (Wakami-cho)

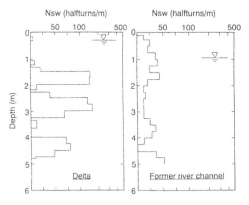

Figure 7. Examples of the penetration test results (Azuma-mura)

weights (5, 10, 10, 25, 25, 25 kg) as shown in Figure 5. The load on the penetrometer is gradually increased to 1 kN (100 kgf) by increasing the number

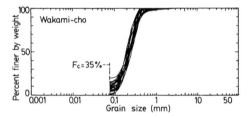

Figure 8. Grain size distribution curve (Wakami-cho)

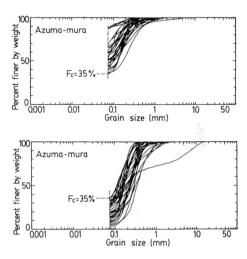

Figure 9. Grain size distribution curve (Azuma-mura)

of weights. When the penetrometer does not penetrate any further with the load of 1 kN, it is rotated and the number of half turns per 0.25 m of penetration is recorded and reported as N_{sw} (halfturns/m). The main advantages of the Swedish weight sounding test are portability, easy operation and enough penetration ability.

The in-situ tests were carried out at the points indicated by black dots in Figures 1 and 2 where the size of small mesh is 100 m * 100 m. The ground water levels were also measured at the respective points. Soil samples of about 50 g to 100 g were taken from the various depths of sounding pits by using a plate spring sampler. They were used for grain size analyses.

Examples of the Swedish weight sounding test results are shown in Figures 6 and 7 for Wakami-cho and Azuma-mura respectively. The ground water levels are also indicated in these figures. Figures 8 and 9 shows the grain size distribution curves obtained in these sites. As seen in Figure 6, the penetration resistance, N_{sw}, generally increased with depth in the clean sand deposits as in Wakami-cho. However, the penetrometer sometimes penetrated without rotation (N_{sw} =0) at certain depth in Azuma-

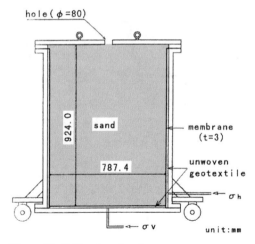

Figure 10. Calibration chamber

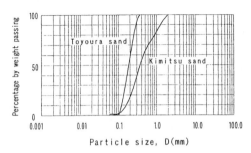

Figure 11. Grain size distribution of Kimitsu sand

mura. Cohesive soil was often found to be attached to the rod. It was also found that the soil samples obtained from the depths of $N_{sw} = 0$ mostly contained fine-grained soils (<75μm) more than about 35%. Therefore, the ground which showed $N_{sw} = 0$ was estimated to be composed of cohesive soils which have low liquefaction potential, even if the gradation of the soil was not obtained from the procured sample (NLDTC 1991, Kuwano et al. 1992, Kuwano 1995).

3 LABORATORY INVESTIGATION

Calibration chamber tests were carried out to investigate the relationships between the penetration resistance, N_{sw}, and the properties of the model grounds.

The calibration chamber used in this study is shown in Figure 10. It is a steel tank with the inner diameter of about 800 mm and the inner height of about 930 mm. The inside of the tank is covered with rubber membranes. Nonwoven geotextiles are inserted between the tank and the membrane and they

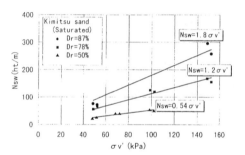

Figure 12. Variation of N_{sw} with σ_v'

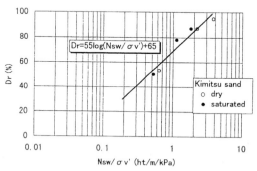

Figure 13. Relationship between Dr and N_{sw} / σ_v'

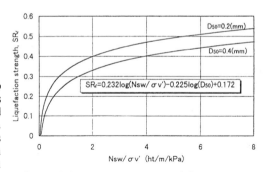

Figure 14. SR_l versus N_{sw} / σ_v' and D_{50}

are saturated. Both vertical and horizontal stresses can be applied to the model ground independently. Ko condition can be achieved by closing the valve of the side pressure bag. In this study, Ko condition was selected to simulate the level ground.

The model ground of Kimitsu sand with Dr of 50 to 95% was made by air pluviation. The grain size distribution curve of the sand is shown in Figure 11. It seems to be similar to that of Wakami-cho. The model ground was dry or saturated. The Swedish weight sounding tests were performed under the effective vertical stress of 49, 98 and 147 kPa.

Relationships between N_{sw} and σ_v' is shown in

Figure 12 for the saturated Kimitsu sand with Dr of 50, 78, 87%. It is seen that Nsw increases linearly with σ_v' for the respective relative density. From the slopes of the lines obtained as above, Figure 13 was made to correlate Dr (%) with Nsw / σ_v' (halfturn/m/ kPa). Data for the dry sand are also plotted in the same figure. Whether the model ground was saturated or dry, the following relationship was obtained:

$$Dr = 55\log(Nsw/\sigma_v') + 65 \qquad (1)$$

Tatsuoka et al. (1980) proposed the following equation to evaluate the liquefaction strength of sand.

$$SR_l = 0.0042Dr - 0.225\log(D_{50}/0.35)$$

$$\text{for } 0.04 \le D_{50} \le 0.6 \text{ mm} \qquad (2)$$

in which $SR_l = (\sigma_d/2\sigma_c')_{Nc=20}$ is the stress ratio to cause liquefaction at the number of loading cycles of 20. Substitution of equation (1) into equation (2) gives

$$SR_l = 0.232\log(Nsw/\sigma_v') - 0.225\log D_{50} + 0.172 \quad (3)$$

By using this equation, liquefaction strength of sand can be evaluated from the penetration resistance of the Swedish weight sounding test. The relationships of equation (3) is shown in Figure 14 for D_{50} of 0.2 mm and 0.4 mm respectively.

4 LIQUEFACTION POTENTIAL EVALUATION

Liquefaction potentials were evaluated for the investigated sites in Wakami-cho and Azuma-mura based on the site investigation data and equation (3).

The liquefaction resistance factor, F_L, is given as follows (Tatsuoka et al. 1980):

$$F_L = SR_l / L \qquad (4)$$

in which L is the maximum stress ratio during an earthquake motion with the maximum ground surface acceleration, $\alpha_{max}^{\cdot}$. L is determined by the following equation:

$$L = \left(\frac{\tau}{\sigma_v'}\right)_{max} = \frac{\alpha_{max}}{g} \cdot \frac{\sigma_v}{\sigma_v'} \cdot (1 - 0.015z) \qquad (5)$$

α_{max} was estimated roughly to be 200 gal for both sites in this research. Examples of F_L evaluations are shown in Figures 15 and 16 for Point Nos. 2-1 and 2-2 in Wakami-cho site. It is seen in Figure 15 that the values of F_L are greater than 1.0 throughout the depth below the ground water level at Point No. 2-1 where the liquefaction damage was not found. On the other hand, as seen in Figure 16, F_L is less than or

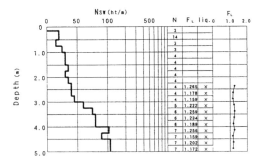

Figure 15. Variation of F_L with depth at the site of no liquefaction damage (Wakami-cho, No. 2-1)

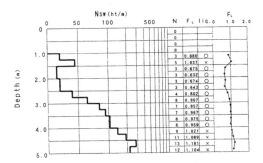

Figure 16. Variation of F_L with depth at the site of liquefaction damage (Wakami-cho, No. 2-2)

close to 1.0 at Point No. 2-2 where the liquefaction damage was observed.

From the ground water level, the gradation of soil and the penetration resistance, Nsw, the thickness of the surface unliquefiable layer, H_1, was determined. As the penetration depth was up to 5 m, the thickness of liquefiable layer, H_2, was simply taken as $H_2=5-H_1$ (m). The data obtained as above are plotted in Figures 17 and 18 for two investigated sites. The curve drawn in the figure is the boundary which discriminates between occurrence and non-occurrence of ground rupturing due to liquefaction (Ishihara 1985). Although 17 points out of 21 points of Wakami-cho are properly plotted in the figure in terms of the liquefaction damage, only 20 points out of 28 investigation points of Azuma-mura are in the proper area in the figure maybe due to the more complicated geological conditions as compared with Wakami-cho. However, the results of the evaluations seem to show reasonably good agreement with the actual liquefaction damage of each point in the sites for the first estimation of liquefaction hazards in housing sites.

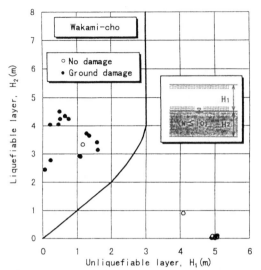

Figure 17. The thickness of the surface unliquefiable layer, H_1, versus the underlying liquefiable layer, H_2 (Wakami-cho)

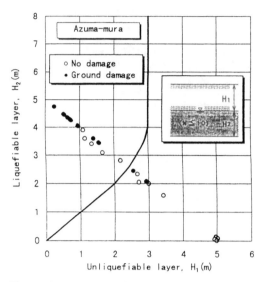

Figure 18. H_1 versus H_2 (Azuma-mura)

5 CONCLUSIONS

The investigations were made for the liquefaction hazard evaluation by the Swedish weight sounding test. The following results were obtained.

1. The following relationship among the relative density, Dr (%), Nsw (ht/m) and the effective overburden pressure, σ_v' (kPa) was obtained from the calibration chamber test results:

$$Dr=55log(Nsw \, / \, \sigma_v')+65$$

2. Liquefaction strength of sand deposit is connected with $Nsw \, / \, \sigma_v'$ and D_{50} (mm) as follows:

$$SR_l=0.232log(Nsw/\sigma_v')-0.225log(D_{50})+0.172$$

3. A chart was prepared to discriminate between occurrence and non-occurrence of ground rupturing due to liquefaction. Results of the evaluation showed reasonably good agreement with the actual liquefaction damage of the sites.

REFERENCES

Broms, B.B. & U. Bergdal 1982. The weight sounding test (WST): State-of-the-art-report. *Proc. 2nd European Symposium on Penetration Testing, Amsterdam*: 203-213.

Ishihara, K. 1985. Stability of natural deposits during earthquakes. *Proc. 11th ICSMFE, San Francisco*, 1: 321-376.

Iwasaki, T., F. Tatsuoka, K. Tokida & S. Yasuda 1978. A practical method for assessing soil liquefaction potential based on case studies at various sites in Japan. *Proc. 2nd Int. Conf. on Microzonation, San Francisco*, 2: 885-896.

Kotoda, K., K. Wakamatsu & S. Midorikawa 1988. Seismic microzoning on soil liquefaction potential based on geomorphological land classification. *Soils and Foundations*, 28(2): 127-143.

Kuwano, J., K. Ozaki & S. Nakamura 1992. Simplified method of liquefaction hazard evaluation for subsurface layers. *Proc. 10th World Conf. on Earthquake Eng., Madrid*, 3: 1431-1434.

Kuwano, J. 1995. Simple method of liquefaction hazard evaluation for houses. *Proc. Infrastructure Development in Civil Eng., Bangkok*: 289-298.

National Land Development Technology Center 1991. *Research report on countermeasures to liquefaction damage of small buildings utilizing liquefaction potential maps*. Tokyo: NLDTC.

Seed, H.B. & I.M. Idriss 1971. Simplified procedure for evaluating soil liquefaction potential. *Proc. ASCE*, 97(SM9): 1249-1273.

Tatsuoka, F., Iwasaki, T., Tokida, K., Yasuda, S., Hirose, M., Imai, T. & Kon-no, M. 1980. Standard penetration tests and soil liquefaction potential evaluation. Soils and Foundations, 20(4): 95-111.

Tokimatsu, K. & Y. Yoshimi 1983. Empirical correlation of soil liquefaction based on SPT N-value and fine content. *Soils and Foundations*, 23(4): 56-74.

Wakamatsu, K. 1991. Microtopographic map of Wakami-cho and Azuma-mura (unpublished).

Wakamatsu, K. 1991. *Maps for historic liquefaction sites in Japan*. Tokyo: Tokai University Press.

Geotechnical Hazards, Marić, Lisac & Szavits-Nossan (eds) © 1998 Taylor & Francis, ISBN 90 5410 957 2

An iterative method for 2D time harmonic elastodynamics in infinite domains

M. Premrov, A. Umek & I. Špacapan
Faculty of Civil Engineering, University of Maribor, Slovenia

ABSTRACT: A new method for solving the time harmonic elastodynamic problem in infinite domains is presented in this paper. The goal is to satisfy a radiation condition, which asserts that at infinity all waves are outgoing. Satisfying the radiation condition is thus one of the basic properties of each good computational soil model. In the presented method an iterative solution of this problem is obtained. An infinite domain is first truncated by introducing an artificial finite boundary (β), on which some boundary conditions must be imposed. The finite computational domain in each iteration is subjected to actual boundary conditions and with different (Dirichlet or Neumann) fictive boundary conditions on β.

1 INTRODUCTION

In solving dynamic soil-structure interaction problems the main problem is computing a dynamic stiffness of a soil. The Soil is namely in a measure of a structure of large or infinite domain and thus can be treated as a halfspace. The basic problem of elastodynamics in infinite domains is to satisfy Sommerfeld radiation condition - the boundary condition at infinity. This asserts that at infinity all waves are outgoing and no energy is radiated from infinity towards the structure. A radiation condition is satisfied automatically as a part of the fundamental solution in the boundary element method. Unfortunately, the fundamental solution is not always available. Although the boundary-element method is regarded as the most powerful procedure for modeling the unbounded medium, it requires a strong analytical and numerical background.

More flexible is the finite element method. The infinite domain is first truncated by introducing an artificial finite boundary (β). On this boundary some boundary conditions must now be imposed. This is the critical step because they must totally eliminate all reflected waves and they must be simple enough. The simplest and most usual boundary condition is the classical 'plane-wave' (PW) damper:

$$u_n(x) = \frac{\partial u}{\partial n}(x) = i\,k\,u(x), \quad on\ \beta. \tag{1.1}$$

Here $u(x)$ is the unknown scattered field, n is the outward normal and k is the wave number. This condition is of the same form as the Sommerfeld radiation condition, which is exactly correct when imposed at infinity but only approximately correct when imposed at a finite boundary β. As a consequence the use of (1.1) leads to the spurious reflection of waves from β.

In order to diminish the spurious reflection, various authors have devised improved local boundary conditions on β. Engquist & Majda 1977 have expressed u_n exactly as a pseudodifferential operator applied to u on β and then approximating this operator by the local differential operators. Bayliss & Turkel 1980 used the asymptotic expansion of u far from the scattered one to obtain similar approximate local boundary conditions. Feng 1983 obtained an exact non-local condition involving an integral over β of u multiplied by a Green's function. Then he approximated it by various local conditions. However, all these local boundary conditions still lead to spurious reflection. Keller 1989 and Givoli & Keller 1990 obtained exact non-reflecting boundary conditions on β, which totally eliminate all reflections. Porat & Givoli 1995 did the same for elliptic artificial boundary β. Pinsky & Thompson 1992 used the approximate local boundary conditions of Bayliss & Turkel in the finite element formulation for the two-dimensional time-dependent structural acoustic problem. The same did Thompson & Pinsky 1993 for three-dimensional problem. Aiello et al. 1994 presented a new iterative procedure for solving electrostatic problems in infinite domains. Different Dirichlet and Neumann boundary conditions are obtained by using the Green's function for a treated problem.

2 ITERATIVE PROCEDURE FOR SOLVING EXTERIOR PROBLEMS

In the presented method an iterative solution for solving elastodynamics in infinite domain is obtained. The infinite domain, which represents exterior boundary value problem, is first truncated by introducing an artificial finite boundary (β). So the finite computational domain Ω_f - interior boundary value problem (bounded with obstacle-actual boundary Γ and artificial boundary β) is obtained - Figure 1.

a)

b)

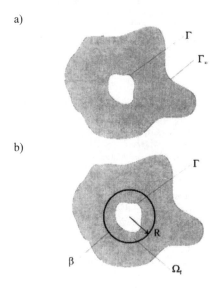

Figure 1. a) Exterior boundary value problem.
b) Interior boundary value problem.

The halfspace outside of the artificial boundary can be represented with DtN (Dirichlet to Neumann) operator (S_N):

$$\frac{\partial u}{\partial n} = S_N \cdot u \quad on \ \beta. \tag{2.1}$$

DtN operator can be in a non-local or local form and can be obtained from Bayliss & Turkel. The finite computational domain Ω_f is subjected in each iteration to actual boundary conditions on Γ and to different fictive boundary conditions on β. The wave equation in each odd iteration is thus:

$$\nabla^2 u + k^2 u = 0 \ ;$$
$$u = u_0 \ on \ \Gamma, \tag{2.2}$$
$$u = DBC \ on \ \beta$$

where DBC represent fictive Dirichlet boundary conditions on artificial boundary. Tractions on β as a result of Dirichlet boundary conditions must now be computed. The solution is in the form:

$$\frac{\partial u}{\partial n} = \sum_{m=0}^{\infty} A_{1m} \cdot H_m^{(1)}(kR) + A_{2m} \cdot H_m^{(2)}(kR) \tag{2.3}$$

where constant A_{1m} represents the amplitude of reflecting waves and A_{2m} the amplitude of outgoing waves. The wave equation in each even iteration is in the form:

$$\nabla^2 u + k^2 u = 0 \ ;$$
$$u = u_0 \ on \ \Gamma, \tag{2.4}$$
$$\frac{\partial u}{\partial n} = NBC \ on \ \beta$$

where NBC represent fictive Neumann boundary conditions on artificial boundary. Displacements on β as a result of Neumann boundary conditions must now be computed. The solution is in the form:

$$u = \sum_{m=0}^{\infty} B_{1m} \cdot H_m^{(1)}(kR) + B_{2m} \cdot H_m^{(2)}(kR). \tag{2.5}$$

The line $y^{(i)}$ which connects the two obtained values is then projected on the DtN operator (S_N), which represents the halfspace outside of the artificial boundary. The new Dirichlet and Neumann fictive boundary conditions on β for the next two iterations are so obtained (Figure 2).

$\partial u/\partial n$ on β

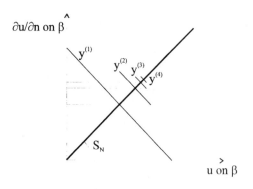

u on β

Figure 2. Iterative procedure for determining new fictive boundary conditions.

344

3 AXISYMMETRIC PROBLEM

Consider an axial symmetric problem with a hole on which constant Dirichlet boundary conditions (u_0) are prescribed (Figure 3).

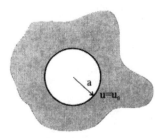

Figure 3. Halfspace with a hole.

3.1 *Exact solution*

The exact solution is well known:

$$u(r) = u_0 \cdot \frac{H_0^{(2)}(kr)}{H_0^{(2)}(ka)} \tag{3.1}$$

3.2 *Approximate solution as an interior boundary value problem*

We solve the problem with the presented method as an interior boundary value problem. By using the exact non-local DtN operator:

$$DtN_{ex} = -k \cdot \frac{H_1^{(2)}(kR)}{H_0^{(2)}(kR)} \tag{3.2}$$

the solution after the first two iterations is in the form:

$$u = u_0 \frac{J_0(kR) - i \cdot Y_0(kR)}{J_0(ka) - i \cdot Y_0(ka)} = u_0 \frac{H_0(kR)}{H_0(ka)} \tag{3.3}$$

This is the exact solution. A very important fact is that this solution is independent of the location of the artificial boundary (R).

By using asymptotic local operators from Bayliss & Turkel 1980 spurious reflections are obtained. The operators are in the form:

$$S_{3/4} = -ik \ , \tag{3.4}$$

$$S_1 = -ik - \frac{1}{2R} \ , \tag{3.5}$$

$$S_2 = -\frac{1}{2R} - ik + \frac{1 - ikR}{8R\left(1 + k^2 R^2\right)} \ , \tag{3.6}$$

$$S_3 = -ik - \frac{1}{2R} + \frac{2}{R}\left(\frac{46 - i73kR + 16k^2 R^2 - i16k^3 R^3}{529 + 1568k^2 R^2 + 256k^4 R^4}\right) \tag{3.7}$$

For the case a=1.0 and k=1.0 the absolute values of the constant A_1 which represents the amplitude of reflecting waves are presented in Table 1.

Table 1. Absolute values of the constant A_1 (after 2. iteration).

R/a	$S_{3/4}$	S_1	S_2	S_3
1.0	0.1805	0.1321	0.0882	0.0940
1.5	0.2266	0.0277	0.0017	0.0023
2.0	0.1698	0.0165	0.0015	0.0014
3.0	0.0989	0.0071	0.0015	0.0012
4.0	0.0930	0.0136	0.0014	0.0014
6.0	0.0526	0.0012	0.0011	0.0011
10.0	0.0347	0.0028	0.0016	0.0012

4 NONSYMMETRIC PROBLEM

Consider a nonsymmetric problem with a hole on which Dirichlet boundary conditions are prescribed in the form:

$$u(ka) = u_0 \cdot cos(n\varphi), \ n = 0,1,... \tag{4.1}$$

4.1 *Exact solution*

The exact solution is well known and it is in the form of an infinite series:

$$u(kr) = u_0 \cdot \sum_{n=0}^{\infty} \frac{H_n^{(2)}(kr)}{H_n^{(2)}(ka)} cos(n \cdot \varphi) \tag{4.2}$$

The infinite series in (4.2) can be replaced with a finite one:

$$u(kr) = u_0 \cdot \sum_{n=0}^{N-1} \frac{H_n^{(2)}(kr)}{H_n^{(2)}(ka)} cos(n \cdot \varphi) \tag{4.3}$$

where N is the maximal number of harmonics. For tractions the exact solution is in the form:

$$\frac{\partial u}{\partial r} = u_0 \cdot \sum_{n=0}^{N-1} \frac{\left[H_n^{(2)}(kr)\right]'}{H_n^{(2)}(ka)} cos(n \cdot \varphi) \tag{4.4}$$

$\mathrm{Re}\ \dfrac{\partial u}{\partial r}$

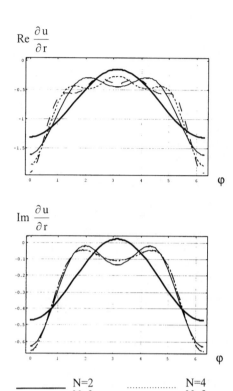

$\mathrm{Im}\ \dfrac{\partial u}{\partial r}$

| ———————— | N=2 | ⋯⋯⋯⋯ | N=4 |
| ———————— | N=3 | --------- | N=5 |

Figure 4. Real and imaginary part of tractions.

The convergence of the solution by using the first N terms is good. Solutions for a=1.0 and R=2.0 are presented in Figure 4.

4.2 Approximate solution as an interior boundary value problem

We solve the problem with the presented method as an interior boundary value problem. For prescribed Dirichlet boundary conditions on r=R the general solution for tractions is in the form:

$$\frac{\partial u}{\partial r} = \sum_{n=0}^{N-1} \left(\begin{array}{c} A_{1n} \cdot \left[H_n^{(1)}(kr) \right]' + \\ + A_{2n} \cdot \left[H_n^{(2)}(kr) \right]' \end{array} \right) \cdot \cos(n\varphi) \qquad (4.5)$$

where A_{1n} is the amplitude of reflected waves and A_{2n} the amplitude of outgoing waves. For prescribed Neumann boundary conditions on r=R the general solution for displacements is:

$$u(r) = \sum_{n=0}^{N-1} \left(B_{1n} \cdot H_n^{(1)}(kr) + B_{2n} \cdot H_n^{(2)}(kr) \right) \cdot \cos(n\varphi)$$

$$(4.6)$$

4.3 Numerical example for N=2

Consider a=1.0, R=2.0 and k=1.0. The solution for tractions after the third iteration with using the $S_{3/4}$ operator is presented in Figure 5.

For all numerical examples it will be denoted
———————— exact solution,
———————— approximate solution.

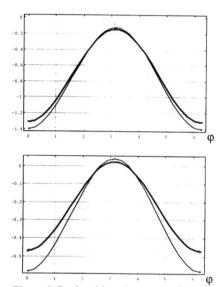

Figure 5. Real and imaginary part of tractions.

Table 2. Absolute values of constants A_{10}, $A_{1.1}$, $B_{1.0}$ and $B_{1.1}$ by using $S_{3/4}$ operator.

	$A_{1,0}$	$A_{1,1}$	$B_{1,0}$	$B_{1,1}$
2.iterat.	0.1560	0.2287	0.1713	0.2089
3.iterat.	0.1687	0.1930	0.1666	0.1873
4.iterat.	0.1686	0.1901	0.1661	0.1814
5.iterat.	0.1702	0.1921	0.1619	0.1768

By using S_1 the results are better. The amplitudes of reflected waves for the first two harmonics are presented in Table 3.

Table 3. Absolute values of constants A_{10}, $A_{1.1}$, $B_{1.0}$ and $B_{1.1}$ by using S_1 operator.

	$A_{1,0}$	$A_{1.1}$	$B_{1,0}$	$B_{1,1}$
2.iterat.	0.0211	0.0655	0.0520	0.1015
3.iterat.	0.0162	0.0496	0.0228	0.0450
4.iterat.	0.0156	0.0512	0.0249	0.0373
5.iterat.	0.0155	0.0515	0.0209	0.0419

The solution for tractions after the projection on S_1 operator after the first iteration is presented in Figure 6.

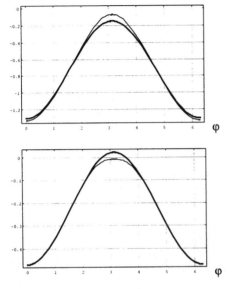

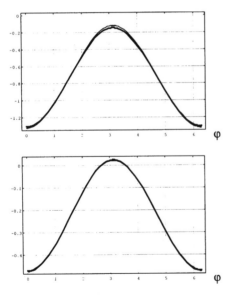

Figure 6. Real and imaginary part of tractions after first iteration.

The amplitudes of reflected waves by using S_2 and S_3 operators for the first two harmonics are presented in Tables 4 and 5. The both operators are in the forms (3.6) and (3.7) independent of the tangential coordinate (φ) and thus simple enough.

Table 4. Absolute values of constants $A_{1.0}$, $A_{1.1}$, $B_{1.0}$ and $B_{1.1}$ by using S_2 operator.

	$A_{1.0}$	$A_{1.1}$	$B_{1.0}$	$B_{1.1}$
2.iterat.	0.0106	0.0804	0.0347	0.1159
3.iterat.	0.0018	0.0653	0.0049	0.0606
4.iterat.	0.0025	0.0667	0.0068	0.0532
5.iterat.	0.0025	0.0672	0.0041	0.0572

Table 5. Absolute values of constants $A_{1.0}$, $A_{1.1}$, $B_{1.0}$ and $B_{1.1}$ by using S_3 operator.

	$A_{1.0}$	$A_{1.1}$	$B_{1.0}$	$B_{1.1}$
2.iterat.	0.0120	0.0794	0.0348	0.1156
3.iterat.	0.0022	0.0647	0.0057	0.0601
4.iterat.	0.0029	0.0660	0.0078	0.0527
5.iterat.	0.0021	0.0660	0.0056	0.0568

It can be seen from Table 4 and Table 5 that there is practically no difference in results between the operators S_3 and S_2. When using the operator S_1 the results are a little worse. The results obtained with the simplest operator $S_{3/4}$ are not good for this distance of the artificial boundary. It can be also perceived that there is practically no difference in results between the fourth and fifth iteration. The results are a little worse for the second as for the first harmonic.

The results obtained for other locations of the artificial boundary (R) after the third iteration are presented in Tables 6 and 7.

Table 6. Absolute values of the constant $A_{1.0}$.

R/a	$S_{3/4}$	S_1	S_2	S_3
1.0	0.0899	0.2247	0.1837	0.1895
1.5	0.2202	0.0323	0.0047	0.0046
2.0	0.1687	0.0162	0.0018	0.0022
3.0	0.0907	0.0054	0.0064	0.0064

Table 7. Absolute values of the constant $A_{1.1}$.

R/a	S_0	S_1	S_2	S_3
1.0	0.1450	0.3609	0.3438	0.3543
1.5	0.2987	0.0820	0.1109	0.1104
2.0	0.1930	0.0496	0.0653	0.0647
3.0	0.0993	0.0319	0.0384	0.0382

Figure 7. Real and imaginary part of tractions after third iteration.

The results are strongly dependent on the locations of the artificial boundary and they are better for the bigger locations. As the location is bigger the difference between the operators is smaller. It can be seen that the results after the third iteration are very good for R/a>1.5 and by using S_1, S_2 and S_3 operators. To obtain good results with the simplest operator $S_{3/4}$, the location must be bigger.

5 CONCLUSIONS

1. With the presented method elastodynamic problems in infinite domains can be solved. The method is based on an iterative variation of fictive boundary conditions on the artificial boundary. By introducing an artificial boundary an interior elliptic problem is obtained and the method is thus convergent. So we do not need a fundamental solution as needed in boundary element method or the Green's function as used in Aiello et al. 1994. The DtN operator is not introduced in a finite element equation as in Pinsky & Thompson 1992 and in Thompson & Pinsky 1993, so we do not have problems with continuous conditions between finite elements. Any finite elements can be used in the analysis.

2. The method in this paper is tested on simple axial symmetric and nonsymmetric problems but it is also applicable for other problems.

3. By using a non-local exact DtN operator for axial symmetric problems an exact solution after second iteration can be obtained and accuracy is independent of the location of the artificial boundary.

4. By using asymptotic local DtN operators the accuracy of the method depends on the selected operator and on the location of the artificial boundary. There is practically no difference in results obtained with S_2 and S_3 operator. By using the simplest $S_{3/4}$ operator the results are not good for the smaller locations of the artificial boundary.

5. The method is iterative but the convergence is relatively fast. There is practically no difference in results already between the fourth and fifth iteration.

REFERENCES

Engquist, B. & Majda, A. 1977. Absorbing boundary conditions for the numerical simulations. *Math. Comp.* 31 (139):629-651.

Bayliss, A. & Turkel, E. 1980. Radiation Boundary Conditions for Wave-Like Equations. *Communications on Pure and Applied Mathematics*, Vol. XXXIII:707-725.

Feng, K. 1983. Finite Element Method and Natural Boundary Reduction. *Proceedings of the International Congress of Mathematicians, Warszawa, 16.-24. August.*

Keller, J.B. 1989. Exact Non-reflecting Boundary Conditions. *Journal of Computational Physics.* 82:172-192.

Givoli, D. & Keller, J.B. 1990. Non-reflecting Boundary Conditions for Elastic Waves. *Wave motion.* 12:261-279.

Porat, G.B. & Givoli, D. 1995. Solution of Unbounded Domain Problems Using Elliptic Artificial Boundaries. *Communications in Numerical Methods in Engineering.* Vol. 11: 735-741.

Pinsky, P.M. & Thompson, L.L. 1992. Local high-order radiation boundary conditions for the two-dimensional time-dependent structural acoustics problem. *Journal for Acoustical Society of America*, 91 (3):1320-1335.

Thompson, L.L. & Pinsky, P.M. 1993. New space-time finite element methods for fluid-structure interaction in exterior domains. *AMD*-Vol. 178. *Computational Methods for Fluid/Structure Interaction*:101-120.

Aiello, G. & Alfonzetti, S. & Coco, S. 1994. Charge Iteration: A Procedure for the Finite Element Computation of Unbounded Electrical Fields. *International Journal for Numerical Methods in Engineering.* Vol. 37:4147-4166.

Geotechnical Hazards, Marić, Lisac & Szavits-Nossan (eds) © 1998 Taylor & Francis, ISBN 90 5410 957 2

Site effects of an M = 7.3-7.5 earthquake, and its tessellated synthesis

L. Sirovich & M. Bobbio
Osservatorio Geofisico Sperimentale, (O.G.S), Opicina Trieste, Italy

F. Pettenati
National Group for Defence against Earthquakes, National Council for Research of Italy, at O.G.S., Opicina Trieste, Italy

ABSTRACT: The regional patterns of the macroseismic intensity of this earthquake of the seventeenth Century are almost insensitive to the gross lithological and topographical characteristics prevailing at the sites. (Each intensity "point" datum always corresponds, however, to the response of a large number of buildings over a relatively large and inhomogeneous area). Rather, source-effects seem to dominate the macroseismic regional patterns, and their geophysical inversion is able to give information on the earthquake source. In fact, an overall consistency between i) the regional patterns of macroseismic intensity, and ii) the synthetically back-predicted intensities has been obtained for this earthquake using a new source-dependent kinematic model (Bull. Seism. Soc. Am., 86, 4, 1019-1027). The focal mechanism found is compatible with the seismotectonics in the area (transcurrent, dextral, sub-vertical fault, 60 km long, with 45° of normal component, strike angle of 28°; nucleation point at 7 km depth, Lat.=37°.08, Long.=14°.93, seismic moment M_0=4.69e26 dyne cm).

1. INTRODUCTION

1.1 *The event studied*

The studied area in SE Sicily (Fig. 1) has not been struck by destructive earthquakes in the instrumental era. Thus, any information about strong earthquakes in the region during past centuries is particularly valuable. However, the only information available for earthquakes of the past is on their damage distribution, which can be retrieved from historical documents. Such information is available for one of the strongest shocks which ever hit the Central Mediterranean area in historical times: the Jan. 11, 1693 earthquake, which caused between 54,000 (Boschi et al. 1995) and 60,000-93,000 casualties in SE Sicily (Barbano & Cosentino, 1981). The city of Catania alone suffered 12,000 deaths (63% of the resident population at the time) and "was almost completely destroyed" (Boschi et al. 1995, page 294).

Information on the damage due to this earthquake is available from the catalogue prepared by Boschi et al. (1995), and recently updated (Boschi et al. 1997). Damage is there expressed in the Mercalli-Cancani-Sieberg, MCS, scale.

Geophysical inversions of the intensity fields of the earthquakes of Jan. 9, 1693 (M=6-6.5); Jan. 11, 1693 (M=7.3-7.5); and Dec. 13, 1990 (M=5.4), and the seismotectonics of NE Sicily have been pre-

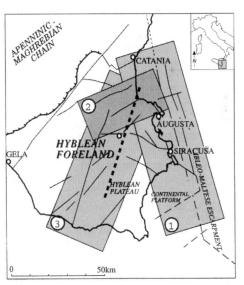

Figure 1. SE Sicily. The patterned stripes indicate the approximate trends of the three series of possible rupture planes taken into consideration. The thick-dashed segment shows the source proposed in this study.

sented more thoroughly elsewhere (Sirovich & Pettenati 1998a, b).

1.2 Source-, path-, and site-effects

Seismic motion (and damage) at a site is the result of the effects due to the source, the path followed by the seismic waves, and the local conditions at the site. Peak ground acceleration (PGA) is sensitive to the geomorphological, geotechnical, and geophysical characteristics of the sites. The first statistical evidence of the amplification of PGA at European sites where relatively soft sediments - with thickness less than about 20 m - lay directly over the bedrock was presented by Chiaruttini & Siro in 1981; this finding recently entered the Italian seismic code (Ministerial Decree of Jan. 16, 1996). Regarding the so-called macroseismic intensity (denoted by I in the following), we presented a case history of small microzones, within towns and hamlets struck by an earthquake in Southern Italy, which experienced amplifications and de-amplifications of 1 to 2 I degrees. These peculiar seismic responses were associated with peculiar geomorphological, geotechnical, and geophysical characteristics of the various microzones within the studied sites (Siro 1982); this case history was included in an international guide for zonation on seismic geotechnical hazards (Anonymous 1993). Note, however, that the aforementioned amplified responses occurred in different parts of the inhabited centres. If the response of each of those centres were to be expressed by one value of I only, at each centre the result would have probably been something close to a "mean" I value (here and after the inverted commas indicate an improper use of the term).

Studying eight earthquakes in California-Nevada, we recently showed that by inverting the macroseismic field of some earthquakes it is sometimes possible to retrieve information on their seismic sources (Sirovich 1996a, b, 1997). Implicitly, this means that in the case of those earthquakes the regional damage distribution (macroseismic intensity, I) was conditioned more by the radiation from the source than by the paths followed by the waves, or by the "mean" (i. e. prevalent) local characteristics of the sites. This observation does not necessarily contradict the previous ones (Siro 1982). In fact, in one case, we are speaking about different responses of small homogeneous microzones of a hamlet or town, while in the other case the I data refer to the "mean" response of each centre. In other words, when one uses I data banks which give only one macroseismic datum per inhabited centre, one must be aware that it is the result of the estimation of damage for the whole town, where different soils etc. are found. This holds, for example, both for the catalogue managed by the American U.S.G.S. for the Californian earthquakes studied (Sirovich 1996a, b,

1997), and for the Italian catalogue by Boschi et al. (1995, 1997).

In this paper we show i) that the regional patterns of I of this seventeenth Century earthquake are rather insensitive to the local lithological and topographical characteristics, and ii) that by inverting them it is possible to retrieve information on the source.

2. THE DATA

The catalogue by Boschi et al. (1995, 1997) (which uses the MCS scale) has 179 I estimates for 179 cities, towns, and villages. These data were used in the geophysical inversion shown in Figure 10.

Regarding the search for site effects, unfortunately we only know the gross lithological and topographical features for 105 sites of those studied by Boschi et al. (1995). But Barbano & Cosentino (1981) also studied the I values produced by the Jan. 11, 1693, earthquake, and the gross lithological and geomorphological features of 12 of their sites are also available to us. They used the Medvedev-Sponheuer-Karnik scale of 1964 (MSK-64), but, fortunately, there is a substantial equivalence between the MCS and the MSK-64 scales (Murphy & O'Brien 1977; Monachesi & Stucchi 1996). Thus, in this work we treated MSK-64 and MCS information from Sicily without adjustments, and obtained a data set of 117 localities. We used this data set to test if there is statistical evidence for a dependence of I on the gross local features of the sites.

Lithological and topographical characteristics of the sites used in this work come from the cross-evaluation of geological studies, maps, and aerial photographs, rendered available in the framework of the SCENARIO project of the European Commission DGXII (Anonymous 1996).

3. SEARCHING FOR SITE EFFECTS

According to current practice, we used linear regressions (epicentral distance versus I) to search for site effects in the sample. Due to the intrinsically descriptive nature of macroseismic scales, the term "pseudo-intensity" is used on the numerical axis of the linear regressions (see also Peruzza 1996). This matter is discussed in more detail elsewhere (Sirovich et al. 1998).

Figure 2 presents the whole data set of 117 sites (all soils and all topographical environments). The distance dispersion is typical for this type of data. Figures 2-8 also show the confidence bands, within which the regression lines have a 95 per cent probability to lie.

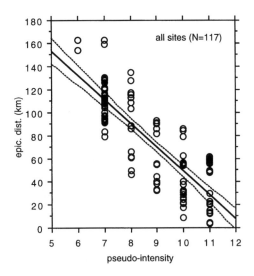

Figure 2. The 117 sites data set used.

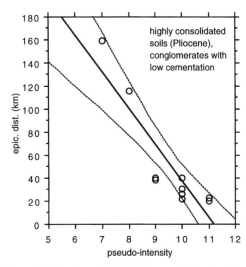

Figure 4. The sites with highly consolidated soils (all grain sizes) of Pliocene and Early-Middle Pleistocene, and conglomerates with low cementation.

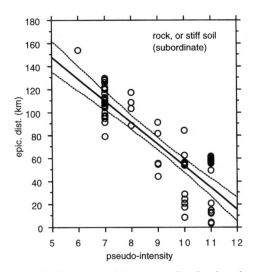

Figure 3. The sites with outcropping hard rock. Sites with stiff soils (subordinate) are included.

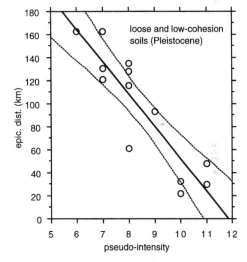

Figure 5. The sites with soils of low consolidation (all grain sizes) of Early-Middle Pleistocene, Middle-Late Pleistocene, and with loose alluvial deposits (sands and gravels).

Figure 3 shows only the sites with outcropping hard rock (limestone and marly limestone, dolomite, lavas, pyroclastic rocks, well-cemented conglomerates and breccia deposits), and those where hard rock prevails and stiff soils are subordinate (Flysch, clayey marl, pre-Pliocene arenaceous deposits).

Figure 4 shows the data for sites with highly consolidated soils (all grain size) of Pliocene and Early-Middle Pleistocene.

The data for sites with soils of low consolidation (all grain size) of Early-Middle Pleistocene, of Middle-Late Pleistocene, and with loose alluvial deposits (sands and gravels), are presented in Figure 5.

Regarding the topographical situation of the studied localities, Figure 6 groups towns and villages which are in the plain, while Figure

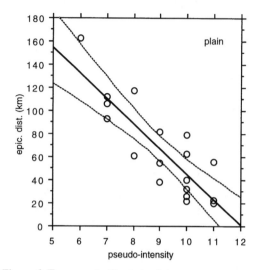

Figure 6. Towns and villages in plain.

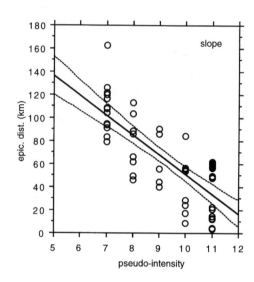

Figure 8. Towns and villages built on slopes.

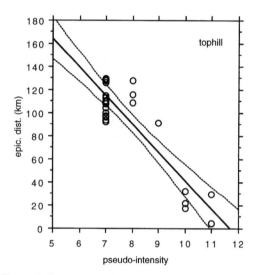

Figure 7. Towns and villages which stand on top of hills.

7 shows those on hilltop, and Figure 8 those which are built on slopes.

The differences between the regressions in Figures 2, 3, and 6-8 are statistically insignificant. The group of data belonging to sites with highly consolidated - all grain size - soils of Pliocene and Early-Middle Pleistocene (Fig. 4) is too small to draw reliable conclusions. Note in this figure, however, the two values (degree 7 and 8) with epicentral distances greater than 110 km (too high for our purposes; see later). The only group of

data which could have a different, statistically significant, behaviour from the others is shown in Figure 5. For example, the 95% confidence band in Figure 3 (rock, and stiff soils) overlaps that in Figure 5 (loose sediments) between $I=11$ and $I=7$-8; but the two bands separate at lower intensities I / higher distances, where our model is less effective (see later).

It is worth noting that regression lines and 95% confidence bands for the towns and hamlets built on hilltop (Fig. 7) overlap the regression lines and confidence bands of those built in the plain (Fig. 6). Thus, at this regional scale (one I value per town), the hilltop scenario seems to be almost transparent to seismic waves. It should be noted, however, that the inhabited centres usually also span over areas around the hilltop, or close to the valley borders; most often the I data refer, thus, to more complex topographical environments than those summarized in the rough classification rendered necessary by the scale of our study. As already said (see 1.2), seismic amplification of homogeneous microzones should really be examined during seismic microzoning, which is beyond the scope of the present work.

We also attempted various aggregations of data according to the lithology and the topography of the sites, but none was able - at this scale - to give acceptable statistical evidence for a systematic dependence of I on the "mean" lithological and topographical characteristics of the sites.

4. INVERTING THE MACROSEISMIC FIELD

The scarce relevance of site effects on the regional I

patterns allows an attempt at geophysical inversion to retrieve the source characteristics.

To do so, a quantitative and objective technique for representing the results of I surveys, and for finding the synthetics that best fit the observations is needed; otherwise one is obliged to compare qualitatively the size and the shape of the experimental I field with those of synthesized pseudo-intensity fields. Note that in current practice the observed macroseismic field is contoured manually to obtain the so-called isoseismals. In this paper we propose to improve on this qualitative practice by i) adopting the Voronoi polygons tessellation technique (Preparata & Shamos 1985), and by ii) computing statistical tests at the sites and over the whole tessellated plane, to judge the quality of the fit obtained with each source tentatively adopted.

Note that the fundamental property of these convex polygons, which circumscribe every control point, is that they honour the data unequivocally. The idea of Voronoi polygons arose from a geometric problem of proximity over the plane, which may be intuitively taken as a problem of partitioning of planes (Preparata & Shamos 1985). These questions are discussed in more detail elsewhere (Sirovich et al. 1998).

Computing statistical tests in the tessellated plane offers some advantages: 1) concentrations of areas (more neighbouring tesseras) with positive, or negative, residuals may suggest errors in the synthesizing procedure, which depend both on azimuth and distance. This is not seen if traditional statistical tests of residuals, azimuth- or distance-dependence are performed; 2) applying statistical tests to the tessellated plane allows the possibility of weighting the sum of the residuals within each tessera as a function of its dimension. In turn, this gives 3) the opportunity of weighting the contribution of the residuals of small tesseras more than the residuals of large tesseras. Note that emphasizing the contribution of small tesseras means emphasizing the importance of the best fits between the observed and calculated I values within areas where many observation points (small tesseras) are available. Note also that small tesseras are usually found close to the epicenter, where high I values are found, and that best fitting these small tesseras by synthesis is more important from an engineering point of view than best fitting, for example, the $I=3$ tesseras. Finally, emphasizing the contribution of the small tesseras decreases the contribution of the large tesseras (with low I) at distances from the source where our model, which is based upon the asymptotic approach (Sirovich 1997), is less effective.

4.1 The tessellated observed field

Figure 9 shows the macroseismic intensities of the Jan. 11, 1693, earthquake in SE Sicily, which were interpreted from historical documents by Boschi et al. (1995). The dots are the "surveyed" sites to which the I values of the catalogue refer. We simply tessellated the data by Boschi et al. (1995) with Voronoi polygons. Note in the figure: i) the irregular shape of the relatively large area of degree 10-11 close to the SE coast of Sicily; ii) that, in its southern part, the $I=11$ area does not reach the coast; contrariwise, it is separated from the coast by two tesseras with $I=10$, and two with $I=9$ and 8 respectively; iii) the extremely fast decay of I with epicentral distance (strictly speaking, the epicentre of this earthquake is not known) from $I=10$ to $I=8$; and iv) the large area with $I=8$ and $I=7$ alternating. Note also the high density of observations (small polygons) close to the epicentral area, and, viceversa, the large $I=6$ polygons in W Sicily and S Calabria.

4.2. The choice of the structures for the inversions

A more comprehensive overview of the regional seismotectonic characteristics of SE Sicily is given elsewhere (Sirovich & Pettenati 1998a); with all this information available, we here decided to adopt three possible families of sources for constraining the I field inversions of the main shock in 1693 (see Fig. 1).

Note that different lengths of rupture along-strike and anti strike, various dip and rake angles, shear waves velocities, and rupture propagation velocities were tentatively used in the model (Sirovich 1997) and the best fit found by trial-and-error.

(In the convention adopted here, the angle of strike is measured clockwise from north, with the plane dipping to the right side. The rake angle is seen on the fault plane from the roof of the fault and measured counterclockwise from 0° to 360°; refer to Figure 1 in: Sirovich 1997).

As said, we did comparisons objectively using statistical tests. The starting principal characteristics of the families of sources were:

1) IBL-MAL: normal, steeply dipping towards ENE (with 0 to 50% dextral or sinistral strike-slip components), faults belonging *latu sensu* to the Ibleo Maltese Escarpment megastructure. In the light of the available seismotectonic information, both offshore and onshore faults in the coastal area were considered;

2) SCO-LEN: normal faults associated with the Simeto-Scordia-Lentini grabens, dipping N and NNW and trending from E-W to NE-SW;

3) SCICLI-EBT78: strike-slip faults, in part associated with the Scicli-Ragusa-Monte Lauro fault system; sub-vertical, trending from N to

Table 1. Source parameters of the best sources for the three areas of Figure 1.

Best SOURCE Parameter	Ibleo-Maltese Escarpment (area 1)	Scordia-Lentini Graben (area 2)	Scicli-EBT78 (area 3)
Latitude	37°.32	37°.28	37°.08
Longitude	15°.28	15°.05	14°.93
Depth (km)	7	7	7
Length (km)	30+50	30+10	35+25
Strike angle	350°	250°	028°
Dip angle	75°	60°	80°
Rake angle	300°	300°	225°
Mach N°	0.60	0.60	0.55
V_S (km/s)	3.5	3.5	3.5
Seismic Moment (dyne cm)	4.65e26	4.68e26	4.69e26

NNE, with right-lateral movements. In particular, EBT78 is a proposed blind strike-slip sub-vertical dextral transfer structure, striking approximately NNE. Note that this proposed structure would lie on the eastern flank of the promontory formed by the Bouguer gravity, and magnetic anomalies in the area of the Hyblean Plateau. This promontory strikes 20°-30° (Ben-Avraham & Grasso 1990; see their Figure 8) as does EBT78. The acronym indicates that the upward vertical projection of this rupture plane almost coincides with the Eastern Border of the Transfer seismogenic zone N° 78; this was one of the seismogenic zones recently adopted by the experts of the National Group against Earthquakes (GNDT) for calculating the seismic hazard of Italy (GNDT 1996). According to their seismotectonic interpretation, zone N° 78 works as a transfer, dextral, zone between two opening ridges (see more details in Sirovich & Pettenati 1998a).

Table 1 shows - for each tentative seismogenic area 1, 2, and 3 in Figure 1 - the kinematic parameters of the sources which were able to give the relatively better fits of the observed *I* field shown in Figure 9.

4.3 *The kinematic function KF*

Our algorithm incorporates simple information on the sources, considers only S_H- and S_V-waves, and uses a simple half-space medium. In spite of these simplistic assumptions, a previous version of KF has already given intriguing results, both while trying to retrieve information on the source of an earthquake which occurred in the pre-instrumental era in the Southern Apennines, Italy, (Chiaruttini & Siro 1991), and best fitting observed intensities in some zones of California and of Nevada, USA, (Sirovich 1996a, b, 1997). The description of the algorithm in question, and the graphical convention used have been already presented (Sirovich 1996a, 1997) and, for saving space, are not repeated here.

This algorithm has since been improved by analysing 1720 "point" *I* data for five earthquakes in the Los Angeles, California, area (Sirovich et al. 1998).

Based upon this data set, a series of statistical analyses gave the linear multiregression (4.1).

Thus, our algorithm now calculates "point" pseudo-intensities i (integer numbers) according to:

$$i = 9.241 + 3.358 \cdot \log_{10}(KF) \\ + (8.04e - 27) \cdot M_0 \tag{4.1}$$

5. RESULTS OF THE INVERSION

5.1 *The proposed source*

Figure 10 shows the bestfit tessellated synthetic pseudo-intensity field of the earthquake. It was calculated adopting the source that we called EBT78; this is the best source that we obtained once the data observed by Boschi et al. (1985) were taken as experimental reference. EBT78 lies in area N° 3 of Figure 1. The kinematic characteristics of the source which produced Figure 10 are in Table 1, fourth column. In our convention, the angle of rake of 225° indicates a dextral transcurrent fault with 45° normal component. The length of rupture is 35 km in the sense of strike, and 25 km antistrike. The rupture velocity is 1.93 km/s.

The synthetics obtained with all other tentative families of rupture planes, compatible with the seismotectonic environment, give worse statistical tests. Even worse statistics are generally obtained if a traditional isotropic "attenuation law", such as equation (5.2), is used (see Table 2). This table shows the residuals (calculated-minus-observed)[2] corresponding to the syntheses obtained with the relatively best sources belonging to the following seismogenic areas (refer to Figure 1):

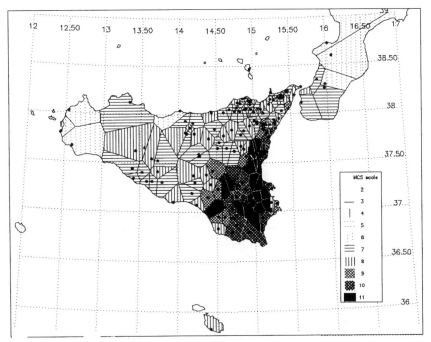

Figure 9. Tessellated observed macroseismic intensities *I* of the Jan. 11, 1693, M=7.3-7.5, earthquake in SE Sicily. The dots are the "surveyed" sites (*I* data from Boschi et al. 1995).

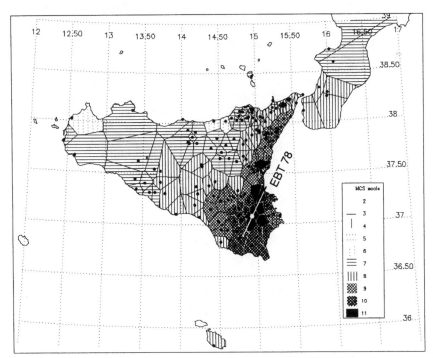

Figure 10. Synthetic pseudo-intensity best fit of the tessellated observed *I* of the Jan. 11, 1693, earthquake (Figure 9). Source: NNE segment of EBT 78 (see Figure 1 and Table 1).

355

Table 2. Residuals (calculated-minus-observed)^2 of the tessellated synthetics obtained with the three sources in Table 1, and of the tessellated I field predicted by the "attenuation law" (equation 5.2).

Best SOURCE Computation of residuals	Ibleo-Maltese Escarp. (area 1)	Scordia-Lentini Graben (area 2)	Scicli-EBT78 (area 3)	"Attenuation law", equation (5.2)
At the I "points"	400	291	181	661
V-V test	1509	2134	1083	1024
C-V test	78,073	43,639	44,193	316,312

i) the Ibleo-Maltese Escarpment (area 1);
ii) the Scordia-Lentini Graben (area 2);
iii) the Scicli-EBT78 (area 3).

Table 2 also shows the residuals of the tessellated pseudo-intensity field predicted by the "attenuation law" in equation (5.2).

Note that we chose the sum of squares instead of the sum of absolute values of residuals because we wanted to emphasize relevant differences (≥ 2)

Only the synthesis achieved with the best Scicli-EBT78 source is shown in this paper.

The key for reading the "computation of residuals" column in Table 2 is as follows:

• «at the I "points"»: the sum of squared residuals was performed directly at the 179 sites surveyed by Boschi et al. (1995) (without tessellation);

• «V-V test»: the fields of both observed I and synthetic pseudo-intensity were tessellated, and the value of the datum observed at each site (or calculated there) was assigned to all its polygon. Then, the sum of the squares of the residuals was calculated over the whole polygon. The total sum obtained over each polygon was weighted with the inverse of the area of the polygon, and then all the weighted sums of all polygons were added together. This kind of analysis of the residuals gives low importance to residuals obtained from isolated sites (large polygons, low I) far from the source.

• «C-V test»: in the third test, the tessellated observed I field is compared directly with the whole synthetic field (which is a continuous function). In this case also, we looked for some logical choice of the weights. They ought overall to reward the successes, or emphasize the misfits, obtained close to the source, in areas with many surveyed sites, where high I were observed. Thus, we chose to calculate the mean distance md between all surveyed sites. Then, formula (5.1) of weight-scaling within each polygon was adapted to this mean distance md.; pd is the distance between each infinitesimal element, of each polygon, and its site.

$$weight = e^{\frac{-\log 2}{md}} \cdot pd \qquad (5.1)$$

In other words, at the periphery of the mean polygon the weight of the residual is close to 0.5. The residual of the hypothetic infinitesimal polygon

would have weight equal to one. The most peripheral part of the largest polygon gives a contribution close to zero to the sum of the residuals' squares. On the other hand, the most peripheral parts of small polygons contribute significantly to the sum.

Regarding Table 2, it is worth adding that i) the source Scicli-EBT78 scores markedly the minimum number of residuals ≥ 2 at all sites, and that ii) the good result (1024) of equation (5.2) in the V-V test is mostly due to the fact that it fits well the tesseras of degree 11.

5.2 Results produced by a traditional "attenuation law"

The residuals produced by a traditional I over log D "attenuation law" are also shown in the Table. We adopted the "attenuation law" which was obtained for SE Sicily by the GNDT Working Group (GNDT, 1996), which recently produced the seismic hazard map of Italy for the Ministry of Civil Defence. The GNDT experts obtained the mentioned "attenuation law" by applying the model of Grandori et al. (1987) to the I data of the Jan. 11, 1693, earthquake, which had been presented by Barbano & Cosentino in 1981; its equation is:

$$I_0 - I_i = \frac{1}{\ln \psi} \ln \left[1 + \frac{\psi - 1}{\psi_0} \left(\frac{D_i}{D_0} - 1 \right) \right] \qquad (5.2)$$

where ln is the natural logarithm, I_0 is the epicentral pseudo-intensity, I_i is pseudo-intensity at the i-th site, $\psi = 0.952$, $\psi_0 = 0.830$, D_i is the epicentral distance of the i-th site from the "macroseismic epicentre" of Boschi et al. (1995), and D_0 represents the radius of a circular source. The purpose of Table 2 is also to show the degree of accuracy with which the intensities of the Jan. 11, 1693, earthquake are back-predicted by one of the best available models of "attenuation" currently in use in the engineering seismology.

6. CONCLUSIONS

6.1 Site effects

The I catalogue used shows the seismic response of each town or village using one value only of I. The statistical analysis performed here shows that, at this scale, the regional I patterns for this earthquake of the XVII Century are rather insensitive to the gross site characteristics prevailing in each urban area, either lithological or topographical. Only a slight tendency for loose-soil-sites (Fig. 5) to amplify I (i. e. damage) beyond 120 km from the source (for I ≤7) is seen.

6.2 The source of the Jan. 11, 1693, earthquake

It should be noted that applying our KF model to a historical earthquake involves more difficulties than treating contemporary, and instrumentally documented earthquakes. In particular, in the study case, the aforementioned (paragraph 4.1) four sites that separate the I=11 area from the SE coast of Sicily become crucial; in fact, the four Voronoi polygons generated by those point data constrain the best source inland. And it is a weak constraint, which is confirmed by Barbano & Cosentino (1981), however. Another constraint - weak because dependent on one "point" only - is given by the good fit of the intensity at Malta. Note, however, that, using a different I catalogue, a source very close to that obtained here is again achieved (Sirovich & Pettenati 1998a).

From the point of view of the regional seismic damage distribution, no Ibleo-Maltese source is able to give a strong I in the Hyblean Plateau, and, contemporaneously, relatively lower I along the SE coast of Sicily, as seen in the real data (Fig. 9). But a causative source on the Escarpment (in area N° 1 of Figure1) cannot be discarded on this evidence only.

Sirovich & Pettenati (1998a) also discuss the results of the modelling of the tsunami of Jan 11, 1693 performed by Piatanesi et al. (1998). Those studies give credence to a rupture plane tangential to the promontories adjacent to Augusta and Siracusa and belonging to the Ibleo-Maltese Escarpment (area N° 1 of Figure 1).

Within the intrinsic limitations of this kind of study, the result presented here constitutes the first quantitative seismological hypothesis available to now on the possible source of one of the most destructive earthquakes which ever hit the Central Mediterranean area in historical times.

ACKNOWLEDGEMENTS

The SCENARIO project group of the European Commission DGXII (in particular: G. Valensise, M. Mucciarelli, and V. Bosi) is thanked for having kindly provided the lithological and topographical characteristics of the studied sites. M. Bobbio solved some computational problems and developed the computer graphics. Supported by the G.N.D.T. of the Italian C.N.R.; grants 96.02963.PF54 and 97.00536.PF54. Our colleague Peter Guidotti revised the English manuscript.

REFERENCES

Anonymous 1993. *Manual for Zonation on Seismic Geotechnical Hazards*. Technical Committee for Earthquake Geotechnical Engineering, TC4 (K. Ishihara, chairman), of ISSMFE. The Japanese Society of Soil Mechanics and Foudation Engineering, Tokyo.

Anonymous 1996. (SCENARIO project group, 1996. *SCENARIO: Time dependent hazard estimates using a multiparametric geophysical observatory*. Final report of the contract EV5V-CT94-0404 to the European Commission DGXII.

Barbano, M. S. & M. Cosentino 1981. Il terremoto siciliano dell' 11 gennaio 1693. *Rend. Soc. Geol. It.* 4: 517-522.

Ben-Avraham, Z. & M. Grasso 1990. Collisional zone segmentation in Sicily and surrounding areas in the Central Mediterranean. *Annales Tectonicæ*, Spec. Issue, 6(2): 131-139.

Boschi, E., Ferrari, G., Gasperini, P., Guidoboni, E., Smriglio, G. & G. Valensise 1995. *Catalogo dei forti terremoti in Italia dal 461 a. C. al 1980*. Istituto Nazionale di Geofisica - SGA storia geofisica ambientale. Roma, CD and paper versions (in Italian).

Boschi, E., Guidoboni, E., Ferrari, Valensise G. & P. Gasperini 1997. *Catalogo dei forti terremoti in Italia dal 461 a. C. al 1990; N°2*. Istituto Nazionale di Geofisica - SGA storia geofisica ambientale. Roma, CD and paper versions (in Italian).

Chiaruttini, C. & L. Siro 1981. The correlation of peak ground horizontal acceleration with magnitude, distance and seismic intensity for Friuli and Ancona, Italy, and the Alpide Belt. *Bull. Seism. Soc. Am.* 71(6): 1933-2009.

Chiaruttini, C. & L. Siro 1991. Focal mechanism of an earthquake of Baroque age in the 'Regno delle Due Sicilie' (southern Italy). *Tectonophysics* 193: 195-203.

GNDT 1996. *Modalità di Attenuazione dell'Intensità Macrosismica* (by L. Peruzza). Report available on the Web site http://emidius.itim.mi.cnr.it/GNDT/home.html, pp. 8.

Grandori, G., Perotti, F. & A. Tagliani 1987. On the attenuation of macroseismic intensity with epicentral distance. In A. S. Cakmak (ed.), *Ground Motion and Engineering Seismology.* Developments in Geotechnical Engineering, 44: 581-594. Elsevier, Amsterdam.

Monachesi, G. & M. Stucchi 1996. *DOM4.1, un data base di osservazioni macrosismiche di terremoti di area italiana al di sopra della soglia del danno.* Report available on the Web site http://emidius.itim.mi.cnr.it/DOM/ home.html.

Murphy, J. R. & L. J. O'Brien 1977. The correlation of peak ground horizontal acceleration amplitude with seismic intensity and other physical parameters. *Bull. Seism. Soc. Am.* 67: 877-915.

Peruzza, L. 1996. Attenuating intensities. *Annali di Geofisica* 34(5): 1079-1093.

Piatanesi, A, Tinti, S., Maramai, A. & E. Bortolucci 1998. Revisione del terremoto tsunamigenico della Sicilia orientale del 1693: simulazione numerica del maremoto. *Proc. Nat. Conf. GNGTS-CNR, Rome, 1996*, in press.

Preparata F. P. & M. I. Shamos 1985. *Computational geometry: an introduction.* New York: Springer Verlag.

Siro L. 1982. Emergency microzonations by Italian Geodynamics Project after November 23, 1980 earthquake: a short technical report. *Proc. 3rd Intern'l Conf. on Microzonation, Seattle USA, June 28 - July 1, 1982*, (3): 1417-1427.

Sirovich, L. 1996a. A simple algorithm for tracing out synthetic isoseismals. *Bull. Seism. Soc. Am.* 86(4): 1019-1027.

Sirovich, L. 1996b. Synthetic Isoseismals of two Californian Earthquakes. *Nat. Hazards* 14(1): 23-37.

Sirovich, L. 1997. Synthetic isoseismals of three earthquakes in California-Nevada. *Soil Dynamics and Earthquake Engineering* 16: 353-362.

Sirovich, L. & F. Pettenati 1998a. Seismotectonic outline of South-Eastern Sicily: an evaluation of available options for the scenario earthquake fault rupture. *Journal of Seismology* in press.

Sirovich, L. & F. Pettenati 1998b. Kinematic information about the source of a 60,000-deaths earthquake of the XVII Century in SE Sicily (Italy) from the inversion of its macroseismic field. Submitted to an international journal.

Sirovich, L., Pettenati, F., Cavallini, F. & C. Chiaruttini 1998. Tessellation, synthesis, and objective evaluation of macroseismic intensity. Submitted to an international journal.

Geotechnical Hazards, Marić, Lisac & Szavits-Nossan (eds) © 1998 Taylor & Francis, ISBN 90 5410 957 2

Construction of data base system of earth dam and its application to earthquake disaster prevention

Shigeru Tani & Masanori Nakashima
National Research Institute of Agricultural Engineering, Tsukuba-shi, Japan

ABSTRACT: Recent large-scale earthquakes have damaged many agricultural facilities such as canals, farm roads, and earth dams. Earth dams are especially important in terms of disaster prevention since the provide irrigation water and their destruction could cause secondary damage to nearby urban districts. The Hyogo-ken Nanbu Earthquake damaged 1,222 earth dams and prompted the construction of a database for preventing damage to earth dams. We developed a database of earth dams, the contents and application of which are described in this paper. This database contains the following information: (1) structures of earth dams (location, height, etc.), (2) visual information (topographical maps, photographs, and drawings), (3) types of previous damages, (4) Digital National Land Information of the Geographical Survey Institute of Japan (geology, elevation, public facilities, etc.), (5) information on active faults and data from the Automated Meteorological Data Acquisition System (AMEDAS).

1. INTRODUCTION

Recent earthquakes have damaged a number of agricultural facilities, such as canals, farm roads, and especially earth dams, which are important for irrigating farm lands. There are over 100,000 earth dams in Japan, 80% of which are 10 m or lower in height and 75% were constructed more than 100 years ago. The history of damage to earth dams began with the construction of earth dams. Traces of damage by the earthquake that occurred in 742 can be observed in an old earth dam, Sayama Ike (Sayama, Osaka Prefecture).

Earth dams damaged by earthquakes in the past are shown in Table 1. The Hyogo-ken Nanbu Earthquake damaged 1,222 earth dams. While only eight of these were completely destroyed, they could have caused secondary damage to nearby urban districts. The earthquake demonstrated the need to create a database for earth dams.

A prototype database system was created in 1991 to prevent damage to earth dams and was tested by entering data samples (Tani and Yoshino, 1991). In 1992, a database was constructed for earth dams in Hokkaido of height 15 m or more, and problems regarding use of the database were studied (Tani, 1993). This system was further modified in 1994 to run on MS-Windows.

Table 1. Earth dam damage caused by several earthquakes

Name of Earthquake	Occurrence Time	Maguni-tude	Number of earth dams damaged
(1)Kita-Tango	Mar 7,1927	7.5	90
(2)Oga	May 1,1939	7.0	74
(3)Niigata	Jun. 16,1964	7.5	146
(4)Matsushiro	Aug.1965~ Dec.1970	Max. 5.4	57
(5)Tokachi-Oki	May 16,1963	7.9	202
(6)Miyagi-ken-Oki	Jun.12,1978	7.4	83
(7)Nihon-kai-Chubu	May 26,1983	7.7	238
(8)Chiba-ken Toho-Oki	Dec.17,1987	6.7	9
(9)Hokkaido Nansei-Oki	July 12,1993	7.8	18
(10)Noto-hanto-Oki	Feb.7,1993	6.6	21
(11)Hyogo-ken-Nanbu	Jan.17,1995	7.2	1222
(12)Ishikari-Hokubu	May 23,1995	5.6	1
(13)Kagosima-Hokuseibu	May 13,1997	6.1	1

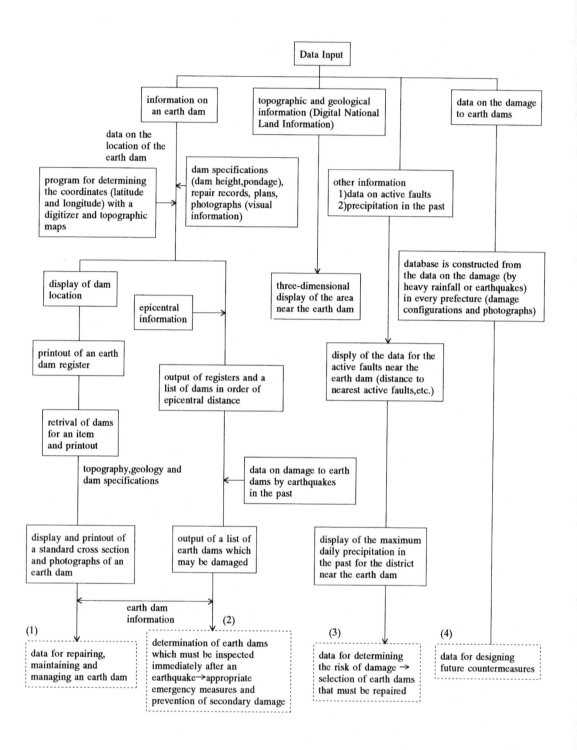

Fig. 1 Conceptual diagram of the database for earth dams

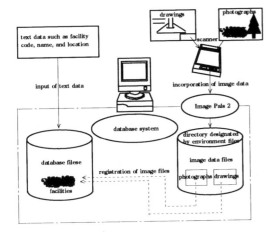

Fig.2　Relationship between the database
and image data

After the Hyogo-ken Nanbu Earthquake, we modified the old system to incorporate data on active faults and to handle images. This paper describes the database we developed and discusses its application for preventing earthquake damage.

2. CONCEPT OF THE DATABASE FOR EARTH DAMS

There are over 100,000 earth dams in Japan, which are mainly used for agricultural purposes. To efficiently maintain and repair these dams, a database should be designed for the earth dams and nearby ground conditions.

Databases are generally classified into those for general use that are created without any special purpose, and those for special purposes. The former type of databases has a complicated structure to suit various purposes but may not be practical for special purposes. Therefore, a database should be constructed by first designing a special-purpose database, investigating and solving problems regarding its utilization, incorporating other special-purpose databases, and finally creating a database for general use.

Although our database was constructed to prevent damage to earth dams, it contains the Digital National Land Information and image data and may be applied for other agricultural facilities. The contents and scope of application are shown in Fig. 1. The main purposes of the database are:

(1) to provide information for daily repair, maintenance, and administration works of earth dams

(2) to identify earth dams that must urgently be inspected after earthquakes and to provide information on such dams, and

(3) to evaluate the risk of damage for each earth dam and to provide information for selecting earth dams that should be repaired.

There is also a method for entering data on damage in the database, and for accumulating damage information for each earth dam. Fig. 2 shows the relationship between the text files and the image data files of the database. The system reads image data by using software for processing images and a scanner, and registers images by linking such image files with a text file.

3. CONFIGURATION OF THE DATABASE FOR PREVINTING DAMAGE TO EARTH DAMS

3.1 *Operational environment of the database*
The standard hardware configuration of the system is as follows:

- Machine: NEC PC-9821Xa12/C8 (PentiumTM CPU, 120 MHz)
- Expanded RAM: 32 MB (RAM of the machine: 16 MB)
- Black-and-white printer: CANON LBP-A404F (300 dpi)
- Color printer: (360 dpi)
- Scanner: (400 dpi)
- External hard disk: (4 GB)
- Optical magnetic disk system:

The basic software is MS-DOS (Ver. 6.2) and Windows (Ver. 3.1, Microsoft). External utility software packages are Image AXIS (CD Co. Ltd.), Object Reive (Bunka Orient Co. Ltd.), Image Pals2 (Unlead System Inc.), Super Print (Unlead System Inc.), Ichitaro Ver. 6 for Windows (Just Systems Co. Ltd.), and Scan II (Seiko Epson Co. Ltd.). The development language was Visual BASIC. The Digital National Land Information, data on active faults, and data from the Automated Meteorological Data Acquisition System (AMEDAS) are used.

3.2 *Contents of the database*

This system consists mainly of 1) information on each earth dam, 2) information related to earthquakes, and 3) geographical and meteorological information, as described below.

3.2.1 *Information on earth dams*

The data of each earth dam are entered into the database in Japanese, numbers, and codes. The data that are contained in the Digital National Land Information, such as topography and geology, are automatically input according to the coordinate axis of the earth dam. The class of seismic intensity and designation by the Prime Minister's Office (for promoting rural development and predicting, earthquakes typhoon areas, etc.) are also automatically input according to the coordinate axis. Approximately 180 data items are controlled for each earth dam with a facility code. A facility code comprises 9 characters, 2 of which represent a prefectural code, 3 a municipal code, and 4 the reference number in the municipality. Entered data are classified into the following three categories:

(1) character and numerical data

Earth dam register (1):
name, facility code, coordinate axis, administrator, year of construction, purpose, designation by the government

Earth dam register (2):
specifications of the dam, pondage, surface ground, topography (Digital National Land Information), materials of the dam,

Earth dam register (3):
Assumed damage
intake works, data on administration

Earth dam register (4):
deterioration of the dam and/or peripheral facilities, sedimentation

Earth dam register (5):
repair history

Earth dam register (6):
damage history (precise data are contained in the damage information)

(2) Damage information

This database also contains precise information related to damage besides the damage history in the earth dam register (6), and the damage histories of several earth dams can be integrated. The following damage information as shown in Fig. 3 is contained:

Damage register (1): name, location, specifications of the dam

Damage register (2): cause of damage, damage configuration, visual information such as photographs of the damage

Damage register (3): outline of repair works

Damage data and about 3,000 photographs of 1,222 earth dams that were damaged by the Hyogo-ken Nanbu Earthquake are now included in the damage information.

Analyses of these data have revealed the relationships between the damage and the dam height, type of ground, and distance from an active fault, as will be reported in the next paper.

(3) Visual information

Visual information comprises of cross sections of the dams, topographical maps for the peripheral regions, and photographs. Color images are saved as JPEG

earth dam register (damage configuration)

name	Saraike		reference number	286830011001
damage configuration investigated on				
cause of damage	earthquake			
name of cause of damage	Hyogo-ken Nanbu Earthquake			
hour and day of damage	5:00 on January 17, 1995	estimated hour and day if unknown		
precipitation	AMEDAS station		station number	
	daily precipitation before and after the damage	3 days before the damage 2 days before the damage 1 day before the damage	the day of the damage 1 day after the damage 2 days after the damage	
earthquake-related information	intensity in magnitude	7	epicentral distance	km
water level when the dam was damaged		note		
degree of damage	B	reason for the judgment	cracks of 1-cm width	
dam body	maximum settlement	cm	maximum crack	width:1.0cm,depth:0.30m
	vertical cracks	crest		
		upstream	downstream	

Fig. 3 Damage register output

files, and black-and-white images as GIF files.

Fig. 4 is a view of an earth dam that was output from the image file. In our conventional system, the data on boring was input as JACIC-type and was output to a floppy disk. The new database uses an image scanner for incorporating images, which is easier to operate.

3.2.2 Information related to earthquakes

This database aims principally to prevent damage by earthquakes. Simply by entering the location of the hypocenter and the intensity of an earthquake, the database predicts earth dams that could be damaged by the earthquake based on past earth-dam damage and outputs a list of such earth dams and, for each municipality, the ratio of estimated dam damage.

Fig. 5 plots the magnitude (M) of an earthquake against distance from the epicenter to the farthest earth dam damaged (the limit epicentral distance). The following regression line that expresses the relationship between M and the limit epicentral distance $\triangle d$ (km) was derived from the data on previous earthquakes.

$$\log \triangle d = 0.858M - 4.28 \quad (6.1 \leqq M < 8)$$
$$\triangle d = 11 \text{ (km)} \quad (5 < M < 6.1)$$

The solid line in the figure represents an approximate limit of damage. The area above the line is the area of no damage, and the area below the line is where damage may occur. Therefore, there is a higher probability of damage in regions nearer to the epicenter for a given intensity.

Fig. 6 plots the mean epicentral distance of each municipality against the percentage of earth dams damaged . The solid line in the figure should show the maximum percentage of earth dams damaged estimated from the mean epicentral distance (estimated maximum damage ratio). However, this relationship could be affected by the conditions of

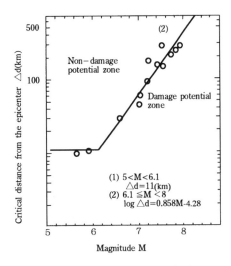

Fig.5 Relation between magnitude and limit epicentral Distance

the ground and embankment. For a comparison with an earthquake of a different intensity, Fig. 7 plots damage ratio against mean epicentral distance for the 1988 Nihon-kai Chubu Earthquake (M=7.7). For each of the Tokachi-oki Earthquake, Niigata Earthquake, Nihon-kai Earthquake, and Hyogo-ken Nanbu Earthquake, the relationship between the maximum estimated damage ratio and the mean epicentral distance was derived from the actual data and is shown in Fig. 8. Since the dam conditions were not considered, the maximum damage ratio was adopted to make a safer judgment (damage was estimated to be slightly severer). By studying the relationship between the ground conditions and damage on earth dams in more detail, it may be possible to more accurately predict damage. An epicentral distance of damage ratio = 0 is equal to the limit epicentral distance.

Fig.4 Visual information of the database

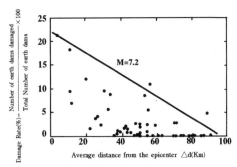

Fig. 6   Mean epicentral distance and the ratio of earth dams damaged (by the Hyogo-ken Nanbu Earthquake)

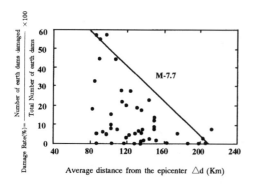

Fig. 7   Mean epicentral distance and the ratio of
          earth dams damaged by the Nihon-kai
          Chubu Earthquake

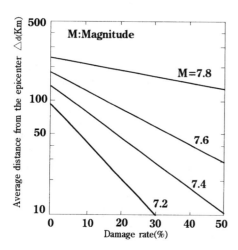

Fig. 8   Relationship between mean epicentral
          distance and estimated maximum damage
          ratio for each district

3.2.3 *Geographical and meteorological information*
The geographical and meteorological information of
this system comprises part of the digital national land
information of the Geological Survey Institute
The following digital information is included:

- FD maps (issued by the Japan Map Center):
    1/200,000 maps

- Elevation for each 250 m × 250 m area (issued by

the Japan Map Center)
- Locations of public facilities (issued by the
Geographical Survey Institute): data of public
facilities in the Digital National Land Information
- Locations of dams (issued by the Geographical

Survey Institute):   data on the locations of dams in
the Digital National Land Information
- Topography and surface geology (issued by the
Geographical Survey Institute):   the topography and
surface geology of each three-dimensional unit

The digital national land information is the oldest
database in Japan and contains a vast amount of data
on various items.   This information is described in
more detail in the Appendix (Geographical Survey
Institute, Ministry of Construction, 1992).
The certainly and activity of every active fault in
Japan are included in our database.   The information
on active faults, the copyright of which is held by
Tokyo University Press, was digitized by Geosupply
Co. (Geosupply,1994).   We were permitted to use
some of the data for our database.

4. DATA INPUT, DISPLAY AND RETRIEVAL IN THE DATABASE

An image displayed by this database is shown in Fig.
9.   The   principal functions of the database are 1)
registration and   modification of data, 2)   display
of data, and 3) retrieval.   Registration   and
modification of data means the input of new data and
modification of old data.   The database uses a file
system that properly updates only the data necessary
for maintaining earth dams, such as maintenance
records, damage information, and deterioration data.
For display, the database displays maps that are based
on the digital information at any magnification or
reduction, and data on active faults.   Figure 8
displays the locations of earth dams together with
data on active faults.
Data items related to earthquakes and active faults
can be retrieved, such as dam specifications (dam
height, poundage, etc.) and location.   The system
outputs a list of dams retrieved and shows their
locations on a map.   Earthquake-related information
is retrieved by entering an epicentral distance and
intensity.   The system displays and outputs earth
dams in order of epicentral distance and their data
such as height and location.   The system rapidly
identifies earth dams that must be inspected
immediately after an earthquake and also provides
the location and other data of the dams.   The
locations of the dams are also displayed on 1/10,000
maps.   The system also displays the maximum
damage ratio (creates a hazard map) for each
municipality with the method described in Section
3.2.1.   Active faults are retrieved by entering values
for fault certainty and activity and a distance from
such faults;

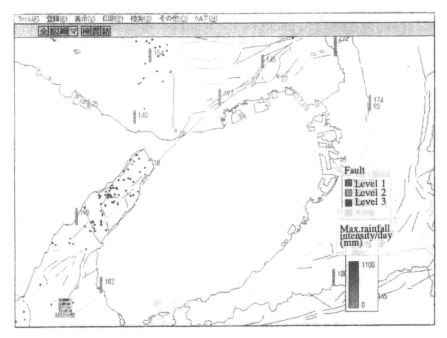

Fig.9　Display of earth dam location, active fault and AMEDAS data

all earth dams that satisfy these conditions are identified. The relationship between active faults and damage is debatable, but most earth dams that are located within 1 km from an active fault were damaged. Identification of such dams is useful for repairing and strengthening earth dams.

5. CONCLUSION

The Hyogo-ken Nanbu Earthquake of 1995 damaged many agricultural facilities, especially to earth dams. Since there are over 100,000 earth dams in Japan, we constructed this database to provide information on each earth dam immediately after an earthquake, to identity and inspect earth dams that may be damaged, and to prevent secondary damage. We finished entering the data for 70,000 earth dams in September 1996. The database was first applied to the 1996 Miyagi-ken Hokubu Earthquake. The information derived by the database was very useful for determining the earth dams which must be inspected immediately. Earth dams that should be improved in terms of earthquake resistance can be identified by using this database. We will make the database more useful for preventing damage by increasing the accuracy of maps and adding more data items.

REFERENCE

Tani, S. (1996) , Damage to Earth dams, Soils and Foundations, pp. 71-72.

[APPENDIX]

{Digital National Land Information}
In 1974, the Geographical Survey Institute of Japan started to compile digital geographical information and completed it in 1979. The institute continues to update and improve the accuracy of some of this data. The digital geographical information expresses locations in latitudes and longitudes and divides the land area of Japan into primary regional sections. Each of these sections corresponds to a 1/200,000 geographic map, and is further divided into 64 equal secondary regional sections (each corresponding to a 1/25,000 map). Each of these smaller sections is further divided into 10 equal third regional sections of 1 square meter). The topographical and geological information in the Digital National Land Information is given in units of this grid.
Recently, an information medium that integrates various geographical information for each administrative unit, such as municipalities, has become necessary. An integrated file of small-scale

365

maps (1/200,000 maps, FD maps) was therefore constructed. The FD maps contain data such as administrative borders, coast lines and roads and are provided on a total of 88 floppy disks, for the whole of Japan. This database was permitted to include some of this data.

{AMEDAS}

The Automated Meteorological Data Acquisition System (AMEDAS) is a well known meteorological database, which automatically and constantly measures the temperature, precipitation, sunshine, and the direction and speed of wind for each of 1,316 stations with automatic meteorological equipment, otherwise known as regional meteorological monitoring systems. For 214 stations of high snowfall, the system also measures the amount of snow cover with automatic snow measurers. The data monitored every hour are transmitted in five minutes to the Regional Meteorological Monitor Center (Otemachi, Tokyo), which automatically edits and statistically analyzes the data, and immediately transmits the data to meteorological observatories and information media. Precipitation is monitored for each 17 km2, and other factors are monitored for each 21 km2. The system is one of the most detailed meteorological networks in the world and was inaugurated in November 1974. AMEDAS data for the past were determined from the data monitored by 1,316 automatic meteorological facilities for every hour. The data are provided on magnetic tapes by the Meteorological Service Support Center.

Geotechnical Hazards, Marić, Lisac & Szavits-Nossan (eds) © 1998 Taylor & Francis, ISBN 90 5410 957 2

A study on the ground flow due to liquefaction behind quaywalls

Susumu Yasuda & Tomohiro Tanaka
Tokyo Denki University, Saitama, Japan

Hiroyuki Nomura
Tokyo Engineering Co., Ltd, Japan

ABSTRACT : Liquefaction associated ground flow behind quaywalls was studied based on case studies, shaking table tests and analyses. Damage to bridges and other structures due to the ground flow during three earthquakes was introduced. The mechanism of the ground flow was demonstrated by the shaking table tests. The quaywalls moved towards the sea or river due to the liquefaction of the ground and inertial force of the wall, then the ground flowed with structures. A simple prediction method to estimate the ground flow was derived from case studies. Effectiveness of the measures by installing sheet piles or improving the ground was demonstrate by the shaking table tests.

INTRODUCTION

The study on liquefaction induced ground flow started after the 1983 Nihonkai-chubu earthquake in Japan because many ground flows occurred on gentle slopes of sand dunes. There are two types of the liquefaction induced ground flow : ①ground flow on gentle slopes ② ground flow behind quaywalls. Studies on ground flow had been mainly focused on the first type before the 1995 Hyogoken-nambu (Kobe) earthquake. However, as the second type of ground flow occurred at many sites during the Hyogoken-nambu earthquake, the authors have studied on the second type of ground flow by conducting several shaking table tests, case studies and analyses.

DAMAGE TO STRUCTURES DUE TO LIQUEFACTION ASSOCIATED GROUND FLOW

About ten years ago, the liquefaction associated ground flow caused by the 1964 Niigata earthquake was measured by pre- and post earthquake aerial survey (Hamada et al., 1986), and clarified that extremely large ground displacements, up to eight meters, occurred along Shinano River as shown in Fig. 1. River shores had been protected mainly by sheet piles. The sheet piles were pushed towards the center of the river and fell down on the bottom of the river. And the ground behind the sheet piles flowed towards the river. The maximum distance of the flowed ground was about 300 meters. Many structures, such as road bridges, a railway bridge,

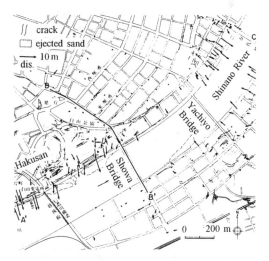

Fig. 1 Horizontal movement on the ground surface in Niigata City during the 1964 Niigata earthquake (Hamada et al. 1986)

buried pipes and buildings, were severely damaged due to the ground flow.

The 1991 Terile-Limón earthquake brought severe damage due to liquefaction at many locations in low land area in Costa Rica. Especially three bridges of Route 36, which runs along the Caribbean Sea, collapsed due to liquefaction and liquefaction-associated ground flow. The Río Vizcaya bridge, a 3-span prestressed concrete I-beam bridge, collapsed completely. One internal support is missing and was supposed to settle down due to liquefaction. The

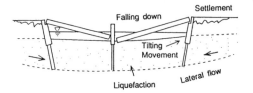

Fig. 2 Mechanism of the damage of bridges due to the ground flow of the river bank during the 1991 Terile-Limón earthquake in Costa Rica (Mora et al. 1994)

abutment rotated 8 degrees and was pushed towards the center of the river due to the ground flow. The Río Bananito bridge and the Río Estero Negro bridge, both are 2-span prestressed concrete I-beam bridges, had similar damage as the Río Vizcaya bridge. The mechanism of the collapse of the bridges is supposed to be as follows (Fig. 2). The ground at one river bank or both river banks liquefied down to the depth of the pile tip or more, and the ground flowed towards the river center, pushing and /or sweeping off abutments and causing down of the span.

Many quaywalls and reinforcements moved toward the sea and settled during the 1995 Hyogoken-nambu (Kobe) earthquake in and around Kobe City in Japan. For example, the average horizontal and vertical displacements of quaywalls in Port Island were 2.7 m and 1.3 m, respectively. Photo 1 shows the moved quaywalls and flowed ground at Rokko Island which is an artificially reclaimed island. The ground behind quaywalls liquefied and flowed toward the sea due to the movement of the quaywalls. The ground flow brought severe damage to many structures, such as bridges, buildings and lifelines. The area to which flow expanded was almost 100 to 200 m behind quaywalls.

Photo 1 Moved quaywall in Rokko Island during the 1995 Hyogoken-nambu (Kobe) earthquake

SEVERAL TESTS TO DEMONSTRATE THE MECHANISM OF GROUND FLOW

Several shaking table tests on two types of quaywall models were carried out to demonstrate the mechanism of the liquefaction associated ground flow behind quaywalls. Figures 3 and 4 show the models for a caisson type quaywall which simulates the quaywall in Kobe and a sheet pile type revetment which simulates the river revetment in Niigata, respectively.

The soil container used was 2200 mm in length, 500 mm in depth and 450 mm in width. The height of the model ground was 300 mm. A model quaywall of 110 mm in height and 57 mm in thickness, and a model sheet pile of 180 mm in length and 1.2 mm in thickness were placed as shown in Figs. 3 and 4, respectively. A model foundation with 4 piles and a footing was placed in the first type of tests to demonstrate the effect of the ground flow on the pile foundation. Very clear Toyoura sand with a mean diameter of 0.175 mm was used. Relative density of a liquefiable layer was adjusted as 40 %. Several piezometers, several accelerographs, several pairs of strain gauges and two displacement transducers were installed in the ground or put on the models. Shaking motion was applied for ten seconds in one direction parallel to the horizontal axis in the figures at a frequency of 3 Hz with several amplitude of acceleration.

Figures 5 and 6 shows time histories of the displacement of the model caisson and the model sheet pile with different amplitude of shaking motion, respectively. As shown in the figures; final displacement of the models and velocity of movement increased with the amplitude of shaking motion in both cases. However, the velocity of movement for the caisson was faster than that for the sheet pile because the inertial force for caisson is stronger. Figure 7 shows a recorded pore pressure in the case for caisson. It is interesting that a negative pore pressure was observed in the ground just behind the caisson during the process of movement of the caisson. This implies that the caisson moved due to not only the force by lateral earth pressure which increased with liquefaction, but also due to the inertial force. On the contrary, the sheet pile must moved due to the increased lateral earth pressure only, because no negative pore pressure was observed in the ground behind the wall during the movement of the wall.

Relationships between the maximum and residual displacements of the model pile foundation and the shaking acceleration in the case of caisson type quaywall are shown in Fig. 8. Both the maximum and residual displacements increased with the shaking acceleration. It is noted that some amount of residual displacements were observed due to the ground flow. This type of residual displacement, up to about 1 m,

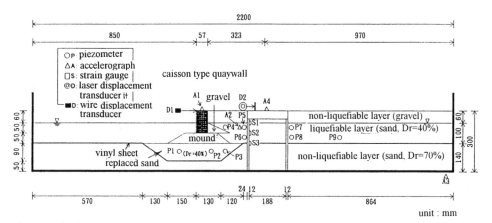

Fig. 3 Model for caisson type quaywall and foundation

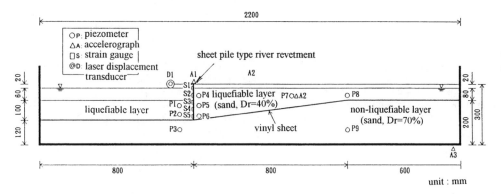

Fig. 4 Model for sheet pile type river protection

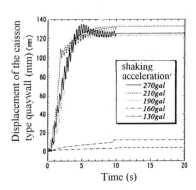

Fig. 5 Time history of the displacement of the caisson type quaywall

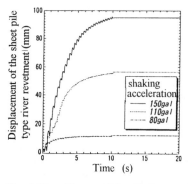

Fig. 6 Time history of the displacement of the sheet pile type river revetment

was observed on the foundations for bridges, buildings and other many facilities during the Hyogoken-nambu earthquake.

Figure 9 explains the mechanism of the liquefaction associated ground flow behind quaywalls which was estimated based on the damage during the past earthquakes, the shaking table tests mentioned before and other tests and analyses conducted by other researchers. A quaywall, a river revetments or a sea revetments moves and settles because of the increase of lateral earth pressure due to liquefaction of the ground behind the wall, and the decrease of strength

369

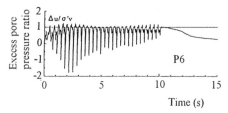

Fig. 7 Time history of the pore pressure near the caisson type quaywall

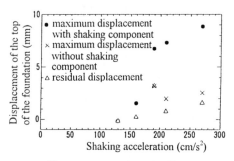

Fig. 8 Relationship between displacement of the foundation in the case of caisson type quaywall

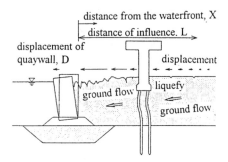

Fig. 9 Mechanism of liquefaction induced ground flow and its damage to a pile foundation

of replaced sand under the wall due to the increase of pore pressure. In case of caisson type wall, inertial force also causes the movement of the wall. Then the liquefied ground behind the wall flowed towards the sea or the river. Lateral force due to the ground flow acts on structures such as pile foundations and causes severe damage.

DEVELOPMENT OF A SIMPLE DESIGN METHOD FOR THE GROUND FLOW

There are three approaches to consider the effect of ground flow into the design of pile foundation or buried pipes:

(1) evaluate the deformation of structures with the deformation of the ground simultaneously,

(2) evaluate the value of pressure which acts on the structures due to the ground flow at first, then evaluate the deformation of structures, and

(3) estimate the displacement of the ground at first, then evaluate the deformation of structures

The first approach is logically correct, but is difficult even though by the recent effective response analysis methods because of the difficulty of estimation of large ground displacement and interaction between the ground and structures.

The second approach is simple and was introduced already in the new specification for highway bridges in Japan in 1996 (The Japanese Road association, 1996). In the new design code, the following method was derived based on back analyses of the damaged highway bridges during the Hyogoken-nambu earthquake :

a) Area to be consider is within 100 m behind a quaywall which is higher than 5 m. Thickness of liquefiable layer is more than 5m.

b) The following forces, shown in Fig. 10, are applied to foundation :

$$q_{NL}=C_s C_{NL} K_p \gamma_{NL} X \qquad (0 \leqq X \leqq H_{NL}) \qquad (4)$$
$$q_L=C_s C_L \{ \gamma_{NL} H_{NL}+ \gamma_L (X-H_{NL}) \}$$
$$(H_{NL} < X \leqq H_{NL}+H_L) \qquad (5)$$

where, C_s: correction factor for the distance, S, from quaywall, 1.0 for $S \leqq 50$ m, 0.5 for $50 < S \leqq 100$ m

C_{NL}: correction factor in non-liquefy layer, 0 for $P_L \leqq 5$, $(0.2P_L-1)/3$ for $5 < P_L \leqq 20$, 1 for $P_L > 20$

C_L: correction factor in liquefy layer, =0.3

K_p: coefficient for passive pressure

X: depth (m)

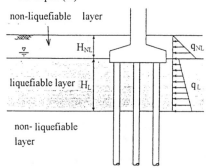

Fig. 10 Earth pressure considered in the specification for highway bridges (The Japanese Road Association 1996)

In the third approach, the displacement of the ground must be estimate at first. There are three grades of estimating methods to estimate the liquefaction associated ground flow :

1) estimate the ground displacement by an empirical formula,
2) estimate the ground displacement by a residual deformation analysis, and
3) estimate the ground displacement by an effective stress response analysis.

A simple method by the first grade has been developed by Ishihara, Yasuda and Iai (Ishihara et al., 1997, Yasuda et al., 1997) based on the following case studies :

In Japan many quaywalls and revetments have been damaged due to liquefaction during past earthquakes. Among them, data on the damage to quaywalls or revetments during the recent six earthquakes, 1964 Niigata, 1978 Miyagiken-oki, 1983 Nihonkai-chubu, 1993 Kushiro, 1993 Hokkaido-naisei-oki and 1995 Hyogoken-nambu (Kobe) earthquakes were reviewed. Based on the collected data, relationships among the displacement of quaywalls and three factors, type of quaywalls, shaking intensity and soil condition, were summarized by Iai as shown in Table 1. The ratio of horizontal displacement to the height, T, was increased with the shaking intensity and thickness of liquefied loose layer. Figure 11 shows relationships between the distance of influence of ground flow, L,

and the ratio of displacement of quaywalls to normalized SPT N values, D/N_1, studied by Yasuda. Though the data are scattered, it can be said that the distance of influence increased with the displacement of quaywalls and decreased with SPT N-values. The mean relationship was $L=250(D/N_1)$. Distributions of displacement on ground surface behind quaywalls were measured at several cross sections on reclaimed lands in Kobe after the Kobe earthquake by Ishihara. Figure 12 summarized the relationships between the ratio of lateral displacement to displacement of quaywalls, U/D, and the ratio of distance from the waterfront to distance of influence, X/L. Lateral displacement on the ground surface decreased with the distance from the waterfront. Simple estimation of the lateral displacement and design of pile foundation against the ground flow can be done based on the study mentioned above. The procedure is as follows: ① estimate the displacement of quaywalls or revetments by Table 1, ② estimate the distance of influence by Fig. 11, ③ estimate the lateral displacement on the ground surface at the site of foundation by Fig. 12, and ④ analyze the bending moment on the pile with an appropriate small soil spring constant.

Table 1 Summery of displacement of quaywalls during past earthquakes (Ishihara et al. 1997, Yasuda et al., 1997, data by Iai)

Type of quaywalls	Intensity of shaking	Depth of liquefied layer	T(%)
Gravity	Level 1 (designed shaking)	Shallower than the bottom of quaywall	5 to 10
		Deeper than the bottom of quaywall	10 to 20
	Level 2 (severe shaking)	Shallower than the bottom of quaywall	10 to 20
		Deeper than the bottom of quaywall	20 to 40
Sheet pile	Level 1 (designed shaking)	Shallower than the bottom of quaywall (ground around anchorage was not liquefied)	5 to 15
		Shallower than the bottom of quaywall (ground around anchorage was liquefied also)	15 to 25
		Deeper than the bottom of quaywall (ground around anchorage was liquefied also)	25 to 50

T: ratio of horizontal displacement of quaywall, D, to the height of quaywall, H

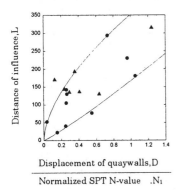

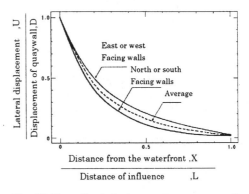

Fig.11 Relationship between distance of influence and D/N_1. (Ishihara et al. 1997, Yasuda et al. 1997, data by Yasuda)

Fig. 12 Normalized displacement versus normalized distance (Ishihara et al. 1997, Yasuda et al., 1997, data by Ishihara)

The authors and their colleague has tried to analyze the deformation of the ground behind quaywalls or river revetments by the second grade. Figure 13 shows a analyzed result for the river revetment in Niigata. In the analysis, the finite element method (FEM) was applied twice as follows:
a) In the first stage, distribution of stress in the ground was calculated by the FEM using the shear modulus before the earthquake.
b) Then, holding the stress constant, the FEM was conducted again using the decreased shear modulus due to liquefaction. The rate of modulus decrease was decided as 1/1000 based on the cyclic torsional shear tests.
c) The difference in deformation measured by the two analyses was supposed to equal the residual deformation.

The appropriate value of the rate of modulus decrease under many conditions has been studied (Yasuda et al., 1995) and the applicability of this method for caisson type wall is studying.

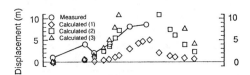

(a) Calculated displacements and measured displacements

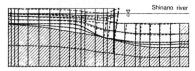

(b) The retaining wall has been installed below the bottom of the liquefied layer

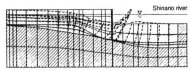

(c) The retaining wall has been installed up to the bottom of the liquefied layer

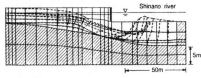

(d) The retaining wall has been installed only up to the middle of the liquefied layer

☐ Liquefied Layer
▨ Not Liquefied Layer
❙ Retaining Wall

Fig. 13 Residual deformation analyses for Shinano River (Yasuda et al. 1992)

TESTS ON COUNTERMEASURES

There are four categories of countermeasures against the damage to structures due to the liquefaction associated ground flow as shown in Fig. 14. Most reliable measure is improve the ground in all area to prevent the liquefaction. However this measure is uneconomical and can not be applied under or near existing structures. The second measure is strengthening structures, for example strengthen by additional piles, to prevent damage even though the flow of the ground occurs. The third measure is strengthen the ground with walls or sand piles, densification of a small area of the ground to prevent the ground flow even though liquefaction occurs. The fourth measure is strengthen quaywalls not to occur the ground flow. The authors has studied the effectiveness of the third measures by conducting several shaking table tests.

The soil container and the model quaywall used for the study is shown in Fig. 15. Dimensions of the soil container is the same as shown in Figs. 3 and 4. A model quaywall was placed 200 mm from the right wall of the container. A model pile made of acrylic plate of 10 mm in thickness and 50 mm in width was positioned 250 mm behind the quaywall. Toyoura sand was used. Relative density of liquefiable layer was adjusted as 50 %. Several piezometers,

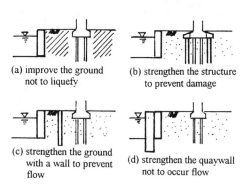

(a) improve the ground not to liquefy

(b) strengthen the structure to prevent damage

(c) strengthen the ground with a wall to prevent flow

(d) strengthen the quaywall not to occur flow

Fig. 14 Countermeasures against the damage due to liquefaction induced ground flow

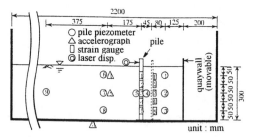

Fig. 15 Model for the shaking table tests on countermeasures

accerelographs, strain gauges, pressure transducers and displacement transducers were installed in the ground or attached on the structures. Shaking motion was applied for ten seconds at a frequency of 3 Hz to induce liquefaction. About 1 to seconds after the stopping the shaking, the model quaywall was moved quickly towards the right side of the container.

Three types countermeasures, install a sheet pile, compact a small area of the ground and cement a small area of the ground, were selected in the study as show in Fig. 16. The model sheet pile was installed in front of the model pile (between the pile and the quaywall) or behind the model pile. Figure 17 shows relationships between displacements of the top of the pile and thickness of the sheet pile. As show in this figure, both the maximum and residual displacements of the pile could reduce by installing the sheet pile.

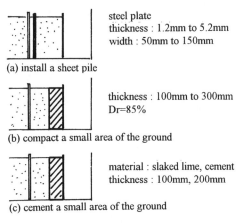

steel plate
thickness : 1.2mm to 5.2mm
width : 50mm to 150mm

(a) install a sheet pile

thickness : 100mm to 300mm
Dr=85%

(b) compact a small area of the ground

material : slaked lime, cement
thickness : 100mm, 200mm

(c) cement a small area of the ground

Fig. 16 Studied countermeasures

The displacements decreased with the thickness of sheet pile. And the installation in front of the pile was more effective than the installation behind the pile.

The density of the compacted ground was 85 %. In the cementation of the ground, Portland cement or slaked lime was mixed with the sand. Figure 18 shows relationships between the maximum and residual displacements of the top of the pile and thickness of the compacted area. Figure 19 shows the displacements of the top of pile with different type of cementation. Both the maximum and residual displacements of the pile could reduce by the densification or cementation of a small area of the ground. The displacements decreased with the thickness of the improved area.

Based on these tests and some analyses by the residual deformation method, the authors are convinced that the measures by the installation of some wall, or the densification or cementation of a small area of the ground, must be effective to prevent the damage to existing structures.

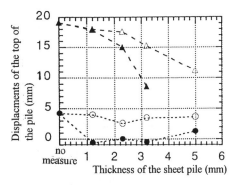

△ : maximum displacement, sheet pile is installed behind the pile
▲ : maximum displacement, sheet pile is installed in front of the pile
○ : residual displacement, sheet pile is installed behind the pile
● : residual displacement, sheet pile is installed in front of the pile

Figure 17 Relationships between displacements of the top of the pile and thickness of the sheet pile

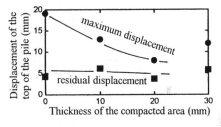

Figure 18 Relationships between the maximum and residual displacements of the top of the pile and thickness of the compacted area

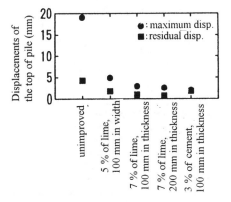

Figure 19 shows the displacements of the top of pile with different type of cementation

373

CONCLUSIONS

Case studies during past earthquakes, several shaking table tests and some analyses on liquefaction induced ground flow behind quaywall, were carried out to study the mechanism, prediction methods and appropriate countermeasures. The following conclusions were derived:

(1) Quaywalls and river or sea revetments move because of the increase of lateral earth pressure due to liquefaction of the ground behind the walls, the decrease of strength of replaced sand under the walls due to the increase of pore pressure, and inertial force of the walls. Then the liquefied ground behind the walls flowed towards the sea or the river.

(2) There are three grades of estimating methods to estimate the liquefaction induced ground flow. The first grade of methods derived from case studies and the residual deformation method which belongs to the second grade methods, introduced in the paper must be useful.

(3) The measures by the installation of some wall and the densification or cementation of a small area of the ground, must be effective to prevent the damage to existing structures.

ACKNOWLEDGMENTS

Shaking table tests were carried out with many researchers and funds by the Association for the Development of Earthquake Prediction, Japan Lime Industry Association, Sumitomo Metal Industries, Ltd. and Nippon Steel Corporation.

REFERENCES

Hamada, M., S. Yasuda, R. Isoyama and K. Emoto 1986. Study on liquefaction induced permanent ground displacement. Association for the Development of Earthquake Prediction.

Ishihara, K., S. Yasuda and S. Iai 1997. A simple method to predict liquefaction induced ground flow. Proc. of the 24th JSCE Earthquake Engineering Symposium : 541-544 (in Japanese).

Mora, S., K. Ishihara, H. Watanabe, S. Yasuda and N. Yoshida 1994. Soil liquefaction and landslides during the 1991 Terile-Limón , Costa Rica, Earthquake. Special Volume for the 13th International Conference on Soil Mechanics and Foundation Engineering : 41-48.

The Japan Road Association 1996. Specifications for Highway Bridges.

Yasuda, S., H. Nagase, H. Kiku and Y. Uchida 1992. The mechanism and a simplified procedure for the analysis of permanent ground displacement due to liquefaction. Soils and Foundations. Vol.32, No.1.: 149-160.

Yasuda, S., N. Yoshida, T. Masuda, H. Nagase, H. Kiku and K. Mine 1995. Stress-strain relationships of liquefied sands. Proc. of the First International Conference on Earthquake Geotechnical Engineering, : 811-816.

Yasuda,S., K. Ishihara, and S. Iai 1997. A simple procedure to predict ground flow due to liquefaction. Abstract Volume for 8th SDEE. :156-157.

Geotechnical Hazards, Marić, Lisac & Szavits-Nossan (eds) © 1998 Taylor & Francis, ISBN 90 5410 957 2

On comparative seismic displacements of rigid retaining walls

Yingwei Wu & Shamsher Prakash
Department of Civil Engineering, University of Missouri-Rolla, Mo., USA

ABSTRACT: A critical review is presented on analysis and design of rigid retaining walls subjected to earthquakes based on limited displacements. Since these walls experience both sliding and rotation, a realistic analysis must consider both motions. A comparison of displacements of a typical wall for several backfills and base soils have been made by the two available methods. The displacements have been compared with Eurocode (1994). Recommendations have then been made for further work.

INTRODUCTION

Conventional static design of rigid retaining walls requires estimating the earth pressure based upon earth pressure theories of Coulomb and Rankine and choosing the wall geometry to satisfy specified factors of safety against sliding, overturning and bearing capacity failure. The factors of safety will be decreased due to increased dynamic pressure under earthquake loading. The rigid walls experience significant sliding and rocking displacement during earthquake. However the wall movement cannot be predicted by the conventional design method (Okabe 1926, Mononobe and Matsuo 1929) and these displacements need to be considered in their design. Prakash and Wu (1997) have prepared a listing of damage of walls during Hokkaido-Nansi-Oki, Northridge and Kobe earthquakes, in which the walls failed by both sliding and rotation.

Richard and Elms (1979) methods considered only sliding of the walls. Rafnsson and Prakash (1994) and Prakash *et.al.* (1995 a, b) developed a solution to predict horizontal movement at the top of a retaining wall under dynamic loading due to simultaneous sliding and rocking motion. Also design charts had been developed for permissible displacement of walls 4m - 10m high and for 21 base soil and backfill combinations. By using this method, Prakash and Wu (1996), Wu and Prakash (1996) compared the performance of two rigid walls during earthquakes with their computations and found good agreement. Also, Wu and Prakash (1997) used only some

provisions of Eurocode-8 Ch7 (1994) to study performance of retaining walls during earthquake with dry or submerged conditions.

A critical review of the behavior of the retaining walls during earthquakes, available methods of analysis and design, including Eurocode (1994) and a comprehensive displacements of a typical wall for several backfills and base soils have been made and recommendations made for further work.

BEHAVIOR OF RETAINING WALLS DURING EARTHQUAKES

Several investigators have recently studied displacements of rigid walls (Fig. 1 and Fig. 2).

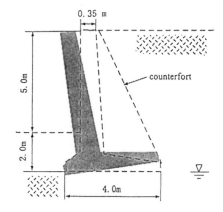

FIG. 1 Failure of cantilever retaining wall after Kobe earthquake, 1995 (Iai, 1998)

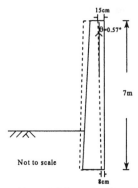

Not to scale

Observed sliding and rocking displacement

FIG. 2 Displacements of wingwalls during the 1981 earthquake in Greece (Rafnsson, 1991).

Based on similar observations, the following behavior of walls is presented below.

Fig. 3 shows a retaining wall (a) subjected to ground motion (b) and its response (c). During the time interval, $0 - t_2$ (Fig. 1b), the wall moves away from backfill from $O_1 b_1$ (Fig. 1c). From time t_2 to t_4, the ground motion and forcing function reverses in direction. But the wall may still have residual velocity and continue to move away from the fill. There may or may not be a partial recovery of the displacement of the wall to c_1 or c_2, because of the large passive resistance to wall movement towards the fill. It will thus be seen that the wall has moved out by c_2c_2' (Fig. 1c) from the original equilibrium position (under one cycle of ground motion). With additional significant pulses of ground shaking, the wall will keep moving away from its static equilibrium position. Hence, the stability of walls will be dependant on the amount of displacement during earthquakes (Prakash, 1981).

It is therefore necessary that a rational design method includes some "*permissible displacement*". Hence, method to compute displacements of rigid walls are also needed.

In practice, the walls experience both sliding and rotation. Therefore, realistic methods must account for both types of motions.

STATE OF ART FOR ANALYSIS AND DESIGN

There are a few design methods which consider permissible displacements.

Richards and Elms Method (1979)

A simplified method for dynamic design of rigid retaining walls had been proposed by Richards and

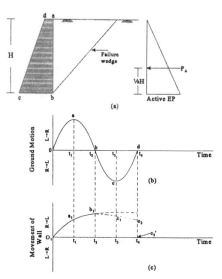

(a)

(b)

(c)

FIG.3 Response of retaining wall to ground shaking. (a). Retaining wall (b). Ground motion and (c). Wall displacement (Prakash, 1981)

Elms (1979). This method is based on Newmark's sliding block analysis (1965) and Franklin and Chang's (1977) solution for upper bound permanent displacements for several natural and synthetic ground motions. This approach determines the wall dimensions based on *permissible displacement*. A brief description of this method is given below:

1. Select permissible displacement (d). *In this method, a permissible displacement is preselected.*
2. Determine A_a and A_v acceleration-coefficients for a given seismic zone (Applied Technology Council, 1978).
3. Determine coefficient of cut off acceleration α_h (Eq. 1).

$$\alpha_h = A_a \left[\frac{0.2A_v^2}{A_a d} \right]^{0.25} \quad (1)$$

Where d is permissible displacement in inches.

4. Compute dynamic active lateral earth pressure behind the wall by using Mononobe-Okabe method for α_h computed in Equation 1.
5. Compute weight of wall by using inertia force of the wall and considering force equilibrium.
6. Apply a factor of safety to the calculated weight. A value of 1.5 is recommended and wall dimensions are then determined

For details, see Richards and Elms (1979), Wu (1995) and Prakash and Wu (1996).

Only sliding motion is considered in this method.

376

Richards and Elms (1979) do not suggest how to determine a permissible displacement for the wall.

Rafnsson's Analysis (1991)

A solution for simulating the response of rigid walls during an earthquake has been proposed by Rafnsson (1991). This model consists of a rigid wall resting on the surface of the soil and subjected to horizontal ground motion. Both material and geometrical damping in sliding and rocking motions have been considered (Rafnsson 1991, Rafnsson and Prakash 1994). Mathematical model in Fig. 4 represents the displacements in active case. Nonlinear behavior of soil is included in defining properties, both at the *base* as well as the *backfill*: Soil stiffness in sliding and rocking; Geometrical damping in sliding and rocking; Material damping in sliding and rocking. For details see Rafnsson (1991).

In Rafnsson (1991) work, the wall dimensions are determined for given factors of safety under static condition and cumulative displacements and rotation of wall are then computed for different loading cycles (magnitude of earthquake) for a given ground motion (sinusoidal motion). Nonlinear soil modulus and strain dependant material damping are used in this solution are shown in Fig. 5 and 6 respectively.

Salient conclusions from this study are:
Effect of horizontal acceleration

Displacements increase with greater horizontal accelerations because higher ground acceleration causes larger exciting force.

Effect of earthquake magnitude

The earthquake magnitude can be represented by number of cycles of ground motion (Seed *et al.*, 1983). A larger earthquake magnitude has larger number of cycles. More cycles of earthquake motion cause larger displacements of retaining wall.

Effect of base soil

The displacements decrease with stiffer soils because the stiffnesses of the base soil is larger.

Effect of wall height

The displacement increases in a non-linear fashion with increasing height of wall. This is obvious because actuating moments increase with square of wall height.

Effect of backfill

The cumulative displacement increases slightly as the backfill becomes loose. However, the increase in

displacement is small and, for all practical purpose, this difference is of no practical significance.

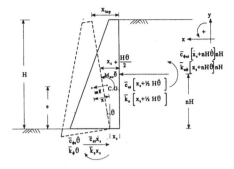

FIG. 4 Mathematical model for stiffness and damping constants for the active case (Rafnsson 1991, Rafnsson and Prakash 1994).

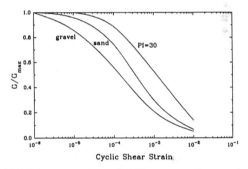

FIG.5 Average values of G/G_{max} versus shear strain (γ) for different soils (After Seed and Idriss 1970, for sand; Seed, Wong, Idriss and Tokimatsu 1986, for gravel; Vucetic and Dobry 1991, for clay with PI=30).

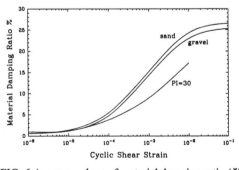

FIG. 6 Average values of material damping ratio (ζ) and shear strain (γ) for different soils (After Seed and Idriss 1970, for sand; Seed, Wong, Idriss and Tokimatsu 1986, for gravel; Vucetic and Dobry 1991, for clay with PI=30).

377

The merit of the above study lies in the fact that:

1. Sliding and rocking motion was considered in a general way for the first time in an analysis.

2). Non-linear soil properties of both the base soil and the backfill have been considered.

3). The computed displacements were of the *same order* as measured displacement in case of retaining wall of 23ft height in Greece (Rafnsson, 1991).

Prakash, Rafnsson and Wu's (PRW) method

By using Rafnsson's model, a complete design procedures and a computer program (Prakash *et al.*, 1996) had been developed. A complete displacement based design procedure was developed which includes:

1. Displacements computed under given equivalent earthquake motion.
2. Safety evaluation of walls by comparing the computed displacement and permissible displacement.
3. Recommendation and revised design (if necessary).

Furthermore, displacements for walls, 4m - 10m high, subjected to ground motion of 0.1g - 0.5g, earthquake magnitude of M5.25 - M8.5 and 3 backfills and 7 base soils (Table 1) had been chosen for detailed study from 3675 possible combinations.

Based on the computation, design charts were prepared for permissible displacements of 1% to 10% height of walls.

However, a discussion on permissible displacements is necessary at this point.

Permissible Displacement

There are no guidelines on the permissible displacements of rigid retaining walls in the field except in Eurocode. Free standing retaining walls are used as wing walls along the spillway structures and as retaining structures along highways. The magnitude of displacements of free standing retaining walls that may be tolerated depends upon:
1. its effect on nearby structures, 2. economy and 3. "*psychological*" effect of a wall displacement.

A wall that is sloping away from the fill after an earthquake may perhaps pose no danger, but it may be perceived to have failed.

Eurocode (1994) introduced permissible displacement for the first time, for the free standing wall as:
1. permissible displacement = $300 \times \alpha$ (mm)   (2)
   Where α is the design seismic coefficient.

The permissible displacement is considered as function of design acceleration only.

Eurocode further recommends consideration of :
2. Non-linear behavior of base soil and backfill.
3. Wall movements in sliding and rocking.

COMPARISON OF RESULTS BY THE TWO METHODS

Parameters Used for Comparison

For comparison of displacement by the two methods, a typical wall of 6m high was studied. The basis of comparison is as follow:

Table 1. Soil properties of base soil and backfill used for analysis (Prakash *et al.* 1995a).
BASE SOIL (BS)

	soil type	γ_d kN/m³	ϕ deg	δ deg	void ratio	ν	c kN/m²	PI	w%
BS1	GW	21.07	37.5	25.0	0.25	0.3	-	-	6
BS2	GP	19.18	36.0	24.0	0.36	0.3	-	-	6
BS3	SW	18.00	35.0	23.3	0.46	0.3	-	-	8
BS4	SP	16.82	34.0	22.7	0.56	0.3	-	-	10
BS5	SM	16.51	33.0	22.0	0.68	0.3	-	4	11
BS6	SC	15.25	30.0	20.0	0.95	0.3	-	13	14
BS7	ML	14.15	32.0	21.3	0.85	0.3	9.57	4	14

BACKFILL (BF)

	soil type	γ_d kN/m³	ϕ deg	δ deg	void ratio	ν	c kN/m²	PI	w%
BF1	GM	19.6	33.0	22.0	0.35	0.3	-	-	10
BF2	GP	18.9	34.0	22.7	0.40	0.3	-	-	8
BF3	SP	15.6	34.0	22.7	0.50	0.3	-	-	8

* All properties for backfill are determined from 90% of the "Standard Proctor" test.

Fix the section by RE's recommendations; then compute the displacements by PRW method and compare with RE permissible displacement.

In *RE* method, permissible displacements must be known to determine the wall section. Therefore, a permissible displacement of 2% of wall height, ie 0.12m, was adopted. The values of A_v (0.4) and A_a (0.4) for specific location, e.g. Orange County (CA.), have been adopted (ATC 1978). Other parameters used in this analysis are listed in Table 2. Cutoff acceleration for the parameters in Table 2, computed from Eq 1 is 0.144g. This value depends upon neither (a) the section of the wall nor (b) on the magnitude of earthquakes.

For a displacement of 0.12m, height of 6m and soil properties as in Fig.7, the wall section was fixed according to *RE* recommendations. Base width used in this study are computed by *RE* method considering full base friction angle (ie. $\delta = \phi$).

A plot of cumulative displacements with number of cycles is shown in Fig. 8. In this figure, three accelerations have been used.

M6.75 Earthquakes

Plot (a) in Fig. 8 is displacement computed by *RE's* cutoff acceleration of 0.144g. According to *RE's* procedure, there is no displacement of the wall for ground motion smaller than cutoff acceleration. However, by *PRW* method, the displacement in 10 cycles (M6.75) is 4.39cm. In *RE's* method, the soil is assumed rigid plastic. Therefore, these displacements cannot be determined and are neglected.

Plot (b) in Fig. 8 is for ground motion of 0.4g. The displacement for 10 cycles (M6.75) is 28.15cm. This is larger than the permissible displacement of *RE's* (12cm) which is the basis of fixing the wall section.

Plot (c) in Fig. 8 is for ground motion of 0.256g (the difference of 0.4g and 0.144g). The displacement for 10 cycles (M6.75) is 14.56cm. This is also larger than *RE's* displacements of 12cm.

Rotation of the wall for 0.4g ground motion and earthquake of M6.75 is 1.29° which results in 13.51cm displacement at the top. Therefore, displacements in sliding only are 14.64 cm for M6.75 earthquake. These values are much closer to the permissible displacement of 12cm used in *RE's* design method. In fact sliding displacement (14.64 cm) for M6.75 is extremely close to the permissible displacement of 12cm adopted here.

Table 2. Parameters used in the analysis.

Parameter	Range of Values
Height of wall (m)	6
Location used for analysis	Orange County, CA
A_a	0.4
A_v	0.4
Permissible displacement (m)	0.12 (2% of height of wall)
Ground Motion (g)	0.4
Frequency (Hz)	1
Magnitude of earthquake	**M6.75 and M7.5**
Number of cycles	10 and 15
Soil properties for base soil and backfill	See Table 1

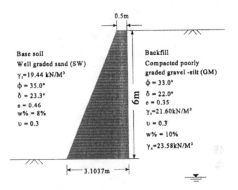

FIG. 7. Section of wall for permissible displacement of 12cm and soil combination used for analysis (BF1-BS3).

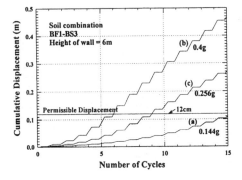

FIG. 8 Cumulative displacements of proposed wall section by using three ground motions (BF1-BS3).

It will thus be seen that:

1. There are significant displacements of this wall for the cutoff acceleration of 0.144g. In real soils, which may be not exhibit rigid-plastic behavior, neglecting those displacements will be unsafe.
2. For design earthquake acceleration of 0.4g, the displacements of the walls is 2.35 times the permissible displacement for earthquake of M6.75 as compared to *RE's* displacements.
3. *RE* method predicted closely only sliding displacements for this wall for M6.75.
4. This wall experiences comparable displacements in rocking also. These cannot be computed by *RE* method.

To continue this comparison further, seven types of base soils and three types of backfills were selected (Table 1). Wall height and top width were 6m and 0.5m respectively throughout. Cumulative displacements are computed for 3 ground motions, ie. 0.144g, 0.4g and 0.256g, and for M6.75 (10 cycles). Fig. 9 shows the range of displacements at these three ground motions. Table 3 lists sliding displacements, rocking degrees and total displacements at the top of the wall. A critical study of Table 3 shows that:

1. The total displacement for cutoff acceleration of 0.144g varies from 4.39cm to 20.84cm. This is neglected in *RE* method.

Table 3. Sliding, and cumulative displacements and rocking degrees computed by various ground motions for M6.75 (10 cycles) by *PRW* method (After Wu and Prakash, 1996).

Soil comb.	base[1] width (m)	Displacement (cm) and rocking (°)[2] in 10 cycles								
		0.144 (g)			0.256 (g)			0.4 (g)		
		Sliding	Rock.	Total	Sliding	Rock.	Total	Sliding	Rock.	Total
BF1-BS1	2.5927	3.12	0.36	6.72	8.40	1.07	19.61	14.97	1.94	35.29[3]
BF2-BS1	2.2636	3.20	0.56	9.20	8.62	1.45	23.80	14.67	2.52	41.06[3]
BF3-BS1	1.7759	4.18	1.20	16.78	8.73	2.53	35.22	14.16	4.10	57.10[3]
BF1-BS2	2.8891	3.83	0.40	8.03	10.77	1.11	22.39	18.10	2.10	40.09[3]
BF2-BS2	2.5321	4.44	0.58	10.44	10.82	1.47	26.21	18.49	2.55	45.19[3]
BF3-BS2	1.9970	5.26	1.16	17.26	10.83	2.49	36.91	17.55	4.05	59.96[3]
BF1-BS3	3.1037	2.30	0.20	4.39	7.54	0.67	14.56	14.64	1.29	28.15
BF2-BS3	2.7266	2.46	0.28	5.39	7.79	0.89	17.11	14.33	1.65	31.61
BF3-BS3	2.1572	3.07	0.56	8.93	7.88	1.48	23.38	13.68	2.59	40.80[3]
BF1-BS4	3.3338	2.97	0.23	5.38	9.69	0.74	17.44	18.14	1.40	32.80
BF2-BS4	2.9350	3.29	0.32	6.64	9.88	0.98	20.14	17.74	1.77	36.28[3]
BF3-BS4	2.3288	3.96	0.62	10.45	9.74	1.58	26.29	16.81	2.75	45.61[3]
BF1-BS5	3.5812	4.45	0.30	7.59	13.20	0.90	22.62	23.99	1.64	41.16[3]
BF2-BS5	3.1591	4.88	0.42	9.28	13.34	1.17	25.59	23.61	2.07	45.29[3]
BF3-BS5	2.5134	5.41	0.79	13.68	12.96	1.86	32.44	22.25	3.20	55.76[3]
BF1-BS6	4.4515	10.19	0.44	14.80	25.26	1.12	36.99	43.93	1.95	64.35[3]
BF2-BS6	3.9476	10.41	0.59	16.59	24.36	1.39	38.92	42.21	2.41	67.45[3]
BF3-BS6	3.1627	10.32	0.97	20.84	22.81	2.12	45.01	38.10	3.56	75.38[3]
BF1-BS7	3.8481	3.99	0.21	6.19	10.65	0.60	16.93	19.43	1.10	30.95
BF2-BS7	3.4010	4.06	0.29	6.99	10.59	0.77	18.66	18.73	1.36	32.97[3]
BF3-BS7	2.4126	4.23	0.49	9.36	10.09	1.18	22.45	17.18	2.02	38.33[3]

[1] Base widths are determined based on *RE* method   [2] Displacements are computed by *PRW* method.
[3] Displacements larger than 5% height of the wall and may represent failure.

2. Displacements in sliding only for 0.4g ground motion is from 14.16cm to 43.93cm. Only in 12 of the 21 cases in this study, the sliding displacements were within 150% of the *RE's* permissible displacements.
3. Total displacements for 0.4g ground motion are from 28.15cm to 32.80cm, neglecting cases with displacements greater than 30cm (5% of the height of the wall) which may constitute failure.

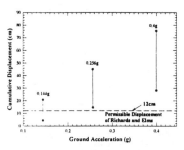

FIG. 9 Displacements computed from different ground accelerations for M6.75.

M7.5 Earthquake

To continue these comparison further cumulative displacements for M7.5 (15 cycles) are presented.
1. In Fig. 8 for cutoff acceleration 0.144g, cumulative displacement by PRW method is 10.86cm.
2. For design acceleration 0.4g, cumulative displacement by PRW method is 46.70cm.
3. For acceleration 0.256g, cumulative displacement by PRW method is 27.12cm.

Fig. 10 shows the range of displacements at these three ground motions. Table 4 lists sliding displacements, rocking degrees and total displacements at the top of the wall. A critical study of Table 4 shows that:
1. The total displacement for cutoff acceleration of 0.144g varies from 10.86cm to 36.40cm. This is neglected in *RE* method.
2. Total displacements for 0.4g ground motion are from 46.70cm to 114.72cm
3. Displacements at ground motion 0.256g are from 21.23cm to 67.93cm.
4. None of the cases in this study show the sliding displacements were within 150% of the *RE's* permissible displacements.

Two more cases were analyzed further as:
1. Increase the weight factor in RE method from 1.5 to 2. The permissible displacement remains the same (12cm). New wall section (Fig 11) is

determined by RE method again. The cumulative displacements computed by PRW method with three acceleration are shown in Fig 12.
2. Increase the permissible displacement to 3 % of height of wall (18cm). The cutoff acceleration is determined by RE method as 0.13g. The new wall section is shown in Fig. 13. The cumulative displacements computed by PRW method with three acceleration are shown in Fig 14.

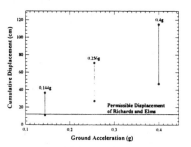

FIG. 10 Displacements computed from different ground accelerations for M7.5.

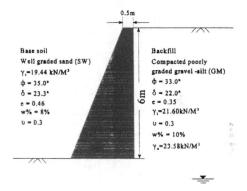

FIG. 11 Section of wall with weight factor of 2 and soil combination used for analysis (BF1-BS3).

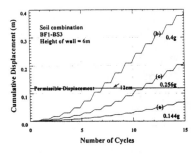

FIG 12 Cumulative displacements of proposed wall with new wall section (weight factor =2).

381

Table 4. Sliding, and cumulative displacements and rocking degrees computed by various ground motions for M7.5 (15 cycles) by *PRW* method.

Soil comb.	base[1] width (m)	Displacement (cm) and rocking (°)[2] in 10 cycles								
		0.144 (g)			0.256 (g)			0.4 (g)		
		Sliding	Rock.	Total	Sliding	Rock.	Total	Sliding	Rock.	Total
BF1-BS1	2.5927	6.62	0.85	15.52	14.26	1.83	33.42	23.61	3.07	55.76
BF2-BS1	2.2636	6.88	1.17	19.13	13.90	2.39	38.93	22.68	3.91	63.63
BF3-BS1	1.7759	7.13	2.05	28.60	13.35	3.89	54.09	21.23	6.16	85.47
BF1-BS2	2.8891	8.52	0.88	17.74	18.00	1.88	37.69	30.95	3.05	62.89
BF2-BS2	2.5321	8.66	1.19	21.13	17.48	2.41	42.72	28.54	3.95	69.90
BF3-BS2	1.9970	8.81	2.00	29.75	16.60	3.85	56.92	26.26	6.10	90.14
BF1-BS3	3.1037	5.76	0.49	10.86	14.13	1.24	27.12	24.19	2.15	46.70
BF2-BS3	2.7266	6.11	0.68	13.23	13.82	1.58	30.37	23.22	2.68	51.28
BF3-BS3	2.1572	6.36	1.19	18.82	13.04	2.46	38.80	21.35	4.06	63.87
BF1-BS4	3.3338	7.33	0.57	13.30	17.39	1.34	31.42	29.64	2.29	53.62
BF2-BS4	2.9350	7.60	0.76	15.56	17.05	1.70	34.85	28.34	2.85	58.19
BF3-BS4	2.3288	7.92	1.28	21.32	15.97	2.60	43.20	26.06	4.28	70.88
BF1-BS5	3.5812	10.38	0.70	17.71	22.85	1.55	39.08	38.34	2.62	65.78
BF2-BS5	3.1591	10.59	0.92	20.22	22.49	1.97	43.12	37.07	3.26	71.21
BF3-BS5	2.5134	10.48	1.50	26.19	20.83	2.99	52.14	33.98	4.90	85.29
BF1-BS6	4.4515	20.11	0.89	29.43	41.16	1.83	60.32	67.93	3.02	99.56
BF2-BS6	3.9476	19.57	1.12	31.30	39.36	2.26	63.03	65.00	3.73	104.07
BF3-BS6	3.1627	18.49	1.71	36.40	35.77	3.34	70.75	57.96	5.42	114.72
BF1-BS7	3.8481	8.36	0.47	13.28	18.55	1.06	29.65	31.36	1.78	50.00
BF2-BS7	3.4010	8.38	0.61	14.77	18.04	1.32	31.86	29.92	2.19	52.85
BF3-BS7	2.4126	8.15	0.96	18.20	16.52	1.94	36.84	27.00	3.16	60.09

[1] Base widths are determined based on *RE* method   [2] Displacements are computed by *PRW* method.

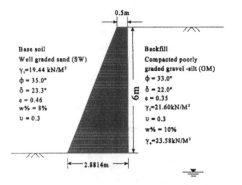

FIG. 13 Section of wall and soil combination used for analysis with permissible displacement of 18 cm (BF1-BS3).

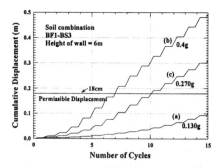

FIG. 14 Cumulative displacements of proposed wall with permissible displacement (18cm).

Table 5. Comparison of displacements with two approaches.

		Weight factor = 1.5 Permissible displ. =12cm			Weight factor = 2.0 Permissible displ. =12cm			Weight factor = 1.5 Permissible displ. =18cm		
Base width (m)		3.1037			4.2657			2.8814		
		Displacement (cm) (Fig. 8)			Displacement (cm) (Fig. 12)			Displacement (cm) (Fig. 14)		
		10 cyl.	15 cyl.		10 cyl.	15 cyl.		10 cyl.	15 cyl.	
Acc^n (g)	cutoff	0.144	4.39	10.86	0.144	3.20	7.15	0.130	3.93	9.90
	max. acc^n	0.400	28.15	46.70	0.400	22.05	38.91	0.400	30.26	49.47
	difference	0.256	14.56	27.12	0.256	9.95	21.10	0.270	17.47	31.06

Typical results for the above 2 cases along with those of Fig.8 are summarized in Table 5 as:
1. By increasing the weight of wall, the displacement at cutoff acceleration of 0.144g at 15 cycles decreases from 10.86cm to 7.15cm; at 0.4g acceleration, it decreases from 46.70 to 38.91cm and at 0.256g acceleration decreases from 27.12cm to 21.10cm.
2. By increasing the permissible displacements, the maximum displacements at 0.4g acceleration and 15 cycles increase from 46.70 to 49.70cm. This is due to the fact that in RE's method, a larger permissible displacement gives a smaller base width.

In both cases, the displacements are too large. For further details see Wu (1998).

Comparison with Eurocode

In Eurocode, the permissible displacement in both sliding and rocking is only 12cm (Eq.2). In all these cases, the displacements are larger than this value.

CONCLUSIONS

Based on solution developed to predict the displacements of rigid retaining walls subjected to both sliding and rocking motions, computed displacements have been compared with Richards and Elms method for typical wall and a wide variety of soil combinations.
1. It is seen that for earthquake of maginitude M6.75, computed sliding displacements are close to RE permissible displacements. For M 7.5, the displacements are very large as compared to permissible displacement. A reasonable design of wall cannot be prepared for RE displacements.
2. Eurocode introduced permissible displacement in the design criteria. The permissible displacement has been related to ground motion only. This needs further evaluation.

3. In all these analysis, a real ground motion has been represented by equivalent number of sinusoidal ground motion. The displacements analysis need be carried out with real ground motion (see Wu, 1998).

REFERENCES

Applied Technology Council, (1978), "Tentative Provisions for the Development of *SeismicRegulations for Building*", Prepared by Applied Technology Council, Associated with the structure Engineers Association of CA.
EUROCODE 8 (EUROPEAN PRESTANDARD 1994) "Design Provisions for Earthquake Resistance of Structures- Part 5: Foundations, Retaining Structures and Geotechnical Aspects", The Commission of the European Communities.
Franklin, A. G., and Chang, F. K., (1977), :Earthquake Resistance of Earth and Rockfill Dams; Report 5, "Permanent Displacements of Earth Embankments by Newmark Sliding Block Analysis", U.S. Army Engineer Waterways Experiment Station, Soils and Pavements Laboratory, Vicksburg, Miss., pp. 1-38.
Iai, S. (1998), "Rigid and Flexible Retaining Walls During Kobe Earthquake", Proc. of the Forth International Conference on case histories in Geot. Engg., ST. Louis, MO., Mar. 8-12, CD-ROM, SOA-4, pp.108-127.
Mononobe, N. and Matsuo, H. (1929), "On the Determination of Earth Pressure During Earthquakes", World Engineering Congress Proceedings, Vol. IX, Tokyo, pp. 177-185.
Newmark, N. M. (1965), "Effects of Earthquakes on Dams and Embankments", The Institution of Civil Engineers, The Fifth Rankine Lecture, Geotechnique, Vol. 15, No.2, January, pp.137-161.
Okabe, S. (1926), "General Theory of Earth Pressure," *Journal of the Japanese Society of Civil Engineers*, Tokyo, Japan, Vol. 12, No. 1.

Prakash, S. (1981), "*Soil Dynamics*" McGRAW-HILL Book Co., New York, N.Y., Reprint, Shamsher Prakash Foundation, Rolla MO.

Prakash, S.,Wu, Y. and Rafnsson, E. A., (1995, a), "On Seismic Design Displacements of RigidRetaining Walls", Proc. Third International Conference on Recent Advances in Geotechnical Engineering and Soil Dynamics, ST. Louis Vol. III, pp. 1183-1192.

Prakash, S.,Wu, Y. and Rafnsson, E. A., (1995, b), "*Displacement Based Aseismic Design Charts For Rigid Walls*", Shamsher Prakash Foundation, Rolla MO.

Prakash, S. and Wu, Y., (1996), "Displacement of Rigid Retaining Walls During Earthquakes ",11th World Conference on Earthquake Engineering, June, 23-28, CD-ROM.

Prakash, S., Wu, Y. and Rafnsson, E. A., (1996), "DDRW-1 - Soft Ware to Compute Dynamic Displacements of Rigid Retaining Walls" Shamsher Prakash Foundation, August.

Prakash, S. and Wu, Y. (1997), "Retaining Structures Under Earthquake Loading", 16th Central PA Geotechnical seminar "Excellent in Geotechnical Engineering" Harrisburg, PA, Oct. 22, 24.

Rafnsson, E. A. and Prakash, S. (1991), "Stiffness and Damping Parameters for Dynamic Analysis of Retaining Walls", Proc. 2nd International Conference on Recent Advances in Geotechnical Earthquake Engineering and Soil Dynamics, Vol. III, pp. 1943-1952.

Rafnsson E. A. and Prakash, S., (1994), "Displacement Based Aseismic Design of Retaining Walls", Proc. XIII Inter. Conf. SMFE, New Delhi, Vol 3, pp. 1029-1032.

Richards, R. and Elms, D. G., (1979), "Seismic Behavior of Gravity Retaining Walls", *J Geot. Engg.Dn.*, ASCE, Vol. 105, No. GT4, April, pp. 449-464.

Seed, H. B. and Idriss, I. M., (1970), "*Soil Moduli and Damping Factors for Dynamic ResponseAnalysis*", Report No.EERC 70-10, Earthquake Engineering Research Center, Dec., pp 1-40.

Seed, H. B., Idriss, I. M., and Arango, I., (1983), "Evaluation of Liquefaction Potential Using Field Performance Data", *J. of Geot. Engg.*, Vol. 109, No. 3, March, pp. 458-482.

Seed, H. B., Wong, R. T., Idriss, I. M. and Tokimatsu, K., (1986), "Moduli and Damping Factors for Dynamic Analysis of Cohesionless Soils", *J Geot. Engg.*, Vol. 112, No. 11, November, pp. 1016-1032.

Vucetic, M. and Dobry, R., (1991), "Effect of Soil Plasticity on Cyclic Response", *J Geot. Engg.*, ASCE, Vol. 117, No. 1, Jan., pp. 89-107.

Wu, Y. (1995), "*Displacement Based Seismic Design Charts for Rigid Retaining Walls*", M.S. Thesis, Univ. of Missouri-Rolla.

Wu, Y. and Prakash, S (1996) "On Seismic Displacements of Rigid Retaining Walls" "Analysis and Design of Retaining Structures Against Earthquakes", ASCE Geot. Spec. Pub. No.60, pp. 21-37.

Wu, Y. and Prakash, S (1997) "Eurocode Based Aseismic Design of Retaining Walls" XIV International Conf. On Soil Mechanics and Foundation Engineering, Hamburg, Germany, Sep. 6-12, Vol I, pp.747-750.

Wu, Y., (1998), "Response of Rigid Retaining Walls to Real Earthquakes", Ph.D. Thesis, Univ. of Missouri-Rolla, USA (Uunder preparation).

Geotechnical Hazards, Marić, Lisac & Szavits-Nossan (eds) © 1998 Taylor & Francis, ISBN 90 5410 957 2

Flow failure – Some data on onset conditions

S. Zlatović
Faculty of Civil Engineering, University of Zagreb, Croatia

K. Ishihara
Tokyo Science University, Japan

ABSTRACT: Landslides of flow failure type mostly occur in very fine sands or non-plastic silty soils that are deposited in a very loose state. It has been shown that the residual strength of these soils depends significantly on the soil fabric, but it is very difficult or even impossible to take undisturbed specimens from these soils. This is why in situ tests are very important if combined with the understanding of the state boundary between dilative and contractive behaviour. To contribute to this understanding, a series of laboratory data is shown in terms of I_{Bu} versus r_c obtained on various sands and non-plastic silts and using three distinctive methods of specimen preparation. For the tested materials, contractive behaviour can be expected if the consolidation pressure (under conditions of isotropic consolidation) is at least 2 times larger than the mean effective stress at the quasy-steady state line.

1 INTRODUCTION

Both natural slopes and artificial earth structures are often found in a very loose state. When the ground is subjected to a shaking as it is in an earthquake, the pore pressure increases and deformations develop. If the gravity shear stresses are larger than the shear strength at the developed deformations, the deformation process develops even after the shaking is terminated. In loose sands and silts this may easily occur resulting in displacements of an order of meters or tens of meters. Such a phenomenon is usually named flow failure.

The testing of silty soils, specially in loose state, is rather difficult, and taking undisturbed specimens is sometimes even impossible. This is why not much data is available in the literature and why series of laboratory tests on reconstituted specimens is presented here.

Because of the difficulties with sampling, the ideal way of testing silty ground for the possibility of flow failure occurrence are in situ tests. Ishihara (1993) proposed a procedure based on understanding obtained on element tests in the laboratory. To contribute to this procedure and to the understanding of silty soil behaviour in undrained shearing, results obtained from a wide range of soils (10 soils, 1, 2 or 3 methods of specimen preparation) are presented.

2 MATERIALS AND PROCEDURES

The tested materials ranged from fine sand to sandy silt. The grain size distribution is shown in Figures 1 and 2 and some properties in Table 1. Nevada sand was Nevada No. 120 sand from USA. Tia Juana silty sand and Lagunillas sandy silt were natural soils taken from dikes close to Lake Maracaibo in Venezuela. Dagupan silty sand was provided from the area liquefied in earthquake on 16 July 1990 in Dagupan, the Philippines. Mochikoshi silt was a tailing material excavated from a liquefied pond deposit in Mochikoshi gold mine on the Izu peninsula, Japan; failure occurred in the Izu-Ohshima-Kinkai earthquake on 15 January 1978. Toyoura sand is a river sand commercially available in Japan. It was crushed in a ball mill into a silt which was then mixed with the original sand.

Three methods of specimen preparation were chosen to produce distinctively different fabrics and to follow approximately some important sedimentation processes in the field. These were: moist placement (MP), dry deposition (DD) and water sedimentation (WS). The specimens, 5 cm in diameter and 10 cm in height, were saturated, consolidated to initial consolidation pressures varying from 0.05 to 0.5 MPa, and axially compressed in undrained conditions. Details on specimen preparation methods and test procedures are given elsewhere (Zlatović 1994, Zlatović & Ishihara 1997).

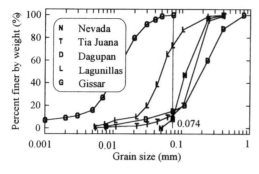

Figure 1. Grain size distribution of natural materials.

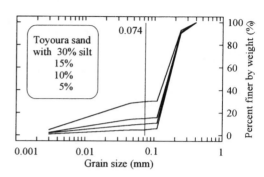

Figure 2. Grain size distribution of Toyoura sand with silt.

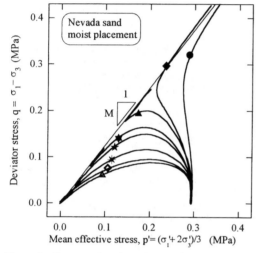

Figure 3. Stress paths of specimens of Nevada sand prepared by moist placement.

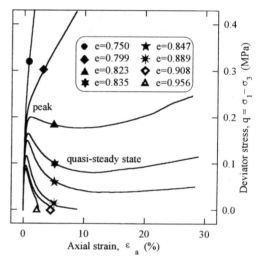

Figure 4. Stress~strain curves of specimens of Nevada sand prepared by moist placement.

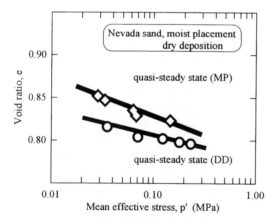

Figure 5. Quasi steady state of Nevada sand.

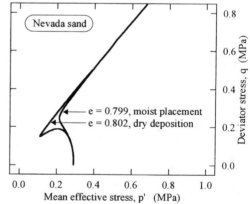

Figure 6. Stress paths of two specimens with same void ratio and confining stress but different fabric.

Table 1 Soil properties

material	D_{50} (mm)	ρ_s (g/cm³)	e_{min}	e_{max}	w_L (%)	I_P (%)
Nevada sand	0.1	2.67	0.511	0.887		
Tia Juana silty sand	0.16	2.68	0.620	1.099		
Dagupan silty sand	0.2	2.825	0.700	1.454		
Lagunillas sandy silt	0.05	2.69	0.766	1.389	26.9	4
Mochikoshi silt		2.70			32.4	13
Gissar silt	0.015	2.755	0.49	1.772		
Toyoura with silt 5%	0.18	2.65	0.566	1.	21.7	1
10%	0.175	2.65	0.532	1.04		
15%	0.17	2.65	0.498	1.08		
30%	0.15	2.65	0.519	1.2		

3 TEST RESULTS AND DISCUSSION

The method of moist placement was used to prepare Nevada sand into a series of specimens with a wide range of initial void ratio. Specimens were saturated and isotropically consolidated at a confining pressure of 0.3 MPa. Results of undrained shearing are shown in Figures 3 and 4: stress paths and stress~strain curves are given in terms of mean effective stress, $p' = (\sigma_1' + \sigma_2' + \sigma_3')/3$, deviator stress, $q = \sigma_1 - \sigma_3$, and axial strain, ε_a; void ratio is shown as e.

For specimens with void ratio under 0.8, deviator stress increased monotonically until the end of the test. However, for looser specimens, after reaching a peak at around 1% of axial strain, both mean effective stress and deviator stress decreased until their minimum, which is represented by the sharp elbow in the stress paths and the very large expanse in the stress~strain curves. The stress state remained constant temporarily, while the strain rate was constant, thus this was the quasi-steady state.

For all the specimens prepared using the method of moist placement, the relationship of mean effective stress at the quasi-steady state, p_s', and void ratio, e, is plotted in Figure 5 as squares. The circles show the quasi-steady state of specimens prepared by dry deposition method. Two distinctive quasi-steady state lines are determined.

Figure 6 shows stress paths of two specimens of Nevada sand with identical values of initial void ratio and consolidation pressure, but prepared with two different methods. Resulting undrained behaviour was quite different: dry deposited specimen had dilative behaviour, while moist placed specimen showed drop in shear strength corresponding to flow failure.

Some test results are given in Table 2, with reference to the specimen preparation method:
$$M = q_s/p_s'$$
where q_s is deviator stress and p_s' is mean effective stress at the quasi-steady state,
$$M_p = q_p/p_p',$$
where q_p is deviator stress and p_p' is mean effective stress at the peak shear strength,
the third parameter
$$p_p'/p_c'$$
is ratio of mean effective stress at the peak and consolidation pressure.

Table 2 Test results

method of specimen preparation		M= q_s/p_s'			M_p= q_p/p_p'			p_p'/p_c'		
		MP	DD	WS	MP	DD	WS	MP	DD	WS
material	silt content									
Nevada	8%	1.25	1.25		0.67↗	0.86		0.70	0.70	
Tia Juana	12%		1.22	1.22		0.73	0.73		0.67	0.55
Dagupan	15%	1.25	1.25		0.83	0.83		0.60	0.60	
Lagunillas	74%	1.24	1.24	1.24	0.625	0.625	0.625	0.62	0.62	0.62
Mochikoshi	92%		1.275	1.275		0.88	0.88		0.52	0.52
Gissar	100%	1.25			0.74			0.63		
Toyoura with silt	5%	1.25			0.7			0.65		
	10%	1.25	1.25	**	0.81	0.81	**	0.625	0.57	
	15%		1.25	**		0.72	**		0.60	**
	30%		1.25	*		0.68			0.60	
average value			1.25			0.8			0.6	

** no flow; * on the boundary between contractive and dilative behaviour

It is interesting to note that these parameters are practically fabric independent (details on fabric effects are given in Zlatović & Ishihara, 1997).

4 SUSCEPTIBILITY TO FLOW FAILURE

Large axial strains in element tests correspond to the occurrence of flow failure in field. It is important, therefore, to be able to detect susceptibility of a soil to contractive behaviour.

The test results of Nevada sand prepared by moist placement and dry deposition show once again that the fabric of soil as well as void ratio and effective stress substantially influence the undrained behaviour and residual strength. Unfortunately, in sandy soils it is quite difficult to obtain undisturbed specimens, and in silty soils often impossible. Imitating the fabric in the laboratory, on the other hand, may be not reliable.

This is why Ishihara (1993) developed a procedure that allows estimation of susceptibility of soil to flow failure on the basis of in situ tests, SPT and CPT. The criterion to decide on the possible onset of flow failure was the boundary between contractive and dilative behaviour, expressed through the initial state ratio, r_c. This is the ratio of the consolidation stress of the soil element and the corresponding effective stress at the quasi-steady state:
$$r_c = p_c'/p_s'$$
One way of representing the level of contractiveness is through brittleness index, I_{Bu}, defined by Bishop (1967):
$$I_{Bu} = (q_p - q_s)/q_p$$

In Figure 7, the brittleness index is plotted versus initial state ratio for all the sands and silts shown, and data from Sladen et al. (1985, 1989) are repeated as small black squares. Since M and stress state at peak in a very loose state are fabric independent (Zlatović & Ishihara 1995, 1997), I_{Bu} is related to r_c:

$$I_{Bu} = 1 - (M p_s')/(M_p p_p')$$

$$= 1 - (M/M_p)(1/ p_p'/p_c')(1/ p_c'/p_s')$$

$$= 1 - (M/M_p)(1/ p_p'/p_c')(1/r_c)$$

As the values of M, M_p and p_p'/p_c' are fabric independent, it is possible to determine them for any given soil and the boundary value of r_c can be estimated.

For the wide range of sands and silts tested, these parameters are almost the same

$$M = 1.25,$$

$$M_p = 0.8,$$

$$p_p'/p_c' = 0.6.$$

This leads to the relationship

$$I_{Bu} = 1 - (1.25/0.8)(1/0.6)(1/r_c)$$

$$I_{Bu} = (r_c - 2.5)/r_c$$

shown as curve in Figure 7. The plotted data fit well with the computed curve. This relationship proves that the boundary value of initial state ratio is 2.5, in other words something larger than 2.

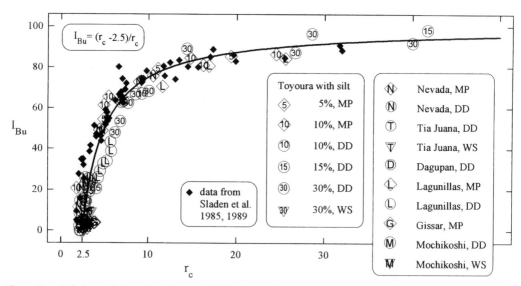

Figure 7.   Brittleness index plotted versus initial state ratio.

5 CONCLUSIONS

Triaxial undrained compression tests were performed on a series of sands and non-plastic silts in loose state, comparing two or three fabrics. Results showed that flow failure can be expected for those initial states in which r_c, initial state ratio, that is the ratio of consolidation pressure and the corresponding effective stress at the quasi-steady state, is larger than some boundary value. Although residual strength is very fabric dependent, this value of r_c is based on three fabric independent parameters of the undrained path of very loose soils, M, M_p and p_p'/p_c' and can be determined for any soil. For usual soils, this value is around 2.5, in other words something larger than 2.

REFERENCES

Bishop,A.W. 1967. Progressive failure - with special reference to the mechanism causing it. Panel discussion, *Proc. Geotechnical Conference, Oslo, Norway* 2: 142-150

Ishihara,K. 1993. Liquefaction and flow failure during earthquakes. *Géotechnique* 43(3): 351-415

Sladen,J.A., R.D.D'Hollander, J.Krahn 1985. The liquefaction of sands, a collapse surface approach. *Canadian Geotechnical Journal* 22: 564-578

Sladen,J.A., J.M.Oswell 1989. The behaviour of very loose sand in the triaxial compression test. *Canadian Geotechnical Journal* 26: 103-113

Zlatović,S. 1994. *Residual strength of silty soils*, D.Eng. Thesis, University of Tokyo, Tokyo

Zlatović,S., K.Ishihara 1995. Minimal peak undrained strength of silty soils. *10th Danube-European Conference on Soil Mechanics and Foundation Engineering*, Mamaia, Romania 2: 471-476

Zlatović,S., K.Ishihara 1997. Normalized behavior of very loose non-plastic soils - effects of fabric. *Soils and Foundations* 37 (4)

Environmental geotechnics

Geotechnical Hazards, Marić, Lisac & Szavits-Nossan (eds) © 1998 Taylor & Francis, ISBN 90 5410 957 2

Stabilisation of the Pliva riverbed in the town of Jajce

Z. Barbalić, Z. Langof, M. Goluža, Z. Šteger, D. Martinović & M. Lasić
Conex, Mostar, Bosnia and Herzegovina

ABSTRACT: In the town of Jajce which has the outstanding natural beauties and a special historic and tourist importance, Pliva riverbed has been eroded along a section of cca 2,0 km. The erosion process is in progress and there is also possibility for the catastrophic consequences to appear. Detailed analysis of the causes of erosion processes, of the current condition related to the riverbed stability and of the possible alternative stabilisation solutions have been done. A stabilisation concept made by combining measures which include structures for the flow velocity reduction, energy dissipation and increase the riverbed stability in some sections was proposed. A special attention was paid to the application of the environmental friendly solutions.

1 INTRODUCTION

More than 600 ago the town of Jajce was built on the Pliva and Vrbas rivers banks. The Pliva river with its riverbed engraved in tuff and with its waterfall at the mouth of the Vrbas River, represents a special features of this old royal town.

From time to time during the last hundred years, some works to protect the Pliva riverbed of erosion and to maintain the beautiful water fall have been undertaken. It is estimated that the riverbed was deepened for about 45 m during the last 1000 years. In the period from 1992 to 1996 there was a sudden worsening of situation - the riverbed was deepened significantly (even up to 6,0 m) while the waterfall was so damaged that there is no more any overflowing.

The critical condition of the very attractive waterfall which attracted a higher public attention, while only few pieces of information have been given about the upper riverbed which is also in a very bad condition especially in its length of about 2,0 km.

The paper presents an attempt to describe briefly a very critical situation in the Pliva River valley upstream of the waterfall as well as a general concept of solutions proposed by the Conex - Zagreb and Conex - Mostar, according to the design made in 1997.

The general hydrographic situation is shown in Figure 1 where the position of Jajce town is marked as well as the water power plant "Jajce I" which makes an important influence upon the hydrologic regime at the part of the Pliva river where the works

for the riverbed stabilisation are to be undertaken urgently.

The characteristics of the Pliva river hydrologic regime have been determined on the bases of the long-rage observations of the water level and flow measurements. The following data have been representative:
- average annual flow: $35,0 \text{ m}^3/\text{s}$
- max. daily flow (probability 0,10): $11.0 \text{ m}^3/\text{s}$
- max. daily flow (probability 0,01): $220.0 \text{ m}^3/\text{s}$.

Average flow duration curve is shown in Figure 2.

A natural hydrologic regime for the observed section was significantly disturbed by construction of the water power plant.

The water power plant, built in 1895, had the installed discharge of $16,0 \text{ m}^3/\text{s}$. The new water power plant, built in 1957, had a maximum discharge of about $75,0 \text{ m}^3/\text{s}$. That is why, as shown in Figure 2, the Pliva flow is only $3,0 \text{ m}^3/\text{s}$ for 352 days which is the minimum outlet for the water power plant Jajce I. Higher flow appears only during 13 days in one year (3,5 % of time).

After construction of the water power plant Jajce I, a changed hydrologic regime with a very low water flow, has caused a faster tuff degradation in the major part of the riverbed. That caused also decomposition of tuff so that tuff was transformed into sand. In addition, due to decomposition (disintegration) of tuff, natural cascades were demolished, and the slope of the Pliva river bed considerably increased which caused also the water velocity to be increased.

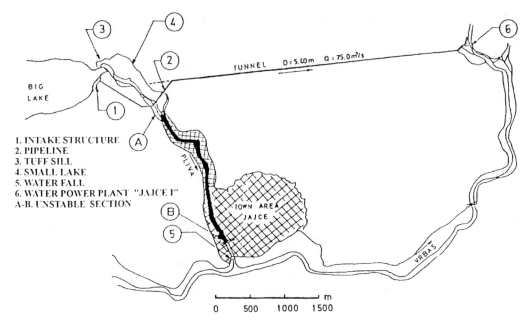

Figure 1. Jajce area-map

1. INTAKE STRUCTURE
2. PIPELINE
3. TUFF SILL
4. SMALL LAKE
5. WATER FALL
6. WATER POWER PLANT "JAJCE I"
A-B. UNSTABLE SECTION

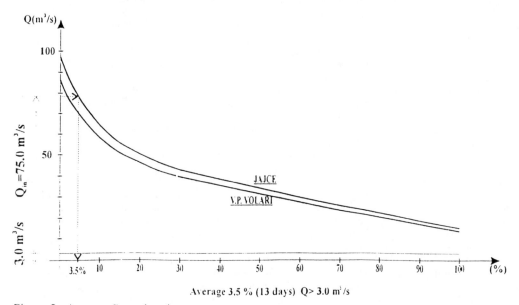

Average 3,5 % (13 days) Q> 3.0 m³/s

Figure 2. Average flow duration curve

An intensive erosion is in process as the slope of the Pliva river bad is about 1,2% today which is adequate for torrent hill flows. The material in the riverbed is of the size being characteristic for large plain rivers. It was found by monitoring the riverbed that the erosion undermined the riverbed in some parts in Jajce even more than 6,0 m during the period of 15 years (Figure 3).

Due to large and in last years even accelerated undermining of the riverbed, the river banks were destroyed in many places and the nearby structures in the river valley are endangered. One bridge, a part of the road long 100 m and some earlier built structures for riverbed stabilisation have been demolished up to now.

The appearance of the Pliva riverbed has become very ugly. Deep gullies are in there and a relatively small quantity of flow is noticed almost during the whole year. Features of the environment in the whole valley became very unpleasant especially in the area of the Jajce town which had been well-know by its beauty and where the tourism is expected to become a very important business field.

Due to a significant lowering of the riverbed, the river banks became even steeper and at several places the landslides were induced By further possible undermining of riverbed and slopes of river banks even bigger landslides could be induced as well as large-scale damages.

Possibilities for the increase of erosion process are rather big today. This can be caused also by a possible break of work in the water power plant, that is even before the high flow in the Pliva river basin occur. It was registered during the war that the erosion process was very fast, while the water power plant Jajce I was not working and all quantities of water flow through Jajce, which caused also great damages.

It is obvious that catastrophic effects can occur and that urgent works for the riverbed stabilisation should be undertaken.

2 RIVERBED IMPROVEMENT CONCEPT

While choosing the concept for the Pliva riverbed improvement of the observed section, the following main aspects were taken into consideration:
– the adequate level of stability of the Pliva river banks to be achieved along the whole section;
– maximum possible adaptation to the Pliva riverbed (to get the works to the minimum);
– minimum occupation of the coastal area in the very narrow Pliva river valley in the town area,
– preservation and improvement of the environmental features;
– the riverbed maintenance expenses should be acceptable;
– the stimulation of the tuff development in the riverbed;
– the remediation of all coastal landslides;
– construction in stages;
– the adequate flood protection level of the coastal area;
– minimum change of coastal underground water level to avoid negative effects for buildings construction in the valley.

The analysis of several possible alternative solutions have been made within the framework of two general concepts:
a) The stabilisation of the Pliva riverbed by creating a series of several lakes;
b) The construction of manmade cascades in order to reduce the riverbed slope.

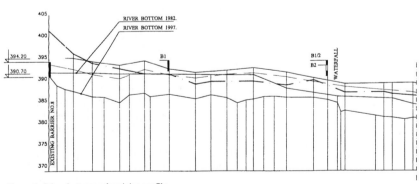

Figure 3. River bottom undermining profile

2.1 *The Plive riverbed stabilisation made by creating a series of several lakes*

This concept includes the construction of a number of dams in order to create the lakes and to reduce the river water velocity. The resulting velocity should be even smaller than the critical velocity by which the sand at the bottom and on banks would be carried away. Undoubtedly, the new lakes with their small waterfalls could improve considerably the features of the environment which were badly aggravated by erosive processes during these last years.

Analysis have showed that tuff sand in lakes could not be moved if the water velocity is lower than 0,77 m/s at the bottom, that is 0,29 m/s at steep banks curves.

It was concluded that for safe riverbed stabilisation, made by velocity reduction in the natural riverbed without any banks and bottom strengthening, a very high structures (dams) should be constructed. Several dams, which would flood a considerable part of the Pliva valley, should be constructed. Besides, due to the large presence of small size grain materials, the foundations and filtration protection below and around the structure should need extremely high funds.

Based on morphology elements analysis, the smallest needed depths of water in the Pliva riverbed are established in order to get the middle profile velocity 1.0 m/s and flow 220.0 m³/s (maximum flow probability 0,01).

Depths of water in lakes, where the water velocity is below 1,0 m/s, could be realised by the construction of 6 dams with nominal height of 6.9 m to 10.0 m.

The dams heights, depending on needed foundation depth and on solutions for energy dissipation, are surely higher for at least 3 meters. Consequently, 6 dams with heights of 10.00 to 13.00 m should be constructed to limit the water velocity in the riverbed at 1.00 m/s.

In the present case, the execution of the water tightness around and below dams, as well as protection against piping, requires an extremely high financial support due to the presence of disintegrated tuff and big caverns in the remaining compact tuff. Such works are rather expensive because grouting, piling and execution of cut off walls in tuff material are very complex and expensive.

On the right bank of the Pliva riverbed, almost on its whole length, the bank is rather low. In order to provide the water velocity less than 1.0 m/s, the flooding or endangering of a number of facilities would be necessary. Besides, the conditions for drainage of coastal areas, which are urbanised to a rather high extent, would be aggravated.

2.2 *Pliva riverbed stabilisation made by manmade cascades*

In the case where it is necessary to build stabilisation structures (for bank erosion protection and on some places even the bottom protection), it is logical to define maximum water velocity which such structures can support. In that way the required dimensions of a new riverbed are smaller which yields the reduction of costs as well as more free space in the coastal area. Based upon the analysis of several options, the value of slope of the riverbed 0.004 is chosen.

Following the analysis of possible options it was concluded that the standard trapezoidal cross section (width of 15.00 m at the bottom) and the slope of the river bank 1:2 is the best solution (Figure 4).

The existing riverbed has some very sharp curves and the layout with adequate curvature needs a lot of excavations in some sections. The accepted criteria is that the freeboard of about 0.40 m be provided in the case of high flow with probability 0.01, while maximum 1000 year flow can occur without freeboard (so that the riverbed can be completely filled up to the top).

Figure 5 shows curves Q = f(h) and v = f(Q) for the adopted new Pliva riverbed. At the water flow minimum Q=3.00 m³/s, the water depth in the riverbed is 0.20 m, and the velocity is 0.95 m/s. These are favourable conditions for development of tuff (the water velocity higher than 0.80 m/s and good aeration) so the gradual bottom strengthening can be expected.

2.3 *Choice of an improvement concept*

On the basis of detailed considerations, the examination of earlier projects for the Pliva river, the analysis of morphologic elements as well as upon hydraulic estimates, the conclusion was made that the combined concept for riverbed improvement should be applied:

a) Riverbed construction of the standard trapezoidal cross section with the slopes protected from erosion should be applied at sections where there is mainly small size grains disintegrated tuff at the bottom. At parts of the bottom without any strong tuff the stabilisation should be made by sills along with bank revetment with toe protection at the depth of 2.0 m in order to limit the bottom erosion The slope 0.004 is achieved (limited) by construction of reinforced concrete sills - cascades.

b) The reduction of slope to 0.004 is also foreseen for sections with mainly strong tuff at the bottom and where the riverbed is branched (with a number of branches) and it will be done by construction of reinforced concrete sills -

cascades. At this section the riverbed has very different forms and dimensions. At some spots even higher local velocities are expected. Almost the whole left bank is made of noncohesive materials with a steep slope and great height. The undermining of the left bank at several spots endangers buildings and road which is at the top of the steep slope. That is why the bank revetments should be constructed almost along the whole length on the left bank. At the right bank, a revetment is necessary only at a short section.

c) The construction of 6 cascades with stilling basin for energy dissipation is foreseen (5 being of nominal height of 2.00 m and one of 1.0 m). These cascades reduce the total fall for 11 meters and reduce the velocity to the magnitude which could be resisted by the strong tuff and by structures for riverbed protection. As for the downstream section, the solution with construction of a chute having the length of 115 m is adopted as well as construction of a stilling basin for energy dissipation.

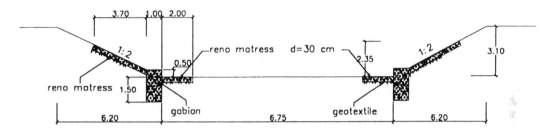

Figure 4. Standard cross section

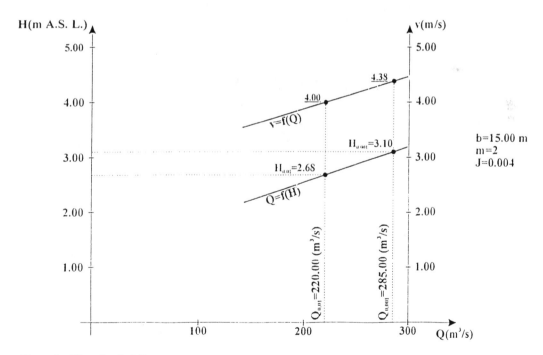

Figure 5. Pliva riverbed flow curve

397

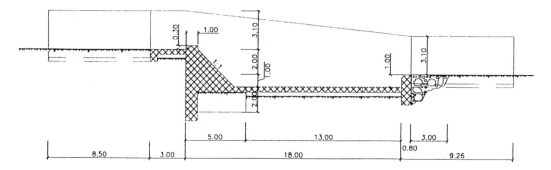

a) Cross section

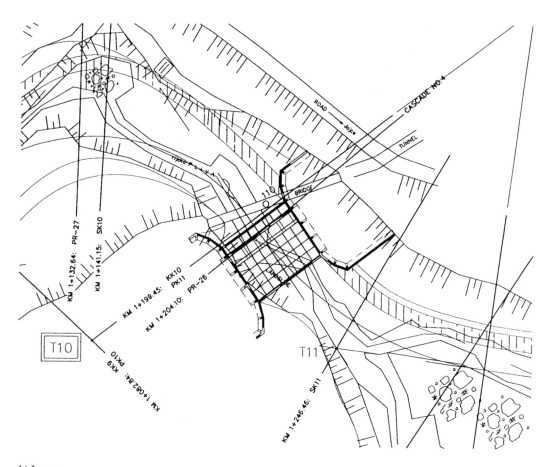

b) Layout

Figure 6. Cascade

Cascades must be safe. Demolishing of one of them can cause also demolishing of others ("dominoes effect") producing a big damaged at the longer section of the riverbed. Reinforced concrete cascades are foreseen (Figure 6).

Several alternatives for construction of riverbed protection structures have been considered (stones, concrete slabs, Reno mattresses, gabions and their combinations).

Based upon the analysis of several options for the slope stabilisation, the solution with revetments construction by application of "Reno mattresses" (basket made of net filled with stones) combined with geotextile leaning on gabions (Figure 7) is chosen. This solution has the following characteristics:

1. It is not necessary to use big pieces of stone and the stones can be found in the vicinity;
2. The structure is adaptable even if some smaller displacements (undermining and other) occur;
3. Rather high water velocities in the riverbed can be tolerated (short-term with nominal value of 6.40 m/s);
4. The construction in stages can be applied depending of the available financial resources;
5. The area is favourable for a possible development of tuff maker;
6. In the cases where a strong tuff is found the solution can be combined easily;
7. After growth of vegetation, it adapts easily to surroundings;
8. The local working force can be used.

Reno mattresses and gabions are adaptable to the environment. It is possible to achieve with them the so-called "environmental friendly solutions".

Pieces of tuff are not good for revetments construction and the Reno mattresses and gabion fillings due to their small density and exposure to disintegration. They can be used for the recesses at the riverbed bottom to be filled in.

The construction of reinforcement structures is foreseen to stabilise the bottom and these structure should be dug in for at least 1.00 m. Their max. distance is 125.0 m.

3 CONCLUSION

The slope of the Pliva river is rather high at the observed section having the length of about 2.0. km. Mainly small grain size materials (disintegrated tuff) are present in the riverbed. In case of higher flow further continuation of the intensive river bottom erosion and especially of the slopes of river banks could be expected. Further riverbed undermining can cause block landslides in the valley flanks with associated destruction of many facilities. The level of risk is very high and it is necessary to start immediately the works for the riverbed stabilisation.

Several possible solution concepts have been analysed. Taking into the consideration all general and special conditions as well as emphasised criteria for preserving specific surroundings characteristics, the solution with six cascades and one chute with basins for energy dissipation, is proposed. In section between these structures the protection of the riverbed by applying the Reno mattresses and gabions combined with geotextile is proposed. Depending on the ratio of presence of strong and disintegrated tuff, the application of the adequate solution for each section of the riverbed is proposed.

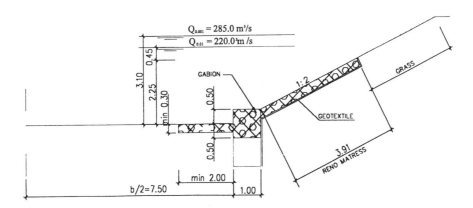

Figure 7. Revetment

4 REFERENCES

Mikulić, S. 1959. *Water Power Plant Jajce* I. Sarajevo: Hydro-energy plants - Energoinvest (in Croatian).

Miljković, E. 1959. *The Pliva river waterfall and riverbed regulation in Jajce*. Sarajevo. *"Our antiquities"* (in Croatian).

Miljković, E. 1959. *Regulation of the Pliva riverbed and the waterfall in the town of Jajce*. Sarajevo. Hydro-energy plants - Energoinvest (In Croatian).

Miljković, E. & Mikulic, S. 1957. *Regulation of the Pliva river and waterfall in Jajce - Main project*. Sarajevo: Elektorprojekt (in CRoatian).

Mitrinović, M., Meler, Z., Katalinic, I. *Regulation of the Pliva riverbed - Design*. Sarajevo: Energoinvest (in Croatian).

Pavletić, Z. 1967. *Report on biology researches done for the Pliva river lowest flow in October 1967* (in Croatian).

Geotechnical Hazards, Marić, Lisac & Szavits-Nossan (eds) © 1998 Taylor & Francis, ISBN 90 5410 957 2

Migration and extraction of heavy metal from contaminated natural clay by electrokinetics

Haik Chung
Geotechnical Engineering Division, Korea Institute of Construction Technology, Kyunggi Ilsan, Korea

Byunghee Kang
Department of Civil Engineering, Inha University, Incheon, Korea

ABSTRACT: This paper presents a experimental study on an electrokinetic extraction for removal of contaminant from natural clay spiked with lead. A series of laboratory experiments including variable conditions such as operating duration, applied electrical current, concentration of lead, and applications of three different chemicals, was performed. The experiment results showed that the amounts of hydrogen ions and electroosmotic water flow transported from anode to cathode increased with increasing in the operating duration and the applied electrical current, and with decreasing in the initial concentration of lead. The efficiency of heavy metal removal from the contaminated soil was increased with increasing in the operating duration, the applied electrical current, and the initial lead concentration. The chemicals used as deprecipitating, complexing, and solubilizing agents were efficient in removal of lead from the soil.

1. INTRODUCTION

Electrokinetics is the movement of water, ions, and charged solid particles between two electrodes under the influence of an electrical field. Electrokinetics can be applied to the fields such as consolidation and stabilization of soft soil and sludge, stabilization of slopes, dewatering in construction sites, remediation of pollutants, barrier system in clay liners, sealing and leak-detection system of geomembrane, injection of chemical, microorganisms as well as nutrients into subsoil strata, and diversion schemes of a contaminant plumes in civil and environmental engineering(Acar et al., 1990; Acar et al., 1993; Mitchell, 1991; Ugaz, 1992; Yeung, 1993).

From the late 1980s, interest in electrokinetic technology for remediation has increased. A study on the electrokinetic remediation has been conducted by several universities, public and private institutes and companies, such as Louisiana State University, University of Texas A & M, MIT, Lehigh University, and Electro-Petroleum Inc. in United States, Geokinetics in The Netherlands, and Leeds University and University of Cambridge in United Kingdom(Acar et al., 1990; Acar et al., 1993; Yeung, 1993; Reed et al., 1995; West et al., 1996; Penn et al., 1996).

Electrokinetic remediation is an emerging in situ technology developed to extract organic and inorganic contaminants from fine grained soils such as clay. Two mechanisms, electroosmosis and electromigration, are the primary driving forces to extract contaminants from the electrokinetic remediation processing(Acar et al, 1990; Alshawabkeh et al, 1992; Yeung et al, 1997). Numerous electrochemical reactions and soil contaminant interactions, such as electrolysis, sorption and desorption of contaminants onto and from a clay particle surface, acidification of soil by the transport of the hydrogen ion, precipitation of inorganic species, occur simultaneously(Acar et al, 1993; Rodsand et al, 1996; Yeung et al, 1997).

In Korea, the industry is commonly located near the seashore, therefore some marine deposits at the seashore are potentially contaminated. Thus this paper presents a study on an electrokinetic remediation for removal of lead from contaminated natural marine clay spiked with lead. For this purpose, a series of laboratory experiments including variable conditions such as operating duration, applied electrical current, concentration of lead, and applications of three different chemicals, was performed. Investigated are pH values of catholyte and anolyte, volume of water extracted from the soil specimen, the changes in pH and electrical conductivity of the contaminated soils, distribution of electrical potential in the soil between the two electrodes. The efficiency of lead removal by the proposed method was evaluated under various operating conditions including the use of three chemicals.

2. ELECTROKINETIC EXPERIMENTS

2.1 *Materials*

Natural marine clay was sampled at Kimpo reclaimed land in the west seashore of Korea. Table 1 and 2 contain the physical and the chemical properties of this marine clay used in this experiment. The index properties of this soil were specific gravity 2.58, liquid limit 32.8%, plastic limit 22.5%, percent finer than #200 sieve 90%, specific surface area 5,249cm^2/g, coefficient of uniformity 4.5, activity 0.114, and coefficient of permeability 2×10^{-7}cm/sec. The compaction properties were maximum dry density 1.73g/cm^3, optimum water content 18.2%. The chemical properties were organic matter content 6.74%, initial pH 6.5 ~7.13, and initial electrical conductivity 1,500μs/cm. The mineralogical composition was as follows: SiO$_2$ 69.20%, Al$_2$O$_3$ 13.97%, Fe$_2$O$_3$ 4.30%, CaO 0.95%, MgO 1.61%, K$_2$O 2.41%, Na$_2$O 2.17%, and TiO$_2$ 1.35%. The study soil was artificially contaminated with lead using an aqueous lead solution. Lead solutions were obtained by stock solution prepared with Pb(NO$_3$)$_2$ and deionized water (Chung et al, 1995; Chung et al, 1997).

2.2 *Testing programs*

The electrokinetic remediation system used in this study was mainly consisted of electrokinetic cell, anolyte and catholyte reservoir, and power supply. The electrokinetic cell was a cylindrical tube with 100mm in length and 100mm in diameter. The end reservoir solutions were circulated constantly by pump to check the anolyte and catholyte pH. Graphite electrodes were used to prevent production of corrosion and two sheets of filter papers were placed at both ends of the specimen to stabilize the soil specimen. Two electrodes were not in direct contact with the soil. To allow venting of gas produced by electrolysis, holes were drilled into the top of each end reservoir.

In this paper, twelve tests were conducted by using one-dimensional electrokinetic test apparatus for lead removal without and with any enhancement. Parameters such as operating duration, current density, concentration of lead, and enhancement methods were chosen in these experiments. Constant current conditions were used in all tests to minimize complicated electrical boundary conditions and to keep the net rates of the electrolysis reactions constant at all testing times(Hamed, 1991). To investigate the

behaviour of the contaminant transport with the variation of each parameter, the currents used in this test were 10, 30, 50, 100mA(current density 0.1, 0.6, 1.3mA/cm^2 each), the operating durations were 5, 15, 30 days, and the lead concentration were 0, 500, 5,000, 50,000mg/kg. In order to enhance the removal of the metals that tend to accumulate and precipitate in the soil, three control methods were chosen.

Table 1. The physical properties of test soil.

Items	Values
Specific gravity	2.58
Atterberg limits (%) Liquid limit Plastic limit Plasticity index	32.8 22.5 10.3
Percent finer than #200 sieve	90.0
Specific surface area (cm^2/g)	5,249
Coefficient of uniformity	4.5
Activity	0.114
Proctor compaction parameter Maximum dry density (g/cm^3) Optimum water content (%)	1.73 18.2
Hydraulic conductivity (cm/sec)	2×10^{-7}

Table 2. The chemical properties of test soil.

Items	Values
Organic matter content (%)	6.74
pH	6.5 ~ 7.13
Electrical conductivity (μs/cm)	1,500
Chemical constituents (%) SiO$_2$ Al$_2$O$_3$ Fe$_2$O$_3$ CaO MgO K$_2$O Na$_2$O TiO$_2$	 69.20 13.97 4.30 0.95 1.61 2.41 2.17 1.35

Three different chemicals like nitric acid as a deprecipitating agent, ethylenediamine as a complexing agent, and acetic acid as a solubilizing agent were used for the enhanced electrokinetic test(Pamukcu et al., 1994; Rodsand et al., 1996; Chung, 1996). A few drops of pure nitric acid were put into lead solution to reduce lead precipitation in soils. Ethylenediamine was injected into the anode chamber of a lead contaminated soil specimen and it penetrates the soil specimen by electroosmotic flow. The amount of ethylenediamine injected was 10 to 1 ratio of moles of ethylenediamine to lead in soil to ensure complexation of all the lead available. Acetic acid of 0.01M was introduced at the cathode chamber to depolarized the cathode reaction.

2.3 *Testing procedure*

Kimpo marine clay was dried, crushed, pulverized and mixed with lead solution. The mixture was cured in a bowl more than 24 hours. The mixture spiked with lead was compacted in an electrokinetic cell. For each clay sample that was prepared, the water content was optimum moisture content, the dry density was 95% of maximum dry density, and the degree of saturation was 100%. The cell was assembled to the electrokinetic test system.

A constant current was applied to the cells and the voltage was measured on a daily basis. pH measurement was made in the anode and cathode reservoir. The time dependant water movement through the soil due to electroosmosis was measured on the outflow by graduated cylinder. After an electrokinetic test, the cell was disassembled from the electrokinetic system and the soil specimen was extruded and sliced in ten sections. The value of pH and the concentration of lead were measured in each slice to assess the change of chemical property in the soil specimen and the efficiency of the process in lead removal.

3. RESULTS AND DISCUSSION

3.1 *Anolyte and catholyte pH*

Plots of anolyte(inflow) and catholyte(outflow) pH for tests with different concentration levels of lead are shown in Figure 1. Anolyte pH was measured at the inflow reservoir and catholyte pH was measured at the outflow reservoir during an electrokinetic remediation test. The anolyte pH dropped to values of 2~3 and the catholyte pH rose to value of 11~13 upon the start of the electrokinetic test regardless of different electrical currents. Subsequently, the pH remained relatively constant with further processing. This results from the electrolysis of water. Therefore, generation of oxygen gas and hydrogen ion at the anode lowers the pH, and generation of hydrogen gas and hydroxide ion at the cathode increases the pH.

In all the tests for different operating durations, electrical currents, contamination levels, and enhancement methods except ethylenediamine injection, the anolyte pH decreased and catholyte pH increased on the start of current application regardless

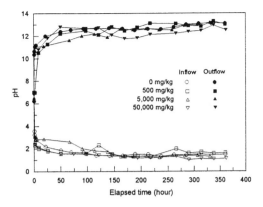

Figure 1. The change of anolyte and catholyte pH by different contamination levels.

of operating durations, applied currents, concentration levels.

3.2 *Soil pH*

The profiles of soil pH across the contaminated specimen for tests with different electrical currents at 15 days of processing and 5,000mg/kg of contamination level are shown in Figure 2. This demonstrates that the acid front generated at the anode advances steadily towards the cathode, while base front generated at the cathode remains in the cathode. The acid front progressed to the cathode by advection, migration, and diffusion and neutralized the base at the cathode. The transient acid front movement in soil was beneficial for metal desorption and dissolution, which in turn contributes to the removal process. The fronts have met at a normalized distance of approximately 0.65 from the anode at a current of 10mA, approximately 0.78 at a current of 50mA, and approximately 0.85 at a current of 100mA.

The final distributions of pH value across the soil specimen decreased with an increase in the operating duration and applied current, because transport of the acid generated at the anode increased with an increase in the duration and current. The final pH value across the soil specimen increased with an increase in the concentration of lead. That reason might be the superposition of electroosmotic flow on electromigration(West et al., 1996). In lower concentration, electroosmotic flow was large, so transport was by electroosmosis and electomigration. But in higher concentration, electroosmotic flow was small, so transport was mainly by electomigration. The final pH value across the soil specimen decreased in the

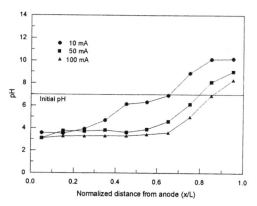

Figure 2. Final pH distribution across the soil specimen by different electrical currents.

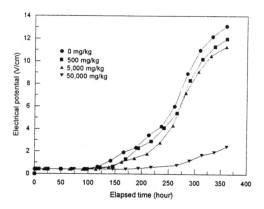

Figure 3. Increments of electrical potential with time by different contamination levels.

enhancement tests such as introducing nitric acid in the soil specimen and acetic acid in the cathode compartment, due to acidification of the soil specimen by applied nitric acid and acetic acid. From the all test results, we can recognize that the quantities of hydrogen ions transported from anode to cathode increased with increasing in the operating duration and the applied electrical current, and decreasing in the initial concentration of lead.

3.3 Electrical potential gradients

The variation of electrical potential across the soil specimen for tests with different contamination levels at operating duration of 15 days and applied electrical current of 50mA is presented typically in Figure 3. The final electrical potentials were 13.1V/cm in the concentration of 0mg/kg, 12.0V/cm in the concentration of 500mg/kg, 11.3V/cm in the concentration of 5,000mg/kg, and 2.5V/cm in the concentration of 50,000mg/kg each due to transport capacity of charge.

From the all test results, here arises that the electrical potential differences across the electrodes increased with an increase in the operating duration and applied current, and decreased with an increase in the concentration of lead, also decreased with an introduction of nitric acid, acetic acid, and ethylenediamine in the soil specimen. The increase of electrical potential in the tests for the different operating duration and electrical current is probably associated with formation of a lead hydroxide, which will reduce ionic strength and soil pore space. The decrease in electrical potential in the tests of different contamination levels is probably related with transport capacity of electrical charge due to difference ionic concentration existed in soil.

3.4 Electroosmotic flow

The electroosmotic flow was directed towards the cathode from the anode in all tests. An example of the electroosmotic flow over the different electrical current applied in an electrokinetic tests with 50mA current and 5,000mg/kg concentration level is presented in Figure 4. These results demonstrate that the electroosmotic flow increased with an increase in the current. The accumulated quantity of flow after 360 hours was 20ml, 120ml, 670ml in 10mA, 50mA, 100mA current each due to variations in the density of applied electrical current. In high current application, electroosmotic flow towards the cathode began almost immediately upon current application. But in the low current application, electroosmotic flow began a little later on the magnitude of applied current.

In the tests with different operating durations, the electroosmotic water flow at the cathode increased with an increase in the operating duration. There was no measurable flow within the first 40 hours of current application. Subsequently, the flow increased rapidly up to 115ml by 300 hours and then increased slowly up to 130ml by 720 hours. In the cases of enhancement tests, the electroosmotic flow was increased in the enhanced case compared with unenhanced cases. The enhanced test introduced acetic acid at the cathode reservoir in all the enhanced tests exhibited highest electroosmotic flow of 400ml after 15 days and unenhanced electrokinetic test exhibited lowest flow of 120ml after 15 days.

The total electroosmotic flow with concentration levels at the end of 360 hours in the application of 50mA was approximately 420ml at a concentration of zero, 390ml at a concentration of 500mg/kg, 120ml at

404

a concentration of 5,000mg/kg, and 25ml at a concentration of 50,000mg/kg. The quantity of flow through uncontaminated and contaminated with up to concentration of 500mg/kg was similar, whereas that

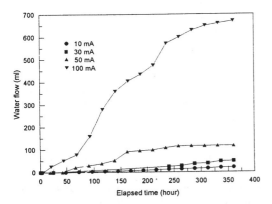

Figure 4. Variation of electroosmotic flow by different electrical currents.

contaminated with 50,000mg/kg was less. These results could be explained by zeta potential(West et al., 1996). In the concentration level of 500mg/kg or less, the zeta potential was more negative. On the other hand in the concentration level of 50,000mg/kg, the zeta potential was less negative. Thus the electroosmotic flow proportioned directly to zeta potential is reduced with higher concentration. Flow rate began to rapidly reduce after about 170 hours for the tests with concentration level of 5,000, 50,000mg/kg, on the other hand flow rate began to slightly reduce after about 300~350 hours for the tests with contamination level up to 500mg/kg.

3.5 Efficiency of lead removal

The normalized lead concentration ratio across the soil specimens determined at the conclusion of electrokinetic experiments for tests with 5, 15, 30 operating durations at an electrical current of 50mA and contamination level of 5,000mg/kg are presented as a function of normalized distance from the anode in Figure 5. This figure shows that lead was progressively advanced from the anode to the cathode with elapsed time. The average concentration ratio of lead in an entire specimen was 0.87 after 5 days, 0.12 after 15 days, and 0.05 after 30 days. This means that the average removal rate of lead in the soil specimen was 13% after 5 days, 88% after 15 days, and 95% after 30 days. Any removal of the adsorbed lead should involve

its desorption into the pore fluid by cation exchange of the hydrogen ions advancing across the soil specimen and its subsequent flushing to the cathode by transport and advection. For the soil specimens where remediation process was operated for a relatively short period, the lead was completely removed from the section near the anode, but accumulated to some extent near the cathode. On the other hand where remediation process was executed for a long period, the lead was evenly removed throughout the soil specimen.

The normalized lead concentration profiles with different electrical currents at the operating duration of 15 days and contamination level of 5,000mg/kg are plotted in Figure 6. The test results shows that the efficiency of extracting lead from a soil specimen increased with an increase in the applied currents. The average concentration ratio of lead was 0.65 in the current of 10mA, 0.12 in the current of 50mA, and 0.06 in the current of 100mA. Thus, the average removal rate of lead was 35%, 88% and 94% in the current of 10mA, 50mA and 100mA. When an electrical field of low currents was applied to the soil specimens in the electrokinetic remediation tests, the lead was completely removed from the anode compartment, but accumulated to some extent near the cathode compartment. But when high currents were applied, the lead was evenly removed across the entire soil specimen.

In the case of electrokinetic remediation tests with different concentration levels after 15 days of test duration at an electrical current of 50mA, it is noted that the efficiency of lead extraction was highest in the highest contaminated soil and lowest in the least contaminated soil. At low level of concentration, the transport of lead was slower because almost all the lead was adsorbed to the soil particle and required a replacement by other cations such as hydrogen ion H^+. But at high level of concentration, the transport of lead was faster because most of the lead was free in the pore and not adsorbed to the soil particle, so relatively easily and quickly moved to the cathode on current application.

Injection of three chemicals into the soil specimen contaminated with lead was tried to enhance the electrokinetic remediation. All enhancement of experiments presented significant lead transport and removal through the soil specimen. The chemicals such as nitric acid, ethylenediamine, and acetic acid used as deprecipitating, complexing, and solubilizing agents were efficient in removal of lead from the soil. But the removal efficiency would be variable with concentration and quantity of chemical agents applied in electrokinetic test. Lead removal efficiency more

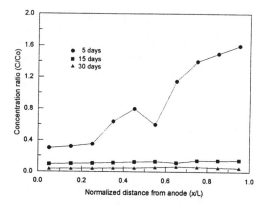

Figure 5. Distribution of final lead concentration across the soil specimen by different operating durations.

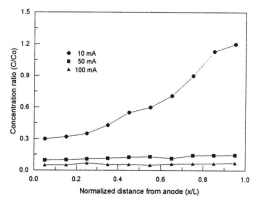

Figure 6. Distribution of final lead concentration across the soil specimen by different electrical currents.

than 88% was achieved in all enhancement experiments conducted in this study. This would result in prevention of hydroxide precipitation and solubilization of the species in transport and it would be rendered efficient removal of species(Rodsand et al., 1996; Pamukcu et al., 1994; Cline et al., 1995).

4. CONCLUSIONS

In this paper, laboratory bench scale experiments demonstrated the removal efficiency of lead from contaminated natural clay. Electrokinetic remediation can be used in extraction of heavy metals from high plastic marine clay. The test results showed that the anolyte pH decreased and catholyte pH increased on the start of current application regardless of operating durations, applied electrical currents, concentration levels. The pH across the soil specimen increased with increasing in the operating duration and the applied electrical current, and decreasing in the initial concentration level of lead. Electroosmotic water flow transported from anode to cathode increased with decreasing in the initial concentration of lead and increasing in the operating duration and the applied electrical current.

The lead was completely removed at the anode section and accumulated near the cathode section in the experiments of low currents and short durations . But the lead was evenly removed throughout the soil specimen in the experiments of high currents and long durations. The extract efficiency of heavy metal from the contaminated soil was increased with increasing in the operating duration, the applied electrical current, and the initial lead concentration. The chemicals such as nitric acid, ethylenediamine, and acetic acid used as deprecipitating, complexing, and solubilizing agents were found efficient in removal of lead from the soil. The addition of the selected chemicals into the contaminated soil proved to expedite the remediation process.

REFERENCES

Acar, Y. B., Gale, R. J., Putnam, G. A., Hamed, J. and Wong, R. L., 1990. Electrochemical processing of soils: Theory of pH gradient development by diffusion, migration, and linear convection. Journal of Environmental Science and Health, A25(6): 687-714.

Acar, Y.B. and Alshawabkeh, A.N., 1993. Principles of electrokinetic remediation. Environmental Science and Technology, 27(13): 2638-2647.

Acar, Y.B. and Alshawabkeh, A.N., and Gale, R.J., 1993. Fundamentals of extracting species from soils by electrokinetics. Waste Management, 13: 141-151.

Alshawabkeh, A. N. Acar, Y. B., 1992. Removal of contaminants from soils by electrokinetics: A theoretical treatise. Journal of Environmental Science and Health, A27(7): 1835-1861.

Chung, H., Lee, Y. and Woo, J., 1995. A study on the remedial technologies of contaminated soil and groundwater. Korea Institute of Construction Technology, Final Research Report, KICT 95-GE-1101-2, pp.5-195.

Chung, H., 1996. Removal of heavy metals from contaminated ground by electrokinetic remediation, Inha University.

Chung, H. and Kang, B., 1997. Removal of lead from contaminated Korean marine clay by electrokinetic remediation technology. Geoenvironmental Engineering Contaminated Ground: fate of pollutants and remediation, Thomas Telford, pp. 423-430

Cline, S. R. and Reed B. E., 1995. Lead removal from soils via bench-scale soil washing techniques. Journal of Environmental Engineering, ASCE, 121(10): 700-705.

Hamed, J. and Acar, Y. B., 1991. Decontamination of soil using electro-osmosis. PhD dissertation, Louisiana State University.

Hamed, J., Acar, Y.B. and Gale, R.J., 1991. Pb(II) removal from kaolinite by electrokinetics. Journal of Geotechnical Engineering, ASCE, 117(2): 242-271.

Ministry of Environment , 1996. A handbook for conservation of soil environment. Korean Administrative Publication No. 12000-67630-67-9613, pp. 1-54.

Mitchell, J.K., 1991. Conduction phenomena: from theory to geotechnical practice. Geotechnique, 41(3): 299-340.

Pamukcu, S. and Wittle, J.K., 1994. Electrokinetically enhanced in situ soil decontamination. Remediation of Hazardous Waste and Contaminated Soils, Marcel Dekker Inc., pp. 245-298.

Penn, M. and Savvidou C., 1996. Centrifuge modeling of the removal of heavy metal pollutants using electrokinetics. Environmental Geotechnics, pp. 1055-1060.

Rodsand, T. and Acar, Y.B., 1996. Electrokinetic extraction of lead from spiked Norwegian marine clay. Geoenvironment 2000, pp. 1518-1533.

Reed, B. E., Berg, M. T., Thompson J. C. and Hatfield J. H., 1995. Chemical conditioning of electrode reservoirs during electrokinetic soil flushing of Pb-contaminated silt loam. Journal of Environmental Engineering, ASCE, 121(11): 805-815.

Segall, B. A., O'Bannon C. E. and Mattias J. A., 1980. Electro-osmosis chemistry and water quality. Journal of Geotechnical Engineering, ASCE, 106(GT10): 1148-1152

Ugaz, A., Puppala, S., Gale, R.J. and Acar Y.B., 1992. Complicating features of electrokinetic remediation of soils and slurries: saturation effects and the role of the cathode electrolysis. Department of Chemistry and Civil Engineering, Louisiana State University.

West, L.J. and Stewart, D. I., 1996. Effect of zeta potential on soil electrokinetics. Geoenvironment 2000, pp. 1535-1549.

Yeung, A.T., 1993. Electro-kinetic flow processes in porous media and their applications.

Yeung, Y. B., Scott, T. B., Gopinath, S., Menon, R. M. and Hsu, C., 1997. Design, fabrication, and assembly of an apparatus for electrokinetic remediation studies. Geotechnical Testing Journal, GTJODJ, 20(2): 199-210.

Geotechnical Hazards, Marić, Lisac & Szavits-Nossan (eds) © 1998 Taylor & Francis, ISBN 90 5410 957 2

Land marine reclamation and environmental problems

R.Ciortan
The Design Institute for Road, Water and Air Transports (IPTANA), Bucharest, Romania

ABSTRACT: The development strategy of Constantza Port in Romania on medium and long timescale, plans construction of wide reclaimed land that requires major filling volumes. A great port complex such as Constantza Port originates a large amount of solid waste, figuring some 35,000 tons/year. To drop costs, a plan has been drawn up to use said waste in reclaiming new port areas. Using solid waste as filling material requires but special actions to protect environment. There were, thus, set dedicated building designs, checked out by theoretical studies and *in situ* researches, to meet demands for limiting pollution and to suit the Port development plan.

1 INTRODUCTION

Constantza Port located west to the Black Sea, 179 NM away from Bosphorus Straits and 85 NM away from Sulina Branch, the waterway that Danube outlets by to Sea, features a millenary existence.

The development of modern Constantza Port, designed as a unitary concept, effectively started in 1896 and has been extended along some major stages that feature the following traits (Table 1):

Table 1. Characteristic features of Constantza Port

Item	North Port	South Port	Final
Total Area (ha)	722	2,500	3,222
• land	404	1,300	1,704
• sea	318	1,200	1,518
Breakwaters (km)	6.77	11.45	17.77
Quays (km)	13.4	50	63.4
Berths (no.)	78	200	278
Depths (m	7.2-14.5	7.0-22.5	7.0-22.5
Traffic (million tons)	60	180	240
Max. Vessel (000 dwt)	80	250	250

Constantza Port is today a couple of great enclosures: Constantza-North that operates all kind of cargo, gas excepted; Constantza-South able to provide future facilities owing to its greater area (Figure 1).

The Port covers some 10 km of sea shore and advances seaward for 5.5 km; its area (some 3,200 ha) includes the outlet to sea of the Danube-Black Sea Canal that connects to Cernavoda Port on Danube.

It really stands as the greatest port complex in the Black Sea area and one of the most important European ports.

Maximum throughput raised to 62,342 thousand tons in 1988, but it decreased these times down to 38,836 thousand tons. Some 60% of it stand for solid cargo. Traffic forecast shows a continuous increase up to 50 millions tons/year during the next 10 years.

Some 4,000-5,000 sea going vessels call for the Port every year. Some 18,000 personnel in more that 200 companies deal with port services.

2 REQUISITE FOR RECLAIMED LAND IN PORT

Constantza Port has been extended seaward, thus making necessary significant earth filling volumes. Total area of reclaimed land shall figures some 1,700 ha; there are some 900 ha accomplished until present i.e. some 100 million cubic meters of filling. In this purpose there have been used materials originated from dredging, borrow pits, rubble. As for Constantza-South Port most of filling materials is spoil from the Danube-Black Sea Canal.

3 DYKES THAT BORDER ENCLOSURES

To restrict scattering of filling materials about basins there have been previously designed and built enclosures bordered by dykes, complying with the general design for Port development. Dykes are rubble mould, featuring a core of quarry run and armours of stone blocks. Core material was also lime spoil from the Danube-Black Sea Canal bed, while stone blocks came from a quarry 50 km away of Constantza. Economical reasons decided on "S slope" design, previous tested on model and proved *in situ* to be true.

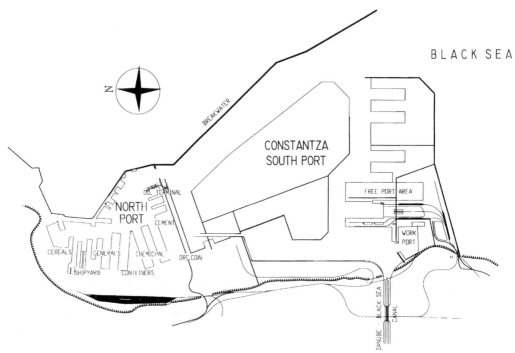

Figure 1. Constantza Port. General layout

4   PROCESSING THE FILLINGS

To transport and dump filling materials there have been used road trucks, conveyors and barges. In respect of working area, materials transported to the port were clay, sandy clay and weathered lime.

Fillings have been accomplished up to level $+5.0 \div +6.0$ m above sea level without any special compaction measures provisioned in this stage.

Further *in situ* measurements, laboratory tests and calculations, have estimated the time behaviour, stating settlements down to 50 cm. In respect of types of future constructions to be erected it shall be decided the design for bottoming and there will be further suggested required strengthening measures.

5   SOLID WASTE COME FROM PORT ACTIVITY

Port run generates some 50,000 tons/year solid waste with the following composition:

- metals   6%
- paper, wood   14%
- domestic   30%
- cement, chemicals   15%
- spoil, rubble   25%
- miscellaneous   10%

To estimate future solid waste, it has been correlated with the traffic forecast. It revealed a solid waste to cargo traffic ratio about 0.25%. Future recover of some metal, wood and paper shall reduce to some 35,000 tons/year the amount to be dumped in designated disposal areas, i.e. decrease to 0.175% of the above said percentage of solid waste in the yearly solid cargo traffic.

6   EXPERIMENTAL STUDIES INTENDED TO USE SOLID WASTE FOR FILLING

To drive waste outside the port area it is required by regulations to use the disposal area outside the Town, 30 km away, thus inducing significant costs with transport means, taxes etc.

Considering rather great quantities of solid waste coming from port run, it has been searched dumping of said materials in landfills intended to reclaim land, thus reducing transport distances to routes inside port only.

Table 2. Permissible Levels for Soil Pollution inside Port Areas
(excerpt from COPSSEC Recommendations issued by IAPH, 1993; Ports &Harbors vol. 38)

Contaminant/Level	Soil [mg/kg dry product]			Underground Waters [ppm]		
	A	B	C	A	B	C
Metals						
Cd	1	5	20		2.5	10
Cr	100	250	800		50	200
Cu	50	100	500		50	200
Mn	900	1,500	2,700	400	800	1,300
Ni	50	100	500		50	200
Pb	50	150	600		50	200
Zn	200	500	3,000		200	800
Inorganic Compounds						
NH_4					1,000	3,000
F		400	2,000		1,200	4,000
CN	1	10	100	5	30	100
S	2	20	200	10	100	300
Br	20	50	300		500	2,000
PO_4					200	700
Pesticides						
Organic-chloride (Individual)		0.5	5	0.01	0.2	1
Organic-chloride (Total)		1.0	10	0.1	0.5	2
Total Hydrocarbons		1,000	5,000	50	200	600

NOTE:

A: Normal values *clean ground* *no external pollutants*
B: Maximum permissible value *low contaminated soil* *it requires proper monitoring*
C: Critical limit *contaminated soil* *it requires cleaning operations*

Table 3. Interpreting Concentration Limits in Case of Soils inside Port Areas

Concentration of Pollutants	Range of Concentrations	Value of Pollutants	Description of Soil	Estimation of Risk/Action to be taken
0	0 . . . A]	normal values	clean area	no risk
A	(A . . . B]	average values	low contaminated area	no cleaning required, risk into normal allowed limits
B	(B . . . C]	maximum permissible values	maximum permissible contaminated area	area shall have to be monitored, risk depending on type of activities and contaminants transport
C	(C . . .	abnormal values	contaminated area	intervention keenly required, risk is unacceptable

It has been previously considered that reclaiming new land requires significant quantities and a long time to accomplish.

To design proper technologies to use solid waste in reclaiming land inside the port there have been run full scale *on site* tests. It has been used solid waste originated from port run along with spoil from the Danube-Black Sea Canal in proportion 1/6÷1/8.

Ten years later, there have been run soil and water tests in said areas. Estimation of resulted soil pollution has been judged as compared with permissible levels of soil contamination by different agents (Table 2) and construed as stated in Table 3.

On base of physico-chemical and microbiologically analyses program, run for soil and underground water samples and of limits provisioned in COPSEC/IAPH Recommendations, it can be stated about said area:
- main chemical pollutants varies from A to B;
- the degree differs from *weak* to *persistency* in case of hydrocarbon residues inside soil and water;
- organic-chloride compounds in soil and water samples varies from A to B;
- most of heavy metals varies from A to B.

411

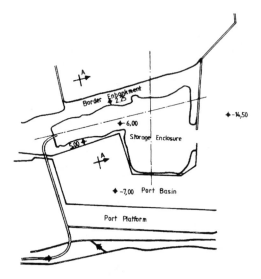

Figure 2. Enclosure intended to reclaim land.

7 THE CONCLUDED DESIGN TO DISPOSE OF PORT SOLID WASTE

On base of previous studies it has therefore been suggested that solid waste shall, latter to a prior sorting, be used to reclaim land in addition to other solid materials intended to fillings.

It is intended to use one existing enclosure inside the Port, bordered by provisional dykes.

Area of said enclosure (Figure 2) figures 120 ha and marks 6.0 m depth average. In this location lime ground at level -14.0 m is covered by a 5.0 m thick dense clay, then a 3.0 m thick silt and a last layer of dredging originated material and other spoil (Figure 3).

Final level of reclaimed land in aforesaid area shall be +5.0 to +6.0 m, thus accounting for 10 million m³

The first stage will feature execution of filling works up to level +1.50 m, then a cover, 0.50 m thick, of sorted material will be dumped and leveled off. The final stage shall feature a second cover, 3.0 m thick, of sorted material to be dumped and leveled off.

8 THEORETICAL ANALYSIS OF CONTAMINANT TRANSPORT INSIDE PORT BASIN

To estimate the effect on port water, there were developed studies on contaminant transport and time domain evolution of pollution degree of water.

The mathematical modeling applied has considered the following:

- Contaminant transport has been measured for different time intervals (Figure 4). It is thus revealed that during the first month the dispersion front of pollutants into water extended to some 440 m from dumping area. In 100 m away of said area the concentration figured 20% of the initial one. During one month time the contaminant in the enclosure did not transport beyond the bordering dyke;
- During the first six months the dispersion front extended to 740 m away of dumping area and concentration figured 10% at 310 m away;
- During the first year dispersion front extended beyond dykes by some 230 m and maximum concentration marked (theoretically) 10% of initial;
- During the first five years the dispersion front reached the extremity of the enclosure close to barges access area. It extends 450 m beyond dikes, but concentration figured utmost 30-40% of the initial one at 60-100 m away of dike;
- Researches revealed contaminants dilution by 40 inside port basin. It therefore renders insignificant concentration of exfiltrated pollutant.

Wave agitation makes dispersion to increase while concentration turns to homogenous and dilution; it shall finally get to even lesser concentrations but mores dispersed over longer contours.

There will be a gap in the enclosure border, 100 m breadth and 5.0 m depth, intended to allow for barges carrying spoil from port maintenance dredging. In the last end, a dyke, to separate filling and disposed material from port basin, shall obstruct this access.

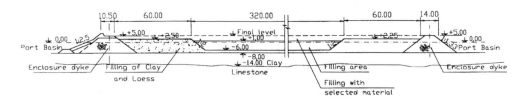

Figure 3. Cross section of submerged dykes and enclosure intended to reclaim land by solid waste filling.

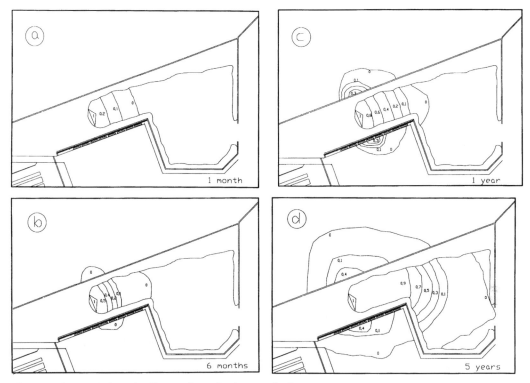

Figure 4. Time evolution of pollutants dispersion into port basin

9   CONCLUSIONS

To dispose of solid waste originated from Constantza Port run, is suggested to use said waste, in addition to other solid materials intended to fillings, in reclaiming new land for port use.

The designs drawn up considered both the economical aspects and the environmental protection ones, especially protection of port waters.

It has therefore been designed building of enclosures bordered by rubble mould dykes of sorted stone that features low permeability. Inside these enclosures can be dumped great quantities of different filling materials.

Both *in situ* tests and theoretical models and calculations helped in identifying and accounting pollution effects and ways to restrict them to permissible levels.

REFERENCES

Ciortan, R.   1996. *Evolution in the concept of arrangement of the Constantza Port.* International Seminar, 12-13 of September, 1996, Constantza, România.

The Official Gazette of Romania, Part 1, No. 303 bis, 6 of October 1997

Ports & Harbors Review, 1993. Issued by The International Association of Ports and Harbors. Vol 38

Geotechnical Hazards, Marić, Lisac & Szavits-Nossan (eds) © 1998 Taylor & Francis, ISBN 90 5410 957 2

Repair of the Pliva river waterfall at Jajce

D.Čorko & D.Lovrenčić
Conex, Zagreb, Croatia

P.Stojić
Faculty of Civil Engineering, University of Split, Croatia

Z.Šteger
Conex, Mostar, Bosnia and Herzegovina

ABSTRACT: The Pliva river waterfall at Jajce is a calc-sinter (tuff calcarious) barrier formed at the confluence of the Pliva and Vrbas. Waterfall was originally located at the very confluence of the two rivers. Some 50 years ago it collapsed and moved upstream about 30 m to its present position. The new position of the overflow profile was protected by certain interventions (construction of overflowing barriers). At the end of 1996 a part of the waterfall collapsed again. That was the reason for new intervention and remedial works. The paper gives a short survey of events preceding the collapse of the part of the waterfall as well as the description of the proposed remedial works. A part of these works was executed preventively and final remediation will be undertaken in the forthcoming period.

1 INTRODUCTION

The Pliva river waterfall is a natural phenomenon and has been the symbol of the city of Jajce for the past 600 years of its existence. The waterfall consists of a calc-sinter barrier formed at the confluence of the Pliva into the Vrbas.

The total length of the Pliva river is about 22 km. In 1895 the first hydroelectric power plant with capacity of 16 m^3/s was constructed 3.5 km upstream of the waterfall. At the same time the Niagara Falls (Westinghouse - Tesla) power plant was constructed. This power plant is considered to be the first modern power plant in the world. In 1956 hydroelectric power plant Jajce I was constructed with considerably greater capacity (Q_i = 60,8 m^3/s). Following the construction of these power plants serious problems realted to river bed deepening and erosion were initiated. In the last 1000 years the river bed deepened by some 45 m but during the period 1952 - 1957 river bed was deepened by 5 m.

The first repair measures on the top of the waterfall were undertaken even before the World War I. In 1947 the waterfall collapsed and the position of the overflow moved upstream by approx. 30 m. In the period from 1956 to 1958 barriers were constructed along the river bed and the overflow profiles were strengthened.

The waterfall suffered another major damage in 1969/1997 and a part on the right side of the waterfall collapsed. At the lowest flow rate (which after the construction of electric power plants lasts over 11 months a year) water does not even reach the waterfall - it sinks into the river bed about 10 m before reaching the overflow.

The new condition is extremely unfavourable and may cause the complete collapse of waterfall partition profile, which would result in further (exceptionally fast) erosion of the river bed by some 20 m. and destruction of city banks. Lower parts of the city would vanish and the remaining parts would be left without road connection with the surrounding areas.

2 GEOTECHNICAL CHARACTERISTICS

The whole river bed of the Pliva from the Great Lake and Small Lake to the mouth of the Vrbas consists of calc-sinter, i.e calcium carbonate. Calc-sinter is formed by mosses and algae from calcium bicarbonate-rich waters under certain conditions (lack of pollution, exposure to light, temperature, velocity, alkalinity, water hardness, etc.). Pavletić (1957) explained in detail the phenomenon and conditions of calc-sinter formation in the Pliva river and its characteristics.

Geotechnical investigation performed for the purpose of waterfall repair indicated that the bedrock composition at the current waterfall profile is typical for materials behind the calc-sinter barriers. During the formation of the calc-sinter barrier the weir of moss and algae is formed on the front side, gradually changing into solid calc-sinter. Muddy calc-sinter, weathered and decomposed calc-sinter and alluvial river deposits consisting of sand, silt, fine-grained gravel, etc. are found at the back of the barrier. In the waterfall zone seven exploration boreholes and two refraction seismic profiles were executed. Investigation results indicate exceptional heterogeneity in rock composition with no regularity.

It is obvious that the current waterfall profile was formed behind the earlier calc-sinter barrier with material of considerably lower characteristics. There is quite a lot of calc-sinter on the front side of the waterfall, which was partly formed after the latest repair of the waterfall in the '50s. It created an impression that geotechnical situation was far more favourable than it actually was.

3 CAUSES OF WATERFALL COLLAPSE

Regardless of the relatively bad geotechnical conditions in the calc-sinter barriers, after interventions in 1956-1958, the waterfall existed for 40 years without sustaining major damage. Causes of waterfall collapse were analysed and several contributing factors were established:

- consequences of war actions during the period 1992 - 1995, when a large number of explosions took place on the very bank in the waterfall zone
- major flood during 1995/1996 when flow rate of over 200 m³/s was recorded (much above the 118 m³/s for which earlier repair was intended)
- earthquake which was felt at that time in Jajce
- higher river water pollution which resulted in decomposition of calc-sinter and destruction of the calc-sinter maker.
- hydroelectric power plant damage inflicted by war actions, which prevented proper water management and controlled drainage into the river bed, i.e. to the waterfall.

The analyses showed that impossibility of proper water management in the hydroelectric power plant facility (which is a direct consequence of damages caused by war actions) could result in a 1000 year high flow rate at any moment. River bed, structures in the river and waterfall overflow profile could not withstand such flows. Consequently significant damage along the whole river bed and on the waterfall itself resulted.

A river barrier (up to 10 m high) upstream of the waterfall was destroyed, one bridge collapsed and two were seriously damaged, parts of bank protections were destroyed, landslides which damaged city roads resulted, etc. The waterfall itself (which consists of three waterfalls - left, right and middle waterfall) was damaged. The right waterfall was totally destroyed. In addition to that a landslide appeared on the right flank, downstream from the waterfall, threatening an important road leading from the city. The terrain in the landslide zone is exposed to imminent waters which at higher flows can overflow the collapsed part of the waterfall, aggravating the damage.

At the moment (at flow rates of 2 to 3 m³/s) water does not even reach the waterfall and sinks into a chasm which opened some 10 m upstream.

The following photographs illustrate the earlier and present condition of the waterfall.

Fig. 1   Original appearance of the waterfall

Fig.2   Waterfall after repair in the '50s

Fig. 3   Waterfall after the last damage - no water
          on the waterfall

Fig. 4   Waterfall after intervention repair - water is
          flowing again

4 DESCRIPTION OF REPAIR WORK

In the preparatory phase of the repair project geodetic survey of the Pliva river bed and the wider area of the waterfall was carried out. Engineering geology, geomechanical and geophysical investigation were also performed, as well as the detailed hydraulic calculations and analyses of assumptions and previous remediation solutions (riverbed and waterfall).

An important result of investigation activities and analyses is the conclusion that rational or environmentally acceptable interventions cannot protect the waterfall from flow rates with minor probability. Flow rates above 118 m³/s would be above acceptable limits for general safety of the immediate waterfall zone. As an illustration of the magnitude of this flow rate let us look at the Pliva flow rates: 2-year high - 116 m³/s; 20-year high - 177 m³/s and 100-year high - 220 m³/s (1000-year high is 285 m³/s). A particular menace for the waterfall area lies in the fact that unproper

management in hydroelectric power plant Jajce I can cause catastrophic flood which corresponds to high waters with low probability.

In view of the circumstances it was proposed to construct a 150 m long tunnel which would direct all waters above the acceptable 118 m³/s limits beside the waterfall into the Vrbas and to protect the waterfall for flow rate up to 118 m³/s.

Calculations indicate that the existing reinforced concrete overflow construction in the zone of the left and middle waterfall is satisfactory, but should be additionally strengthened. For that purpose a reinforced concrete slab is to be constructed at the bottom between the two waterfalls (left and middle) with stilling basin. The slab should be protected in the edge zone of the waterfall by micro-piles, and in the inner part by jet grouting columns. A system of drainage channels with appropriate outlets in the concrete body of the overflow barrier will be provided for drainage of possible waters which might appear below the concrete plate. River bank in the area of left waterfall shall be protected from water dynamic and erosion impact by appropriate protective wall.

Similar interventions are planned in the river bed in front of the right and middle waterfall as well as the reconstruction of the overflow construction (in a new location). The ground-plan of the right waterfall is formed according to the conditions in the field and corresponding hydraulic analyses. Foundations of the overflow construction are also protected by micro-piles.

Weak points in the vertical face of the waterfall will be protected by reinforced concrete anchor slabs. Drilled drains will be provided for drainage of possible water which might appear behind those slabs.

Natural stilling basin at the heel of the waterfall and actually at the river Vrbas level will be protected by stones as well as the re-shaped landslide mass.

Reinforced concrete barriers on the right flank where the landslide occurred will be protected by prestressed ground anchors.

All reinforced concrete barriers in the river bed upstream of the waterfall zone will be additionally strengthened by prestressed ground anchors to increase their (too low) resistance to sliding.

Appropriate bank protection will be provided for proper direction of the river flow. Figure 5 shows the ground-plan of the waterfall with location of relief tunnel and Figure 6 schematically presents the reconstruction work and protection of the right waterfall.

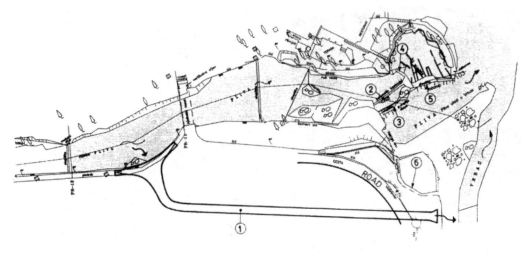

① Tunnel　　　　　　　　　　　　　④ Intermediate waterfall
② Precipe　　　　　　　　　　　　 ⑤ Left waterfall
③Right waterfall (demolished)　　　 ⑥Slip surface

Fig. 5　Ground-plan of the waterfall and relief tunnel

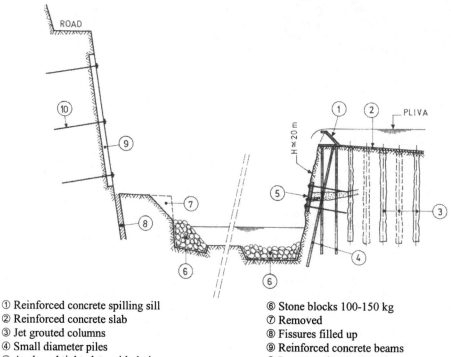

① Reinforced concrete spilling sill　　　⑥ Stone blocks 100-150 kg
② Reinforced concrete slab　　　　　　⑦ Removed
③ Jet grouted columns　　　　　　　　⑧ Fissures filled up
④ Small diameter piles　　　　　　　　⑨ Reinforced concrete beams
⑤ Anchored tight plate with drainage　　⑩ Prestressed ground anchors

Fig. 6　Schematic presentation of reconstruction work and protection of the collapsed right waterfall

5 CONCLUSION

The Pliva river waterfall is a natural gift, unfortunately exposed to destruction. Human activity (electric power plant operation, war actions, pollution, etc.) primarily and partly consequences of erosion are to be blamed for it. Repair work will try to prevent the disappearance of the waterfalls, although it must be clear that after human intervention it becomes less a natural phenomenon and more an artificial formation. Unfortunately, that is necessary, and we must strive to protect the nature as much as possible.

Favourable conditions will facilitate the formation of calc-sinter which will in time hide from view human intervention and the natural phenomenon will continue. Experiences gained from earlier repair work undertaken in the '50s show that calc-sinter has already covered the constructions, continuing the work of nature where it was interrupted.

One thing is certain - such problems are not frequently encountered and experiences are relatively small, which made our work in trying to find the solution difficult. In addition to find the technical solution, preconditions for continuation of formation of calc-sinter as the vital element of this exceptional phenomenon had to be created. Time will show to what extent the executed repair works are successful.

REFERENCES

Design solutions of remedial works on riverbed and waterfall of Pliva river in 1952 (in Croatian).

Pavletić, Z. 1967. Report on biological testing of the lower Pliva river (in Croatian).

Energoinvest 1967. Remedial works on Pliva river and its waterfall in Jajce - Condition control after 10 years (in Croatian).

Spajić, S. 1997. Report on waterfall destruction (in Croatian).

Institute of Civil Engineering, University of Mostar 1997. Report on geotechnical investigation works on the Pliva riverbed and waterfall in Jajce (in Croatian).

Conex - Mostar and Conex - Zagreb 1997. Hydraulic analysis and remedial work projects on Pliva riverbed and waterfall in Jajce (in Croatian).

Geotechnical Hazards, Marić, Lisac & Szavits-Nossan (eds) © 1998 Taylor & Francis, ISBN 90 5410 957 2

Verhalten einer Schmalwand aus zementhaltiger Suspension in kontaminiertem Boden

O. Henögl & S. Semprich
Institut für Bodenmechanik und Grundbau, Technische Universität Graz, Austria

R. Völkner
Grün + Bilfinger Ges.m.b.H., Wien, Austria

KURZFASSUNG: Zur Sicherung der Altlast 'Gailitzspitzdeponie' auf dem Werksgelände der ehemaligen Bleiberger Bergwerksunion in Arnoldstein, Kärnten, war eine Umschließung mit einer doppelten, i.M. ca. 7,5 m tiefen, erosionsstabilen Rüttelschmalwand mit einem k-Wert von < 1.10^{-9} m/sec vorgesehen. Da das Grundwasser hydratationsschädliche Sulfatgehalte erwarten ließ, wurde pro m^3 Schmalwandmasse eine bereits bei ähnlichen Bauaufgaben bewährte Rezeptur eingesetzt. Für ungestört entnommene Proben aus der hergestellten Schmalwand wurde im Alter von 28 Tagen jedoch ein Durchlässigkeitsbeiwert von nur 1.10^{-8} m/s gemessen. Auch hinsichtlich des Festigkeitsverhaltens der Proben mußte festgestellt werden, daß die Proben selbst nach 3 Monaten kaum erhärteten und unter dem als erosionsstabil geltenden Wert lagen. Deshalb wurden Versuche zur Erosionsstabilität durchgeführt, bei denen die Durchströmung einer Schmalwand realistischer als bei der herkömmlichen Untersuchung mittels Pinhole-Test simuliert wurde. Die Ergebnisse zeigten, daß bei den vorort vorhandenen Korngrößenrelationen selbst eine weichplastische Schmalwand keine erhöhte Erosionsgefährdung aufweist. Um die Ursache des Nichterhärtens herauszufinden, wurden im Labor unterschiedliche Dichtwandmassen-Rezepturen unter Beigabe von vorort anstehendem Boden und einem Modellboden untersucht. Nur die mit Boden aus der Umgebung der Schmalwand angereicherten Proben erhärteten nicht. Untersuchungen des Bodens ergaben außergewöhnlich hohe Blei- und Zinkgehalte, die im Grundwasser nicht vorhanden sind. Da hohe Blei- und Zinkgehalte bekanntermaßen das Abbindeverhalten von Zement behindern, kann davon ausgegangen werden, daß diese die Ursache für das verzögerte Erhärten der mit Boden angereicherten Schmalwandmasse sind. Zusammenfassend kann für vergleichbare Bauaufgaben empfohlen werden, den Boden vor Beginn der Baumaßnahmen chemisch zu untersuchen, untergrundspezifische Eignungs- und Voruntersuchungen unter Berücksichtigung der Filterkriterien durchzuführen und diese Ergebnisse dann in Ausschreibung und Qualitätsanforderungen einfließen zu lassen.

1 PROJEKT GAILITZSPITZDEPONIE

Die Gailitzspitzdeponie liegt auf dem Werksgelände der ehemaligen Bleiberger Bergwerksunion in Arnoldstein, Kärnten. Auf dieser Deponie wurden im wesentlichen Industrierückstände deponiert. Zur Sicherung dieser Altlast wurde 1995 und 1996 eine Umschließung mit einer doppelten Rüttelschmalwand ausgeführt (Abb.1). Der Umfang dieser im Mittel 7,5 m tiefen Umschließung betrug für die innere und äußere Dichtwand je ca. 550 m, so daß eine Gesamtbauleistung von ca. 8000 m^2 zu erbringen war (Abb.2). Die Ausschreibung sah eine erosionsstabile Schmalwandumschließung mit einem k$_f$-Wert < 1 . 10^{-9} m/s unabhängig vom Alter vor. Die Kenntnis der Baugrundverhältnisse im Bereich der Deponie basierte auf den Ergebnissen von 5 Kernbohrungen, 18 Schürfen und 4 schweren Rammsondierungen (Abb.2). Unter einer Überlagerung aus bis zu mehreren Metern mächtiger Schlacke folgen Grobkiese bis Feinsande in Wechsellagerung, gefolgt von ebenfalls wechselgelagerten

Grob-/Mittelkiesen bis Mittel-/Feinsanden (Abb.3). Der Grundwasserspiegel wurde örtlich in Tiefen von ca. 2 bis 7 m unter Geländeoberfläche angetroffen. Da erhöhte Sulfatgehalte im Grundwasser zu erwarten waren, wurde zur Herstellung der Dichtwand eine bereits bei anderen Bauvorhaben bewährte Schmalwandmasse (Walhalla-Kalk, Regensburg 1995) mit folgender Rezeptur pro m^3 eingesetzt:

125 kg Protomix C der Fa. Walhalla, Regensburg
525 kg Steinmehl Carolith FM, Feldbach, Stmk.
760 kg Leitungswasser der Gemeinde Arnoldstein

In dem Merkblatt der Fa. Walhalla-Kalk werden als Sollwerte nach 28 Tagen ein Durchlässigkeitsbeiwert k$_{10}$ ≤ 1. 10^{-9} m/sec und eine einaxiale Druckfestigkeit q$_u$ ≥ 70 N/cm^2 genannt.
Die erwarteten hydratationsschädlichen Sulfatgehalte bestätigten sich in später durchgeführten chemischen Untersuchungen des Grundwassers.

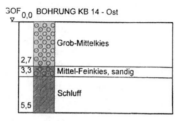

Abb. 3. Ergebnisse der Erkundungsbohrungen KB 13-Ost und KB 14-Ost

Abb.1. Herstellung der Rüttelschmalwand

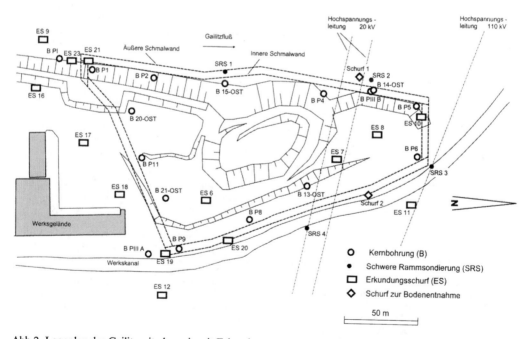

Abb.2. Lageplan der Gailitzspitzdeponie mit Erkundungsprogramm und geplanter Dichtwand

Abb. 4 zeigt die dabei ermittelten Sulfatgehalte in Beziehung auf ihre angreifende Wirkung auf Hochofenzement.

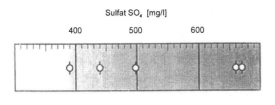

Abb.4. Sulfatgehalt des Grundwassers

Im Oktober 1995 wurden von der ausführenden Firma Grün + Bilfinger wesentliche Abschnitte der westlichen, südlichen und östlichen Deponieumschließung hergestellt. Die Durchlässigkeitsversuche an Rückstellproben der Schmalwandmasse zeigten, daß nach 28 Tagen der geforderte Grenzwert von 1.10^{-9} m/sec erreicht war (Abb.5). Die 28-Tage Druckfestigkeit dieser Proben ergab hinreichende Werte von ca. 80 N/cm^2.

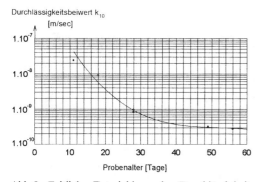

Abb.5. Zeitliche Entwicklung der Durchlässigkeit von Rückstellproben der verwendeten Schmalwandmasse

2 DURCHLÄSSIGKEIT UND DRUCK-FESTIGKEIT DER HERGESTELLTEN SCHMALWAND

Mit Beginn der Arbeiten wurde eine auf der Westseite gelegene Prüfkammer hergestellt, in der im Zeitraum vom 16.10. - 20.10.1995 ein Auffüllversuch durchgeführt wurde und dessen Ergebnisse bezüglich der Systemdichtigkeit der Kammer zufriedenstellend ausgefallen sind. Darüberhinaus wurden zur Über-

prüfung der Qualität der hergestellten Dichtwand Schürfen ausgeführt.

Hierbei zeigte sich, daß ihre Festigkeit augenscheinlich nicht den Erwartungen entsprach. Daraufhin wurden ungestörte Proben aus der Wand entnommen und auf ihre Durchlässigkeit und einaxiale Druckfestigkeit nach den einschlägigen Regelwerken untersucht (ÖNORM B 4421-1, B 4415, Empfehlungen des AK "Geotechnik der Deponien und Altlasten"). Für 28-Tage alte Proben wurde ein Durchlässigkeitsbeiwert von nur 1.10^{-8} m/s gemessen. Untersuchungen nach weiteren 3 Monaten zeigten keine wesentliche Änderung. Nach ca. 8 Monaten erreichten die Proben aber dann doch Durchlässigkeiten von i.M. ca. 5.10^{-9} m/sec (Abb.6).

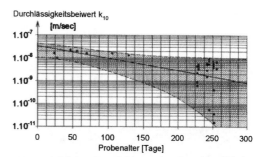

Abb.6. Zeitliche Entwicklung der Durchlässigkeit von ungestörten Proben aus der hergestellten Schmalwand

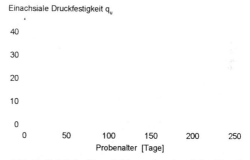

Abb.7. Zeitliche Entwicklung der einaxialen Druckfestigkeit von ungestörten Proben aus der hergestellten Schmalwand

Aber hinsichtlich des Festigkeitsverhaltens der Proben mußte festgestellt werden, daß die Proben selbst nach dieser Zeit kaum erhärteten (Abb.7).

3 UNTERSUCHUNGEN ZUR ERO-SIONSSTABILITÄT

Wenngleich auch die Ausschreibung keine Mindestdruckfestigkeit für die hergestellte Dichtwand forderte, ist die Erosionsstabilität zur langfristigen Sicherstellung ihrer Funktion eine unumgängliche Notwendigkeit. Schmalwandmassen wurden bisher generell ab einer einaxialen Druckfestigkeit von $q_u = 15 - 20$ N/cm^2 als erosionsstabil bezeichnet (Kirsch & Rüger 1976, Martin et.al. 1986). Da die Ergebnisse der Druckfestigkeitsuntersuchungen an den aus der Dichtwand entnommenen Proben diese Mindestwerte nicht erreichten, wurde die Ingenieurgemeinschaft für Geotechnik Plankel, Pelzl & Partner, Lauterach, Anfang 1996 beauftragt, Untersuchungen zur Erosionsstabilität der hergestellten Schmalwand durchzuführen. Dafür stellten Plankel, Pelzl & Partner Untersuchungen an, bei denen die Durchströmung einer Schmalwand hinsichtlich einer möglichen Erosion (Abb.8) realistischer simuliert wurde, als es bei der herkömmlichen Untersuchung zur Erosionsstabilität mittels Pinhole-Test nach British Standard BS 1377: Part 5: 1990 der Fall ist.

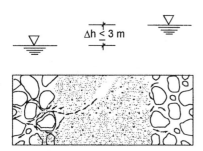

Abb.8. Schematische Darstellung von Erosionserscheinungen

Es wurden zunächst ungestörte Proben aus der hergestellten Schmalwand in einen Durchlässigkeitsprüfzylinder derart eingebaut, daß die verhältnismäßig weichen Proben satt an der Außenwand des Prüfzylinders anlagen. Es erfolgte keine Verdichtung, sondern nur ein sehr vorsichtiges Andrücken der Proben beim Einbau in den Prüfzylinder. Die Schmalwandproben wurden etwa bis zur halben Höhe des Prüfzylinders eingebaut. Unmittelbar darüber wurde der in der jeweiligen Tiefenstufe anstehende Boden eingebaut. Anschließend erfolgte die Durchströmung der Probe von unten nach oben. Unter der Annahme, daß die Schmalwand an jeder Stelle eine Mindeststärke von 10 cm aufweist sowie der Annahme, daß eine maximale Wasserspiegeldifferenz von $\Delta h = 3{,}0$ m an der Schmalwand noch zu keinen Schäden führen darf, ergibt sich ein hydraulisches Gefälle von $i = 30$.

Bei den Versuchen wurde das hydraulische Gefälle so lange gesteigert, bis ein deutlicher Anstieg der Durchlässigkeit bzw. eine Trübung des durchgeströmten Wassers festgestellt wurde. Durch den fehlenden Seitendruck wurde auch die Möglichkeit einer Umströmung der Probe nicht unterbunden. Erst bei einem Gradienten von $i = 70$ nahm die Durchlässigkeit überproportional zu und es zeigte sich gleichzeitig eine Trübung.

Zusätzlich wurden Versuche in der triaxialen

Durchlässigkeitszelle mit einem Zelldruck von 0,1 bar über dem Durchströmungsdruck durchgeführt. Der Probeneinbau und der Versuchsablauf erfolgte wie oben. Selbst bei einem extrem hohen Gradienten von $i = 200$ konnte keine Trübung des Wassers und keine Durchlässigkeitserhöhung festgestellt werden. Mit den Versuchen wurde somit nachgewiesen, daß unter den bei der Gailitzspitzdeponie vorort vorhandenen Korngrößenrelationen selbst eine weichplastische Schmalwand keine erhöhte Erosionsgefährdung aufweist (Plankel, Pelzl & Partner, 1996) (Abb.9).

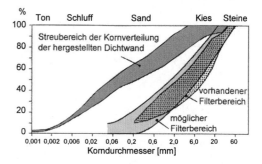

Abb.9. Nachweis der Erosions- und Suffosionsstabilität durch Gegenüberstellung der Korngrößenverteilungen

4 URSACHE FÜR DAS VERZÖGERTE ERHÄRTEN DER HERGESTELLTEN SCHMALWAND

Um die Ursache des Nichterhärtens herauszufinden, wurde an aus der Schmalwand entnommenen Proben der Anteil des im Zuge der Herstellung unvermeidbaren Bodeneintrags bestimmt. Der Masseanteil für Korndurchmesser $d > 0{,}063$ mm betrug zwischen 50 und 65 %. Danach wurden unterschiedliche Dichtwandmassen-Rezepturen untersucht, bei denen alternativ Grazer Leitungswasser oder aus mehreren Brunnen entnommenes und gemischtes Deponiegrundwasser verwendet wurde. Für die Mischprobe des Deponiegrundwassers hat sich als einzig auffallender Meßwert der Sulfatgehalt zu 498 mg/l ergeben. Dieser Wert übersteigt den für Trinkwasser im Österreichischen Lebensmittelbuch, Kap. B1, III. Aufl., mit 250 mg/l geforderten Wert der zulässigen Höchstkonzentration. Alle anderen untersuchten Parameter einschließlich Blei- und Zinkgehalt lagen unterhalb der dort angegebenen zulässigen Grenzwerte. Außerdem wurden Versuche ohne und mit Beigabe von aus der Umgebung der verzögert erhärtenden Schmalwand stammendem Boden (Schurf zur Bodenentnahme, Abb.2) sowie einem Modellboden aus Quarzkörnern durchgeführt. Nur die mit Boden aus der Umgebung der Schmalwand angereicherten Proben erhärteten nicht (Abb.10).

Daher schied das Grundwasser als Ursache für das verzögerte Erhärten der Dichtwand aus. Die Ursache war also im anstehenden Baugrund zu suchen:

Untersuchungen haben außergewöhnlich hohe Blei- und Zinkgehalte des Bodens bis zu ca. 1000 mg/l im basischen Eluat bzw. bis zu 0,7 % in der Trockensubstanz ergeben. Da diese bekanntermaßen das Abbindeverhalten von Zement behindern, kann davon ausgegangen werden, daß die Blei- und Zinkanteile die Ursache für das verzögerte Erhärten der mit Boden angereicherten Schmalwandmasse sind (Lieber 1967). Auch Wruss bestätigt in seiner Stellungnahme (Wruss 1995), daß hohe Anteile von Blei und Zink das Abbindeverhalten des Zementes verzögern. Wahrscheinlich werden die an die Bodenkörner adhäsiv gebundenen Blei- und Zinkanteile bei Veränderung des pH-Wertes von neutral auf basisch im Zuge der Einbringung der Schmalwandmasse mobilisiert.

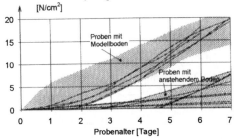

Abb.10. Zeitliche Entwicklung der einaxialen Druckfestigkeit von Proben mit verschiedenen Schmalwandmassenrezepturen unter Verwendung des anstehenden Bodens bzw. eines Modellbodens

5 RESÜMEE

Für vergleichbare Bauaufgaben kann folglich empfohlen werden, den Chemismus des Bodens über die gesamte relevante Tiefe vor der Bauausführung im Zuge der Planung zu erkunden. Damit werden anschließend untergrundspezifische Voruntersuchungen sowie Eignungsprüfungen der zu verwendenden Schmalwandmasse möglich. Die Forderung nach Erosionsstabilität sollte zusätzlich die Untersuchung der vor Ort vorhandenen Korngrößenverteilungen unter Einbeziehung der bodenmechanischen Filterkriterien vorsehen. Schließlich sollten die daraus gewonnenen Erkenntnisse in die Ausschreibung der Ausführungsarbeiten und deren Qualitätsanforderungen einfließen.

LITERATUR

ESW Consulting Wruss 1995. Stellungnahme bezüglich A9 - Gailitzspitzdeponie (unveröffentlicht).

Kirsch & Rüger 1976. Die Rüttelschmalwand - Ein Verfahren zur Untergrundabdichtung. *Vorträge der Baugrundtagung in Nürnberg:* 439-476.

Deutsche Gesellschaft für Erd- und Grundbau.

Lieber, W. 1967. Einfluß von Zinkoxyd auf das Erstarren und Erhärten von Portlandzement. *Zement-Kalk-Gips* Heft 3: 91-95.

Martin, A. et.al. 1986. Schmaldichtwände am Lech. *Wasserwirtschaft* 76 Heft 12: 545-612.

Plankel, Pelzl & Partner 1996.Geotechnischer Bericht - Gailitzspitzdeponie Arnoldstein - Erosionsstabilität Schmalwand (unveröffentlicht).

Walhalla-Kalk, Regensburg 1995. Technisches Merkblatt "Walhalla Protomix-C".

Geotechnical Hazards, Marić, Lisac & Szavits-Nossan (eds) © 1998 Taylor & Francis, ISBN 90 5410 957 2

Reuse of petroleum contaminated soil as a road material

H. Koyuncu, A. Tuncan, M. Tuncan & Y. Güney
Anadolu University, Eskişehir, Turkey

ABSTRACT: In this study, petroleum contaminated soil were stabilized by mixing pozzolanic fly ash, lime and cement in order to use as a sub-base material. This stabilization produced physically, mechanically and chemically stabilized new mixtures. Unconfined compressive strength, permeability, California bearing ratio (CBR), freeze/thaw (durability), pH, electrical conductivity, cation exchange capacity, total metal content and leachate experiments were conducted on stabilized mixtures. The best result in all of the tests conducted is obtained with petroleum contaminated soil stabilized with Lime (20%) + Fly ash (10%) + Cement (5%). This mixture can be effectively and safely used as sub-base material for the road construction.

1 INTRODUCTION

The petroleum industry can present high risks to the environment. Oil well drilling activities produce large quantities of wastes in the form of drilling muds and cuttings. These wastes are disposed of in open pits (mud-pits) and large quantities of unstabilized wastes remain at a number of sites. To reduce the high risks of environmental impacts, effective and economical alternatives are to stabilize the wastes and reuse them as sub-base material in the road construction. Anadolu University and Turkish Petroleum Corporation (TPAO) have been cooperating to determine the feasibility of reuse of petroleum drilling wastes as sub-base after stabilizing with pozzolanic materials. Turkish Petroleum Corporation constructs roads between petroleum tower and the nearest village. Therefore, stabilized mixtures can be economically and effectively used for the road construction.

Soil stabilization has been used for decades in geotechnical and civil engineering constructions. Stabilization is the improvement of physical, chemical and mechanical properties of wastes, encapsulation of pollutants and reduction of solubility of toxic substances. An economical and fast solution is to stabilize petroleum contaminated drilling wastes by mixing with additives. The stabilization method is applied to petroleum drilling wastes to bind the oil and metals in a structure formed by cementing pozzolanic materials to produce chemically and physically stable

and mechanically handable mixtures. The stabilization of petroleum drilling wastes is accomplished by pozzolanic reactions which include both chemical and physical interactions between the wastes and additives. Stabilization with lime, fly ash and cement is the simplest and inexpensive technique. Lime, fly ash and cement are used to increase pH, to contribute surface area and to retain hydrocarbons, respectively. The main purpose of stabilization is to increase the strength of sub-base material to support road traffic, to maintain integrity of the material during repeated freezing/thawing, to reduce the leachability of the waste constituents. Therefore, it is provided to have a substantial strength and environmental durability. Mixtures used in the laboratory and their grain size distributions are given in Table 1 and Figure 1, respectively.

This paper reports the results of a comprehensive laboratory study on physical, mechanical and chemical properties of stabilized mixtures in order to use as sub-base material in the road construction.

2 SOLIDIFICATION/STABILIZATION (S/S) TECHNIQUES

Solidification/stabilization of contaminated soil or waste is a relatively new technique. Solidification and stabilization techniques are to prevent or minimize the release of a contaminants to the environment by producing a solid mixture, improving handling characteristics, decreasing surface area for contaminant transport, reducing mobility of the contaminant when exposed to leaching fluids and bonding the contaminant into a non-toxic form (Poon, 1989). Advantage of S/S is that the additives are widely available and relatively inexpensive. Disadvantages of it are the volume of treated material may increase with the addition of additives. S/S of organic and inorganic wastes can be done by the addition of industrial by products such as fly ash, lime and portland cement, or a combination of both materials which often results in a pozzolanic reaction (U.S. EPA, 1992). The purpose is to neutralize the sludge, produce a soil like, compactable mixture, and entrap hydrocarbons in a structure formed by cementing the clay into continuous monolith. A pozzolan is defined as a material that does not exhibit cementing ability when it is used by itself. However, when it is used with portland cement or lime, cementitious reactions occur. A pozzolan is defined in ASTM C595 (1985) as a siliceous or siliceous and aluminous material which in itself has little or no cementious value of mixture, chemically react with calcium hydroxide at ordinary temperature to form compounds having cementious properties.

Table 1. Mixtures used in the laboratory experiments.

Mixture	
Mixture 1	Petroleum Drilling Waste (PDW)
Mixture 2	PDW+ L (%15)+ C (%5)
Mixture 3	PDW+ L (%15)+ C(%5)+ FA (%5)
Mixture 4	PDW+ L (%20)+ C (%5)+ FA (%10)
Mixture 5	Standard sub-grade material used in Turkey

L: Lime, C: Cement, FA: Fly Ash

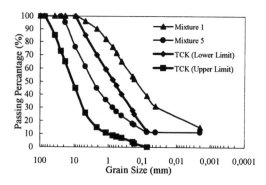

Figure 1. Grain size distributions of mixtures 1 and 2 and the upper and lower limits of Turkish General Directorate of Highways (TCK).

2.1 Solidification technique

Solidification techniques eliminate the free liquid, increase the bearing strength, decrease the surface area of the wastes and produce a monolithic solid product of high structural integrity. Solidification may involve encapsulation of fine waste particles (microencapsulation) or large blocks of waste (macroencapsulation). Therefore, solidification encapsulates the waste in a monolithic solid of high structural integrity (Pamukcu et al., 1989). The major advantage of encapsulation process is that the waste never comes in contact with water and very soluble chemicals such as sodium chloride which can be successfully encapsulated. Encapsulation is to isolate the waste from the environment. The contaminants do not necessarily interact chemically with additives, but are mechanically locked within the solidified matrix called microencapsulation. The size of the microencapsulated particles may range from very small values to larger agglomerates visible under the light microscopy or even to the naked eye. Wastes can also be macroencapsulated by the additives.

2.2 Stabilization technique

Stabilization is the process of reducing the hazardous potential of waste material by converting the contaminants into their least soluble, mobile or toxic form. Stabilization involves reduction in the free moisture and bulk mobility of the waste and therefore, improves the mechanical properties of waste. Stabilization of wastes with lime and fly ash is the simplest and least expansive technique. Because lime provided pH control and fly ash contributed surface area. Lime and fly ash give the microencapsulating the sludge droplets inside a fine, stiff and porous structure. Portland cement is added to improve the interparticle binding. Mechanical properties are also improved greatly. Addition of lime decreases the water content, thus reducing the thickness of diffuse double layer of water surrounding clay particles. The solubility of hydrocarbons was also dramatically reduced (Martin et al., 1990).

3 MATERIALS

3.1 Petroleum drilling waste (PDW)

Petroleum drilling waste contains drilling muds and cuttings which consist of both liquid and solid phases. Drilling muds are used to cool, clean or lubricate bits, to carry cuttings up from the bottom and to control subsurface pressure. Drilling muds are normally classified as oil-base and water-base mud (Rogers, 1963). Oil-base mud contains a continuous oil phase consisting of usually diesel oil, bentonite clay and chemicals. Water-base mud contains fresh or salt water phase, bentonite clay and chemicals. Water-base mud is preferred for its low cost. However, certain situations require an oil-base mud for drilling operations in formations sensitive to water-base fluids. Drilling cuttings are the solid cuttings which are brought to the surface from the well during borings. In this study, water-base mud was used. This mud consists of chloride, barite, bentonite clay, caustic and carboxy methly cellulose (CMC). In addition to these, wastes also consist of diesel oil and rock fragments.

Petroleum contaminated soil (PCS) in New Jersey is considered hazardous waste if they exceed 3% total petroleum hydrocarbons (TPH) (Bell et al., 1989). Constituents of drilling wastes are given in Table 1.

3.2 Additives

In this study, fly ash, lime and cement are used as additives. Some properties of the additives are given in Tables 2, 3 and 4.

One of the most popular pozzolans in the industrial by product is fly ash. Most fly ash is a by-product of coal burning power plants. Fly ash is a fine, silt size material consisting spherical, sometimes hollow, glassy particles and is primarily composed of silica, alumina and various oxides and alkalies. It is pozzolanic in nature and can react with hydrated lime to produce cement-type products. For this reason, lime/fly ash mixtures can be used for stabilization of highway bases and sub-bases. It is also used in combination with lime, portland cement and aggregate as a road base and as a backfill. Fly ash acts as a conditioner in the waste to absorb liquid and to coat waste globules. Fly ash used in this study is Type C fly ash according to ASTM C618 (1993). Type C fly ash is self cementing due to the presence of lime or other chemical compounds. Type C fly ash has self-hardening characteristics. Because it contains large quantities of calcium oxide. Therefore, the addition of water alone produces a cementing reaction without the addition of lime. This kind of process is suitable for S/S techniques. Total amount of $SiO_2 + Al_2O_3 + Fe_2O_3$ is 61.76 %. Minimum requirement is 50 % to be a type C (Connor, 1990).

Lime has been used as a practical soil stabilizing agent for a long time. The most common alkali used in the stabilization is lime. The hydrated form is more common because it is easier and safer to store and handle. It decreases the plasticity and oil content of wastes (Van Keuren et al., 1987). It increases workability and improves the strength and deformation properties of wastes. If the lime is mixed with iron and aluminium bearing minerals, it will form a crude form of cement. Many types of coal power plant fly ash contain at least some amounts of iron and aluminium. If fly ash is mixed with lime, then the mixture can react with the wastes in a manner analogous to portland cement. Lime and lime/fly ash mixtures tend to be less expensive than cement. Lime can be used for the treatment of acidic wastes due to its high acid buffering capacity. As a result, the pH of the stabilized product remains high for a longer period of time minimizing the mobility of metals. Lime is widely used in the stabilization of the petroleum contaminated soils. Addition of lime increases the reaction rate of the pozzolanic material to form a pozzolanic cement. An increase in the lime content started to decrease strength because the mixture became brittle and friable therefore, cohesion of the mixture decreases.

Portland cement was used to improve the interparticle binding and mechanical properties of wastes. It reacts with water and binds to the aggregate within a short time. Cement is used as a stabilizing material for the construction of highway bases and sub-bases. It increases the strength of soil and its strength increases with curing time.

Table 2. Some properties of additives.

Properties	PSA	Lime	Fly Ash	Cement
PH	8.06	12.60	12.34	12.63
CEC (meq/100gr)	25.21	16.43	9.46	21.74
EC (miliSimens/cm)	11.17	6.29	10.38	4.97
Moisture Content (%)	83.12	0.50	0.30	1.90
Organic Matter (%)	6.74	24.38	1.05	2.95
Spesific Gravity	2.98	2.45	2.44	3.19
Bulk Density (gr/cm^3)	1.80	0.53	0.94	1.01
Oil Content (%)	4.00	----	----	----
Opt. Moisture Content (%)	9.60	----	----	----
Dry Unit Weight (gr/cm^3)	2.06	----	----	----
Grain Size Analysis				
Gravel (%)	10	----	----	----
Sand (%)	55	5	15	6
Silt (%)	20	75	72	58
Clay (%)	15	20	13	36

4 EXPERIMENTAL RESULTS

Each experiment was conducted on at least there duplicate samples and the average of those results are reported in this study. The percentages both in the case of oil and additives were based on the dry weight of the soil. All of the specimens were mixed at optimum water content before addition the oil and the additives.

Table 3. X-ray analysis of additives.

Elements	PSA	Lime	Fly Ash	Cement
SiO_2	22.63	2.72	45.68	20.35
Al_2O_3	5.13	0.37	9.04	5.19
Fe_2O_3	2.49	0.25	7.04	3.34
CaO	6.66	63.88	15.20	64.56
MgO	1.74	4.00	5.85	1.52
P_2O_3	0.08	0.01	0.83	0.07
K_2O	2.40	0.06	1.46	0.70
Na_2O	0.87	0.00	2.85	0.05
SO_3	7.41	0.02	8.10	2.04
Cl	0.116	0.004	0.000	0.003
TiO_2	0.35	0.03	0.51	0.25
SrO_2	0.18	0.00	0.03	0.03
Mn_2O_3	0.06	0.00	0.10	0.07
Loss of ignition	14.79	27.48	1.51	1.26
Total	64.91	98.82	98.19	99.48
CO_2	5.04	10.52	0.71	4.86

Table 4. Total metal contents of additives (mg/kg).

Total metals	PSA	Lime	Fly Ash	Cement
Lead (Pb)	430	106.1	80.1	120
Chromium (Cr)	146	70.4	144	111
Zinc (Zn)	177	43.1	88	77
Nickel (Ni)	137	34.7	369	24
Cobalt (Co)	35	42	42	21
Copper (Cu)	95	52.5	213	473
Cadmium (Cd)	<0.5	----	----	----
Arsenic (As)	<5	----	----	----
Molybdenum (Mo)	148	115	99	86

4.1 *Moisture content, pH, organic matter content, electrical conductivity (EC), oil content and cation exchange capacity (CEC)*

Organic wastes increase the water content of a soil. The water holding capacity of a soil is directly related to its bulk density and soil texture. In this study, optimum moisture content is used to prepare the mixtures.

Cation exchange capacity refers to the total amount of cations which are held exchangeably by a unit mass or weight of a soil (meq/100gr soil). The CEC of a soil depends on its clay, humus and organic matter content.

The extent of accumulation and migration of salt measured by electrical conductivity is strongly influenced by climate. Sites located in arid areas have the highest EC values in the surface soil ranges from 3

to 10 milisimens/cm, whereas sites locates in humid areas typically had EC values less than 1 milisimens/cm (Ryan et al., 1983).

Moisture content, pH and organic matter content of mixtures are given in Table 5. It can be seen from Table 5 that the addition of lime significantly increases the pH of wastes. However, it slightly decreases organic matter content of waste. Electrical conductivity, oil content and cation exchange capacity of mixtures are given in Table 6. Addition of lime, fly ash and cement decreases EC, CEC and oil amount of waste. These are because of the encapsulation of petroleum drilling wastes by the additives.

Table 5. Moisture content, pH and organic matter content of mixtures.

Mixtures	Moisture Content (%)	PH	Organic Matter Content (%)
Mixture 1	9.60	8.06	6.74
Mixture 2	9.60	12.40	6.02
Mixture 3	9.60	12.45	6.52
Mixture 4	9.60	12.55	6.60
Mixture 5	7.70	7.50	0.76

Table 6. Electrical conductivity, cation exchange capacity and oil amount of mixtures.

Mixtures	Electrical Conductivity (miliSimens/cm)	Cation Exchange Capacity (meq/100g)	Oil Content (%)
Mixture 1	11.17	25.21	4.00
Mixture 2	8.56	19.01	1.74
Mixture 3	8.55	18.78	1.43
Mixture 4	8.01	18.56	1.09
Mixture 5	2.14	13.12	----

4.2 *Unconfined compressive strength (UCS)*

Several geotechnical parameters are used to determine the performance of S/S wastes. Among others, the UCS is most frequently used. Other parameters include permeability, compressibility, environmental durability and chemical leachability for highly toxic wastes. UCS may be an acceptable indicator of the S/S waste's resistance to leaching. A sensitivity analysis of pozzolanic activity was performed by using unconfined compressive strength which is an indicator of strength gain from the fly ash and hydrated lime mixtures. Addition of cement to pozzolans is to support heavy highway traffic and to maintain integrity of the material

430

under repeated freezing/thawing. Strength and durability values depend on waste type, water content, mix ratio, curing time and temperature. Permeability, durability and strength of most S/S wastes are interrelated. Strength measurements is able to give an idea of the permeability and durability characteristics of the S/S waste. Permeability is inversely proportional to strength. Durability increases with strength which increases with the amount of cement used in the processing.

UCS values of 50 psi for 28 days stabilized petroleum contaminated wastes have been reported. UCS of 50 psi is required for solidified wastes to support the final cover or to ensure chemical stabilization of wastes containing free liquids. UCS may also be used as a design of standard for mechanical stability and UCS of 50 psi is often specified for stabilization of hazardous wastes (Martin et al., 1990). UCS value of 200 psi has been reported for various processes (Poon, 1989). For purposes of comparison, unconfined compressive strengths of lime stabilized natural soils range from 80 psi to 1100 psi, depending on the amount of pozzolanic materials present and curing period (Pamukçu et al., 1989). UCS values of the treated waste ranged from 170 psi to 720 psi. These values are above EPA's minimum recommendation of 50 psi for UCS for stabilized wastes to be disposed of in a hazardous waste landfill (U.S. EPA, 1992). However, these UCS values are well below the minimum ASTM/ACI standard of 3000 psi for use in construction of concrete sidewalls (ASTM, 1991). High UCS values indicate that a high degree of contaminant immobilization has occurred. Additions of lime, fly ash and cement significantly increase the strength of waste given in Table 7. When the amount of lime and fly ash increases, the strength of wastes also increases. Therefore, mixture 4 is the best working material with petroleum drilling waste. This value is in the range of limits published in the literature.

4.3 California bearing ratio (CBR) test

California bearing ratio test was developed to classify the suitability of a soil for use as a subgrade or base course material in highway construction (AASHTO T193-63, 1987 or ASTM D1883-73, 1978). The CBR test measures the shearing resistance of a soil under controlled moisture and density conditions. The CBR number is obtained as the ratio of the unit load required to effect a certain depth of penetration of the penetration piston into a compacted specimen of soil at some water content and density to the standard unit load. The CBR number is used to rate the performance

of soils primarily for use as bases and subgrades beneath pavements of road. It can be seen from Table 8 that addition of lime, fly ash and cement gives a good idea whether waste can be used as a sub-base material for road construction or not. Mixture 4 can be safely used as a sub-base material for the road construction.

Table 7. The variation of unconfined compressive strength values of mixtures.

Mixtures	Unit Weight (gr/cm^3)	Unconfined Compressive Strength (kg/cm^2)			
		Fresh	7 Day	14 Day	28 Day
Mixture 1	2.062	1.85	1.90	1.89	1.95
Mixture 2	1.890	2.14	5.66	7.58	10.01
Mixture 3	1.910	2.38	7.04	7.75	10.53
Mixture 4	1.840	2.69	8.78	9.68	14.15
Mixture 5	2.228	3.27	3.50	4.11	4.35

Table 8. California bearing ratio values of mixtures.

Mixtures	Unit Weight (gr/cm^3)	CBR (No.)	Rating/Uses (CBR No.)
Mixture 1	2.43	12.1	Fair/Sub-Base (7-20)
Mixture 2	2.08	22.0	Good/Base, Sub-Base (20-50)
Mixture 3	2.28	31.5	Good/Base, Sub-Base (20-50)
Mixture 4	2.33	44.5	Good/Base, Sub-Base (20-50)
Mixture 5	2.54	34.9	Good/Base, Sub-Base (20-50)

4.4 Permeability (hydraulic conductivity)

Stabilized mixture's permeability is used as a measure of its capability to physical isolation of contaminants as well as an indicator of internal movement of water. The permeability of most cementious materials various with additives used, curing time and conditions. It is generally known that the permeability of the S/S material decreases with increasing amount of cement used in the processing. Characterization of immobilization is generally done with leaching and permeability tests.

The permeability of solidified wastes falls in the range of 10^{-5} to 10^{-7} cm/sn depending on the type of measuring (Connor, 1990). Wide range of permeabilities of 10^{-4} and 10^{-7} cm/sn has been reported by different organizations (Poon, 1989). Most solidification processes result in material permeability of 10^{-5} to 10^{-6} cm/sec (Pamukcu et al., 1989). Low permeability prevents the movement of fluids in the pores and shows the rate of leachate generation. The variation of permeability values of stabilized mixtures is given in Table 9. Addition of lime, fly ash and cement decreases the permeability of wastes as expected.

In this study, triaxial permeability tests are used to evaluate the hydraulic conductivity of the stabilized mixtures. 14 psi of cell pressure, 1 psi of backpressure and 2 psi of hydraulic head were applied for permeation of rain water through the mixtures. Water samples were collected for chemical analysis of the influent and effluent from the stabilized mixtures. Total metal concentrations were also determined. The Permeability of mixtures is given in Table 8. Permeability of mixtures is in the range of 10^{-6} which is agreed with the literature.

Table 9. The variation of permeability test results of mixtures.

Mixtures	Unit Weigth (kg/cm^3)	Permeability kx10^{-6} (cm/sn)	
		fresh	28 days
Mixture 1	2.062	3.61	6.70
Mixture 2	1.890	3.43	4.90
Mixture 3	1.910	3.10	2.59
Mixture 4	1.840	2.97	1.33
Mixture 5	2.228	1.50	1.66

4.5 *Freeze/thaw durability*

Another factor would significantly affect the leaching potential especially long-term of the S/S waste is durability. Durability can be defined as the ability of the material to retain its mechanical and dimensional properties under different weathering conditions. Stabilized subgrade material of roadway is subjected to environmental effects. Over time, these environmental stresses cause cracking, peeling and crumbling of stabilized mixtures. Therefore, environmental durability is one of the most important parameters for stabilized mixtures. In principle, the damage due to freezing/thawing is closely related to the availability of water. In the freezing, ice lenses forms which causes frost heave. Upon thawing of the ice lenses, the excess water changes the mixture into a mud with no strength. Frost heave and loss of subgrade strength are the most detrimental forms of highway damage. Although there is a direct relationship between strength and durability, this does not hold in the case of frost damage.

The method to evaluate the environmental durability of the solidified waste is freezing/thawing test (ASTM D560-82, 1985). Samples exposed to freezing for 24 hours at a temperature of less than -20°C, then thawing

in water for 24 hours. Freeze/thaw durability test results are shown in Table 10. The average cumulative mass loss was least for the mixture 4 which survived 12 cycles of freeze/thaw.

Table 10. Freeze/thaw durability test results of mixtures.

Mixtures	Moisture Content (%)	Number of Cycles	Mass Loss (%)
Mixture 1	9.6	2	total deterioration
Mixture 2	9.6	12	30.34
Mixture 3	9.6	12	21.21
Mixture 4	9.6	12	21.32
Mixture 5	7.7	2	total deterioration

4.6 *Total metal contents*

pH control is a necessary for metal fixation in most fixation systems. High pH is desirable because metal hydroxides have minimum solubility in the range of pH 7.5 and 11. The solubility of chromium, zinc, nickel and lead is in the pH range of 7.5 and 9. Solubility increases rapidly above pH 9 (Connor, 1990). Soil pH and CEC are important to immobilize of metals. The soil pH should be maintained between 6 and 8 to minimize the immobilization and degradation processes. Most metals in petroleum industry wastes exist not very water soluble compounds. Metals such as arsenic, barium, cadmium, chromium, cobalt, copper, lead, mercury, molybdenum, nickel, zinc and vanadium are the toxic metals according to federal or state agencies in the USA. The metals have been precipitated with lime or other alkali to produce metal hydroxides, sulphides or other compounds that have low solubility under the conditions of precipitation usually in the pH range of 6 to 8 and a dilute water medium. The hydroxides of heavy metals such as arsenic, cadmium, chromium, lead and zinc exhibit minimum solubility in the pH range of 7.5 to 10 and solubility increases rapidly as pH increases. Solidification with portland cement provides good fixation of cadmium, chromium and nickel (Connor, 1990).

Lead occurs in soils, rocks and minerals. It is also found in substantial amounts in cement and lime. Petroleum industry wastes contain both inorganic and organic lead compounds. When the pH level dropped below 8, it began to leach significantly. Lead leaching also increases very rapidly above pH 12. Lime additions of 10 to 15 % reduced lead leaching in the EPT test by a factor of 185, but excess lime above 15% resulted in high leaching rates. Allowable lead concentration in the petroleum drilling waste is 100 mg/kg (Mucsy et al.,

1984). EP toxicity limit is 300 mg/l. The leachable lead is less than 5 mg/l based on the EP toxicity test. Lead concentrations of 3 different mixtures are significantly decreased under the allowable limits after stabilization.

Drilling muds contain both soluble and difficult soluble chromates for corrosion control and chromium lignosulfonates. Most of the drilling mud is disposed of as a waste after using. Other uses of chromium compounds are in batteries, magnetic tapes, chemicals, ceramics, pyrotechnics, electronics and fungicides. Allowable chromium concentration in the petroleum drilling waste is 100 mg/kg (Mucsy et al., 1984). Chromium concentration in petroleum drilling waste is around the allowable limit. Chromium concentrations of 3 different mixtures are significantly decreased under the allowable limit after stabilization.

Cadmium and its compounds are highly toxic. The primary treatment method is precipitation with alkali, usually lime. The pH of mixture is the most important factor controlling cadmium solubility. Solubility of cadmium decreases as pH increases. Between pH 8 and 11 about 99% of the cadmium is speciated as the solid carbonate (Conner, 1990). Allowable cadmium concentration in the petroleum drilling waste is 5 mg/kg (Mucsy, et al., 1984). Kaolinite clay, fly ash and sawdust has cadmium amount of 0.05 mg/g, 0.22 mg/g and 0.11 mg/g, respectively. Cadmium contents of soil in non-polluted areas are usually below 1 ppm (Bolt, 1976). Cadmium concentration in petroleum drilling waste is under the allowable limit.

Zinc is an aquatic toxin according to EPA. Zinc is not toxic to humans. Zinc is found in all natural waters and soils, and is an important nutrient in plant and animal life. Normal soils contain 10-30 ppm. Allowable zinc concentration in the petroleum drilling waste is 300 mg/kg (Mucsy et al., 1984). Zinc concentration in petroleum drilling waste is under the allowable limit. Zinc concentrations of 3 different mixtures are significantly decreased under the allowable limit after stabilization.

Copper is an aquatic toxin according to EPA. Copper is abundant in minerals and soils. Allowable copper concentration in the petroleum drilling waste is 100 mg/kg (Mucsy et al., 1984). Copper concentration in petroleum drilling waste is around the allowable limit. Copper concentrations of 3 different mixtures are significantly decreased under the allowable limit after stabilization.

Allowable nickel concentration in the contaminated soil is 100 mg/kg (Saarela, 1992). Nickel concentration in petroleum drilling waste is under the allowable limit.

Total metal concentrations of stabilized mixtures are shown in Table 11.

Table 11. Total metal concentrations of mixtures.

Mixture	pH	EC	Total Metals (mg/kg)					
		(mS/cm)	Pb	Zn	Cr	Cu	Cd	Ni
PDW[&]	----	----	100*	300*	100*	100*	5*	100[+]
Soil[&]	----	4	100[#]	300[#]	100[#]	100[#]	3[#]	50[#]
Mixture 1	8.06	11.17	430	177	146	95	<0.1	137
Mixture 2	12.40	8.56	218	158	123	57	<0.1	116
Mixture 3	12.45	8.55	222	145	126	54	<0.1	89
Mixture 4	12.55	8.01	269	112	108	41	<0.1	53
Mixture 5	7.50	2.14	57	185	15	13	<0.1	23

[&] Allowable Limits; *Mucsy et al., 1994; [#] Turkish Standards, 1994; [+] Saarela, 1992.

4.7 Leachate analysis

One of the best indicators of the stabilization efficiency is the result of a leaching test on the stabilized mixture. This test involves permeating the sample with water or another permeant and analyzing the effluent for mobilized contaminants. Leachate is the major long-term concern in the stabilization. In this study, rain water was used as a permeant. The leachates collected from triaxial permeability test and analyzed for total metals, pH and electrical conductivity. These leachates were collected after 24 hours of steady state water permeation through the saturated matrix of each sample. Table 12 shows the results of pH, total metals and electrical conductivity of the leachate collected from the stabilized mixture. It can be seen that after 24 hours, addition of lime/fly ash/cement to the petroleum drilling waste significantly increases pH values of stabilized mixtures, as expected. pH is an important parameter which affects the immobilization of metals (Tuncan et al., 1995). With proper pH management, which is provided in this case, metals are effectively immobilized. High pH is desirable because metal hydroxides have minimum solubility in the range of pH 7.5-11.0. Electrical conductivity of stabilized mixtures are decreased due to the additives in the mixtures. The concentrations of total metals are under the allowable limits before and after stabilization, in general.

Table 12. Leachate test results of mixtures after 28 days.

Property	pH	EC	Total Metals (mg/l)					
		(mS/cm)	Pb	Zn	Cr	Cu	Cd	Ni
U.S. EPA	----	----	5	----	5	----	1	----
RW	7.30	0.025	<0.01	4.27	0.02	<0.01	<0.01	<0.01
Mixture 1	7.72	46.2	0.69	1.20	1.07	1.67	<0.01	1.71
Mixture 2	11.9	27.7	0.28	0.15	0.76	1.35	<0.01	0.90
Mixture 3	12.0	26.5	0.23	0.21	0.43	0.77	<0.01	0.50
Mixture 4	11.9	20.5	0.31	0.38	0.80	0.49	<0.01	0.36
Mixture 5	7.18	1.26	0.07	0.21	0.03	0.05	<0.01	<0.01

RW: Rain Water

5 CONCLUSIONS

The following conclusions have been drawn from this study.

1. Additions of fly ash, lime and cement significantly increase pH value, decrease cation exchange capacity. However, oil content of stabilized PDW is decreased to around 1% which is under the limits.

2. UCS of stabilized PDW are increased up to around 200 psi. CBR value of stabilized PDW indicated that mixtures 2, 3 and 4 are used a sub-base material.

3. Permeability values are around in the range of 10-6 which is between the limits. After the leachate test, total metal concentrations are under the limits. After stabilization, concentrations of zinc, nickel copper, chromium and cadmium are reduced under the allowable limits, except the concentration of lead.

The best results in all of the tests conducted is obtained with mixture 4. This mixture can be effectively and safely used as sub-base material for the road construction.

REFERENCES

AASHTO T193-63, 1987. Standard spesifications for transportation materials and methods of sampling and testing. The American Association of State Highway and Transportation Officials.

ASTM 1991. *Annual book of ASTM standards*. ASTM Philadelphia, PA.

ASTM C595. 1985. *Standard Specification for Fly Ash and Other Pozzolons for Use With Lime.*

ASTM D560-82. 1985. *Methods for freezing -and-thawing tests of compacted soil-cement mixtures.*

ASTM C618. 1993. *Standard specification for fly ash and raw or calcined natural pozzolan for use a mineral admixture in portland cement concrete.*

ASTM D1883-73. 1978. *Test method for bearing ratio of laboratory-compacted soils.*

Bell, E. & Kostecki, P.T. & Calabrase, E. J. 1989. An update an national survey of state regulatory policy: cleanup standards. *Petroleum Contaminated Soils. Proc. of the Conference on Hydrocarbon Contaminated Soils.* Vol 3. pp. 49-72.

Bolt, G.H., Bruggenwert, M.G.M.1976, *Pollution of soil.* Soil Chemistry A. Basic Elements, Elsevier Scientific Pbl. Co., pp.192-263.

Conner, J.R. 1990. *Chemical fixation and solidification of hazardous wastes.* Van Nostrand Reinhold, New York, 692 p.

Martin, J.P. & Biehl, F.J. & Browning, J.S. & Van-Keuren E.L. 1990. Constitutive behaviour of clay and pozzolon-stabilized hydrocarbon refining waste. *Geotechnics of Waste Fills Teory and Practice*, ASTM STP 1070, Philadelphia, pp.185-203.

Mucsy, G. & Kranicz-Pap E. & Urbany, G. 1984. Disposal of oil-containing sludges on farmlands. *Hazardous and Industrial Waste Management and Testing*, 3rd Symposium, ASTM STP 851, Philadelphia, pp 135-151.

Pamukçu, S. & Hijazi, H. & Fang, H.Y. 1989, Study of possible reuse of stabilized petroleum contaminated soils as construction material. *Petroleum Contaminated Soils, Proc. of the Conference on Hydrocarbon Contaminated Soils,* Vol 3., pp. 203-214.

Poon, C.S. 1989. A critical review of evaluation procedures for stabilization/solidification processes, *4th International Hazardous Waste Symposium on Environmental Aspects of Stabilization/Solidification of Hazardous and Radiactive Wastes*, ASTM STM 1033, Philadelphia, pp.114-124.

Rogers, W.F. 1963. *Composition and properties of well drilling fluids.* Gulf Publishing Company, 678 p. Houston, Texas, USA.

Ryan, J.R. & Hanson, M.L. & Loehr, R.C. 1983, Land treatment practices in the petroleum industry. *Land Treatment, Hazardous Management Alternative, Water Resources Symposium*, No.13, pp.319-345.

Saarela, J. 1992. Effects of tailings spread around three mines in Finland, *Proceedings of the Mediterranean Conference on Environmental Geotechnology*, Çeşme, Turkey. pp. 483-489.

U.S. EPA. 1992 *Silicate Technology corporation's solidification/stabilization technology for organic and inorganic contaminants in soils.* Applications analysis report. Environmental Protection Agency. EPA/540/AR-92-10.

Tuncan, A. & Tuncan, M. & Koyuncu, H. 1995. Physical and chemical stabilization of petroleum drilling wastes, *The Second International Conference on the International Coastal Environment*, MED-COAST, October 24-27, Tarragona, Spain.

Turkish Standards. 1994. Ministry of Environment, Solid waste management. R.G. No:20814.

Van Keuren, E. & Martin, J. & De Falco, A. 1987. Pilot field study of hydrocarbon waste stabilization. *Proceedings of the 19th Mid-Atlantic Industrial Waste Conference*, pp. 330-341, Technomic Publishing Company.

Geotechnical Hazards, Marić, Lisac & Szavits-Nossan (eds) © 1998 Taylor & Francis, ISBN 90 5410 957 2

Impervious barriers for landfills in karst

D. Kovačić, B. Kovačević-Zelić & M. Vrkljan
Faculty of Mining, Geology and Petroleum Engineering, University of Zagreb, Croatia

D. Znidarčić
Department of Civil, Environmental and Architectural Engineering, University of Colorado at Boulder, Colo., USA

ABSTRACT: The aim of the research project is to prove that the soil materials from flysch deposits, which are locally available on the Istrian peninsula at low cost, can be used as the components of the liner layers in municipal landfills. The results of the soil investigation presented in the paper are a part of this project which is still under way. On several sites on the Istrian peninsula the samples of marly material were collected and tested in the laboratory. The results show that all samples could be classified as clayey material of low plasticity. The results of X-ray analysis proved that all samples are composed of cca 30 - 40% clay minerals. Permeability tests performed in the oedometer cell prove that tested soil material can achieve permeability less than 1×10^{-7} cm/s.

1 INTRODUCTION

One of the most serious problems facing urban communities today is the efficient and longterm disposal of municipal solid waste. In recent years a number of municipal solid waste landfills are planned either as new sanitary landfills or as remedial works on existing waste dump sites in Croatia. Almost half of them need to be located in the karst region along the Adriatic coast.

The most critical parts of a sanitary landfill are impervious layers in the bottom liner system and in the final cover. The most important characteristic of the materials for the construction of these layers is its impermeability. The research project has been initiated with the aim to prove that the soil materials from flysch deposits, which are locally available at low cost, can be used as the components for the liner layers.

The territory of Croatia consists mostly of sedimentary rocks which cover more than 95% of its surface. Almost half of this surface belongs to the karst region where mesozoic and paleogene carbonate rocks prevail. According to the classification used by Garašić (1991) the whole region can be subdivided into Inner Karst region, Middle Karst region and Outer Karst region (Figure 1). The latter region extends along the Adriatic coast. Its area totals about 10 000 km².

The flysch formations in the mentioned karst regions are generally acting as hydraulic barriers indicating their sealing capabilities. Thus it is logical to search within the flysch layers for the materials suitable for the construction of the impervious liners for the sanitary landfills. The flysch formations are mainly situated within the Outer Karst region in Istria and in the areas of Rijeka, Zadar, Split and Dubrovnik (Velić et al. 1983).

Within the framework of the paper the attempt has been made to prove that the clayey components of the flysch formation could be used for the construction of the impervious layers.

Figure 1. Croatian karst areas.

2 GENERAL PROPERTIES OF THE FLYSCH FORMATIONS IN THE OUTER KARST REGION

A number of geological and geotechnical investigations of the flysch formations in the Outer Karst region have been conducted so far.

Generally shales, marls, siltites and sandstones are considered to be the most significant and the most frequent lithologic members of the flysch formations in the mentioned region (Magdalenić et al. 1980).

The results of X-ray difraction analyses on flysch samples from Istria indicate that the dominant clay minerals are illite and hydro-mica along with septe-chlorite, kaolinite and montmorillonite (Magdalenić 1972).

Šestanović (1990) reported similar results for Split region where illite, illite-muscovite and montmorillonite are most frequent clay minerals.

Geotechnical characteristics of flysch deposits were investigated in various projects. Jašarević et al. (1987) presented the results of mineral analyses of the flysch material from Rijeka region and Dubrovnik region. The proportion of phyllosilicates, typical clay minerals, in the weakly cemented (geologically loosely bound) flysch samples (clayey silt, silty clay) is 50% or more. The particle size distribution data show the proportion of clay size particles to be between 15% and 50%.

The results of the Atterberg limits tests on the flysch samples taken in Istria (Novosel et al. 1972) classified the material as clay of intermediate to high plasticity. Similar results were obtained for the samples from Dubrovnik region (Lovinčić 1992).

A recent research (Miščević & Roje Bonacci 1997) proved that marl component from the flysch deposits in Split region satisfies the permeability criterion for bottom liners in municipal landfills.

The investigation presented in the paper has been focused on the Istrian peninsula.

3 LINER MATERIAL SPECIFICATIONS

Clayey soil is the most common natural lining material. The main factors which affect the quality of clay liners are permeability, degree of compaction, moisture content, clay composition, field placement technique and liner thickness. The criteria for choosing a liner material is primarily based on the recompacted permeability achievable under field conditions. A soil material that can be compacted to obtain a low permeability (1×10^{-7} cm/s or less) when compacted to 90-95% of the maximum Proctor's dry density at wet of optimum moisture content is often chosen for landfill liner construction. In order to satisfy this requirement the plasticity index, liquid limit and some minimum requirements regarding grain size distribution should be specified.

According to Bagchi (1990) a soil with the following specifications would prove suitable for liner construction:

Liquid Limit	≥ 30% (25-30%)
Plasticity Index	≥ 15% (10-15%)
Percentage Fines	≥ 50% (40-50%)
Clay Fraction	≥ 25% (18-25%)

The minimum requirements recommended by Daniel (1991) are:

Percentage Fines	> 25%
Plasticity Index	> 10%
Percentage Gravel	< 50%
Max. Particle Size	25 to 50 mm

In a study of possible sources of natural materials for landfill liners in Great Britain, Jones et al. (1995) introduced the term "suitable" material. They used the suitability criteria specified by the National Rivers Authority (NRA). The NRA defines suitable materials as those clays with the following characteristics:

Liquid Limit	< 90%
Plasticity Index	< 65%
Clay Fraction	> 10%

In a review of German landfill lining systems Bishop et al. (1995) state that the chosen soil material must fulfill the following requirements.

Percentage Fines	> 20%
Organic Content	< 5%
Carbonate Content	< 15%
Dry Density	> 95% of Proctor max.

It is obvious that there is no standard and widely accepted criteria for choosing suitable clayey materials. The criteria presented by Bagchi (1990) are clearly the most stringent ones with respect to fines and clay content. However it is also clear that percentage of fines, clay fraction, plasticity index PL and liquid limit LL are appropriate criteria in selecting the suitable liner materials.

4 THE AREA UNDER INVESTIGATION

4.1 Stratigraphic and palaeogeological characteristics

Istria belongs to the NW part of the Adriatic Carbonate Platform. According to Velić et al. (1995) deposits in Istria can be divided into four sedimentary units or megasequences as shown in Figure 2. Since only unit IVb is related to the present research its characteristics are briefly described.

Transitional beds consist of clayey limestone, calcitic marls and marls, composed of fine-grained carbonate and a siliciclastic matrix with planktonic foraminifera and bioclasts of benthic organisms. They were deposited in the significantly deeper environments of the Middle Eocene age.

Flysch deposits of the Middle and Upper Eocene age outcrop in the Pazin, Labin and Plomin basins, Mountain Učka and partially on Mountain Ćićarija. They are characterised by an alternation of marl and carbonate sandstone beds.

The appearance of the flysch deposit in various stages of weathering at site Paz is presented in Figure 3.

Figure 3. Marl exposure at site Paz.

The description of the samples is shown on Table 1.

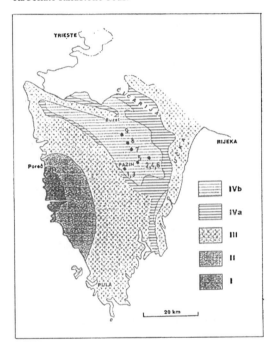

Legend:
I) Bathonian - Lower Kimmeridgian
II) Upper Tithonian - Upper Aptian
III) Upper Albian - Lower Campanian
IV) Palaeocene - Eocene
a) Foraminifera limestones
b) Transitional beds and flysch

Figure 2. Surface distribution of outlined megasequences in Istria (Velić et al. 1995).

4.2 *Samples collection*

Several sites with flysch outcrops were visited in the region called "Central Istria" and samples for laboratory testing were collected as indicated on Figure 2. The numbers correspond to numbers of samples given in Table 1.

Table 1. Sites and sample description.

No.	Site	Description
1	Pazin I	Grey crushed marl with yellow grey silty filler, low grain strength
2	Paz	Calcified marl, yellow-grey, high grain strength
3	Pazin II	Grey mud, very soft
4	Paz	Yellow-grey silty material with marl fragments
5	Cerovje	Grey-green marly clay
6	Paz	Calcified marl, yellow-grey, pieces with sharp edges, high strength
7	Kovačići	Large fragments of marl with silty cementd filler
8	Oslići	Calcified marl, yellow-grey, pieces with sharp edges
9	Krušvari	Large fragments of marl with silty cementd filler

5 GEOTECHNICAL CLASSIFICATION

Grain size distribution by sieve and hydrometer analysis as well as the determination of Atterberg limits were performed in soil mechanics laboratory. In addition, the activity A as defined by Skempton (1953) was calculated on the basis of the test results. Relevant parameters are given in Table 2.

Table 2. Soil parameters.

Sample No.	Percentage Fines (%)	Clay Fraction < 2μm (%)	LL (%)	PL (%)	A
1	81	11.0	35	17	1.56
2	64	9.0	26	10	1.15
3	90	10.7	32	14	1.30
4	69	7.5	27	13	1.71
5	71	7.9	44	24	3.06
6	67	5.1	30	13	2.58
7	76	8.2	39	21	2.59
8	59	7.4	28	11	1.57
9	78	9.0	42	23	2.50

The results show that there is no significant difference among samples. Sample No. 5 has the highest plasticity corresponding to the largest amount of clay minerals as shown later.

These results indicate that the flysch samples satisfy the liner material criteria presented above except that the clay fraction is somewhat lower than the generally accepted values. It is particularly noted that all but one sample are classified as active clays defined as clays with activity greater than 1.25 (McCharty, 1988). This finding is verified by the mineral composition as shown later.

6 SEDIMENTATION ANALYSIS AND MINERAL COMPOSITION

6.1 Sedimentation analysis on insoluble residue

In addition to the conventional soil tests the granulometric analysis was performed by an alternative sedimentation (pipette) method (made by D. Aljinović, Faculty of Mining, Geology and Petroleum Engineering). First, the carbonate cement in the sample is dissolved by treating it with dilute 5% hydrochloric acid. In the specimens No. 2 and No. 5 the $CaCO_3$ content of 50,6% and 24,0% respectively, is removed.

The pipette method is then used on the insoluble residue. The samples are soaked for 10 days and ammonia solution is added as a deflocculating agent. Clay fraction, defined in this case as the particle size less than 4 μm, is 23% and 33% for samples No. 2

and No. 5, respectively. On the basis of these results the material is classified as silty marl by the criterion proposed by Konta (1973).

Clearly, by this method the content of particles smaller than 2μm would also be higher than the values indicated in Table 2. As such the material would be more in compliance with the criteria given above in section 3.

6.2 Mineral composition

Mineral composition has been determined on four samples by X-ray powder diffraction (XRD). The method of cation exchange capacity (CEC) has been used for the determination of expandable component. XRD patterns were taken using a Philips diffractometer (graphite monochromator, $CuK\alpha$ radiation, proportional counter). Diffraction patterns of original samples were recorded as well as the patterns of insoluble residue following the removal of carbonates by means of diluted hydrochloric acid (pH ≈ 3). After soaking in water and after sieving, the particles less than 63 μm are separated and dispersed by ultrasonic dispergator. The particles less than 2 μm are then separated by sedimentation. XRD patterns were also taken of samples with preferentially oriented grains (draied in air, treated by ethylene glycol and heated at 650° C).

Cation exchange capacity was determined by Kjeldahl method (Šušterčić 1969) with prior replacement of exchangeable cations using ammonium acetate. Exchangeable cations are placed within layered clay minerals and their position (interstitial between the layers) are not fixed. CEC was determined on insoluble residue of air dried samples and recalculated for the original sample. By matching the CEC results and proportions of carbonates and clay minerals according to the intensity of diffraction lines (for clay minerals) the approximate proportions of clay minerals and expandable component are determined (Table 3). (The term smectite is used as synonym for montmorillonite).

Expandable component is found in randomly interstratified illite-smectite (samples No. 1 and No.5) and also as separate smectite in samples No. 2 and No. 4.

The mineral composition indicates the clay content of about 30% with majority of the minerals being montmorillonite or illite. This finding reinforces the activity results by which all the samples were classified as active. The grain size distribution indicates that some of the clay minerals are contained in the fraction > 2μm size particles, which is further confirmed by the clay content difference obtained by the two dispersion methods.

438

Table 3. The results of semi-quantitative X-ray analysis (made by Prof. D. Slovenec, Faculty of Mining, Geology and Petroleum Engineering).

Sample No.	Mineral Type	Mass Ratio (%)	Expand. Comp. Mass Ratio (%)
1	calcite	45	
	dolomite	<5	
	quartz	12	
	plagioclase	5	
	clay minerals		
	illite-smectite* +		
	illite	25	9
	chlorite	5	
2	calcite	55	
	quartz	8	
	plagioclase	<5	
	clay minerals		
	illite-smectite* +		
	smectite + illite	25	10
	chlorite	<5	
4	calcite	60	
	quartz	8	
	plagioclase	<5	
	clay minerals		
	illite-smectite* +		
	smectite + illite	20	9
	chlorite	<5	
5	calcite	20	
	quartz	38	
	plagioclase	5	
	K. felsdspar	few	
	clay minerals		
	illite-smectite* +		
	illite	35	17
	chlorite	<5	

*randomly interstratified minerals

7 COMPACTION TESTS

7.1 *Processing of soil*

It was decided to use marl material from Paz (sample No.6) in order to test compaction and permeability properties.

The soil obtained from the field was processed to form the suitable soil portion. A soil was prepared by crushing it mechanically. Two samples were made out of the crushed soil. The first sample consisted of the soil that passed through a sieve with 4.0 mm openings and the second sample consisted of fraction less than 2.5 mm. The soils were then mixed carefully with water to achieve the desired water content. The moistened soils were stored for about 3 days to allow time for the soil particles to hydrate.

7.2 *Compaction*

Two compaction tests were performed using standard Proctor procedures (ASTM D 698). The results are given in Table 4.

Table 4. Compaction parameters.

Fraction mm	Dry density g/cm^3	Opt. water content %
0 - 4	1.95	12.9
0 - 2.5	1.93	13.3

The respective compaction curves indicate rather narrow range of water content in which maximum dry density can be obtained. The values of optimum water content are lower than expected for the clayey material of similar index properties.

8 PERMEABILITY TESTS

The coefficient of permeability k of two specimens prepared for the compaction tests was measured in the laboratory. The tests were performed in oedometer cell (modified to allow direct measurement of k). Hydraulic gradient of 50 was used in both tests. Water was used as the permeant. The tests lasted approx. 24 hours for each loading step. The results of the permeability tests on two samples are shown in Table 5. σ is the vertical pressure, e is void ratio and k is coefficient of permeability.

Table 5. Permeability tests.

Fraction mm	σ kN/m^2	e	k cm/s
0 - 4	100	0.378	2.46×10^{-8}
	200	0.358	1.82×10^{-8}
	400	0.341	1.52×10^{-8}
0 - 2.5	100	0.366	3.57×10^{-8}
	200	0.349	2.70×10^{-8}
	400	0.329	2.14×10^{-8}

Both samples satisfy the required permeability criteron.

9 CONCLUSIONS

The results presented in the paper demonstrate some unusual characteristics for the clayey component of the flysch formation as they pertain to the construction of impervious barriers for landfills. First, the results indicate that by using the conventional compaction technique, samples with permeability of less than 10^{-7} cm/s can be created in the laboratory. As such, the material is suitable as the construction material for impervious liners in landfills. This result is obtained despite relatively low clay content and low plasticity. However, the result is reasonable since the material is classified as clay of high activity and the clay minerals are predominantly montmorillonite and illite.

Clearly the clayey component of flysch formation is a good candidate for constructing impervious liners for landfills. Before such material can be used in construction, it is necessary to investigate its behaviour under field compaction condition and to verify that the field permeability test will produce similar favourable results. It is also important to evaluate the potential difficulties in processing large quantities of the material, especially with regard to the compaction water content control. While only field experiments will be able to give the final answer to these questions, our experience in the laboratory so far has been only positive. In conclusion, the presented results confirm the hypothesis that the flysch formations are good source of material for the construction of impervious barriers for landfills.

ACKNOWLEDGEMENT

The work described in this paper is funded by the U.S. - Croatian Joint Board on Scientific and Technological Cooperation. The grant designated No. JF150 was awarded in November 1995 for the proposed 3 year project. This support is gratefully acknowledged.

REFERENCES

Bagchi, A. 1990. *Design, construction, and monitoring of sanitary landfill.* New York: John Wiley & Sons.

Bishop, D.J. & Carter, G. 1995. Waste disposal by landfill - German landfill lining systems. In Sarsby (ed.), *Proc. Int. Symp. on Geotechnics Related to the Environment - GREEN 93*, Bolton: 127-134, Rotterdam: Balkema.

Daniel, D.E. 1991. Compacted Clay and Geosynthetic Clay Linings. In *XV Ciclo di Conferenze di Geotecnica di Torino*, Torino 19-22 Nov. 1991.

Daniel, D.E. & Benson, C.H. 1990. Water Content - Density Criteria for Compacted Soil Liners, *Journal of Geotechnical Engineering*, A.S.C.E., Vol. 116, No. 12, 1811-1830.

Garašić, M. 1991. Morphological and hydrogeological classification of speleological structures (caves and pits) in the Croatian karst area. *Geološki vjesnik (A Journal of the Institute of Geology Zagreb and Croatian Geological Society)*, Vol. 44, 289-300, Zagreb.

Jašarević, I., Jurak, V. 1987. Stability of the flysch coastal slopes of the Adriatic Sea in the static and seismic conditions. In *Proceedings of the 6th International Congress on Rock Mechanics*, Vol. 1, 411-418.

Jones, R.M., Murray, E.J., Rix, D.W. & Humphrey, R.D. 1995. Selection of clays for use as landfill liners. In Sarsby (ed.), *Proc. Int. Symp. on Geotechnics Related to the Environment - GREEN 93*, Bolton: 433-438, Rotterdam: Balkema.

Konta, J. 1973. *Kvantitativni system rezidualnih hornin, sedimentu a vulkanoklastickych usazenin.* Praha: Univeza Karlova Praha (in Czech).

Lovinčić, M. 1992. Analysis of flysch ground properties. *Ceste i mostovi (Roads and Bridges)*, Vol. 38, No. 3-4, 61-64 (in Croatian).

Magdalenić, A., Crnković, B., Jašarević, I. 1980. Problemi vezani za radove u flišu. In *5. Jugoslavenski simpozij za mehaniku stijena i podzemne radove*, Vol. 2, 93-109 (in Croatian).

Magdalenić, Z. 1972. Sedimentology of central Istria flysch deposits. *Acta Geologica VII/2*, Zagreb, 71-96 (in Croatian).

McCharty, D.F. 1988. *Essentials of Soil Mechanics and Foundations: Basic Geotechnics.* Egglewood Cliffs, New Jersey: Prentice Hall.

Miščević, P., T. Roje Bonacci 1997. Possibility of using marl in the preparation of sealing coats. *Građevinar (The Journal of the Croatian Society of Civil Engineers)*, Vol. 49, No. 2: 87-93 (in Croatian).

Novosel, T., Ilijanić, N. 1972. Estimate of the flysch complex by engineering-geological research of the area of the accessing roads for the tunnel "Učka", In *2. Jugoslavenski simpozij o hidrogeologiji i inžinjerskoj geologiji*, Vol. 2, 243-250 (in Croatian).

Skempton, A.W. 1953. The colloidal activity of clays. *Proc. 3rd Int. Conf. Soil Mech. Found. Eng.* (Switzerland), Vol. I, p.57.

Šestanović, S. 1990. *Osnove geologije i petrologije*, Zagreb: Školska knjiga (in Croatian).

Šušterčić, N. 1969. *Pogonske analize.* Zagreb: Školska knjiga (in Croatian).

Velić, I., Velić, J. 1983. Petrografska karta Jugoslavije, *Šumarska enciklopedija 2*, Zagreb: JLZ (in Croatian).

Velić, I., J. Tišljar, D. Matičec & I. Vlahović 1995. A Review of the Geology of Istria. In I. Vlahović & I. Velić (eds), *Excursion Guide-Book - Proc. 1st Croatian Geological Congress*, Opatija: 5-30. Zagreb: Institute of Geology.

Geotechnical Hazards, Marić, Lisac & Szavits-Nossan (eds) © 1998 Taylor & Francis, ISBN 90 5410 957 2

Cyclic direct simple shear testing of OII landfill solid waste

Neven Matasović & Thomas A. Williamson
GeoSyntec Consultants, Huntington Beach, Calif., USA

Robert C. Bachus
GeoSyntec Consultants, Atlanta, Ga., USA

ABSTRACT: As a part of pre-design activities for landfill closure Cyclic Direct Simple Shear (CyDSS) testing was conducted on reconstituted solid waste specimens from the Operating Industries, Inc. (OII) landfill in southern California. The testing was conducted in an on-site laboratory using a large-diameter (460 mm) CyDSS device. The results of the testing program indicate: (i) the tested solid waste behaves as a hysteretic material under CyDSS conditions; (ii) Masing rules apply for the tested solid waste material; and (iii) significant modulus reduction and an equivalent viscous damping increase occur at cyclic shear strains beyond approximately 0.1 percent.

1 INTRODUCTION

1.1 *Site conditions*

The Operating Industries, Inc. (OII) landfill is located in a relatively arid area of southern California, approximately 16 km east of downtown Los Angeles (Figure 1).

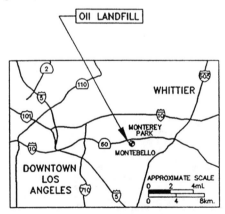

Figure 1. OII landfill site location map.

The site is divided by California State Road 60 (the Pomona Freeway) into a relatively small and level north parcel and a steeply sloping 58-hectare south parcel. The south parcel has been in the process of being closed since 1985 under the United States Environmental Protection Agency (EPA) Superfund program. Pre-design activities for final closure of the south parcel were initiated in 1993.

The OII landfill site was formerly a sand and gravel quarry pit. The approximately 60-m deep pit was filled with solid waste over a 40-year operating period. The site accepted residential, commercial, and industrial solid wastes. In addition, liquid wastes were accepted at the landfill at times, primarily at the west end of the south parcel. It has been reported that waste was disposed without any separation or compaction. Daily soil cover was placed on the side slopes of the landfill and on the decks and waste faces as part of routine operations as the landfill rose above grade.

The landfill last received waste in 1984, at which time interim soil cover was placed on top of the landfill. The interim cover soils appear to have been derived, like the daily cover soils, primarily from local borrow sources and typically vary in classification from silty clay to silty sand. The interim cover side slope inclination varies from 3 horizontal to 1 vertical (3H:1V) to 1.5H:1V.

1.2 *Laboratory testing background*

The OII site is located in an area of high seismicity. The design earthquake established for final closure is a moment magnitude 6.9 event capable of generating a peak horizontal acceleration of 0.61 g in a hypothetical bedrock outcrop at the geometric center of the site. The site is instrumented by a pair of strong motion instruments which recorded strong ground shaking in several earthquakes. Back analyses of these data by Matasovic et al. (1995) and Idriss et al. (1995) provided insight into the response of OII solid waste in the intermediate strain range (cyclic shear strains ranging from 0.001 to 0.1 percent). However, preliminary forward analyses indicated that cyclic shear strains could exceed 1 percent in the design earthquake.

In order to characterize the behavior of the OII solid waste at cyclic shear strains in excess of 0.1 percent, a cyclic laboratory testing program using equipment capable of generation of large cyclic shear strain (0.1 to 6 percent cyclic shear strain) was conducted. This program consisted of a series of large-scale cyclic direct simple shear (CyDSS) tests on reconstituted specimens. The results of these tests are presented herein. Additional information regarding development of the modulus reduction and damping curves for OII solid waste from the results of these tests is presented in Matasovic and Kavazanjian (1998).

2 TESTING EQUIPMENT AND TESTING PROCEDURES

2.1 CyDSS device

The CyDSS laboratory testing program described herein involved the design and fabrication of a CyDSS device capable of testing a large-diameter sample of solid waste. The CyDSS device designed and fabricated specially for the OII landfill project is shown on Figure 2.

The CyDSS device measures 2 m in overall length and 2 m in overall height. The device is designed to test a 460-mm diameter cylindrical specimen. The specimen height can be varied from 25 to 460 mm using the series of stacked teflon-coated stainless-steel rings which confine the test specimen. Each ring is 12 mm thick. The bottom ring is placed on a trolley attached to a low-friction roller bearing block which is attached in turn to a stainless steel bearing shaft. The trolley is connected to the shaft of a servo-hydraulic actuator assemble through a shear arm.

Top and bottom platens contain multiple steel pins which provide an interlock between the specimen and the platens. The bottom platen also contains drainage channels for drainage of pore fluid. The top platen is connected to a plunger that moves through vertically mounted low-friction roller bearings. These roller bearings allow vertical movement of the plunger during shear while restricting rotation of the top platen. The plunger includes a centrally located load cell that can measure the applied vertical load.

Shear load is applied through a closed loop servo-hydraulic actuator that is controlled either manually or by a personal computer (PC). The actuator can control both monotonic and cyclic loading tests that can either be displacement or stress controlled. Vertical normal loads are applied to the specimen by a hydraulic ram which bears on the top plunger.

2.2 Data acquisition

During the CyDSS test, vertical and horizontal displacements are measured using two linear variable differential transformers (LVDTs). The horizontal LVDT was linked to a servo-controller via the data acquisition system (Figure 3) to provide feedback and to control the cyclic loading.

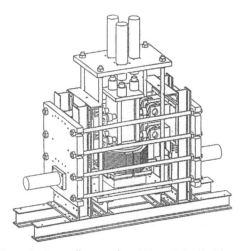

Figure 2. Large-diameter (i.e., 460-mm) Cyclic Direct Simple Shear (CyDSS) Device.

Figure 3. Data acquisition system (PC not shown).

Data acquisition was performed using a PC with a data acquisition controller. A digital/analog board was used to send control signals to the servo-hydraulic actuator and to receive the signals from the test instruments. A view of the large-diameter CyDSS device during a test is presented in Figure 4.

Figure 4. Large-diameter CyDSS test in progress.

2.3 *Testing program and procedures*

There were several health and safety concerns associated with testing and handling large volumes of solid waste. Accordingly, all laboratory testing was conducted in an on-site laboratory established at the OII landfill. Due to the difficulties with recovery and handling of large-diameter intact samples, testing was conducted on reconstituted specimens.

The specimens consisted of reconstituted bulk samples of solid waste recovered from the landfill using a 840-mm diameter bucket auger. Laboratory tests on the reconstituted specimens were conducted at the field moisture content and in-situ unit weight.

Table 1 identifies the borings and depths from which bulk specimens used in the CyDSS testing program were recovered, approximate age of the waste specimens as determined from dated materials recovered during sampling, testing stress levels, moisture content, shear wave velocity, in-situ unit weight of the specimens, and information regarding visual classification of the solid waste and soil or soil-like constituents of the bulk specimens. More details of OII waste characterization program can be found in GeoSyntec (1996).

A total of nine displacement-controlled (strain-controlled) staged CyDSS tests were conducted on the reconstituted solid waste specimens. Each test was conducted using one of the following two test scenarios.

Scenario (A): Five loading cycles at four different shear strain levels, approximately 0.1, 0.3, 1, and 3 percent, followed by monotonic loading. The sample was loaded at 0.1 Hz during cyclic loading and at a shear strain rate slower than 0.5 mm/min during monotonic loading.

Scenario (B): Twenty-five loading cycles at approximately 0.5 percent shear strain at 0.1 Hz followed by monotonic loading as described in Scenario A.

3 TEST RESULTS AND INTERPRETATION

3.1 *Test results*

Six strain-controlled tests on Monterey 0/30 sand were initially conducted to evaluate the system friction. The results of the CyDSS tests conducted on the Monterey 0/30 sand are plotted on Figure 5. Also shown on Figure 5 are the CyDSS testing results on a standard sand with the same grain size distribution carried out using the Norwegian Geotechnical Institute-type (NGI-type) device (Matasovic and Vucetic, 1993). The NGI-type device tests a 45-mm diameter specimen confined by a wire-reinforced rubber membrane and is believed to have negligible system friction.

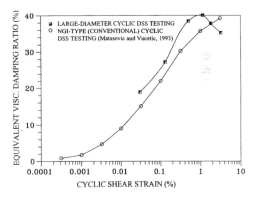

Figure 5. Evaluation of large-diameter CyDSS device system friction.

Figure 5 indicates that the equivalent viscous damping ratio of the large-diameter prototype CyDSS device is on the order of 4 percent.

The results of the CyDSS testing program on reconstituted specimens of OII solid waste are presented in Figures 6a to 6i. The results are presented as plots of shear stress versus shear strain. On all of the figures, the shear stress is presented normalized by the initial vertical consolidation stress.

Table 1. Large-diameter cyclic direct simple shear testing program for OII landfill solid waste

Test No.	Boring No.[a]	Sampling Depth	Age of Sample[b]	Normal Stress Applied	Moisture Content[c]	Shear Wave Velocity[d]	Unit Weight[e]	Scenario[h]
1, 2	BA-1	3.4-6.1 m	-	100.3; 95.8; 94.1 kPa	15.3%	122 m/s	16.0 kN/m³	A, B
3	BA-3	3.4-6.1 m	1980	97.4 kPa	16.6%	162 m/s	16.9 kN/m³	B (modified)
4,5	BA-1	9.2-12.2 m	-	176.2; 176.4 kPa	33.5%	231 m/s	18.5 kN/m³	A, B
6	BA-3	30.5-33.5 m	1964	511.4 kPa	24.7%	195 m/s	12.0 kN/m³	A
7, 8	BA-2	15.2-18.3 m	1983	292.1; 292.3 kPa	41.2%	231 m/s	16.5 kN/m³	A, B
9	BA-1	15.2-18.3 m	1984	315.1 kPa	25.5%	148 m/s	18.1 kN/m³	A

Test No.	Waste Composition[f]						Soil and Soil-Like Materials Composition[g]
	Paper and Cardboard	Plastics and Rubber	Wood	Metals	Glass	Textiles and Misc.	
1, 2	1.6%	7.0%	2.3%	0.6%	2.5%	1.2%	84.8% (SM; 10% Gravel; 50-60% Sand; 30-40% Fines)
3	1.7%	1.1%	0.4%	0.8%	0.4%	0.9%	94.6% (CH; 10% Gravel; 15% Sand; 75% Fines)
4, 5	6.9%	11.8%	4.0%	0.8%	5.6%	5.3%	64.6% (CH; 5% Gravel; 5% Sand; 90% Fines)
6	10.3%	0.4%	2.9%	0.9%	0.1%	0.1%	85.4% (Fine Grained)
7, 8	1.2%	3.7%	3.4%	4.9%	0.4%	1.5%	84.8% (ML; 10% Gravel; 30-40% Sand; 50-60% Fines)
9	9.5%	3.2%	3.7%	2.4%	1.6%	2.2%	77.5% (CL; 10-15% Gravel; 30-40% Sand; 50-55% Fines)
Average	5.2%	4.5%	2.8%	1.7%	1.8%	2.0%	82.0%

[a] BA = 840 mm-diameter Bucket Auger (BA) borehole.
[b] Age of the solid waste part of the sample estimated, where possible, from dated newspapers contained in the solid waste.
[c] Moisture content of soil and soil-like materials measured prior to the test.
[d] Shear wave velocity measured in the 460-mm consolidation apparatus for same initial density and overburden pressure.
[e] In-situ unit weight. The unit weight of the reconstituted sample in the CyDSS apparatus was ±2 percent of this value.
[f] Percentage developed on weight basis.
[g] Unified Soil Classification System by visual classification: Soil-like materials refer to the materials that may be a product of waste decomposition.
[h] Refer to text for description of testing scenarios.

444

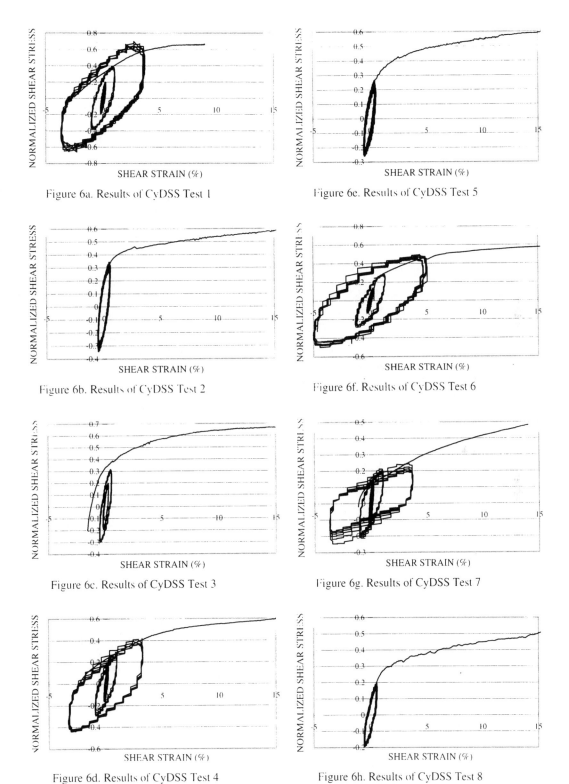

Figure 6a. Results of CyDSS Test 1

Figure 6e. Results of CyDSS Test 5

Figure 6b. Results of CyDSS Test 2

Figure 6f. Results of CyDSS Test 6

Figure 6c. Results of CyDSS Test 3

Figure 6g. Results of CyDSS Test 7

Figure 6d. Results of CyDSS Test 4

Figure 6h. Results of CyDSS Test 8

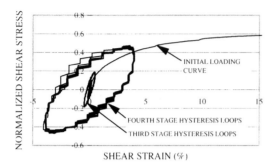

SHEAR STRAIN (%)

Figure 6i. Results of CyDSS Test 9.

Test results shown on Figures 6a to 6i indicate that the response of OII solid waste to strain-controlled cyclic loading is non-linear and hysteretic. The results also indicate that the cyclic loading slightly densified the specimens tested and thus increased their resistance under subsequent monotonic loading.

3.2 *Data interpretation*

In geotechnical earthquake engineering, the non-linear hysteretic behavior of soils and soil-like materials subjected to cyclic loading is commonly modeled b an equivalent-linear model (Seed and Idriss, 1970 The equivalent-linear model consists of a shea modulus reduction curve and a material dampin curve. The modulus reduction curve is evaluate from the slope of the line connecting the tips of th cyclic stress-strain loops (the secant modulus, G_s The G_s values are commonly normalized by the initi shear modulus, G_{mo}, which is typically evaluated fro the (laboratory) measured unit weight and shear wav velocity. The damping curve is evaluated from th area of cyclic stress-strain loops assuming that material damping can be characterized by the strain-dependent equivalent viscous damping ratio defined as a ratio of damping energy to the equivalent strain energy (see, e.g., Ishihara, 1996).

Recently, a variety of non-linear constitutive models have also been used to model the behavior of soil and soil-like materials under cyclic loading. Non-linear constitutive models are more complex than the equivalent-linear model and use specific "rules" to describe the stress-strain behavior during cyclic loading. The most common rules employed in geotechnical earthquake engineering to characterize the stress-strain behavior during the cyclic unloading and reloading are known as Masing (1926) rules. The first Masing rule postulates that the tangent shear moduli at the reversal points of the unloading and reloading branches of the loop are identical to G_{mo}. The second Masing rule states that the shape of the

reloading branch is the same as that of the positive part of the backbone curve enlarged by the factor of two, and similarly, that the unloading branch has the shape of the negative part of the backbone curve enlarged by a factor of two.

The Masing rules were originally proposed by Masing for the cyclic stress-strain behavior of brass and were later applied to granular materials (Newmark and Rosenblueth, 1971). Finn et al. (1977) demonstrated that the Masing rules are applicable for fully saturated sands while Idriss et al. (1978) demonstrated their applicability for clays.

To investigate the validity of the Masing rules as applied to OII landfill solid waste, the results of one characteristic test (Test 9) are crossplotted in Figure 7. Figure 7a shows the initial loading curve superimposed on the third-stage hysteresis loops (cyclic loading to approximately 0.5 percent strain) for Test 9. Figure 7b shows the initial loading curve superimposed on the fourth-stage hysteresis loops (cyclic loading to approximately 4 percent strain) for Test 9. Also indicated on Figures 7 are the initial shear moduli at the reversal points of the unloading and reloading branches of the cyclic loops (Figure 7a), and the initial loading curve enlarged by the factor of two (reloading curve; Figure 7b).

Figures 7a and 7b indicate that OII landfill solid waste, like the other soil and soil-like materials, follows the Masing rules during cyclic loading. Consequently, the results of monotonic loading (the initial loading curve) can also be used to evaluate the modulus reduction and equivalent viscous damping of OII landfill solid waste.

Figures 8 and 9 show the results of interpretation of OII landfill solid waste testing results in terms of the equivalent-linear model parameters, i.e., solid waste modulus reduction and damping.

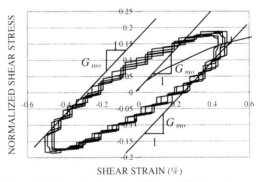

SHEAR STRAIN (%)

Figure 7a. Investigation of validity of Masing rules - Test 9, third-stage hysteresis loops.

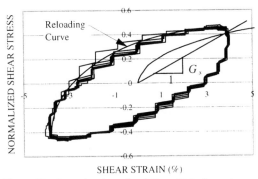

Figure 7b. Investigation of validity of Masing rules - Test 9, fourth-stage hysteresis loops.

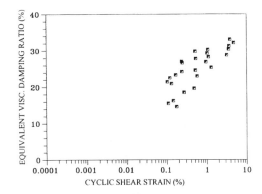

Figure 9. OII landfill solid waste equivalent viscous damping ratio evaluated from CyDSS test results.

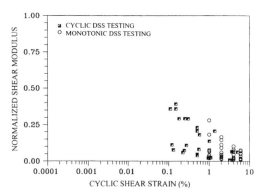

Figure 8. OII landfill solid waste modulus reduction evaluated from monotonic and CyDSS test results.

Figure 8 shows the results of both cyclic and monotonic testing (for cyclic shear strains of 1 percent and above). As expected, based upon the evaluation of applicability of the Masing rules, the G_s/G_{mo} values evaluated from the results of monotonic testing show no discernible difference from the G_s/G_{mo} values evaluated from the results of cyclic testing.

Figure 9 shows the equivalent viscous damping ratio values evaluated from the results of CyDSS testing. The damping values shown in Figure 9 include a correction for system friction which was developed using the series of six CyDSS tests on Monterey 0/30 sand discussed previously. Demonstrating the validity of the Masing rules, evaluation of the OII solid waste damping values from monotonic testing results is theoretically possible. However, evaluation of the damping values requires (numerical) integration of the initial loading curve along the entire strain range of interest. This integration may be affected by the robustness of the measurements in the small strain range. An attempt to evaluate equivalent viscous

damping from the monotonic test results is currently underway in GeoSyntec Consultants.

4 SUMMARY AND CONCLUSIONS

As a part of pre-design studies for landfill closure, a laboratory testing program using strain-controlled CyDSS testing was conducted on OII landfill solid waste. The testing was conducted on-site using a large-diameter (i.e., 460-mm) CyDSS device constructed specially for this project.

Six bulk solid waste specimens were obtained from depths ranging from 3.4 to 33.5 m, where the in-situ vertical consolidation stresses ranged from 94.1 to 511.5 kPa. The six specimens tested were fibrous and had a content of soil and soil-like materials ranging between 65 to 95 percent. Staged cyclic loading on specimens reconstituted to in-situ unit weight and the field moisture content was used to evaluate the stress-strain behavior of the solid waste under cyclic shear strains ranging from 0.1 to 6 percent.

The results presented herein may be used to characterize OII landfill solid waste in the large strain range (cyclic shear strain ranging from 0.1 to 6 percent). In the large strain range, the solid waste tested behaves as hysteretic material. The Masing rules apply to the material tested. Significant modulus reduction and an equivalent viscous damping increase occur at cyclic shear strains beyond approximately 0.1 percent cyclic shear strain. Both the secant shear moduli and damping were found to relatively unaffected by material moisture content and composition.

Modulus reduction and damping for solid waste may depend on numerous factors that have not yet been explored, including waste composition and age, waste soil content and type, loading path, and specimen size.

Therefore extrapolation of the results presented in this paper to other conditions and materials must consider these uncertainties.

ACKNOWLEDGMENTS

The authors wish to express their sincere appreciation to New Cure, Inc. (NCI), in particular, Les LaFountain and Ken Hewlett, which represents the potentially responsible parties for design and construction of the final cover at the OII landfill, for authorizing the work described herein. Gary Schmertmann, Dennis Vander Linde, and William Mazanti, all of GeoSyntec, also played key roles in the design and fabrication of the CyDSS device. We also wish to thank Edward Kavazanjian of GeoSyntec who reviewed the manuscript and provided many valuable suggestions.

REFERENCES

Finn, W.D.L., Lee, K.W. and Martin, G.R. (1977), "An effective stress model for liquefaction," *Journal of the Geotechnical Engineering Division*, ASCE, Vol. 103, No. GT6, pp. 517-533.

GeoSyntec (1996), "Waste mass field investigation, Operating Industries, Inc. landfill, Monterey Park, California," *Report No. SWP-2*, GeoSyntec Consultants, Huntington Beach, California.

Idriss, I.M., Dobry, R. and Singh R.D. (1978), "Nonlinear behavior of soft clays during cyclic loading," *Journal of the Geotechnical Engineering Division*, ASCE, Vol. 104, No. GT12, pp. 1427-1447.

Idriss, I.M., Fiegel, G., Hudson, M.B., Mundy, P.K. and Herzig, R. (1995), "Seismic response of Operating Industries landfill," In: *Earthquake Design and Performance of Solid Waste Landfills*, ASCE Geotechnical Special Publication No. 54, pp. 83-118.

Ishihara, K. (1996), "Soil behavior in earthquake geotechnics," Clarendon Press, Oxford, United Kingdom, 350 p.

Masing, G. (1926), "Eigenspannungen und Verfestigung beim Messing." Proc. *2nd International Congress on Applied Mechanics*, Zürich, Switzerland, pp. 332-335.

Matasovic, N. and Vucetic, M. (1993), "Cyclic characterization of liquefiable sands," *Journal of Geotechnical Engineering*, ASCE, Vol. 119, No. 11, pp. 1805-1822.

Matasovic, N., Kavazanjian, E., Jr. and Abourjeily, F. (1995), "Dynamic properties of solid waste from field observations," Proc. *1st International Conference on Earthquake Geotechnical Engineering*, Tokyo, Japan, Vol. 1, pp. 549-554.

Matasovic, N. and Kavazanjian, E., Jr. (1998), "Cyclic characterization of OII landfill solid waste," *ASCE Journal of Geotechnical and Geoenvironmental Engineering*, Vol. 124, No. 3.

Newmark, N.M. and Rosenblueth, E. (1971), "*Fundamentals of earthquake engineering*," Prentice-Hall, Inc., Englewood Cliffs, New Jersey, pp. 162-163.

Seed, H.B. and Idriss, I.M. (1970), "Soil moduli and damping factors for dynamic response analyses," *Report No. EERC 70-10*, Earthquake Engineering Research Center, University of California, Berkeley, California, 40 p.

Geotechnical Hazards, Marić, Lisac & Szavits-Nossan (eds) © 1998 Taylor & Francis, ISBN 90 5410 957 2

Sanierung einer Altlast durch Umlagerung und Errichtung einer Deponie im Bergbaugebiet

K. Schippinger & S. Hohl
Dipl.-Ing. Dr. Schippinger und Partner, Ziviltechniker Ges.m.b.H., Graz, Austria

H. Hannak
Lafarge Perlmooser AG, Wien, Austria

KURZFASSUNG: In den Jahren 1965 bis 1984 wurde der im Werk Peggau (Steirische Montanwerke AG) bei der Zementproduktion angefallene Elektrofilterstaub auf der Deponie Eichberg nördlich von Graz abgelagert. Durch Sickerwasseraustritte am Dammfuß der Deponie kam es im Jahr 1990 zur Beeinträchtigung eines Brunnens einer Wassergenossenschaft. Daraufhin wurden zur Sanierung mehrere Projektsvarianten ausgearbeitet, wobei jene Variante realisiert wird, die eine schrittweise Räumung des Altlastmaterials und eine Umlagerung auf eine hydrogeologisch geeigneten Fläche vorsieht.

1 EINLEITUNG

In der Altlast "Elektrofilterstaubdeponie Eichberg" sind im wesentlichen Filterstäube aus den Filteranlagen der Schachtöfen des ehemaligen Zementwerkes in Peggau der Perlmooser Zementwerke AG abgelagert worden. Der Filterstaub ist feinkörnig und weist eine vorwiegend anorganische Zusammensetzung auf. Als Hauptbestandteile sind Alkaliverbindungen (im wesentlichen Kaliumsulfat) zu nennen. Diese Anteile sind relativ leicht wasserlöslich und bewirken einen hohen pH-Wert. Bei den Neben- und Spurenelementen sind vor allem Aluminium, Fluorid, Nitrit, Arsen und Chrom (vorwiegend in sechswertiger Form) zu nennen.

Nach Ende der Deponietätigkeit wurden die betroffenen Flächen rekultiviert und landwirtschaftlich genützt.

Durch Sickerwasseraustritte am Dammfuß der Deponie kam es zu einer Verklausung des Zitollbaches und in Folge zu einer Beeinträchtigung des Brunnens der Wassergenossenschaft Zitoll.

2 PROJEKTSVARIANTEN

Sicherung mittels Entwässerungsmaßnahmen durch die Herstellung einer Tiefendrainage, Verbindungs- und Ableitungskanalisation sowie von Brunnenfassungen und Ableitung der Quellwässer. Weitergehende Untersuchungen ließen Zweifel hinsichtlich der Wirksamkeit des vorgelegten Sicherungsprojektes aufkommen. Die Ausarbeitung einer Variantenstudie ergab als einzig zielführende Maßnahme die schrittweise Räumung des Altlastmaterials und Umlagerung auf eine hydrogeologisch geeignetere Fläche, die als Deponie entsprechend auszubilden und abzudichten ist.

3 AUSFÜHRUNGSVARINTE

Das im Jahr 1995 erstellte Projekt sieht die vollständige Räumung der Altlast Zitoll mit Abtransport des Materials nach Retznei vor. Im Bereich des Tagbaugeländes des Zementwerkes Retznei erfolgt eine Konditionierung des angelieferten Elektrofilterstaubes mit hydraulischen Bindemitteln und danach der Einbau in die eigens zur Übernahme des Altlastenmaterials errichtete Sanierungsdeponie.

4 DEPONIESTANDORT - RETZNEI

Durch die Beendigung des Abbaues in einem Teil des Tagbaues Retznei der Lafarge Perlmooser AG, Retznei ergab sich eine Fläche, die sowohl von der Topographie, als auch von der geologisch/ hydrogeologischen Eignung zur Übernahme des Altlastmaterials geeignet war. Die Entfernung des Deponiestandortes zum Ablagerungsort des Elektrofilterstaubes beträgt auf der Straße ca. 90 km.

4.1 Geologie

Das Tagbauareal des Zementwerkes der Lafarge Perlmosser AG in Retznei gehört im geologischen Sinne dem sogenannten Leithakalk an, der im Bereich der mittelsteirischen Schwelle in mehrere, räumlich voneinander isolierte Vorkommen aufgeteilt ist. Im Raum Retznei handelt es sich um Algenschuttkalke und verschiedene mergelige Kalke, welche im Tagbau gewonnen und zur Zementherstellung verwendet werden. Diese Leithakalkentwicklung wird im Bereich des Tagbaues vom Steirischen Schlier, einer mächtigen Abfolge von Tonmergeln, unterlagert.

Das Hangende der Kalke bilden ebenfalls dunkle, harte Tonmergel und sandige Mergel. Bisweilen sind reine Sandlagen eingeschaltet. Das Leithakalkriff von Retznei ist somit nahezu "allseitig" von tonigmergeligen Gesteinen eingeschlossen. Dies hat zur Folge, daß die zusitzenden Wässer aus dem Tagbautiefsten ausgepumpt werden müssen, da ein Abströmen durch die undurchlässigen Mergeltone verhindert wird.

4.2 Hydrogeologie

Die Lagerstätte ist mittels Bohrungen gut untersucht. Nach den Auskünften der Werksleitung zeigen die Profile, daß der Riffkörper überall auf den Tonmergeln aufliegt und im Hangenden von tonigmergeligen-sandigen Gesteinen überlagert wird. Dadurch ergibt sich in vertikaler Richtung eine natürlich dichte Barriere gegen den Untergrund.

4.3 Bodenmechanik
4.3.1 Geländestabilität

Vom Niveau der Deponieaufstandsfläche wurde im Zuge der Aufschließungsmaßnahmen eine Kernbohrung abgeteuft. Die aus der Mergelschicht entnommenen Bodenproben haben im Scherversuch einen mittleren Reibungswinkel von $\varphi' = 24\,°$ und eine Kohäsion von im Mittel $c = 100$ kN/m². Bei der gewählten Deponieform (Haldendeponie) und den vorliegenden Untergrundverhältnissen sind demnach keine Stabilitätsprobleme zu befürchten.

4.3.2 Setzungsverhalten

Die trockenen, kompakten Mergel besitzen infolge ihrer festen Konsistenz ein sehr günstiges Setzungsverhalten. Unter der Voraussetzung, daß die Aufstandsfläche in diesen Schichten zu liegen kommt und daß die Überschüttungshöhe ca. 12 m beträgt, kann davon ausgegangen werden, daß die ursprüngliche Vorbelastung durch die zwischenzeitlich abgebauten Kalke keinesfalls erreicht wird, sodaß mit keinen nennenswerten Setzungen zu rechnen ist.

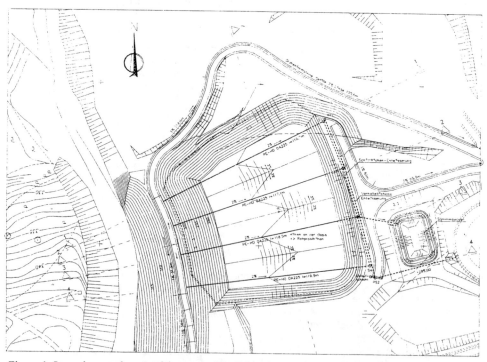

Figure 1: Lageplan aus dem Ausführungsprojekt Schippinger & Partner (1996)

4.3.3 Durchlässigkeit

Die abgeteufte Kernbohrung (Bohrloch 3) liegt etwa in der Mitte des vorgesehenen Deponieareals und wurde vom bestehenden Niveau (ca. 292 m ü.A.) bis in eine Tiefe von 15,5 m niedergebracht. Die aus dieser Bohrung in einer Tiefe von ca. -3,8 m und -8,3 m entnommenen ungestörten Bodenproben wurden im geotechnischen Labor des Institutes für Bodenmechanik und Grundbau der TU Graz hinsichtlich ihrer Durchlässigkeit und ihrer Kornverteilung untersucht. Die im Labor in der Triaxialzelle ermittelten Durchlässigkeiten liegen für beide Proben bei einem k-Wert $< 10^{-10}$ m/s. Die Sieblinie weist für die Mergelformation einen tonigen Schluff mit geringem Sandanteil (ca. 7,5 %) auf. Zur Abschätzung des Einflusses eventuell vorhandener Klüfte durch die jahrelangen Sprengarbeiten wurden im Niveau der bestehenden Abbauterrasse Versickerungsversuche durchgeführt. Es wurde eine Insitu-Durchlässigkeit für die Versickerungsgrube mit $k = 7.10^{-9}$ m/s ermittelt.

5 BAUAUSFÜHRUNG

Die nach den Plänen der Dipl.-Ing. Dr. Schippinger & Partner, Ziviltechniker Ges.m.b.H., neu errichtete Deponie weist folgende Kenndaten auf (Fig.1):

Die gesamte abzudichtende Fläche beträgt 14.400m², wobei 5.800 m² auf die Basis und 8.600 m² auf die Böschungen entfallen.

5.1 Aufstandsfläche, Rohprofilierung

Die Aufstandsfläche der Deponie wurde mit der projektsgemäß vorgegebenen Neigung von 3 % in Quer- bzw. 2 % Längsrichtung profiliert.

Diese Fläche wurde entsprechend den geotechnischen Vorgaben mit einer Einbaudichte von $D_{pr} \geq 99$ % hergestellt um dem Auftreten von Setzungen entgegenzuwirken. Die beim Einbau erzielte Verdichtung wurde mittels Lastplattenversuchen überprüft.

5.2 Dichtungsmaßnahmen

Sowohl die Deponiebasis (Fig.2), als auch die Böschungsflächen wurden mit einer dreilagigen mineralischen Dichtschicht entsprechend dem Stand der Technik versehen.

Mindestwerte:
- k-Wert $\leq 10^{-9}$ m/s
- Mindeststärke 60 cm
- Herstellung in mindestens 3 Lagen
- organische Beimengung < 5 %
- Einbaudichte ≥ 97 % Proktordichte
- ev. Beimengung von dichtigkeitserhöhenden Zusätzen lt. Eignungsprüfung.

Die mineralische Dichtung wurde aus aufbereiteten lokalen Mergelschichten gewonnen. Die prinzipielle Eignung dieses Materials wurde durch eine Eignungsprüfung des Institutes für Bodenmechanik und Grundbau der TU-Graz (1996), nachgewiesen.

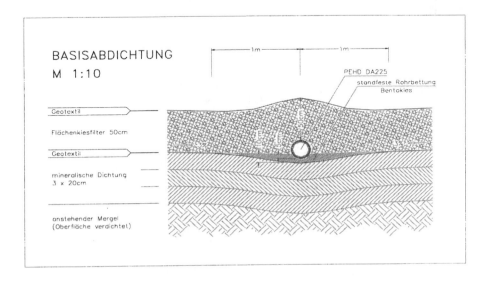

Figur 2: Basisabdichtung aus dem Ausführungsprojekt Schippinger & Partner (1996)

5.3 Drainageschicht

Um ein Abfließen der Sickerwässer in freier Vorflut zu ermöglichen, wurde auf der Basis vollflächig ein Drainagekörper aus Kies mit der Körnung 16/32 in einer Stärke von 50 cm aufgebracht. Die Trennung dieses Flächenkiesfilters von der mineralischen Dichtung bzw. erster Abfallage erfolgte durch Geotextilien, wobei das obere Geotextil UV-stabilisiert sein mußte und eine höhere Durchlässigkeit besitzt.

5.4 Drainageleitungen

Zur Erfassung der anfallenden Wässer innerhalb der Deponiefläche wurden in einem Maximalabstand von 30 m Drainagerohre parallel verlegt. Diese Leitungen wurden aus PEHD, DA 225 mm hergestellt und sind über 2/3 ihres Umfanges gelocht. Die Drainagen gehen vom Beginn der Dammdurchdringung an in geschlossene Leitungen über, die weiter zu den Sickerwasserschächten führen. An der gegenüberliegenden Seite (Böschung) wurden die Drainagerohre zu Wartungszwecken aus der Deponie geführt und mit Spülöffnungen versehen.

5.5 Sammelleitungen

Die zuvor erwähnten Sickerwasserschächte (S1 - S5) sind durch Rohrleitungen miteinander verbunden. Vom Schacht S 2 führt ein Ableitungskanal zum Sammelbecken. Dieser Kanal weist ein Gefälle von 10 ‰ und DA 250 mm auf.

5.6 Wartungsschächte

a) Sickerwassersystem
Die zur Wartung der Sickerwasserleitung erforderlichen Schächte wurden aus Fertigteilen mit einem Innendurchmesser von 1,5 m hergestellt. In diesen Schächten ist durch Ausbauen eines Verbindungsstückes die Trennung zwischen Regenwasser und Sickerwasser möglich. Des weiteren besteht die Möglichkeit, mittels Schieber die Drainageleitungen abzusperren.
b) Regenwassersystem
Die Schächte für die Wartung des Regenwasserkanals wurden ebenso in Fertigteilbauweise errichtet, weisen jedoch aufgrund der im Gegensatz zum Sickerwassersystem nicht erforderlichen Einbauten lediglich einen Innendurchmesser von 1,0 m auf.

5.7 Sammelbecken

Die aus dem Sickerwassersammelsystem ankommenden Wässer werden in dem östlich der Deponie situierten Auffangbecken zwischengespeichert. Dieses Becken ist als Erdbecken ausgeführt und mit einer dreilagigen mineralischen Dichtung versehen. Als Schutz vor Erosionen wurde auf der Dichtung vollflächig ein Geotextil verlegt. Um ein Aufschwimmen dieses Geotextils zu verhindern wurde am Beckenboden eine 20 cm starke Kieslage aufgebracht.

6 SANIERUNG

6.1 Arbeitsablauf

Als wesentliche Punkte des Sanierungablaufes sind folgende Arbeitsschritte zu nennen:
- Abbau des Altlastenmaterials in der E-Filterstaubdeponie Eichberg
- Transport des Materials von der Altlast am Eichberg zum ca. 90 km entfernten Tagbaugelände in Retznei
- Beprobung des Altlastenmaterials zur Festlegung der Mischungsrezeptur
- Behandlung des Altlastenmaterials durch Vermischung mit hydraulischen Bindemitteln und eventuell erforderlichen Zuschlagstoffen
- Beprobung des Altlastenmaterials nach Konditionierung im Tagbau Retznei
- Verdichteter Einbau des behandelten Altlastenmaterials in der Sanierungsdeponie
- Beprobung des Verfestigungsproduktes nach dem Einbau

6.2 Räumung und Transport

Das Altlastenmaterial ist mit einer unterschiedlich mächtigen Schicht Humus und Boden überdeckt, die vorweg abgeschoben wird. Danach wird das Material abgebaut und gesiebt, um es von Störstoffen (Gesteinsbrocken, Bauschutt udgl.) zu befreien, die später eine Beschädigung des Mischers zur Folge haben könnten. Das Überkorn wird danach mittels einer Brecheranlage auf ein Größtkorn <50 mm zerkleinert.
Der Transport des Ausgangsmaterials erfolgt mittels LKW bis zu der im Betriebsareal des Bergbaues errichteten und der Behandlungsanlage angeschlossenen Zwischenlagerfläche.

6.3 Beprobung der Altlast

Zur Entscheidung, ob Altlastenmaterial vorliegt, wird die Durchführung einer im folgenden beschriebenen Schnellelution durchgeführt. Diese Analysemethode basiert im wesentlichen auf der Tatsache, daß der E-Filterstaub auf Grund der hohen löslichen Salzanteile (besonders Kaliumsulfat) bei Elution mit

Wasser eine sehr hohe elektrische Leitfähigkeit aufweist und sich deshalb von inertem Material (geringe Leitfähigkeit bei Elution) leicht unterscheiden läßt.

Die Schnellelution wird wie folgt durchgeführt: In einem Kübel mit rund 12 Liter Fassungsvermögen wird genau 1 kg Material, welches aus einer Mischprobe aus gefördertem Aushub entnommen wird, eingewogen. Anschließend werden genau 10 Liter Leitungswasser zugegeben. Das eingebrachte Deponiematerial wird anschließend für 20 Minuten durch mehrmaliges Umrühren eluiert. Nach dieser Zeit werden pH-Wert und elektrische Leitfähigkeit im Eluat bestimmt (die Messung erfolgt mit einem pH-Meßgerät mit Glaselektrode und einem Leitfähigkeitsmeßgerät mit Temperaturkompensation).

Zusammenfassend wurde als Vorgangsweise für die Entscheidung, ob das gelöste Altlastenmaterial zur Behandlung abtransportiert werden muß, folgende Vorgangsweise gewählt:
- Schnellelution an einer Mischprobe; der Abtransport hat zu erfolgen, wenn folgende Grenzwerte überschritten werden:
pH-Wert: 11,5
elektr. Leitfähigkeit: 2 mS/cm
- zusätzliche optische Kontrolle auf verdächtig erscheinende Bestandteile.

Bei der Festlegung dieser Grenzwerte wurden die Grundwerte des zur Elution verwendeten Leitungswassers miteingerechnet. Unter dieser Berücksichtigung entsprechen die angeführten Grenz-werte den vorgegebenen Grenzwerten für eine Bodenaushubdeponie (pH-Wert: 6,5 bis 11 bzw. geogen bedingt 6,5 bis 12; elektr. Leitfähigkeit: 1,50 mS/cm bzw. geogen bedingt 2,50 mS/cm).

6.4 Beprobung im Sohlbereich

Im Übergangsbereich vom Altlastenmaterial zum natürlichen Untergrund reicht zur Unterscheidung, inwieweit noch kontaminiertes Material vorliegt, die Schnellelution allein nicht aus. Es muß nämlich zusätzlich festgestellt werden, ob Schadstoffgehalte aus dem E-Filterstaub in den Untergrund eingebracht wurden.

Wird also im Zuge der Altlastenräumung der natürliche Untergrund erreicht, so werden vorerst wie-

der Schnellelutionen am geförderten Material durchgeführt. Werden die für pH-Wert und elektr. Leitfähigkeit vorgegebenen Grenzwerte unterschritten, muß zusätzlich eine Elution wie zur Ermittlung der Mischungsrezepturen im (nachfolgenden Kapitel) beschrieben wird, durchgeführt werden. Das untersuchte Material muß dann nicht auf die Deponie im Tagbau Retznei gebracht werden, wenn im Eluat folgende Grenzwerte für die kritischen Parameter des Filterstaubes eingehalten werden:

Tabelle 1: Einzuhaltende Grenzwerte

Parameter	Grenzwert (mg/l)
Aluminium Al	2,0
Fluorid F	3,0
Nitrit NO_2	1,0
Arsen As	0,1
Chrom gesamt Cr	0,1

6.5 Beprobung des Materials zur Festlegung der Mischungsrezeptur

Die Beprobung des aus der E-Filterstaubdeponie angelieferten Altlastenmaterials erfolgt in Retznei. Von der Zwischenlagerfläche wird je 500 t Altlastenmaterial eine Mischprobe durch Entnahme an fünf verschiedenen Stellen zusammengestellt und in einem Behälter durchmischt.

Im Zuge der Voruntersuchungen des Institutes für Baustofflehre und Materialprüfung der Leopold - Franzens - Universität Innsbruck unter Leitung von a.Univ.Prof. Dr. Walter Lukas zur Verfestigung des E-Filterstaubes aus der Altlast wurde festgestellt, daß das Altlastenmaterial in unterschiedlich vermischter Form im Deponiekörper vorliegt. Daraus resultieren auch unterschiedlich hohe Konzentrationen im Eluat bei Auslaugung des Filterstaubes. In Abhängigkeit der Eluatkonzentrationen sind unterschiedliche Mischungsrezepturen zur Konditionierung und Verfestigung des E-Filterstaubes erforderlich. Die für die Anwendung der drei Mischungsrezepturen maßgeblichen Eluatkonzentrationen sind wie in der Tabelle 2 angegeben, definiert:

Tabelle 2: Mischrezepturen

	Rezeptur 1	Rezeptur 2	Rezeptur 3
Aluminium Al	< 50 mg/l	50 - 100 mg/l	> 100 mg/l
Fluorid F	< 150 mg/l	150 - 300 mg/l	> 300 mg/l
Nitrit NO_2	< 15 mg/l	15 - 50 mg/l	> 50 mg/l
Arsen As	< 1 mg/l	1 - 3 mg/l	> 3 mg/l
Chrom gesamt Cr	< 1,5 mg/l	1,5 - 3,5 mg/l	> 3,5 mg/l

6.6 Mischrezepturen

Die Mischrezepturen wurden ebenfalls vom Institut für Baustofflehre und Materialprüfung der Leopold - Franzens - Universität Innsbruck, unter Leitung von a.Univ.Prof. Dr. Walter Lukas entwickelt und setzen sich wie folgt zusammen:

Mischrezeptur 1: 64 Masse-% Filterstaub (1800 kg) 16 Masse-% Bindemittel (150 kg PZ 275 C Retznei, 300 kg Flugasche) 20 Masse-% Wasser (560 kg inklusive Eigenfeuchte des E-Filterstaubes)

Mischrezeptur 2: 56 Masse-% Filterstaub (1560 kg) 16 Masse-% Bindemittel (150 kg PZ 275 C Retznei, 300 kg Flugasche) 8 Masse-% Zuschlag (220 kg) 20 Masse-% Wasser (560 kg inklusive Eigenfeuchte des E-Filterstaubes)

Mischrezeptur 3: 48 Masse-% Filterstaub (1340 kg) 16 Masse-% Bindemittel (150 kg PZ 275 C Retznei, 300 kg Flugasche) 16 Masse-% Zuschlag (450 kg) 20 Masse-% Wasser (560 kg inklusive Eigenfeuchte des E-Filterstaubes)

Die in Klammer angeführten Massenangaben beziehen sich auf die in der Deponie im Tagbau Retznei aufgestellte Mischanlage mit einem Fassungsvermögen von 1,5 m³.

6.7 Behandlung des Altlastenmaterials

Die Behandlung der angelieferten E-Filterstäube erfolgt in der im Tagbaugelände des Zementwerkes Retznei aufgestellten Mischanlage. Die Konditionierung erfolgt unter Zugabe von hydraulischen Bindemitteln, Zuschlagstoffen und Anmachwasser nach den zuvor beschriebenen Rezepturen. Das Material ist nach der Behandlung im wesentlichen der Eluatklasse Ib nach ÖNORM S 2072 zuzuordnen. Ausnahmen stellen lediglich die Parameter Leitfähigkeit (Eluatklasse IIIa) bzw. pH-Wert (Eluatklasse II b) dar.

6.8 Beprobung nach Konditionierung

Nach Verarbeitung des Altlastenmaterials entsprechend den vorgenannten Mischungsrezepturen wird das Mischgut mittels Förderband in die Deponie im Tagbau Retznei eingebracht. Aus diesem Mischgut, wird durch Absieben auf die Körnung bis 32 mm Probenmaterial direkt aus der Deponie entnommen. Die Herstellung der Proben erfolgt durch dreilagigen Einbau in Zylinderschalungen (PE-Rohre mit 100 mm Durchmesser und 120 mm Höhe), wobei jede Lage mit einem Proctorgerät (Fallkörper 2,5 kg) mittels 15 Schlägen verdichtet wird. In analoger Weise wurden auch im Rahmen der Voruntersuchungen die Probekörper hergestellt. Die hergestellten Probekörper werden in Retznei zwischengelagert

und in einem Alter von ca. 14 Tagen zur weiteren Untersuchung verschickt.

Für die vorgeschriebenen Untersuchungen an den verfestigten Proben (Druckfestigkeit, k_f-Wert, Elution nach ÖNorm S 2072, 2,25 Tage Elution) ist die Herstellung von 5 Probekörpern erforderlich, wobei eine Probe als Rückstellprobe erhalten bleibt.

Die Herstellung der Proben erfolgt je 500 t verarbeitetes Altlastenmaterial, was entsprechend der im vorigen Kapitel angeführten Mischungsrezepturen einer Mischgutmenge von rund 1000 t entspricht. Es wird also je 1000 t Mischgut eine Probenserie hergestellt.

Zusätzlich zu den aus dem auf die Deponie geförderten Mischgut entnommenen Proben werden je 10.000 t verdichtet eingebautem Mischgut 5 Proben mittels Ausstechzylinder entnommen und in analoger Weise wie die mit dem Proctorgerät hergestellten Probekörper untersucht.

6.9 Bauzeit

Die Deponiebauarbeiten wurden im Dezember 1996 begonnen und im Juni 1997 fertiggestellt. Danach begann die Anlieferung und Behandlung der E-Filterstäube. Mitte Dezember 1997 wurden die Konditionierungsarbeiten aufgrund der niedrigen Temperaturen (< -5° C) eingestellt.

Die Sanierungsarbeiten werden voraussichtlich 24 Monate dauern und Ende 1998 abgeschlossen sein.

6.10 Untersuchungsergebnisse

Wie die Gegenüberstellung der Ergebnisse der ersten Untersuchungen von Prof. Lukas am unverfestigten Ausgangsaltlastenmaterial (Tab. 3) und am verfestigten Altlastenmaterial (Tab. 4) zeigt, werden die wasserlöslichen Nebenelemente wie Chrom, Arsen, Fluorid und Nitrit durch das Konditionierungsverfahren erheblich verringert.

Auch die Vorgaben hinsichtlich der Mindestdruckfestigkeit und der maximalen Durchläßigkeit konnten erfüllt werden.

7 FINANZIERUNG

7.1 Kosten

Die Gesamtkosten für die Sanierung (Räumen der Altlast am Eichberg, Materialtransport vom Eichberg nach Retznei, Errichtung der Deponie in Retznei, Konditionierung und Einbau der Filterstäube in die neue Deponie) belaufen sich auf ca. 90 Mio ATS.

Tabelle 3: Elution des unverfestigten Altlastenmaterials

Probe vom	pH	LF	NO_2	F	Al	As	Cr_{ges}
		mS/cm	mg/l	mg/l	mg/l	mg/l	mg/l
27.06.1997	13,11	24,5	26,5	3,2	1,5	0,022	0,44
08.07.1997	12,50	5,85	5,20	0,3	1,8	0,286	0,09
22.07.1997	12,37	5,94	12,0	0,5	2,8	0,078	0,22
24.07.1997	11,37	1,89	2,42	0,3	2,8	0,023	0,03
29.07.1997	12,04	3,33	4,55	1,6	2,2	0,130	0,08
12.08.1997	12,02	4,48	10,3	2,1	5,2	0,037	0,12

Tabelle 4: Elution des verfestigten Altlastenmaterials

Probe vom	pH	LF	NO_2	F	Al	As	Cr_{ges}
		mS/cm	mg/l	mg/l	mg/l	mg/l	mg/l
27.06.1997	12,10	1,73	0,94	0,12	0,47	0,007	<0,01
08.07.1997	11,70	1,01	0,88	0,15	0,39	0,009	<0,01
22.07.1997	10,90	0,87	0,80	0,10	0,32	0,013	<0,01
24.07.1997	11,50	1,11	0,92	0,40	0,22	0,022	0,02
29.07.1997	10,41	0,86	0,85	0,37	0,20	0,018	<0,01
12.08.1997	9,52	0,22	0,20	0,42	0,15	0,001	<0,01

7.2 Förderung

Oben genannte Kosten werden von der Österreichischen Kommunal Kredit AG gefördert. Der Restbetrag wird von der Lafarge Perlmooser AG aufgebracht.

LITERATUR

Institut für Bodenmechanik und Grundbau der TU-Graz, 1996. Attest über die Eignungsprüfung im Labor von Material aus dem Tagbau Retznei zur Verwendung als mineralische Abdichtung. Graz, 1996.

ÖNORM S 2072, 1990. Eluatklassen (Gefährdungspotential) von Abfällen. Wien: Österreichisches Normungsinstitut, 1990

Dipl.-Ing. Dr. Schippinger & Partner 1996. Sicherung der Altlast "Steirische Montanwerke" E-Filterstaubdeponie Retznei - Einreich-und Ausführungsprojekt. Graz, 1996

Geotechnical Hazards, Marić, Lisac & Szavits-Nossan (eds) © 1998 Taylor & Francis, ISBN 90 5410 957 2

Engineering properties of compacted power industry wastes

R. Steckiewicz & K. Zabielska-Adamska
Bialystok Technical University, Poland

ABSTRACT: Compaction of fly ash is an effective way to improve the properties of the embankment material, such as: strength improvement, compressibility reduction, and loss of permeability. The samples were compacted by Standard Proctor and Modified Proctor methods at various moisture content in order to define dry density-water content relationship and determine the correlation between both dry density and water content, then values of the shear strength, the penetration resistance and the permeability.

1 INTRODUCTION

The Polish net building programme of limited-access highways and other highways, and the modernization of inferior in class highways cause usability of power industry wastes in earth constructions. Application of these wastes as a material for linear embankments and area macrolevelling is a very simple way of utilizing them.

The frequent obstruction in the progress of power engineering wastes utilization is lack of corresponding research results which are necessary to power-usability evaluation and which should be made for wastes from every power station as a prospective source of embankment material. Because power industry wastes are used in the earth constructions, it is necessary to investigate their properties depending on the compaction. Compaction of fly ash is an effective way to improve the properties of the embankment material, such as: strength improvement, compressibility reduction, and loss of permeability.

1.1 *Tested material*

The properties of compacted wastes were tested on the example fly ash/slag mixture (called fly ash) from dry storage yard of the Thermal-Electric Power Station in Bialystok. To date, over 1,880,000 tons of fly ash and slag have been accumulated in waste dumps, where slag is 10÷15% of the whole waste production.

2 RESEARCH RESULTS

2.1 *Compaction*

The samples were compacted by Standard Proctor and Modified Proctor methods at various moisture content in order to define dry density-water content relationship and determine the correlation between both dry density and water content, then values of the shear strength, the penetration resistance and the permeability.

It was found that recompacted samples were not representative. When working with recompacted samples, the maximum dry densities obtained were higher and optimum moisture contents were lower than those achieved with virgin samples, in the same compaction conditions, which are shown on Figure 1. It is necessary to say, that laboratory establishing of the compaction parameters according to Polish national standard (for mineral soils), admitting to re-use of the same sample, leads to the incorrect estimate of power industry wastes compactibility effects. The every point of compaction curve (the moisture-density relationship) ought to be determined for separate prepared fly ash sample.

The increase of maximum dry density at decreasing optimum moisture content is connected with grain size reduction and increase of specific surface of recompacted fly ash, what is opposite to compaction effects of mineral soils. This phenomenon can be explained by better compaction of fly ash caused of improvement of grain-size distribution by partially crumbling of dynamically

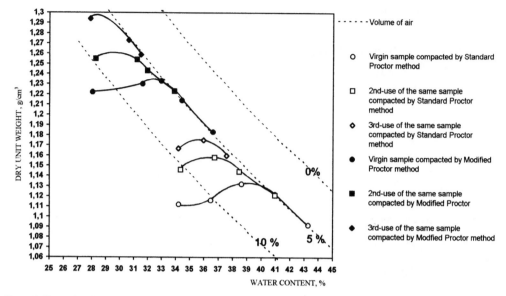

Figure 1. Comparison between the compacting of virgin fly ash samples and re-used samples obtained in Standard and Modified Proctor tests for example sample of fly ash

Table 1. The oedometer test results for fly ash samples compacted at optimum water content

$$M = \frac{\Delta\sigma}{\varepsilon}, \text{kPa}$$

Compactive method	Loading σ, kPa				Unloading σ, kPa	Reloading σ, kPa
	12.26-49.03	49.03-98.07	98.07-196.13	196.13-294.20	196.13-12.26	12.26-196.13
Standard Proctor	3890	7150	10,970	21,260	50,690	42,930
Modified Proctor	6660	9490	14,340	28,450	69,000	51,810

Table 2. The swell results (made with load 2.44kPa)

Parameter	Compactive method	
	Standard Proctor	Modified Proctor
Average value of swell, %	0.0104	0.0286
Range of compaction water contents, %	33.63 ÷ 43.26	27.14 ÷ 37.44

compacted grains. It should be added, that graining of the investigated fly ash changed after compaction only insignificantly.

2.2 Compressibility and expansion

Fly ash compressibility depends on compaction of tested material. The oedometer test results are shown

in Table 1. These results relate to specimens made up at optimum moisture content compacted to standard and modified maximum dry density (Proctor tests).

The swell measurements were determined using CBR moulds; received results are shown in Table 2. Fly ash was compacted both dry and wet of optimum (about 5%) by standard and modified compactive energy.

It should be stated, that compacted fly ash from the Thermal-Electric Power Station in Bialystok does not swell at minimum load 2.44kPa, just as other power engineering wastes (Pachowski 1976).

It was observed that fly ash expansion increases with the compactive effort. Fly ash moisture, tested after soaking of samples (at maximum expansion), practically does not depend on compaction water content of fly ash.

2.3 *Shear strength*

Shear strength parameters were obtained using unconsolidated undrained (UU) triaxial compression tests and direct shear tests. The UU test was applied due to its quickness, and because values of shear strength parameters obtained from this test are usually lower, so safer in estimation of fly ash strength than from other triaxial tests. A number tests were conducted on freshly compacted samples at low confinement loads to determine the mechanical parameters values at conditions which are similar to those existing in road constructions.

Fly ash specimens were compacted using two different compactive energies: standard and modified.

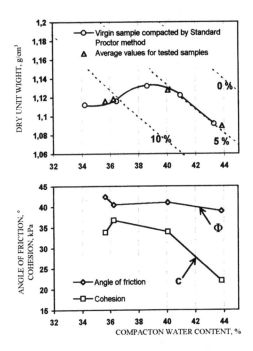

Figure 2. Shear strength parameters obtained in direct shear apparatus for samples compacted by Standard Proctor test

a) Compaction - Standard Proctor method

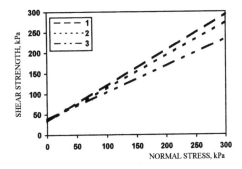

b) Compaction - Modified Proctor method

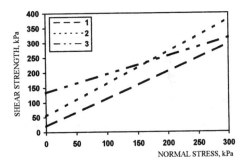

Figure 3. Shear strength of fly ash samples obtained using:
1. direct shear test - samples were compacted in Proctor moulds and cutted out, 2. direct shear test - samples were compacted directly in box of apparatus,
3. triaxial compression test (UU);
fly ash samples were compacted at water content closed to moisture water content by methods:
a) Standard Proctor, b) Modified Proctor

The shear strength results of fly ash samples compacted wet and dry of optimum ($\pm$ 5%) were obtained in direct shear apparatus.

It was found, that fly ash shear strength τ_f is practically independent on the initial water content and dry unit weight. Shear strength is only a function of normal stress σ. Obtained graphs of a function $\tau_f(\sigma)$ are closed to each other for samples compacted dry of optimum and at optimum water content; lower shear strength of fly ash were obtained for samples compacted wet of optimum. Figure 2 shows the correlation between friction angle or cohesion and compaction water content, obtained for fly ash samples compacted by Standard Proctor method. The less values of friction angle were obtained when compaction water content was greater. The greatest cohesion was obtained when the fly ash was compacted dry of optimum, if the fly ash was compacted wet of optimum - cohesion was insignificantly decreasing. The similar relationship was determined for fly ash samples compacted by Modified Proctor method.

The estimation of fly ash shear strength depends not only on the measuring instruments which are used for investigations, but also on the sample preparation procedure. Cutting out the samples, particularly those compacted using Modified Proctor test, with the aid of cutting ring from Proctor moulds damages the structure of tested fly ash. It causes the reduction of shear strength, and especially the value of cohesion. The comparison between strengths of the fly ash samples with the same dry density and water content (closed to optimum), compacted using two different compactive efforts was made, what is shown on Figure 3. These results were obtained using triaxial compression test (UU) and direct shear test (for samples compacted directly in box of apparatus or cutted out from Proctor moulds).

It should be stated, that lower values of friction angle and higher values of cohesion were obtained using triaxial compression test, independently of compactive energy. The most extensive study illustrating the influence of measuring instruments on shear strength parameters was published by Zabielska-Adamska (1997).

2.4 California Bearing Ratio

The investigations of California Bearing Ratio (CBR) values were carried out on unsoaked and soaked (for 96 hours) fly ash samples compacted both dry and wet of optimum ($\pm 5\%$).

Laboratory results of California Bearing Ratio are shown on Figure 4.

It should be noted, that CBR value varies according to compaction conditions. The CBR values depend not only on the water content, but also on the dry unit weight that is reached; it does not depend on water content and dry unit weight of fly ash samples after its maximal expansion (for soaked samples). The greatest CBR value of unsoaked fly ash is obtained when fly ash is compacted dry of optimum or at optimum water content; the greatest CBR value of soaked samples is determined for compaction at optimum moisture content. After soaking CBR results are decreasing in comparison with tested directly after compaction, on the average 7.9% for samples compacted by Standard Proctor method and 29.8% by Modified Proctor method.

The soaking tests on fly ash produce a large reduction of the CBR values for fly ash samples

- - - - - Volume of air

———○——— Virgin sample compation by Standard Proctor method

———●——— Virgin sample compacted by Modified Proctor method

◇   Average values of tested samples compated by SP method (without soaking)

◆   Average values of tested samples compacted by MP method (without soaking)

□   Average values of tested samples compacted by SP method and soaking

■   Average values of tested samples compacted by MP method and soaked

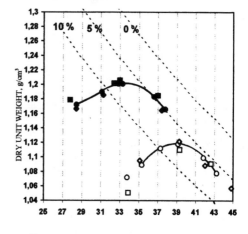

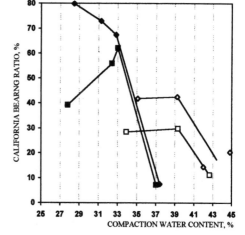

———◇——— CBR value for samples compated by Standard Proctor method (without soaking)

———◆——— CBR value for samples compacted by Modified Proctor method (without soaking)

——□—— CBR value for samples compacted by Standard Proctor method and soaked

——■—— CBR value for samples compacted by Modified Proctor method and soaked

Figure 4. Variation of the CBR values with compaction

compacted dry of optimum, and a small for those compacted at above optimum and on the wet side of optimum, just as on mineral cohesive soils (Rico Rodriguez et al. 1988).

2.5 *Permeability*

The permeability of a compacted samples, like the other mechanical properties, depends on the dry density and degree of saturation.

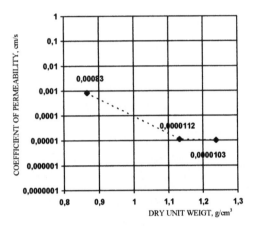

Figure 5. Differences in the permeability of fly ash samples in the loosest state and compacted by Standard Proctor method and Modified Proctor

Figure 5 gives the changes in permeability reached for fly ash in the loosest state and compacted by two different efforts, Standard Proctor and Modified Proctor, at moisture equalled optimum moisture content. The permeability coefficients obtained from laboratory tests of compacted fly ash samples were several times lower than uncompacted samples (in the loosest state).

3 CONCLUSIONS

The graining of tested fly ash corresponds to a silt, but it will, if properly compacted, give a very satisfactory fill material. The relatively high values of shear strength and penetration resistance were obtained.

Compaction have the significant effect on values of mechanical parameters and permeability of fly ash. The oedometric modulus increased up to 25%, the shear strength up to 30% and the CBR values doubled for fly ash samples compacted using Modified Proctor method in comparison with those compacted by Standard Proctor method.

ACKNOWLEDGMENT

The research described herein is a part of thesis presented to Bialystok Technical University, in 1997, by Zabielska-Adamska, K. *Geotechnical properties of power engineering wastes from Bialystok Thermal-Electric Power Station in aspect of its usability for structural fill*, in partial fulfillment of the requirements for the degree of Doctor of Philosophy. The work was supported by Dean's Research Grant from Bialystok Technical University.

REFERENCES

Pachowski, J. 1976. *Fly ashes and their use in road building*. Warsaw: Wydawnictwa Komunikacji i Łączności (in Polish).
Rico Rodriguez, A., H. del Castillo & G. F. Sowers 1988. *Soil mechanics in highway engineering*. Clausthal-Zellerfeld: Trans Tech Publication.
Zabielska-Adamska, K. 1997. *Geotechnical properties of power engineering wastes from Bialystok Thermal-Electric Power Station in aspect of its usability for structural fill*. PhD thesis, Bialystok Technical University, Bialystok (in Polish).

Geotechnical Hazards, Marić, Lisac & Szavits-Nossan (eds) © 1998 Taylor & Francis, ISBN 90 5410 957 2

The effect of petroleum hydrocarbons on geotechnical properties of kaolinite and Na-bentonite clays

A.Tuncan, M.Tuncan, Y.Güney & H.Koyuncu
Anadolu University, Eskişehir, Turkey

ABSTRACT: Most petroleum activities cause significant petroleum pollution into the ground and ground water. The presence of petroleum products affects the physical, mechanical and chemical processes in the soil after a spill is completed. Geotechnical properties of petroleum contaminated clays are important for offshore, nearshore and under water structures. Engineers must understand the behavior of clays influenced by petroleum products to provide proper foundation designs of oil stroge reservoirs and analysis of slopes in the nearshore and to understand the environmental damage to harbor and underwater structures. In this study, the effects of petroleum hydrocarbons such as crude oil, No. 6 fuel oil and kerosene on geotechnical properties of kaolinite and bentonite clays were investigated. It can be observed that the addition of crude oil, No. 6 fuel oil and kerosene decreases Atterberg limits, compressibility and strength of bentonite clay. Addition of kerosene increases of strength of kaolinite clay. While addition of crude oil, fuel oil and kerosene decreases Atterberg limits of kaolinite clay, they increases compressibility.

1 INTRODUCTION

The rapid development of industry has resulted in the release of large quantities of industrial waste containing a variety of organic and inorganic pollutants. Most industrial wastes contain varying amounts of total petroleum hydrocarbons, organic chemicals and heavy metals. Among them, petroleum hydrocarbon contamination is one of the most important problems for the scientists to solve. Petroleum hydrocarbon contamination may occur through a variety of sources such as the following: tanker accidents, discharge from coastal facilities, petroleum production facilities, pipelines and natural seepage. Oil spills and leakages have resulted in the contamination of large volumes of soil and ground water. When the oil is spilled on the soil, it tends to move downward under the gravity force. The rate of movement depends not only on the chemical and physical characteristics of the soil but also on the characteristics of the petroleum product (Tuncan and Pamukcu, 1992).

Petroleum is mostly composed of hydrocarbons which is immiscible with water and cemically inactive compounds. Although the petroleum products are immiscible with water, this does not mean that the various constituents are absolutely insoluble in water. Solubility differences are observed among the many petroleum products (Baehr, 1984). The lighter the product, the greater its relative content of soluble constituents. For example, gasoline and kerosene exhibit a higher total solubility than crude oil and fuel oil. Hydrocarbons are basically compounds of carbon and hydrogen. A component of petroleum is heavy molecular weight crude oil. Clay particles can adsorb large quantities of hydrocarbons which increases with salinity. Through this process oil wets the clay particle surfaces and also agglomerates the particles. Oil can penetrate deep into the soil column and remain in the soil for many years after a spill. A conceptual representation of how oil gets incorporated in the soil is shown in Figure 1. Petroleum and its products enter soil and undergroundwater through accidents. Similar to water, petroleum products which are released in bulk quantities at the soil surface can penetrate through the soil surface. If the released quantity is large, downword migration occurs with all soil pores being saturated with petroleum products. Petroleum products remain attached to soil particles via capillary forces (Dragun, 1989). This petroleum is immobile and is known as residual petroleum. If the migration of petroleum reaches the water table, it will begin spreading laterally over the water table. This layer of petroleum assumes the shape of a pancake and is commonly known as the pancake layer or free floating petroleum.

It is widely recognized that many of the soil characteristics, such as shear strength, compressibility and permeability are influenced by the pore fluid. The changes in the chemistry of the pore fluid within the soil matrix are likely to result in significant internal changes in their engineering behavior.

In this study, the effects of petroleum hydrocarbons on geotechnical properties of kaolinite and Na-bentonite clays were investigated. Crude oil, No.6 fuel oil and kerosene were used as a petroleum hydrocarbons. Some geotechnical properties such as consistency limits, organic matter, shear strength and consolidation characteristics were determined.

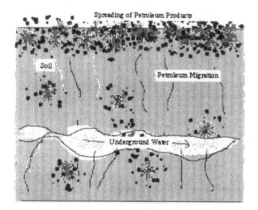

Figure 1. A conceptual representation petroleum products in the soil and underground water.

2 INVESTIGATION

An extensive laboratory work was conducted to investigate the mechanisms of interaction between clays and petroleum hydrocarbons. The laboratory experiments were conducted using artificial clay specimens compacted with crude oil, No.6 fuel oil and kerosene with tap water. The materials used and the experiments conducted are presented below along with the pertinent results.

2.1 Materials

The kaolinite and Na-bentonite clays were obtained commercially in powder form. Some of the measured properties of clay minerals used in this study are given in Table 1. The crude oil, No.6 fuel oil and kerosene were obtained from Turkish Petroleum Corporation (TPAO). Crude oil is immiscible with water and chemically inactive. No.6 fuel oil is also immiscible with water but lighter than crude oil. Kerosene is lighter than No.6 fuel oil. Some properties of crude oil, No.6 fuel oil and kerosene are given in Table 2.

2.2 Experiments and testing

A laboratory testing program was implemented to evaluate the influence of crude oil, No.6 fuel oil and kerosene on the consistency limits, organic matter content, shear strength and the compressibility of

kaolinite and Na-bentonite clays. The amount of contaminating agent was measured as weight percentage of the dry clay and premixed at optimum water contents of 25% and 40% for kaolinite and Na-bentonite clays, respectively. The proportion of these contaminants were selected as 5%, 10% and 20% by dry weight of the clay. This mixture were then statically compacted. After compacting, samples were tested for unconfined compression strength and consolidation.

Table 1. Particle size distribution, Atterberg limits and specific gravity of clay minerals.

Clay Minerals	Particle Size Distribution(%)			Atterberg Limits (%)			Specific Gravity
	Sand	Silt	Clay	LL	PL	PI	Gs
Na-Bentonite	2	46	42	395	106	289	2.60
Kaolinite	5	55	40	40	28	12	2.63

Table 2. Selected properties of crude oil, No.6 fuel oil and krosene.

Properties	Crude Oil	No.6Fuel Oil	Kerosene
API Gravity (oC)	18	15	6
Viscosity (oC)	30	22.5	10.6
Pour Point (oC)	-6	-1.1	-----
Flash Point (oC)	1.1	34	57.8
Freeze Point (oC)	3.4	3.5	0.08
Sulphur Content (% by weight)	6	0.1	-----

3 EXPERIMENTAL RESULTS

3.1 Atterberg limits

Atterberg limits depend on the type and amount of clay fraction, exchangeable cations and pore water chemistry. The effects of crude oil, No.6 fuel oil and kerosene on properties of Na-bentonite and kaolinite clays are given in Tables 3, 4 and 5. Whereas addition of crude oil and No.6 fuel oil decreased the plasticity, addition of kerosene increased the plasticity of kaolinite and bentonite clays, which may be attributed to loss of cohesion in the first case. Kaolinite clay became non-plastic when mixed with crude oil and No.6 fuel oil as shown in Tables 3 and 4. Addition of crude oil and fuel oil increases the organic matter content of Na-bentonite and kaolinite clay. While addition of kerosene increases the organic matter content of kaolinite clay, it decreases that of bentonite clay.

3.2 Ignitable organic matter content

Organic matter was determined by the ignition loss method. Samples were predried at 105 °C for 24 hours. These samples were then heated at 550 °C for 2 hours. The loss of weight as a result of the second heating was assumed to be due to loss of organic matter only. Organic matter increased with addition of crude oil and No.6 fuel oil for both kaolinite and Na-bentonite clays. Organic matter may cause high compressibility, low permeability and low strength. Increased organic content also causes an increase in the water content which may be due to entrapment of the water inside oil-clay or organic matter-clay domains.

Table 3. Atterberg limits and ignitable organic matter content of Na-bentonite and kaolinite clays contaminated with crude oil.

Mixture	Atterberg Limits (%)		Ignitable Organic Matter Content (%)
	LL	PI	
B+TW+CO(0%)	395	289	6.80
B+TW+CO(5%)	330	201	8.28
B+TW+CO(10%)	321	195	8.94
B+TW+CO(20%)	312	189	12.75
K+TW+CO(0%)	40	12	3.23
K+TW+CO(5%)	44	NP	11.00
K+TW+CO(10%)	36	NP	14.69
K+TW+CO(20%)	30	NP	21.00

B: Na-Bentonite, K: Kaolinite, TW: Tap Water, CO: Crude Oil, NP: Non-Plastic

Table 4. Atterberg limits and ignitable organic matter content of Na-bentonite and kaolinite clays contaminated with No.6 fuel oil

Mixture	Atterberg Limits (%)		Ignitable Organic Matter Content (%)
	LL	PI	
B+TW+FO(0%)	395	289	6.80
B+TW+FO(5%)	405	292	6.60
B+TW+FO(10%)	410	327	7.60
B+TW+FO(20%)	365	294	11.00
K+TW+FO(0%)	40	12	3.23
K+TW+FO(5%)	46	NP	7.91
K+TW+FO(10%)	34	NP	12.12
K+TW+FO(20%)	32	NP	18.09

B: Na-Bentonite, K: Kaolinite, TW: Tap Water, FO: No.6 Fuel Oil, NP: Non-Plastic

3.3 Shear strength

The shear strength of clays are influenced by void ratio, water content, interparticle forces or cohesive bonds and pore fluid chemistry. It is shown in Table 6 that the shear strength of petroleum contaminated clays were determined by unconfined compressive test. Shear strength of Na-bentonite clay decreased with increasing the amount of crude oil, No.6 fuel oil and kerosene. This is mainly because of the reduced cohesion among the clay particles. Decreases in shear strength are attributed to a higher repulsive forces. While the addition of crude oil and fuel oil decreased the shear strength of kaolinite, addition of kerosene increased the shear strength of kaolinite. It appears that kerosene promotes a higher angle of shearing resistance rather than cohesion. This is because of the presence of a highly cemented samples. This is attributed to the increase of attractive forces and decrease of repulsive forces.

Table 5. Atterberg limits and ignitable organic matter content of Na-bentonite and kaolinite clays contaminated with kerosene.

Mixture	Atterberg Limits (%)		Ignitable Organic Matter Content (%)
	LL	PI	
B+TW+KE(0%)	395	289	6.80
B+TW+KE(5%)	300	205	5.50
B+TW+KE(10%)	355	269	4.80
B+TW+KE(20%)	415	340	4.20
K+TW+KE(0%)	40	12	3.23
K+TW+KE(5%)	41	10	5.03
K+TW+KE(10%)	44	21	5.24
K+TW+KE(20%)	48	24	6.10

B: Na-Bentonite, K: Kaolinite, TW: Tap Water, KE: Kerosene

Table 6. Unconfined compressive strength of Na-bentonite and kaolinite clays contaminated with petroleum hydrocarbons.

Mixture	Unconfined Compressive Strength UCS (Pa)			
	Tap Water	Crude Oil	Fuel Oil	Kerosene
B+TW+(0%)	0.22	-----	-----	-----
B+TW+(5%)	-----	0.20	0.19	0.21
B+TW+(10%)	-----	0.15	0.15	0.21
B+TW+(20%)	-----	0.09	0.11	0.15
K+TW+(0%)	0.05	-----	-----	-----
K+TW+(5%)	-----	0.02	0.02	0.15
K+TW+(10%)	-----	0.01	0.01	0.16
K+TW+(20%)	-----	0.00	0.00	0.20

B: Na-Bentonite, K: Kaolinite, TW: Tap Water

3.4 *Consolidation*

Consolidation parameters of petroleum contaminated kaolinite and Na-bentonite clays were determined by odometer test apparatus. Compression index, C_c, is given in Table 7. While addition of crude oil and No.6 fuel oil slightly decreased the compression index of Na-bentonite clay, addition of kerosene decreased the compression index. However, addition of crude oil, No.6 fuel oil and kerosene increased the compression index of kaolinite clay.

Table 7. Consolidation characteristics of Na-bentonite and kaolinite clays contaminated with petroleum hydrocarbons.

| Mixture | Compression Index (Cc) | | | |
	Tap Water	Crude Oil	No.6 Fuel Oil	Kerosene
B+TW+(0%)	0.37	-----	-----	-----
B+TW+(5%)	-----	0.31	0.40	0.47
B+TW+(10%)	-----	0.38	0.21	0.45
B+TW+(20%)	-----	0.30	0.30	0.42
K+TW+(0%)	0.18	-----	-----	-----
K+TW+(5%)	-----	0.17	0.17	0.12
K+TW+(10%)	-----	0.23	0.24	0.11
K+TW+(20%)	-----	0.33	0.38	0.42

B: Na-Bentonite, K: Kaolinite, TW: Tap Water

3.5 *Predicted mechanisms between petroleum hydrocarbons and clay*

Clay minerals adsorb petroleum hydrocarbons or petroleum hydrocarbons adsorb clay particles on their surfaces. Adsorption increases with salinity because salinity concentrates the clay particles in the oil phase. When salt is added to water-oil mixed system, cations dissolve in the oil phase and chlorine ions remain in the water phase. This causes oil phase to be positive with respect to water subsequently, adhesion occurs between the negatively charged clay particle surface and the positively charged oil surface shown in Figure 2. Therefore the clay particle surface become wetted by oil. The large specific surface area of the Na-bentonite and kaolinite clays and its compositions also affect the oil retention on the clay particle surfaces.

In spherical agglomeration oil wets the clay particle surfaces promoting flocculation and agglomeration of the particles. The adsorption of the oil components on the clay particles causes the wettability to change from water-wet to oil-wet. The adsobed oil layers increase the viscosity of the oil near the surface. This adsorbed layer can not be replaced or removed by water. This resulting interaction between clay and oil in water has been characterized as "spherical agglomeration" (Puddington and Sparks 1975). In the spherical agglomeration process, clay particles in one liquid suspension are treated with a bridging liquid which wets the clay particles and is immiscible with the first liquid. An important aspect of the spherical agglomeration process is the relative wettability of the suspended clay particles by the two immiscible liquids.

When the bridging liquid is water, it is adsorbed on the surface of the clay particles. When two or more suspended clay particles collide, adhesion occurs having the formation of water bridges between the particles. The bonding force between the clay particles is due to the interfacial tension of the liquids involved. Large spherical particles need more bridging liquid for their formation. The amount of bridging liquid required for the agglomeration depends on the geometry of the particles. Flat surfaces require less bridging liquid to have a very strong adhesive bond than spherical surfaces. If water is the suspending medium, the most suitable bridging liquids that would cause spherical agglomeration are organic compounds such as oil, which is insoluble in water. The surface of the solid must be hydrophobic to form an agglomeration. Clay minerals are hydrophobic and therefore will form spherical agglomeration in presence of oil in a water suspension. The basic mechanism is given in Figure 3, where at equilibrium:

$$Tco - Tcw = Tow \cos\theta$$

Tco, Tcw and Tow are the surface tensions at the interface of the clay and oil, clay and water, and oil and water, respectively. θ is the contact angle measured in the liquid. Spherical agglomeration occurs when contact angle, θ, is greater than 90°. In this case, the clay particles migrate into the oil phase. Therefore, clay particles are absorbed by the oil.

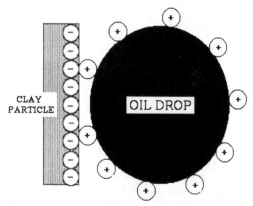

Figure 2. The mechanism of adhesion between oil drop and clay particle.

Spherical agglomeration observed in the sedimented specimens of marine clay photomicrographs can be seen in Figure 4. The clay particles in water immiscible liquids are believed to be randomly arranged and the agglomerations are more or less spherical in shape. When oil drops and clay particles are mechanically agitated as in the case of preparing oil mixed water suspensions of clay in the settling columns, the total energy of interaction between the double layers together with the forces due to particle and oil drop motion is responsible for bringing the oil drops and clay particles into sufficiently close proximity.

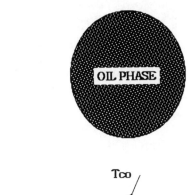

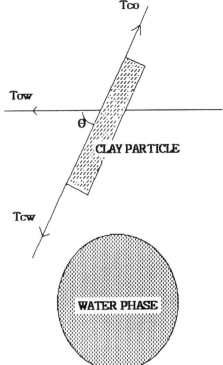

Figure 3. The mechanism of spherical agglomeration.

An interesting product of this process is observed in Figure 4, where inside of a spherical shell of clay is exposed when the oil drop was removed by some

means. The spherical agglomeration of the clay particles along with adhesion and coating of these particles on an oil drop surface is evidenced in this feature.

Figure 4. SEM micrograph of clay specimens (magnification: 1000X). S: Spherical agglomeration.

As a result of these mechanisms different interactions were observed. One is the clay particles coating the surfaces of oil drops as shown in Figure 5. The average diameter of these spherical drops ranged from a fraction of a micron to several microns.

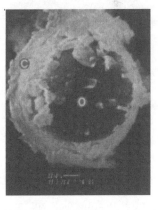

Figure 5. SEM micrograph of crude oil mixed clay specimens (magnification: 9000X).
O: Oil drop, C: Clay particles.

Second interaction is the adhesion of clay particles to oil layers. The coating of oil surfaces with clay particles is evident in the micrograph in Figure 6.

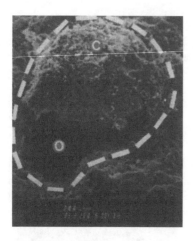

Figure 8. SEM micrograph of crude oil mixed clay specimens (magnification: 100X).
O: Oil surface, S: Spongy structure.

Figure 6. SEM micrograph of crude oil mixed clay specimens (magnification: 300X).
O: Oil surface, C: Clay surface.

4 CONCLUSIONS

Another mechanism is when the oil is incorporated in the sediment as an irregular shaped mass or a thin layer of material, as shown in Figure 7. There is adhesion between these oil surfaces and the clay particles surrounding them. Such oil surfaces act as blockhades and prevent water from flowing through the soil voids, therefore result in the net effect of lowered permeability.

The following conclusions can be drawn from this study.
1. Petroleum hydrocarbons reduces the plasticity of Na-bentonite and kaolinite clays significantly.
2. Petroleum hydrocarbons reduce the UCS of Na-bentonite clay. Crude oil and No. 6 fuel oil reduces the UCS of kaolinite clay. However, kerosene increases the UCS of kaolinite clay significantly.
3. Crude oil and fuel oil reduces the compression index of Na-bentonite clay slightly. However, kerosene increases the compression index of Na-bentonite clay. Petroleum hydrocarbons increase the compression index of kaolinite clay.

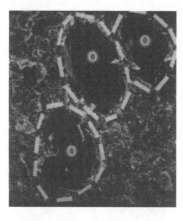

Figure 7. SEM micrograph of crude oil mixed clay specimens (magnification: 200X). O: Oil surface.

Crude oil mixed marine clay appeared to have "spongy" structure in general as observed in Figure 8. This is probably due to the agglomeration caused by oil. Finally, visual observation and handling of the crude oil mixed marine clay specimens indicated that crude oil reduced cohesion of the clay. most probably due to the agglomeration of clay into flocs and significant reduction of the effective specific surface area of the material.

REFERENCES

Baehr, A.L. 1984, Immiscible containment transport in soils with an emphasis on gasoline hydrocarbons, Ph.D., University of Delaware, p. 347.
Puddington, I.E. and Sparks, B.D. 1975, Spherical agglomeration processes, Mineral Science Engineering, Vol. 7, No. 3, pp. 282-288.
Tuncan, A. and Pamukcu, S. 1992, Predicted mechanisms of crude oil and marine clay interactions in salt water, Proc. of the Mediterranean Conference of Environmental Geotechnology, pp.109-121 Cesme, Turkey.
Tuncan, A., Tuncan, M. and Ozmen, E. 1993, The effect of petroleum hydrocarbons on geotechnical properties of kaolinite clay , 6 th National Clay Symposium, KIL'93, pp.197-207 Istanbul, Turkey.
Dragun, J., 1989, Recovery techniques and treatment technologies for petroleum and petroleum products in soil and groundwater, Petroleum Contaminated Soils, Vol.1, eds. Kostecki, P.T. and Calabrese, E.J., pp. 211-217 .

Geotechnical Hazards, Marić, Lisac & Szavits-Nossan (eds) © 1998 Taylor & Francis, ISBN 90 5410 957 2

Possibility for reassigning the mining stock-piles on the basis of the physical-mechanical characteristics

S.A.Živković, J.Nuić & D.Krasić
Faculty of Mining Geology and Petroleum Engineering, University of Zagreb, Croatia

ABSTRACT: The refuse stock-piles formed as a product of the coal exploitation represent an "ideal" place for laying off communal waste. Flexible mining technology enables depositing of toxic material as well if safety measures have been performed. The examinations of the physical-mechanical features of the discarded rock material and of hydrogeological conditions have shown that even larger masses of the waste might be deposited without endangering mining technology. With the technology adapted to the discarding of rock masses, the conditions for continuous laying off communal waste have been created. The example of the waste depot formed on the mining object is given by the brown coal mine Mostar, P.K. "Vihovići" where during the Native war an uncontrolled depot of waste occurred and which asks for an urgent sanation.

INTRODUCTION

A coal deposit in the immediate vicinity of the town of Mostar, lays in a series of lake sediments of the Tertiary Age. The series lays transgressively over massive limestones of the Cretaceous Age and is covered by quatenary formations.

From the beginning of the 20th century till the sixties, the exploitation of coal had been performed in underground pits and after that time till 1992, the coal was exploited on the surface pit P.K. "Vihovići" (note: P.K. = surface pit)

During a period of exploitation of the surface pit, 3500000 t of coal was obtained and 170000 m³ of solid waste mass was laid off on the stock-pile. The discarded masses take the area of 270000 m², with the thickness of 50-60 m. The result of this exploitation was a pit crater with the volume of 4300000 m³.

The war activities (1991-93) caused, in the beginning, an uncontrolled, without any preparatory works, discarding of the city waste in the pit crater, and then on the stock pile.

This extremely hard situation which the town of Mostar was confronted with (the existence of the waste depot in the city nucleus) demands urgent measures of sanation. As a first step, before sanation works have been started on the waste depot, or simultaneously with these works, the discarded waste (about 300000 m³) should be made harmless and then, following a plan, laid off on the location of the existing waste depot.

STRUCTURAL COMPOSITION OF THE DISCARDED ROCK MATERIAL ON THE STOCK-PILE

The sediments on the site of the P.K."Vihovići" may be divided into three groups:
  -solid (limestone basis) deposits;
  -plastic (bottom clay under coal layer) deposits;
  -elastoplastic (coal deposits with the refuse overlay) deposits.

The stock-pile is formed with elastoplastic layers consisting of:
  -sandstones, in the nearby roof of the coal layer interstratified with marl and coal;
  -black marl, penetrated with limonite slugs (inserts);
  -dark grey marl, penetrated with mangano-limonite slugs;
  -light grey marls, which surfaces (multi-layered) are filled with mangano-iron oxides of brown colour;
  -grey marls, which are the greatest in thickness, make the youngest level of the marl;
  -deposited matter of the alluvial detritus, as a product of superficial alluvions.

The refuse material was laid off according to the priorities, depending on the technological requirements of the coal exploitations, and makes a heterogeneous discarded mass of an irregular geometric form, unprepared for the drainage of surface waters and subjected to the erosion coming from the inclined slopes and cuts.

Fig. 1 Discarded masses on the stock-pile of the P.K. "Vihovići"

TECHNOLOGY OF DISCARDING THE MASSES

Destruction, i.e. mining, is used as a technology for obtaining the refuse overlay. A characteristic feature of this method is that rock (mined) material of heterogeneous composition and of various granulation is being obtained.

The transport of the masses was performed by trucks, and discarding by means of tilting down from higher terraces to the lower ones, at the height of 15-20 m.

By geomechanical examinations carried out on the waste pile there were achieved medium values of the parameters essentials for the calculation of the stability of the inclinations (the samples taken one year after the waste-depot was consolidated):

- cohesion $c = 10$ kN m^{-2}
- angle of the inner friction $\varphi = 39,0°$
- volume mass $V = 14,7$ kN m^{-3}

The calculation of the inclined slopes stability was done by means of the E. Hoek and J.W. Bray method for a circular cylindrical breaking surface passing through the slope foot, in a homogenous soil (rock material discarded ten years ago).

The investigations performed on the unsettled waste depot show a satisfactory stability of the inclinations, assuming that watering x=4H till x=∞. The height of discarded refuse masses, in the limits of 50-60 m, with the inclination slope of α≈30°, has a prescribed safety coefficient F=1,3, required for the permanent inclinations of the surface pits. The result has bean given in form of a diagram on depandance of the hight and inclination of the stock-pile slope for more water-filled conditions and for safety factors F=1,3.

(The above stated conclusion confirms a situation of the waste depot where there is no sliding of masses and where no secondary movements have been noticed.)

NEWLY RESHAPED WASTE DEPOT

Through variant solutions and computer modelling, taking into consideration the following criteria:
- fitting into the environment;
- maximal stability in the frame of possible techno-economical solutions;
- correct depositing of already discarded waste material;
- using the future area for economical necessities

the sanation-exploitation works have been proposed. Redistribution of the masses on the waste depot along with new masses (the exploitation of the most qualitative coal zones), makes possible a reshapment of the waste depot, formation of the plateaus and the sanation of the slopes to have stable inclinations which do exclude the sliding of the final and terrace slopes.

As a basic protection from the erosive movement of the material, a controlled drainage of surface waters should be carried out, as from the territory watered from the river basin, so also from precipitations which directly come to the waste depot.

For achieving satisfactory factors of the inclinations stability, the proposed parameters are as follows:

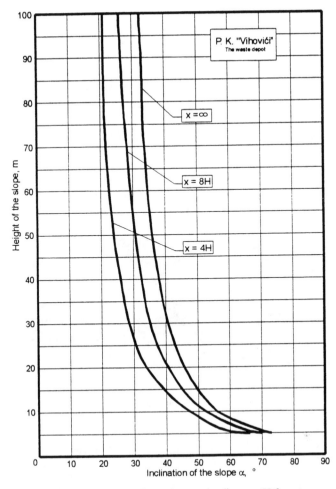

x - level of the water on the slope plane at the distance H from terrace margin.

Fig. 2   Diagram of the slopes stability on the waste depot of the P.K. "Vihovići"

-height of the waste depot terrace   15 m
-width of the terrace level   10 m
-final inclination of the waste depot slope   <30°
-inclination of the waste depot terrace slope   30°

well as the distance from underground waters (K40-K60) guarantee that laid off waste shall not have any influence on nearby or distant environment.

PLACING OF THE WASTE IN THE NEWLY RESHAPED WASTE-DEPOT

By means of reshaping the waste depot with the measures provided against the sliding of the discarded material and by a controlled drainage of the surface waters, there have been created conditions for storage of already uncontrolled discarded waste.

Water impermeability of discarded rock material as

TECHNOLOGICAL MEASURES FOR IMPROVING PHYSICAL-MECHANICAL FEATURES OF THE MATERIAL ON NEWLY RESHAPED WASTE DEPOT

Previous chapters present the results of geomechanical examinations performed on the working environment of the discarded waste material.

A given proposal asking that during war

471

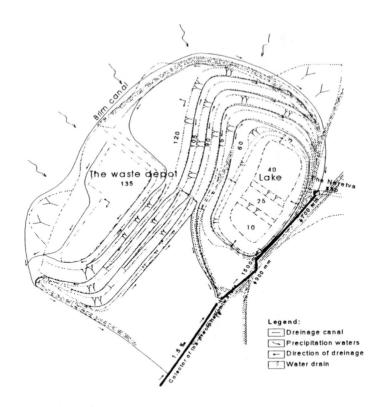

Fig.3 Newly reshaped waste depot

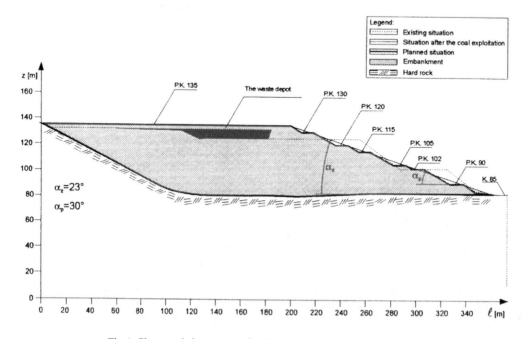

Fig.4 Characteristic cross-section through a newly reshaped waste depot

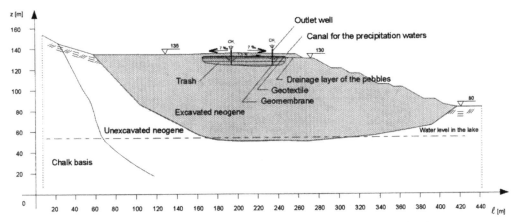

Fig.5 Characteristic cross-section through the waste depot located at the Kl35

activities the uncontrolled discarded waste in the pit crater and on the waste depot should be definitively placed on the existing location, demanded additional safety measures for stabilising the working environment. These measures would be as following:

-discarding of new masses as well as redistribution of already discarded masses to be effected in the layers of 0.5 m with a recurrent forcing down;

-to reduce inclination of the slopes on the proposed declivity;

-to fill all the holes and to level surfaces with the inclination towards drainage canals;

-to sanify biologically the terrace slopes and plateaus with cultures corresponding to climatic conditions.

CONCLUSION

The war activities interrupted the work of the mine and brought to the inundation of the pit crater. Discarding the communal, industrial and hospital refuses into the pit crater and on the unprepared waste depot, represents, and it may be said with certainty, an ecological bomb.

The investigations carried out with the aim to sanify the resulting circumstances, show that there is a possibility to discarding the uncontrolled laid off waste on the site but under strictly defined preliminary actions and work technology.

The stated proposal depends, in many aspects, on physical-mechanical features of the working environment. Elastoplastic discarded rock material is suitable to be reshaped, it is stable on the terraces' slopes, may be pressed and it is poorly impermeable to water. Thus it is possible to form in it, under certain conditions, a waste depot of a definite volume, without being nocuous or dangerous to the environment.

REFERENCES

Kaiser,W. (1993.): Rekultivierung von Berg und Tagebauflächen. Stuttgart. Deutscheland.

Koprić,F. (1972.): The analysis of stability of the external refuse stock-pile on the surface pit "Vihovići". The brown coal mine Mostar. Mostar. B&H.

Nuić, J., Živković,S., Grabowsky,K. et al (1996.): A study of sanation, exploitation and recultivation of the surface pit "Vihovići" and of the reorganisation of the Mostar mine. EU-administration Mostar, Zagreb.

Nuić,J., Živković,S., Krasić,D. et al. (1997.): Sanation of the mining sites near urbane areas. 27th International Conference on Safety in Mines Research Institutes, New Delhi, India.

Olschovy,G. (1993.): Rekultivierung durch Landschaftspflege und Landschafsplanung. Bergbau und Landschaft. Hamburg-Berlin. Deutscheland.

Stević,M. (1991.): The mechanics of the ground and rocks. The faculty of geology and mining Tuzla. Tuzla. B&H.

Sunarić,D. (1976.): The elaboration of geomechanic conditions for the coal exploitation in the north-west part of the surface pit "Vihovići". The faculty of geology and mining of the University Belgrade. Belgrade. FR Yugoslavia.

Šalović,M. (1983.): The elaboration of geomechanic examinations in the area of surface pit "Vihovići". The institute for mining researches Tuzla. Tuzla. B&H.

Foundations and soil improvement

Geotechnical Hazards, Marić, Lisac & Szavits-Nossan (eds) © 1998 Taylor & Francis, ISBN 90 5410 957 2

Shallow foundations – Experimental study under cyclic loading

S. Amar
Laboratoire Central des Ponts et Chaussées, Paris, France

F. Baguelin
Terrasol, Montreuil, France

Y. Canepa
Laboratoire régional de l'Est Parisien, Melun, France

ABSTRACT: This paper provides a detailed description of cyclic loading tests performed on a large square shallow foundation (1 m x 1 m) installed on two natural soils (a silt and a sand). The main results are detailed and compared with those obtained from static loading tests. Practical informations to estimate the behaviour of shallow foundations under cyclic loading are given.

1. INTRODUCTION

In order to update and validate the French foundation design rules, the Laboratoires des Ponts et Chaussées, have undertaken a major programme of research dealing with shallow foundations. (Amar et al, 1987, Canepa et al, 1990). This programme consisted in particular of on-site loading tests of large foundations.

118 tests were performed. Most of them were conducted to failure, in order to examine the influence of different parameters on the bearing capacity of soils (for example embedment of foundations, eccentric or inclined loading, foundations at the crest of slopes, etc.). These failure tests were performed by gradually increasing the load in step. Some series of tests were carried out in order to examine long term creep and the behaviour of a foundation under cyclic loading.

It is actually important, in the context of new limit state justification rules (Fascicule 62, 1993, Eurocode.7, 1996), to distinguish the behaviour of foundation between different types of loading and to have a good understanding of the effect of the type of

loading for some structures which are subjected to very severe conditions, for example because of wave action (Laue, 1997).

This paper describes the 2 series of cyclic loading tests conducted on silt and on sand. Table 1 summarizes the main geotechnical parameters of the two experimental sites.

2. DESIGN OF THE TEST SET-UP

Figure 1 shows the experimental set-up used for these tests.

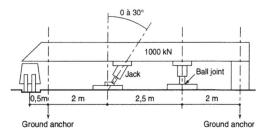

Figure 1. Design of the experimental set-up

Table 1. Summary of geotechnical results

Site	Soil type	w (%)	W_L	I_P	γ_d (kN m³)	c_u (kPa)	φ_u (°)	c' (kPa)	φ (°)	p_l (kPa)	E_M (kPa)	q_c (kPa)	q_d (kPa)	N SPT
Jossigny	Silt	20-25	38	14	16	38	0	12	32	500	6200	1200	1800	8
Labenne	Fine sand	4-5	-	-	16	-	-	0	32	910	7800	4000	3800	15

The reaction device consisted of a beam held by very deeply bonded ground anchors.

The load was applied by a jack. The device could exert a maximum force of 1 000 kN.

For the cyclic loading tests, the hydraulic power unit was controlled by a microcomputer which was also responsible for the real time acquisition and processing of measurements.

A diagram of the device is shown in figure 2.

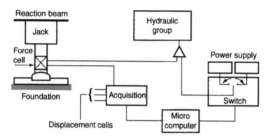

Figure 2. Diagram of the control device for cyclic tests

3. THE PROGRAM TESTS

The following tests were performed at each site in order to study the effect of loading cycles:

- A reference static load test carried out to failure

- Four or five cyclic tests.

Figure 3 and 4 show the locations of the different tests carried out on the two expérimental sites.

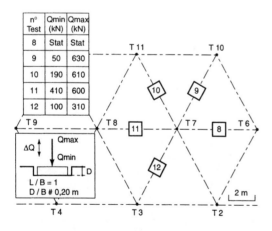

Figure 3. Cyclic tests at the Labenne site.

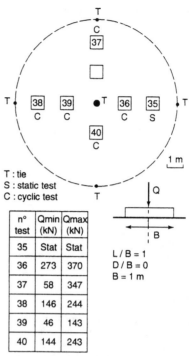

T : tie
S : static test
C : cyclic test

n° test	Qmin (kN)	Qmax (kN)
35	Stat	Stat
36	273	370
37	58	347
38	146	244
39	46	143
40	144	243

L / B = 1
D / B = 0
B = 1 m

Figure 4. Cyclic tests at the Jossigny site.

4. LOADING PROCEDURE

The same procedure was used for all the cyclic loading tests. Figure 5 shows the loading procedure and the notations used. It can be seen that all the tests began with static loading before the cycles. Table 2 recapitulates the characteristics of each cyclic test.

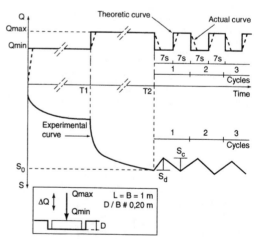

Figure 5. Loading procedure for "cyclic" tests.

478

Table 2. Characteristics of the cyclic tests

Test	Q_{min} (kN)	Q_{max} (kN)	Qr (kN)	ΔQ (kN)	$\dfrac{\Delta Q}{Q_{max}}$	$\dfrac{Q_{max}}{Qr}$	$\dfrac{\Delta Q}{Qr}$	No. of cycles	T1 (min)	T2 (min)
Jossigny	Qr = 411 kN (reference static test N° 35)									
36	273	370		97	0.26	0.90	0.24	1131	10	20
37	58	347		289	0.83	0.84	0.70	557	10	20
38	146	244		98	0.40	0.59	0.24	10003	10	20
39	46	143		97	0.68	0.35	0.24	1501	10	20
40	144	243		99	0.41	0.59	0.24	1224	30	60
Labenne	Qr = 869 kN - (reference static test N° 8)									
9	44	625		581	0.93	0.72	0.67	1079	30	60
10	191	605		414	0.68	0.70	0.48	498	30	60
11	406	601		195	0.32	0.69	0.22	1101	2	
12	104	309		205	0.66	0.36	0.24	7537	165	195

The loading procedure for the static tests (figure 6) was conventional, i.e.:

- loads applied by steps,

- equal increments of loading,

- duration of a loading step: 30 minutes

5. EXPERIMENTAL RESULTS

5.1 Behaviour under static loading.

Figures 7 and 8 show typical behaviours during a static load test. Figure 7 shows the raw loading curve for test 35 at Jossigny. Conventionally, the static limit load Q_r is that which gives rise to a settlement s of 10 cm (= 10 %B). Its value in this case was 411 kN.

Figure 8 shows the change in settlement under constant loading during a static test. Settlement against the log of time produces an almost linear plot. The creep is given by the gradient As of these stabilization lines.

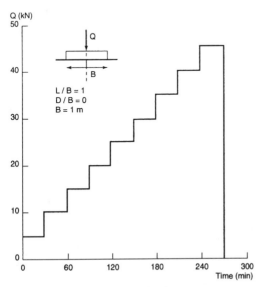

Figure 6. Loading procedure for "static" test N° 35 (Jossigny)

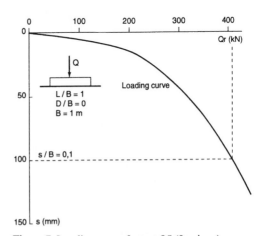

Figure 7. Loading curve for test 35 (Jossigny)

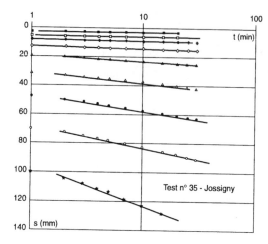

Figure 8. Typical stabilization lines

Figures 9 and 10 compare for the Jossigny and Labenne sites respectively change in the gradients (As) of the stabilization lines obtained during the reference static test and during the static loadings performed at the beginning of each cyclic test.

It can be seen that for a particular site there is no significant difference in the measured "static" creeps.

5.2 Behaviour under cyclic loading

5.2.1 Observed phenomenons

Figure 11 shows the raw results from 3 tests on the Jossigny site (silt). The graphs show, for each test respectively, the settlement obtained for Q_{max} and Q_{min} versus the number of cycles. Generally, the settlement of the foundation increases with the number of cycles and for a particular value of ΔQ the rate of settlement is greater the higher the value of Q_{max}.

5.2.2 Analyse of "cyclic creep"

When the settlement of foundations is plotted versus the log of the number of cycles (figure 15), the measured settlement after about 100 loading cycles is practically distributed along a straight line. The gradient Ac of this straight line has been used to characterize the "cyclic creep".

Table 3 shows the mean values of Ac which were measured in the course of different tests and Figures 12, 13 et 14 indicate respectively the influence of Q_{max}/Q_r, $\Delta Q/Q_r$ and $\Delta Q/Q_{max}$ on the values of Ac.

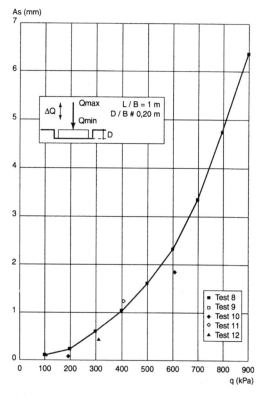

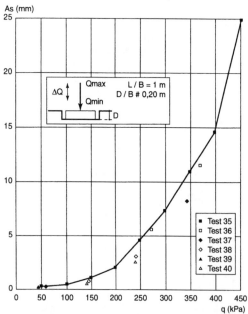

Figure 9. "Static" creeps measured on the Jossigny site

Figure 10. "Static" creeps measured on the Labenne site.

Table 3. Static and cyclic creep

Test	As	stb	sta	tb	ta	Ac	As/Ac
		mm	mm	min	min		
Jossigny							
36	12.5	83.5	64	282	19.9	16.96	1.36
37	11	122.4	64.3	133	28.9	87.68	7.97
38	4.33	38.6	21.4	2391	13.8	7.72	1.78
39	1.04	11.2	9.3	357	27.0	1.69	1.62
40	4.32	24.2	18.4	300	21.1	5.02	1.16
Labenne							
9	2.54	93.1	50.5	244	12.4	32.84	12.93
10	2.34	63.1	41.2	126	11.3	20.88	8.92
11	2.29	57.6	50.3	257	51.2	10.36	4.52
12	0.63	15.5	13.5	1925	11.5	0.90	1.43

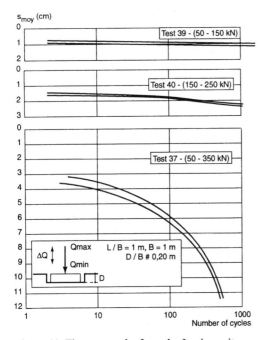

Figure 11. The raw results from the Jossigny site.

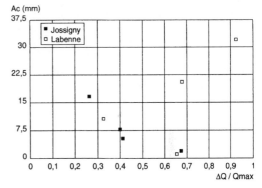

Figure 12. Ac versus $\Delta Q/Q_{max}$.

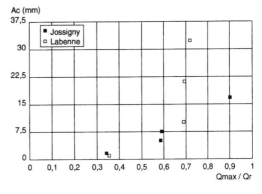

Figure 13. Ac versus to Q_{max}/Q_r.

In spite of the dispersion of the results, the following trends can be detected (see table 3 and figures 12 and 13):
- Ac increases with dQ when the maximum load Q_{max} is constant (see test 36 and 37 or test 9, 10 and 11).
- Ac is little altered by Q_{max} when the amplitude of

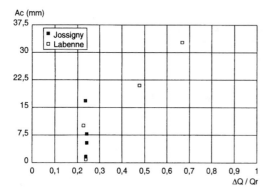

Figure 14. Ac versus $\Delta Q/Q_r$.

5.3.2 Comparison between static creep As and cyclic creep Ac.

Figures 16, 17 and 18 show the influence of $\Delta Q/Q_{max}$, Q_{max}/Q_r and $\Delta Q/Q_r$ on the Ac/As ratios. The trends observed during investigation of Ac seem to be confirmed. On the basis of these results

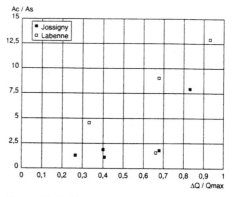

Figure 16. Ac/As versus $\Delta Q/Q_{max}$.

the loading cycle ΔQ is constant and $Q_{max} < 0.5\ Q_r$ (see tests 38 and 39).

5.3 Comparison with static creeps

5.3.1 Effect of cycles

When the settlements obtained during a cyclic loading test are compared with those from static loading with an intensity and duration equivalent to the maximum cyclic load (see Figure 15) the deleterious effect of the cycles is clearly apparent.

This result applies for all the cyclic tests and the differences between the two curves increase the greater the ΔQ of the cycles and the nearer the maximum load Q_{max} for the cycle was to the static limit load Q_r.

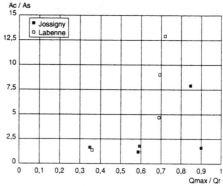

Figure 17. Ac/As versus Q_{max}/Q_r.

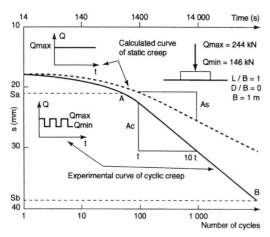

Figure 15. Effect of cycles. Test N° 38 (Jossigny).

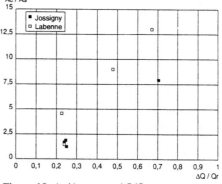

Figure 18. Ac/As versus $\Delta Q/Q_r$.

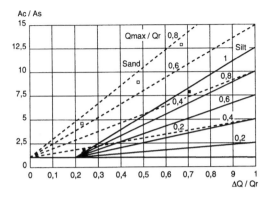

Figure 19. Chart for determining Ac/As

realisés sur sites par les LPC (1978-1990), *Rapport interne LCPC*, 1.17.02.0

Eurocode 7 (1996), Partie I : Calcul géotechnique, projet de norme européenne ENV 1997-1

Fascicule 62 - Titre V (1993), Règles techniques de conception et de calcul des fondations des ouvrages de génie civil. *Cahier des clauses techniques générales applicables aux marchés publics des travaux*, Ministère de l'équipement.

Laue J. (1997), Settlements of shallow foundations subjected to combined static and repeated loadings. Proceedings of the Fourteeth International Conference on Soil Mechanics and Foundation Engeeniring, Hambourg.

we have proposed the chart in Figure 19 for the silts and sands we tested.

This chart can directly provide values for Ac/As according to $\Delta Q/Q_r$ and Q_{max}/Q_r.

CONCLUSIONS

The following can be learnt from this study:

- for the tested soils (silt, sand) for the same level Q_{max} of loading, cyclic loading is more deleterious than a static load,

- this adverse effect increases the greater the amplitude ΔQ of the cycles and the nearer the maximum load Q of the cycles is to the static limit load Q_r,

- the effect of cycles can be estimated in practice, on the one hand, from the characteristics of the cycle (ΔQ, Q_{max} and the period of cycles) and, on the other hand, from the behaviour of the foundation when subjected to static loads (Q_r and A_s).

Lastly, although this study has provided some data it is a long way from covering all soils and all the parameters which can play a part in the problem (frequency of cycles etc.). For this reason further research, which will also deal with a wider range of soils, is required in order to gain a better understanding of the behaviour of foundations under this type of loading.

REFERENCES

Amar S., Baguelin F., Canépa Y. (1987), Comportement des fondations superficielles sous charges cycliques, *Bulletin de liaison des LPC*, 149, p. 23-28.

Canépa Y., Depresles D. (1990), Catalogue des essais de chargement de fondations superficielles

Geotechnical Hazards, Marić, Lisac & Szavits-Nossan (eds) © 1998 Taylor & Francis, ISBN 90 5410 957 2

The use of the jet-grouting method for the stabilization of the underground

L. Ballarin & F. Forti
Geokarst Engineering S.r.L., Area Science Park, Trieste, Italy

ABSTRACT: This work analyzes the jet-grouting technique, identifying the main technologies and their fields of application. The study and the analysis of the machineries employed and the realization of the jet-grouting technique as well as the general organization of a site are particularly important.

1 INTRODUCTION

This work analyzes the jet-grouting technique through the different fields of usage, different methodology, and in particular examines the various equipments employed in sites for the potential application of the system.

2 TECHNIQUE AND OPERATIVE FIELDS

The jet-grouting technique was born approximately at the end of the 50s and beginning of the 60s, when some Pakistan companies used it for the stabilization of the ground in some civil works. However, between the end of the 60s and the beginning of the 70s, the Japanese gave a clear impulse to this methodology, developing it and widening its range of action in the various fields of special foundations.

Today this technique has asserted itself also in many European countries and in Italy in particular, with numerous successful applications. The jet-grouting technique has been recently exported also to the countries of East Europe, as Croatia and Slovenia, always obtaining very good results.

The jet-grouting finds its best field of employment in the stabilization of soils, or in lithoid grounds with characteristics similar to soils (mylonitizated areas, soft limestones, broken up marls).

Before starting the real jet-grouting work, a series of preparatory works as a careful investigation of the area of intervention, with the carrying out of a geological survey together with a series of drillings with continuous coring, with the extraction of undisturbed samples to determine those parameteres

able to qualify the ground from a geotechnical point of view (grain size distribution, shear resistance, uniaxial compression and free expansion, water content, Atterberg limits) as well as tests in hole (SPT, permeability) must be realized. These parameters are necessary to calculate the type of jet-grouting to be carried out.

At present three methods are mostly used and are the monofluid system, that with 2 fluids and that with 3 fluids.

In the monofluid system the first phase of the work envisages the drilling of a borehole, with distruction of the core, with rotation or roto-percussion, with or without lining, according to the type of ground and to its degree of cohesion, up to the depth established by the project. In the second phase the drill rods are extracted at a speed comprehended between 30-50 cm/min., and at a rotary speed comprehended 10-20 rev./min. with intervals every 3-8 cm and the contemporary injection of a cement mixture at very high pressure, from 300 to 800 bars, with an average flow of 300 l/min., which acts both as breaking up and binding element. The ratio concrete/water can vary according to the ground to be treated and must be always comprehended within 1:1 to 1.5:1. Therefore, columns of stabilized ground, which may have variable diameters from 30 to 100 cm, are obtained. The different diameters are obtained operating on different times of reascent of the set of drill strings, on the rotary speed, on the diameters and the number of nozzles and on the pressure of injection. As for the calculation of the range of action the following formula may be used:

$Ra = f.(P.t.To.\mu u.A)$, where Ra = range of action, P = pressure of injection, t = time of reascent, μu =

diameters of nozzles and A = density of the boiacca of concrete. Before starting the jetting work, in an adjacent field with the same lithological and stratigraphical characteristics, some test columns, made with different parameters calculated during the project, must be carried out. After the hardening of cement (2-3 days), the whole ground around the columns must be excavated, for a depth of 2-4 m, the average diameter which has been obtained must be measured and compared with the calculated data; then, whether some corrections are to be made must be considered. In order to be able to verify the integrity of the column, also in depth, some coring with double tube core barrel and, within the hole, some tests of permeability with parker may be carried out; or by the echo method or sonic coring, some log, also along the hole made during the coring, may be carried out. Moreover, when the columns are used as supporting elements, some load tests analogous to those carried out on the injected piles are necessary.

Some values of the universal compression and free expansion on the samples of treated ground are reported below:
Clays - Clayey silts 12 - 40 kg/cm^2;
Silts - Sands 30 - 120 kg/cm^2;
Sands - Gravels 60 - 200 kg/cm^2.

The columns obtained have high values of the uniaxial compression with free expansion, whereas they present scarce rates in the tangential stresses. In order to further improve the resistance of the treated ground, once the phase of injection is over, a scaffold or an iron pipe, which further increases the values of resistance, is inserted by the method of vibration.

Any soil can be treated varying the parameters previously described; as grounds with presence of aquifer and percolating movements up to 10-2 - 10-1cm/sec., where, using particular mixture, the hardening of cement is obtained in a few minutes.

The obtained columns may never be compared with an injected pile, but must be always considered as stabilized ground.

The 2-fluid system gives better performance with respect to the monofluid system and is used in more cohesive undergrounds, where superior diameters of treated ground are to be obtained.

The operativness is similar to the one of the monofluid system as far as the drilling is concerned, while, during the phase of injection, besides the cement injected at high pressure, there is also a nozzle for the exit of compressed air at 12 bars which favours the breaking up process of the ground. In this way some columns with diameters of 80-140

cm may be obtained.

In the grounds with high cohesive values or for columnar treatments which require big diameters, the 3-fluid jetting system is used. The carrying out of the first phase is analogous to the systems previously analyzed; in the reascending and injection phases, instead, are a high pressure jet of water (400-500 bars), and a flow of at least 300 l/min., which penetrates into the mass of ground, breaking it up; a second jet of compressed air at 12 bars completes the process of separation, acting on finer materials and finally, the third jet, with the injection of the boiacca of concrete at a pressure of 50-70 bars and a flow of about 1000 l/min, penetrates the mass of ground, by now deprived from the finest component, impregnating it in a uniform way. By this system some columns, whose diameter may reach 2 m, are obtained.

The executive schemes may be carried out in the most different ways according to the type of work to be realized.

A fundamentally important point, for a good result of the columnar treatment, is that, once a column has been completed, in no case a new column adjacent the one just finished must be carried out, since the high pressure of injection would completely break up the column previously realized, destroying the work done and, therefore, also the mechanical resistance. Before carrying out a new column, a complete hardening of the cement must be waited for, not less than 2-3 days. The engineer, the head of the site or the expert in jetting have to organize the work in the site, programming and planning the displacements, in order to obtain the best results with the least consumption of time and energies.

The fields of application of the jet-grouting technique may be the most various, both for temporary and definitive works. However, the immagination and the expertise of the planner, in collaboration with the geologist for the advice on the type of ground, give the best solutions for the employment of this innovative technique. This system, thanks to its great verastility, will be able to be operative in further new fields in the future, since this techinque, still young, may be further on developed and applicated in many fields of civil engineering and particularly for the use of special foundations.

3 EQUIPMENTS

The first plants and machines for the works of jet-grouting have come from the transformation and the

486

adaptation, made by the technicians of each company, of machineries designed and constructed for other purposes. Today the major construction companies have, in their catalogues, complete equipments for the carrying out of such works. In spite of the vast range of equipments, still today the transformations which enable to operate in particular working situations are frequent.

A complete plant for the monofluid jet-grouting is made of a silos for the storage of cement, a mixer and an agitator for the preparation of cement, a high pressure pump and a drilling rig properly transformed for the works of jetting.

All the single equipments are analyzed below.

The silos has a capacity of 25 m³, below it a screw conveyor is coupled to send the cement to the mixer and agiatator, mounted on slide, where the mixture cement/water is prepared and sent to the high pressure pump. These plants are available containerized and are completely automatic both for the type of mixture which is to be used, and for the quantity of cement necessary for the injection of the single pile. The productive capacity must not be inferior to 500 l/min. The automatization allows a strong reduction of hand work limited only to the control and to the works of daily cleaning and maintenance.

The high pressure jetting cement pump is made of a diesel engine from 350 to 500 HP to which a 2 or more speed gear box is coupled. By a Cardan shaft and a chain trasmission, the movement is transferred to a triplex pump with plunging pistons, with a lubricator for pistons. The whole is mounted on a slide- like- chassis for easy transport and positioning in the site. The pump complex may be containerized in soundproofing cabins, which make the personeel's work less hard.

The triplex pumps are built with pistons of different diameters and combined to a different rotary speed (blow frequency per minute). They give different pressure and working flows (pressure 300-800 bars, flow from 40 to 700 l/min.). All the modern pumps are equipped with an automatized control panel for the determination of the number of blows and pressure, and are provided with safety devices in the case of overpressures, which automatically stop the movement in order to avoid damages or explosions of pipes or other parts.

A particular attention must be paid to the periodic maintenance of the triplex pump. Since the pistons work in contact with a very abrasive material, as cement is, all their seals must be periodically changed, in order to have always the maximum efficiency and to avoid damages to mechanical parts.

The valves and the valve holders and relative seals must be treated in an analogous way. It is always opportune to have in the site a large supply of these parts in order not to face a stoppage of the work.

Anyhow, all the pumps recently built have revision intervals of about 500 - 700 hours of work.

The connection for the sending of cement between the pump and the drilling rig, occurs through a high pressure pipe with a diameter 1" or 1 1/2", with a SAE 100 R9 or superior quality rubber pipe with the point of explosion of at least 1600 bars; it must be placed so as not to make curves with a radius inferior to 300 - 400 m, in order to avoid explosions or damages.

As for the two-fluid system an air compressor with a capacity of 6000 l with a working pressure of 12 bars is also necessary. It is opportune to couple it to a tank, having no return valves, in order to have always an air reservoir available.

As for the three-fluid system, besides the air compressor and the high pressure pump, which is used for the injection of water at 400 - 600 bars, a pump for cementized injections must be available, with a low pressure of 50 - 70 bars, but with a flow of at least 1000 l/min.

As for the drilling rigs, the constructors have generally oriented themselves in the usage of machines for surveys and micropiles, properly transforming the mast and the rotary head, as well as equipping them with some necessary accessories for the jetting, as the timer for the programmed reascent of the rods, and a high pressure manometer, placed on the delivery pipe of cement, possibly near the control panel, so as the operator can check the pressure during the phases of injection and, therefore, verify the continuity of arrival of cement.

The machines on crawler are always preferred because, for the small displacements to be made in the sites, often on very bad grounds, this type of machines offers very good capacities of movement.

The power of a diesel or electric engine must be comprehended between 50 and 150 HP in order to be able to have a torque of 400-1000 kg/m at the rotary head.

The mast of the drilling rig can be placed in any position, from vertical to horizontal, with lateral displacement on two sides or rotate by 360° around a turntbale, this for a higher verstility of use. The mast must be properly streched with an extension which supports and drives the whole continue set of drill rods. The rotary head must be a hydraulic drill string chuck; the clamps of the drill string chuck are of the exact dimension of the type of rods employed.

There are also special rigs for works in gallery. They

are made of an undercarriage mounted on crawlers on which an electric or diesel engine, the oleodynamic part and the relative tanks are placed. On the drilling rig the mast is positioned and is horizontally supported by some iron beams and oleodynamic pistons which enable both the inclined movement with respect to the horizontal and the rotation along the longitudinal axis of the machine so that it can cover the whole arch of the gallery. It is used in works of pre-stabilization and stabilization inside tunnels.

The rods for the monofluid system must have an external diameter of about 60 mm, especially if machines with a high rotation torque are used.

At the beginning of the set of rods is the tool, a roller or drag bit with steps, chosen on the base of the type of ground to be drilled. It is screwed at the monitor or nozzle holder, through a connection of the same diameter of the rods. On the monitor the two nozzles with the same diameters of the holes, which can vary from 1.2 to 3 mm, are screwed. During the phase of injection, not to let the cement go out through the tool, once the drilling phase is over, a metal sphere, falling in a seat placed between the connection of the tool and the monitor, is placed. The high pressure of injection guarantees the setting of the sphere in the seat, doing so, the cement goes out through the nozzles and not through the tool. The couplings between the rods must be kept very cleaned and greased, because a minimum leak of cement would go and "cut" the thread irreparably. On the top of the set of rods the rotary swivel is screwed, on which, in its turn, a high pressure pipe for the sending of the cement mixture coming from the high pressure pump is screwed.

Once the pit is drilled, the whole set of rods must be washed by very high pressure water (400 - 600 bars) so that any trace of cement present inside the rods is sent out. A particular care must be taken during the removal of the nozzles for their cleaning or replacemnet, paying attention to the pressure inside the rods which must be zero, otherwise damages to the operator may occur.

In the case of the 2 or 3-fluid systems the diameters of rods are bigger, respectively of 70 and 90 mm, because inside them there are 2 or 3 different passages for each single fluid. The monitor is equipped with more nozzles, one for fluid; an analogous scheme for the rotary swivel, on which two or three pipes are screwed, is necessary.

As for the organization of a site for jet-grouting, almost 4 people have to be employed and they are distributed as follows: one person for the cementation plant, one for the high pressure pump, an operator for the drilling rig and his assistant. A very important point is that of the positioning of the various equipments as fixed plants (silos, mixer - agitator and cement pump) placed so as not to obstacolate drilling operations, but to be easily reached by trucks for cement supply and in a place such as to enable to stretch the pipe sending cement without making too many curves, dangerous for the resistance of the pipe itself.

4 SUMMARY

The jet-grouting technique or columnar treatment has been developed since the end of the 50s and the beginning of the 60s and it greatly developed in some Japanese companies. In Europe and in Italy some working methodologies, which have been able to enlarge the field of application of this innovative technique, have been further on developed. This method consists in the stabilization of the ground by an injection of high pressure cement.

The most used techniques are: that with one fluid, an injection of only high pressure cement, that with two fluids, an injection of compressed air at 12 bars and one of high pressure cement (400 bars) in order to obtain diameters of about 1 m also in very cohesive grounds, and that with three fluids, an injection of high pressure water at 400 bars, one of air at 12 bars and finally an injection of low pressure cement (30-40 bars), in order to obtain columnar treatments with diameters of more than 1 m.

The machineries employed are: a drilling rig transformed for jet-grouting works, a cement pump for high pressure injections and, finally, a mixing plant made up of a silos for the storage of cement connected to a mixer and an agitator. If the two or three-fluid systems are used, a big compressor of 21,000 l and a working pressure of 12 bars must be available.

In order to carry out the jet-grouting correctly, various parameters must be carefully considered in order to be able to calculate in advance both the diameter of the column to be obtained and the quantity of cement to be injected in the underground. Besides, the arrangement of columns must be carefully studied, considering the type of foundations to be obatined and their realization.
The general organization of the whole site is of fundamental importance in order to be able to carry out the work without problems.

Geotechnical Hazards, Marić, Lisac & Szavits-Nossan (eds) © 1998 Taylor & Francis, ISBN 90 5410 957 2

Ein vermeintlicher Schadensfall bei Unterfangung mittels HDI-Säulen

P. Bilz
Dresden, Germany

ABSTRACT: Sections of a representative old building, partly damaged by war, situated in direct neigbourhood of a construction pit for a new building, must have secured by underpinning. This measure was carried out by means of jet grouting. After uncovering the excavation lining heavy cracks developed in a short time, pointing out to lacks of quality. However, detailed investigations demonstrated that it was solely a question of shrinkage which up to now were not observed in such an extent.

1 EINLEITUNG

Das Düsenstrahlverfahren zur Bodenverfestigung geht im Ursprung auf britische und japanische Anwendungen in den 60er Jahren zurück. In Deutschland kam das Verfahren erstmals Ende der 70er Jahre zur baupraktischen Ausführung. Die Anwendungspalette ist inzwischen breit gefächert (vgl. u.a. Hilmer und Rizkallah 1990; Schnell und Vahland 1997).

Während bei der klassischen Injektion im Ingenieurbau stets das Einpressen eines die Verfestigung bewirkenden (mehr oder weniger flüssigen) Mittels in vorhandenen Hohlräumen (Poren, Spalten, Klüfte) des Untergrundes bei Drücken bis 20 bar erfolgt, wird bei der Anwendung der Düsenstrahltechnik mit Drücken bis 800 bar die vorhandene Bodenstruktur zerstört, ehe anschließend ein Vermörtelungseffekt zum Tragen kommt.

Dieses Verfahren wurde von verschiedenen Herstellern mit unterschiedlichen Bezeichnungen belegt (Kluckert 1996):
- Hochdruck (Düsenstrahl)Injektion (HDI)
- Soilcrete
- Jet grouting
- Rodinjet

Auf eine ausführliche Beschreibung des Verfahrens wird bewußt verzichtet, die angegebene (und dort weitere) Literatur liefert ein umfassendes Bild. Nach Kluckert (1996) werden folgend vier Verfahrensvarianten unterschieden:

Verfahren 1   Hochdruckschneiden mit Zementsuspension

Verfahren 2   Hochdruckschneiden mit Zementsuspension + Luftummantelung des Schneidstrahls

Verfahren 3   Hochdruckschneiden mit Wasser + Niederdruck-Verfüllen mit Zementsuspension

Verfahren 4   Hochdruckschneiden mit Wasser + Luftummantelung des Schneidstrahls + Niederdruck-Verfüllen mit Zementsuspension

2 URSACHEN FÜR QUALITÄTSMÄNGEL UND SCHÄDEN

In seinem Beitrag zur Baugrundtagung 1996 in Berlin hat Kluckert in sehr eindrucksvoller Art und Weise acht herstellungsbedingte Fehlergruppen beschrieben, die als Qualitätsmängel zu bezeichnen sind und ihrerseits zur Beeinträchtigung der Funktion der betreffenden Baumaßnahme (Baugrubenumschließung, Gründungsveränderung bzw. -sanierung, Abdichtung u.a.) oder zu Schädigungen an Bestandsbauten führen können. In gestraffter Form wird zu den ursachen zitiert. Planungs- bzw. Entwurfsmängel sind hier nicht behandelt.

2.1 Der Säulendurchmesser ist zu klein

Das bedeutet, die Säulen
- können im Durchmesser über die ganze Länge zu klein sein

- können Einschnürungen haben, d.h. der ∅ ist partiell zu klein
- können durch Düsschatten Fehlstellen haben, die durch Steine oder vorher verfestigte Zonen entstehen

Durch zu kleine Säulen entstehen im HDI-Körper Fehlstellen.

- Bei Unterfangungen und Lastabschirmungen sind kleinere Fehlstellen ungefährlich,
- bei größeren Fehlstellen kann dies sehr gefährlich sein, sie sind daher zu sanieren

Bei druckwasserhaltenden HDI-Körpern

- hat in der Regel schon eine einzige Fehlstelle katastrophale Folgen, dies gilt besonders für rollige und fließgefährdete Böden. Diese Böden fließen in der Regel schlagartig durch die Fehlstellen, wenn diese vorher nicht bemerkt und saniert werden. Die Gefahr des Einfließens von Böden wird sehr oft unterschätzt.

2.2 Der Säulendurchmesser ist zu groß

Wenn die Primärsäulen so groß werden, daß sie den Ansatz der Sekundärsäulen (Schließer) überdecken, entstehen Düsschatten und somit Undichtigkeit in der Wand oder der Sohle.

2.3 Die Säulen sind oben oder unten zu kurz

Wenn der Düsvorgang durch einen Meßfehler zu hoch begonnen wird, werden die Säulen unten zu kurz. Wenn der Düsvorgang zu früh gestoppt wird, dann werden die Säulen oben zu kurz. Und wenn die vermutete Fundamentunterkante nach oben vorspringt, ohne daß dies bemerkt wird, werden die Säulen ebenfalls oben zu kurz. Auch durch Versickern der Suspension in den Porenraum rolliger Böden kann der Suspensionsspiegel soweit absinken, daß oben ein Stück der Säule fehlt. Meßfehler können nur durch sorgfältiges Arbeiten und Einsatz von qualifiziertem Personal verhindert werden. Das Absinken der Suspension muß durch Kontrolle und ständiges Nachfüllen mit Zementsuspension verhindert werden.

2.4 Die Säulenachsen weichen von der Sollneigung ab

Die Abweichungen können durch

- Meßfehler entstehen
- oder durch Abweichungen des Bohrgestänges aus der Sollachse. Dies geschieht, wenn das Gestänge zu dünn ist oder das Bohrgestänge an Bohrhindernissen abweicht.

2.5 Abweichungen der Ansatzpunkte im Grundriß

Durch Fehler beim Einmessen oder wenn die tatsächliche Höhenkote des Bohrplanums von der geplanten Höhenkote abweicht, verschiebt sich der Bohransatzpunkt relativ zur Fundamentvorderkante und somit auch die Säule, die das Fundament tragen soll. Die Säulen ragen einerseits zuweit in die Baugrube hinein und müssen kostspielig abgestemmt werden. Andererseits fehlt auf der Rückseite der Säule ein Stück des statisch erforderlichen Unterfangungskörpers. Der umgekehrte Fall ist auch möglich.

2.6 Der HDI-Körper hat zu geringe Festigkeit

Dieser Fehler kann folgende gründe haben:

- zu hoher W/Z-Faktor schon beim Anmischen
- zu hoher resultierender W/Z-Faktor durch falsches Verhältnis von Schneidwasser zu Suspensionsmenge bei Verfahren 3 und 4
- zu schnelle Ziehzeiten in bindigen Böden
- chemische Einflüsse durch organische Böden, wie Faulschlamm oder Torf oder durch kontaminierte Böden und Grundwässer

Bei Unterfangungen bedeutet dies, daß

- die Vertikallasten aus Gebäuden nicht aufgenommen werden können
- die Horizontallasten aus Erd- und Wasserdruck nicht aufgenommen werden können
- bei Verankerungen die Pressungen unter den Ankerplatten nicht aufgenommen werden können

Bei reinen Dichtungskörpern spielen Festigkeiten in der Regel eine untergeordnete Rolle.

2.7 Der HDI-Körper hat zu hohe Festigkeit

Dies kann bedingt sein durch

- zu niedriger W/Z Faktor
- zu geringe Ziehgeschwindigkeit für den anstehenden Boden

Hinweis: Die Festigkeit läßt sich nach oben nur schwer begrenzen und sie ist bodenabhängig. Gezielte Maßnahmen, die zuverlässig wirken und nicht ins Gegenteil umschlagen, sind sehr schwierig zu steuern.

Zu hohe Festigkeit des HDI-Körpers ist für die statische Funktion ohne Einfluß. Lediglich das Entfernen des Vorwuchses oder das Durchörtern von sehr harten HDI-Körpern kann sehr kostenträchtig werden.

2.8 Fehler, die durch die HDI-Produktion zu Gebäudeschäden führen

Bei der HDI-Produktion können auch Fehler gemacht werden, die zu Schäden an dem zu unterfangenden Bauwerk oder an benachbarten Bauwerken führen. Diese Schäden werden durch Hebungen oder Senkungen verursacht. Senkungen können entstehen, wenn zuviele frische Säulen zu dicht nebeneinander unter einem Fundament hergestellt werden. Die frische Zementsuspension übernimmt ja zunächst keine Lasten. Durch Gewölbewirkung im Fundament muß die Gebäudelast auf die benachbarten Bodenabschnitte neben der frischen Säule übertragen werden. Umgekehrt können bei nicht ordnungsgemäßem Rücklauf der Überschußsuspension Hebungen des Untergrundes und der darauf befindlichen Gebäude entstehen. Diese Hebungen können mehrere Dezimeter betragen. Es ist bei HDI-Arbeiten also immer auf einen geregelten Rückfluß der Über-schußsuspension zu achten. Bei tiefen Bohrungen in nicht standfesten Böden kann es erforderlich werden, die Übertragungsstrecke oberhalb der eigentlichen HDI-Säule mit Standrohren zu versehen. In anderen Fällen kann es notwendig werden gesonderte Entlastungsbohrungen abzuteufen, über die sich der Druck abbauen kann. Es können auch automatische Schlauchwaagensysteme oder elektronische Überwachungssysteme installiert werden.

Bild 1.   Altbau mit HDI-Unterfangung

3 UNTERSUCHUNGSOBJEKT

Ein in exponierter Lage am Dresdner Elbhang gelegenes Areal, welches vor der Kriegszerstörung im Wesentlichen durch Gebäude einer ehemaligen Brauerei bebaut war, ist in den vergangenen 3 Jahren zu einem Wohn- und Geschäftskomplex umgestaltet worden. Dabei sind denkmalgeschützte Altbausubstanz rekonstruiert und Neubauten unterschiedlichster Form und Größe errichtet worden. Bei der Rekonstruktion sind insbesondere die Fassade des ehemaligen Hauptgebäudes, umfangreiche Keller- und Geschoßgewölbe sowie ehemalige Stallgebäude erhalten worden.

Für die Errichtung der Tiefgeschosse eines an die erhaltenswerte Altbausubstanz unmittelbar angrenzenden Neubaues war der Aushub einer zwischen 9,0 und 9,4 m tiefen Baugrube erforderlich. Die säulen- und wandpfeilerartigen Stützen der Altkellergewölbe sind auf 0,5 bis 0,75 m unter Gelände reichenden Blockfundamenten gegründet, so daß deren Unterfangung erforderlich wurde. Geplant und ausgeführt wurde diese Unterfangung in Form überschnittener rückverankerter HDI-Säulen, deren Grundrißanordnung im Untersuchungsbereich aus Bild 2 hervorgeht. Die hieraus resultierende Dicke der Unterfangungkörper

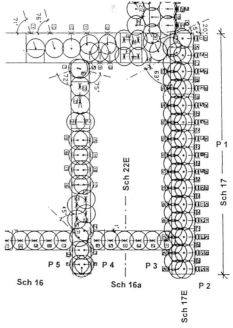

Bild 2.   Grundriß der HDI-Säulen mit Schnitten und Prüfstellen

491

detiefe von 1,5 m lag die Gesamthöhe der HDI-Säulen (senkrecht gemessen) im untersuchten Baugrubenbereich zwischen 5,9 und 6,5 m. Im Bild 3 ist die Situation im Bereich des Schnittes 17 dargestellt.

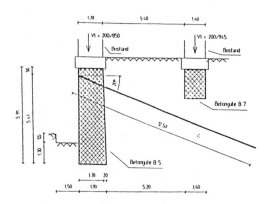

Bild 3.   Schnitt 17 als Beispiel

Die Herstellung der HDI-Säulen ist im September erfolgt, der Baugrubenaushub im Bereich des untersuchten Bereiches fand Mitte Januar des Folgejahres (also ca. 4 Monate später) statt. Ende Januar/Anfang Februar zeigten sich an den halbseitig freigelegten HDI-Säulen Risse, die sich zunehmend ausweiteten und verbreiterten. Ende der ersten Februardekade war optisch ein derart kritisches Rißbild entstanden, daß durch den Prüfstatiker Einsturzgefahr des baugrubenseitigen teiles des Altbaues sowie akute Gefährdung der Arbeiter in der Baugrube signalisiert wurden. Die Bilder 4 und 6 vermitteln einen Eindruck vom Rißbild zu diesem Zeitpunkt. Die Risse waren ausgeprägt vertikal und horizontal orientiert. Zwischen benachbarten Säulen zeigten sich zum Teil Ablösungserscheinungen (Bilder 4 und 6).

Obwohl die Rißbilder nicht den aus der Betonzerstörung durch übermäßigen (Druck-)Beanspruchung bekannten (d.h. überwiegend schrägen) Verlauf zeigten, wurden entsprechend der im Abschnitt 2 zusammengefaßten Fehlerquellen und -ursachen Überprüfung der Säulengeometrie vorgenommen. Die in Tabelle 1 zusammengestellten Ergebnisse weisen zum Teil erhebliche Mängel in dieser Hinsicht aus. Insbesondere waren die Säulendurchmesser in den Wandbereichen (Schnitt Sch 17 /18, 16a) sowie in dem Endbereich Sch 17E bis zu ca. 0,3 m zu gering. Lediglich im Endbereich Sch 22E lag der Durchmesser über den geforderten. Hier stimmte -wie auch im Endbereich Sch 17E- jedoch die Anzahl der Säulen nicht: statt 5 bis 7

waren nur 2 hergestellt worden, wobei im Bereich Sch 22E die zweite unvollständig war (vgl. Bild 6).

Eine Nachrechnung der Unterfangung des am stärksten von den Solldaten abweichenden Mittelpfeilers (Sch 22E), der auch gemäß Statik die höchste Belastung erfuhr (V = 1440 kN) ergab weder bezüglich der Druckspannungen (vorh. $\sigma_D \approx 0,9$ N/mm^2, an den zugehörigen Prüfstellen 4 und 5, aber auch kleiner als das Minimum aller Prüfwerte 2,0 N/mm^2 gemäß Tabelle 2) noch bezüglich der Tragfähigkeit im aus Mittelsand bestehenden Untergrund Versagensfälle.

In Auswertung der Geometriediskrepanzen an den HDI-Säulen sowie der optisch beeindruckenden Rißbildungen war schließlich in Übereinstimmung mit der Denkmalspflege entschieden worden, den im Bild 1 sichtbaren Teil des Altbaues aus Sicherheitsgründen vollständig abzutragen, zumal die Schutzwürdigkeit dieses Bauteiles (ein Seitenflügel der rekonstruierten Hauptfront) nicht unumstritten war.

Tabelle 1.   Ergebnisse der Geometrieüberprüfung

Bereich	Anzahl der Säulen		Durchmesser der Säulen[m]		Einbindetiefe [m]	
	Soll	Ist	Soll	Ist	Soll	Ist
Sch 17/			1,10	0,80	1,30	
18			(-0,2[1])	0,85	1,45	1,53
Sch 16a			1,10	0,78	0,72	1,50
Sch 17E	7	2	1,10	0,80	1,30	1,53
Sch 22E	5	1+1[2]	1,10	1,45	1,50	>1,50

[1] Überschneidungsmaß
[2] nicht voll ausgebildet (vgl. Bild 6)

4 MATERIALPRÜFUNG

Bei dem zwischenzeitlich erfolgten Abriß der vermeintlich geschädigten HDI-Säulen sind Probestücke entnommen und daraus 25 Prüfkörper mit ~ 50 mm Durchmesser hergestellt worden. Die Ergebnisse der Dichte -und Druckfestigkeitsprüfung nach DIN 1048 (in einer staatlichen Materialprüfanstalt) haben die in Tabelle 2 zusammengefaßten Ergebnisse gebracht. Die Umrechnung von β_D auf β_{w200} erfolgt gemäß DIN1045 (Ziffer 7.4.3.5.3.). Da es bisher noch keine speziellen Regelungen für HDI-Materialien gibt, ist der Bezug auf Beton in der Praxis üblich. In Anlehnung an DIN 4093 wird die zulässige Druckspannung σ_D zu 0,2 β_{w200} angesetzt (vgl. letzte Spalte in Tabelle 2).

Tabelle 2.   Prüfergebnisse

Nr.	Maße d [mm]	d/h	Dichte ρ [g/cm³]	Druckfestigkeiten β_c	β_w200 [N/mm²]	zul.σ_D
1/1	49,1	0,96	1,46	15,0	13,5	2,7
/2	49,1	0,95	1,47	14,0	12,6	2,5
/3	49,1	0,96	1,46	11,2	10,1	2,0
/4	49,1	0,97	1,43	14,2	12,8	2,6
/5	49,1	0,95	1,42	13,3	12,0	2,4
2/1	49,2	0,96	1,62	28,8	26,9	5,4
/2	49,2	0,98	1,58	22,3	20,1	4,0
/3	49,4	1,00	1,65	22,5	20,3	4,1
/4	49,3	0,97	1,60	20,0	18,0	3,6
/5	49,3	0,96	1,64	19,4	17,5	3,5
3/1	49,3	0,98	1,50	21,8	19,6	3,9
/2	49,3	0,96	1,66	28,5	25,7	5,1
/3	49,2	0,96	1,50	20,0	18,0	3,6
/4	49,2	0,95	1,67	30,6	27,5	5,5
/5	49,2	0,96	1,52	21,8	19,6	3,9
4/1	49,3	0,97	1,73	15,5	14,0	2,7
/2	49,4	0,98	1,73	17,8	16,0	3,2
/3	49,3	0,97	1,71	14,3	12,9	2,6
/4	49,2	0,96	1,74	16,2	14,6	2,9
/5	49,3	0,99	1,77	21,9	19,7	3,9
5/1	49,3	1,01	1,79	31,0	27,9	5,6
/2	49,3	0,96	1,97	47,5	42,8	8,6
/3	49,4	0,95	1,79	30,5	27,5	5,5
/4	49,3	0,97	1,99	52,5	47,3	9,5
/5	49,3	0,98	1,94	44,2	39,8	8,0

1/1 = Probekörper-Nr./Prüfkörper-Nr.

Demgemäß weisen die untersuchten Bereiche unterschiedliche Dichten ρ und Festigkeiten β_c bzw. β_w200 auf:

Abschnitt 17/Wandbereich:
ρ   = 1,42 ... $\overline{1,45}$ ... 1,47 g/cm³
β_w200 = 10,1 ... $\overline{12,2}$ ... 13,5 N/mm²
zul. σ_D = 2,0 ... $\overline{2,4}$ ... 2,7 N/mm² (< 5 N/mm²)

Abschnitt 17/Ende
ρ   = 1,50 ... $\overline{1,59}$ ... 1,67 g/cm³
β_w200 = 17,5 ... $\overline{21,3}$ ... 27,5 N/mm²
zul. σ_D = 3,5 ... $\overline{4,3}$ ... 5,5 N/mm² (≤ 5 N/mm²)

Abschnitt 16a / 16 (Mittelpfeiler)
ρ   = 1,71 ... $\overline{1,82}$ ... 1,99 g/cm³
β_w200 = 12,9 ... $\overline{26,3}$ ... 42,8 N/mm²
zul. σ_D = 2,6 ... $\overline{5,3}$ ... 9,5 N/mm² (≤ 7 N/mm²)

Die Klammerwerte geben die Forderungen der statischen Berechnung an. Während die mittleren, gemessenen Dichtewerte in der bei Rizkallah/Hilmer (1989) angegebenen Größenordnung (1,6...2,0 g/cm³) liegen, erreichen die Mittelwerte der zulässigen Druckfestigkeiten die geforderten Werte nicht

Die statische Berechnung weist für die HDI-Säulen z.T. beträchtliche Schubbeanspruchungen aus, die in Tabelle 2 für die untersuchten Abschnitte zusammengestellt sind. In Anlehnung an DIN 1045 (Tabelle 13) sind zulässige Schubspannungen lim τ_0 durch Extrapolation der Tabellenwerte für < B 15 abgeschätzt und in Tabelle 3 aufgenommen worden. Gleichfalls sind in dieser Tabelle (letzte Spalte) die nach DIN 4093 für den Gebrauchslastzustand ermittelten Schubfestigkeiten eingetragen (τ_G = 0,2 • σ_D). Während die nach DIN 1045 ermittelten Grenzwerte mit Ausnahme des Bereiches Schnitt 16a erheblich unter den erforderlichen liegen, erfüllen die nach DIN 4093 bestimmten Größen die Forderungen prinzipiell alle.

Tabelle 3   Schubfestigkeiten

Bereich	erf. τ [1]	lim. τ_0 [2]	τ_G [3]
		[kN/m²]	
Sch 17	250	130 (Platte)	245
Sch 17 E	250	190 (Balken)	425
Sch 16 a	83	130 (Platte)	525
Sch 16	536	220 (Platte)	525
Sch 22 E	421	260 (Balken)	525

[1] gemäß Statik   [2] nach DIN 1045   [3] nach DIN 4093

5 SCHWINDVERHALTEN

Ließ bereits das beschriebene Rißbild Zweifel aufkommen, daß die Ursache hierfür in ungenügender Festigkeit des verfestigten HDI-Materials zu suchen sind, so wurden diese Zweifel bei Betrachtung der Säulenbruchstücke gestützt: Die diametrale Rißtiefe betrug nur einige cm. Hier muß selbstkritisch eingeräumt werden, daß durch Entnahme von Bohrkernen die Rißtiefe schnell und einfach hätte überprüft werden können. Allerdings muß auch bemerkt werden, daß übereinstimmend alle Beteiligten Gefahren bei der Ausführung jeglicher Arbeiten unmittelbar an den „geschädigten" HDI-Säulen sahen.

Konsultationen bei Baustoff- bzw. Betonspezialisten führten schließlich auf die Beurteilung des Schwindverhalten des Materials hin, welches sowohl

Bild 4.   Beispiel für Rißbildungen

Bild 6.   Unvollständige HDI-Säule

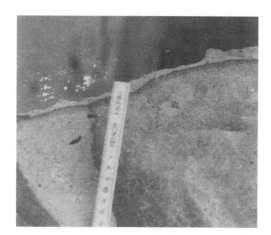

Bild 5.   Randzone der Verfestigung

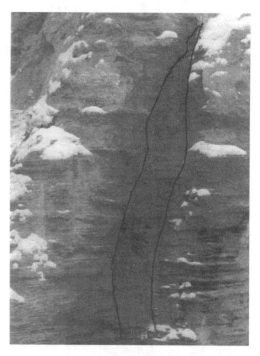

aus dem relativ hohen Zementanteil in diesem aus einem gleichförmigen Sand hergestellten „Beton" (der eigentlich einen Zementmörtel darstellt) als auch aus dem nach dem baugrubenseitigen Freilegen der HDI-Säulen überwiegend einseitigen Austrocknung erklärbar ist. Ein Schwindmaß von 0,05 bis 0,1% ist als real anzusehen und führt bei Säulendurchmessern von 0,8 bis 1,4 m zu möglichen Rißbreiten von 2,5 bis 4,4 mm, die am Objekt in dieser Größenordnung sichtbar gewesen sind (vgl. Bilde 4). Rißtiefen sind nicht gemessen worden; an den Probestücken war jedoch eine Verfärbungszone von 10 bis 15 mm erkennbar, die als Rißtiefe gedeutet wird (vgl. Bild 5).

Bild 7.   Nicht verfestigter Keil zwischen einer senkrechten und einer schrägen HDI-Säule

6 SCHLUßFOLGERUNGEN

Die Untersuchung dieses vermeintlichen Schadensfalles durch den Autor bestätigt einerseits die Feststellungen von Kluckert (1996), daß Geometrieabweichungen sowie unvollständige Ausbildung von HDI-Körpern in der Praxis häufiger vorkommen. Andernseits waren Schwinderscheinungen an HDI-Körpern zwar prinzipiell bekannt (einschließlich der betonanalogen Abhängigkeit vom Zementanteil), aber selbst bei renovierten Spezialtiefbaufirmen in den hier beobachteten Ausmaßen nicht bekannt. Von ausschlaggebender Bedeutung ist nach Auffassung des Autors im vorliegenden Fall die vergleichsweise späte und einseitige Freilegung der HDI-Wände und -säulengruppen. Der Schwindprozeß verlief in den nahezu 3 Monaten zwischen Herstellung und Freilegung im naturfeuchten sandigen Milien langsam und gleichmäßig. Dann schloß sich die einseitige Freilegung in wenigen Tagen an und der Schwindvorgang -unterstützt durch die direkte Sonneneinstrahlung von Osten her auf die freiliegende Materialseite- wurde forciert. In der Baupraxis werden HDI-Körper, die als Baugrubenwände dienen, in der Regel (dem allgemeinen Termindruck geschuldet) wenige Tage nach der Ausführung der HDI freigelegt, so daß offenbar der Schwindprozeß eine untergeordnete Rolle spielt.

LITERATUR

Hilmer, K. 1991. Schäden im Gründungsbereich. Berlin: Ernst und Sohn.

Hilmer, K. u. V. Rizkallah 1990. Bauwerksunterfangung und Baugrundinjektion mit hohen Drücken (Düsenstrahlinjektion). *Veröff. des LGA-Grundbauinstitutes Nürnberg,* Heft 55.

Kluckert, K.D. 1996. 20 Jahre HDI in Deutschland -Von den Fehlerquellen über die Schäden zur Qualitätssicherung. *Vorträge der Baugrundtagung 1996 in Berlin,* 235-258. Essen: Deutsche Gesellschaft für Geotechnik e.V.

Rizkallah, V. 1986. Underpinning and Foundations near old Buildings. *X. Int. Conf. On Soil Mech. And Found. Engineering, Stockholm,* Vol. 3, 153 ff.

Schnell, W. und R. Vahland 1997. Verfahrenstechnik der Baugrundverbesserungen. Stuttgart: B. G. Teubner.

Geotechnical Hazards, Marić, Lisac & Szavits-Nossan (eds) © 1998 Taylor & Francis, ISBN 90 5410 957 2

Fundamentverstärkung und Fundamentsicherung mit SOILCRETE-jet grouting bei der Rekonstruktion Wenzelsplatz 33 in Prag

R.Cuda & Z.Boudik
Keller Speciální Zakládání Praha, Czech Republic

ABSTRACT: Die Anforderungen und Wünsche des Bauherrn, die Vorschläge der Projektanten und die Vorgaben der Behörden und Nachbarn des prominenten Platzes verlangen umfangreiche Sicherungsarbeiten an den Fundamenten der betroffenen Bauwerke. Denkmalgeschützte Gebäudeteile bleiben bestehen, das Gebäudeinnere wird rekonstruiert, neue Belastungsverhältnisse auf die bestehenden Fundamente werden geschaffen. Die Errichtung einer Tiefgarage ist Voraussetzung zur Erfüllung zeitgemäßer Nutzungsansprüche an das Gebäude. Die angrenzenden Bauwerke, errichtet in äußerst unterschiedlichen Zeitepochen, müssen dafür unterfangen werden. Der rekonstruierte Gebäudeteil wird auf SOILCRETE im tragfähigen Boden gegründet. Die Aushubarbeiten für die Tiefgarage reichen mehr als 10 m unter die Geländeoberkante, während die Nachbarfundamente auf SOILCRETE lasten. Ein Verbindungstunnel zwischen der Tiefgarage und späterer unterirdischen Zufahrt vom Wenzelsplatz ist vorbereitet. Mit SOILCRETE - jet grouting gelingen die anspruchsvollen grundbautechnischen Sicherungsarbeiten.

1 EINFÜHRUNG

Die grundlegenden politischen Änderungen, die in der Tschechoslowakei im November des Jahres 1989 eintraten, zogen auch bedeutende ökonomische Veränderungen nach sich. Das Eigentum, das bis zum erwähnten Umsturz in staatlicher Verwaltung war, kam nun allmählich in private Hände.

Dieser Trend verläuft kontinuierlich seit dem Jahre 1989 und bezieht sich selbstverständlich auch auf Immobilien. Die Privatisierung ist besonders markant an attraktiven Orten in den Städten der Tschechoslowakei und deren Nachfolger: der Tschechischen Republik . Natürlich gibt es in der Hauptstadt Prag eine Vielzahl solcher Fälle.

So wie die Immobilien im Zentrum Prags ihre neuen Besitzer finden, so ändern sich auch die Ansprüche seitens der Nutzung der existierenden Gebäude. Dort wo es früher bloß zweitrangige Wohnungen gab und sogar schlechtere, wünschen die neuen Besitzer einen kompletten Umbau in Geschäftszentren mit Büroräumen und Wohnungen deren Standard der außerordentlich attraktiven Lage im Stadtzentrum entspricht.

Zu einem solchen Komplex gehören natürlich auch Tiefgaragen und weitere Untergeschosse in denen die verschiedensten technologischen Einrichtungen untergebracht werden, die zur Versorgung der ganzen Objekts nötig sind.

Ähnlich war die Aufgabenstellung auch beim Objekt am Wenzelsplatz Nr.33, das sich im wahren Herzen Prags befindet. Auch hier hat sich der Investor für einen Umbau des Objekts zu einem Geschäftszentrum mit Büroräumen, modernen Wohnungen und Tiefgaragen entschieden. Die Kellergewölbe des Objekts stammen aus dem 15. Jhd., aus der Zeit der Gotik, das Gebäude aus Natursteinmauerwerk aus dem 18. Jhd. Das ganze Objekt steht unter Denkmalschutz.

Zur Vorbereitung der weitgehenden Rekonstruktionen der Zubauten und Neubauten wurden in der ersten Phase folgende Arbeiten durchgeführt:
- Demolierung einiger Objekte im Hintertrakt
- Sicherung der Baugrube
- Aushub der Baugrube
- Errichtung eines neuen Stahlbetonskellets im Bereich des Hintertrakts.

Die anknüpfenden Tätigkeiten sind so geplant, daß das Objekt i.J. 1998 dem Bauherrn übergeben werden kann.

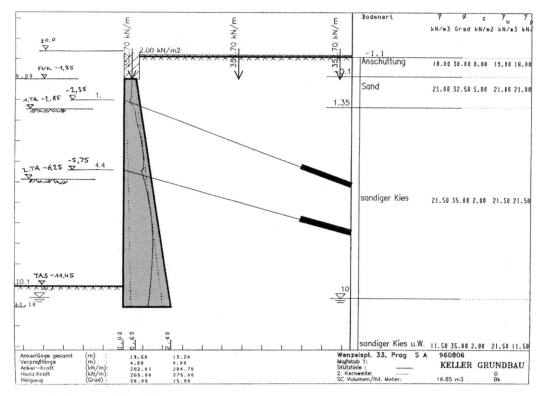

Bodenart	γ kN/m3	φ Grad	c kN/m2	γ_u kN/m3	γ kN
Anschüttung	18.00	30.00	0.00	19.00	18.00
Sand	21.00	32.50	5.00	21.00	21.00
sandiger Kies	21.50	35.00	2.00	21.50	21.50
sandiger Kies u.W.	11.50	35.00	2.00	21.50	11.50

Ankerlänge gesamt	(m) :	13.60	13.20
Verpreßlänge	(m) :	4.00	4.00
Anker-Kraft	(kN/m):	262.01	284.70
Horiz.Kraft	(kN/m):	265.00	275.00
Neigung	(Grad) :	20.00	15.00

Wenzelspl. 33, Prag S A	960806	
Maßstab 1:		KELLER GRUNDBAU
Stützlinie :	——	
2. Kernweite:	......	0
SC Volumen/lfd. Meter:	16.65 m3	Bk

Bild 1. Typischer statischer Schnitt

2 SPEZIALTIEFBAUARBEITEN

Das Baugelände wurde in zwei Teile aufgegliedert. Der Bereich der abgerissenen Bauten wurde Neubau benannt. Der zweite Teil trägt die Bezeichnug Altbau. Er ist ungefähr identisch mit dem Grundriß des ursprünglichen Gebäudes, das zur Rekonstruktion bestimmt ist.

Für den Neubau war es nötig die Baugrube zu sichern und die Fundamente der Nachbarobjekte zu unterfangen.

Im Teil Altbau wurde die Baugrube gesichert und die Fundamente unterfangen. Weiters mußten an einigen Stellen die ursprünglichen Fundamente für die Belastungsverhältnisse verstärkt werden.

Im ursprünglichen Entwurf waren Schlitzwände zur Sicherung der Baugrube vorgesehen. Der Zustand der Nachbargebäude und Probleme mit der Zufahrt zum Baugelände waren der Grund weshalb eine andere Technologie gewählt wurde. Man entschied sich für das SOILCRETE - jet grouting Verfahren (Hochdruckbodenvermörtelung HDBV) mit verschiedenen Ankerhorizonten und Aussteifungen.

3 GEOLOGIE

Das Baugelände befindet sich im Zentrum Prags. Das Stadtviertel „Neustadt" breitet sich rechts vom Mol- daufluß aus. Es wurde im 14. Jahrhundert vom tsche- chischen König Karl IV. gegründet, ist also seit dieser Zeit intensiv bewohnt.

Die Archivdokumentation ergab folgende geologische Struktur:

a) vom Geländeniveau bis zur Tiefe 2 bis 4m: heterogene Anschüttung verschiedener Art und Herkunft - - primäre Kulturschicht

b) Bis zur Tiefe cca 7 m sandige Flußsedimente, dann bis zu 16 - 17 m Sandkies und Kies

c) Ab 16 bis 17 m Tonschiefer

Um genauere Angaben zu erhalten, wurden 3 Aufschlußbohrungen im Gebiet des künftigen Bauplatzes ausgeführt. Das Resultat dieser im Frühjahr 1994 durchgeführten Bohrungen bildete dann zusammen mit dem geotechnischen Gutachten eine der Unterlagen für den Entwurf der Baugrubensicherung.

Die genauere geologische Struktur ist im folgenden wiedergegeben:

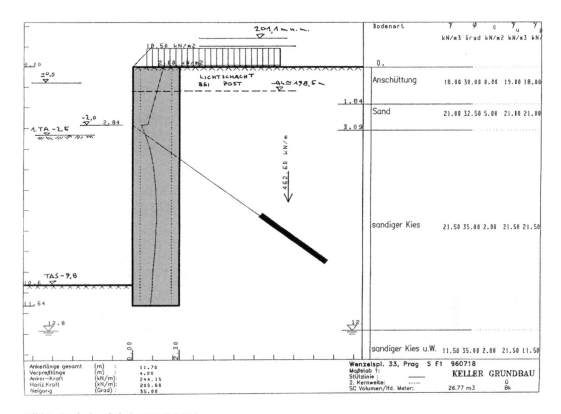

Bodenart	γ	φ	c	γ_u	γ
	kN/m3	Grad	kN/m2	kN/m3	kN/
0.					
Anschüttung	18.00	30.00	0.00	19.00	18.00
Sand	21.00	32.50	5.00	21.10	21.00
sandiger Kies	21.50	35.00	2.00	21.50	21.50
sandiger Kies u.W.	11.50	35.00	2.00	21.50	11.50

Ankerlänge gesamt	(m) :	11.70	
Verpreßlänge	(m) :	4.00	
Anker–Kraft	(kN/m):	244.15	
Horiz.Kraft	(kN/m):	200.00	
Neigung	(Grad) :	35.00	

Wenzelspl. 33, Prag S F1 960718
Maßstab 1: ———
Stützlinie :
2. Kernweite: ....
SC Volumen/lfd. Meter: 26.77 m3

KELLER GRUNDBAU
0
Bk

Bild 2. Statischer Schnitt TELECOM

a) Bis zur Tiefe 1,3 bis 1,6m heterogene Anschüttung (Bauschutt, Erde, Sand und Kies)

b) Sedimente-schluffiger Sand von einer Mächtigkeit 1,2 bis 3,2m, sandiger Kies - Mächtigkeit 10,3 bis 12,2.

c) Felsartiger Tonschiefer wurde in der Tiefe von 14 bis 15 m erreicht, die Oberfläche dieser Schicht war zerschert und zerlegt.

Der Grundwasserspiegel befand sich 13,2 m unter dem Geländeniveau.

Bei lang andauernden Niederschlägen kann er um 1m steigen. Das Grundwasser wurde als nicht aggressiv klassifiziert. Für Betonkonstruktionen, die dauernd mit dem Grundwasser in Berührung sind, ist Portlandzement ausreichend.

4 NEUBAU

Im Teil Neubau wurde eine tiefe Baugrube gewählt mit zwei Untergeschossen und lokal auch drei, die bis 10,4 m unter das existierende Geländeniveau reichen.

Die Nachbargebäude hatten verschiedene tragende Systeme unterschiedlichen Alters z.B. Ziegelgewölbe, Natursteinmauerwerk, Stahlbetonskelett u. a. Die Fundamentunterkante der angrenzenden Objekte bewegte sich von cca 0,00 bis 6,4 m unter Geländeniveau.

Um die unterschiedlichen Bedingungen zu berücksichtigen wurden insgesamt 46 statische Schnitte untersucht und beurteilt (Bild 1).

Die größte Aufmerksamkeit wurde der Baugrubensicherung in dem Bereich gewidmet, der an das Gebäude der Firma Telecom angrenzt, weil dort sehr strenge Bedingungen für zulässige Erschütterungen vorlagen. Zur Ermittlung der wirklichen, durch das Bohrgerät hervorgerufenem dynamischen Erschütterung wurden im voraus spezielle Messungen durchgeführt.

Gewölbe in den angrenzenden Bauten wurden zusätzlich mit Spannankern versehen. Diese Arbeiten wurden vor dem Ausheben der Baugrube durchgeführt.

Wie schon oben erwähnt sind die Nachbarbauten unterschiedlichen Alters (vom 15. bis zum 20. Jahrhundert) und über die Fundamentunterkante lagen meistens nur ungenaue Angaben vor. Zur Ermittlung der tatsächlichen Fundamentunterkante wurde eine große Anzahl von Schürfen gegraben und die ermittelten Daten in den statischen Berechnungen berücksichtigt.

Auf diese Weise war die verläßliche Sicherung der angrenzenden Bauten gewährleistet.

Die einzelnen Ankerhorizonte wurden im Laufe der Aushubarbeiten entsprechend der Lage der Keller in den Nachbarobjekten durchgeführt.

Die gesamte Kubatur der Baugrube beträgt 13 000 m^3. Im Objekt Neubau wurden insgesamt 2 390 lfm Hochdruckbodenvermörtelung durchgeführt und 1 365 lfm Anker verlegt.

Bei der Vorspannung der Anker wurden 10 Druckmeßdosen zur Kraftmessung und Beobachtung der Ankerkräfte eingebaut.

5 ALTBAU

Für das Objekt Altbau ist eine weitgehende Rekonstruierung vorgesehen um das Objekt den neuen Forderungen anzupassen. Vor dem Objekt sind Tiefgaragen unter dem Wenzelsplatz geplant, die zu einem späteren Zeitpunkt errichtet werden sollen. Deshalb wurde schon in dieser Vorbereitungsphase eine Verbindungskommunikation unter der existierenden ebenerdigen Hauseinfahrt in einer Tiefe von 8 - 10,5m gebaut.

Diese 4 m breite Rampe bildet auch die Zufahrtsverbindung zu den Tiefgaragen im Neubau. Für die Sicherung der entsprechenden Baugrube wurde das SOILCRETE Verfahren und 3 Aussteifungshorizonte angewendet.

Solange die Tiefgaragen unter dem Wenzelsplatz noch nicht durchgeführt sind, werden die Räumlichkeiten der Rampe vorübergehend andern Zwecken dienen (z.B. als Lagerräume).

Die zusätzlichen Belastungen, die durch die Rekonstruktion entstehen, machten es nötig die existierenden Fundamente zu verstärken und an einigen Stellen neue Fundamente zu errichten. Auch diese Aufgaben wurden mit dem SOILCRETE Verfahren gelöst, wobei man allerdings in räumlich sehr beschränkten Bedingungen arbeiten mußte.

6 RISIKEN

Bei Bauarbeiten im dicht besiedelten Stadtzentrum muß man mit verschiedenen spezifischen Risiken und Problemen rechnen.

In unserem Falle waren die angrenzenden Objekte verschieden Alters und im Laufe der Jahrhunderte wechselten sie auch ihre Besitzer. Deshalb gab es von vielen Objekten keine Dokumentation, oder fehlten Angaben über Umbauten, die im 20. Jhd. in älteren Objekten durchgeführt wurden.

Es war daher nötig noch vor Beginn der eigentlichen Bautätigkeit eine Beweissicherung der umgehenden Objekte durchzuführen.

Zur Überprüfung der Fundamentunterkanten wurden sowohl im Altbau als auch im Neubau Schürfe gegraben, denn es gab von insgesamt 7 Objekten bloß für 4 Objekte diesbezügliche Angaben.

Bei der Beweissicherung der angrenzenden Objekte ergab es sich, daß einige sehr baufällig sind, andere unter Denkmalschutz stehen und daß darum besondere Maßnahmen nötig sind.

Einige Objekte wurden mit Spannankern und Spritzbeton gesichert. In der statischen Berechnung werden entsprechend vorsichtige Erddruckansätze gewählt.

Ein weiteres, sehr kompliziertes Problem stellte die Sicherung des Grundstücks der Firma Telecom an der nordöstlichen Seite der Baugrube dar. Die Firma Telecom verlangte hier eine Begrenzung der dynamischen Vibration, die bei der Hammerbohrung, vor allem beim Durchgang der Unterfangung bzw. durch den Beton beim Bohren für die Spannanker entstehen.

Der offensichtlich schwierigste Augenblick des ganzen Bauvorhabens war die Feststellung, daß das Nachbarobjekt der „Stiftung für Architektur und Bauwesen" mit seinem Stahlbetonskelett eine Fundamentunterkante besitzt die bloß 1,5 unter dem Geländeniveau liegt, obwohl der anschließende Kellerfußboden auf einer Minuskote von 3,5 - 4 m liegt, also um 2 - 2,5 m tiefer. Diese Tatsache rief eine große Anzahl von Besprechungen aller Teilnehmer hervor und Diskussionen wie das Nachbarobjekt zu sichern wäre.

Die beteiligten Fachleute schlugen 2 Varianten vor:
- Sicherung mit Mikropiloten
- zusätzliche SOILCRETE - Sicherung

Schließlich wurde die Sicherung durch SOILCRETE als technisch verläßlicher gewählt, obwohl gewisse Erschwernisse damit verbunden waren, wie z.B. Betriebsbegrenzung in den Werkstätten des Objekts. Auch mußte das Kellermauerwerk, fraglicher Beschaffenheit und Güte, mit einer Verschalung versehen werden, um das mögliche Durchdringen der Zementsuspension in den Keller zu verhindern.

7 ZUSAMMENFASSUNG

Der Bau von Geschäftszentren in dicht besiedelten Stadtteilen kommt heutzutage ohne spezielle Tiefbauarbeiten nicht umhin. Unterschiedlich von Bauarbeiten im freien Gelände sind diese Arbeiten mit einigen spezifischen Problemen und Risiken verbunden:

- gedrängte Arbeitsbedingungen
- beengte Zufährtskommunikationen zum Bauplatz
- erhöhte ökologische Forderungen (Lärm, Staub)
- beschränkte Arbeitszeit (keine Nachtarbeit)
- Arbeit an Objekten, die unter Denkmalschutz stehen
- Bauarbeiten, unter Aufsicht der Archäologen
- die angrenzenden Objekte können sehr baufällig sein
- Forderung des Bauherrn, die Bauzeit so kurz wie möglich zu halten

Der Firma KELLER ist es gelungen bei der Realisierung des Bauauftrags für das Objekt am Wenzelsplatz Nr.33 in Prag die oben erwähnten anspruchsvollen Bedingungen zu erfüllen und dem Investor ein technisch hochwertiges Bauwerk in den geforderten Terminen zu übergeben.

Damit hat Firma KELLER bewiesen anspruchsvolle Projekte auch im historischen Zentrums Prags zu realisieren.

Geotechnical Hazards, Marić, Lisac & Szavits-Nossan (eds) © 1998 Taylor & Francis, ISBN 90 5410 957 2

Investigation of the behavior of propped diaphragm walls in deep excavations

A. Edinçliler
Union of the Municipalities of the Marmara Region, İstanbul, Turkey

E. Güler
Department of Civil Engineering and Protection of Environment and Improvement Research Center, Boğaziçi University, İstanbul, Turkey

ABSTRACT: Deep excavations cause displacement and settlement in surrounding soil during and after excavations. In this study, the effect of a propped diaphragm wall is investigated. An 8 meter deep excavation has been modelled by using a finite element program called CRISP which is based on Critical State Theory. The excavation has been modeled by using 225 nodes and 392 elements. The excavation has been carried out in four stages and it has been assumed that the excavation is propped with struts. When the whole excavation is carried out in one stage and all the struts are placed after the completion of the excavation, larger lateral displacements are observed when compared to placing the struts after each excavation stage. In the case of using a 10.7 meter deep wall, it has been observed that when the whole excavation is completed, the lateral displacements in the excavation base are reduced.

1. INTRODUCTION

Increased need of deep excavations and retaining structures because of new technological developments in civil engineering and increasing value of the land use increased the importance of the retaining structure design. Stability problem of deep excavations can be solved by using cantilever and supported retaining structure. The factor of safety of an excavation against failure can be calculated. Retaining structure, however, must also be designed to prevent excessive soil deformations, to increase the stability and to carry the vertical loads.

Deep excavations cause excessive lateral deformations and displacements at the soil surface during and after excavations. The most important parameters to check the deformations of the propped diaphragm walls are the rigidity of the wall and determination of the strut spaces according to the type of surrounding soil. In addition, deformations of the supported system depend on the excavation depth, the size of the excavated area, excavation rate and time before construction of the supporting structures (Potts and Fourie 1984; Fleming 1985).

The aim of this study is to investigate the effect of designing the propped diaphragm walls extending deeper than excavated depth on displacements, settlements and stresses of the soft soil.

In order to find the best solution, several analyses have been made by using a finite element program called CRISP which is based on Critical State Theory (Britto and Gunn 1987). With the help of preparing finite element model, stress and deformation values can be determined at each step of excavation and consolidation.

2. FINITE ELEMENT MODEL REPRESENTING DEEP EXCAVATION

Observations have been made by creating a model 60 m. wide and a 16 m. deep. As seen from Fig. 1a, because of symmetry, the half of the system which has 225 nodes and 392 elements has been considered.

The excavation considered has a depth of 8 m. and a width of 10 m. (Fig. 1b). The soil has been modeled by Modified Cam Clay Soil which simulates elasto-plastic soil behavior. The following numerical values have been used to represent the soil: $\kappa = 0.05$, $M = 1.00$, $\gamma = 20$ kN/m^3, $\lambda = 0.30$, $\nu = 0.30$, $k_x = 1.9$ E-9 m/s, $e_{cs} = 2.95$, $\gamma_w = 10$ kN/m^3, and $k_y = 0.1$ E-9 m/s. Diaphragm wall is assumed to behave as a elastic material. The properties are: $E_x = 2.E7$ kPa, $\nu_x = 0.16$, $G_{hv} = 8.6.E7$ kPa, $E_y = 2.E7$ kPa, and $\gamma = 20$kN/m^3.

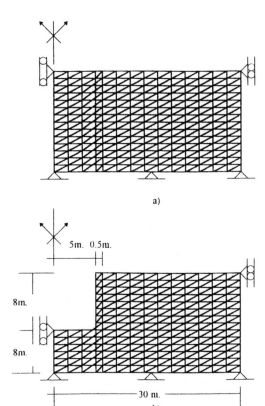

a)

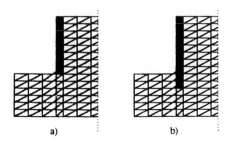

a)   b)

Figure 2. Placing diaphragm wall
a) Analysis 1 b) Analysis 2

Figure 1. Finite Element Model representing deep excavation a)Boundary conditions b) Dimensions

The appropriate fixity codes are specified for all the element sides along the boundaries. The lateral and vertical displacements are restricted along the bottom boundary. On each side horizontal fixities have been used (Fig.1a). At a depth of 16 m., isotropic stress of 169 kPa and an excess pore pressure of 160 kPa are defined. It is assumed that the ground water level is at the ground level. A preconsolidation pressure of 320 kPa is chosen at the same depth. To observe the effect of the depth of the diaphragm wall on deformations and stresses, firstly, a 8 m. deep wall is considered (Fig.2a). After that, the depth of the wall has been increased to 10.7 m. (Fig.2b). Bar elements representing struts has been placed after excavation. Struts having a length of 1 m have been assumed. To prevent then bars supporting 10 m., 1m. long bar elements were chosen. Therefore, to obtain the same lateral deformations, Elasticity Modulus of steel has been used by dividing the Elasticity Modulus by 10. The following physical parameters have been used to represent the struts by

using bar elements: E = 3.E6 kPa, ν = 0.31, and A = 0.001 m^2.

3. FINITE ELEMENT ANALYSIS

To observe soil behavior for the cases of modeling the diaphragm walls designed at the excavation depth and deeper than excavation depth, two analysis have been carried out. While Analysis 1 simulates the use of an 8 m. deep wall, Analysis 2 represents placing a 10.7 m deep wall for the same excavation depth. In the Analysis, struts were placed after each excavation step.

To compare the different design techniques, analyses have been carried out in nine increment blocks, consisting of 30 increments. Excavation and placing the struts has been defined in 8 increment blocks. To define the plastic behavior of the soil properly, each increment block was divided into small loading steps. In the first increment block, 2 m. excavation was performed and the strut has been placed in the second increment block. In the following six increment blocks, the same principle was followed. In the ninth increment block, consolidation has been assumed to last for 3 months, consisting of 14 increments. At this stage, it is assumed that the soil is subjected to one way drainage and excess pore water pressure boundary was defined. Deformations that occur at the end of each 2 m. excavation are shown in Fig. 3.

4. NUMERICAL RESULTS

After the whole excavation is completed and struts are placed, it is seen that vertical and horizontal deformations is reduced in the case of using the diaphgram wall that is longer than excavation depth. Fig.4 shows the critical points subjected to excessive

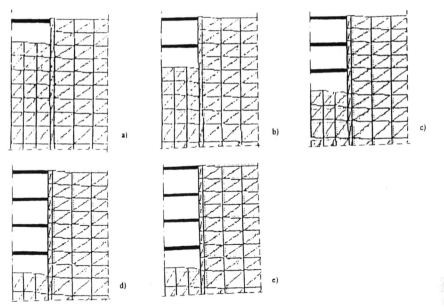

Figure 3. Soil Deformations a) after 2 m. excavation, b) after 4 m. excavation, c) after 6 m. excavation
d) after 8 m. excavation, e) after consolidation

deformations. In Table 1, displacements in x and y directions for two cases after excavation and consolidation can be seen. Also, horizontal and vertical stresses calculated in the center of Element No. 6 are given in Table 2.

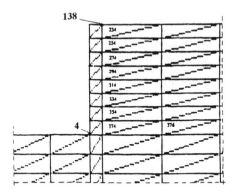

Figure 4. Critical points subjected to excessive deformations

5. RESULTS

The analyses revealed the following results:

1. The arrows in Fig.5 show the direction and relative magnitude of the displacements of various points after consolidation. In the case of using a 10.7 meter deep wall, it has been observed that when the whole excavation is completed, the lateral displacements in the excavation base are reduced.

2. Stress paths for the elements No.374-376 just behind the wall has been given in Fig. 6. As seen from Figure, in the case of designing the wall depth equal to excavation depth (Analysis 1), the soil at the defined location reaches yielding conditions. Comparing stress paths of Analysis 2, it is clear that designing the wall longer than the excavation depth prevents this yielding of soil.

3. When observing lateral displacements at Element

Table 1. Displacement values after excavation and consolidation

Type of Analysis	Element No. 4				Element No. 138			
	After excavation		After Consolidation		After excavation		After Consolidation	
	x (m.)	y (m.)	x (m.)	y (m.)	x (m)	y (m.)	x (m.)	y (m.)
Analysis 1	0.101	0.015	0.124	0.018	0.040	0.014	0.040	0.015
Analysis 2	0.068	0.012	0.071	0.012	0.036	0.010	0.036	0.011

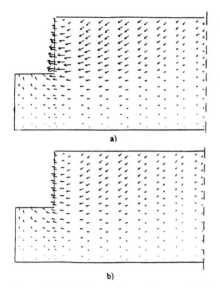

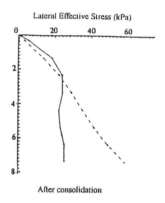

Figure 5. Soil deformations after consolidation
a) Analysis 1 b) Analysis 2

Table 2. Stresses in Element 6

Type of Analysis	Stresses after excavation (kPa)		Stresses after Consolidation (kPa)	
	x	y	x	y
Analysis 1	169	28	142	14
Analysis 2	140	28	125	17

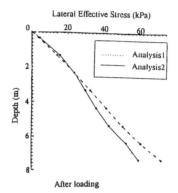

After loading

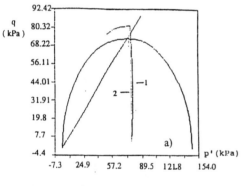

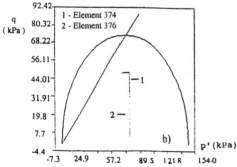

Figure 6. Stress paths at Element No. 374 and 376
a) Analysis 1 b) Analysis 2

After consolidation

Figure 7. Lateral soil stresses behind the wall

No. 234, 254, 274, 294, 314, 354, and 374, it is seen that using Analysis 2 namely the wall longer than the excavation depth also decreased lateral deformations at the points close to ground surface (Fig. 7).

REFERENCES

Britto, A. M. & Gunn, M.J. 1987. *Critical State Soil Mechanics via Finite Elements*, Ellis Horwood, Chichester, England.

Fleming, W.G.K. 1985. Diaphragm Walls, *Derin*

Kazılar ve Iksa Methodları Sempozyumu, ZMTM
Türk Milli Komitesi, Bogaziçi, University:61-71.

Potts, D.M. & Fourie, A.B. 1984. The Behavior
of a Propped Retaining Wall: Results of a Numerical
Experiment, *Geotechnique 34*, :384-404.

Geotechnical Hazards, Marić, Lisac & Szavits-Nossan (eds) © 1998 Taylor & Francis, ISBN 90 5410 957 2

Reconstruction of demolished bridge across Sava river near Županja

Z. Ester
Faculty of Mining, Geology and Petroleum Engineering, University of Zagreb, Croatia

M. Vrkljan
Civil Engineering Institute of Croatia, Zagreb, Croatia

G. Brlek
Scandkop, Zagreb, Croatia

ABSTRACT: This paper presents a unique method that was utilized underwater to hash up the reinforced concrete roadway and steel construction of demolished bridge by using explosives. The bridge construction was hashed up into pieces of favourable dimensions to be taken out by crane, so it would not cause any damage to the surrounding objects, especially piers number. 11 and 12.

1 INTRODUCTION

The Motorway Bridge across the Sava River near Županja was constructed in 1968. Its total length was 791,75m and width was 10,0m. The main part of the construction over the water flow had three arcs (85,0m, 134,0m and 85,0m). Steel piers were set 6m apart from each other, with approximate height of 3,5m (min. 2,35m, max. 5,5m) and they were holding a reinforced concrete deck slab 0,18m thick. The transversal steel stiffeners were set 6,07m apart in end sections, and 6,09m in the middle sections.

During the war in Croatia, in 1991, for the strategic purposes, the collapse of the main bridge construction into Sava River was performed by explosives. The middle arc 120m long, fell into the water, as well as a part of the right bank arc 36m long. With the exception of small pieces next to each of the piers (ca. 4m), the hole collapsed part of the bridge sunk 7m deep into the water regarding the lowest river level.

The reconstruction of the bridge began in 1996. It was decided that the destroyed part of the bridge should be replaced by a new construction using the existing piers in Sava River, which have not been blasted during demolition. Since the collapsed part of the bridge was located in the very bridge axes, it had to be removed first.

There were several possibilities discussed for removal of collapsed material. Finally, it was decided to use explosive to demolish the bridge remains into pieces, which could be lifted by a floating crane. Blasting had to be performed in the safest way possible, primarily without damaging the existing piers in the river, living quarters and other buildings in Orašje, IFOR's pontoon bridge and a cable ferry on Sava River (the only connection between the Republic of Croatia and the Federation of Bosnia and Herzegovina at that time).

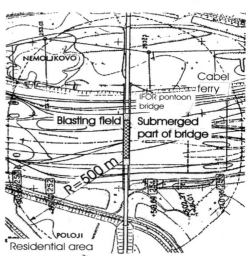

Figure 1: Situation map, with safety radius from Blasting field

The objects mentioned were located within 500m from the blasting point (Figure 1).

2 THEORETICAL PART

Underwater explosive detonation causes primary and secondary effects. The primary effect is a shock wave which is a result of four main waves: direct wave, surface-reflected wave, bottom-reflected wave, and bottom-refracted wave (Figure 2).

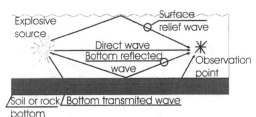

Figure2: Main waves in underwater blasting

The highest pressure is achieved by a direct wave, and it can be approximately defined as a function of time (Joachim, C.E. & al. 1997):

$$P_{(t)} = P_m e^{-\left(\frac{t-t_a}{\Theta}\right)} \qquad (1)$$

where P_m stands for maximal pressure value (kPa):

$$P_m = 52.400(\lambda)^{-1.13} \qquad (2)$$

Θ is a time constant (s):

$$\Theta = 9.2 \times 10^{-5} Q^{\frac{1}{3}} (\lambda)^{0.18} \qquad (3)$$

t_a stands for time of shock wave arrival at the observation point (s).

$$t_a = \frac{r}{1.524} \qquad (4)$$

λ is the calculated value of explosive ($m/Q^{1/3}$), Q is explosive mass, r distance between blasting point and observation point (m), 1,524 velocity of expansion of a shock wave in water (m/s). Impulse of a shock wave is:

$$I_t = 5.76 \times Q^{\frac{1}{3}} (\lambda)^{-0.89} \qquad (5)$$

Figure 3 shows the relation P_m and r/r_0, where r_0 represents a radius of blasting cartridge.

The secondary effects of underwater detonation represents a pulsating gas products after explosion (successive expansion and compression), that depends on the type of explosive, and it can be approximately calculated by the equation (Ostrovskii, A.P. 1962):

$$P = P_0 \left(\frac{r_o}{r}\right)^m = P_0 \left(\frac{v_o}{v}\right)^{\frac{m}{3}} \qquad (6)$$

where P_o is average value of pressure from explosion products (kPa), r_o radius of blasting cartridge (m), m time constant, r radius of bubble (m), v bubble volume (m³), v_o initial volume of a bubble (m³).

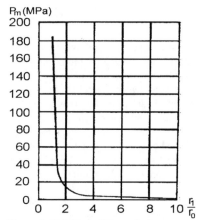

Figure 3: Relation of P_m and r/r_o

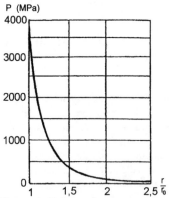

Figure 4: Relation between pressure P and bubble radius r/r_o for first pulsation

510

Figure 4 shows the relation between pressure P and relative values of the bubble radius r/r_0 for the first pulsation.

As shown in Figure 3, the pressure of a shock wave P_m decreases rapidly with a distance from detonation point, while on the other hand the pressure of bubble which is carrying explosion products, although 10 times smaller in values, it reaches even 4 times larger impulse because of the much longer time of pulsation.

Characteristics of the effects of underwater blasting have been of great importance for choosing the explosive that was used to blast the reinforced concrete deck slab.

The chosen explosive should have the following blasting-technical characteristics: density ρ (kg/dm³) much over 1, good water-resistance, plastic consistence, relatively good sensitivity - possibility of initiation with detonation fuse, extreme energy of blow - at least 5MJ/kg (Persson, P.A. et al. 1994).

On the other hand, for the security of the piers in Sava River and other objects, we were looking for an explosive with higher shock wave energy compared to the energy of a bubble which is carrying explosion products.

In table 1 there is a list of blasting and energy characteristics for most of the commercial explosives.

The explosive Gelamon-40 was chosen for that it best fits the required blasting and security conditions.

Table 1: Blast and energy characteristics for some commercial explosives

Expl. type	Density (kg/dm³)	Water resis.	Sens. to det.	Blow energ. (MJ/kg)	Bubble energ. (MJ/kg)
Gelamon	1,45	Excellent	Excellent	1,77	3,29
ANFO	0,98	Bad	Bad	1,28	3,40
TNT	1,64	Excellent	Good	1,44	3,07
Emulsion	1,5	Excellent	Poor	1,66	3,53
Heavy ANFO	1,1	Poor	Poor	1,03	3,02

3 PERFORMANCE OF BLASTING OF BRIDGE

Due to a very poor visibility in the waters of Sava River (only 15cm) and high water velocity a complete preparation of blasting cartridges was carried out on a floating object. Sections 6.0m × 6.07m were blasted one at the time. Section included an area between the main and transversal stiffeners.

A frame of concrete steel $\varnothing$ 18mm, 5.0m × 5.0m was constructed. The explosive Gelamon-40 in cartridges with diameter of $\varnothing$ 60mm and mass of 2.0kg was attached to the frame. A necessary concentration of blasting cartridge for breaking reinforced concrete deck slab 0.18m thick is 7.2 kg/m', therefore the explosive was placed in two rows. The total mass of explosive per section was 144 kg. Initiation was performed with double detonating cord C-20. An additional weight - a steel frame was put on the blasting cartridge and thus the best contact with the concrete deck slab was achieved which enabled a transition of shock waves to the deck slab with minimum energy loss.

Figure 5 shows the construction of a blasting cartridge for blasting of a section.

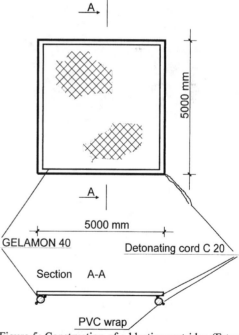

Figure 5: Construction of a blasting cartridge (Ester, Z. & Krsnik J. 1996)

The complete construction was lifted by a crane and placed into a position by sliding along the steel wires that divers fixed to the section to be blasted. This way the divers work was reduced to a minimum, as well as a possibility for mistakes while placing the blasting cartridges into position for blasting (Figure 6 & Figure 7).

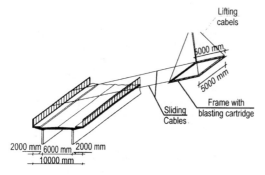

Figure 6: Placing blasting cartridge on section

Figure 7: Placing blast cartridge into water

Before activating the blasting cartridge the floating object was moved about 100m away from the blasting position. Initiation of the blasting cartridge was performed on the floating object by electrical detonator mounted to a detonating cord.

The totals of 26 blastings were performed with no damage done to either IFOR's pontoon bridge neither to apartment building in the nearby Orašje. Relatively high water above the blasting place disabled dispersal of blasted material and also damped the shock wave.

Figure 8 shows the moment of blasting one of the sections.

Eventual damage of the bridge piers in Sava River and their functionality for reconstruction of the bridge was determined by special tests.

Figure 8: Moment of blasting

4 TESTING DAMAGE ON EXISTING PIERS

Field and laboratory testing was conducted according to the program agreed with the bridge-repair designer (IPZ - Zagreb) with the purpose of determining the level of damage inflicted on piers 11 and 12. Prior to the commencement of works, top parts of the piers projecting above the water level were removed, and pier surfaces were flattened.

According to the program, two boreholes 10,0 m in depth and one borehole 15,0 m in depth, were drilled at each pier.

The core drilling was performed in the period from 16 to 26 July 1997. The cored samples were extracted from the boreholes, disposed in boxes, described and photographed, and transported to laboratory for testing. Some of these samples are presented in Figure 9.

Figure 9. Samples from the central borehole on pier 12

Aggregate particles of less than 16 mm are dominant in the concrete samples although, considering the size of concreting segments, a higher proportion of coarser particles could have been expected. Some grains in the aggregate are weathered but there is a good link between the aggregate and the cement paste. A somewhat weaker structure is locally visible.

The inspection of the cored samples has shown that, from the standpoint of drilling, the samples are of a very good quality. The quality of concrete was determined quantitatively by means of CR and RQD indices which are normally used to determine quality of the rock mass. The CR (core recovery) index amounts to about 90%, while the RQD index is very close to the CR index. An average RQD index amounts to 87.2% for all samples.

The frequency of crack occurrence as determined by FF index is quite low and ranges from 2.2 to 4.4 which means that an average spacing between the cracks varies from 23 to 45 cm. It can be concluded from the shape of the cracks that most of them were formed during or due to the drilling process, while only a few of them resulted from concrete segregation or other local heterogeneity's present at the time of concreting.

None of the cracks observed during the sample inspection point to the concrete failure due to shear or tensile stresses.

To perform the compressive strength testing, an equal number of samples, uniformly arranged by height, were taken from all three boreholes on individual piers, and this from each concreting segment 4 m in height.

Compressive strengths of individual samples vary by depth but, as a rule, values are somewhat higher in the lowest concreting segment, when compared to those registered in upper segments. The minimum value of all tests is 17.4 MPa, the maximum value is 59.8 MPa, while the mean value is 40.61 MPa.

In order to test the static modulus of elasticity for concrete, one sample was taken from each concreting segment and this from the central borehole of each pier. In total, 8 samples were tested, and the values of elasticity modulus range from 27000 to 37600 MPa.

In addition, geophysical borehole logging was performed in four boreholes in order to determine in the continuous manner - by depth - the porosity, density and natural radioactivity of samples.

This logging was performed using the digitized instrument RG PORTALOG III manufactured by the English Company Robertson Geologging. This instrument uses the processor PC 386 and the

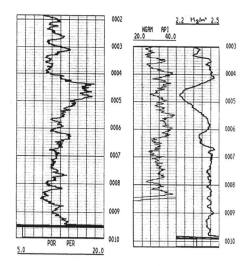

Figure 10: Geophysical logging results for porosity, density and natural radioactivity

MS-DOS operating system. The data are registered on thermal paper. The maximum depth of measurement is 500 meters.

The porosity was determined by means of the probe DNNS, DUAL NEUTRON, while the density and natural radioactivity was measured using the probe FDGT, gama-gama and gama.

An example of the geophysical logging readout is presented in Figure 10.

This geophysical logging has shown that the porosity and density of concrete varies by depth. The density varies in a relatively narrow range from 2.32 to 2.51 g/cm3 except at two points where the densities of 2.20 g/cm3 and 2.23 g/cm3 were registered. The porosity values vary from 8.0 to 18.0%. The intervals with somewhat more pronounces fall in density, i.e. with an increase in porosity, may be explained by the presence of smaller or larger nests in concrete which are due to technology used during concreting.

5 CONCLUSION

The overall testing results show that the piers 11 and 12 were realised in an appropriate way, with the concrete of satisfactory quality. The use of explosives to divide bridge remains into smaller sections did not damage the piers. No cracks in concrete give reasons to believe that the concrete failure at piers is due to either shear or tensile stresses.

REFERENCES

Ester, Z. & Krsnik J. 1996. Projekt razbijanja eksplozivom armirano betonske kolovozne ploče srušenog mosta preko rijeke Save kod Županje. Zagreb: Hydrocon.

Ester, Z. & Cirković, G. 1997. Demolition of reinforced concrete bridges during reconstruction of the motorway Zagreb-Krapina. *Proceedings of 21st Int. Conf. Explosives and Blasting Technique, Las Vegas, 2nd-5th February 1997:133-144. Cleveland ISSE.*

Joachim, C.E. & C.R. Welch 1997. Underwater shocks from blasting. *Proceedings of 21st Int. Conf. Explosives and Blasting Technique, Las Vegas, 2nd-5th February 1997:525-536. Cleveland ISSE.*

Ostrovskii, A.P. 1962. Deep-hole drilling with explosives. New York: Consultants Bureau.

Persson, P.A.., Holmberg R. & L. Jaimin 1994. *Rock Blasting and Explosives Engineering.* Boca Raton: CRC Press.

Geotechnical Hazards, Marić, Lisac & Szavits-Nossan (eds) © 1998 Taylor & Francis, ISBN 90 5410 957 2

Laboratory tests of the bearing capacity of spherical anchors in sand

L. Frgić & K. Tor
Faculty of Mining, Geology and Petroleum Engineering, University of Zagreb, Croatia

F. Verić
Faculty of Civil Engineering, University of Zagreb, Croatia

ABSTRACT: The paper discusses ultimate bearing capacity of spherical anchors in cohesionless soil under vertical pulling force. For that reason, several representative anchor diameter and embedment depth ratios were laboratory tested. The anchors were pulled out by given their vertical displacements, and the corresponding force intensity was measured. Regularities were observed in limit force increase with growing anchor depth and diameter, as well as the convergence of pulling force to certain limit value. The pulling force computations for the test cases were performed by Finite Elements Method (FEM). The problem was treated as an axial-symmetrical case. The vertical displacements of anchor are given in steps. Opening of cracks was modelled by exclusion of the elements subject to tensile stresses, thus providing a clear insight into the development of tension zones as the anchor displacement increased. The relationships between the anchor displacement and the pulling force as obtained by experiments and by numerical method have been statistically analyzed.

1 INTRODUCTION

In modern geotechnical structures the anchors are ever more frequently used as bearing elements for transfer of tensile forces from the structures to the soil. The soil and/or rock resistance against pulling is used to take over tensile forces. The frequency of anchors application in the engineering practice demands to study the complex problem of anchor/soil interaction in transferring the forces from the structure to the soil.

The problem of transferring tensile forces to the soil and the anchors bearing capacity involve numerous questions such as the soil failure conditions and shape of the sliding area, distribution of tensile stresses along the anchoring section, stress/strain relationships within the anchor/soil system depending on anchor space orientation, rheological processes in the soil etc.

An anchor, as a bearing element within the structure, could play a crucial role in structure safety.

The paper discusses behavior of anchors (i.e. spherical anchor bearing capacity) in cohesionless material exposed to vertical pulling force.

The laboratory tests were performed in order to get the relationships governing the analyzed problem.

2 LABORATORY TESTS

The laboratory tests were conducted for 14 representative relationships of the spherical anchor diameter D to embedment depth H (embedment coefficient $1 < \lambda = H/D < 4$).

The tests were conducted in a specially constructed tester consisting of a sand box (1), steel frame (2), measuring devices (3 and 7) and model anchor (9). The tester scheme is represented in Figure 1. The sand box measuring 70 cm x 77 cm x 48 cm is placed on the lower part of the frame made of rigid steel shapes. The steel frame deformations during the testing may be neglected because the frame is sufficiently stiff. The anchors were pulled out by using 16 mm dia screws (4) with rectangular thread and 80 mm dia rotating disk (3). On the disk is engraved an angular graduation (one grade for displacement of 1/20 mm). One complete rotation lifts the screw for 2 mm. The side dowel (10) prevents screw rotation in block (5) during the operations. The plate and the block through which the screw penetrates are fitted with precise bearing rings (6). The disk has six holes on the sides. A rod of the same diameter provides precise rotation resulting in desired displacement even for higher pulling forces. The rotating plate with the screw is connected with the anchor (9) by a dynamometer (7) and a steel rod with Φ 6 mm (8). The dynamometer and the rod are fitted with joints, which transfers only longitudinal forces, namely the forces in the direction of the rod carrying the anchor. The rod elongations (8) during testing can be neglected since they amount 2/1000 mm for the extreme force of 250 N. The anchor body and rod are connected by a 20 mm long thread. The turned wooden balls, diameters D= 25 mm, 50 mm, 75 mm and 100 mm, are used as spherical anchors.

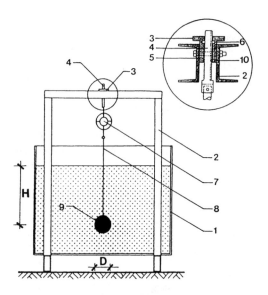

Figure 1. Outline of the laboratory tester for anchor pulling out

The tests were performed using the sand from the Botovo sand pit near Drava River, dried in laboratory driers to zero moisture content and sieved on 2 mm mesh sieve.

The dry sand was placed in the box in layers. Since reliable data can be obtained only for soil of constant and reliable mechanical characteristics, special attention has been paid to the model preparation.

Compaction uniformity of test models has been checked by determining the sand density in the containers of the known volume. Penetrometer measurements were also used for the control of uniform compaction of sand in the model, so their significance is

only comparative. The measurements were performed by a modified penetrometer needle.

Vertical position of anchors during sand filling was controlled by vertical guides set in two directions perpendicular to each other.

The result was a dry filled model which was submerged in water. Water was drained through the perforated bottom and valves mounted on the side walls of the box, near the bottom. Since the sand permeability is high, the model has been left idle for 24 hours.

The anchors were pulled out, one by one, by manual rotation of the disk following the engraved gradation, i.e. by presetting vertical movement of the anchor. The force intensity was measured by precision ring dynamometer.

The tests of relationships between the pulling force and the anchor displacement in natural soil were conducted for the cases of shallow foundation at which total soil failure takes place. The failure surface extends from the anchor up to the soil surface. The following cases were tested, designated as D_λ: 100_1; 100_2; 100_3; 75_1; 75_2; 75_3; 75_4; 50_1; 50_2; 50_3; 50_4; 25_2; 25_3 and 25_4. At least three tests were run for each of the above relationships. The extraction was continued until maximum force was achieved, as indicated by the absence of the force increment, namely decrease in force value during further displacement increase.

The measurement results (groups of data pairs for anchor displacement d and adequate values of pulling force P), are shown in the form of diagram expressed by the below expression

$$P = f(d) \tag{1}$$

and they have a characteristic shape.

The regularities were observed in limit force increase with growing anchor depth and diameter, as well as convergence of pulling force to certain limit value.

3 STATISTICAL ANALYSIS OF MEASUREMENT RESULTS

The limit pulling forces are determined by the approximation of obtained measurement results using exponential functions.

As the approximation curve a curve with a following shape has been adopted

$$P = a \times [1 - b \times e^{-\alpha \times d} - (1-b) \times e^{-\beta \times d}] \tag{2}$$

516

The exponential curve has an asymptote which presents limit pulling force a=P_{gr} for large displacements (d=∞). The unknown parameters P_{gr}, b, a and b may be determined by an iterative method only.

The method is based on the optimization of parameter values, where the minimum deviation of the squares of differences between the measured values and the corresponding values of the approximation curve is requested (Turk 1978).

The measurement data and the corresponding exponential function of approximation for anchor 50_2 are plotted in Figure 2.

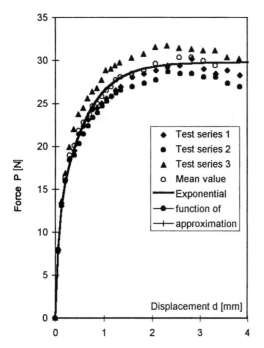

Figure 2. Approximation curve of the measurement data for 50_2 (D_λ) anchor

The Figure 3 shows approximation curves for other embedment depths for anchor diameter D = 50 mm, while the Figure 4 shows approximation curves for the measurement results with embedment coefficient λ=H/D=2.

Special program was developed for this statistical analysis. Values of approximation exponential function parameters are given in Table 1.

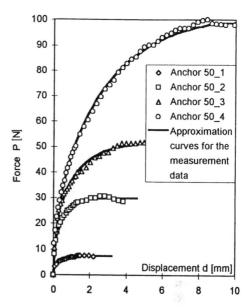

Figure 3. Approximation curve of the measurement data for anchor with diameter D = 50 mm, for different embedment depths

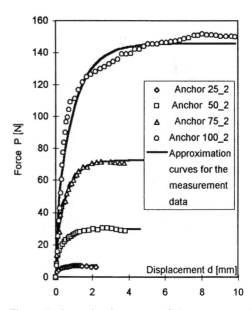

Figure 4. Approximation curve of the measurement data for embedment coefficient λ=H/D=2

Table 1. Paramters of the measurement results approximation functions

Anchor design.	Depth H (mm)	P_{gr} (N)	b	α	β
25_2	50.	7.28	0.629	2.641	28.478
25_3	75.	12.23	0.609	1.566	16.886
25_4	100.	21.31	0.557	0.769	8.309
50_1	50.	7.42	0.602	2.057	22.200
50_2	100.	29.80	0.665	1.747	19.347
50_3	150.	51.81	0.627	0.783	8.451
50_4	200.	102.00	0.749	0.349	3.758
75_1	75.	27.41	0.638	1.636	17.634
75_2	150.	72.60	0.739	1.544	16.628
75_3	225.	135.49	0.568	0.708	7.640
75_4	300.	227.36	0.728	0.164	1.768
100_1	100.	50.72	0.618	4.947	53.396
100_2	200.	145.74	0.764	0.995	10.698
100_3	300.	301.50	0.784	0.153	1.654

4 NUMERICAL MODELLING

The limit pulling force computations for the test cases were performed by Finite Elements Method (FEM). The anchor extraction from the soil was treated as an axial-symmetrical case. The displacements at the nodal points of the anchor are given per increments. The finite element mesh for calculation of the pulling force for anchor 50_2 is represented in Figure 5.

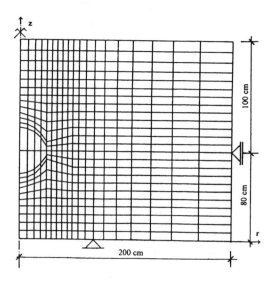

Figure 5. Finite elements mesh for calculation of the 50_2 anchor pullout force

Since we are dealing with an axial-symmetrical problem (axis z is the axis of symmetry), the symmetry of displacement and load is ensured by vertically moving supports in the nodes in axis of symmetry and on the right boundary of the respective area. The fixed supports are set in the nodes of the lower boundary of the observed area.

The soil is modelled as a homogenous isotropic elastoplastic material. The parameters of shearing stress c and φ are determined by standard testing of sand samples. The material properties used in calculation are specified in Table 2.

Table 2. Physical properties of the soil

Property	Value
Unit weight	15.0 kN/m³
Young's modulus	10,000.0 kN/m²
Poisson's ratio	0.3
Cohesion	0.5 kN/m²
Friction angle	28.0°

The elastic material model with a hundred times higher modulus of elasticity was adopted for anchor elements in order to get a relatively undeformable body.

To eliminate the effect of adhesion between the anchor and soil, and accelerate the process of relaxation, the first series of elements under the anchor has been modelled using reduced modulus.

Following the concept of initial stresses in Gaussian points, the gravity load is added to the total stress vector in the first calculation increment.

Since the planes of soil failure (discontinuity) have not been known in advance, it was not possible to use sliding elements. The finite elements program was completed with subprograms for changing moduli of individual elements. The exclusion of the elements subjected to tensile stresses was performed by gradual reduction of modulus, thus enabling tensile strains with insignificant transfer of force.

Thus modelling of cracking and sliding was ensured and a clear insight was gained into development of tension zones as the anchor displacement increased.

5 STATISTICAL ANALYSIS OF NUMERICAL CALCULATION RESULTS

The same statistical method was applied to the analysis of numerical calculation results.

The curve of the same shape was used for approximation curve:

$$F = F_{gr} \times [1 - b \times e^{-\alpha \times d} - (1 - b) \times e^{-\beta \times d}] \qquad (3)$$

since the numerical result diagrams are similar in form to the measurement diagrams.

The convergence of calculation force to the limit value was obtained, which is in concordance with the measurement results.

The Figure 6 shows approximation curves of the numerical computation results for the anchor D=75mm, while the Figure 7 shows approximation curves of the numerical computation results for the embedment coefficient λ=3.

The values of the limit pulling forces obtained by approximation of the numerical calculation results F_{gr} are given in Table 3 and Figure 8 for comparison purposes, together with the values of the limit pulling forces obtained by approximation of the laboratory test results P_{gr}.

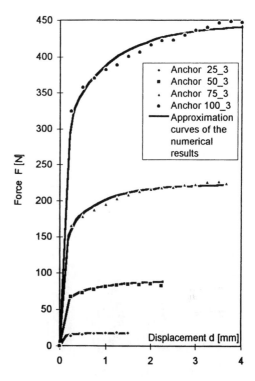

Figure 7. Approximation curve of the numerical calculation results for embedment coefficient λ=H/D=3

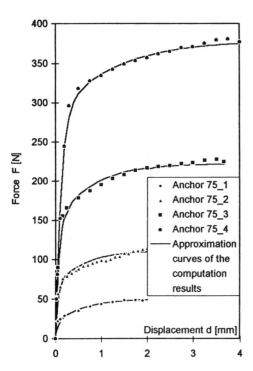

Figure 6. Approximation curve of the numerical calculation results for anchor with diameter D = 75 mm, for different embedment depths

Table 3. Limit pulling force values obtained by approximation of the laboratory testing and numerical calculation results

Anchor designation D λ	Depth H (mm)	Laboratory tests Force P_{gr} (N)	Numerical calculation Force F_{gr} (N)	Ratio $\frac{P_{gr}}{F_{gr}}$ -
25_2	50.	7.28	12.22	0.60
25_3	75.	12.23	17.64	0.69
25_4	100.	21.31	26.45	0.81
50_1	50.	7.42	20.71	0.36
50_2	100.	29.80	46.04	0.65
50_3	150.	51.81	89.11	0.58
50_4	200.	102.00	138.30	0.74
75_1	75.	27.41	50.57	0.54
75_2	150.	72.60	112.03	0.65
75_3	225.	135.49	222.49	0.61
75_4	300.	227.36	379.96	0.60
100_1	100.	50.72	95.47	0.53
100_2	200.	145.74	224.08	0.65
100_3	300.	301.50	443.60	0.68

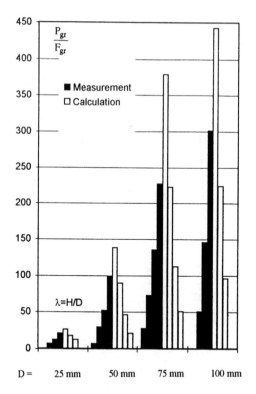

Figure 8. Limit pulling force values presented by approximation of measured data and numerical calculation results

6 CONCLUSION

The chosen approximation curves describe well the observed phenomenon.

Comparison of the values of limit forces obtained by the FEM analysis and the laboratory measurement results shows qualitative analogy.

We also performed some field tests and received the same relationships.

The overview of the results of approximation by exponential data curves for laboratory tests and the FEM analysis results indicates that the approximate limit forces from the measurement data are in average 62% of the calculation values.

The differences can partly be attributed to the unreliable measurements of soil mechanical characteristics, and only partly to the imperfection of the mathematical model.

REFERENCES

Das, B.M. 1990. *Earth Anchors.* Elsevier Scientific publishing Co.: Amsterdam.

De Borst, R. & P.A. Vermeer 1984. Possibilities and limitations of finite elements for limit analysis. *Geotechnique,* Vol. 34, No. 2, 199-210.

Desai, C.S. & J. T. Christtian 1977. Numerical Methods in Geotechnical Engineering. New York: McGraw-Hill.

Frgić, L., M. Magerle & K. Tor 1997. Influence of Boundary Conditions on Stress and Strain Results near Underground Rooms. *Proc. of 2nd Congress of Croatian Society of Mechanics,* 157-164. Supetar, Croatia.

Gudehus, G. 1977. *Finite Elements in Geomechanics.* John Wiley & Sons: London.

Heternyi, M. 1954. *Handbook of Experimental stress analysis.* New York: John Wiley & Sons.

Hinton, E. & Owen, D.R. 1977. *Finite element programing.* London Academic Press, London.

Hinton, E. & Owen, D.R. 1779. *An introduction to finite element computations.* Pineridge Press Limited, Swansea.

Murray, E.J. & J.D. Geddes 1987. Uplift of anchor plates in sand, *J. Geotech. Engin. Div.,* ASCE, Vol. 113, No. 3, 202-215.

Murray, E.J. & J.D. Geddes, 1989. Resistance of pasive inclined anchors in cohesionless medium, *Geotechnique,* Vol. 39, No. 3, 417-431.

Naylor, D.J., G.N. Pande, B. Simpson & R. Tabb 1981. *Finite elements in geotechnical engineering,* Pineridge Press, Swanse.

Owen, D.R. & Hinton, E. 1980. *Finite elements in plasticity: Theory and practice.* Pineridge Press Limited, Swansea.

Saeedy, H.S. 1987. Stability of circular vertical earth anchor. *Canadian Geotechnical Journal,* Vol. 24, No. 3, 452-456.

Sarač, Dž. 1989. The uplift capacity of shallow buried anchor slabs. *Proc. 12th ICSMFE,* Vol. 2, 1213-1216, Rio de Janeiro.

Sarač, Dž., F. Verić & K. Horvat 1976. Dimenzioniranje temelja dalekovodnih stupova prema vlačnoj sili. *VI savjetovanje o prenosu električne energije u SFRJ, Cavtat,* 1-27. Liber Zagreb.

Turk, S. & L. Budin 1978. *Analiza primjenom računala.* Školska knjiga, Zagreb.

Zienkiewicz, O.C. 1977. *The finite element method* McGraw-Hill Book Co., U.K.

Geotechnical Hazards, Marić, Lisac & Szavits-Nossan (eds) © 1998 Taylor & Francis, ISBN 90 5410 957 2

Origins of mass deformations of building in Norilsk region

V. I. Grebenets & A. G.-o. Kerimov
Research Institute of Bases and Underground Structures, Norilsk, Russia

V. T. Paramonov
Mining Management, Norilsk Plant, Russia

ABSTRACT: Observations conducted in the 90-s provided evidence of increasing number of deformed structures. It has been determined that thawing of permanently frozen soils enclosing end-bearing piles leads to a drastic decrease in piles stability owing to the action of bending moments and shearing forces.

A thorough examination of the current construction norms and specification has revealed that the most commonly used pile types in Norilsk region (0,4x0,4 m and less) have not been specified by the state construction standards. Violations of maintenance rules for geotechnic systems (service cellars, collectors etc.) and distinctive features of structures heat release result in thawing of frozen layers. Nonuniform thawing soil settlement, presence of frozen lenses and rise of ground water enhance local variations of unfavorable impacts on deep emplaced piles.

DIE THESEN: Die Naturbeobachtungen in Nordensibirien in das Ende des XX Jahrhundertes haben gezeigt, daß die Anzahl der deformierten Objekte Ständig größer wird. Es ist festgestellt, daß das Tauen des die enthaltenden stähnden Pfähle Permafrostes bei der Degradation ewiger Frostboden zur helfige Standfestigkeit der Pfähle durch den Einfluß der Krümmungsmomente und der Verschiebenskräfte herbeigeführt wird. Die Analyse der existierenden Dokumente auf diesem Gebiet hat gezeigt, daß die verbreitesten in Nordsibirien Typen der Pfähle (0,4x0,4 M und weniger) durch die Nationallalbaunormative nicht geregelt wird. Die Verletzung des Betriebes der Geotechniksysteme (die speuillen im Winter belüfteten Keller, die Kollektoren für der Ingenieurkommunikation u.a.), die Besondersheiten der Wärmeaussonderungen vor Anlagen rufen das Tauen des Permafrostes heraus. Die ungleichmäßigen Senkunden der tauenden Gründe, Die Erhalung der Schichten des Frostbodenes, die technogenische Bewässerung verstärken die Lokalverschiedenheiten der negativen Einflüsse auf die tiefen Pfähle.

1 INTRODUCTION

The issue of foundation engineering on an industrial scale in permafrost regions, as a whole, has been resolved. The results of observations conducted in the 80-90-s in Norilsk and other North Siberian cities however demonstrated the increasing number of deformed structures. Well-known features [*Danilov, 1978*] of permafrost are there frost-facial variations along dip and strike. The foundation soil even of one structure varies in composition, structure, density, temperature, ice-content, frost conditions. Consequently, permafrost warming and thawing phenomena manifestations vary, which results in different foundation displacements causing structure deforma-

tions. Furthermore, different loads on particular supports enhance differentiation of foundation dislocations. Currently about 25 % of large structures in Norilsk are situated on foundations beds with disturbed frost conditions and about 250 buildings has been deformed. For the last 5 years foundation beds with technogenetically minimized bearing capacity have numbered 5 times as great as for the 50 previous years. It has been established [*Grebenets et. al., 1994*] that a general tendency in big cities of the region is towards permafrost degradation. Yet local permafrost warming up and thawing manifest themselves more pronouncedly and have more adverse economical and environmental impact than general warming up with a stable temperature balance.

DISCUSSION AND NEW OBSERVATIONS

Deformations of structures caused by changed permafrost conditions have originated for the most part from the stages of construction development of the region. Unindustrialized construction and lack of expertise and experience account for actual disregard of permafrost phenomena in the early stages of the Arctic regions development. Foundation engineering conceptions suitable for southern regions were mechanicaly adopted. The most frequently laid foundations were the ones of strip and pier types resting on rock bed or on lightly thaw-compressed soil. And that imposed particular requirements on construction sites of big industrial structure. Buildings and structures built on permafrost got damaged in numbers. Their service life seldom exceeded 7-10 years.

Beginning in the mid-50s pile foundations found application. Their specific feature was wide use of so called "floating piles", the bearing capacity of which was ensured by their freezing-in in permafrost. The findings of the engineering survey and permafrost exploration determined the pile field design (primarily the depth of emplacement and the number of piles): the lower was the temperature of permafrost and coarser the texture, the less was the depth of the emplacement. Clearly, permafrost degradation led the to impermissible deformations of buildings and structures whereas load on a structure remained unchanged adfreezing forces freezing together a pile and the soil decreased as the temperature rose. It was the permafrost degradation that caused mass deformation of buildings and structures in the 90-s.

At present in foundation engineering in permafrost areas the following lines of investigation have been brought forward:

- a technology advancement of different designs of pile foundation (drilled-sunken, drilled-cast-in-situ, drilled-driven, combined ones)
- specific requirements for pile composition brought about by underground water phase transformations, temperature fluctuations, technogenic excess moisture content of concrete, underground water corrosiveness etc.
- a solution to the arisen problem of reducing stability of frozen-in pile foundations under changing geothermal conditions
- a considering of impairing geocryological conditions in the twenty first century under the effects of technogenesis and the global tendency toward warming-up.

Nowadays, the most extensively employed technique is sinking of a pile into a predrilled hole, larger in diameter, followed by grouting-in the hollow (so-called sunken piles). The bearing capacity of such piles is secured by the strength of adfreezing these piles and the enclosing soil together (floating piles) or by resting a pile end on a hard, mostly rock bed (standing piles). In either case, the piles are restrained and counteract bending moments and horizontal loads within permafrost. Similar conditions are required for butt-joined drilled-driven piles of deep emplacement. They are sunk into predrilled-through-permafrost holes, and after that driven through underlaying soft (unfrozen) soils to a rock bed. Consequently, the maintenance of soils in a permanently frozen state has acquired great importance. In actual practice, however, this task presents a complicated problem. There are some special situations possible, in which piles lose their stability, i.e. this type of pile foundations is rather insecure. Within recent years soil thawing has caused damage of several big buildings, and construction deformations of 25 nine and five-storeyed residental buildings have reached a critical stage. These buildings may get damaged within the next few years.

It has been believed for a long time that end-bearing piles are the most reliable ones, their bearing capacity ensured by pile ends resting on an uncompress rock bed. As of now, quite a number of structures have been erected on end-bearing pile foundations (piles of deep emplacement including), and practically in all instances permafrost maintenance has to be provided.

Russian state construction norms and specifications [SNiP, 1986] do not list adequately defined specifications of end-bearing piles. At the design stage the accuracy of calculation can be affected by the type of pile foundation design and the definition of design parameters. The problem of the calculation accuracy becomes even more complicated in the designing of deep emplaced end-bearing piles, when a side surface (or its part) of a pile has proved to be frozen into a lens of permafrost.

Russian state construction norms [SNiP, 1986] specify as one of the basic types drilled reinforced concrete piles, which are installed in the course of drilling of the holes with or without bottom widening, grouting of sand-cement grout into them and sinking of cylinder or prism shaped units sectionally continuous, 0,8 m and more in sides or diameter. Thus, the current construction norms do not specify pier piles of rectangular section, 0,4x0,4 m in size and less, which have found wide use in construction in Norilsk industrial region. The existing recommendations for calculation, choice of geometrical parameters (reduced length, radius of ineytia of cross-section) can not be readily followed in designing of piles 0,4 m and less in sectional size. Thus, the state constructions norms lay the groundwork for possible future decrease in stability of structures.

Obviously, the imperfection of the preparation process of making a hollow in the rock is one more cause for deformations as it is no possible to completely drain mud remains out of the well bottom.

In many cases of constructing mud remains (0,2 - 0,5 m in height) freeze before installation of a sunken end-bearing pile and in the process stratify (sediment and elutriation of particles) to produce an ice horizon in the upper part; so settlement initiation of the loaded end-bearing piles is aided by disappearance of the ice horizon if grounds thaw. A further reason for decrease in reliability of the structures built on end-bearing piles is the impossibility to define the surface strength of rocky grounds with assurance where the pile end is set according to the drilled-and-driven technique of sinking the upper permafrost layer by a leading well and driving the end-bearing pile through the lower thawed layers down to the bearing horizon. In this case water penetration and filtration over the rock surface destroy the compacted ground core formed under the pile plane end upon driving.

In the design of load-bearing elements to buildings and structures the pile foundations relative to the building frame elements are considered as fixed bearings. In connection with great permafrost facies changeability of grounds upon thawing settlements and other displacements of deep-seated end-bearing piles are differential and in consequence impermissible (up to destructions) structure deformations evolve.

It should be pointed out that due to the technogenic effect-violation of maintaining cold ventilated subfloor spaces, heat release of underground collectors for utility lines, mechanized movement of snow masses, use of artificial well-filter bedding course, technogenic inundation and salinization of a seasonally thawed layer and etc. - the geocryological conditions at building and structure foundations in Norilsk region are not stable, a tendency for permafrost degradation is observed through the urbanized territory [Grebenets et.al., 1994].

Pile joints are one more problem on the way to provide reliability of the deep-seated end-bearing piles which are not designed on bending moment influence and shearing force effects may take place in the ununiform thawing process of the permafrost grounds where foundations are installed in.

On clear example of failure deformation one of the greatest structures built on end-bearing piles in Norilsk - Peak Boiler House of Norilsk Heat Power Station No.1 supplying the city and industrial zones with the heat - might be called. The foundation of the structure built at the beginning of the '70s and consisted of a boiler room and a pump room are constructed from end-bearing piles, in this case loads put on them are in the range from 900 kN to 1800 kN and bending moment is up to 70 tm.

Pile clusters (of 3 - 4 piles) run along the extreme axes of building frame spans. Other loads are taken by isolated piles with cross-section 40 x 40 cm arranged in the form of a grid (6 m). In particular cases cylinders with round cross-section (720 mm in diameter) are installed, 3 % being in the total quantity.

Technical and geocryological conditions of the site before building were reasonably easy: gabbrodolerit bedding under permafrost morainic loams. Actual surface of rocky grounds where end-bearing piles jammed is in a depth range of 16-17 m to 23-24 m as to the surface, under these conditions the least depth of the rock surface bedding is located at the area close to the building geometrical centre in plan. The building layout relative to the rock surface as well as eolation of the upper horizons of the gabbrodolerit enhance the overall tendency toward the structure unstability when grounds thaw. Before building the depth of seasonally loam thawing was not over 1,4 m; temperature at the level of zero annual amplitudes was - 3,5 °C. The building was constructed and maintained in terms of "SNiP" [SNiP, 1990] that meant no preservation of the permafrost state of foundation bed grounds.

Permafrost grounds were in the process of destroying during maintenance of the structure that released heat and water to the basement in large amount. According to the studies [Grebenets et. al., 1996] depth of foundation bed ground thawing at the top over 20-year period of maintenance ranged from 4-5 m to 8-11 m and pressure filtration of the technogenic water discharge of Norilsk Heat Power Station No.1 situated distance 300 m was over rocky ground surface.

Thawing and elevation of the permafrost temperature within the foundation placement was found as a serious danger to the structure stability. As permafrost grounds thawed the eccentrically loaded and jointed piles were provided with a possibility to bend in the wells of a larger diameter than the piles have. Over 20 years unmanaged hot water discharges occurred at the basement of Peak Boiler House of Norilsk Heat Power Station No.1 as well as migrated through some horizons from the main building of Peak Boiler House of Norilsk Heat Power Station No.1. The water discharges and the conductive heat from the structure to the basement destroyed permafrost and caused ground settlements under the building with producing cool subfloor space in places not be designed that resulted in concrete destruction of piles and a socle floor of great moister. Full-scale investigations have shown that all floor girders within the pump room were subject to concrete destruction, grid foundation beams of the boiler room were deformed at the rate about 20 %, protective concrete layer was completely destroyed in places, there were numerous cracks over the grid foundation beams as well as over the pile heads. Reinforced concrete de-struction of the grid foundation beams and the socle floor resulted in an actual loss of "reinforcing disk " that combined pile heads, that was, neutralization and redistribution of horizontal loads and moments that took place upon piles eccentrically loaded were not available.

The study of the original design and engineering concepts as well as executive documentation materials revealed that the execution of pile foundation design was not sufficiently correct. During object installation standard documentation on building con-struction within permafrost zone was not strictly followed, particularly, floor waterproofing was not done, emergency water discharges penetrated through the basement and wetted structures, ground floor was not made as a monolithic reinforcing disk. According to the scheme of strength calculation in foundation design the end-bearing piles restraint length was rated to be equal to 8 diameters i.e. 3,2 m as a pile diameter is 40x40 cm. It was assumed that a pile should be restrained in the frozen ground at a 3 m depth (active layer and thermal effect from a building were not even taken into consideration). It should be noted that no special measures to preserve the ground in frozen state were provided by the de-sign and this resulted in ground thawing in course of the object operation. Naturally, under the influence of ground thawing process the restraint depth of end-bearing piles was changed. Therefore the rated length of piles should be increased. It is well known that there is a direct relationship between the stability of end-bearing piles and their rated length. Taking into account basement thawing it is necessary to use the full length of piles in calculations of their stability.

Three-dimensional models according to the scheme "basement ground - end-bearing pile - grid foundation - building frame" were developed and calculated for the evaluation of ground thawing criti-cal depth at which end-bearing piles lose their stabil-ity in the basement of Peak Boiler House (Norilsk Heat Power Station No.1). The calculation was per-formed by the computer programme based on the finite-element method. Actual length of end-bearing piles, existing strength parameters of the system ele-ments material defined by non-destructive test meth-ods analysis and compression tests of concrete sam-ples, as well as all possible loads applied on founda-tions and ceilings were used in the calculation scheme.

Calculation analysis of the basic model taking into account changed permafrost-ground state where de-formation mainly takes place in the direction of end-bearing piles row with maximum length and larger external loads, showed that three-dimensional rigidity of the system is insufficient.

Taking into consideration, that end-bearing piles stability depends on their cross-section geometric parameters the situation with enlarged size of the cross-section up to 8000 mm (cast-in-place end-bearing piles) was simulated. Calculations showed that this ensures the reliability of a building. The cross-section enlargement of the existing joined piles is an extremely labourintensive process comprising post installation around each pile in 24 m deep wells and in this case deformation of individual structures (even collapse) may occur, because a stripped pile, prepared for reconstruction (40x40 cm in cross-section) is turned to bend significantly (wittin the pit limits) under the influence of applied eccentric loads and moments.

As a result of different models calculation it was revealed that with ground thawing up to 12 m depth the building loses its stability.

To ensure reliable operation of Peak Boiler House (Norilsk Heat Power Station No.1) special measures were developed. To ensure spatial rigidity of a building, a supporting frame consisting of a large diameter (up to 1000 mm) cast-in-place piles was designed and realized. It was installed along the first row where prohibitive heel deformation of the entire structure occurs. The following stabilization meas-ures were proposed: prevention of hot water pene-tration to the basement through the floor and instal-lation of water drainage system for emergency dis-charge; restoration of the floor reinforcing disk and grid foundation beams by means of their splicing; additional reinforced concrete frame construction at basement surface level; restoration of basement fro-zen ground conditions and formation of peculiar "frozen reinforcing disk" joining all piles and ensur-ing their common and function due to earth-sheltered ventilation cavities. It is necessary to perform floor hydroinsulation with modern materials, pile heads and grillage strengthening with reinforced concrete casing and united together by a system of cross-over beams at the level of pile heads bottom.

CONCLUSIONS

Many similar surveys, calculations and develop-ment of certain engineering considerations were performed in the region in 1990s for 15 large struc-tures built on deep seated end-bearing piles subjected to deformation due to geocryological and hydro-geological changes.

While evaluation of deep-seated end-bearing piles stability within changing geocryological conditions the following issues are still under question:
- the possibility to classify drilled-and-sunken piles of 40x40 cm or less cross-section (this type is widely used in Norilsk region) installed in a big diameter

drill holes as end-bearing piles according to national "SNiP" standards, 1986;

- the necessity to consider basement thawing ground as a linear deformation medium and to evaluate its strength to deformation occurring in piles under horizontally applied loads;

- quantitative analysys of the affect of change in geocryological conditions on foundation bearing capacity.

Taking into account the progressively worsening tendency in geocriologic conditions of buildings and structures basements in Norilsk Industrial region, the lack of proper and detailed regulations on calculation, design and installation of deep-seated end-bearing piles, it is expedient to perform a scientific research aimed at the development of recommendations (at Federal and regional construction regulations level) concerning their design, calculation and operation within possible changes in geocryological conditions.

REFERENCES

Danilov I.D. 1978. *Polar lithogenesis.* Proc. "Nedra". Moscow. *[in Russian].*

Grebenets, V.I., Fedoseev, D.B. & Lolaev, A.B. 1994. *Technogenesis influence on the frozen ground.* Proceeding influence on the 7th Congress International Association of Geology, Lisboa, Portugal. Editor R.Oliveira, A.A.Balkema, p.p. 2533-2536. Rotterdam.

SNiP 2.02.03-85. (Russian Building Codes and Rules). 1990. *Piles foundation.* Gosstroy USSR, Moscow. (in Russian).

SNiP 2.02.04-88. (Russian Building Codes and Rules). 1990. *Bases and foundations in permafrost.* Gosstroy USSR, Moscow. (in Russian).

Grebenets, V.I., Kerimov, A.G., Melnik, M.F., Khlopuk, L.Y. 1996. *Reliability of building Norilsk Heat Power Station No.1.* Proceeding of the regional Scientific conference "The extreme North 96". Norilsk, Russia. p.p. 47-49, *[in Russian].*

Geotechnical Hazards, Marić, Lisac & Szavits-Nossan (eds) © 1998 Taylor & Francis, ISBN 90 5410 957 2

Tunnelling in poor ground – Choice of shield method based on reliability

M.T. Isaksson
Department of Civil and Environmental Engineering, Royal Institute of Technology, Stockholm, Sweden

ABSTRACT: This study deals with a possible methodology for the choice of shield excavation method, based on an economic model. By using a geological/constructability model the robustness of the method can be defined. The robustness of the method can be increased by certain measures. A developed probability-based decision model has been used to estimate expected costs for various shield methods of excavating. Clearly formulated decision criteria are defined: the choice of method should not be based on the lowest standard cost but on the lowest expected total cost. A simulation has been carried out in a case study with three different excavation methods in the Grauholz tunnel in Switzerland. The result shows that the method can be made more robust by implementing additional measures which give less probability of major cost increases.

1 INTRODUCTION

The correct choice of excavation method is decisive for maintaining the cost ceilings and time frames of a tunnel project. Therefore, it is essential to select a method that is robust, and can cope with any type of geology that may occur along the whole extent of the tunnel. The right decision about the tunnelling method should be taken as early as possible in the planning phase. Factors influencing this decision could be geology, length of tunnel, number of tunnels, time schedule, space in front of the tunnel adit, buildings above, availability of trained personnel etc.

To facilitate the decision about the cheapest method of excavation, decision theory can be applied. The decision criterion for excavation is to minimise the risk of cost increases. The willingness of the construction parties to accept risks must also be regarded.

The choice of alternatives for construction is standardly based on earlier projects under similar geological and other conditions. A proper decision analysis, including main tree, fault tree, maximum cost and so on, is very often not carried out due to lack of time and funds. In the absence of a large bank of experience and skilled personnel, it is almost impossible to compete for and realise certain design solutions. In such cases, expensive solutions may be the only reasonable alternative.

This study deals with a possible methodology for the choice of shield method. Robustness, the probability of unwanted events and measures to increase robustness are examined. Certain excavation methods can be excluded directly due to external circumstances. Others are worth analysing further.

Research into choice of excavation method has been conducted since the 1970s at Massachusetts Institute of Technology in USA (Einstein & Vick 1974). A simulation program called Decision Aids in Tunnelling (DAT) has been developed and applied to the choice of excavating system by estimating project costs and time (Einstein et al. 1991, 1996), (Salazar 1983).

This article is part of a scientific project at the Division of Soil and Rock Mechanics at the Department of Civil and Environmental Engineering, Royal Institute of Technology (KTH) in Stockholm, Sweden.

2 THEORETICAL APPROACH

2.1 Simulation model

A probability-based decision model is used to estimate expected costs for drilling with different shield machines. The advance rate varies depending on constructability state, that is, ability to cope with the geology, which affects the construction time when estimating the costs. Constructability state I means that problems arise, state II means standard operation and state III means that work is progressing very smoothly.

The simulation is based on an economic model. The costs of every excavation method in every geological zone and for every constructability state consist of standard costs and a risk factor; see equation (1). To obtain a probable cost, including the risk of the method working poorly, the product of the costs and the probability of a certain constructability state is estimated in a given zone; see equation (2). Then the probable costs for every zone are added together. The parameters included in the cost model, for instance, advance rate, are expressed as stochastic variables which are modelled as triangularly distributed. A Monte Carlo simulation of the total implementation costs displays the distribution of costs.

The robustness of the method can be increased through measures. If the method is more robust when applied to a certain geological formation, the probability of a problem arising is decreased.

Expected costs are estimated according to these formulas:

- The cost at a given constructability state in a given zone

$$C_{zone.i} = C_n + P_f * C_f \qquad (1)$$

- The total cost of the whole tunnel

$$C_{tot} = \sum_{all.zones} \sum_{all.states} \left(C_{zone.i} * P(state) \right) \qquad (2)$$

where C_n = standard cost for one zone, P_f = probability of an unwanted event, C_f = cost consequences of this event and $P(state)$ = probability of the constructability state.

The decision criteria for the choice of method is, consequently, not only the minimum standard costs min $\sum_{all.zones} C_n$ but the minimum expected total cost

min $\sum_{all.zones} \sum_{all.states} \left(C_{zone.i} * P(state) \right)$.

The following parameters are necessary for the simulation of the costs:

- advance rate
- standard cost for excavation and support
- probability of an unwanted event in the given constructability state
- cost consequence of this event
- probability for geological/constructability state

The methodology to determing these parameters is described below.

2.2 Methodology to calculate advance rate for shield machines

A study of tunnelling operations and how these are affected by disturbances has been carried out to provide a basis for calculating the advance rate. This can be done if the geology is constant throughout a section of the tunnel (zone) and is obtained from a subjective estimation and follow-up data from earlier projects. The advance rate is then calculated according to the following formula:

$$y = \frac{T * \lambda}{\left(tb + tf + tk + td \right)} \qquad (3)$$

where T = working hours per day, λ = round length which means the length at maximal extended cutter head, tb = time for excavation, tf = time for installing concrete segments, tk = time for change of cutter, td = various down time. These time parameters express the numbers of hours for one round of length λ.

$$tb = \frac{\lambda}{\left(v * rpm * 0{,}06 \right)} \qquad (4)$$

where rpm = rounds per minute of the cutter head, v = penetration rate per round.

$$tf = ts * seg + to \qquad (5)$$

where ts = the time to install a segment in the ring, seg = the number of segments per ring, and to = time for changeover between previous operation and commencing the installation.

$$tk = MTTR * \frac{\pi^2 d\lambda}{(MTTF * 4n)} \qquad (6)$$

where $MTTR$ = mean time to repair cutters, $MTTF$ = mean time to fail, and n = number of cutters on the cutter head.

An estimation of down time (td) can be made with fault tree analysis. In a main tree, the principal factors that cause down time are defined. Reasons for down time in the respective main trees are illustrated with sub trees. As a basis for estimating the probabilities in the case of each initial event in the following example, data from the Grauholz tunnel has been used: (Aebersold 1994).

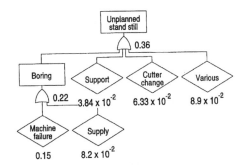

Figure 1. Main tree for unplanned down time

2.3 Methodology for estimation of standard costs

Standard costs are estimated for different methods of shield excavation. The costs depends of the tunnel diameter and consists of:
- Fixed costs
- Time-dependent costs

The following equation is used to estimate standard costs for the entire tunnel using shield excavation.

$$Cn = \sum C_{fix} + \frac{\sum C_{tid} * i * \lambda}{y * dm} \qquad (7)$$

where $Cfix$ = fixed costs, $Ctid$ = time-dependent costs, i = number of rounds along the whole tunnel, λ = round length, y = advance rate and dm = number of days per month.

The fixed costs consists of:
- site establishing and de-establishing
- material costs according to the following formula:

$$Cfix_2 = \left[\frac{\pi * d^2}{4}(E + S + B) + \pi * d *(I * tj * 1,3 + M * t)\right] * i * \lambda \qquad (8)$$

- shield and back-up system according to the following formula:

$$Cfix_3 = (1 + R)^N * d * \left(\frac{Re + A}{2}\right) \qquad (9)$$

where d = tunnel diameter, E = running costs, S = wear costs, B = bentonite consumption, I = grouting between segment and rock, tj = space between segment and rock, M = costs for support, t = segment thickness, A = purchase price for shield machine and back-up system, Re = scrap value of shield and backup, R = annual interest, N = depreciation time.

The time-dependent costs include:
- site office with personnel
- machine costs
- wages for workers

2.4 Methodology for estimating probability of constructability state

Probabilities for various constructability states depend on the geology and on the robustness of the method, that is to say its sensitivity to variation in geological factors such as water and clay content in the soil or the occurrence of boulders.
- *State I* means that the method cannot handle the problems. The probability of unwanted events is high.
- *State II* is the optimal method under given circumstances. Low probability of unwanted events.
- *State III* means that excavation with a cheaper method could be conducted without increasing the risk that problems will occur.

To be able to estimate the probability for constructability state, the probability of geological factors must be assessed. The geological factors could be designated A, B and C. A geological factor of type A could, for example, signify a high water content, B a high frequent occurrence of boulders, and C soil with a high clay content. The probability of a certain factor occurring could be estimated starting from, for example, pre-investigation results.

Table 1. Examples of probability of the occurrence, or non-occurrence of a geological factor

Geological factor	Occurrence	Non-occurrence
	+	-
A	0.9	0.1
B	0.8	0.2
C	0.5	0.5

The excavation method can be rendered less sensitive to interference and the costs lowered if a method is chosen that is not sensitive to variation of the parameters that could vary over the entire length of the tunnel. Preventive steps can also be taken to reduce the sensitivity and thus, the probability of an unwanted event. The steps, or combinations of steps, that are the optimum choice in an actual case depend on a number of factors. Knowledge of technical possibilities and ability to judge how the rock will behave is necessary for the correct steps to be taken.

529

Table 2. Examples of steps to increase robustness in shield machines

Cause	Steps to increase robustness in shield machines
Boulder occurrence	Integrate stone crusher
Inflow of water	De-watering measures
High clay content	Centrifuges/separate centre head

When there is a risk of disturbance, it is very important to implement robust excavation systems. The diagram below illustrates an example of how costs can rise drastically when a geological factor increases.

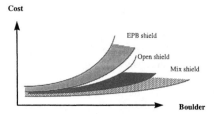

Figure 2. Example of disturbance sensitivity of various types of machine to boulder frequency in the soil.

Table 3. Examples of how occurrence of geological factors will give the probability of the constructability state for a certain method of excavation with or without steps taken to increase the robustness.

A	B	C	State of constructability without steps taken	State of constructability with steps taken	P(state)
+	-	-	III	III	0.09
+	+	-	II	II	0.36
+	+	+	I	I	0.36
-	+	+	I	II	0.04
-	-	+	I	II	0.01
-	-	-	III	III	0.01
-	+	-	II	II	0.04
+	-	+	I	II	0.09

By means of table 3 probabilities for different states can be estimated. It is evident that the occurrence of factor C during excavation with a certain shield machine without measures involves problems. With measures, however, the geological factor C could occur without major problems.

By adding up the probability figures in table 3 for different states of constructability, the figure below will be obtained.

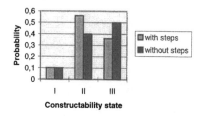

Figure 3. Example of probability for different states of constructability, both with and without steps taken for certain shield method.

2.5 Methodology for risk estimation

The risk of unwanted events can be expressed as the product of probability and consequence according to the following:

$$R = Pf * Cf \qquad (10)$$

where Pf =probability of the unwanted event, and Cf = cost consequence of the event.

By using a rough analysis and fault trees, it is possible to structure causes, probabilities and consequences of unwanted events. The probability of disturbance could also be obtained by using experience values from earlier events. According to the definition of constructability states, a simplified probability assessment can be performed (See table 4).

Table 4 Assessment of probabilities and consequences of unwanted events for different constructability states. This is valid for all zones and methods.

Constructability state	Pf	Consequence class
I	0.9	2-3 (4)
II	0.1	2
III	0.05	1

The cost consequences due to damage when excavating the tunnel can be calculated as the sum of the costs of down time and extra measures taken e.g.support, according to the formula below.

$$Cf = \sum C_{time} * t + Ext \qquad (11)$$

where ΣC_{time} = the sum of time-dependent costs, t = down time and Ext = extra costs for support, for example.

The cost consequences can be divided into different consequence classes according to table 5.

Table 5. Definition of cost consequences

Conse quence class	Consequence description	Cost [% of tender sum]	Example
4	Extensive damage	>100	Method failed
3	Considerable damage	50-100	Face collaps
2	Limited damage	10-50	Machine failure
1	Minor damage	<10	Segment fault

Table 6. Unwanted events in the case of shield excavation. Cause statistics based on 15 projects (The consequence class is based upon a subjective assessment).

Type of problem	Description	Class of conse quence	No.
Geology			33
Problems at the face	Face collapse	2-3	9
	Collapse to the surface	3	2
	Collapse and water inflow	3	2
	Jamming	3	6
	Vertical deviation	2	3
	No contact cutter head- rock	2	2
Other	Methane gas leakage	3	1
Function failure	Transport syst not suitable for geology	2-3	1
(TBM	Cutters not suitable for geology	2-3	2
	Grippers not suitable for geology	4	4)
Problems at the perimeter	Jamming of segments	1	1
Problems at surface	Sink hole	1	2
	Settlements	1-3	2
Mechanics			7
Cutter head	Main bearing failed	1	2
	Screw conveyor failed	1	2
Shield break	Head / tail	1	1
	Main component out of order	2	2
Misce- llaneous			2
Material failure	Hydraulics	1	1
	Ventilation fans	1	1

To gain a picture of possible unwanted events and their related cost consequences in the case of shield excavation, table 6 has been compiled, based on literature data.

The events that occurred can be divided into geological, mechanical and other events. The geological events are often caused by a combination of unpredicted geology and human error in the form of shortcomings in information flow or in the decision-making process. Mechanical failure are often a result of faulty component, and other events may be due to erroneous dimensioning or manufacturing faults.

3 CASE STUDY

The developed theory for choice of method has been applied to the Grauholz Tunnel in Bern, Switzerland. The double-track railway tunnel has a diameter of 11.6 m and a length of 5550 m. Construction began in 1990 and was completed in 1993.

The geology of the area can be roughly divided into three zones (Rohrer 1994). Zone 1 was composed of moraines with a relatively high clay content and occasional boulders. The water pressure rose in this zone to 0.37 MPa. Zone 2 consisted of sandstone with a maximum overburden of 120 m. The last zone was made up of moraines and boulders above groundwater level.

The following three types of shield machines were considered for the construction of the tunnel (Rohrer 1994):

- Mixshield with the option of rebuilding from hydroshield to TBM.
- EPB shield in which the soil mixture ahead of the cutter head absorbs soil and water pressure. The excavated material is transported by a screw conveyor.
- Open shield which does not have the possibility to support groundwater pressure at the face. This means, that the shield can be operated in groundwater if the soil is not impermeable, or reinforced with grouting or ground freezing.

The advantages and disadvantages of different types of machine have been summarised in table 7.

Table 7. Advantages and disadvantages of different types of machine

Machine type	Advantage	Disadvantage
Mix shield	• Little influence on groundwater	• Overpressure at face necessary for change of cutters • Bentonite and separation expensive
EPB shield	• No separation necessary • Different layering can be overcome	• Difficult to plan in inhomogenous soil
Open shield	• Low investment costs • Possible to operate in different soil/rock hardness	• No occurence of water acceptable in non-cohesive soil • Clay content <6%requires face support • No system for supporting face

Source: (Barbendererde 1991), (Maidl 1994)

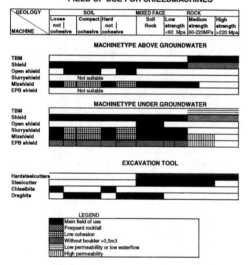

Figure 4. Geological field of use for different shield machines

As can be seen in figure 4, only a relatively small number of geological and hydrogeological parameters are required to make a preliminary choice of machine type. The large overlapping areas, however, show that the final choice of machine type is a complex process, encompassing both technical parameters and economic, environmental and safety factors.

The decision is further influenced by local negotiation strategies and claims management.

4 RESULT

The probability-based decision model (described in the Section 2) for assessing expected costs for excavation with different shield methods has been applied in the Grauholz Tunnel. In figure 8, a probable cost for excavating the tunnel has been calculated for the three types of machine. In order to judge the distribution of the total implementation costs for the different machine types and thereby facilitate a decision, the tree has been evaluated using a Monte Carlo simulation.

Figures 5 till 7 show the results of a Monte Carlo simulation of the total costs of the whole tunnel, i.e. the sum of all three zones. In the figures, the costs with and without measures and the standard costs have been marked. It is clear from the figures that the standard costs can vary considerably, which renders them less efficient as the basis of a decision. The figures also show that the method can be made more robust for different geological factors with the help of additional measures.

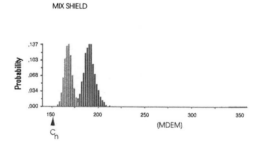

Figure 5. Mix shield with and without centrifuges for separation

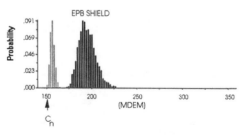

Figure 6. EPB shield with and without capability to crush boulder (technique not possible today)

The result of the simulation shows that:

1 The mixshield technique is suitable for excavation both in soil under water pressure and in dry sandstone, which was present in the Grauholz Tunnel project. In very clayey soils, however, considerable separation of fines is necessary.

2 The EPB shield is sensitive to rocks. The cost of this type would likely be less than for mix shield if it were able to crush rocks. However, this is not possible with present-day technology. This method involves no separation

3 The open shield requires major drainage and support measures for excavation of large-diameter tunnels in soil under high water pressure. Thus using this method entails very high costs.

5 CONCLUSIONS

The developed model can be used in the choice of shield machine. It is important to have clearly formulated decision criteria, i.e., not only to choose the method with the lowest standard cost estimate, but to select the one that gives the lowest expected total costs depending on the cost increases that may occur as a result of incorrect geological assumptions. By assessing the probability of the method succeeding or failing because of variations in the geology, the robustness of the method can be studied.

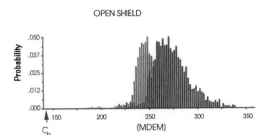

Figure 7. Open shield with and without comprehensive drainage measures

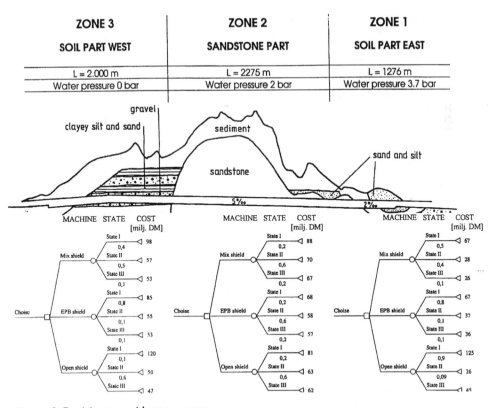

Figure 8. Decision tree without measures

A further refinement of the methodology for calculating the robustness of an excavation method is planned, since a robust method is less sensitive to different disturbances. The manufacturing industry has been carrying out different forms of robustness calculation for many years (the Taguchi method).

REFERENCES

Aebersold, W. 1994. Betrieb und erzielte Leistung mit dem Mixschild. In K. Kovari (ed) *Referate der Studientagung vom 26. Mai 1994 in Schönbühl, Fachgruppe für Untertagebau*, SIA Dokumentation D0116, Grauholztunnel II

Babendererde, S. 1991. Kritische Betrachtungen zum Einsatz von Hydroschilden. In K. Kovari & R. Fechtig (eds) *Sicherheit und Risiken bei Untertagebauwerken, ETH- Symposium, 21/22 März 1991* Zürich.

Einstein, H.H., J.P. Dudt, V.B. Halabe, & F. Descoeudres 1996. Geologic uncertainities in Tunnels. *Uncertainty in the Geologic Environment: From Theory to Practice.*

Einstein, H.H., F. Descoeudres, J.P. Dudt, & V. Halabe, Oktober 1991. Entscheidungshilfen für den Tunnelbau. *Schlussbericht über die im Auftrag des Bundesamtes für Verkehr ausgeführten Arbeiten.*

Einstein, H.H. & S.G. Vick 1974. Geological model for a tunnel cost model. *Proc. Rapid Excavation and Tunneling Conf., 2nd, II*:1701-1720

Maidl, B. 1994. *Handbuch des Tunnel- und Stollenbaus Band 1: Konstruktion und Verfahren*, Zweite Auflage Essen: Glückauf GmbH.

Rohrer, U. 1994. Überlegungen des Unternehmers in der Offertphase. In K. Kovari (ed) *Referate der Studientagung vom 26. Mai 1994 in Schönbühl Fachgruppe für Untertagebau*, SIA Dokumentation D0116, Grauholztunnel II

Salazar, G.F. 1983. Standastic and Econific Evaluation of Adaptability in Tunnelling Design and Construction. *Ph.D. Thesis, Department of Civil Engineering, Massachusetts Institute of Technology,* Cambridge, MA

Geotechnical Hazards, Marić, Lisac & Szavits-Nossan (eds) © 1998 Taylor & Francis, ISBN 90 5410 957 2

Onshore structures critical load and selection of design solution

K. Ivandić & F. Verić
Faculty of Civil Engineering, Zagreb, Croatia

ABSTRACT: In the paper the analysis of the kinetic energy absorption of the ship while putting to the onshore structure is presented in a variant when the total energy is absorbed by elastic fenders and in a variant when the total ship kinetic energy is absorbed by fenders and onshore structure through their deformations without any effect of soil and respecting the deformability of soil, that is, the energy delivered into the soil. Great differences between the mentioned approaches are evident concerning the distribution of displacements and internal forces in the structure which is particularly expressed at the computation of pile ultimate bearing capacity. A rationalization of the force reduced by 30% approx. as compared to the standard solutions provides an absorption of the ship kinetic energy while putting ashore at a speed greater than permited.

1 INTRODUCTION

The onshore structures used for putting ashore of large ships with deep draught are constructed on the shore out into the sea and founded on bored piles. For such structures two types of load are characteristic: vertical and horizontal. Vertical loads are produced by the dead load of a structure, by the cranes and vehicles driving in the dock area and by the useful load i.e. the containers temporary stored on the dock. Horizontal loads occur due to the impact of a ship when putting ashore, the ship pressure or by the tensile force acting on bollards due to wind action on the anchored ship as well as by the horizontal acceleration of the mass during quake.

A special type of horizontal load is acting on piles due to frequently major vertical load in the storage open areas for bulk cargo which are placed in the immediate vicinity, behind the onshore structures.

The ground under the shore plateau is inclined from the area behind the onshore structures up to the bottom of the sea in front of the onshore structure. The difference in elevation is mostly ranging from 12 to 20 m. Due to the inclined ground the pile free length (when a pile appears from the soil and enters the cap structure) is changed which has caused certain problems at the transmission of horizontal loads. The superstructure connecting all piles into a whole, requires equal horizontal displacement of all piles within a frame. Therefore the piles with minor free length take over a considerably greater portion of horizontal load than the piles with major free length. In the direction perpendicular to the shore the

onshore structures are the frames with two, three or more verticals i.e. piles that mostly reach the same depth but have varying free lengths.

At the transmission of load into the soil there are entirely contradictory requirements with respect to the mentioned two types of load. Vertical loads require that the central piles within frame have major diameter because they transmit major loads. Horizontal load requires matching of the pile dia and the pile free length. Accordingly, by reducing free lengths of the piles also their diameters should be reduced in order to provide equal utilization of all piles. Namely, at a horizontal displacement of few centimeters or even ten centimeters in the piles with major free length a minor fixed-end moment is being activated while at the piles with minor free length the activated bending moment can cause breaking of the pile. It should be particularly observed that the major horizontal and vertical loads are not acting simultaneously.

In the paper the above mentioned problems are considered and some solutions suggested. In particular, the problem of the horizontal load activated on the occasion of a ship coming alongside the onshore structure is discussed when the ship kinetic energy during impact with the onshore structure should be annuled by the potential energy of the deformed fenders, onshore structure, piles and the soil.

Also in the paper the application and certain restrictions of Mindlin's solution for the concentrated force acting within the elastic halfspace are considered.

2 THEORETICAL BASIS

The analysis of the horizontally loaded onshore structures founded on piles is approached by using the solution of the problem of horizontaly loaded piles in deformable soil. Depending on the type of soil modelling (linear, nonlinear model), selection of particular model parameters and the computation method (analytical, numerical) , the more or less differing solutions of the problem are obtained (Tomlinson 1991). In the paper the soil is modelled as a lineary elastic halfspace while the computations have been made by using the finite difference method.

2.1 *Main Assumtions of the Analysis*

The mathematical model of the horizontally loaded piles is given by the differential equation:

$$EI \frac{d^4 y}{dz^4} = -p + f \tag{1}$$

where E = modulus of pile elasticity, I = moment of inertia of the pile cross-section, y = unknown function of the pile horizontal displacement, p = unknown function of the soil reactive pressure, f = known function of the pile external load.

The method of the problem solution is dependent on the additional ratio of the unknown function of displacement y and the reactive pressures p. If the soil is considered as an elastic halfspace there are two different solutions, analytical solution for a beam on elastic support (Hetenyi 1946) or numerical solution obtained by girder quantization (Poulos 1980) after the Mindlin formula for the concentrated force acting within the elastic halfspace (Mindlin 1936). The Equation (1) is written in the terms of differential:

$$\frac{EI}{h^4} [B] \{y\} + [D] \{p\} = \{f\} \tag{2}$$

where $[B]$ = matrix of differential equations coefficients, $[D]$ = diagonal matrix, $\{y\}$ = unknown vector of displacement, $\{p\}$ = unknown vector of reactive pressure, $\{f\}$ = vector of external load, EI = pile stiffness, h = differential step.

Compatibility condition:

$$\{y\} = \{y_s\} = [U] \{p\} \tag{3}$$

where $[U]$ = matrix of influential coefficients of soil displacement, $\{y_s\}$ = vector of soil displacement.

The matrix elements of the influential coefficients $[U]$ are obtained by integrating the Mindlin formula

upon the rectangular plane for horizontal concentrated force acting within the elastic space. When the values of matrix elements are known the folowing can be written:

$$\frac{EI}{h^4} [B] [U] \{p\} + [D] \{p\} = \{f\} \tag{4}$$

$$(\frac{EI}{h^4} [B] [U] + [D]) \{p\} = \{f\} \tag{5}$$

$$\frac{EI}{h^4} [B][U] + [D] = [C] \tag{6}$$

$$[C] \{p\} = \{f\} \tag{7}$$

Equation (7) is the system of linear algebraic equations which can be easily solved to get values of reactive pressures and displacements along the pile and deformation energy of the pile and the soil.

2.2 *Review of the application of Mindlin's solution*

At the analysis of stress in homogeneous, elastic and isotropic halfspace two standard solutions are mostly applied: Boussinesq theory (1885) when a force acts on the limit plane of a halfspace and Mindlin theorem when a force (horizontal or perpendicular) acts within this halfspace. In the application of the mentioned solutions there is a significant difference; the force on the limit plane can be actually applied without disturbing the whole of this halfspace while at the actual application of force within the halfspace its whole is disturbed from the limit plane to the place of force application. Because of this fact the Mindlin's solution must be applied with certain restrictions and corrections.

The analysis of the stress on the surface of a semisphere under a force concentrated on the limit plane of a halfspace (Boussinesq) gives a representation of the force application in halfspace. A similiar analysis can be also made for the force which is applied to the halfspace at a particular depth (Mindlin). If a sphere is described around the application point and the stresses on the sphere envelope are calculated a theoretical representation of that force application is obtained.

Accordingly when a vertical force is applied (Figure 1) it is practically being divided into a part which extends the halfspace part beyond the depth of force application and a part which compresses the halfspace part below the depth of force application. More precisely, the force at greater depth is divided in two equal parts: 50% goes to the tensile part and 50% to the compressive part. By reducing the depths of force application the tensile part is also reduced while the compressive part is increased. In the case when the force rises to the halfspace surface the solution becomes Boussinesq solution.

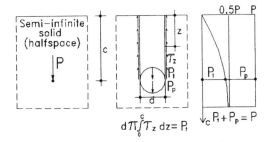

Figure 1. Vertical force in the interior of the semi-infinite solid (halfspace).

It is particularly important that the shear stresses on the cylinder envelope beyond application force hold equilibrium to the tensile part of the total force. Thus the Mindlin's solution becomes acceptable when force is applied into the soil by a pile so that 50 % of the total load is transmitted by friction on the mantle and 50% by the load on the pile point.

At the application of horizontal force into the halfspace depth (Figure 2) the pattern of this application can be analyzed in the same way. By analyzing the stresses on the sphere around the place of force application a similiar situation to that of the perpendicular force application is obtained. Namely, a part of the horizontal force is employed for the compression of the halfspace before force application and a part for the extension of the halfspace behind the force. Hence certain adaptations need to be done while using Mindlin solution because the horizontal force can be physically brought into the soil only by compressive stress.

By reducing the halfspace modulus of elasticity to 50% of the value at the application of the Mindlin solution for the whole halfspace a displacement is obtained which corresponds to the full value of the modulus of elasticity for the transmission of total load, however, by the compressive component only which is shown in Figure 3.

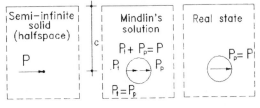

Figure 2. Horizontal force in the interior of the semi-infinite solid (halfspace).

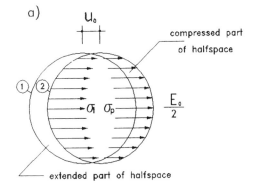

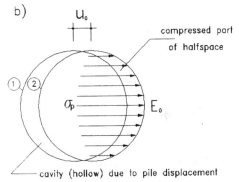

Figure 3. Influence of the a) computed and b) realistic horizontal load transfer in to the soil on the choice of the value of the modulus of elasticity where ① Initial position of the pile and ② Position of the pile after the movement.

2.3 Forming of the system rigidity matrix

The cap structure connecting the piles is considered to be absolutely rigid so that there is no change of the angle between the pile points and the cap.

The horizontal rigidity of the system as well as the fixed-end moment required to obtain the unit horizontal displacement and zero angle of rotation of the point of the pile embedded in the deformable soil are indirectly determined from the system flexibility matrix. By the generalized unit forces acting on the point of the pile the generalized displacements are obtained forming the system flexibility matrix (Figure 4 and Figure 5). By the inversion of the matrix and by multiplying the obtained matrix with the unit displacement and the zero angle of rotation the relevant horizontal force and the fixed-end moment on the point of every pile is obtained. The total horizontal rigidity of the system is now sum of the horizontal forces on the point of the pile.

$$[F] = \begin{bmatrix} \Delta_{H=1} & \varphi_{H=1} \\ \Delta_{M=1} & \varphi_{M=1} \end{bmatrix} \tag{8}$$

$$[F]^{-1} = [K] \qquad (9)$$

$$\left\{ \begin{array}{c} K_H \\ M \end{array} \right\} = [K] \left\{ \begin{array}{c} 1 \\ 0 \end{array} \right\} = \left[\begin{array}{cc} k_{11} & k_{12} \\ k_{21} & k_{22} \end{array} \right] \left\{ \begin{array}{c} 1 \\ 0 \end{array} \right\} = \left\{ \begin{array}{c} k_{11} \\ k_{21} \end{array} \right\} \qquad (10)$$

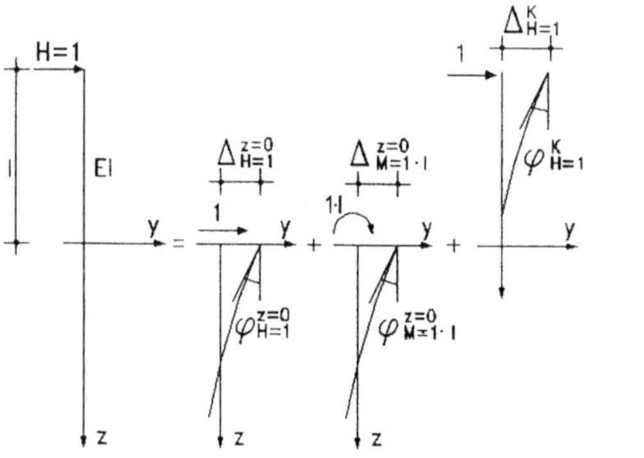

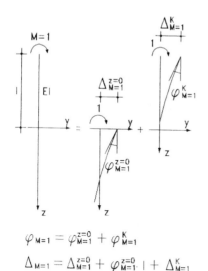

$$\varphi_{H=1} = \varphi_{H=1}^{z=0} + \varphi_{M=1\cdot1}^{z=0} + \varphi_{H=1}^{K}$$

$$\Delta_{H=1} = \Delta_{H=1}^{z=0} + \varphi_{H=1}^{z=0}\cdot1 + \Delta_{M=1\cdot1}^{z=0} + \varphi_{M=1\cdot1}^{z=0}\cdot1 + \Delta_{H=1}^{K}$$

Figure 4. General displacements of the pile due to the unit horizontal force on the point of the pile.

$$\varphi_{M=1} = \varphi_{M=1}^{z=0} + \varphi_{M=1}^{K}$$

$$\Delta_{M=1} = \Delta_{M=1}^{z=0} + \varphi_{M=1}^{z=0}\cdot1 + \Delta_{M=1}^{K}$$

Figure 5. General displacements of the pile due to the unit moment on the point of the pile.

When the rigidity of each pile is known and thus also the total rigidity of the system the displacements and forces in every pile can be computed as well as the deformation energy of the particular members of the carrying system. The effect of the ship striking

of the ship striking against the onshore structure is computed by the ship acting kinetic energy as obtained from the ship mass and the speed of the ship movement. In the mentioned cases the problem is to determine the equivalent static force because kinetic energy appears to be an input value. The problem is solved by the energy equilibrium of the problem i.e. equality of the ship kinetic energy (E_{kin}) and the deformation energy of the elastic fenders (E_{ef}), onshore structure (E_{st}), piles (E_p), and the soil (E_s), in which the piles are placed (Equation 11). In other words, action of the internal forces upon displacements in the direction of that forces shall be equal to the kinetic energy of the ship.

$$E_{kin} = E_{ef} + E_{st} + E_p + E_s \qquad (11)$$

When the deformation energy of the fenders solely provides equilibrium for the kinetic energy (the structure is considered to be absolutely rigid) the equivalent static force is greater than in the case when the deformability of the structure has been also taken into consideration. The represented diagrams of the ratio of the deformation and the equivalent static force and the energy acting on fender have been given by various fender manufacturers (Committee for Waterfront Structures 1986). By taking into consideration also the deformability of the the soil in which the piles are placed the true distributions of forces and structure displacements can be obtained and accordingly the total deformation energies of the onshore structure and the soil as well.

3 ANALYSIS OF THE RESULTS

In the paper the results of the analysis of an onshore structure composed of a cap structure i.e. beam and the three 40 m long reinforced-concrete piles with a diameter of 1.8 m; 1.5 m and 1.0 m (Figure 6) are discussed. By the soil parameters variation (modulus of elasticity) the effect of soil deformability on the magnitude of rigidity and the fixed-end moments on the piles points is represented. Also the displacements and the deformation energy of the pile for the horizontal force acting on onshore structure are given.

By increasing the stiffness of the deformable soil the state of the soil absolute rigidity is being approached which has been represented in Figure 7. The computed rigidity K of the structure for the modulus of soil elasticity $E=10^7$ kN/m^2 amounts approx. 90% of the stiffness K_u for the undeformable soil. It is quite obvious that the values of the modulus of elasticity of the natural soils cannot be as great. For the real moduli of soil elasticity the obtained pile stiffnesses have been many times smaller as compared to the stiffnesses for the undeformable soil which means that the assumption on the soil absolute rigidity is not realistic and it gives an incorrect distribution of forces and structure displacements.

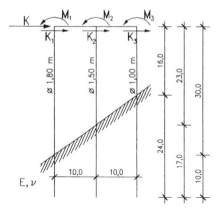

Figure 6. Geometry of the analysed construction in the deformable soil.

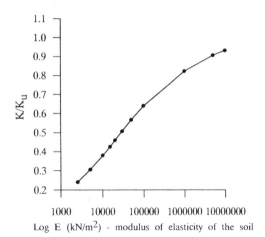

Figure 7. Relative rigidity of the system for different values of the modulus of elasticity of the soil.

In Figure 8 the relationship of the relative rigidity (K_i/K_{ui}) of particular piles and the change of modulus of soil elasticity is represented where K_i = rigidity of the pile for deformable soil and K_{ui} = rigidity of the pile for undeformable soil.

The change of structure rigidity results also in the change of activated forces and horizontal displacements on the pile point. The system with lesser stiffness will experience greater displacements and at the equal acting force the deformation energy of such systems will be greater. Thus the equivalent force becomes smaller which will provide a more economical solution provided that the requirement on

frame maximum horizontal displacement has been satisfied. On the other hand, there arises a question about the distribution of force and displacement depending on the modulus of soil elasticity in the case of a constant deformation energy. It may be seen in Figure 9 that the displacement and the force magnitudes very much differ for the case of the real moduli of soil elasticity as compared to the case of the undeformable soil.

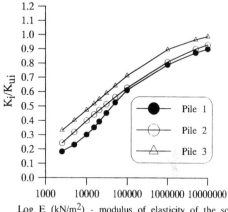

Figure 8. Relative rigidity for each pile for different values of the modulus of elasticity of the soil.

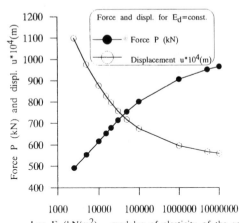

Figure 9. Horizontal forces and displacements of the construction for the constant value of the deformation energy.

Figure 10 shows the effect of the modulus of soil elasticity on the distribution of forces and moments on the pile point. The ratio of the moment and force is constant for the case of the undeformable soil which means that it is not dependent upon the pile rigidity and amounts half of the pile free length. However, when taking into consideration the deformability of the soil this ratio is not constant any more. The greatest difference as compared to the case of the undeformable soil is found at piles with major free lengths. Hence in this case the greatest difference was found for pile 1 (major free length L_f) where this ratio was reduced for 50% of the value compared to the case of undeformable soil. By all means the distribution of force and moment shall be analyzed in every pile because this distribution is being changed from pile to pile with every change of the soil modulus.

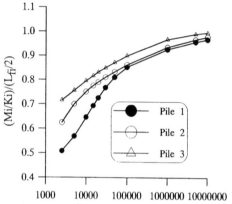

Figure 10. Ratio of the moment and the force on the cap of the piles and their half of the free lenghts for different values of the modulus of elasticity of the soil.

The system deformation energy is the force action upon the displacement and it can be written $E_d = P\delta/2$. However $\delta = P/K$ and hence the expression for energy $E_d = P^2/2K$. This is square function of P with constant stiffness K as represented in Figure 11.

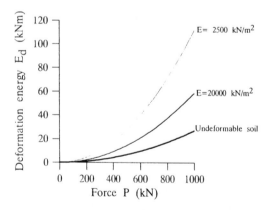

Figure 11. Deformation energy of the system versus horizontal force for different values of the modulus of elasticity of the soil.

The range of the deformation energy values is very large. The quantity of energy for the case of the undeformable soil is the least while the quantity of energy for the identical force is the greatest for the least soil model.

4 CONCLUSIONS

1. The halfspace modulus of elasticity must be reduced to 50% of the value at the application of the Mindlin solution for the whole halfspace. The displacement corresponds in this case to the full value of the modulus of elasticity for the transmission of total load by the compressive component only.

2. Only by taking into account the deformability of the soil in which the onshore structure has been founded a true distribution of horizontal forces and displacements on the point of a pile can be established which at the same time also affect the quantity of the deformation energy of each pile as well as the entire structure. The differences between the cases of varying soil stiffnesses can be considerably great as illustrated by the analysis and that can result in wrong analyses and breaking of the pile due to the wrong insight into the true distribution of forces and displacements in piles.

3. Respecting the deformation energy of the onshore structure, piles and the deformable soil in which the piles are placed more economical solutions than common analyses can be attained because a greater quantity of deformation energy is achieved in the structure and hence also minor equivalent static force, however, provided that the requirement on the structure maximum displacement has been met. Also reduced ekvivalent force about 30% approx. (for the usual dimensions of the construction) to standard solution increases the reserve for the absorption of the kinetic energy while the ship putting ashore at a speed greater than permited.

REFERENCES

Boussinesq, J. 1885. *Applications des potentielles a l'etude de l'equilibre et du mouvement des solides elastiques*. Gauthier Villars, Paris.

Committee for Waterfront Structures 1986. *Recomndations of the Committee for Waterfront Structures EAU*. Ernst&Sohn.

Hetenyi, M. 1946. *Beams on elastic foundations*. Ann Arbor: Univesity of Michigan Press.

Mindlin, R.D. 1936. Force at a point in the interior of a semi-infinite solid. *J. Appl. Physics 7. No5*:195-202.

Poulos, H.G. & E.H. Davis 1980. *Pile foundation analysis and design*. New York: Wiley.

Tomlinson, M.J. 1991. *Pile design and construction practice*. E. and FNSPON.

Geotechnical Hazards, Marić, Lisac & Szavits-Nossan (eds) © 1998 Taylor & Francis, ISBN 90 5410 957 2

Untersuchungen zur Sanierung der Donaubrücke Russe – Giurgiu

J. Jellev & G. Stefanoff

Universität für Architektur, Bauwesen und Geodäsie, Sofia, Bulgaria

ZUSAMMENFASSUNG: Die einzige Donaubrücke zwischen Bulgarien und Rumänien verbindet die bulgarische Stadt Russe mit der rumänischen Giurgiu. Es handelt sich um eine der längsten kombinierten (Straßen- und Eisenbahn-) Brücken Europas. Die Brücke ist in den Jahren 1952-54 errichtet worden. Die Strohmpfeiler sind mittels Luftdrucksenkkästen bei einer Gründungstiefe von 5 bis 9 m in Mergel eingebunden. Die restlichen Pfeiler übertragen die Lasten mittels 10 bis 14 m langen Stahlbetonpfählen auf eine Sand-Kies-Schichte, die einige Meter über dem gewachsenen Kalkstein liegt. Im Laufe der Jahre hat sich verschiedenes geändert: 1. Die Hydrologie des Flußes und die damit zusammenhängenden hydraulischen Beanspruchungen - eventuelle teilweise Bloßlegungen der Pfeilergründungen. 2. Die Belastung ist viel intensiver (und mit höheren Lasten) geworden. 3. Im Jahre 1977 verursachte das Erdbeben von Vrancea (Rumänien) erhebliche Schäden, inkl. in der Stadt Russe, weshalb auch bestehende Konstruktionen auf Erdbebensicherheit überprüft werden müssen. Dies alles gab im Jahr 1992 der gemischten Bulgarisch-rumänischen Kommission, die mit der Überwachung der Brücke betreut ist, den Anlass eine dringende Untersuchung der Stabilität der Pfeilergründungen zu verordnen. Diese Untersuchungen, die dabei erzielten Ergebnisse und die vorgeschriebenen Maßnahmen werden in diesem Aufsatz dargestellt.

1 EINLEITUNG

Die einzige Donaubrücke zwischen Bulgarien und Rumänien, die die bulgarische Stadt Russe mit der rumänischen Stadt Giurgiu verbindet, ist in Forschungs- und Projektierungsinstitute der gewesenen Sowietunion projektiert und während der Jahre 1952 und 1954 erbaut worden. Sie ist eine der größten kombinierten (Eisenbahn- und Straßen-) Brücken in Europa und erstreckt sich über 37 Felder: eine Öffnung zu 86 m, 2 Durchlaufträger über je 2 Felder zu 160 m, 4 Durchlaufträger über je 2 Felder zu 80 m und 24 Öffnungen zu 34.40 m mit einer Gesamtlänge von ca 2224 m. Die Strohmpfeiler sind mittels Druckluftsenkkästen in 5 bis 9 m Tiefe auf Mergel gegründet (Fig.1).

Die anderen Pfeiler übertragen die Brückenlasten mittels 10 bis 14 m langen Pfählen auf eine Sand-Kies-Schichte bis auf einige Meter über den gewachsenen Fels (Fig.2).

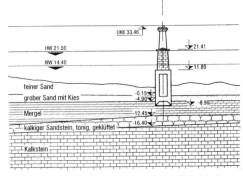

Abb. 1. Lithologischer Aufbau des Baugrundes bei Pfeiler Nr. 21

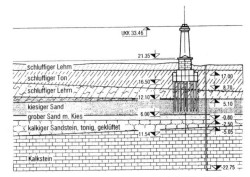

Abb. 2. Lithologischer Aufbau des Baugrundes bei Pfeiler Nr. 25

Der technische Zustand der Brücke wurde erhalten gemäß der Vereinbarung (1954) und des Reglements (1954), sowie der Vorschriften der Beschlüsse der jährlichen Sitzungen der Bulgarisch-rumänischen Kommission.

In den Jahren 1991-1992 ergab sich ein erschwertes Belastungsregime des Brückenüberbaues, welches noch andauert, durch den langzeitigen Aufenthalt von schweren LKW, öfters in 2 Kolonnen nebeneinander. Deshalb mußten theoretische und experimentelle Untersuchungen aller Brücken- elemente durchgeführt werden: des Flußbettes, der Gründungen der Stützenkörper (Pfeiler und Wiederlager) und des Überbaues; gleichzeitig wurden durch präzise geodätische Messungen die eventuellen Verschiebungen der Stützenkörper verfolgt. Von der Bulgarisch-rumänischen Kommission wurde die Aufgabe als eine dringende "Untersuchung der Stabilität der Brückenstützenkörper für die veränderten Belastungsbedingungen durch Verkehrslasten, seismische Beanspruchungen und Änderungen der Flußsohle" formuliert. Diese Untersuchungen, deren Ergebnisse und die vorgeschriebenen Sanierungsmaßnahmen stellen den Inhalt dieses Aufsatzes dar. Dabei wird nur die südliche (bulgarische) Brückenhälfte betrachtet.

2 LASTANNAHMEN

Die rechnerischen Untersuchungen der Brückenstützen wurden mittels Belastungen, die dem reellen erschwerten Nutzungsregime für die Verkehrslasten (Eisenbahnzüge und Autokolonnen) Rechnung tragen, durchgeführt. Für die anderen Belastungen wurden die Vorschriften vieler bulgarischer, russischer und Euronormen (Eurocode 1, 7 und 8) eingehalten. Die Besonderheit der Belastungen erforderte es strenger vorzugehen als es die Normen vorsehen. So z.B. wird gewöhnlich die Windbelastung auf die auf der Fahrbahn befindlichen Autos (LKW und PKW) nicht in acht genommen; wegen dem langzeitigen Aufenthalt der Autokolonnen (hauptsächlich TIR) mußte diese Windbelastung auch berücksichtigt werden.

Die Belastungen auf die Brückenstützkörper wurden in folgende drei Gruppen unterteilt:

I. Hauptkräfte.

A. Bleibende Lasten.

1) Eigengewicht des Überbaues.

2) Eigengewicht der Stützenkörper bei HW und bei NW.

3) Erddruck vom Eigengewicht des Bodens bei HW und bei NW.

4) Hydrostatischer Druck (Auftrieb) bei HW und bei NW.

5) Hydrodynamischer Druck bei HW und bei NW.

B. Verkehrslasten.

6) Lotrechte Lasten.

a) Eisenbahnzug - besteht aus einer Diessellokcmctive "06-00" und vierachsigen Lastenwagons mit einer äquivalenten Belastung $q = 61.8$ kN/m.

b) Autokolonne - besteht aus 5-achsigen LKW "TIR" von einem Gewicht 465 kN (1x65 + 4x100) in einer Kolonne mit einem Zwischenabstand von 1 m, mit einer äquivalenten Belastung $q = 28.2$ kN/m.

7) Erddruck von Verkehrslasten.

II. Zusatzkräte.

8) Windbelastung - quer zur Brücke besteht sie aus:

a) Windbelastung vom Brückenüberbau; sie wurde unter folgenden Voraussetzungen ermittelt;

- durch eine Autokolonne belastete Brücke mit spezifischer Windbelastung $w = 1.5$ kPa,

- durch einen Eisebahnzug belastete Brücke mit spezifischer Windbelastung $w = 1.5$ kPa,

- die Windlast auf die Autokolonne und den Eisebahnzug wirkt 1.5 m über Straßenfahrbahnoberkante und Schienenoberkante,

- Belastung der dem Wind ausgesetzten Flächen des Überbaues;

b) Windbelastung auf die Stützenkörper selbst mit spez. Windbelastung $w = 1.5$ kPa bei HW, MW und NW.

9) Brems - und Zugkräfte.

10) Reibungskräfte in den beweglichen Lagern.

III. Besondere Kräfte.

11) Stoß durch schwimmende Fahrzeuge. Laut Beschluß der Bulgarisch-rumänischen Kommission wurde angenommen:

a) frontal zum Stützenkörper (in Strohmrichtung)- 3000 kN,

b) quer zum Stützenkörper-1250 kN;

diese Kräfte greifen in einer Höhe von 2 m über MW (erste Variante) oder HW (zweite Variante) an.

12) Seismische Beanspruchungen. Bei den Untersuchungen wurden die Stützenkörper als senkrecht eingespannte Kragarme angenommen - mit der zugehörigen Masse vom Überbau (entsprechend dem real existierenden Nutzungsregime), ihre Eigenmassen und die dazugehörende Wassermassen bei HW und MW.

13) Belastung durch Eis. Wurde bei HW-Eisgang und Eisdicke von 40 cm ermittelt.

Wegen der Einzigartigkeit des Bauwerkes und seiner Bedeutung für die Verkehrssysteme beider Länder (Rumänien und Bulgarien) in Nord-Süd-Richtung wurden die Stützenkörper und ihre Gründungen auf die ungünstigsten Kombinationen von Belastungen und Einwirkungen untersucht, wobei in manchen Fällen strenger vorgegangen wurde, als es die bestehenden Verordnungen vorschreiben.

3 STATISCHE UNTERSUCHUNGEN

Es wurden folgende Nachweise eingebracht:

a) Nachweis des Angriffspunktes der Resultierenden aus den Belastungskräften in der Fundamentsohle bei HW und bei NW unter Einhaltung folgender Erfordernisse:

-für die Pfeiler

$$\frac{e_{oi}}{r_i} \leq 0.1 \text{ bei bleibenden Lasten (A);}$$

$$\frac{e_{oi}}{r_i} \leq 1 \text{ bei bleibenden (A) und Verkehrs-(B) Lasten}$$
$$(A+B);$$

- für die Widerlager

$$\frac{e_{oi}}{r_i} \leq 0.5 \text{ bei bleibenden Lasten (A);}$$

$$\frac{e_{oi}}{r_i} \leq 0.6 \text{ bei bleibenden und Verkehrslasten}$$
$$(A+B);$$

dabei bedeuten:

e_{oi} - Exzentrizität der Resultierenden zur Längsachse (x) und zur Querachse (y) der Sohlfläche,

$r_i = W_i/A$ - Kernweite der Sohle,

A - Sohlfläche,

W_i - Widerstandsmoment der Sohlfläche gegenüber den Richtungen x und y.

b) Nachweis der Sohlspannungen der Senkkastengründungen, als frei aufliegend berechnet (ohne Berücksichtigung der elastischen Einspannung im Baugrund). Für die verschiedenen Belastungskombinationen wurden die Sohlspannungen nach Navier berechnet

$$\sigma_{min}^{max} = \frac{V_i}{A_i} \pm \frac{M_x \cdot y_i}{W_{xi}} \pm \frac{M_y \cdot x_i}{W_{yi}} \ .$$

Es wurde nachgewiesen ob folgende Bedingungen erfüllt sind:

$$\sigma = \frac{V_i}{A_i} \leq R_d \text{ und}$$

$$\sigma_{max} \leq 1.2 R_d \ ;$$

hierin ist

σ die mittlere Sohlpressung,

σ_{max} - die größte Eckspannung,

R_d - Bemessungswert des Sohldruckwiderstandes des Baugrunds (Mergel, Sandstein) für die entsprechende Belastungskombination. Auf Grund der Ergebnisse von Laboruntersuchungen auf einaxialen Druck in natürlichem Zustand und nach Wassersättigung der Mergel und Sandsteine und der statischen Probebelastungen wurde der Bemessungssohldruckwiderstand bei den Belastungskombinationen wie folgt angenommen:

- für Hauptbelastungen (I) $R_d = 1.8$ MPa,
- für Haupt-und Zusatzbelastungen (I+II) $R_d = 2.0$ MPa.

c) Nachweis der Pfahlbelastung der Pfahlgründungen. Nachdem das Rostwerk zusammen mit dem Grüngungskörper als unendlich steif angenommen wurde, wurde die Pfahlbelastung nach der Formel für exzentrischen Druck ermittelt

$$N_{min}^{max} = \frac{V_i}{n} \pm \frac{M_x \cdot y_i}{y_i^2} \pm \frac{M_y \cdot x_i}{x_i^2} \ .$$

Die maximale Belastung wurde mit der Bemessungstragfähigkeit R_{cd} verglichen. R_{cd} wurde auf folgende Arten ermittelt:

- von Tabellen mit empirischen Werten für Spitzenwiderstand und Mantelreibung,
- mittels Rammformeln (während des Rammens),
- von 6 statischen Pfahlprobebelastungen, die während der Gründung durchgeführt wurden.

Für einen Sicherheitsgrad 1.4 wurde die Bemessungstragfähigkeit wie folgt festgelegt:

für Druckpfähle $R_{cd} = 600$ kN,

für Zugpfähle $R_{cd} = 200$ kN.

4 ANALYSE DER ERHALTENEN UNTERSUCHUNGSRESULTATE

4.1 Bei den Druckluftsenkkastengründungen

- Für Haupt- und Zusatzbelastungen (I+II) sind die maximalen Eckspannungen in der Gründungssohle geringer als die Bemessungssohlpressungen im Mergel, bzw. im Sandstein. Nur bei einem Pfeiler befindet sich die Kräfteresultierende außerhalb des Kernes, jedoch klaffender Fuge bleibt die maximale Sohlpressung unterhalb der zugelassenen Bemessungsspannung.

4.2 Bei den Pfahlgründungen

- Für Hauptbelastungen (I) ist der Sicherheitsgrad bedeutend größer als 1.
- Für Haupt- und Zusatzbelastungen (I+II) ist der Sicherheitsgrad größer als 1.3. Nur für einen Pfeiler ist er kleiner als 1. An dieser Stelle befinden sich jedoch günstige Umstände: die Pfahlspitzen reichen bis in ein festes Konglomerat und außerdem ist ringsum die Pfahlgründung eine Spundwand eingerammt.
- Für Haupt-, Zusatz- und besondere Belastungen (I+II+III) ist der Sicherheitsgrad, mit Ausnahme des Widerlagers (Nr. 37), geringer als 1. Wenn man bei den Zusatzkräften, wegen der gleichzeitigen Kombination mit den besonderen Belastungen, den Wind vernachlässigt, wächst der Sicherheitsgrad an, bleibt jedoch bei HW bei 6 Pfeilern, und bei NW bei weiteren 3, unter 1.

5 HYDROLOGISCHE UNTERSUCHUNGEN, FESTSTELLUNGEN UND EMPFEHLUNGEN

Seit der Errichtung der Brücke (1954), bis zum Beginn der durchgeführten Untersuchungen (1992), hat sich die Flußsohle im Bereich der Brücke verändert. Bei der neuen Gestaltung der Flußsohle in Nähe der Pfeiler ist ihre Standsicherheit gesunken. Deshalb mußten im Bereich der Brücke neue hydraulische Untersuchungen rechnerisch und an Modellen durchgeführt werden. Diese Untersuchungen enthielten:
- Wahl der analytischen Verfahren und Modell-Untersuchungen;
- Erbauen eines aerodynamischen Modells der Donau im Brückenbereich;
- Modelluntersuchungen für zwei charakteristische Wassermengen;
- Hydraulische und morphologische Berechnungen zur Bestimmung der Auskolkung der Flußsohle und der Strohmdynamik bei und nach der Brücke.

Die Untersuchungsergebnisse ermöglichten es die Entwicklungstendenz der Flußbettgestaltungsprozesse in Nähe der Brückenstützkörper zu erfassen, folgende Rückschlüsse zu ziehen und Empfehlungen zu machen:
- Im allgemeinen weist das Donauflußbett im Brückenbereich eine relative Stabilität der Flußsohle und der Ufer auf. Teilweise wesentliche Verformungen der Flußsohle sind nur seitlich und hinter zwei der Strohmpfeiler (bei Pfeiler Nr. 21 auch vorn) zu verzeichnen (Fig. 3).

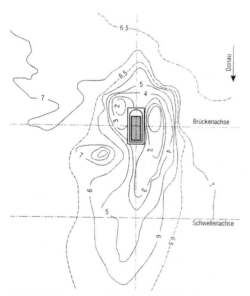

Abb. 3. Auskolkungen der Flußsohle bei Pfeiler Nr. 19

- Die Auskolkung der natürlichen Flußsohle bei den Brückenpfeilern ist ein unvermeidlicher Vorgang wegen der Umströhmung mit verhältnismäßig größeren Geschwindigkeiten, die Änderungen in der Strohmrichtung hervorrufen; es bilden sich aufwärtsgerichtete sekudäre Wirbelströhmungen, die Partikeln des Sohlalluviums erfassen und sie außerhalb des Brückenbereiches transportieren. Der etwas schräg zu den Pfeilern gerichtete Strohm erleichtert die Vertiefung der Kolke.
- Bei der vorhandenen Kornverteilung der aktiven Schichte der Flußsohle sind Auskolkungen um die Pfeiler bis zur absoluten Kote ±0 möglich, was eigentlich bei Pfeiler Nr. 21 zu beobachten ist, wo jedoch dieser Prozess schon abgeklungen sein dürfte. Für die anderen Pfeiler, wo die Flußsohlbildung noch nicht abgeschlossen ist, kann man eine ähnliche Entwicklung prognostizieren, besonders nach Ausführung von Maßnahmen zur Flußsohlensanierung bei anderen Pfeilern.
- Es besteht auch eine Ungewißheit der Entwicklung der Ströhmungsprozesse um die Pfeilern, besonders bei außergewöhnlichen Situationen, wenn z.B. in der Donau größere als die maximalen Bemessungswassermengen durchfließen.

Unter den soeben beschriebenen Umständen erweist es sich als notwendig die Wiederherstellung der ursrünglichen (vor Errichtung der Brücke) Flußsohle im Brückenbereich. Dies wird die statische Funktion der Senkkastengründungen, besonders bei seismischen Einwirkungen, verbessern, da die elastische Einspannung der Pfeiler wirksamer würde. Die letzten Erdbeben in Rumänien und Bulgarien haben gezeigt, daß bei einer Gründungstiefe D ≥ 0.25H (Höhe des Bauwerkes) das Bauwerk besser die seismischen Einwirkungen verträgt und es weniger zu Schäden kommt.

6 MASSNAHMEN ZUR SANIERUNG DER BRÜCKENGRÜNDUNGEN

Die statischen und hydraulischen Untersuchungen, sowie ihre Deutung, führten zu folgenden Beschlüssen:

1) Durch entsprechende Organisation des Verkehrs über die Brücke muß die Bildung von ununterbrochenen Autokolonnen unterbunden werden. Dadurch werden die Ecksohlpressungen, bzw. die maximalen Pfahlbelastungen, unter den angegebenen Bemessungswerten verbleiben. Diese Maßnahme ist schon erfüllt.

2) Für zwei der Strohmpfeiler (Nr. 19 und Nr. 21) mußten Befestigugsmaßnahmen entworfen werden, die die weitere Auskolkung der Flußsohle im Brückenbereich verhindern sollen.

In Zusammenhang mit 2) wurde ein Projekt zur Wiederherstellung der Flußsohle ringsum die Pfeiler bis zum ursprünglichen Niveau entworfen, welches im zentralen Teil der Brücke einer absoluten Kote + 7 m und hinter der Brücke + 6.5 m entspricht. Dies wird durch die Herstellung einer Schwelle in der Flußsohle, 50 m hinter den Pfeilern Nr. 19 und Nr. 21, erreicht. Der Raum zwischen den Pfeilern und der Schwelle, inkl. der Auskolkungen, wird mit Flußkies ausgefüllt (Fig. 4).

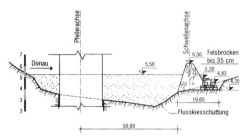

Abb. 4. Sohlschwelle 50 m flußabwärts von Pfeiler Nr. 19

Um den hydraulischen und hydromorphologischen Einfluß der Entwurfslösung für die Befestigung der gefährdeten Pfeiler auf die Stabilität der Flußsohle festzustellen, wurden hydraulische Untersuchungen an Fragmentmodellen (1997) der Pfeiler Nr. 19 und Nr. 21 durchgeführt. Die erhaltenen Ergebnisse bewiesen die Wirksamkeit der gewählten Lösung. In Kürze steht die Ausführung des Entwurfes zur Befestigung der Flußsohle und der Pfeiler bevor. Bis dahin werden systematische Beobachtungen und Messungen durchgeführt, die die weiteren Verformungen der Flußsohle im Bereich der Strohmpfeiler feststellen sollen.

Die an der Donaubrücke Russe-Giurgiu durchgeführten Untersuchungen beweisen noch einmal von welcher ausschlaggebenden Bedeutung für die Bauwerke Beobachtungen während ihrer Nutzung sind, um rechtzeitig die drohenden Gefahren zu erkennen und die entsprechenden Schutzmaßnahmen zu treffen.

SCHRIFTTUM

Vereinbarung (1954) zur Erhaltung und Betreuung der Eisenbahn- und Straßenbrücke an der Donau zwischen Russe und Giurgiu, Giurgiu (nicht veröffentlicht).

Reglement (1954) für die technische Erhaltung und die Betreuung der Eisenbahn- und Straßenbrücke an der Donau zwischen Russe und Giurgiu, Giurgiu (nicht veröffentlicht).

Hydraulische und theoretische Modelluntersuchungen (1993) an der Donau im Bereich der Brücke Russe Giurgiu. Archiv BDZ (Bulgarische staatliche Eisenbahnen) (nicht veröffentlicht).

Hydraulische Untersuchungen an Fragmentmodellen (1997) der Pfeiler Nr. 19 und Nr. 21 der Donaubrücke bei Russe-Giurgiu. Archiv BDŽ (Bulgarische staatliche Eisenbahnen) (nicht veröffentlicht).

Geotechnical Hazards, Marić, Lisac & Szavits-Nossan (eds) © 1998 Taylor & Francis, ISBN 90 5410 957 2

Analysis of an example of a nailed wall in soft clayey soil

Predrag Kvasnička & Leo Matešić
Faculty of Civil Engineering, University of Zagreb, Department of Geotechnical Engineering, Croatia

Bojan Vukadinović
Geotehnički Studio d.o.o., Zagreb, Croatia

ABSTRACT: The objective of the paper is to analyse the behaviour of a nailed wall in soft clay deposits constructed for a temporary protection of an excavation pit in the centre of Zagreb in the year 1996. It has been the first attempt to build a nailed wall in Croatia. The paper deals with actions, displacements and a failure mechanism of the soil nailing structure. The analysis of the evolution of displacements has been carried out based on soil parameters and settlements measured during construction. It has been performed by using program FLAC, provided by Itasca Consulting Group, Inc. Minneapolis, Minnesota. By means of FLAC an attempt has been made to model the influence of the rise of groundwater level on stability of the nailed wall. The numerical analyses showed that the nailed wall in Mandrovićeva Street should stay stable under normal conditions and that the instability of a section of the wall was probably caused by the unexpected raise of groundwater.

1 INTRODUCTION

To date, the technique of stabilisation of excavations and slopes by soil nailing has been successfully developed and applied in soils such as cemented sands, clayey and sandy silts, and similar soil materials. Because of the low values of an internal friction angle and the possibility of significant clay creeping in the areas of higher stress concentration around the nails, it is generally accepted that soil nailing may not be successfully applied in clays. Consequently, very little research has been done on the subject, and there is practically no field experience on soil nailing in deposits of clays of medium to high plasticity.

The main feature of soil nailing is that the existing natural soil is reinforced as opposed to backfill reinforcement. The inclusions, commonly called nails, are installed as the excavation proceeds by utilising a "top-down" construction procedure, unlike reinforced earth walls constructed from the bottom up. This allows soil retention in areas where little space is available for the excavation. The soil nailing concept is to reinforce the soil with passive inclusions so that the nailed soil mass behaves as a composite unit similar to a gravity retaining wall supporting a soil backfill (Recommendations Clouterre, 1991; Juran & Elias, 1991; Mitchell & Villet, 1987). In that sense, soil nailing also differs from the conventional tie-back excavation support, since the soil nails are not prestressed, i.e. their resistance can be mobilised only by the movement of soil mass or the face of the excavation to which the nails are fixed. Therefore, the nails develop tension only when the lateral deformation as a result of excavation takes place. Lack of pretensioning of the nails is the cause of higher wall movements and settlements, which may cause damage to buildings behind walls if precautionary measures were not taken.

Soil nails are relatively thin elements around which concentration of stresses may easily develop. In soft soil, which have high creep characteristics, such stresses may cause large deformation of the soil in time and corresponding displacement of the nails, and ultimately failure of the structure. Such large movements of nails through soft soil are often referred to as "combing". The geology of Croatia is such that large parts of the most populated and developed regions are on deposits of medium to soft clays. Therefore, for truly successful and cost-effective introduction of the soil nailing construction technology to Croatia a critical issue of the applicability of soil nailing to soft clayey deposits must be addressed.

2 GENERAL COMMENTS

The applicability of soil nailing technology in clay deposits was tested on a soil nailed wall in Zagreb. The

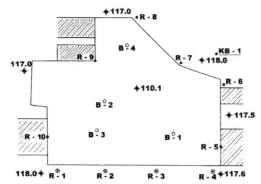

Figure 1 Plan view of the excavation pit in Mandrovićeva Street in Zagreb.

wall is located along the excavation pit in Mandrovićeva Street. The depth of the excavation is approximately 7.4 m (from 117.5 down to 110.1 m above the see level). A plan view of the excavation pit is presented in Figure 1. As there are some houses next to the excavation pit, it was necessary to provide protection not to cause loss of bearing capacity, settlements or lateral movements of the surrounding houses. A soil retention was constructed in a slightly different way from standard nailed walls. Before excavation, steel I-beams of 200 mm in thickness and 10 m in length were pushed into ground spaced 1.8 m. As excavation proceeded, by utilising a "top-down" construction, two anchors were pushed into soil next to each I-beam at an acute angle. The space between I-beams was filled with reinforced concrete (a steel net as a rebar) instead with shortcrete facing. Foundations of the houses were supported with "step-taper" piles that were pushed in soil before excavation. During excavation the space between the piles was filled with concrete and later "nailed" like other sides of the excavation pit. The nails consisted of steel round tubes of 50 mm in diameter and 6 m in length.

Soil is soft to medium silt and clay. It was explored in a standard way; the soil investigation was carried out by means of borings and test pits. The location of the borings (B-1, B-2, B-3, B-4 and KB-1) and monitoring points (R-1 to R-10) may be seen in Figure 1. The depths of borings range from 10.0 to 16.0 m. SPT tests were carried out in borings. The groundwater level was measured with piesometers and found that it was approximately at a depth of 4 m below the soil surface. Anyhow, the soil was almost saturated up to the soil surface. Disturbed and undisturbed samples were obtained to identify and classify the soil and to carry out standard laboratory tests. Triaxial tests on undisturbed samples were carried out in order to provide stress-strain relationships of soil. Three pullout

tests on grouted anchors served for determination of the bearing capacity of the nails. Comprehensive review of the soil investigation, the laboratory test results and the pull out test diagrams are presented in the companion paper by Vukadinović et al. (1998).

For this study, particularly interesting was the northern part of the excavation pit (near the monitoring points R-7 and R-8) where the extensive settlements occurred behind a part of the soil nailed wall. The state of this part of the wall was obviously less stable than the other parts and close to failure. The researchers believe that this state was initiated by an unexpected rise in the groundwater level at this part of the wall. It seemed that the rise in the groundwater level was a result of the water inflow in places caused by, for example, a water pipe burst. Water was monitored as coming out from nail holes throughout this section of the wall. This event was used in a numerical analysis to simulate the wall failure with a rise in the groundwater level.

3 NUMERICAL ANALYSIS

3.1 Basic concepts

The analysis of displacements leading to failure is carried out on the basis of the assumed loads and soil data and evaluated based on recorded settlements of soil surface behind the wall. The analysis is performed using program FLAC, provided by Itasca Consulting Group, Inc. Minneapolis, Minnesota. FLAC is a two-dimensional explicit finite difference code primarily intended for geotechical engineering applications which simulates the behaviour of structures built of soil, rock or other materials that may undergo plastic flow when their limit is reached.

Key features of this numerical analysis include:

- five excavation stages,
- emplaced support components at each excavation stage, and
- development of soil settlement behind the wall as the result of actions of soil pressure and seepage forces.

As always in modelling, the simplest possible model, consistent with the reproduction of physical processes that are important for the problem at hand is used. Soil is modelled as isotropic horizontal layers that behave as an elastic, perfectly plastic Mohr-Coulomb material. The reinforced concrete (facing) is assumed to be homogeneous linearly elastic material represented as beam elements. The nails are assumed to be homogeneous isotropic linearly elastic material represented as cable elements with the limited bond stiffness and bond strength. The problem is originally 3D with regularly spaced reinforcement. It is reduced to 2D by averaging the reinforcement effect in three dimensions over the distance between the reinforcements according to suggestions made by Donovan et al. (1984). Material properties are linearly scaled in order to distribute discrete effect of the reinforcement over the distance between reinforcement in a regularly spaced pattern. Reinforcing effects of nails are calculated for one-meter width of the soil-nailed wall.

3.2 Soil parameters

In geomechanics, when performing an analysis the problem always involves a data-limited system and, because of the high uncertainty in the property data base, the selection of soil parameters is often the most difficult in the generation of a model. For the excavation under consideration, material properties are derived from *in situ* and laboratory testing program; therefore, the use of simple material model such as Mohr-type, seemed to be the only reasonable option.

Mohr-type material model in FLAC assumes an isotropic material behaviour in the elastic range described by two elastic constants, bulk modulus (K) and shear modulus (G), and strength parameters: cohesion (c) and angle of internal friction (φ). In our case values of the moduli were attained indirectly by obtaining at first Young's modulus E_i and assuming the values of Poisson's ratio, taking into consideration the type of analysis (drained or undrained). Young's modulus was calculated on the basis of triaxial tests as recommended by Duncan & Chang (1970):

$$E_i = Kp_a \left(\frac{\sigma_3}{p_a} \right)^n \; ; E_0 = Kp_a \; ; E_i = E_0 \left(\frac{\sigma_3}{p_a} \right)^n$$

in which E_i is the initial tangent modulus, σ_3 is the minor principal stress, p_a is atmospheric pressure, K and n are a modulus number and the exponent determining the rate of variation of E_i with σ_3.

K and n for two clay specimens from the boring log KB1 (from depths of 3.50 to 3.90 m and 5.00 to 5.40 m) were derived from the results of undrained triaxial tests by plotting the values of E_i against σ_3 on log-log scales and fitting a straight line to the data, as shown in Figures 2 and 3. With the well known relations, from E_i and Poisson's coefficient ν, values of K_i and G_i were derived for every characteristic soil layer. Considering that the drainage conditions were close to undrained state (due to relatively short excavation time and taking into consideration small clay permeability),

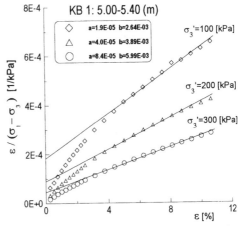

Figure 2. Transformed stress-strain curve of triaxial test results for specimen KB1: 5.00 to 5.40m.

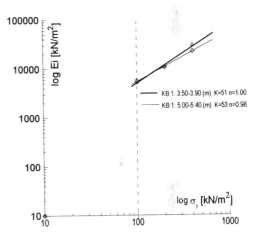

Figure 3. Variation of initial tangent modulus with confining pressure under undrained triaxial test conditions.

average value of $\nu = 0.4$ was assumed and values for K_i and G_i calculated (Table 1). Strength parameters - cohesion c_i and angle of internal friction φ_i, were estimated on the basis of the results presented in Vukadinović et al. (1988). Lower bound values were chosen as input parameters to FLAC.

Approximate values of undrained shear strengths were obtained indirectly by using the "step-taper" piles and small piles ($\varnothing$ 245 mm, $\varnothing$220 mm and $\varnothing$130 mm) as cone-penetrometers and back calculating the shear strengths according to Sanglerat (1972). The results are presented in Table 2.

Table 1. Input values (in FLAC) for the moduli K_i and G_i and the strength parameters c_i and φ_i and undrained cohesion c_{ui}.

height above the see level [m]	σ_v [kN/m²]	K_i [MN/m²]	G_i [MN/m²]	c_i [kN/m²]	φ [°]	c_{ui} [kN/m²]
113.5-117.5	30.7	2.66	1.11	4.0	25	30.0
112.5-113.5	67.2	5.82	2.46	2.0	23	30.0
108.0-112.5	90.0	7.65	3.21	4.0	25	30.0
106.0-112.5	95.2	8.25	3.82	8.0	28	40.0

Table 2. The undrained shear strengths in kN/m², back calculated from the results of the step-taper pile penetration.

height above the see level (m)	"step-taper" piles, group 1	"step-taper" piles, group 2	small "step-taper" piles
113	-	28.7	-
112	23.3	31.0	24.0
111	24.0	32.3	28.0
110	25.0	33.0	30.0
109	25.5	33.6	31.0
108	25.8	34.0	32.0
107	26.1	34.0	-

3.3 Numerical modelling

In the numerical analysis, actual field excavation and construction of a nailed wall is simulated by five phases performed by first compacting the soil mass under gravity to establish equilibrium in situ conditions and then sequentially excavating to various levels and introducing support elements. As the inflow of water was the cause of high deformations on the northern part of the wall special attention was paid to the effect of water induced stresses. Generally, conditions existing long after construction are critical and control the safety factor of the trust for which the retaining wall must be designed. It can be evaluated by an effective stress analysis using pore pressures determined by natural ground water conditions. For the long term structures it is not necessary to determine the state of safety immediately after excavation, but as the result of standstill in the excavation process, nailed wall in Mandrovićeva Street was the case in which both situations (drained and undrained) could be critical.

Therefore, the following project situations were analysed:

- total stress analysis, with shear strengths of soil

equal to undrained strength (with input data for moduli and undrained cohesion as in Table 1),
- drained stress analysis with groundwater at a depth of 4.0 m below the soil surface and
- drained stress analysis with the groundwater at the soil surface in the background (with input data for moduli and drained strength parameters as in Table 1),

The criteria for the successful modelling was the compliance of the calculated settlements and displacements with the measured ones. For the numerical analysis a system with 40 x 40 grid was generated. The so-called null model was used to simulate excavation; beam and cable elements were introduced to simulate concrete facing and grouted nails respectively. In FLAC, soil to nail interface obeys elastic-perfectly plastic behaviour. The grid was configured for the groundwater flow in drained analysis to get the distribution of pore pressures. The pore pressure distribution was important because it was used in the computation of the effective stresses at all points of the system.

4 RESULTS AND CONCLUSION

The results of calculated as well as of the measured settlements are divided into two groups - with stable and unstable behaviour of the nailed wall structure. Calculated and measured settlements are presented in diagrams in Figures 4 and 5.

In Figure 4, the settlements derived by numerical modelling were compared with the measured settlements for the stable section of the wall (monitoring points: R-1, R-3, R-4, R-5, R-6 and R-10). In calculation, stable behaviour of the wall was obtained for the undrained-type analysis and the drained-type analysis with the groundwater level at a depth of 4 m. Although the drained analysis was carried out at real time, the settlements which were the result of calculation could not be matched directly to settlements measured during the progress of excavation. In modelling, it was not possible to anticipate the number and duration of standstill periods that occurred during excavation. Therefore, the development of the calculated settlements in time was adjusted to the measured ones which can be visible as strait horizontal lines in settlement-time diagrams in Figures 4 and 5.

The settlements for the unstable section (monitoring points: R-7 and R-8) are presented in Figure 5. In calculation, comparable settlements were obtained for the drained analysis supposing that the groundwater level was at the soil surface. A point in Figure 5 marked with an arrow corresponds to the time when a support to the wall was provided by means of an embankment to prevent its further deformation.

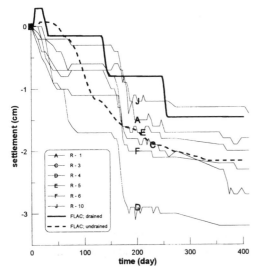

Figure 4. Diagram of the measured and calculated settlements of the soil behind the nailed wall in the drained analysis for the stable portion of the nailed wall (in the calculation groundwater is on the top of the soil surface behind the wall).

The pattern of the failure mechanism of the unstable section developed by FLAC is presented in Figure 6. The effect of the nailing is that the Rankine's failure surface which would otherwise develop (without nails), is impeded, and the other failure surface with higher safety factor becomes critical. It confirms the supposition that the nailed soil mass acts like a composite unit.

The failure surface was evidenced even *in situ* by monitoring the soil surface and damage to small huts behind the wall.

Attention is drawn to the fact that in spite of the use of a simple elastic-perfectly plastic constitutive relations, calculated and measured settlements corresponded well for both types of behaviour.

The calculated forces in nails as regards stable behaviour were below the measured pull-out forces for the test nails (approximately 50 kN, Vukadinović et al., 1998); but as for unstable behaviour, forces were close or even equal to pull-out values.

The numerical analyses showed that the nailed wall

in Mandrovićeva Street should stay stable under normal conditions and that the failure of a section of the wall was probably caused by the unexpected groundwater flow. Generally speaking, by using experience gained with nailed wall in clay in Mandrovićeva Street is positive. However, it should be taken into account that the expected settlements and horizontal deformations are more extensive with the nailed walls then with retaining structures with prestressed anchors. This fact could be decisive factor when making decision on the type of the retaining system of excavation to be used.

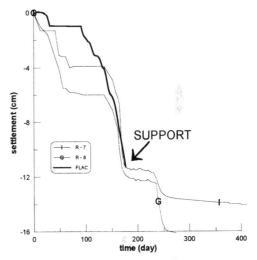

Figure 5. Diagram of the measured and calculated settlements of the top of the facing in the drained analysis for the failed section of the nailed wall (in the calculation groundwater is on the top of the soil surface behind the wall).

5 ACKNOWLEGEMENT

The investigation of nailed walls in soft clays is an integral part of the ongoing research by Croatian and U.S. investigators. It represents a continuation of existing research collaboration in the field of earth reinforcement supported by the U.S.-Croatian Science and Technology Joint Fund. The authors wish to thank Mladen Vucetic, associate professor at Civil and Environmental Engineering Department, University of California, Los Angeles, for his encouragement and support. Special thanks to students Đurđica Keglević and Otto Krasić who made efforts to prepare soil parameters for computations in FLAC.

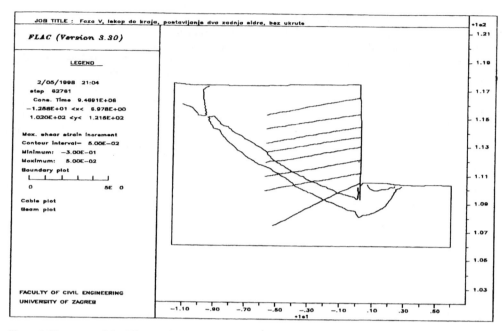

Figure 6. The pattern of the failure mechanism for the case with groundwater level on the soil surface (from FLAC).

REFERENCES

Bruce, D.A. & Jewell, R.A. 1987. Soil Nailing: Application and Practice - Part 1 and 2. *Ground Engineering*: 10-15, 21-33.

Donovan, K., Pariseu, W.G. & Cepak, M. 1984. Finite Element Approach to Cable Bolting in Steeply Dipping VCR Stops. *Geomechanics Application in Underground Hardrock Mining*. New York: AIME: 65-90.

Duncan, J.M. & Chang, C.-Y. 1970. Non-linear analysis of stress and strain in soils. *Journal of the Soil Mechanics and Foundation Division* 96 (SM 5): 1629-1653.

Gassler, G. & Gudehus, G. 1983. Soil nailing statistical design. *Proc. 5th Eur. Conf. Soil Mech. and Found. Eng.* 3, Helsinki: 491-494.

Juran, I & Elias, V. 1991. Ground anchors and soil nails retaining structures. *Foundation Engineering Handbook, Chapter 26*, New York. Edts Van Nostrand Reinhold: 868-905.

Mitchell, J.K. & Villet, W.C.B. 1987. *Reinforcement of earth slopes and embankments*, TRB-NCHRP, Report No. 290.

Recommendations clouterre 1991 for designing, calculating, constructing and inspecting earth support systems using soil nailing. Washington: FHWA.

Sanglerat, G. 1972. *The penetrometer and soil exploration*. Amsterdam: Elsevier Publishing Co.

Vukadinović, B., Sorić, I. & Verić, F. 1998. Foundation pit protection in soft clay, applying modified soil nailing method, *XI Danube - European Conference on Soil Mechanics and Foundation Engineering*, Poreč.

Geotechnical Hazards, Marić, Lisac & Szavits-Nossan (eds) © 1998 Taylor & Francis, ISBN 90 5410 957 2

A numerical method for calculation of propped retaining wall

Miloš Lazović
Faculty of Civil Engineering, Belgrade, Yugoslavia

ABSTRACT: The developments of diaphragm wall construction have led to the frequent use of free or support by ground anchors embedded cantilever retaining walls. The method is suitable for urban sites where disturbance of the soil beneath existing foundations close to the margins of the excavation are to be avoided. Ground anchors may be placed at a single level, or in deep excavations at several levels, installed at each successive stage of excavation. The design of cantilever or propped retaining walls is currently based on approximate limit equilibrium calculations, or on finite element method. In this Paper, applied numerical solution is based on finite element method and on nonlinear equivalent frame method. To model the soil behavior an hypoelastic constitutive law is used.

1 INTRODUCTION

The recent developments of slurry trench methods of in situ diaphragm wall construction and secant piles have led to the frequent use of free or propped embedded cantilever retaining walls for retained cuttings and cut end cover tunnels in urban environments where land use is restricted. This is a suitable method of construction for sites where insufficient space is available around the excavation to slop back the sides. The method is also suitable for sites where considerations of noise and vibration of the soil beneath existing foundations close to the margins of the excavation are to be avoided. The rigidity of the diaphragm wall in combination with support by pre-loaded ground anchors followed by strutting with the floors of the permanent structure can reduce the inward deflection on the structure end hence maintain stability and excessive soil movement. The magnitude of the soil movements will depend on the geometry of the cutting and wall, the method of the wall construction, the properties of the wall and the soil, the position of the prop or props and the initial stress prevailing in the soil. Earth pressures should be calculated at the various stages of construction of an anchored wall. Thus at stage 1 (Fig.3) the wall is acting as vertical cantilever. At stage 2, when the top level ground anchors are stressed the pressure on the back of the wall will be determined by the stresses induced in the anchors and it may at some stage be intermediate between the "at rest" and passive condition

depending on the amount of reverse movement of the wall. After final excavation, the wall acts in stage 3 as a vertical beam cantilevering above the anchors and restrained by yielding supports at anchor level and at the embedded portion below excavation level. The depth of the toe bellow final excavation level must be such as to provide the required passive resistance. At the end of construction, a check should be made to ensure that the wall can withstand "at rest" earth pressure and any hydrostatic pressure on the rear face. Also, in all types of ground expect massive rock some inward yielding of the sides of the strutted excavation will take place. Thus in soft silts and clays there is the additional risk of upward heaving of the bottom of the deep excavation accompanied by major settlement of the ground surface.

This paper considers design aspects of embedded, propped and unpropped wall formed by excavation in front of the wall. Embedded wall may be used in either temporary or permanent works situations. For permanent construction critical design condition occur in the long term after all excess pore pressures, developed during the retaining wall construction, have dissipated. For such long-term stability, effective stress soil parameters are used. In the temporary works situation excess pore water pressure may not dissipate fully during the life of the structure. In such cases design is carried out in terms of total stresses. When designing embedded retaining walls the main criterion is to prevent unserviceability. To ensure overall stability, cantilevers are often designed using the fixed earth support method in which the wall is assumed to rotate about a point near to its toe. The required embedment depth is then obtained by taking moments about the toe, and this depth is increased by an empirical 20%. For design, the moment equation is used to ensure that restoring moments exceed overturning moments by a prescribed safety factors. A propped retaining wall are designed using the fixed earth support method in which the wall is assumed to rotate about a prop point. The required prop force is then obtained by taking horizontal equilibrium.

In the finite element analyses the soil and diaphragm wall are in a state of plane strain. The excavation of a row of element takes place over several load steps. The equivalent nodal forces applied by the elements to be excavated are first calculated and then removed over several increments of the analysis. During these increments the excavated elements are treated as ghost elements with very low stiffness. The accuracy of the analysis depends on the finite element mesh, constructive soil models and number of increments employed. It can be concluded that nonlinear incremental finite element analysis is very complex and not suitable for practical design. In this Paper, the proposed numerical method is based on linear finite element method and on nonlinear equivalent frame method.

2. THE PROPOSED NUMERICAL METHOD

As is mentioned before, design of embedded cantilever or propped retaining wall is very complex, and must follow the constracting stages. The soil is assumed to be fully drained. The water pressure is hydrostatic below the water table. The incremental and iterative method of analysis should be applied because of the nonlinear stress-strain relations.

As in most geotechnical engineering problems, a knowledge of in-situ state of stress is necessary. The in situ state of stress in soil is defined in terms of the current values of effective vertical stress (σ'_{vo}) and effective horizontal stress (σ'_{Ho}). For horizontal, level ground, the in situ vertical stress is

$$\sigma_{vo} = \sum \gamma_i h_i \qquad (1)$$

However, the horizontal stress is more difficult to evaluate. The stress ratio K_0, which is the at rest coefficient of horizontal soil stress, is defined as $\sigma'_{Ho}/\sigma'_{vo}$. For normally consolidated soil, the simplified Jaky equation provides reasonable estimates for K_0, as is given below:

$$K_0 = 1 - \sin \varphi \qquad (2)$$

Many factor affect the in situ state of stress in soil, including: overconsolidation, aging, chemical bonding, etc. Overconsolidation is probably most influential for the majority of soils. For the overconsolidated soils, the general relationship for K_0 is often expressed as:

$$K_0 = (1 - \sin \varphi) \cdot OCR^n \qquad (3)$$

As suggested by Schmith (3), the exponent n may be expressed as a function of:

$$n = \sin \varphi_{tc} \qquad (4)$$

In some cases, close to the margins of the excavation there are existing structures. The foundation pressures generate additional stresses in the soil. The stresses in an elastic and isotropic half-space produced by a uniform vertical load, over a flexible strip foundation, may be written in the following form:

$$\sigma_z = \frac{p}{\pi} \left(arctg \frac{b-x}{z} + arctg \frac{b+x}{z} \right) - \frac{2pbz(x^2 - z^2 - b^2)}{\pi \left[(x^2 + z^2 - b^2)^2 + 4b^2 z^2 \right]} \qquad (5)$$

$$\sigma_x = \frac{p}{\pi} \left(arctg \frac{b-x}{z} + arctg \frac{b+x}{x} \right) + \frac{2pbz(x^2 - z^2 - b^2)}{\pi \left[(x^2 + z^2 - b^2)^2 + 4b^2 z^2 \right]} \qquad (6)$$

A diaphragm wall is constructed by excavation in an trench which is temporarily supported by a bentonite slurry. So, there isn-t change in the horizontal stress in soil. But, there is some change in the vertical stress in dependence of the trench width. In this paper a correction coefficient is used to take into account the change in vertical stress behind the wall end and as well in the front of the wall. The change in vertical stresses generates additional settlement on existing structures.

Before excavation, diaphragm wall and the soil are making statically an equilibrium system. This statical system can be represented by an equivalent frame as is shown in Fig.1.

The diaphragm wall is discretized by twonodded beam elements. The influence of soil is substituted by the horizontal springs(boundary elements) at nodal points, at both sides of diaphragm wall. The

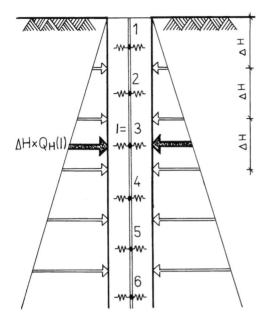

Fig 1. Equivalent frame

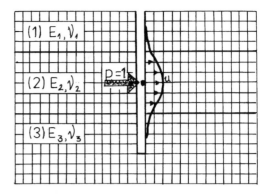

Fig. 2 Finite element mesh

hyperbola in conjunction with the relation between initial modules and confining pressure proposed by Janbu. The following expression for the tangent modules can by obtained as

$$E_t = K \cdot \left(\frac{\sigma'_3}{p_a}\right)^n \left[1 - \frac{R_f(\sigma_1 - \sigma_3)(1 - \sin\varphi)}{2(\sigma_3 \sin\varphi + c\cos\varphi)}\right]^2 \quad (8)$$

In that expression the Mohr-Coulomb failure criterion is incorporated. For unloading the initial modulus is used. To correct the evaluated horizontal nodal displacements a diagonal matrix D is formed. The coefficients in D, are ratios between modulus of elasticity and tangent modulus at every nodal points. The stiffness of the boundary elements may be evaluated as a ratio between nodal forces and corrected nodal displacements.

$$K_i = \frac{\sigma_H(i) \cdot \Delta H}{U_i^*} \quad (9)$$

The excavation was simulated by sequentially removing the thin soil layers-slices in front of the wall. Removal of slices was simulated by first calculating the equivalent nodal forces arising from the stresses acting within these slices and then applying those which acted on slices remaining in their opposite sense as boundary conditions for further increments of the analysis. Correct account was taken of both the initial stresses and those stress changes which occurred during the excavation process. At the base of the excavation the soil is subjected to passive stress relief.

As a result of applying nodal forces on the equivalent frame, horizontal displacements of the nodal points towards the excavation are obtained. The consequence of this displacements are changes of horizontal stresses in the soil. In the front of the

system is subjected to horizontal ground pressure at rest and hydrostatic pressure below ground water level. This loading is in equilibrium. In this paper the finite element method is employed to calculate tangential stiffness of the boundary elements. The soil and diaphragm wall are in state of plane strain. It is sufficient to consider only a slice between two sections separated by a distance of unit length. The mesh of finite elements is shown in Fig.2.

It was assumed that soil behaves as linearly elastic material. Every soil layer is defined with two parameters E_s and v_s, determined for the stress level in the middle of every layer. Using finite element formulation the soil flexibility matrix F_s is numerically evaluated.

The coefficients in F_s, are horizontal nodal displacements due to external applied unit horizontal nodal forces. By using the principle of superposition the horizontal displacements of the nodal point, due to horizontal soil pressure at rest, may be written in the matrix form as:

$$U = F_s \cdot P \quad (7)$$

To model the soil behaviour an hypoelastic Duncan-Cheng model is used. The hypoelastic concept can provide simulation of constitutive behaviour in a smooth manner, and hence can be used for hardening or softening geologic materials. Use of the hyperbola for representing stress-strain curves for soil, was proposed by Kondner. To incorporate this aspect, Duncan and Chang used the

Table 1. Soil parameters

4Layer	h	g'	φ	c	K	n	R_f	E
(1)	3.5	17.2	28.0	0.	75.960	0.7	0.80	8000
(2)	2.8	18.2	32.0	0.	78.760	0.7	0.85	15000
(3)	5.0	9.5	10.0	48	63.150	0.8	0.90	24000
(4)	3.5	9.8	14.0	57.0	90.500	0.8	0.92	34000

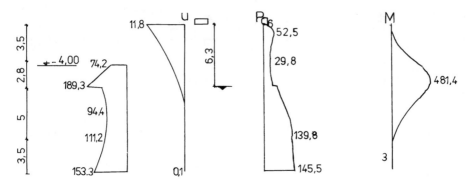

Phase 1: Excavation to Level -4m.

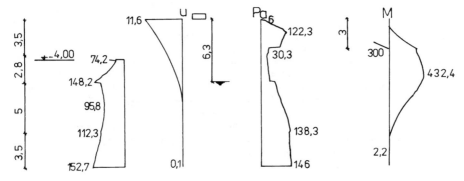

Phase 2: Prestressing up to 300 kN by ground anchor at level -3m

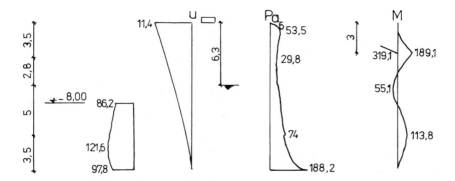

Phase 3. Excavation to Level -8m.

Fig. 3. Earth pressure, wall displacement and bending moment

wall, passive site, the pressure in the soil increases and is given by,

$$\sigma_{Hp}(I) = \hat{\sigma}_{Hp}(I) + \frac{P_p}{\Delta H} - \Delta\sigma_{Ho}(I) \qquad (10)$$

where are,

σ_{Hp} =horizontal stress in nodal i, on the passive site;

$\hat{\sigma}_{Hp}$ =stress in the same nodal, before applying incremental load;

P_p =force in the boundary element on passive site;

$\Delta\sigma_{Ho}$ =decrease of the horizontal stress due to excavation.

On the back of diaphragm wall, active site, the pressure in the soil decreases end is given by

$$\sigma_{Ha}(I) = \hat{\sigma}_{Ha}(I) + \frac{P_a}{\Delta H} \qquad (11)$$

where are

σ_{Ha} = horizontal stress in nodal i, on the active site;

$\hat{\sigma}_{Ha}$ = stress in the same nodal, before applying incremental load;

P_a =force in the boundary element on the active site.

After all needed calculation are performed, it is necessary in all nodal points, on both sides of diaphragm, to calculate the safety factors according to next expression:

$$F_s = \frac{2(\sigma_3 \sin\varphi + C\cos\varphi)}{(\sigma_1 - \sigma_3)(1 - \sin\varphi)} \qquad (12)$$

If this factor is larger than previously calculated (before excavation slice), it means that this node have undergone unloading. In this nodes the initial modulus have to be used. Also it is necessary to check whether the stresses in the nodes are greater than active, are smaller than passive. In the nodes where it is not satisfied, the boundary elements are removing, and replacing bay active and passive pressure. This procedure is used iterattively within every incremental loading to monitor the plastic zone development at the interface of diaphragm and soil. All nodal stresses and deformations obtained at the end of iteration process, within the considered increment, are stored. The local, safety factors in the soil are also saved. This procedure is repeated for the next increment of excavation, and obtained results are added to the already stored from previous one. If the anchors or supports are designed, they can be also incorporated in the calculation process. The prestresed force in the anchor is applying incrementally, and in every increment above explained iterations are performing. After that in the considered node the boundary element is prescribed with the stiffness equal to anchor stiffness. The procedure is repeated for the next level of excavation, and if it is needed the new anchors are incorporated in the model. If the excavation takes under the level of ground water table, the increments of hydrostatic pressure are applying.

According to above explained procedure the computer program is made. Using this program efficient calculation can be easily performed.

3. NUMERICAL EXAMPLE

As a numerical example the results of calculation for a propped retaining wall in Moscow is shown. The 0,80m tick and 14.80m deep concrete retaining wall is assumed to be linearly elastic with a Young's moduls E of 30 GPa and Poisson's ratio 0.18. To model the soil behaviour an hypoelastic constitutive law is used, with parameters shown in table 1. Passive soil pressure, horizontal wall displacements, active soil pressure and bending moment in the diaphragm wall, for phases 1,2 and 3 are shown in Fig. 3. This results are in very good agreement with the values obtained by measurements.

4. CONCLUSIONS

The proposed numerical method is very efficient for practical design of cantilever or propped retaining walls. The calculation can be performed using standard commercial computer programs. The influence of construction stages, initial stresses and constitutive law of soil, can be correctly taken in to account.

5. REFERENCES

Potts, D.M. & Fourie, A.B.(1986) A numerical study of the effects of wall deformation on earth pressures. *Int. J. for Numer. and Analyt. Methods in Geomech.* 10, 383-405.

Lazović, M. (1995). *Prilog proračunu podupretih potpornih konstrukcija*, SANU, Beograd.

Schmidt, B.,(1966). Discussion of "Earth pressures at rest related to stress history", *Canadian Geotechincal Journal*, Vol. 3. No. 4, Nov, pp. 239-242

Burland, J.B., Potts, D.M. & Walsh, N.M. (1981). The overall stability of free and propped embedded cantilever retaining walls. *Ground. Engng.* 14, No. 5, 28-37.

Fourie, A.B. & Potts, D.M. (1989). Comparison of finite element and limiting equilibrium analyses for an embedded cantilever retaining wall. *Geotechinique* 39, No. 2. 175-188.

Geotechnical Hazards, Marić, Lisac & Szavits-Nossan (eds) © 1998 Taylor & Francis, ISBN 90 5410 957 2

Reliability assessment of tunnel design

M. Marenče
Verbundplan, Office Salzburg, Austria

ABSTRACT: Introduction of the probabilistic safety determination based on partial safety factors defined in the European Code represents, a difficult problem for geotechnical structures. Because of the incompleteness of the subsurface investigations, the underground heterogeneity and specificity in comparison to the other construction materials, it is not possible to describe the reality of the underground environment.
This paper aims to introduce the reliability assessment as alternative or complementary safety concept in the tunnel design. The reliability assessment calculation with the computational model is outlined on the practical example.

1. INTRODUCTION

Traditionally, evaluations of the structural adequacy have been expressed by safety factor. A safety factor can always be expressed as a ratio of capacity to demand. The use of precisely defined single values is in an analysis known as a deterministic approach. By the geotechnical projects, as a rule, we have to work with problems which cannot be fully defined by comprehensive and objective concepts. A primary deficiency is that the parameters (material properties, strength, loads, etc.) cannot be assigned as single precise values, and are biased by assumptions, qualitative statements, opinions and their relations. Together with the calculation model which represents an idealisation of the natural behaviour, the deterministic safety definition concept is overloaded by engineer's judgement, and ambiguous, uncertain data.

Another approach, the probabilistic approach, extends the safety concept to explicitly incorporate uncertainty in the parameters. The uncertainties can be quantified through statistical analysis of existing data or judgementally assigned. Even if judgementally assigned, the probabilistic results will be more meaningful than the deterministic analysis because the engineer provides a measure of uncertainty of his judgement in each parameter.

The reliability assessment analysis is suggested in European code as an alternative method to partial safety factors. The reliability assessment analyses combine what is known about a structure with reasonable limits for the unknowns to assess reliability. The structure reliability is dependent on the adequacy of design, quality of construction, and condition of the structure. The reliability assessment of a structure will depend on the evaluator's knowledge of these factors. Reliability analyses can be made with limited information, and during evaluation extended by the unknown or insufficient adjusted factors to improve the analysis. The presented method introduces the reliability assessment as an alternative or complementary safety concept in the tunnel design. The concept is shown and discussed on one of the possible tunnel failure mechanisms.

2. BASIC PRINCIPLES OF RELIABILITY ANALYSIS

In the probabilistic approach the parameters are treated as random variables. Random variables assume a range of values in accordance with a function termed a probability density function or probability distribution. Although the value of a parameter is uncertain or variable, the probability density function quantifies the likelihood that the value of the random variable lies in a given interval. When parameters are defined as random variables, functions of these parameters such as the safety factor also become random variables and can be expressed in probabilistic terms.

An engineering reliability analysis determines the probability of unsatisfactory performance, $P(u)$, defined as the probability that the value of a function which characterises the performance of the system exceeds some limit state.

The reliability, R, i.e., the probability that the unsatisfactory performance will not occur, is given by

$$R = 1 - P(u) \qquad (1)$$

A concept used in the calculation of reliability is the safety ratio. Important parameters should be defined as random variables, then the total capacity, C (resistance), and the total demand, D (load), are also random variables. The safety ratio, SR, is the quotient of the capacity and demand. The probability of unsatisfactory performance could then be expressed as

$$P(u) = P(SR < 1) = P[(C/D) < 1] \qquad (2)$$

In many applications of reliability analysis, and also in European code (EVN 1991-1, 1996), the probability of unsatisfactory performance is discarded in favour of the reliability index, ß, which is a measure of how much the expected average value of the safety ratio exceeds the limit state. Expressing reliability in terms of the reliability index is preferred because the reliability index can be calculated knowing only the expected value, μ, the standard deviations, σ, and the coefficient of variation, V, of the variables which have a measurable and easily understood meaning. That the expected value, the standard deviations, and the coefficient of variation are interdependent, knowing any of two the third is defined. The expression of capacity divided by demand, C/D, in terms of the distributed random variables is the performance function and is estimated by lognormal distribution. The reliability index, ß, is defined as a distance between the expected mean performance valueof the performance, $\mu_{\ln(C/D)}$, and the limit state, expressed in normalised units of standard deviation of performance, $\sigma_{\ln(C/D)}$, and is given by

$$\beta = \frac{\mu_{\ln(C/D)}}{\sigma_{\ln(C/D)}} \qquad (3)$$

A ß=3.0 implies that the expected value of the performance function lies three standard deviations above the limit state. The limit state is defined as a ratio of capacity to demand approaching unity, i.e., $\ln(C/D) \Rightarrow 0$. In the figure 1 the shaded area under the curve on the left of the limit state is probability of unsatisfactory performance. The US Army Corps of Engineers (1992) defines based on the reliability index and probability of unsatisfactory performance the expected performance level (Table 1)

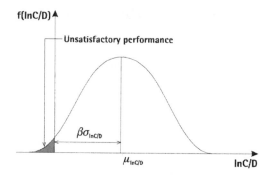

Figure 1. Lognormal distribution of capacity and demand and reliability definition

Table 1. Expected performance level

Expected performance level	Beta	Probability of Unsatisfactory Performance
Hazardous	1.0	0.16
Unsatisfactory	1.5	0.07
Poor	2.0	0.023
Below average	2.5	0.006
Above average	3.0	0.001
Good	4.0	0.00003
High	5.0	0.0000003

3. STRUCTURAL FAILURES

The tunnelling experience shows that the most hazardous phases are during the excavation. The criteria for safety assessment can be drawn from following sources (Dudeck, 1993):
- experience and observations from previous similar underground works
- geologic and geotechnical site exploration of the ground conditions
- in-situ monitoring during excavation
- geotechnical numerical analyses

Each of these items has a special range of significance during the design and construction phases and the risk assessment should take into considerations all four sources of information or should at least try to combine them as far as possible by interpreting their relative contributions to the safety aspects.

4. PROBLEM DESCRIPTION

The reliability assessment of tunnel design is shown on the design of the railway section between Kumanovo and Bulgarian border in Republic of Macedonia. The railway route, to be newly constructed or modified, has a total length of 71 km with 31 tunnel sections. Due to the pressed time schedule the tunnel design has to start with only few investigations carried out. Hence, it was necessary to check the range of the geotechnical parameters, chose the excavation method and support measurements, and estimate the construction safety.

The first geotechnical information used for design was based on the geological mapping and the seismic refraction. The geotechnical parameters from laboratory and „in-situ" tests in following project stages improved the first approximations. Selected design method together with the numerical simulation describing the excavation process has to fulfil following requirements:
- Model is defined on the existing data which should be updated during the design process.
- Scatter of the geological data and assumptions should be included in the calculation to estimate the design safety.
- To get acceptance of the owner the valuation of geological data should be done based on the international recognised classification system.
- Application should be simple and transparent.
The design procedure based on the known theories but combined to fulfil the project requirements have been developed and is presented in figure 2.

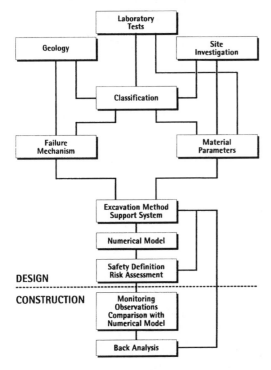

Figure 2. Design procedure

Numerical simulation of the excavation and the support system by the two-dimensional finite element method is selected, and is carried out by

program system FINAL (Swoboda, 1997). The selected calculation method is defined for fix input parameters and is supplemented by reliability theory to include the variation of the input parameters. Each parameter variation is associated with new calculation which is together with results interpretation a time consuming process.

The Taylor Series - Finite Difference (TSFD) estimation process (US Army Corps of Eng. 1992) is selected for reliability analysis. The TSFD estimation is practical for engineering problems, especially complicated numerical calculations, because it requires only $2n+1$ deterministic calculations of the safety ratio with n representing the number of random variables of influences.

5. IDENTIFICATION AND CHARACTERI-SATION OF RANDOM VARIABLES

5.1 Rock mass characterisation

Based on the geological field data, seismic refraction and geotechnical test the RMR classification of rock mass (Bieniawski, 1989) is performed. The geological data along the railway route is divided in the geological units which are dependent on disturbance divided in geological homogeneous regions. For each region the RMR classification is performed. The parameters influencing the classification, such as rock strength, RQD, spacing and conditions on discontinuities, ground water and orientation of discontinuities, are defined based on the observations, data form tests and estimation. These values are not characterised by a single value and the probability distribution is defined for each parameter. The rock mass rating (RMR) is defined as a sum of the six single parameter ratings. These six parameters with their probability distributions are combined by the Monte-Carlo-method to get final rating (Edlmair et al. 1996). Based on the large number of simulations the expected mean value and the standard deviation of RMR is defined for each homogenous region.

For the selected calculation the section in Mica schist characterised as a homogeneous region III (poor rock) is selected. The middle RMR value, $\mu_{RMR}=38$ and the standard deviation of $\sigma_{RMR}=7$ ($V_{RMR}=18.4\%$) is calculated. For the ratings $\mu_{RMR}-\sigma_{RMR}=31$, $\mu_{RMR}=38$ and $\mu_{RMR}+\sigma_{RMR}=45$ the strength based on the Hoek-Brown strength criterion (1988) and elasticity modulus based on the work of Serafim & Pereira (1983) are defined. The strength curves are presented on figure 3, and the elastic and strength parameters of rock mass are specified in table 2.

Table 2: Rock mass parameters

	RMR	σ_c[MPa]	m[-]	s[-]	E_r[Gpa]
$\mu-\sigma$	31	30	0.0724	0.00001	3.35
μ	38	30	0.1193	0.00003	5.01
$\mu+\sigma$	47	30	0.2269	0.00015	8.41

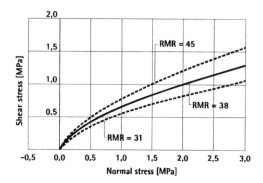

Figure 3. Hoek-Brown strength envelope

5.2 *Primary state of stress*

The primary state of stress represents a stress condition in the ground under the undisturbed conditions, before the tunnel is excavated. No measurements of the primary state of stress have been realised. The vertical stress was assumed as the weight of overburden because no special geotectonical conditions are expected,. The overburden is statistically defined by middle overburden height and standard deviation from the assumed geological profile. The middle overburden height is defined by μ_h=48 m and σ_h=9.0 m (V_h=18.8%). Unit weight of rock is defined with γ=26kN/m³.

The horizontal stress is defined as a portion of the vertical stress, defined by the horizontal stress coefficient. The middle value of the horizontal stress coefficient k, defined as a function of the Poisson's ratio, is taken as μ_k=0.33. No data for the standard deviation exists and the σ_k=0.10 (V_k=33.3%) is assumed.

5.3 *Initial stress relief factor*

The three-dimensional excavation process in the two-dimensional calculation is usually modelled by initial stress relief factor (Swoboda et al., 1993). The initial stress relief factor is influenced by ground conditions, excavation method, size and speed of excavation.

The stiffness reduction model is used. Based on the experience the initial stress relief factor is defined and used as the expected mean value μ_α=0.4. The standard deviation of σ_α=0.1 (V_α=25.0%) has been foreseen for calculation.

5.4 *Support system*

Tunnel cross-section together with primary support system is shown in figure 4. Primary support system consists of the systematic bolting in crown and shotcrete lining thickness of 15 cm in crown and 10 cm in sides without invert. The systematic bolting with 4.0 m long SN rock bolts is used to increase the strength of the surrounding rock mass and suppress local instabilities. Action of rock bolts is assumed by increase of the cohesion in the bolted region (Marenče, 1993). Shotcrete with quality C25 was foreseen. The compressive strength tests of shotcrete gives mean of μ_{fc}=3.1kN/cm² and standard deviation of σ_{fc}=0.5kN/cm²(V_{fc}=16.1%). For the reliability assessment ultimate strength values are used

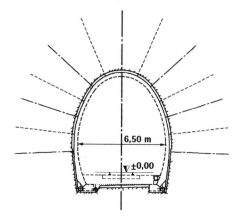

Figure 4. Tunnel cross-section and estimated primary support system

6. IDENTIFICATION AND INTEGRATION OF PERFORMANCE FUNCTION

The performance function is selected to numerically characterise a mode of performance of the structure. For each performance mode one or more performance functions may be considered. For the example, the performance modes considered for tunnel design are: failure in surrounding rock mass, shotcrete failure and footing failure. The problem of the face stability cannot be discussed on the 2D excavation model.

The performance function of shotcrete failure is defined simply as a ratio of the ultimate shotcrete strength and stress calculated in each specified calculation. The results of the calculation with used values for random variables are presented in Table 3. For five random variables eleven deterministic calculations are necessary. One analysis (μ-run) is made using the mean value of each of five random variables. The remaining analyses are grouped into pairs of analyses for each random variable. In each pair, the values of one standard deviation above and below the mean are used in combination with the mean values of remaining variables. The difference in the ratio of capacity to demand for that pair of analyses determines the effect that the variable has on the performance function. The difference divided by two is tabulated as C/D-σ for each random variable. The square root of the sum of the squares of these values is tabulated as a standard deviation under the column headed C/D. The values of C/D-σ make easy to find out which parameters are significant to the results of the analyses.

Table 3. Performance function of shotcrete failure

Run	RMR	h	k	α	C	D	C/D
	[-]	[m]	[-]	[-]	[kN]	[kN]	[-]
1	31	48.0	0.33	0.4	4650	1618	2.87
2	45	48.0	0.33	0.4	4650	694	6.70
3	38	39.0	0.33	0.4	4650	852	5.46
4	38	57.0	0.33	0.4	4650	1387	3.35
5	38	48.0	0.23	0.4	4650	1098	4.23
6	38	48.0	0.43	0.4	4650	1135	4.10
7	38	48.0	0.33	0.3	4650	1113	4.18
8	38	48.0	0.33	0.5	4650	1003	4.64
9	38	48.0	0.33	0.4	3900	1110	3.51
10	38	48.0	0.33	0.4	5400	1110	4.87
μ	38	48.0	0.33	0.4	4650	1110	4.19
σ	7	9.0	0.10	0.1	750		2.30
C/D-σ	1.91	1.05	0.07	0.23	0.68		

where the random variables are:
- RMR is rock mass rating
- h   is the height of the overburden
- k   is the horizontal stress coefficient
- α   is initial stress relief factor, and
- C   is capacity defined as shotcrete strength, and
- D   is calculated normal force in shotcrete lining

7. RELIABILITY INDEX AND SAFETY FACTOR

The reliability index, β, is calculated by the equation (3) where

$$\sigma_{\ln C/D} = \sqrt{\ln\left[1 + (\sigma_{C/D}/\mu_{C/D})^2\right]} = 0.514$$

and

$$\mu_{\ln C/D} = \ln\mu_{C/D} - \frac{\sigma_{\ln C/D}^2}{2} = 1.301$$

The reliability index of β=2.53 is calculated, which in mathematical explanation means, probability of the unsatisfactory performance of approximately P=6‰. The performance function is presented in figure 5.

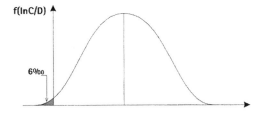

Figure 5: Performance function of shotcrete failure

Based on the European code (EVN 1991-1, 1996) the aspired reliability index for exploitation period is β_{EVN}=3.8. The standard deterministic calculation for mean parameters gives safety factor in shotcrete of 1.6, which is higher than prescribed 1.3 (ÖNORM 4700, 1995). For the calculation with 95% probable rock mass parameters (Edlmair et al., 1996), the deterministic safety factor is only 1.1, which is insufficient.

The calculation method can also be used as a parameter sensitivity analysis, because the so called difference, C/D-σ, gives importance of each random variable. This example shows that the most important parameter, as expected, is the quality of the rock mass. To improve the design, additional investigations during the excavation works have been prescribed.

Both analyses, reliability assessment and deterministic safety calculation show that for the rock mass with rating lower than RMR=35 the safety against shotcrete failure will be insufficient and additional measures have to be used to increase safety. In this case an additional support class with modified support system is used. The modification is foreseen in form of adopted systematic bolting. The new deterministic safety calculation shows increase of the safety from 1.1 to 1.7 for 95% probable rock mass.

8. CONCLUSION

The tunnel design procedure is commonly based on the deterministically defined safety concept. An alternative or complementary safety concept by the reliability assessment is suggested. The method gives possibility to find probability of the unsatisfactory performance defined by random variables, the distribution of which is quantified through statistical analysis of existing data or just judgementally assigned. The method should be sufficiently accurate to rank the relative reliability of various structures and components, but it is not an absolute measure of probability. The possible unsatisfactory performances, defined as failure conditions, have to be considered by the design engineer. The results of the reliability analysis and the results of calculations have to be critically discussed and the numerical model is only one item of the tunnel safety aspects. The presented example is concerned to the shotcrete failure only, and in the praxis the other possible failure mechanisms have to be also considered.

To minimise the calculation effort to find reliability index, the Taylor Series - Finite Difference method is suggested. The method enables definition of the reliability index using only $2n+1$ calculation, where n is a number of random variables and is practical for engineering problems

The calculated reliability index gives an indicative value for the possibility of the unsatisfactory performance, and represents an additional help in design procedure. The Taylor Series - Finite Difference method gives also possibility for the parameter sensitivity analyses, defined through the so called difference, C/D-σ, for each random variable.

The numerical methods used for the tunnel design in deterministic or probabilistic manner give answer on the possible difficulties during the execution, and the structure reliability, but cannot substitute the safety control, in form of measurements, observations and back-analysis, during the excavation works.

REFERENCES

Bieniawski, Z.T. 1989. *Engineering Rock Mass Classification.* Wiley & Sons: New York.

Dudeck, H. 1993. Safety analysis and risk assessment for underground structures. In Ribero e Sousa & Grossmann (eds), *Eurock 93*: 787-793, Rotterdam: Balkema.

Edlmair, G., Marenče, M. & Jung, G. 1996. Bewertung geotechnischer Risiken beim Eisenbahntunnelprojekt in Mazedonien. *Felsbau: 332-336.*

Hoek, E. & Brown, E.T. 1988. The Hoek-Brown failure criterion - a 1988 update. *Proceedings of the 15 Canadian Rock Mechanics Symposium:* 31-38.

Marenče, M. 1993. *Numerical Model of Rockbolts.* Dissertation on University of Innsbruck.

ÖNORM EVN 1991-1. 1996. *Eurocode 1: Grundlagen der Tragwerksplannung und Einwirkungen auf Tragwerke. Teil 1: Grundlagen der Tragwerksplannung.* Wien, Österreichisches Normungsinstitut.

ÖNORM EVN 1997-1. 1996. Eurocode 7: *Entwurf, Berechnung und Bemessung in der Geotechnik, Teil 1: Allgemeine Regeln.* Wien: Österreichisches Normungsinstitut.

ÖNORM 4700. 1995. *Stahlbetontragwerke, EUROCODE-nahe Berechnung, Bemessung und konstruktive Durchbildung.* Wien:Österreichisches Normungsinstitut.

Serafim, J.L. & Pereira, J.P. 1983. Consideration ot the geotechnical classification of Bieniawski. Int. symp. *On engng. geology and underground construction,* Lisbon.

Swoboda, G., Marenče, M. & Mader I. 1993. Finite element modelling of tunnel excavation. *Engineering Modelling:* 51-64.

Swoboda G. 1997. *Programsystem FINAL - Finite element analysis of linear and nonlinear structures under static and dynamic loading, Version 7.0.* University of Innsbruck.

US Army Corps of Engineers 1992. *Reliability Assessment of Navigation Structures.* Engineering Technical Letter No. 1110-2-532, Washington DC.

Geotechnical Hazards, Marić, Lisac & Szavits-Nossan (eds) © 1998 Taylor & Francis, ISBN 90 5410 957 2

Erosion in harbours caused by ship's propulsors

B. K. Mazurkiewicz

Technical University of Gdańsk, Poland

ABSTRACT: Today most ships try to reach or depart from their final berthing position in a harbour by using their own manoeuvring equipment. To these equipment belong the main propellers in combination with bow and stern thrusters or waterjet propulsors. They induce waterjets with velocities of water particles reaching 10 m/s. These jets are meeting the berthing structure and the bottom at the structure, causing liquefaction of the bottom soils and intensive erosion leading to considerable reduction of the stability of the structure.

The paper presenting the interaction between propulsor waterjet, quay wall and erodible bottom, gives the detail characteristics of the whole phenomena and ways of reduction of the bottom current velocities caused by propulsor waterjets. This gives the possibility of proper protection of the harbour bottom and thus the stability of the berthing structure. Finally an example of protection structure is given together with description of its failure which took place during operation.

1 INTRODUCTION

Modern ships, mainly ferry and container ships, carry out their berthing monoeuvres mostly independent from costly and time consuming foreign assistance like harbour tugboats. Instead they use bow and stern thrusters together with their main propeller or propellers and rudder or rudders. For instance the modern passenger ship „Century" is equipped with 3 bow thrusters, 2 stern thrusters and two main propellers connected to engines of total power 46,530 kW. The main changes concern container ships of the fifth and higher generation listed in Table 1 (Mazurkiewicz 1997). These ships have bow thrusters of 30 tons capacity at 2000 kW power and stern thrusters of 20 tons capacity at 1400 kW power. This is the main reason that manoeuvring deep draft ships might cause severe erosion and scouring in front of marine civil engineering structures founded on erodible bottom. This concerns such structures as piers, jetties, quay walls, etc., errected e.g. on sandy soils. The result of the erosion is the danger of loss of stability of the quay wall structure, particularly if the scouring takes place for a significant distance along the quay wall, and if together with scouring a deep liquefaction of the bottom soil is observed.

Ships equipped with 8 propellers, which might rotate independently can cause propeller waterjets in which the velocities of water particles often reach 10 m/s. In addition, the observed increase of draught of the new types of ships, can cause a significant reduction of the underkeel clearance at the quay wall or in the harbour basin, particularly at low water levels. Underkeel clearance is defined as the minimum gap between the keel at its lowest point and between the bottom of the harbour basin.

In the last years the problem became more complicated due to the introduction of high-speed crafts i.e. so called fast ferries. In comparison with conventional ferries, the new high-speed crafts are operating about twice as fast. This means that the generated wake wash will include waves with significantly higher wave lengths while the wave height still will be of the same magnitude as for other ships. The main problem associated with the wake wash from high-speed ferries occurs when the waves are generated in shallow water or are propagating into shallow water. The wave length becomes shorter and the wave amplitude will increase with approaching the shore. This peaking phenomenon will be more pronounced for the waves generated by the high-speed ferries and they will be affected by the bottom earlier than short waves.

In high-speed ferries the waterjet propulsion units are generally used. In comparison with conventional propellers, the water flow rate through the waterjet pump is smaller than the flow through

Table 1. Main parameters of container ships of the newest generations

Ship Parameters	APL	New	New	New	New	New
TEU-capacity	4,800	5,400	6,400	7,200	8,000	8,024
Length $_{OA}$,m	275.0	277.0	318.2	325.0	338.0	335.0
Length $_{pp}$,m	262.0	262.0	-	310.0	323.0	321.0
Beam, m	40.0	40.0	42.8	43.0	46.0	46.0
Height, m	24.3	24.3	26.2	26.5	26.5	26.7
Draught, m	14.0	14.0	14.0	14.0	14.0	13.0
Deadweight, tons	66,300	67,000	84,000	89,000	100,000	110,000
Speed, kn	24.6	24.0	25.0	24.0	25.0	23.0
Power, kW	48,840	49,000	54,900	52,000	70,000	70,000

the propeller disk, but the water in the jet will experience a higher peak pressure. This means an increase of the amount of surface water which is processed by the waterjets.

Another characteristic feature of waterjet propulsors is the reversing buckets. During reversing and astern manoeuvres the water jet is deflected downwards at an angle of 30-40 degrees, and in combination with high flow rate and jet velocity, bottom interaction effects may be more significant compared with conventional propellers.

The actions, which are needed to protect the harbour structure against the direct attack of the propulsor waterjet, are dependent principally on this fact, if one considers an existing structure or a newly designed structure which has to be constructed in the future. In the first case principally protection of the harbour bottom being in direct contact with the protected structure is made, namely through the placement of a special bed protection structure either in the form of a layer of stones having the required single stone diameter or by precast joined concrete blocks laid on a geotextile filter, or by mattresses of geotextile pillows filled with concrete (Fig.1).

The parameters of the protection structure elements are dependent mainly on the velocity of the propulsor waterjets at the harbour bottom. The same concerns the estimation of the range of the protection, i.e. distance from the protected quay wall to the outside edge of the bed protection structure. Taking into consideration the mobilisation of the passive earth pressure in the front of the harbour structure, it might be assumed that the width of the protection bed, perpendicular to the structure, should guarantee the mobilisation of at least 90 % of the maximum passive earth pressure for a soil wedge resisting in undisturbed state of the soil.

The calculation of the velocities of propulsor

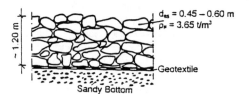

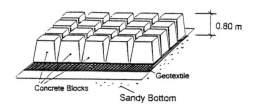

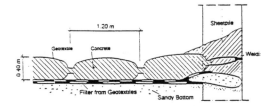

Figure 1. Types of bottom protection: a-riprap, b-mattress of concrete blooks on geoextile filters, c-mattress of geotextile pillows filled with concrete (Zimmermann et.al.1995)

waterjets and induced bottom water currents allows comparison between the above current velocities and limiting velocities for the soil from which the harbour bottom is built, i.e. for the estimation of the danger of occurence of bed erosion. This means that a possibility exists to protect fully the stability of the harbour structure if the type and range of this protection will be adequate to the actual valid dependencies concerning diameter of the protection system elements as well as the range of the protection structure.

In the case of new quay wall structures the alternative to scour protection may be an increased structural foundation depth depending on the anticipated liquefaction and scouring depths. Additional alternatives may be special devices deflecting the propulsor waterjets to the surface or placing the vertical wall backwards behind a row of piles creating room for conversion of the waterjet energy into turbulence with smaller effects on the structures foundation. In all cases it is necessary to have the velocities and velocity distributions induced by the various propeller and waterjet propulsor configurations, the ship positions and the effects from various structural modifications at or in front of quay wall structures.

From the performed investigations it is evident that for instance for container ships of the fourth generation the liquefaction of the sandy soils at the quay wall reached the depth of 4.0 m below the harbour bottom level, which could mean that such depth should be taken into consideration when estimating the design depth of the quay wall. One can of course perform calculations taking into consideration the known formula of Hjulstrøm (Rodatz 1991). It seems, however, that due to the random character of the manoeuvres of berthing and deberthing of ships and the existence of natural bottom currents (river, basin, etc.) causing also defined bottom particles movement the assumption of an estimated value of 4.0 m is in this case correct.

The further directions of actions, concerning new erected structures, result from the analysis of interaction of propulsor waterjets with the harbour structure and the bottom at the structure. In this respect two phenomena can be considered:

a) Influence of the friction of the surface of the installed protection structure on the reduction of the velocity of bottom currents induced by propulsor waterjets.

b) Reduction of the propulsor waterjets velocities through the deflection of the waterjet to the water surface converting their energy into unharmfull turbulence.

Detailed consideration of the above phenomena allow to state that a possibility exists through the increase of the roughness of the protection surface, through the arrangement on the harbour structure of special deflectors of the waterjets and through a special form of the harbour structure, to reduce considerably the bottom currents acting at the quay wall. It leads to a considerable reduction of the weight and range of the protection system.

The applied solution however, does not give the answer to the question, how the outside edge of the protection system has to be constructed to avoid a failure of the whole protection layer and thus the protected structure. In the following a case is presented in which a protection system is shown which started to fail due to inadequate solution of the edge of this system.

2 ESTIMATION OF THE EDGE OF THE BOTTOM PROTECTION AREA

The considerations in the eventual explanation of the phenomena occuring during manoeuvring of ships has to start with the estimation of the velocities of bottom currents on the area on which the berthing and deberthing of ships take place. Taking into consideration all possible alternatives of manoeuvres for occuring wind and currents, one can state that the maximum bottom velocities occur on much larger areas than estimated for protection, while it has to be assumed that the action of propulsor waterjets at a certain distance from the quay wall is about 1.2 times larger than the action at the quay wall. As result of certain simulation tests it was obtained that the maximum bottom velocities occur on the whole water area on which the manoeuvres take place. For the analysed case the approach manoeuvres are starting at a distance of 400 to 500 m from the quay wall while the maximum velocities were registered at a distance of 150 m from the mooring line which could mean that the edge of the protection area should be at least at a distance where current velocities are smaller than allowable from the point of view of grain size of bottom soils. This is however, not acapted from many aspects and thus an edge at smaller distance should be considered.

In addition it has to be mentioned that for waterjet propulsors local current velocities in the order of 5 m/s may occur at water depths of 15 to 20 m when a large waterjet is reversed with the jet directed 40 degrees from the horizontal.

Taking into consideration the estimated bottom velocities due to the action of propulsor waterjets of the magnitude of 7.0 m/s, it is possible to calculate the required diameters, d_r, of stones or blocks from the following formula (EAU 1996):

$$d_r \geq \frac{\left(v_b^{max}\right)^2}{B^2 \cdot g \cdot \left(\frac{\rho_s - \rho_w}{\rho_s}\right)} \tag{1}$$

567

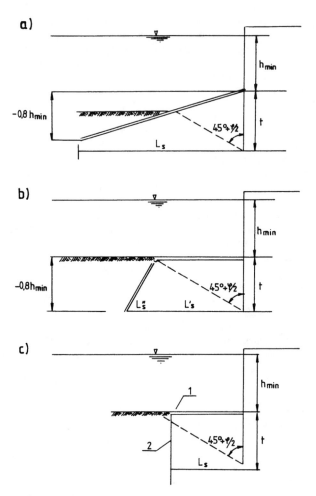

Figure 2. Possible solutions assuring the stability of the edge of the bottom protection system: 1-bottom protection structure, 2-sheet wall

where: v_b - water particle velocity at bottom, m/s,
B - stability coefficient = 1.25 for a ship with centrally situated rudder,
g - acceleration of gravity, 9.81 m/s,
ρ_s - stone density = 2,650 kg/m³,
ρ_w - water density = 1,000 kg/m³.

Thus for v_b^{max} = 7.0 m/s one obtains the required diameter of single stone $d_r \cong 1.94$ m and the weight of this stone 10,126 kg. This means that the edge of the protection structure made of such stones should be stable with an additional requirement, i.e. that scouring which could occur behind the protected area, should not cause its damage due to falling of stones in scour holes etc.

The solution assuring the stability of the

protection edge could be based on following assumptions (Fig.2) (Mazurkiewicz 1997):

a) The stone fill (riprap protection), or mattresses of concrete blocks on geotextile filter, or mattresses of geotextile pillows filled with concrete, are placed with an inclination to such a depth at which the reduction of the waterjet is so large that the stability of the edge is guaranteed.

b) The edge of the protection structure made for example in the form of a mattress of geotextile pillows filled with concrete, is connected through a hinge to an elastic mattress which is placed in the harbour bottom at large inclination to a depth at which the danger of scour does not exist or is significantly reduced.

c) The edge of the protection structure is

568

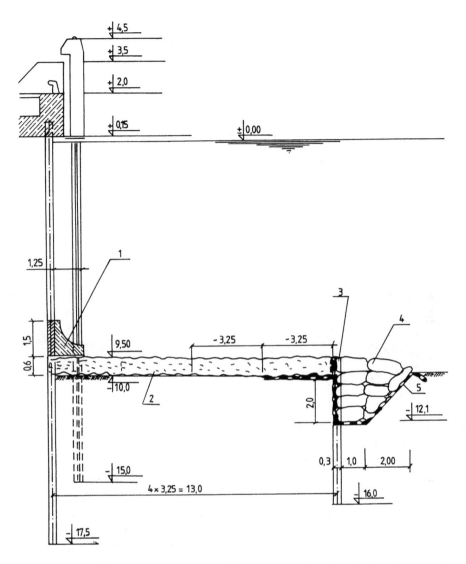

Figure 3. Cross-section of the recommended quay wall bottom protection: 1-prefabricated concrete element protecting the connection between the bottom mattress and the sheet wall, 2-mattress of geotextile pillows filled with concrete, 3-sheet wall l = 6.5 m, 4-trench filled with sand bags 0.5 x 1.2 x 1.5 m, 5-geotextile filter.

connected to a vertical or inclined sheet wall, while in the front of the sheet wall a trench filled with stones or blocks is made.

The introduction of a proper solution must be based on the estimation of the bottom currents on the assumed depth, in relation to the axis of the propulsor induced waterjet. Existing methods do not consider the damping effects of the water and of the liquefied bottom soils.

Taking into consideration the above assumptions,

as well as the results of experience gained during the operation of ferry boat terminals, it is recommended that further solutions of the protection system for all modern ships should be made in the way shown in Figure 3.

The bottom protection system can be designed taking into consideration the velocities calculated using the following formulae for the maximum velocity v_m behind the propeller:

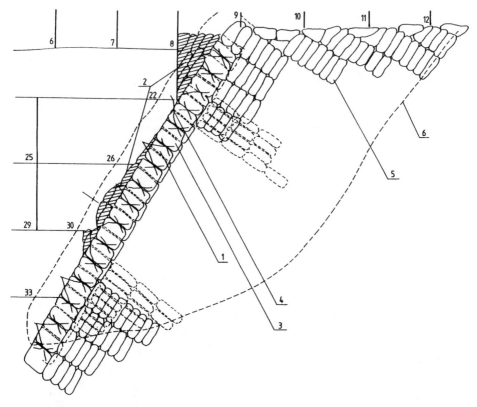

Figure 4.   Repair of the failed quay wall bottom protection: 1–area of the washed out scour hole under the mattresses filled with porous concrete, 2 –filling free spaces with sand bags 4.0 x 0.9 m, 3–sand bags 3.0 x 3.0 x0.3 m, 4–nylon ropes, 5–sand bags 4.0 x 1.0 m, 6–protection area border line.

$$v_m = C_p \left(\frac{P_c \cdot k}{\rho_w \cdot D} \right)^{1/3} \qquad [m/s] \qquad (2)$$

where:

    D - propeller diameter,
    P_c - total energie power,
    k - reduction factor of the engine power during berthing manoeuvres, usually k = 0.5,
    C_p- coefficient = 1.48 for a free propeller (not in a tunnel),
    ρ_w - water density, kg/m³.

and for the maximum bottom velocity v_b^{max}:

$$v_b^{max} = v_m \cdot E \cdot \left(\frac{h_p^{min}}{D} \right)^a \qquad (3)$$

where:

E = 0.42 for central position of the rudder,
a = - 0.28 for two propellers of the main drive,
$h_p^{min} = z + R_t^{min}$,m,
z = the position of the propeller axis above the basic (keel) line, m,
R_t^{min} = minimum clearance under the keel, m.

3 EXAMPLE OF A PROTECTION STRUCTURE

The described case concerns a quay wall for ferry boats , which are characterised by the total power of one engine P_c = 7,920 kW, and the diameter of the main propeller D = 4.3 m. The maximum velocity v_m behind the propeller according to (2) is:

$$v_m = 1.48 \sqrt[3]{\frac{7920 \cdot 0.5}{1 \cdot 0 \cdot 4.3^2}} = 8.85 \ m/s$$

570

The maximum bottom velocity v_b^{max} according to (3) and for:

$z = 2.35$ m, $R_t^{min} = 0.25$ m $h_p^{min} \cong 2.6$ m

is: $v_b^{max} = 8.85 \cdot 0.42 \left(\dfrac{2.6}{4.6} \right)^{-0.28} = 4.28$ m/s.

The change of the water particle velocity v_b^z with depth z is, for the analysed case:

$$v_b^z = 5.59 \left(h_p^z \right)^{-0.28}$$

which means that an allowable velocity could be reached at the depth of 112 m, and means that a special protection of the edge which will guarantee a required stability of the quay wall in all loading conditions, i.e. for velocities which are higher than allowed for the bottom soil, is needed.

The analysed quay wall was protected by mattresses of geotextile pillows filled with concrete. The thickness of the mattress was 0.48 m. The width of the mattresses perpendicular to the quay wall was 30 to 45 m while the edge was protected by a layer of reinforced concrete slabs of dimension $4.0 \cdot 2.0 \cdot 0.2$ m (weight in dry = 3,840 N).

After one year of operation, at the edge of the protection layer a deep scour hole was observed to which the protection concrete slabs were sliding. In addition, under the concrete mattress an erosion hole was formed causing the breaking of the mattress itself and of the filling concrete.

The repair works concerned filling with a porous concrete of the gaps washed out under the mattresses and placement of geotextile protection bags of dimension $3.0 \cdot 3.0 \cdot 0.3$ m filled with sand. These bags were placed in layers and connected between them and with the concrete mattresses by nylon ropes of 12 to 14 mm dia. The erosion holes behind the protected area were filled with ballast bags of 4.0 m length and 1.0 m dia filled with sand and placed cross-wise. Totally over 100 bags $3.0 \cdot 3.0 \cdot 0.3$ m and above 500 bags of $4.0 \cdot 1.0$ m (Fig.4) were placed.

During some months of operation all bags were broken away from the protection mattresses and displaced behind the protection area. The difference between the top of the mattresses and the bottom of the erosion hole was equal to 1.8 m. For the time being further damage of mattresses is occurring, requiring their permanent repair. This means that the applied system as is now clear, has not fulfilled the assumed conditions

4 CONCLUSIONS

As result of the above repair works it can be concluded that the main task of the bottom protection system, required for quay walls at which modern ships are mooring, is a proper solution of the edge of this system. The presented proposal of the bottom edge structure seems to guarantee a stable protection of harbour structures against the erosion of harbour bottom caused by modern ship's propulsors.

5 REFERENCES

EAU-*Arbeitssausschuss für Ufereinfassungen 1996.* Kolkbildung und Kolksicherung vor Ufereinfassungen besonders infolge schiffsbedingter Erosionseffekte (E 83). Wilhelm Ernst & Sohn

Mazurkiewicz,B. 1997a. Failure of Quay Wall Bottom Protection Structure. *Proc.Int.Conf. on Foundation Failures*, Singapore, 12-13. May 1997: 229-236

Mazurkiewicz,B. 1997b. Influence of new generation of ships on the design parameters of quay walls. *Proc. 4th Int. Seminar on Renovation and Improvements to Existing Quay Structures.* Gdańsk 26-28.May 1997: 153-160

Rodatz,W. 1991 Gründungen im offenen Wasser.*Grundbau Taschenbuch, Teil 2,*261-338, Wilhelm Ernst & Sohn.

Zimmermann,C., H. Schwarze, N. Schulz& S.Henkel 1995. Ships propeller induced velocities near quay structures. *Proceedings of the Int. Conf. on Coastal and Port Engineering in Developing Countries,* Rio de Janerio, Brazil, 22-29. Sept.1995.

Geotechnical Hazards, Marić, Lisac & Szavits-Nossan (eds) © 1998 Taylor & Francis, ISBN 90 5410 957 2

Improvement of the bearing of the soils by using plastic-rubbish matters

T. Messas, R. Azzouz, C. Coulet & M. Taki
L2M, IUT A Génie Civil, Lyon 1, France

ABSTRACT : In order to study the effect of plastic material on the bearing capacity of superficial foundation, a series of bearing tests on reinforced soils by plastic materials were carried out. A small scale reproduction experimentation was made in laboratory, on Taylor-Scheneebeli analogical material that responds to Coulomb critirion. The influence of a big number of experimental parameters on soil bearing was observed such as the plastic quantity by layer, geometrical form of the plastic pieces, the depth of the first layer, layers number, the distance between layers and the way the continuous layers are fixed.

1 INTRODUCTION

The "in situ" tests on real scale are experimental tests under ideal conditions for a better understanding of the behaviour of a structure. However, there are many disadvantages associated with these tests, among which the main ones are the following:

1. To have the necessary settlement or rotation corresponding to the real conditions may take a very long time to be made. The experiments must be done at a faster pace than usual.

2. The changes in weather conditions (rain, frost, dry conditions) result in variations of hydrostatic pressures and densities.

3. The experimentation on real scale is costly.

This is why their use is limited. This brings us to use analogous media for reduced models. The tests on the Scheneebelli model have certain advantages as compared to experiments on real structures :

1. The speed and small cost.

2. The study of the behaviour of the structures at rupture.

3. The control and the study of the relative importance of the various parameters at work.

The study presented in this article contributes to the improvement of the bearing capacity of the soils by using plastic matter. We have shown the advantage of the influence of a number of experimental parameters on the bearing capacity of the soils such as :

1. The quantity of plastic pieces per layer.

2. The depth of the layer.

3. The geometric form of the plastic pieces.

4 The number of layers.

5. The spacing out between the layers.

6. The arrangement of the plastic matter.

7. The fixing mode.

An interpretation of the improvement of the soil bearing capacity shall be given. A comparison will be made between reinforcement by discontinuous layers and reinforcement by continuous layers. This comparison will enable us to quantify the tensile strength due to discontinuous reinforcement.

2 THE MATERIALS AND EQUIPMENT

2.1 Materials used

2.1.1 Analogous soil

This soil is represented by the analogous model of Scheneebelli (1957). This is obtained by the piling up of stainless steel duralumin rollers with four different diameters : 2, 3, 4 and 5 mm which are distributed randomly. The corresponding weight ratios are 29 %, 38 %, 21 % and 12 % respectively. Such a medium possesses some of the fundamental properties of the soils (homogenity, isotropy) and obeys Coulomb's law for pulverulent granular media. The mechanical characteristics of this soil are determined by the shear box test. The shear box has the dimensions of 200 x 100 x 60 mm^3. This test gave a value of angle of internal friction equal to 22° and zero cohesion.

2.1.2 Plastic reinforcing material

The reinforcement of the layer of analogous soil, is performed by using two types of wastes :
1. White and translucent polyethylene sheeting with a surface mass of 125 g/m² and tensile strength at rupture equal to 1.8 kN/m.
2. Soft plastic wastes.

2.2 Test material (figure1)

2.2.1 Description of the frame

Scheneebelli's wheel rollers are arranged in a frame with the help of metallic frames having a large rigidity in order to dump out all the vibrations.
The stacked rollers make a massif 150 cm wide and 86 cm hight. These dimensions are chosen for all the lines of rupture to be able to develop without interference from the edges or the bottom of the frame. The lateral sides of the frame present several slots to allow the insertion of plastic layers (figure1).

2.2.2 Foundation

The foundation is formed of a 20 cm long, 6 cm wide and 15 cm high steel plate. This height has been chosen in order to prevent the soil from flowing back at the time of penetration. The wheels have been joined to the steel flate with the wheels of the same type so as to make them rough (figure1).

2.2.3 System of loading

This system comprises of a mechanism fixed to the foundation. The mechanism screwed to a ring which enables the transmission and measurement of the forces which are applied to it according to the penetration. The penetration is measured with the help help of a comparateur which can read up to $1/100^{th}$ of a mm and which is fixed to the frame (figure1)

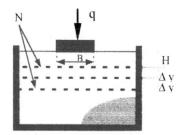

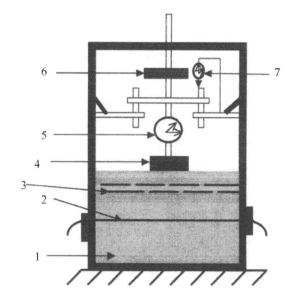

1 : Embankment (massif) of rollers
2 : Continuous layer
3 : Discontinuous layer
4 : Steel of the foundation
5 : Ring
6 : System of loading
7 : Comparateur
B : Width of the steel
N : Number of layers
Δy : Spacing between the sheets
P : Penetration
M : Mass of plastic matter per layer/sheet
S : Area of plastic piece
H : The depth of the 1st sheet/layer
R. B. C. = q/q_0 : Ratio of bearing capacity
q : Bearing capacity of the reinforced soil
q_0 : Bearing capacity of the non-reinforced soil

Figure1 : Diagram of the experimental model

3 MODE OF OPERATION OF A TYPICAL TEST

The mode of preparation of Scheneebelli's cylindrical wheels is the same for all the tests. These rollers are piled up by hand and wabe up 6 cm layers They are put into place by means of a right. The rollers can be rearranged and the environment made homogeneous The plastic layers of 6 cm width and variable length are placed into positions which depend upon the test.

The embankment (massif) is of 150 x 86 cm^2 and has an apparent density of 6.6. The loading tests are performed at each 0.5 mm. The test is stopped at a penetration of 80 mm. Then the frame is unloaded and the wheels are moved to avoid the formation of heterogeneous zones for the next test.

4 RESULTS AND DISCUSSIONS

The curves drawn are the curve of bearing capacity and penetration. In addition to these a third curve the so called bearing capacity ratio is also drawn.

4.1 The effect of the quantity of plastic matters per layer on the gaining of bearing strength

To point out the effect of the mass of plastic matters, the bearing capacity test is performed. In this test, the number of layers (N= 1), the layer depth (H = 4 cm) is kept the same while the mass of the plastic matter per layer is changed. The size of the plastic pieces is also kept constant at 6 x 30 cm^2. The results are presented in the figures 2 and 3. The following remarks are made : the bearing capacity depends on the mass of plastic matter. This phenomenon can be explained as follows : As the quantity of the plastic matters increases, the overlap between the plastic pieces also increases. Hence, there is a better coutinuity between the plastic pieces. Their behaviour, therefore, tends increasingly towards that of plastic matter in the form of continuous layers. They can consequently support tensile forces. This continuity is made by the overlapping and friction between the plastic pieces embedded in the soil.

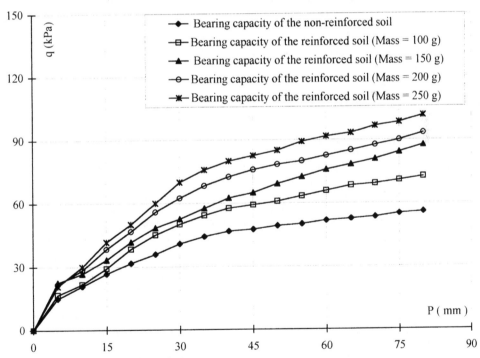

Figure 2 : Effect of the quantity of plastic matters per layer on the gaining bearing strength (N = 1 , H/B = 20 % , S = 6 x 30 cm^2)

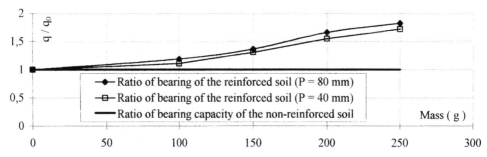

Figure 3 : Evolution of the ratio of bearing capacity (R. B. C) as a function of the quantity of plastic matters per layer (N = 1 , H/B = 20 %, S = 6 x 30 cm²)

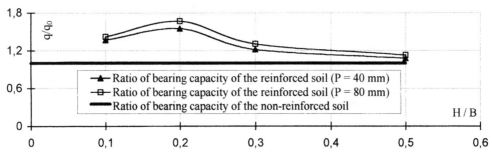

Figure 4 : Variations of the ratio of bearing capacity (R. B. C. = q/q₀) as a function of the depth of the depth of the plastic layer (N = 1 , M = 70 g , S = 30 x 6 cm²)

4.2 The effect of the depth of the plastic layer on the bearing strength

A series of tests are performed with the aim of investigating the above-noted effect. During these tests, following variables are kept constant while varying the depth H. The layers of the plastic matter (N = 1). The quantity of the plastic matter (M = 70 g). The size of the plastic pieces 30 x 6 cm². The arrangement of the plastic pieces staggered. To relate the variable H to the steel of foundation, the ratio H/B is used. Figure 4 shows three curves. These show the variation of the ratio of thebearing capacity (or relative bearing capacity) as a function of H/B. The curves are for reinforced soil, non-reinforced soil and soil without plastic pieces. The following remarks are made :

1. The experiments showed the maximum of the coefficient of improvement is for H/B equal to 20%.

2. The results also reveal that beyond a certain value of depth H/B greater than 50 %, the plastic matter has no influence whatsoever on the bearing strength.

4.3 The effect of the shape of the plastic pieces

A series of tests are performed in order to investigate the effect of the shape of the plastic pieces on the bearing capacity. In this test, the same number of layers of plastic pieces (N = 1) are placed at the same depth (H/B = 20 %) with the same quantity of plastic matter (M = 70 g).

The results obtained are presented in figure 5. It is observed that the bearing capacity increases with the increase in size of the plastic pieces. The greater the size of the plastic pieces, the greater the recovery is.The continuity between the plastic pieces is ensured by the recovery and the frictional effects involving the plastic pieces.

4.4 The effect of the number of layers of plastic matter on the bearing capacity

The punching tests carried out were performed in order to investigate the variation of the bearing capacity as a function of penetration and study the evolution of the coefficient of the bearing capacity as a function of the number of layers of the plastic pieces. The following variables are kept constant

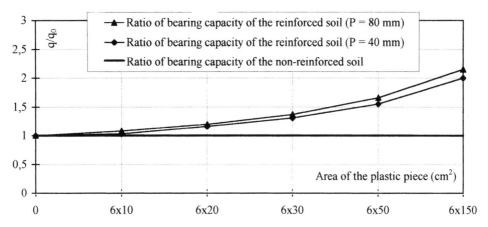

Figure 5 : Evolution of the ratio of bearing capacity (R. B. C. = q/q_0) as a function of the shape of the plastic pieces (N = 1 , H/B = 20 % , M = 70 g)

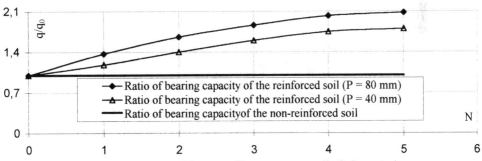

Figure 6 : Evolution of the ratio of bearing capacity (R. C. B. = q/q_0) as a function of the number of layers of plastic matter on (H/B = 20 % , Δy = 4 cm , M= 150 g)

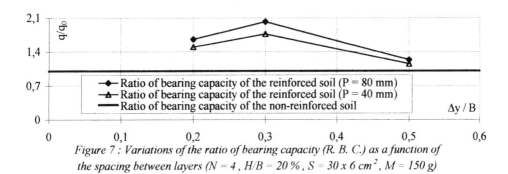

Figure 7 : Variations of the ratio of bearing capacity (R. B. C.) as a function of the spacing between layers (N = 4 , H/B = 20 % , S = 30 x 6 cm^2 , M = 150 g)

while varying the number of layers of the plastic pieces. The depth of the 1st layer (H/B = 20 %), the quantity of the plastic matter per layer (M = 150 g), the shape of the plastic pieces (30 x 6 cm^2), the staggered arrangement and the spacing between the layers (Δy = 4 cm). The conclusion is that the bearing capacity registers an increase with the increase of N up to 4 layers. Beyond this figure, the increase in the bearing capacity remains constant (figure 6).

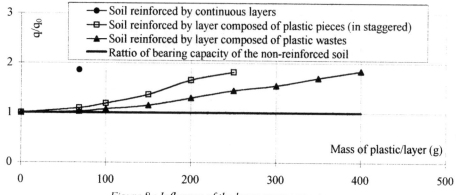

Figure 8 : Influence of the layer arrangement

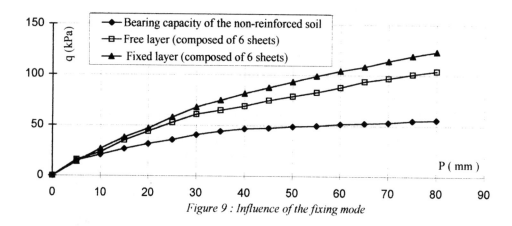

Figure 9 : Influence of the fixing mode

4.5 The effect of the layer spacing out

The influence of this parameter has been analysed for three different spacings : $\Delta y = 2$ cm, $\Delta y = 4$ cm, $\Delta y = 10$ cm. To relate this variable to the width of the steel flange, the coefficient $\Delta y/B$ has been used while keeping the following parameters constant:

1. The number of layers (N = 4).
2. The depth of the 1st layer (H/B = 20 %).
3. The quantity of the plastic matter per layer.
4. The size of the plastic pieces and the staggered arrangement of the plastic pieces.

The results are shown in figure 7. The following remarks are made :

1. An improvement in the bearing capacity of the soil is observed when the penetration of the steel flange becomes important.
2. The maximum of the coefficient of improvement $\Delta y/B$ is equal to 30 %.
3. The more the deep layers recede from the zone of influence, the more the carrying capacity of

the soil approaches the case of the reinforcement by a single layer «the 1st layer»

4.6 The effect of the layer arrangement

The effect of this parameter is studied for three types of arrangement of plastics :

1. Continuous layer of 6 cm wide and 150 cm long continuous layer.
2. Discontinuous layer in pieces cut into small rectangles of 6 x 30 cm and arranged in staggered pattern.
3. Discontinuous layer composed of plastic wastes.

The quantity of plastic is varied for the three cases of reinforcement. The approximate evolution of the bearing capacity of the soil according to the mass of plastic matter is presented in figure 8. It is observed that the soil bearing capacity varies with the quantity of plastic matter. This can be explained because the quantity of plastic chips increases, the overlapping between the chips increases in the same way and hence an enhanced continuity is observed. Their behaviour increasingly approaches the one of the soils having continuous layers of plastic matters. The latter can support tensile stresses. It is also observed that beyond a certain quantity of plastic, the bearing strength of the soil does not increase anymore.

4.7 The effect of the mode of fixation of the plastic layers

Two different conditions are imposed upon the layers to study this effect : free layer and fixed layer.

All the tests are made with the same layer depth H = 4 cm. The results are presented in figure 9. According to these results, it can be maintainedthat the inclusion of the plastic increases the bearing capacity of the soils. This increase is quite distinct when the plastic is fixed. This difference results from the effective strength in the tension from the fixed layer, especially for large punching displacements.

5 CONCLUSION

The object of this study is to contribute to the improvement of the soil bearing capacity when plastic matters are added to it. The distinguishing characteristics feature of this experimental study have been to relate the increase in bearing capacity of the soil reinforced by discontinuous layers to the one of reinforcement by continuous layers. The lab tests have been made on an embankment formed of a Taylor-Scheneebelli type analogous material. This material obeys Colomb's law and has an angle of internal friction equal to 22°.The principal findings of the study are restated as under :

1. The more the mass of plastic matters increases, the more importantthe overlapping effect and consequently their load-resisting behaviour approaches more closely that of continuous plastic matters.

2. It appears that the maximum effect would be obtained with plastic matters in the form of continuous bandes/strips.

3. However, with the increase in the size of the plastic pieces there is a corresponding increase in the overlapping between the pieces which leads to a greater bearing capacity.

4. The more the deeper layers are away from the zone of influence, the more the bearing capacity approaches the case of reinforcement by a single layer (the 1st layer of the plastic pieces). The

increase in strength is maximum for a depth of H/B equal to 20 %.

5. An optimal spacing of 4 cm between the layers enables to obtain the maximum improvement in the soil's bearing capacity.

6. The gaining in the bearing capacity of the soil with the number of layers is observed but beyond N equal to 4 the increase remains constant.

7. The bearing capacity for fixed layers is observed to be greater than that for free layers.

This study shows that the addition of plastic material to the soil whether in the form of continuous or discontinuous strips result in increasing the soil's bearing capacity. Furthemore, this improvement is more significant for large value recess.

REFERENCES

Binquet, J. & Lee, K. 1975. *Bearing capacity tests on reinforced earth slabs.* Journal of geotechnical engineering division, Vol. 101, N° GT12, p. 1241-1256.

Coulet, C. & Rakotondramanitra, J. D. & Bacot, J. 1987. *Soil reinforcement making use of waste plastic materials study with large shear box machine.* VIIIth Nat. Conf. on Soil Mecanics and Foundation Engineering, Wroclaw, Pologne.

Omine, K. & Ochiai, H. & Katot, T. 1996. *Effect of plastic wastes in improving cement treated soils.* Environmental Geotechnics, Kamon 1996 Balkema, Rotterdam. Vol. 2, p. 875-880.

Messas, T. & Azzouz, R. & Coulet, C. & Curtil, L. 1997. *Soil renforcement by using layers of soft or discontinuous plastic wastes.* GREEN2 - 2nd International symposium on géotechnics and the environment 8-11 September 1997, KRAKOW, POLAND.

Geotechnical Hazards, Marić, Lisac & Szavits-Nossan (eds) © 1998 Taylor & Francis, ISBN 90 5410 957 2

Bodenverbesserung mit Kiesstopfsaülen beim Autobahnbau in der Slowakei

G. Mosendz
Keller špeciálne zakladanie, spol sr.o., Bratislava, Slovakia

ABSTRACT

Der rasche Ausbau der Verkehrsverbindungen zählt zu den vordringlichen Infrastrukturprogrammen der Slowakei. Dazu gehört auch die Fortzetzung der Autobahstrecke im Raum Bratislava sowie die Ost-West Verbindung Bratislava-Žilina-Košice. Die Wahl der Trassenführung gelingt nicht immer im problemlosen Baugrund. Insbesondere im Randbereich der Städte verlangt der Strassenaufbau oder die Dammschüttung nach Bodenverbesserungen. Die Kellersche Rüttelstopftechnik schafft die Voraussetzung für einen sicheren Strassenbau in verschiedenen weichen und lockeren Böden oder auch im Bereich ehemaliger Deponien und Anschüttungen mit unterschiedlicher Zusammensetzung. Berücksichtigt bleiben dabei die Anforderungen des Projektes nach vollständiger Veermeidung eines Bodentausches oder einer teuren Verfuhr des Deponieinhaltes. Ausgeführt wird eine wirtschaftliche, +in die Tiefe und Raster anpassungsfähige Bodenverbesserung für den Aufbau der Dämme der Verkehrsflächen.

1. PLANUNG

Die Tiefenverdichtung stellt eine Bodenverbesserung dar. Unter anderm wird die Tiefenverdichtung auch zur Bodenverbesserung in Anschütungen und Deponien verwendet. Die Ausführungsplanung der Tiefenverdichtung macht die Grundbaufirma und der Bodengutachter (Planer) welche dieses Verfahren ausführt. Für die Ausführungsplanung der Tiefenverdichtung ist notwendig einzuholen einen Lageplan, Querschnitte und zukünftige Belastungen von dem prijektiertem Bauobjekt. Ebenso ist notwendig die Kentniss der Geologie und der Hydrogeologie. Auf Grund dieser Unterlagen, sowie auch auf Grund der Besichtigung der Baustelle, der Konsultation mit dem Statiker und dem Projektanten, wird der Ausführungsplan der Tiefenverdichtung auf Grund der Erfahrungen ausgeführt. Der Ausführungsplan beinhaltet die Austeilung der Punkte der Tiefenverdichtung und einen oder mehrer Querscghnitte mit Bezug auf den Baugrund.

2. MASCHINEN, GERÄTE

Zu der Tiefenverdichtung sind folgende Maschinen und Geräte erforderlich:

- ein geeigneter Typ des Tiefenrüttlers für die Art der Tiefenverdichtung und den anstehenden Boden
- ein geeignetes Traggerät für den Rüttler und das entweder eine Tragraupe der Firma Keller, oder ein Baggerkran. Der Tiefenrüttler ist auf der Tragraupe vertikal geführt, mit der Möglichkeit der Aktivierung durch Druck vom Eigengewicht der Tragraupe. Am Baggerkran hängt der Tiefenrüttler lose. Der Tiefenrüttler auf dem Baggerkran gehängt wird meistens verwendet für die Tiefenverdichtung im Kies-Sand, während der Tiefenrüttler montiert auf dre Tragraupe wird verwendet für Verdichtung mit Luftspülung (Rüttelstopfverdichtung). Die Leistung mit Einsatz der Tragraupe ist grösser als beim Einsatz des Baggerkranes. Es werden verschiedene Tiefenrüttler, je nach Boden verwendet. Im Snad-Kies ist der S-Rüttler und im bidigem Boden der Schleusenrüttler und andere Rüttlertypen.
- für die Tiefenverdichtung mit Luftspülung ist notwendig ein geeigneter Kompressor, für Wasserspülung eine Wasserpumpe und eine Wasserentnahmemöglichkeit,
- der Elektromotor im Tiefenrüttler wird angetrieben mittels eines geeignetem Stromagregat oder Netzanschluss,

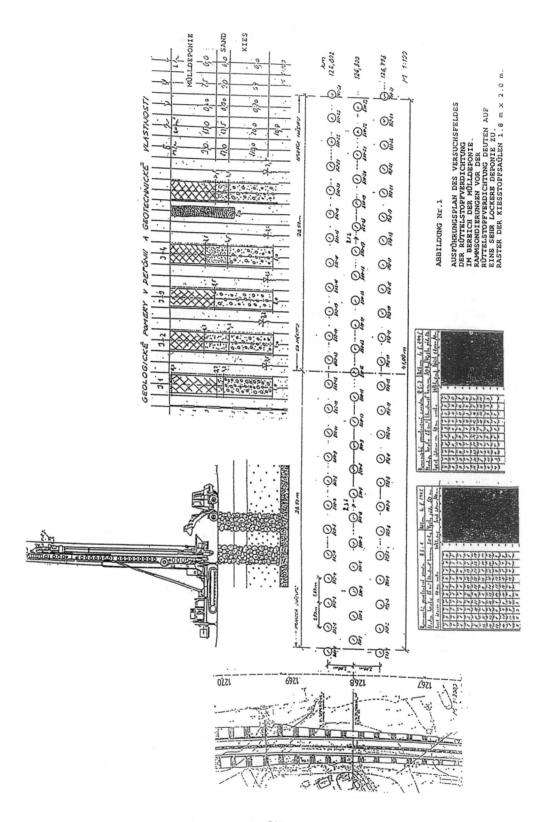

ABBILDUNG Nr.1

AUSFÜHRUNGSPLAN DES VERSUCHSFELDES
DER RÜTTELSTOPFVERDICHTUNG
IM BEREICH DER MÜLLDEPONIE.
RAMMSONDIERUNGEN VOR DER
RÜTTELSTOPFVERDICHTUNG DEUTEN AUF
EINE SEHR LOCKERE DEPONIE ZU.
RASTER DER KIESSTOPFSÄULEN 1.8 m × 2.0 m.

582

- die Zugabe des Kiesmateriales erfolgt mittels eines Radladers mit Schafelinhalt ca 1,0 - 1,5 m3,
- Werstadt- und Ersatz- Kontainer,
- Personaltagesaufenthaltkotainer.

3. ARBEITSABLAUF

Laut Ausführungsplan werden ausgesteckt die einzelnen Verdichtungspunkte am Arbeitsplanum der Baustelle. Die Punkte werden ausgesteckt mit Stahlstäben 40 cm lang. Der Tiefenrüttler wird genau auf den ausgesteckten Punkt gesetzt. Mittels Vibrationen, Wasser- oder Luft- Spülung und des Eigengewichtes und der Aktivierung beim Einsatz der Tragraupe, dringt der Tiefenrüttler in die notwendige iefe ein. Die notendige Tiefe stellt dar meistens den tragfähigen Untergrund auf welchen die Kiesstopfsaüle gestützt wird. Das erreichen des tragf+higen Untergrundes macht sich bemerkbar durch erhöten Wiederstand zun Eindringen des Rüttlers. Nach Erreichen des tragfähigen Untergrundes wird die Rüttelstopfsaüle von unten nach oben aufgebaut. Zu diesem Zweck wird Kies zur Rüttlerspitze eingeführt. In Schritten ca 0.5 m und beim Einbau des geeigneten vorher abgestimmten Kiesmateriales, wird die Kiesstopfsaüle bis zur Oberfläche oder der gewünschten Tiefe ab Arbeitsplanum hergestellt. Die Herstellung der Kiesstopfsaüle verfolgt in der Kabine des Traggerätes der Rüttlerfahrer. Dabie wird eine gewisse Amperaufnahme des Tiefenrüttlers für die Festgkeit der Kiesstopfsaüle oder des Bodens je nach Tiefe erreicht.

4. HERSTELLUNGSPARAMETER

Die Ampereaufnahme in Abhängikeit der Tiefe der Rüttlerspitze und der Zeit werden graphisch aufgezeichnet während der Kiesstopfsaülenherstellung. Damit steht zur Verfügung von jeder einzelnen Kiesstopfsaüle eine graphisch Darstellung der Herstellung. Ein weiterer wichtiger Parameter ist der Kiesverbrauch in die Saülenherstellung. Der Verbrauch des Kiesmaterials wird ermittelt von der zahl der Kiesschaufeln der Tragraupe, wobei der Inhalt einer Kiesschaufel bekannt ist, welche notwendig ist für eine Kiesstopfsaüle.
Auf diese Art ist es möglich zu ermitteln den durchschnittlichen Kiesverbrauch pro 1 lfm Kiesstopfsaüle und den Durchmesser der Saüle.
Die Kontrolee der Qualität der Tiefenverdichtung (Rütteldruckverdichtung) in nichtbindigen Böden - Kies und Sand ist möglich mittels der

Rammsondierung vor und nach der Tiefenverdichtung.
In bindigen Böden, Deponien und Anschüttungen welche mittels Vibrationen nicht zu verdichten sind, verbessern die Rüttelstopfsaülen der Rüttelstopfverdichtung diese Böden und die Efektivität der Rüttelstopfverdichtung läst sich nachweisen mittels Berechnungen, wobei es notwendig ist zu kennen die physikalisch-mechanischen Eigenschaften des Bodens, welcher verdichtet wird und des Einbaumateriales der Kiesstopfsaülen, den Durchmesser der Kiesstopfsaülen und das Grundwasserniveau, sowie die Belastung des mit Kiesstopfsaülen verbesserten Bodens. Möglich sind auch Belastungsversuche, welche jedoch ziemlich aufwendig sind. Der Erfolg der Rüttelstopfverdichtung hängt vo Raster und Durchmesser der Saülen ab. Je grösser der Durchmesser und kleiner der Abstand umso höher ist die Bodenverbesserung.

5. AUTOBAHN D2/D61 km 126,6 - 127,0 BRATISLAVA - PETRZALKA

5.1. GROSSVERSUCH

Auf dieser Baustelle befindet sich eine ca 30 Jahre alte Komunalmülldeponie (Schlakke, Asche, Textilfetzen, Glass, Papier, Knochen, Gummi und Bauschutt) und es wurde beschlossen hier eine Bodenverbesserung des Untergrundes des Autobahndammes welcher später ca 8 m hoch sein wird.
Den eigentlichen Tiefenverdichtungsarbeiten ging voran ein Grossversuch. Das Versuchsfeld 41 x 6 m mit einem Raster der Kiesstopfsaülen 2,0 x 1,8 m wurde im Bereich der Deponie gewählt. Auf Grund der Resultate des Grossversuches und dessen Auswertung wurde gewählt der definitive Raster der Kiesstopfsaülen. Zu dem Zweck wurde ausgearbeitet ein Ausführungsplan der Rüttelstopfverdichtung im Bereich des Versuchsfeldes (Abbildung Nr. 1). Vor der Rüttelstopfverdichtung wurden 2 Rammsondierungen gemacht zur Bestätigung der Tiefe der Anschüttung und Lagerungsdichte des Kieses aus der nahen Aufschlussbohrung J1.
Die Resultate der Rammsondierung haben bestätigt die angenommene Geologie und es wurde festgestellt das die Deponie reicht bis zur Tiefe 3,5-4,0 m. Die Deponie selbst ist sehr locker gelagert und weist im Durchschnitt 1 Schlag auf 10 cm Eindrigtiefe mit der Schweren Rammsonde, d.h. Fläche Rammspitze 15 cm2, Rammbär 50 kg, Fallhöhe 50 cm. Der Untergrund der Deponie besteht vom mitteldicht bis dicht gelagertem Kies, wie das beschrieben wurde in den Aufschlussbohrungen J1-J4.

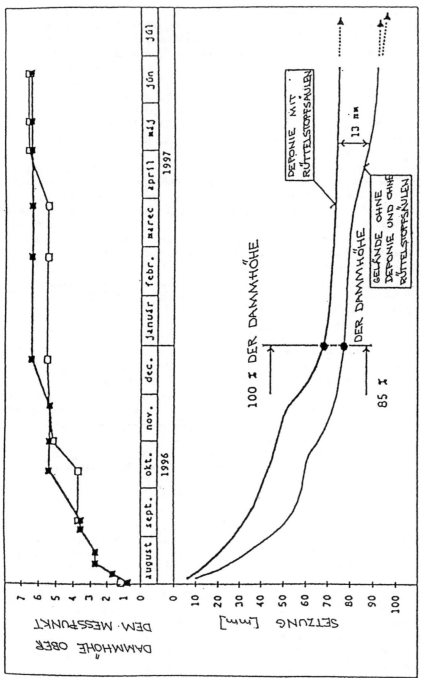

SETZUNG DER FUNDAMENTUNTERKANTE DES AUTOBAHN DAMMES. AUF DER AUTOBAHN
D2 / D61 BRATISLAVA - PETRŽALKA KM 127,000 UND KM 127,030

ABBILDUNG Nr. 2

Der Tiefenrüttler unterstützt durch Luftspülung, Eigengewicht und Aktivierung durch das Gewicht der Tragraupe, drang leicht in den tragfähigem Kiesuntergrund welcher festgestellt wurde auch mittels Aufschlussbohrungen und Rammsondierungen ein. Der Tiefenrüttler drang in die tragfähige Kiesschicht auf eine Tiefe von 0,5 - 1,0 m hinein, was durch einen grösseren Wiederstand bemerkbar war und die damit zusammenhängende hohe Ampereaufnahme auf einen Wert ca 150 Ampere. Durch den Rüttler entstand ein Holraum im Boden. In den Hohlraum zylindrischer Form wurde Kiesmaterial der Körnung 8 - 32 mm eingebaut. Der Röttler brachte das Kiesmaterial nach unten und seitlich in den Boden ein. Für die Zugabe des Kiesmaterales wurde ein Radlader mit einer Schaufel 1,5 m3 verwndet. Für jede einzelne Kiesstopfsaüle wurde festgehalten die Anzahl der Kiesschaufeln an der Tragraupe. Der durchschnittliche Verbrauch vom Kies auf 1 lfm Kiesstopfsaüle war 0,40 m3. Von diesem Verbrauch wurde errechnet der Durchmesser der Kiesstopfsaüle von ca 74 cm. Die Aufnahme des Verdichtungstromes lag zwischen 90 Ampere und 120 Ampere. Diese Werte sind für eine Deponie zureichend, jedoch mit Hinsicht auf die Heterogenität und den speziellen Charakter der Deponie, auf welche bekantlich auch der Abfall von der Gummifabrik ausgeführt wurde, sind möglich auch kleinere Werte der Stromaufnahme dargestellt in Ampere. Bei manchen Kiesstopfsaülen wurden festgestellt auch extreme Tiefen des tragfähigen Untergrudes 10,4 - 11,0 m. Die Ursache kann eine weiche Auffülung eines Flussarmes, oder eine spezifisch lockere Lagerung vom Kies - Sand sein.

5.2. SETZUNGSBERECHNUNG

Die Arbeiten auf dem Versuchsfeld ergaben die notwendigen Parameter für eine Setzungsberechnung des Deponieuntergrundes verbessert mittels Kiesstopfsaülen. Ermittelt wurde die Tiefe der Kiesstopfsaülen, deren Durchmesser, der Verbrauch des Kieses, die Ampereaufnahme und die Schichtleistung. Bestätigt wurden auch die geologischen Verhältnisse wie schon beschrieben von den Aufschlussbohrungen J1 - J4, mit Ausnahme 3 Kiesstopfsaülen bei welchen der Untergrund festgestellt worden ist in Tiefen grösser als 10 m. Es wurde gemacht eine Setzungsberechnung nach Priebe für einen Raster 1,8 x 2,0 m, mit einer Belastung 150 kN/m2, was entspricht einer Dammhöhe von 8 m. Die geotechnischen Werte sind ersichtbar von der Abbildung Nr.1, der Saülendurchmesser wurde eingesetzt 70 cm. Errechnet wurde eine Gesamtsetzung von 11,5 cm nach der Rüttelstopfverdichtung.

5.3. AUSFÜHRUNG

Die Resultate des Grossversuches mit der Setzungsberechnung waren zufriedenstellend und es wurde durchgeführt die Bodenverbesserung mit Tiefeverdichtung im Bereich der Komunalmülldeponie, d.h. Länge 330 m x Breite 41 m = Fläche 13.350 m2 im Raster der Punkte der Kiesstopfsaülen von 1,8 x 2,0 m. Insgesamt wurden hergestellt 24.837 lfm Kiesstopfsaülen. Es wurde gearbeitet mit 2 Verdichtungseinheiten in Tag und Nacht Schicht. Jede Verdichtungseinheit bestand aus 1 Tiefen-Rüttler, 1 Rüttler in Reserve, 1 Tragraupe, 1 Stromaggregat 150 kVA, 1 Kompressor 6m3/min, 1 Radlader. Die Manschaft pro Schicht bestand von 1 Rüttler-Tragraupenfahrer, 1 Verdichtungsarbeiter und 1 Radladerfahrer, also insgesamt 3 Mann. Die gesamte Baustelle verfügte weiters über 1 Bauführer und 1 Bauleiter. Die Arbeiten wurden ausgeführt im Zeitraum Juni - August 1996.

5.4. SETZUNGSKONTROLLMESSUNG

Zur Kontrolle der Setzungsberechnung der Gesamtsetzung von 11,5 cm nach der Bodenverbesserung durch Tiefenverdichtung und dem zeitlichem Verlauf der Konsolidation des Untergrundes der Dämme (90% der Konsolidation in 14 Monaten) wurden im Untergrund des Dammes errichtet 2 Messprofile zur Messung der Setzung. Ein Messprofil wurde errichtet im Bereich der Komunalmülldeponie verbessert mit Kiesstopfsaülen im km 127.000. Das zweite Messprofil, als Referenzmessprofil, wurde errichtet auserhalb dem Berich der Komunalmülldeponie im geologisch ursprünglichem Naturzustand im km 127.030. Im Messprofil in der durch Kiesstopfsaülen verbesserten Komunalmülldeponie im km 127.000 war die Setzung dank der Bodenverbesserung und dem intensiven Ausbau des Autobahndammes zum Juni 1997 praktisch abgeschlossen, wie dargestellt in Abbildung Nr.2. Sichtbar ist eine Diferenz zwischen dem zeitlichem Verlauf der Setzung des Profiles im km 127.000 und km 127.030. Die Differenz ist klein und nach 18 Monaten ab Beginn des Autobahndammausbau ist sie 13 mm zu Ungunsten des Messprofiles, welches auser der durch Kiesstopfsaülen bodenverbesserten Komunalmülldeponie liegt.

Die tatsächlich gemessene Gesamtsetzung des Autobahdammuntergrundes im Bereich der durch Kiesstopfsaülen verbesserten Mülldeponie war nur 7,5 cm, also weniger wie durch die Setzungsberechnung ermittelt.

6. WEITERE AUTOBAHNBAUSTELLEN

Ein ähnliches Problem wie das obenbeschriebene trat auch an anderen Bereichen des Autobahnausbaues auf. Mülldeponien befanden sich in der Trasse der Autobahn. Mit dem Verfahren der Bodenverbesserung durch Tiefenverdichtung mit Kiesstopfsaülen wurde der Autobahnuntergrund verbessert auf folgenden Baustellen :

Autobahnbereich Bratislava - Jarovce. Eine ca 20 Jahre alte Industriemülldeponie bestehend hauptsächlich aus Gummiabfall wurde vrbessert mit ca 13.000 lfm Kiesstopfsaülen im Zeitraum 18.11.1996 - 4.2.1997.

Autobahnbereich Nove Mesto nad Vahom wo Naturweichböden nahe einem Bach anstanden wurde mit 2.864 lfm Kiesstopfsaülen verbessert im Zeitraum 10.2. - 4.3.1997.

Autobahnbereich Nemsova - Ladce. Eine Komunalmüll- und Industriemüll- Deponie mit Abfall von der Holzindustrie, Farben und Lacke, Spitalmüll, Hausabfall, alles relativ frisch wurde mit 40.329 lfm Kiesstopfsaülen verbessert im Zeitraum 3.3.- 16.6.1997.

7. SCHLUSSWORT

Die Tiefenverdichtung nach dem Rüttelstopfverfahren durch einrütteln von Kiesstopfsaülen in den Untergrund von Autobahnen und Srassen wo wenig tragfähiger Boden ansteht stellt dar eine wirtschaftliche und schnelle Lösung dieses Problemes. Die Tiefenverdichtung als Alternative zum Bodentausch erspart die Notwendigkeit der Errichtung einer neuen Deponie und die Verfuhr der Altdeponie was kostspielig ist. Die Kiesstopfsaülen sind fähig sich mit der Deponie partiell zu verformen, wobei die Deponie und die Kiesstopfsaüle sich gleichmäsig setzen und somit nicht bedrohen die Autobahnkonstruktion mit durchstnzen sei es zum Beispiel mit Betonpfählen. Der Untergrund der Autobahn wird homogenisiert, rasche Konsolidierung und geringe Setzungen treten ein.

LITERATUR

Priebe H., Die Bemessung von Rüttelstopfverdichtungen, Die Bautechnik, 72. Jahrgang, Heft 3/1995

Sidak N., Sohldichtung mittels Tiefenverdichtung Wasserkraftwerk Gabcikovo - Nagymaros, Wehr am Umlauf, Sonderdruck Keller, Bratislava 1993

Báslik R., Jakubis I., Higway Landfill Settlement, Konferenzbuch der 3.Geotechnischen Konferenz veranstaltet durch die Slowakische Technische Universität, Bratislava 1997

Mosendz G., Improvement of Foundation Soil by Gravel Vibro Columns, Konferenzbuch der 3.Geotechnischen Konferenz veranstaltet durch die Slowakische Technische Universität,Bratislava 1997

Geotechnical Hazards, Marić, Lisac & Szavits-Nossan (eds) © 1998 Taylor & Francis, ISBN 90 5410 957 2

Abdichtungsinjektionen mit Feinstbindemitteln

Hanno Müller-Kirchenbauer & Jürgen Rogner
Universität Hannover, Germany

Carsten Schlötzer
Ingenieurgesellschaft Grundbauinstitut Hannover Dr.-Ing. Karl Weseloh - Prof. Dr.-Ing. Hanno Müller-Kirchenbauer und Partner mbH, Germany

ABSTRACT: Zur Untergrundabdichtung von Kornfraktionen des Mittelsandbereichs mit Feinsandanteilen mittels Permeationsinjektion werden die Injektionsmaterialien auf der Basis von Wasserglaslösungen aufgrund ökologischer Bedenken in der Bundesrepublik Deutschland derzeit nur begrenzt eingesetzt. Als Alternative zu diesen chemischen Lösungen werden die neuartigen Feinstbindemittelsuspensionen gesehen, für deren Einsatz derzeit jedoch Bemessungskriterien fehlen. Auf der Basis von Laboruntersuchungen und von Erfahrungen aus baupraktischen Einsätzen konnten erste Ergebnisse zum Ausbreitungsverhalten dieser neuartigen Injektionsmassen gewonnen werden.

1 EINLEITUNG

Durch das Einpressen von Injektionsmitteln in den Untergrund können Böden verfestigt oder abgedichtet werden. Bei sogenannten Permeationsinjektionen wird der anstehende Porenraum des Bodens durch das Verpreßmittel weitgehend aufgefüllt (Kutzner 1991, Müller-Kirchenbauer 1968, 1996). Dabei wird die Hohlraumstruktur des Bodens aus Klüften im Festgestein oder Poren in Lockergesteinen allgemein nicht verändert. Nachfolgend soll auf Verpreßarbeiten mittels Permeationsinjektion zur Abdichtung von nichtbindigen Lockergesteinen mit neuartigen Suspensionen eingegangen werden.

Für die Auswahl des jeweiligen Verpreßmittels werden neben den technischen Anforderungen an das verpreßte System Kennwerte des anstehenden Bodens maßgebend, insbesondere die Porengrößen und deren Verteilungen. Hierfür stellen die Korngrößenverteilung beziehungsweise der sogenannte wirksame Korndurchmesser d_{10} die bodenmechanisch maßgeblichen Parameter dar. Nach allgemeinen Kriterien wird die Injizierfähigkeit eines Bodens deshalb auch nach der Korngrößenverteilung oder der Durchlässigkeit abgeschätzt (Kutzner 1991).

Rollige Böden, die überwiegend aus Kiesen bestehen, lassen sich mit konventionellem Zementmörtel oder –suspensionen verpressen (Bild 1). In Feinkiesen und Grobsanden ist eine Verfüllung des Porenraums dagegen nur noch mit Suspensionen auf der Basis von Injektionszementen möglich. Für die Ver-

pressung von Mittelsanden bis herab in den Bereich der Feinsande kamen bisher Silicatgellösungen aus Wasserglas und einem Reaktiv zum Einsatz, die aufgrund derzeitiger ökologischer Bedenken für Untergrundabdichtungen nur noch zurückhaltend zugelassen werden. Der Übergangsbereich zu den bindigen Böden stellt auch die Grenze zum Anwendungsbereich der vergleichsweise kostenintensiven Kunststoffinjektionen dar (Bild 1).

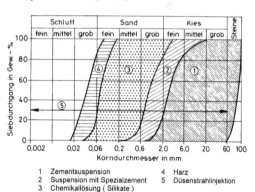

1 Zementsuspension
2 Suspension mit Spezialzement
3 Chemikallösung (Silikate)
4 Harz
5 Düsenstrahlinjektion

Bild 1. Anwendungsbereiche von Injektionsmitteln (Vorl. Merkblatt 1993).

Als Alternativverfahren zur Permeationsinjektion mit chemischen Lösungen wurde in letzter Zeit oftmals das sogenannte Düsenstrahlverfahren eingesetzt, welches in der einschlägigen Literatur auch als Hochdruckinjektion (HDI) bezeichnet wird. Durch den Schneidstrahl wird bei diesem Verfahren im Unterschied zur Permeationsinjektion die Bodenstruktur jedoch weitgehend aufgelöst. Außerdem kommt es beim Düsenstrahlverfahren zu einem vergleichsweise hohen Suspensionsrücklauf und damit –verbrauch.

Mit dem Ziel, auch für den Korngrößenbereich der Mittelsande mit einem begrenzten Feinsandanteil mineralische und damit im Hinblick auf die Grundwasserbelastung als vergleichsweise unbedenklicher geltende Verpreßmaterialien für eine Permeationsinjektion zur Verfügung zu stellen, wurden von der Baustoffindustrie sogenannte Feinstbindemittel als Suspensionsgrundstoffe entwickelt. Nachfolgend sollen erste Erfahrungen mit diesen Materialien im Labor sowie auch zur Herstellung von Sohlabdichtungen für Baugruben dargestellt und die sich daraus ergebenen Probleme sowie erste Lösungsansätze zusammenfassend erläutert werden. Darüber hinausgehend resultiert aus dieser innovativen Technik derzeit noch ein umfassender Forschungsbedarf, der abschließend ebenfalls charakterisiert werden soll.

2 FEINSTBINDEMITTEL

Seitens der Baustoffindustrie wurden in letzter Zeit sogenannte Feinstbindemittel entwickelt, mit denen feststoffhaltige Suspensionen für Permeationsinjektionen in Mittelsanden mit begrenzten Feinsandanteilen aufbereitet werden können. Je nach wirksamem Korndurchmesser der Suspensionspartikel beziehungsweise nach der Feinheit der Feststoffe des Feinstbindemittels sollen Sande mit Feinsandanteilen bis zu 50 Massen% verfüllt und dadurch abgedichtet werden können (Tausch und Teichert 1990).

Bei den Feststoffen der Feinstbindemittel handelt es sich nach Herstellerangaben (Dyckerhoff Zement GmbH 1996) im wesentlichen um sehr fein gesichtete mineralische Komponenten wie Tonmehle und Hochofenschlacke. Zusätzlich können die Rezepturen die Hydratation verzögernde und verflüssigend wirkende Komponenten wie auch Stabilisatoren enthalten. Die Hochofenschlacke reagiert latent hydraulisch, so daß für die Einleitung der Abbindereaktion ein Katalysator erforderlich wird. Deshalb enthalten die Feststoffe in geringen Anteilen auch besonders fein gemahlenen Portlandzementklinker, dessen Hydratationsprodukte die Reaktion der Hochofenschlacke einleiten können.

In den bisherigen baupraktischen Anwendungen zur Herstellung von Injektionssohlen wurden überwiegend Feinstbindemittel eingesetzt, dessen Feststoffpartikeldurchmesser d_{95} bei einem Siebdurchgang von 95 Massen% unterhalb von 0,016 mm lagen (Bild 2). Hieraus lassen sich nach den Produkt-

informationen der Hersteller sedimentationsstabile wasserreiche Suspensionen mit Wasser-Bindemittelwerten zwischen 5 und 6 aufbereiten (Dyckerhoff Zement GmbH 1996). Für einen guten Aufschluß der Feststoffe im Dispersionsmittel Wasser ist eine energiereiche Aufbereitung mit einem Kolloidalmischer erforderlich. Die Verarbeitungszeit der Suspensionen entspricht mit etwa 60 Minuten denen konventioneller Silicatgellösungen. Das Fließverhalten der Suspension kann mit der Marshzeit als Summenparameter aus der Wichte, der Viskosität und der Fließgrenze charakterisiert werden. Für Wasser wird im allgemeinen eine Marshzeit von etwa 28 s und für chemische Injektionslösungen eine von etwa 30 s bestimmt. Für die Feinstbindemittelsuspensionen ergibt sich unmittelbar nach deren Aufbereitung eine Marshzeit von rund 32 s, die im Verarbeitungszeitraum nur unmaßgeblich ansteigt (Dyckerhoff Zement GmbH 1996). Darüber hinaus ist nach Herstellerangaben im Verarbeitungszeitraum eine Sedimentation der Suspensionspartikel nicht zu erwarten.

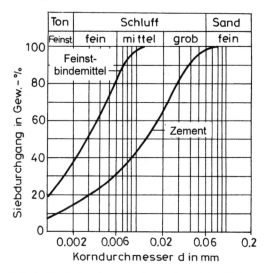

Bild 2. Typische Kornverteilung eines Feinstbindemittels im Vergleich zu einem Normalzement (Vorl. Merkblatt 1993).

3 ANFORDERUNGEN

Die Anforderungen an eine Feinstbindemittelsuspension zur Abdichtung eines nichtbindigen Lockergesteins lassen sich wie folgt zusammenfassen:
- Das aufbereitete Verpreßmittel muß letztlich so fein sein, daß das anstehende Porensystem wirksam penetriert, aufgefüllt und damit abgedichtet werden kann. Dies setzt zunächst fein gesichtete Feststoffe

voraus, die im Verarbeitungszeitraum insbesondere nach einem Kontakt mit Wasser nicht wesentlich agglomerieren sollten (Bild 3 beziehungsweise Schulze 1992), wodurch die Penetration erschwert beziehungsweise sogar verhindert würde.

- Im Verarbeitungszeitraum sollten die suspensionsrheologischen Parameter weitgehend konstant bleiben, so daß mit der Verpreßsuspension eine möglichst große Endreichweite erzielt werden kann.

- Die eingebrachte Suspension sollte nach der Injektion in der Porenmatrix wirksam abbinden, um die dichtenden Eigenschaften sicherstellen zu können.

- In Orientierung zu den allgemeinen Anforderungen an Abdichtungsinjektionen sollte die Durchlässigkeit k des verpreßten Gesamtsystems in einer Größenordnung unterhalb von $5 \cdot 10^{-7}$ m/s liegen (Kutzner 1991).

- Für reine Abdichtungszwecke ist die Eigenfestigkeit der verpreßten und anschließend abgebundenen Suspension zunächst von allgemein untergeordneter Bedeutung. Die einaxiale Druckfestigkeit q_u sollte jedoch hinsichtlich einer Erosionsstabilität gegenüber den zu erwartenden hydraulischen Gradienten bei mindestens etwa 200 kN/m² liegen.

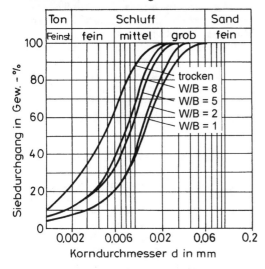

Bild 3. Einfluß des Wasserbindesmittelwertes W/B auf die Korngrößenverteilung von Feinstbindemittelsuspensionen (Vorl. Merkblatt 1993).

4 INJEKTIONSKRITERIEN

Bei einem Einsatz der Injektionstechnik zur Abdichtung von Porensystemen wie auch zu deren Verfestigung wird zunächst anhand gängiger Injektionskriterien abgeschätzt, inwieweit ein Verpreßmittel gegen-

über einem bestimmten Kornsystem verpreßbar ist. Hierfür stehen die allgemein empirischen Kriterien auf der Basis der Korngrößenverteilungen beziehungsweise auch der Durchlässigkeiten des zu verpressenden Erdstoffes zur Verfügung (Bild 1). Darüber hinaus lassen sich auch aus den sogenannten Filterkriterien Hinweise auf die Verpreßbarkeit einer Porenmatrix ableiten, indem diese Kriterien gedanklich umgekehrt werden. Sie orientieren sich an dem Korndurchmesser d_{85} aus der Korngrößenverteilung des Verpreßmittels, der bei einem Siebdurchgang von 85 Massen% abgegriffen wird, und dem Korndurchmesser D_{15} aus der Korngrößenverteilung des Bodens, der verpreßt werden soll. Demnach stellt die Größe d_{85} ein Maß für das Größtkorn des Injektionsmittels und D_{15} ein Maß für die Porengröße des Bodens dar. Eine Injektion wird allgemein dann als möglich angesehen, wenn diese beiden Parameter im Verhältnis

$$N = \frac{D_{15}}{d_{85}} \geq 24 \qquad (1)$$

stehen. Liegt das Verhältnis N unterhalb eines Wertes von 11, ist eine wirksame Verpressung der feststoffhaltigen Suspension in das Porensystem erfahrungsgemäß nicht mehr möglich. Im dazwischen liegenden Bereich ist die Injektion nicht in jedem Fall möglich. Somit wird hierfür ein entsprechender experimenteller Nachweis erforderlich.

Gemäß den Hinweisen und Empfehlungen des Vorläufigen Merkblatts für Einpreßarbeiten mit Feinstbindemitteln im Lockergestein (1993) wird für den Nachweis der Injizierfähigkeit folgende Vorgehensweise empfohlen (Bild 4): In eine wassergesättigte Bodensäule mit einem Durchmesser d von mindestens 50 mm und einer Höhe, die mindestens dem Vierfachen des Durchmessers entspricht, wird die Suspension von unten eindimensional verpreßt. Dabei sollte eine möglichst geringe und konstante Beaufschlagungsrate eingestellt werden.

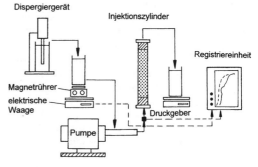

Bild 4. Prinzipskizze der Versuchsanordnung (Vorl. Merkblatt 1993).

Nach den Hinweisen im vorangehend genannten Merkblatt gilt der jeweils geprüfte Boden mit der gewählten Feinstbindemittelsuspension dann als verpreßbar, wenn

- eine Suspensionsmenge den Boden durchströmt hat, die dem dreifachen Porenvolumen entspricht, und
- gleichzeitig ein Verpreßdruck von 0,6 MPa nicht überschritten wurde.

5 ÜBERTRAGBARKEIT KONVENTIONELLER ANSÄTZE AUF FEINSTBINDEMITTELINJEKTIONEN

Hinsichtlich der Fragestellung, inwieweit die klassischen Injektionskriterien gemäß Abschnitt 4 unmittelbar auf das Problem der Feinstbindemittelinjektionen übertragen werden können, besteht nach ersten Ergebnissen aus Laborversuchen sowie baupraktischen Anwendungen zur Herstellung von Sohldichtungen noch ein gewisser Forschungsbedarf. Insbesondere gilt dies auch für die Versuchsmethodik, um aus den Ergebnissen der Laborversuche die Verpreßfähigkeit einer Feinstbindemittelsuspension in das Porensystem einer nichtbindigen Porenmatrix abzuschätzen.

Die klassischen Injektionskriterien (Bild 1) sind zunächst allgemein für konventionelle Verpreßmittel anwendbar, was durch eine Vielzahl von Praxisfällen auch weitgehend nachgewiesen ist. Feinstbindemittelsuspensionen sollen als wasserreiche feststoffhaltige Verpreßmittel einen Teil des Anwendungsbereiches abdecken, für den bisher vorzugsweise chemische Injektionslösungen auf Silicatgelbasis verwendet wurden. Bei den wasserreichen Feinstbindemittelsuspensionen kann es jedoch bereits unmittelbar nach deren Aufbereitung zu Veränderungen kommen, die sich auf die Injizierfähigkeit auswirken können (Schulze 1992, Vorläufiges Merkblatt 1993). Aufgrund der hohen Mahlfeinheit der Feinstbindemittel verfügen diese auch über eine vergleichsweise hohe spezifische Oberfläche. Da die Hydratationsreaktion vorwiegend als topografische Oberflächenreaktion abläuft, ist von einer vergleichsweise hohen Reaktivität der Feststoffe bei einem Wasserkontakt auszugehen. Die Folge davon können Agglomerationen sein, so daß sich die Korngrößenverteilungslinien der Feinstbindemittelsuspensionen, je nach Wasser-Bindemittelwert, entsprechend verschieben können (Bild 3). Allgemein beginnt die Hydratationsreaktion sofort nach einem Wasserkontakt der reaktiven Feststoffpartikel, was letztlich ebenfalls durch die Wasseranlagerung und die Entstehung der Hydratationsprodukte auf ein Partikelwachstum führt. Beiden

vorangehend genannten Phänomenen kann jedoch in begrenztem Rahmen durch eine entsprechende Abstimmung der Rezeptur für das Feinstbindemittel entgegengewirkt werden, in dem die Hydratation verzögernde sowie auch verflüssigend wirkende Komponenten zugesetzt werden.

Darüber hinausgehend können sich auch sogenannte Feststoffverlagerungen (Schlötzer und Müller-Kirchenbauer 1995) ungünstig auf die Verpreßbarkeit von Feinstbindemittelsuspensionen auswirken. Zur wirksamen Durchführung einer Permeationsinjektion ist vorauszusetzen, daß die Suspension das Porensystem penetrieren kann. Eine filterkuchenbildende Oberflächenfiltration ist von vornherein auszuschließen, da in solchen Fällen naturgemäß keine Auffüllung der Porenmatrix zustande kommen kann. Grundsätzlich sind allerdings auch Filtrationen im Innern des Porensystems nicht völlig auszuschließen. Bei diesen sogenannten Tiefenfiltrationen oder Kolmationen können, ausgehend von Porenengstellen, einzelne Fließwege mit ausfiltrierten Feststoffen verstopft werden, während weiterhin Wasser aus der Suspension ausgepreßt wird. Damit ist eine weitere Durchströmung zumindest örtlich nicht mehr möglich. Als Folge davon kann sich schließlich das Porensystem bis hin zu einer Stagnation der Verpressung mehr und mehr zusetzen.

Zusammenfassend läßt sich bereits aus den vorangehenden Erläuterungen ableiten, daß eine Übertragung der klassischen Injektionskriterien wie auch der konventionellen Bemessungsansätze für Permeationsinjektionen auf das Problem der Feinstbindemittelinjektionen nicht ohne weiteres möglich scheint.

6 DERZEITIGER ERFAHRUNGSSTAND

Neben umfangreichen Versuchsreihen zur Aufbereitung verschiedener Feinstbindemittelsuspensionsrezepturen und zu deren Rheologie wurden am Institut für Grundbau, Bodenmechanik und Energiewasserbau (IGBE) der Universität Hannover auch eine Vielzahl von Verpreßversuchen durchgeführt. Dabei wurden einerseits in Anlehnung an das Vorläufige Merkblatt für Einpreßarbeiten mit Feinstbindemitteln in Lockergestein (1993) eindimensionale Verpreßversuche durchgeführt. Andererseits wurden auch größermaßstäbliche Laboruntersuchungen zur räumlichen Ausbreitung der Suspensionen durchgeführt.

Wesentliches Ergebnis dieser Untersuchungen war zunächst, daß bei eindimensionaler Verpressung in ein nichtbindiges Korngerüst mit Feinsandanteilen von etwa 20 Massen% Reichweiten im Sinne von Durchströmungslängen von etwa 0,7 m erzielt werden konnten (Müller-Kirchenbauer & Schlötzer

1994, Müller-Kirchenbauer et al. 1996 sowie auch Schulze 1994). Bei räumlicher Ausbreitung lagen die zum Teil jedoch auch versuchsbedingt begrenzten Reichweiten bei etwa 0,15 m und 0,50 m (Bild 5).

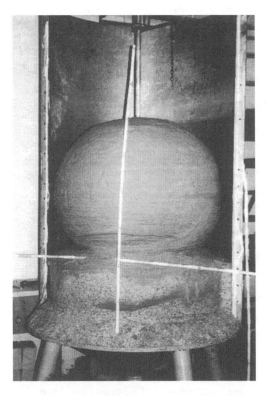

Bild 5. Verpreßkubatur eines großmaßstäblichen Laborversuchs (d ≅ 100 cm).

In situ wurden bisher in verschiedenen Projekten vor allem im Berliner Raum Feinstbindemittelsuspensionen zur Herstellung sohlgedichteter Baugruben verpreßt. Die Injektionsraster und -parameter hinsichtlich der Verpreßmengen, -raten und -drücke wurden bei der Planung dieser Maßnahmen im allgemeinen aus den Kenntnissen und Erfahrungen für Silicatgelinjektionen übertragen. Somit ergibt sich beispielsweise für einen Raster aus gleichseitigen Dreiecken mit Seitenlängen von etwa 1,5 m eine erforderliche Reichweite von rund 1,0 m. Bei einigen dieser Projekte mußte jedoch festgestellt werden, daß zumindest örtlich die erforderlichen Reichweiten offenbar nicht erzielt wurden und somit eine durchgehende Abdichtungswirksamkeit der Injektionssohle nicht voll erreicht wurde.

Nach derzeitiger Kenntnis ist nicht auszuschließen, daß die Bemessung einer Feinstbindemittelinjektion auf der Basis
- konventioneller Verpreßraster und -parameter mit konstanter Verpreßrate zwischen allgemein 6 l/min und 8 l/min sowie
- von Ergebnissen eindimensionaler Laborverpreßversuche gemäß dem Vorläufigen Merkblatt (1993) nicht in jedem Fall auf eine ausreichende Abdichtungswirksamkeit führt. Dabei ist zunächst von einem Unterschied hinsichtlich der eindimensionalen und der räumlichen Ausbreitung des Verpreßmittels im Porensystem auszugehen. Bei konstanter Verpreßrate q und eindimensionaler Ausbreitung ist die mittlere effektive Geschwindigkeit v_s im Porenkanal über den Fließ- oder Penetrationsweg weitgehend konstant. Demgegenüber vermindert sich bei quasi kugelförmiger Ausbreitung im isotropen Baugrund die mittlere effektive Geschwindigkeit v_s mit zunehmendem Abstand r von der Verpreßquelle:

$$v_s = \frac{v_f}{n} = \frac{4 \cdot q}{\pi \cdot r^2 \cdot n} \; . \tag{2}$$

Gleichzeitig verlängert sich mit der Entfernung von der Verpreßquelle die Aufenthalts- und damit auch die Reaktionszeit der hydraulisch aktiven Feinstbindemittelpartikel in den Poren. Hierbei kann es bereits zu Veränderungen bestimmter Eigenschaften sowie zur Affinität zu den silicatischen Porenkanalwandungen kommen. Aufgrund der abnehmenden Transportgeschwindigkeit des reaktiven Bindemittelpartikels im Porensystem können Einflüsse aus
- Sedimentationen in Richtung der Gravitationskraft,
- zunehmender Aktivierung der Partikeloberflächen durch die initiale Hydratation, woraus zunehmend Adhäsions- beziehungsweise Agglomerationseffekte resultieren können,
- einem Quellen einzelner Partikel und
- Tiefenfiltrationen beziehungsweise Kolmationen, die ausgehend von Porenengstellen das System mehr und mehr zusetzen und damit verstopfen,
nicht ausgeschlossen werden. Hierauf weisen auch die im Laborversuch festgestellten rechnerischen Differenzen zwischen dem eingebrachten Verpreßvolumen der Feinstbindemittelsuspension und dem dabei erzielten Volumen der letztendlich entstandenen kugelförmigen Verpreßkubaturen hin. Diese lagen für Kugeln mit einem Durchmesser von 0,15 m bei etwa 87 % und mit einem Durchmesser von etwa 0,45 m bei rund 70 %, bezogen auf einen theoretisch verpreßbaren Porenraum von 70 % des Gesamtporenraums.

7 ZUSAMMENFASSUNG UND AUSBLICK

Um zumindest mittelsandige Böden mit begrenzten Feinsandanteilen, die bisher allgemein nur mit chemischen Silicatgellösungen verpreßbar waren, durch weitgehend mineralische Materialien abzudichten, wurden seitens der Baustoffindustrie sogenannte hydraulisch abbindende Feinstbindemittel entwickelt. Diese sollen sich, je nach ihrer Mahlfeinheit, für Verpreßarbeiten in Böden mit bis zu 50 % Feinsandanteilen einsetzen lassen.

Feinstbindemittelsuspensionen wurden bereits verschiedentlich zur Herstellung sohlgedichteter Baugruben, vor allem im Berliner Raum, verpreßt. Dabei wurden die gängigen Injektionskriterien und die Injektionsraster wie auch die –parameter der konventionellen Injektionstechnologie weitgehend übernommen. Allerdings mußte hierbei festgestellt werden, daß die erforderliche durchgehende Abdichtungswirksamkeit nicht in allen Fällen ohne weiteres erreicht werden konnte.

Für weitergehende Laboruntersuchungen zur Verpreßbarkeit von Feinstbindemittelsuspensionen in rolligen Böden liegt derzeit der Vorschlag vor, diese anhand eindimensionaler Verpreßversuche abzuschätzen (Vorläufiges Merkblatt 1993).

Grundsätzlich ergibt sich bei Injektionen in situ aus einer punktförmigen Quelle heraus eine räumliche Ausbreitung. Dabei entstehen aufgrund der Baugrundanisotropie im allgemeinen ellipsoide bis hin zu kugeligen Kubaturen bei weitgehend isotropen Verhältnissen. Die Fortschreitungsgeschwindigkeit des penetrierenden Verpreßmittels nimmt über die Länge des Fließweges mehr und mehr ab. Hierbei kann es aus den obengenannten Gründen zu einer zumindest lokalen Verstopfung des Porensystems kommen. Auf diese Phänomene weisen neben Erfahrungen aus baupraktischen Anwendungen bereits die Ergebnisse von Laboruntersuchungen hin. Dabei wurde unter anderem festgestellt, daß die tatsächlich erzielbare Verpreßkubatur kleiner ist, als es der theoretisch rechnerisch aus dem Verpreßvolumen ableitbaren Kubatur entsprechen müßte.

Somit muß einerseits davon ausgegangen werden, daß die Injektionskriterien, die Injektionsraster und auch die Verpreßparameter für das neu entwickelte System der Feinstbindemittelinjektionen anzupassen sind. Aus heutiger Sicht könnte die Ausführungssicherheit beispielsweise durch die Wahl verdichteter Verpreßraster und eine Bemessung der Maßnahmen mit darauf abgestimmten Verpreßvolumina erhöht werden, wobei diese jedoch gleichzeitig auch die im Versuch erhaltenen Verlustvolumina beispielsweise aus sedimentations- und filtrationsbedingten Einflüssen erfassen. Weiterhin liegen Hinweise dafür vor,

daß erhöhte zeitliche Verpreßraten zu verbesserten Reichweiten führen könnten. Die Empfehlungen zur Abschätzung der Reichweitenentwicklung im Rahmen erforderlicher Grundsatz- beziehungsweise Eignungsprüfungen sollten insoweit überarbeitet beziehungsweise modifiziert werden, so daß hierbei auch die offensichtlich allgemein maßgebenden räumlichen Effekte berücksichtigt werden können.

Aus den vorangehenden Erläuterungen läßt sich im Hinblick auf die Durchführung von Abdichtungsinjektionen mit neuartigen Feinstbindemitteln noch ein umfassender Forschungsbedarf ableiten. Die hierzu erforderlichen experimentellen Untersuchungsprogramme und theoretischen Betrachtungen sind Gegenstand der aktuellen wie auch der geplanten Forschungen am Institut für Grundbau, Bodenmechanik und Energiewasserbau (IGBE) der Universität Hannover. Die Ergebnisse werden später ebenfalls veröffentlicht.

LITERATUR

Kutzner, C. (1991): Injektionen im Baugrund. Ferdinand Enke Verlag. Stuttgart.

Müller-Kirchenbauer, H. (1968): Zur Theorie der Injektionen. Veröffentlichungen des Institutes für Bodenmechanik und Felsmechanik der Universität Fridericana Karlsruhe. Heft 32.

Müller-Kirchenbauer, H. & Schlötzer, C. (1994): Neue Forschungsergebnisse auf dem Gebiet der Injektionstechnik im Hinblick auf Abdichtungs- und Verfestigungsmaßnahmen in der Praxis. Vortrag im Rahmen des Lehrgangs "Injektionen mit Feinstbindemitteln in der Geotechnik." 01./02.02.1994. Technische Akademie Esslingen.

Müller-Kirchenbauer, H., Wichner, R., Friedrich, W. & Schlötzer, C (1996): Zur Bemessung vertikaler und horizontaler Dichtelemente sohlgedichteter Baugruben. Vorträge der Baugrundtagung 1996 in Berlin. Deutsche Gesellschaft für Geotechnik e.V.. Seite 99 - 113.

Schlötzer, C. & Müller-Kirchenbauer, H. (1995): Filtrationsverhalten von Dichtsuspensionen für Einphasenschlitzwände. Bautechnik 72. Heft 9. Seite 600 - 607.

Schulze, B. (1992): Injektionssohlen. Theoretische und experimentelle Untersuchungen zur Erhöhung der Zuverlässigkeit. Veröffentlichungen des Institutes für Bodenmechanik und Felsmechanik der Universität Fridericana Karlsruhe. Heft 126.

Schulze, B. (1994): Der Einfluß der Rezeptur auf das Eindringverhalten von Injektionssuspensionen in der Praxis. Vortrag im Rahmen des Lehrgangs

"Injektionen mit Feinstbindemitteln in der Geotechnik." 01./02.02.1994. Technische Akademie
Esslingen.

Tausch, N. & Teichert, H.-D. (1990): Injektionen mit
Feinstbindemitteln - Zum Eindringverhalten von
Suspensionen mit Mikrodur in Lockergestein. 5.
Christian-Veder-Kolloquium. "Neue Entwicklun-
gen in der Baugrundverbesserung."
26./27.04.1990. Graz.

Dyckerhoff Zement GmbH Wiesbaden (1996):
Dyckerhoff-Kundenhandbuch Tiefbau/ Umwelt-
technologie.

Vorläufiges Merkblatt für Einpreßarbeiten mit
Feinstbindemitteln in Lockergestein (1993). Bautechnik 70. Heft 9. Seite 550 - 560.

Geotechnical Hazards, Marić, Lisac & Szavits-Nossan (eds) © 1998 Taylor & Francis, ISBN 90 5410 957 2

Controlling the soil instability by excavating an underground gallery

G. Nicola, L. Udrea & R. J. Bally
STIZO Enterprise, Department for Special Foundations and Drilling, Bucharest, Romania

ABSTRACT: Self hardening stable cement-bentonite suspensions were successfully applied by the authors to underground cavities filling, foundation blocks consolidation and their monolith adherence to the surrounding soil, hampering the movement of fluids through the ground. Their penetration in ground occurs by filling the preexisting voids or by hydraulic fracturing. The latter allows the suspension penetration and hardening in soil with fine pores, where grouting is not efficient, resulting in a considerable extension of groutable soils. The case histories mentioned in the paper are obvious examples of the successful application of grouting by hydraulic fracture in sandy silt with permeability coefficient lower than $10^{-5} \div 10^{-6}$ cm/s. Some less usual parameters of the suspension and resulting hardened material are mentioned: the cohesion according Lombardi plate, nonlinear (hyperbolic) constitutive law and plastic deformation.

1 CAVITIES AND WEAK ZONES IN THE GROUND

The problem to control cavities and weak zones in the ground in frequent in the authors' enterprise activity. Figures 1 and 2 outline usual cases of incidents at the contact between the construction infrastructure and the surrounding ground.

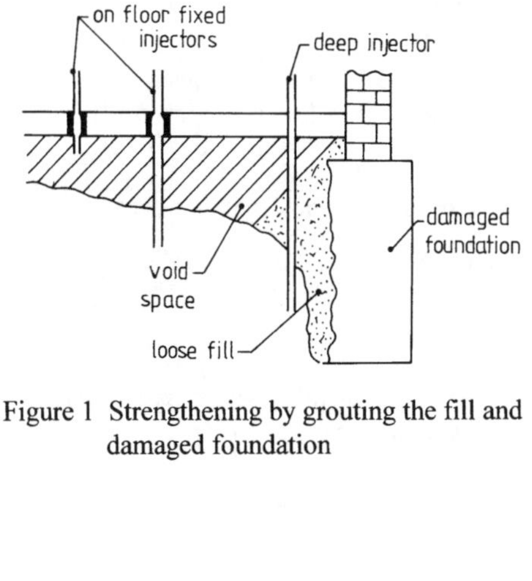

The first refers to holes and loose soil around the foundation and underground walls or beneath pavements and floors.

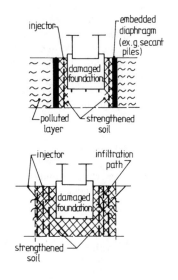

Figure 1 Strengthening by grouting the fill and damaged foundation

Figure 2 Foundation protection in polluted space

The second presents the consequences of aggressive fluids filtration resulting in foundations and soil corrosion.

Both situations were successfully controlled by grouting stable cement bentonite suspensions. Its effect implied:
- the filling and cementation of voids and mineral particles;
- generation of a honeycomb structure of hard lenses due to the hydraulic fracturing of the soil, resulting in ground strengthening and filtration hampering (the so-called "lense grouting") (Al-Alusi, 1995).

Figure 3 presents a loess sample including a hardened cement-bentonite lense due to suspension fracturing grouting blocking of the aggressive solutions traject.

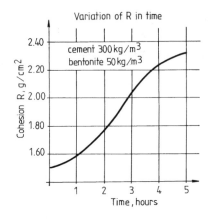

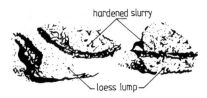

Figure 3 Lumps of grouted loessial soil

Besides the usual parameters applied to characterize or to control the suspension along all the three specific phases (fluid, setting, hardening) the cohesion determined by the plate method proposed by Lombardi (1985) for cement suspensions was tested also for cement-bentonite suspension (dipping into suspension and then pulling out a metal rough plate: the quantity of suspension trained up by the plate gives an estimation of the suspension cohesion).

Graphs on figure 4 emphasize the method ability to quantitatively express the suspension cohesion variation in time (the setting evolution) or in terms of the ratio between ingredients.

2 CONTROLLING THE SANDY SILT HYDRO-DYNAMIC INSTABILITY DURING THE EXECUTION OF AN UNDERGROUND GALLERY

Two examples of the use of the stable self-hardening cement-bentonite suspensions will be given in the following, both referring to incidents occurring during the execution of an underground gallery.

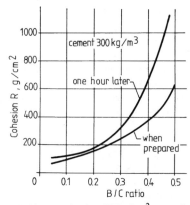

Figure 4 Slurry cohesion "R" g/cm^2 according to Lombardi plate method

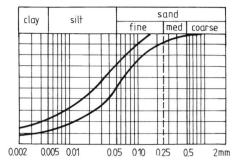

Figure 5 Sandy silt grain size distribution

The 4.5 m outer diameter gallery passed through a highly liquefiable sandy silt (Fig. 5) lying under a loessial soil completely wetted at its base (Fig. 6).

The gallery bottom was at 14.00 m from ground surface, 7.00 m bellow the underground water table.

596

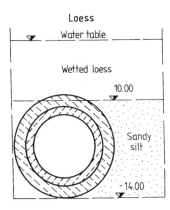

Figure 6 Geotechnical profile along the gallery

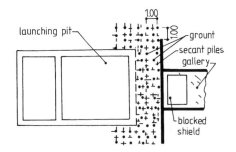

Figure 8 Control of the unstable interval between the gallery front and a launching pit

Frequent eruptions of liquefied sandy silt, up to 80 m³, generated substantial difficulties for the excavation and also large settlements on ground surface, extending 30 ÷ 40 m both sides of the canal traject and severely disturbing some small houses.

The authors' enterprise was asked to propose and apply adequate solutions to control the soil instability and permitting to resume the excavation works.

Two main solutions both including self-hardening stable cement-bentonite suspensions were adopted and applied separately or together:

- embedded walls of secant piles including in the joints a sleeve tube permitting, if necessary, an ulterior grouting to improve the watertightness of the contact or to extend, by lense grouting, the stabilized soil zone (Fig. 7);
- ground stabilization by lense grouting.

A first intervention zone (Fig. 8) was situated between the shield position, blocked by soil eruptions and a pit intended to bring out the existing shield and launch a new one.

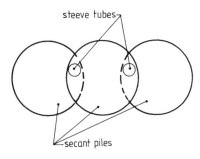

Figure 7 Watertight diaphragm

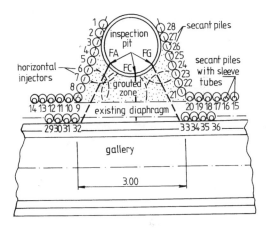

Figure 9 Assuring the passage from a visiting pit to the gallery

A second zone implied the link between a visiting pit and the gallery (Fig. 9).

The grouted self-hardening suspension was composed, for 1 m³, from cement 300 kg, bentonite 50 kg, sodium silicate 10 ℓ (density 1.2; Marsh viscosity 40", stability 98-100%, gel time 3-4 h, $R_{2\mu} > 1000$ kPa).

Laboratory determinations emphasized a nonlinear stress-strain relation and a plastic behaviour before rupture of the hardened slurry (Fig. 10 and Fig. 11).

Resuming, after grouting, the excavation works, the stabilized gallery front appeared like a succession of hardened lenses and compacted sandy silt layers (like "a layered cake") (Fig. 12).

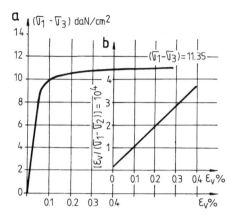

Figure 10 Triaxial compression of a hardened slurry sample; hyperbolic assumption:

a) $\sigma_1 - \sigma_3 = \dfrac{\varepsilon_v}{10^{-5} + 9.06 \cdot 10^{-2} \cdot \varepsilon_v}$

$(\sigma_1 - \sigma_3)_{lim} = 11.37 \ \text{daN/cm}^2$

b) linearised expression:

$$\dfrac{\varepsilon_v}{\sigma_1 - \sigma_3} = 10^{-5} + 9.06 \cdot 10^{-3} \cdot \varepsilon_v$$

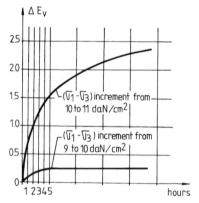

Figure 11 Plastic behaviour before rupture (sample presented in Fig. 10)

3 CONCLUSIONS

Self-hardening stable cement-bentonite suspensions were successfully used for lense grouting or secant piles including joint grouting sleeve tubes to solve difficult problems as instability of highly liquefiable sandy silts.

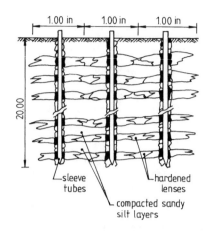

Figure 12 View of the lense grouted gallery front ("a layered cake")

Lombardi plate for slurry cohesion and hyperbolic stress-strain relation proved to be efficient interpretation methods.

REFERENCES

Al Alusi, H. R. 1995. Lense grouting in geotechnical engineering. *XI Afric CSMFE*, Cairo, 374-380.
Lombardi, G. 1985. The role of cohesion in cement grouting of rock. *XI ICOLD*, Lausanne, vol. 3, Q58, R13, 236-261.

Geotechnical Hazards, Marić, Lisac & Szavits-Nossan (eds) © 1998 Taylor & Francis, ISBN 90 5410 957 2

Flaws occurring in sewer construction – Solutions for consolidation

Augustin Popa
Department of Civil Engineering, Technical University of Cluj, Romania

Florin Lăcătuş, Valer Rebeleanu & Tibor Tokes
CONSAS Company, Cluj, Romania

ABSTRACT: The construction of a 4,30 m diameter sewer imposed the crossing of some areas of saturated sand soils. The water bearing sand inside the sewer shield to the acurence of soil sinking resulting in damaging the construction within those areas. The size and extension of these flaws imposed some novel technologies to be adopted in shield construction procedures. The choise was for a solution consisting of consolidating an earth are by injecting some binding substances, and through which the shield construction could advance. The paper presents a technological method of earth consolidation and control of construction execution.

1. INTRODUCTION

The main sewer, built in 1910 of the town of Brăila has undergone many demages due to the settlement of the foundation ground. Consequently a new sewer was started to be built, 1.6 km long. Out of this length, 0.5km crossed a package of fine , dusty sands. The site is also full of one-story houses with a low rigidity. All along the route of the sewer flaws and defects appeared, such as road deck and yard falls, cracks and falls of the walls of some houses.

2. DESCRIPTION OF WORK

The main sewer is built with a partially mechanized shield, in its open rear being mounted 7 (seven) reinforced concrete prefabricated keystone fitted with hook bolts (Figure 1).

A normal operation requires an inner lining with reinforced, plastered torcrete able to protect against aggressive waters collected. The raft of the drain is situated at a depth of -13.00 m ÷ - 14.50 m and the inner diameter measures 3,60 m, while the outer diameter is of 4.30 m.

Difficulties became obvious from the first 200 m ditch digging, when the surface lauers began deforming. Besides the compression of the layers, the land around settled. In order to do with these flaws, a set of two gel-concrete cast walls, which bordered the front shield area , was built. The tow cast or woulded walls 0.60 m thick were introduced

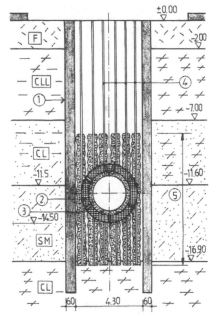

Figure1. Cross section 1. Gel concrete walls; 2. Reinforced concrete prefabricated; 3. Torcret ; 4. Injected pylons; 5. Grouting zone.

in the layer of clayey dust at −16.90 m.

Though the cast walls reduced the difficulties in advancing with the shield, the other aspects remained still unacceptable as the settlement of the surrounding ground went on demaging the buildings along the route.

In view of the above, extra measures were to be adopted in order to consolidate in due time all the flowing-in layers. The consolidation works aimed at stabilizing the soils masses in the shield front all along the width of the area delineated by the cast walls where instability levels were higher.

3. THE SOIL PROFILE

The soil through which the sewer was built is caracterized by a succesion of layers of the following form : The earth surface is covered by a landfill, 0.30 – 1.80 m thick, followed by a package of loessial layers, of hard plastic silty clays 4 - 6 m thick , then by layers whose thickness varies from 2.70 to 3.90 m. There follows a 7 – 8.50 m thick a layer of fine silty sands-sandy silt whose relative density increases with depth from a loose to an average. The basic layer contains silty clay , of a viscous, plastic consistency.

The level of the underground water is at – 11.5 – 14.5 m deep and at foundation of the fine sands a layer of water under pressure was met. Using dynamic penetration tests (DPT) the relative density was pointed out all along the experimental sector (see Figure 2).

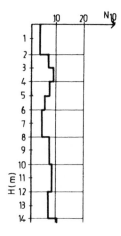

FIGURE 2. Characteristic DPT

It was established that loessial layers of up 7 – 8 m thick do not flow, are not in a soft plastic state and that they are followed by fine loose – averagely loose saturated sands. The sewer passes through the silty clay, soft loessial – fine silty loose saturated layers.

4. CONSOLIDATION SOLUTIONS

The consolidation solutions presented below were inspected and investigated in the experimental site developed along the sewer route :
a) consolidation with sleeve tube injection grouting, where suspension of cement – bentonite were injected at pressures varying from 3 to 5 bars;
b) consolidation with quick lime piles;
c) consolidation with vertical drains;
d) in depth compacting with some pylons made by pressure injections with a tough enough mortar as not to penetrate the soil and as to only exert lateral pressures upon it.

Among the consolidation methods mentioned above, the best results were obtained with the expanding grouting pylon compaction. The construction technology comprised the following stages (Lăcătuş, 1996):
• a network of pylons as in Fig.3 was drawn;

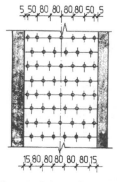

Figure3. Column disposition plan

• pylons of 50 mm diameter were drilled up to 16.00 m thick and bentonitic mud was used
• taking out of the casing and replacing it with the lancet-shaped injection seal, reaching a depth of up to 16 m.
• grouting in sections of 0.50 – 1 m long, upwards, with a viscous cement mortar of a pressure of 15 – 20 bars, up to the maximum fill – up of the section.

The grouting mortar used contained : cement - sand : 1: 2~1:4; with 5 - 20% bentonide and 3 - 4% sodium silicate added to it. Its toughness measured with the standard funnel was 300 - 700 CP.

Workability was assured with 2 – 4 % plastifier, which reduced the w/c ratio to 0.30 – 0.40.

The consolidation was obtained by the effect of densification of the package of silty clay loessial saturated sands, due to the horizontal stresses brought in by the pylons acting upon the surrounding ground. Besides a phenomenon of filling in some voids in the fine sands was seen. The negative effect appeared during the shield progress as small quantities of water were gathered and flowed in from the compact sand package.

The control of the consolidation works was made by :

- samples were prelevated and tested to compression after 7 and 28 days;
- recording of the vibropercution spear penetration time in the 1 m length of consolidated ground;
- CPT, DPT, SPT tests.

After the consolidation works were carried out, the soil bearing capacity increased 2 – 7 times while the lancet penetration value increased 7 – 30 times.

Increased sand density between the pylons was estabilished with CPT tests and the formula (Lunne, Chistoffersen, 1983) :

$$D = 0,34 \ln\left[\frac{q_c}{61\left(\sigma_z^{'}\right)^{0.71}}\right] \tag{1}$$

5. CONCLUSIONS

The consolidation of the soil made up of saturated loose earth with injected cement pylons presents the following advantages :

1. The lowest cost per 1 m^3 of consolidated ground;
2. The highest rate of building due to the large building site;
3. The possibility of correcting less consolidated layers;
4. The possibility of compacting saturated earth layer with grouted cement pylons leads to front side which keeps its vertical position in front of the shield.

REFERENCE:

Lunne, T.H.P. Cristoffersen 1983 : *Interpretation of cone penetrometer for offshore sands.* Prac. Offshore Technology Conf Houston : 181 – 188.

F. Lăcătuş, V. Rebeleanu 1996 : *Consolidarea stratelor nisipoase saturate traversate de scut la canalul colector Brăila. Raport tehnic.* Cluj Napoca (unpublished).

Geotechnical Hazards, Marić, Lisac & Szavits-Nossan (eds) © 1998 Taylor & Francis, ISBN 90 5410 957 2

Seasonal movements of building on the expansive subsoil

O.J. Puła & M. Stachoń
Technical University of Wrocław, Poland

ABSTRACT: Many clay soils in Poland, especially in the central and western part, posses a large potential for volume change. The buildings of low-rise housing are particularly vulnerable since the substrata is not heavily loaded and the structures themselves do not posses a great deal of stiffness. A considerable number of small towns around Wroclaw has of this type of buildings. The paper is concerned with to the case of building of that type which undergo cracking of the division walls. During two year observations the vertical displacement were measured at the eight bench marks.

1 INTRODUCTION

Expansive soils occur extensively through the world and have been reported in numerous countries. Many clayey soils in Poland, especially from the central and south part of the country, are also sensitive to contact with water. The existing zones of the pliocen clays territories of shallow depositions of drifts (in many cases just below the ground surface) are denoted in Figure 1.

LEGEND:
- RANGE OF EXPANSIVE CLAY
- EXPANSIVE CLAY OUTCROPPINGS
  IN   POLAND (0 - 25 m)

Fig. 1 Outcropings of expansive clay in Poland

The shrinkage and swelling processes in soils can result in damage to the structures and, as it was pointed by Burland (1984), the buildings of low - rise housing are particularly vulnerable since the substrata is not heavily loaded and the structures themselves do not posses a great deal of stiffness.

A considerable amount of small towns from south-west part of Poland (specially around Wroclaw) are comprised of this type of buildings. They were constructed 50 - 60 years ago and were founded on strip footings at depth of 0.7 m to 1.5 m below the ground level.

2 TECHNICAL DATA OF THE BUILDING

The five storeys apartment building was built at the beginning 1940. The total length of the building is 75.0 m and its width 8.75 m. The building consist of three parts. Each part of building is shifted in relation to previous, about half width of building. Between segments there are no expansion joint. The basement structure was designed below ground floor .

The foundations were shallow strip footings which extended to a depth of between 1.50 m, 1.80 m and 2.0 m below ground level.

The building structure is very light as is made as a steel skeleton one .

External and internal walls are made from hollow brick. The intermediate floors were constructed as a flat floor structure full filed by brick wall. Thickness of this external wall was 42 cm.

3 GEOTECHNICAL CONDITIONS

In 1994 a conventional six shell-and-auger borings were drilled up to five meters below ground level. Three of the boreholes were situated in front of the building and three others at the back side of it. Results of drilling and archives data shows that the

Boreholes 4, 5 and 6 were situated at the back side of the building. The specimens in the boreholes were collected from different depths ranging from 2.2 m to 3.5 m below the ground level.

Laboratory tests were carried out on undisturbed specimens of the Tertiary greyish-yellow silty clay. Laboratory tests included: index tests, swelling and shrinkage tests and unconsolidated undrained (UU) tests.

Unit weight of the specimens from the boreholes of numbers 1 and 3 ranges from 19 kN/m^3 to 21 kN/m^3 and respectively for specimens from the boreholes of numbers 4 to 6 is 19 kN/m^3, similarly to those

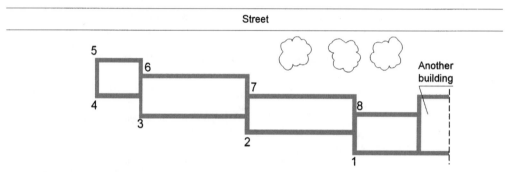

Fig. 2 The general layout of the building

area is underlain by very thick (more than 12 m) layer of clays from Teritary and Pleistocene drift .

The following crossection may be described near building deposits starting from the top. The upper layer consist of waste (clayly soil with crushed bricks and stones) of thickness of 1.5 - 2.0 m. Just below the upper layer silty and sandy clay were carried out till 5.0 m below surface.

Borehloes indicated that the upper part of clay layer is soft I_L = 0.29 - 035, below 3,0 m (measuring from the ground level) the firm state was observed I_L = 0.01 - 0.19.

4 LABORATORY TESTS

Specimens for laboratory tests were obtained from six boreholes. Two boreholes (number 1 and 3) were situated at the front side of the building with neighbouring old trees. The borehole number 2 was performed in the cellar of the building.

specimens from the borehole of number 2. Index test results are summarized in Table 1.

Table 1. Atterberg Limits

Borehole number (1)	Water content w_o [%] (2)	Liquid limit w_L [%] (3)	Plastic limit w_P [%] (4)	Plasticity index I_P[%] (5)
1 and 3	22	67	26	41
2	37	58	22	36
4 to 6	30	68	22	46

4.1. *Swelling and shrinkage tests*

The swelling tests were carried out to get the values of the swell pressure. The tests were conducted at no volume changes of soil possible during the swell process. Results of the swelling tests of specimens from different boreholes are presented in Table 2. The swell tests finished after 24 hours in the case of every specimen.

The Shrinkage Ratio is generally known and is defined as the ratio between the volume change, expressed as a percentage of the final dry volume, and the change in water content, expressed as percentage of the dry weight of the soil varied from 0.246 to 0.469 for tested specimens.

Results of the swell test are summarised in Table 2.

Table 2. Results of the swell test

Borehole number	Initial water content w_n [%]	Final water content w_k [%]	Swell pressure p_c [kPa]
(1)	(2)	(3)	(4)
1	22	24	66
2	36	50	0,0
3	22	25	242
4	31	32	24
5	29	30	48
6	30	31	16

Results of these tests in total values of stress are presented in Figure 3.

4.2. *Triaxial tests*

Triaxial UU testing was carried out on specimens at the natural water content. Measurements of the degree of saturation S_r show that it values ranged for tested specimens between 0.855 and 0.990.

Results of these tests in total values of stress are presented in Figure 3.

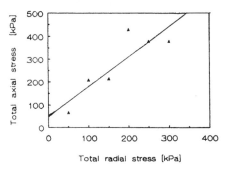

Fig. 3 UU Triaxial tests

Results of triaxial tests .analysis show rather low values of the strength parameters. The estimated total of friction angle is $\phi_u = 8^0$ and total value of cohesion $c_u = 22$ kPa.

The huge value of sell pressure in borehole number 3 closely correspond with big amount of $CaCO_3$.

5 CASE HISTORY

One of the most notable characteristics of some clay soils from the engineering point of view is their susceptibility to slow volume changes due to swelling or shrinkage. Such volume changes can give rise to ground movements which may result in severe damage buildings. Low-rise buildings are especially vulnerable to ground movements as they usually do not possess the weight or strength to resist.

Differences in the period and magnitude of precipitation and evapotranspiration are the major factors influencing the swell-shrink response of a clay soil beneath a building. Poor surface drainage or leakage from underground pipes can produce concentrations of moisture in clay. Trees with high water demand and unisulated hot process foundations may dry out clay causing shrinkage. The depth of active zone in expansive clays (i.e. the zone in which swelling and shrinkage occurs in wet and dry seasons respectively) varies.

The first cracking of the structures were reported ten years ago. The recent drought conditions, shrinkage of the soil led to different settlement of some parts of the structure, which results severe vertical and diagonal cracking of walls. The cracks ran from roof to floor and occurred above doors, and cornices had pulled away from walls. These cracks were several millimetres wide. In 1994 a conventional shell-and-auger borings and soil sampling were performed. Also vertical displacements of the building have been monitoring started to be monitored (since November of 1993 till to August 1995). The vertical displacements were measured at the eight bench marks which were installed on external walls. The place of location of each bench mark is shown in Figure 2. Bench marks 5 to 8 were installed on external wall in the vicinity of a street. A row of old trees is growing by this side of the building.

The observations were started in November 1993 (point 0 on the time axis). On the Figure 4 displacements of bench marks during first two hundred days observations are presented.

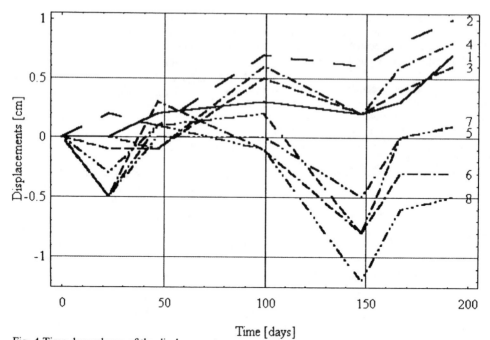

Fig. 4 Time dependency of the displacements

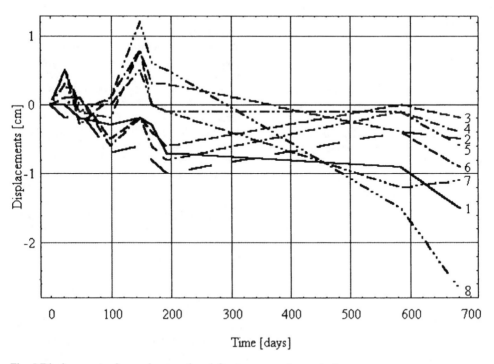

Fig. 5 Displacements observed versus time (after two years observation).

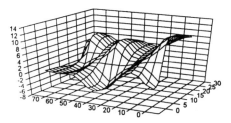

Fig.6 Isometric view, settlements of building from November 1993 to March 1994.

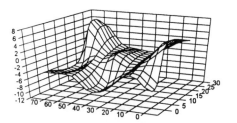

Fig.7 Isometric view, settlements of building from November 1993 to May 1995

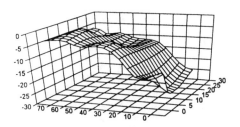

Fig.8 Isometric view, settlements of building from November 1993 to August 1995

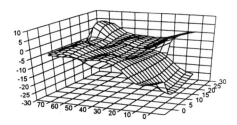

Fig.9 Isometric view, - upper mash settlements of building from November 1993 to May 1994; lower to August 1995

Seasonal vertical movements (up and down vertical) of a range more than 30 mm of the building were reported.

Total displacements (from November 1993 till August 1995) are presented in Figure 6.

The vertical displacements of each part of the building are closely connected to a season of year and the distance to the row of trees. In the case of increasing humidity of soil bench marks on a wall went up (usually after a winter in March - Figure 6). In May soil getting dry and displacements were smaller (Figure 7). Each point of the building rise up or settle on his own way and this causes a real danger for the structure. Many craks in external and internal walls are caused by such non-uniform movements of each point of building, during period of last few years. The result of non - uniform displacements are presented in Figure 6 - 9.

6 CONCLUSION

Soils with a capacity for swelling cover nearly a half of the territory of Poland and are serious cause of foundation problems. The full economic impact of the damage to structures is difficult to quantify precisely, but it is comparable with the costs of natural disasters. However, since the effects are most seriously concentrated on individual small structures, there has been relatively limited co-ordinated action. The case records that have been presented in this paper show the range of values of soil from clay sites where desiccation has occurred, and where there has been consequential building damage. Some of these records show that the desiccation caused by trees can be comparable with other reasons for example the shrinkage process caused by high temperature at the summer season. For analysed building settlements caused in the front part of building were two times higher then in the back part of building (where there was no trees).

This is particularly dangerous if, at the beginning a small tree, is planted quite near (even a few meters) from the new building. After forty years roots of big tree can completely will change geotechnical conditions above foundation of the building.

7 REFERENCES

Bell,F.G., J.C.Cripps,& M.G.Culshaw 1993. Volume changes in weak rocks: Prediction and measurement. *Proc.Gotech.Engieering of*

Hard Soils-Soft Rocks: 925-932. Rotterdam: Balkema.

Bell,F.G.,R.R.Maud 1994 Volume changes of soils and their effects in Natal, South Africa. *Proc.7th Intertional Congress International Association of Engineering Geology:*771-779. Lisboa:Balkema

Chandler,R.J.,M.S.Crilly,&G.Montgomery-Smith 1992. A low-cost method of assessing clay desccation follow-rise buildings. *Proc. Inst. Civ. Engineers* 92: 82-89

Driscoll, R.1983. The influence of vegetation on the swelling and shrinking of clay soils in Britain. *Geotechnique 33*:93-105

O'Neil , M.W. & A.M. Poormoayed 1980. Methodology for foundations on expansive clays. *Proc. American Society Civ.Engineers Journal Geotech. Engineering Div.*:1345-1367.

Przystański J. 1991. Posadowienie budowli na gruntach ekspansywnych. Politechnika Poznańska. *Rozprawy nr 244* (in polish).

Pula O. , Engineering problems associated with expansive clays in Poland. *Proc.7th Int. Congress International Association of Engineering Geology:* 627-633. Lisboa: Balkema

Van der Merwe, D.H.1964. The prediction of heave from the plasticity index and the precentage clay friction. *The Civ.Engineering in South Africa 6*, No; 103 - 107.

Geotechnical Hazards, Marić, Lisac & Szavits-Nossan (eds) © 1998 Taylor & Francis, ISBN 90 5410 957 2

Grundwasserschutz und Abdichtungsmaßnahmen mit Soilcrete-jet grouting

K. Saufnauer
Keller Grundbau, Wien, Austria

INHALT: Eine wichtige Aufgabe des Grundbaues ist der Schutz des Grundwasserhaushaltes vor unzulässiger Beeinflussung durch Bauvorhaben, sowie auch der Schutz der Bauten und ihre Nutzer vor den Gefahren des Wassers.
Der Bericht beschreibt den Stand der Technik des Soilcrete-jet grouting Verfahrens und zeigt Anwendungsbeispiele im Grundwasserschutz und für Abdichtungsmaßnahmen auf Baustellen in Österreich und den angrenzenden Nachbarländern.
Begleitende Meß- und Kontrollmaßnahmen gewährleisten eine sichere Ausführung der Aufgabenstellung.

1 EINLEITUNG

1.1 *Das Soilcrete Verfahren*

Die weiterhin wachsende Bedeutung von Abdichtungen im Untergrund, als Umschließung von tiefen Baugruben zur Vermeidung weitreichender Grundwasserabsenkung, als Sicherung von Grundwasserhorizonten, als wesentliche Begleitmaßnahme im Flußbau, im Dammbau und im Deponiebau fordert eine ständige Weiterentwicklung an grundbautechnischen Lösungen.

Die Soilcrete Technik ist ein von der Firma Keller Grundbau auf dem europäischen Baumarkt eingeführtes Verfahren und erfüllt in zahlreichen Anwendungen auch Grundwasserschutz- und Abdichtungsanforderungen.

Das Soilcrete Verfahren ist eine, die herkömmlichen Bodeninjektionen erweiternde Methode der Bodenverfestigung. Bei herkömmlichen Bodeninjektionen wird das Injektionsgut durch Druck in vorhandene Hohlräume des Untergrundes (Poren, Fugen, Klüfte) eingepreßt. Im Gegensatz dazu wird bei der Soilcrete Technik das Korngefüge des Baugrundes von einem energiereichen Flüssigkeitsstrahl aufgelöst, gleichzeitig werden die Bodenteile mit Zementsuspension vermischt und der erfaßte Raum damit ausgefüllt.

Zur Erzeugung des energiereichen Strahles werden Pumpendrücke in der Größenordnung von 300 bis 500 bar eingesetzt. Diese bewirken die hohen Austrittsgeschwindigkeiten des Flüssigkeitsstrahles durch eine oder mehrere Düsen.

Damit sich die aufgebrachte Druckenergie möglichst weitgehend in Geschwindigkeitsenergie umsetzen kann, darf nicht in ein geschlossenes System gepumpt werden, sondern ist durch Entlastungsöffnungen, gewöhnlich der Bohrlochringraum, der Druck außerhalb der Düsen auf den hydrostatischen Druck zu begrenzen.

Die Erosionsweite des Düsenstrahles richtet sich nach Bodenart und Verfahrensart.

Im Gegensatz zu den herkömmlichen Injektionsverfahren, bei denen das Injektionsgut in die Hohlräume des Bodens eindringen muß, ist die Soilcrete Technik nicht durch die Bodendurchlässigkeit begrenzt. Neben Kiesen und Sanden kann praktisch der gesamte Bereich des Schluffes und unter Berücksichtigung der wirtschaftlichen Grenzen auch tonreiche Böden mit Soilcrete bearbeitet werden.

1.2 *Soilcrete Eigenschaften*

Soilcrete wirkt im Baugrund je nach Aufgaben-stellung als Verfestigungs- oder Abdichtungskörper. Die Soilcrete Festigkeit wird von Art und Menge des Zementanteiles sowie den verbleibenden Bodenanteilen in der Soilcrete Masse bestimmt. In gut abgestuften Kies/Sandböden können Festig-keiten bis 25 N/mm² angegeben werden, mit zu-nehmenden Schluff/Tongehalt nehmen die Druck-festigkeit deutlich ab.

Die Soilcrete Abdichtungseigenschaft wird durch geeignete Suspensionsrezeptur erreicht. Die übliche Wasser/Zementsuspension wird dabei durch ent-sprechende Füller und Bentonit ergänzt.

1.3 *Herstellungsablauf*

Die Geräteausrüstung besteht aus einer an geeigneter Stelle auf dem Baugelände installierten Misch- und Pumpanlage. Von dort laufen die Schlauch- und Steuerleitungen zum Soilcrete Bohr-gerät am Einbauort. Die örtlichen Bedingungen bestimmen die Größe des Bohrgerätes. Die Auf-rüsthöhe reicht von 2,0 m für Kellerräume oder Arbeitsschächte bis zu Lafettenhöhen von 35 m für tiefe Bohrungen im Freien.

Bohren

Ein Bohrgestänge von 10 cm Durchmesser wird mit dem Düsenhalter und der Bohrkrone drehend abgeteuft. In der Regel unterstützt ein Spülstrom aus Soilcrete Suspension den Vorgang und hält den Ringraum um das Gestänge für den Abfluß der Bohrspülung offen. Für Durchbohrungen von Mauerwerk und Beton werden spezielle Bohrkronen verwendet.

Schneiden

Das Auflösen des Korngefüges mit einem im Winkel von ca. 90° zur Bohrachse austretenden energiereichen Flüssigkeitsstrahl beginnt an der tiefsten Stelle des Soilcrete Elementes. Über-schüssiges Wasser-Boden-Zement-Gemisch fließt über den Bohrlochringraum zutage.

Soicretieren

Gleichzeitig mit dem Erodieren des Bodens wird Zementsuspension unter Druck zugeführt und durch verfahrensbedingte Turbulenzen im unmittelbaren Produktionsbereich optimal eingemischt.

Erweitern

Soilcrete Elemente jeder Form lassen sich sowohl frisch in frisch als auch frisch gegen fest miteinander verbinden und kombinieren. Die Herstellungsreihenfolge wird auf die gegebene Aufgabe abgestimmt.

1.4 *Ausführungsformen*

Die geometrischen Formen von Soilcrete werden durch zielgerichtete Bewegungen des Bohrgestän-ges erzeugt:

Ziehen des Gestänges ohne Rotation ergibt Lamel-len.

Ziehen und Schwenken des Gestänges ergibt Teil-säulen (Viertelsäule, Halbsäule).

Ziehen mit Rotation des Gestänges ergibt Säulen.

Die Soilcrete Grundformen können beliebig anein-ander gereiht, ineinander übergehend oder kombi-niert angeordnet werden, sodaß verschiedene Körperformen herstellbar sind.

1.5 *Ausführungskontrolle*

Bei der Herstellung von anspruchsvollen Soilcrete Kubaturen, insbesondere auch für Abdichtungs-maßnahmen, werden oft aufwendige Meß- und Kontrollmaßnahmen verlangt.

Dazu gehört die exakte lage- und höhenmäßige Vermessung der Bohrungen, sodaß Bohransatz-punkte und Bohrneigungen plangemäß stimmen. Ebenso wie der Bohransatz ist die Genauigkeit des Bohrzielpunktes entscheidend. Die Kontrolle der Bohrtiefe ist durch Laserunterstützung sehr genau möglich oder erfolgt durch Einsatz von Instru-menten zur automatischen Identifikation des Produktionstiefe.

Die maßgeblichen Herstellungsparameter wie Drücke, Durchflüsse, Produktionszeiten und -geschwindigkeiten können für jeden Produk-tionspunkt elektronisch aufgezeichnet werden, um Störungen noch während der Herstellung zu erken-nen, die Überwachung lückenlos zu gewährleisten und allenfalls notwendige Korrekturen gezielt vor-zunehmen.

Die fertiggestellte Soilcrete Kubatur ist mit Hilfe von Kontroll-Schürfen oder Kontroll-Bohrungen überprüfbar.

Zu beachten bleibt, daß es trotz jahrzehntelanger Praxis weiterhin vorkommt, daß für besondere Bodenschichten keine ausreichenden Erfahrungen verfügbar sind. Hier bringt einzig die vorherige

Ausführung und Überprüfung von Probesäulen die erforderliche Sicherheit bei der Auswahl der Herstellungsparameter.

Der Einsatz der beschriebenen Meß- und Kontrollmaßnahmen drückt sich in den zusätzlichen Kosten des Gesamtprojektes aus. Gerade bei Abdichtungsaufgaben ist jedoch eine hohe Herstellsicherheit dringend geboten.

2 SOILCRETE AUSFÜHRUNGSBEISPIELE FÜR ABDICHTUNGSAUFGABEN

2.1 *Baugrubensicherung mit Grundwasserabdichtung*

Eingriffe in den Grundwasserhaushalt bringen insbesondere im innerstädtischen Tiefbau rechtliche, technische und ökologisch Probleme.
Die Forderung nach Grundwasserabdichtungen anstelle von Grundwasserabsenkungen ist vielfach maßgebend für die Baugrubenplanung.
Dazu kommt der Wunsch das Bauherrn und des Architekten nach maximaler Platznutzung des Grundrisses und der Wunsch der Bauausführung nach unkompliziertem Bauablauf, auch für die Tiefgeschoße, innerhalb einer sicheren und trockenen Baugrube.
Die statisch erforderliche Unterfangung mit Soilcrete und die Verlängerung einer Soilcrete Schürze bis zum Stauhorizont erfüllen diese hohen Ansprüche.
Die Unterfangung außerhalb der Baufluchtlinie läßt dem Bauwerk maximalen Platz, sodaß die wasserdicht ausgeführten Kellerwände getrennt und unabhängig von der Baugrubensicherung, errichtet werden können.
Die Soilcrete Schürze, die keine Gebäudelasten mehr berücksichtigen muß, kann wirtschaftlich unterhalb der statischen Beanspruchung der Baugrubensicherung bis zum Stauhorizont verlängert werden.

2.1.1 Wohnhaus Bräuhausgasse, Wien

Der Baugrubenaushub für die Kellergeschosse des neuen Wohnhauses reichte ca. 2 m unter den Grundwasserhorizont. Der Grundwasserträger ist sandiger, schluffiger Kies, den Stauer bildet „Wiener Tegel" in einer Tiefe von ca. 4 m unter der Baugrubensohle.
Die Unterfangungsstatik für die Nachbargebäude und die Gehsteigfronten verlangte eine verankerte

Soilcrete-Wand, die Abdichtung bis zum Stauhorizont erfüllte eine Soilcrete Lamellenwand unterhalb der statischen Kubatur.
Während des Baugrubenaushubes und der Bauarbeiten wurde eine leicht bewältigbare Restsickerwassermenge abgepumpt, womit die Baugrube nach dem Stand der Technik als dicht angesehen werden konnte.

2.1.2 Geschäftshaus Vysoka Ulica, Bratislava

Eine große Herausforderung an den Grundbau stellte die Baugrube Vysoka Ulica, Bratislava. Die Baulinie war begrenzt von Häusern, die zum Teil unter Denkmalschutz stehen. Auf dem Bauplatz selbst gab es Altbestandsobjekte mit Untergeschossen aus mehreren zeitlichen Epochen, unterirdische Hohlräume wurden befürchtet. Der Baugrubenaushub für 3 Untergeschosse ging bis 13 m unter das Geländeniveau und damit ca. 7 m unter den Grundwasserspiegel.
Ein durchgehender Stauhorizont in wirtschaftlich erreichbarer Tiefe war nicht gegeben. Die Geologie unterhalb des oberflächennahen Kieshorizontes wechselte in Lagen von schluffigen Sanden bis tonigen Schluffen.
Die Baugrubensicherung erfolgte durch die statisch erforderliche Soilcrete Kubatur und der Vertiefung als Soilcrete Schürze zur Verlängerung des Grundwasserströmungsweges.
Dadurch konnte der Grundwasserspiegel innerhalb der Soilcrete-Umschließung mit Schachtbrunnen abgesenkt werden und die Kellergeschosse mit herkömmlicher Außenisolierung den Grundriß der Baugrube maximal ausnutzen.

2.1.3 Bürohaus Na Belidle, Prag

Die Baugrube im Prager Zentrum in geringer Entfernung zur Moldau war unmittelbar beeinflußt vom Wasserstand des Flusses mit einem Bemessungswasserspiegel bis ca. 5 m über den tiefsten Aushubbereichen. Voruntersuchungen der Nachbargebäude zeigten außerdem einen sehr schlechten Zustand der Fundamentmauern.
Aufgabenstellung war die Ausführung eine dichten Baugrubensicherung für zwei Untergeschosse bis ca. 9 m Tiefe. Grundwasserträger ist ein sandiger Kies, der Stauhorizont, ein unterschiedlich stark verwitterter Schiefer, liegt in ca. 13 m Tiefe.
Die Soilcrete Technik erfüllte alle Ansprüche. Schon während der Herstellung der einzelnen Säulen erfolgte auch eine Vermörtelung und damit

Verbesserung der Nachbarfundamente. Die Soil-crete Säulen bereiteten damit in äußerst schonender Weise die Fundamente als Bestandteil der Unter-fangung vor und bildeten gleichzeitig auch die dichte Umschließung der Baugrube.

Die nachfolgenden Bauarbeiten erfolgten bei Ab-pumpen einer geologisch bedingten unvermeid-lichen Restsickerwassermenge ohne Komplika-tionen.

2.2 Abdichtungsaufgaben im Flußbau

Planung und Bau von Wasserkraftwerken verlangen auch die Rücksichtnahme auf den Grundwasser-haushalt sowohl im Unterwasser als auch im Oberwasser des Krafthauses.

Das Unterwasser wird eingetieft, der Grundwasser-spiegel wird kommunizierend mitgezogen. Verti-kale Dichtungselemente sollen den Einfluß auf das Grundwasser verhindern.

Der Stauwasserspiegel im Oberwasser wird deutlich über den bisher natürlichen Grundwasserspiegel und auch über das Geländeniveau der angrenzenden Ufer gehoben. Uferbegleitende Dichtungselemente unterbinden eine Anhebung des Grundwasser-spiegels und unkontrollierte Schadstoffbelastungen. Sie verhindern weiters eine Durchnässung bzw. Überschwemmung des Hinterlandes.

Hauptanwendungsgebiete für Dichtungselemente mit Soilcrete sind Abdichtungen unter schwierigen Randbedingungen. Dazu zählen innerstädtische Bereiche, Produktionsstätten, Bereiche unterhalb von Gebäuden, Industrieanlagen oder Straßen mit den typischen Behinderungen im Arbeitsraum und im Baugrund.

Hier ermöglicht die Soilcrete Technik mit seiner geometrischen Flexibilität und der Option für den Einsatz auch sehr kleiner Bohrgeräte die Realisierung der Abdichtung.

2.2.1 Kraftwerk Freudenau,Wien

Das Dichtungselement im Ufer des Oberwassers bildet eine Schlitzwand bis ca. 20 m Tiefe durch Dammschüttung und Donaukiesschichten bis in den schluffig, tonigen Stauer der tertiären Boden-schichten. Im Bereich des Heizkraftwerkes Donaustadt kreuzen die unterirdischen Einbauten für die Kühlwasserentnahme und den Kühlwasser-rücklauf die Trasse der Dichtwand.

Die Schlitzwand endete hier im definierten Sicher-heitsabstand von den Rohrleitungen, die offene Lücke wurde mit einer Soilcrete Lamellenwand

geschlossen. Die Bohrungen konnten zielgenau neben den gesicherten Rohrleitungen abgeteuft werden, die Soilcrete Lamellen wurden unter den Rohren ausgeführt.

Der kraftschlüssige, rundum dichte Anschluß erfolgte durch Verfüllung des Sicherungsschlitzes mit Dichtwandsuspension.

Eine weitere Schlüsselstelle bildete das Anschluß-detail der Schlitzwand an das Pfeilerfundament der Rohrbrüche. Hier sorgten Soilcrete Säulen als Zwickeldichtung zwischen Schlitzwand und Funda-mentbeton für eine durchgehende Dichtung.

2.2.2 Donauufer Medvedov, Slowakei - Ungarn

Die Dichtungsschlitzwände am Donauufer im Raum Medvedov an der slowaksich - ungarischen Grenze dienen dem Hochwasserschutz für den Ausbau des Wasserkraftwerkes Gabcikovo. Die außergewöhn-liche Dicke des sandig, kiesigen Grundwasser-trägers ermöglicht keine Einbindung der Dichtwand in den Stauhorizont, sodaß die Schlitzwände bis ca. 25 m tief als unterströmte Dichtwände ausgeführt werden.

Auch hier mußten Schlitzwandlücken dort offen bleiben, wo das Arbeitsumfeld für die schweren Baggergeräte ungeeignet war, oder unterirdische Einbauten das Öffnen eines Schlitzes verhinderten.

Unter anderem querte die stark frequentierte Zufahrtsstraße zum Grenzübergang die Dichtwand-trasse. Der Lückenschluß erfolgte durch eine Soilcrete Lamellenwand, ausgeführt mit fächerför-migen Bohrungen vom Straßenrand. Der Verkehr konnte ungehindert aufrecht erhalten bleiben.

Dazu waren Soilcrete Bohrungen bis ca. 30 m Tiefe und Neigungen bis ca. 70° von der vertikalen Achse erforderlich.

2.3 Baugruben - Sohlenabdichtung

In Abhängigkeit der Geologie des Baugrundes ist es technisch oft nicht möglich, die seitliche Umschließung in einen Stauhorizont einzubinden.

Um in weiterer Folge eine Unterströmung der Baugrubenwände zu verhindern, wird eine dichte Sohle ausgeführt.

Die Soilcrete Technik bietet dafür verschiedene Vorteile gegenüber den konventionellen Injektions-verfahren. Vor allem die weitgehende Unabhängig-keit von Schichtwechseln und der Einsatz von ausschließlich anorganischen Materialien wie Zement, Bentonit und Kalksteinmehl macht Soilcrete heute sehr erfolgreich in diesem

Anwendungsgebiet. Auch die zielgenaue Möglichkeit zur dichten Anbindung der Sohle an die seitlichen Umschließungswände sorgt für ein hohes hochgehaltene Sicherheitsniveau.

Soilcrete Sohlen werden vor allem als tiefliegende Sohlen mit Bemessung nach der geforderten Auftriebssicherheit hergestellt. Sie können aber auch, mit entsprechenden Druckfestigkeitseigenschaften, gleichzeitig als Dichtungselement und als statisch wirksames Element, wie z.B. als Sohlaussteifung oder als tragfähige Gründungssohle, genützt werden.

2.3.1 Reichstag, Berlin

Die Umgebung des Reichstagsgebäudes in Berlin wurde in der Fachwelt mit dem Titel der größten Baustelle Europas ausgezeichnet.

Vorübergehend wirkte das Gelände rund um den Reichstag wie eine fast endlose Baugrube.

Der Baugrubenaushub führte im Berliner Feinsand/Mittelsand weit unter den Grundwasserspiegel.

Die Dimension der Arbeiten und die befürchteten Auswirkungen auf den Grundwasserspiegel ließ die Behörde sehr strenge Forderungen für ein Grundwassermanagement festlegen.

Mehrere 10.000 m² auftriebssichere Sohlenabdichtung bis ca. 25 m Einbautiefe wurden mit Soilcrete ausgeführt.

Die hohen Qualitätsansprüche erforderten ein ebensolches Maß an Qualitätssicherung.

Dazu gehörte eine strenge interne Disziplin schon beginnend bei der Planung, bei der Genauigkeit der Herstellung und Dokumentation sowie bei den begleitenden Untersuchungen und Kontrollen.

Mit großem Aufwand und Kosten konnten schließlich auch hier die Anforderungen gewährleistet werden.

2.3.2 Kraftwerk Gabcikovo, Slowakei

Die Baugrube für die Wehranlage im Bereich Cunovo hatte eine Grundrißfläche von ca. 12.000 m². Die Geologie des Donauufers zeigte Kiese und Sande ohne Stauhorizont in nutzbarer Tiefe, der Grundwasserspiegel war unmittelbar von der nahen Donau beeinflußt und forderte entsprechende Grundbautechnik für die Baugrube.

Die seitliche Umschließung wurde durch Einphasenschlitzwände mit eingestellten Spundwänden hergestellt.

Die Sohle wurde mit einer 3,5 m dicken, hochliegenden Soilcrete Sohle abgedichtet. Sie mußte Auftriebssicherheit bis 6,5 m Wasserüberdruck gewährleisten und Festigkeitsansprüche für die Gründung des Wehrbauwerkes erfüllen.

Durch Unterteilung des Baugrubenfeldes in 10 abgeschlossene Kassetten konnte das Ausführungsrisiko und die terminliche Ablaufplanung optimiert werden.

3 ZUSAMMENFASSUNG

Das Soilcrete Verfahren wird in zahlreichen Ausführungsformen für Abdichtungs- und Grundwasserschutzmaßnahmen angewandt. Die Ausführungskontrolle begleitet den gesamten Herstellungsablauf und gewährleistet die geforderten Qualitätseigenschaften. Beispiele zeigen die Anwendungspraxis für verschiedene Aufgaben bei der Baugrubenumschließung und im Wasserbau.

LITERATUR

Berg J., Samol H. 1986. Soilcrete - ein Bodenverbesserungsverfahren im Grund- und Wasserbau
Sonderdruck: Wasser + Boden

Raabe E.-W., Toth S. 1987. Herstellung von Dichtwänden und -sohlen mit dem Soilcrete Verfahren.
Fachaufsatz

Keller 1997. Das Soilcrete Verfahren
Prospekt

Keller. Baustellenberichte

Geotechnical Hazards, Marić, Lisac & Szavits-Nossan (eds) © 1998 Taylor & Francis, ISBN 90 5410 957 2

Wirtschaftliche Fundierung durch Bodenverbesserung

Erich Schwab & Martina Zinsenhofer
Österreichisches Forschungs- und Prüfzentrum Arsenal Ges.m.b.H., Wien, Austria

Robert Gotic
Geoengineering KEG, Wien, Austria

ZUSAMMENFASSUNG: Im Bereich der Tiefenverdichtungsverfahren nehmen Bodenverbesserungs-maßnahmen mittels Rüttelstopfverdichtung, aufgrund der nahezu generellen Einsetzbarkeit in Bezug auf die Untergrundverhältnisse schlecht tragfähiger Böden, den größten Raum ein. Im Jahre 1997 wurde dieses Verfahren erstmals in Kroatien für die Fundierung einer Kathetrale erfolgreich eingesetzt. Einsatzbereiche, Berechnungsmethoden und Grenzen der Methode werden beschrieben. Der erfolgreiche Einsatz der Rüttelstopfverdichtung wird anhand von zwei Fallbeispielen aufgezeigt.

1. EINLEITUNG

Durch Anpassung an die Untergrundverhältnisse und optimale Ausnützung der vorhandenen Boden-eigenschaften stellen die verschiedenen Bodenver-besserungsmaßnahmen heute (1998) einen fixen Bestandteil im Spezialtiefbau dar. Bereits vor 3000 Jahren wurden Bodenverbesserungsmaßnahmen bei der Errichtung babylonischer Tempel eingesetzt und in China wurden Holz, Bambus und Stroh zur Bewehrung des Bodens verwendet, van Impe (1989). Der Einsatz dieser Maßnahmen beruhte aus-schließlich auf Erfahrungswerte. In der modernen Geotechnik werden, aufgrund genauer Kenntnisse des Verformungs- und Festigkeitsverhaltens von Böden, Bodenverbesserungsmaßnahmen in immer größerem Ausmaß auch bei komplexen Untergrund-erhältnissen und komplizierten Fundierungs-problemen erfolgreich eingesetzt. Diesem Thema wurde daher bereits 1981 bei der XI. Internationale Konferenz für Bodenmechanik und Grundbau in Stockholm eine Hauptsitzung gewidmet. Mitchell (1981) streicht in seinem *State of the Art* Bericht u.a. die Einführung von vibrierenden Methoden zur Ver-dichtung kohäsionsloser Böden als die bedeutendste Entwicklung in den letzten 50 Jahren heraus.

Der wesentlichste Vorteile von Bodenverbesse-rungsmaßnahmen liegen in ihrer Wirtschaftlichkeit durch individuelle Anpassung an die jeweiligen lokalen Bodenverhältnisse (Flexibilität), Ausnutzung der vorhandenen Bodenfestigkeit und in ihrer Umweltfreundlichkeit.

Van Impe (1989) gibt eine komplette Beschreibung der gängigsten Bodenverbesserungsmaßnahmen und teilt diese ein in:
- *temporäre Maßnahmen*, wie z.B. Grundwasser-haltung, Bodenvereisung, Elektroosmose
- *permanente Maßnahmen ohne Materialzugabe*, wie z.B. Oberflächenverdichtung, Tiefenver-dichtung (mittels Fallmassen, vibrierend,) thermische Maßnahmen und
- *permanente Maßnahmen mit Materialzugabe*, wie z.B. Kalk- oder Zementstabilisierung, Schottersäulen, Bodentausch, bewehrte Erde, etc.

Bei Baugrundverbesserungen durch Tiefenrüttler wird der anstehende Boden bis zur erforderlichen Tiefe so verbessert, daß die Bauwerkslasten sicher und innerhalb der geforderten Setzungsmaße abgeleitet werden. Bei konventionellen Pfahl-gründungen hingegen, wird der nicht tragfähige Boden durch Tragelemente überbrückt und die Lasten in die tiefer gelegenen tragfähigen Schichten eingebracht. Die Bodeneigenschaften werden dabei grundsätzlich nicht verändert und die Ableitung der Lasten erfolgt nur partiell über Reibung in den schlecht tragfähigen Schichten.

In diesem Beitrag wird auf die Boden-verbesserungsmaßnahme durch Tiefenrüttler näher eingegangen, da diese Methode, wie bereits erwähnt, ein breites Anwendengsgebiet hat und 1997 erstmals in Kroatien erfolgreich eingesetzt wurde.

2. BAUGRUNDVERBESSERUNG DURCH TIEFENRÜTTLER

In zahlreichen Veröffentlichungen (z.B. D'Appolonia, 1954; Plannerer, 1965; Brown, 1977; Kirsch, 1979, 1993; Mitchell, 1981; van Impe, 1989; van Impe et al., 1997) wurde der Aufbau und die Wirkungsweise von Tiefenrüttlern bereits ausführlich beschrieben. In den folgenden Ausführungen wird daher nur auf die Merkmale jener Tiefenrüttler näher eingegangen, die bei den nachfolgend beschriebenen Fallbeispielen zum Einsatz gekommen sind.

Bei den eingesetzten Tiefenrüttlern werden durch Exzentermassen an der Rüttlerspitze horizontale Schwingungen erzeugt, die mehr oder weniger ungedämpft auf den umgebenden Boden einwirken. Die Exzenter rotieren mit konstanter Geschwindigkeit. Daraus folgt, daß bei größerem Bodenwiderstand die aufgebrachte Energie, ausgedrückt durch die Amperaufnahme des Motors, umso größer sein muß. Die Amperaufnahme ist somit ein direktes Maß für den Bodenwiderstand und den Verdichtungserfolg. Die wesentlichsten Dimensionen und Merkmale von den heute am häufigsten eingesetzten Tiefenrüttlern sind:

–Durchmesser	30 cm bis 50 cm
–Länge	3 m bis 4 m
–Gewicht	16 kN bis 25 kN
–Motorleistung	5 kW bis 150 kW
–Drehzahl	1800 bis 3600 UpM
–Amplitude	3 mm bis 25 mm
–Schlagkraft	130 kN bis 230 kN

Grundsätzlich muß zwischen der Bodenverbesserung von grobkörnigen Böden (Rütteldruckverdichtung, Vibroflotation, Vibro Compaction) und der Bodenverbesserung von feinkörnigen Böden (Rüttelstopfverdichtung, Vibro Replacement) unterschieden werden.

2.1 Bodenverbesserung von grobkörnigen Böden (Rütteldruckverdichtung)

Natürliche oder geschüttete grobkörnige Böden mit einer Dichte kleiner als die maximale Dichte, können durch Schwingungen oder Vibrationen verdichtet werden. Bei der Versenkung des Tiefenrüttlers bildet sich in der Regel an der Oberfläche ein Setzungstrichter. Füllmaterial wird von der Oberfläche beigegeben. Diese Methode ist bis zu einer Tiefe von ca. 50 m noch wirtschaftlich. Der Verdichtungserfolg hängt neben den Rüttlerparametern in erster Linie von der Korngrößenverteilung und der Kornform des Bodens sowie vom effektiven Überlagerungsdruck ab. Durch Variation des Rasterabstandes kann heterogener Untergrund in einen gleichmäßig tragfähigen Baugrund umgewandelt werden. Bei einer Erhöhung der Lagerungsdichte wird der Reibungswinkel und der Steifemodul erhöht und somit die Tragfähigkeit und Erdbebensicherheit verbessert. Grobkörnige Böden mit einem Feinkornanteil (Korngröße < 0,06 mm) von mehr als 15 % können in der Regel nicht mehr effektiv durch Vibration verdichtet werden.

2.2 Bodenverbesserung von feinkörnigen Böden (Rüttelstopfverdichtung)

Wie bereits oben erwähnt können feinkörnige Böden nicht mehr durch Schwingungen verdichtet werden. Auch bei der Oberflächenverdichtung werden in solchen Fällen keine vibrierenden Walzenzüge mit Glattmantel sondern Walzen mit knetenden Bandagen (z.B. Schaffußwalze) eingesetzt. Die Bodeneigenschaften feinkörniger Böden können jedoch auch mittels Tiefenrüttler, durch Einrütteln von Kiessäulen, verbessert werden. Ursprünglich wurde die Methode der Rüttelstopfverdichtung nur in Schluffen, sandigen Schluffen und schluffigen Tonen geringer Plastizität eingesetzt. Wie jedoch das eine Fallbeispiel zeigt kann die Methode auch erfolgreich in plastischen schluffigen Tonen breiiger bis steifer Konsistenz erfolgreich und wirtschaftlich eingesetzt werden.

Der Rüttler wird vibrierend bis zur Endteufe versenkt und erzeugt dabei in der Regel, aufgrund der Kohäsion des Bodens, einen standfesten zylindrischen Hohlraum. In diesen Hohlraum wird Fremdmaterial (ein rolliger Kies) entweder von oben chargenweise oder mittels Schleuse eingebracht und verdichtet (gestopft). In feinkörnigen Böden bewirkt die Rüttelstopfverdichtung eine Erhöhung der Tragfähigkeit sowie eine Verminderung der Setzungen und Setzungsunterschiede durch die hohe Wasserdurchlässigkeit der Kiessäulen aber auch eine Beschleunigung der Konsolidierungszeit und somit eine raschere sukzessive Zunahme der Scherfestigkeit des natürlichen Bodens. Der Erfolg der Bodenverbesserungsmaßnahmen hängt in feinkörnigen Böden nicht so sehr von den Rüttlerparametern, sondern mehr von den Bodeneigenschaften, den geotechnischen Eigenschaften des Zugabematerials und dem geometrischen Abstand der Schottersäulen ab.

In wassergesättigten feinkörnigen oder organische Böden breiiger bis weicher Konsistenz können nur selten homogene Kiessäulen hergestellt werden. Hier hat sich erfolgreich der Einsatz von Geotextilien bewehrt. Der Tiefenrüttler wird vor dem Versenken mit einem „Strumpf" aus Vlies ummantelt und das Zugabematerial wird leicht in diesen Strumpf gestopft. Hier ist die Zugabe mittles Schleuse unbedingt erforderlich. Eine gleichmäßige, homogene Stopfsäule kann so hergestellt werden und das Vlies verhindert nicht nur das Ausdrücken des Kieses in den breiigen Boden sondern auch die Eintragung von Feinteilen aus dem Porenwasser in die Kiessäulen während der Konsolidierung.

2.3 Baugrundbeurteilung für Bodenverbesserungsmaßnahmen

Zur Beurteilung des Baugrundes für Bodenverbesserungsmaßnahmen werden in erster Linie die Resulate von statischen und dynamischen Sondierungen herangezogen. In Sanden, Schluffen und tonigen Böden werden Drucksondierungen sowie Sondierungen mit der schweren Rammsonde und der Standardsonde (SPT) bevorzugt eingesetzt, während in grobkönigen Böden fast ausschließlich Sondierungen mit der schweren Rammsonde verwendet werden. In breiigen bis weichen feinkörnigen Böden kann auch die undränierte Scherfestigkeit, ermittelt z.B. mit dem Scherflügelgerät oder der Drucksonde zur Beurteilung herangezogen werden.

In Österreich ist der Einsatz von Rammsondierungen zur Beurteilung des Baugrundes weit verbreitet. Biedermann (1979) gibt in seinem *State of the Art* Bericht umfangreiche Korrelationen und Bodenkennwerte der verschiedenen Sondiermethoden, die für Berechnungen und Baugrundbeurteilungen herangezogen werden können.

In Abb.1 sind die Körnungsbereiche für optimalen Einsatz von Bodenverbesserungsmaßnahmen mit Tiefenrüttlern dargestellt.

Die tatsächliche Beurteilung der Notwendigkeit zur Verbesserung eines Baugrundes hängt natürlich in erster Linie von den Bauwerkslasten und von den zulässigen Setzungen ab. Es kann somit kein allgemein gültiges Kriterium für die Notwendigkeit allein über den Bodenzustand (Eindringwiderstand bei Rammsondierunge, Spitzenwiderstand bei Drucksondierungen oder undränierte Scherfestigkeit)

gegeben werden. Im nächsten Abschnitt werden einige ausgewählte Beispiele von Sondierresultaten vor und nach einer Bodenverbesserung von konkreten Bauvorhaben gezeigt. Die Untergrundverhältnisse können für gewisse Baumaßnahmen zufriedenstellend sein, für die geplanten Bauvorhaben war jedoch eine Bodenverbesserung erforderlich.

2.4 Beurteilung der Verbesserung des Untergrundes nach einer Tiefenverdichtung

Der Erfolg einer Bodenverbesserung kann direkt nach der Durchführung der Arbeiten grundsätzlich nur nach einer Rüttelsdruckverdichtung durch Sondierungen durchgeführt werden. Bei Rüttelstopfverdichtung ergeben Sondierungen direkt nach der Durchführung der Arbeiten nur einen ungefähren Hinweis auf den Erfolg. Tatsächlich greifen in diesen Böden die Verbesserungsmaßnahmen erst während und nach der Belastung durch das Wechselspiel zwischen Kiessäulen und umgebenden Boden.

Abb.2. zeigt das Bodenprofil und die Resultate von Drucksondierungen vor und nach einer Rütteldruckverdichtung mit Zugabe von 10 % natürlichem Sand des verdichteten Volumens.

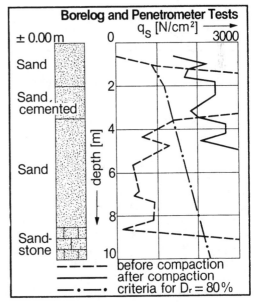

Abb.2.Resultate von Drucksondierungen vor und nach einer Rüttelstopfverdichtung.

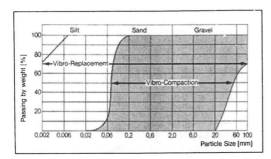

Abb.1. Korngrößenbereiche für Rüttelstopfverdichtung (Vibro-Compaction) und Rüttelstopfverdichtung (Vibro-Replacement).

Für eine Industrieanlage in einer seismisch aktiven Zone mußte der Untergrund, bestehend aus Sand und Feinsand mit Schlufflagen, verbessert werden. Es wurde eine Rüttelstopfverdichtung gewählt. Der Verdichtungserfolg kommt bei diesen Untergrungverhältnissen deutlich durch die Zunahme des Eindringwiderstandes zum Ausdruck, Abb.3.

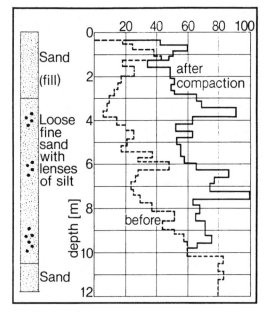

Abb.3. Resultate von Rammsondierungen vor und nach Verdichtungsmaßnahmen mit Tiefenrüttlern.

3. BEMESSUNG DER RÜTTELSTOPFVER-DICHTUNG

Für feinkörnige Böden hat Greenwood (1970) als Erster ein brauchbares Bemessungsdiagramm vorgestellt. Die Verbesserung oder die relative Setzung des mit Stopfsäulen verbesserten Bodens wird in Abhängigkeit vom Abstand der Stopfsäulen und der undränierten Scherfestigkeit dargestellt. Im deutschsprachigen Raum wird die Bemessung einer Rüttelstopfverdichtung in der Regel nach Priebe (1976, 1978) durchgeführt. Die Bemessungsdiagramme nach Priebe stimmen gut mit den Resultaten der Berechnungen nach der Methode der finiten Elemente von Baalam & Poulos (1977) überein (Kirsch, 1979).

Die Tragfähigkeit von einzelnen Kiessäulen in feinkörnigen Böden kann mit der undränierten Scherfestigkeit des anstehenden Bodens (c_u) und den Festigkeitsparametern der Kiessäule (c, φ) abgeschätzt werden, van Impe & De Beer (1983), van Impe et al. (1997). Auch für das Verformungsverhalten von mit Stopfsäulen verbesserten feinkörnigen Böden gibt van Impe (1989, 1997) einfache Lösungen an. Dazu ist neben den geometrischen Abmessungen der Stopfsäulen noch die Kenntnis der elastischen Parameter, die Kompressions- und Festigkeitseigenschaften des anstehenden Bodens erforderlich.

4. FALLBEISPIELE

In zwei typischen Fallbeispielen wird der erfolgreiche Einsatz der Rüttelstopfverdichtung einmal bei relativ homogenen und einmal bei sehr heterogenen Untergrundverhältnissen kurz beschrieben.

4.1 Bodenverbesserung für eine Kathetrale in Kroatien

In Karlovac ist eine Kathetrale zu errichten. Der Untergrund am Bauplatz besteht aus sandigen, tonigen Schluffen breiiger bis halbsteifer Konsistenz und locker bis mitteldicht gelagerten schluffig, kiesigen Sanden. Die Rammsondierungen ergaben Schlagzahlen von 0 bis 8 Schlägen pro 10 cm Eindringung. Die Zahl der Schläge bei der Sondierung mit der Standardsonde (SPT) lag zwischen 3 und 7.

Laut Bodengutachten des Planungsteam wurde der Untergrund als nicht geeignet beurteilt, die relativ hohen Lasten der Einzelstützen (bis zu 1200 kN) innerhalb der zulässigen Setzungsdifferenzen aufzunehmen. Es wurde eine Fundierung mit Ortbetonpfählen vorgeschlagen. Geoengineering wurde beauftragt eine Variante zum Gründungsvorschlag auszuarbeiten. Von Geoengineering wurde eine Rüttelstopfverdichtung mit einem Raster von 1,4 m x 1,1 m mit einer durchschnittlichen Säulenlänge von ca 8,5 m vorgeschlagen. In Abb.4 ist der Grundriß der Kathedrale mit dem vorgeschlagenen Stopfraster gezeigt. Die Rüttelstopfverdichtung war um ca. 30 % billiger als die Fundierung mit Bohrpfählen.

Da das Verfahren in Kroatien bisher noch nicht eingesetzt worden war, wurde vom Bauherrn eine Probeverdichtung gefordert. Dabei wurden die ersten Erfahrungen auf dieser Baustelle gesammelt und die Bauherrnschaft von der technischen und wirtschaftlichen Durchführbarkeit überzeugt.

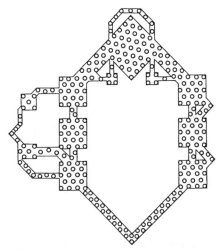

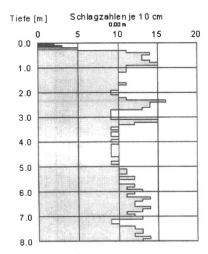

Abb.4. Grundriß und Austeilung der Stopfsäulen sowie der Fundamente der Kathetrale.

Abb.6. Resultate der Rammsondierungen nach der Herstellung der Rüttelstopfsäulen.

Es waren insgesamt 262 Stopfsäulen herzustellen. Die Arbeiten wurden in einem Zeitraum von 2 Wochen durchgeführt. Die Herstellung wurde anhand der Ampere - Tiefen - Schreiber überprüft. Weiters wurden nach Fertigstellung der Arbeiten Kontrollsondierungen mit der schweren Rammsonde durchgeführt. In Abb.5 und 6 sind typische Sondierdiagramme vor und nach der Rüttelstopfverdichtung dargestellt. Die Kontrollsondierungen wurden jeweils genau zwischen zwei Stopfsäulen durchgeführt. Aufgrund des relativ hohen Sandanteils konnte ein guter Verdichtungseffekt sofort nach

Fertigstellung der Stopfsäulen erzielt und durch Sondierungen nachgewiesen werden.

Weiters wurden in Abständen zwischen 5 m und 20 m vom Ansatzpunkt einer Stopfsäule Schwingungsmessungen durchgeführt. Die Messungen wurden während der Abteufung und während des Stopfens durchgeführt und zeigen, daß die resultierende Schwinggeschwindigkeit bei den gegebenen Untergrundverhältnissen Werte erreicht, die weit unterhalb von zulässigen Werten liegen, die Schäden an Gebäuden hervorrufen, Abb.7.

4.2 Bodenverbesserung für eine Lagerhalle in Österreich

Der Projektstandort liegt in Oberösterreich im

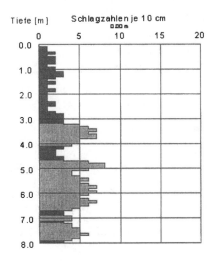

Abb.5. Resultate der Rammsondierungen vor der Herstellung der Rüttelstopfsäulen.

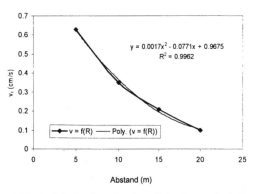

Abb.7. Maximalwerte der Schwinggeschwindigkeiten in verschiedenen Abständen vom Ansatzpunkt des Tiefenrüttlers.

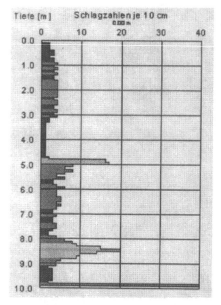

Abb.8.Resultate einer typischer Rammsondierungen mit Kaverne

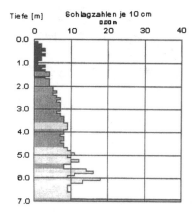

Abb.9. Resultate einer typischer Rammsondierungen im Bereich steifer Konsistenz

Bereich der s.g. Traun- Enns- Platte und der Untergrund wird von quartären, glazialen und fluvialen Ablagerungen und deren Verwitterungsprodukte gebildet.

Die Bodenaufschlüsse ergaben, daß unter einer gering mächtigen kiesigen Deckschicht (Anschüttung) ein unterschiedlich mächtiger Lehmhorizont ansteht. Das Liegende bildet im wesentlichen eine Konglomeratschicht, aber auch Schichten aus einem Gemenge aus Sanden, Kiesen und Steinen oder verlehmten Kiesen (jüngere Deckenschotter) wurden

angetroffen. Im Bereich der stellenweise angetroffenen Konglomeratpakete, welche erst im Bereich der gering verwitterten Deckenschotter größere Mächtigkeit aufwiesen, aber auch im oberflächennahen Verwitterungslehm angetroffen wurden, wurde örtlich auf Kavernen unterschiedlicher Größe gestoßen.

In Abb.8 und 9 werden typische Resultate von Rammsondierungen des anstehenden Bodens gezeigt. Die Heterogenität des Baugrundes kommt dadurch deutlich zum Ausdruck

Aufgrund der hohen Lasten der Einzelstützen (bis zu 1528 kN) und der Inhomogenität des angetroffenen Untergrundes in seiner Zusammensetzung, in seinem Zustand bzw. in der Lagerungsdichte und somit in seinem unterschiedlichen Setzungsverhalten, wurde vorgeschlagen, den Untergrund durch eine Rüttelstopfverdichtung zu verbessern. Konventionelle Setzungsberechnungen und Berechnungen nach der Methode der finiten Elemente haben ergeben, daß durch die Verbesserungsmaßnahme bei einer Belastung von 86 kN/m² bis 177 kN/m² für die Fundamente bzw. 45 kN/m² für den Hallenfußboden die Setzungen 3 cm nicht überschreiten werden und die Setzungsunterschiede im Bereich von 1 cm liegen werden. Höhere Werte wurden mit einer Berechnung nach Priebe (1993) ermittelt.

Die Anpassung der Verbesserungsmaßnahme an die Belastungs- und Untergrundverhältnisse wird hier als wesentlicher Vorteil der Rüttelstopfverdichtung angesehen. Der Verdichtungsraster kann entsprechend der vorhandenen Lasten und eine Fundierung direkt am gewachsenen Boden, da zum Teil keine ausreichende Grundbruchsicherheit gegeben ist und die Setzungsunterschiede mehr als 5 cm betragen können, den Untergrundverhältnissen angepaßt werden. Auch Abweichungen der Untergrundverhältnisse im Meterbereich können bei dieser Methode durch entsprechende Wahl des Rasterabstandes berücksichtigt werden.

Die Herstellung der Schottersäulen erfolgte mit einem Schleusenrüttler auf einer Tragraupe. Damit ist gesichert, daß das Zugabematerial zweifelsfrei bis zur Rüttlerspitze gelangt und lotrechte Säulen hergestellt werden.

Ausgeführt wurden Rüttelstopfsäulen mit einem Durchmesser von ca. 600 mm bei einem Durchmesser des Schleusenrüttlers von ca. 500 mm. Der Abstand zwischen den Rüttelstopfsäulen liegt im Fußbodenbereich zwischen 1,10 m und 2,50 m im Fundamentbereich je nach Fundamentgröße und abzutragenden Gesamtlast, zwischen 1,00 m und 1,95 m. Der vorgeschriebene Mindestabstand der Rüttelstopfsäulen von 1 m wird sowohl im Fußbodenbereich als auch im Fundamentbereich eingehalten.

Die Prüfung der Tiefenrüttelung ist durch Kontrolle der geometrischen Größen erfolgt (Abstand, Tiefe, Durchmesser der Kiessäulen,

Menge des Zugabematerials, Herstellungszeit der Säule, Leistungsaufnahme des Rüttlers mittels selbstschreibenden Tiefenschreiber), weiters durch Setzungsmessung während und nach der Lastaufbringung und durch Probebelastungen von Einzelsäulen und Säulengruppen.

Zur Beurteilung und zur Kontrolle der Säulenherstellung wurde ein Zeit - Weg - Diagramm und ein Zeit - Ampere - Diagramm herangezogen. Der Eindringwiderstand und der „Stopfwiderstand" wird durch Amperaufnahme beschrieben und wurde im Zuge der Probestopfungen für das gesamte Baulos festgelegt.

Für diese Baustelle wurden folgende Einbaukriterien festgelegt:

Ein Anstieg der Amperaufnahme läßt auf einen größeren Eindringwiderstand (Verdichtungsgrad, Konsistenz) des anstehenden Bodens oder des eingebauten Kiesmaterials schließen und ist daher ein gutes Kriterium für die ordnungsgemäße Herstellung der Stopfsäule. Selbst beim Abteufen können dadurch dichter gelagerte Schichten erkannt werden.

Kurz vor Erreichen der Endteufe, steigt der Stromverbrauch drastisch an und fällt dann wieder ab. Die geringe Ampereaufnahme resultierte aus dem geringen Widerstand des aufgeweiteten Loches. Der Rüttler hebt sich aufgrund seines goßen Spitzenwiderstandes durch die aufgewendete Energie selbst aus.

Das gleiche Phänomen läßt sich beim Stopfen erkennen, wenn der eingebrachte Kies genug verdichtet ist.

Hat der Rüttler die maximale Tiefe erreicht (das Trägergerät hebt teilweise vom Boden ab), wird dieser einige Male angehoben und wieder unter voller Aktivierung abgesenkt um wirklich sicher zu gehen das die tatsächlich maximale Tiefe erreicht ist. Danach erfolgt eine überwiegend aufwärts gerichtete Bewegung die mit der sogenannten „Stopfbewegung" des Rüttlers (auf- und abwärts) kombiniert ist, wodurch die Verdichtung des Kiesmaterials erfolgt.

Eine weitere Überprüfung der Säulenproduktion erfolgt über die Kiesmenge, die mit Hilfe des Rüttelsäulenprotokolles den einzelnen Tiefen zugeordnet und der Verdichtungsaufwand in der jeweiligen Tiefe ermittelt werden kann. Dadurch kann auch das Vorhandensein von Kavernen ersichtlich gemacht werden.

Wie zuvor angeführt kann zur Überwachung der ordnungsgemäßen Herstellung und Verdichtung der Stopfrüttelsäulen die Schreiberprotokolle, die den Eindringwiderstand, die Verdichtungsverhältnisse und die Kiesmenge beschreiben, herangezogen werden. Zu ordnungsgemäßen Überwachung werden Kontrollparameter an Produktionssäulen im Bereich der ausgewählten Versuchsflächen festgelegt.

Eine Abweichung der Kontrollparameter von den Einbaukriterien, muß aber nicht bedeuten, daß die Herstellung der Rüttelstopfsäulen nicht korrekt verlief. Dieser Umstand kann sich aufgrund der unterschiedlichen Bodenverhältnisse ergeben. Es werden daher zur Klassifizierung und Beschreibung der Rüttelstopfsäulen deren Kontrollparameter von den Einbaukriterien abweichen, 4 Prototypen festgelegt, die die geringe Ampereaufnahme bzw. die geringe Eindringtiefe erklären bzw. erläutern.

Zur Überprüfung der Tragfähigkeit und Setzungen des Fußbodens und eines Fundamentes wurden vier

Fundament - exzentrisch belastet

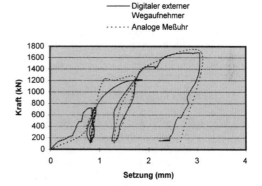

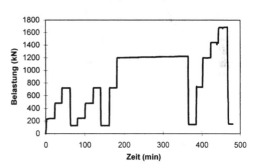

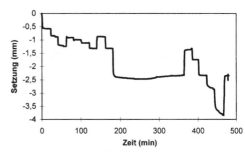

Abb.10. Kraft - Setzungskurve, sowie Belastungs - Zeit und Zeit Setzungskurve für den undränierten Fall.

Probebelastungen durchgeführt. Ein Versuch wurde auf der Rüttelstopfsäule und ein zweiter Versuch zwischen den Rüttelstopfsäulen durchgeführt. Die Belastungsversuche wurden im Fußbodenbereich durchgeführt. Die restlichen zwei Probebelastungen wurden auf einem Fundament durchgeführt.

Es wurden wie geplant die Belastung bis zu 140 %-Gebrauchslast von 63,0 kN/m² bzw. 1.680 kN durchgeführt. Die Tragfähigkeit der Kiessäulen war bei maximaler Belastung bei weitem nicht erschöpft. Die Setzungen waren überraschend niedrig, ca. 3 mm bei Maximallast. Abb. 10 zeigt ein charakteristisches Ergebniss der Probebelastung für den undränierten Fall mit kraftgesteuerten Belastung.

Prognosen für den Langzeit - Setzungsverlauf lassen erwarten, das die Kriech- oder Sekundärsetzungen, infolge des Abbaus von Schubspannungen, 60 % der durch die konventionelle Setzungsberechnung erhaltenen Werte erreichen werden. Für den Bettungsmodul wird eine Verbesserung um ca. 30 % erwartet.

5. SCHLUSZFOLGERUNGEN

Die vorhergegangen Fallbeispiele haben gezeigt, daß eine Bodenverbesseung mittels Rüttelstopfsäulen sowohl in homogenen Böden als auch in inhomogenen Böden eine technisch einwandfreie, umweltfreundliche sehr wirtschaftliche Methode im Vergleich zu anderen Fundierungsmaßnahmen darstellt. Durch die maßgeschneiderte Anordnung der Rüttelstopfsäulen im Stopfraster an die abzutragenden Lasten und die optimale Anpassung an die lokalen Untergrundverhältnisse zeigt dieses Verfahren wesentliche technische und ökonomische Vorteile.

Dies war auch einer der Gründe warum diese Art der Bodenverbesserung erstmals in Kroatien zur Anwendung gekommen ist. Vielleicht ist gerade mit diesem Projekt der Weg für diese umweltfreundliche Methode geebnet worden.

LITERATUR

Van Impe, W.F. 1989. *Soil Improvement Techniques and their Evolution.* Rotterdam: A.A.Balkema.

Mitchell, J.K. 1981. *Soil Improvement - State of the Art Report.* Stockholm. XI ICSMFE 4: 509-565.

Kirsch, K. 1979. *Erfahrungen mit der Baugrundverbesserung durch Tiefenrüttler. Geotechnik.* 1/79: 21-32.

D'Appolinia, A. 1954. Loose Sands - *Their Compaction by Vibroflotation. Spec. Techn. Publ.* ASTM 156.

Plannerer, A. 1965. *Das Rütteldruckverfahren. Mitt. Inst. Grundbau und Bodenmechanik.* TU Wien: Heft 6

Brown, R.E. 1977. *Vibroflotation Compaction of Cohesionless Soils. J.* Geot. Engng. Div. ASCE 103: 1437-1451.

Van Impe, W.F., F. De Cock, J.P. Cruyssen & J. Maertens 1997. *Soil Improvement Experiences in Belgium:* Part II. *Vibrocompaction and Stone Columns.* Ground Improvement 1: 157-168.

Biedermann, B. 1979. *Dynamische und statische Sonden und ihre praktische Bedeutung in der Bodenmechanik.* Symposium: Sondierungen und in situ Messungen, ÖFPZ Arsenal: 5-85.

Greenwood, D.A. 1970. *Mechanical Improvement of Soils Below Ground Surface.* Proc. Ground Engng. Conf. Inst. of Civ. Engeneers. London: 11-22.

Priebe, H. 1976. *Abschätzung des Setzungsverhaltens eines durch Stopfverdichtung verbesserten Baugrundes.* Die Bautechnik 53: 160-162.

Priebe, H. 1978. *Abschätzung des Scherwiderstandes eines durch Stopfverdichtung verbesserten Baugrundes.* Die Bautechnik 55: 281-284.

Priebe, H. 1995. *Die Bemessung von Rüttelstopfverdichtung.* Bautechnik 3: 183-191.

Balaam, N.P. & H.G. Poulos 1977. *Settlement Analysis of Soft Clays Reinforced with Granular Piles.* The University of Sidney. School of Engineers

Van Impe, W.F.& E.E. De Beer, 1983. *Improvement of Settlement Behaviour of Soft Layers by Means of Stone Columns.* Proc. 8th Europ. Conf. S.M. Found. Engng. Helsinki. Balkema 1:309-312

Geotechnical Hazards, Marić, Lisac & Szavits-Nossan (eds) © 1998 Taylor & Francis, ISBN 90 5410 957 2

Investigation of interaction of foundations with basement with the help of modelling installation

K. Sh. Shadunts, V. V. Ramensky & S. I. Matsiy

Kuban State Agrarian University, Krasnodar, Russia

ABSTRACT: Many hazards of the buildings and the constructions take place due to differential settlement of the foundations. The interaction of the basement being deformed with the foundation and the constructions plays an important role. The interaction of the system has been considered: the construction of the structures, the foundation and the basement on the installation which includes the elements which model the operation of the system components. The model of the basement is assembled with the help of a number of bellows with alternating rigidity being regulated. The load device includes the levers with the suspensions and the loads, and a frame with rigidity functions as the foundations and the part over the foundation.

1 INTRODUCTION

The normative documents of Russia suggest to carry out the calculations of the basements according to the deformations taking into consideration the conditions of the co-operative operation of the system: the construction and the basement. Such calculation should take into account rigidity of the foundations and the construction above the foundation and their ability to re-distribute stresses under non-uniform compressibility of the basement which, for example, take place due to local steeping. The analytical solution is rather complicated, especially when three-dimensional interaction of the system elements is considered.

We have worked out an installation for the investigation of a foundation with a basement including the models of a basement and a foundation, the load gauges and the measuring devices. "Structure (building) - foundation - basement" system model suggests a consideration of the mutual influence of the components which have their own rigidity and deformation characteristics which frequently differ greatly from one another.

During modelling it is impossible to take into account all the factors of the processes being investigated, that's why only the main factors determining the essence of re-distribution of the stresses are taken into account.

In the paper we have considered the results of modelling of the interaction of flexible foundations with the soil basement; the deformations of these foundations influence the stress-strain state of the basement when loading takes place; as a results of it, re-distribution of the efforts between the parts of the construction takes place.

For such type of the foundations the bearing capacity of the whole system is determined by rigidity and resistance of the constructions to the cracks, but not the deformations and bearing capacity of the basement which is characteristic for rigid foundations. Average pressure on the basement is often lower than its design resistance.

Besides, the deformation of structurally unstable (slump, swelling, thawing out soils, etc.) and heterogeneous (made grounds, underworking and karst soils, etc.) soils of the basements is of eventual (possible) character and is arranged on the contact surface. The deformation of larger part of the contact surface of the basement takes place in the linear stage, and local parts which differ greatly in their characteristics are deformed in the plastic stage and re-distribute power influences on the nearest parts of the basement.

In case of flexible foundations the stress - strain state inside the compressed thickness of the soil does not have the determining value, and "construction - foundation - contact surface of basement" system interaction should be considered.

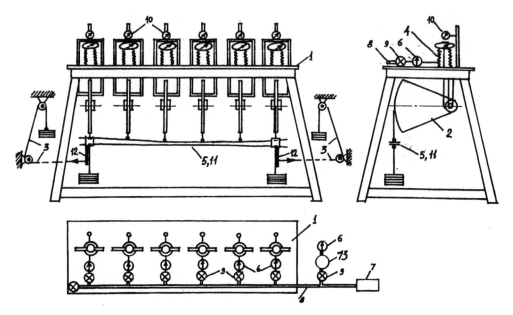

Fig. 1 General view of the device for modelling

2 DEVICE FOR MODELLING

A device for modelling has been worked out and developed at the chair of basements and foundations of the civil engineering faculty of the Kuban State Agrarian University (city of Krasnodar) (Shadunts, 1992). It helps to investigate various constructions which interact with the basement: beam and frame foundations of the buildings and the constructions (locks, pipings, spillways, sewage works, tanks, fragments of box - shaped foundations, etc.), buildings as a whole reduced to a beam with equivalent rigidity, piles undertaking horizontal loads, thin - wall retaining walls of the angle type, etc.

The general view of the device is given in Fig.1. The device consists of the foundation model installed on the operation table 1 in the form of bellows 2 with the mechanism of regulation of their rigidity including compressor 3, connecting pipes 4, pressure gauge 5, valves 6 and compensating capacity 7. The load device consists of levers 8 with suspensions 9 and loads 10. The model of foundation 11 can be made of a beam - strip (metal, wood, plastic, etc.) or of two strips with a regulated gap between them in order to change rigidity and can have the cantilevers for realization of various boundary conditions (bending moments and turning angles) 12 and load block fixtures 13.

The device has the following principle of operation. In bellows 2 having cylindrical form and corrugated side surface compressor 3 creates the air pressure which is measured with the help of pressure gauge 5 and regulated with the help of valves 6. The pressure value is chosen and distributed on the bellows in accordance with the planned distribution of basement rigidity in accordance with the conditions of similarity theory and the purposes of the experiment.

A load is transferred on each bellows via lever 8. With the help of load block fixtures 13 the moment loads are transferred which reproduce the peculiarities of the construction being investigated (for example, the influence of the supports of the frame foundation, bending moment or transverse force influencing the pile crown, etc.) and re-distribute the efforts between levers 8. Vertical displacements of the bellows are registered by displacement indicators 14, and the basement model response is registered by dynamometers 15 (if the bellows are calibrated, it is not necessary to install the dynamometers).

When modelling of the distribution change processes of the basement stiffness coefficient, for example, in case of slump soil steeping, air pressure is regulated in the bellows in accordance with the specified program.

3. SIMILARITY CONDITIONS

Rigidity choice of the bellows, the material, the gaps and the geometrical sizes of strips 11 as well as the efforts applied to levers 8 and load block fixtures 13 is determined on the grounds of the similarity theory from the criteria analysis of the equations of the deformation characteristics of the basement and the differential equation of a bend of a beam.

$$q = -EI \, d^4w/dx^4 \qquad (1)$$

$$q = dQ/dx \qquad (2)$$

$$q = d^2M/dx^2 \qquad (3)$$

$$\Theta = dw/dx \qquad (4)$$

The contact conditions can be represented by the most widely spread models of the soil basement:

Vinkler's model

$$S = q/Kb \qquad (5)$$

or linear flexible half-space model

$$S = q \, (1-v^2) \, \omega / E_0 \qquad (6)$$

The connection between K and E_0 can be represented as:

$$K = E_0 /(\omega \, F^{1/2} \, (1-v^2)) \qquad (7)$$

where q, Q, M - average pressure on the basement when the test with the help of a press tool takes place, linear reactive pressure of the basement, transversal force and bending moment in the beam;

Θ - turning angle of beam axis;

EJ - beam rigidity (E and J - modulus of elasticity of material and moment of inertia of cross section of the beam, respectively);

b, h - width and height of cross section of the beam, respectively;

x - axis of the points of the neutral axis of the beam;

w, S - deflection of the beam and settlement of the basement;

$E_0 \ v$ - modulus of deformation of elastic half-space (basement soil) and Poisson's ratio;

K - basement bed ratio;

ω - ratios without sizes depending on the form and area of charge.

The connection between the corresponding values of the phenomenon being investigated in nature (N) and the model (M) is determined by means of similarity ratios $C = M/N$.

Having accepted $C_b = C_x$, $C_v = 1$, and noting $C_w = C_S$, $C_N = C_Q$ we shall have the following similarity ratios:

$$C_q C_x^3 / C_E C_h^3 C_w = 1 \qquad (8)$$

$$C_q C_x / C_Q = 1 \qquad (9)$$

$$C_q C_x^2 / C_M = 1 \qquad (10)$$

$$C_\Theta C_x / C_w = 1 \qquad (11)$$

$$C_w C_k C_x / C_q = 1 \qquad (12)$$

$$C_w C_E / C_q = 1 \qquad (13)$$

C_b, C_x, C_h, C_N, C_q, C_Q, C_M, C_Θ, C_w, C_S, C_k, C_E, C_v are the scales of linear dimensions (width, length, thickness of a beam), concentrated loads, reactive pressure of the soil, force, bending moment, turning angle, beam deflection, basement settlement, bed ratio, modulus of deformation and Poisson's ratio of the basement, respectively.

The scale of lengths C_x is selected for the constructive reasons and is determined by the dimensions of a test rig. As w, S, q, M are the values determined in the process of the experiment, accuracy and convenience of the performance of the investigations will depend on the value of C_w and C_q greatly ($C_m = C_q C_q^2$; $C_w = C_S$).

When modelling of the basement is performed with the help of the system of the flexible supports, reactive pressure is also determined via the support settlement values

$$p = q/b = KS; \qquad q = KSb \qquad (14)$$

Here ii is more convenient to use the value of compliance of the supports which is inverse to the bed ratio value:

$$C = 1/KF \qquad (15)$$

$$Q = Sb/CF \qquad (16)$$

$$C_q = C_w C_x / C_c C_x^2 = C_w / C_C C_x \qquad (17)$$

As the ratios C_Q and C_m are determined when the values of C_q and C_x are known, it is enough to consider the dependencies (8, 11, 12) in order to ensure similarity taking into account (17).

The dependence (13) expresses the connection of the main deformation characteristic of the basement (E_0) determined during the engineering and geological survey with bed ratio (K).

The dependence (11) is used when the foundation construction operation is modelled taking into consideration the formation of fissures and the character of destruction. Using the relation (11) and the expressions determining the connection of the longitudinal deformation of the beam depending on the position in relation to the neutral axis (y) with curvature ($\mathit{æ}$)

$$\varepsilon_x = \mathit{æ}y; \qquad \mathit{æ} = d\theta/dx \qquad (18)$$

we shall get

$$C_\varepsilon = C_w C_h / C_x^2 \qquad (19)$$

Let us imagine three dependencies which provide similarity (taking into consideration the condition $C_N = C_q C_x$)

$$C_N = C_E C_h^3 C_w / C_x^2 \qquad (20)$$

$$C_C = C_w / C_N \qquad (21)$$

$$C_h = C_\varepsilon C_x^2 / C_w \qquad (22)$$

As it has been noted, the condition (22) is necessary when the formation of the fissures and the foundation construction destruction character are modelled which is required not always. If this requirement is not taken into consideration, the fulfilment of the correlation's (20) and (21) is enough.

4. MODELLING FACTORS CLASSIFICATION

All the factors of beam modelling on the flexible supports can be conditionally subdivided into constructive, main, derivative and auxiliary factors.

The constructive factors are conditioned by the purposes of the experiment and the construction of the test rig, and the preliminary linear dimensions (l, b) and modulus of deformation of the beam material (E) can be accepted.

The main factors characterize accuracy and convenience of the performance of the experiment: transversal deformations of the beam axis (w), settlement of flexible supports (S), compliance of flexible supports (C), concentrated loads (N), beam cross section height (h), longitudinal deformations of the external fibres of the beam (ε).

The derivative factors are determined analytically from the main ones: bending moments (M) and transversal forces (Q), linear distributed loads (q).

The auxiliary factors carry out the connection of the characteristics of flexible basement of the model and the nature: bed factor (K) and modulus of deformation of the basement (E_0).

Similarly one can differentiate the similarity factors. The latter can be selected in accordance with several schemes.

For example, when the preliminary selection of factors C_x (it is determined by the dimensions of the test rig) and C_E (metal, wood, plastic, reinforced cement, etc. are the most convenient materials for modelling) takes place, one of the factors is appointed as the main one and the rest factors are calculated with the help of variation of the factors which determine accuracy (C_w) or convenience of the experiment performance (C_N). The selection of the factors is considered to be the most expedient when C_w is preliminary selected proceeding from the required accuracy of the experiment performance. The nomographs are compiled for the convenience of the

selection of the values of the similarity factors depending on their combination (Fig. 2).

The use of a lever load fixture simplifies the performance of the experiments and widens the possibilities of modelling. Let us consider two principle schemes of modelling of the flexible foundations on the soil basement represented by flexible supports: a) the model of the foundation is in contact with flexible basement, b) the loads from the model to the flexible supports are transferred with the help of a lever, for example, with the relations of the arms 1:10.

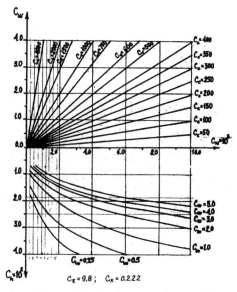

Fig. 2 Nomographs for the convenience of the selection of the values of the similarity factors

5. SCHEMES OF THE FOUNDATIONS MODELLING

When the settlements $S_1 = S_2$ and loads $N_1 = N_2$ are equal, the compliance of the flexible supports expressed by the dependence via bed factor

$$C_i = S_i / N_i = 1 / K_i b_i l_i \qquad (23)$$

will be as follows according to the specified schemes:

$$C_1 = S_1 / N_1 \qquad (24)$$

$$C_2 = S_2 / N_2 = 0.1\, S_2 / 10\, N_2 = 0.01\, S_2 / N_2 = 0.01\, C_1 \qquad (25)$$

$$C_2 = 0.01\, C_1 \qquad (26)$$

Table 1

Comparison of compressibility factors of flexible supports during modelling according to schemes (a) and (b)

Compressibility factors of flexible supports according to scheme (a)	Variants of possible compressibility factors of flexible supports according to scheme (b)					
	1	2	3	4	5	6
C_2	$0.01\,C_1$	C_1	C_1	$0.1\,C_1$	C_1	$10\,C_1$
S_2	S_1	S_1	$100\,S_1$	S_1	$10\,S_1$	S_1
N_2	N_1	$0.01\,N_1$	N_1	$0.1\,N_1$	$0.1\,N_1$	$0.001\,N_1$

It means that for modelling of the same process according to the scheme (b) an effort which is smaller by (a) factor of 100 or a support which rigidity is smaller by a factor of 100 is required; it simplified greatly the performance of the experiment or increases its accuracy. Several possible relations of compressibility factors of flexible supports according to two specified schemes which give the same results are given in Table 1.

In the general form the relation between the specified schemes of loading during modelling is expressed by the dependence:

$$S_2/C_2N_2 = 100\ S_1/C_1N_1 \qquad (27)$$

which demonstrates the advantage of the application of a lever load fixture during modelling (scheme b).

beam when the frame constructions are considered is carried out when equality of the bending moments and turning angles is achieved in the junction units of the frame post with the foundation beam when geometrical similarity takes place (the forms of nature and model are similar) and the corresponding unit of the device. Modelling of the boundary conditions for each concrete type of the construction is planned from the analysis of differential equations of bending of the rods which comprise to given construction. The correspondence of the boundary conditions is provided by the application of the calculated efforts to cantilever rods fixed to the foundation model. For example, in case of the frame foundation the boundary conditions are provided when a free edge of the cantilever rod is brought under the influence of a horizontal effort to a position which corresponds to its zero displacement.

6. APPLICATION

The advantage of the methods being suggested will tell on modelling of the constructions with high rigidity and extent, such as, for example, the buildings on the flexible foundation if they are considered as the beams with equivalent rigidity. According to the usual scheme of modelling (a) such an experiment could be performed on a test rig of the large dimensions with power facilities of a large power, it is expensive and laborious. The device being worked out allows to investigate such constructions on the small models using the usual weights for loading.

For the convenience of modelling of the constructions of flexible foundations which operate in the flexible stage a universal model of the beam is worked out made of two plates with a regulated gap and a fixed one (fixation is carried out with the help of the bolts). Rigidity of such a beam is expressed by the following equation depending on the size of the gap between the plates:

$$EI_{II} = 2EI_1[1+3(z/h+1)^2] \qquad (28)$$

where z - thickness, EI_1 - rigidity of one plate.

Modelling of the boundary conditions, for example, of the upper frame response on the foundation

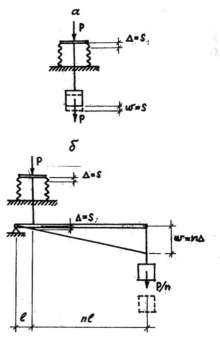

Fig. 3 Schemes of modelling

627

7. CONCLUSIONS

An installation for "soil basement - foundation – overhead construction of buildings and structures" system operation modelling allows to investigate a wide range of the tasks connected with the determination of stress strain state of the elements of the system if the analytical solutions are difficult to be obtained. The modelling methods being worked out are especially effective when the possible deformations of the buildings erected on the structurally unstable soils are evaluated under the local changes of their compressibility during operation.

Geotechnical Hazards, Marić, Lisac & Szavits-Nossan (eds) © 1998 Taylor & Francis, ISBN 90 5410 957 2

The excessive movements and sliding at the ore deposit at the Port of Koper

I. Sovinc
Ljubljana, Slovenia

ABSTRACT: In the Port of Koper (Slovenia) at the terminal of bulk cargo, the process of extensive movements and sliding was detected in 1986. After the review of the survey data of soil movements, field and laboratory investigations were carried out as well as in situ "inclinometer" and "pore water" measurements. The admissible height of the ore deposit was chosen, with control of the ore density (checking the compressibility and water content). The ore deposits have been in full exploitation since 1986, and no significant movements have been observed after the results of the stability calculations. The data are published here for the first time.

1 INTRODUCTION

There are two general rules that are to be taken into consideration when designing a foundation:
- settlement of foundation must be in permissible limits,
- the allowable bearing capacity must not be exceeded.

At Port of Koper the subsoil consists mainly of normally consolidated sea sediment, deposited there by rivers, streams and torrents. As far as grain size is concerned, various coherent and non-coherent materials are included, rocky stone and gravel as well as sand, silt and clay. The coarse grained sediments have been rounded and polished by the sea and its wave motion. The fine silty and clayey grains deposited in suspensions by fresh waters gradually subsided in a loose honeycomb and flake-like structure. Under pressure of the new sediments the layers of this hardly permeable silty-clayey sediments consolidated and compacted. However, as the natural increase in the load was gradual, the initial structure was more or less preserved. Thus a few meters thick layers of recent and very loose coherent materials have been formed in the process which is still continuing (Sovinc & Vogrinčić 1994).

Due to the changes that occurred in the beds of the rivers, streams and torrents as well as in their mouths, and owing to the changes in the waters flowing into the sea in different weather conditions, as well to the effects of the wave motion, layers of organic materials and sand intercalated between these porous and hardly permeable coherent materials. If these permeable intercalations reached the surface, they speeded consolidation. In places where was no possibility of such consolidation, the primary phase of consolidation has probably not been completed yet. The fine sandy sediments have in most cases settled rather loosely, particularly in places where grains of earth and silt as well as organic particles were mixed with them. The porosity of such loosely settled sand is often great, and their structure is unstable. This is the reason why the settlement of foundation on this sandy floor is greater than expected from experience we had with sand

The investigations carried out for various construction projects in Koper after World War II was rather extensive: a review of the available data has shown that it is Koper Bay where the silty-clayey sediments are the loosest.

The solid rocky substratum of the town of Koper and its environs consists of Eocene flysch, which can be on the town's outskirts seen on the surface. The space immediately above the solid flysch is covered with a layer of moulded flysch.

In 1982 the Port of Koper's management decided to build a bulk cargo terminal with a suitably strengthened surface layer and a possibility to transport ores from there by rail (position of tracks is shown in Figures 1, 2).

However, a year after the terminal was opened, some large movements of the railway tracks appeared as well as signs of slides. The latter occurred particularly due to the fact that the bulk

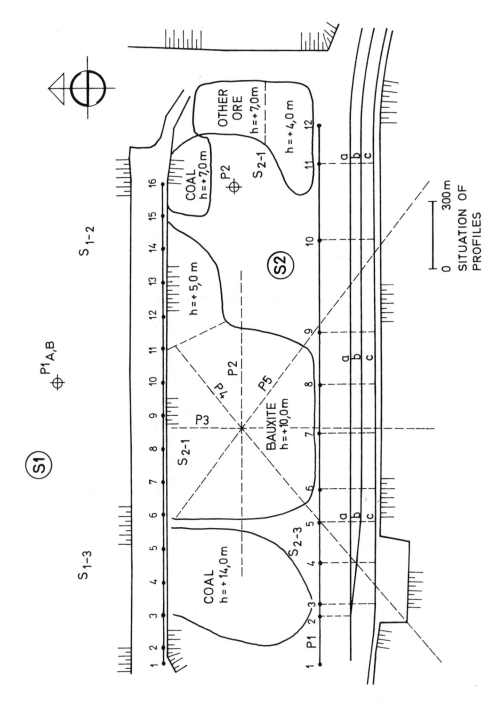

Figure 1. Site plane with locations of piezometers P1 and P2, inclinometers A, B and C, and measurement points.

density of ores shipped to Koper was underestimated and the compressibility was ignored and because even the height of the deposited ore had not been controlled

2 STABILITY OF ORE DEPOSITS

In the stability assessment the analyzed part of the ground is considered to be a rigid body or a number of such bodies. Failure lines are of different forms: straight, circular, logarithmic spirals, etc.

If the ground is homogenous and isotropic, then the rupture line usually occurs as a part of a circle. In natural conditions, a special attention must be devoted to the layers of either greater or smaller strength. In such cases, the presumed failure line must follow the boundaries between the layers. Where a solid layer occurs in the analyzed sections, it is advisable that the three-dimensional rupture lines are taken into account. The stability assessment can be made only when all the geotechnical characteristics of the ground are known. In the ground of low anisotropy, circular forms of rupture lines are usually chosen (Sovinc, 1987) and common analytical methods, so-called methods of slices, are used with the application of various computer programs (e.g. Bishop, Spencer, Janbu).

In the calculations of deformation of the ground or tracks in the immediate vicinity of ore deposits, the settlements occurring due to the weight of the pier were taken into account.

The changes in hydrological characteristics were decided to be monitored by recording the level of groundwater, while the movements of the surrounding soil were to be followed by inclinometers. Their number and distribution were given at longitudinal and transverse sections that must be made before ore deposits were built.

2.1 Control of stability for spirals failure lines

The ground stability under ore deposits was analyzed initially by choosing spiral failure surface. The position of the spiral was adjusted to the place and inclination of the cracks , the depth of the place where and the zones of the lowest shear strengths that were recorded, and the place of the expected dislodge edge. For the bulk density of bauxite with surface height of +20.00 m, the value of 15.3 kN/m^3 was accepted, as measured "in situ" which gave the pressure of 273 kN/m^2 . Assuming an average angle along the slide the safety factor obtained was 1 (F = 1 at zero cohesion). We repeated this analysis considering the pore pressures measured and found out that with the angle of internal friction $\phi'=12.5°$ the necessary cohesion, c', was 3.5 kPa.

2.2 Control of stability for combined rupture lines

Considering that the above described and analyzed spiral sliding surface run probably too deep in the area of TRT 86/3 boring, the section was analyzed also with the sliding surface, initially composed of a circular curve. For the rupture line selected in this way, we looked for unstable conditions with a number of parameters of shear strength, given in effective values. First of all we state the results of the analyses as per c' = 0, $\varphi' \neq 0$ method. The analyses were valid for such ore materials, the friction angle of which is 34° and which are separated, the same as was the head of bauxite at the time when some great deformations and slides occurred at the   ore deposits and railway tracks.

3 RESULTS

Figure 1 shows the situation with railway tracks a, b and c. The P1 and P2 were the places where pore water pressure was measured, and A, B and C were positions of the inclinometers. The state of the surface after rupture is presented. The height of the stored ore is given as well. Figure 3 shows the inclinometer readings at the inclinometer C. Figure 4 shows the horizontal movements in the inclinometers A, B and C. Loads on the ground are shown above. Figure 5 presents the movements of track at various time intervals.

4 CONCLUSIONS

On the basis of the results obtained by laboratory and field geotechnical investigations of the ground, as well as of the analyses of survey data and of geomechanical estimation of the conditions of stability and bearing capacity of the ground, we state herewith valid allowable soil loads as per separate parts of the ore deposits:
1) for the area of the northern ore deposits S1:
    $p_{allowable}$ = 153 kN/m^2 (100% planned load);
2) for the area of the southern ore deposits:
    sections 1 & 3:
    $p_{allowable}$ = 135 kN/m^2 (90% planned load);
    section 2:
    $p_{allowable}$ = 115 kN/m^2 (75% planned load).

Conditions under which the above stated loads could be allowed were:
1) Consent by the Water Management Institute Ljubljana.
2) Steady control of allowable load of the soil through regular measuring of bulk density (the compressibility and water content effect) and on

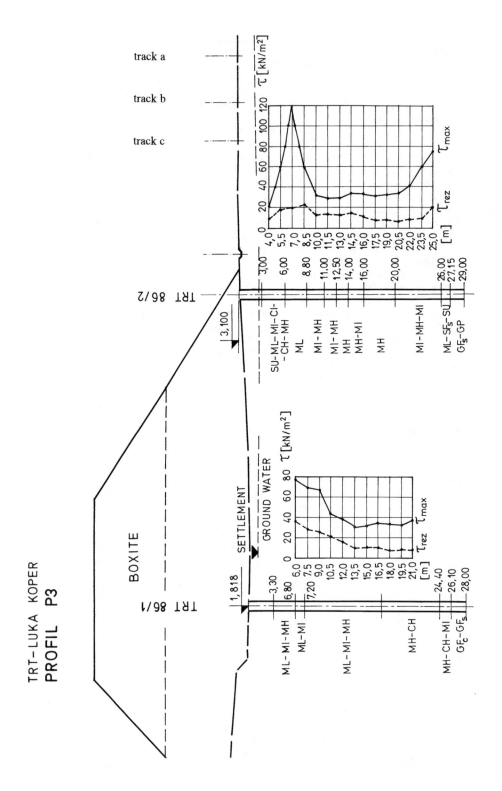

Figure 2. Profile 3 with characteristic soil profile.

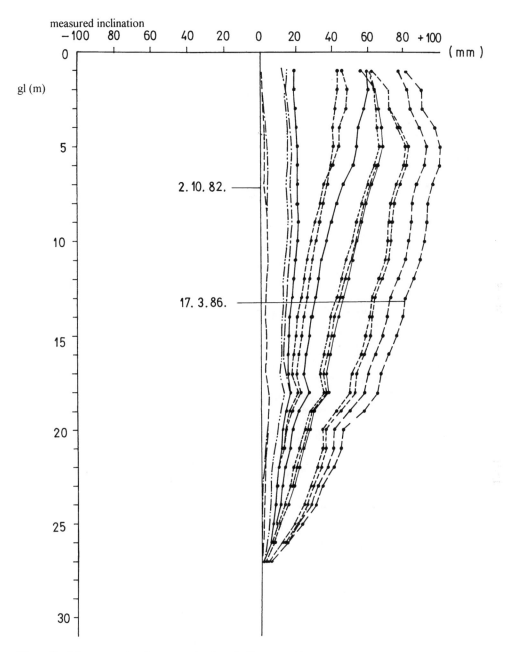

Figure 3. Measurements of movements along inclinometer C.

the basis of these data calculated allowable heights of deposited material.

3) Occasional geotechnical measurements at the terminal according to the program as given by Laboratory for Soil Mechanics at IMFM in Ljubljana.

4) At least half-yearly analysis of the data obtained by measurements as given under item 3. The aim of the analyses is to ascertain the time when the values stated herewith can be increased.

5) To deposit the loads as even as possible.

6) To prepare a project of improvement of deformed surface especially in the area of the deep rupture line.

VERTICAL SETTLEMENT OF TRACTION (a)

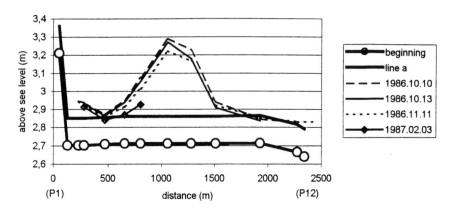

Figure 4. Vertical movements of railway track at a different time intervals

HORIZONTAL MOVEMENT - DEPTH 4 METERS

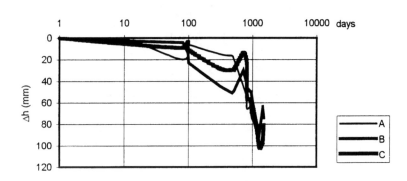

Figure 5. Horizontal movements in the inclinometers A, B and C.

7) To continue with the lengthy and long-term laboratory investigations of the soil samples, with the purpose as given under item 4.

ACKNOWLEDGMENT

This work was kindly supported by the Institute for Mathematics, Physics and Mechanics at the University of Ljubljana (IMFM). The author wishes to express his gratitude to Dr. G. Vogrinčić and to Dr. J. Liquor for their stability analyses.

REFERENCES

Sovinc, I. 1987. Excessive movements at ore depot TRT at Port of Koper. *Report. IMFM, University of Ljubljana (in Slovenian)*.
Sovinc,I., G.Vogrinčić 1994.Geotechnical Properties of Marine sediments from Koper Bay. *Proceedings of 13th ICSMFE, New Delhi*.

Geotechnical Hazards, Marić, Lisac & Szavits-Nossan (eds) © 1998 Taylor & Francis, ISBN 90 5410 957 2

Geosynthetics use in protection of exposed pipeline sections when crossing water barriers

Y. I. Spector & Y. D. Perezhogin
Ufa State Petroleum Technological University, Russia

ABSTRACT: The fork network of mains on the territory of Russia that are crossing the numerous water barriers produced a great number of underwater crossings. One of the main reasons of the underwater pipelines failure is river bed reforming. When repairing the exposed underwater pipeline sections , the flexible bracings based on geosynthetics use appear to be mostly fitted to bottom deformations. The investigations of filtration ability, strength, geosynthetics durability, deformability of protective coat models, rules of pipe models baring, influence of bracing models parameters upon baring protection efficiency were carried out. The results obtained are assumed as a basis for constructive decisions on repair of the 300 - 700 mm dia. gas lines crossing the rivers.

1 FILTRATION ABILITY

When protecting the exposed underwater pipeline sections from erosion by water and while repairing them, the use of geosynthetics as the basic material for constructive development of protection, in fact as the underlying material under the facing and in kind of shells for different fillers is increasing.

When using the geotextile for soil suffosion protection and drainage such material characteristics as filtration and water conducting ability as well as sufficient porosity that excludes its colmatage are taken into account.

The investigations results showed the filtration coefficient of not woven geotextile being in operation for a definite time and partly sand contaminated to be: $k=(1.0-5.0)*10^{-5}$ m/sec. The residual geotextile porosity $n=0.32-0.74$ provides the sufficient filtration for a long time.

The geotextile permeability and filtration coefficient depend, to a great extent, on the load normal to the material surface. The loading provokes the geotextile compression. The most intensive material compression is observed with loads up to 0.2 MPa. The further load increase insignificantly changes the hydraulic characteristics of the material. With the load up to 2 MPa the coefficient of the geotextile lateral filtration can be decreased by 2-2.5 times.

2 STRENGTH AND DURABILITY

It is known that the geotextile strength significantly decreases by the solar radiation influence. Experiments showed that the tensile strength of multifibrous woven and not-woven materials on the basis of polyesters, polyamides, polypropylenes and others, applied in exposed constructions, sharply decreases (up to 25-35%) during the first year of use, then decreases according to linear dependence and in 20 years reaches for some materials 20% of the original strength. The more durable are the geosynthetics on the basis of polyethylene and polypropylene, the less durable are polyester and polyamide geosynthetics.

In constructions protected from the solar ultraviolet radiation the geotextile is more durable. During the first year of its use the geotextile strength can be decreased by 10-17.5%. One of the factors influencing the durability of the geotextile is its possible deformation value in the process of use. The material "Terfil 2" was long-term investigated in conditions of creep and water influence.

The tests were carried out on test benches where the specimens were placed into special bathes filled with water, then the specimens were loaded with a long acting load that was 30, 40, 50% and etc. of the breaking load when short-term tensile testing. The test data analysis allows to make the following conclusions: the continuous loading of the materials investigated leads to the gradual deformations development. The process of deformations increase provokes the material failure. The initial load up to 80-90% of the material tensile strength produces the significant specimen deformation with its following failure during several minutes. The relatively low stresses do not produce the failure, however the deformations are continuously increasing up to their fixed stabilization. Figure 1 shows the deformations against time at different continuous loads (from 37 up to 75% of the breaking load). Creep test was carried out within 3600 hours (5 months), the specimens did not fail. The equilibrium state in the material for the specimens loaded up to 37% of the breaking load was achieved within 72 hours (curve 4). With the load increase much more time was needed for the equilibrium state achievement.

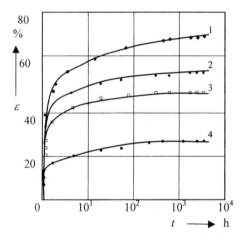

Figure 1. Geotextile deformation (ε) against time (t) at different loads: 1 - 75% of breaking load; 2 - 70% of breaking load; 3 - 60% of breaking load; 4 - 37% of breaking load.

In coastal protection structures the geotextile is subjected to the variable temperature, influence, especially in the zone of strengthening above the constructional water level. When geotextile laying near the day surface, for instance in combination with the thin reinforced concrete covering, the interval of the influencing temperatures can be within -40° and +40° C. In order to study the temperature influence on the geotextile strength the specimens were tested in the above interval.

The tests were carried out on the unit ZMGIT according to standard method. All tests showed strength minimums within the temperatures -10° C and +40° C.

Table 1. Results of geotextile chemical resistivity investigation .

Material name	Chemical reagent	Breaking load with the specimen width 5 cm, N	Strength variation,%
Material with lavsan polyester fibre prevailed	Salt solution	467	-16.4
	Kerosene	586	+4.8
	Acid medium	559	
	pH=5	531	-5.0
	Alkali medium		
	pH=9	584	+4.5
Material with carbon fiber prevailed	Salt solution	1574	+2.5
	Kerosene	1513	-1.0
	Acid medium	1529	
	pH=5	1429	-4.6
	Alkali medium		
	pH=9	1365	-10.7
Material from Polyethylene	Alkali medium	126 110	-12.6

The geotextile being under the water for a long time should be chemically resistant. The table above contains data of different materials strength variation.

3 COATS DEFORMABILITY

One of the most acceptable protection designs is the two-layered geotextile mat in kind of sewed together sections when joint intervals are filled with the stabilized soil. With the dense filling of the individual voids with the soil a system of shells close to cylindrical by form is formed. The shells divide the coat into separate sections.

A series of laboratory model tests was carried out with the aim of the filled shell deformability study. The shells with the diameter 88 mm made of geotextile "Terfil" on the basis of polypropylene and filled with the sand were used. Specimens of different length for two schemes corresponding to the real service conditions of flexible coat were tested, i.e. the scheme of shell supporting at two supports and the scheme of cantilever type bending. It was assumed that the tensile strength of the flexible shell depends on the longitudinal force T and its dead weight. The linear load per unit of the shell length was 88.2 N/m. In each test the length L and the maximum deflection h of the shell were measured (Figure 2).

For approximation of test data an equation of flexible fibre was used. The dependence, approximating test data, is the following:

$$h = a \cdot D \left[ch\left(b \cdot \frac{L}{D} \right) - 1 \right]$$

where h- shell maximum deflection, m; a, b-ratios depending upon the longitudinal force T, a =10.9; 13.7; 17.5 and b = 0.09; 0.07; 0.06 at T = 0; 30; 50 N respectively; D - shell diameter, m; L - shell length, m.

4 INVESTIGATIONS OF LOCAL EROSION BY WATER AND STABILIZATION OF THE BOTTOM IN THE UNDERGROUND PIPELINE RANGE

The tests were conducted on the hydraulic laboratory unit with a closed system of water circulation. The unit included a centrifugal pump, pressure and suction pipes, overflow tank. The experimental part of the unit was a rectangular tray on the bottom of which a levelled layer of sand was laid. The level of the tray filling was controlled with a set of plates serving as a weir. The water consumption was controlled with a valve in the pressure line and measured with a flowmeter. As a pipeline model a section of a plastic pipe with a diameter d=20 mm was used that was placed whether directly in the bottom or with the half diameter deepening (Figure 3). The most part of sand composition were the particles of small fractions (0.14 - 0.315 mm) - 38.7% and medium fractions (0.315 - 0.65 mm) - 59.9%.

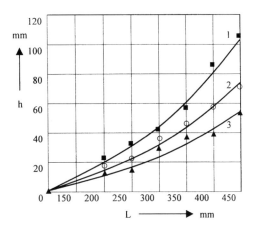

Figure 2. Graph of maximum deflection (h) against length (L) with the shell bending: 1 - T= 0; 2 -T= 30N; 3 - T= 50N.

The average diameter of sand particles was 0.38 mm. The tests were conducted with the following values of the flow hydraulic parameters: rate Q = 4.5 - 22.5 m^3 /h; depth H = 0.035 - 0.06 m; speed V = 0.18- 0.53 m/s; Reynolds number Re = 19600 - 9830; Froude number Fr = 0.094 - 0.48; H/d = 1.75 -3.0. The value of speed that did not lead to erosion by water, V_H, was used as the characteristics of soil stability against erosion.

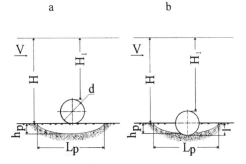

Figure 3. Scheme of the local erosion by water in the pipeline range bottom: a – non-deepened pipeline; b – half-deepened pipeline.

The investigations of the local erosion by water in vicinity of the underwater pipeline considered two characteristic conditions of the flow and soil base interaction.
- $V < V_H$, when there was no natural bottom erosion due to drifts displacements, and the local erosion

was determined only in the pipeline presence;
- $V > V_H$, when the local erosion of the bottom in the pipeline region was increased due to drifts displacement.

The tests of non-deepened and half-deepened pipes were conducted with the same hydraulic parameters, this provided their comparison. The analysis of test data showed the essential influence of the pipeline deepening l/d on the character of erosion. In case of half-deepened pipeline and with the relative flow velocity $V/V_H=0.96$ - 1.03 although the soil erosion by water under the pipe up- and down -stream was observed, however it appeared to be less than the value of deepening.

The tests confirmed the case of the bottom erosion in the region of non-deepened pipeline provided the flow speed was essentially less than the speed that did not lead to the erosion by water.

The increase of the flow average speed caused the increase of the bottom local erosion parameters in the pipeline region h_p/d, L_p/d, h_p/L_p despite the pipeline deepening. In case of half-deepened pipeline the parameters values were essentially less. The bottom erosion by water along the lateral sides of half-deepened pipeline, provided the hydrostatic pressure gradient being essentially less than its critical value, confirmed the conclusion of some researchers concerning the erosion by water up- and down-stream as the consequence of vortices appeared in this region.

The flow hydraulic parameters were described with the value $Fr_D\,H/d$, that considered the flow kinetic energy and the pipe relative dimensions. The most researchers characterize the bottom ridges that are formed when drifts displacement with the parameter Fr. The local erosion sink can be considered as a single ridge, the formation of which depends upon the pipeline presence.

Figures 4 - 6 show the results of test data processing, they are applicable when $V/V_H = 0.66$ - 1.03 and $Fr_D > 0.25\ H_l/H$. With H_l decrease the depth increases and the erosion length decreases. Test data analysis shows the decrease of erosion depth with the pipe deepening increase. In this case the relation h_p/H is characterized with more stable values compared with the relation h_p/d.

The test values of the erosion relative depth h_p/H in vicinity of pipeline appeared to be 1.5- 2.0 times more than the ridges height of the given relief.Thus, the increased flow turbulence because of pipeline availability in vicinity of the bottom provokes the increased erosion compared with the natural conditions. When the flow speed exceeds one leading to erosion by water, the role of the natural erosion is increased compared to the role of

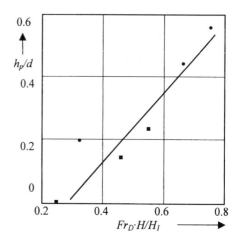

Figure 4. h_P/d dependence on H/H_l
• l/d=0;
▪ l/d=0.5

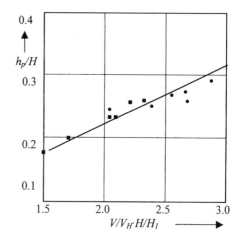

Figure 5. h_P/d dependence on $V/V_H·H/H_l$ when $V/V_H>1$
• l/d=0;
▪ l/d=0.5.

the pipeline itself.

The aim of the next series of experiments was the determination of the ways of protection against the local erosion in vicinity of the exposed pipeline with the help of the geotextile diaphragm that excludes the flow influence upon the bedding being eroded by water and prevents from loss of particles from under the pipeline. Three schemes of protection according to different problems that can occur

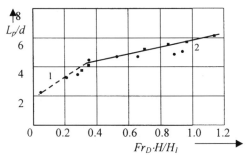

Figure 6. L_p/d dependence on $Fr_D \cdot H/H_l$:
1 – when $V/V_H < 1$;
2 – when $V/V_H > 1$.

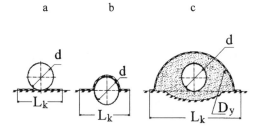

Figure 7. Schemes of bottom stabilization in the region of the local erosion in vicinity of pipelines with the geotextile application: a - scheme 1; b - scheme 2; c - scheme 3.

during the service life of crossings were considered (Figure 7):

Scheme 1 - the placement of the geotextile onto the bottom under the non-deepened pipeline in order to prevent the following erosion and deflection of the pipeline. The scheme can be applied when running repairs as a temporary measure for the pipeline stabilization;

Scheme 2 - the placement of the geotextile onto the top in the region of the half-deepened pipeline in order to stabilize the bottom marks and to prevent the following pipeline baring. The scheme can be applied in complex with the partial soil discharge as the more long-term measure compared to scheme 1;

Scheme 3 - the geotextile placement along the embankment near the exposed pipeline. The scheme can be applied both when running repairs and major maintenance of the crossing as a long-term measure of pipeline protection from the outside influences.

The aim of the experimental investigations was the evaluation of the protective geotextile coat critical length L_K, at which there was no local bottom erosion in vicinity of the pipeline. The investigations have been carried out with the hydraulic conditions used in former tests. The value L_K increased with the flow speed increase analogous to an increase of a sink length of the local erosion L_P for the non-deepened and half-deepened pipelines. Figures 8 and 9 show the experimental results for schemes 1 and 2. In all tests L_K appeared to be less than Lp.

When scheme 3 application, the length of geotextile coat depends essentially on the dimensions of pipe embankment.

For fully bare pipeline the height of the embankment should be not less than $h_0 = d + 0.5$ m according to Codes being currently in force. The protective geotextile coat should cover the whole surface of the embankment. Having taken conditionally its form as a circumference part with the diameter D_y, chord

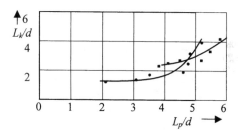

Figure 8. Charts of relations L_K/d and L_P/d dependence:
• $l/d = 0$;
▪ $l/d = 0.5$.

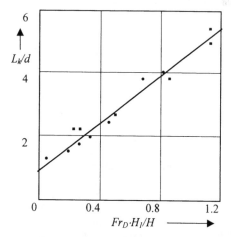

Figure 9. Charts of L_K/d dependence on $Fr_D \cdot H/H_l$ when $V/V_H > 1$;
• $l/d = 0$;
▪ $l/d = 0.5$.

639

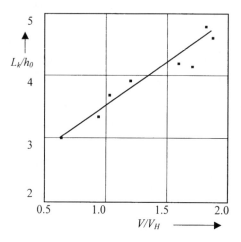

Figure 10. Dependence of L_K/h upon V/V_H .

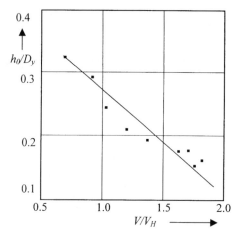

Figure 11. Dependence of h_0/D upon V/V_H .

length L_K and segment height h_0 when $d = 0.5$- 1.2 m, one can determine $L_K/h_0 = 4.0$ -5.2, $h_0/d = 0.15$ - 0.18.

Thus, the embankment can be conditionally considered as a part of some cylindrical fill with the diameter D_y that projects above the bottom surface up to the relative height $h_0/D_y = 0.11$ - 0.20. According to investigations data of soil erosion around the cylinder buried into the bottom, the local erosion is not observed when the height of the projected part is not less than $0.3D_y$. When tests conducting the relation L_K/d was changed from 3.6 to 5.7 and the relation L_K/h_0 was changed from 3.0 to 4.75 and increased with the flow speed increase (Figure 10). The corresponding values of the h_0/Dy relation were changed

with the flow speed increase from 0.3 - 0.26 when $V/V_H<1$ to 0.15 when $V/V_H = 1$-8 (Figure 11).

When designing some protective scheme for the real object one should be guided by the experimental values of the parameters Fr and V/V_H corresponding to the maximum values of flows that are forming the river beds at flood. As the analysis of hydrogeological conditions shows the relation V/V_H is less than 3.0 for the most plain rivers and accordingly the protective coat length L_K does not exceed $6,5h_0$ by scheme 3.

5 CONCLUSIONS

The experimental tests show the geosynthetics to have the sufficient filtration ability, strength, durability and deformability. The mechanism of the bottom local erosion by water in the area of the exposed pipeline is studied. The erosion by water takes place when the flow speed is essentially less than that not leading to erosion. Three constructive schemes are suggested of geosynthetics use for the protection against the local erosion by water in vicinity of the exposed pipeline. Recommendations on the geometrical dimensions of protective coats determination are given. The results of the investigations for the repair of gas lines with a diameter of 300-700 mm at river crossings are used. The results obtained allow to work out the technology of works, to organize test sections in situ with their subsequent regular investigation.

REFERENCES

Spector, Y.I., L.A. Babin & M.M. Valeev 1996. *New technologies in pipeline construction based on technical soil reclamation.* Moscow :Nedra.

Goldin, E.R. 1988. *Flexible protective coats with geotextile.* Moscow : Transport.

Earth reinforcement 1996. *Proceedings of the International Symposium , Fukuoka, Japan, 12-14 November 1996.* Rotterdam: Balkema.

Geotechnical Hazards, Marić, Lisac & Szavits-Nossan (eds) © 1998 Taylor & Francis, ISBN 90 5410 957 2

Tiefenverdichtungs-Methoden zur Gründung von Industriehallen in Ungarn

G. Strauch
Keller Grundbau Ges.m.b.H., Wien, Austria

Zs. Böröczky
Keller Grundbau Mélyépitö Kft Budapest, Hungary

ABSTRACT:Die Tragfähigkeit von Böden kann durch Tiefenverdichtungsgeräte wesentlich erhöht werden. Der Typ des angewandten Rüttelgerätes und das Verfahren hängt vom Boden und dem zu erreichenden Ziel ab. Die Anwendung eines umweltfreundlichen, die Konsolidation beschleunigenden, die Setzungen und das Erdbebengefahr vermindernden Verfahrens ist sowohl im Hoch- als auch im Tiefbau erfolgreich und wirtschaftlich. Dazu einige Beispiele aus Ungarn.

1. Die geotechnischen Gefahren

Tragfähiger Baugrund ist heute eine knappe Ware. Man wendet sich zu den Grundstücken, wo früher eine Bautätigkeit unmöglich war, da der Baugrund weich, das Grundwasser zu hoch oder die Erdbebengefahr zu groß ist.

Ist die Tragfähigkeit des Bodens ungenügend, bestehen für die zu errichtenden Bauten diverse Gefahren. Der Geotechniker hat die Aufgabe, diese Gefahren zu erkennen, den Boden zu untersuchen und geeignete, wirtschaftliche Methoden für eine Gründung vorzuschlagen. Eine dieser Methoden ist die Bodenverbesserung.

2. Bodenverbesserung mittels Tiefenverdichtung

Seit 1933 hat sich die Firma Keller Grundbau GmbH mit den Problemen der Baugrundverbesserung mittels Tiefenrüttler beschäftigt. Die Verfahren haben zum Ziel, durch Eindringen eines Tiefenrüttlers in den Boden dessen Tragfähigkeit zu erhöhen. Dabei können Böden oder Schüttungen großer Mächtigkeit mit oder ohne Grundwasser so verbessert werden, daß sie unter statischen oder dynamisch beeinflußten Belastungen sehr geringe und gleichmäßige Setzungen erleiden und gleichzeitig eine hohe Grundbruchsicherheit gewährleistet ist.

Der Tiefenrüttler ist ein schlankes, zylindrisches Gerät, das durch motorangetriebene Unwuchten quer zu seiner Längsachse Schwingungen erzeugt und infolge Eigengewicht, der Schwingungen und Aktivierung in den Boden abgesenkt wird.

Die Tiefenrüttelverfahren werden in 3 Gruppen eingeteilt: in Rütteldruckverfahren, in Rüttelstopfverfahren und in Sonderverfahren.

2.1 *Die Rütteldruckverdichtung*

Tiefenrüttler hängen am Seil eines Baggers oder werden an einer Tragraupe geführt und dringen infolge ihres Eigengewichtes, der Vibration und der beim Versenkvorgang verwendeten Wasserspülung an der Rüttlerspitze in den Boden ein. Dieser Versenkvorgang ist bei beinahe allen rolligen Böden einschließlich geringer Einlagerungen bindiger und organischer Schichten möglich.

Der Widerstand, den der Boden dem eindringenden Rüttler entgegensetzt, das Eigengewicht des Rüttlers, seine Schwingungsparameter und die Stärke der Wasserspülung sind die hauptsächlichen Variablen, welche die Versenkzeit beeinflussen.

Rollige Böden, die natürlich abgelagert oder künstlich geschüttet sind und nicht ihre größtmögliche Lagerungsdichte besitzen, werden unter dem Einfluß von Schwingungen mit Tiefenrüttler umgelagert. Gegenüber dem Ausgangszustand wird das Porenvolumen vermindert und dadurch Steifezahl und Reibungswinkel der behandelten Böden erhöht. Mit dem von Keller entwickelten Tiefenrüttler können Böden im oder außerhalb des Grundwassers bis in jede gewünschte Tiefe verdichtet werden. Die derzeit baupraktisch angewandten Tiefen liegen in der Regel zwischen ca. 3 und 20 m, in Einzelfällen werden bis zu 55 m erreicht.

Es wird beim Tiefenverdichtungsvorgang die Reibung der Bodenkörner zeitweise aufgehoben, wobei diese sich infolge der Schwerkraft zwanglos und spannungsfrei in eine sehr dichte Lagerung einordnen.

Die durch die Erhöhung der Dichte des Bodens

bewirkte Volumensverminderung führt zu einer Setzung der Geländeoberfläche. Die Zugabe von örtlich anstehendem oder angeliefertem Material gleicht die Verminderung des Porenvolumens aus. Die Zugabe einer gegebenenfalls im anstehenden Boden nicht vorhandenen Fehlkörnung während der Rüttelarbeiten ist möglich und dient zur weiteren Reduktion des Porenvolumens.

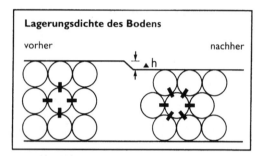

Lagerungsdichte des Bodens

vorher · nachher

Für die Tiefenrüttelung eignen sich grob- und gemischtkörnige Böden mit Steinen bis zu einem Korndurchmesser von 100 mm. Die Feinkörnigkeit des Bodens grenzt die verdichtbaren von den durch Rüttelstopfverdichtung zu behandelnden Böden ab.

Der Grenzgehalt des Bodens an Schluff- und Tonteilen, bei denen eine Umlagerung mit den geeigneten Rüttlern noch möglich ist, liegt im allgemeinen in der Größenordnung von ca. 3 %.

In der Regel ist eine Gefährdung bestehender Bauwerke durch Schwingungen des Tiefenrüttlers bei einem Horizontalabstand von 10 m nicht mehr gegeben. Bei ordnungsgemäßer Gründung und geringen Verdichtungstiefen (unter 10 m) kann der horizontale Abstand zwischen Rüttler und Gründungssohle des benachbarten Bauwerkes auf das Maß der Rütteltiefe unter Gründungstiefe des Nachbarprojektes verringert werden. Die Zulässigkeit der angegebenen Mindestabstände von bestehenden Bauwerken großer Schwingungsempfindlichkeiten muß durch Messungen oder sorgfältige Beobachtungen zu Beginn der Arbeiten überprüft werden. Im allgemeinen sind die Erschütterungen, die durch einen geeigneten Tiefenrüttler in den Boden eingeleitet werden, nicht vergleichbar mit den sehr viel stärkeren Erschütterungen, die beim Rammen von Spundwänden, Stahlträger oder Rohren aber auch bei Verdichtarbeiten mit Rüttelwalzen auftreten. Es gibt eine große Anzahl von Beispielen von Verdichtungen unmittelbar neben bestehenden Bauwerken.

Der Durchmesser der verdichteten Bodensäule liegt zwischen rund 1,5 m bei sehr feinkörnigem Sand und etwa 2,5 m bei leicht sandigem Kies. Durch Aneinanderreihen der Bodensäulen läßt sich jedes gewünschte Bodenvolumen mit hoher Lagerungsdichte herstellen.

Die nach dem Rütteldruckverfahren verdichteten Böden sind ohne Vorspannung umgelagert. Die erzielte Lagerungsdichte ist bleibend und wird auch nicht durch wechselnde Grundwasserstände oder dynamische Einwirkungen beeinflußt.

Dies wird gezielt bei Bauwerken mit hohen dynamischen Lasten eingesetzt (Erdbebengebiete, Papiermaschinen, Rotationsdruckmaschinen, etc.)

2.2 *Die Rüttelstopfverdichtung*

Bei der Ausführung der Rüttelstopfverdichtung wird in die vom Rüttler geschaffene Bodenöffnung abschnittsweise Grobmaterial (Schotter oder Kies) eingebaut und vom Rüttler verdichtet und verdrängt. Durch die Entwicklung des Schleusenrüttlers können heute in Verbindung mit speziellen Traggeräten Böden bis hin zu Weichböden technisch einwandfrei verbessert werden, wobei beim Schleusenrüttler das dem Baugrund zuzugebende Material über ein seitlich am Rüttler angebrachtes Materialrohr zur Rüttlerspitze geführt wird und dort unter Druckluftbeaufschlagung austritt. Das Traggerät kann dabei aktiviert werden, d. h. sein Eigengewicht kann zum Teil als vertikale Last auf den Rüttler aufgebracht werden. So ist es möglich, zusätzlich zur Rüttelenergie eine vertikale Kraft an der Rüttlerspitze einzuleiten und bei jedem Einfahren und bei jedem Stopfvorgang eine Vorbelastung auszuüben und eventuelle härtere Zwischenschichten leichter durchfahren zu können

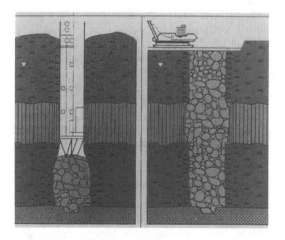

Ausgeführt wird dieses Verfahren der Baugrundverbesserung für praktisch alle vorkommenden Grundbauaufgaben in Hoch- und Tiefbau, Verkehrswegebau, Industriebau, Wasserbau, etc.

Vorteile des Rüttelstopfverfahrens:

Da die Baugrundverbesserung durch Rüttelstopfverdichtung in nahezu allen Bodenarten bis in große Tiefen auch unterhalb des Grundwasserspiegels anwendbar und ausführbar ist und damit die erdbautechnischen Anforderungen an den Untergrund auch bei tiefer liegenden feinkörnigen Böden mit ggf. erforderlichen Verbaumaßnahmen und Grundwasserhaltungen vorgenommen werden kann, stehen schnelle, wirtschaftliche Möglichkeiten zur Verfügung, die sich den jeweiligen örtlichen Gegebenheiten (z. B. Tieflage und Mächtigkeit) schnell und flexibel anpassen können.

Durch entsprechende Rasterung (Abstand der Rüttelstopfsäulen in Quer- und Längsrichtung) kann das System der Baugrundverbesserung in jedem Fall den einzelnen Anforderungen angepaßt werden.

Durch die Baugrundverbesserung mittels Rüttelstopfverdichtung werden in erdbau- und grundbaulicher Hinsicht folgende Verbesserungen erreicht:

- Verdichtung des Untergrundes und damit Erhöhung der Lagerungsdichte,
- Verdübelung des Untergrundes mit den tieferen tragfähigen Bodenschichten,
- Verbesserung der Scherfestigkeit des Untergrundes zur Erhöhung der Standsicherheit,
- Masseerhöhung des Untergrundes und dadurch Verbesserung seines Schwingungsverhaltens.

2.3 Die Sonderverfahren

Seit der Entwicklung von Tragraupe und Schleusenrüttler in den 70er Jahren stehen für die Baugrundverbesserung durch Stopfverdichtung und die abgeleiteten Verfahren, die Vermörtelten Stopfsäulen und die Betonrüttelsäulen, standardisierte Geräte und eine standardisierte Ausführung seit über 20 Jahren zur Verfügung.

2.3.1. Vermörtelte Stopfsäulen und Fertigmörtelstopfsäulen

Bei organischen Zwischenschichten im zu verbessernden Bodenbereich oder in sehr weichen, breiigen Böden, die auf Dauer keine seitliche Stützung erwarten lassen, kann die äußere Stützung des Säulenmaterials durch einen innere Bindung des Stopfmaterials ersetzt werden. Dies wird durch Zugabe einer Zementsuspension während der Herstellung ermöglicht, die dann mit dem Zuschlagstoff einen Verbundkörper bildet. In Fortentwicklung dieses Verfahrens wird heute überwiegend ein vorgefertigter Mörtel (Kiesbeton) über den Schleusenrüttler im Baugrund verstopft (Fertigmörtelstopfsäule).

Die Anwendung, Herstellung und Bemessung von Vermörtelten Stopfsäulen und Fertigmörtelstopfsäulen ist über eine allgemeine bauaufsichtliche Zulassung des Institutes für Bautechnik, Berlin, geregelt.

2.3.2. Betonrüttelsäulen

Eine sehr leistungsfähige Variante zu den Vermörtelten Stopfsäulen sind die Betonrüttelsäulen, bei denen Pumpbeton anstelle von Zugabematerial mit Suspension in dem vom Rüttler erzeugten Raum verpumpt und so ein pfahlartiger Gründungskörper im Baugrund erstellt wird.

Die Grundgeräte sind ein Tiefenrüttler mit seitlichem Betonrohr mit Anschluß an eine Betonpumpe sowie ein Traggerät mit Aktiviermöglichkeit. Der Tiefenrüttler wird bis zur vorgesehenen Tiefe eingefahren. Nach Erreichen der Endtiefe wird begonnen, Beton über das Zugaberohr zu verpumpen. Gleichzeitig mit dem Betonaustritt werden durch Anheben und Wiederversenken des Rüttlers Stopfvorgänge im Fußbereich vorgenommen, die dazu dienen, einen Fuß auszustopfen und bei Kiessanden auch die Lagerungsdichte zu erhöhen. Danach wird der Rüttler unter weiterem Verpumpen von Beton gezogen.

Die Betonrüttelsäulen werden unarmiert hergestellt und unmittelbar unter den Lastbereichen angeordnet, wobei der kleinste Pfahlabstand ca. 1,50 m nicht unterschreiten sollte. Die Betonrüttelsäulen sind in der Regel Aufstandspfähle, die infolge des Fußes und der Verdichtung tragfähiger Schichten bei geringer Einbindung hohe Gebrauchslasten gestatten. Lasten bis über 800 kN sowie Einbau von Anschlußbewehrungskörben ist möglich.

2.3.3. Kombinierte Verfahren mit "Teilvermörtelten Stopfsäulen"

Teilvermörtelte Stopfsäulen bestehen aus einer Kombination von Rüttelstopfsäulen im Bereich der grobkörnigen Böden und vermörtelten Stopfsäulen im Bereich der Weichschichten.

Für die Tragfähigkeit der Rüttelstopfsäule ist die Stützkraft des umgebenden Bodens erforderlich. Wenn die Stützkraft des umgebenden Bodens (z. B. bei organischen Weichschichten) durch Entwässerung oder Verrottung abnimmt, kann sich die Rüttelstopfsäule unter Auflast verformen und in den umgebenden Boden ausweichen, wodurch Setzungen entstehen können (fehlender Umschließungsdruck).

Um diese Verformungen zu vermeiden, wird bei dem System "Teilvermörtelte Stopfsäule" die Rüttelstopfsäule in diesen Bodenschichten verfestigt, indem während des Einbaus des Kieses entweder eine Zementsuspension zugegeben oder ein stampffähiger Beton als Zugabematerial (Fertigmörtelstopfsäule) eingebaut wird. Oberhalb der Weichschichten wird dann die Rüttelstopfsäule bis zur Arbeitsebene fortgesetzt.

Zur Ableitung der Spannung in den tragfähigen Untergrund unterhalb der durch vermörtelte Stopfsäulen verbesserten Bodenbereiche wird ein verdichteter Kiesfuß zur Vergrößerung der Aufstandsfläche und zur Erhöhung der Tragkraft ausgebildet.

Dieses Verfahren eignet sich besonders für Verkehrsdämme in weichen Bodenzonen.

2.4. *Die Geräte*

Die Tiefenrüttler verdrängen beim Einfahren in den Baugrund den Boden durch horizontale, umlaufende Schwingungen im Bereich ihrer Spitze. Die Keller'sche Tiefenrüttler sind mit Elektro-Motoren ausgerüstet, die eine Unwucht mit 1800 UpM bis 3600 UpM antreiben. Ihre Schlagkraft reicht von 131 bis 221 kN. Die Schwingweiten betragen zwischen 4,6 und 18,5 mm.

Vibrator	Amperaufnahme kW	Drehzahl min-1	Schlagkraft kN	Amplitude mm	Gene ratio
T	35-50	3000	150	4	1
M	50	3000	160	7	2
A	58-80	3000-2000	160	14	2
S	120-150	1800	220-290	16-25	3
L	100	3600	200	5	3

Der für eine Tiefenverdichtung ausgewählte Rüttler soll
• den zu verbessernden Boden mühelos durchfahren,
• rollige Böden optimal verdichten und
• bei bindigen Böden das Einbaumaterial zuverlässig einbringen und in den Boden schlagen.

Angesichts der verschiedenen Bodenarten und ihrer Reaktion auf die Rüttlerschwingungen müssen mehrere Rüttlertypen und Verfahrensweisen zur Auswahl stehen.

3. Tiefenverdichtung gegen Erdbebengefahr

Beim Einsatz von Tiefenrüttlern wird der Boden gezielt Schwingungen ausgesetzt, um die Bodenteilchen umzulagern und damit eine größere Lagerungsdichte zu erreichen. So gesehen sind Böden, die keine entsprechende Wirkung zeigen und deshalb mit Stopfverdichtung verbessert werden, nicht empfindlich in bezug auf Erdbebenerschütterungen.

Der Verflüssigungswiderstand eines Bodens hängt hauptsächlich von der Kornverteilung ab. Robertson und Campanella geben für kritische Sande mit D_{50} (mittlerer Korndurchmesser) Beziehungen zwischen dem zyklischen Spannungsverhältnis und dem normierten Spitzendruck der Drucksonden an.

Durch Stopfverdichtung kann die Gefahr einer Verflüssigung des Bodens erheblich vermindert oder sogar völlig gebannt werden. Bisher gab es jedoch keine Ansätze für eine Bemessung.

Seed und Booker untersuchten, inwieweit die Drainfähigkeit der Stopfsäulen die Gefahr der Verflüssigung herabsetzt. Dementsprechend basieren die Ansätze auf Formeln für Drains. Die ermöglichen eine Bemessung in Abhängigkeit von der Durchlässigkeit und damit indirekt von der Feinkörnigkeit des behandelten Bodens. Die Durchlässigkeit darf allerdings nicht zu gering sein. Durch diese Einschränkung fallen weitgehend die Böden aus der Bemessungsmöglichkeit heraus, in denen Stopfverdichtung vorgesehen wird, weil die Rüttlerschwingungen allein sie nicht ausreichend verdichten.

Neben der Drainwirkung ist bei einer Stopfverdichtung die Bodenbewehrung durch Säulen mit einem hohen Reibungsanteil von ausschlaggebender Bedeutung. Die sich daraus ergebende Bodenverbesserung im Hinblick auf Verflüssigung ist dann besonders günstig, wenn sie von einem Zugabematerial getragen wird, welches absolut erdbebenunempfindlich ist.

Das von Seed und Idriss eingeführte Bemessungsverfahren anhand von Feldversuchen, welches von Robertson und Campanella auf die zuverlässigeren Drucksondierungen abgestimmt wurde, beruht auf dem Vergleich des zyklischen Spannungsverhältnisses, welches im Erdbebenfall möglicherweise auftritt, mit dem, welches auf Grund vorhandener Sondierwiderstände vermutlich vom Boden ertragen wird. Das zyklische Spannungsverhältnis wird dabei aus der mittleren Scherspannung aus zyklischer Belastung in einem horizontalen Schnitt im Boden und der anfänglich wirksamen Überlagerung in eben diesem Schnitt gebildet. Die anfänglich wirksame Überlagerung hat unmittelbar keine physikalische Bedeutung, sondern relativiert den Bezug zur Tiefe und liefert im übrigen einen dimensionslosen Verhältniswert. Die mittlere Scherspannung ergibt sich aus der bewegten Masse, im Normalfall also aus der gesamten Überlagerung.

Die Wirkung einer Stopfverdichtung beruht auf Lastumlagerung. Auf Grund ihrer größeren

Steifigkeit übernehmen die Säulen im Belastungsfall erheblich mehr Last als ihrem Flächenanteil entspricht. Eine derartige Umlagerung wird schon während der Säulenherstellung durch seitliche Verdrängung von Zugabematerial erzwungen.

Im Erdbebenfall wird vor Beginn einer Verflüssigung die maximal mögliche Umlagerung erreicht; denn Verflüssigung kann nur von den Feldmitten des jeweiligen Rasters ausgehen, weil die Stopfsäulen aus grobkörnigem Material nicht verflüssigen und der umgebende Boden wegen der Drainung zu den Säulen hin nicht dazu neigt.

Die Sicherheit, die eine Stopfverdichtung gegen Verflüssigung gewährt, hängt also davon ab, wieviel Last von den Stopfsäulen übernommen wird. Der verflüssigungsgefährdete Boden dazwischen wird nur noch von der Restlast beansprucht. Es genügt somit, für diese Restlast den Nachweis nach dem Verfahren von Seed und Idriss bzw. von Robertson und Campanella zu führen.

Von Priebe wurde der Lastanteil ermittelt, den die Stopfsäulen übernehmen. Die gesuchte Restlast auf dem Boden zwischen den Stopfsäulen ergibt sich aus einer einfachen Umformung der von ihm angegebenen Formeln. Zur Berücksichtigung der durch die Stopfverdichtung erzielten Verbesserung wird die maßgebliche Scherspannung und damit das zyklische Spannungsverhältnis, welches möglicherweise zu Verflüssigung führt, entsprechend reduziert.

Die Rüttelstopfverdichtung ist besonders in erdbebengefährdeten Gegenden eine geeignete Gründungsmaßnahme, weil sie einerseits eine gewisse Flexibilität besitzt und andererseits Bodenverflüssigung weitgehend verhindert. Die stabilisierende Wirkung beruht auf dem Reibungswiderstand der Stopfsäulen, die einen hohen Anteil der äußeren Lasten und des Bodeneigengewichts übernehmen, und auf ihrer Fähigkeit, Porenwasserüberdruck im Boden - zumindest in ihrer unmittelbaren Umgebung - augenblicklich abzubauen.

4. Bemessung der Stopfverdichtungen

Die Bemessung der Rüttelstopfverdichtung wird durch die Setzungsabschätzung und einer eventuellen Grundbuchuntersuchung überprüft, wobei Berechnungsansätze und Methoden bei Priebe (1988) eingehend beschrieben sind.

Die Bemessungsmethode bezieht sich auf die verbessernde Wirkung von Stopfsäulen in einem Baugrund, der ansonsten gegenüber dem Ausgangszustand unverändert ist. Zunächst wird dabei ein Wert oder ein Faktor ermittelt, um den die Stopfsäulen den Boden gegenüber dem Ausgangszustand verbessern. Um diesen Verbesserungswert erhöht sich z. B. rein rechnerisch die Steifezahl des unbehandelten Bodens oder

vermindert sich eine für den unbehandelten Boden ermittelte Setzung. Alle weiteren Bemessungsschritte greifen letzten Endes auf diesen Ausgangswert zurück.

5. Beispiele für Tiefenverdichtungsanwendungen in Ungarn

Seit 1982 ist Keller in Ungarn präsent. Diverse kleine und mittlere Projekte wurden mit Erfolg ausgeführt. Die bedeutendsten Großprojekte, bei denen das Stopfverdichtungsverfahren eingesetzt war, sind der Flughafen Budapest Ferihegy 2, die Opel-Werke in Szentgotthard und das Einkaufzentrum Cora in Törökbálint. Nachstehend wird über die jüngsten dieser Großprojekte berichtet.

5.1. OPEL Szentgotthard

Die Adam Opel AG plante im Werk Szentgotthárd den Neubau einer Produktionshalle. Wegen der später in der Halle arbeitenden Hochpräzisionsgeräte war es nötig, einen stets setzungsarmen Baugrund herzustellen und dafür das geeignete Gründungskonzept zu bestimmen.

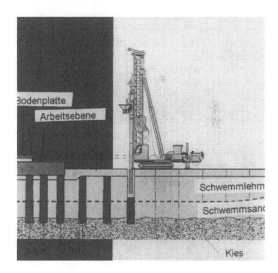

Das Baugelände liegt am Westrand des Westpannonischen Beckens. Der Schichtenaufbau besteht aus Schwemmlehm des Flusses Raab, unterlagert von Schwemmsand und Kies. Die Stützenfundamente wurden auf Vermörtelten Stopfsäulen gegründet, die in die sehr tragfähigen Kiese reichen. Die Bodenplatte wurde auf einem Flächenanteil von ca. 70%, auf dem die Mächtigkeit des Schwemmlehms mindestens 3 m betrug, ebenfalls über vermörtelten Schottersäulen im Kies gegründet. Im restlichen Teil erfolgte eine Flachgründung nach Austausch von Teilen des Schwemmlehms gegen lagenweise eingebrachten und mit Oberflächenrüttlern verdichteten Kiessand. Die Abgrenzung der beiden Methoden zur Gründung des Hallenbodens erfolgte während der Ausführung durch den Baugrundsachverständigen. So war die Optimierung von Ausführungszeit und Wirtschaftlichkeit unter Beachtung der Setzungskonformität möglich. Der Gutachter stützte sich auf umfangreiche Baugrunduntersuchungen und die Aufzeichnungen der laufenden Bodenverbesserungsmaßnahmen.

Die Ausführung

Der gesamte Bereich der Stützenfundamente und 70 % des Hallenbodens wurden auf ca 3300 Stück vermörtelte Stopfsäulen gegründet. Für das Arbeitsplanum und den späteren Bodenaufbau wurden ca 1,3 m der bindigen Bodenschichten abgetragen und durch Kiessand ersetzt.

Technische Daten der Herstellung der vermörtelten Stopfsäulen:
- Einbauraster im Bereich der Bodenplatte ca 2,0 x 2,0 m
- Säulenlängen bis 7 m
- Zulässige Belastung je Säule 350 kN

Die Kontrolle: Automatische Amperetiefenschreiber zeichneten für jede Säule Versenktiefe, Energieaufnahme und Herstellzeit auf. Der Suspensionsverbrauch wurde mittels Durchflußmessung ermittelt, die Splitteinbaumenge pro Säule protokolliert.

Zur Ausführung wurden 2 Geräte eingesetzt, die die Leistung in Tag- und Nachtschicht in 7 Wochen zur vollsten Zufriedenheit des Bauherren, der Bodengutachter und des Auftraggebers erbracht haben.

5.2. CORA Törökbálint

In Törökbálint, einer westlichen Nachbargemeinde von Budapest wird auf einem 93 ha großen Areal die Shopping City Budapest in mehreren Stufen realisiert. CORA, einer der größten Hypermarketbetreiber in Frankreich baute seinen ersten ungarischen Supermarket in Törökbálint auf.

Die 380 m x 120 m große Halle aus Stahlbetonfertigteilen wurde auf einem stark sackungsgefährdeten Boden gegründet. Die Bodenverhältnisse wurden aus 10 Bohrprofilen und etlichen Sondierungen ermittelt. Da die einzelnen Bodenarten in unregelmäßiger Form vorlagen, teilte man sie in 3 Schichtgruppen ein:
• Löß - Deckschicht in verschiedener Dicke und Mächtigket. Der Boden neigt über dem Grundwasserspiegel zum Einsacken, unter dem Wasser ist er locker.

• Mittlere und fette Tone in unregelmäßiger Lage und Tiefe. Sie kommen sowohl über als auch unter Grundwasser vor, es bilden sich Linsen, ihre Konsolidation dauert lange.
• Nichtbindige Schichte, deren Lage kaum nachvollziehbar ist.

Der Grundwasserhorizont wurde bei ca 8 m unter Gelände erbohrt.

Die Belastungen: Einzelstützen bis 1300 kN
                     Fußbodenplatte   20 kN/m2

Das gewählte Gründungsverfahren war die Keller'sche Rüttelstopftechnik. Die Lasten der Einzel- und Streifenfundamente wurden durch Schottersäulen in den tragfähigen Baugrund abgeleitet. Der unterschiedlichen Lage des tragfähigen Bodens entsprechend wurden die Schottersäule bis in die erforderlichen Tiefen (bis zu 13 m) hergestellt. Die Tragfähigkeit der einzelnen Säulen wurde auf 260 kN ausgelegt. Für die Einzelfundamente wurden 3 bis 5 Schottersäulen hergestellt. Aufgrund der unterschiedlichen Bodenverhältnisse wurde der Fußboden ebenfalls mit Schottersäulen gegründet.

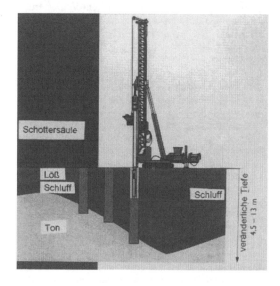

Die Setzungsberechnung ergab Setzungsdifferenzen zwischen Einzelfundamenten und Fundamentplatte unter 2 cm.

Die über 4.000 Schottersälen wurden mit Keller'schen Schleusenrüttlern und Tragraupen hergestellt. Zur Ausführung wurden bis zu 3 Geräteeinheiten eingesetzt, die die Leistung in Mehrschichtbetrieb in knapp 9 Wochen erbracht haben.

Die Qualitätskontrolle der Stopfverdichtung umfaßte die Leistungen:

- Ausführungsplan mit Punktnummern, Dokumentation, Vermessung, tägliche Berichte.
- Kontinuierliche Aufzeichnung und Kontrolle der Bodenverbesserungsparameter: Versenktiefe, zeitlicher Verlauf von Versenken und Verdichten, Kontinuität des Verdichtungsvorganges und damit der Verdichtungssäule, Amperaufnahme bei Versenken und Verdichtvorgang.
- Zugabematerial wurde laufend kontrolliert und der Verbrauch dokumentiert.
- Durchführung von 2 Belastungsversuchen mit der 1,5 -fachen Gebrauchslast.
- Durchführung von Rammsondierungen.

Alle Messungen und Kontrollen brachten für alle Beteiligten zufriedenstellende Ergebnisse.

6. Schlußbemerkung

Die Rüttelstopfverdichtung nimmt den größten Raum innerhalb der Tiefenverdichtungsverfahren ein. Durch die große Flexibilität der Geräte und Verfahren können damit viele Grundbauprobleme technisch, wirtschaftlich und rasch gelöst werden. Diverse EDV-Programme stehen für die Setzungsabschätzungen zur Verfügung. Jedoch müssen die Eingangsparameter durch Bodenaufschlüsse und deren Begutachtung vorliegen. Wie Sie sehen konnten, können mit diesen Methoden die unterschiedlichsten Probleme, wie auch Gefahren durch Erdbeben oder dynamische Einwirkungen von Maschinen beherrschbarer gemacht werden.

Literatur

(1) Priebe, H.: Die Bemessung von Rüttelstopfverdichtungen, *Die Bautechnik*, 72. Jahrgang. Heft 3/1995

(2) Kirsch, K.: Die Baugrundverbesserung mit Tiefenrüttlern, 40 Jahre Spezialtiefbau: 1953-1993, *Festschrift*, Werner-Verlag GmbH, Düsseldorf, 1993

(3) Priebe, H.: The prevention of liquefaction by vibro replacement, *International Earthquake Conference Berlin*, June 1989

(4) Robertson, P.K. / Campanelle, R.G.: Liquefaction Potential of Sands Using the CPT, *Journal of the Geotechnical Engineering Division*, ASCE. Vol. 111, 1985

(5) Seed, H.B. / Booker, J.R.: Stabilization of Potentially Liquefiable Sand Deposits Using Gravel Drains, *Journal of the Geotechnical Engineering Division*, ASCE. Vol. 103, 1977

(6) Seed, H.B. / Indriss, I.M.: Simplified Procedure for Evaluating Soil Liquefaction Potential, *Journal of the Geotechnical Engineering Division*, ASCE. Vol. 97, 1971

(7) Stockhammer, P. / Strauch, G. / Böröczky, Zs.:

Talajjavítás mélytömörítő vibrátorral, *Közúti közlekedés- és mélyépítéstudományi szemle* XLVII. 1997

Geotechnical Hazards, Marić, Lisac & Szavits-Nossan (eds) © 1998 Taylor & Francis, ISBN 90 5410 957 2

On the calculation methods of Green's functions for the layered half-space

Andrej Štrukelj

Faculty of Civil Engineering, University of Maribor, Slovenia

ABSTRACT: The scope of work presented is the derivation of the three-dimensional Green's function for the layered half-space. In that case the half-space is loaded with a vertical point load and the method of potentials is used for the solution. Partial differential equations occurring in this problem are made ordinary ones through the Hankel integral transform. The integral that represents the inverse Hankel transform of solution contains a singularity of first order. It is made regular by extracting the singularity from the integral of inverse transform. The appearance of Stonely waves are also discussed. Finally, the contour integrals are evaluated by substitution of Bessel function occurring in the integrand with the more suitable one and by closing the integration contour on the upper half-plane by the infinite half-circle. The two different integration contours have been used.

1 INTRODUCTION

The frame of this paper is the development of methods and procedures for the description of the motion of an arbitrary shaped foundation. Since the infinite half-space cannot be properly described by a model of finite dimensions without violating the radiation condition, the basic problems are infinite dimensions of the half-space as well as its non-homogeneous nature. Consequently, an approach to solve this problem indirectly by developing Green's function in which the non-homogeneity and the infiniteness of the half-space has been investigated. When the Green's function is known, the next step will be the evaluation of contact stresses acting between the foundation and the surface of the half-space through an integral equation. The equation should be solved in the area of the foundation using Green's function as the kernel. The derivation of three-dimensional Green's function for the homogeneous half-space (Kobayashi and Sasaki 1991) was made using the potential method. Partial differential equations occurring in the problem were made ordinary ones through the Hankel integral transform. The general idea for obtaining the three-dimensional Green's function for the layered half-space is similar. But in that case some additional phenomena may occur. One of them is the possibility of the appearance of Stonely surface waves in the contact surfaces of lay-

ers. Their contribution to the final result is in most cases important enough that they should not be neglected.

2 THE DERIVATION OF GREEN'S FUNCTION

The surface of the horizontally layered half-space is loaded by the concentrated load $P \cdot H(t)$. As the wave motion generated on such way is axially symmetric, the cylindrical co-ordinate system is introduced. The local co-ordinate systems having their origins on the top surface of the each layer are also defined. The governing equation for the each layer is the known equation of motion:

$$\mu \cdot \nabla^2 \vec{U} + (\lambda + \mu) \cdot \vec{\nabla}\vec{\nabla} \bullet \vec{U} = \rho \cdot \ddot{\vec{U}} \tag{1}$$

which can be separated into two parts:

$$\frac{\mu}{\rho} \cdot \nabla^2 \vec{u} + \frac{\lambda + \mu}{\rho} \cdot \vec{\nabla}\vec{\nabla} \bullet \vec{u} = -\omega^2 \cdot \vec{u} \quad \text{and}$$

$$\ddot{T} = -\omega^2 \cdot T \tag{2}$$

The displacement vector can be written in potential form as:

$$\vec{u} = \vec{\nabla}\varphi + \vec{\nabla} \times \vec{\psi} \tag{3}$$

The two non-zero components of the above defined displacement vector are:

$$u = \frac{\partial \varphi}{\partial r} - \frac{\partial \psi}{\partial z} \qquad \text{and}$$

$$w = \frac{\partial \varphi}{\partial z} + \frac{1}{r} \cdot \frac{\partial (r \cdot \varphi)}{\partial r} \qquad (4)$$

The relationships between the stress components and displacements are therefore:

$$\sigma_z = (\lambda + 2 \cdot \mu) \cdot \frac{\partial w}{\partial z} + \frac{\lambda}{r} \frac{\partial (r \cdot \psi)}{\partial r} \qquad \text{and}$$

$$\tau_{rz} = \mu \cdot \left(\frac{\partial u}{\partial z} + \frac{\partial w}{\partial r} \right) \qquad (5)$$

After introducing the Eq(3) into the first of Eq(2) and obeying the well known expressions for the wave numbers in longitudinal and transversal direction ($k_T = \omega/c_r$ and $k_L = \omega/c_L$) two wave equations for two potentials have been obtained. In cylindrical co-ordinate system they have the following form:

$$\frac{\partial^2 \varphi}{\partial r^2} + \frac{1}{r} \cdot \frac{\partial \varphi}{\partial r} + \frac{\partial^2 \varphi}{\partial z^2} + k_L^2 \cdot \varphi = 0 \qquad \text{and}$$

$$\frac{\partial^2 \psi}{\partial r^2} + \frac{1}{r} \cdot \frac{\partial \psi}{\partial r} + \frac{\partial^2 \psi}{\partial z^2} - \frac{\psi}{r^2} + k_T^2 \cdot \psi = 0, \qquad (6)$$

Partial differential equations Eq(6) can be translated into ordinary ones by using the Hankel integral transform $r \rightarrow \xi$, which is defined as:

$$H^n(f(r)) = \bar{f}^n(\xi) = \int_0^\infty f(r) \cdot J_n(\xi \cdot r) \cdot r \cdot dr \qquad (7)$$

Corresponding inverse transform is:

$$f(r) = \int_0^\infty \bar{f}^n(\xi) \cdot J_n(\xi \cdot r) \cdot \xi \cdot d\xi \qquad (8)$$

Hankel transform of Eq(4), Eq(5) and Eq(6) are:

$$\bar{u}^1 = -\xi \cdot \bar{\varphi}^0 - \frac{d\bar{\psi}^1}{dz} \qquad \text{and}$$

$$\bar{w}^0 = \frac{d\bar{\varphi}^0}{dz} + \xi \cdot \bar{\psi}^1 \qquad (9)$$

$$\bar{\sigma}_z^0 = \mu \cdot \left[(2 \cdot \xi^2 - k_T^2) \cdot \bar{\varphi}^0 + 2 \cdot \xi \cdot \frac{d\bar{\psi}^1}{dz} \right] \qquad \text{and}$$

$$\bar{\tau}_{rz}^1 = -\mu \cdot \left[2 \cdot \xi \cdot \frac{d\bar{\varphi}^0}{dz} + (2 \cdot \xi^2 - k_T^2) \cdot \bar{\psi}^1 \right] \qquad (10)$$

$$\frac{d^2 \bar{\varphi}^0}{dz^2} - (\xi^2 - k_L^2) \cdot \bar{\varphi}^0 = 0 \qquad \text{and}$$

$$\frac{d^2 \bar{\psi}^1}{dz^2} - (\xi^2 - k_T^2) \cdot \bar{\psi}^1 = 0 \qquad (11)$$

The fundamental solutions can be written in following form:

$$\bar{\varphi}^0 = \Phi_1 \cdot e^{\sqrt{\xi^2 - k_L^2} \cdot z} + \Phi_2 \cdot e^{-\sqrt{\xi^2 - k_L^2} \cdot z} \qquad \text{and}$$

$$\bar{\psi}^1 = \Psi_1 \cdot e^{\sqrt{\xi^2 - k_T^2} \cdot z} + \Psi_2 \cdot e^{-\sqrt{\xi^2 - k_T^2} \cdot z} \qquad (12)$$

Since the layered half-space has to be treated as a continuous media, four continuity conditions have to be introduced on each contact surface. On the contact of two layers the equal normal stresses, shear stresses and the displacements in both directions are demanded. Considering that the layered half-space consists of n parallel layers resting on a homogeneous half-space there are $4 \cdot n$ of continuity conditions and two boundary conditions on the top of the first layer:

$$\tau_{rz}\big|_{z=0} = 0; \qquad \sigma_z\big|_{z=0} = -\frac{P \cdot H(t) \cdot \delta(r)}{2 \cdot \pi \cdot r}. \qquad (13)$$

On the other hand the fundamental solutions for the potentials φ and ψ for each layer have four integration constants while the fundamental solutions for the underlying half-space, where the radiation conditions should be introduced, have another two of them. For evaluating the values of the $4 \cdot n + 2$ unknown constants there are thus $4 \cdot n + 2$ equations. The matrix of this system is band matrix with the bandwidth of maximum 8 terms. The right side of the system is a column matrix where each one except the first term is equal to zero. As the point of interest is only the evaluation of the surface motion, only the solution for the first four integration constants is needed. Potentials $\bar{\varphi}^0$ and $\bar{\psi}^1$ have now the following form:

$$\bar{\varphi}^0 = C_1 \cdot e^{\sqrt{\xi^2 - k_L^2} \cdot z} + C_2 \cdot e^{-\sqrt{\xi^2 - k_L^2} \cdot z};$$

$$\bar{\psi}^1 = C_3 \cdot e^{\sqrt{\xi^2 - k_T^2} \cdot z} + C_4 \cdot e^{-\sqrt{\xi^2 - k_T^2} \cdot z} \qquad (14)$$

Referring to the Hankel transform of Eq(7) and taking into account that on the surface z=0 the relation for the vertical component of Green's function in transformed domain ($\bar{w}^0$) is obtained. Its inverse transform can be obtained by putting it into Eq(8):

$$w = \int_0^\infty \xi \cdot \overline{w}^0 \cdot J_0(\xi \cdot a) \cdot d\xi \qquad (15)$$

where a assigns the dimensionless frequency defined as: $a = r \cdot \omega / c_T$. The product $\xi \cdot \overline{w}^0$ does not vanish in infinity, but it converges to a constant value $-(1-\nu)$, as can be proved. Therefore this constant value should be subtracted from the integrand. After this and excluding the singularity $1/r$ from the integral, Eq(15) has the following form:

$$w_H = \frac{(1-\nu)}{r} \cdot \left(1 - \frac{a}{1-\nu} \cdot I_{(a)}\right) \qquad (16)$$

where:

$$I_{(a)} = \int_0^\infty \left(\xi \cdot \overline{w}^0 + 1 - \nu\right) \cdot J_0(\xi \cdot a) \cdot d\xi \qquad (17)$$

It can be seen from the structure of the solution in transformed domain that each layer contributes two couples of conjugate branch points. These branch points as well as Rayleigh and Stonely poles lie along the same lines, which slopes depend on the value of material dumping ratio ϕ. In order to make the solution single valued, the appropriate branch cuts have to be introduced. For the limit case, where $\phi \to 0$ materials of the layers are linear elastic and all branch points and poles are lying on the real ξ axis. The usual way to calculate the integral Eq(17)

is to close the integration contour in the complex ξ plane, so the following steps are introduced:

$$2 \cdot J_0(\xi \cdot a) = \frac{2}{\pi} \cdot \int_0^\pi \cos(\xi \cdot a \cdot \sin\zeta) \cdot d\zeta$$

$$= \frac{1}{\pi} \cdot \int_0^\pi \left[e^{-i\cdot\xi\cdot a\cdot\sin\zeta} + e^{i\cdot\xi\cdot a\cdot\sin\zeta}\right] \cdot d\zeta \qquad (18)$$

$$= h(-\zeta \cdot a) + h(\xi \cdot a)$$

The integral $I_{(a)}$ becomes than $\frac{1}{2} \cdot I_{(h)}$, where:

$$I_h = \int_{-\infty}^\infty \left(\xi \cdot \overline{w}^0 + 1 - \nu\right) \cdot h(\xi \cdot a) \cdot d\xi \qquad (19)$$

The integration contour can now be closed by an infinite semicircle lying in the upper half of the ξ plane and the residue theorem can be used for the evaluation of the integral. $I_{(h)}$ It can be proved that the contour integral over the semicircle vanishes as the radius R reaches infinity. The integration problem is now reduced on the integration along the branch cut and evaluating of the residues in singular points enclosed by the integration contour. In the Fig.1 the example of integration path for the dumped half-space consisting of one layer on the homogeneous half-space is shown. The solution of similar problem for the non-damped half-space will be discussed later in this section.

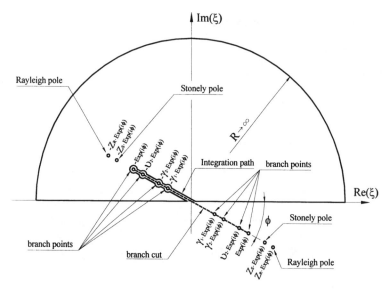

Figure 1. Integration path for the dumped one-layer half-space (The meaning of symbols γ_1, γ_2 and υ_2 will be clearified later in the section.)

Rayleigh waves appear always when the free surface exists. On the other hand the appearance of the Stonely waves depends on the density and shear modulus ratios of the neighbouring layers. To illustrate this statement the system of two coupled but different non-dumped half-spaces have been investigated (Fig.2).

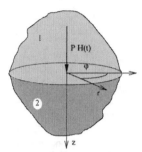

Figure 2. Model of two coupled half-spaces

Obeying the radiation condition for both half-spaces and introducing the substitution $\xi = k_{T1} \cdot \eta$ the fundamental solutions for the potentials can be written in following form:

$$\overline{\varphi}_1^0 = C_{11} \cdot e^{2 \cdot \pi \cdot \sqrt{\eta^2 - \gamma_1^2} \cdot \zeta} = C_{11} \cdot e^{2 \cdot \pi \cdot \alpha_1 \cdot \zeta}$$
$$\overline{\psi}_1^1 = C_{12} \cdot e^{2 \cdot \pi \cdot \sqrt{\eta^2 - 1} \cdot \zeta} = C_{12} \cdot e^{2 \cdot \pi \cdot \beta_1 \cdot \zeta}$$
$$\overline{\varphi}_2^0 = C_{21} \cdot e^{-2 \cdot \pi \cdot \sqrt{\eta^2 - \gamma_2^2} \cdot \zeta} = C_{21} \cdot e^{-2 \cdot \pi \cdot \alpha_2 \cdot \zeta} \qquad (20)$$
$$\overline{\psi}_2^1 = C_{22} \cdot e^{-2 \cdot \pi \cdot \sqrt{\eta^2 - \upsilon_2^2} \cdot \zeta} = C_{22} \cdot e^{-2 \cdot \pi \cdot \beta_2 \cdot \zeta}$$

where ζ represents the ratio of the z co-ordinate and the wavelength of the transverse waves in the upper half-space. The symbols γ_1, γ_2 and υ_2 represents the ratios between k_{L1} and k_{T1}, k_{L2} and k_{T1} and k_{T2} and k_{T1} respectively. In this case four continuity conditions are needed. The first two of them require the equal values of the normal (Eq.(21)) and shear stresses (Eq.(22)) on the contact surface:

$$(2 \cdot \eta^2 - 1) \cdot C_{11} + 2 \cdot \eta \cdot \beta_1 \cdot C_{12} - \frac{\mu_2}{\mu_1} \cdot (2 \cdot \eta^2 - \upsilon_2^2) \cdot C_{21}$$
$$+ \frac{2 \cdot \mu_2}{\mu_1} \cdot \eta \cdot \beta_2 \cdot C_{22} = -\frac{P}{2 \cdot \pi \cdot \omega^2 \cdot \rho} \qquad (21)$$

$$2 \cdot \eta \cdot \alpha_1 \cdot C_{11} + (2 \cdot \eta^2 - 1) \cdot C_{12} + 2 \cdot \frac{\mu_2}{\mu_1} \cdot \eta \cdot \alpha_2 \cdot C_{21}$$
$$- \frac{\mu_2}{\mu_1} \cdot (2 \cdot \eta^2 - \upsilon_2^2) \cdot C_{22} = 0 \qquad (22)$$

The second pair of continuity conditions demands the equality of vertical (Eq.(23)) and horizontal (Eq.(24)) displacements in the contact surface:

$$\alpha_1 \cdot C_{11} + \eta \cdot C_{12} + \alpha_2 \cdot C_{21} - \eta \cdot C_{22} = 0 \qquad (23)$$

$$\eta \cdot C_{11} + \beta_1 \cdot C_{12} - \eta \cdot C_{21} + \beta_2 \cdot C_{22} = 0 \qquad (24)$$

From the obtained system of equations the four unknown integration constants, which are in fact functions of η, can be obtained. The real zeros of the system determinant represent so-called Stonely poles defined as the ratios of the first half-space shear wave front velocity and Stonely wave velocities. In the plane, where first of the axes represents the ratio of densities (ρ_1/ρ_2) and the second one represents the ratio of the shear modulus (μ_1/μ_2), two curves (A and B) define the boundaries of the real solutions region, as can be seen in the Fig.3. The curve A is connecting all points for which the Sonely wave velocity (c_s) coincide with shear wave velocity of the first half-space (c_{T1}) and the curve B represents the boundary for which the Sonely wave velocity (c_s) coincide with shear wave velocity of the second half-space (c_{T2}).

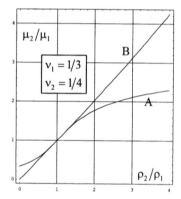

Figure 3. The region of real solutions of the system determinant (between curves A and B)

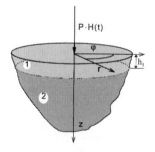

Figure 4. Model of the one layer half-space

In the case of the elastic layer resting on a homogeneous half-space (Fig.3) the fundamental solutions for the potentials look like:

$$\overline{\varphi}_1^0 = C_{11} \cdot e^{2 \cdot \pi \cdot \alpha_1 \cdot \zeta} + C_{12} \cdot e^{-2 \cdot \pi \cdot \alpha_1 \cdot \zeta}$$
$$\overline{\psi}_1 = C_{13} \cdot e^{2 \cdot \pi \cdot \beta_1 \cdot \zeta} + C_{14} \cdot e^{-2 \cdot \pi \cdot \beta_1 \cdot \zeta}$$
$$\overline{\varphi}_2 = C_{21} \cdot e^{-2 \cdot \pi \cdot \alpha_2 \cdot \zeta} \tag{25}$$
$$\overline{\psi}_2 = C_{22} \cdot e^{-2 \cdot \pi \cdot \beta_2 \cdot \zeta}$$

Two boundary and four continuity conditions can be used to build the system of equations for evaluation of the six unknown integration constants. The first of the boundary condition requires the equilibrium of the normal stresses on the free surface of the half-space.

$$(2 \cdot \eta^2 - 1) \cdot (C_{11} + C_{12}) + 2 \cdot \eta \cdot \beta_1 \cdot (C_{13} - C_{14})$$
$$= -\frac{P}{2 \cdot \pi \cdot \omega^2 \cdot \rho} \tag{26}$$

The second boundary condition requires the equilibrium of the shear stresses on the free surface:

$$2 \cdot \eta \cdot \alpha_1 \cdot (C_{11} - C_{12}) + (2 \cdot \eta^2 - 1) \cdot (C_{13} + C_{14}) = 0 \tag{27}$$

The remaining four equations are created on the basis of continuity conditions. To make the solutions more universal another normalisation has been introduced. The layer thickness has been written as the product of the wavelength of the shear waves of the layer and the non-dimensional coefficient Z_1. The first continuity condition demands the equilibrium of normal stresses in the contact plane:

$$(2 \cdot \eta^2 - 1) \cdot (C_{11} \cdot e^{2 \cdot \pi \cdot \alpha_1 \cdot Z_1} + C_{12} \cdot e^{-2 \cdot \pi \cdot \alpha_1 \cdot Z_1})$$
$$+ 2 \cdot \eta \cdot \beta_1 \cdot (C_{13} \cdot e^{2 \cdot \pi \cdot \beta_1 \cdot Z_1} - C_{14} \cdot e^{-2 \cdot \pi \cdot \beta_1 \cdot Z_1})$$
$$- \frac{\mu_2}{\mu_1} \cdot (2 \cdot \eta^2 - \upsilon_2^2) \cdot C_{21} \cdot e^{-2 \cdot \pi \cdot \alpha_2 \cdot Z_1} \tag{28}$$
$$+ 2 \cdot \frac{\mu_2}{\mu_1} \cdot \eta \cdot \beta_2 \cdot C_{22} \cdot e^{-2 \cdot \pi \cdot \beta_2 \cdot Z_1} = 0$$

The requirement of the equilibrium of shear stresses in the contact plane defines the second continuity condition:

$$2 \cdot \eta \cdot \alpha_1 \cdot (C_{11} \cdot e^{2 \cdot \pi \cdot \alpha_1 \cdot Z_1} - C_{12} \cdot e^{-2 \cdot \pi \cdot \alpha_1 \cdot Z_1})$$
$$+ (2 \cdot \eta^2 - 1) \cdot (C_{13} \cdot e^{2 \cdot \pi \cdot \beta_1 \cdot Z_1} + C_{14} \cdot e^{-2 \cdot \pi \cdot \beta_1 \cdot Z_1})$$
$$+ 2 \cdot \frac{\mu_2}{\mu_1} \cdot \eta \cdot \alpha_2 \cdot C_{21} \cdot e^{-2 \cdot \pi \cdot \alpha_2 \cdot Z_1} \tag{29}$$
$$- \frac{\mu_2}{\mu_1} \cdot (2 \cdot \eta^2 - \upsilon_2^2) \cdot C_{22} \cdot e^{-2 \cdot \pi \cdot \beta_2 \cdot Z_1} = 0$$

In the contact plane the compatibility of displacements in both vertical (Eq.(30)) and horizontal (Eq.(31)) direction is also necessary:

$$C_{11} \cdot \alpha_1 \cdot e^{2 \cdot \pi \cdot \alpha_1 \cdot Z_1} - C_{12} \cdot \alpha_1 \cdot e^{-2 \cdot \pi \cdot \alpha_1 \cdot Z_1}$$
$$+ \eta \cdot C_{13} \cdot e^{2 \cdot \pi \cdot \beta_1 \cdot Z_1} + \eta \cdot C_{14} \cdot e^{-2 \cdot \pi \cdot \beta_1 \cdot Z_1} \tag{30}$$
$$+ C_{21} \cdot \alpha_2 \cdot e^{-2 \cdot \pi \cdot \alpha_2 \cdot Z_1} - \eta \cdot C_{22} \cdot e^{-2 \cdot \pi \cdot \beta_2 \cdot Z_1} = 0$$

$$\eta \cdot C_{11} \cdot e^{2 \cdot \pi \cdot \alpha_1 \cdot Z_1} + \eta \cdot C_{12} \cdot e^{-2 \cdot \pi \cdot \alpha_1 \cdot Z_1}$$
$$+ C_{13} \cdot \beta_1 \cdot e^{2 \cdot \pi \cdot \beta_1 \cdot Z_1} - C_{14} \cdot \beta_1 \cdot e^{-2 \cdot \pi \cdot \beta_1 \cdot Z_1} \tag{31}$$
$$- \eta \cdot C_{21} \cdot e^{-2 \cdot \pi \cdot \alpha_2 \cdot Z_1} + C_{22} \cdot \beta_2 \cdot e^{-2 \cdot \pi \cdot \beta_2 \cdot Z_1} = 0$$

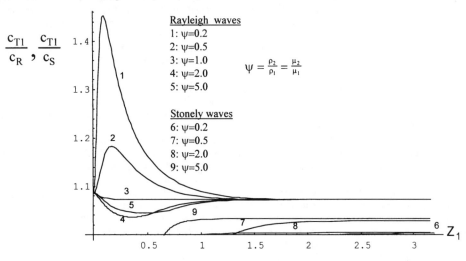

Figure 5. Dispersion curves for one layer half-space

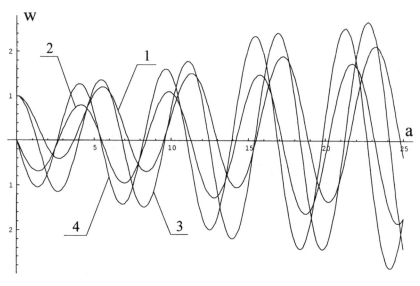

1-Real part of Green's function for the surface of the layer resting on an elastic half-space

2-Imaginary part of Green's function for the surface of the layer resting on an elastic half-space

3-Real part of Green's function for the surface of the homogeneous half-space

4-Imaginary part of Green's function for the surface of the homogeneous half-space

Figure 6. Green's function for the non-dumped homogeneous and one-layer half-space

The determinant of created equation system has four zeros. Two of them represent Stonely surface wave and the other two the Rayleigh wave appearing because of the free surface. The appearance and velocity of the Stonely waves depends now not only on the properties of the neighbouring two materials but also on the thickness of the top layer. In the limit when the thickness of the layer go to infinity the velocity of the Stonely waves becomes more and more similar to the values from the previous example and the velocity of the Rayleigh waves reaches the values for the homogeneous half-space. On the Fig.5 the ratios of the shear velocities and the velocities of Rayleigh and Stonely waves (c_{TI}/c_R and c_{TI}/c depending on the ratio of the layer thickness and the wave length of the shear wave in the layer (Z_1) is shown. The curves on Fig. 5 define therefore the positions of Rayleigh and Stonely poles in the complex ξ-plane or in the case of non-damped half-space the position of the poles on the real ξ-axis. In the following example the vertical component of Green's function for the elastic layer resting on the elastic half-space is calculated. The ratio between the wavelength of shear waves in the layer and the layer thickness is considered to be $2 \cdot \pi$. No material dumping is considered. Poisson's ratios of the layer and half-space are 1/3 and 1/4, respectively. The density of the half-space is 1.5 times greater and the shear module is 1.6 times greater than corresponding

characteristics of the layer. All of the evaluation steps in transformed domain have been made analytically, only the contour integration for the evaluation of the inverse transform was obtained with numerical algorithms. The results are plotted on Fig.6.

The main advantage of results presented is their accuracy because all essential steps of Green's function evaluation except of the contour integration along the branch cut are made analytically. On the other hand the disadvantage of this method is that the mathematical effort for obtaining the Green's function is increasing drastically with the increase of the number of layers. The above-described semi-analytical method is therefore very useful for testing the efficiency and accuracy of numerical methods. Very successful algorithm for solving this problem has been developed on the basis of previous-discussed analytical steps. Only the order of the steps should be changed. As the integration constants are in fact the functions of variable η the system of equations has to be solved for each value of η separately. The result is therefore discrete functions that enter into further analysis. As in that case the evaluation of system determinant zeros is very difficult and the problems occur also by the evaluation of residues, the idea was to change the integration path so, that all singularities are outside the contour. One of the possible solutions that give very good results is presented in Fig.7.

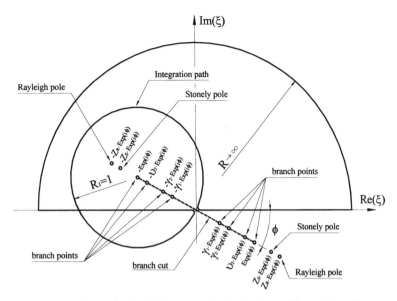

Figure 7. Integration path suitable for numerical evaluation of Green's function

As the integral of inverse Hankel transform along the infinite half-circle is equal zero it can be seen that the result of contour integration along the whole real axis is equal to integral along the unit circle. There is no need to calculate the zeros of system determinant and the evaluation procedure can be performed automatically.

3 CONCLUSIONS

Until now various numerical solutions for Green's functions of the layered half-space have been given. They are more or less accurate depending on the way of introducing of the radiation and continuity conditions. Since the results obtained by the semi-analytical and numerical method practically coincide and the above-derived method is very stable the future work will therefore be directed in developing the software that should cover the problems of evaluating the Green's functions and evaluation of the contact stresses between the soil and the foundation.

REFERENCES

Auersch, L.: *Wave Propagation in Layered Soils: Theoretical Solution in Wavenumber Domain and Experimental Results of Hammer and Railway Traffic Excitation.* Journal of Sound and Vibration, 1994, Vol. 173, No.2, pp. 233-264
Ewing, W.M. & Jardetzky, W.S. & Press, F.: *Elastic waves in Layered Media*, Mc. Graw-Hill Book Company, 1957
Kobayashi, T. & Sasaki, F. (1991): *Evaluation of Green's Function on Semi-infinite Elastic Medium.* Kajima Technical Research Institute, KICT Report, No.86
Luco, J. E. & Apsel, R. J.: *On the Green's Function for a Layered Half-Space.* Part 1. Bulletin of the Seismological Society of America, 1983, Vol.73, No.4, pp. 909-929
Luco, J. E. & Apsel, R. J.: *On the Green's Function for a Layered Half-Space.* Part 2. Bulletin of the Seismological Society of America, 1983, Vol.73, No.4, pp. 931-951

Geotechnical Hazards, Marić, Lisac & Szavits-Nossan (eds) © 1998 Taylor & Francis, ISBN 90 5410 957 2

Results of static and dynamic loading tests on driven steel-pipe piles

Geza Vogrinčič
FMF, University of Ljubljana, Slovenia

Gorazd Strniša
SLP Ljubljana, Slovenia

ABSTRACT: This paper describes the results of maintained load tests conducted on steel-pipe piles and the evaluation of ultimate bearing capacity determined from static and appropriate dynamic measurements and numerical analysis. For prediction of optimal pile driving conditions the GRLWEAP[TM] computer program was used. During driving and redriving dynamic measurements were performed with a Pile Driving Analyzer® (PDA). CAPWAP® program was applied for simulation of static load test based on dynamic measured data. The closed ended steel-pipe piles with outer diameter of 508 mm and length of more than 40m were driven with a diesel hammer through normally consolidated recent silty-clayey sediments into a bearing strata of sandy-gravely soils. Several redirving redriving dynamic tests were made after static load test. The results of the static loading test were then compared with the pile bearing capacity obtained through PDA measurements and CAPWAP® analysis.

1 INTRODUCTION

In accordance with the development plan for Port Koper, the cattle port terminal is now being constructed. The quay platform 43.5m long and 24.7m wide, is composed of a concrete slab over prefabricated reinforced concrete beams supported by a total of 50 vertical and 10 inclined closed-ended steel pipe piles. The piles were welded together on site from 12m long spirally welded tube sections of 508 mm outside diameter and 8.8 mm wall thickness. The piles were placed in a grid of 4.2m x 5.25m as shown in Figure 1. A preliminary analysis of pile behavior during driving, optimal selection choice of driving equipment and evaluation of ultimate bearing capacity of the driven pile was carried out using the widely used computer program GRLWEAP[TM] (Goble Rausche Linkins and Associates, Inc.). The test driving of three selected working piles followed,

determining the driving criteria and the bearing capacity estimated from measurements obtained during driving using PDA. Twenty days later, one pile at position F15 was statically load tested. These static results were then compared with the pile bearing capacity values obtained through PDA measurements and CAse Pile Wave Analysis Program (CAPWAP®). In addition, at the end of pile installation, five piles were checked by dynamic testing (for three of them the operational driving hammer and for the remaining two a free falling drop weight were used to generate the dynamic load).

2 SUBSOIL CONDITIONS AT THE SITE OF THE TEST PILES

A typical soil profile (Table 1) was indicated by borehole No. L9, five meters away from the location of the static load test pile F15.

Table 1. Typical soil profile

Layer No.	Elevation (m)	Description	Properties
1	0-23	recent clayey-silty sediments	$w = 47$ %, $w_P = 24$ %, $w_L = 57$ %, $I_P = 33$ %, $I_c = 0.30$, $\gamma = 17.2$ kN/m^3, $c_u = 17$ kPa from CPTU: $0.4 < q_c < 0.7$ MPa, $1.2 < M < 1.5$ MPa, $c_u \cong 23$ kPa
2	23-26	sandy silt	from CPTU: $1.0 < q_c < 1.3$ MPa, $1.6 < M < 2.7$ MPa, $c_u \cong 41$ kPa
3	26-49	silty&clayey gravel	$N > 50$ blows from CPTU on the level −29.0m: $q_c > 26$ MPa, $\varphi => 33°$
4	>49	Eocene flysh	$N > 100$ blows

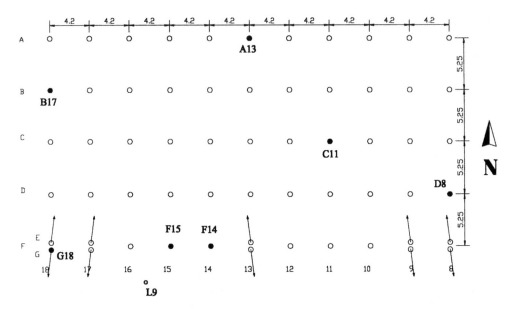

Figure 1: Plan showing the location of the piles and borehole No.L9.

Sub-surface stratigraphy and soil materials were identified and their geotechnical properties estimated using five borings with standard penetration tests (SPT), vane shear tests, a cone penetration test with pore pressure measurement (CPTU) plus laboratory tests.

Approximately thirty meters from the borehole No. L9 the CPTU-1 (ISMES, with thrust capacity 200 kN) was performed (penetration reached the depth of 29m below the seafloor) with pore pressure measurement. The dissipation or decay of pore pressure with time after stopping the penetration was measured at five different depths.

The interpretation of CPTU data evaluated after Mitchell and Gardner 1975; Sanglerat 1972; Meigh 1987; Meyerhof 1976 are given in Table 1.

From the pore pressure dissipation measurements we estimated (after Levadoux & Baligh 1986) the average value of soil permeability for upper layers of cohesive soils k = 5.4 . 10^{-10} m/s.

3 PRELIMINARY ESTIMATION OF PILE BEARING CAPACITY AND SELECTION OF AN OPTIMAL DRIVING SYSTEM

Before driving the test pile a wave equation computer analysis was made using GRLWEAP^{TM}, a computer program which simulates the behavior of an impact driven pile. The program contains mathematical models that describe hammer, driving system, pile and surrounding soil during a hammer blow.

For the preliminary analysis it was assumed that the spirally welded steel pipe pile had a 508 mm outside diameter, 8 mm wall thickness, 0.203 m^2 gross section, 125.66 cm^2 steel cross sectional area. Driving of the pile was analyzed between penetrations 25m to 47m.

The soil was modeled with following limit soil resistance values during driving: skin friction: 2.7 kPa for the top layer composed of very soft silty-clayey soil, 25 to 65 kPa for the underlying silty&clayey and sandy gravel layer ; toe resistance up to the level −21.0m was neglected, below that it was assumed to linearly increase from 5 MPa to 8 MPa.

Two different diesel hammers (Delmag D46-13 and Delmag D36-13) were analyzed. Driveability graphs for hammer D46-13 were given in the Figure 4. The results obtained show that piling is possible with either of the analized hammers to a depth of 35.0 m. For deeper penetrations the hammer D46-13 should be used because the blow count (BLCT) using the D36-3 would then exceed 500 blows/m (< 2mm/blow) and therefore driving would be very slow and uneconomical.

Expected ultimate bearing capacity at the end of driving (ED) with pile toe penetration at − 39.0m would be approximately 2600 kN with blow count BLCT≈300 blows/m.

We expect that, because of soil set-up effects, the ultimate bearing capacity would be at least 3200 kN within one month after ED.

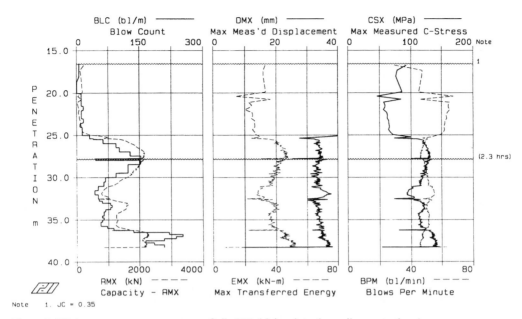

Figure 2: PDA measurements - summary of pile F15 driving data along pile penetration.

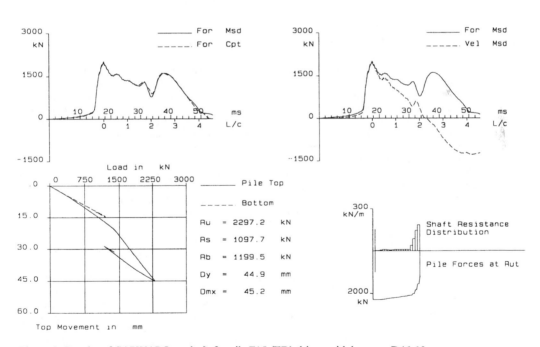

Figure 3: Results of CAPWAP® analysis for pile F15 (ED) driven with hammer D46-13.

4 TEST PILE DRIVING RESULTS

Pile F15 was selected for static axial load test and piles F14 and F16 as reaction piles. During driving of piles F14 and F15, dynamic measurements were taken with the PDA, allowing for an estimate of bearing capacity by the Case Method, dynamic compression and tension stress values in the pile,

659

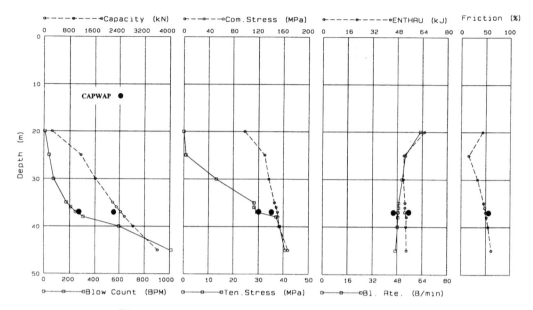

Figure 4: GRLWEAP™ driveability graph for pile JK508/8mm, hammer D46-13 and CAPWAP® results for F15 (ED).

energy transferred to the pile, blows per minute for hammer check and pile integrity. The main goal of these measurements was to determine pile penetration which assure final ultimate bearing capacity 3200 kN and so factor of safety F = 2 to the maximum designed load. Measurements were performed during driving of pile F14 from penetration -16.5m to –38.25m and for pile F15 from -16.0m to -38.46m. Figure 2 shows plotted summary of pile F15 driving data versus pile penetration.

The results show that the bearing capacity increased first near –27.0m and secondly, after a temporary decrease, at –37.0m. For the last meter of penetration, the set per blow was 4.8 mm, the Case Method capacity (RMX using Jc=0.35) indicated an ultimate bearing capacity of approximately 2300 kN. Pile F14, located only 4.2m away from F15 showed at nearly the same level an ultimate bearing capacity of 2600 kN and penetration 4.0mm/blow. The maximum compressive stress caused by impact was less than 160 MPa.

Recorded dynamic data for a selected hammer blow at the end of driving of test pile F15 were analyzed with CAPWAP® /CAse Pile Wave Analysis Program/ to obtain ultimate bearing capacity and simulation of static load test if performed immediately after installation of the pile. In principle CAPWAP® is an analytical method combining field recorded data and wave equation solution in order to compute the ultimate bearing capacity of the pile. In the course of CAPWAP® analysis a model of the pile

and soil is initialized with one of the measured value /force F(t) or velocity v(t)/. The result is the calculated response of the pile-soil model. If calculated values are the same or nearly the same as second measured quantity then the assumed model is accepted. As a result we obtained soil segment resistance forces, end bearing, soil quakes, damping values and ultimate bearing capacity.

With this model it becomes possible to perform a simulated Static Loading Test. Procedure and method is standardized ASTM D4945-89 and documented as a Recommendation for dynamic pile load test by Technical Commitee 5 of the German Geotechnical Society.

The CAPWAP® results for pile F15 were graphically presented in Figure 3.

The graph on the upper right of Figure 3 shows measured force and velocity for analyzed impact, the upper left depicts the "match" or agreement between measured and calculated force values, the lower right represents the shaft resistance distribution along the pile and the lower left presents the simulated static load test curve or load- set plot.

Figure 3 indicates that the CAPWAP® ultimate bearing capacity (Ru = 2297 kN) is in the same range as determinate from PDA measurements. Comparisons between the GRLWEAP™ driveability analysis and a measurement on the site with CAPWAP® is shown in the Figure 4. The agreement at the level –38.0m is acceptable.

5 STATIC LOAD TEST RESULTS

Static loading test was performed on test pile F15 driven to -38.2m. The total length of the test pile was 41.0m with 34.0m embedded in to the soil. Only last 17.7m of that embedment was in silty-gravelly soil layer.

The two reaction piles, F14 and F16 in Figure 1, were connected by a steel welded box girder serving as support for 4 reversible hydraulic jacks which allowed for a controlled application of the test load to the load test pile. To diminish the possibility of extraction, the tension piles were filled with concrete. The so-called "maintained loading test" procedure was chosen to determine the axial bearing capacity of pile F15. According to this method, the piles were gradually loaded and the settlements of the pile top was mesured and recorded for each particular loading step, either until it ceased in accordance with ASTM-D 1143-81, or at least for 2 hours. The vertical pile penetrations were measured at three measurement points in one cross-section (about 2.0m below the pile head) by means of micrometers and precise leveling instrument. At the same time, the displacements of the reaction piles were also observed. The loading test started 20 days after initial driving. During the first phase a load of 1776 kN was applied in four increments. This was followed by gradual unloading of the pile and then by another sequence of loading which was terminated when reaction pile F14 reached its uplift capacity. Because of this limitation, the pile F15 could only be subjected to a maintained load of 3108 kN for two hours. During that time pile F14 was extracted approximately 80 mm. An attempt to increase the load on the test pile in next loading step failed due to the accelerated upward movement of tension pile F14. With maximum applied axial force of 3108 kN the failure of the soil was not achieved. Graphic presentation of the results (load-settlement plot) is given in Figure 5.

The evaluated ultimate bearing capacity values according to both DIN 4026 standard and Davisson's method were 3340 kN and 3500 kN, respectively, using extrapolated load settlement curve.

6 DYNAMIC MEASUREMENTS DURING REDRIVING

Test pile F15 was PDA tested during a restrike 65 days after installation using hammer D46-13. The CAPWAP® analysis was also performed. We find out that only approximately 3000 kN static soil resistance could be activated with this hammer. The set during redriving was only 2.3 mm/blow.

Additionally, five piles in the positions G18, D8, C11, B17 and A13 (Figure 1) were tested with the

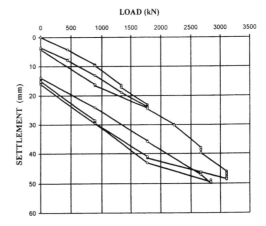

Figure 5: Static load test (load settlement diagram of the test pile F15).

PDA. Four of them were tested during redriving (elapsed time after installation about two weeks). In order to fully activate the static soil resistance in the last two redriving tests (piles B17 and A13), a free fall weight of 100 kN was used. With free fall height of 1.5m on both tested piles permanent penetrations between 6.5 to 7.5 mm/blow were achieved and so soil resistance were fully mobilized. CAPWAP® analyses were performed for all controlled tested piles. The results confirmed that using hammer D46-13 impact only up to 2966 kN static resistance could be activated. Piles tested with free fall weight proved that ultimate bearing capacity of the piles exceeded 3200 kN.

As an example, CAPWAP® analysis results for pile A13, tested with the free fall weight, are presented in Figure 6. It can be seen from force and velocity versus time curves in the upper right of Figure 6 that the skin resistance in the lower part of the pile and toe bearing were very high.

7 CONCLUSIONS

It is very useful for large piling projects that calculations include preliminary estimation of pile behavior during driving using a wave equation computer program such as GRLWEAP™. Dynamic measurements with the Pile Driving Analyzer® during driving and restrike are essential and should be combined with CAPWAP® analysis to assess not only the static bearing capacity and its distribution, but also to check hammer efficiency, pile stresses during driving and pile integrity.

In the case presented, the preliminary estimates of dynamic pile behavior, pile dynamic measurements

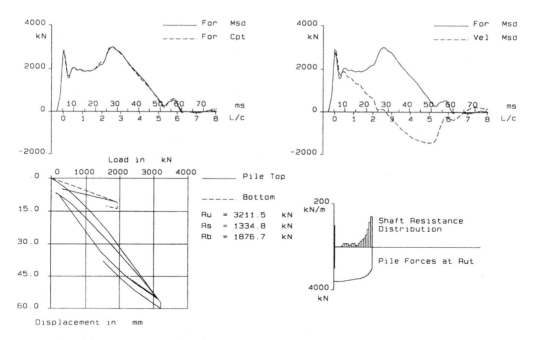

Figure 6: CAPWAP® results for restrike drop weight test of pile A13.

and static load results are all in good agreement. Unfortunately during static loading the ultimate bearing capacity was not achieved because of reaction pile failure. Using pile dynamic measurements and especially using a properly selected free fall weight for pile impact testing it was possible to prove the required static soil resistance which was also the ultimate static bearing capacity of the pile.

REFERENCES

CAPWAP® Manual – Case Pile Wave Equation Analysis Program (1997-1), GRL, USA.

GRLWEAP™ Manual - Wave Equation Analysis of pile driving (1997-2), GRL, USA.

Likins,G.E. & M.Hussein, F.Rausche 1988. Design and testing of pile foundations, *3rd Int. Conf. on application of stress-wave theory to piles*, Ottawa 25.-27.5.1988, Canada, 644-658.

Levadoux, J.N. & M.M. Baligh 1986. Consolidation after undrained piezocone penetration. I. Prediction, II. Interpretation. *Journal of geotechnical Engineering*, ASCE, Vol.112, No.7: 707-745.

Meyerhof, G.G. 1976. Bearing capacity and settlement of pile foundations. *Proc. Am. Soc. Civ. Engrs .-J. Geotech. Engng. Div.*, Vol.102(GT3): 195-228.

Meigh, A.C. 1987. Cone penetration testing – Methods and interpretation. *Ground Engineering report: In situ Testing*,London, CIRIA.

Mitchell, J.K. & W.S. Gardner 1975. In situ measurement of volume change characteristics. *Proceedings of the ASCE Specialty Conference on In Situ Measurements of Soil Properties*, Raleigh, North Carolina, 2: 279-345, ASCE.

Rausch,F. & G.G Goble, G.E. Likins 1985. Dynamic determination of pile capacity, *Journal of the soil mechanics and foundation division, ASCE*, Vol. III:367-383.

Sanglerat, G. 1972. The penetrometer and soil exploration, Amsterdam: Elsevier.

Vogrinčič,G. & G.Strniša 1991. Bearing capacity of large diameter steel pipe pile determined by static and dynamic testing, *4th international DFI Conference*, Stressa, Italy, Vol.I:659-664.

Geotechnical Hazards, Marić, Lisac & Szavits-Nossan (eds) © 1998 Taylor & Francis, ISBN 90 5410 957 2

Foundation pit enclosure in soft clay applying modified soil nailing method

Bojan Vukadinović
Geotehnički Studio d.o.o. Zagreb, Croatia

Igor Sorić
Karst d.o.o. Zagreb, Croatia

Franjo Verić
University of Zagreb, Faculty of Civil Engineering, Geotechnical Department, Croatia

SUMMARY: The excavation enclosure method by nailing of soil is customary applied around the world, while the number of so implemented protective structures in Croatia is quite small. The paper presents the description of a case executed in 1995/96 in Mandrovićeva Street in the centre of Zagreb. The excavation enclosure was accomplished by application of modified elements of soil nailing, while along the existing buildings it was done by jacked piles. A description of soil characteristics and structural elements, as well as of realized displacements is given. In the course of works at one side of the foundation pit undesirable deformations occurred due to unexpectedly high ground water table and errors in co-ordination in advancement of works.

1. GENERAL

The described operation was protecting an excavation 7 to 8 m deep, made as interpolation along the existing city street, at its southern side, adjacent to residential houses and courtyard areas along other sides.

New building has in its basement two storeys of undergound garages, which extend accross the whole plot area, making thus excavation immediately adjoining the existing buildings conditional.

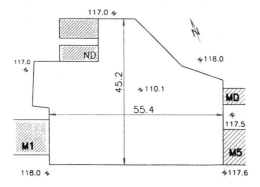

Figure 1 - Construction site

The protective structure is made of nails (soil anchors) 6 m long, tied to H-sections. In between of H-sections a concrete lining wall was built, 20 cm thick. Behind the wall the drains and filtering layer

for relaxation of hydrostatic pressures were built in.

The existing residential premises along which the excavation was made were additionally protected by jacked piles.

The foundation pit has a circumference of approx. 200 m. Altogether 109 rolled H-sections were driven in, 28 piles were jacked and 1400 nails driven in.

2. SITE DESCRIPTION

2.1 Siting

The plot geometry and ground levels are shown in Figure 1. The existing ground lays at elevation $117 \div 118$ m, while the elevation of final excavation is 110.1 m.

Buildings along the street (southern side), marked with M1 and M5, are residential houses with semi-cellar and two floors high plus attics. The building marked with MD is a ground-floor family house. All the buildings are founded upon concrete strips, buildings M1 and MD at elevation 115.6 m and building M5 at elevation 114.8 m.

2.2 Soil Properties

Investigations at the subject site were made by nine test bore holes and piezometric observations. An additional picture of soil properties was also obtained through measurements in the course of piles jacking and nails testing.

Figure 2 - Detail of the north-western side of excavation

The site ground is composed of consolidated clay fill up to 2 m thick, with underlying silty clay down to the depth of 10 m. Beneath clayey layers lie silty sand and gravel.

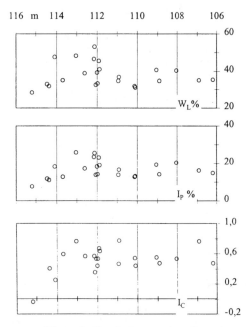

Figure 3 - Physical properties of tested samples

The consistency state of the clay layer is medium, while between depths of 2 and 4 m (elevations 116 and 114 m) it is soft to very soft in places. The consistency index, as well as axial strength are within this area essentially lesser, while the strength parameters at direct shearing are within the framework of figures obtained for the other samples.

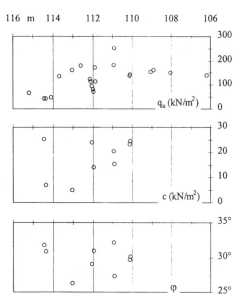

Figure 4 - Mechanical properties of tested samples

Competent indices of clayey materials properties are shown in Figures 3 and 4. It should be mentioned that within the area of very soft material sampling was aggravated, so that the successfully taken samples present somewhat better picture of the environment characteristics than these really are.

2.3 Ground Water Table

The ground water table was in the course of investigation operations established at elevation of approx. 116 m. As here we have the case of poorly permeable materials, it was assessed that it will be possible to lower it by drainage.

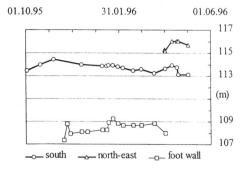

Figure 5 - Observations of ground waters table

For the purpose of lowering the ground water table

in the course of works performance, as well as for ground waters pumping during the exploitation period of premises, horizontal drains and drainage along the whole excavation surface were built in.

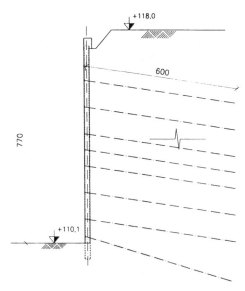

Figure 6 - Characteristic section of the enclosure

The driven in horizontal drains were executed using perforated pipes, which were also used for nails, but without grouting.

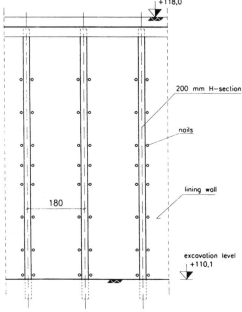

Figure 7 - The enclosure view

The drainage was also executed using flexible perforated plastic pipes, 10 cm in diameter and a layer of drainage geotextile. It was built in behind the lining reinforced concrete wall.

In the course of performance of works the ground water table was observed through built in piezometers. In this process three piezometers were plugged, so as to show piezometric table of water in the foot wall layer of gravel, while seven of them were executed in the open variant, so as to show the highest table of water in its profile.

The observation results are shown in Figure 5. Three tables of water were noted. Permanent water in the foot wall layer of gravel at elevation ≈ 108 m, as well as surface seepage waters at elevation ≈ 116 m on the northern side of the intervention at elevation ≈ 114 m.

The heightened table of water on the northern side was noted only afterwards, in the course of operation, when it was noted that increased deformations of structure occur in this part. Such difference in water tables of northern and southern sides was impossible to foresee, while the cause of this phenomenon remained unknown.

3. DESIGN

The stability control and structure designing were executed following recommendations of l'Ecole Nationale des Ponts et Chaussees / US Department of Transportation, as well as some approaches described in Vučetić & al. reference (1994). Moreover, the conventional Rankine method made a part of wall stability analysis. Finally, the German method of nailed wall failure was used, while some other approaches could be seen in Kvasnička & al. reference.

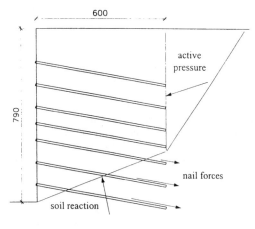

Figure 8 - The nailed wall falure modelling; German bilinear method

4. OPERATION DESCRIPTION

The first phase in performance of works consisted of driving in the 200 mm steel H-sections along the whole circumference of the pit, excepting lines along the existing buildings. Sections were driven in at interspaces of 1.8 m and to the depth of up to 8.5 m. Subsequently, excavation in vertical phases 1 m high begun, accompanied with a successive nails driving in and constructing of the lining wall.

Figure 9 - View of building M1

The entire time of works from the commencement of excavation until the concreting of floor slab was approximately 9 months. In this process an excavation down to approximately 4 m was made from the mid-October to the end of November, while in January there were no excavation works. By the end of March the excavation was in parts done down to the final depth, while at the end of April first part of the floor slab was concreted. The entire floor slab was finished by June.

4.1 Soil Nailing

Nails were made of steel pipes of the 49 mm outer diameter. Length of nails is 6 m, while the inner 3 m are perforated. Nails were driven in at an inclination of 8° and 12° against the horizontal. They are tied in twos to each vertical H- section, so that the medium horizontal distance is 90 cm. The vertical distance of rows is 75 to 100 cm.

After driving in, grouting through the pipe body was performed. Altogether 55 m³ of grout was used, $70 \div 100$ dm³ per nail in upper rows, and $15 \div 30$ dm³ in lower rows. In the process the terminal pressure of grouting was retained until the commencement of grout setting. By driving out of the grout through pipe perforations a forming of anchoring zone in the soil was accomplished.

4.2 Testing of Nails Against Extraction

Testing of nails was performed until the failure, maintaining a constant force.

Results of nail testing are shown by diagram in Figure 10. General data on tested nails are given in the Table 1.

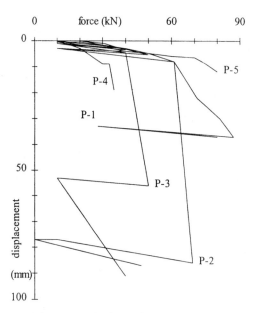

Figure 10 - The force/displacement diagram of nails tested

Nails 1, 2 and 5 demonstrated good properties. After the failure loading, unloading and repeated loading, nails retained their loadbearing properties. As repeated loading was performed after long term period and the material properties are even less than $I_c = 0.5$, such a soil behaviour could be treated as akind of self remedy phenomenon.

The test nail 4 was not grouted, which resulted in essentially lesser carrying capacity. Nail 3 failed at the force of 40 kN, and after its failure it behaved similarly to nail 4. It is speculated that a failure of contact between the grout and the steel pipe occurred.

Table 1 - Data on test nails

Test nail	P-1	P-2	P-3	P-4	P-5
Elevation of lead-in point of nail (m)	115	114	115	111	111
Quantity of grouted mix (dm³)	100	100	60	0	15
Terminal pressure of grouting (MN/m²)	0.4	0.7	0.6	-	1.0
Period from grouting to testing of nail (days)	13	10	12	-	55

On the basis of results obtained from performed tests it can be assessed that the extraction force of grouted nails is approx. 50 kN, while the magnitude of undrained soil strength is $25 > c_u > 45$ kN/m².

4.3 Jacked Piles

Buildings M1, M5 and MD were additionally protected by jacked piles, 133 to 245 mm in diameter and 8.5 m long.

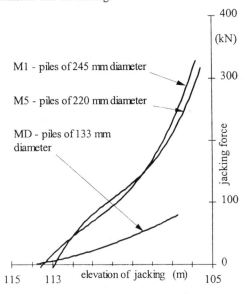

M1 - piles of 245 mm diameter

M5 - piles of 220 mm diameter

MD - piles of 133 mm diameter

400

(kN)

300

jacking force

100

0

115   113   elevation of jacking (m)   105

Figure 11 - Forces of piles jacking

Piles were executed by jacking of joined steel elements 50 cm long. After jacking of the last element, the hollow space within the pile was filled up with concrete, while braces enabling direct transmission of force from the old foundation to the pile were built in.

Figure 11 shows piles jacking measured forces. On the basis of measured values it was assessed that the magnitude of undrained soil strength is $25 > c_u > 45$ kN/m².

It is interesting that in execution of piles having diameter of 245 mm at three piles jacking was stopped before building in of braces. Piles remained

unloaded in duration of 15 to 40 hours. When the jacking was continued, an essential increase of resistance against jacking was observed on all three piles, by a force of 100 kN.

5. SETTLEMENT ALONG THE EXCAVATION PERIMETER

Settlements during the excavation were continuously monitored at reference points. Reference points were put on buildings along the excavation perimeter and on the ground, at the southern side, at a distance of approx. 2 m from the edge of excavation.

A leaping advancement of settlement was noticeable, caused by interruption of works during January and April.

Settlements of the following magnitudes were noted:
- southern side (towards the street) $\approx 2 \div 3$ cm
- buildings on jacked piles $\approx 1 \div 2$ cm
- northern side (courtyards and sheds) $\approx 14 \div 16$ cm

Figure 12 - View of the eastern side of foundation pit and building MD

At the northern side of intervention, vertical displacements of 15 cm magnitude were measured (Figure 13). The horizontal displacement is estimated to be 10 to 15 cm. Cracks opened in the ground at a distance of $4 \div 5$ m from the enclosure structure. Buildings being situated in this part along the perimeter of excavation, which are small and poorly built, originally not being foreseen for protection, suffered considerable damage.

The high table of ground water which was confirmed by observation from subsequently installed piezometers caused such large deformations. It transpired that the water table on the north-eastern side of the intervention remains approx. 2 m higher than in the rest of the intervention. Here, the retaining structure was streng-

667

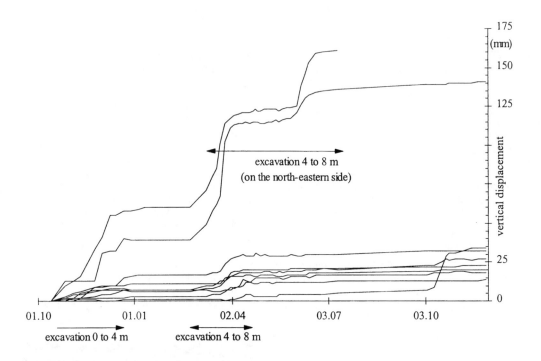

Figure 13 - Settlements along the excavation perimeter

thened by additional 16 m long drilled anchors.

Settlement of building M1 on the side facing the foundation pit was 15 mm, which resulted in hair-like cracks in the lower storey. Here appeared an opening of 15 mm to the adjacent building. It is assessed that this is the magnitude of horizontal displacement of building towards the foundation pit.

Buildings MD and M5 have shown settlements of the 15÷20 mm magnitude on the side facing the foundation pit and hair-like cracks appeared here too. Between the building M5 and the adjacent street building in row appeared an interspace of 1÷2 mm.

6. CONCLUSION

The enclosure of excavation by soil nailing can be successfully performed in the environment of soft clays.

The process demonstrates that a particular attention is required in the course of works, because errors can have quite unpleasant consequences.

According to the experience gained, immediate settlements at excavation are considerably pronounced in soft clays. If the clay had better deformation properties (e.g. Ic > 0.6), immediate deformations would be considerably lesser. In such a case the classical approach to the soil nailing would

also be an acceptable engineering solution. To neutralize immediate deformations of soft clays, H sections and stiff lining wall were introduced.

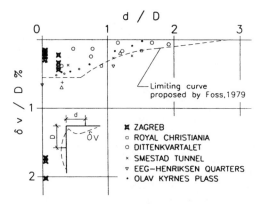

Figure 14 - Immediate settlements behind anchored retaining structures in soft to medium stiff clay (Athanasui, X ECSMFE, 1991)

At the north-eastern side displacement overstepped the expected ones, due to the fact that low buildings along the edge of site were not planned to be protected. Besides, on the same side increased tables of ground water were recorded, so the retaining structure was stabilized by execution of additional anchors.

The performed observations of displacements show that ground settlements at the perimeter of excavation of the remaining part of the pit are within the magnitude of 0.2÷0.4% of the excavation height. This is within the framework of acceptable values recommended for such kind of materials (Foss, 1979). The measured displacements were compared with measured values at five other sites, and these values are entered into the diagram according to Athanasui (X ECSMFE, 1991), shown in Figure 14.

REFERENCES

Athanasui C. M & Schram A, 1991. *Back-calculation of Case Records to Calibrate Soil-structure Interaction Analysis by Finite Element Method of Deep Excavation in Soft Clays*, X ECSMFE, Firenza 1991, Associazione Geotecnica Italiana

French National Research Project Clouterre, 1991. *Soil nailing Recommendations-1991*. Federal Highway Administration with the permission of the Presses de l'Ecole Nationale des Ponts et Chaussees

Kvasnička P. & Matešić L. & Vukadinović B. 1998. *Analisys of an Example of a Nailed Wall in Soft Clayey Soil*, XI Danube - European Conference on Soil Mechanics and Foundation Engineering

Vučetić, M. & al. 1994. *Analysis of Soil-Nailed Excavations Stability During the 1989 Loma Prieta Earthquake in California*, USGS Professional paper

Geotechnical Hazards, Marić, Lisac & Szavits-Nossan (eds) © 1998 Taylor & Francis, ISBN 90 5410 957 2

Polish experiences on founding and underpinning of buildings by means of jet grouted piles

Z. Żmudziński & E. Motak
Cracow University of Technology, Geotechnical Institute, Poland

ABSTRACT: This paper presents Polish experiences with application of piles executed by jet grouting method for underpinning the existing and founding of the new buildings. The authors elaborated a method of evaluation of bearing capacity of the jet grouted piles as presented in the paper, which has enabled solving of the designing problems. In the paper some examples of underpinning and founding the new, mainly industrial buildings, together with the results of test loadings and laboratory testing of soilcrete strength are presented.

1. INTRODUCTION

The method of jet grouting, invented in Japan in 1973 found during subsequent years a broad application in many countries of Western Europe. That method consists in cutting the soil and mixing its particles with cement grout, flowing out of a nozzle in the drilling rod at a pressure of 20 to 70 MPa at a speed over 100 m per second. The soil particles surrounded with cement grout create a mixture, which after hardening constitutes a building material: soilcrete. The method of jet grouting allows to form elements such as piles, pillars and walls of different dimensions in dependence on the kind and state of the soil in the substrate and on the pressure of the grout flowing out of the nozzle. Forming of soilcrete elements is possible as well in cohesive as non cohesive soils, situated in any position, and even of elements of changeable dimensions. This method finds actually broad application in building as well as in civil and water-engineering thanks to its numerous technical and ecological advantages. The use of cement for forming elements assures a constantly chemically neutral behaviour of the hardened soilcrete in the underground. Executing of soilcrete elements does not cause vibration, shocks and disturbances in the vicinity of the place of work performance. In foundation practice that method is used for underpinning the existing and also for founding new objects.

For a couple of years piles executed by the jet grouting method have been used in Poland as the effect of activity of two enterprises „Geocomp" in Cracow and „Geoservice" in Wrocław. Introduction of that kind of piling enabled cooperation between the mentioned firms and specialists from the Geotechnical Institute of Cracow University of Technology; they worked out a method of evaluation of bearing capacity for jet grouted piles, elaborated a series of projects of new foundations and also of underpinning existing buildings, and subsequently performed the author supervision over the execution of piles, carried out test loadings and tested resistance of soilcrete material of the piles. This paper presents some of these realisations.

2. PROPOSED METHOD FOR COMPUTATIONAL EVALUATION OF BEARING CAPACITY OF JET GROUTED PILES (ŻMUDZIŃSKI & MOTAK 1995).

During designing the first foundations on jet grouted piles, the authors came across a serious difficulty, which was lack of a method enabling evaluation of bearing capacity of such kind of piles. It was not inserted in the Polish Standards PN-83/B-02482 and the authors could not find any information about it in technical publications. In that situation the authors decided to elaborate an adequate method in their own scope. Very important information was provided in the publication of Bustamante. & Gianeselli (1994) where the results of measurements of deformation distribution along the pile axis during test loadings were inserted. These proved, that over 90% of total bearing capacity of the pile loaded with axial compressive force causes friction resistance on its lateral surface. An analogous assumption was made by the authors of the paper in the proposed method. Use was also made of partial safety coefficients according to PN-83/B-02482 which should assure the global safety coefficient F=2,0 for piles loaded with compressive force and F=2,2 for piles loaded with pulling out force.

According to PN-83/B-02482 the condition of the limit state of bearing capacity for a jet grouted pile is described by the formula:

$$Q_r \leq m \cdot N \qquad (1)$$

where: $Q_r = \gamma_f \cdot Q_n$ - computational (designed) axial load of the pile, kN

$\gamma_f > 1$ - coefficient of loading

Q_n - characteristic load, kN

m - correction coefficient; m=0,7 -compressed pile, m=0,65 -pulled out pile,

N - computational bearing capacity N_t of compressed pile or N^w of pulled out pile.

Computational bearing capacity of the compressed pile is given by the formula:

$$N_t = 1,1 \ \Sigma \ t_i^{(r)} \cdot A_{si} \qquad (2)$$

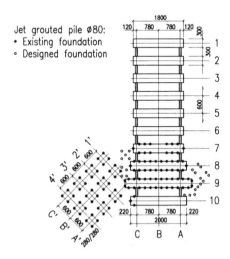

Jet grouted pile ⌀80:
• Existing foundation
○ Designed foundation

Figure 1. Underpinning with jet grouted piles of foundations in Rzeszów.

where: $t_i^{(r)} = \gamma_m \cdot t_i^{(n)}$ - computational value of unit friction resistancte on lateral surface of the pile in the geotechnical layer „i", kPa

$\gamma_m \leq 0,9$ - material coefficient, evaluated according to the Polish Standards PN-81/B-03020

$t_i^{(n)}$ - characteristic value of unit friction resitance, kPa

A_{si} - lateral surface of the pile within the geotechnical layer „i", m²; $A_{si} = \pi D \cdot h_i$

h_i - depth (thickness) of geotechnical layer „i", m

Computational bearing acapacity of pulled out pile is designated with the formula:

$$N^w = \Sigma \ t_i^{(n)} \cdot A_{si} \qquad (3)$$

where the indications are the same as in formula (2). The values of $t_i^{(n)}$ were accepted according to PN-83/B-02482 and Bustamante & Doix (1985).

3. SELECTED EXAMPLES OF UNDERPINNING OF EXISTING BUILDINGS OR FOUNDING OF NEW ONES ON JET GROUTED PILES.

3.1 *Underpinning of existing foundations in Rzeszow (Żmudziński & Motak et al. 1994)*

Underpinning executed in 1993 concerned two buildings, the foundations of which, after realization of the buildings in unfinished state, were for seven years loaded only with 40% of the full foreseen load. The elaboration of design was preceeded by geotechnical survey of subsoil, which indicated, that below the foundation level silty soils occured, showing the liquidity index I_L=0,20 to the depth of 6,0m and I_L= 0,30 at greater depth. Having analysed the load of each foundation, underpinning of a part of strip foundations of building no 1 with 67 piles and underpinning of all footings of building no 2 with 48 piles and additionally execution of 23 new piles was designed. There were together 134 grouted piles 0,8m in diameter and 4,0 to 6,0m in length realised, their situation is presented in Fig.1. The piles were grouted at pressure of 45 MPa. The compression tests of soilcrete samples showed mean value of its resistance R_{c28}=5,57MPa. The piles were executed by enterprise Geoservice, Wrocław (*Rybak, Borys, Noga 1993*).

3.2 *Test loadings of jet grouted piles in Kostrzyn (Żmudziński & Motak 1995).*

In 1994 a foundation on jet grouted piles in Kostrzyń

for a building of framework reinforced construction, 55m of length, 15m of breadth was realized. It consisted of 132 piles 0,40/0,55m in diameter and 9,0m in length, reinforced axially with steel pipe $\varnothing$ 127/95mm. The piles were carried out by enterprise Geoservice within four weeks. The authors performed test loadings for two from these piles (that was for the first time in Poland). The soil conditions and results of the loadings are presented in Fig. 2 a, b. The piles were executed of grout c/w=1,0 made of cement brand 35, at jet pressure 40-45 MPa. Test loadings were executed after 31-32 days following grouting. The admissible load of tested piles, settled as result of test loadings was 1004 and 897 kN and it exceeded the designed values 700 and 650 kN. The

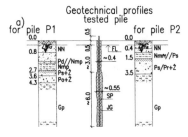

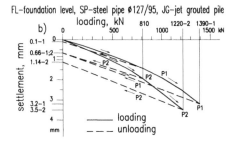

Figure 2. Results of pile load testing in Kostrzyń.

compression strength of soilcrete samples, taken out of the tested piles, was equal $R_{c28}=16,7 - 19,5$ MPa. By the end of 1995 and during 1996-1997 further designs were elaborated and jet grouted piles were executed for underpinning of existing buildings and construction of new objects of the big industrial works in Cracow in the area of occurrence of silty-clayey soils of loess origin.

3.3 Underpinning of an industrial hall and foundations of a trestle (Żmudziński & Motak 1997).

The hall of steel construction with a reinforced concrete ground floor, built in 1962, was founded on footings resting on Franki piles. As a result of reconstruction of the hall, connected with increase of loads acting on the ground, the underpinning of the existing footings by means of jet grouted piles was decided.

The subsoil of the hall consisted of silty embank-ments of mean thickness 4,1m, below which a layer of soil of loess origin 8,6m thick occurred; its upper part made of silt, mean $I_L=0,16$ and its lower part of clayey silt, mean $I_L=0,24$. Beneath a sand layer 1,7m thick appeared, under which a layer of organic clay 5,4m thick occurred, showing mean $I_L=0,26$. It was underlayed by a series of non cohesive soils: sands and gravels of a great depth. A stabilized groundwater level was stated at the depth 16,0 - 17,6m under the terrain surface.

The underpinning concerned 30 footings and was realized by means of 111 jet grouted piles, 0,50 and 0,60m in diameter and 10 and 11m in length. The required admissible bearing capacity of piles was equal 800 to 950 kN. It was computed by the method elaborated by the authors (Żmudziński & Motak 1995). The required compression strength of soil-crete was $R_{c28}=7,0$ MPa. For limiting the deformation of the pile body, the piles were reinforced axially with a steel pipe $\varnothing$ 102/82, made of steel R35. The underpinning was executed by enterprise Geocomp Cracow in winter time, within 2,5 months under supervision of the authors. The grout c/w=1,2, made of cement brand P-45 was jeted with pressure 40 MPa. Mean expenditure of cement for 1m of pile was 303 kG. As example the underpinning construction of one footing is presented in Fig. 3a, b.

In the vicinity of the described hall a new designed trestle of steel construction was built. It was supported with seven footings, resting on 48 jet grouted piles 0,70m in diameter and 12,0m in length, vertical or inclined at 1:5 or 1:10 to the vertical. The piles were reinforced with steel pipes $\varnothing$ 108/80 mm. made of steel St18G2A. The required admissible bearing capacity of the piles was 850 kN. The soil conditions were approximately the same as described before, as well as the technological requirements concerning the jet grouted piles. From the number of 159 piles for the hall and the trestle as a whole, four were test loaded, according to the Polish Standards

PN-83/B-02482. All the piles were in the first cycle loaded to the designed load, then unloaded and in the second cycle they were loaded until they reached 1,64÷1,83 overloading coefficient, however one of the piles was in consequence of a break down of the hydraulic jack overloaded only to 1,49 overloading coefficient. In Fig. 4 the graphical presentation of results of one test loading is shown.

The interpretation of test loading results was also performed by the analytical method (Żmudziński & Wolski 1997) which allowed to confirm, that at the designed load the settlement of piles was equal to 1,51 to 3,65mm, mean 2,49mm. The admissible of piles were also computed by the analytical method, the value of which proved to be at 13,7 to loads 48,6% (mean 29%) higher than the designed ones.

The tests of physical properties and compressive strength of soil-crete, carried out on 57 samples, taken by the method of core drilling from four piles at distances of 1,5m along the pile axis, gave following mean results: water content 42,75%, density 1498 kg/m^3, compressive strength of samples of slenderness λ=2 R_{C28}=8,80 MPA, modulus of elasticity E=2908 MPa.

3.4 *Underpinning of a water container of volume 400m^3 (Żmudziński & Motak 1997)*

A cylindrical steel container was supported by means of a reinforced concrete rim and an overground wall of brick 0,38m thick, 2,1m high, resting on a cylindrical concrete wall 0,6m which went down stepwise into a foundation ring 8,74m in outer diameter and 5,94m in inner diameter, founded at the depth 3,40 m under the terrain surface.

The container was built in 1960. The subsoil of the container foundation consisted of: to the depth 0,8m silty embankments, from 0,8 to 1,8m silt I_L=0,25, from 1,8 to 2,9m silt I_L=0,40, from 2,9 to 5,0m silt interbedded by silty clay I_L=0,59, from 5,0 to 6,3m silty clay I_L=0,40, from 6,3 to 6,7m clay I_L=0,40, from 6,7 to 12,8m clay I_L=0,24, from 12,8 to 14,8m watered medium grained sand I_D=0,60.

Settlement measurements of the container, continued during the period 1961-1995 by levelling of 4 benchmarks, showed its mean settlement s_m=10,66cm and inclination θ=0,054. These values exceeded the admissible ones according to the Polish Standards PN-81/B-03020, equal s_m=7,0cm and θ=0,003. In container with 9 jet grouted piles, 0,70m in

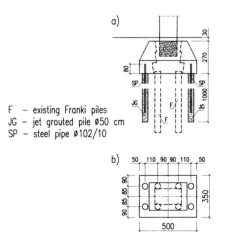

F – existing Franki piles
JG – jet grouted pile ⌀50 cm
SP – steel pipe ⌀102/10

Figure 3. Underpinning of one of the hall footings in Cracow.

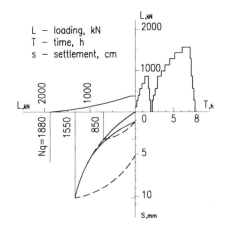

L – loading, kN
T – time, h
s – settlement, cm

Figure 4. Load testing results for one of jet grouted piles.

diameter and 10,0m in length, presented in Fig. 5 was elaborated. It assumed taking over by the piles 80% of the whole load of the container with water, according to evaluation the bearing capacity of piles by the method of the authors (Żmudziński & Motak 1995). The required compression strength of soilcrete was accepted at R_{C28}=7 MPa. For limiting their deformation the piles were reinforced with steel pipes ⌀ 102/82mm. The piles were executed after having drilled through the concrete foundation wall holes 110mm in dia, inclined at 8° to vertical. The upper ends of the pipes were connected by concreting with the wall on the length of 0,80m. Execution of underpinning took place at the end of 1995 by the Enterprise Geocomp, Cracow.

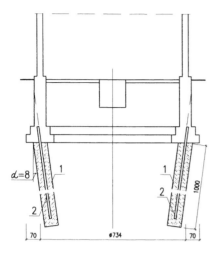

1 — jet grouted pile ⌀70cm
2 — steel pipe ⌀102/82mm

Figure 5. Underpinning of a container with jet grouted piles.

3.5 *Other realizations*

At the beginning of 1997 underpinning of two neighbouring buildings in connection with building an additional storey was realized (Żmudziński & Motak 1997). One of the buildings three storey, 60,6m long, 11m broad, built of bricks, partially provided with cellars, was founded on streap foundations at a depths -1,5, -3,4 and -4,1m. The second, four storey building, 36m long 15,3m broad, without cellars had a mixed construction: outer walls and walls at the staircase built of bricks were founded on streap foundations at a depth 1,5m, the inner columns were supported by footings. After 25 years of exploitation these buildings showed many cracks and fissures. Examinations of the subsoil of the buildings proved occurrence of a layer of weak silty soils (I_L=0,59÷0,44) 4,2m thick. Underpinning of the buildings was realized by means of jet grouted piles 0,50m in diameter, 7-8m in length, the number of which amounted to 48 for the first building and 27 for the second. Inclined piles were executed trough the core drilling holes in the walls from the terrain surface or from inside of the cellars.

In 1997 underpinning of two gable-walls of an industrial hall with jet grouted piles was executed. These walls showed great uneven settlements and it proved that they were founded on soft silty soil. Underpinning of each wall was realized with 16 piles 0,40m in diameter and 5,0 to 7,0m in length. Inclined piles were executed by means of holes drilled through the walls on both sides.

CONCLUSION

The Polish experiences so far have indicated a possibility of replacement of classical piles by jet grouted piles taking into account the range of the achieved bearing capacity and settlement, with simultaneous shortening of realization time and reduction of cost. Advantageous features of jet grouting technology are clearly revealed in difficult geotechnical problems, in particular in underpinning foundations of municipal buildings (for instance of historic buildings), in rooms of small height. The results of Polish realizations and testing of jet grouted piles proved to be better than foreseen and not worse than presented in publications.

REFERENCES

Bustamante, M & B.Doix 1985. Une méthode pour le calcul des tirants et des micropieux injectées. *Bull. Liaison Labo P. et Ch.140. 1985*

Bustamante, M. & L.Gianeselli 1994. Portance d'un groupe de colonnes de sol traité par „jet grouting" sous charge verticale axiale *Bull. Liaison Labo P. et Ch.189. 1994.*

Rybak, C., R.Borys & L.Noga 1993. Jet grouting - modern technology of strengthening underground and building foundations. *Inżynieria Morska i Geotechnika no 4/1993, Gdańsk (in Polish).*

Pietrzyk, K., Z.Żmudziński & E.Motak 1997. Selected problems of application of jet grouted piles. *Proc. of XLIII Scientific Conference of Civil and Water Engineering Committee of Polish Academy of Sciences, Krynica 1997, vol. VIII (in Polish).*

Żmudziński, Z., Motak, E., Noga, L., Łukasiński, S. & L.Ziemiański 1994. Unconventional methods of evaluation and increase of bearing capacity of existing foundations. *Proc. of Scientific Conference.„Building Damages" Szczecin-Międzyzdroje, 1994 (in Polish).*

Żmudziński, Z & E. Motak 1997. Application of jet grouted piles for underpinning of buildings in area of Cracow. Examples of realization. *Proc. of*

Scientific Conference. „Building Damages"
Szczecin-Międzyzdroje, 1997 (in Polish).

Żmudziński, Z & E. Motak 1995. Application of jet
grouted piles in building foundation. *Proc. of XLI
Scientific Conference of Civil and Water
Engineering Committee of Polish Academy of
Sciences, Krynica 1995, vol. VIII (in Polish).*

Żmudziński, Z & E. Motak 1995. Computational
evaluation of bearing capacity of jet grouted piles.
*Monograph no 194, Cracow University of
Technology, Cracow 1995 (in Polish).*

Żmudziński, Z & B.Wolski 1997. Bearing capacity of
jet grouted piles in analytical method of
interpretation of test loading results. *Proc. of
XLIII Scientific Conference of Civil and Water
Engineering Committee of Polish Academy of
Sciences, Krynica 1997, vol. VIII (in Polish).*

Geotechnical parameters

Geotechnical Hazards, Marić, Lisac & Szavits-Nossan (eds) © 1998 Taylor & Francis, ISBN 90 5410 957 2

Small strain stiffness of Venetian soils from field and laboratory tests

S.Cola, G.Ricceri & P.Simonini
University of Padova, Italy

ABSTRACT: This paper presents selected results of an extensive in-situ and laboratory geotechnical investigation on Venetian quaternary deposits, which formed part of a project on protection from flooding of the old city of Venice. In particular, the evaluation of the very small strain shear stiffness of cohesive and granular formations by very accurate in-situ and laboratory testing is considered and discussed.

1 INTRODUCTION

The city of Venice shows a rather precarious equilibrium, whose margin of security is being eroded annually at an ever increasing rate.

The rate of deterioration is being accelerated by the increasing frequency of the flooding of the old city, by the increase of pollution of both the lagoon and the atmosphere and by the reduction of the freeboard of the city as a result of the eustatic rise in sea level, coupled with a regional man-induced subsidence of the general lagoon area, which was important in the period 1950-1970 (Ricceri & Butterfield, 1974).

In order to protect the old city and the Venetian lagoon against flooding, the Italian Government decided to finance an important project regarding the design of special mobile barriers, located at the three inlets of the lagoon.

The design of the barrier foundations requires an accurate knowledge of the geotechnical parameters of Venetian soil formations.

To this purpose a comprehensive geotechnical investigation was carried out in two phases.

The first phase was considered as a preliminary investigation (to be used for the initial design of the barriers) and was performed in relation to the three lagoon inlets using some standard forms of investigation.

The second phase was designed in order to characterise the Venetian subsoil more accurately.

Before starting with the final stage of soil investigation, a special geotechnical test site had been selected to this effect at the Malamocco inlet - the so

called "Malamocco Test Site" (MTS) (Figure 1) - where, in a limited area, deep boreholes together with piezocone, dilatometer, selfboring pressuremeter, screw plate and cross/down hole tests were carried out on contiguous verticals. In addition, in order to obtain high quality undisturbed specimens, a new large diameter sampler was also used.

The aim of the MTS is to check the possibility of establishing reliable correlations and/or to validate available interpretation methods between in-situ and laboratory test results, thus leading to a more extensive use of the less expensive in-situ tests in the design of mobile barrier foundations.

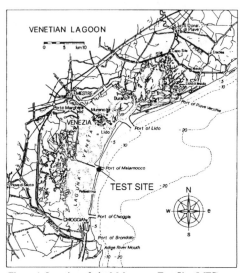

Figure 1. Location of the Malamocco Test Site (MTS).

This paper is particularly concerned with certain aspects in determining the maximum stiffness by in-situ and laboratory tests carried out within the MTS.

The design company Consorzio Venezia Nuova co-operated with ISMES and the University of Padova in the execution of these geotechnical investigations (Ministero dei Lavori Pubblici, 1994).

2 BRIEF GEOLOGICAL HISTORY OF THE VENETIAN SOIL

The quaternary deposits of the Venice Lagoon, reaching a depth of approximately 900-950 m, were formed throughout the Pleistocene period and

are composed of a complex system of interbedded sands, silts and silty clay sediments. Their accumulation took place in different phases, during which marine regression and transgression alternated with the rivers transporting fluvial materials from the nearby Alps.

In the twelfth century, when the first future citizens of Venice settled on the islands, the rivers Brenta, Sile, Piave and others discharged waters and sediments into the Venetian lagoon. In order to prevent this, the rivers were diverted into extensive canals around the lagoon's periphery.

After the eighteenth century no further hydraulic works were carried out and so the lagoon was not subjected to any significant alteration.

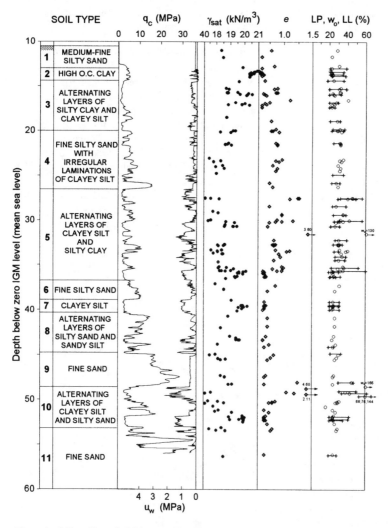

Figure 2. Soil profile at the Malamocco Test Site.

3 SOIL PROFILE OF THE MALAMOCCO INLET

Due to such a complex geological history, the sediments exhibit a great non-homogeneity with variation of particle size distribution even in a sample a few centimetres long. Hence it was very difficult to devise a scheme on a soil profile where the different formations (cohesive/granular) could be clearly distinguished.

Figure 2 depicts a tentative soil profile, up to 60 m below zero IGM level ($\approx$ mean sea level), determined from a borehole log and compared with the results of a piezocone test (q_c = tip resistance; u_w = pore pressure), carried out on a nearby vertical. Eleven basic formations were selected on the basis of the in-situ testing results. Figure 2 also shows the variations in the depth of the bulk density (γ_{sat}), the void ratio e, Atterberg limits (LL, LP) and natural water content (w_o) determined from the samples drawn up through two contiguous boreholes, the standard diameter borehole and the large diameter one.

It can be noted that the great majority of samples, other than sands or silty sands, fall into the category of silt ($PI \leq 10\%$, $LL \leq 35\%$) and very silty clay ($10 \leq PI \leq 25\%$, $35 \leq LL \leq 50\%$). Granular formations are composed mainly of medium-fine sand and fine silty sand ($D_{50} < 0.3$ mm). Some peaty layers - not depicted in the profile of Figure 2 - are embedded between formations 5 and 10.

4 IN-SITU STATE OF STRESS

The prediction of an in-situ stress state is usually a complicated problem and in the case of the Venetian soils is even more so, because they have undergone a complex stress history of unloading and reloading which has been shown to be very difficult to reconstruct with precision.

The trend of effective overburden stress σ'_{vo} with depth is reported in Figure 3, which also shows the values of preconsolidation stress σ'_p determined from oedometric tests using Casagrande's method.

Calculated laboratory values of $OCR = \sigma'_p/\sigma'_{vo}$ were plotted on the same figure and compared with those estimated using the results of a dilatometer test (DMT). An appreciable decrease of OCR was clearly observed. The high OCR values (>10) in the formation 2 are characteristic of the well known *caranto*, an high o.c. clay on which most historical venetian buildings are founded. The deeper formation are usually n.c. or slightly o.c., but there are some lowers with $OCR \approx 2$: for these soils, as for *caranto*, the overconsolidation is caused by superficial oxidation during glacial periods.

The coefficient of pressure at rest K_o, determined by the DMT and from the uniaxial reconsolidation stage in computer controlled CK_0D/U triaxial tests, is also plotted in Figure 3: K_0 decreases strongly with depth starting from values much higher than unity above 20 m and approaching 0.5 at greater depth. Note that the triaxial K_o values are less sensitive than DMT ones to overconsolidation states: this may be due to the stress relief particularly evident in silty soils and caused by the sampling disturbance.

5 MAXIMUM STIFFNESS FROM LABORATORY TESTS

The determination of small strain shear stiffness requires laboratory tests to be carried out with apparatuses and/or devices capable of measuring strain levels below 10^{-5}, which has been shown to be approximately a threshold value for the elastic reversible behaviour of most soils (e.g. Jardine, 1992; Tatsuoka and Shibuya, 1992).

Within this investigation, three types of apparatus:
- triaxial cell with local strain tranducers;
- triaxial cell with bender elements;
- resonant column.

5.1 *Triaxial tests with local strain tranducers*

Some very accurate undrained triaxial compression and extension tests (TX) were performed on 70 mm diameter and 140 mm height specimens, trimmed from large undisturbed samples. These tests were carried out using an automated stress/strain path

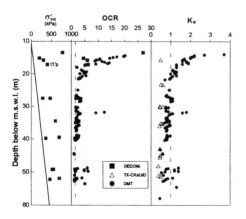

Figure 3. Soil stress history at the Malamocco Test Site.

triaxial system, recently designed and set up at the University of Padova.

In our case the triaxal cell was modified to measure internal strains on soil specimens (Ricceri et al., 1997). The local measuring system is composed of non contact proximeters with a resolution of 1 μm and targets glued onto the membrane: with 4 transducers employed for the vertical strain and 2 for the radial .

Figure 4 shows the maximum shear stiffness determined from those triaxial tests only carried out on the cohesive samples, plotted against vertical effective consolidation stress.

No general trend may be observed: this is due to

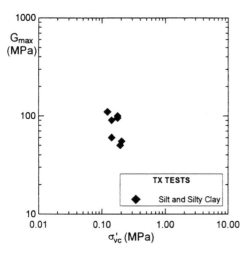

Figure 4. Maximum shear modulus in TX tests.

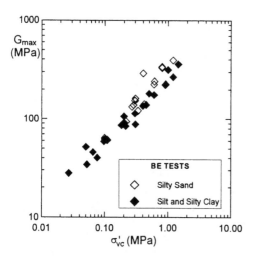

Figure 5. Maximum shear modulus in BE tests.

the limited range of consolidation stresses adopted in these tests.

5.2 *Triaxial tests with bender element system*

The Bender element (BE) system has been used in several triaxial tests, carried out both on sandy and clayey silty formations. In our case, piezoceramic elements (transmitter and receiver) are embedded in the base pedestal and in the top plate of the triaxial apparatus.

Shear wave velocity measurements have been recorded on 50 mm diameter and 100 mm height specimens at increasing K_o stress levels during the consolidation stage.

Figure 5 depicts the values of G_{max} as a function of vertical stress. The trend of shear modulus is quite regular with slightly higher values for the sandy samples, especially at greater stress levels.

5.3 *Resonant column tests*

Resonant column (RC) tests have been performed with a "fixed-free" apparatus on both granular and cohesive soils. The torsional vibrations are applied at different consolidation levels on 50 diameter and 100 height specimens. The range of consolidation stress was selected between 0.5 and 1.0 of the geostatic vertical effective stress.

The trend of G_{max} versus p'_c, plotted in Figure 6, is similar to that obtained from BE tests: no particular difference can be appreciated between the two types of formations.

Note that in the resonant column test the stress state imposed on the soil before shearing is isotropic: this may affect comparison with the other tests where a K_o stress state acts in the soil.

6 STIFFNESS FROM INSITU TESTS

Another way to estimate soil stiffness at very small strain levels can be provided by the seismic in-situ tests. With these, the disturbance inherent to the operations of sampling, trimming and setting up the specimens for laboratory testing could be avoided. On the contrary, when the soil properties vary significantly with depth (i.e. the case of Venetian soils) it seems relatively difficult to relate the exact measured velocity to the layer interested by wave propagation.

At the Malamocco Test Site, two different seismic testing techniques were used, i.e. cross hole (CH) and down hole (DH) test.

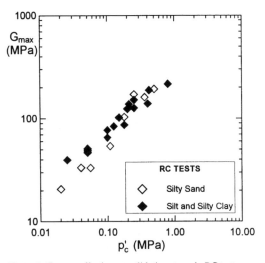

Figure 6. G_{max} vs. effective consolidation stress in RC tests.

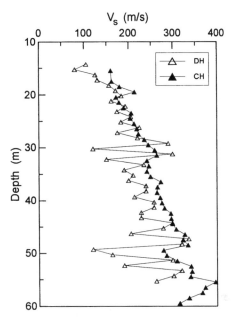

Figure 7. Profiles of V_s with depth from in-situ tests.

6.1 Cross hole test

Shear wave velocities V_s were measured at depth intervals equal to 1 m in cross hole tests performed on three verticals located at the corners of an equilateral triangle circumscribing the core of the MTS. The distance among the three verticals is approximately 5 m.

Average values of wave velocity have been plotted in Figure 7 against depth. A general increase of the wave velocity with depth is observed in very small velocity variations among the different soil formations.

6.2 Down hole test

In order to perform down hole (DH) wave measurements a seismic cone was utilized. The tests consisted of measuring the time span required by the signal sent by the wave actuator to reach two geophones installed at the tip of the piezocone and spaced 1.0 m apart.

At the MTS, 61 down hole tests were carried out during the cone penetration. Wave velocity recording was taken at depth intervals equal to 1 m. In Figure 7 the velocity measurements up to 60 m were plotted against depth.

A comparison with CH values shows that the DH velocity is generally lower than those of CH. In addition, large variations of shear wave velocity were measured especially at depths between approximately 29-34 m and 45-52 m below m.s.w.l. This may be due to the presence of thin, deformable peaty layers -

not reported for the sake of simplicity on the profile shown in Figure 2 - which could influence the propagation of shear waves across the horizontal soil layering.

In order to calculate the field shear modulus G_f from both CH and DH shear wave velocities the following equation:

$$G_f = \frac{\gamma}{g} V_s^2 \qquad (1)$$

was utilized, where g is the gravity acceleration and γ is the average bulk modulus determined from laboratory tests on undisturbed samples.

Figure 8 shows the trend of G_f vs. the geostatic vertical effective stress: note the significant data scatter for the DH tests due to the large variations of wave velocity discussed above.

7 COMPARISON AMONG VARIOUS TEST RESULTS

The particular trends of G_{max} and G_f plotted in Figures 4-8 suggested the performance of a statistical analysis of data based on the use of some widely accepted semi-empirical equations.

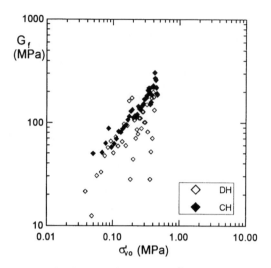

Figure 8. G_f vs. vertical effective stress in CH and DH tests.

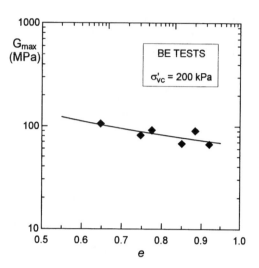

Figure 9. G_{max} vs. void ratio at constant σ'_{cv} in BE tests.

It is recognized that the value of shear wave velocity and, therefore, of maximum shear modulus is influenced by the mean effective stress p', by the overconsolidation ratio OCR, and void ratio e. In a simple form, G_{max} measured in the laboratory may be represented by the following expression (Hardin and Black, 1969; Vucetic and Dobry, 1991):

$$G_{max} = A \cdot f(e) \cdot \left(\frac{p'}{p'_r}\right)^n (OCR)^k \quad (2)$$

where:
A = dimensional material constant;
$f(e)$ = void ratio function;
p'_r = reference stress;
k, n = exponents.

It has been observed that the effect of OCR and of the horizontal effective stress is negligible for lightly overconsolidated cohesive and granular soils, that is the soils showing small variation of K_0 with depth (Afifi and Richart, 1973; Jamiolkowski et al. 1994; Shibuya and Tanaka, 1996).

Therefore, the maximum modulus can only be related to e and σ'_v, which are obtainable in ordinary site investigation practice. Thus, the above expression can be expressed in a simpler form:

$$G_{max} = A \cdot f(e) \cdot \left(\frac{\sigma'_v}{\sigma'_{vr}}\right)^n \quad (3)$$

Some empirical relationships have been proposed in the past to express the function f(e) in eq. (2) or in eq. (3) (see Table I).

Table I. Suggested void ratio functions.

$f(e)$	Reference
$(2.97 - e)^2 / (1 + e)$	Hardin and Black (1969)
$0.67 - e / (1 + e)$	Shibata and Solearno (1975)
$e^{-(1.1+1.5)}$	Jamiolkowski et al. (1994)
$e^{-1.5}$	Shibuya and Tanaka (1996)

Equation (3) was used in the interpretation of our results introducing :

$$f(e) = e^m \quad (4)$$

where m is an exponent and the reference stress is assumed equal to 1 MPa.

Due to the reliable estimate of G_{max}, e and σ'_v obtained with the bender element tests, eq. (3) was applied in two steps in the analysis of their results.

The first step considered the influence of void ratio on G_{max} at constant σ'_v. Figure 9 shows the trend of G_{max} vs. e at σ'_v =200 kPa: the exponent m in function (4) turned out to be -1.07 (r^2=0.560).

Assuming that the value of m, the fitting of eq. (3) on all BE data gave the following expression:

$$G_{max} = 181 \cdot e^{-1.07} \left(\frac{\sigma'_v}{\sigma'_{vr}}\right)^{0.58} \text{ (MPa)} \quad (r^2=0.977) \quad (5a)$$

for silts and silty clays and:

$$G_{max} = 218 \cdot e^{-1.07} \left(\frac{\sigma'_v}{\sigma'_{vr}}\right)^{0.64} \text{ (MPa)} \quad (r^2=0.920) \quad (5b)$$

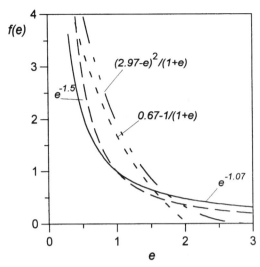

Figure 10. Trend of various *f(e)* functions used in current geotechnical practice.

for all soils.

For the sake of comparison the function $f(e)=e^{-1.07}$ is drawn on Figure 10 together with the functions reported above. The trend of the new function is very similar to that proposed by Jamiolkwski et al. (1994) and Shibuya and Tanaka (1996) but less sensitive to the void ratio variation for $e<1.0$.

Now it is interesting to compare the maximum shear modulus determined from the five types of tests - four dynamic and one static - carried out in the MTS investigation.

Figure 11 collects all the experimental determinations reported and illustrated in the preceding section. G_{max} and G_f are divided by the function $f(e)=e^{-1.07}$ and plotted against σ'_v. Note that in the case of in-situ data the vertical effective stress is the geostatic one whereas for laboratory tests it is the vertical consolidation stress (e.g. in the BE tests) ranging from an overconsolidated to a normally consolidated stress state.

Due to the relatively large variation of in-situ void ratio profile (Figure 2), one of the most difficult problems to solve in order to compare the field data with those from laboratory tests is the selection of representative values of e_o to be introduced into the function $f(e)$. Since CH and DH velocity measurements were taken at 1.0 m intervals, a value of e_o averaged within each meter of depth was utilized.

Both $G_{max}/f(e)$ and $G_f/f(e)$ show an approximate linear increase with vertical effective stress in the log-log scale. More particularly, CH and DH $G_f/f(e)$ values seem to represent an upper bound with respect to those values determined in the laboratory.

The expression (5b) was superimposed on the laboratory and field experimental data in Figure 11. In addition, the same data (150 points) were fitted using expression (3) with $f(e)=e^{-1.07}$. The following expression was therefore determined:

$$G_{max} = 216 \cdot e^{-1.07} \left(\frac{\sigma'_v}{\sigma'_{vr}} \right)^{0.61} \text{ (MPa)} \quad (r^2=0.740) \quad (6)$$

which differs only slightly from eq. (5b).

CONCLUSIONS

Selected results of an extensive site and laboratory investigation, carried out on the soils forming the seabed at one of the three inlets of the Venetian lagoon, have been presented in this study. More particularly, the estimate of the small strain shear stiffness carried out by using five types of test- four dynamic and one static - has been examined and discussed.

From the main results of this experimental investigation the following conclusions can be drawn:

1. Due to a complex geological history, the soils at the MTS exhibit a great non-homogeneity with variation of particle size distribution even in centimetric scale. Therefore it is relatively difficult to distinguish clearly between cohesive and granular formations;

2. Excluding the upper highly overconsolidated clay the cohesive soils under approximately 18 m are slightly overconsolidated or normally consolidated with K_o ranging from about 0.4 to 1.0;

3. Among the laboratory tests considered in this study, the triaxial ones equipped with bender elements have shown to be a versatile method of estimating the maximum stiffness. They can provide many simple determinations of G_{max} at different stress levels in K_o condition. In addition, this test enabled a reliable estimate of the current void ratio as a function of vertical consolidation stress to be carried out;

4. Comparing the cross hole and down hole test results the CH shear wave velocities are generally higher than those of DH, the latter being much more influenced by soil stratification especially in the presence of very deformable peaty layers;

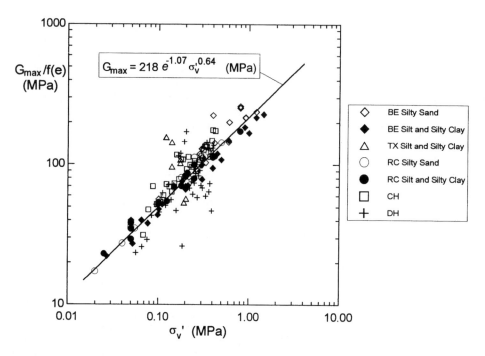

Figure 11. Correlation between G_{max}, e and σ'_v derived from both laboratory and in-situ tests.

5. Due to the reliable estimate of G, e and σ'_v obtained with the bender element tests, these values have been selected and successfully used to fit a semi-empirical three-parameter equation relating the maximum shear modulus to the void ratio and the vertical effective stress;

6. Finally, the same equation has been applied to the analysis of all data from field and laboratory investigations. In spite of the rough estimate of void ratio profile with depth used in the interpretation of field results, the general fitting showed very little difference with that determined by the bender element method.

REFERENCES

Afifi, S. E. A. and Richart, F.E. Jr (1973). Stress-history effects on shear modulus of soils. *Soils and Foundations*, Vol. 13, No. 1, 77-95.

Hardin, B. O. and Black, W. L. (1969). Vibration modulus of normally consolidated clay. *J. of SMFE Div.*, Proc. ASCE, Vol. 95, No. SM1, 33-65.

Jardine, R. J. (1992). Some observation on the kinematic nature of soil stiffness. *Soils and Foundations*, Vol. 32, No. 2, 111-124.

Jamiolkowski, M, Lancellotta, R and Lo Presti, D. C. F. (1994). Remarks on the stiffness at small strains of six Italian clays, *Prefailure behaviour of Geomaterials*, Shibuya, S, Mitachi, T. and Miura, S. (Eds.), Vol. 2, 817-836. Rotterdam: Balkema.

Ministero dei lavori pubblici - Magistrato alle acque (1994). Nuovi interventi per la salvaguardia di Venezia. Interventi alle bocche lagunari. Campagna di indagini geognostiche, prove geotecniche e prove di laboratorio. Bocca di Malamocco.

Ricceri, G. and Butterfield, R. (1974). An analysis of compressibility data from a deep borehole in Venice, *Geotechnique* 24, No. 2, 175-192.

Ricceri, G., Simonini, P. and Cola, S. (1997). Stiffness of clayey silts of the Venetian quaternary basin from laboratory tests. *Proc. XIV Int. Conf. On Soil Mech. Found. Eng.*, Rotterdam: Balkema.

Shibata, T. and Solearno, D.S. (1975). Stress strain characterstics of clays under cylcic loading. *J. of JSCE*, No. 276, 101-110 (from Shibuya and Tanaka, 1996).

Shibuya, S. and Tanaka, H. (1996). Estimate of elastic shear modulus in Holocene soil deposits. *Soils and Foundations*, Vol. 36, No. 4, 45-55.

Tatsuoka, F. and Shibuya, S. (1992). Deformation Characteristics of Soil and Rocks from Field and Laboratory Tests. Report of the Inst. of Industrial Science, Serial No. 235, University of Tokyo.

Vucetic, M. and Dobry, R. (1991). Effect of soil plasticity on cyclic response. Journal of Geotechnical Engineering, Vol. 117, No. 1, 89-107.

Geotechnical Hazards, Marić, Lisac & Szavits-Nossan (eds) © 1998 Taylor & Francis, ISBN 90 5410 957 2

Small-strain testing in an NGI-type direct simple shear device

M. Doroudian & M. Vucetic
Civil and Environmental Engineering Department, University of California, Calif., USA

ABSTRACT: A simple shear device named the Double Specimen Direct Simple Shear (DSDSS) device was designed to investigate static and cyclic properties of soils at small strains, such as the cyclic stress-strain loops, maximum shear modulus, G_{max}, secant shear modulus, G_s, and equivalent viscous damping ratio, λ. Cyclic shear strains between 10^{-4} % and 4% were successfully applied and measured in a single test on the same specimen. The measurement of very small strains was achieved by completely eliminating the friction of load-transfer mechanism and by reducing the effects of mechanical compliance of the device to practically zero. This was facilitated by using a special configuration of two parallel specimens.

1 INTRODUCTION

The cyclic behavior of soils is nonlinear and hysteretic, and consequently the shear moduli and damping of soils are strain-dependent. This dependency is shown in Figure 1, where a cyclic loop with all pertinent parameters is presented. In the figure, γ_c = cyclic shear strain amplitude, τ_c = cyclic shear stress amplitude, G_{max} = maximum shear modulus at strains approaching zero, and G_s = secant shear modulus corresponding to γ_c and τ_c. The hysteretic damping of the soil can be expressed by the equivalent viscous damping ratio

$$\lambda = \frac{\Delta w}{2\pi \ \tau_c \ \gamma_c} \tag{1}$$

where Δw = the area of the stress-strain loop (Jacobsen 1930). The figure also includes the initial loading backbone curve which is constructed by extending the initial loading curve into the negative domain.

In soil dynamics, it is customary to represent the cyclic behavior shown in Figure 1 in the normalized form in terms of the G_s/G_{max} versus γ_c and λ vs. γ_c curves plotted in a semi-logarithmic coordinate system. These two curves are presented jointly in Figure 2. Such representation of the initial loading cycle was originally introduced by (Seed & Idriss 1970). It can be seen that the horizontal logarithmic scale in Figure 2 allows for more precise presentation of the small-strain behavior than that in Figure 1. Figure 2 shows accurately how in the small-strain

range the damping increases with the cyclic strain amplitude, γ_c, while the normalized modulus is being reduced. Such nonlinear behavior necessitates precise evaluation of the cyclic stress-strain properties in the range of small strains for the analysis of soil dynamics problems such as the ground motions during earthquakes, machine foundation vibrations, traffic vibrations, behavior of the foundations of offshore structures subjected to ocean wave loads, etc.

To date, it has been common practice to obtain design curves such as those presented in Figure 2 by combining two types of laboratory tests. For the properties at very small strains, tests based on the wave propagation through the specimen have been used, while the medium and large-strain properties have been tested on separate specimens by cyclic triaxial, cyclic simple shear or cyclic torsional tests. More recently, however, considerable efforts have been made to develop testing devices which can directly measure cyclic properties of a single specimen over the entire range of small and large strains. As a result of these efforts, several triaxial and torsional devices for small-strain testing were designed and successfully used (Kim et al. 1991; Ampadu & Tatsuoka 1993; Tatsuoka et al. 1994; Stokoe et al. 1995).

In this paper, a recently developed simple shear device for small-strain testing is described and some typical results are presented. The configuration and basic idea of the newly developed device is shown in Figure 3. Its first version was developed and tested in 1995 (Doroudian & Vucetic 1995). In the meantime the device has been improved to test even smaller

687

strains and apply in the same test large strains almost to the failure. The reliable and accurate measurement of properties at very small strains was made possible by eliminating the friction of load-transfer mechanism and by reducing the effects of mechanical compliance and other sources of errors to practically zero. This was facilitated by employing a specially designed configuration of two parallel specimens, instead of one specimen configuration that is traditionally employed in direct simple shear and other geotechnical testing devices. Hence the new device was named the Double Specimen Direct Simple Shear (DSDSS) device.

2 DESCRIPTION OF THE DEVICE

In the past, the measurement of soil stress-strain properties at very small strains in direct simple shear (DSS) devices has been deemed practically impossible for two principal reasons. First, a typical DSS device has a considerable equipment compliance, and second, the friction of the load-transfer mechanism interferes with the load transmitted onto the soil specimen..

The friction of the load transfer mechanism was eliminated in the DSDSS device by introducing a double specimen concept shown in Figure 3. With the use of two soil specimens the load is applied directly via a load cell onto the soil. Besides that, the new DSDSS device was built from thick stainless steel components to minimize the mechanical compliance. As indicated in Figure 3, the shear stress in the new DSDSS device is calculated simply by dividing the horizontal shear force detected by the load cell by two times the area of the specimens.

In the DSDSS apparatus vertical load is applied pneumatically by a piston. The magnitude of the vertical load is adjusted by an air pressure regulator and is measured by the vertical load cell mounted between the top cap and a pneumatic piston shaft. During consolidation, the vertical settlement of the specimens is measured either by a dial gauge or an LVDT bridged between the top plate of the device and the rod which is connected directly to the top cap of the specimen (Doroudian & Vucetic 1995).

Horizontal cyclic load is measured by a load cell connected directly to the middle cap. The horizontal load can be applied either by means of a pneumatic piston, or manually. In the tests treated in this paper the horizontal load was applied manually. The pneumatic or a hydraulic cyclic loading systems were not considered because they introduce vibrations that are too large for the type of the small-strain testing conducted.

The horizontal displacement of the middle cap with respect to the top and bottom caps is measured by a proximity transducer bridged between one of these two caps and the middle cap. For better results, two proximity transducers can be mounted independently, one to the bottom and the other to the top cap, and their measurements averaged. The particular non-contact transducer used in this study is capable of position measurements of 0.00003 mm.

The above-described system for measurement of displacements bypasses all connections and parts which are the source of false deformations in the standard DSS device, except for the connections of the porous stones to the caps and the shear deformability of the caps and the porous stones themselves. To minimize these false deformations, the bottom, middle and top caps were manufactured from hard stainless steel and the porous stones from brass. The brass porous stones were firmly pressed in, glued, and in addition to that screwed to the caps. Based on simple calculations of the deformability of steel caps and brass porous stones, it was concluded that the effects of possible mechanical compliance caused by this deformability are negligible.

The specimens setup is designed in such a way that after the completion of consolidation the specimens can be aligned vertically by means of an adjusting system built into the base of the apparatus. Following such alignment, the top and bottom caps can be firmly fixed to the top and bottom plates of the device with the help of threaded rods mounted in the plates. Accordingly, such system of threaded rods enables the application of the classic NGI constant volume equivalent undrained DSS conditions proposed by Bjerrum & Landva (1966).

In the testing described here, the device was connected to a data acquisition system so that measured data could be automatically recorded and stored. In fact, the stress-strain curves generated during the testing were monitored and controlled on the computer screen in real time.

To enable the measurements of very small forces and displacements, in addition to the elimination of the load-transfer mechanism friction and reduction of the false deformations to negligible levels, the new device is also equipped with instruments of high precision. The measuring instruments proved to be capable of reproducing consistent results when subjected to the same loading and deformation conditions during their calibration. The horizontal displacement proximity transducer was calibrated using a high-precision micrometer. As already mentioned, its precision was determined as somewhat smaller than 0.00003 mm, which for the height of the specimen of 20 mm corresponds to the shear strain of around 10^{-4}%. The calibration curves were linear and did not show any hysteretic behavior. In addition to that, the output voltage signal of the proximity transducer was greatly

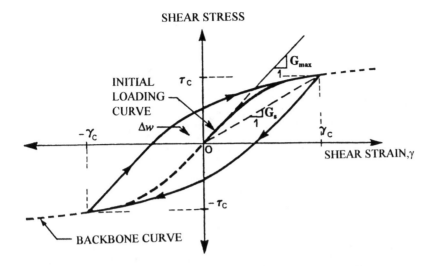

Figure 1. Idealized cyclic hysteretic stress-strain loop.

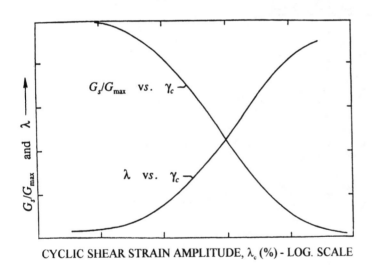

Figure 2. Secant shear modulus reduction curve and damping curve.

amplified. The amplification was such that the increment of displacement corresponding to a digitization voltage step of the data acquisition system was less than 0.000001 mm. If this displacement increment is divided by the height of the specimen, which was typically around 20 mm, it would corresponds to a shear strain well under 10^{-5} %. The precision and sensitivity of horizontal load cell was 0.0001 N, and it was calibrated for loading and unloading conditions of interest. The calibration did not show any output voltage shifting or hysteretic behavior.

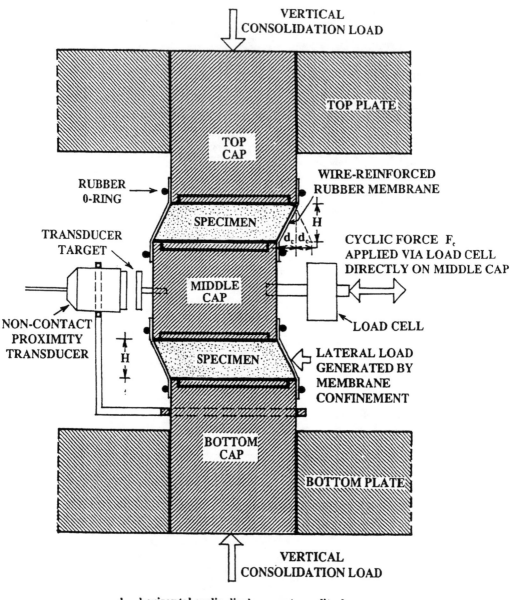

VERTICAL
CONSOLIDATION LOAD

TOP PLATE

TOP
CAP

RUBBER
0-RING

WIRE-REINFORCED
RUBBER MEMBRANE

SPECIMEN

H

TRANSDUCER
TARGET

d_c d_c

CYCLIC FORCE F_c
APPLIED VIA LOAD CELL
DIRECTLY ON MIDDLE CAP

MIDDLE
CAP

LOAD CELL

NON-CONTACT
PROXIMITY
TRANSDUCER

H

SPECIMEN

LATERAL LOAD
GENERATED BY
MEMBRANE
CONFINEMENT

BOTTOM
CAP

BOTTOM PLATE

VERTICAL
CONSOLIDATION LOAD

d_c = horizontal cyclic displacement amplitude
F_c = horizontal cyclic shear force
H = constant height of specimen
A = area of specimen
$\gamma_c = d_c/H$ = horizontal cyclic shear strain amplitude
$\tau_c = F_c/2A$ = horizontal cyclic shear stress amplitude

Figure 3. Specimens configuration, loading conditions, and definition of test parameters in DSDSS test.

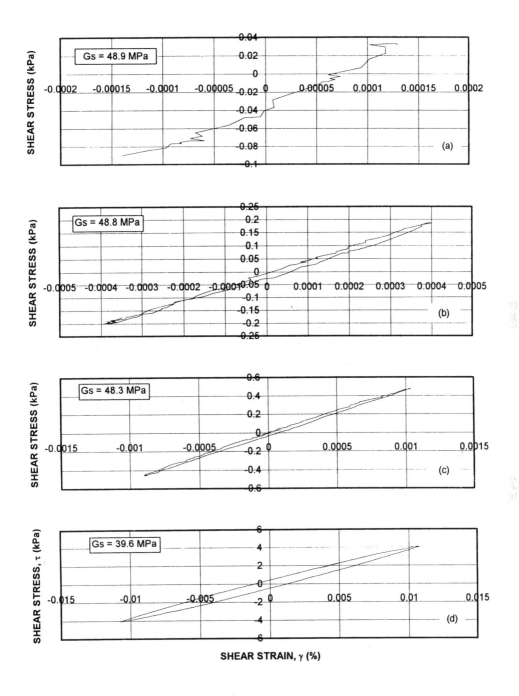

Figure 4. Stress-strain loops and corresponding secant shear moduli, G_s.

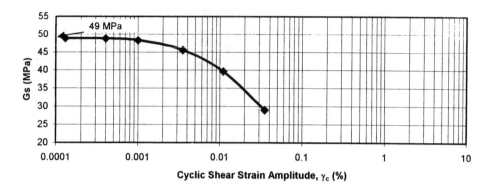

Figure 5. Estimation of maximum shear modulus, G_{max}, from the secant modulus reduction curve.

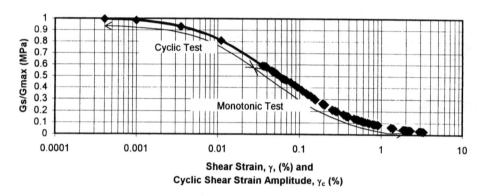

Figure 6. Normalized secant shear modulus, G_s/G_{max}, reduction curve.

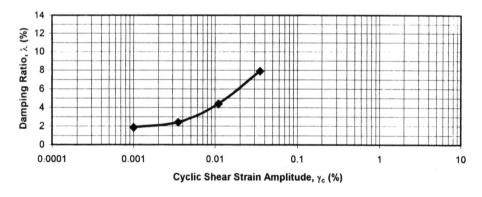

Figure 7. Variation of the equivalent viscous damping ratio, λ, with cyclic shear strain amplitude, γ_c.

3 EXAMPLE OF TEST RESULTS

To show the capabilities of the DSS device, the results of a series of 6 cyclic tests followed by a monotonic test on the same specimen of a clay are presented in Figures 4 through 7. The clay had the following characteristics after the consolidation under the vertical effective stress $\sigma'_{vc} = 95$ kPa and before the shearing: Plasticity Index, $PI = 6.1$, water content, $w = 20\%$, void ratio, $e = 0.59$, and the degree of saturation, $S = 89.3\%$. It was retrieved from a site in Los Angeles affected by the 1994 Northridge Earthquake from a depth of 5 m. The 6 cyclic tests were conducted at the following cyclic strain amplitudes: $\gamma_c = 0.00013\%$, 0.0004%, 0.001%, 0.0035%, 0.011% and 0.035%. After the last cyclic test conducted at $\gamma_c = 0.035\%$, the two DSDSS specimens were sheared monotonically to the shear strain $\gamma = 3.4\%$.

The stress-strain curves obtained at $\gamma_c = 0.00013\%$, 0.0004%, 0.001% and 0.011% are presented in Figure 4. As shown in Figure 4a, the stress-strain loops in this range of $\gamma_c = 0.0001\%$ could not be recorded but the secant shear modulus, G_s, could still be evaluated. Beyond $\gamma_c = 10^{-3}\%$, the cyclic loops were clearly recorded and both G_s and the equivalent viscous damping ratio, λ, could be easily and quite accurately determined.

To evaluate the usefulness of the DSDSS device for seismic site response and other dynamic analyses, the values of the secant shear moduli, G_s, are plotted versus cyclic shear strain amplitude, γ_c, in Figure 5. The best fit curve was drawn through the data points and easily extrapolated to shear strain of $10^{-4}\%$ to estimate the maximum shear modulus G_{max}. The value of G_{max} obtained in this way was 49 MPa. In Figure 6, the curve from Figure 5 is presented in the normalized form with respect to G_{max}, along with the results from the monotonic part of the test. The values of the equivalent damping ratio, λ, obtained from cyclic tests performed at 6 different γ_c amplitudes are plotted versus γ_c in Figure 7.

4 CONCLUSIONS

In the DSDSS device presented in this paper, the soil stress-strain curves were determined reasonably well in the range of cyclic shear strains, γ_c, from $10^{-3}\%$ to large shear strains. The secant shear moduli, G_s, were determined for cyclic shear strains as small as $\gamma_c = 10^{-4}\%$. The equivalent viscous damping ratio, λ, was determined with confidence for γ_c as small as $4 \times 10^{-4}\%$. Also, by extrapolating the G_s versus γ_c curve into the domain of very small strains of $10^{-4}\%$, it was possible to estimate the maximum shear modulus, G_{max}. In conclusion, the DSDSS device can be used for constructing both the G_s / G_{max} versus γ_c and λ versus γ_c design curves covering very small and large shear strains.

5 ACKNOWLEDGMENTS

The research described was partially supported by the National Science Foundation (NSF) Small Grant for Exploratory Research No. 9012975, and the Contract # 014493;01 from the Southern California Earthquake Center (SCEC).

REFERENCES

Ampadu, S. K. & F. Tatsuoka 1993. A Hollow Torsional Simple Shear Apparatus Capable of a Wide Range of Shear Strain Measurement. *Geotechnical Testing Journal, GTJODJ:* 16 (1): 3-17.

Bjerrum, L. & A. Landva 1966. Direct Simple-Shear Test on a Norwegian Quick Clay. *Geotechnique:* 16(1): 1-20.

Doroudian, M. & M. Vucetic 1995. A Direct Simple Shear Device for Measuring Small-Strain Behavior. *Geotechnical Testing Journal, GTJODJ:* 18(1): 69-85.

Jacobsen, L. S. 1930. Steady Forced Vibrations as Influenced by Damping. Transactions. *ASME:* 52(15): 169-181.

Kim, D., K. H. Stokoe & W. R Hudson 1991. Deformational Characteristics of Soils at Small to Intermediate Strains from Cyclic Tests. *Report No. 1177-3:* University of Texas: Austin.

Seed, H. B. & I. M. Idriss 1970. Soil Moduli and Damping Factors for Dynamic Response Analyses. *Report No. EERC 70-10:* University of California: Berkeley.

Stokoe, K. H., S. K. Hwang, J. N. K. Lee & R. D. Andrus 1995. Effects of various parameters on the stiffness and damping of soil at small to medium strains. *Proc Int. Symp. on Pre-failure Deformation Characteristics of Geomat:* 785-816. IS-Hokkaido: Sapporo. Japan. Rotterdam: Balkema.

Tatsuoka, F., S. Teachavorasinskun, J. Dong, Y. Kohata & T. Sato 1994. Importance of Measuring Local strains in cyclic triaxial tests on granular soils. *Dyn. Geotech. Testing II, ASTM STP 1213:* Ebelhar, R. J., V. Drnevich & B. L. Kutter. eds. 288-302.

Geotechnical Hazards, Marić, Lisac & Szavits-Nossan (eds) © 1998 Taylor & Francis, ISBN 90 5410 957 2

Evaluation of quick multistage oedometric relaxation test

E. Imre

Technical University of Budapest, Hungary

ABSTRACT: The 'quick' multistage oedometric relaxation test is a new type of the oedometric tests. Its duration is only one day since the stages are interrupted after 10-20 minute elapsed times except the last one that is longer than the 99% dissipation time. The points of the compression curve can be determined on the basis of the solution of inverse problems. These solutions may be non-reliable sometimes for short stages. In this case the parameters can be estimated with the use of the coefficient of consolidation identified for the last long stage using a special section of the merit function.

1 INTRODUCTION

The multistage oedometric relaxation test (MRT) is characterized with stepwisely increasing displacement load. The load imposition is not instantaneous since the volume of the sample - being constant during the stages - can be decreased between two stages only.

A research has been initiated about the multistage oedometric relaxation test. Model was elaborated and, inverse problem solution method was suggested. Various versions of the model were validated (Imre 1990, 1993, 1995a, 1995b, 1996).

A new multistage testing procedure, the 'quick' multistage oedometric relaxation testing procedure was applied that is treated in here. In this procedure the stages are interrupted before the 'end' of the pore water pressure dissipation. Points of the compression curve are determined by the solution of inverse problems.

Results of the evaluation of about twenty quick multistage oedometric relaxation tests are presented.

Model parameters were determined and compared with the ones identified from conventional compression test data. The reliability of the inverse problem solution was analyzed.

It follows from the first encouraging results that the applicability of the quick testing procedure is worthy to be examined for the conventional multistage compression test, too.

2 LABORATORY TESTS

2.1 *Methods*

Geonor type automatic swelling pressure apparatus h-200 A was used. The pore water pressure at the bottom, the total stress and the displacement at the top of the sample were measured.

The displacement load was increased in equal steps of 0.1 mm. The maximum load was between 0.6 and 2.0 mm. The rate of average strain was of 10^{-3} to $5 \cdot 10^{-2}$ %/s during the load increase.

Table 1. Soil physical parameters

Soil	1	2	3	4	5	6	7	8	9	10	11	12	13	14	15
I_P [%]	10.0	16.0	17.0	19.2	20.5	22.2	22.8	23.3	26.7	29.9	31.7	37.0	37.8	41.0	62.8
w_L [%]	32.4	37.2	41.7	42.2	52.5	44.3	57.9	53.7	65.1	63.0	56.1	64.1	63.6	72.0	119
I_L [%]	0.3	0.2	0.1	0.2	0.1	0.2	0.1	0	0.1	0	0.1	0	0.2	0.03	0.1

Quick multistage oedometric relaxation tests and conventional multistage compression tests (MCT) were made on twenty duplicate samples with various plasticity. Results concerning the silty-clayey soils are treated only (Table 1).

Most of the samples were taken from depths of 5 to 20 m (beneath the water table) in Szeged and Szolnok areas. The geotechnical features of these areas are described by Réthati et al (1978) and Baranya (1985) respectively.

2.2 *Stress and displacement histories*

The history of the total stress, pore water pressure at the sample bottom, their difference - the effective stress at the sample bottom - and, the history of the controlled displacement load are shown on the example of soil 7 in Figure 1.

In general, the relaxation behaviour was characterized by three subsequent phenomena that were as follows in chronological order : (i) time delay with constant total stress in the first seconds (that is not apparent in Figure 1), (ii) fast total stress decrease in the first seconds-minutes, (iii) total stress variation with time dependent rate.

The initial stress drop - with about constant stress rate - was significant if the displacement load was greater than about 1 % mean strain. The model does not concern this period. Therefore, two time variables (t' and t) are used throughout this work (t' elapsed time from the end of the imposement of the displacement load, t elapsed time from the end of the stress drop).

The total stress and the pore water pressure at the sample bottom basicly showed decreasing tendency from the end of the stress drop.

2.3 *Instability*

The displacement load was difficult to be controlled during the fast stress decrease. Overloading then partial unloading of the sample in terms of the displacement occurred in several cases that is indicated by symbol x in the $v(t',0)$ history.

This instability occurred if the load was larger than a threshold value v_I and, was followed by a rebound in the total stress if the load was greater than a second threshold value v_{II} being larger than v_I. Both threshold values were different for the various samples.

No rebound was observed for many samples and no instability occurred for some samples. The overloading generally increased with the stage number.

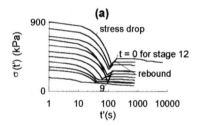

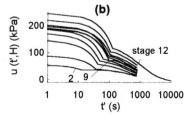

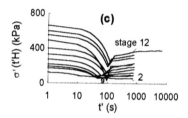

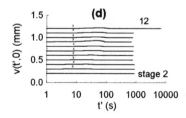

Figure 1. Time histories for (a) total stress, (b) pore water pressure, (c) effective stress, (d) displacement

3 MODELLING

A joined model was elaborated for the evaluation of the oedometric relaxation test. It consisted of a coupled consolidation part-model and, and empirical relaxation part-model. The solution of the model for the total stress σ and the pore water pressure u (Imre, 1993) are as follows :

$$\sigma(t) = -\sum_{i=1}^{\infty} \varphi_i \, e^{-i^2 \pi^2 T} + \sigma_\infty - \Delta\sigma^r(t) \tag{1}$$

$$u(t,y) = -\sum_{i=1}^{\infty} \varphi_i [\cos(\frac{i\,\pi}{H}y) - 1] \cdot e^{-i^2\pi^2 T} \tag{2}$$

where σ_∞ the instant total stress, $T = c_v t/H^2$ time factor, c_v the coefficient of consolidation, φ_i Fourier coefficients, t time, y distance from the sample top.

The relaxation part-model :

$$\Delta\sigma^r(t) = s \cdot \sigma(0) \cdot \frac{1}{1-sb} \cdot \log(\frac{t+t_1}{t_1+t_3}); t > t_3 \tag{3}$$

where s the coefficient of relaxation, t_1 delay time, t_3 the pause of relaxation, $b = \log((t_1 + t_3)/t_1)$. Values of t_3 and b are zero if no partial unloading occurs.

Parametric initial pore water pressure functions

$$u(0,y) = A \cdot |y|^3 + B \cdot |y|^2 + C \cdot |y| \tag{4}$$

$$u(0,y) = G \cdot (1 - e^{-\frac{|y|}{F}}) \tag{5}$$

relate to models 'H' and 'E', respectively.

4 INVERSE PROBLEM

4.1 Formulation of the inverse problem

The solution of the model is $v(t,p)$: $\Re^1 \times \Re^M \to \Re^L$ where t time and $p \in \Re^M$ parameter vector. The hipervector $v \in \Re^{L \times N}$ contains N measured/computed values for the model solution v^i (i = 1..L) at discrete values of time $t^1...t^N$. The map $v(p)$: $\Re^M \to \Re^{L \cdot N}$ is constructed in such a way that times $t^1...t^N$ are substituted in $v(t,p)$. The 'real-life/noise-free merit function' $R(p)$: $\Re^M \to \Re^1$:

$$R(p) = \| h(p) \| \tag{6}$$

is a norm of the error vector $h(p) = v(p) - v$ that is used in a normalized form in here. The global minimizer $a = [a^1...a^M]$ of the merit function is the weak solution of the inverse problem.

A noise-free data vector and, a 'follower' noise free merit function can be constructed for the global minimizer vector a. The global minima of the real-life and the follower merit functions coincide.

4.2 Subminimisation

The inverse problem was solved with the aid of a subminimisation the theory of which is as follows. A section of the merit function $R(f(p_1), p_1)$ - called minimal section - is defined where $p_2 = f(p_1)$: $\Re^{M-J} \to \Re^J$ is a map that renders one or more p_2 for a specified p_1 such that each p_2 is the solution of the following conditional inverse problem:

$$R(p_2, p_1) = \min!, \quad p_1 = p_{1,*} \tag{7}$$

It follows that the minimal section $R(f(p_1), p_1)$ and the merit function $R(p)$ have the same global minima. Therefore, if the minimal section is minimised instead of the merit function then the original inverse problem with M parameters can be reduced to an inverse problem with M-J parameters.

The notion of parameters with linear and non-linear dependence is introduced. Let us assume that the parameter vector is split into two parts :

$$p = (p_1, p_2), \quad p_1 \in \Omega^{(M-J)}, \quad p_2 \in \Omega^{(J)} \tag{8}$$

and, the solution $v(t,p)$ is written as follows :

$$v(t,p) = p_2 \cdot B(t, p_1), \quad B \in \Omega^{(M-J) \cdot L}, \quad v \in \Omega^{(L)} \tag{9}$$

If p_2 is maximal then $v(t,p)$ depends linearly on p_2 and, non-linearly on p_1.

If the solution depends linearly on p_2 then solution of the linear inverse problem (7) can uniquely be determined with the SVD algorithm (Press et al 1986).

4.3 Solution of the inverse problem

The measured data were evaluated in such a way that the minimal section of the merit function related to the non-linearly dependent parameters was minimised instead of the merit function.

The minimisation was made in such a way that a one dimensional minimal section was geometrically represented and the deep minima were selected. For this the (7) was solved and the merit function was evaluated along a coordinate mesh generated in the subspace of the non-linearly dependent parameters.

4.4 Reliability of the solution

The solution of the inverse problem is reliable if it is unique and, its error (e.g. confidence region) does not exceed the range of the parameter.

The solution is unique if (i) the map $v(p)$ is locally one-to-one type and, (ii) only one solution is physically admissible.

If the map $v(p)$ is locally one-to-one type then the global minima of the merit function are not degenerated. If the error of a parameter does not exceed the range of a parameter then the global minima of the merit are not quasi-degenerated.

Deep minima of the merit function were geometrically represented using the minimal sections of the merit function concerning a non-linearly dependent parameter.

Confidence intervals of the parameters were analytically determined with the covariance matrix method and, the Tshebyseff inequality.

5 MODEL IDENTIFICATION

Models H and E were validated in such a way that they were fitted on measured data and, the reliability of the solution was tested (Imre, 1995a).

The solution of the inverse problem needed excessive computational effort since these models contained 8/7 parameters, 4/5 out them were with non-linear dependence (Table 2).

It turned out that parameter t_3 was reliable in some cases only, parameter t_1 was not reliable. Therefore, some reduced model-versions were produced so that the number of the superfluous non-linearly dependent parameters could be decreased.

5.1 *Reduced model-versions*

In the reduced model-versions the relaxation part-model was modified or deleted as follows.

In model-version HCR/ECR value of t_3, t_1 are specified, a modified relaxation parameter (s_k)

$$s_k = s \cdot \sigma(0) \tag{10}$$

is identified, where term $\sigma(0)$ is as follows

$$\sigma(0) = \sigma_\infty + 2 \cdot A \cdot H^3 + \frac{4}{3} \cdot B \cdot H^2 + C \cdot H \tag{11}$$

$$\sigma(0) = \sigma_\infty + G(1 - e^{-\frac{H}{F}}) \frac{1}{H} \left(\frac{H}{1 - e^{-\frac{H}{F}}} - F \right) \tag{12}$$

for model-versions HCR/ECR, respectively. The solution depends linearly on s_k.

Model-versions HCRT/ECRT are the same as model-versions HCR/ECR except that parameter t_3 is identified, value of b is set to zero (i.e. the partial unloading effect is approximatively described).

The relaxation is neglected, only the consolidation is described in model-versions HC/EC.

5.2 *Model discrimination*

The reduced model-versions were evaluated in a such a way that they were fitted on measured data. The reliability of the solution of the inverse problem was checked and, the fitting error related to the various model-versions were compared (Imre, 1995b).

The fitting error was about the same for every model-version where the relaxation was not neglected. For the consolidation models HC/EC the fitting error was twice as large, these models were less precise.

Table 2. Model parameters and mesh points

Model	Parameters		Number of mesh points
	with linear dependence	with non-linear dependence	
H	A, B, C, σ_∞	$c_v, \sigma_\infty, s, t_3, t_1$	125,280
E	$G, \sigma_\infty,$	$F, c_v, \sigma_\infty, s, t_3, t_1$	1,250,280
HCRT	$A, B, C, \sigma_\infty, s_k$	$c_v, t_3,$	1044
ECRT	G, σ_∞, s_k	$F, c_v, t_3,$	10,440
HCR	$A, B, C, \sigma_\infty, s_k$	c_v	36
ECR	G, σ_∞, s_k	F, c_v	360
HC	A, B, C, σ_∞	c_v	36
EC	G, σ_∞	F, c_v	360

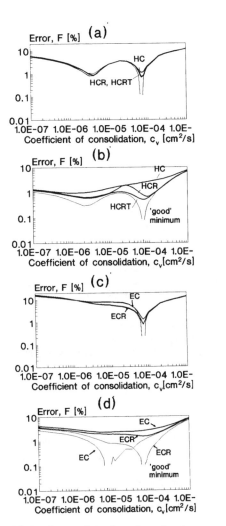

Figure 2. c_v minimal sections for long stages. (a), (b): Model family H with soils 13 and 10; (c) (d): Model family E with soils 13 and 10.

The c_v minimal sections for the various model-versions are shown in Figure 2 where real-life and follower noise-free merit functions are in thick and thin lines, respectively.

Generally two minima occurred for model family H (the one with the larger c_v value was physically admissible) and, one minimum for model family E.

The inverse problem solution was sometimes uncertain for models HC/EC (e.g. see Figs 2b, 2d). In this case approximate solution of the inverse problem was chosen using the c_v minimal section and a c_v value determined with a model-version containing a relaxation term.

6 EVALUATION OF THE SHORT STAGES

6.1 *Minimal sections*

The c_v minimal section related to the stages of a multistage oedometric relaxation test are shown in Figure 3 where the last long stage is in thick line, the short stages are in thin lines.

Two minima were encountered not only for model family H but also for model family E.

These minima were not necessary distinct in some cases yielding a single quasi-degenerated global minimum as it can be seen on Figures 3b and 3d.

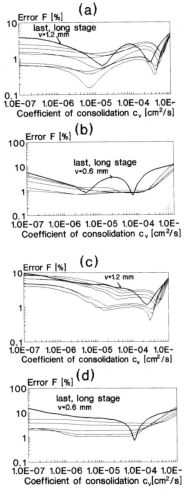

Figure 3. c_v minimal sections (a), (b): Model-version HCRT with soils 7 and 13; (c), (d): Model- version ECR with soils 7 and 13.

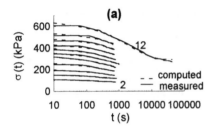

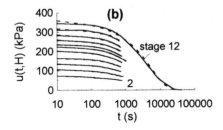

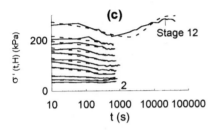

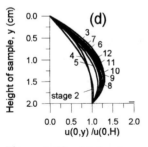

Figure 4. Simulated and measured stresses for soil 12: (a) total stress, (b) pore water pressure, (c) effective stress, (d) initial condition.

The slight deviation of the experimental and the theoretical curves can be attributed to the negligence of the relaxation modelling.

The stresses within samples are shown in Figures 5 and 6 for a clay and a silt sample. The effective stress basicly decreases at the upper, increases at the lower part of the sample with time for the clay. The effective stress is negative at some sample points for the silt sample indicating that the linear model is not acceptable.

6.3 *Compression curves*

The compression curve is a relation between the displacement load $v(0,0)$ and, the effective stress of the sample 'after' dissipation $\sigma_{>99\%}$. According to the experiences, this curve is time dependent since the constitutive law is time dependent.

The simulated final effective stress is equal to parameter σ_∞ for the consolidation models HC/EC. The identified value of parameters σ_∞ and, c_v are related to the time dependent compression curve.

For those model-versions that include the relaxation the simulated final effective stress is time dependent:

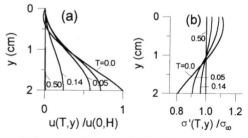

Figure 5. Stresses within sample 7: (a) pore water pressure, (b) effective stress

If the minima were not distinct then approximate solution of the inverse problem was chosen on the basis of the c_v minimal section using the c_v value determined for the last long stage.

6.2 *Simulated stresses*

Measured and computed MRT stress histories and the identified initial condition for soil 12, model-version HC are shown in Figure 4.

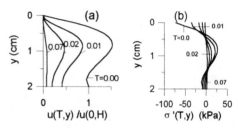

Figure 6. Stresses within sample 1: (a) pore water pressure, (b) effective stress

$$\sigma_{>99\%}(t) = \sigma_\infty - \Delta\sigma^r(t) \qquad (13)$$

The identified parameters σ_∞ and, c_v are time independent and are related to the 'instant' compression curve since the time dependency is described with the relaxation part-model.

It follows that the identified parameters σ_∞ and c_v have different meaning for those model-versions that neglect or include relaxation.

The MRT compression curves related to model HC are shown with the conventional compression curve in Figure 7.

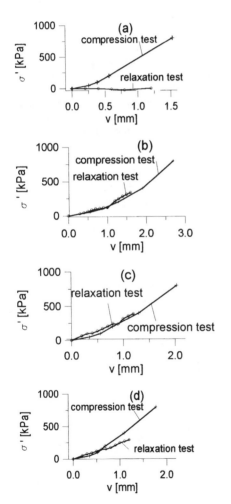

Figure 7. Compression curves for soils 1 (a), 6 (b), 7 (c) and 12 (d).

According to the results, the compression curve related to the multistage oedometric relaxation test is about identically equal to zero for the silt.

The compression curves determined from multistage oedometric relaxation test and multistage compression test are similar for soils with plasticity index of 20-23 % except that the MRT compression curves are generally non-monotonous.

The conventional compression curve is generally intersected by the multistage oedometric relaxation test compression curve for more plastic soils.

6.4 *Parameters*

Typical parameter values determined from multistage oedometric relaxation test data using model HC and multistage compression test data using the Terzaghi's model are tabulated in Tables 3, 4.

The coefficients of consolidation identified form MRT data are somewhat greater than the ones identified from MCT data. This indicates that the soil samples are probably in overconsolidated state during MRT stages.

7 DISCUSSION

7.1 *Stress drop*

Fats initial stress drop - that is possibly the dual counterpart of the initial settlement that occurs during multistage compression tests - is reported by Whitman (1957) and, Kondner & Stallknecht (1961) related to soil relaxation tests with fast partly drained load imposition.

The load imposition applied herein was partly drained and fast, too. The strain rate was larger than 0.001 %/s that is the upper validity boundary of the Leroueil's relation (1985).

7.2 *Parameters from MRT and from MCT*

The multistage oedometric relaxation test compression curves were not monotonous due to the small load increments reflecting the microstructural redistribution of the grain structure (Bagi, 1991).

The slight deviation of parameters identified from

Table 3. Soil 14, Terzaghi model, MCT results

load σ (kPa)	50	100	200	400	800
c_v (cm^2/s)	3e-5	1e-5	5e-5	9e-5	1e-4

Table 4. Soil 14, model-version HC, MRT results

v [mm]	$R(a)$ [%]	A [kPa/cm³]	B [kPa/cm²]	C [kPa/cm]	c_v [cm²/s]	σ_∞ [kPa]
0.1	1.28	9.44	-65.46	137.42	5.0E-4	103.75
0.2	1.49	10.67	-91.98	213.88	5.5E-4	141.55
0.3	1.59	18.05	-142.04	311.16	5.5E-4	181.26
0.4	1.45	30.39	-204.87	416.18	5.5E-4	225.30
0.5	1.90	26.33	-222.41	498.28	6.0E-4	277.51
0.6	3.82	123.01	-561.32	822.71	4.0E-4	308.83

multistage compression test data and multistage oedometric relaxation test data for clays can be explained by the differences of loading regimes.

For silts, large difference in the compression curves and negative simulated MRT effective stresses were observed. Both phenomena are possibly dependent on the initial stress drop as follows.

The effective stress

$$\sigma'(t,y) = \frac{1}{H} \cdot \int_0^H u(t,y)\,dy \ - u(t,y) + \sigma_\infty \qquad (14)$$

may be negative if σ_∞ is "too small", since the first two terms of the right hand side of Equation (15) is negative in some sample parts.

The initial total stress includes parameter σ_∞ :

$$\sigma(0) = \frac{1}{H} \cdot \int_0^H u(0,y)\,dy \ + \sigma_\infty \qquad (15)$$

The total stress drop detemines the value of the initial total stress.

8 CONCLUSION

According to the first results, the quick multistage oedometric relaxation test can be used for the determination of the compression curve and the coefficient of consolidation of the soil.

The deep minima of the merit function related to the short stages can be quasi-degenerated in some cases. This problem seems to be solvable using the c_v minimal section of the merit function and the c_v value determined for the last long stage.

The difference of the parameters identified from multistage oedometric relaxation test data and multistage compression test data is small for clays and, may be large for some loose silty soils.

REFERENCES

Bagi, K. (1991). Numerical analysis of granular assemblies. Doctoral Thesis. TU of Budapest.

Baranya, T. (1985). Szolnok altalajnak statisztikai értékelése. Mélyép. Szemle, 35:49-53.

Imre, E. (1990). Multistage oedometric relaxation test. Proc. of the 9th Nat. CSMFE Cracow, 1990. Vol. 1. 171-179.

Imre, E. (1993). Analytical studies of stress changes in soil under forced displacement load. Ph. D. Thesis, TU of Budapest.

Imre, E. (1995a). Evaluation of oedometric relaxation test data. Proc. of XI. ECSMFE, Copenhagen, Vol.3, 95:100.

Imre, E. (1995b). Model discrimination for the oedometric relaxation test data. Proc. of 8th. Baltic Conf. pp. 55-60.

Imre, E. (1996). Inverse problem solution with a geometrical method. Proc. of 2nd Inter. Conf. Inverse problem solution in engineering.

Kondner, R. L. & Stallknecht, A. R. (1961). Stress Relaxation in Soil Compaction. Proc. of Highway Research Board. Vol. 40, 617-630.

Lacerda, W. A. & Houston W. N. (1973). Stress Relaxation in Soils. Proc. of the 8th ICSMFE Vol.1. No.1. pp. 221-227.

Leroueil, S.; Kabbaj, M.; Tavenas, F.; Bouchard, R. (1985). Stress-strain rate relations... Geotechnique, Vol.35., No.2. pp. 159-175.

Mitchell, J. K. (1976). Fundamentals of Soil Behavior, Wiley, pp. 422.

Press, W.H.; Flannery, B.P.; Teukolsky, S.A.; Wetterling, W.T. (1986): Numerical Recipes. Cambridge Univ. Press, Cambridge

Rétháti, L. & Ungár, T. (1978). Nagyobb települések talajfizikai jellemzoinek statisztikai értékelése. Építés-Építészett. X. Kötet, 1-2.

Whitman, R. V. (1957). The Behaviour of Soils Under Transient Loading. Proc. of the 3rd ICSMFE Vol. 1., 207-210.

Geotechnical Hazards, Marić, Lisac & Szavits-Nossan (eds) © 1998 Taylor & Francis, ISBN 90 5410 957 2

Tonenklassification durch Korrelations-Regressionsanalyse

W.A. Iwanova, G.I. Trapov & P.Ch. Stoeva
Lehrschul 'Untertagebau', Universität für Bergbau und Geologie 'St. Iwan Rilski', Sofia, Bulgaria

ZUSAMMENFASSUNG: Der über der Kohle liegende Komplex des Ost-Maritza Lignittertiärbeckens (Bulgarien) ist aus Tonen und Sandtonen mit einer verschiedenen Mineralzusammensetzung aufgebaut. In dem mehrschichtigen (bis 120m) mächtigen Komplex sind vier lithologische Hauptabarten abgesondert.

Die Komplexenlaborangaben charakterisieren die senkrechte und laterale Inhomogenität der Tone, was auch durch die niedrigen Korrelationskoeffizienten zwischen den einzelnen Parametern bewiesen wind. Das bedingt die Ausnutzung einer mehrwertmäßigen Korrelations-Regressionsanalyse. Die Ergebnisse davon geben die Möglichkeit zur theoretischen und praktischen Anwendung der Bindungen zwischen den Eigenschaften der lithologischen Abarten.

Der Ost-Maritza Lignittertiärbecken (Bulgarien) ist aus Tonen, Sandtonen und Lignitkohle aufgebaut. Im Becken wird die Lignitkohle in drei großen Tagebauen gewonnen. Der mehrschichtige mächtige (120m) über der Kohle liegende Komplex ist den Eigenschaften, dem Mineralbestand und dem strukturellen Zustand nach mannigfaltig sowohl im senkrechten als auch in lateraler Hinsicht. Dieser Komplex stellt ein großes Interesse dar von Standpunkt der Standfestigkeit der Arbeits- und Standböschungen sowie der Grabbedingungen und der Leistung der großen Schaufelradbagger.

Der über der Kohle liegende Komplex besteht aus vier lithologischen Hauptabarten: schwarzen organischen Tonen, mit oder ohne Kohleneinschlüssen; graugrünen Tonen mit Karbonatstoff unter verschieener Form (dispergierte Kalkkomponente, Kalklecken unregelmäßiger Form, Kalkkonkretionen mit verschiedenen Ausmaßen); blaugrünen sandigen Oxidtonen und Sandtonen mit verschiedenen der Fläche nach Eigenschaften, mit einem unterschiedlichen Verbundenheitsgrad.

Zur Erhaltung einer komplexen Charakteristik der Tonabarten sind verschiedene Reviere aus dem über der Kohle liegenden Komplex an den Stellen abprobiert, wo die Leistung der Bergbaumaschinen gemessen ist. Im Labor sind die physikalischen Eigenschaften untersucht: Volumendichte (Ro); natürlicher Wassergehalt (Wn); Verlaufgrenze (WL); Korngrößenzusammensetzung: Tonfraktion (Mc), Sand (S), Staubfraktion (P); Festigkeitseigenschaften: Kohäsion (C), Innenreibungswinkel (Fi), Einachsendruckfestigkeit (Rn); Ankleben der Tone

im natürlichen Zustand auf Metallflächen (Pol) und Mineralzusammensetzung: Montmorillonit (Mm), Quarz (Qv), Kalkstoff (Cv), Sulphate (SO), Pyrit (Pr) und organischer Stoff (Ov).

Bei der Eigenschaftsanalyse wird die Klassifizierung der lithologischen Abarten in den oben angeführten vier Gruppen und entsprechend in den Untergruppen bestätigt.

Gruppe I: dazu gehören die schwarzen organischen Tone mit oder ohne Kohleneinschlüssen, mit einem Tonenkomponentegehalt von 54 bis 87%. In der Tonfraktion überwiegt der Montmorillonit, der von 34 bis 47% variert. Diese Tone enthalten Sand nur im punt 1 (Anprobepunkt) – S=15%. In den schwarzen Tonen ist eine bestimmte Quarzmenge (2÷8%) festgestellt, die dimensionell der Staubfraktion (P=13÷46%) (Tabelle 1) entspricht. In allen Untersuchungen bis jetzt für den Ost-Maritza Becken wird der Quarz an der Sandfraktion verbunden. Im Falle stellt der Quarz nach Ausmaßen eine Staubkomponente dar. Das wird durch den langen Transport der klastischen Stoffe aus den speisenden Provinzen erklärt.

Die wichtigsten Klassifikationsdaten, welche Gruppe I in ·vier Untergruppen teilen, sind: makroskopische Besonderheiten; Quarzgehalt; Plastizität und genauer WL; organischer Stoff; Ankleben; Standfestigkeit an Einachsendurch.

Untergruppe I.1. Sie wird mikroskopisch als schwarze organische Tone und eine Schicht aus Tonkohle charakterisiert.

Untergruppe I.2[a]. Sie ist von 70% Ton und 30% Holzkohlenschicht vertreten.

Tabelle 1. Gruppe I. Schwarze organische Tone

Anprob ierungs punkt	Beschreibung	Mc %	Mm %	S %	P %	Qv %	Ro g/cm³	Wn %	WL %	Ov %	Pol g/cm²	C MPa	Fi /°	Rn MPa
					Untergruppe I.1									
1	Schwarze Tone mit Sand und Tonkohle	65	34	15	20	8	1.64	42	84	13	20	0.203	15	0.660
					Untergruppe I.2ᵃ									
5	Schwarze Tone mit	64	47	0	36	3	1.56	48	90	11	28	0.204	8	0.570
6	Holzkohle (30%)	66	47	0	34	3	1.56	48	90	11	28	0.204	8	0.570
					Untergruppe I.2ᵇ									
11	Schwarze Tone mit Holzkohle (40%)	64	46	0	36	4	1.76	39	95	7	16	0.249	19	0.820
					Untergruppe I.3									
23	Schwarze	87	37	0	13	8	1.81	36	97	9	52	0.340	35	1.420
25	organische	66	40	0	34	3	1.53	61	96	26	50	0.160	10	0.380
26	Tone	54	42	0	46	3	1.71	47	102	20	52	0.312	28	0.960

Untergruppe I.2^b. Sie ist von 60% Ton und 40% Holzkohlenschicht vertreten.

Untergruppe I.3. Schwarze organische Tone – 100%.

Die oben angefürte Kategorisierung in Untergruppen beruht auf folgendem: der Prozent-quarzgehalt wird von der ersten zur letzten Untergruppe (von 8 bis 3%) kleiner; die Verlaufgreze (WL) wächst (von 84 bis 102%) an; der organische Stoff wächst auch an und in der dritten Untergruppe erreicht er bis 26%.

Die schwarzen organischen Tone werden mit dem höchsten Wn und der höchsten Plastizität (Tabelle 1) charakterisiert. Wichtige Charakteristik für diese Gruppe ist das Ankleben auf Metall, das von 20 bis 52g/cm² variiert und wächst von der Untergruppe I.1 zu der Untergruppe I.3 an. Die Einachsendruck-festigkeit und die Kohäsion wachsen in Richtung von der Untergruppe I.1 zur Untergruppe I.3 an. Das Rn variiert von 0.57 bis 1.42 MPa , und die Kohäsion – von 0.20 bis 0.34 MPa. Die ziemlich hohe festgestellte Festigkeit in der Untergruppe I.3, absehen davon, daß Kohlenschichte und Ingradiente fehlen, wird durch den Tixotropeneffekt der Wieder-herstellung der kolloidaktiven Bindungen in diesen Tonen im Zerschneidungsprozeß erklärt.

Gruppe I besitzt eine große makroskopische Ungleichartigkeit und Unterschiede in allen Eigen-schaften, was Probleme beim Tontypisieren schafft. Um eine verallgemeinerte Charakteristik gemacht zu werden, führte man eine Korrelations-Regression-sanalyse der Eigenschaften durch, die die Möglich-keit gab, bestimmte Abhängigkeiten mit theoreti-schem und praktischem Aspekt zu finden.

Die erchaltene Korrelationsmatrix zeigt die Bindung aller Paare von den untersuchten Eigen-schaften.

Als Abrechnungskriterium der Bindungsenge sind Korrelationskoeffiziente über 0.5 angenommen. Auf diesem Grund ist die Auswahl der für jede Gruppe ausschlaggebeuden Eigenschaften durchgeführt. So z. B. für die schwarzen Tone ist der organische Stoff eine spezifische Komponente und wird mit Wn verbunden; Wn beeinflußt die Festigkseitseigen-schaften (Kohäsion, Innenreibungswinkel, Einachsen-druckfestigkeit); das Ankleben auf den Metall-konstruktionen, was das Gewinnen und Transport der Tone erschwert, wird mit der Verlaufgrenze verbunden; Wn übt einen Einfluß auf die Volumen-dichte und den organischen Stoff usw. aus.

Gruppe II. Zu dieser Gruppe gehören die graugrünen plastischen Tone mit einem hohen Tonfraktionsprozent – von 51 bis 88%. Der Montmorillonit im Ton variiert von 19 bis 30%. Die Standfraktion ist von 1 bis 5% (Tabelle 2). Auch diese Gruppe besitzt Quarz (10÷29%), mit der Staubdimension und wird nicht von dem Prozentgechalt der Sandfraktion beeinflußt, wie der Fall mit P.8 – Qv=27% bei P=45% und S=0%, oder im P.19 – Qv=28%, P=31% und S=0% ist. Der Gehalt an organischem Stoff ist kleiner im Vergleich zu diesem in Gruppe I.

Die hohe Verlaufsgrenze dieser Tone (37÷90%) wird durch das hohe Montmorillonitprozent (19÷42%) bedingt. Die erhöhte Menge der Staub-fraktion – bis 46% bedingt die relative Verkleinerung des Wn (26÷46%) gegenüber der ersten Gruppe.

Diese Tone werden auch mit einem hohen Anklebegrad auf dem Metall (Pol=20÷80g/cm²) charakterisiert. Die Einachsendrukfestigkeit wird in

Tabelle 2. Gruppe II. Graugrüne Tone

Anprobierungspunkt	Beschreibung	Mc %	Mm %	S %	P %	Qv %	Cv %	SO %	Ro g/cm³	Wn %	WL %	Ov %	Pr %	Pol g/cm²	C MPa	Fi /°	Rn MPa
Untergruppe II.1																	
3	Graugrüne plastische Tone ohne Kalkstoff	62	30	1	37	19	0	5	1.89	28	73	0	3	21	0.108	31	0.508
Untergruppe II.2																	
8	Graugrüne plastischeTone mit	55	19	0	45	27	1	0	2.00	24	37	1	1	37	0.183	18	0.642
10	dispergiertem Kalkstoff	53	28	5	42	19	3	1	2.01	26	65	1	0	24	0.300	36	1.530
Untergruppe II.3ᵃ																	
13	Graugrüne	72	28	4	24	20	11	0	1.90	26	75	2	1	30	0.120	24	0.385
14	plastischeTone mit	66	28	0	34	16	13	0	2.00	25	70	3	1	40	0.272	24	0.655
15	Kalkflecken	88	29	0	12	10	0	3	1.68	46	90	18	0	61	0.140	7	0.628
Untergruppe II.3ᵇ																	
19	Graugrüne plastische Tone mit Kalkkonkretion mit verschiedenen Ausmaßen	69	19	0	31	28	0	0	1.97	26	45	1	0	80	0.217	8	0.675
24		51	42	3	46	14	5	0	1.87	37	86	3	0	20	0.150	13	0.534

den Grenzen von 0.385 bis 1.533MPa verändert. Diese Veränderung entspricht völlig der Veränderung der Kohäsion – von 0.108 bis 0.300 MPa und des Innenreibungswinkels – von 7 bis 36°.

Die wichtigsten Klassifikationsangaben, die die Gruppe II in 4 Untergruppen einteilt, sind: makroskopische Beschreibung; Quarzgehalt, bzw. Staubkomponente; Anwesenheit von Kalkstoff und Kalkkonkretionen; Ankleben.

Untergruppe II.1. Graugrüne plastische Tone ohne Kalkstoff.

Untergruppe II.2. Graugrüne plastische Tone mit dispergiertem Kalkstoff.

Untergruppe II.3ᵃ. Graugrüne plastische Tone mit Flecken unregelmäßiger Form und Flechen aus dem staubartigen Kalkstoff.

Untergruppe II.3ᵇ. Graugrüne plastische Tone mit Kalkkonkretionen verschiedener Ausmaßen.

Die oben angeführte makroskopische Charakteristik ist auf folgende Art und Weise verbunden: Untergruppe II.1. – Cv=0%; Untergruppe II.2. – Cv=1÷3%; Untergruppe II.3ᵃ. – Cv=11÷13%; Untergruppe II.3ᵇ – Cv – bis 5% plus Kalkkonkretionen, die bei der Untersuchung des Mineralgehalts nicht abgerechnet werden können.

Der Prozentgehalt des Quarzes und der Staubkomponente sind mit einer allgemeinen Erhöhungstendenz von Untergruppe II.1 zur Untergruppe II.3ᵇ verbunden. Mit der selben Tendenz ist auch die Anklebeerhöhung – von Untergruppe II.1 (21g/cm²) zur Untergruppe II.3ᵇ (bis 80g/cm²) verbunden.

Für die Festigkeitseigenschaften kann keine gesetzmäßige Veränderung abgerechnet werden, da sie von der Beteiligung des Kalkstoffes oder der Konkretionen abhängen. Die allgemeine Tendenz beim Zerschneiden dieser Tone ist, daß der Kalkstoff Widerstand beim Zerstören, unabhängig von seiner Art, ausübt. Die niedrigeren Festigkeitseigenschaften werden mit dem Mangel an Kalkstoff oder mit der Anwesenheit von Litifikations- und tektonischen Strukturfehlern (Tabelle 2) verbunden.

Die Korrelationsmatrix gibt eine engere Verbindung zwischen dem organischen Stoff und der Tonfraktion; der Volumendichte und dem natürlichen Wassergehalt, und dem natürlichen Wassergehalt mit der Tonfraktion, dem Montmorillonit, der staubartigen Quartfraktion und der Volumendichte. Die Einwirkung der Volumendichte und des natürlichen Wassergehalts auf die Werte des innenreibungswinkels und der Kohäsion ist offensichtlich. Das Ankleben ist eng mit der Tonkomponente und dem Montmorillonit verbunden, seine Eigenschaften waren oben angeführt. In dieser Gruppe sind die Korrelationskoeffiziente zwischen dem Ankleben und dem organischen Stoff 0.39, wovon ersichtlich ist, daß der organische Stoff keine entscheidende Rolle für diese Tone spielt und nur für den natürlichen Wassergehalt, die Volumendichte und die Tonfraktion aus schlaggebend ist.

Gruppe III. Blaugrüne Tone, sandige, kalkhaltige, mit oder ohne karbonateinschlüssen.

Der prozentgehalt der Tonkomponente in diesen Tonen variert in sehr breitigen Grenzen – von 16 bis

73%. Das hängt von der Stelle des Abprobierens der litolhogischen Abart im Massiv ab, der im Ganzen in bezug sowie auf Tongehalt als auch auf Sandgehalt inhomogen ist.

Der Montmorillonit ist in Übereinstimmung mit dem Gehalt der Tonfraktion, aber er ist verhältnismäßig kleiner als die I und II Gruppe.

Der Prozentgehalt an Quarz wächst in manchen Punkten bedeutend an, was auch hier von der beteiligten Staubfraktion abhängt und weniger (P.17, P.22, P.34 und P.42) von dem großen Sandprozent (P.17 – S=37%; P.22 – S=43%) (Tabelle 3).

Der Kalkstoff erreicht 38% und ist in größer Menge aus allen Gruppen beteiligt.

Die große Menge Standfraktion und die Anwesenheit eines hohen Kalkstoffprozentes beeinflußen die niedrigeren Werte von Wn (13÷24%) und die Verlaufsgrenze (WL=28÷72%).

Das Ankleben hängt von allen bisher beschreibenen Gesetzmäßigkeiten ab und erreicht bis 42g/cm². Die Proben, wo kein Ankleben vorhanden ist, werden als Kalktone oder Tonsande charakterisiert.

50%, p.17) und mit Kalkstoff (bis 38%, P.18).

In dieser Untergruppe sind die Tonkomponente (16÷33%) und der Montmorillonit (6÷14%) in geringer Menge, und die Staubfraktion – 32÷47%. Der natürliche Wassergehalt (15÷19%) und die Verlaufsgrenze (33÷41%) sind auch niedrig. Die Festigkeitseigenschaften für die drei Anprobepunkte sind auch verhältnismäßig niedrig (von 0.360MPa bis 0.532MPa).

Untergruppe III.2. Sie wird mit einem hohen Gehalt an Tonkomponenten (36÷73%) und entsprechend Montmorillonit bis 30% charakterisiert, die Staubfraktion sinkt bis 19% (Tabelle 3).

De Kalkstoff bewegt sich in den Grenzen von 0 bis 20% und hängt von dem Anprobieren im Massiv ab. Der natürliche Wassergehalt in dieser Untergruppe variiert von 13 bis 31%. Für diese Untergruppe hat der organische Stoff keinen Einfluß und stellt kein Interesse dar.

Beim Zerschneiden leistet der Kalkstoff Widerstand. Die Einachsendruckfestigkeit erreicht den Wert von 1.055MPa (P.30). Die niedrigeren

Tabelle 3. Gruppe III. Blaugrüne Tone

Anprobierungs punkt	Beschreibung	Mc %	Mm %	S %	P %	Qv %	Cv %	Ro g/cm³	Wn %	WL %	Ov %	Pol g/cm²	Rn MPa
					Untergruppe III.1								
17	Sand, grob, tonartig	16	6	37	47	50	2	1.97	15	33	0	0	0.452
18	Ton mit Sand und Karbonateinschlüssen	33	2	35	32	23	38	2.10	16	38	0	0	0.360
22	Ton mit Sand und Karbonateinschlüssen	19	19	43	38	31	18	1.86	19	41	3	8	0.532
					Untergruppe III.2								
30	Kalkton	62	26	4	34	16	18	2.03	23	68	2	7	1.055
32	Kalkton	73	30	4	23	14	20	1.82	24	63	2	9	0.433
33	Kalkton	70	30	4	23	14	20	1.82	24	60	2	9	0.433
34	Sand grob, tonartig mit Karbonateinschlüssen	20	20	46	34	39	0	2.19	14	40	0	0	0.679
35	Ton, staubartig (Emersion)	36	26	10	54	33	2	2.18	14	50	0	42	0.898
38	Kalkton	55	26	2	43	12	14	1.88	31	72	0	25	0.442
42	Sand, verschiedenenkörnig, tonartig	38	12	43	19	19	0	2.01	13	28	0	0	0.788

Die Einachsenfestigkeit ist bei den Kalktonen (P.30 – 1.055MPa, P.35 – 0.898MPa) und in den Tonsanden (P.42 – 0.788MPa) am höchsten.

Unabhängig von der großen Eigenschaftsmannigfaltigkeit und der makroskopischen Beschreibung kann diese Gruppe nach einigen Merkmalen in zwei Untergruppen eingeteilt werden:

Untergruppe III.1. Blaugrüne Tone mit Sand (bis

Werte des Rn – 0.433MPa werden von den Litifikations-, Mikro- und Makrospalten bedingt.

Die Korrelationmatrix der Gruppe hat einen spezifischen Charakter, da sie summar die Verbindungen zwischen den Parametern zweier sich nach Eigenschaften unterscheiden der Untergruppen wiederspiegelt. In diesem Fall schenkt man Aufmerksamkeit der Verbindung zwischen dem natürlichen

Wassergehalt und der Tonfraktion, dem Montmorillonit, dem Sand, dem Quarz und der Volumendichte, oder dieser zwischen dem Quarz und den Sand- und Staubfraktionen.

Gruppe IV. Dazu gehören Sande mit verschiedenen Strukturbindungen. In diesen lithologischen Abarten wird ein niedriges Prozent an Tonkomponente (12÷17%) und ein entsprechend niedriger Montmorillonitgehalt (6÷14%) (Tabelle 4) festigestellt. Der Sandfraktionsgehalt ist von 48 bis 71%, und an Quarz – 30÷67%. Die Staubfraktion ist 14÷35%. Der Quarzgehalt entspricht sowohl der Sand- als auch der Staubfraktion. Damit kann die Tatsache erklärt werden, daß der Sand einerseits aus Quarz besteht, und andererseits, in manchen Fällen, entspricht der Quarz dimensionsgemäß der Staub-

Untergruppe IV.2. Quarzsand, großkörnig, mit großen anisometrischen Körnern, chaotisch verteilt, mit Ton (Wasserkolloidbindungen).
Untergruppe IV.3. Quarzsand, schwach verbunden.
Untergruppe IV.4. Quarzsand, fest verbunden.
Die oben angeführte Kategorisierung der Untergruppen beruht auf dem erhöhten Gehalt an. Sandfraktion und Quarz gegenüber allen bis hierher beschriebenen Gruppen, sowie auf dem verkleinerten Prozent an Wassergehalt, was der gesetzmäßigen Reduzierung der Tonfraktion folgt.
In der Untergruppe IV.1 wird Sand in bedeutenden Mengen (S=48÷70%) festgestellt, summar varieren Sand und Staub von 83 bis 84%. In dieser Untergruppe ist Quarz in der geringsten Menge (30÷41%)

Tabelle 4. Gruppe IV. Sande mit verschiedenen Strukturbindungen

Anprob ierungs punkt	Beschreibung	Mc %	Mm %	S %	P %	Qv %	Ro g/cm^3	Wn %	WL %	Rn MPa
					Untergruppe IV.1					
2	Quarzsand, verschiedenkörnig,	16	14	70	14	41	1.86	17	40	0.300
4	schwachverbunden mit Ton	16	14	70	14	41	1.86	17	40	0.300
7	(Wasserkolloidbindungen)	17	10	48	35	30	1.89	13	32	0.497
9		17	10	48	35	30	1.89	13	32	0.497
					Untergruppe IV.2					
21	Quarzsand, verschiedenkörnig mit groben anesoidischen Körnern, chaotisch verteilt, mit Ton	16	6	54	30	67	1.93	8	17	0.840
					Untergruppe IV.3					
28	Quarzsand, schwachverbunden	12	6	71	17	59	1.85	12	31	0.485
29		12	6	71	17	59	1.85	12	31	0.485
					Untergruppe IV.4					
37	Quarzsand, festverbunden	12	6	71	17	59	1.85	12	31	0.485
40		12	6	71	17	59	1.85	12	31	0.485

fraktion – z.B. im P.28 (Qv=59%, S=71%, P=17%).
Die Anwesenheit eines hohen Sandprozentes ist mit der Erniedrigung des natürlichen Wassergehaltes (8÷17%) und der Verlaufsgrenze WL (31÷40%) verbunden. Es ist interessant zu bemerken, daß die Einachsedruckfestigkeit in nahen Grenzen von 0.300 bis 0.497MPa erhalten bleibt.
Die wichtigsten Klassifikationsdaten, die Gruppe IV in vier Untergruppen einteilt: makroskopische Besonderheiten; Sandgehalt, bzw. Quarz; Tonfraktionsgehalt, bzw. montmorillonit; natürlicher Wassergehalt und plastische Eigenschaften.
Untergruppe IV.1. Quarzsand, mit verschiedenen Korngrößen, schwach mit Ton verbunden (Wasserkolloidbindungen).

vorhanden. Die Tonkomponente ist 16÷17% mit Gehalt entsprechend an Montmorillonit 10÷14%. Leztere bedingen die Wasserkolloidbindungen in dieser Untergruppe.
Untergruppe IV.2. Der Sandgehalt und Staub ist 84%, aber die Anwesenheit großer anisometrischer Quarzkörner bedingt das relativ höchste Quarzprozent, den niedrigsten natürlichen Wassergehalt (8%) und die niedrigste Verlaufsgrenze (WL=17%). Mit den betrachteten Eigenschaften ist auch die erhöhte Einachsendruckfestigkeit (Rn=0.840MPa) verbunden.
Die Untergruppen IV.3 und IV.4 haben ähnliche Eigenschaften. Der Unterschied dazwischen beruht auf der makroskopischen Charakteristik und dem verschiedenen Verbindungsgrad.

Bei dem Analysieren der Korrelationsbindungen zwischen den verschiedenen Eigenschaften in der Matrix bemerkt man in dieser Gruppe, daß ein wichtiger Parameter der Sand darstellt. Er verbindet sich mit dem Quarzgehalt der Volumendichte und der Einachsendruckfestigkeit. Die Volumendichte ist auch ein wichtiger Parameter und verbindet sich mit der Tonfraktionsgehalt, dem Sand, dem Staub und der Einachsendruckfestigkeit. Die Tonkomponente und der Montmorillonit sind ausschlaggebend nur für die erste (IV.1) und zweite (IV.2) Untergruppe, abgesehen davon, daß sie enge Verbindungen mit einer Reihe Eigenschaften besitzen. Hier überwiegen die Wasserkolloidbindungen. In diesem Fall werden diese Korrelationsabhängigkeiten analysiert und nur in Zonen ausgenutzt, wo die angeführten zwei Untergruppen überwiegen. Die Montmorillonit-anwesenheit, was die Wasserkolloidbindungen bedingt, beeinflußt die Einachsendruckfestigkeit, aber gleichzeitig damit auch die Bindungsenge zwischen dem natürlichen Wassergehalt und der Verlaufs-grenze.

Die durchgeführte Analyse für die Eigenschaften der lithologischen Abarten im über der Kohle liegenden Komplex des Ost-Maritza Beckens wurde bei der Zusammenstellung mathematischer Modelle ausgenutzt, die die Regressionsverbindung zwischen der Stundenleistung (Qt) der Schaufelradbagger der Grubenarbeiter und den ausschlaggebenden physika-lisch-mechanischen, technologischen u.a. Eigen-schaften und der Mineralzusammensetzung wieder-spiegeln. Die erhaltenen Modelle sind adequate und geben die Möglichkeit die Einwirkung der aus-schlaggebenden Eigenschaften nach Bedeutung abgestuft zu werden (Abb.Abb.1, 2, 3, 4).

Model fitting results for: Qt

Independent variable	coefficient
CONSTANT	7176.900182
S	−0.631293
WL	−15.321586
Pol	−12.550389
Fi	20.670955
Kf*	−403.505132

* Kf – Spezifische Grabwiderstand

Analysis of Variance for the Ful Regression

Sourse	Mean Square	P-value
Model	260603.	.1144
Error	6141.59	

Plot of Qt

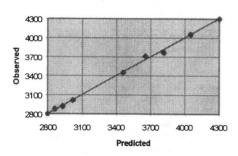

Abb.1. Zusammenhang zwischen der betrachteten und theoretischen (vom Regressionsmodell) Stunden-leistung (Qt) [m^3/h] für Gruppe I.

Model fitting results for: Qt

Independent variable	coefficient
CONSTANT	3782.535071
SO	118.206664
Wn	18.969099
Pr	7.537466
Fi	5.02048
Kf	−26.614963

Analysis of Variance for the Ful Regression

Sourse	Mean Square	P-value
Model	108100.	.3922
Error	59553.7	

Plot of Qt

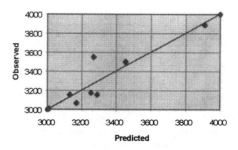

Abb.2. Zusammenhang zwischen der betrachteten und theoretischen (vom Regressionsmodell) Stunden-leistung (Qt) [m^3/h] für Gruppe II.

Model fitting results for: Qt	
Independent variable	coefficient
CONSTANT	2679.874557
Mc	0.313034
S	1.574251
Qv	13.31335
Cv	14.081748
Rn	−11.259603
Kf	−80.610395

Model fitting results for: Qt	
Independent variable	coefficient
CONSTANT	934.182603
Mc	105.773465
Mm	97.02816
Wn	1.502613
Kf	−29.153878

Analysis of Variance for the Ful Regression

Sourse	Mean Square	P-value
Model	114197.	.0836
Error	18798.6	

Analysis of Variance for the Ful Regression

Sourse	Mean Square	P-value
Model	697574.	.0106
Error	45101.4	

Plot of Qt

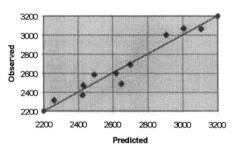

Plot of Qt

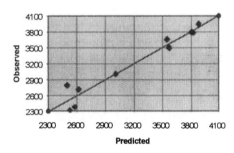

Abb.3. Zusammenhang zwischen der betrachteten und theoretischen (vom Regressionsmodell) Stundenleistung (Qt) [m^3/h] für Gruppe III.

Abb.4. Zusammenhang zwischen der betrachteten und theoretischen (vom Regressionsmodell) Stundenleistung (Qt) [m^3/h] für Gruppe IV.

Als Schluß kann gesagt werden:

Die durchgeführte Korrelationsanalyse und die erhaltenen Korrelationsbindungen zwischen den physikalischen und Festigkeitseigenschaften und der Mineralzusammensetzung für alle Gruppen beweisen die Wichtigkeit dieser Analyse für die Klassifikationsziele und die praktische Anwendung. Wegen der Inhomogenität der Pliozänsedimente muß zur Erhaltung einer genaueren Information in den Gruppen, wo diese engen Verbindungen fehlen, eine Korrelationsanalyse für jede einzelne Untergruppe gemacht werden.

Die praktische Anwendung der Korrelations-Regressionsanalyse wird zur Zeit für die Verbindung zwischen der Stundenleistung der Schaufelradbagger (Qt) und den ausschlaggebenden Eigenschaften der lithologischen Abarten ausgenutzt.

Stoeva P., W. Iwanova, & G. Trapov u.a. 1990, 1991, 1992, 1993, 1994. Kategorisierung der über der Kohle liegenden Tone in den Tagebauen „Trojanovo-Nord" und „Trojanovo 3" nach Grabbedingungen. Wissenschaftliche Berichte, *Archiv der Universität für Bergbau und Geologie*, Sofia.

Комаров И., Н. Хайме, А. Бабенышев 1976. *Многомерный статистический анализ в инженерной геологии*. Москва, Недра.

LITERATUR

Iwanov I. & K. Scheyretov 1969. Über die Abhängigkeit zwischen den Festigkeitsdaten der Tone und ihrem Grabwiderstand. *Zeitschrift Waglista*, No.5.

Geotechnical Hazards, Marić, Lisac & Szavits-Nossan (eds) © 1998 Taylor & Francis, ISBN 90 5410 957 2

Underground exploitation of dimension stone in the region of Istrian peninsula

B. Kovačević-Zelić & S. Vujec
University of Zagreb, Faculty of Mining, Geology and Petroleum Engineering, Croatia

ABSTRACT: Throughout Croatia one can find a lot of buildings which testify a long tradition of a stone application in civil engineering and architecture. Dimension stone was extracted mostly by surface and only rarely by underground methods. Lately, underground methods have been used more frequently in the world, because of environmental, organisational and economical reasons. In the Kanfanar quarry in the region of Istrian peninsula an existing active surface exploitation had to be transferred underground because of economical and environmental constraints. Regular room and pillar method was chosen for the trial underground exploitation. In the paper the conventional pillar design procedure is presented and compared to the numerical results made by the usage of finite difference code FLAC. It was concluded that in a complex geological environment, numerical analyses can better describe the behaviour of a rock mass. Therefore, the parameters for the trial underground exploitation were chosen according to the numerical results.

1 INTRODUCTION

There is a long tradition of dimension stone exploitation in Croatia. Dimension stone was extracted mostly by surface and only rarely by underground methods. Nevertheless, there is one abandoned quarry of rudist limestone called "Sv. Stjepan" near the town of Buzet on the Istrian peninsula (Fig. 1). More than a hundred years of mining activity make this quarry exceptionally significant. The dimensions of the underground working space are 150 x 150 m. The height of this space is 5-7 m and there are 29 randomly placed support pillars. The subsidence has not been recorded at the surface until these days. The quarry "Sv. Stjepan" was declared unprofitable and closed by the end of 1965 (Gabrić et al. 1995).

Croatia is full of historical monuments, which testify stone application especially in civil engineering and architecture dating back to the Roman period up to these days. Some of them are: the amphitheatre in Pula; the Diokletian Palace in Split; the basilica Eufrasiana in Poreč; the cathedrals in Trogir, Šibenik, and Zagreb; the old town of Dubrovnik. Throughout Croatia one can find a lot of public buildings, temples, houses and palaces made from local stones. Dimension stone from Istria was also intensively used for the construction and restoration of Italian town of Venice. In recent

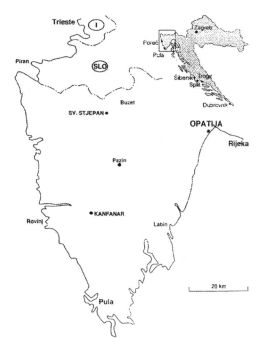

Figure 1. Map of Istrian peninsula.

Table 1. Material properties for the numerical analyses.

Material type	Height of layer (m)	Density (kgm⁻³)	Bulk modulus (MPa)	Shear modulus (MPa)	Cohesion (MPa)	Friction angle (°)	Tensile strength (MPa)
Upper roof	5.0	1800	8.5	3.9	0.05	20	0.0
Immediate roof	8.9	2630	7600	4340	17	50	5.3
Exploitable layers	6.8	2635	9345	5878	13	55.5	4.4
Immediate footwall	5.0	2635	9500	5900	15	60	5.0

times, modern architectural designs have been composed from dimension stone: town squares in Zagreb and Rijeka, National Library in Zagreb, hotel Sheraton in Zagreb, Istrian National theatre in Pula etc. (Cotman 1995).

Underground exploitation of dimension stone has been spreading lately in the world for three main reasons: economy, organisation and environment. The latter aspect is the most important in the region of Istrian peninsula in Croatia which is well known as natural protected area and an area of tourism.

The other advantage of underground exploitation is the fact that created subsurface openings can be used later. They have been already used world-wide for many purposes such as for storing of liquids, food, industrial products, strategic materials; for creating protected areas for military and civil use (archives, museums, research centres); for storing of industrial and toxic waste; or as cheese storehouses and wine cellars (Fornaro & Bosticco 1995).

Underground exploitation of ornamental stone is, from technological point of view, different from surface quarrying only in the first stage, namely in the removal of top slice; descending slices are worked as in conventional quarries. But, in this case stability problems require adequate studies and stability checks in order to avoid expansive artificial support measures (Pelizza et al. 1994, Fornaro & Bosticco 1995).

In the Kanfanar quarry, in the region of Istrian peninsula (Fig. 1), an existing active surface exploitation had to be transferred underground because of economical and environmental constraints. The analysis of room stability and pillar design was made by numerical modelling with the 2-dimensional finite difference code FLAC. The results of these analyses are presented in the paper.

2 GEOLOGICAL AND GEOTECHNICAL DATA

Croatia has a significant number of quarries producing dimension stones mostly of sedimentary origin. The region of Istria is characterised by

dimension stone deposits which are situated in limestones and clastites of Upper Jurassic including Lower and Upper Cretaceous till Eocene. In the Kanfanar quarry the limestones of Lower Cretaceous, which are known under commercial name "Giallo d'Istria" or "Istrian yellow", are exploited (Crnković & Jovičić 1993).

The characteristic geological profile, schematically presented in Figure 2, was created according to previously made geological research. The position of the dominant discontinuities, that could affect the results of numerical analyses, was also determined by the in-situ prospection. Three horizontal discontinuities (bedding planes), that are placed in the immediate roof, were obvious at the outcrops. Two sets of vertical discontinuities (fractures of tectonic origin), that are orthogonal to each other, were also observed.

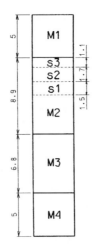

Legend: M1 - Upper roof; M2 - Immediate roof; M3 - Exploitable layers; M4 - Immediate footwall; s1, s2, s3 - Bedding planes.

Figure 2. Schematic presentation of geological profile - Kanfanar.

Physical and mechanical properties of intact rock materials (Table 1) were determined by the laboratory tests. The discontinuities were introduced into the analyses with the following properties:

cohesion, $c = 0$ MPa,
friction angle, $\varphi = 34°$,
normal stiffness, $k_n = 2.7$ GPa/m,
shear stiffness, $k_s = 1.0$ GPa/m.

Shear strength parameters of discontinuities were determined by direct shear test. The values of normal and shear stiffness were taken from the literature data for limestones (Bandis et al. 1983).

3 PILLAR DESIGN

The main function of pillars in room-and-pillar mining is to maintain the stability of the overlaying strata. In many cases, pillars are left as permanent support and must perform their task throughout the life of the mine.

In general, pillars consist of mineable ore. In order to reduce ore losses, it is essential to fully utilise the load-bearing capacity of pillars. On the other hand, overestimation of pillar strength, leading to pillar failure, can result in production losses and substantial costs for remedial action (Krauland & Soder 1987).

The conventional pillar design procedure essentially consists of estimating the pillar strength and the load on pillars and linking the two trough a proper safety factor. But this approach is based on some presumptions, which are not always satisfied. Moreover, no single pillar strength formula has been accepted as the most reliable one, even for coal mine pillars which have received the most research attention (Bieniawski 1984). Many of the pillar strength formulae are similar and nearly all are of one of the two following forms (Galvin et al. 1996):

Linear form:

$$S_p = k_1 \cdot \left[A + B \cdot \left(\frac{w}{h} \right) \right] \qquad (1)$$

Power form:

$$S_p = k_2 \cdot \frac{w^a}{h^b} \qquad (2)$$

where: S_p = pillar strength; k_1, k_2 = strength of a cube; w = pillar width; h = pillar height; and A, B, a, b = field or laboratory derived dimensionless parameters.

As an example, the formulae by Obert-Duvall (1967) which has a historical significance, takes the following form:

$$S_p = k_1 \cdot \left(0.778 + 0.222 \cdot \frac{w}{h} \right) \qquad (3)$$

Pillar strength can also be determined using the Hoek-Brown criterion which is defined by (Hoek & Brown 1980):

$$\sigma_1 = \sigma_3 + \left(m \cdot \sigma_c \cdot \sigma_3 + s \cdot \sigma_c^2 \right)^{1/2} \qquad (4)$$

where: σ_1 = major principal stress; σ_3 = minor principal stress; σ_c = uniaxial compressive strength; and m, s = empirical constants.

In the case of slender pillar, the stress state across the centre of the pillar is very close to uniaxial stress conditions in which $\sigma_1 = S_p$ and $\sigma_3 = 0$ (Hoek & Brown 1980). By the substitution of these relation into (4), the pillar strength can be expressed in the following form:

$$S_p = \left(s \cdot \sigma_c^2 \right)^{1/2} \qquad (5)$$

Pillar loading can be determined by calculation or by stress measurements. Above others, the following calculation methods can be used: tributary area approach and numerical methods (Krauland & Soder 1987).

According to tributary area approach, the average pillar stress is expressed in the following manner:

$$\sigma_p = \frac{A_t}{A_p} \cdot \sigma_v \qquad (6)$$

where: σ_p = average pillar stress; σ_v = virgin vertical stress; A_t = area supported by one pillar; and A_p = area of the pillar.

For the case of rectangular pillars, equation 6 takes the following form:

$$\sigma_p = \frac{(w+c) \cdot (l+c)}{w \cdot l} \cdot \sigma_v \qquad (7)$$

where: w = pillar width; l = pillar length; and c = room width i.e. pillar span.

Factor of safety F_s is defined as the ratio between pillar strength and pillar stress:

$$F_s = \frac{S_p}{\sigma_p} \qquad (8)$$

The assumptions made in the formulation of this approach lead to conservative estimate of the pillar load. Measurements have shown that this approach overestimates the pillar load by about 40%. This approach is still used in practical mining applications due to its simplicity (Bieniawski 1984).

On the contrary, numerical methods allow one to simulate complex geologic structures and geometric configurations (Krauland & Soder 1987). Therefore, the finite difference code FLAC (Itasca 1991) was chosen for numerical calculations in the Kanfanar quarry. These analyses will be presented hereafter. Pillar load was also calculated by the tributary area approach, for the comparison purposes only.

4 NUMERICAL ANALYSES

Regular room and pillar method was chosen for the trial underground extraction of dimension stone. Two basic cases with different pillar dimensions had to be examined (Fig. 3, Table 2). These cases were acceptable from the technological point of view.

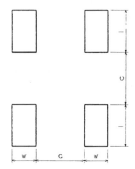

Figure 3. Room and pillar method

Table 2. Types of the analyses.

Dimension	Case 1	Case 2
w (m)	2.8	3.3
l (m)	5.0	5.0
c (m)	5.5	5.5

It is well known that the mechanical response of rock mass depend far more on the behaviour of discontinuities rather than on the strength of rock material itself. Numerical methods offer a very powerful and effective tool in the analysis of jointed rock structures.

Therefore, the analysis of room stability and pillar design was made by means of numerical modelling with 2-dimensional finite difference code FLAC (Itasca 1991). The Mohr-Coulomb's plasticity model was chosen for intact rock materials with the properties assigned according to data in Table 1. Special attention was focused on the influence of discontinuities with different position in

the roof-strata. Because of the nature of discontinuities in the Kanfanar quarry, they were introduced into the models using interface elements.

Four different models were analysed, according to the available geological data:

Model 1 - roof without discontinuities (competent rock),

Model 2 - roof with three parallel horizontal discontinuities,

Model 3 - roof with one vertical discontinuity in the middle of the room,

Model 4 - roof with one vertical discontinuity near the pillar.

It was anticipated to adjust pillar layout with respect to variations in geology i.e. the distribution of the discontinuities, which should be mapped continuously with the mining progress. Therefore, the numerical models with the discontinuities placed inside pillars were not considered. Instead of that, the models with the enlarged pillar span to 10.0 m were also examined for both cases of pillar dimensions. These models represent the situation of the necessity of pillar adjustments in the case of discontinuity occurrence in the pillars. Moreover, they represent the crossings too.

5 DISCUSSION OF NUMERICAL RESULTS

Numerical results are presented in terms of stress and strain components. Special attention is focused on the maximum compressive stresses in pillars σ_{max}, the appearance of tensile stresses inside the pillars or in the roof-strata and maximum vertical displacements in rooms δ_{max}.

The results of the numerical analyses for models with the pillar spans of 5.5 m and 10.0 m are presented in tables 3 and 4, respectively.

Table 3. Results of numerical analyses – c = 5.5 m.

	Case 1		Case 2	
	σ_{max} (MPa)	δ_{max} (mm)	σ_{max} (MPa)	δ_{max} (mm)
Model 1	1.211	0.46	1.132	0.420
Model 2	1.223	0.47	-	-
Model 3	1.224	0.50	1.139	0.454
Model 4	1.275	0.65	1.178	0.585

Table 4. Results of numerical analyses – c = 10.0 m.

	Case 1		Case 2	
	σ_{max} (MPa)	δ_{max} (mm)	σ_{max} (MPa)	δ_{max} (mm)
Model 1	1.871	0.98	1.676	0.904
Model 2	-	-	-	-
Model 3	1.881	1.00	1.694	0.985
Model 4	2.526	2.40	2.314	2.240

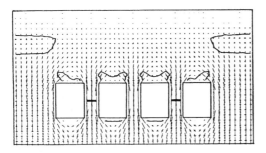

a) Case 1

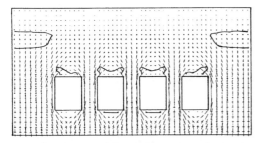

b) Case 2

Figure 4. Stress distribution for model 1 and c = 5.5 m.

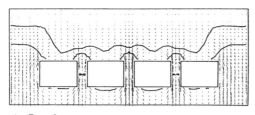

a) Case 1

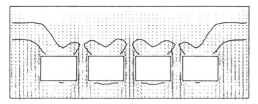

b) Case 2

Figure 5. Stress distribution for model 1 and c = 10.0 m.

Model 2 was not used for all the cases because the results of the case 1 (Table 3) have indicated that the horizontal discontinuities did not very much affect the results in comparison to the model 1, especially

if we consider the magnitudes of displacements. This could be explained by the very good quality of intact rock material and a relatively high position of these discontinuities in the roof.

If we compare the results presented in tables 3 and 4, we can make the following conclusions:

- Compressive stresses are higher for pillar width w = 2.8 m (case 1) than for pillars of w = 3.3 m (case 2), as it was expected. For the pillar span of c = 5.5 m this difference is between 7-8%, and for the pillar span c = 10.0 m it is from 9-11%.

- The influence of the pillar span increase from 5.5 m to 10.0 m on the magnitude of compressive stresses is much more pronounced. The difference between compressive stresses is 54-98% for the case 1 and 48-96% for the case 2.

- The magnitude of maximum compressive stresses in pillars is almost equal in all models if we are looking to the same case and the same pillar span. Only for the models 4, where the vertical joint is placed near the pillar, one of the pillars suffers higher loading. Moreover, the vertical displacements are also much larger for these models.

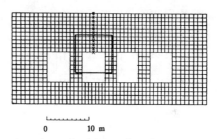

0   10 m

a) Basic mesh with interface element

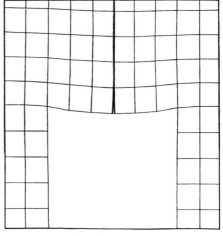

b) Detail of deformed mesh enlarged 1000 times

Figure 6. Model 3 – case 1.

Model 1 is chosen for the presentation of numerical results by means of stress distribution. Figures 4 and 5 represent the results for both cases of analyses and for two different pillar spans. The results of the numerical analyses presented by the stress distribution lead to the following conclusions:

- There were only compressive stresses inside the pillars in all the models of the case 2.

- The zone of mixed stress state (one principal stress is tensile and the other is compressive) inside the pillars was observed in all the models of the case 1.

- The zones of mixed stress state were also observed above the opened rooms in the cases 1 and 2 and the pillar span c = 5.5 m.

- In the case of pillar span increase to 10.0 m, the zones of the pure tension were also obtained beside the zones of mixed stress state. The zones of mixed stress state were enlarged in comparison to the same models for the case of c = 5.5 m. Moreover, they were connected together for model 4 in case of pillar width w = 3.3 m, and for all the models in case of pillar width w = 2.8 m.

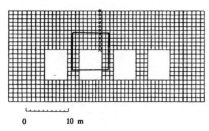

```
0          10 m
```

a) Basic mesh with interface element

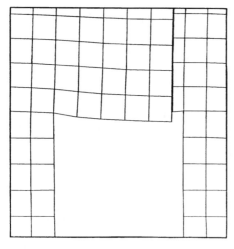

b) Detail of deformed mesh enlarged 1000 times

Figure 7. Model 4 – case 1.

However, it should be emphasized that the plasticity indicators were never obtained for the intact rock materials.

The influence of the vertical discontinuities, modelled with interface elements, was much more pronounced. The interface element placed in the middle of the room caused the development of a tension crack. The length of tension crack was approximately 1 m for pillar span of 5.5 m (Fig. 6) and approximately 2 m for pillar span of 10.0 m. It caused only local instabilities and not the general roof failure.

Interface element near the pillar caused larger vertical displacements (Fig. 7). It did not induce shear failure on interfaces for room width of 5.5 m. But, for the room width of 10.0 m, shear failure and slippage along discontinuity could occur and cause a general roof failure. Therefore, the crossings should be designed with special attention and in conformity with the measured joint distribution.

6 CONVENTIONAL PROCEDURE

The results obtained by the conventional procedure will be presented hereafter for the comparison purposes only.

By the substitution of σ_c=88.6 MPa and s=0.1 (rock mass of very good quality) into (5), pillar strength Sp=28.0 MPa is obtained. Average pillar stress is determined by using the equation (7), and safety factor according to equation (8). Table 5 gives the results of this analysis in terms of average pillar stresses, σ_p, and safety factor, F_s.

Table 5. Tributary area approach.

Pillar span	Case 1		Case 2	
c (m)	σ_p (MPa)	F_s	σ_p (MPa)	F_s
5.5	2.02	13.9	1.81	15.5
10.0	4.44	6.3	3.92	7.1

Considering these results, we can conclude the following: The magnitudes of average pillar stresses given by tributary area approach are much higher than that obtained by numerical analyses for model 1 (Tables 3-4). The difference between the results of these two approaches is 67% (case 1) and 60% (case 2) for pillar span c = 5.5 m. This difference is much bigger for the pillar span c = 10.0 m. and come to 137% (case 1) and 134% (case 2). The values of safety factors seem to be extremely high.

It is evident that in this example, the results obtained by the conventional procedure are not realistic, because of the assumptions on which this procedure is based and because of the fact that the effect of discontinuous nature of a rock mass cannot be introduced into this type of calculation.

7 CONCLUSIONS

In the paper, the conventional pillar design procedure was compared to the results of numerical analyses. The conventional approach to the pillar strength and pillar load determination can give misleading results in complex geologic environment. Numerical models, on the contrary, can incorporate the effects of different geological materials and discontinuities.

In the case of the Kanfanar quarry, it was recognised that the discontinuities will have the dominant influence on the results. Therefore, the parameters for the trial underground exploitation of dimension stone were chosen according to numerical calculations.

It was decided to start the exploitation with the pillars of w = 3.3 m. The occurrence of the zones of mixed stress states inside the pillars of w = 2.8 m was considered to be unacceptable in this phase of pillar design.

The stability of rooms is always satisfactory beside the fact that the zones of pure tension were obtained for the crossings (c = 10.0 m). But, the discontinuities which can occur in the roof-strata will always require special care and probably adequate safety measures.

It should be mentioned here that all the results were obtained in two-dimensional analyses. In reality, the stability problem of room-and-pillar method is three-dimensional. Due to the lack of measurements in this phase of the pillar design, presented models are in this respect more qualitative and should be interpreted as such. Adequate in-situ measurements and monitoring is necessary during the trial phase. Simple methods of monitoring like convergence measurements and mapping of pillar fracturing should be adopted as mining progresses.

Special attention should be given to the permanent observation of the distribution of discontinuities. If it is necessary pillar layout should be adjusted with respect to variations in geology.

As soon as mining development takes place, the design refinement should be undertaken. Further research is expected.

REFERENCES:

Bandis, S.C., Lumsden, A.C. & Barton, N.R. 1983. Fundamentals of Rock Joint Deformation. *Int. J. Min. Sci. & Geomech. Abstr.* 20(6): 249-268.
Bieniawski, Z.T. 1984. *Rock Mechanics Design in Mining and Tunneling*. Rotterdam: Balkema.
Cotman, I. 1995. Stone production in Croatia. *Stone World*, July/95: 50-56.
Crnković, B. & Jovičić, D. 1993. Dimension Stone Deposits in Croatia. *The Mining-Geological-Petroleum Engineering Bulletin* 5: 139-163.
Fornaro, M. & Bosticco, L. 1995. Underground Stone Quarrying in Italy-Its Origins, The Present Day and Prospects (Part 4). *Marmomacchine International* 9: 64-87.
Gabrić, A., Galović, I., Sakač, K. & Hvala, M. 1995. Mineral Deposits of Istria - Some Deposits of Bauxite, Building-Stone and Quartz Sand (Excursion C). In I. Vlahović & I. Velić (eds), *Excursion Guide-Book – Proc. 1st Croatian Geological Congress, Opatija, 18-21.10.1995:* 111-137. Zagreb: Institute of Geology. (in Croatian)
Galvin, J.M., Hebblewhite, B.K. & Salamon, M.D.G. 1996. Australian Coal Pillar Performance. *ISRM News Journal* 4(1): 33-38.
Hoek, E. & Brown, E.T. 1980. *Underground Excavations in Rock*. London: The Institution of Mining and Metallurgy.
Itasca Consulting Group, Inc. 1991. *FLAC Users Manual – Version 3.0*. Minneapolis, Minnesota.
Obert, L. & Duvall, W.I. 1967. *Rock Mechanics and the Design of Structures in Rock*. New York: John Wiley & Sons, Inc.
Pelizza, S., Mancini, R., Fornaro, M., Peila, D. Cardu, M. & Bosticco, L. 1994. Design Criteria to transfer Underground Ornamental Stone Quarries. *Proc. XVI. World Mining Congress, Sofia, 12-16 Sept. 1994*: 425-434.

Geotechnical Hazards, Marić, Lisac & Szavits-Nossan (eds) © 1998 Taylor & Francis, ISBN 90 5410 957 2

Possibility of hammer tunnelling in complex geological conditions

T. Kujundžić, S. Dunda & S. Vujec
Faculty of Mining, Geology and Petroleum Engineering, University of Zagreb, Croatia

ABSTRACT: The paper presents the in situ testings of the hydraulic hammer productivity upon boulder breaking. There are also factory data regarding hydraulic hammer productivity upon tunnelling and the published data about the hydraulic hammer productivity on concrete tasks. The article contains preliminary research on fracture toughness of rocks, conducted according to the suggested methods by the International Society for Rock Mechanics at the Faculty of Mining, Geology and Petroleum Engineering in Zagreb. The aim of the research is to precisely determine the factors of the resistance to the appearance and advance of fractures within the rock, which should decrease dilemma upon decision when to apply hydraulic hammers upon tunnelling.

1 INTRODUCTION

The research presented in this paper is aimed at the decrease of the geotechnical hazard by the application of the relatively new tunnel excavation technology by means of hydraulic hammers.

Hydraulic hammers can be employed for a wide variety of uses e.g.:
- boulder breaking after production blasting
- ditch excavation for pipelines
- through cut and side hill cut excavation in road construction
- various excavations, demolition of ruinous buildings and excavation of the underground structures in urban areas, where blasting is not allowed
- scaling after blasting in tunnel and road constructions
- underground construction works (excavations, boulder breaking, etc.)
- excavation of mineral row materials on the open pit
- cleaning of moulds in metallurgy.

They can also be employed for scaling and tunnelling. Upon the application of the classical excavation method in tunnelling (drilling and mining) scaling is necessary to remove the unstable pieces from the mined contour in case of insufficient breakage. For such works mini-excavators with hydraulic hammers are usually used.

This procedures have been applied in Croatia by the Viadukt company upon the excavation works of the Tuhobić tunnel on the highway Karlovac-Rijeka

and their application is continued. However, a heavy hydraulic hammer, as the primary excavation machine in tunnelling has not been applied in Croatia so far.

Tunnelling by means of such machines started in Italy in 1985, where a tunnel was excavated in this way for the first time. The total length of the tunnel was 1200 m and the surface of the cross-section was 98 m^2. The tunnel was excavated in layered limestone with dolomite. Daily advance was between 6.0 and 8.0 m. The Rammer hydraulic hammer S86, which was attached to Caterpillar hydraulic excavator CAT 235, was used (Lawton, 1994). Since then, such a way of tunnelling has been spreading outside Italy, especially in Japan.

According to the data of the factory Rammer-Finland, the tunnel length of 32.56 km of different cross sections, mostly in limestone was excavated in Italy from 1985-1994. Table 1 shows the tunnels where heavy hydraulic hammers of the Rammer Factory were used as primary excavation machines (Rammer, 1994).

2 PRODUCTIVITY OF HYDRAULIC HAMMERS

In mining industry hydraulic hammers are mostly used for breaking of boulders at open cast mining of mineral row materials after production blasting.

In order to determine the factors which affect the productivity of hydraulic hammers, the in situ testings of productivity of hydraulic hammers on boulder breaking were conducted (Kujundžić, 1997).

Table 1 - Rammer hammer tunnelling in Italy

NAME / LOCATION	YEAR	ROCK MASS	EXCAVATED CROSS SECTION (sq.m)	TOTAL LENGTH (m)	LINEAR PROGRESS (m/day)	HAMMER	EXCAVATOR
Malborghetto	1985	Layered limestone with dolomite	98	1200	6,0-8,0	Rammer S86	CAT 235
Val di Flungo	1986	Layered limestone with flintstone	80	950	1,5-2,0	Rammer S86	LIEBHER 962
Rieti	1987	Moist and silliceous limestone	80	2100	7,0-8,0	Rammer S86	LIEBHER 962
Sanremo	1988	Sedimentary formation with sandstone, marl, limestone marl and clayed schist	98	1800	10,0	Rammer S86	FIAT FE45
Palermo	1988	Hard limestone with joints and carstifications	66	1760	3,0	Rammer S86	CAT 235
Sanremo	1988	Sedimentary formation with limestone, marl sandstone and claystone	95	1800	7,0	Rammer S86	FIAT FE45
Siracusa	1989	Weathered and fractured basalt	66	1800	3,0	Rammer S86	CAT 235
Sanremo	1989	Hard limestone pudding stone tough schist	65	7300	8,5	Rammer S86	CAT 235
Vicenza	1990	Dolomite limestone layered limestone	120	3250	5,0	Rammer G120	CAT 245 LIEBHER 974
Morgex	1990	Gneiss and micaschist with calc-schist	120	930	4,5	Rammer S84	CAT 235
Imperia	1990	Layered limestone	98	1570	6,0	Rammer S86	CAT 235
Casina	1991	Siltstone and claystone	65	1000	4,0	Rammer S86	LIEBHER 954
Porto Vado	1991	Fractured limestone	98	2000	7,0	Rammer S86	CAT 235
S. Bernardo	1992	Gneiss and micaschist	98	1200	6,0	Rammer G120	CAT 235
Val Brembana	1994	Layered limestone	98			Rammer G100	FIAT FH400
Taormina	1994	Layered marl	86	1300	4,0	Rammer G100	RH 20
Lastra a Signa	1994	Layered limestone	98	2600	5,0	Rammer G100	CAT 235

The testing were done on 4 quarries (Vukov Dol, Očura, Široko brdo, Belaj) for three rock types (marble, dolomite, limestone), with 5 different hammers (Montabert BRH 125; Krup HM 960, HM 900, HM 720-1 CS and HM 600). The measurements in situ comprised volume defining of selected boulders and time of their breaking to particular granulation. On the grounds of the given data the productivity on a particular boulder was calculated and the graphic logarithm correlative model was made separately for each measured location. The graphs that present the inter-dependence of productivity upon the block volume are in Figures 1. and 2. The correlation curve (Figure 1.) is presented by the logarithm curve, since it is considered that the productivity can in no way be equal zero. In Figure 2. the log correlative model is repeated whereby the rare boulder sizes have been excluded.

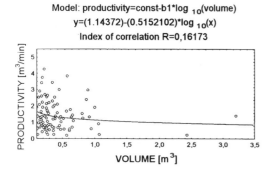

Model: productivity=const-b1*log $_{10}$(volume)
y=(1.14372)-(0.5152102)*log $_{10}$(x)
Index of correlation R=0,16173

Figure 1. - Results of the productivity measurements for the hydraulic hammer Krupp HM 600 in the quarry Očura.

The results of the testings for all rock types on all the locations have shown that the productivity does not depend (or depends insignificantly) on the boulder volume within the volume range, characteristic of engineering and geological conditions at the particular location. The comparison of the hammer productivity to different technical characteristics in the same boulder type, upon boulder breaking to equally determined granulation, has shown that the productivity directly depends on the power of the applied hammer. Figures 3 and 4 present the dependence of the used impact energy upon the boulder size for the applied hammers in the same rock type. According to the diagram, both hammers used an equal amount of impact energy for the boulder breaking of the same volume. For example, the hammer HM 960 of 1600 kilo mass, used up 1490.60 kJ of impact energy on the boulder

of 1 m^3 volume and the hammer BRH 125 of 345 kilo mass used 1925.50 kJ. The fact that the more powerful hammer (HM 960) used less energy per block of the same volume seems to be unlogical, but it can be explained by the way of the hydraulic hammer application upon the boulder breaking. Through light lifting of the front part of the excavator i.e. by pressing the hammer against the working surface, the additional compressive stress on the boulder is initiated, which increases the productivity of the boulder breaking (Anderson, Papineau, 1989).

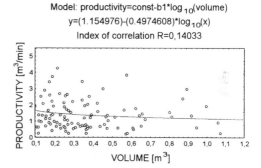

Model: productivity=const-b1*log $_{10}$(volume)
y=(1.154976)-(0.4974608)*log $_{10}$(x)
Index of correlation R=0,14033

Figure 2. Selected results of the productivity measurements for the hydraulic hammer Krupp HM 600 in the quarry Očura.

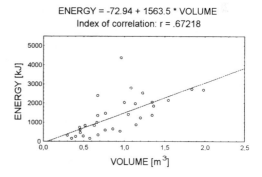

ENERGY = -72.94 + 1563.5 * VOLUME
Index of correlation: r = .67218

Figure 3. Impact energy used up upon the application of the hydraulic hammer Krupp HM 960 on the quarry Vukov Dol.

Out of the relation between the masses of the hydraulic hammers and the relation between the masses of the excavators is to be concluded that upon the application of the hydraulichammer HM 960 the compressive stress was higher, which caused

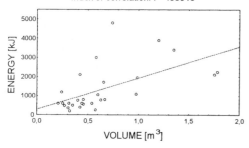

ENERGY = 287.85 + 1637.6 * VOLUME
Index of correlation: r = .58516

Figure 4. Impact energy used up upon the application of the hydraulic hammer Montabert BRH 125 on the quarry Vukov Dol.

decreased usage of impact energy.

Besides the characteristics of the rocks and the power of the hammer (number of strikes and energy of the particular strike) the hammer productivity upon such works depends on general causes such as: experience of the operator, type of the tool (moil, chisel, blunt tool, etc.) excavator mass and the condition of the machine, needed granulation of breaking, work organization and alike.

In order to determine the productivity of hydraulic hammers upon tunnelling (Rammer, 1994),the Finnish factory Rammer provides for all their hammer types orientation diagrams showing the dependence of the productivity upon the type and structural fabric of the rock. According to them the rocks fall in fourteen categories and rock massive structural fabrics in six. Regarding these data a diagram in the Figure 5 showing the dependence of the productivity upon the type and structure of the rock for the hammer S86 has been modified (simplified). Here, the rocks are ranged in 6 categories according to quality and in 4 categories according to their structural fabric. The first rock group comprises: limestone, micaschist, dolomite, biotitic granite, amphibolic gneiss. The second group are andesites and the third one are granitic-gneiss and clay slate. The fourth group contains: granite, calcareous sandstone, concreted shingle, gabro, feldspathic sandstone and weathered basalt. The fifth group consists of: augitic diorite, biotitic gneiss, riolite, diorite, amphibolic granite, amphibolic schist and fresh basalt and the sixth one contains: weathered diabase, sandstone, piroxenic quartzite and unweathered diabase.

Such diagrams serve only as sketchy orientation and can be used for comparisons of different hammer types, but not for determination of the hammer

application on the particular location. Therefore it is valuable to quote the scientific paper by H. Fukuda (1997) with data about the productivity of hydraulic hammers, realized on a specific task, namely upon the construction works of the Shin-Shin-Hisoyama Tunnel. At this hydraulic tunnel, having a length of 2963 m and an excavation cross section of 95 m^2, excavation was initially planned to be performed by the NATM by drill and blast methods with a short bench. However, because of the fact that the tunnel portal was close to the planned site of a nuclear power station and the fact that any effect on the two existing parallel tunnels must be kept to a minimum, it became clear that the work would become subject to restrictions in the use of explosives. Thereupon, after a comparative study with the controlled blasting method and referring to examples of construction overseas, an excavation method using a large model hydraulic breaker weighing 3.8 tons as the main equipment was employed for the first time in Japan. Tunnel excavation was performed by dividing cross section into three sections: heading, bench and invert.

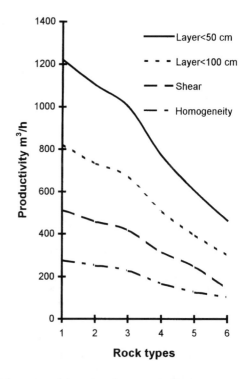

Figure 5. - Orientation diagram of the dependence of the hammer productivity upon rock type (factory data).

Figure 6 shows the average values of productivity for representative types of rock which appeared in this tunnel divided by excavation section. The maximum advance in this tunnel was realized in tuff upon bench excavation 32.3 m³/h, heading excavation 26.0 m³/h and invert excavation 11.7 m³/h.

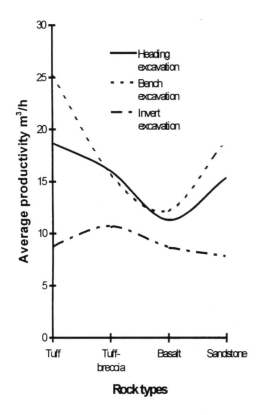

Figure 6. - Average results of the hydraulic hammer realized upon excavations of the Shin-Shin-Hisoyama Tunnel.

From the same point of view the work by Biligin, Kuzu, Eskikaya and Ozdemir (1997) is useful, since they have compared the productivity of the roadheaders and hydraulic hammers in Istanbul Metro drivages. The comparison comprised Alpine ATM 75 and Eickhoff ET 250 transverse type roadheader and hydraulic hammers Krupp HM 185, HM 720 and Montabert BRH 250. Alpine ATM 75 and Eickhoff ET 250 type roadheaders were only used in small part of Taksim Tunnels. Hydraulic hammers Krupp HM 185-720 and Montabert BRH

250 attached to Hitachi 200 Excavators were used mainly in the major parts of the metro tunnels. Drilling and blasting method was used when andesite and diabase were encountered along the tunnel route.

Overall performance of hydraulic hammers:
- Average net productivity
  (cutting rate), m³/h   11.9
- Daily advance, m   1.8
- Best daily advance, m   3.6
- Average weekly advance, m   10.8
- Average monthly advance, m   43.2

The authors have offered the empirical formula according to which the machine performance can be predicted prior to starting a tunnel project that will definitely define the tunnel drivage economy. The formula was published by Biligin, N., Yazici, S., and Eskikaya, S., in 1996.

$$ICR = 4,24 \cdot (RCMI)^{-0,567} \qquad (1)$$

In these equation:
ICR = Net productivity (instantaneous), m³/h
RCMI = Rock mass cuttability index, MPa

$$RCMI = \sigma_c \, (RQD/100)^{2/3} \qquad (2)$$

σ_c = Uniaxial compressive strength, MPa
RQD = Rock quality designation.

3 THEORETICAL DISCUSSION

From this it follows that the productivity of hydraulic hammers and their application as primary machines for tunnelling mostly depends on structural fabric of the rock massive and physical and mechanical characteristics of the rocks.

Rock masses as geological environment, are heterogeneous and fractured i.e. discontinuous. If planar discontinuitis are within the rock mass and if the distance among them is such as to build a thick fracture-net, which divides the rock into smaller irregular shapes, the tunnelling by hydraulic hammering becomes easier. The fragmentation of the rock massive as the consequence of tectonic disturbances and even more the disintegration caused by genetic processes make the application of hydraulic hammers suitable. Jointing within intrusive and effusive magmatic rocks causes clearly visible fractures along which the stone mass is divided into pieces of different shapes and sizes such as thin and thick plates, three- to six-edged prism, cube, globe and completely irregular shapes. The genetic characteristic of the sedimentary rocks such as layerness and those of the metamorphic rocks such as layerness and schistosity make the application of hydraulic hammers suitable.

The suitability of the hydraulic hammer

application upon tunnelling also depends on physical and mechanical characteristics of the rocks such as: strength, hardness, toughness, abrasivity, etc. There is no generally accepted unique rock classification, which would satisfy all the needs and be useful for all the engineering works in rock masses (Vujec, 1997).

At the conference of Rapid excavation and tunnelling, which was held in New York in 1987, Bienawski has presented major rock mass classifications and their applications. None of them can directly be applied for the determinations of the productivity, i.e. applicability of hydraulic hammers upon tunnelling, which is due to their way of operation.

4 TESTING OF FRACTURE TOUGHNESS OF THE ROCKS

In view of the prior in situ measurements (Kujundžić, 1997) and taking into consideration the procedure of a great number of boulder breaking, almost the same breakage mechanism was observed, which was especially impressive on the boulders of large volume.

Upon the contact between the block and working tool there is intensive crushing and pulverization of rock material. Soon after that a fracture (or fractures) appear within the boulder with visible advance. When the fracture reaches free space of the boulder, it breaks apart. From this it follows that the productivity of the hammer depends on the speed of fracture appearance and its advance. Therefore, the rock characteristic, upon which the productivity of hydraulic hammer depends mostly, is the resistance of the rock to the appearance and advance of the fractures, i.e. fracture toughness.

The fracture toughness measurement method has been introduced by the International Society of Rock Mechanics (ISRM, 1989). In order to precisely determine resistance indicators to appearance and advance of rock fractures, i.e. to classify the rocks according to fracture toughness, a measure instrument for fracture toughness testings of rock samples was constructed according to the demands of ISRM at the Faculty of Mining, Geology and Petroleum Engineering in Zagreb (Kujundžić, 1997). On the instrument there were conducted laboratory testings of fracture toughness of the rock samples from the above mentioned quarries, in which the productivity of hydraulic hammers was measured. Geometry of samples and testing procedure were realized according to the demands and principles of the ISRM.

The testing was conducted on two levels. During the first level, only the maximum force was measured, while in the second one the force and quantity of the displacement during at least 4 cycles of loading and unloading of the sample were continuously observed. The results of the fracture toughness testing on one sample are presented in Figure 7.

Due to the fact that the number of the tested samples is small, the results of the fracture toughness of rocks cannot be directly connected with the measurements of the productivity of hydraulic hammers.

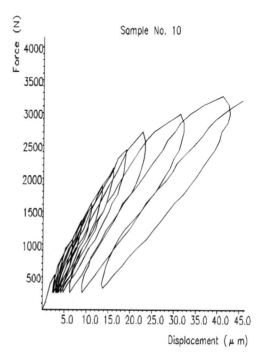

Figure 7. - Results of the fracture toughness testing on the sample no. 10.

5 CONCLUSION

The underground construction work is a very complicated and sometimes also a risky engineering task. In case, when the effects of mining upon tunnelling can cause damages on the surrounding objects (urban areas, closeness of the pipe of another tunnel,...) hydraulic hammers are the alternative solution of the machine tunnelling.

The traditional drill and blast method is still preferred in compact and hard rocks. However, in geologically complicated conditions such as fractured and layered rocks, the drilling of blast holes and their charging with explosives becomes more difficult and slowed down, and the efficiency of explosives is reduced. This results in slower advance and inevitable increase of excavation costs. In such complicated geological conditions in which primary (genetic) and secondary (post-genetic) fractures divide the rock into pieces of different shapes and sizes, the application of hydraulic hammers is desirable.

The productivity of hydraulic hammers depends on frequency and mutual space relation of natural discontinuities, but also on physical and mechanical characteristics of rocks, especially on their fracture toughness.

The completion of the rock classification according to fracture toughness will offer more precise indicators about appearance and speed of the fracture advance within the specific rock type, which will make easier the decision as when to apply hydraulic hammers upon tunnelling.

This publication is based on two projects. The first project No. JF279 was financed by U.S.-Croatian Science Technology Joint Fund in cooperation with Bureau of Mines. Ministry of Science and Technology of the Republic of Croatia financed the second project No. 195007 "Ecologically directed excavation of mineral raw materials".

REFERENCES

Anderson, J.T., Papineau, W.N. (1989): Rock excavation with boom mounted hydraulic impact hammers. *Foundation engineering: Current principles and practices*, pp. 418-431, Evanston, IL, USA.

Bieniawski, Z. T. (1987): Rock mass classification as a design aid in tunnelling. *Proc. Rapid excavation and Tunnelling Conference*, AIME, New York.

Biligin, N., Kuzu, C., Eskikaya, S. and Ozdemir, L. (1997): Cutting performance of jack hammers and roadheaders in Istanbul Metro drivages. *Tunnels for People, Golser, Hinkel & Schubert*, pp. 455-460, Balkema, Rotterdam.

Fukuda, H. (1997): Large cross section hard rock tunnel excavated by hydraulic breaker for first time in Japan. *Tunnels for People, Golser, Hinkel & Schubert*, pp. 343-348, Balkema, Rotterdam.

ISRM (1988): Suggested methods for determining the fracture toughness of rock (F. Ouchterlony coordinator) *Int. J. Rock Mech. Min. Sci. & Geomech. Abstr.*, Vol. 25, pp. 71-96.

Kujundžić, T. (1997): Research of the hydraulic hammers productivity according to the physical and mechanical rock characteristics. *Master's Paper*. Faculty of Mining, Geology and Petroleum Engineering in Zagreb, Zagreb

Lawton, D. (1994): Hydraulic hammers for no-blast applications. *Tunnels & Tunnelling*, January, pp. 40-41.

Rammer (1994): Hammer tunnelling in short (unpublished).

Vujec, S. (1997): Rock mass classification. *Technical Encyclopaedia 13 Ter-Ž*, pp. 205-206, Leksikografski Zavod Miroslav Krleža, Zagreb.

Geotechnical Hazards, Marić, Lisac & Szavits-Nossan (eds) © 1998 Taylor & Francis, ISBN 90 5410 957 2

The behavior of electroosmotic radial drainage consolidation of soft clay

Kwang Shin Lee

Department of Geotechnical Engineering, WooDae Engineering Consultants Co., Korea

ABSTRACT : The purpose of this paper is to study the effect of electrode geometry and to explore the feasibility of a soil improvement method incorporating electroosmotic treatment with a vertical drain. A series of conventional consolidation tests and electroosmotic consolidation tests were conducted under a vertical sand drain with a overburden pressure. The electroosmotic consolidation tests were carried out using specially designed consolidation cells to simulate field condition as much as possible. For predicting the degree and rate of consolidation of a fine grained soil under loading and electroosmosis condition with vertical sand drain, two types of electroosmotic cells, circular ring type and plate slice type consisting of four, six or eight electrodes, were used and consolidation settlement, porewater pressure, and discharge from cell were monitored during the test.

1. INTRODUCTION

In recent years, a large number of huge construction site for housing and industrial facilities by dredging and reclamation are in progress in the western and southern seaside parts of Korea. It is a matter of primary concern to provide soil of sufficient strength in the weak ground thereby increasing the strength of soils in effective and economical way. Conventional ground improvement techniques, such as pack drain, sand drain and plastic drain method, are currently used in combination with surcharge.

The research is focused on the preliminary study for the utilization of electroosmosis in field. Even though electroosmotic process as a method of improving the shear strength property of fine grained soils has been successfully applied for many years, researches on the effect of electrode geometry on electroosmotic radial drainage consolidation and its application have been little studied. The purpose of this paper is to study the effect of electrode geometry and to explore the feasibility of electroosmosis in ground improvement method, a series of conventional radial consolidation tests and electroosmotic radial drainage consolidation tests were performed with overburden pressure. In these applications, the electroosmotic radial drainage consolidation apparatus which is modified from the conventional Rowe cell consolidation test device was used. The apparutus consists of specially designed consolidation cells with rectangle, hexagon and octagon types of electrode dispositions to simulate field condtion as much as possible because in lab scale the anode is a circular ring whereas in field installation the anodes are vertical rods spaced about the cathode.

Consolidation settlement, pore pressure, and discharge from cell were monitored during the tset.

2. ELECTROOSMOTIC RADIAL DRAINAGE CONSOLIDATION APPARATUS

Figure 1 shows the electroosmotic radial drainage apparatus for measuring consolidation from radial drainage under vertical load. Figure 2 shows a schematic diagram of electroosmotic radial drainage consolidation test apparatus. It consists of a electroosmotic radial drainage cell, loading frame, a power supply, displacement and pore pressure measuring devices, water level observation panel and other devices.

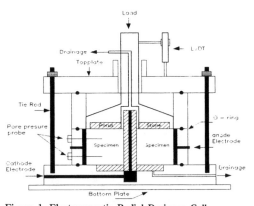

Figure 1. Electroosmotic Radial Drainage Cell.

The electroosmotic radial drainage cell is made of plexiglass in which the electrodes are placed directly at center of the soil sample and side of cell. The electrodes used in this test are copper plates to increase electrical conductivity where the electrode interfaces with soil sample. The soil sample has a diameter of 10cm and is 5cm in height. Water that is squeezed or drained out of the soil sample during the consolidation process is collected in the water level measuring panel through the drainage valves attached the top and bottom of the electroosmotic radial drainage cell.

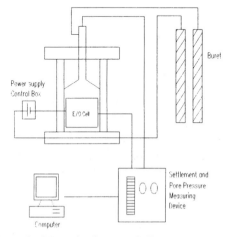

Figure 2. Schematic diagram of Electroosmotic Radial Drainage Consolidation Apparatus.

3. EXPERIMENTAL RESULTS

The samples used in this study were reconstituted with marine silty clay sampled from the western seaside of Korea. The soil properties of the samples used in this study are described in Table 1.

Table. 1 Soil properties of the sample.

Specific Gravity	water content (%)	Plastic Limit (%)	Liquid Limit (%)	Classification of Soil
2.65	43.5	21.2	31.5	CL

A total of 10 tests were performed under applied load of $3.92\,N/cm^2$ and applied voltage (0V, 3V, 6V, 8V, 12V, 16V). Variation of electroosmotic radial consolidation tests are shown in Table 2.

Table 2. Variation of electroosmotic radial drainage consolidation tests.

No voltage	0V	Overburden pressure : $3.92\,N/cm^2$ Cell type : Ring cell
Variation of voltage	3V	
	6V	
	8V	
	12V	
	16V	
No. of electrode	4	Overburden pressure : $3.92\,N/cm^2$ Voltage : 12 V
	6	
	8	
	Ring cell	

When electroosmotic pressure is applied to the fine grained soils in the field with surcharge load, excess porewater pressure by surcharge load and negative porewater pressure by electroosmosis are developed concurrently in soils. Therefore, it is necessary to analyze the compositive characteristics of consolidation both by loading and electroosmosis. In this study, radial drainage consolidation test without voltage and electroosmotic radial drainage consolidation test with voltage variation are both performed to investigate the effect of electroosmosis and electrode pattern on the consolidation characteristics of the fine grained soils. The electroosmotic radial drainage consolidation tests with ring cell under applied load of $3.92\,N/cm^2$ were carried out according to the variation of voltage and the variation of water content between anode and cathode obtained from those are compared with each other in Figure 3. The effect by electroosmotic pressure is comparatively evident and water content decrease with the increase of voltage gradient.

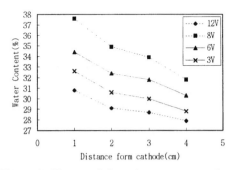

Figure 3. The variation of water content between anode and cathode.

The electroosmotic radial drainage consolidation tests with ring cell under applied load of $3.92\,N/cm^2$ were performed with the variation of voltage and the e/e_0-log t curves obtained from those are compared with each other in Figure 4. here, e is variation of void ratio and e_0 is initial value of void ratio. In Figure 4, the variation of void ratio according to the increase of voltage is normalized with the initial value of void ratio and the amount of settlement increases with the increase of voltage gradient.

Also, the electroosmotic consolidation tests with cells of rectangle, hexagon, octagon type under applied load of $3.92 \, N/cm^2$ were performed according to the variation of voltage and the e/e_0-log t curves obtained from those are compared in Figure 5. By these results, the amount of settlement increases with the increase of electrode number.

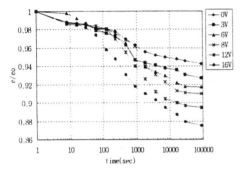

Figure 4. e/e_0-log t curves according to the variation of voltage gradients(Ring cell).

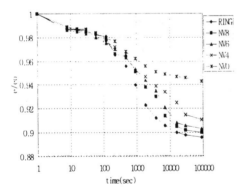

Figure 5. e/e_0-log t curves according to the variation of electrode pattern(12V).

The comparison of the value of settlement according to the effect of electrode geometry such as rectangle, hexagon, octagon pattern are shown in table 3. The measured values of settlement through electroosmotic consolidation tests with cell of octagon, hexagon, rectangle pattern are 0.943, 0.914, 0.848 times the value of settlemnt through that with cell of ring pattern. By

those results, a correction must be made for the fact that in laboratory the anode is a circular ring whereas in field installation the anodes are vertical rods spaced about the cathode.

Table 3. Comparison of e/e_0 according to electrode pattern.

Cell type	Ring	Octagon	Hexagon	Rectangle
e/e_0	0.895	0.901	0.904	0.911
Relative %	1	0.943	0.914	0.848

The variation of pore pressure according to the increase of voltage for 1 hour are shown in Figure 6, 7.

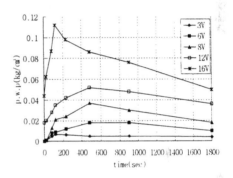

Figure 6. The curve of pore pressure and time according to voltage variation.

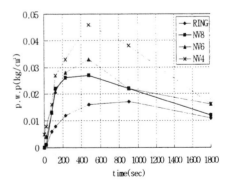

Figure 7. The curve of pore pressure and time according to electrode pattern.

During the test to measure the pore presure, a large amount of gas was generated for a short time after the application of voltage. and for the next 24 hours, it continued to generate. After that, the amount of gas generated was decreased.

The variations of electroosmotic discharge according to the increase of voltage are shown in Figure 9, 8. The electroosmotic discharge increase with the increase of voltage and electrode number.

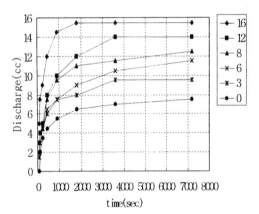

Figure 8. The curve of discharge and time according to voltage variation.

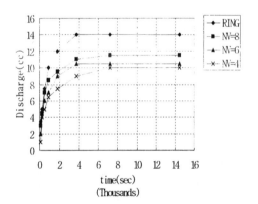

Figure 9. The curve of discharge and time according to electrode pattern.

4. CONCLUSIONS

The electroosmotic radial drainage consolidation tests were performed for clayey soils with variation of voltage and with electrode pattern under overburdon. Based on the experimental results presented herein, the following conclusions can be made.

1) The value of settlement was increased during the electroosmotic radial drainage consolidation test with the increase of voltage gradient which also shows electroosmotic effect can reduce the consolidation time of soft soils.

2) As shown in Table 3, a correction for the electrical radial drainage case is provided for field installations where the anode of octagon, hexagon, rectangle pattern surround a central cathode.

This research was performed and electroosmotic radial drainage apparatus was developed for the development of ground improvement technology in R&D Center, SunKyung Engineering and Construction Ltd by the author.

5. REFERENCES

Bjerrum, L. & O. Eide 1956. Stability of strutted excavations in clay, *Géotechnique.*

Donald Bjerrum, L., Moun, J., and Q. Eide 1967. Application of Electroosmosis to Foundation Problem in a Norwegian Quick Clay, *Géotechnique 17 : 214-235.*

Esrig, M. I. and Gemeinhardt J. P. 1967. Electro kinetic stabilization of Illitic clay, J. of the SMFE, Div., Proc. of the ASEC, Vol.93, No. SM3, pp.109-128.

Gray, D. H. 1970. Electro Chemical Hardening of Clay Soils, *Géotechnique 20 : 81-93.*

Gray, D. H., J. K. Mitchell 1967. Fundamental Aspects of Electro-Osmosis in Soils, J. of Soil Mechanics and Foundation Div. Proc.of the ASCE, Vol.93, SM6.

Johnston I. W. and Butterfield R. 1977. A laboratory investigation of soil consolidation by electroosmosis Australian Geotechnics Journal.

Mitchell, J. K. 1967. Conduction phenomena : from theory to geotechnical practice, *Géotechnique 41(3): 299-340.*

Morris, D. V. and Hills, S. F. and Caldwell, J. A. 1985. Improvement of Sensitive Silty Clay by Electroosmosis Can. Geotech. J. 22 : 17-24.

Robert L. Nicholls and Rene L. Herbst, Jr. 1961. Consolidation under Electrical Pressure Gradients, J. of the SMFE, Div., Proc. of the ASEC, Vol.93, No. SM5 : 139-150.

Geotechnical Hazards, Marić, Lisac & Szavits-Nossan (eds) © 1998 Taylor & Francis, ISBN 90 5410 957 2

Geotechnical investigations and design in permo-carboniferous clastites: An example

B. Majes & J. Logar
University of Ljubljana, Faculty of Civil and Geodetic Engineering, Slovenia

Z. Popovič
Institute for Geology, Geotechnics and Geophysics, Ljubljana, Slovenia

ABSTRACT: A development of the geological model of the Golovec tunnel designed in permo-carboniferous clastites is presented. The results of the first stage of geotechnical investigations served as the basis for tunnel design. During the earthworks in the southern portal area a landslide occurred. The remedial measures were designed based on additional investigations. As the construction continued additional movements deep under the surface started. Only the extensive third investigation stage gave an explanation of previous events.

1 INTRODUCTION

A section of the eastern Ljubljana ring motorway runs along the region consisting of permo-carboniferous clastites (combination of slate and sandstone) which are considerably folded and disturbed due to intensive tectonic movements in the geological history. The soft rock is covered by a several meters thick layer of clayey weathered material and weathered sandstone gravel. A part of this motorway section comprises also a 550 m long double tube tunnel with maximum overburden of 90 m. Each tunnel tube has a cross-sectional area of 130 m^2 enabling the construction of three traffic lines.

Due to insufficient geotechnical investigations of the lythologically heterogeneous soft permo-carboniferous rock, considerable difficulties occurred during the construction of the cutting in front of the southern portal of the Golovec tunnel. In November 1995, at the beginning of the earthworks for the cutting in front of the tunnel, a 10 to 15 m thick layer of weathered rock and soil started to slide in an area of 20.000 m^2. Since the landslide affected the portal area of the future tunnel, the remedial measures had to be adopted. The purpose of the rehabilitation was to stabilise the landslide and enable further works in the tunnel. At that stage it had already been clear that additional engineering geological mapping, in-situ and laboratory testing was needed. Based on these investigations a new geological profile was established and strength parameters were revised based upon laboratory tests and back calculations.

As a result of additional testing and considering the aim of remedial works the landslide masses were redistributed and an anchored pile wall, which supported the remaining landslide material, was constructed. In front of the wall the whole landslide material was removed. Such a measure should have provided a safe continuation of the tunnel works. However, with further works a deep slide in the tectonically disturbed rock appeared up to the depth of 55 m. In order to stop the slide of such large dimensions and to re-design the tunnel, another set of additional geotechnical investigations was performed and a third engineering geological model was determined. The deep landslide was stabilised by a 70,000 m^3 embankment at the toe of the landslide, and the entry into the tunnel was designed as a portal structure of large diameter piles. However, as the tunnel works in the right tunnel tube which runs almost parallel to and in the direct vicinity of a major fault continue, the contractor is still facing serious problems.

In the design of the southern portal structure of the left tunnel tube the forecast geological structure was not taken into account properly.

The introduction to the paper clearly shows three major sources of hazard with regard to geotechnical investigations and interpretation of test results:

- Lack of quality data (not enough borings, borings not deep enough, poor quality of drilling and consequently poor quality samples for laboratory tests, absence of additional exploration methods such as geophysical testing) which leads to unreliable interpretation. The latter can be caused by many factors, e.g. positive previous experience in similar conditions, time and financial limitations.
- Misinterpretation of the acquired data, which leads to inadequate geological model.
- Poor understanding of the geological and geotechnical data and interpretation stated in reports.

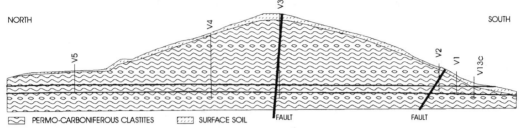

⌇⌇ PERMO-CARBONIFEROUS CLASTITES ⌇⌇ SURFACE SOIL FAULT FAULT

Figure 1. Geological model presented in the report after the 1st investigation stage for the left tunnel tube

2 INVESTIGATIONS IN INDIVIDUAL STAGES

2.1 Investigations prior to the construction – 1st stage

In addition to preliminary investigation works (review of the available data, geological mapping, analysis of aerial photographs) six boreholes were made with depths ranging from 23 to 90 m. The locations can be seen in Figure 1. In these boreholes pressuremeter tests, permeability measurements and down-hole test were performed. The samples taken from the boreholes were examined in laboratory in order to determine strength parameters, unit weight of rock and mineralogical and petrographical composition. Detailed test results are not presented in this paper. The angle of internal friction was found to vary between 23° and 34° without any significant difference between weathered rock/soil and better rock samples taken from slate and siltstone. Sandstone samples were tested in uniaxial compression (q_u=40 MPa) and uniaxial extension (tensile strength σ_t=8 MPa).

The rock was found to belong to permo-carboniferous clastites of poor and very poor quality. Slate represent 15% of the entire rock mass, siltstone 40% and sandstone 45%. A small quantity of conglomerate was also encountered. The rock mass was reported to be tectonically damaged and locally transformed into tectonic clay. Slate was found to be extremely sensitive to atmospheric influences. The sandstone formation was reported to be only partially damaged by tectonic activities. The rock is covered by up to 5 m thick clayey layer mixed with sandstone gravel.

At this stage the rock mass was considered nearly impervious. Only minor local water inflows were expected during the tunnelling.

Two faults were identified in the area in NW-SE direction and are shown in the longitudinal cross-section (Fig. 1) of the tunnel together with the geological model which was presented in the first geological report. According to this report 25% of the tunnel would be bored in the most difficult conditions (slates and siltstone, tectonic clay and low overburden), 15% in tectonically damaged siltstone with higher overburden, 25% in tectonically damaged sandstone and 35% in sandstone with lower degree of tectonic damage.

2.2 Investigations after the first landslide – 2nd stage

The earthworks for the cutting in front of the southern tunnel portal had caused a landslide with a velocity 20-40 cm/day. Several fissures were observed of approx. 1 m in width with the height difference more than 1 m. Landslide had progressively reached its final area of 20.000 m^2.

A new stage of the in-situ investigation followed immediately. Simultaneously, the works started to achieve temporary stable conditions by moving 25.000 m^3 of the surface soil from the top of the landslide to the toe of the landslide. A set of 12 new boreholes was made and equipped with inclinometer casings. These investigations were carried out in order to identify the landslide characteristics and to find the permanent solution to the southern portal area in such a way that a safe continuation of earthworks and tunnelling would be possible. With regard to the depth of this new set of boreholes it was decided that they have to penetrate at least 5 m into better rock. Figure 2 presents a cross-section along the right tunnel tube in the area of the southern portal with a new and more elaborated geological model. The permo-carboniferous rock on the model was divided into the upper weathered and wet slate layer resting on predominantly dry and firmer rock. Due to considerable thickness of the surface soil layer it was difficult to find rock outcrops and nearly impossible to determine the strike and dip of rock discontinuities. The surface layer, a weathered clayey soil with weathered sandstone gravel, preserved its structure. The foliation planes were dipping inside the slope, therefore a favourable structure was expected.

The underground water was observed at the bottom of the surface soil layer and the underlying slate layer was wet. This fact was considered one of the major causes for sliding as slip surface was found within the weathered slate layer (Fig. 2).

A new set of laboratory shear strength tests showed an even larger scatter in the angle of internal friction ($18°\leq\varphi'\leq40°$). Small values of cohesion were measured on some samples.

A back analysis was made in order to find the average angle of internal friction along the slip surface resulting in φ'=19°. The value lies within the

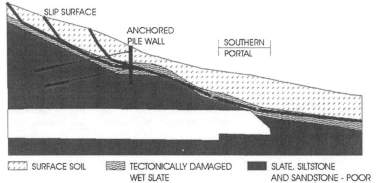

SLIP SURFACE

ANCHORED
PILE WALL

SOUTHERN
PORTAL

SURFACE SOIL   TECTONICALLY DAMAGED
WET SLATE

SLATE, SILTSTONE
AND SANDSTONE - POOR

Figure 2. Geological model in the slide area presented after the 2nd investigation stage - right tunnel tube

laboratory test results and was used in the subsequent calculations.

As permanent solution an anchored wall of large diameter piles was constructed. Between the wall and the tunnel portal the entire surface layer and a layer of weathered rock was removed. Such temporary surface should ensure safe start of tunnel excavation.

2.3 Final stage of investigations – 3rd stage

A few months after the completion of all previously described activities, in Autumn 1996 with more rainfall, an increase in anchor forces was observed and simultaneously small movements of the wall. Four additional deep inclinometers were installed in the area between the wall and the tunnel portal. A new and a much deeper slip surface was located 14 m beneath the bottom of the wall, i.e. within the tunnel tube.

At this stage after 18 boreholes were made it was obvious that the geological model of the southern Golovec slope was far from being reliable.

A new set of deep boreholes was designed and performed. The majority of these boreholes was equipped with inclinometer casings and a few of them served as piezometers. In addition to the investigations on the southern slope northern slope was investigated too, in order to avoid similar problems on the location of northern portals. 14 new boreholes with depths ranging between 20 and 115

m were bored at the southern slope and 7 boreholes up to 50 m deep at the northern slope. These additional boreholes were performed with high quality equipment and the core was carefully studied by specialists in structural geology and tectonics. Hydrogeological conditions were also studied in more detail. The result of this investigation phase is shown as a geological model in the longitudinal section of the right tube of the Golovec tunnel in Fig. 3. From this model it became clear that pre-determined shear planes existed dipping toward south. The northern slope is globally less susceptible to sliding. However, a large area of tectonically damaged rock was found in the location of the northern portals.

The behaviour of the underground water was found to be largely influenced by heterogeneous lythological structure of the rock mass. Water can be found in fissured sandstone with higher permeability, whereas weathered slate layers and especially faults act as hydro-geological barriers. Two faults were identified at this stage too, but the direction was NE-SW, i.e. nearly perpendicular to the one stated in the first report. After all construction activities were terminated also the rock structure could be observed in the cutting. The foliation planes were dipping nearly parallel to the slope.

The southern slope was stabilised by loading the toe of the slope with 70,000 m^3 of gravely material with high shear resistance, which was easy to grout

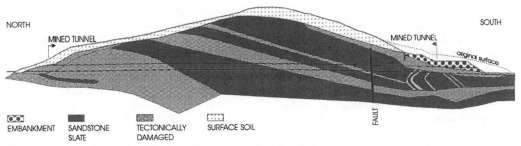

NORTH   MINED TUNNEL

MINED TUNNEL   SOUTH

original surface

FAULT

EMBANKMENT   SANDSTONE
SLATE

TECTONICALLY
DAMAGED

SURFACE SOIL

Figure 3. Final geological model of the Golovec tunnel (right tube).

above the tunnel tube. The grouted gravel formed a firm arch thus minimising additional pressures on the shallow tunnel tube beneath. The southern portals were constructed from large diameter bored piles in order to stabilize the portal area.

3 PRESENT SITUATION

The first 30 m of the right tube of the Golovec tunnel were excavated in sandstone and siltstone without major problems using NATM principles. Major problems were encountered when the work was carried out in slate, which was extremely weak and tectonically damaged. In contact with water this material soon turns into soft clay. It was then decided to continue the excavation under previously prepared pipe-roof.

Although major deformations were observed at the surface due to tunnelling, now there are no signs of global sliding. The pile wall from the first rehabilitation stage had undergone large settling caused by tunnelling without significant lateral deformations. Both portal structures (on the left and right tubes) are stable, too.

At the northern side portal structures were constructed similar to the ones at the southern side. At the moment preparations are being carried out to start the tunnel excavation from northern side as well in order to make up for some of the lost time.

4 DISCUSSION

Such severe difficulties as those encountered at the construction of the Golovec tunnel are usually caused not by a single but by many unfavourable factors. The major sources of problems as can be seen now are as follows:

- inadequate quality and quantity of investigations with regard to rock conditions on site, especially at both portal areas,
- failure in constructing a reliable and a more detailed geological model from the existing investigation data,
- inadequate design of portal cuttings in the original project,
- too fast excavation in the vicinity of the portal,
- fairly good previous experience with tunnelling in permo-carboniferous rock in Slovenia,
- extremely fast weathering of permo-carboniferous slate when exposed to atmospheric influences.

Looking back at the 2nd investigation stage from the present state of knowledge one would require at least a few deeper boreholes and the best drilling equipment.

In the 3rd investigation stage many issues were improved, the most important being:

- drilling up to the ultimate depths that can be of technical importance (up to 20 m beneath the tunnel in our case)
- using the best available drilling equipment,

- involving the most experienced experts in structural geology, tectonics and hydrogeology.

After tunnelling 90 m in the right tube and 10 m in the left one the last geological model was found to coincide fairly well with the actual geology.

5 CONCLUSIONS

Permo-carboniferous clastite formations in Slovenia are among the most unfavourable for construction purposes. They are usually folded, tectonically crushed, weathered and in most cases exposed to previous sliding. In such conditions the construction of cuttings is the most difficult. The slope stability is mostly governed by pre-existing shear planes, by dip and strike of foliation planes and by the percentage of softer rock (slate, schist) in the whole rock mass. Some of these factors are difficult to discover in advance with routine investigations. Therefore careful design and elaboration of different types of investigation procedures should be used to obtain as much information as possible. It is advisable to perform investigations in several stages.

The presented geological models elaborated after individual investigation stages clearly show that only quality geotechnical investigations can lead to a reliable geological model. It is obvious however, that exact geological model can not be elaborated in advance at a reasonable cost (see Fig. 4). Therefore a careful construction with good monitoring is essential. In the case of tunnelling a pre-drilling has to be considered.

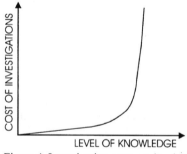

Figure 4. Investigation costs vs. level of knowledge

From the Golovec landslide and some other experiences in permo-carboniferous rock formations in Slovenia we have learned that the fact that natural permo-carboniferous slopes are mostly in a state of limit equilibrium has to be taken into account. The design of cuttings is therefore only possible with stiff retaining structures in order to limit the deformations that can lead to the reduction of shear strength to its residual value.

Tunnel portals have to be designed in a way which minimizes pre-cuts.

Geotechnical Hazards, Marić, Lisac & Szavits-Nossan (eds) © 1998 Taylor & Francis, ISBN 90 5410 957 2

Effect of drying and wetting on mechanical characteristics of Eocene flysch marl

P. Miščević

Faculty of Civil Engineering, University of Split, Croatia

ABSTRACT: The Eocene flysch in the region of Split (Dalmatia, Croatia) is characterized by the presence of layers with different characteristics. It includes thin-layered marls, clayey marls, calcareous marls, clastic layered limestones, calcarenites and breccias. The effect of drying and wetting on the deterioration of strength is investigated on an unweathered marl. The changes of strength are measured with the point load test on four marl samples with a different carbonate content. The specimens are prepared with five different paths of wetting and drying. The deterioration of analysed soft rock progresses quickly by repetition of drying-wetting. Experimental results show that the strength is reduced radically when the soft rock is wetted after the desiccation. The strength reduction level in this process depends on the desiccation degree and degree of wetting after desiccation. From the results it can be concluded that wetting-drying is an important process in the weathering of marl. For the practical use this material should not be allowed to desiccate and wet on free surface in a structure because of the significant strength reduction.

1 INTRODUCTION

The Eocene flysch in the Split region (Croatia) mainly includes: marly clay, clayey marls, marls, calcareous marls, calcarenites, limestones and calcareous breccias. There are fragments of calcareous breccias and the breccia-conglomerates and of calcarenites and detrial limestones bound by biocalcarenite and calcite cement (Jašarević 1993). The marls are mainly composed of calcite and clay minerals with some quartz, feldspars and plagioclase.

The majority of marly layers can be classified according to their physical-mechanical properties as soft rock. The use of these materials in geotechnical structures showed that they are susceptible to a change in properties due to weathering processes. In this paper the weathering of the soft rock as a construction material is defined as "degradation or deterioration of natural structural materials under the direct influence of the atmosphere, hydrosphere and man's activities within the engineering time scale." The term "engineering time scale" (i.e. a period ranging from a few years to a few decades) is used to distinguish this type of weathering from the weathering processes of hard rocks on a geological time scale (Fookes 1988).

One of the most important changes of property due to the weathering process, is the deterioration of strength (Maekawa 1991, Yamaguchi 1993). According to investigations by Miščević (1995 & 1997), the drying and wetting process has great influence on the changes of properties. In the paper the effect of drying-wetting on the deterioration of strength is investigated on unweathered samples of Eocene flysch marl from the Split region. The strength is measured with the point load test on irregular lumps with the different degree of desiccation and wetting.

2 LABORATORY TEST

2.1 *Material used for samples*

Samples were collected from four different places around the town Split in the Dalmatian region, Croatia. The characteristics of each sample are presented in Table 1 with carbonate content (C), slake durability index (I_{D2}) and natural moisture content (w_o) at the moment of the excavation.

Table 1. Characteristics of used samples

sample	carbonate content C (%)	slake durability index I_{D2} (%)	natural moisture content w_o (%)
S1	83.23	99.09	1.39
S2	64.47	95.28	2.42
S3	68.68	96.27	3.16
S4	81.88	98.67	0.92

All samples were "fresh" samples proceeded immediately after a deep excavation, in order to prevent any kind of weathering.

2.2 Analysis procedure

The strength of samples is measured with the point load test method on irregular lumps. Irregular lumps for test specimens were chosen to avoid any deterioration during the preparation of specimens. Preparation of a regular shapes mostly demands cutting with use of water, which is in fact wetting of samples. The influence of the wetting is one of the investigation points, so for the purpose of the investigation it was avoided. Also, cutting of investigated materials without water is almost impossible because the heat developed during the cutting process crushes these mostly thinly layered materials.

Tested samples were from 5 cm to 10 cm long, 4 cm to 8 cm wide (W) and 3 to 6 cm high (D), nearly shaped as blocks.

Uncorrected point load strength (I_S) is calculated with Eq. (1)

$$I_S = \frac{P}{D_e^2} = \frac{P}{\frac{4WD}{\pi}} \tag{1}$$

where P is applied force. Correction on the standard 50 mm size sample is made with Eq. (2)

$$I_{S(50)} = F * I_S = \left(D_e / 50\right)^{0.45} * I_S \tag{2}$$

where "the size correction factor F" is obtained from ISRM (1985) "Suggested method for determining point load strength". The accuracy of the correction factor applied for the analysed material is not investigated because the applied value does not influence the correlation with strength change. All values are multiplied with the same factor and the aim of the analysis was to investigate the change of the strength, not to establish the value of the material strength.

The mean value of the corrected point load strength ($I_{S(50)}$) is calculated by deleting the highest and lowest value from 15 tests and calculating the mean of the remaining values.

The strength is measured for each sample from Table 1. for five different paths of wetting and drying. The paths are described in Table 2.

Drying of specimens is performed in the drying oven at the temperature of 105° C. "Dry state" is defined as a state when the change of water content is less than 0.20 % after last 24 hours of drying. The specimens were not dried to a completely dry state in order to avoid any kind of transformation of calcium carbonate in material during the drying process.

Saturated state of a specimens is reached by immersion in water until there is no change of specimen mass in last 24 hours.

Table 2. Applied wetting-drying paths

path	description of the wetting-drying path
I	specimens with natural moisture after excavation
II	specimens dried from natural moisture
III	specimens wetted from natural moisture to saturated state
IV	specimens desiccated nearly to dry state from natural moisture, and then wetted to moisture close to natural
V	specimens desiccated nearly to dry state from natural moisture, and then wetted to saturated state

2.3 Results

The changes of the strength for analysed samples are presented in Tables 3 and 4 with the mean value of the corrected point load strength ($I_{S(50)}$), for drying-wetting paths described in Table 2.

Table 3. Changes of the strength represented by mean value of the corrected point load strength $I_{S(50)}$ for samples S1 and S2

sample	S1		S2	
path	moisture content (%)	$I_{S(50)}$ (MPa)	moisture content (%)	$I_{S(50)}$ (MPa)
I	1.39	2.19	2.42	0.67
II	1.01	2.07	0.76	1.43
III	3.01	1.49	5.51	0.78
IV	2.10	1.82	2.98	1.02
V	2.74	0.71	6.80	0.22

Table 4. Changes of the strength represented by mean value of the corrected point load strength $I_{S(50)}$ for samples S3 and S4

sample	S3		S4	
path	moisture content (%)	$I_{S(50)}$ (MPa)	moisture content (%)	$I_{S(50)}$ (MPa)
I	3.16	0.99	0.92	1.25
II	1.13	1.11	0.48	1.90
III	4.66	1.31	2.73	2.23
IV	1.61	1.31	1.31	1.02
V	4.83	0.47	2.87	0.48

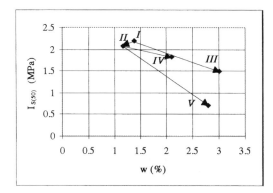

Figure 1. Mean value of the corrected point load strength $I_{S(50)}$ plotted against water content w for sample S1, with the wetting- drying paths

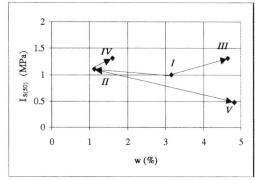

Figure 3. Mean value of the corrected point load strength $I_{S(50)}$ plotted against water content w for sample S3, with the wetting- drying paths

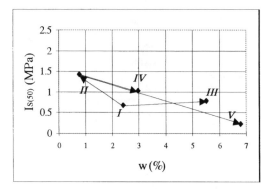

Figure 2. Mean value of the corrected point load strength $I_{S(50)}$ plotted against water content w for sample S2, with the wetting- drying paths

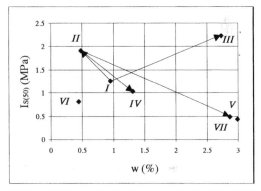

Figure 4. Mean value of the corrected point load strength $I_{S(50)}$ plotted against water content w for sample S4, with the wetting- drying paths

The results from Table 3 are plotted with the wetting-drying paths in Figure 1 (sample S1) and Figure 2 (sample S2). The results from Table 4 are plotted with the wetting-drying paths in the Figure 3 (sample S3) and Figure 4 (sample S4).

The moisture at the end of paths III and V is not exactly the same because of the opened cracks inside the samples after desiccation and moistening. The cracks are filled with water influencing significantly the measured water content. The amount of new cracks is not predictable and the water inside the cracks is not possible to remove without significant deterioration of samples.

The specimens were not dried to a completely dry state in order to avoid any kind of transformation of calcium carbonate in material during the drying process. That is the reason for slight differences in a water content in the "dry" state.

3 STRENGTH CHANGES

Figures 1-4 show that the strength is changed with the first desiccation or wetting from a natural moisture state. The strength is mostly increased.

The most significant changes of the strength (decrease) occur after the samples were dried and then moistured. This change is shown with paths IV and V in Figures 1-4. The value of the decrease depends on the moisturizing degree after desiccation. Significant change occurs when the moisture is again close to or higher than the natural moisture. It is the highest when the state of saturation is reached. The loss of the strength compared to value immediately after excavation is more than 50%.

Samples susceptible to slaking, loose the strength significantly with the increase of a moisture content to the saturated state. Slaking process is only a part of complex weathering process causing deterioration of material into smaller pieces with development of cracks along the sample. But for quick changes of the moisture content, it is almost the most significant process, so the analysed samples are correlated with a slake durability index.

With the repetition of the wetting-drying cycles the decrease of a strength continues. Samples with a carbonate content lower than cca. 65 % and with a slake durability index smaller than cca. 95 %, are completely deteriorated after a few cycles.

Figure 5. with path VI presents the change of strength after the second cycle of wetting-drying for the sample S3. The second cycle is performed with desiccation after path V was finished and moistening again to a saturated state .

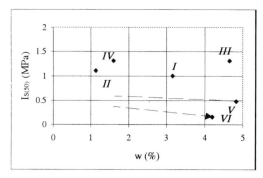

Figure 5. Change of strength after the second wetting drying cycle of sample S3 (path VI)

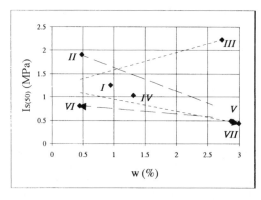

Figure 6. Change of strength after second cycle of wetting drying for sample S4 with two different paths (paths VI and VII)

The decrease of strength with second cycle of wetting-drying for the sample S4 is shown in Figure 6. Path VI is wetting to saturated state and then desiccation, after the path II was finished. Drying and

then wetting to saturated state, after the path III was performed, is designated as path VII.

All changes of strength are measured after a relatively quick wetting-drying process (few days of moistening or desiccating). In reality the changes, meaning decrease of strength, need significantly more time.

The main purpose of the performed analysis was to emphasize the problem of relatively quick strength decrease beginning. The importance of the process is in the fact that for some kinds of the marls the decrease is more than 50% even after first cycle of wetting-drying.

The total decomposition of marls due to weathering in the engineering time scale, for the engineering work is a very well known problem. However, treating marl after excavation as a soft rock in the engineering work usually leads to solutions that do not take into account the problem of quick beginning of the deterioration process and loss of strength. As a result, a few months after the work is finished, because of incorrectly treated problem, some usual problems occur:
- erosion of unprotected slope surfaces after excavation;
- unpredicted amount of settlement of the embankment made of a crushed marl;
- unexpected slides of a slopes cut in marl deposits (Šestanović 1994).

4 CONCLUSIONS

Wetting-drying process is an important process influencing the behavior of marl. One of the important effects is a initially significant strength decrease. The first cycle of material desiccation or moistening from the natural moisture has usually a small influence on the strength change. But every further step of the wetting, after the material was desiccated, results in a significant decrease of strength. The first wetting, close to the saturated state, after the samples were desiccated, resulted in a decrease of strength of more than 50% compared with the strength of sample with natural moisture immediately after the excavation.

Every further cycle of desiccation and moistening results in a further decrease of strength. The analysed samples with a carbonate content less than 70%, after only few wetting-drying cycles were so deteriorated that it was not possible to perform the point load strength test on them (example in Figure 5, path VI). Marl is more susceptible to change (decrease) of the strength with the increase of clay component content and decrease of the carbonate component content.

Finally, it can be concluded that marl used in every kind of engineering works should be protected from wetting and drying immediately after the excavation. Unprotected marl can relatively quickly loose its strength, depending on the velocity of wetting-drying cycles. The loss of strength is significant so it cannot be neglected in engineering solutions. The process begins at the surface but in the course of time it continues deeper in the mass and can jeopardize the stability or functionality of engineering work.

REFERENCES

Fookes, P.G., C.S. Gourley, & C. Ohikere 1988. Rock weathering in engineering time. *Quar. J. of Engin. Geology.* 21:33-57.

Jašarević, I., M. Lovinčić & M.S. Kovačević 1993. Some correlations between mineralogical-petrografical composition and geotechnical properties of flysch layers. Greece. *Geotech. Enging. of Hard soils-Soft rocks.* Eds. A. Anagnostopoulos. Rotterdam: Balkema. 159-168

Maekawa, H. & K. Miyakita 1991. Effect of repetition of drying and wetting on mechanical characteristics of a diatomaceous mudstone. *Soils and Found.* 31(2): 117-133.

Miščević, P. & T. Roje-Bonacci 1995. Prediction of additional settlement part caused by weathering of fill material. *Int. J. Engin. Mod.* 8(1-2): 37-44.

Miščević, P. 1997. The investigation of weathering process in the flysch terrains by means of index properties. *Proc. of Int. Sym. on Eng. Geology and Environment. Athens, Greece.* 1: 273-277. Rotterdam:Balkema.

Olivier, H.J. 1979. Some aspects of the influence of mineralogy and moisture redistribution on the weathering behaviour of mudrocks. *Proc. of the 4th Int. Con. on Rock Mech.,* ISRM /Montreux/ Vol. 3, Theme 1/.

ISRM 1985. "Suggested methods for determining point load strength",. *Int. J. for Rock Mech. Min. Sci. & Geo. Abstr.* 22: 51-60.

Šestanović, S., N. Štambuk & I. Samardžija 1994. Control of the Stability and Protection of Cut Slopes in Flysch. *Geologia Croatica.* 47(1): 139-148.

Yamaguchi, H., I. Kuroshima & M. Fukuda, 1993. Settlement-swelling characteristics of tertiary mudstone induced by effect of drying-wetting. *Proc. of the Int. Symp. Geotech. Eng. of Hard Soils-Soft Rocks.* Eds. A. Anagnostopoulos. Rotterdam: Balkema. 1343-1351.

Geotechnical Hazards, Marić, Lisac & Szavits-Nossan (eds) © 1998 Taylor & Francis, ISBN 90 5410 957 2

Plasticity index – Indicator of shear strength and a major axis of geotechnical modelling

Ž. Ortolan
RNK-Geomod, Zagreb, Croatia

Z. Mihalinec
Civil Engineering Institute of Croatia, Zagreb, Croatia

ABSTRACT: The plasticity index of a material stands out as the most significant indicator of peak and residual friction angle of coherent soils and soft rocks. The highest values of plasticity index, but also the liquid limit, correspond to lowest expected, and practically confirmed, values of friction angle. This fact opens the possibility of proceeding a new aproach to exact geotechnical modeling. An ultimate aim of modeling is a consistent correlational geotechnical column in a relatively limited area of investigation, and its structure requires a consistent application of the RNK (reference level of correlation) method. Definitions are provided for "correlational geotechnical column" and "reference level of correlation".

1 INTRODUCTION

Exact geotechnical modeling procedures aspire after the full correlation of formations, within a single geotechnical column. By using the possibilities of detailed correlation of formations, which is actually quite common in geological sciences, it has been demonstrated (Ortolan, 1990) that all results obtained by geotechnical testing, in a relatively limited area of investigation, may be interpolated in the corresponding part of vertical sequence in the correlational geotechnical column. This is achieved by means of one or several reference layers, one of which is selected as the principal or reference layer. Ortolan (1996) proposed that this layer be called the reference level of correlation - RNK, while defined it as a clearly recognizable bedding or any other reference plane, with respect to which it is possible to exactly determine the height position of all studied profiles with individual results obtained by point testing for any kind of material, in a single vertical lithostratigraphical or geotechnical sequence (correlational engineering-geological and/or geotechnical column). This enables us to obtain a consistent engineering-geological soil model in which it is possible to logically attribute concrete numerical parameters to each individual layer along the entire height of the studied vertical layer sequence what leads to obtaining a geotechnical model. From the correlational column devised in such way it is possible to differentiate, in a general way, zones of minimum residual shear-strength parameters, with

their thicknesses and continuities. The correlational geotechnical column of a studied area is the "key" through which overall geotechnical relations are explained in the required number of profiles and for two-dimensional and thre-dimensional modeling.

The thoroughly developed method of geotechnical modeling is called the RNK method, or the reference level of correlation method. At that, it has been demonstrated that the presence of visually reco-gnizable reference layers is not required. In fact, any reference plane can be selected as RNK provided that laboratory tests are of sufficient density. Then the position of formations (layers) may be determined by unequivocal positioning of the plane through three points, even in zones where this was not possible by measuring layer position on the basis of visible outcrops.

When analyzing numerous known results obtained by testing residual shear strength of different types of soils and soft rocks, Ortolan (1996) determined in which way the plasticity index of materials can become the most significant sign of the presence of zones with minimum residual parameters of resistance to shear, and he suggested a method for the wider application of this index.

Authors will attempt to demonstrate in this paper that the plasticity index is significant, not only for the residual friction angle of materials, but also for the peak friction angle. In addition, an emphasis will be placed on a good responsiveness of liquid limit to these parameters.

Table 1. Laboratory testing results for samples of natural materials (residual friction angle).

No	LL	PL	Pi	CF	ΦR	SOIL TYPE	REFERENCES
	(%)		(%)		(°)	STRATUM / AGE	
1	71,0	30,0	41,0		12,0	London Clay	(Chandler &
2	44,0	20,0	24,0		26,0	Cowden Till	Hardie 1989)
3	33,0	13,0	20,0		28,5	Falkirk Till	
4	23,0	17,0	6,0	42,0	31,0	Very Sensitive Clays	(Kenney 1967)
5	34,0	21,0	13,0	44,0	29,2		
6	37,0	22,0	15,0	43,0	26,6		
7	32,0	22,0	10,0	55,0	27,5		
8	67,0	31,0	36,0	70,0	28,8		
9	31,0	18,0	13,0	42,0	27,9	Clay - Shales	
10	71,0	28,0	43,0	80,0	14,0		
11	121,0	40,0	81,0	58,0	9,1		
12	117,0	35,0	82,0	50,0	6,3		
13	145,0	42,0	103,0	56,0	5,7		
14	96,0	25,0	71,0	58,0	5,7		
15	59,0	32,0	27,0		6,3	Cucaracha Shale	
16	106,0	35,0	71,0	70,0	10,2	Clay Interbeds in a	
17	106,0	40,0	66,0	68,0	9,1	Limestone Series	
18	62,0	18,0	44,0	52,0	15,6		
19	59,0	37,0	22,0	72,0	15,1	Kaolinite	
20	67,0	21,0	46,0	57,0	8,5	Materials	(Kenney 1977)
21	113,0	22,0	91,0	52,0	5,1	Containing	
22	72,0	29,0	43,0	57,0	9,6	Montmorillonite	
23	66,0	24,0	42,0	53,0	9,1		
24	62,0	24,0	38,0	64,0	7,4		
25	25,0	18,0	7,0	45,0	16,7	Materials Containing	
26	31,0	19,0	12,0	44,0	18,3	Hydrous Mica	
27	41,0	21,0	20,0	38,0	30,1		
28	71,0	28,0	43,0	55,0	7,1	Bury Hill Etruria Marl	(Lupini et al. 1981)
29	59,0	28,0	31,0	43,0	8,1	Clay Gouge from	
30	57,0	24,0	33,0	50,0	9,4	Carboniferous Shales	
31	26,0	20,0	6,0	32,0	10,1	and Mudstones	
32	31,0	19,0	12,0	32,0	12,1		
33	65,0	32,0	33,0	52,0	8,7	Overconsolidated	
34	82,0	33,0	49,0		11,1	Clays	
35	59,0	30,0	29,0	50,0	9,2		
36	59,0	28,0	31,0		8,6		
37	63,0	26,0	37,0	51,0	7,3		
38	95,0	34,0	61,0	59,0	9,4		
39	82,0	28,0	54,0	57,0	8,4		
40	66,0	24,0	42,0	53,0	8,0		
41	62,0	26,0	36,0	46,0	8,2		
42	62,0	26,0	36,0	46,0	7,8		
43	85,0	27,0	58,0	50,0	6,6		
44	58,0	26,0	32,0	52,0	10,7		
45	59,0	23,0	36,0	51,0	7,1		
46	93,0	32,0	61,0	60,0	7,0		
47	46,0	25,0	21,0		28,8	Soft Clays	
48	94,0	34,0	60,0	50,0	12,6		
49	41,0	21,0	20,0	38,0	28,7		
50	94,7	23,6	71,1	44,0	8,4	Pleistocene clays	(Ortolan 1996
51	59,8	18,3	41,5	26,0	20,5		
52	83,6	24,4	59,2	53,0	8,6		
53	68,8	24,7	44,1	37,0	10,8		
54	73,7	27,0	46,7	52,0	10,3		
55	57,0	21,6	35,4	36,0	19,3		
56	92,0	39,8	52,2	52,0	8,3		
57	87,5	39,0	48,5	40,0	15,1		
58	80,5	35,2	45,3	32,0	8,0		
59	81,8	33,8	48,0		9,6		
60	34,4	18,3	16,1	14,0	25,8		
61	34,4	18,3	16,1	14,0	26,0		
62	42,6	14,8	27,8	17,0	24,4		
63	92,9	16,4	76,5		6,7		
64	59,3	20,3	39,0	26,0	19,9		
65	57,8	16,5	41,3	21,0	13,3		
66	49,4	21,3	28,1	27,0	23,0		
67	70,0	15,8	54,2	39,0	6,2		
68	43,0	15,8	27,2	19,0	21,5		
69	57,5	13,6	43,9	36,0	10,1		
70	81,5	30,3	51,2	44,0	6,7		
71	85,0	26,9	58,1	31,0	5,4		
72	78,0	25,6	52,4	34,0	6,8	Pliocene and	
73	85,6	23,3	62,3	44,0	9,0	Plioquaternary	
74	159,6	35,9	123,7	75,0	7,6	Clays	
75	132,7	29,9	102,8	70,0	8,5		
76	79,1	21,0	58,1	50,0	10,5		
77	75,3	27,1	48,2	61,0	11,4		
78	87,5	28,2	59,3	61,0	9,0		
79	74,4	22,4	52,0	52,0	6,2	Pliocene and	
80	80,5	27,6	52,9	40,0	7,1	Plioquaternary	
81	70,1	28,4	41,7	44,0	21,7	Clays	
82	74,4	29,7	44,7	35,0	13,8		
83	73,5	22,0	51,5	36,0	13,8		
84	71,7	26,6	45,1	36,0	14,5		
85	73,1	23,7	49,4	34,0	13,4		
86	68,8	27,3	41,5	51,0	14,6		
87	49,7	24,6	25,1	8,0	22,8	Upper	
88	25,5	8,6	16,9		26,6	Quaternary	
89	31,0	21,2	9,8		28,8	Clay	
90	25,5	14,5	11,0		26,3		
91	51,0	20,5	30,5	42,0	24,4		
92	39,9	17,2	22,7	37,0	26,9		
93	76,6	28,2	48,4	70,0	9,3	Terra Rossa	
94	59,0	21,4	37,6	40,0	19,2	Eocene	
95	54,8	22,2	32,6	48,0	18,6	Residual	
96	53,7	21,1	32,6	52,0	22,6	clay	
97	60,4	23,6	36,8	19,0	20,3	Miopliocene	
98	66,0	19,3	46,7	21,0	9,9	Residual Clay	
99	45,0	15,9	29,1	30,0	12,5	Triassic clay	
100	33,0	16,8	16,2	7,0	19,3		
101	34,9	17,6	17,3	8,0	21,5		
102	44,0	18,4	25,6	16,0	13,4		
103	31,5	15,0	16,5	13,0	22,5		
104	29,0	15,2	13,8	7,0	21,7		
105	31,2	13,2	18,0	10,0	20,3		
106	27,8	12,5	15,3	13,0	18,7		
107	33,2	16,0	17,2	12,0	22,9		
108	94,5	40,8	53,7	45,0	7,6	Miocene Clay	
109	53,5	24,3	29,2		26,6	Clay Shales	
110	105,5	62,9	42,6	25,0	15,8		
111	47,5	29,9	17,6		25,0		
112	57,0	27,0	30,0	70,0	12,8	Upper Carboniferous	(Skempton 1985)
113	60,0	27,0	33,0	52,0	12,1	Etruria Marl	
114	64,0	28,0	36,0	52,0	9,9	Upper Lias	
115	75,0	29,0	46,0	58,0	11,1	Atherfield	
116	80,0	29,0	51,0	55,0	11,8	London Clay	
117	165,0	119,0	46,0	65,0	39,0	Allophane	(Wesley 1977)
118	95,0	65,0	30,0	76,0	35,0	Halloysite	
119	101,0	44,0	57,0	83,0	24,5	Halloysite	
120	213,0	167,0	46,0		39,0	Allophane	

2 OVERVIEW OF RESULTS AND DISCUSSION

Over the past decade, parallel laboratory tests of residual shear strength, Atterberg plasticity limits and granulometry, have served to numerous authors for making correlations of residual or peak friction angle with: clay fraction content - CF (Skempton 1964, Lupini et al. 1981, Chandler 1984, Skempton 1985, Mesri & Cepeda-Diaz 1986), liquid limit - LL (Mesri & Cepeda-Diaz 1986), plasticity index - PI (Gibson 1953, Voight 1973, Kanji 1974, Wesley 1977, Lupini et al. 1981, Chandler 1984) or with all of these three indicators combined through experiments (Collotta et al. 1989).

Kenney (1967, 1977) investigated the influence of mineralogical composition on the residual strength of natural soils and mineral mixtures.

Lupini et al. (1981) investigated numerous other influences on residual shear strength of cohesive soil, and they have also described in detail the residual shear mechanism and the reasons of possible deviations from determined principles.

Skempton (1970) pointed to the relevance of peak shear strength, fully softened shear strength and

Table 2. Laboratory testing results for samples of natural materials (peak friction angle).

No	LL	PL	PI	CF	Φ_P	SOIL TYPE STRATUM / AGE	REFERENCES
	(%)			(%)	(°)		
1	127,0	36,0	91,0	77,0	21,0	Klein Belt Ton	(Gibson 1953)
2	123,0	36,0	87,0	61,0	23,0	Shelhaven Clay	
3	107,0	31,0	76,0	62,0	23,5		
4	98,0	30,0	68,0	61,0	22,0		
5	74,0	25,0	49,0	50,0	20,5	London Clay	
6	45,0	25,0	20,0	57,0	30,0	Massena Clay	
7	73,0	28,0	45,0	50,0	22,0	Illite Clay	
8	55,0	26,0	29,0	62,0	19,0	Chicago Clay	
9	63,0	38,0	25,0	78,0	21,5	Kaolinite	
10	47,0	22,0	25,0	23,0	26,0	Wiener Tegel	
11	32,0	16,0	16,0	38,0	31,5	Horten Clay	
12	48,0	25,0	23,0	54,0	31,0	Boston Clay	
13	0,0	0,0	0,0	0,0	31,0	Stone Court Sand 10% Mica	
14	0,0	0,0	0,0	0,0	32,0	Stone Court Sand	
15	0,0	0,0	0,0	0,0	33,0		
16	41,4	19,4	22,0	18,0	27,7	Pleistocene Clay	(Mihalinec &
17	38,0	18,0	20,0	27,0	27,3		Ortolan 1993)
18	81,5	30,3	51,2	44,0	21,0		(Mihalinec et
19	47,5	29,7	17,8	19,0	26,6		al. 1988)
20	42,0	26,4	15,6	17,0	27,3		
21	48,0	23,0	25,0		22,8		(Nonveiller 1964)
22	25,0	23,0	2,0		34,2		
23	38,0	32,0	6,0		31,8		
24	42,0	22,0	20,0		29,2		
25	38,0	20,0	18,0		27,9		
26	60,0	33,0	27,0		27,9		
27	43,0	20,0	23,0		24,7		
28	44,0	19,0	25,0		25,6		
29	43,0	22,0	21,0		27,5		
30	58,0	21,0	37,0		22,8		
31	62,4	21,1	41,3	36,0	23,2	Pleistocene Clay	(Ortolan &
32	62,7	23,5	39,2	29,0	22,7		Mihalinec 1995)
33	56,6	20,9	35,7	31,0	20,8		
34	60,6	22,7	37,9	32,0	22,2		
35	50,9	19,9	31,0	27,0	23,7		
36	62,6	20,5	42,1	25,0	22,7		
37	55,0	20,9	34,1	22,0	22,7		
38	39,0	25,7	13,3	11,0	27,0		(Ortolan &
39	91,0	34,9	56,1	46,0	21,6		Sapunar 1997)
40	92,0	39,8	52,2	52,0	21,6		
41	49,5	29,1	20,4	15,0	26,6		
42	51,0	25,7	25,3	19,0	26,9		
43	41,0	23,1	17,9	10,0	28,1		
44	39,0	24,3	14,7	9,2	27,7	Pliocene or	(Ortolan 1997)
45	52,6	27,9	24,7	18,0	29,8	Pleistocene Clay	
46	40,2	23,6	16,6		29,0		
47	52,3	30,3	22,0	13,8	29,7		
48	44,0	22,0	22,0	36,0	25,0	Walton's Wod	(Skempton 1964)
49	26,0	13,0	13,0	17,0	32,0	Selset	
50	70,0	27,0	43,0	47,0	22,0	Jari	
51	82,0	29,0	53,0	55,0	20,0	London Clay	
52	53,0	28,0	25,0	69,0	21,0	Walton's Wod	
53	83,0	30,0	53,0		20,0	Kensal Green	
54	95,0	31,0	64,0		19,5	Brown London Clay	(Skempton &
55	95,0	33,0	62,0		20,7		La Rochele 1965)
56	93,0	30,0	63,0		21,3		
57	57,0	26,0	31,0	69,0	21,0	Carbonif. Mudstone	(Skempton &
58	83,0	32,0	51,0	55,0	20,0	Brown London Clay	Petley 1967)
59	75,0	29,0	46,0	57,0	18,0	Atherfield Clay	
60	69,0	29,0	40,0	47,0	24,0		
61	53,0	25,0	28,0	45,0	23,0	Upper Siwalik Clay	
62	58,0	27,0	31,0	52,0	22,0		
63	69,0	26,0	43,0	58,0	20,0	Blue London Clay	
64	165,0	119,0	46,0	65,0	40,0	Allophane	(Wesley 1977)
65	95,0	65,0	30,0	76,0	38,0	Halloysite	
66	213,0	167,0	46,0		40,0	Allophane	

Voight (1973) and Chandler (1984) have provided some explanations for deviations of residual friction angles of some soft rock varieties from the general trend, in correlation with the plasticity index (mudstone, claystone, shale), and the same was done by Wesley (1977) for allophane and halloysite clays.

Despite some imperfections, a successful correlation between the plasticity index and the residual friction angle can clearly be noted in most of the above mentioned papers. However, this fact is not sufficiently exploited in exact geotechnical modeling (Ortolan 1996).

In order to demonstrate validity of the above assertions, results have been collected from numerous analyses of peak and residual shear strength of natural materials and some of their mixtures. These results are presented in Tables 1 and 2. Symbols used in Tables 1 and 2, in all figures and in the text, are explained as follows:

- LL = liquid limit (%)
- PL = plasticity limit (%)
- PI = plasticity index (%)
- CF = clay fraction content (%)
- Φ_R = residual friction angle (°)
- Φ_P = peak friction angle (°).

Interdependencies of individual parameters from Tables 1 and 2 are presented in Figures 1-3. It can be noted that, for presented data, the plasticity index (Fig. 1) and the liquid limit (Fig. 2) are not very firmly related to the clay fraction content.

Perhaps this links could have been somewhat better, but their present status is due to the fact that various authors use different procedures for preliminary preparing soil samples for hydrometric analysis of grain size distribution. The result is that the estimated clay content may vary by as much as 200 - 300% (Mulabdić et al. 1994).

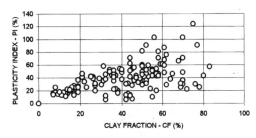

Figure 1. Dependence of plasticity index on the clay fraction content (data from Tables 1 and 2).

residual shear strength in some concrete cases.

Mesri & Cepeda-Diaz (1986) have drawn attention to the link between the residual friction angle and the fully softened friction angle, or the friction angle at critical state (Skempton 1970).

On the other hand, when drawn on plasticity chart, data from Tables 1 and 2 are concentrated in relatively narow zone (Figure 3), almost paralley with A-line, except for allophane and halloysite clays.

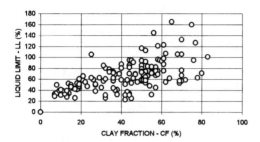

Figure 2. Dependence of liquid limit on the clay fraction content (data from Tables 1 and 2).

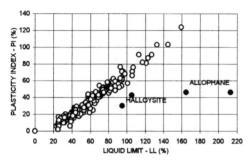

Figure 3. Dependence of plasticity index on liquid limit - Plasticity chart (data from Tables 1 and 2).

We should state at this point that such a clear deviation of samples from general principles, in the plasticity chart, may also point to the unsuitability of correlations for such materials, as discussed below in more detail.

The example presented in figure 4 shows in how extremely narow is the zone of all data points on plasticity chart may fall (almost a line parallel with A-line), for the relatively limited investigated area, if materials belong to the same lithostratigraphical unit.

Let us now present, for samples given in Tables 1 and 2, the dependence of the peak and residual friction angle on: clay content (Figure 5), liquid limit (Figure 6) and plasticity index (Figure 7). As was expected, the diagram of friction angle dependences on clay content gives an excessively wide range of results, while dependence of friction angle on liquid limit and plasticity index are obvious, and the results are quite similar. It should be noted that the dispersion of results is greater for friction angle versus liquid limit than versus plasticity index.

According to expectations, the dispersion of results in the diagrams showing the dependence of the residual (Figure 8) and peak (Figures 9 and 10) friction angle on plasticity index is even less pronounced, in cases of relatively limited areas with similar or identical conditions of sedimentation.

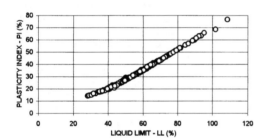

Figure 4. Dependence of plasticity index on liquid limit (Plasticity chart) for the Danube bridge in Novi Sad (Ortolan et al. 1990, unpubl.).

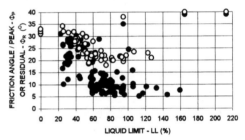

Figure 6. Dependence of the peak and residual friction angle on the liquid limit (data from Tables 1 and 2).

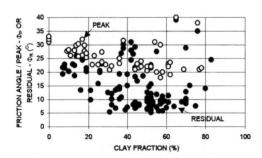

Figure 5. Dependence of the peak and residual friction angle on the clay fraction content (data from Tables 1 and 2).

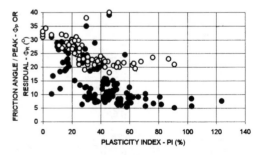

Figure 7. Dependence of the peak and residual friction angle on the plasticity index (data from Tables 1 and 2).)

In Figure 8, a subparallel sequence of results for individual types of materials should be noted. The similar principle can be observed in papers by Voight (1973) and Kanji (1974).

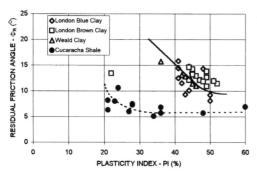

Figure 8. Dependence of the residual friction angle on plasticity index for sediments in relatively limited zones (Bishop et al. 1971).

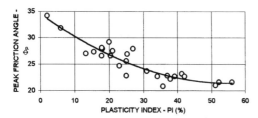

Figure 9. Dependence of the peak friction angle on plasticity index for the Pleistocene clays in the southern foothills of Medvednica (data from Tables 1 and 2).

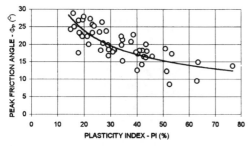

Figure 10. Dependence of the peak friction angle on the plasticity index for the Danube bridge in Novi Sad (Ortolan et al. 1990, unpubl.)

Ortolan (1996) determined that, depending on the part selected from the same sample of the Pleistocene clay from the southern foothills of Medvednica (these clays are known for their heterogeneity), the plasticity index of material may vary up to 18%. In order to

obtain even better correlations, sample selection for plastic limit testing should be limited to a same zone of sample that is subjected to friction angle testing. In fact, the best solution would be to create a data base containing plasticity indices determined on samples taken from narrow zones of shear surfaces that are formed during friction angle testing in laboratory.

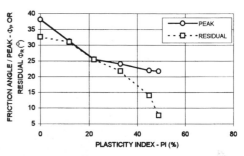

Figure 11. Dependence of the peak and residual friction angle on the plasticity index for the Happisburgh til, London clay and their mixtures (Lupini et al. 1981).

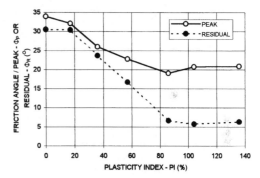

Figure 12. Dependence of the peak and residual friction angle on the plasticity index for the sand, bentonite and their mixtures (Lupini et al. 1981).

Figures 11 and 12 show results obtained by testing dependence of residual friction angles on the plasticity index (Lupini et al. 1981) for mixtures of natural materials. These results correspond well with the general trend shown in Figures 7 and 13.

Figure 13 presents results, from Tables 1 and 2, for correlation of the residual and peak friction angle with the plasticity index. Except for allophane and halloysite clays, the peak friction angle shows a significant dependence on the plasticity index. The residual friction angle also shows a separate grouping of results for siltstones, alumina and shales, to the left of the line of general dependence on plasticity index. All these deviations have already been partly explained in available literature.

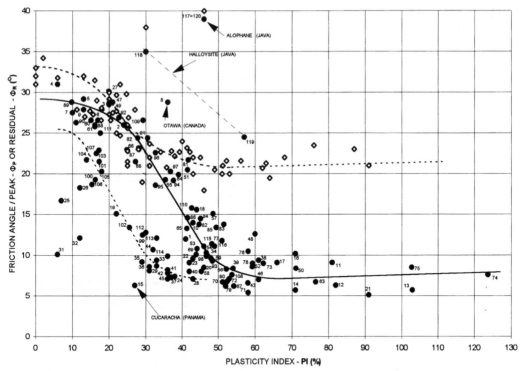

Figure 13. Correlation of the peak and residual friction angle with the plasticity index of natural materials (Tables 1 and 2).

However, despite the observed deviations, the common trend of individual restricted zones is in compliance with general principle, although it should be noted that dependence curves, on locations subjected to a great number of tests, are subparallel to the general orientation of the basic curve showing dependence of the friction angle on the plasticity index (family of subparallel curves).

3 PRACTICAL APPLICATION

As a result of analysed correlations the defined principle (Figure 13) can be used in the exact geotechnical modeling of relatively limited spaces, in the manner suggested by Ortolan (1996). The practical application of this will be shown on three examples where the full correlations of formations were made to obtain a single correlational geotechnical column (Figures 14, 15 and 16).

The results are self explanatory, and in all three cases (landslides) the positions of slip planes were determined in an unequivocal way. Over the last decade, similar methodology has been applied quite successfully on a number of landslides. A lot of them have subsequently been improved, so theoretical assumptions have been confirmed in the practice, through supervision of such landslide improvement works.

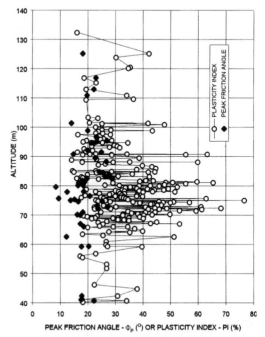

Figure 14. Variations of the peak friction angle and the plasticity index in the correlational geotechnical column for the Danube bridge in Novi Sad (Ortolan et al. 1990, unpubl., Stanić & Mihalinec, 1991).

748

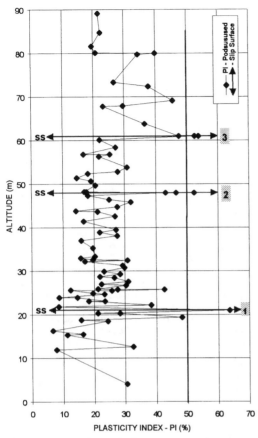

Figure 15. Variations of the plasticity index in the correlational geotechnical column for the Podsused landslide (Ortolan 1996).

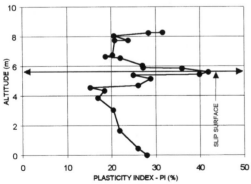

Figure 16. Variations of the plasticity index along the depth of the correlational geotechnical column for the Frkanovec landslide (Ortolan 1997, unpubl.).

The above presentations clearly point to the possibilities for exact geotechnical modeling in soils and soft rocks, which should in the future result in more accurate geotechnical calculations and in a safer design, but also in cost optimization.

It should be noted that the entire methodology would not make any sense if samples from longer intervals are tested. All samples have to be taken from intervals that must be as short as possible (up to 10 cm in length). That is because longer intervals of materials with identical characteristics are very rare in nature, and the zones containing materials with the lowest shear resistance could be missed. Ortolan (1996) has determined that two (slip planes 2 and 3, Figure 15) out of three slip planes in the "Podsused" landslide near Zagreb range in thickness from below 1 mm to several mm, in an area extending to about 1 square kilometer. The accuracy of the positions of slip planes, for this landslide, were checked by several procedures. The best results were obtained by the non-stereotype interpretation of displacement vectors for individual sliding bodies, through photogrammetric measurements (Ortolan & Pleško 1992, Ortolan et al. 1995). Three-dimensional analyses of stability (Mihalinec & Stanić 1991) have also confirmed accuracy of the spatial geotechnical model of the "Podsused" landslide.

4 CONCLUSION

The plasticity index has been proven as the most significant indicator of strength for coherent materials. The principle of general reduction of angle of internal friction with an increase of plasticity index is valid for the peak friction angle as well a for the residual friction angle. A similar principle, with a somewhat greater dispersion of results, is also valid for the dependence of friction angle on liquid limit. The clay particle content is not an appropriate indicator of the strength of materials, and should therefore be avoided. These are the reasons why large-scale testing of plastic limits for samples taken from specified depths is recommended for geotechnical modeling. The way is thus traced to a more exact geotechnical modeling and much safer design practices. The reference level of correlation method (RNK method; Ortolan 1996) should consistently be applied in data analysis.
Results presented in this paper have been thoroughly checked in practice.

REFERENCES

Bishop, A.W., G.E. Green, V.K. Garga, A. Andresen & J.D. Brown 1971. A new ring shear apparatus and its application to the measurement of residual strength. *Geotechnique* 21(4): 273-327. London.

Chandler, R.J. 1984. Recent European experience of landslides in over-consolidated clays and soft rocks. *Proc.IV Int. Sump. on Landslides* 1: 61-81. Toronto.

Chandler, R.J. & T.N. Hardie 1989. Thin-sample technique of residual strength measurement. *Geotechnique* 39(3): 527-531. London.

Collotta, T., R. Cantoni, U. Pavesi, E. Ruberl & P.C. Moretti 1989. A correlation betwen residual friction angle, gradation and the index properties of cohesive soils. *Geotechnique* 39(2): 343-346. London.

Gibson, R.E. 1953. Experimental determination of the true cohesion and true angle of internal friction in clays. *Proc. 3rd Int. Conf. on Soil Mech. and Found. Eng.* 1: 126-130. Zurich.

Kanji, M.A. 1974. The relationship betwen drained friction angles and Atterberg limits of natural soils. *Geotechnique* 24: 671-674. London.

Kenney, T.C. 1967. The influence of mineral composition on the residual strength of natural soils. *Proc. Geotech. Conf. on Shear Strength Prop. of Nat. soils and Rock* 1: 123-129. Oslo.

Kenney, T.C. 1977. Residual strengths of mineral mixtures. *Proc. 9th Int. Conf. on Soil Mech. and Found. Eng.* 1: 155-160. Tokyo.

Lupini, J.F., A.E. Skinner & P.R. Vaughan 1981. The drained residual strength of cohesive soils. *Geotechnique* 31(2): 181-213. London.

Mesri, G. & A.F. Cepeda-Diaz 1986. Residual shear strength of clays and shales. *Geotechnique* 36(2): 269-274. London.

Mihalinec, Z. & Ž. Ortolan 1993 (unpubl.). Regional road Velika Gorica - Pokupsko, improvement projekt for landslide in km 17+750 (in Croatian). *Civil Eng. Inst. of Croatia.* Zagreb.

Mihalinec, Z. & B. Stanić 1991. Three-dimensional slide analysis procedure (in Croatian). *Građevinar* 9: 441-447. Zagreb.

Mihalinec, Z., B. Stanić, Ž. Ortolan & D. Cvijanović 1988 (unpubl.). Urban design for Pantovčak - Zelengaj area, geotechnical investigations (in Croatian). *Civil Eng. Inst. of Croatia.* Zagreb.

Mulabdić, M., S. Sesar & Z. Matacun 1994. Hydrometric analysis of cohesive soil (in Croatian). Hidrometrijska analiza granulometrijskog sastava tla. *Geotehnika prometnih građevina - saopćenja* 1 *(procedings 1)*: 153-158. Zagreb.

Nonveiller, E. 1964. The landslide in stiff fractured clay at Prekrižje in Zagreb (in Croatian). *Građevinar* 16(2): 58-64. Zagreb.

Ortolan, Ž. 1990. The role of correlation methods in determining zones of minimum parameters of the shear resistance. *Proc. 6th Int. IAEG Congress:* 1675-1679. Rotterdam: Balkema.

Ortolan, Ž. 1996. The creation of a spatial engineering-geological model of deep multi-layered landslide on an example of the Podsused landslide in Zagreb (in Croatian). *PhD. Thesis.* Zagreb.

Ortolan, Ž. 1997 (unpubl.). Geotechnical investigations and improvement measures for Frkanovec landslide (in Croatian). *RNK-Geomod.* Zagreb.

Ortolan, Ž. 1997a (unpubl.). Geotechnical investigations and improvement measures for Perjavica landslide (in Croatian). *RNK-Geomod.* Zagreb.

Ortolan, Ž. & Z. Mihalinec 1995 (unpubl.). Geotechnical investigations and improvement measures for Grmoščica landslide (in Croatian). *Civil Eng. Inst. of Croatia.* Zagreb.

Ortolan, Ž., Z. Mihalinec, B. Stanić & J. Pleško 1995. Application of repeated photogrammetric measurements at shaping geotechnical models of multi-layer landslides. *Proc. 6th Int. Symp. on Landslides:* 1685-1691. Rotterdam: Balkema.

Ortolan, Ž. & J.Pleško 1992. Repeated photogrammetric measurements at shaping geotechnical models of multi-layer landslides. *Rudarsko-geološko-naftni zbornik* 4: 51-58. Zagreb.

Ortolan, Ž. & N. Sapunar 1992 (unpubl.). Prekrižje extension - geotechnical investigations (in Croatian). *Civil Eng. Inst. of Croatia.* Zagreb.

Ortolan, Ž., B. Stanić & Z. Mihalinec 1990 (unpubl.). Landslide on the right ride of Danube at Bridge Sloboda region in Novi Sad (in Croatian). *Civil Eng. Inst. of Croatia.* Zagreb.

Skempton, A.W. 1964. Long term stability of clay slopes. *Geotechnique* 14: 77-101. London.

Skempton, A.W. 1970. First-time slides in over-consolidated clays. *Geotechnique* 120: 320-324. London.

Skempton, A.W. 1985. Residual strength of clays in landslides, folded strata and the laboratory. *Geotechnique* 35(1): 3-18. London.

Skempton, A.W. & P. La Rochele 1965. The Bradwell slip: a short-term failure in London clay. *Geotechnique* 15(3): 221-242. London.

Skempton A.W. & D.J. Petley 1967. The strength along structural discontinuities in stiff clays. *Proc. Geotech. Conf. on Shear Strength Prop. of Nat. soils and Rock* 2: 29-46. Oslo.

Stanić, B. & Z. Mihalinec 1991. Bridge Sloboda - Novi Sad, landslide on the right ride of Danube (in Croatian). *Prvi znanstveni kolokvij mostovi:* 323-330. Brijuni.

Voight, B. 1973. Correlation betwen Atterberg plasticity limits and residual shear strength of natural soils. *Geotechnique* 23: 265-267. London.

Wesley, L.D. 1977. Shear strength properties of haloysite and allophane clays in Java, Indonesia. *Geotechnique* 27(2): 125-136. London.

Geotechnical Hazards, Marić, Lisac & Szavits-Nossan (eds) © 1998 Taylor & Francis, ISBN 90 5410 957 2

Rheologisches Verhalten von verschieden strukturierten Gesteinsabarten

P.Ch.Stoeva
Lehrschul 'Untertagebau', Universität für Bergbau und Geologie 'St. Iwan Rilski', Sofia, Bulgaria

ZUSAMMENFASSUNG: Es sind im Labor verschiedene Gesteinsabarten (Steinsalz, Tone, Lignitkohle) von den bulgarischen Bergbaumassiven untersucht. Es ist das spezifische rheologische Gesteinsverhalten festgestellt:
– Die Verformungsentwicklung in der Zeit mit sich veränderten Kriechgeschwindigkeiten bei ständigem Gewicht wird von der Kristallgröße (in dem Steinsalz und von der Anwesenheit von Makroingradienten (in Tonen mit Kohleeinschlüssen).
– Die Geschwindigkeitsverformung bzw. der Viskosität für verschiedene lithologische Abarten ist charakteristisch und hat eine theoretische und praktische Bedeutung bei der Interpretation des rheologischen Phänomens. Es sind die entsprechenden rheologischen Modelle vorgeschlagen.

Das rheologische Verhalten der Gesteine hängt von einer Reihe Faktoren und Bedingungen ab: von dem Platz der Gesteine im Massiv, von ihrer Zusammensetzung. Aber die größte Bedeutung haben die Struktur und die Größe der Teilchen und die sie aufbauenden Ingradiente. Diese Unterschiede beeinflußen den ganzen Verformungsprozeß in der Zeit. Die Deutung des rheologischen Phänomens bei den verschiedenen Gesteinen findet immer größere Anwendung bei der Wahl der Berechnungsparameter beim Grubenbau und anderen Einrichtungen, sowie das Abklingen der Verformungen und ihre Aktivisierung bei allen gleichen Bedingungen (gleichem gespanntem Zustand, gleicher Feuchtigkeit und konstanter Struktur). Einer der größten Probleme der Durchführung eines normalen technologischen Prozeßes in bestimmten Bergwerken ist die Standfestigkeit der Einrichtungen:
– unterirdische Kammern, Zwischenkammerpfeiler und Bohrungen in der Salzlagerstätte „Mirovo", Prowadja (Bulgarien) und
– die Arbeits – und Nichtarbeitsböschungen in den Kohlentagebauen des größten Pliozänsbeckens in Bulgarien – des Ost-Maritzas Beckens, Stara Sagora.
In einzelnen Zonen der beiden abzubauenden Bergwerke entstehen langfristige Verformungsprozesse mit großen Ausmaßen der summaren Verformungen, die die Kammerwänd deformieren, schließen oder dehermetisieren die Bohrungen oder rufen Rutschungserscheinungen in den Tagebauen hervor. Alles das ist mit großen materiellen Verlusten verbunden, Störung der Umwelt und Gefahr vor menschliehen Opfern.

Die im Labor durchgeführten rheologischen Untesuchungen mit langer Dauer beweisen, daß die Gesteine mit verschiedenen Strukturen verschiedenes Verhalten besitzen und sich von der klassischen Definitionen für die Verformungen in der Zeit unterscheiden. Das spezifische Verhalten gibt eine Möglichkeit zur Auslegung der Prozesse und zum Prognosieren der verformungen im Massiv.
Bei Untersuchung von 100% Salz wird festgestellt, daß sie bei allen Spannungen von 0.1 von der zerstörenden einachsigen Spannung (σ_n) bis zu der Grenze der langfristigen Festigkeit kriecht. Eine charaktere Besonderheit bei der langfristigen Verformung ist, daß sich bei verschiedenen Fällen eine Mobilisation zwischen den Kristallen – ein typischer makrorheologischer Prozeß, vollzieht. Nur bei den Riesenkristallen bei der Grenze der langfristigen Festigkeit werden die makromechanischen Fehler der kristallinen Splitterung mobilisiert.
Das großkristalline Salz 100% bei Spannung $\sigma_p > 0.3\sigma_n$ wird stufenförmig mit einigen Kriechgeschwindigkeiten (BC – CC^I – C^IC^{II}) (Abb.1a) verformt, die asymetriach anwachsen oder abnehmen. Die stufenartige Äußerung des Verformungsprozesses wird von der ungleichmäßigen und unorientierten Unterbrechung der Makrostruktur vorausbestimmt. Außerdem schafft die verschiedene Anordnung der Makrokristalle in der Struktur der lithologischen Abart energetische lokale Baragen mit großer Trägheit für eine relative Versetzung oder Verdrehung. In dem feinkristallinen Salz 100% ist der Verformungsprozeß verhältnismäßiger regel-

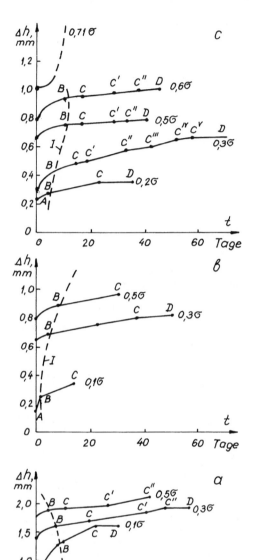

Abb. 1. Abhängigkeit der absoluten Verformung (Δh) und der Zeit (t) – Δh=f(t) für Salz 100%: a – grobkristallin; b – feinkristallin; c – nadelförmig, orientiert.

mäßiger, aber mit einer unterstrichenen kinematischen Mobilisation der zwischenkristallinen Kontakte (Abb.1b). Das nadelartige fluidal orientierte Salz 100% kriecht mit sich wechselnden abnehmenden und anwachsenden Geschwindigkeiten $(BC - CC^I - C^IC^{II} - C^{II}, C^{III} - C^{III}C^{IV} - C^{IV}C^V)$ (Abb. 1c). Hier mobilisiert der Kriechprozeß aufeinanderfolgend die Makrofehler in den imvoraus verformten Kristallen bis zum Erreichen einer langfristigen Festigkeit. Ein Analogyphänomen ist von Ter Stepanjan (1988) bei Untersuchung der makroporigen Diatomitgesteine beobachtet.

Die Veränderung der Verformungsgeschwindigkeit in den verschiedenen Revieren ist sehr vielsagend bei dem nadelartigen Salz. Es nimmt seine Geschwindigkeit im Intervall BB^I ab, was der Retardation und der klassischen Verformung entspricht. Das Abnehmen der Geschwindigkeit bis C^I ist mit den unregelmäßigen Verdichtungsprozessen (Abb. 2a) verbunden, wonach die Geschwindigkeit wegen Öffnen und Schließen der Splitter, lokalen Mikro- und Makrobaragen, Verdichten und Entdichten der Struktur anwächst und abnimmt, Im betrachteten Fall kann die Existenz einer postzerstörende Phase nicht angenommen werden, da der Prozeß mit Abklingen der Verformungen nach 54 Tagen endet. Die Abhängigkeiten: v=f(lgt) (Abb. 2a), dε=f(lgt) und η=f(lgt) (Abb. 2b) haben eine Bedeutung bei der Erklärung des Mechanismus des Verformungsprozesses als Ganzes und der Feststellung der quantitativen Parameter, besonders für den Viskositätskoeffizienten (η) bei den Prognosen, Berechnungsschemen zur Einschätzung des plastisch-viskosen Zustandes des Gesteinsmassivs.

Das geschaffene makrorheologische Modell der Salzabarten (Stoeva, Paraschkevov u.a., 1991) gibt eine Möglichkeit mit folgender Zustandsgleichung beschrieben zu werden.

Für den ersten Intervall:

$$\varepsilon(t) = \sigma/E_1 + \int_0^{t_{ret(\lim)}} t_{ret}(\varepsilon,\sigma)dt + \Sigma\{\eta\sigma/E_2 +$$

$$+ [\varepsilon_{i(pl)}+\eta_i] + \dots +[\varepsilon_{in(pl)}+\eta_{in}]\} \qquad (1)$$

Für den zweiten Intervall:

$$\varepsilon(t) = \sigma/E_1 + \sigma/E_{p1}(t) + \int_0^{t_{ret(\lim)}} t_{ret}(\varepsilon,\sigma)dt +$$

$$+ \Sigma\{\eta\sigma/E_2 + [\varepsilon_{i(pl)}+\eta_i] + \dots +[\varepsilon_{in(pl)}+\eta_{in}]\} \quad (2)$$

Für den n-ten Intervall:

$$\varepsilon(t) = \sigma/E_1 + \sigma/E_{p1}(t) + \sigma/E_{p2}(t) + \dots$$

$$+ \sigma/E_{p(n-1)}(t) + \int_0^{t_{ret(\lim)}} t_{ret}(\varepsilon,\sigma)dt + \dots + \Sigma\{\eta\sigma/E_2 +$$

$$+[\varepsilon_{i(pl)}+\eta_i] + \dots +[\varepsilon_{in(pl)}+\eta_{in}]\} \qquad (3)$$

wo: i=1, 2, 3...n; E_1, E_2 – elastische Moduls; t_{ret} – Retardationszeit; $\varepsilon_{i(pl)}$ – plastische Verformung sind.

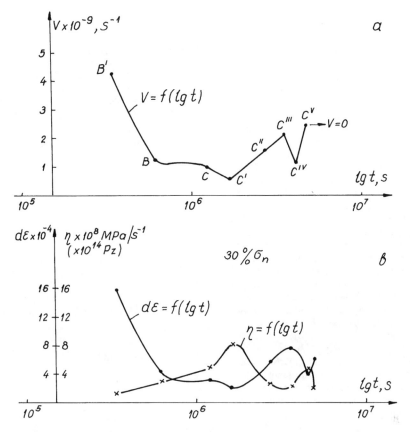

Abb. 2. a – Abhängigkeit ν=f(lgt); b – dε=f(lgt) und η=f(lgt), für Salz nadelförmig, orientiert.

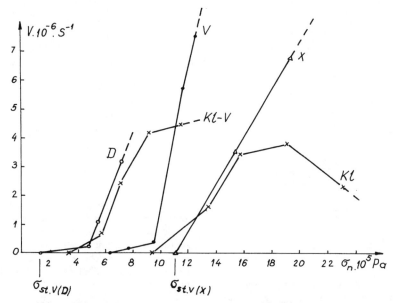

Abb. 3. Abhängigkeit ν=(σ$_n$) für verschiedene Kohlenlithotype.

Die Lignitkohle im Ost-Maritza Pliozänbecken ist auch bei einem Einachsendruck bei verschiedenem Prozent der Zerstörungsspannung untersucht. Bei den verschiedenen Kohlenlithotypen ist die Kriechgeschwindigkeit verschieden und hängt von ihrer Spezifik ab. Der Xylenlithotyp (X) wird mit einer ständigen Kriechgeschwindigkeit und hohen Kriechschwelle (nach Maslov, 1968) charakterisiert (Abb. 3), was die Systemelastizität beweist, die von der Festigkeit und Zähigkeit der Strukturbindungen bedingt ist. Das Xylen verringert seine Festigkeit mit 70%, was außer mit der Elastizität der Strukturbindungen noch mit der Abschichtung der Holzstruktur erklärt wird. Die übrigen Lithotype, wie Düren (D), Vitren (V), Klaren-Vitren (Kl–V), besitzen verschiedene Kriechgeschwindigkeiten in den verschiedenen Revieren. Ein Interesse stellt aber das Klaren (Kl) dar, das ein besonderes Verformungsverhalten hat, welches außer von den Strukturbindungen, noch von seiner großen Inhomogenität, von der Anwesenheit von Mineraleinschlüssen, und von der verschiedenen Verteilung der körnigen Struktur im allgemeinen Volumen abhängt. Es wird festgestellt, daß jede Kriechgeschwindigkeit kleiner als die vorhergehnde ist. In diesem Fall kann man von „abklingenden Intervallen der Kriechgeschwindigkeit" bei ständiger Belastung (Abb. 3) sprechen. Da das Kohlenflöz aus den obenangeführten Lithotypen besteht, kann die Verformung eines mehrkomponenten Massiv durch ein superponiertes rheologisches Modell dargestellt werden, aufgebaut auf Grund der Zusammenwirkung einer jeden Komponente (Abb. 4). Das summiert einerseits das Verformungsverhalten der Komponente und andererseits – beeinflußt die Umverteilung der Spannung und „das ins Gleichgewichtbringen" des allgemeinen Spannungs-Verformungszustandes. Dieses Problem ist in geringem Grade untersucht. In der Literatur ist ein superponiertes Modell von Florin-Wjalov [1978], von Bolzmann [1978] u.a. angewiesen. Das superponierte Modell sieht wie auf Abb. 5 aus und wird mit der gleichung:

z. B.:

$$d[\varepsilon''_{11(p)}]/dt = \sigma/\eta'''_{11} \qquad (4)$$

$$\varepsilon''_{11(p)} = \int_{t_{z.anfang}}^{t_{z.end}} (\sigma/\eta'''_{11})dt \qquad (5)$$

wovon

$$\varepsilon_{cr} = \int_{t_{z.anfang}}^{t_{z.end}} [\sigma/(K_5/\eta'''_{11} + K_1/\eta'''_3 + K_4/\eta'''_{14} + \cdot$$

$$+ K_3/\eta'''_{10} + K_2/\eta'''_6)]dt \qquad (6)$$

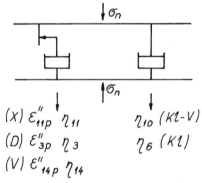

Abb. 5. Superponiertes rheologisches Modell für die Kriechverformung.

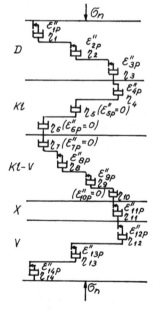

Abb. 4. Modell der Kriechverformung verschiedener Kohlenlithotype: Düren (D), Klaren (Kl), Klaren-Vitren (Kl–V), Xylen (X), Vitren (V).

Das Modell ist von den Untersuchungsangaben für jeden Lithotyp durch Computerbearbeitung und durch Schaffen einer superponierten kinetischen Abhängigkeit geprüft, was einem „mittelmäßigen Kohlenflöz" entspricht, das auf Grund der Bilanzverteilung der Lithotype in der Kohlensohle beschrieben ist.

Die schwarzen über der Kohle liegenden Tone mit Kohleneinschlüssen werden auch mit mehreren

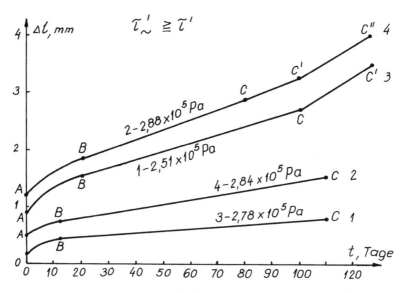

Abb. 6. Schwarze organische Tone (Kurven 1 und 2) Schwarze organische Tone mit Kohleneinschlüssen (Kurven 3 und 4).

Geschwindigkeit (Abb. 6) verformt. Auch hier, wie bei der Tonkohle, ist jede folgende Geschwindigkeit größer als die vorangehende. Das ist den Kohleningradieten zu verdanken, die ein Inhomogenität nicht nur in physikalischer sondern auch in mechanischer Hinsicht schaffen. Die tatsächliche Kriechgeschwindigkeit wird mit bedeutender Versetzung erhalten – etwa 80–100 Tage. Das wird mit der Trägheit der kohlenteilchen erklärt, die viel langsamer ihre maximale relative Versetzung erreichen, wegen ihrer Makroausmaßen gegenüber den Tonteilchen wegen der allgemeinen größen Beweglichkeit der sie einfügenden Tonmasse.

Die vorgeschlagenen Modelle und Ideen zur Beschreibung heterogener Massive nach der Bilanzverteilung und einem superponierten Modell finden eine große Anwendung bei der Einschätzung der Standgestigkeit der Bergwerkeiwichtungen in den Untertage- und Tagebauen. Die theoretische Bedeutung beruht auf der Analyse der vereinbarten Kriechprozesse verschiedener lithologische Abarten und der Prozeßbeschreibung detailliert und als Ganzes. Auf Grund des superponierten Verformungsprozesses wird der Verformungsmechanismus erklärt, der einem entsprechenden den Schema entspricht, wonach die Stabilitäts berechnungen durchgeführt werden.

Маслов Н. Н. 1968. *Длительная устойчивость и деформация смещения подпорных сооружений*. Энергия, Москва.

Stoeva P., R. Paraschkevov, E. Totzev. 1991, Makrorheologisches Modell von Salzstrukturen. *Internationales Salzsymposion*, St. Petersburg.

Тер-Степанян Г. И. 1988. Явление скачкообразной перестройки структуры грунтов при деформирования. *Инж. геол.*, 3.

LITERATURVERZEICHNIS

Вялов С.С. 1978. *Реологические основы механики грунтов*. Высшая школа, Москва.

Geotechnical Hazards, Marić, Lisac & Szavits-Nossan (eds) © 1998 Taylor & Francis, ISBN 90 5410 957 2

SPT results reliability

Gorazd Strniša & Ivan Lesjak
SLP d.o.o., Slovenia

ABSTRACT: This article deals with reliability issues of SPT /Standard Penetration Test/ results. Intention was to find out differences between actual or transferred and theoretical energy and necessary calibration coefficients for different SPT configurations on the basis of measured transferred energy in accordance with EUROCODE 7 and ASTM D4633 recommendations. More than 2000 blows on different sites and on different SPT systems were recorded by PDA /Pile Driving Analyzer®/. We found big differences and disparity in transferred energy on different SPT driving systems. We tested all SPT systems commonly encountered in Slovenia. Because of the extreme variability of transferred energy and therefore the unreliability of the associated SPT N-values, we recommended to exclude some SPT driving systems from geotechnical practice.

1 INTRODUCTION

Intention of this research was to find out reliability and repeatability of SPT test results (Standard Penetration Test), which is one of the most common in-situ geotechnical test used to determine soil geotechnical characteristic of noncohesive soils. In fact, the SPT test is practically the only in-situ test used in Slovenia for noncohesive soils. Geotechnical engineers use the "N" values from the SPT for numerous correlations.

The number of blows required to drive the last 300 mm (one foot) of a 450 mm sampler penetration is the "N-value" and indicates soil strength. The SPT N-value is used for many geotechnical evaluations and influences the engineer's design. Unfortunately, the N-value depends not only on soil strength – driving resistance, but also upon SPT hammer energy input. Given a low reliability of the in-situ test, the design must be very conservative to reduce risk. Reliable N-values result in lower and more economical safety factors.

For the present research, 29 transferred energy measurements were made on eleven (11) different SPT systems which could be divided in five (5) typical groups. These five categories considered differences in release and lifting systems.

The original design of the Standard Penetration Test is generally attributed to Terzaghi and Peck in the 1920s and 1930s. The test procedure starts with lowering the drive rods with sampler or cone to the bottom of the clean borehole. The hammer of 63.5 kg mass, the hammer fall guide and the anvil make up the drive system assembly whose weight is initially supported by the drive rods. An initial or seating drive of 150 mm penetration is applied with a hammer fall of 760 mm and the number of blows is recorded. The sampler or cone shall then be driven in the same manner an additional 300 mm. The total number of blows required for the 300 mm penetration after the seating drive is termed the penetration resistance number N which is recorded in the field. Basic prescription for SPT test are: hammer mass is 63.5 kg +- 0.5 kg; hammer fall is 760 mm +- 10 mm. Maximum mass of hammer assembly on the top of the first driving rod segment should be less than 115 kg and driving rods diameter from 35 to 60 mm. Borehole diameter should be less than 150 mm. As stated, the SPT test should actually be performed with a sampler (diameter 54 mm) to recover disturbed soil samples. However, in Slovenia the mere use of a cone with diameter 50 mm and angle of 60° is very common.

Diameter and mass of driving rods are not prescribed strictly with any standard. For good energy transfer, rod splices should be tight and driving rod diameters should be equal over the whole drive rod length.

In accordance with ENV 1997-3 automatic hammer drop mechanism is required while ASTM D1586 permits different dropping systems.

2 FIELD SPT NUMBER CORRECTIONS

Energy losses are induced by the hammer assembly

due to frictional and other effects, which cause the hammer velocity at impact to be less then the free fall velocity. Further losses of energy are originated by the impact on the anvil, depending on its mass and other characteristics. The type of machine, skill of the operator and other factors can also influence the energy delivered to the drive rods.

If the length of rods is less than 10 m, the energy reaching the tip is reduced and the correction factors (λ) shown Table 1 should be applied.

Table 1: Correction factors due to rod length.

Rod length below anvil (m)	Correction factor λ
> 10	1.00
6 – 10	0.95
4 – 6	0.85
3 – 4	0.75

In case cone is used instead of a sampler, a correction factor (κ) should be used according to Nonveilller/Eiler (1955) $\kappa = 0.75$.

The effect of the overburden pressure and therefore the density index may be taken into account by applying to the measured N-value the correction factor C_N. For an effective overburden pressure of 100 kPa, the correction factor becomes $C_N = 1$. A detailed description of this correction is given in ENV 1997.

One of the basic corrections of the field N-value uses the k_{60} factor. This factor adjust for the variation in hammer or hammer system efficiency and reflects the assumption that 60 percent of theoretical SPT hammer energy represents the historical average energy transfer on which many empirical relations have been based.

For general design and comparison purposes, the N-values shall be adjusted to a reference energy ratio of 60 %.

Energy correction factor (k_{60}) is defined as relationship between actually transferred energy from the hammer to driving rods ($E_{measured}$) and 60% of theoretical energy of the driving system (E_{60}).

$$k_{60} = E_{measured} / E_{60}$$

The following value for final reference blow count N_{60} would be obtained including all correction factors:

k_{60} ... energy correction factor
κ ... sampler/cone correction factor
λ ... correction factor due to rod length
C_N ... overburden pressure correction factor

The final adjusted SPT blow count after all corrections is: $N_{60} = N * k_{60} * \kappa * \lambda * C_N$

In accordance with ENV 1997 and for corrected or final blow count for $C_N = 1$ (effective overburden pressure 100 kPa) is:

$$(N_1)_{60} = N * k_{60} * \kappa * \lambda$$

3 ENERGY MEASUREMENTS

Similar to pile driving, the SPT installation procedure is governed by stress wave propagation. One dimensional wave mechanics can be used to analyze these measurements and evaluate energy transfer. The energy transmitted can be determined from the work done from the expression:

$$E = m * v^2 / 2 \quad or \quad E = F * h$$
$$v = (2*g*h)^{1/2}$$

E ... kinetic or potential energy
v ... hammer impact velocity
F ... ram weight
g ... earth gravitational acceleration

Theoretical kinetic energy (E_{100}) for SPT is :

$$E_{100} = m * v^2 / 2 = 63.5 * 2 *g * h/2$$
$$= 63.5 * 2 * 9.85 * 0.76 / 2 = 0.475 \ kJ$$

and theoretical potential energy (E_{100}):

$$E_{100} = F * h = 63.5 * 9.85 * 0.76 = 0.475 \ kJ$$
$$E_{60} = E_{100} * 0.6 = 0.475 * 0.6 = 0.285 \ kJ$$

Transferred energy can be calculated from measured force, F(t), and velocity, v(t), induced by the hammer impact in the top of the driving rod:

$$E(t) = \int (F(t) * du(t)) = \int (F(t) * v(t) * dt)$$

F(t) .... force at rod top due to impact
v(t) .... rod top velocity due to impact
dt .... time increment
du .... displacement increment

Therefore, force and velocity should be measured at the top of the drive rod while it is impacted by the hammer and then the transferred energy can be calculated. Usually, instead of velocity, acceleration is measured and converted to velocity by integration over time.

A specially instrumented reusable rod section inserted into the drill string during SPT hammer operation measures strain (force) and acceleration. The authors have used a Pile Driving Analyzer® (PDA) measured by Pile Dynamics, Inc. for force and acceleration measurements and for the energy calculation. The PDA obtains immediate results for each hammer blow. It conforms to the ASTM

D4633 specification for SPT energy measurement. Typical F (t) and v (t) records are shown in Figure 1. Numerical results for each blow are presented in diagrams generated with the PDAPLOT program; an example of this presentation is in Figure 2.

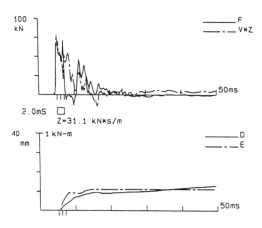

Figure 1: Typical F (t) and v (t) records.

4 SPT SYSTEMS

In Slovenia drilling contractors use five typical SPT systems. The vary in their lifting and dropping mechanisms:

A Automatic drop system (TRIN). Hammer is lifted in special frame with a drop mechanism that assures a constant drop height. Hammer and drop mechanism are lifted by cathead and rope.

B Hands drop system (KV+SKL). Hammer guide is a steel rod; hammer is lifted with the same steel cable and clutch which are also used for lifting of drilling rods. Drop height is marked with chalk on guide rod. Operator lifts hammer using clutch and releases it when hammer reaches mark.

C Hands drop system (VRV+VIT). The same system as in B except that hammer is lifted with rope (not steel wire) which is wrapped once or twice around cathead. Operator pulls rope to lift hammer and releases it for drop.

D Automatic drop system (VERIGA). This system is actually a super heavy dynamic penetrometer (DPSH). Hammer is lifted with chain and automatic drop mechanism. In this case rod diameter is always 35 mm.

E Automatic drop system (KRONA). The same as system A, except hammer with hole is guided by a 35 mm rod. Automatic drop mechanism on the top of the hammer assures constant drop height. Anvil is very small.

Actual hammer masses and actual drop heights on site were not controlled.

5 FIELD MEASUREMENTS AND RESULTS

Measurements on 29 sites were made. More then 2000 blows were recorded.

Table 2: Measurement locations and used SPT systems.

No.	Location	B.hole	Machine	SPT group
1	TE-TO LJ	P33/15	LINK BE.	A
2	TE-TO LJ	P55/15	LINK BE.	A
3	SHELL JESENICE	V1	FRASTE	C
4	SHELL JESENICE	V2	FRASTE	C
5	MOST LUKA KP	N1	JANEZ	B
6	MOST LUKA KP	N2	JANEZ	B
7	AC ŠENT.-BLAG.	NB1	FRASTE	C
8	AC ŠENT.-BLAG.	NB1	FRASTE	C
9	AC BLAG.-VRAN.	6-3/V4	SIMCA	C
10	AC BLAG.-VRAN.	6-2/V2	J. 4/GZL	C
11	NB5 AC ŠENT.-BL.	NB5	FRASTE	C
12	NB16 AC ŠENT.-BL.	NB16	KANAR.	C
13	6-1 AC BL.-VRAN.	V19	SIMCA	A
4	6-1 AC BL.-VRAN.	V19	SIMCA	A
15	PAC PODNANOS	NV7	JANEZ	A
16	PAC PODNANOS	NV7	JANEZ	A
17	PAC PODNANOS	NV7	JANEZ	A
18	AC VRAN.-BLAG.	6/6/14	JANEZ	B
19	AC PESN.-SLIV.	V83	JANEZ	B
20	AC PESN.-SLIV.	V83	JANEZ	B
21	SŠC KRŠKO	V2	KANAR.	C
22	SŠC KRŠKO	V2	KANAR.	C
23	SŠC KRŠKO	V2	KANAR.	C
24	MOST MOKRON.	DP1	SPT-DP	D
25	MOST MOKRON.	DP1	SPT-DP	D
27	MOST MOKRON.	DP2	SPT-DP	D
28	MOST MOKRON.	DP2	SPT-DP	D
29	CRP. PETROL, MB	V2	KRONA	E

With measurements we define maximum transferred energy (Emax), minimum transferred energy (Emin), averaged transferred energy (Eavg), standard deviation on averaged transferred energy (Estd/avg) and energy correction factor k_{60} for different SPT groups or SPT systems due to the differences in the lifting and dropping mechanisms.

Typical records of transferred energy versus blow number are presented in Figure 2 for different SPT measurements. Final results are in Table 3.

Figure 3 summarize averaged transferred energies and corresponding energy correction factors for site tests with different SPT systems. Very big difference

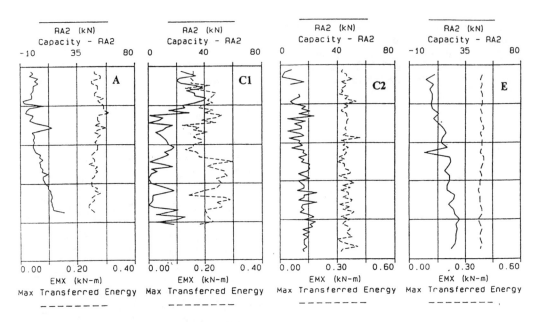

Figure 2: Transferred energy versus number of blows.

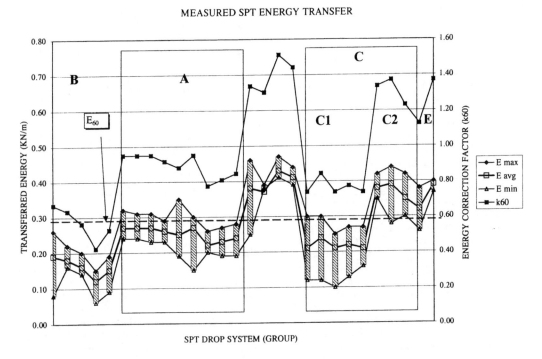

Figure 3: Averaged transferred energies and corresponding energy correction factors for site tests with different SPT systems.

Table 3: Energy correction factor k_{60} for different SPT groups (E in kJ).

SPT	E_{max}	E_{min}	E_{avg}	$E_{std/avg}$	$k_{60\ avg}$
A	0.35	0.15	0.25	0.02	0.88
B	0.26	0.06	0.16	0.03	0.56
C	0.44	0.1	0.28	0.08	0.98 *
C1	0.3	0.1	0.22	0.02	0.77
C2	0.44	0.26	0.36	0.03	1.27
D	0.47	0.25	0.40	0.03	1.41
E	0.4	0.37	0.39	0.01	1.37

* ... Unreable statistical average value for group C (C = C1 + C2) – very big deviations.

in using SPT without automatic drop mechanism (SPT group C1) and with automatic drop mechanism (SPT group E) could be seen from transferred energy graphs in Figure 2.

6 CONCLUSIONS

On the basis of performed measurements it is possible to conclude that very big differences in transferred energy existed among different SPT system groups. As can be seen from Table 3, transferred energy values varied from 0.06 kJ to 0.47 kJ leading to energy correction factors on averaged transferred energies (k_{60}) are from 0.56 to 1.41 in other words, the field recorded N-values would have been high by up to 41% (low transferred energy) or low by up to 44% (high energy) if the mesurements would not have been made or used for correction.

As an illustration of this problem an example is presented in Table 4. It shows how a field N-value of 30 would change for the different SPT systems.

Table 4: N_{60} for different SPT groups (N_{filed}=30).

SPT GROUP	k_{60-avg}	N_{60} ($N_{field} = 30$)
A	0.88	26
B	0.56	17
C1	0.77	23
C2	1.27	38
D	1.41	42**
E	1.37	41
$N_{60\ max}$		41
$N_{60\ min}$		17

** needs additional correction because of rod friction

With out energy correction factor design engineer adopts the same soil characteristics for all cases although from N_{60} could be seen significant differences. With such low SPT reliability, the design

must be very conservative to reduce risk.

Using SPT without automatic drop mechanism also results in big variations within a single test - as could be seen from Figure 2 – (c). Deviations on transferred energy from blow to blow using hand drop SPT systems are so extreme that it should be excluded from geotechnical practice.

The final conclusion is that the k_{60} correction factor of each SPT equipment has to be known, if the N-values are to be used for the quantitative evaluation of foundations or for the comparison of results.

REFERENCES

EUROCODE ENV 1997-3/1995

ASTM D1633-84 and ASTM D 1586-84

ASTM D4945-89, *Standard Test Method for High-Strain Dynamic Testing of Piles* (1989)

Goble,G.G & Hasan Abou-matar, 1994, *MESUREMENTS ON THE STANDARD PENETRATION TEST*, GRL and Associates

GRLWEAP™ *Manual - Wave Equation Analysis of pile driving (1997-2)*, GRL, USA.

Hussein,M. & F.Rausche, *Pile design and construction control by dynamic methods-Case history*, Geotechnical News, Vol. 8, No.4, december 1990, pages 24-28 (1990)

Likins,G.E. & M.Hussein, F.Rausche 1988. Design and testing of pile foundations, *3rd Int. Conf. on application of stress-wave theory to piles*, Ottawa 25.-27.5.1988, Canada, 644-658.

Strniša,G. & I.Lesjak, *Metoda "CAPWAP®" kot alternativa klasični statični obremenilni preizkušnji*, Gradbeni vestnik, Ljubljana 1987 (36), pages 64-68 (1987)

Geotechnical Hazards, Marić, Lisac & Szavits-Nossan (eds) © 1998 Taylor & Francis, ISBN 90 5410 957 2

Small strain soil stiffness in foundation settlement predictions

A. Szavits-Nossan & M. S. Kovačević
University of Zagreb, Faculty of Civil Engineering, Croatia

R. Mavar
Civil Engineering Institute of Croatia, Zagreb, Croatia

ABSTRACT: Overprediction of settlements of foundations on stiff clays based on stress-strain measurements on soil samples in the laboratory is often met in practice. This was recently attributed to deficiencies in measurements of soil stiffness at small strains in conventional laboratory tests. Shear stiffness of stiff clays, gravels and sands at very small strains is readily measured by modern geophysical non-destructive field methods. The paper presents a comparison of predicted and measured settlements obtained by series of plate load tests on a well compacted gravely road embankment. Predictions of settlements were based on shear wave velocity profiles measured by the SASW method, and they compared well to the measured values.

1 INTRODUCTION

Calculation of settlements of shallow foundations on stiff clays based on stiffness testing of intact soil samples by conventional laboratory equipment usually leads to overprediction. Overprediction is now attributed to the inability of conventional laboratory and field equipment to measure reliably small strains which are relevant for settlement predictions of shallow foundations (e.g. Jardine et al. 1986, Burland 1989, Tatsuoka & Kohata 1995). The situation with sands and gravels is even worse because of the difficulties in recovering intact soil samples and their mounting onto appropriate laboratory apparatus. Settlement predictions based on correlations with various field tests remain the most reliable alternative (e.g. Burland & Burbridge 1985 for correlation with SPT).

Recent laboratory research with small strain measuring devices directly mounted on soil samples in the triaxial apparatus has shed much light onto the mechanical behaviour of stiff clays and sands at small strains. Highly non-linear stress-strain behaviour and higher stiffness than measured by conventional laboratory equipment were found at small strains (Burland & Symes 1982, Clayton & Khartush 1986, Goto et al. 1991). There are also indications that shear modulus reduction curves for static loading correspond well to shear modulus reduction curves for dynamic loading (Atkinson & Sallfors 1991). It was also found that static shear stiffness at very small strains measured in the laboratory corresponds very well to the dynamic stiffness obtained by geophysical methods in the field

(Tatsuoka & Kohata 1995). Fundamental links between static and dynamic deformation properties were established (provoking even an enthusiastic paper title, Burland 1989). These links opened up the way for wider application of geophysical measurements for the determination of *in situ* soil stiffness. A hypothesis that settlements on stiff clays, gravels and sands may be predicted by the use of shear wave velocity profiles of the ground follows naturally from these findings. This hypothesis has become particularly attractive after the development of the spectral analysis of surface waves (SASW) testing technique. SASW is a quick, cost effective, and non-destructive testing technique that works entirely from the soil surface. It can accurately determine shear wave velocity profiles of horizontally layered sites up to depths of 10 m and more.

The present paper describes an investigation in which settlements of uniformly loaded circular plates 30 cm in diameter were predicted using results of SASW. Predictions were compared with measured settlements of the plate load test (PLT) thus testing the above hypothesis.

2 FIELD MEASUREMENTS

2.1 *The test site and the testing program*

A motorway embankment under construction was used as a convenient test site for the comparison of settlements predicted by SASW and those measured by PLT. The embankment was constructed over level natural ground by placing and compacting 80 cm thick layers of well-graded gravel up to the height of approximately 10 m. After each layer was compacted, SASW testing was performed. At the location of the impact in the SASW test, PLT was performed.

2.2 *The SASW test*

The SASW method provides detailed profiles of shear wave velocities in horizontally layered strata working entirely from the surface (e.g. Nazarian & Stokoe 1983, Addo & Robertson 1992, Stokoe at al. 1994, Hiltunen & Gucunski 1994). A vertical impact at the surface generates transient Reyleigh waves that propagate at different speeds in layered strata. Vibration transducers are located at known distances on the surface of the stratum. The dependency of surface wave velocities on wave frequency for the specific site, known as the dispersion curve, is computed by performing the spectral analysis of recorded signals at known distances. A theoretical dispersion curve for a horizontally layered site, which matches the measured curve, is sought by an iterative trial and error algorithm by adjusting layer thicknesses and layer shear wave velocities. Soil layer density and Poisson ratio have to be assumed in this algorithm.

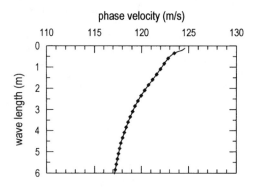

Figure 1. A typical measured dispersion curve (discrete symbols) and the corresponding theoretical dispersion curve (full line) for the test site.

The required testing equipment consist of: surface impact generators (falling weights), vibration sensors (geophones), an analog-to-digital converter and a computer supplied with adequate software. Each of these elements has to satisfy certain requirements to obtain meaningful and reliable results. Details of the testing procedure and equipment employed in the investigation covered in this paper are described elsewhere (Szavits-Nossan et al. 1998). Impact location and vibration transducers were placed along a line, about 15 m long, on the embankment surface. Geophones were

placed at distances from 10 cm up to 15 m enabling the determination of stiffness profiles up to the depth of at least 10 m. The SASW testing was performed by the Geotechnical Laboratory of the Faculty of Civil Engineering of the University of Zagreb. Figure 1 shows a typical pair of the measured dispersion curve and the corresponding theoretical curve at the test site, while Figure 2 shows the corresponding shear wave velocity profile. Poisson's ratio $v = 0$ was assumed due to the unsaturated state of the embankment. A different assumption may slightly influence the results. Soil density of $\rho = 2$ Mg/m^3 was assumed for all embankment layers.

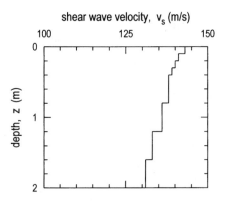

Figure 2. Shear wave velocity profile for the theoretical dispersion curve of Figure 1.

2.3 *The PLT*

The use of PLT in predicting foundation settlements in practice is considered controversial, since it covers too shallow soil depths. Its use is decreasing (Ladd et al. 1977) and is mostly limited to controlling the stiffness of road embankment layers. The PLT was performed by pressing a stiff round steel plate, 30 cm in diameter, vertically into the ground. The location of the plate was at the location of the impact in the SASW test. Vertical plate displacements were measured against a non-moving horizontal bridge. Micrometers were used at three equally distant points on the plate. Readings were made at average vertical pressures of 150, 250, 350 and 450 kPa. For a given vertical pressure the settlement of the plate was defined as the average of the three micrometer readings. Figure 3 shows the results of three micrometer readings for a typical PLT at the testing site. The settlements in Figure 3 do not approach the bearing capacity, not even for highest contact pressures. It is estimated that the highest contact pressure of 450 kPa is somewhere between 30% and 50% of the bearing capacity.

764

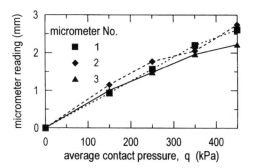

Figure 3. Typical PLT results at the site.

3 PREDICTIONS AND MEASUREMENTS

3.1 Prediction of settlements

Settlement s of a circular plate having a diameter of $D = 30$ cm was calculated by Equation 1, assuming a uniform contact pressure q and Poisson's ratio of zero. The vertical stress distribution $\Delta\sigma_v(z)$ was calculated assuming a homogeneous isotropic elastic halfspace. The integral in Equation 1 was numerically evaluated under the centre of the circular plate up to the depth of $d \approx 10$ m. Poisson's ratio of zero was assumed for two reasons. First, the embankment was in an unsaturated state inferring drained conditions. Secondly, the influence of the Poisson's ratio is considered small regarding the expected accuracy of settlement predictions. In other circumstances different values of the Poisson's ratio may be assumed.

$$s = \int_0^d \frac{\Delta\sigma_v(z)}{E(z)} \, dz \qquad (1)$$

The vertical distribution of the Young modulus $E(z)$ was calculated by Equation 2 from the vertical distribution of shear wave velocity, $v_s(z)$, obtained from SASW testing.

$$E(z) = 2G(z) = 2\rho \, v_s^2(z) \qquad (2)$$

Settlements were calculated for each test location. Due to the assumptions of linear elasticity, the calculated settlements are proportional to the uniform contact pressure at the ground surface.

3.2 Comparison with measurements

The computed settlements s were compared with measured settlements by PLT. The results of the statistical analysis for 24 locations are shown in

Table 1. The analysis was made separately for each contact pressure intensity q as well as cumulatively for all contact pressures. The analysis shows a remarkably good agreement with predictions. Measured settlements were in average 11% larger than the predicted settlements. No significant differences are apparent for various contact pressures. Figure 4 shows the comparison between predicted and measured stiffness, defined as q/s.

Table 1. Average ratio r of measured (s_{PLT}) vs. predicted (s_{SASW}) settlements

Average contact pressure	$r = s_{PLT} / s_{SASW}$	
(kPa)	Mean	St. deviation
150	1.15	0.19
250	1.14	0.17
350	1.10	0.16
450	1.06	0.16
150-450	1.11	0.17

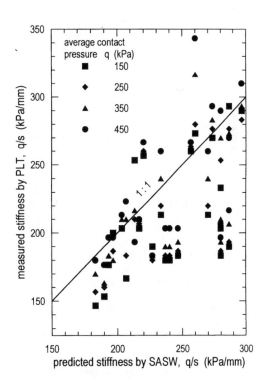

Figure 4. Measured stiffness by plate load test (PLT) with plate diameter of 30 cm compared to predicted stiffness by theory of elasticity and SASW.

765

An average underprediction of settlements by about 11% indicates an average reduction of the shear modulus to 90% of its original value at very small strains. Larger local reductions are expected at depths in the ground where higher concentrations of shear strains are found.

4 CONCLUSIONS

1. Measured settlements of circular plates of the plate load test on well-graded and well-compacted gravels are only about 11% larger than settlements predicted by the theory of elasticity. The ground stiffness was obtained from shear wave velocities measured by SASW. Circular plate of 30 cm in diameter was loaded up to the contact pressure of 450 kPa.

2. The SASW testing technique may successfully be used to measure ground stiffness for the prediction of settlements of shallow foundations on gravels. Measured ground stiffness should be reduced by about 10% to allow for shear modulus reduction due to shear strains induced by the foundation load when this load is in the range of 30% to 50% of the bearing capacity.

3. Further research is required for testing the relevance of SASW results for the prediction of settlements of shallow foundations when larger foundations are used, and different foundation soils are encountered.

REFERENCES

Addo, K. O. & P. K. Robertson 1992. Shear-wave velocity measurement of soil using Rayleigh waves. *Canadian Geotechnical Journal.* 29(4): 558-568.

Atkinson, J. H. & G. Salfors 1991. Experimental determination of stress-strain-time characteristics in laboratory and in situ tests. *Proc. Tenth European Conf. Soil Mechanics and Foundation Engineering.* Rotterdam: Balkema. III: 915-956.

Burland, J. B. & M. C. Burbidge 1985. Settlement of foundations on sand and gravel. *Proceedings of the institution of Civil Engineers.* 78(1): 1325-1381.

Burland, J. B. 1989. Ninth Laurits Bjerrum Memorial Lecture: "Small is beautiful"-the stiffness of soils at small strains. *Canadian Geotechnical Journal.* 26: 499-516.

Burland, J. B. & M. J. Symes 1982. A simple axial displacement gauge for use in the triaxial apparatus. *Géotechnique* 32: 62-65.

Clayton, C. R. I. & S. A. Khartush 1986. A new device for measuring local axial strains on triaxial specimens. *Géotechnique* 36: 593-597.

Goto, S., F. Tatsuoka, S. Shibuya, Y. S. Kim & T. Sato 1991. A simple gauge for local strain measurements in the laboratory. *Soils and Foundations* 31(1): 169-180.

Hiltunen, D. R. & N. Gucunski 1994. Annotated bibliography on SASW. In: R. D. Woods (ed.), *Geophysical characterisation of sites*, Volume prepared by XIII ISSMFE Technical Committee # 10 for XIII ICSMFE, 1994, New Delhi, India. Rotterdam: Balkema. 27-34.

Jardine, R. J., D. M. Potts, A. B. Fourie & J. B. Burland 1986. Studies of the influence of non-linear stress-strain characteristics in soil-structure interaction. *Géotechnique* 36: 377-396.

Ladd, C. C., R. Foot, K. Ishihara, F. Schlosser & H. G. Poulos 1977. Stress-deformation and strength characteristics. *Proc. Ninth Int. Conf. Soil Mechanics and Foundation Engineering,* Tokyo, Vol. 3, 421-494.

Nazarian, S. & K. H. Stokoe 1983. Use of spectral analysis of surface waves for determination of moduli and thicknesses of pavement systems, *Transportation Research Record.* No. 954.

Stokoe, K. H., S. G. Wright, J. A. Bay & J. M. Roësset 1994. Characterization of geotechnical sites by SASW method. In: R. D. Woods (ed.), *Geophysical characterisation of sites*, Volume prepared by XIII ISSMFE Technical Committee # 10 for XIII ICSMFE, 1994, New Delhi, India. Rotterdam: Balkema. 15-25.

Szavits-Nossan, A., R. Mavar & M.-S. Kovačević 1998. Experience gained in testing pavements by spectral analysis of surface waves. In: Proc. 1st Int. Conf. Site Characterization, Atlanta, Georgia, Rotterdam: Balkema (in print).

Tatsuoka, F. & Y. Kohata 1995. Stiffness of hard soils and soft rocks in engineering applications. *University of Tokyo, Report of the Institute of Industrial Science* 38(5).

Geotechnical Hazards, Marić, Lisac & Szavits-Nossan (eds) © 1998 Taylor & Francis, ISBN 90 5410 957 2

Soil physical properties of sand–bentonite mixtures

G. Telekes
Ybl Miklós Polytechnic, Hungarian Academy of Sciences, Geotechnical Research Group, Technical University of Budapest, Hungary

B. Móczár & J. Farkas
Technical University of Budapest, Hungary

ABSTRACT: The paper presents the results of the research supported by the National Research Foundation, Hungary (Grant No.: T 007399), carried out in the Department of Geotechnics, Technical University of Budapest, Hungary. Investigations were carried out determining the soil physical properties on sand - bentonite mixtures. The permeability, the compacting factor, the shrinkage was investigated by changing the mixture ratio in 10 % increments. The effect of the mixing methods to the homogeneity of the mixture are also discussed.

1 INTRODUCTION

Developing waste deposit sites it is important to have natural liners for insulation. For the natural liners it is essential to have both small permeability coefficient and suitable density. Small permeability coefficient could be secured by cohesive soils, however the cohesive soils cannot be compacted well and those soils which can be compacted has insufficient permeability coefficient. This makes it reasonable to use mixed soils providing optimal features of the insulation layer. In the reported experiment sand–bentonite mixtures was investigated. Changing the sand – bentonite mixing ratio in 10 % increments the compacting factor, the permeability coefficient and the shrinkage limit was measured. The effect of mixing technology to the homogeneity of the mixtures was also investigated.

2 INVESTIGATION OF THE COMPONENTS OF THE SOIL MIXTURES

Before making the experiment with the soil mixtures the components were investigated. The grain – size distribution curve of the sand used for the mixtures are shown in Figure 1.The uniformity coefficient of the sand U = 3,6 and contain 2 – 3 % MO.

The soil physical parameters of the bentonite are as follows:
Liquid limit: 79.7 – 83.5 %
Plastic limit: 15.1 – 17.8 %
Index of plasticity: 64.6 – 65.7 %

GRAIN-SIZE DISTRIBUTION CURVE

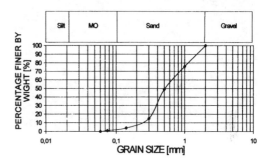

Figure 1. Grain-size distribution curve of the used sand

3 DETERMINATION OF THE MIXING METHOD TO THE HOMOGENEITY

Two mixing methods were investigated to decide which gives better homogeneity. Microscopical pictures were made about both the dry hand mixing and the wet machine mixing samples. The microscopical picture of the dry mixing are shown in Figure 2 and of the wet mixing are shown in Figure 3. The pictures show a 70%-30% sand-bentonite mixtures. It is clearly seen, that dry hand mixing makes better homogeneity, so during the experiment dry hand mixing were applied.

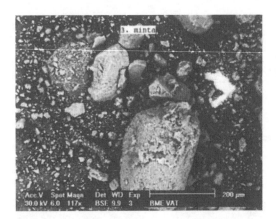

Figure 2. Sand-bentonite mixture by dry hand mixing

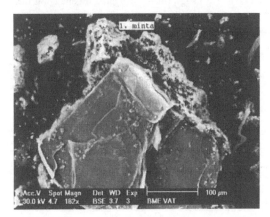

Figure 3. Sand-bentonite mixture by wet machine mixing

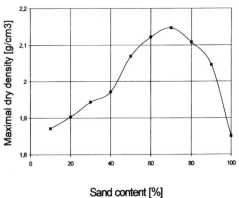

Sand content [%]

Figure 4. The maximal dry density versus the sand content

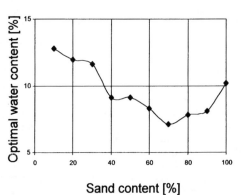

Sand content [%]

Figure 5. The optimal water content versus the sand content

Table I. Result of the two series of Proctor tests

	1st series	2nd series	1st series	2nd series
ρ_{dmax}	2,143 g/cm³	2,112 g/cm³	100 %	98,553 %
Sand content	71,486 %	76,859 %	100 %	107,516 %
w_{min}	7,388 %	7,923 %	100 %	107,241 %
Sand content	73,991 %	81,104 %	100 %	109,613 %

4. INVESTIGATION OF THE DEGREE OF COMPACTION

The degree of compaction of the sand-bentonite mixtures was investigated by changing the mixture ratio in 10 % increments. Each Proctor test was made twice. The maximal dry density versus the sand content of the first series of Proctor tests are shown in Figure 4.

The optimal water content versus the sand content of the first series of Proctor tests are shown in Figure 5

Table I. contains the results of the two series of Proctor tests. The deviation in the results of the two series of compaction tests is less than 10 %. It probably comes from the different routine of the two laboratory assistants, from the different intensity of the dry mixing etc.

The maximal dry density was obtained from the compaction test, when the sand-content in the mixture was around 70%-75%.

5 DETERMINATION OF THE PERMEABILITY

Permeability tests were made at constant water pressure, at varying water pressure and in triaxial pressure cell.

The permeability coefficient in the triaxial apparatus was determined at the degree of compaction of 100 %.

The results of the triaxial permeability tests are shown in Table II.

Table II. Results of the permeability tests

Sand /bentonite mixing ratio	Trρ 100%
100/0	$9{,}20*10^{-5}$
90/10	$8{,}20*10^{-6}$
80/20	$1{,}50*10^{-6}$
70/30	$7{,}70*10^{-8}$
60/40	$5{,}20*10^{-9}$
50/50	$4{,}35*10^{-9}$
40/60	$3{,}85*10^{-9}$
30/70	$3{,}24*10^{-9}$
20/80	$2{,}79*10^{-9}$
10/90	$1{,}72*10^{-9}$

Studying the results of the permeability tests made with different methods there are some deviation comparing with the expectation. The different test methods made by different results. The expectable results comes probably from the triaxial tests. The other two methods the preparation and the saturation of the samples were difficult.

The permeability versus sand content to the degree of compaction 100 % from the triaxial permeability tests are shown in Figure 6.

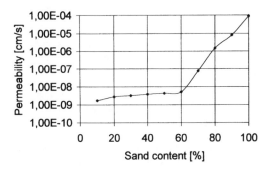

Figure 6 Permeability versus sand content

The permeability coefficient around the sand-bentonite mixture ratio 65-35 % reach the value of 1.00E –08.

6 INVESTIGATION OF THE SHRINKAGE BEHAVIOUR

The shrinkage behaviour of the sand-bentonite mixture was also investigated.

From the sand-bentonite mixture 70/30 %, 60/40 % and of 50/50 % four tests were made in each sample.

The shrinkage limit versus the sand content are shown in Figure 7 and the linear shrinkage versus the sand content of the sand-bentonite mixture are shown in Figure 8.

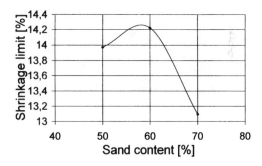

Figure 7. Shrinkage limit versus sand content

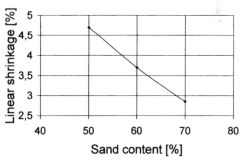

Figure 8. Linear shrinkage versus sand content

The diagrams are made from the average of the four tests.

The maximal shrinkage limit was obtained from the tests, when the sand-content in the mixture was around 60 %.

Figure 8 shows that 10 % changes in the sand content makes around 1 % in the linear shrinkage.

7 INVESTIGATION OF OTHER SOIL PHISYCAL PARAMETRS

Making the shrinkage tests from the sand-bentonite mixture 70/30 %, 60/40 % and of 50/50 % the phase composition, the void ratio and the porosity were also investigated.

The diagrams are made from the average of the four tests.

The phase compositions are shown in Figure 9, the void ratio versus the sand content are shown in Figure 10 and the porosity versus the sand content are shown in Figure 11.

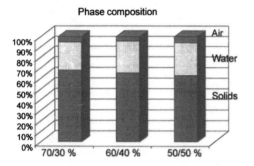

Figure 9. Phase composition of sand-bentonite mixture

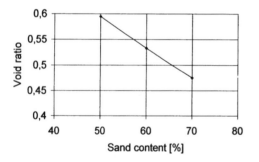

Figure 10. Void ratio versus sand content

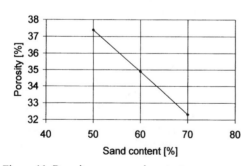

Figure 11. Porosity versus sand content

8 CONCLUSION

The analysis of the mixing technology shows that dry mixing makes better homogeneity than wet mixing.

Analysing our results of the sand-bentonite mixture as well as studying papers from other authors shows that the results depend lesser degree the sand and higher degree from the bentonite. The origin and the mineral matter composition of the bentonite could influence the behaviour of the mixture.

The best compaction could be produced by a sand-bentonite mixture ratio of 70-75 %/30-25 %. The optimal water content belongs to the best compaction is around 7-8 %.

Since the permeability of the mixture is reach the value of 1E-8 cm/s at the sand-bentonite mixing ratio of 60-65 %/40-35 %, consequently adding more bentonite to the mixture is not economical.

When the bentonite content reach around 40 %, and the mixture getting dry small cracks are occurred, however if the mixture getting wet again the cracks are closed.

10 % changes in the sand content makes 1 % changes in the linear shrinkage.

The results help to solve how to make good natural liner for waste depositories.

9 REFERENCES

R. P. Chapius (1990), „ Sand-bentonite liners: predicting permeability from laboratory tests „ Canadian Geotechnical Journal, Vol.:27, No.:1.

R. P. Chapius (1990) „ Sand-bentonite liners: field control methods „ Canadian Geotechnical Journal, Vol.:27, No.:2.

D. Haug & L. C. Wong (1992), „ Impact of molding water content on hydraulic conductivity of compacted sand-bentonite „ Canadian Geotechnical Journal, Vol.:29, No.:2.

T.C. Kenney et al. (1992), „ Hydraulic conductivity of compacted bentonite-sand mixtures" Canadian Geotechnical Journal, Vol.: 29, No.: 3.

B. Králik & J. P. Molnár (1993), „ Soil treatment technology and its results for the Aszód Waste Depositories" Acta Politechnica (in Hungarian)

G. Telekes at al. (1997), „Report on National Research Fund (OTKA) Research, entitled: Investigation of permeability of soil mixtures" (in Hungarian)

Landslides and slope stability

Geotechnical Hazards, Marić, Lisac & Szavits-Nossan (eds) © 1998 Taylor & Francis, ISBN 90 5410 957 2

Aspects concerning instability phenomena in some Romanian oil fields

S. Andrei, S. Manea & L. Jianu
Technical University of Civil Engineering Bucharest, Romania

ABSTRACT: Romanian Sub-Carpathian regions, where a large number of oil fields lie, are faced with instability phenomena, of various types.

Based on complex, multidisciplinary researches, hazard maps have been achieved for a number of oil sites (located in Moldavia, Oltenia and Muntenia). Suitable stabilisation steps were taken regarding the high-risk areas.

The paper is a survey of the principles and methods to achieve hazard maps such as the above-mentioned ones. Also presented are details regarding the horizontal drains, which work well in clayey soils (e.g. Cilioaia landslide, Moldavia). These solutions have been experimented successfully on many unstable sites, by using various technologies.

1 GENERAL

Romanian oil field works lie mainly in the hill areas bordering the Carpathian Mountains, where slope instability occur, either as temporarily stabilised slides, or as active ones.

Because of the large instability phenomena that occurred in North Moldavian oil fields in 1992 (Manea, Andrei, Antonescu 1996), a complex programme focusing on the study of slope stability in these areas was initiated in 1993. The purpose of this multi-disciplinary programme, carried out in cooperation by the PETROSTAR S.A. Design Company - Ploiesti and the Technical University of Civil Engineering - Bucharest, was to establish a methodology that would allow to detect the areas already affected by landslides or in danger to be affected, and consequently to draw a hazard map, as well as to recommend solutions for the prevention and reduction of instability risk.

The methodology achieved, as presented in Figure 1, has been checked so far on 10 oil fields in Moldavia, Oltenia and Muntenia.

An important stress is placed, within this methodology, on the geotechnical laboratory testing of soils. The purpose of the tests is to obtain a description of the soils, as complete as possible, especially of the ones located close to the failure surface. The shear strength parameters of the soils

were obtained by tests carried out on the reversible direct shear apparatus, the rotational shear one and the triaxial one. The methods that were used can apply both to saturated and to unsaturated soils (Andrei, 1997; Andrei, Manea, 1992).

From these tests, both the peak and the residual shear strength parameters were obtained, as well as the strength mobilisation with deformation.

Considering the large mass of results, all the data obtained form the laboratory tests were processed and all the parameters were saved in a database concerning each site based on an original procedure created at the Technical University of Civil Engineering - Bucharest and working with "prints" and "state diagrams" (Andrei, Manea, 1987).

As for the computation stage, the following steps were considered:
– to outline the water presence within the soil masses under study, by means of the hydrodynamic spectra;
– to check the slope stability under different hypothesis, generally connected to the water presence.

The check-up is done by using an original software based on the Janbu method which, a software able to consider the progressive mobilisation of the shear strength (Manea, Andrei, Antonescu, 1996).

STUDYING STAGES

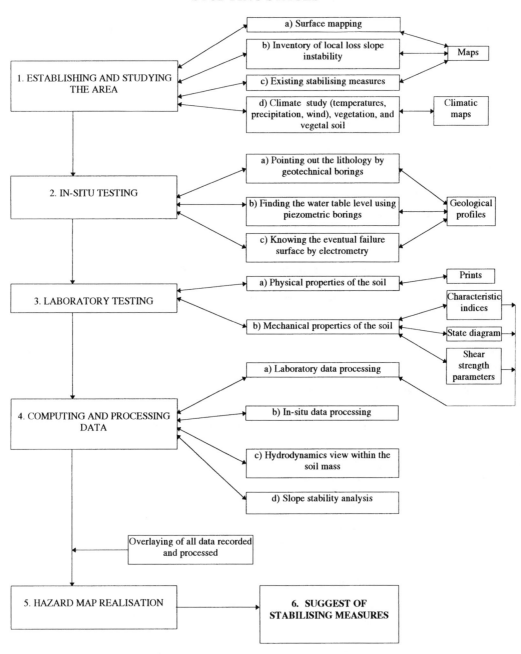

Figure 1. Methodology for creating hazard maps and for suggesting stabilising measures

2 ELABORATION OF HAZARD MAPS

By means of a surface mapping software, three-dimensional hazard maps were achieved, as well as prospective maps regarding the dynamics of instability phenomena.

The hazard map of an oil field presented in Figure 2 defines the risk degrees, from one to four, based on the geomorphological and geotechnical characteristics, as well as the computed safety factors.

The use of this complex methodology in studying 10 oil fields allows for some essential aspects to be pointed out.

Thus, it was found that the morphological zones between 200 and 1,000 m altitude, which display successive parallel ridges and valleys wrinkled by poor water sources, generally face instability problems; partially stabilised slides may be found here and there, but also areas of re-activation, as revealed by tilting tree trunks, water stagnation and hydrophilic vegetation zones.

Bustuchin - West oil works
HAZARD MAP

1 to 4 Risk degrees
·I-I' to X-X' Different failure surfaces taken into account

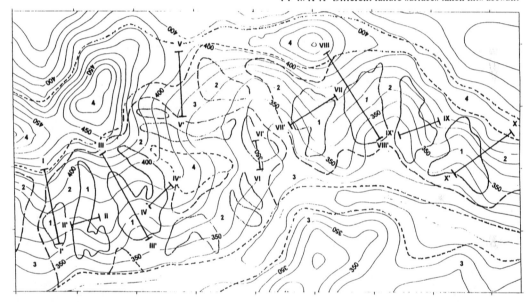

Risk degree	Slide probability	Description	Stability factor	Morphological characteristics
1	Very high	Active slides	≤ 1	- Steep slopes; - Wide zones with surface and underground water; - Saturated soil ($S_r \cong 1$); - Clay with rock fragments in flowing state.
2	High	Temporary stabilized slides, prone to re-activation	1.1 - 1.3	- Variable slopes; - Wary ground surface, stagnant water, hydrophilic vegetation.
3	Medium	Old stabilized slides, possible re-activation	> 1.3	- Medium slopes; - Relatively wary ground surface; - Few surface water accumulations; - Saturation degree $S_r \cong 0.8$.
4	Low	Relatively stabilized zones	> 1.7	- Ridges and aprons of cemented rocks (as sandstone); - Mostly afforested zones.

Figure 2. Hazard map

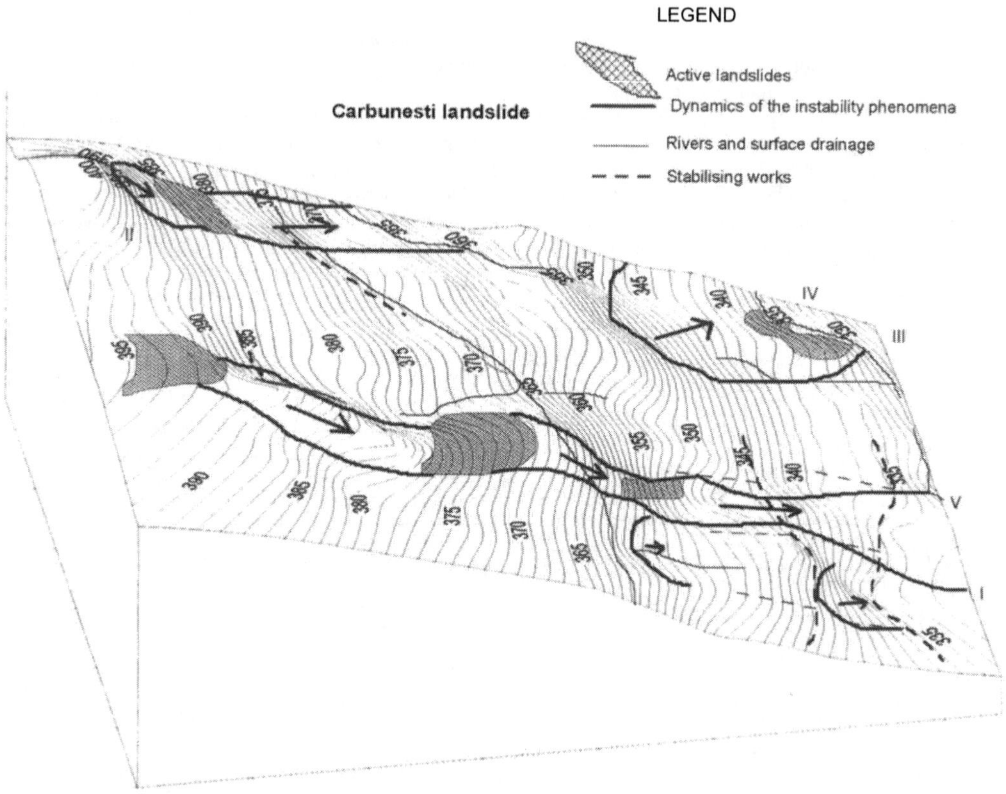

Carbunesti landslide

Figure 3. Dynamics of instability phenomena

The soil layers are predominantly parallel and conform, generally in thicknesses from 2 to 12 m; mainly cohesive sediments are to be found lying on the bed-rock - marl or sandstone. Within these layers, especially where coarser materials occur (such as sand), groundwater may be present, the level of which can undergo important seasonal variations, depending on the precipitations level.

The potential sliding surfaces are generally situated at the contact between the bed-rock and the shallow deposits, where a permanent water layer was often found, which encourages the slide. The slide surfaces are generally deep and parallel with the surface of the bed-rock. Shallow slides of small extent are often caused by local factors such as erosion, cracks, engineering works, etc.

Finally, a map of the forecasted dynamics of instability phenomena was drawn, together with suggestions for stabilising works, as shown in Figure 3.

3 DESCRIPTION OF THE HORIZONTAL DRAINS

Since the main factor leading to loss of stability in all studied areas was water, the steps to be taken in order to prevent and diminish instability focused on this particular factor.

For this purpose, alongside with the classical measures regarding the drainage and collecting of surface water, the use of horizontal drains was also studied, including the case of clayey soils.

This methodology was successfully experimented in two stages (1994 & 1995) in stabilising the large landslide at Cilioaia, Moldavia (Studies, 1994-1997).

At a first stage, the projected drainage system consisted of seven groups made of two horizontal 25-30 m long boreholes, placed at a 30-40 m distance, covering an overall distance of 250 m.

Due to the technical means available to the constructor, the horizontal drains were actually carried out by boring with water, while the boreholes had a final diameter of 76 mm and lengths of 25-40 m with upwards slopes of about 4-5°.

The above-mentioned technology used in carrying out the drains and the high degree of non-homogeneity of the soil led to some deficiencies in the location of drains (sometimes, long stretches of pipe had a tendency to bend downwards, or the failure surface could not be reached, for efficient drainage).

Taking into account the deficiencies noticed, a new drainage system was carried out in 1995 at the same location, with a different technology (dry boring by means of a German, Klemm 605 equipment). Ten drains were carried out, each about 40 m long and 75 mm in final diameter (Figure 4).

Simultaneously, the MODE software based on the finite difference method was used to model the changes in ground water level.

Results obtained according to the values measured in-situ show that in both directions, across and longitudinal, the effect of the drains extends along 8-15 m. Areas influenced by each group of drains are estimated to be around 1,000-2,000 m^2 (Figure 5).

The decrease of the ground water level is directly related to the safety factor of the slope, being observed an increase of about 15-20%.

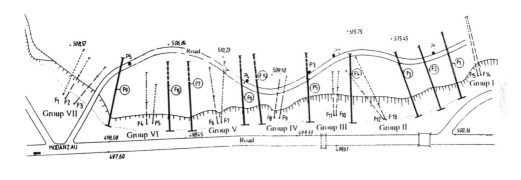

Figure 4.   Drainage system

The design of horizontal drains was done in accordance with the methods proposed by Kenny et al. (1997), Lau & Kenny (1984) and Nonveiller (1991).

To check the efficiency of the drains, a system was set up for the measurement of the water table level and the volume of drained water.

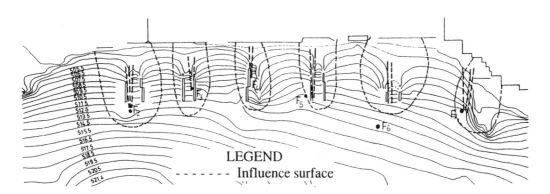

Figure 5. Drainage influence

The above led to the creation of a computer programme that assesses the optimum distance between groups of drains for a desired increase in the safety factor.

It is worth mentioning that by providing a well drained layer at the bottom of the slided mass, a good drainage of the slope is obtained, over time, as the good behaviour of the system has shown since 1994.

4 CONCLUSIONS

A methodology of elaborating hazard maps has been creating for the Romanian oil fields. This methodology shows positive results, therefore there is a will to achieve hazard maps for all oil fields in Romania, considering their slope stability features, in order to increase the security conditions for activities taking place there.

Experiments revealed the applicability of the horizontal drains systems for landslides and, consequently, it was suggested to use this solution for other areas as well, including for clayey soils, by means of an adequate methodology.

REFERENCES

Andrei, S. 1977. *Water in unsaturated soils* (in Romanian). Editura Tehnica, 117 p, Bucharest.

Andrei, S. & Manea, S. 1987. Forecast of moisture and volume changes in unsaturated soils. *Proc. of 9th ECSMFE*, vol. 1, pp 533-536, Dublin.

Andrei, S. & Manea, S. 1992. On soil collapse prediction. *Proc. 7th Int. Conf. on Expansive Soils*. vol. 1, pp 67-72, Dallas, Texas.

Andrei, S., Manea, S., et al. 1996. Hazard mapping for oil fields affected by landslide phenomena. *Proc. of 8th National Conf. for Soil Mechanics and Foundation Engineering*, Iasi, Romania.

Andrei, S., Manea, S., Jianu, L. et al. 1996. On stabilisation on landslides in clay soils by using horizontal drains. *Proc. of 8th National Conf. for Soil Mechanics and Foundation Engineering*, Iasi, Romania.

Kenny, T.C., Pazin, M., Chai, W.S. 1977. Design of horizontal drains for soil slopes. *Journal of Geotechn. Eng. Div.*, GT 11, pp 1311-1323.

Lau, K.C. & Kenny, T.C. 1984. Horizontal drains to stabilise clay slopes. *Canadian Journal*, vol. 21, no. 2, pp 241-249.

Manea, S., Andrei, S., Antonescu, I. 1996. Landslide investigations in Carpathian Mountains. *7th Int. Symposium on Landslides*, Trondheim, Norway.

Nonveiller, E. 1981. Efficiency of horizontal drains on slope stability. *Proc. of ICSMFE*, Stockholm, vol. 3, pp 495-500.

* * * Studies on the soil slope desiccation by using flexible horizontal drains within Zemes and Bustuchin oil works areas. *Technical University of Civil Engineering Bucharest.* 1994-1997.

Geotechnical Hazards, Marić, Lisac & Szavits-Nossan (eds) © 1998 Taylor & Francis, ISBN 90 5410 957 2

Slope stability analysis in an open mining area

Anton Chirica, Rolland Mlenajek, Andrei Olteanu & Cristian Banciu
Technical University of Civil Engineering Bucharest, Romania

ABSTRACT: In a southern Romanian village, the inhabitants have sued juridical action against the local mining company considering that the instability phenomena brought about by the coal quarry and the new sterile dump situated near the village, are the main causes of their buildings damages. Many buildings have been monitorised (129) and 19 geotechnical boreholes were performed. Inclinometric tests were performed during the year 1996 in 15 boreholes spread on four calculus profiles. A complex stability analysis was performed and the results are pointed out in this paper.

1. INTRODUCTION

In 1995 at the request of the "Motru Mining Company" a research team from Technical University of Civil Engineering Bucharest started to study the main causes of the civil buildings damage which occurred in the village situated nearby the open mining area.

The research program was performed until the end of 1996 and it consisted of complex geotechnical surveys including: boreholes (127 m total length), hydrogeological and inclinometric measurements as well as laboratory geotechnical tests. A number of 129 houses were subsequently inspected registering the types of damage, the history of their development and the main features of structures and building foundations. In order to establish the real causes of the civil buildings damage, using the obtained geotechnical data, general ground slope stability analyses were performed. The main results of the field and laboratory tests and their interpretation are further presented.

2. GEOLOGICAL AND HYDRO-GEOLOGICAL ASPECTS

The village named "Valea Manastirii" is located in the southern part of Romania on the northern bank of river Motru valley (Figure 1). In the river Motru alluvial plain, the dump was erected 30 m above the natural ground level having +200 m as compared with the "Black Sea" level. As it can be seen in the Figure 2 the quarry is situated on the hill top plateau. One of coal layers having 5 m thickness which is currently operated can be also seen on the same photo. A part of the sterile ground has been deposited in the dump body. The southern margin of the quarry is situated at more than 200 m from the nearest house, but in general, excavation begins to develop at an average distance of over 400 m from the village limit. As we noted before, the dump is situated in the southern part of studied zone, on the alluvial plain of the river Motru. The minimal distance between the northern dump slope foot and the southern constructed area limit is about 120 m.
Within the village area the bank is characterized by an average slope of 1:5 greater to the superior site and in torrential valley zones.

Our field investigation performed within the village area pointed out the several local zones characterized by typical landslide morphology (waved ground surface, steep free faces, inclined trees). By geological point of view, the cross section based on the performed geotechnical drillings, has shown the bedrock (superior Pliocene as age), represented by succession of clayey and sandy layers, with coal intercalated strata, and Quaternary shallow deposits. The drillings also pointed out alluvial deposits, mainly coarse in the alluvial plane and in the two buried terraces situated at the superior slope side.

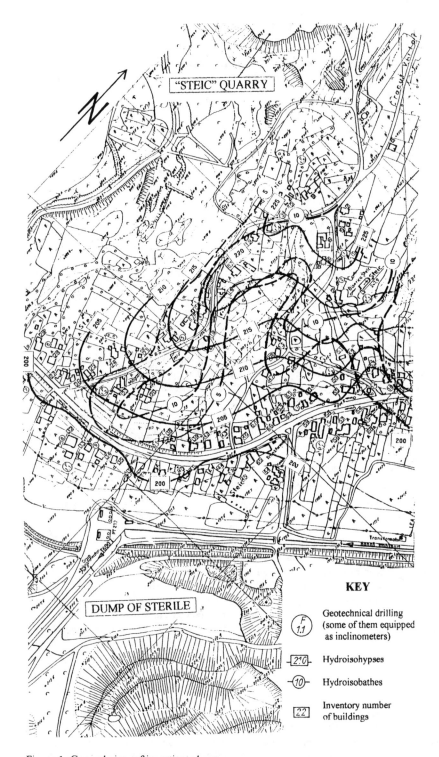

KEY

(F 1.1)	Geotechnical drilling (some of them equipped as inclinometers)
[210]	Hydroisohypses
(10)	Hydroisobathes
[22]	Inventory number of buildings

Figure 1. General view of investigated area

Figure 2. General view of the coal quarry

slope base the ground water levels are strongly influenced by the precipitation regime. Asking the inhabitants about the variations of ground water levels they observed in their domestic wells during the times, we found out big amplitudes of piezometrical level (seasonal variations of $2 \div 4$ m).

The seasonal variations of ground water levels induce changes in the soil physical state and its mechanical behavior. More sensitive in the cycling changes of the saturation degree are the colluvial soils on which the bulk of constructions is funded so that we concluded that this can be one of the causes of registered structure damage.

On 10 September, 1996 we measured the piezometrical level of ground water in 29 domestic wells spread out on the village area. A part of this "water points" is presented in Figure 1. Considering this investigation, we were able to mark the lines of equal elevations of ground water table (hydroisoypses) and the lines of equal depths of ground water table (hydroisobathes). Ground water surface is situated at depths, varying between less than 5 m (usually 2m) at the slope foot (southern side of built area) and more than 10 m to the hill top (northern side of the village). The general flow direction of ground water is from North to South being characterized by an average hydraulic gradient of 0.08. We noted that within the village zone does not exist a continuous aquifer, the ground water being collected by the pervious boundary of bedrock and shallow deposits many times cut off by the

3. GEOTECHNICAL INVESTIGATIONS RESULTS

The complex geotechnical research program begun in 1996 with 19 boreholes situated in four calculus profiles used for stability analysis. The typical geological deposits are the following: clayey silt, silty clay, terrace deposits and alluvial plain deposits. In the Table 1 their main characteristics are presented in detail.

Some samples of the already mentioned clayey silt deposit were studied using specific oedometer tests in order to observe their behavior under water content variation. A swelling pressure of about $p_s = 70$ kPa was obtained. This result shows that the studied clay has volumetric deformations during water content variation (expansive clay).

The lythology of the studied area (pointed out by the drillings performed in-situ) shows the soil

Table 1 Geotechnical Characteristics of shallow deposits

Deposits	I_P	I_C	w	n	γ	γ_S	ϕ'	c'
	(%)	-	(%)	(%)	kN/m^3	kN/m^3	(°)	kPa
Clayey silt	$20 \div 23$	$0.60 \div 0.70$	$19 \div 22$	$39 \div 47$	$16.5 \div 18.2$	$26.6 \div 26.8$	$17 \div 23$	$10 \div 30$
Silty clay	$28 \div 32$	$0.50 \div 0.65$	$20 \div 28$	$38 \div 42$	$17.2 \div 19$	$26.7 \div 27$	$14 \div 17$	$25 \div 36$
Terrace deposits	-	-	$16 \div 18$	$29.5 \div 30.5$	$18.1 \div 18.6$	26.5	$32 \div 36$	0.0
Alluvial plan deposits	-	-	$22 \div 25$	$31 \div 36$	$17.8 \div 18.2$	26.5	$28 \div 33$	0.0

colluvial clayey formations. This fact determines a low discharge of ground water from the slope area to the pervious deposits of the river Motru alluvial plain, and consequently, the tendency of colluvial deposits to be saturated until the levels pointed out by the hydrogeological survey. It is very important to mention that due to the lack of a real drainage to the

capacity to slow flow that induce in the shear plane a great amplitude of the displacements. Consequently the shear strength parameters tend to residual values for cohesion and effective friction angle (c'_r and ϕ'_r). For the clay from the failure plane, the domain of the minimum values for the $\tan\phi_r'$ is $0.18 \div 0.296$, leading to values of $\phi_r' = 6 \div 17°$ (according to the I_p values).

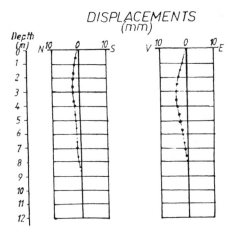

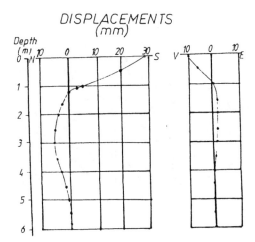

Figure 3. Inclinometric measurements in the drilling F 1.2

Figure 4. Inclinometric measurements in the drilling F 1.4

The back stability analysis results, presented in the next chapter, confirmed these values. Taking into account the just mentioned domain, stability analysis have been performed using for this shear strength parameters the values $\phi' = 6°; 10°; 14°$ and $17°$ and $c' = 0; 5$ and 10 kPa respectively.

A special attention was dedicated to the active sliding surface position. For this reason, 15 drilled - holes were equipped as inclinometric wells, and a set of 5 inclinometric measurements was performed during the interval of May $22 \div$ October 15, 1996.

The deformation registrations were made by SINCO digital device having an accuracy of 0.25 mm /m. A number of 6 inclinometric wells locations are pointed out in Figure 1. A special highlighting refers to the inclinometric F 1.1 and F 2.1 located in the relative plate zone, on the alluvial plain. The representatives of company that made the inclinometers were surprised when we designed the above mentioned locations, because they had used the methodology only in order to survey the ground movement in the slope area. We did not abandon the idea explaining them that it is necessary to place a row of inclinometers parallel with the dump slope foot in order to verify if any eventual influence of the fill in ground movement exists. The lack of deformation registered in the above mentioned devices strongly confirmed that no influences of dump exist in constructed area, even in the case there was no influence of a mining work highlighted by inclinometric measurements. Only two inclinometers (F 1.2 and F 1.4) pointed out some displacements of the shallow part of colluvial deposits (Fig. 3-4).

However, in the zones were inclinometers F 1.2 and F 1.4 were placed, our field investigations localized shallow landslides pointed out by specific morphological features.

4. STABILITY ANALYSIS RESULTS

Studying the general stability of the slope on which the village is located four calculus profiles were considered.

In Figure 1, the positions of the first two sections are presented while in Figure 5 and 6 respectively, the corresponding calculus profiles are shown. (Chirica et al., 1996). The possible slide surfaces were divided into three categories:

1. Extensive sliding surfaces as 1.1, 1.3, 1.4, 2.1, 2.3 etc.
2. Medium length sliding surfaces as 2.2, 2.4 etc.
3. Local sliding surfaces as 1.2, 2.5.

The stability analysis was performed in two steps as follows:

1. Back calculus for all considered sliding surfaces in order to determine the most probable limit values of shear strength parameters in conditions of safety factor having the value 1.00 (Fig. 7 and Fig. 8).

2. Stability calculus for all considered sliding surfaces using the values of shear strength parameters obtained from back calculus on one hand and the values determined by laboratory tests on the other. Finally 240 calculus variants were taken into

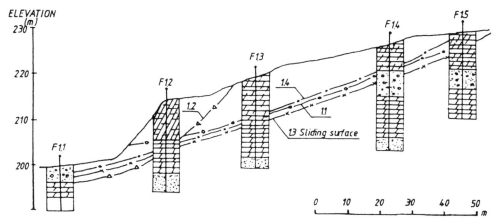

Figure 5. Stability calculus profile 1.1

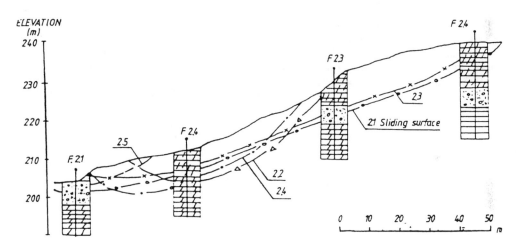

Figure 6. Stability calculus profile 2.2

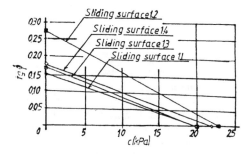

Figure 7 Results of back analysis for the profile 1-1

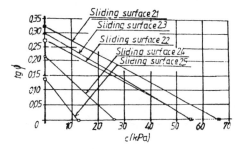

Figure 8 Results of back analysis for the profile 2-2

account, corresponding to the 19 sliding surfaces analyzed in the above mentioned conditions.

Taking into account all the obtained results the most probable values of shear strength parameters are confined in the interval $10° \div 17°$ for the friction angle and $0 \div 10$ kPa for the cohesion. We also observed, from the diagrams presented in Figure 7 (for the calculus profile 1-1) and Figure 8 (for the calculus profile 2-2) that the extensive sliding surfaces, characterized by the smallest values of shear strength parameters, are the most dangerous.

In the two calculus steps a STAB 1 numerical program was used (Chirica, 1982) based on Fellenius method. The main results of the performed stability analysis are presented in Table 2. The analysis clearly emphasizes that the studied slope are characterized by a natural instability potential. The entire zone is affected by temporary stabilized landslides which can be reactivated when the water table level rises or when the slope is overloaded by new built constructions. Though we did not performed stability calculus in dynamic conditions it is also clear that the vibrations induced by the heavy trucks carrying coal has a negative influence on the safety stability factor.

Table 2. Safety factor values for analyzed sliding surfaces

Failure surfaces	SAFETY FACTOR F_S VALUES FOR			
	$\phi'=10°$ c'=0kPa	$\phi'=10°$ c'=5kPa	$\phi'=14°$ c'=0kPa	$\phi'=17°$ c'=0kPa
1.1	0.40	1.40	-	-
1.2	0.70	0.84	0.90	1.06
1.3	0.65	1.30	-	-
1.4	0.60	1.20	-	-
2.1	0.30	0.60	0.75	0.90
2.2	0.40	0.70	0.83	1.00
2.3	0.40	0.70	0.85	1.00
2.4	0.55	1.05	1.11	-
2.5	0.76	-	-	-
3.1	0.40	0.80	0.84	1.01
3.2	0.33	0.67	0.75	0.90
3.3	0.30	0.62	0.70	0.83
3.4	0.30	0.65	0.70	0.83
4.1	0.30	0.54	0.72	0.80
4.2	0.30	0.50	0.60	0.75
4.3	0.40	0.72	0.82	0.99
4.4	0.40	0.72	0.82	1.00

5. BUILDING DAMAGE INVENTORY

A number of 129 houses were investigated in order to establish the constructive features, the damage types and their history. The inventory was completed with photos and videotapes registration. Our investigations pointed out the following observations:

1. A large number of houses are hardly damaged due to the phenomena specific to foundation on unstable ground; the structures cracks were amplified by several design and construction deficiencies and by the low quality of the construction materials.

2. The buildings have, in general, a single floor and have structural walls made of brick masonry with lime mortar, wood or concrete blocks; the slabs are wood made.

3. The bricks are characterized by low quality being hand made and insufficiently burnt; moreover, the masonry is poorly executed, presenting regions with non interlaced joints partially filled with mortar.

4. Even the wood made buildings are the oldest, they have the best behavior, presenting less damage than the others.

5. The cracks are mostly due to the displacements at foundation level, being in continuous evolution; the displacement triggering is represented by the rainfalls.

Several types of damages were identified as following:

1. Cracks in walls, having inclined directions at the corners of the buildings, produced by local settlements of the foundation ground; the cracks are spreading becoming even wider and deeper in the foundations.

2. Breaks of walls and foundations due to global settlement of the building, or as a consequence of differential settlements.

3. Vertical shearing of structural and partition walls favored by the non-existence of required differential settlements joints between the parts of building erected at different times.

6. CONCLUSIONS

The complex research performed in "Valea Manastirii" village, revealed that the closely situated mining works (a coal quarry and a dump of sterile) do not influence the constructed perimeter.

A strong argument supporting the above conclusion is given by the lack of any deformation registered in the inclinometers placed around the village. The house damages, sometime very severe (Fig. 9-11) are due to the following causes:

1. Foundation ground geotechnical special features (expansive clay).

2. Ground displacements, in the past, before the mining activity.

Figure 9. Damages registered at building No. 82

Figure 10. Damages registered at building No. 84

Figure 11. Damages registered at building No. 48

3. Imperfections in the buildings design and construction.

The natural cause leading to the ground displacements is the following one:

The slope where is the village location is affected by temporarily stabilized and shallow active landslides; seasonal variation of saturation degree determine at their turn, the local ground displacements characterized by reduced sliding velocities.

Inspecting 129 buildings the following main design and construction faults were pointed out:

1. Too shallow foundation works at depths smaller than 1 m, improper to the ground geotechnical conditions.

2. Actual tendency to construct brick and concrete blocks made buildings that through their greater weight overload the ground, leading to appreciable settlements and local reactivation of temporary stabilized landslides.

3. The lack of differential settlement joints between the buildings parts erected in different stages.

4. The low quality of construction materials and the superficiality of local builders that have not an adequate professional training.

Our researches emphasized the fact that despite the appearances, the major economical and social problems appeared in the studied zone are not due to the mining works placed in the village vicinity. The damage affecting the majority of buildings situated in "Valea Manastirii" village area emerge from constructive solutions improper to specific features of the foundation ground. So that, every case has to be separately treated after a detailed geotechnical survey in order to select adequate remedial works and constructive solutions for the new buildings. Popular experience determined, in the past, the construction of wood made houses, lighter and more flexible, more suitable for the local geotechnical conditions. The nowadays tendency of replacing the traditional construction materials with heavier materials was proved this time rather improper.

REFERENCES

Chirica, A. (1982) - Stability analysis software based on limit equilibrium method *Proceedings 3rd. National Symposium C.E.I.. Sibiu*: 16 - 26

Chirica, A., Mlenajek, R., Olteanu, A., Banciu, C. (1996) - Ground stability conditions corresponding to the open mining area "Valea Manastirii" and quarry "Steic - Lupoaia". *Research raport no. 66. Technical University of Civil Engineering, Bucharest*: 1-96

Geotechnical Hazards, Marić, Lisac & Szavits-Nossan (eds) © 1998 Taylor & Francis, ISBN 90 5410 957 2

Geotechnical hazards related to marls in Greece

J.C.Christodoulias & H.Ch.Giannaros
Central Public Works Laboratory, Athens, Greece

A.C.Stamatopoulos
Kotzias-Stamatopoulos, Soil Consultants, Athens, Greece

ABSTRACT: This paper presents geotechnical hazards related to the tendency of Greek marls to destabilize. Two case histories have been examined, one concerning a site in the north part and the other in the south part of Greece. In the first case marls of high initial strength (about 400 KN/m^2 or 4 kg/cm^2 in compression) turned into slurry when they were subjected to strong shaking. In the second case, the yearly cycle of prolonged dry period followed by heavy rains, resulted in intensive deterioration and sliding of cut slopes. The causes and mechanisms of failures are analysed, and remedial measures are reviewed.

1. INTRODUCTION

Greek marls include clays and silts with a wide range of carbonate content. Depending on carbonaceous material and admixtures of iron compounds, they are coloured white, off-white, yellow and green. The age of marls is Neocene and low Quarternary (Pliocene, Pliestocene). Their origin is either marine or lacustrine depending on the phase of tectonic movement during deposition. Greece has undergone cycles of downward and upward movements resulting from Neo-Alpine orogeny.

1.1 *Case 1*

In the north part of Greece, near the towns of Ptolemais and Servia, large deposits of lignite are extracted daily, for the needs of a thermoelectric power station (Fig. 1). Marls of high initial strength are sandwiched between the layers of lignite and after the separation, the marly material is transported by conveyor belts to tailing disposal sites.

Maximum haulage distance from open pit mines is about 5 km in Ptolemais extraction site, and only 300 m in the area of Servia. At the begining the marly deposits were transported and dumped by trucks with no visible symptoms of liquefaction, but subsequently trucks were replaced by conveyor belts, running at a speed of 5 m/s. Belts have a width of 1200 mm, and are supported by rotators 1100 mm apart. During transportation, blocks and large chunks of marl progressively disintegrate and arrive at the disposal site in a slurry-like consistency, never recovering any part of the initial strength, but tending to flow. The marls from Servia were fully liquified after a few hundred meters of

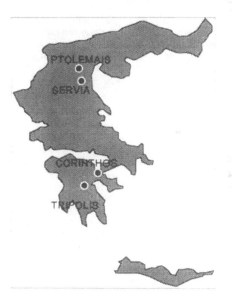

Fig. 1   Map of Greece

transportation, receiving on the way about 4 shocks/m of travel (Kotzias & Stamatopoulos, 1983).

1.2 *Case 2*

In the south part of Greece, Peloponese district (Fig. 1), the new motorway connecting the cities of Corinth and Tripolis, involved a considerable number of excavated slopes. Exposure of hitherto buried marls lead to deterioration, creep and

787

ultimately failure. As it can be seen from the histograms of precipitation for the decade 1980-1990 (Fig.2), the area of construction has a prolonged dry

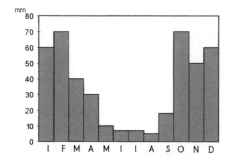

period, followed by an also long period of rainfall. These acute changes result in successive cycles of full saturation and dryness of the marly slope material.

The first winter after exposure, intensive deterioration appeared on the surface, which during the following winter progressively converted to shallow creep, especially in the areas of high saturation. Finally the phenomena of instability evolved to shallow translational slides and flows (Giannaros et al. 1992).

2. GEOLOGY

In the north, during the Neo-Alpine orogeny of the upper Pliocene era (1.5 million years ago), the Bitolji-Servia region was unevenly downfaulted as a graben, relative to adjacent but much older marine limestone, sandstone and flysch. Water sediments such as lignite, marls-boglimes and others, were deposited in the lakes within the graben. Intensive downfaulting throughout the Pliocene differentiated the developing basins by further grabens and horsts (long rock blocks uplifted along faults on both sides). There are locations where the thickness and relatively small depth of lignite layers allows economically profitable exploitation by surface extraction. Interlayers of marl must, however, be removed and taken away (Anastasopoulos & Koukouzas, 1972).

In the south, the area of the Peloponese near the city of Cor'nth, is covered by Pliocene marls, soft limestones and sandstones, and relatively loose conglomerates. Preeminent among these are Pliocene marls of yellow to white, but sometimes also light gray or blueish colour. They contain about 35 - 75% calcium carbonates and also a large percentage of clay minerals. Their grain size distribution is predominantly of the silt sizes. Marls have developed a structure of strong bonds of cementing material linking individual particles, composed of carbonates. Whithin the marly mass

are intercalations of sandstone, spheritic conclomerates, and marly limestone. The deposits are normally brackish and lacustrine. The lacustrine phase predominates in the upper parts and is full of fresh water shells. Sometimes there are thin lignite beds included in the marl deposits of the upper part (I.G.M.E., Geological maps of Greece).

3. LABORATORY TEST RESULTS

3.1 *Gradation*

Results from the sites of the two cases gave remarkably similar results. Percent passing the U.S. sieve no 200 is about 85 and percent of sizes smaller than 0.002 mm is of the order of 30% - 40% (Fig. 3).

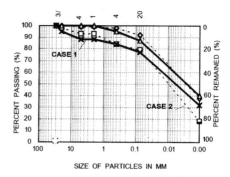

Fig. 3   Grain size distribution for case1 and case2

3.2 *Plasticity*

The plasticity chart shows that the results of Ptolemais and Servia marls, are plotted bellow the line A (fig. 4), falling mostly on the ML area and occassionally on the MH area, thus indicating organic silt. For the Corinth marls, all samples are above line A, indicating mostly CL material and occassionally CH.

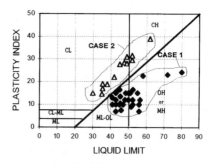

Fig. 4   Plasticity chart

3.3 *Strength*

Unconfined compressive strengths for four sample
from Ptolemais open pit are reported in Table /
Results are for undisturbed samples and also fc
samples remoulded and compacted (Wilson, 1964)
in order to examine how the dry density, strength
and degree of saturation of liquified marl change
with compaction. Five undisturbed soil samples from
the area near Corinth are reported on the same
table.

Table A. Unconfined Compressive Strength of Marls

Sample Number	1	2	3	4	5
Ptolemais / Servia Undisturbed					
strength (kN/m²)	560	372	284	343	
dry density (kg/m³)	1,100	1,080	1,540	1,090	
water content %	47	48	23	51	
Ptolemais / Servia Compacted after Remoulding					
strength (kN/m²)	78	62	83	33	
dry density (kg/m³)	1,180	1,010	1,490	1,190	
water content %	44	48	22	49	
Korinthos Undisturbed					
strength (kN/m²)	101	42	110	78	306
dry density (kg/m³)	1,600	1,500	1,678	1,876	1,782
water content %	23.5	27.3	28.8	21.9	18.2

The above tests indicate a high strength and low
density for the marls of Case 1, and a somewhat
lower strength coupled with a much higher density in
the marls of Case 2. Liquefied material from Case 1,
when compacted in the laboratory gave a strength
of the order of 15% of the strength in the undisturbed
state.

In addition to the unconfined compression tests
samples from the motorway cuts of Case 2 were
subjected to:

a. Quick triaxial tests (UU) which revealed
compressive strength ranging from 22 to 56 Kpa.

b. Consolidated undrained (CUPP) tests which
revealed effective shear strength parameters c' = 9
- 10 Kpa, and φ' = 31 - 35 degrees.

c. Ring shear tests in order to evaluate the residual
strength parameters which, gave c' = 0 and φ' = 21
- 27.5 degrees.

3.4 *Mineral Composition*

Analyses of 14 samples from Ptolemais/Servia
(Case 1) indicated mean values of 88.5% $CaCO_3$
and 2.4% organic material.

In two samples taken from the slope cuts of the
motorway (Case 2) semiquantitative analyses were
conducted, in order to identify the percent of clay
minerals by x-ray diffraction method (Bayliss, 1986).
Results were as follows: smectite 3 - 4% / chlorite 2
- 4% / illite 10 - 19% / quarge 25 - 26% / feldspar 13
- 25% / calcite 33 - 34%.

4. CAUSE OF FAILURE

4.1 *Case 1*

Late Pliocene lacustrine marls of North-western
Greece are a metastable, highly calcareous and
nearly saturated soil-like material. Marls are
generally ML and occasionally MH silt in terms of the
unified soil clasification system. Being lacustrine-
type calcareous deposits, these marls have a
characteristically in situ low density coupled with
high strength. These properties can be explained in
terms of a honeycomb structure (Terzaghi & Peck,
1967) held together by calcareous bonds between
fine grains of calcite. When transported by conveyor
belt they commonly liquify and exhibit only an
insignificant fraction of their original strength. Here,
the cause of failure is, clearly, the destruction of
bonds by mechanical action (strong shaking).

4.2 *Case 2*

Failure appeared in places where the impregnation
of slopes was more pronounced. Failure surfaces
were shallow and elongated, with depth varying
between 1.5 and 3.0 m.

It seems that the water action is the main cause in
the development of landslides. This action comes
from rainwater during winter. The water action is
facilitated by the dense joints in the marls and by the
shrinkage cracks that are opened during the dry
periods. Water action operates as follows: Swelling
with reduction of strength / Increase of pore water
pressure through accumulation in joints and cracks /
Erosion of cut slopes with formation of deep
drainage paths that facilitate further swelling etc. in
greater depth.

In figure 5, an attempt is made to approach the
problem of slope failure due to water impregnation
by stability analyses. The slope fails when water
impregnation exceeds a certain depth z (fig. 5).

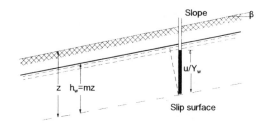

Fig. 5   Failure mechanism of marly slopes

Assuming that the failure surface is a plane (an
assumption which agrees with observations) the
problem reduces to that of an infinite slope, and the
factor of safety F is given by the expression
(Skempton & Delory, 1957):

$$F = \frac{c' + (\gamma - m\gamma_w)\, z \cos^2 \beta \tan \varphi'}{\gamma \cdot z \cdot \sin \beta \cdot \cos \beta}$$

where: z = depth of failure surface, assumed equ
to the depth of the zone of impregnation, i.e., m = ,
and $h_w = z$.

Also:
β = slope of inclination (degrees).
γ = bulk unit weight of marl (20 KN/m³).
γ_w = unit weight of water (10 KN/m³).
c' φ' = effective shear strength parameters of marl
after impregnation.

In table B and figure 6, the variation of factor of
safety versus depth of impregnation is presented for
β = 33°, 35°, 38°.

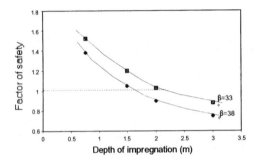

Fig. 6 Factor of safety versus depth of impregnation

As it can be seen in Fig. 6, for factors of safety
about 1, the depths of failure surfaces varies
between 1.5 and 2.5 m, approximately the same on
the depths of the observed failure surfaces of the
slopes that have failed.

Table B. Values of Factor of Safety F.

Depth of Impregnation z (m)	Factor of Safety (F)		
	β = 33°	β = 35°	β = 38°
1.0	1.52	1.51	1.38
1.5	1.20	1.14	1.05
2.0	1.03	0.98	0.91
3.0	0.86	0.82	0.76

5. REMEDIAL MEASURES

5.1 *Case 1*

Catchment ponds were constructed at the bottom of
the tailings, with dykes made of non-liquefiable
material. This measure proved suitable for stopping
the flow of liquefied marl into the nearby farmlands.

5.2 *Case 2*

Creep and sliding was checked by adopting gentler
slopes (about 27° with the horizontal), by
constructing drainage trenches, by applying a
concrete lining at the surface, by planting with
strong and deep rooted grass and bush, and by
installing gabbion walls at the foot of the cuts.

6. CONCLUSIONS

The geotechnical hazards related to Greek marls
include destabilization caused by external factors. In
Case 1 mechanical disturbance by strong shaking
resulted in liquefaction. In Case 2, recent road cuts
that exposed the material to the weather, caused
shrinkage cracks that, in combination with other
joints, allowed rainwater to penetrate deep into the
soil mass. Impregnation resulted in softening of the
soil, and accumulation increased pore water
pressure. There was creep that developed into
sliding. Stability was further impaired by surface
drainage of rainwater that cut deep channels into the
soil.

REFERENCES

Anastasopoulos I. & Koukouzas K. 1972. Economic
    Geology of the Southern part of Ptolemais Lignite
    Basin Macedonia - Greece. Institute of Geology
    and Subsurface Research, Athens, Greece.
Bayliss, P. 1986. Quantitative analysis of
    sedimentary minerals by powder diffraction.
    Powder Diffraction, Vol. 1, No 2, June, pp. 37-39,
    U.S.A.
Giannaros H., Pahakis M. & Christodoulias J. 1992.
    Long term stability problems of marly cutting
    slopes on the Korinth - Tripolis motorway. 2nd
    Greek National Conf. on Geotech. Engineering.
    Thessaloniki, Oct. 21 - 23, Vol. 1, pp 431 - 4381
I.G.M.E. Geological maps of Greece, 1:50000
Kotzias P. & Stamatopoulos A. 1983. Sensitivity of
    very hard Pliocene Marls. Journal of the
    Geotechnical Engineering Division, ASCE, Vol.
    109, N. 12, dEC. 1982, pp 1526 - 1533.
Skempton A. W. & Delory F. A. 1957. Stability of
    natural slopes in London clay. Proc. IV ISSMFE,
    Vol. 2, pp 378 - 381.
Terzaghi K. & Peck R. B. 1967. Soil Mechanics in
    Engineering Practice. John Wiley, N.Y., p 31.
Wilson S. 1964. Suggested Method of test for
    Moisture - Density Relations of soils using
    Harvard Compaction Apparatus. A.S.T.M.,
    Procedures for testing soils, Committee D-
    18, 4th Edition, Dec., p 160.

Geotechnical Hazards, Marić, Lisac & Szavits-Nossan (eds) © 1998 Taylor & Francis, ISBN 90 5410 957 2

Slope stability issues of rainfall induced landslides

B. D. Collins & D. Znidarcic
Department of Civil Engineering, University of Colorado at Boulder, Colo., USA

ABSTRACT: The slope stability issues concerning rainfall induced landslides have been investigated and are presented. The results of numerical analyses performed using a two-dimensional finite element code for unsaturated seepage are used in determining the decrease in soil suction during infiltration as well as the development of positive pore pressures. Infinite slope analysis is used to develop an expression for the stability of long slopes that depends on the pore pressure profile. A comprehensive method of analysis is presented that calculates the time and depth of failure and gives information on the type of triggering mechanism that will occur based on the soil, slope, and rainfall parameters.

1 INTRODUCTION

Rainfall induced landslides are a geotechnical hazard in many areas of the world. Although these failures tend to occur more frequently in tropical areas, the European continent is also greatly affected by these processes (Becht and Rieger, 1997 and Polemio, 1997).

Landslides in soils caused by rainfall are often referred to as rainfall induced debris flows or earthflows, and have traditionally been analyzed as a three step process consisting of initiation, mobilization, and flow stages. In this paper, the initiation stage is analyzed. Because these types of slope failures occur due to the infiltration of water from intense rainfalls, the flow of water through an initially unsaturated soil is emphasized to determine the triggering mechanism for failure.

The results of infiltration analyses using a two-dimensional finite element code show the pore pressure profile that will develop in slopes of varying soil type. Infinite slope stability is used to develop an expression for the critical depth of failure that is dependent on the soil and slope parameters. The critical depth and time for failure is seen to be affected by both negative and positive pore pressures in the pore pressure profile. A method of analysis for understanding the triggering mechanisms is presented by superimposing the failure criterion from stability analysis over a given infiltration trace for specified rainfall and soil parameters.

2 INFILTRATION

Rainfall induced landslides often occur on relatively long slopes where a homogeneous weathering profile has developed, leaving a shallow layer of residual soil overlying a bedrock interface. These characteristics allow the use of infinite slope assumptions to analyze the pore pressure profiles that occur in a slope. For infinite slope analysis, the stress conditions are assumed to be the same anywhere along the slope, thus simplifying the analysis of this problem considerably.

2.1 *Methods of Analysis*

The effect of seepage on slope stability is typically addressed in most analyses by calculating the factor of safety or critical depth for an infinite slope subject to seepage parallel to the slope surface. This type of analysis assumes that steady state flow is taking place over a given fraction of the soil depth (Craig, 1992), but in order to simplify the analysis, it is often assumed that the phreatic surface coincides with the slope surface and that the slope is completely saturated. For initially saturated slopes, infiltration is not possible and rainfall will have no effect on slope stability.

For slopes that are initially unsaturated, the effect of rainfall at the slope surface will have a dramatically different effect. The pore water pressure pattern that develops in the soil will occur as a transient process as the infiltrating water moves downward into the soil profile. Several factors not taken into account in saturated analyses are needed to perform the unsaturated flow and slope stability analyses properly. The shear strength of the soil mass will depend on the degree of suction according to the relationships outlined by Fredlund and Rahardjo, (1993). In addition, the transient change of the pore water pressure profile must be calculated using the equations for the flow of water through an unsaturated soil. These equations are based on the unsaturated characteristic curves for the particular soil which will be shown to affect the type of triggering mechanism for failure.

2.2 *Infiltration Analyses*

To study the infiltration of water from rainfall into an unsaturated soil, infiltration analyses were performed using a two-dimensional finite element code (SEEP/W) for unsaturated seepage (GEO-SLOPE, Int. Ltd., 1994). Because of the assumptions made using infinite slope analysis concerning homogeneity, one-dimensional column infiltration analyses were performed for varying unsaturated soil parameters. A hydrostatic suction distribution is assumed for the initial condition, which is a reasonable estimate of in-situ conditions prior to rainfall in many cases.

To simulate the maximum infiltration rate at the top surface, the boundary condition was selected to have zero pore pressure. This boundary allows for continuous water infiltration at the surface without ponding. It is important to note that ponding will not occur on slopes.

The infiltration analysis results for a typical coarse grained soil are presented in Figure 1. The

infiltration front is seen to move downwards starting from the initial unsaturated hydrostatic pressure head line. The results indicate that the development of positive pore pressure heads occur as the infiltration front progresses downward. This is due to the dramatic decrease in the hydraulic conductivity in the unsaturated zone (zone under suction) of a coarse grained soil. With such a small hydraulic conductivity and continuous infiltration from above, large hydraulic gradients are formed and positive pressure heads develop.

Infiltration analysis results for a typical fine grained soil are presented in Figure 2. Again, the infiltration front moves downwards beginning at the initial unsaturated hydrostatic pressure head line. However, in this case positive pressure heads are not formed during the infiltration process. The unsaturated hydraulic conductivity for a fine grained soil does not change as drastically with increased suction as for the coarse grained soil so that infiltration progresses without the formation of large hydraulic gradients. It should be noted that due to the much smaller saturated hydraulic conductivity of the fine grained soil, infiltration occurs much more slowly compared with the coarse grained soil as the time steps indicate.

The differences in the development of the pore pressure profiles that occur in slopes as a result of rainfall infiltration will be seen to govern the type of triggering mechanism for slope failure. The presence of both negative and positive pressure heads in the slopes mandates a stability analysis that takes both into account.

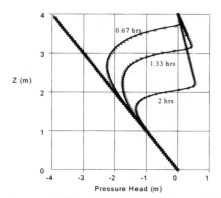

Figure 1. Infiltration Results for a Coarse Grain Soil

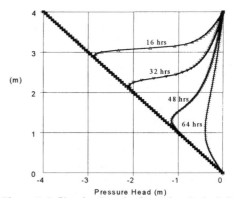

Figure 2. Infiltration Results for a Fine Grain Soil

3 SLOPE STABILITY

In order to investigate the issues related to the stability of landslide prone slopes, infinite slope analysis was performed. As previously mentioned, infinite slope analysis can be used to correctly model the stress state and pore pressure regime in long slopes. Debris flow failures often have small depth to length ratios, and form failure planes parallel to the slope surface, thus substantiating the use of infinite slope analysis (Cascini & Versace, 1988; Reid et al, 1988, Transportation Research Board, 1996).

3.1 *General Overview of Stability Issues*

The infiltration analysis results presented in Figures 1 and 2 indicate dramatically different pore water pressure profiles. This difference will affect the type of triggering mechanism that is likely to cause failure for a particular soil slope. The fine grain soil results show that positive pressure heads do not form during infiltration. Failure will therefore only be able to occur as a result of the loss in shear strength due to the loss of suction as the infiltration front progresses downwards. However, for the coarse grained soil, positive pressure heads do develop. The positive pressure heads indicate a saturated state which will lead to the formation of seepage forces due the elevation head difference along the slope. However as will be shown in the next section, the direction of seepage will not necessary be parallel to the slope during the infiltration process.

3.2 *Infinite Slope Assumptions*

Considering an infinitely long slope with slope angle β, measured from the horizontal, the forces acting on a vertical slice of the slope include the weight of the soil, W, and the normal and tangential reaction forces, N and T, acting on the base of the slice (Figure 3). A formulation for failure of an infinite slope can be developed by recognizing that the failure surface will occur at the critical depth d_{cr}. As is the assumption with infinite slope analysis, side forces on the slice are equal and of opposite direction and cancel. When the slope is subject to infiltration from rainfall, the slice is subject to infiltration from the vertical direction only. Because each slice of an infinitely long slope is subject to the same amount and intensity of rainfall, an individual slice can be treated as a one-dimensional soil column subject to vertical infiltration.

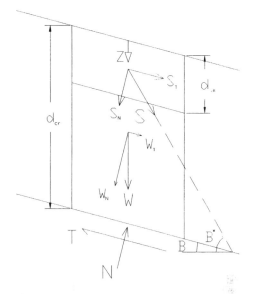

Figure 3. Free Body Diagram of a Typical Infinite Slope Slice

However, because an infinite slope is two-dimensional in nature, infiltrating water in one slice will be at a higher elevation than a slice just down-slope from it. Thus even for identical pressure head profiles, a two-dimensional seepage pattern will develop as a consequence of the elevation head difference between slices. Calling the resultant of this seepage pattern on a slice the seepage force S, results in the free-body diagram of a slice of an infinite slope shown in Figure 3. The seepage force will vary in direction, magnitude, and location of action within the slice.

As initially vertical infiltration takes place, a flow regime will be induced according to the difference in elevation heads of the slices as outlined above. Let β^* be the angle of seepage direction taken from the horizontal which will describe the direction of the seepage force.

The magnitude of the seepage force, S is dependent on the hydraulic gradient, "i" of the seepage which in turn can be calculated by examining the head difference on the surface of the slope in terms of β and β^*: The following formulation can be developed by calculating the head difference on the surface of the slope in terms of β and β^* as is shown in Figure 4. Here, two constant head lines and three flow lines are shown in a typical flow net seepage pattern which would result from infiltration from the slope surface.

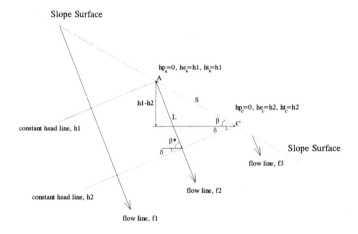

Figure 4. Infinite Slope Profile Showing Relationship Between "i" and β, β^*

The slope angle β, and the direction of seepage, β^* are shown as they have been defined in Figure 3. The angle δ is the angle formed between the horizontal and the constant head line directions.

Choosing two points, A and C at the slope surface to coincide with the total head lines, h1 and h2, the elevation head values at these points can be found. At the surface, the pressure head is zero, so that the elevation heads for points A and C are simply h1 and h2 respectively. The elevation distance between these two points is then simply (h1-h2).

By using trigonometry and some fundamental identities, the hydraulic gradient can be found in terms of β and β^* as

$$i = \frac{\sin\beta}{\cos(\beta - \beta^*)} \qquad (1)$$

It can be seen that at $\beta^* = \beta$, seepage is parallel to the slope and the commonly known solution of $i = \sin\beta$ for seepage parallel to the slope surface is recovered. The seepage force is given by $i \cdot \gamma_w$ where γ_w is the unit weight of water.

3.3 Equilibrium of a Slice

As mentioned in the last section, the location of action of the seepage force will also vary within the slice. The seepage force, S can be decomposed into normal and tangential components. By realizing that it will act only over a certain depth of the slice termed the depth of infiltration d_{in}, the following components are obtained for a slice of width, a:

$$S_N = i \cdot \gamma_w \cdot \cos\beta \cdot [\sin\beta^* - \cos\beta^* \cdot \tan\beta] \cdot a \cdot d_{in} \qquad (2a)$$
$$S_T = i \cdot \gamma_w \cdot \cos\beta \cdot [\cos\beta^* + \sin\beta^* \cdot \tan\beta] \cdot a \cdot d_{in} \qquad (2b)$$

The weight force, W can also be decomposed into normal and tangential components. Because only a portion of the slope will be saturated from infiltration, the unit weight of the slice will change depending on its saturation state. The buoyant (effective) unit weight, γ' is used for the part of the slope above the depth d_{in}, while the total unit weight, γ is used for the rest of the slope. For a slice of width a, the normal and tangential weight components are:

$$W_N = [\gamma' \cdot d_{in} + \gamma \cdot (d_{cr} - d_{in})] \cdot a \cdot \cos\beta \qquad (3a)$$
$$W_T = [\gamma' \cdot d_{in} + \gamma \cdot (d_{cr} - d_{in})] \cdot a \cdot \sin\beta \qquad (3b)$$

Because a debris flow is known to be a translative movement, force equilibrium in the normal and tangential directions is chosen to satisfy the limiting equilibrium equation for failure. Moment equilibrium is automatically satisfied by the redistribution of normal forces at the base of the slice, i.e. by moving the assumed position of the normal force, N.

Solving for force equilibrium in the normal and tangential directions gives:

$$N = [\gamma' \cdot d_{in} + \gamma \cdot (d_{cr} - d_{in})] \cdot a \cdot \cos\beta +$$
$$i \cdot \gamma_w \cdot \cos\beta \cdot [\sin\beta^* - \cos\beta^* \cdot \tan\beta] \cdot a \cdot d_{in} \qquad (4a)$$
$$T = [\gamma' \cdot d_{in} + \gamma \cdot (d_{cr} - d_{in})] \cdot a \cdot \sin\beta +$$
$$i \cdot \gamma_w \cdot \cos\beta \cdot [\cos\beta^* + \sin\beta^* \cdot \tan\beta] \cdot a \cdot d_{in} \qquad (4b)$$

The corresponding normal and shear stresses, σ and τ can be substituted into the Mohr-Coulomb Failure Criterion given by:

$$\tau = c_t + \sigma \cdot \tan\phi' \tag{5}$$

where c_t is the total cohesion defined by Fredlund (1979) as:

$$c_t = c' + u_c \cdot \tan\phi^b \tag{6}$$

c' and ϕ' are the effective cohesion and friction angle respectively, and ϕ^b is the friction angle with respect to matrix suction, u_c.

The critical depth for infinite slope failure can be found using Equations 4a, 4b, and 5 yielding:

$$d_\sigma = \frac{c_t}{\gamma \cdot \cos^2\beta \cdot [\tan\beta - \tan\phi']} +$$
$$\frac{\gamma_w}{\gamma} \cdot d_{in} \cdot \left[1 - \frac{\tan\beta}{1 + \tan\beta \cdot \tan\beta'} \cdot \left(\tan\beta + \frac{1}{\tan(\beta - \phi')} \right) \right] \tag{7}$$

In deriving Equation 7 the assumption that $\gamma' = \gamma - \gamma_w$ is made. This is only valid if γ is the saturated unit weight of the soil which is certainly not true for depths larger than d_{in}. However, the change in the degree of saturation only marginally affects the unit weight of soil and the stated assumption is justified in order to simplify the analysis.

Equation 7 gives the critical depth for failure of an infinite slope subject to seepage at direction β^* and acting over a depth of d_{in}. When there is no infiltration, $d_{in} = 0$, the conventional slope stability equation for the critical depth of an unsaturated slope is recovered. The case for parallel seepage,

where $\beta^* = \beta$, is also seen to be a special case of Equation 7.

3.4 *Critical Depth in Terms of Pore Pressures*

Although the physical meaning of the results of seepage can be more easily understood in terms of the depth of infiltration, d_{in} and the seepage direction, β^*, it is more convenient to work in terms of pore pressures. In addition, by putting the formulation outlined above in terms of pore pressures, the link to the results of the infiltration analyses can be more easily made.

Again, considering a slope at angle β, subject to seepage at direction β^*, the pore pressure head h_p at a given depth, Z can be determined graphically as is shown in Figure 5. The mathematical relationship is developed by considering two points, E and F on the surface of the slope. A constant head line h2 is chosen to pass through point F so that the point on the h2 line directly beneath point E is defined and called point D. The pressure head at this point, h_{pD} is the variable of interest in this analysis. Again, as in Figure 4 the direction of seepage, β^* is the angle between the flow lines and the horizontal direction and the slope angle is defined as β.

The pressure head at point D is given by:

$$h_{pD} = x / \tan\beta^* \tag{8}$$

where x is defined by the geometry in the triangle formed by the sides x and $(Z - h_{pD})$:

$$x = (Z - h_{pD}) / \tan\beta \tag{9}$$

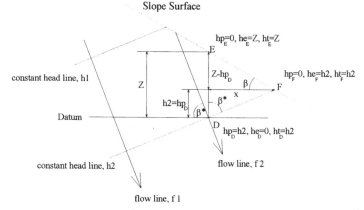

Figure 5. Infinite Slope Profile Showing Relationship between hp and d_{in}, β^*

Thus, the pressure head at a point D, hp_D can now be defined in terms of β, β^*, and Z:

$$h_p = \frac{1}{1 + \tan\beta \cdot \tan\beta^*} \cdot (Z) \qquad (10a)$$

Equation 10a can be rearranged so that the slope of the positive pore pressure infiltration results in Figure 1 can be seen to be function of β and β^* only:

$$\frac{Z}{h_p} = 1 + \tan\beta \cdot \tan\beta^* \qquad (10b)$$

Taking this one step further, by examining an infiltration analysis for a given soil and rainfall intensity, the depth/pressure head ratio, Z/h_p at a given depth of infiltration, d_{in} can be converted to a seepage direction angle, β^*. For a slope subject to infiltration, the maximum pressure head at the surface will be zero due to an inability to pond water on a slope. Thus, the Z/h_p trace will be a diagonal line starting at zero and moving downwards into the slope as Figure 1 shows. For the positive pressure head domain that is being examined here, the given depth, Z will coincide with the depth of infiltration, d_{in} so that $Z = d_{in}$, and the Z/h_p ratio can be thought of as a d_{in}/h_p ratio. Either way, β^* is calculated from this ratio, and the β^*, d_{in} pair can be used in Equation 7 to find the critical depth for this form of infiltration. The larger the Z/h_p ratio from the infiltration results, the larger the seepage angle will be. Larger values of β^*, indicating more vertical infiltration will result in larger values of d_{cr} and a larger portion of the soil slope will be mobilized during failure. With flow closer to being parallel to the slope, smaller values of β^* will result in smaller values of d_{cr}. As β^* changes during infiltration, the depth of infiltration, d_{in} will also increase as water penetrates deeper into the slope. The variation of d_{in} will be seen to be inconsequential to the critical depth when pore pressures are examined.

The results for this variation of two parameters, β^* and d_{in} can be realized much more easily by using Equation 10a to restate Equation 7 in terms of pressure head. By substituting for β^* in Equation 7, the depth of infiltration is eliminated and the new formulation for d_{cr} is:

$$d_{cr} = \frac{c_t - \gamma_w \cdot h_p \cdot \tan\phi'}{\gamma \cdot \cos^2\beta \cdot (\tan\beta - \tan\phi')} \qquad (11)$$

The critical depth for infinite slope failure is now a function only of pressure head, h_p (or pore pressure,

$u = \gamma_w \cdot h_p$) and the given material and slope characteristics, c_t (c'), ϕ', γ, and β. It is important to note here that although d_{in} is eliminated from the formulation when pressure heads are examined, this variable still has a valuable meaning. Stated simply, the horizontal component of seepage can be the same for varying d_{in}. Thus, there will be no difference in the horizontal force equilibrium for a given slice subject to seepage. On the other hand, the vertical component of seepage will be different for varying d_{in}. As infiltration becomes deeper, d_{in} increases and the vertical component will grow larger. At the same time, the weight force will be switching from total unit weight to buoyant unit weight which will make the weight force smaller. Thus, the vertical force equilibrium will remain unchanged as well. Since neither the horizontal nor vertical components will change, the critical depth will remain unchanged as well. For an increasing d_{in}, only a change in pore pressure will affect the critical depth as Equation 11 shows.

An equation that depends on both negative and positive pressure heads can now be written by incorporating Equation 6 for the total cohesion in terms of capillary pressure head, h_c into Equation 11:

$$d_{cr} = \frac{c' + \gamma_w \cdot h_c \cdot \tan\phi^b - \gamma_w \cdot h_p \cdot \tan\phi'}{\gamma \cdot \cos^2\beta \cdot (\tan\beta - \tan\phi')} \qquad (12)$$

Equation 12 allows the stability envelope to be defined for both negative and positive pressure heads. This allows a smooth transition to be made in calculating the critical depth for infiltration analyses that move through both the negative and positive regimes as in seen in Figure 1.

4 INFLUENCE OF INFILTRATION ON SLOPE STABILITY

The equation for slope stability presented above can now be used to make quantitative predictions about slope failure by incorporating the results of infiltration presented earlier in this paper. Because the equation was developed in terms of the pressure head, the resulting stability envelope can be superimposed on an infiltration trace for the particular soil being investigated.

4.1 Analysis Results

The stability envelope for slope parameters, $\beta=40°$, $\gamma=20kN/m^3$ and shear strength parameters $c' = 10$ kPa, $\phi' = 20°$, and $\phi^b = 10°$ is shown overlain on the

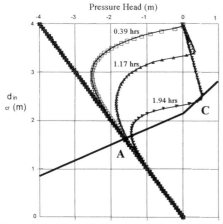

Figure 6. Combined Infiltration / Slope Stability Plot for Coarse Grained Soil

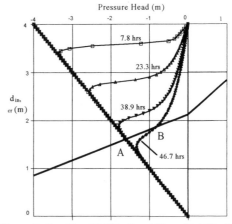

Figure 7. Combined Infiltration / Slope Stability Plot for Fine Grained Soil

infiltration trace for the coarse grain soil (Figure 1) with slightly different time steps in Figure 6. The points marked with letters (A and C) coincide with the intersections of the stability envelope with the infiltration trace. Point A is located on the initial hydrostatic pressure head profile and defines a failure point that will always occur using infinite slope analysis. This is because of the assumption made about the side forces being equal and in opposite direction for infinite slope analysis. Without lateral confinement of the slope, failure will always be predicted to occur at some depth for any given set of slope and soil parameters. Thus, point A does not contribute any further information about the failure mechanisms from infiltration.

Point C indicates that failure of the slope defined by the given parameters will occur at a depth of 1.6 meters at an elapsed time of 1.94 hours after rainfall begins. The triggering mechanism here is due to the development of seepage forces in the slope since failure occurs in the positive pressure head regime.

Similar analysis results are presented in Figure 7 for the same stability envelope as in Figure 6, this time overlain on the fine grained soil infiltration trace (Figure 2) with slightly different time steps. It is clear that the development of seepage forces will not cause failure for this case since positive pore pressures do not develop. Again point A shows the analytical solution for infinite slope analysis and does not explain the triggering mechanisms due to infiltration. Point B shows that failure is predicted to occur at a depth of 2.2 meters after 46.7 hours of rainfall. The failure mechanism is due to the decrease in shear strength that occurs from the loss of soil suction in the negative pore pressure regime.

Failure for the fine grain soil will therefore occur at a greater depth and after a much longer period of time compared with the coarse grained soil.

The results presented here indicate the type of information that can be gained from performing these type of analyses. In addition to the quantitative statements about the time and depth of failure, qualitative generalizations can be made about the type of triggering mechanism that is likely to occur for rainfall induced landslides.

4.2 *Comparison of Results with Field Observations*

The results of the analysis method presented here are at least qualitatively verified by comparison with observations made in the field about rainfall induced landslides. Reid et al., (1997) observed that shallower failure surfaces occur as a result of positive pore pressures in the soil (i.e. failure from seepage), while deeper failure surfaces form without the formation of positive pressures (i.e. reducing suction). De Campos et al., (1991) observed the occurrence of shallow debris flow failure surfaces and concluded that the formation of a seepage pattern in the slope led to failure. Rahardjo et al., (1996) also observed formation of shallow debris flow failure planes, however their analyses led to the conclusion that these failures occurred as a result of a reduction in suction. This report is still consistent with the findings presented here because the soils observed by Rahardjo et al, (1996) were predominantly fine grained soils. It has been shown that positive pore pressures are less likely to occur in fine grained soils, thus dismissing the formation of seepage induced failures in these cases.

5 CONCLUSIONS

The triggering mechanisms of rainfall induced landslides have been investigated and a comprehensive method for predicting the depth and time of failure for long slopes subject to rainfall has been presented. The analysis method allows for both qualitative and quantitative statements to be made regarding the failure mechanisms in relation to the soil and slope parameters. Observations from the field confirm the predicted behavior.

6 ACKNOWLEDGMENTS

The authors wish to acknowledge the financial support of the United States Department of Education's Graduate Assistance in Areas of National Need (GAANN) Fellowship Program which provided funding during the time in which this research was being performed.

7 REFERENCES

Becht, M. & Rieger, D. (1997). Spatial and Temporal Distribution of Debris-Flow Occurrence on Slopes in the Eastern Alps. *Proc. 1st Int. Conf. on Debris-Flow Hazards Mitigation*, San Francisco, California. ASCE, New York, pp 516-529.

Cascini, L. & Versace, P. (1988). Relationship between rainfall and landslide in a gneissic cover. Proc. 5th Int. Symp. Landslides, Lausanne, 565-570, Rotterdam: Balkema.

Craig R. F. (1992). *Soil Mechanics*. 5th ed. Chapman & Hall, London.

De Campos T. M. P., Andrade M.H.N. & Vargas Jr. E. A., (1991). Unsaturated Colluvium Over Rock Slide in a Forested Site in Rio de Janeiro, Brazil. *Proc. 6th Int. Symp. on Landslides*, Christchurch, New Zealand, Vol. 2, pp 1357-1364.

Fredlund, D.G. (1979). Appropriate concepts and technology for unsaturated soils. *Canadian Geotechnical Journal* 16: 121-139.

Fredlund, D.G. & Rahardjo, H. (1993). *Soil Mechanics for Unsaturated Soils*. John Wiley & Sons, Inc.

GEO-SLOPE Int. Ltd. (1994). *SEEP/W User's Guide*. GEO-SLOPE Int. Ltd., Calgary.

Polemio, M. (1997). Rainfall and Senerchia Landslides, Southern Italy. *Proc. 2nd Pan-American Symp. Landslides/ 2nd Brazilian Conf. Slope Stability*, Rio de Janeiro, Brazil. pp 175-184.

Rahardjo H., Chang M. F. & Lim T. T., (1996). Stability of Residual Soil Slopes as Affected by Rainfalls. *Proc. 7th Int. Symp. on Landslides*, Trondheim, Norway. A. A. Balkema, Rotterdam, Netherlands, pp 1109-1114.

Reid, M.E., Nielsen, H.P. & Dreiss, S.J. (1988). Hydrologic Factors Triggering a Shallow Hillslope Failure. *Bull. Assoc. Eng. Geologists* 25 (3): 349-361.

Reid M. E., LaHusen R. G. & Iverson R. M., (1997). Debris-Flow Initiation Experiments Using Diverse Hydrologic Triggers. *Proc. 1st Int. Conf. on Debris-Flow Hazards Mitigation*, San Francisco, California. ASCE, New York, pp 1-11.

Transportation Research Board. (1996). *Landslides Investigation and Mitigation*: Special Report 247. National Academy Press, Washington.

Geotechnical Hazards, Marić, Lisac & Szavits-Nossan (eds) © 1998 Taylor & Francis, ISBN 90 5410 957 2

Landslide hazard assessment in Hong Kong

R.A. Forth
University of Newcastle upon Tyne, UK

ABSTRACT: Landslides induced by torrential rain have been recorded for the past one hundred years in Hong Kong. Those causing fatalities are particularly well documented commencing with a retaining wall collapse in 1925 which killed 75 people. The most notorious incidents occurred in 1972 when two major disasters occurred resulting in 138 deaths. One of these disasters was caused by the failure of a fill slope and in 1976 another fill slope failure in the same area of east Kowloon in which 18 people perished led to the establishment of the Geotechnical Control Office (GCO) in 1977. The GCO initiated the Landslip Preventative Measures programme resulting in the expenditure of millions of dollars to improve the safety of slopes over the past 20 years. Private developments on slopes were controlled by the rigorous enforcement of investigation, design and construction standards empowered by the Building Ordinance. Nevertheless a small number of incidents with fatalities continue to occur up to the present day, again induced by heavy rainfall. This paper examines the relationship between rainfall and landslides during a particularly severe rainstorm in August 1982 when over 572 mm of rain fell in a five day period. This rated as the most severe five day rainfall period in August ever recorded. Over 200 landslide incidents were reported and 5 persons were killed and 3 injured.

1. INTRODUCTION

Hong Kong is a densely populated territory with the majority of dwellings being on low land surrounded by hills rising to an elevation of 1,000 m. As a result of its phenomenal economic growth since 1947, available flat land was soon taken up and construction commenced on the lower slopes of the hills. Much of this construction consisted of the erection of flimsy squatter huts, but many engineered structures of ever-increasing size were also constructed. This development coincided with the growth of the subject of soil mechanics and it is probably fair to say that the early developments were constructed with limited geotechnical appraisal. Consequently, given the torrential rainfall that Hong Kong receives in the wet season, the territory has been subject to landsliding throughout recorded history.

The most notorious incidents occurred on 18th June 1972 when two major landslides occurred at separate locations resulting in the death of 138 people. Following a landslide in 1976, when a further 18 people lost their lives, a Commission of Inquiry was set up and this led to the formation of the Geotechnical Control Office (GCO) in Hong Kong. Since 1978 the GCO has maintained a database of more than 3,000 landslides and a form of risk assessment has been used to rank slopes and allow a systematic investigation and, if necessary, reconstruction of unstable slopes.

1982 was an exceptionally wet year in Hong Kong with two significant rainstorms, the first in May and the second in August. Both these were "top ten" storms (i.e. ranking in the 10 worst storms ever recorded on the criteria of rainfall totals). August 1982 was in fact the wettest August since records began in 1884 and the five-day rainfall between 15th and 19th August, 522.6 mm, was the highest 5-day rainfall for any August on record.

Five people were killed, three injured and more than 1,500 made homeless as a result of some 800 reported incidents of which 204 were landslides. This is a significantly lower number of casualties (22 killed, 26 injured) and reported landslides (550) than the storm in May of the same year.

This paper reviews the events of the 1982 storms and comments on the rainfall-landslide relationship as well as some of the landslide hazard assessment studies carried out in Hong Kong.

2. THE AUGUST 1982 STORM

August 1982 was the wettest August since records began in 1884 and the rainstorm of the five days 15th-19th August was the highest recorded 5-day rainfall for any August on record. The comparison of this storm with others is given in Table. 1, and suggests that (from data at the Royal Observatory) the storm could be expected to occur once every eight years. However, there is a wide variation in rainfall with differences of over 120 mm in the same 24 hour period between areas only 2 km apart. The rainfall distribution is shown in Figure 1.

The storm event was associated with Severe Tropical Storm Dot which landed in South China between Shantor and Xiamen. Dot dissipated over South China early on 16th August but widespread thundery showers and heavy rain affected Hong Kong until 19th August. The rain was heaviest on 16th August and 238.3 mm of rain was recorded during the 8 hour period ending at 1.00 p.m. The total daily rainfall for each of the 5 days as measured at the Royal Observatory is presented in Table 2.

The variation in rainfall across the Territory was recorded on the automatic gauges monitored by the GCO. The highest 4 day rainfall was recorded at Stanley with 571.5 mm between the 16th and 19th August, and the highest one day rainfall occurred on the Peak, 322.0 mm on the 16th August. The maximum rainfall intensity was 71.5 mm per hour recorded at Stanley and Sheko on the morning of 16th August.

3. ANALYSIS OF LANDSLIPS

3.1 *Cut Slopes and Retaining Walls*

Cut slope failures accounted for about two thirds of all the reported landslides, of which the majority were in soil slopes. Most of these were small scale (volume less than 50 m^3). Mixed soil/rock failures were less frequent (19 in all) and damaged roads - one major highway was partially closed for 9 weeks whilst remedial works were carried out. Of the 8 rock slope failures, 5 were major but no fatalities resulted. 19 retaining wall incidents were recorded, of which 15 were complete collapses. Three deaths and one injury were caused by the failure of retaining walls, all in squatter areas. Of the 19

walls, 8 were skin walls where the height to width ratio is very large, and 3 walls were of substantial cross-section with a height-width ratio of 3 : 1.

3.2 *Natural Slope Failures*

Twenty-six failures of natural slopes occurred, seven of which were major, one of which caused two fatalities. The majority of these failures occurred in squatter areas and 91 huts were evacuated.

3.3 *Fill Slopes*

Twelve fill slope failures occurred of which two were classified as major.

Table 1

The August 1982 rainstorm compared to other rainstorms since 1884		
Period	August 1982 rainstorm figures	Ranking of August 1982 rainstorm at that time
3-day	462.2 mm (16th to 18th)	10th highest (Highest = 854.9 mm)
2-day	414.6 mm (16th to 17th)	10th highest (Highest = 841.2 mm)
1-day	334.2 mm (16th)	5th highest (Highest = 534.0 mm)
24 hr	362.4 mm (ending 9 pm on 16th)	10th highest (Highest = 697.1 mm)

Table 2

Rainfall during the rainstorm of August 1982 (Royal Observatory and GCO figures)		
Day and date	Rainfall at RO gauge in Tsim Sha Tsui	Maximum rainfall recorded at GCO gauge in Tsuen Wan (Cho Yiu Estate)
Sun 15 Aug.	29.4 mm	27.5 mm
Mon 16 Aug.	334.2 mm	311.0 mm
Tues 17 Aug.	80.2 mm	148.0 mm
Wed 18 Aug.	47.6 mm	68.5 mm
Thur 19 Aug.	31.2 mm	26.5 mm

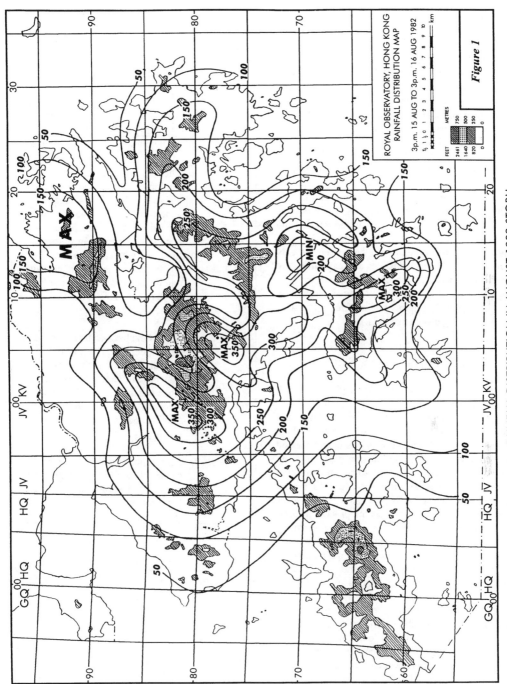

RAINFALL DISTRIBUTION IN AUGUST 1982 STORM

4. THE LANDSLIDE PREVENTATIVE MEASURES (LPM) PROGRAMME

The LPM Programme was established in 1976 following the failure of a fill slope in the Sau Mau Ping area with the loss of 18 lives. This failure occurred near to the disastrous landslide in 1972 when 71 people lost their lives, and prompted the Government to establish a Commission of Inquiry.

The Geotechnical Control Office (GCO) was established to administer the LPM Programme (initially with consultants and subsequently largely in-house). A Catalogue of Slopes was established and each slope was "ranked" largely according to its height, slope angle, and proximity to buildings. High ranking slopes were investigated and subsequently reconstructed (if found to be sub-standard). The programme has been largely successful in reducing landslides and casualties, but occasionally landslides occur such as one in 1994 which led to a review of all procedures adopted by the Geotechnical Engineering Office (formerly GCO), including design methods.

5. LANDSLIDE RISK ASSESSMENT

The ranking list for slopes is a form of risk assessment and indicates the order in which slopes should be investigated. The site investigation and laboratory testing carried out for a particular slope provides data which traditionally has been analysed using conventional limit equilibrium methods. The factor of safety so calculated was used to determine the need for remedial measures, according to a matrix of consequence and risk (Table 3). However it has long been recognised that the process of back saturation in the triaxial cell at least partially destroys the mineral bonding in the weathered igneous rocks and the cohesion values obtained are thus lower than in reality. Various methods such as dead-load testing were employed to overcome this difficulty, with limited success.

More recently quantitative risk assessment has been used drawing on the unique inventory of more than 3000 landslides that have occurred since 1978. Data on geology, slope angle, slope age, groundwater and geometry have been recorded.

A proposed method (Fell et al, 1996) for assessing the probability of sliding of an individual slope PF is to
(a) determine the average probability of sliding (Pa) from the landslide statistics, and

Table 3. Recommended Factors of Safety for the analysis of existing slopes and for remedial and preventive works to slopes for a ten-year return period rainfall (from Geotechnical Manual for Slopes 1984, Hong Kong Government)

	Recommended Factor of Safety against loss of life for a ten-year return period rainfall		
Risk to Life	Negligible	Low	High
	>1.0	1.1	1.2

Note: 1. These factors of safety are minimum values to be used only where rigorous geological and geotechnical studies have been carried out, where the slope has been standing for a considerable time, and where the loading conditions, the groundwater regime and the basic form of the modified slope remain substantially the same as those of the existing slope.

2. Should the back-analysis approach be adopted for the design of remedial or preventive works, it may be assumed that the existing slope had a minimum factor of safety of 1.0 for the worst known loading and groundwater conditions.

3. For a failed or distressed slope, the causes of the failure or distress must be specifically identified and taken into account in the design of the remedial works.

(b) determine the factor (F) by which Pa should be multiplied to assess the probability of the individual slope failing.

Hence Pt = Pa x F.

It is not clear how this approach can be reconciled with conventional factors of safety for slopes. The approach is also only valid where large amounts of raw data is available.

6. REFERENCES

Fell, F., P. Finlay, & G. Mostyn 1996. Framework for assessing the probability of sliding of cut slopes. *Proceedings of the 7th International Conference on Landslides, Trondheim*: 201-208.

Geotechnical Control Office, Hong Kong Government 1984. *Geotechnical Manual for Slopes.*

Geotechnical Hazards, Marić, Lisac & Szavits-Nossan (eds) © 1998 Taylor & Francis, ISBN 90 5410 957 2

Pipeline response to landslide loads

M.Georgiadis, C.Anagnostopoulos & N.Kasimatis
Aristotle University of Thessaloniki, Greece

ABSTRACT: Pipeline damages caused by landslides can create serious environmental hazards and therefore special care is needed in the design. A series of model tests were carried out to study the response of pipelines to landslide induced loads. A model pipeline, instrumented with strain gauges, was used to investigate the effect of pipeline embedment on pipeline bending moments caused by a landslide. Measured pipeline response was compared to predictions made through a numerical analysis in which the pipeline was modelled as an elastic beam on non-linear springs. Predictions made using several spring characteristics underestimated significantly the pipeline bending moments and displacements. A new load-displacement relationship was proposed which produced good agreement between measured and predicted pipeline responses.

1 INTRODUCTION

The continuously increasing need for energy has led to the construction of a large network of onshore and offshore pipelines, which carry oil or gas from the production to the consumption areas. These pipelines are often required to pass through areas of potential landslides and therefore special care is needed in the design, to avoid geoenvironmental hazards. The most secure approach for avoiding such hazards is the selection of the appropriate safe pipeline routing. Unfortunately, this is not always possible since it would increase enormously the pipeline length and the corresponding cost. It is necessary in those cases, to define the landslide characteristics, such as soil properties and slide geometry, and to incorporate them into a soil-pile interaction analysis. This analysis will lead to a safe pipeline system, which may consist of multiple smaller diameter pipelines instead of a single large diameter pipeline or charge from crossing the landslide area at right angles to crossing it parallel to the movement direction.

In this type of analysis, the pipeline and the soil are modelled as an elastic beam and a series of non-linear perpendicular to the pipeline springs, respectively (Fig. 1). The key problem is the evaluation of: (a) the load which is applied by the landslide to the pipeline and (b) the load-displacement relationship of the soil springs.

Providing that this information is known, the pipeline analysis and design is rather simple. Several investigators, based on model tests on ground anchors, laterally loaded piles and short pipe sections, have proposed semi-empirical equations relating the landslide loads and non-linear spring stiffnesses to the pipe diameter and embedment and to the soil properties (Hansen, 1961, Smith, 1962, Ovesen, 1964, Audibert & Nyman, 1977, Akinmusuru, 1978, Trautmann & O'Rourke, 1985, Georgiadis, 1991, Hsu, 1994).

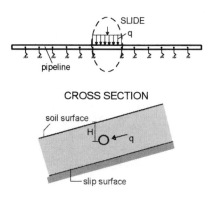

Figure 1. Pipeline model

The non-linear spring behaviour is usually described by a hyperbolic or an exponential equation relating force to spring deformation. The ultimate spring load as well as the uniformly distributed load, which is applied by the landslide to the pipeline, are related to the soil properties and the pipeline embedment. Depending on the type of tests and the experimental conditions on which these correlations have been based, the ultimate loads proposed by several investigators can vary by more than 100%. For example, in the case of a pipeline in sand with $\varphi=35°$ and embedment to diameter ratio $H/D=4$, the ultimate load given by Akinmusuru (1978) is 2,5 times the load given by Ovesen (1964). It is obvious that the large differences in landslide loads and spring characteristics make the accuracy of the computed pipeline bending moments rather questionable.

Due to these uncertainties, a series of model pipeline tests were carried out, in which bending moments along the pipeline and maximum pipeline displacements were measured and compared to numerical predictions obtained using landslide load and spring characteristics relationships proposed by several investigators.

2. MODEL TESTS

A series of model tests were carried out in which the effect of pipeline embedment on bending moments and maximum lateral displacements was investigated. The model consisted of the steel three tanks shown in Figure 2, which were filled with loose sand. The middle tank was resting on rails and was able to advance laterally, simulating the landslide movement, while the other two tanks were fixed to the floor. The length of the fixed tanks was 1400mm, while the length of the middle moving tank was 200mm, equal to the model landslide width.

A 25mm O.D. (23mm I.D.) and 3.2m long plastic tube was running through the middle of the three tanks, along the longitudinal direction, as shown in Figure 2. It's flexural stiffness was $EI=0.023kNm^2$. The tube was instrumented with eight pairs of strain gauges which measured the variation of bending moments and axial forces along the model pipeline. A displacement transducer was used to measure the maximum lateral pipeline displacement which occurred at the middle of the central tank.

The three tanks were filled with loose fine to medium sand (SP) with grain sizes between No.20 and No.100 standard U.S. sieves. The dry unit weight and the friction angle of the sand were $\gamma_d=14kN/m^3$ and $\varphi=32°$, respectively.

Tests were performed at three different embedments (H) of 62.5, 87.5 and 112.5 mm, corresponding to embedment to diameter ratios (H/D) of 2.5, 3.5 and 4.5, respectively. All strain gauge and displacement transducer readings were simultaneously recorded through a multi-channel data-logging system.

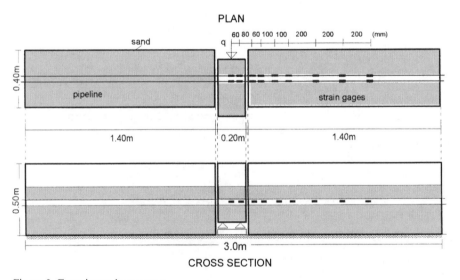

Figure 2. Experimental apparatus

3. NUMERICAL ANALYSIS

The behaviour at the model pipeline was analysed using a simple computer code which was based on the transfer matrix method and an iterative procedure. The pipeline was considered as an elastic beam resting on a series of non-linear springs located outside the landslide area. A uniformly distributed load was applied to the middle 200mm of the pipeline, simulating the landslide earth pressure (Fig. 1). The analysis was performed for various load - displacement spring relationships (*p-y* curves) and distributed landslide loads (*q*), proposed by several investigators.

The distributed landslide load was derived through the equation :

$$q = N_h \cdot \gamma \cdot H \cdot D \qquad (1)$$

where γ is the soil unit weight and N_h is a bearing capacity factor depending on the soil friction angle (φ) and the embedment to pipe diameter ratio (H/D).

The spring stiffness was described with the following hyperbolic relationship :

$$p/p_u = (y/y_u) / [\, a + b\,(y/y_u)\,] \qquad (2)$$

where p is the spring load per unit area corresponding to a displacement y, p_u is the ultimate spring load ($p_u = q$) which is reached at a displacement y_u and a,b are empirical coefficients.

Several investigators have proposed different values of N_h, y_u, a and b, which result to different landslide loads and spring characteristics. Some typical *p-y* curves (Smith, 1962; Audibert & Nyman, 1977; Trautmann & O'Rourke, 1985) derived for H/D=3.5 are presented in Figure 3. On the same Figure is also plotted a *p-y* curve corresponding to the following exponential relationship which proved to represent better the pipeline response:

$$p / p_u = 1 - e^{-K\,y/p_u} \qquad (3)$$

where K is the initial stiffness of the *p-y* curve, which is a function of the pipeline embedment and the sand density. The variation of K with H/D for loose sand is shown in Figure 4. The ultimate spring load is equal to twice the value proposed by Trautmann & O'Rourke (1985). The ultimate loads proposed by Smith (1962) and Audibert & Nyman (1977) are 15 and 30 percent lower, respectively. The numerical analysis was performed using the *p-y* curves of Figure 3 and the corresponding landslide loads *q*.

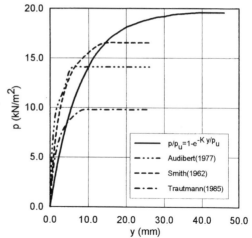

Figure 3. "p-y" curves

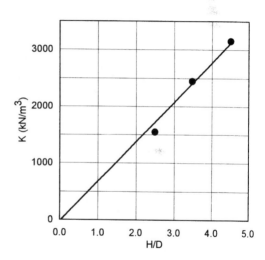

Figure 4. Initial spring stiffness

It provided the bending moments along the pipeline and the pipeline displacements for different *H/D* values.

4. EXPERIMENTAL AND NUMERICAL RESULTS

Results of the experimental and numerical analysis are presented in Figures 5-7. These figures show measured and predicted bending moments along the model pipeline for embedment ratios of 2.5, 3.5 and 4.5, respectively. Predictions were made using

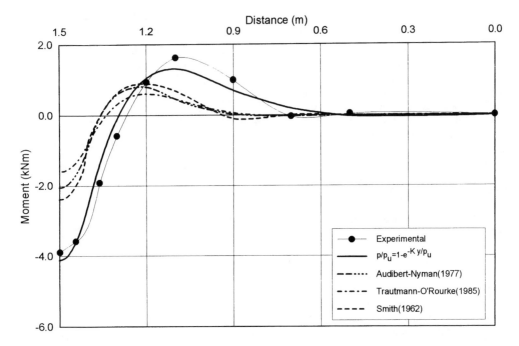

Figure 5. Measured and predicted bending moments for H/D=2.5

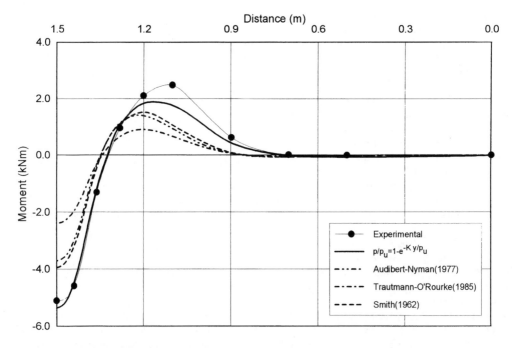

Figure 6. Measured and predicted bending moments for H/D=3.5

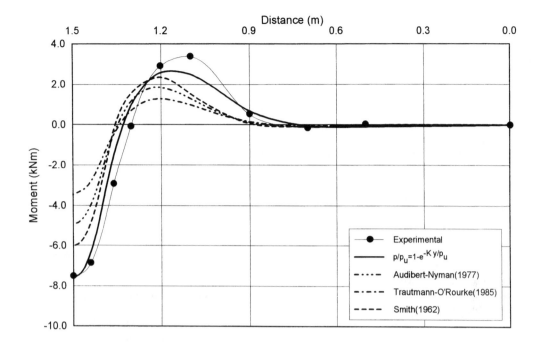

Figure 7. Measured and predicted bending moments for H/D=4.5

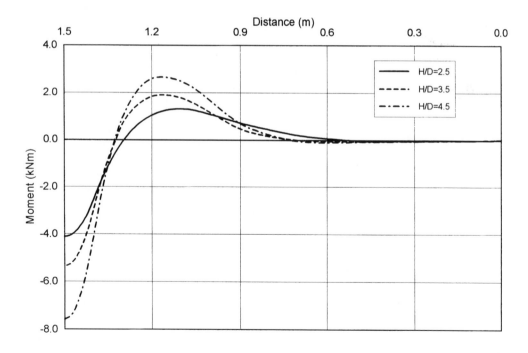

Figure 8. Effect of embedment ratio on predicted bending moments

the *p-y* curves and mudslide loads described in the previous section. Measured and predicted maximum pipeline displacements are presented in Figure 9 as a function of embedment ratio. Measured pipeline axial forces were negligible.

Figures 5-8 demonstrate that the use of equation (2) underestimated significantly both measured bending moments and maximum displacements. Measured bending moments were even 100% larger than some predicted values, while differences in pipeline displacements were even greater. A much better correlation between measured and predicted pipeline responses was obtained by modelling the soil spring characteristics with equation (3).

Figure 8 presents bending moments predicted using equation (3) in the numerical analysis and illustrates the effect of pipeline embedment. It shows that the increase of the embedment depth increases seriously the pipeline bending moments. The maximum bending moment for H/D=4.5 is about 100% larger than the value derived for H/D=2.5.

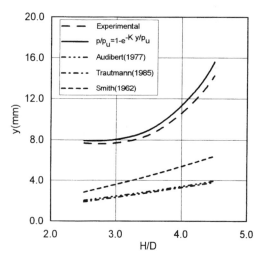

Figure 9. Measured and predicted maximum pipeline displacements

5. CONCLUSIONS

Model pipeline tests and numerical analysis demonstrated that currently used pipeline - soil interaction relationships underestimate seriously the pipeline bending moments. An exponential load - displacement relationship has been proposed, which improved significantly the pipeline response

predictions.

The model test results revealed that the axial pipeline stresses were rather insignificant compared to the pipeline bending stresses.

Both measurements and predictions demonstrated that increase of the pipeline embedment increases significantly the landslide induced bending moments. It is therefore required, in areas of increased landslide potential, to select the minimum possible pipeline embedment.

REFERENCES

Akinmusuru J.O., 1978. Horizontally loaded vertical plate anchors in sand. *J. of Geotech. Engrg. Div.*, American Society of Civil Engineers (ASCE), 104 (GT2): 283-286.

Audibert J.M.E. & Nyman K.J., 1977. Soil restraint against horizontal motion of pipes. *J. of Geotech.Engrg.Div.*, Americal Society of Civil Engineers (ASCE), 103 (GT10): 1119-1142.

Georgiadis M., 1991. Landslide drag forces on pipelines. *Soils & Foundations*, 31 (1): 156-161.

Hansen J.B., 1961. The ultimate resistance of rigid piles against transversal forces. *Danish Geotech.Institute*, Bulletin 12, Copenhagen, Denmark.

Hsu T.W., 1996. Soil restraint against oblique motion of pipelines in sand. *Canadian Geotech. J.*, 33: 180-188.

Ovesen N.K., 1964. Anchor slab, calculation methods and model tests. *Danish Geotech. Institute*, Bulletin 16, Copenhagen, Denmark.

Smith J.E., 1962. Deadman anchorages in sand. *U.S. Naval Civil Engrg. Laboratory*, Tech. Report R-199, Port Hueneme, U.S.A.

Trautmann C.H. & O'Rourke T.D., 1985. Lateral force displacement response of buried pipe. *J.of Geotech.Engrg.*, American Society of Civil Engineers (ASCE), 111 (9): 1077-1092.

Geotechnical Hazards, Marić, Lisac & Szavits-Nossan (eds) © 1998 Taylor & Francis, ISBN 90 5410 957 2

Learning from a large landslide in Northern Italy

G.Gottardi, G.Marchi & P.V.Righi
DISTART, University of Bologna, Italy

ABSTRACT: The village of Corniglio, in the Parma Apennines (Northern Italy), has been involved for the last three years by a huge landslide - currently one of the largest in Europe - causing considerable socio-economic damages. The main slope movement, which has now reached displacements of over 50 m, is a composite landslide, reactivated by its original causes, namely geomechanical characteristics decay due to weathering, high groundwater pressures following periods of intense rainfall and local seismic activity. Strictly correlated, an adjacent slope movement involving the old city centre has also appeared since early 1996, fortunately giving rise only to limited damages so far. The paper aims to present the currently available information on such very instructive case history: in particular it describes the extensive site investigation and monitoring scheme carried out both on the main landslide and, above all, on the city centre, which enabled to outline the stratigraphic profiles, provide the geotechnical characterisation of the main units and develop a preliminary stability analysis of the two interconnected slope movement phenomena. However, due to the unusual landslide size, all possible remedial works are necessarily partial and very expensive.

1. INTRODUCTION

After a long period of quiescence, lasting since 1902, in mid-November 1994 a large ancient landslide - over 3000 m long, 1000 m wide and up to 120 m deep - resumed its activity, striking the village of Corniglio in the Parma Apennines (Northern Italy, see index map in Fig. 1). Corniglio is a municipal chief centre with a total resident population of about 3860 people and is located at an altitude of 688 m a.s.l. on a region with a high density and frequency of various slope movements, essentially due to the tormented geological history which led to the formation of the Apennines and to the widespread presence of weathered clayey and marley soils.

The existence of such a considerable slope movement has been historically recorded since the 17th century (Almagià, 1907) and was previously mentioned in legends going back to the Middle Ages, showing an average reactivating time interval of about one century. In the Seventies, neglecting the presence of the quiescent landslide, the old hamlet of Linari - located at the lower portion of the slope, some hundreds of metres west of the main civic centre (Fig. 1) - underwent widespread urban development, with the construction of about 70 new buildings, including holiday houses and storehouses

for the seasoning of the famous locally produced Parma ham.

The area delimited by the slope movement is shown in Figure 1 and extends from the Mount Aguzzo at South (1150 m a.s.l.) to the River Parma at the Northern boundary (550 m a.s.l.) for about $2 \cdot 10^6$ m^2 and with a slope angle between 8° and 23°. The side boundaries can be identified in correspondence of Rivulet Maltempo at West and Rivulet Lumiera at East. The depth of the surface of rupture is estimated to vary between 30 and 120 m, providing an approximate volume of displaced material of 200 millions of cubic metres. The landslide foot moved up to several tens of metres in the past three years, destroying Linari and inducing its evacuation. Fortunately the old village of Corniglio, located hillside, on the right flank of the main landslide, has been so far interested only by much smaller displacements, continuously monitored.

The latest reactivation of the Corniglio landslide began on November 17th 1994, after a period of intense rainfall which, from September 1st to November 15th, reached 1070 mm, of which 700 mm during September only. The movement took place by means of rotational slips in the crown area: a main large scarp and several other secondary scarps were produced, causing a general

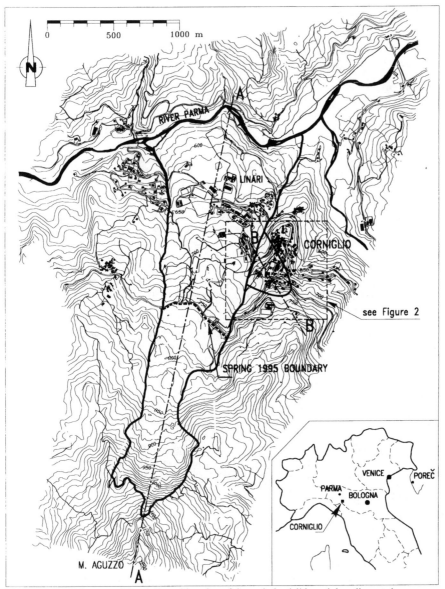

Figure 1. Plan of the area of Corniglio and location of the main landslide and the adjacent slope movement.

retrogression of the landslide uppermost boundary of over 200 m with respect to the previous event of 1902. Related shallow earth flows in the middle landslide portion reached a velocity of over 50 m/day. Subsequently, after a period of frosty and dry weather, the movement gradually slowed down. At the end of Spring 1995 the slope movements had reached the lower boundary shown with a dashed line in Figure 1.

After heavy Summer rainfall and a 3.3 magnitude seismic shock, on January 1996 large displacements started again in the crown area, inducing the whole soil mass to resume movements, this time as far as the Parma riverbed, which was narrowed so that the regular water flow had to be guaranteed by mechanical removal of the depleted material. The final narrowing of the riverbed (April 1996) resulted in about 20 m, i.e. 1/3 of its original width. During February 1996 the landslide moved at an average rate of 30-40 cm/day with peaks of 80 cm/day. The overall soil movement has been greater than 50 m in Linari.

In concomitance with the mass reactivation of the main landslide, ground tension cracks and buildings fissures were also observed in the southern part of the old village of Corniglio. Late in April the buildings and streets of this area had undergone a 20 to 25 cm displacement, showing large cracks, while the gas and water pipelines were damaged in several points.

The significant socio-economic impact induced the Italian Government (*Ministero della ProtezioneCivile*) to appoint a special Committee in charge of the Civil Defence Interventions, so that a rather extensive investigation of such an impressive slope movement became eventually possible. More details on the landslide evolution can be found in Larini et al. (1997) and Gottardi et al. (1998).

2. GEOLOGICAL SETTING

Along the Po Valley side of the Northern Apennines several piled up allochthonous units crop out, defined as Ligurian and Sub-ligurian Units. In the study area the formations belonging to these units are made up of calcareous and arenaceous flysches, often accompanied by thin marly-clayey beds and tectonic and sedimentary mélanges. During the two last glaciations of the Late Pleistocene the region was subject to glacial and periglacial processes (Federici & Tellini, 1983) which favoured intense rock weathering, with the formation of large detrital covers. The lithological situation, morphological evolution and present climatic conditions (the area has a mean precipitation of about 1500 mm/year which can reach 2400 mm/year on the highest surrounding peaks) cause a high density and frequency of various slope movements, up to 9 km in length. The most common types of landslides are intermittent, slow roto-translational slides and earth flows, one of which is precisely the Corniglio landslide.

The geological boundary situation in the study area is described as follows (Cerrina Ferroni ed., 1990):
• landslide crown (altitude of about 1150 m, near M. Aguzzo): highly tectonised Upper Cretaceous clayey-marly-calcareous flysch (locally known as *Flysh di Monte Caio*, FC);
• western margin of the main landslide body and foot area along the Parma riverbed (550 m a.s.l.): Upper Cretaceous-Middle Eocene clay shales with thin calcareous beds and chaotically arranged blocks (*Argille e Calcari* Formation, AC);
• eastern margin, where also the Corniglio civic centre lies: Upper Oligocene arenaceous-pelitic flysch (*Arenaria di Ponte Bratica*, ABR), stratigraphically cropping out on top of AC. The eastern landslide boundary coincides with a fault line juxtaposing ABR with AC, where the Rivulet Lumiera is situated.

3. FIELD INVESTIGATIONS

The geometrical delineation of the main landslide body - once movements resumed - was relatively simple because of the large displacements and the considerable volume of displaced material. As already mentioned, the landslide moved a first time (Winter 1994-95) as far as the slope middle part; following the reactivation of the subsequent Winter, triggered by new important detachments in the crown area, the whole ancient landslide moved as far as the River Parma, with huge amounts of accumulated material emerging at the foot.

Field investigations of a widespread area on complex soil formations are always rather expensive and often incomplete anyway. Earlier available information consisted only of a quite detailed historical report of the previous event (Almagià, 1907), strikingly similar to the recent one. In order to rapidly acquire as many information as possible on the slope movement, late in 1994 a first site investigation of the main landslide body was planned, with special emphasis on the area where the hamlet of Linari was located.

A new aerial photography survey was carried out in order to update existing mapping and keep a check on the movement. From the end of 1994 to May 1995 18 boreholes, up to a maximum depth of 89 m, were drilled in the Linari area. All but two boreholes were later equipped with inclinometer casings, 5 of which being locally slotted in order to provide preliminary information on the piezometric level. Part of the borings were quickly drilled with core-destruction and part with continuous coring. Undistorted samples recovery of such complex soils, often at a considerable depth, was virtually impossible and only a generic qualitative description of main sediments is available.

A quick and efficient way of investigating the subsoil of the whole area was provided by geophysical investigations. In December 1995 the first geophysical survey started within the main landslide body and consisted of seismic refraction measurements over a total length of 10 km, 3 down-hole tests and 2 tomography surveys which used previously drilled boreholes. The seismic refraction survey was deployed along 23 seismic lines covering the whole slope extension in order to better define the side boundaries of the landslide, to acquire a general definition of the subsoil conditions and, above all, to provide some information on the bedrock profiling and on the depth of a possible surface of rupture.

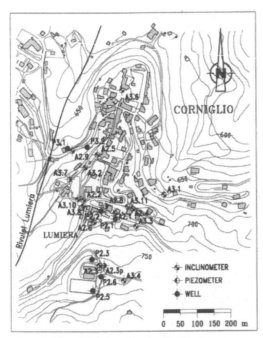

Figure 2. Plan of field investigations in the city centre.

from 8 different boreholes, on which the following laboratory tests were carried out: particle size and calcium carbonate content analysis; Atterberg limits and index properties; direct shear, unconfined compression and point load tests. Samples for the determination of the shear strength parameters were reconstituted from the material passing to sieve n. 40 (0.42 mm), thus reflecting the characteristics of the dominant clayey component.

Direction and intensity of total displacements of the city centre were also constantly monitored through surface topographic surveys. In Summer 1996 a second geophysical survey was undertaken in the Lumiera area, including both seismic reflection and seismic refraction measurements and borehole tests (tomography and down-hole). Finally a detailed geomechanical and structural survey was made on the rocky outcrops on top of which the centre of Corniglio is located.

Such an extensive site investigation programme enabled us to draw a rather detailed stratigraphic profile of the Corniglio subsoil and, despite the soil heterogeneity, define its main geological and geotechnical properties. Both topographic surveys and inclinometric measurements provided a rather accurate idea of the extent and intensity of the slope movements. On the other hand fewer information on the soil permeability and the groundwater pore pressure are available, unfortunately inadequate to outline a reliable scheme of the very complex groundwater circulation.

4. THE MAIN LANDSLIDE

As mentioned, the two different slope movements individuated (the "historic landslide", i.e. the main body resulting from previous movements, and the much smaller displacements involving the old city centre of Corniglio) are inevitably correlated and the stratigraphic units found along a profile drawn along sections A-A and B-B (Fig. 1) rather similar. However, following the two separate and subsequent site investigations campaigns, the two parts are herein separately treated.

4.1 The stratigraphic profile

The main source of data for the "Linari" landslide is provided by the former geophysical survey. In fact the inclinometers initially installed were not enough deep to reach the probable surface of rupture (over 100 m deep at the landslide foot) and rigidly moved together with the soil mass. The stratigraphic profile, essentially derived from the measured values of the seismic velocity v_p, is outlined in Figure 3 which shows a longitudinal section of the whole slope (section A-A in Fig. 1).

Following the landslide resumed activity in February 1996, the first noticeable cracks appeared in the buildings of the southern part of the old village of Corniglio, locally known as "Lumiera". A correlation between the reactivation of the adjacent main landslide and the smaller displacements of Corniglio appeared immediately evident. Due to the greater socio-economic importance of the old city centre, a second extensive site investigation was planned and carried out on the Lumiera area in order to obtain as much detailed information as possible on the extent of the displacements, the characteristics of the relevant soils and the groundwater circulation.

Figure 2 shows the location of 18 of the 21 boreholes drilled between February and May 1996 in two separate periods (respectively labelled A2 and A3): a total of 12 were continuous-coring, the other 9 core-destruction borings. Six boreholes of the series A3 were pushed to a depth greater than 100 m (A3.5 as deep as 175 m!). 18 inclinometer casings, 2 observation wells and 2 open standpipe Casagrande piezometers were subsequently installed. Eight wells were also drilled with the core-destruction method. In addition, in 6 borings, permeability and pressuremeter tests at various depths were carried out.

A total of 36 samples (due to the nature of soils and the widespread presence of rocky fragments in the clayey matrix, again only disturbed samples could be collected) were taken at several depths

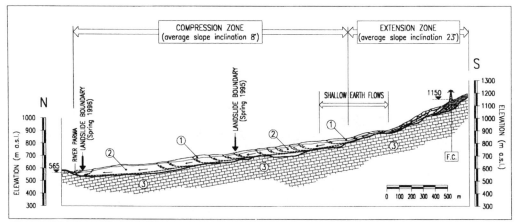

Figure 3. Main landslide longitudinal section and progressive mechanism of failure.

The units 1 (v_p between 600 and 1600 m/s) and 2 (v_p between 1600 and 2600 m/s) represent the landslide debris, i.e. a chaotic accumulation of material derived from ancient or recent slope movements and subsequent weathering processes, which reaches a depth exceeding 120 m. Unit 3 (v_p between 2600 and 3700 m/s) is the bedrock basically made up of the AC formation.

The detrital material is highly heterogeneous and anisotropic, because of the presence of rocky and scaly fragments, and therefore of rather difficult geotechnical characterisation. The relevant main parameters obtained from the limited site and laboratory investigation are as follows: unit weight of 20 kN/m^3 and undrained strength (from SPT and pocket penetrometer tests) of 100 to 200 kPa; nearly all samples fall within the CL group on the Casagrande plasticity chart (w_L = 25 to 38% and PI = 5 to 15%). Particle-size analysis and visual examination show the widespread presence of scales even in the finest fractions.

4.2 The mechanism of failure

On the basis of the landslide historic evolution and the geomorphological surface evidence a possible schematic mechanism of failure has been reconstructed and drawn in Figure 3. At the beginning instability phenomena took place in the top part (FC outcropping area), with retrogressive roto-translational slides which caused the regression and widening of the landslide crown. In Figure 3 the area affected by these movements is designated as "extension zone" and is characterised by a much higher slope inclination (23° compared to the 8° of the lower part). The mobilised mass overloaded the landslide mid-high portion and reactivated the middle part of the ancient slope movement (as far as the Spring 1995 boundary).

New large detachments took place in early January 1996 along rotational failure surfaces near the crown, with propagation of slides in the middle part. This seems to confirm the assumption that only an additional overload on the middle part of the old landslide body could mobilise its lower portion, causing the progressive spreading of the movement both at the surface and in depth.

Progressive failure downstream eventually affected the Parma riverbed, generating a "compression zone" in the landslide lowest portion with prevalent advancing roto-translational movements, which in some points reveal the surfaces of rupture. The main landslide movement has also induced lateral dragging effects and consequent widening along the flanks. Finally, between the "extension zone" and the "compression zone" shallow earth flows have often been observed.

As regards the present state, distribution and style of activity, the "historic landslide" should be rated as a composite (sliding and flowing) multiple landslide, reactivated by its original causes. In particular, it consists of: a) retrogressive landslide in the upper portion (within AC and FC); b) advancing landslide in the lower portion; c) widening landslide on both sides.

4.3 A parametric stability analysis

The above reconstructed failure mechanism was then followed on performing the stability analyses - using the standard limit equilibrium method - first of the upper-middle part and subsequently of the whole slope. The stratigraphic profile of Figure 3 and a surface of rupture coinciding with the interface between units 2 and 3 (i.e. basically the boundary between the landslide debris and the bedrock) were adopted. Due to the substantial lack of information on the subsoil pore pressure and to the poor quality

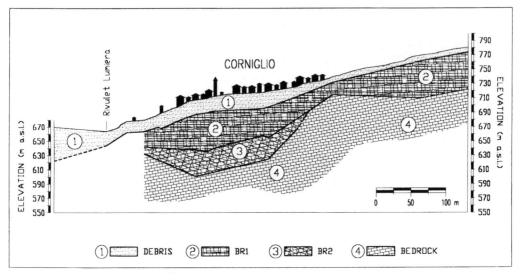

Figure 4. City centre stratigraphic profile along section B-B.

of the shear strength parameters, an extensive parametric study was carried out.

The whole mobilised mass was considered made up essentially of only two different types of soil (i.e. with a separate set of geotechnical parameters), according to the different slope inclination. Therefore, for example, assuming an angle of shear strength $\phi = 31°$ in the upper part, together with a reasonable water level 5 m below surface on the more sloping zone and coinciding with the surface level all over the rest, a peak shear strength angle $\phi_p = 18°$ (plus a cohesion of 50 kPa) and a residual $\phi_r = 13°$ were obtained in the lower part. The parametric study has also shown that, if the shear strength parameters are deduced from the stability analysis of the landslide mobilised as far as the Spring 1996 boundary, only considering a pore pressure acting on the surface of rupture well above the ground surface or, alternatively, the additional action of a local seismic shock, the whole landslide can move as far as the River Parma.

5. THE CITY CENTRE LANDSLIDE

5.1 The stratigraphic profile

From all the available information on the old city centre (geological and geomorphological data, second geophysical survey, borehole logs, laboratory and in situ tests) a rather accurate section of the subsoil profile was deduced (Fig. 4), where the soil intrinsic complexity and heterogeneity has been brought back to four main stratigraphic units, characterised by common geotechnical parameters.

Starting from the ground surface:
- Debris: essentially made of a brown sandy-silty matrix including marly-arenaceous fragments, with a downstream increasing thickness, up to about 20 m in correspondence of the Rivulet Lumiera. The detrital material is highly weathered immediately below surface and progressively tends to exhibit the properties of the underlying unit. Seismic velocities are of the same order of magnitude of the main landslide debris ($v_p = 1200$-1400 m/s).
- Units BR1 and BR2 are both ascribable to the original ABR formation which provides the rocky fragments (from small grains to cobbles) of calcareous marl and sandstone, embedded in either a plastic silty-clayey (BR1) or a blackish clay shale (BR2) matrix. Unit BR1 can be found at depths up to 70 m, whilst BR2 is about 40 m thick - at the most - and tends to disappear at the section ends.
- Bedrock: compact and only moderately fractured (RQD between 90 and 100%) arenaceous-pelitic flysch (ABR). It is characterised by seismic velocities greater than 3000 m/s.

The geotechnical properties of the debris are very similar to those already obtained for the main landslide (see section 4.1). The average shear strength angle was found to be $\phi' = 16°$, but it must be kept in mind that shear tests were performed on the finest fraction and that the operational strength will obviously depend also on the relative importance of the coarser fraction: Figure 5 shows the relevant granulometric distribution of 6 debris samples.

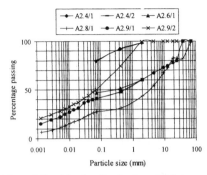

Figure 5. Particle size distribution of debris.

The few available piezometric readings are shown in Figure 6; location and characteristics of the installed piezometers are grouped in Table 1.

Table 1. Piezometers installed in the city centre.

Borehole	Elevation (m a.s.l.)	Piezometer installed Type	Depth (m)
A2.3P	763	Observation well	35.0
A2.8	718	Observation well	80.0
A3.7	710	Casagrande C1	35.0
A3.7	710	Casagrande C2	61.5

5.2 Analysis of displacements

The slope movement involving the city centre appeared only when the whole main landslide (i.e. as far as the River Parma) reactivated in February 1996. A close relationship between the two slope movements was immediately noticed by examining all the displacements measurements within the two areas.

Movements in the Lumiera area were recorded essentially in two separate periods, in concomitance with the large displacements of the main landslide. Figure 7 shows a good example of an inclinometer log from mid-October to December 1996: although small displacements are recorded down to 80 m, a rather clear local surface of rupture appears located at about 20 m of depth. From the analysis of Figure 7 as well as of all the other inclinometer readings, two main movements can be identified: one more superficial and quick, involving the detrital unit and characterised by a well defined surface of rupture; the other, deeper and slower, passing through the unit BR1 and showing more distributed plastic deformations.

Finally in Figure 8 the modulus and the direction of the integral displacements measured by all the inclinometers in the Lumiera area in the same period are represented: the prevalent direction of the slope movement appears to be along the maximum slope inclination.

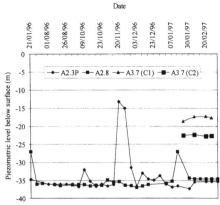

Figure 6. Piezometric readings in the city centre.

6. CONCLUSIONS

The huge landslide developed within the municipality of Corniglio and involving the hamlet of Linari is one of the largest slope movements in Europe: an estimated volume of mobilised mass of about 200 millions of cubic metres, several tens of metres of displacement at the foot, more than 70 buildings and over a hundred people urgently evacuated in February 1996.

The unusual dimensions of the landslide together with the considerable socio-economic impact on the local community enabled to finance and carry out a preliminary site investigation programme in order to characterise the subsoil and monitor the displacement evolution.

Despite the wide extension of the study area and the subsoil intrinsic complexity and heterogeneity, a rather detailed picture of the situation has been drawn: the stratigraphic profile, the mechanism of failure and the instability causes can all be estimated with a good degree of accuracy. A parametric stability analysis has also demonstrated the influence on the overall factor of safety of limited water level fluctuations, especially in the upper and steeper part. Unfortunately only insufficient information on the groundwater pressure have been so far available.

Slope movements recorded in the old city centre area since early 1996, fortunately a couple of orders of magnitude smaller, are clearly very closely related to the main landslide reactivation. Following the subsequent extensive field investigation campaign based on the city centre, a large amount of data on the subsoil parameters and the displacement evolution is now available. Data analysis and interpretation are still in progress, but a more accurate geotechnical characterisation and therefore a more reliable stability analysis are thus possible.

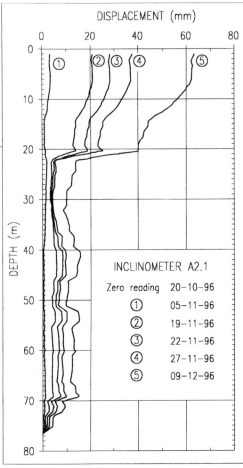

Figure 7. Example of inclinometer logs in the city centre.

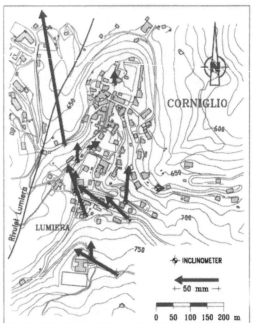

Figure 8. Integral displacement vectors from mid-October to December 1996.

Such a comprehensive investigation should eventually lead to the design of possible remedial measures, in addition to the emergency works which were already carried out, focusing on the preliminary control of shallow and subsurface waters. However all possible interventions to stop or even only keep under control land movements of such proportions appear anyway partial and very expensive.

At the same time little can be done to prevent the reactivating causes to be soon active again. Only prevention and careful planning of the territory seem to be the obvious tools to cope with the correct management of urban slopes.

ACKNOWLEDGEMENTS

The Authors wish to thank Luigi Samorì and Laura Tonni for their invaluable help in the preparation of figures.

REFERENCES

Almagià R. (1907). Studi geografici sopra le frane in Italia. *Mem. Geograf. Ital.*, **13**, Rome.

Cerrina Ferroni A. (ed.) (1990). Geological Map of the Emilia-Romagna Apennines, 1:50,000 scale, Neviano degli Arduini - Sheet 217. *Regione Emilia Romagna, Ufficio Cartografico*, Bologna.

Federici, P.R. & Tellini, C. (1983). La geomorfologia dell'Alta Val Parma (Appennino Settentrionale). *Riv. Geograf. Ital.*, **90**, 393-428, Rome.

Gottardi, G., Malaguti, C., Marchi, G., Pellegrini, M., Tellini, C., & Tosatti, G. (1998). Landslide risk management in large, slow slope movements: an example in the Northern Apennines (Italy). *Proc. Second Int. Conf. Env. Management (ICEM2)*, University of Wollongong, Australia, 10-13 Feb. 1998 (in press).

Larini, G., Marchi, G., Pellegrini, M. & Tellini, C. (1997). La grande frana di Corniglio (Appennino settentrionale, Provincia di Parma) riattivata negli anni 1994-96. *Proc. Int. Conf. "La Prevenzione delle Catastrofi Idrogeologiche: il Contributo della Ricerca Scientifica"*, Alba 5-7 Nov. 1996, 1-12, C.N.R.-I.R.P.I., Turin.

Geotechnical Hazards, Marić, Lisac & Szavits-Nossan (eds) © 1998 Taylor & Francis, ISBN 90 5410 957 2

Stability of Danube embankments in the Slovakian territory

J. Hulla, E. Bednárová & D. Grambličková
Slovak Technical University, Bratislava, Slovak Republic

J. Cábel, D. Janovická & Z. Kadubcová
Water Economy Construction, Bratislava, Slovak Republic

ABSTRACT: After the long-term disastrous flood in 1965, the protective embankments of the Danube in Slovak territory broke in two places. A period of intensive theoretical and experimental studies followed, that dealt with the problems of filtration stability, ensuing the criteria for sealing and drainage elements, particularly in connection with the construction of the water power system of Gabčíkovo-Nagymaros, the design of which was being worked out during those days. Presently a part of the system - the river project of Gabčíkovo - is in operation, thus making it possible to verify the effectivenes of the stability criteria.

1 INTRODUCTION

When the protective embankments broke in two places back in 1965, the water flooded a large part of the fertile area of Žitný ostrov. Thanks to the sample help of the army, some enterprises, and inhabitants from all over the country, no lives were lost. Huge material damages were, however, impossible to avoid. More than 10,000 houses were destroyed or were seriously damaged, there were vast losses in agriculture and the environment.

A period of destructive flood consequences elimination, investigation of reasons of the embankment break, design and implementation of protective measures followed. Experts came to an agreement that the reason of the embankment break was not the embankment itself but the soil lying below it. The protective measures had to correspond with it, a distinctive feature being terrific financial demand. Thus it became completely understood that the protection of our and Hungarian territory against floods got integrated into the construction of water power system Gabčíkovo-Nagymaros.

The destructive floods which attacked Central Europe, especially Germany , Czech Republic and Poland in 1997 caused lots of embankment breaks, loss of life, and giant economic damage. Slovakia and Hungary did not avoid them either, but along the Danube there were no serious problems, due particularly to the protective measures implemented. Engineers were given an opportunity to check the effectiveness of the protective measures, summarize that knowledge and offer it to serve in the reconstruction of damaged embankments in similar conditions within neighbouring countries.

2 GEOLOGICAL CONDITIONS

The geological environment in the Slovakian or Hungarian territory along the river Danube is formed by quarternary alluvium (gravel, sand, silt) which for example near Bratislava reachs the depth of about 10 m, near Gabčíkovo down to the depth of 400 m, near Čičov to the depth of about 100 m, at the state border 15 - 20 m, neogenous sandy and clay sediments are below them.

In the subsoil of the protective embankments there are mainly gravely soils with varying permeability. Figure 1 shows permeability coefficient (Carman-Kozeny) depth function for the newest boreholes between Palkovičovo and Čičov, the values ranging between $k = 10^{-4}$ and 10^{-1} ms^{-1} following no

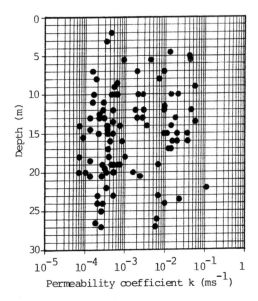

Figure 1: Values of gravely soils permeability coefficients below the embankments between the villages of Palkovičovo and Čičov.

years discharge is 10,600 m^3s^{-1}, the biggest discharge was 10,400 m^3s^{-1} in 1954, in 1965 the value was 9224 m^3s^{-1} and in 1997 "only" 7500 m^3s^{-1}. The destructive character of the flood in 1965 was caused by the long period of its duration, discharges over 6000 m^3s^{-1} lasted for 40 days (Kunsch & Škoda 1995).

3 EMBANKMENT ACCIDENTS IN 1965

It has already been mentioned that the reason of the embankment accidents near the villages of Čičov (Fig. 2) and Patince was not the embankment bodies itself. The experts came to an agreement that the reason was the subsoil. In both cases the break schemes and their consequences were almost the same.

rules. The ratio of horizontal and vertical permeability, gravely soil anisotrophy, values are $\lambda = k_h/k_v = 10$.

Degraded gravely layers, practically with no sand particles, a distinctive feature of which is very high permeability, are often located in a not very deep strata below the embankments, which is due to long-term water level fluctuation. Water can flow at very high speeds with negligible hydraulic losses in such layers, which is very unfavourable to the stability of the impermeable silty covering layers at the outer side of the embankment.

The Danube flows across the area of Slovakia through the alluvium in such manner that even at minimum discharges water flows from the river bed into the surrounding environment, the water level being very close to the surface. The environmental balance conditions that had been created in the past have significantly changed after construction of the water works in Germany and Austria. Gravel supply has stopped and erosion processes of the Danube river bed have started, together with the water level decrease. From 1958 to 1989, the water level under Bratislava decreased by 2 m (Liška 1995).

The average discharge of the Danube in Bratislava is about 2000 m^3s^{-1}, one hundred

Figure 2: Flood broken embankment in 1965 near the village of Čičov.

A good concept about the break stages can be perceived from Figure 3. According to Peter & Šebesta (1965), first of all, springs occur at the outer foot of the embankments, gradually bringing sandy particles to the surface (1). At longer lasting flood states the pressure line gradient gets smaller, the load of the impermeable covering layers by uplift gets bigger, new springs emerge and under the embankment channels get created (2). In the following stages the channels gradually get enlarged (3), get together and form a sort of a tunnel (4), the embankment slumps into it (5) and by intensive hydrodynamic effects gets quickly washed away and thus a break emerges (6). A distinctive feature of such a break is at significant depth, which in our cases reached about 18 to 20 m, being twice or three times bigger height than that of the embankment. The broken sections of the embankments were 80 to 100 m long.

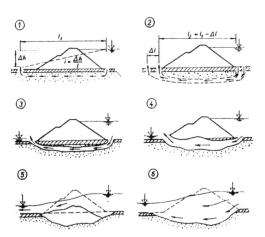

Figure 3: Distinctive stages of embankment subsoil break: 1 - water springs under aerial foot, 2 - formation of channels under the embankment, 3 - enlargement of the channels, 4 - formation of a "tunnel", 5 - embankment slump, 6 - embankment wash away and break (Peter & Šebesta 1965).

The survey after the flood showed that about 2313 water springs with sand being brought to the surface, emerged at our territory, 318 of which were found in the very nearness of the aerial embankment foot (Jakubec 1967). Regulations for the design of protective measures against the influence of seepage from rivers and reservoirs under construction of water works on the Danube (Bažant & Hálek 1966) were quickly worked out.

4 PROTECTIVE IMPACT OF THE GABČÍKOVO WATER SYSTEM

One of the main aims of the Gabčíkovo water system is the protection of the surrounding area against floods. This aim enables to fill in a reservoir with a volume of 194.8×10^6 m^3, a power channel with a discharge capacity of 4500 m^{3}s^{-1} together with the old Danube river bed where with no further problems, 8000 m^{3}s^{-1} can flow (Fig. 4).

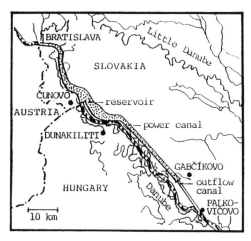

Figure 4: The main parts of the Gabčíkovo project

The protective impact of the Gabčíkovo water system is efficient especially in the upper part of the Žitný ostrov between Bratislava and Palkovičovo. The volume of the reservoir is relatively small, but due to a prompt information system about the development of Danube discharges in the area of Austria it is partially possible to reduce maximum discharges by an appropriate manipulation with water levels even below Palkovičovo, which was put to use during the flood in 1997.

5 UNPROTECTED PART BETWEEN PALKOVIČOVO AND ČIČOV

Due to having stopped the works of the Hungarien side on the construction of the Gabčíkovo - Nagymaros Project the deepening of the Danube river bed under Palkovičovo was not realised. Connected corrections of the embankments between Palkovičovo and Čičov were not realised too. After the flood in 1965, the conditions got better by embankment consolidation, additional fills and tank formation below the outer embankment foot at some locations. Higher discharges have these measures, however, proved to be insufficiently reliable in some locations. Therefore throughout 1995 to 1997 further measures focused particularly on the embankment subsoil were carried out.

At the water embankment foot a sealing wall made of self-hardening suspension down to depth of 20 to 25 m was built as well as the water embankment side was sealed by a PVC foil (Fig. 5). On these locations the gravely soils reach down to the depth of 100 m and the sealing wall was impossible to get fixed into the impermeable clay base not because of the technical but economical reasons as well.

The sealing wall makes water flow in its surrounding in the vertical direction in which the gravely soils are less permeable due to their anisotrophy. In our conditions, however, the interruption of continuity of degraded and very permeable layers is important. Undreground water flow used to concentrate into such layers.

At the place of the sealing wall a hydraulic loss important for the stability of the covering impermeable layers in protected areas takes place. At vertical water flow near the sealing wall the stability of the sandy layers can get endangered. As a result of contact piping, sandy particles can get flown into the pores of the gravely soils. Underground water flow concentrates below the downer edge of the sealing wall where proper attention should be paid to inner piping.

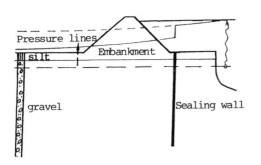

Figure 5: Embankment subsoil protection by sealing walls.

The positive influence of the sealing wall upon the embankment subsoil protection was throughly analysed by with the ose of FEM.

For illustration, Figure 6, shows the time development of the uplift load of the covering layers (under the aerial embankment footing) and of the filtration velocities (under

the embankment base). The results make obvious that due to the sealing wall construction the initial hydrodynamic load intensity of the embankment subsoil (v_f) and uplift load of the covering layers (p) decreases. In addition to that the whole process is from the time development point of view distinctively retarded.

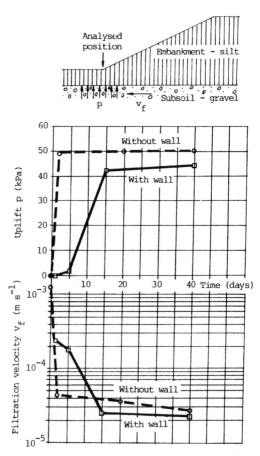

Figure 6: Influence of a sealing wall upon the time development of the uplift and filtration velocity below the aerial embankment foot.

The decrease of the filtration motion intensity and the contemporary increase of the covering layers uplift load is slower.

The maximum uplift of the covering layers for the protected subsoil takes place about 15 days after the begining of flood discharges.

The same for the unprotected subsoil can get activized even after 2 days.

6 FILTRATION STABILITY

The notion of filtration stability expresses a stability of the environment and civil engineering structures at hydrodynamic load. When not fully ensured, filtration defects can take place. In the conditions of the Danube protective embankments especially a break of a less permeable cover by uplift, piping inside gravely soils, contact piping in a layered environment and hydraulic liquefaction of sandy soils can take place. This paper pays attention to the inner piping and the contact piping only.

At a given hydrodynamic load, sandy particles in pores of gravely soils start moving. If they are washed away, permeability increases, resistivity of the medium decreases and the destabilizing effects of the flowing water get more intensive.

Inner piping can be a dangerous factor for the medium and the civil engineering structures if the three conditions are fulfilled together:
- geometrical (there must be fine smaller particles and bigger pores or fissures where the particles can freely move)
- hydrodynamic (the fine particles must be effected by water flow forces getting them to move)
- spacial (there must be free premises for the moving fine particles to get disposed in, or washed away from further).

Sandy particles of a diameter d_S can be washed away from gravely soils, being available from a formula according to Busch & Luckner (1973)

$$d_s \leq 0.243 \sqrt[6]{d_{60}d_{10}} \, d_{17} \qquad (1)$$

where d_{10}, d_{17} and d_{60} being diameters of the characteristical particles.

All of such particles are not subject to the same geometrical conditions of movability, even though the size of some pores are bigger. For smooth grain size curves washing is not possible, since the fine particles are not free.

In our Danube conditions, there are such gravely soils which do not contain some fractions or such fractions are only of small presence thus making motion of the smaller particles easier. Even in such conditions not all the fine particles in the pores are free. A part of them carries geostatic load and is not movable. Results gotten from the formula (1) thus have to be thoroughly analysed.

In order that the particles with the diameter d_S get to motion, hydrodynamic conditions also have to be fulfilled - the stream force must exceed the resistance preventing the particles from motion. Generally these effects are expresed by critical gradient or water flow values, understanding Darcy´s law gradients being easily able to get converted to velocities. Filtration velocities are often the result of calculations in the stage of design or direct measurements when checking effectiveness of seepage measures.

The critical velocities (v_{crit}) are most often expressed dependent on the permeability coefficient k values. Especially formulas by Sichardt ($v_{crit} = \sqrt{k}/15$) and Ťavoda ($v_{crit} = 0.041\sqrt[3]{k}$) ensuring appropriate stability of sandy particles around driven wells (v_{crit} and k being in ms^{-1}) are known.

According to an adjusted formula by Busch & Luckner (1973) the critical velocity is stated by:

$$v_{crit} = 0.6\left(\frac{\gamma_s}{\gamma_w} - 1\right)\left[0.82 - 1.8n + 0.0062\left(\frac{d_{60}}{d_{10}} - 5\right)\right]$$
$$\sin\left(30^o + \frac{\vartheta}{8}\right)\sqrt{\frac{n \, g \, d_s^2 k}{v}} \qquad (2)$$

(γ_s - specific soil weight, γ_w - specific water weight, n - porosity, ϑ - flow direction, g - gravity acceleration, k - permeability coefficient, v - kinematic viscosity).

Use of the mentioned and other criteria given in detail in the paper of Hulla & Cábel (1977) is possible in any case where grain size curves of the gravely soils is available. The possibilities to get such information in practice are restricted, so a statistical approach has been chosen. Within the area between Medveďov and Čičov, all the grain size curves of the gravely soils got at a new monitoring system foundation were chosen. The basis also contained permeability coefficient calculated according to Carman and Kozeny. The preceeding analysis showed that such values can be trusted.

The analysis of the results in Figure 7 implies that critical velocities increase directly with increasing permeability of the gravely soils. The wider spread of the results gained by different criteria is caused by the fact that apart from the permeability coefficients other qualities are taken into account.

From the given virtually not very clear results the lower bound for the critical velocities (depicted by a full line) can be uniquely stated and an equation derived:

$$v_{crit,min} = 0.032 \ k^{0.5} \qquad (3)$$

It is very probable, that for the filtration velocities being below this line the hydrodynamic conditions for piping shall not be fulfilled and the sandy particles in the gravely skelet shall not get into motion.

The upper bound of the critical velocities is in Figure 7 depicted by a broken line and an equation has been derived:

$$v_{crit,max} = 0.4 \ k^{0.45} \qquad (4)$$

One can expect that for higher values of filtration velocities being above the upper bound of critical velocities the sandy particles shall surely get into motion and an examination of the third - spacial condition is inevitable.

Examination of the spacial condition is also actual when the filtration velocities fall between the bounds of critical velocities calculated from formulas (3) and (4). In such cases the motion of free sandy particles is possible, near the upper bound are very probable.

The third condition of the piping origin and the development is given by the presence of the free premises into which free movable particles can be washed. Such free premises originate e.g. after breaking relatively impermeble cover by uplift. The danger is lower, if the lower critical bound of critical velocities is exceeded in bigger depths. The free particles move in gravely soils horizontally, if they can get together and colmatage they can be a barrier against further motion.

Contact piping takes place at underground and seepage water flow through a layered medium when finel silty or sandy particles can be washed into pores of the coarser gravely soils. Sandy layers near vertical sealing walls make water flow in a vertical direction are also subject to contact piping. Washing sandy layers into pores of gravely layers can imply a deterioration or damage of the sealing walls.

A less dangerous situation can occur at territories without impermeable cover at longer distances from protective embankments where as a result of contact piping ground surface can settle down.

To judge the danger of contact piping occurence the same conditions as for inner piping are valid (geometrical, hydrodynamic, spacial).

For the geometrical conditions adjusted criteria for filters can be put to use. Contact piping between sandy and gravely soils can according to Terzaghi´s and Karaulov´s criteria take place when:

$$\frac{D_{15}}{d_{85}} > 5 \ ; \qquad \frac{D_{60}}{d_{40}} > 10 \ ; \qquad \frac{D_{60}}{D_{10}} > 10 \qquad (5)$$

(the symbol D stands for characteristic diamatres of a gravely soil, the symbol d stands for those of a sandy soil).

Hydrodynamic conditions are best to judge based on critical gradients. For gravely soils critical velocities gined by formulas (2) to (4) can provide a good basis for determination of critical gradients implying from Darcy´s law ($i_{crit} = v_{crit} / k$). For sandy soils the values of critical gradients range between $i_{crit} = 0.15$ to 0.35.

7 THE INFLUENCE OF THE FLOOD IN 1997 UPON THE EMBANKMENT SUBSOIL

Having built the sealing walls between Palkovičovo and Čičov filtration velocities at aproximately average discharges in the Danube (about 2000 $m^3 s^{-1}$) have been measured as a basis for the sealing walls effectiveness judgement. The measurements were done by

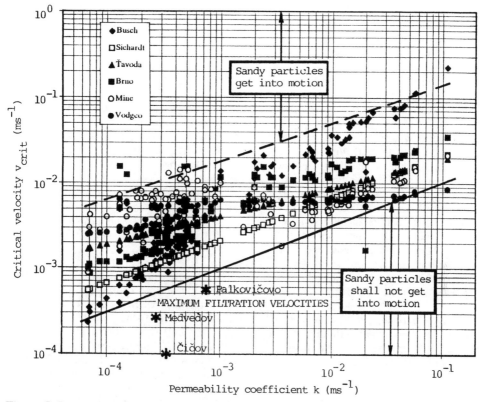

Figure 7: Inner piping critical velocities serve as a function of gravely soil permeability coefficients gained by different criteria and maximum values of filtration velocities during the flood of 1997.

tracer methods based on vertical water flow in borehole monitoring. The results can be characterized by a medium filtration velocity value of $5 \times 10^{-6}\,ms^{-1}$ (Fig. 8).

The effectiveness of the sealing walls has shown to be quite positive, the increase of filtration velocities below the embankments and their near surrounding can be considered negligible.

A check of maximum filtration velocity values taking place in bigger depths thanks to the sealing walls was also important. For the individual locations monitored, maximum filtration velocities were depicted in Figure 7, where they can also be judged from the piping point of view. All the values are below the lower bound of critical velocities, thus sandy particles stability in the gravely soil pores was not threatened.

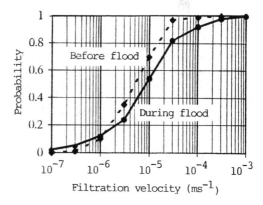

Figure 8: Empirical distribution functions of filtration velocities for average and flood discharges in the Danube.

8 CONCLUSIONS

Geological, hydrogeological and hydrological conditions in Slovak territory have been endangering the Danube protective embankments stability during floods. Embankment breaks during the flood in 1965 were caused by intensive hydrodynamic effects in their subsoil.

The Gabčíkovo waterworks and realised measures on the left side of Danube embankments within the frame construction of the protection measures of the Nagymaros part on the Slovak territory provide a satisfactory protection for both Slovakian and Hungarian territory between Bratislava and Palkovičovo and between Čičov and Hungarian border. Between Bratislava and Palkovičovo we obtained higher flood protection separate of discharge $(10,600 \text{ m}^3\text{s}^{-1})$ to the power channel $(4500 \text{ m}^3\text{s}^{-1})$ and to the old Danube river $(6100 \text{ m}^3\text{s}^{-1})$.

In the section between Palkovičovo and Čičov, where their protective influence does not take place so distinctively, special measures - sealing walls interconected to the water embankment side by a foil - have had to be built.

The sealing walls are in the conditions given an important contribution to embankment subsoil filtration stability ensurement. They decrease the load of impermeable covering layers on the aerial sides of embankments by uplift and keep hydrodynamic effects within acceptable range.

Special attention was paid to checking the criteria for piping in gravely soils. Boundary criteria making possible to distinct main filtration stability or filtration damage stages were stated.

Based on the new knowledge the sealing wall effectiveness under conditions of intensive hydrodynamic load during the flood in 1997 was positively judged.

ACKNOWLEDGEMENTS

The authors of this text thank the heads of the Firm Water Economy Construction, namely the general director Dr. Eng. Július Binder and director Eng. Ján Stanko, for the permanent care to the credibly operation of the Water power system Gabčíkovo, for the effective subvention of the stability problems solution connected by their flood protection.

REFERENCES

Bažant, Z. & V.Hálek 1966. *Regulations for the design of protective measures against the influence of seepage from rivers and reservoirs under construction of water works on the Danube.* Brno: VUT.

Busch, K.F. & L. Luckner 1973. *Geohydraulik.* Leipzig: VEB Deutscher Verlag fur Grundstoffindustrie.

Hulla, J. & J.Cábel 1997. Analysis of criteria for filtration stability. *Inžinierske stavby* 45: 145 - 149.

Jakubec, L. 1967. Engineering geological conditions for the filtration failures at the Czechoslovak part of the Danube during highwater in 1965. In *Highwater in the Danube in 1965.* Bratislava: ČSVTS.

Kunsch, I. & P.Škoda 1995. Highwater in 1965 and their importance between historical highwaters in the Danube. In *The Danube - artery of the Europa:* 9-20. Bratislava: T.R.I. Medium.

Liška, M. 1995 : *Development of the Slovak - Hungarian section of the Danube.* Bratislava: WEC.

Peter, P. & Š.Šebesta 1965. After katastrophy of the Danube embankments. *Inženýrské stavby* 7: 1-8.

Geotechnical Hazards, Marić, Lisac & Szavits-Nossan (eds) © 1998 Taylor & Francis, ISBN 90 5410 957 2

Landslide hazard in the Medvednica submountain area under dynamic conditions

V. Jurak – *Faculty of Mining, Geology and Petroleum Engineering, University of Zagreb, Croatia*

I. Matković – *Civil Engineering Institute of Croatia, Zagreb, Croatia*

Ž. Miklin – *Institute of Geology, Zagreb, Croatia*

D. Cvijanović – *Zagreb, Croatia*

ABSTRACT: In the Zagreb wider area the Medvednica submountain area is considered as a separate geotechnical macro-zone distinguished by a considerable density of unstable slopes and landslides. The mentioned zone is a densely constructed residential area. The respective zone is affected by the known hypocentre in the Medvednica massif in which the earthquakes with a magnitude M > 5,5 may be expected. In the last hundred years there were three strong earthquakes generated in the mentioned massif. The expected response of the slope to the seismic excitation defined by the adopted dynamic parameters can be reflected in two ways: - either as a total failure of the slope previously affected by slow sliding; - or as initial deformations of the slope. Both of these two ways open the access for other factors, meteorological and anthropogenic in particular. The gradation of these results is likely to point out to the necessity of slope instability zoning.

1 INTRODUCTION

The elements of geotechnical hazard in the densely populated submountain zone (Fig. 1) are unstable slopes and landslides on the one hand and strong earthquakes from the registered epicentral area on the other hand. Concerning seismic activity there is one epicentral area which is particularly significant being at a distance of only 16 km from the city centre - area source Kašina. The city area is also affected by the fault zone of the Žumberak - Medvednica - Kalnik fault which can be considered as a line source (Fig. 2).

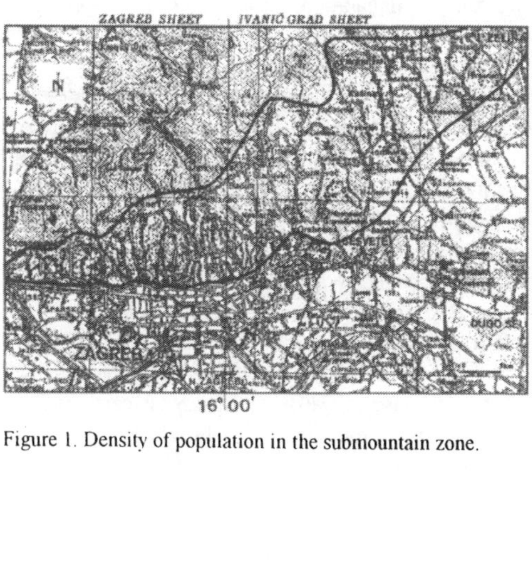

The landslides in the submountain zone were predominantly solved as separate cases although there had been some earlier attempts to carry out systematization and classification (Fijember 1951). Later on, Cesarec & Polak (1986) classified movements on the slopes (according to Skempton & Hutchinson 1969) noting some triggering factors. They pointed at the methods of stability analysis applied to the Zagreb landslides and the monitoring

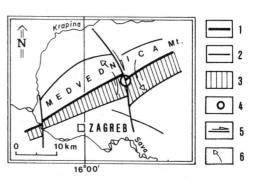

Legend: 1 - Regional fault; 2 - Fault; 3 - Fault zone; 4 - Epicentre of the strongest earthquake; 5 - Fault with designated horizontal movement; 6 - Rotation of tectonic blocks.

Figure 2. Seismotectonic model of Medvednica and the locations of seismic excitation sources (after Cvijanović et al. 1980).

Figure 1. Density of population in the submountain zone.

of some repaired landslides. At that time they studied the landslides under static conditions and their work was meant to be used as a guideline to those who would be engaged in the problem of slope slides in the Zagreb area. In this respect, the regional approach to slope stability under seismic conditions as presented here is related with the mentioned study. Accordingly, in the submountain zone the geotechnical hazard could be defined by the behaviour of the slopes under dynamic conditions although some other incidents may take place in this zone as well.

This work has been developed in the framework of the research projects of the Ministry of Science and Technology: Subsurface-Geologic and Geohazard Explorations in Croatia (in succession of Healthy Living) and the standing project OIGK RH (Basic Engineering-Geological Map of the Republic of Croatia), scale 1:100,000.

2 NATURAL CONDITION OF THE SUBMOUNTAIN ZONE

2.1 Geologic Structure and Relief

The submountain zone extends over the south-east foot of the Medvednica mountain from Podsused to Zelina, parallel with the direction of its massif.

Among the relief elements a submountain valley is being distinguished as well as the nearly perpendicular ridges of the predominant north-south and partly north northwest-south southeast direction with the mountain streams flowing in the same directions. The basic elements of the area relief are represented in Figure 3.

Geologic structure of the submountain zone is distinguished by the neogene deposits overlapping the older rocks of the Medvednica massif. In the area with massive occurrence of unstable slopes and slides three stratigraphic members are prevailing: the youngest members of Miocene - Lower and Upper Pont (M^1_7 and M^2_7) and plioquaternary deposits (Pl,Q) that are covered with loess (lQ_1) (Šikić 1995, Basch 1995). The lower-pont deposits are represented mostly by clayey marls while the upper - pont - Rhomboidea deposits consist of clayey-sandy marls and clays with gradual transition into poorly cemented up to loose silty sands and sandy silts. They form a continuous belt under the plioquaternary sediments having a gentle inclination toward south and south-southeast.

The plioquaternary formations have been sedimented on the eroded base and in general the paleorelief is inclined toward south and south-southeast. They have a heterogeneous composition - predominantly silty clay with lenses of sandy gravels which is a basal member in places. Their origin is believed to be proluvial, proluvial-fluvial, fluvial and fluvial-lacustrine. In general, they become thicker toward south where their maximum depth obtained by boring is 65 m and where they are also described in detail (Polak 1978). It is to be mentioned here that the existence of swelling minerals is rather important for the behaviour of these deposits (Slovenec & Šiftar 1991).

2.2 Seismicity

As to the knowledge so far the city of Zagreb is within the most active zone of the continental part of Croatia (Cvijanović 1983). In the Zagreb epicentral area the earthquakes are the result of the contact between the structures of the Panonian basin and the structures of the "Medvednica - Kalnik range". In Croatia this zone covers the northern hillsides of Bilogora, Kalnik, Ivanščica, Medvednica, Vukomeričke gorice and Žumberačka gora. According to the groups of earthquake hypocentres in Zagreb and its proper area the following localities can be distinguished: Medvednica, Žumberačka gora, Pokuplje, Kalnik, Ivanščica and Zagorje, northern part of Bilogora (Cvijanović et al. 1978).

The submountain zone is characterized by the earthquakes with the epicentre in the Medvednica mountain. These are also the strongest earthquakes occured so far in the Zagreb epicentral area.

Generally speaking, the major number of the

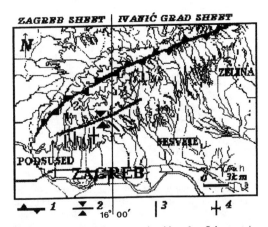

ZAGREB SHEET | IVANIĆ GRAD SHEET

Legend: 1 - Medvednica mountain ridge; 2 - Submountain valley; 3 - Orographic axis; 4 - Location of prevailing slope model.

Figure 3. Basic elements of the Medvednica relief.

earthquakes and the strongest earthquakes refer to the southern hillsides of the Medvednica mountain and some of their hypocentres were in the Podsused - Zelina area. In the seismically most active part of Medvednica, near the villages Kašina and Planina, there was the hypocentre of the strongest earthquake occured on November 9, 1880 (Fig. 2). The maximum intensity in the epicentre and in the proper area was estimated to be IX degree MCS and VIII degree MCS in the remaining part of the city of Zagreb. The magnitude of this earthquake is estimated to be M = 6,0 - 6,5. In the same locality another two strong earthquakes occured: on December 17, 1905 (I_0 = VII - VIII degree MCS, M = 5,6) and on January 2, 1906 (I_0 = VIII degree MCS, M = 6,1). The depth of the hypocentre of the mentioned earthquakes is estimated to be 5 to 10 km. Also, it is very significant that the hypocentre of the strong earthquake on December 17, 1901 was under the city (I_0 = VII degree MCS, M = 4,6) in the Šestine area.

The mentioned quantitative data on seismic activity in the Medvednica mountain could be completed with the following information:

a) along with strong quakes usually also a great number of subsequent weaker aftershocks occur and their cumulative effect may be important as well;

b) since the considered area from Podsused to Zelina is in a fault zone the fractures in the ground caused by displacements along the fault may be also expected.

2.3 Basic Geotechnical Macrozones

The wider area of Zagreb can be clearly divided into three macrozones with specific properties of the

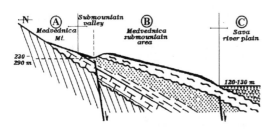

Legend: A - Well petrified neogene and older rocks; B - Slightly petrified and poorly cemented neogene rocks (soils), plioquaternary clays and loess; C - Holocene proluvial and alluvial fans.

Figure 4. Schematic cross-section of the basic geotechnical macrozones in the Zagreb area.

ground condition. Each of these zones is distinguished by the respective geologic structure, relief, geodynamic site conditions and hydrogeologic properties and accordingly they can be referred to as the basic geotechnical macrozones (Jurak & Mihalić 1995). The mentioned zones are completely separated one from the other by faults (Fig. 4).

In the hydrologic zoning of the Medvednica catchment area the submountain zone i.e. the geotechnical macrozone B is also being distinguished as a separate morphohydrographic belt (Rupčić & Žugaj 1982).

3 DISTRIBUTION OF LANDSLIDES

There is a great number of data on the landslides in the area of Zagreb. The data are collected from the Basic Engineering-Geological Map of the Republic of Croatia, scale 1:100,000, "Zagreb" Sheet and "Ivanić-grad" Sheet (Miklin et al. in press). The landslides are represented by dots (area smaller than 1 ha) and by polygons (area bigger than 1 ha). The biggest landslides reach up to 1 sq. km (Fig. 5).

The greatest number of landslides is in the area of miocene (M_7^{1-2}) and plioquaternary deposits (Pl,Q) that form a narrow submountain zone composed of marl and clay. In the "Zagreb" Sheet the plioquaternary deposits prevail and landslides are not so numerous as in the "Ivanić Grad" Sheet where the miocene deposits prevail and the number of landslides is twice as much, however, in total they cover a smaller area (Table 1).

The prevailing model of most slopes of the submountain zone is represented in Figure 6. Characteristic is the hypsometric diferentiation according to the type of instability which also requires a selection of the methods of dynamic stability analysis - one method for the circular slip surface and the other for the infinite slope.

4 CONDITION OF SLOPES AND SEISMIC PARAMETERS

4.1 Condition of Slopes prior to Earthquake

Considering the classification of the causal factors as preparatory factors and triggering factors it can be said that the origin of recent landslides is affected by ground conditions, geomorphological processes (permanent erosion of the slope toe), physical processes with seasonal characteristics (prolonged high precipitation) and particularly significant permanent and cumulative man-made processes. All

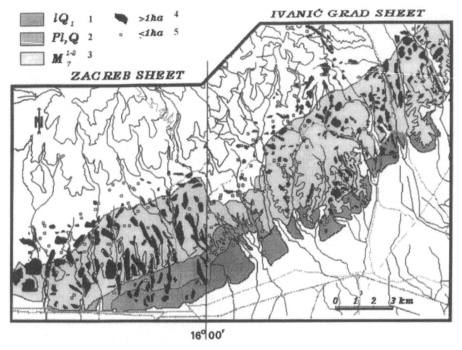

Legend: 1 - Loess; 2 - Plioquaternary formations; 3 - Lower and upper-pont deposits; 4 - Landslide area exceeding 1 ha; 5 - Landslide area smaller than 1 ha.

Figure 5. Distribution of landslides and unstable slopes in the submountain zone.

above mentioned factors are preparatory factors.

As triggering factors two incidents can cause massive occurence of landslides. These are extreme hydrologic occurences - torrential floods from Medvednica caused by the intensive, short - period rainfalls with 1000 year return period and the quakes with the direct and indirect effect which are considered in this paper. Both mentioned incidents enter from the group of physical processes (Popescu 1994). For the illustration, in the summer 1989 severe torrents were registered having all the characteristics of natural disaster (Gajić-Čapka 1990).

The condition of most slopes in the moment of seismic excitation is represented in Figure 7 according to the UNESCO classification - Working Party on World Landslide Inventory (1993) is respected.

4.2 Seismic Coefficient (k_S)

The main parameter for dynamic computation of landslides at seismic excitation is seismic coefficient k_s which is in correlation with the coefficient of seismic intensity K_S. According to the Regulations on Technical Standards for Design and Computation of

Table 1. Landslide distribution in the "Zagreb" Sheet and "Ivanić Grad" Sheet, Basic Engineering-Geological Map of the Republic of Croatia - the state completed with the year 1991.

Medvednica submountain area	Surface area (sq. km)	Landslide endangered areas (sq. km)	Active landslides area (sq. km)	Number of landslides represented by dots	Number of landslides represented as polygon	Landslide Total
"Zagreb" Sheet	127.63	32.77	8.07	49	98	147
"Ivanić Grad" Sheet	202.72	78.28	11.84	81	178	259
Total	330.35	111.05	19.91	120	276	406

830

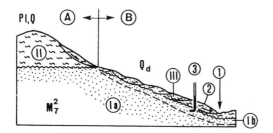

Figure 6. Prevailing model of most slopes in the submountain zone (after Jurak et al. 1996).

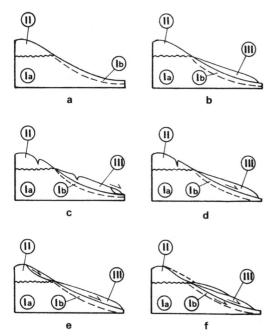

Figure 7. Condition of slopes in the moment of seismic excitation based on the prevailing model (identical designations of geotechnical units as in Figure 6).

Engineering Structures in Seismic Areas (1984) the latter is related to the degree of design seismicity. For the illustration, for VII degree MCS the corresponding factor is $K_S = 0.025$; for VIII degree $K_S = 0.050$ and for IX degree $K_S = 0.10$.

According to the results of engineering-seismological investigations made so far in the Zagreb area, the most probable values of the coefficients for the respective return periods in the area from Podsused to Zelina are represented in Table 2. The values for the 200 and 1000 year return period have been read from the Zagreb seismic microzoning maps (Geotehnika - Geoexpert 1988) and for the 500 year return period adopted from the paper of Marić et al. (1995).

Table 2. Main parmeters for the computation of slope stability under dynamic conditions in the Podsused - Zelina area.

Return period (year)	I_{max} (MCS)	a_{max} (g)	K_s
200	7.5-8.0	0.14-0.24	0.035-0.05
500	8.6	0.21	0.08
1000	8.0-9.0	0.20-0.40	0.05-0.10

It should be mentioned here that the values of the expected maximum acceleration a_{max} (g) refer to the bedrocks while the values of the expected maximum intensity I_{max} (MCS) already comprise the main local site conditions.

Since the parameter of seismic coefficient k_s has entered the dynamic stability computations, the first approximation at this regional approach to dynamic stability of slopes was carried out very carefully by adopting equal values for both coefficients (coefficient of seismic intensity K_S and seismic coefficient k_S).

5 COMPUTATION OF SLOPE STABILITY

The computation of stability was made for the slopes of submountain zone (a ...f) taking into account the prevailing model (Figs 6, 7). A formation of slide body was anticipated in geotechnical unit II - silty clay (predominantly clay of low plasticity - CL after USCS) and, respectively, in unit III - overburden composed of clay-silt-sand-gravel mixture (CL/S/G after USCS) and in unit Ib - weathered zone composed of silty sand (SM after USCS).

For geotechnical unit II the critical circular slip surfaces have been determined by using the SLOPE/W/Ver. 3 computer programme in which the method of limit equilibrium is used to compute safety factor. For units III and Ib the safety factors have been defined according to the computational model of an infinite slope with the site surface parallel to the slip surface and with the given inclination. In both cases dynamic excitation is computed as pseudo-static load i.e. the effect of inertial horizontal force $H = k_S \times slice\ weight$ has been respected.

The selected computational model is represented in Figure 8. The characteristic values of the input parameters are given in Table 3. The values have been adopted according to the data obtained from the documentation on investigation works performed so far. The parameters c_R and φ_R refer to the assumed residual values of cohesion and angle of internal friction.

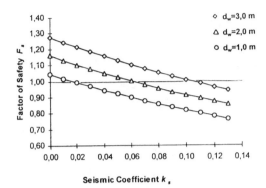

Figure 9. Influence of earthquake and groundwater level to slope stability.

Table 3. The parameters for the computation of slope dynamic stability.

Geotechnic. unit	c Cohesion (kN/m²)	φ Angle of internal friction (°)	γ Unit weight of soil (kN/m³)	z Depth up to slip surface (m)	d_w Depth up to ground water (m)	β Slope inclination (°)	k_s Seismic coeff.	Remark
Ib	15	27	19.0	7.0	Water on surface	25	0.035-0.10	Weathered zone
II	20 c_R=0	20 φ_R=20	20.0	5.0	Water on surface	15-30	0.035-0.10	Overlying formation
III	10 c_R=0	20 φ_R=20	18.0	5.0	Water under pressure	10-20	0.035-0.10	Resedimented material

For circular slip surfaces also a possibility of the occurrence of tension cracks has been taken into consideration. As extreme hydrogeological conditions the following has been adopted: groundwater level on the ground surface for geotechnical unit II and water under artesian pressure for geotechnical unit III. The results of stability computation are represented in Table 4 where the safety factors for the circular slip surfaces have been obtained by Bishop method. The change of safety factor in relation to the groundwater level for the case of infinite slope model is represented in Figure 9.

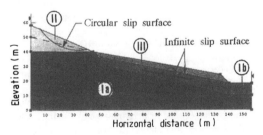

Figure 8. Computational model

6 CONCLUDING REMARKS

The results exhibit sensitivity and changeability of the condition of the existing slopes concerning seismic excitation as well as abrupt changes of pore pressure due to unfavourable hydrologic conditions. In other words, the present condition of slopes (a...f, Fig. 7) can be changed to a higher degree of instability up to the slope failure. By the probability gradation of particular results a basis for the representation of the slide hazard caused by earthquake would be obtained.

Since the microzoning of the Zagreb area so far have not respected the dynamic stability of slopes (Bubnov et al. 1971a, b, Geotehnika - Geoexpert 1988) we recommend a range of activities as follows: - to use the documentation; - to review the existing unstable slopes and landslides and to register the new ones as well as to elaborate their inventory according to the appropriate methodology (Matković et al. 1995); - to make statistical analysis of the particular elements of landslides and unstable slopes; - to reduce the types of landslides and unstable slopes to a few models; - to represent the distribution of

Table 4. Computation results for slope dynamic stability.

Slope condition	Slip surface		c (kN/m²)	φ (°)	γ (kN/m³)	z (m)	d_W (m)	β (°)	k_S	F_S
	Geotech. unit	Slide model								
a	II	Circular slip surface	20.0	20	20.0	5.0	0.0	22	0.035	1.120
									0.100	0.960
	Ib	Infinite slip surface	15.0	27	19.0	5.0	2.0	25	0.035	1.062
									0.100	0.913
b	II	Circular slip surface	20.0	20	20.0	5.0	0.0	22	0.035	1.120
									0.10	0.960
	III	Infinite slip surface	8.0	20	18.0	5.0	1.0	12	0.075	1.006
									0.10	0.919
c	II	Circular slip surface with tension crack	20.0	20	20.0	5.0	In slide body gravity centre	22	0.035	1.326
									0.10	1.137
	III	Infinite slip surface	10.0	20	18.0	5.0	Dry Under art.press.	20	0.0	1.346
									0.10	0.504
d	II	Circular slip surface with tension crack	20.0	20	20.0	5.0	In slide body gravity centre	22	0.035	1.326
									0.10	1.137
	III	Infinite slip surface	0.0	20	18.0	5.0	Dry	20	0.035	0.901
e	II	Circular slip surface	0.0	20	20.0	5.0	Dry	22	0.035	1.001
								In slide body grav.cen.	22	0.035	0.898
	III	Infinite slip surface	0.0	20	18.0	5.0	Dry	20	0.035	0.901
f	II	Collapse - displaced mass	-	-	-	-	-	-	-	-
	Ib	Infinite slip surface	15.0	27	19.0	7.0	2.0	25	0.035	0.892
									0.100	0.762

models by a "Map of Landslide Models"; - to compute the stability of models by the simulation of a strong earthquake from the known epicentre during unfavourable hydrologic conditions; - to elaborate a map of the condition of slopes after the earthquakes which would represent a "Potential Map of Seismically Induced Landslides" in scale 1 : 10,000 or 1 : 5000. The suggested procedure follows to a certain extent the Deterministic landslide hazard analysis as presented in the paper by van Westen et al. (1997).

While considering the landslide - seismic excitation relationship one should also note very complex phenomena during propagation of seismic waves from the hypocentre to the landslides under consideration as well as the direction of orographic axes against the epicentral area. There is, however, a question how to quantify the insufficiently investigated effects of earthquake, particularly the influence of topography, which so far have been recorded and described in some countries (Ishihara 1985, Bard 1995). The position of landslides on the slope is also likely to affect the selection of seismic coefficient (k_S).

The above recommended extensive work could be a firm basis for the evaluation of slope instability hazard in the Medvednica submountain area.

REFERENCES

Bard, P.-Y. 1995. Effects of surface geology on ground motion: Recent results and remaining issues. *Proc.of 10th Europ. Conf. on Earthquake Engin., Vienna, 28 August - 2 Sept. 1994*: 1: 305-323 Rotterdam: Balkema.

Bash, O. 1995. Medvednica Geological Map (In Croatian). In K. Šikić (ed), *Medvednica Geological Guide*: Zagreb.

Bubnov, S., D. Cvijanović, V. Jurak, A. Magdalenić, D. Skoko & D. Vukovojac 1971a. Seismic zoning of Zagreb. Earthquake Enginering. *Proc. of the third Europen Symp. on Earthquake Engin., Sofia, 14-17 Sept. 1970*: 95-102. B.A.S.

Bubnov, S., D. Cvijanović, V. Jurak, A. Magdalenić & D. Skoko 1971b. Preliminary map of the sesmic microzoning of the town of Zagreb (In Croatian). *Proc. of 1st Yugosl. Symp. on Hydrogeol. and Engin. Geology. Herceg Novi, 4-8 May 1971*: 2: 67-73. YCHEG, Belgrade.

Cesarec, M. & K. Polak 1986. Landslides in the Zagreb region (In Croatian). *Proc. of XVI Symp. YSSMFE, Aranđelovac, 5-8 November 1986*: 2: 169-184. YSSMFE.

Cvijanović, D. 1983. Seismicity of the territory of the Socialistic Republic of Croatia (In Croatian). *Integral geotechnical investigations of urban units for the purposes of geotechnical and seismic microzoning. Zadar, 12-14 May 1983*: 1: 13- 40.

Cvijanović, D., E. Prelogović & D. Skoko 1978. Seismic risk in the Zagreb area (In Croatian). *Građevinar*, 30, 2: 33-40. Zagreb.

Cvijanović, D., E. Prelogović, D. Skoko, K. Marić & D. Mišković 1980. Seizmotectonic zoning of Medvednica (In Croatian). *Proc. of 6th Yugsl. Symp. on Hydrogeol. and Engin. Geology, Portorož, 12-16 May 1980*: 2: 13-25. YCHEG, Belgrade.

Fijember, M. 1951. Experiences with remedy of hillside slide of the Zagreb terrace. *Građevinar*, 3, 11-12: 17-35. Zagreb.

Gajić-Čapka, M. 1990. Characteristics of the short-period precipitation during floods in the Zagreb wider area, summer 1989 (In Croatian). *Extraordinary meteorological and hydrological events in the Socialistic Republic of Croatia in 1989*. M6-13: 30-35. Republic Meterological Department of the Socialistic Republic of Croatia, Zagreb.

Geotehnika - Geoexpert 1988. (unpubl.) Seismic microzoning of the town of Zagreb (14 municipal areas) (In Croatian). Professional documents, Zagreb Public Record Office.

Ishihara, K. 1985. Stability of natural deposits during eartquakes. *Proc. of the eleventh intern. conf. on soil mechanics and foundation eningeering, San Francisco, 12-16 August 1985*: (editor Publications Committee of XI ICSMFE), 1: 321-376. Rotterdam/Boston: Balkema.

Jurak, V. & S. Mihalić 1995. Basic geotechnical zoning of the Zagreb region (In Croatian). *Proc. of Second Conf. of the CSSMFE, Varaždin, 4-6 October 1995: Geotechnical Engineering in Cities*, 1: 429-439.

Jurak, V., I. Matković, Ž. Miklin & S. Mihalić 1996. Data analysis of the landslides in the Republic of Croatia: Present state and perspectives. *Proc. of the seventh International Symposium on Landslides, Trondheim, 17-21 June 1996*. 3: 1923-1928, Rotterdam: Balkema.

Marić, B., D. Dujmić & V. Jurak 1995. Determination of the reconstruction condition of some parts of Medvedgrad castle (In Croatian). *Proc. of Second Conf. of the CSSMFE, Varaždin, 4-6 October 1995: Geotechnical Engineering in Cities*, 1: 351-358.

Matković, I., V. Jurak & Ž. Miklin 1995. Inventories of unstable slopes and landslides in the Zagreb region (In Croatian). *Proc. of Second Conf. of the CSSMFE, Varaždin, 4-6 October 1995: Geotechnical Engineering in Cities*, 1: 367-376.

Miklin, Ž. et al (in press) OIGK Sheet "Zagreb" and Sheet "Ivanić Grad" Institute of Geology Zagreb.

1984. Draft regulation on technical standards for design and computation of egineering structures in seismic areas (In Croatian). *Građevinar*, 36, 7: 295-314. Zagreb.

Polak, K. 1978. Einige Merkmale der quartärer Sedimente die auf dem Beispiel des Rutschgeländes Jelenovac bei Zagreb untersucht wurden (In Croatian). *Geološki vjesnik*, 30/1: 151-165. Zagreb.

Popescu, M. E. 1994. A suggested method for reporting landslide causes. *Bull. of IAEG*, 50: 71-74. Paris.

Rupčić, J. & R. Žugaj 1982. Regulation of the streams Medvednica and Vukomeričke Gorice (In Croatian). *Građevinar*, 34, 3: 91- 100. Zagreb.

Skempton, A. W. & J. N. Hutchinson 1969. Stability of natural slopes and embankment foundations. *State of the art volume. Proc. of 7 th ICSMFE, Mexico 1969*: 7: 291-340.

Slovenec, D. & D. Šiftar 1991. Vermiculite and smectite in clastic sediments of the southern slopes of the Mt. Medvednica. *Geološki vjesnik*, 44: 121-127. Zagreb.

Šikić, K. 1995. Structural relations and tectogenesis of the Medvednica wider area (In Croatian). In K. Šikić (ed), *Medvednica Geological Guide*: 31-40. Zagreb.

Westen, C.J.van, N. Rengers, M.T.J. Terlien & R. Soeters 1997. Prediction of the occurrence of slope instability phenomena through GIS-based hazard zonation. *Geol.Rundsch*, 86: 404-414. Springer - Verlag.

WP/WLI 1993. A suggested method for describing the activity of a landslide. *Bull. of IAEG*, 47: 53-57. Paris.

Geotechnical Hazards, Marić, Lisac & Szavits-Nossan (eds) © 1998 Taylor & Francis, ISBN 90 5410 957 2

Experiences on stabilisation of landslide in South Black Forest

H.-G. Kempfert & D. Zaeske
Institute of Geotechnique, University of Kassel, Germany

M. Stadel
Kempfert + Partner GmbH, Kassel, Germany

ABSTRACT: This paper presents the experiences on a redevelopment of a landslide in South Black Forest. A big part of a hillside was sliding with velocities up to 90cm/year downstream and endangered a building located above the creep zone. To reduce the movement of the soil masses, a doweling of the hillside was executed. The redevelopment consists of a serie of piles, that are embedded in the stable underground and anchored backwards at the pile caps. Due to the support at both ends of the pile, the load distribution is advantageous and enables an economic design of the piles. The analytical formulations used to determine the forces in the dowels will be reflected and the success of the redevelopment will be verified by results of measurements and calculations with the FEM.

1 INTRODUCTION

In Spring 1991 a landslide in South Black Forest was realized. The area affected by the landslide is located uphill the country road No.132 between the villages Badenweiler and Sehringen. The moving soil mass on the hillside covers an area about 80 m in North-South and 50 m in West-East direction and creeps in dependence of the intensity of precipitation with creep velocities between 10 and 90 cm/year. The resulting movement of the hillside became so large, that a redevelopment was necessary, in particular, a building on the hillside was in danger to suffer damages.

The concept of the redevelopment provides constructional reinforcements to reduce the creep movement of the bed load on an acceptable degree.

This report describes the landslide that appeared before the redevelopment was executed. The effectiveness of the redevelopment will be analyzed with numerical calculations with the FEM and compared with present measurements.

2 LANDSLIDE

The concerned area is situated between the country road 132 in the west at a height of 497 m NSL and a hospital in the east at 522 NSL. The hillside has an inclination of 13°, the situation is illustrated in Figure 1. Simplifying, the formation has a structure of two layers: a loose soil layer, which forms the creeping soil, and clay stones, which forms the deeper stable underground.

The thickness of the loose soil layer is between 9 and 16 m and is made of clayey soil in the upper 4 to 5 m, which has less coarse grain particles with mainly stiff, partially plastic or semi-solid consistency. The deeper section of the loose soil layer down to the claystone consists of loam and/or debris with fine fractions of clay, sand-clay or sand-silt mixtures. The loam and/or debris is partial soaked and softened due to water supply. The stable underground is formed by clay stone of the opalinus-clay (Braunjura alpha). The stratigraphical sequence is a serie of uniform dark-grey foliated clay stones, which could not be divided by petrography. The clay stone in the upper 3 m is softened to greybrown clay in semi-solid and solid consistency. Several slickensides were determined in the clay stone, which indicate significant tectonic pretensions by the mountain.

The hillwater was found in different levels. It flows above the watertight layer of claystones in the creeping bed load to downstream.

The movement of the bed load is recorded by inclinometer - measurements since October 1991. The measurements indicate that the creeping bed load demarcates itself legible from the stable underground.

The movement of the soil mass takes place on a thin sliding plane, that is located more or less parallel to the surface of the hillside. The depth of the sliding plane for a characteristic profile of the hillside is illustrated in Figure 2.

The boundary line of the creeping soil upstream is

observed by detected gaps in the soil close underneath the area of the hospital. The sliding plane is located at a depth of 11 to 12 m, partially up to 15 m and has an almost constant inclination of 10° to 13°. The area affected by the landslide extends downstream until 200 or 300 m behind the country road. Before the redevelopment was made in Dec. 1995, the recorded creep velocities of the moving soil mass were between $v_o = 10$ and 90 cm/year.

The reason for the increasing and decreasing rates of creep velocities is the seasonal changing of the amount of precipitation, which influences the hill water conditions and the groundwater flux, however this effect can not be quantified exactly. The total lateral displacement of the road has reached a value of about 1,0 m since the measurements started in Oct. 1991. The consequences of the displacements in the area of the road are shown in Figure 3.

3 REDEVELOPMENT MEASURE

3.1 Concept of the measure

The concept of the redevelopment was to stabilize the creeping soil masses by security measures with the destination to reduce the forward movement as far as no danger for the hospital appears in foreseeable future.

The redevelopment consists of 15 tied-back piles with a diameter of 120 cm at spacing of 4 m, each pile has a length of 20 m and is embedded in the stable underground, they are placed about 20 m from the country road in the hillside. The pile caps are tied backwards with two injection anchors, which have an inclination of 25° from horizontal. In ground level, all caps are connected with a girder, that has a total length of 60 m.

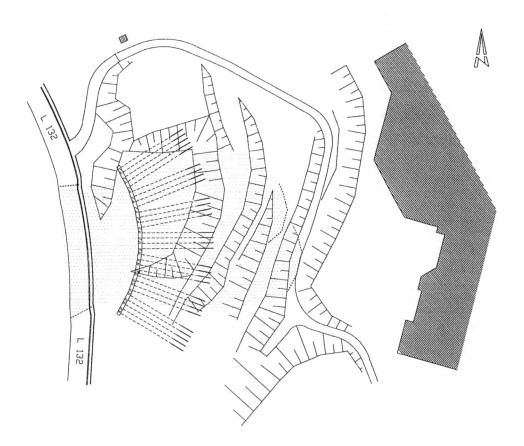

Figure 1. Plan of the redevelopment

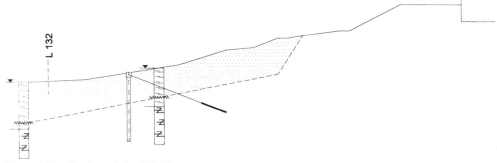

Figure 2. Profile through the hillside

The girder is used for the absorption of the anchoring forces and for compensation of the different bearing capacities of the piles. Due to a better bearing behaviour of the construction and a more suitable integration in the existing nature, the piles were disposed in arched configuration. The anchoring of the pile caps was necessary, because the sliding plane is located in a relative deep horizon, the stressing of only one end restrained piles would become too large for the chosen pile diameter. In the hillside, a drainage-system was also installed to reduce the hydraulic forces caused by the hill-water.

3.2 Design of the redevelopment

To estimate the forces absorbed by the structural elements of the redevelopment, two different analytical formulations were considered. The first assumption is based on the state of equilibrium between acting and resisting forces in the sliding plane. The stability against sliding of the bed load is defined by the ratio of the resisting forces and the acting shearing forces along the slide direction.

$$\eta = \frac{G' \cdot \cos \beta \cdot \tan \varphi + H}{G' \cdot \sin \beta + S} \qquad (1)$$

where G' is the effective gravity load of the soil, β the inclination of the sliding surface, φ the angle of shear friction in the sliding plane, H the tangential resisting forces and S the hydraulic thrust of the hill-water per metre.

Originating from the assumption that the landslide occured before the redevelopment has a safety factor of 1,0 and from the known location of the sliding surface, the decisive angle of shear friction in the sliding plane was determined. Water level located 5 to 7 m above the sliding plane was also taken into account. Using the established angle of shear friction, the necessary resisting force to increase the safety factor from 1,0 to 1,15 was calculated.

The second formulation avoids the difficulties in determination of absolute values for the calculation like the effective angle of shear friction in the sliding plane or the hill-water conditions to simulate the equilibrium state correctly. The required resistance force along the sliding surface to reduce the creep movement can be obtained from a logarithm toughness law for soil (Kreuter&Lippomann, 1993)

$$H = l \cdot \tau_0 \cdot I_{v\alpha} \cdot \ln\left(\frac{v_0}{v_1}\right) \qquad (2)$$

where l is the length of the creeping soil body, τ_o the average shear stress along the slide plane, $I_{v\alpha}$ the viscosity index of the soil, v_o the initial creep velocity and v_1 the aimed creep velocity.

The average shear stress in the sliding plane can be determined by the relation

$$\tau_0 = \gamma_r \cdot h \cdot \sin \beta \cdot \cos \beta \qquad (3)$$

with γ_r = weight of the saturated soil of the bed load and h = depth of the sliding plane.

Assuming, that in future no unforeseen and unfavourable conditions will occur, which will lead to greater creep velocities than 90 cm/year, the target of the redevelopment was to reduce the creep velocity to a value smaller than 0,5 cm/year.

Both formulations lead almost to the same value for the necessary resisting force, that is about 350 kN/m. For the design of the construction, the calculated resisting force was converted to a line load acting on the piles above the sliding surface. In account of the anchored pile caps, the load was assumed to be distributed uniformly along the pile. The part of the piles beneath the sliding plane was idealized as a subgrade reaction with constant modulus of subgrade in horizontal direction. The acting forces in the structural elements of the redevelopment, determined at this system, enabled the design of the pile, the girder and the anchors.

Figure 3. Situation at the road before the redevelopment

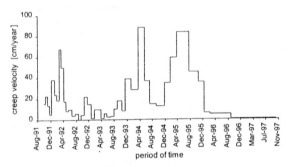

Figure 4. Measured velocities of the creeping soil mass

Figure 5. Redevelopment measure today

4 NUMERICAL EXAMINATIONS

The effectiveness of the redevelopment will be compared with the results of numerical calculations with the FEM. Only a small section of the hillside has been considered in the calculations. Using the properties of symmetry, a section of the construction was extracted and translated in a three-dimensional model for the FEM, see Figure 6.

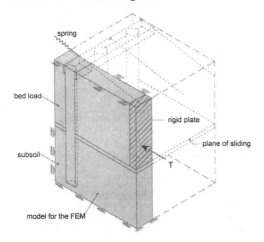

Figure 6. Model for the FEM

The tied-back piles are simulated by only half a pile in the model, whereas the connection between the pile and the girder is assumed to be fixed. The anchor is simulated by a spring with an external load in the size of half of the value of the prestressing. Boundary conditions are defined with free displacement in direction of the creep movement for the bed load and fixed for the stable underground. Between the bed load and the subsoil a thin layer is included to simulate the sliding surface. The FEM-model is build up with second-order hexahedral solid elements. To determine sliding of the soil along the surfaces of the pile and the girder special contact elements are implemented, they allow relative displacements and cracking between soil and pile. The material behaviour for the bed load and the subsoil is assumed to be elastic, hence, the elements in the plane of sliding have viscous behaviour by decreasing shear modulus with time.

The pushing force from the hillside is simulated by a horizontal load acting on a rigid plate at the end of the hill-section. This plate ensures a continuos movement of the bed load, the same as observed by inclinometer measurements. The total shearing force along the sliding surface is calculated on the base of equation (3) and yields the total load T, acting on the rigid plate (Figure 6). To obtain correct values for the time depended shear relaxation modulus, in a

first calculation the pile and the girder were removed by replacing the material parameters of the concerned elements with the values of the surrounding soil. The viscous damping coefficients were calibrated, in such a way, that a creep velocity of 90 cm/year was achieved, which corresponds to the maximum value before the redevelopment. The material parameters are shown in table 1

Table 1. Material parameters for the FEM

section	Young's modulus	Poisson's ratio
bed load	5500 kN/m²	0.35
subsoil	55000 kN/m²	0.30
pile / rig. plate	$3 \cdot 10^6$ kN/m²	0.00

The final Model including the redevelopment was analyzed in two steps. First, the external load is applied instantaneously in a simple static step. In a second calculation step the load was held constant and the shear relaxation effect was applied in 20 time increments, where each increment represents a duration of five days, so a total time of t = 100 days was inspected. Figure 7 shows the deformed mesh after the second calculation step at t = 100 days.

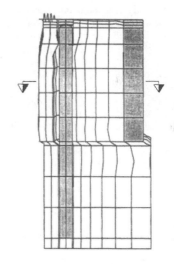

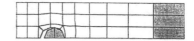

Figure 7. Deformed mesh

839

The displacement of the rigid plate, which represents the movement of the bed load, is illustrated in Figure 8. After the redevelopment was completed, the slope failure decreases due to retardation, which is a time-depend process. To judge the effectiveness of the redevelopment, Figure 8 also shows the creeping displacement in the initial state correspondending to $v_o = 90/365 = 0,247$ cm/day.

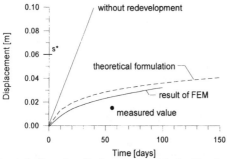

Figure 8. Creeping displacements of the bed load

For the calculation with the FEM, the material behaviour for the bed load and the subsoil is assumed to be linear elastic without any limit of admissible stresses. Factual the stresses increases with strains only until a defined yielding pressure is reached (Winter, 1980). The yielding pressure is given by several authors in the range of $2,5...7,5 \cdot c_u \cdot d$, where c_u is the undrained shear-strength and d the diameter of the pile. The point of time when the yielding pressure is attained depends on the relative displacement between the pile and the surrounding soil. The magnitude of the relative displacement, s^*, necessary to mobilize the yielding pressure can calculated by the equation Winter's (1980):

$$s^* = \varepsilon_u \cdot d \tag{4}$$

ε_u is the limit strain of the soil and is supposed with $\varepsilon_u = 5\%$ for the bed load. The total time t^* until the yielding pressure occurs, can be calculated by (Schwarz, 1987) from

$$t^* = \frac{s^*}{v_o \cdot \ln(v_o / v_1)} \cdot \left(\frac{v_o}{v_1} - 1\right) \tag{5}$$

In this case, after $t^* = 839$ days, the pressure on the pile does not increase further and a constant creeping velocity of $v_1 = 0,5$ cm/year is present. If the pile, connected resistant to bending with the girder, is assumed to be nearly undisplaceable, which agrees with the results of the FEM, the motion of the bed load, shown in Figure 8 displays also the relative displacement between pile and soil. In account of the regarded duration of 100 days for the numerical calculation, the assumption of elastic behaviour for

the bed load is admissible. The increment of displacement before the yielding pressure is reached, can be derived as a function of time t by

$$s = \frac{s^*}{\ln(v_o / v_1)} \cdot \ln\left[1 + \frac{v_o \cdot \ln(v_o / v_1)}{s^*} \cdot t\right] \tag{6}$$

The results of the theoretical formulation are shown in Figure 7.

Both the results of the FEM and equation (6) leads to larger displacements, than indicated by the inclinomter-measurements, where 53 days after the redevelopment was done a movement of only 1,5 cm was detected. Next inclinometer results are presented for this profile after 255 and 487 days and show very small additional displacements less than 0,3 cm. This is tendentiously conform to anticipation, because the supposed value of the initial creep velocity of 90 cm/year is an upper bound of the measured velocities in field (see Figure 4); the factual mean annual is smaller. Due to the pessimistic assumption for the initial conditions, the calculation leads to more untimely results. The seasonal variations of the creep velocity in the initial state are primary caused by the hill-water conditions. The executed redevelopment includes a drainage system, that reduces the streaming potential, thus the shearing force decreases. This effect does not enter into the theoretical calculations of FEM or formulation (6).

CONCLUSIONS

The redevelopment reduces the sliding of the hillside significantly. The measurements after the redevelopment indicate very small movements of the bed load, whereas analytical methods forecast the success of the measure slight untimely. Since the redevelopment was executed two years ago (Dec. 1995), the creep velocity reduces continuously over a time interval of about 839 days and approaches a limit velocity of about 0,5 cm/year to be attained in spring 1998.

REFERENCES

Kreuter, H. & Lippomann, R. 1993. Planung und Überwachung von Verdübelungsmaßnahmen zur Sanierung kriechender Hänge. Tiefbau, Ingenieurbau, Straßenbau 7: 491-493.

Winter, H. 1980. Bemessung von Pfahlgründungen und Hangverdübelungen auf Fließdruck. Vorträge Baugrundtagung Mainz 1980: 563-593

Schwarz, W. 1987. Verdübelung toniger Böden. Veröffentlichung des Institutes für Bodenmechanik und Felsmechanik der Universität Fridericiana in Karlsruhe. Heft 105

Geotechnical Hazards, Marić, Lisac & Szavits-Nossan (eds) © 1998 Taylor & Francis, ISBN 90 5410 957 2

Major landslide triggered by local instability

Rolf Larsson & Elvin Ottosson
Swedish Geotechnical Institute, Linköping, Sweden

Göran Sällfors
Chalmers University of Technology, Gothenburg, Sweden

ABSTRACT: A major landslide occurred at Agnesberg in the Göta River valley in Sweden in 1993. The consequences of the slide were extensive and the risks involved if the slide had spread were even greater. The slide occured in a river valley with deep deposits of soft, normally consolidated, sensitive clays where landslides are recurrent and where adapted methods of investigations and undrained total stress stability analyses had been developed. Previous investigations in the slide area using these methods had yielded satisfactory safety factors. However, erosion from the water flow and traffic in the river had created locally very steep underwater slopes which were not stable in a long term effective stress perspective. A local slide in such a slope with highly sensitive clay resulted in a series of retrogressive slides. An unobserved local erosion and a method of assessment of the stability conditions which did not take such a chain of events into account thereby resulted in a major slide with a hazard to human safety and very great economic values.

1 INTRODUCTION

The landslide at Agnesberg occurred on April 14, 1993 and comprised an area 80 by 30 m in an industrial area along the Göta River. The back-scarp of the slide almost reached a building in the area, which however was undamaged. The slide masses partly filled up the shipping channel in the river, where the depth was reduced by about 2 m.

The community of Agnesberg is located about 10 km upstream of the centre of Gothenburg, the second largest city in Sweden, and about 2.5 km upstream of the fresh water intake supplying the city. Behind the industrial area, the main railway line through Western Sweden and a major highway run parallel to the river in the valley. The residential areas in the community are located in a side valley behind the traffic lines, Fig. 1.

The topography in the area is characterised by a ground surface sloping slightly towards the river. At the river bank, there is a low wall for erosion protection and a water depth of about 1 m. The water depth increases slowly towards the centre of the river and then increases rapidly at the edge of the shipping channel, which is at least 8 m deep. The clay layers in the slide area are about 35 m thick and overlie deposits of coarser soil. The clay layers thin out towards the valley sides, where the ground rises steeply. Because of the topography, there are high artesian water pressures in the coarse bottom layers, with a water head 6 - 8 m above the water level in the river.

The slide activity along the Göta River valley is high, both in the main valley and the side valleys. Most of the

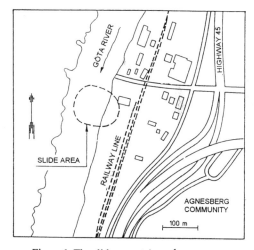

Figure 1. The slide area at Agnesberg.

slides occur in rural areas, where no particular precautions have been taken, but major slides in developed areas have occurred, e.g. in Surte in 1950 and Göta in 1957. The large landslide in Tuve in 1977 occurred in a parallel valley in the same region. The experience gathered shows that a large number of local slides spread progressively and/or retrogressively to finally comprise all the soft clay in the local area within limits set by surrounding firmer ground. The major slides in the 50s led to an extensive investigation into the stability conditions in the valley and a method of stability analysis based on undrained shear strength, primarily determined by field vane shear tests, and also on the safety factors required, (SOU 1962:48). This method has been used for all stability evaluations in the area until recently. The building at the back-scarp of the slide in Agnesberg was constructed as late as 1990 and the stability of the area was then considered satisfactory. However, the investigations and the research carried out after the landslide at Tuve have successively made it clear that undrained analyses are not sufficient even in soft clays. New guidelines for slope stability analyses issued by the Swedish Commission on Slope Stability in 1995.

The landslide at Agnesberg caused a stop in the river traffic for a considerable period of time until stabilising measures had been taken and the water depth in the channel had been restored. This was accomplished mainly by use of lime/cement columns on land and successive excavation and refilling with rock fill in the river. These processes and the fresh water intake had to take place intermittently. The speed on the railway line was reduced and a monitoring and warning system had to be operated until the remedial works were completed. Because of the risk of retrogressive slides, the total risks involved industrial areas and plants along the river, the railway line, the highway, large parts of the community of Agnesberg, interruption of hydroelectric power production in the river, the fresh water supply of Gothenburg city and subsidiary environmental effects of damming the river and chemical pollution.

Parallel to investigations for the remedial works, an investigation into the cause and course of the slide was performed jointly by the Swedish Geotechnical Institute and Chalmers University of Technology, (Larsson et al. 1994).

2 COURSE OF THE SLIDE

The slide was observed visually at 9 o'clock in the morning by the pilot on a passing ship. A 20 m wide strip along about 50 m of the river bank had then disappeared into the river. Three and a half hours later, a retrogressive slide took another strip about 8 m deep and later during the day a number of smaller retrogressive slides followed. A further retrogressive slide, which extended the length of the slide area, followed 1.5 months later. The larger slide events were also recorded in terms of peaks in turbidity at the fresh water intake downstream. Allowing for the flow rate and the distance, a sequence of events with an initial subaqueous slide at about 6 a.m., a major slide at about 9 a.m. and another relatively large slide at about 12:30 p.m. could be deduced, Fig. 2.

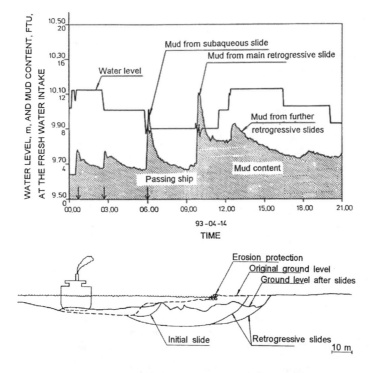

Figure 2. Recorded turbidity and deduced course of the slide.

3 GEOTECHNICAL CONDITIONS

The geotechnical conditions were investigated by gathering all previous investigations in the area, new investigations for the remedial works in and behind the slide area, and supplementary investigations on both sides of the test area. The previous investigations consisted mainly of field vane tests, undisturbed piston sampling and routine testing and oedometer tests in the laboratory. The new investigations comprised the same types of tests supplemented by pore pressure measurements and piezocone tests in the field and triaxial tests in the laboratory. The piezocone tests were performed with special highly sensitive "clay probes" and proved very useful in accurately determining the stratigraphy. The methods of evaluation of shear strength and pre-consolidation pressure proposed by Larsson and Mulabdic (1991) could be verified in a number of parallel investigations and also these parameters could be estimated reliably from the piezocone tests, (Åhnberg 1995). Fig. 3.

The soil profile in the area consists of a 30 to 40 m thick clay layer on coarser soil. The upper 13 m of the clay layer consists of grey, high-plastic clay with infusions of shells and plant remains. The clay below is grey, black-spotted and medium-plastic with infusions of shells. Occasional small pockets of different material have been found. The clay in the entire profile is highly sensitive, particularly the medium-plastic layer, and in most parts of the profile it is classified as a "quick clay".

Before the slide, the ground surface in the area between the railway and the river bank was almost level, with an inclination of only 1:50. A certain amount of fill had been laid out to provide sufficient bearing capacity for storage areas and service roads. The bottom profile in the river before the slide can only be assumed from the sounding which was made in connection with the investigations for the adjacent building in 1990 and the later soundings performed upstream and downstream of the slide area. A common characteristic of these results is that on the side of the river where the slide occurred they show a shallow bottom close to the river bank and a very steep slope at the edge of the shipping channel. Within the slide area, a bottom profile can be assumed with 1 m water depth at the river bank, a 24 m wide shallow area with the water depth gradually increasing to about 2 m and a steep slope with a height of about 6 m and inclination about 1:1.5 down to the shipping channel.

The mean water level in the river is located 0.5 to 1 m below the ground level behind the wall for erosion protection. It varies somewhat with the weather conditions and the sea level at the river mouth, and can be up to 1 m lower. It also varies significantly when large ships are passing. At the time of the slide, the water level corresponded roughly to the mean water level. The free ground water level outside the river is high and only 0.5 to 1 m below the ground surface over the investigated area up to the railway.

The water pressure in the coarse soil below the clay has in all measurements been found to correspond to a water head about 7 m above the mean water level in the river. The pore pressure distributions below the free ground water level have been found to be fairly linear with depth within the clay layer. At each point, there is thus a constant gradient and an upward pore water flow, but this gradient varies over the area, depending on the thickness of the clay layer. It is particularly high in, and close to, the shipping channel in the river, where the clay thickness is reduced, Fig. 4.

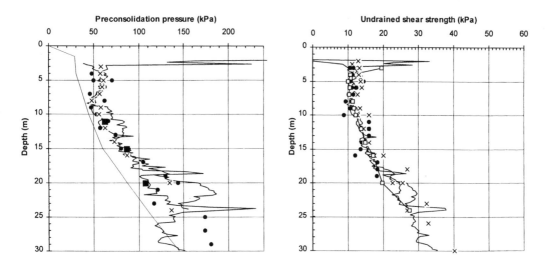

Figure 3. Comparison of pre-consolidation pressures and undrained shear strengths determined by different methods. (Åhnberg 1995)

The clay in the area is only slightly overconsolidated in relation to the in situ effective vertical stress, except for a certain part below the deeply eroded/dredged channel in the river. The undrained shear strength has been determined by field vane tests, piezocone tests and fall cone tests, all of which give consistent and unanimous results. In addition, the results correspond to what can be estimated by empirical relations based on pre-consolidation pressures and consistency limits, (Larsson et al. 1984). Also the results of the triaxial tests follow the expected pattern. The undrained shear strength outside the river area is fairly constant, about 11 kPa down to 10 m depth, and then increases by about 1 kPa per metre depth. Below the river, there are no dry crust effects and the effective stresses are lower. The undrained shear strengths here are about 6 kPa at the river bottom, increasing directly and more rapidly with depth to become more uniform over the area at great depths. Fig. 5.

In this type of clay, the drained shear strength parameters can empirically be estimated as $c' = 0.03\sigma'_c$ and $\phi' = 30°$. This gives a c' value of 1.5 kPa down to a depth of 10 m below the ground surface or the water level respectively.

4 ANALYSIS OF THE STABILITY

The traditional method of estimating stability in the area has been undrained total stress analysis. Using this method, the investigation before construction of the building at the back-scarp in 1990 yielded a safety factor of 1.5, which was considered sufficient. The new guidelines for slope stability analyses demand that a so-called combined analysis should be made, in which not only totally drained or undrained conditions are considered, but also partly drained conditions. For each part of a potential slip surface, both undrained and drained conditions are compared and the most disadvantageous combination is selected. In soft clays, this normally leads to a critical condition, in which the most superficial parts of a slip surface with low effective stresses are drained and the deeper parts remain undrained. The main exceptions are cases where high artesian water pressures exist and where drained parameters may become critical for larger parts of the slip surfaces also in soft clays. For the long slip surfaces considered in the previous investigation, such analyses reduced the safety factor to 1.4 (1.36), which is a fairly typical reduction in soft clays, Fig. 6.

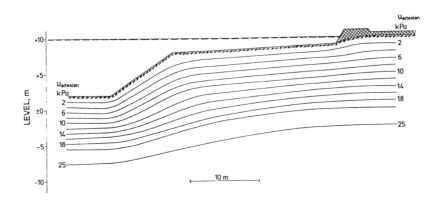

Figure 4. Distribution of artesian pore pressures (pressures above the hydrostatic values).

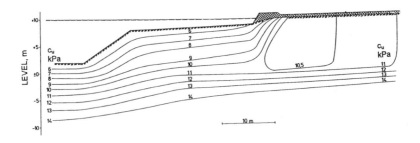

Figure 5. Distribution of undrained shear strength.

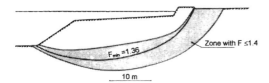

Figure 6. Calculated safety factors for long slip surfaces.

However, when the steep subaqueous slope towards the shipping channel is studied, it is found that this is barely stable. Combined analyses here yield safety factors close to unity for slip surfaces involving considerable volumes of soil, whereas undrained analyses yield safety factors of ≥ 1.25 and drained analyses yield minimum factors of 1.14 for relatively superficial slides, Fig. 7.

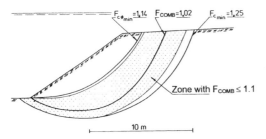

Figure 7. Calculated stability of the subaqueous slope.

A possible slide involving an amount of soil corresponding to what is calculated by the combined analyses would also lead to further consequences. The soil masses are highly sensitive and would end up in flowing water. The masses can consequently be expected to flow out and leave a steep, unsupported back-scarp. A new analysis of the stability of the remaining soil masses shows that also in totally undrained conditions the safety factor for a large slide involving part of the shoreline has now become unity. A retrogressive slide can thus be expected after a short period of time allowing for some creep effects to occur. Also this slide would leave a steep unsupported back-scarp, leading to further retrogressive slides, Fig. 8.

This calculated process is in line with the actual observations. The ultimate end of this process is difficult to predict because it depends on the extent to which the masses in the preceding slides flow away and also on very local variations in the soil conditions. In this case, the slide propagation was probably stopped by a small gradual increase in the undrained shear strength with distance from the river and the forcing upwards of the slide surfaces to shallower depths, because of some remaining support.

5 HAZARDS OF INADEQUATE INSPECTION AND INCOMPLETE METHODS OF STABILITY ASSESSMENT

The cause of the slide at Agnesberg could be attributed to the geotechnical conditions and the high and steep underwater slope. Previously, the underwater slopes in the river had not been recorded in detail and for the method of assessment of stability employed they were not of major concern. In fact, the only check that had been made continuously was to ensure that the water depth within the channel was sufficient for shipping, although there is a certain awareness that new and larger ships equipped with bow propellers may cause considerable erosion along the channel.

The slide at Agnesberg led to a new investigation of the stability conditions for the developed areas along the river. The previous stability assessments in certain areas were then significantly revised and, among other things, very extensive stabilising measures had to be taken for a large chemical plant located about 10 km upstream of the fresh water intake for Gothenburg city.

A new large slide along the same river but outside the investigated area occurred in 1995. This slide was considerably larger than the slide at Agnesberg, but since it occurred in rural country and only arable and pasture land was involved, it caused only minor problems for shipping. Also this slide has been investigated by the Swedish Geotechnical Institute and Chalmers University of Technology. The course of events was different, but also in this case heavy erosion in the river was considered to be a major triggering factor, (Andersson et al, 1997).

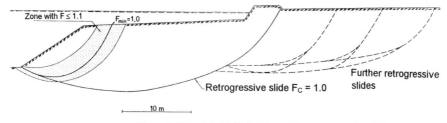

Figure 8. Calculated initial slide and first retrogressive slide.

Recently, in the spring of 1997, a large landslide involving several single-family houses occurred in another part of Sweden, leading to the evacuation and demolition of an entire residential district. Also in this case the soil consisted of soft clay and the assessment of stability during development of the area in the 70s had been made on the traditional lines with undrained analyses. Investigations of the causes of the slide are in progress and from the preliminary results it is obvious that artesian water pressures, which are not accounted for in an undrained analysis, play a significant role.

Insufficient stability assessments in developed areas have been found to be common and a process of identifying developed areas with potential stability problems has been taking place in Sweden for several years. The government has also drawn up a special yearly budget for remedial works in hazardous areas.

6 CONCLUSIONS

Stability assessments should involve all possible aspects. In particular, the ground water conditions should be considered in every case. For slope stability analyses, undrained total stress analyses alone are never sufficient, even in soft clays. For subaqueous slopes and slopes towards watercourses, the topography and possible erosion should be checked regularly. Local slides can spread and involve very large areas, even in areas with almost flat ground.

7 REFERENCES

Andersson, H., Ottosson, E. and Sällfors, G. (1997). The landslide at Ballabo. Report in progress, Swedish Geotechnical Institute, Linköping. (In Swedish).

Larsson, R., Bergdahl, U. and Eriksson, L. (1984). Evaluation of Shear Strength in Cohesive Soils with Special Reference to Swedish Practice and Experience. Swedish Geotechnical Institute, Information No. 3. Linköping. (also in shorter version in ASTM Geotechnical Testing Journal, Vol. 10, No. 3, 1987)

Larsson, R. and Mulabdic, M. (1991). Piezocone tests in soft clay. Swedish Geotechnical Institute, Report No. 42. Linköping.

Larsson, R., Ottosson, E, and Sällfors, G. (1994). The landslide at Agnesberg. Swedish Geotechnical Institute, Report No. 44, Linköping. (In Swedish).

Statens offentliga utredningar (1962). Rasriskerna i Götaälvdalen. SOU 1962:48. Stockholm. (In Swedish).

Swedish Commission on Slope Stability (1995). Guidelines for slope stability investigations, Report 3:95, Linköping. (In Swedish).

Åhnberg, H. (1995). Use of CPT tests in very soft soils. International Symposium on Cone Penetration Testing, CPT'95, Vol. 2, pp. 619-624, Linköping.

Geotechnical Hazards, Marić, Lisac & Szavits-Nossan (eds) © 1998 Taylor & Francis, ISBN 90 5410 957 2

Scales of landslide hazard and risk mapping

S. Mihalić
Faculty of Mining, Geology and Petroleum Engineering, Zagreb, Croatia

ABSTRACT: The preparation of landslide hazard and risk maps is required as a base for rational land use planing and decision-making in landslide prone areas. In this paper, an inventory was made of the currently available methods for landslide hazard and risk zonation, in order to give recommendations for the use of specific methods in the relation of the scale of analysis. A hierarchical set of activities aimed at obtaining landslide-related information for all levels of land use planning in the Republic of Croatia is constructed. This set encompasses: establishing of national landslide inventory at the regional scale (<100,000), statistical landslide hazard analysis of geological-morphological factors at the medium scale (1:25,000) and geotechnical characterisation of slope movements followed by landslide risk analysis at the detailed scale (>1:5000).

1 INTRODUCTION

The term landslides denotes the movement of a mass of rock, debris or earth down a slope (Cruden 1991). Most of terrain in hillside areas has been subjected to landslides at least once under the influence of a variety of causal factors. The number of slope instability grows with increased urbanisation and development in landslide-prone areas. Hence, landslides continue to be one of the most threatening and widespread geohazards.

There is an increasing trend for geohazards to be recognised in planning legislation and guidance, especially in the last decade (the International Decade for Natural Disaster Reduction). Many countries presently have specific planning or development policies aimed to reduce losses due to natural hazards (Schuster 1991, McInnes 1996). Methods vary from guidance documentation alone, through mandatory building codes and finally to insurance or disaster relief schemes (Statham et al. 1995). Zoning and subdivision ordinances are used to divert development into areas where the risks are less and to ensure that, where developments are permitted, appropriate engineering measures are incorporated.

Risk assessment is prerequisite condition for all the methods of hazard prevention and mitigation. It consists of three steps: (1) hazard assessment - identification of past/present landslides as well as prediction of the future occurrence; (2) vulnerability analysis - identification of the location and distribution of population, infrastructure and vital economic activities exposed to a potential or present landslide; and (3) calculation of expected loss (risk) from the hazard and vulnerability. Hazard analysis requires a detailed knowledge of the geo-environmental predisposition factors and initiation events that led to landsliding. This is in domain of earth scientists. Vulnerability and risk evaluation also includes other disciplines then the earth sciences, such as urban planning, social geography, economy, etc. The end result of hazard and risk analysis should be presented in informative documents, usually in the form of various maps that display the spatial distribution of hazard and risk classes. These documents are made use of by decision-makers who have to define a general risk prevention policy.

Many methods and techniques are proposed for landslide hazard and risk mapping over the last 30 years (Brabb 1984, Hansen 1987, Mihalić 1996). Significant progress has been made by establishing the basic definition of terms related to hazard and risk assessment (Varnes 1984). Van Westen's (1993) overview of the available methods is of great importance for improving of a quality as well as for achieving a uniform approach to landslide hazard mapping on international level. On the contrary, examples of landslide risk zonation are still rare because of difficulties in assessing of the probability of landsliding and of the vulnerability of elements at risk. However, the evaluation of risk corresponds to a political, economic and social necessity. Therefore, research of operational risk evaluation methods is in progress (Ragozin 1994, Rezig et al. 1996).

The first fundamental step of the hazard and risk assessment is the identification and mapping of all landslide phenomena, i.e. compiling the landslide inventory. It should be kept in mind that only those factors to which computer access is possible could be analysed (Fernández et al. 1996, Rosenbaum & Popescu 1996). This shows the full importance of landslide cartographic data bases which should be valid nation-wide. In many countries, the development of such databases started in the nineties. The most important are examples of France (Leroi 1996), Germany (Krauter et al. 1996), USA (Brown 1992) and Canada (Cruden 1996). Moreover, there is a tendency of establishing a World Landslide Inventory (Brown et al. 1992). To ensure consistency of data recording, the International Geotechnical Societies' UNESCO Working Party on World Landslide Inventory (WP/WLI), initiated in the 1988 at the 5th International Symposium on Landslides, is suggesting a standard terminology for describing landslides (WP/WLI 1993). GIS technology is also essential for the assembling hazard and risk models as well as for an efficient and rapid information exchange between scientists, engineers, policy makers, and all the people and institutions dealing with landslide hazard.

Before starting any data collection, a number or interrelated things should be clearly defined such as the aim of a study, the scale and degree of precision of presented results, and the available resources in the form of money and manpower. To achieve the optimisation of costs and quality, application of the different data analysis methods at various scales is required.

Accordingly, a concise review of current methods of landslide hazard and risk assessment is presented in this paper. The review is aimed to compare the methods and to propose a logical set of activities related to preparation and implementation of hazard and risk maps in the field of land use planning in the Republic of Croatia. This set comprises all levels of urban planning, from national to the local scale.

2 LANDSLIDE HAZARD AND RISK

The terminology concerning hazards and risk used in this paper conforms to the definitions proposed by Varnes (1984). Evaluation of various risk components (hazard, vulnerability, cost) and of the landslide risk as a whole presupposes that answers are available for the questions as shown in Table 1 (Leroi 1996).

Table 1. Risk components with questions connected to landslide risk assessment.

Component	Question
Hazard	1) Which type of movement is involved?
	2) Where are the potentially unstable areas?
	3) At which moment can the identified phenomenon be triggered?
	4) How far can the phenomenon be propagated?
Vulnerability	5) What are the interactions with the environment, natural or modified by Man?
Cost	6) What is the cost of the caused damage?

Accordingly, landslide research aimed at hazard and risk mapping comprises the aspects of landslides summarised in the following paragraphs.

Since the term of landsliding encompasses all "movement of a mass of rock, debris or earth down a slope" (Cruden 1991), it should be defined which types of movement are present in the studied area. The types of the movement are essentially those defined by the International Geotechnical Societies' UNESCO Working Party on World Landslide Inventory: falls, topples, slides, spreads and flows (Cruden & Varnes 1996).

Location of the potentially unstable areas should be determined by engineering geological mapping of the landslides. The objective would be to record identifiable landslide features and dimensions (IAEG Commission on Landslides 1990).

Evaluation of the probability or timing of the future occurrence is dependent on the probability of occurrence of their triggering factor. Trigger is an external stimulus, such as intense rainfall, that causes a near-immediate response in the form of a landslide by rapidly increasing the stresses or by reducing the strength of slope material (Wieczorek 1996). Hence, it is of primary importance to differentiate the conditions that caused slope instability and the processes that triggered the movement (Popescu 1994).

Landsliding causes damages both in the areas of instability initiation and in the areas of transport and of reception of movements. In order to be able to describe where the landslide is moving, it is necessary to investigate it's activity. UNESCO Working Party on World Landslide Inventory (1993) suggests describing the landslide activity in terms relating to state, distribution and style of activity.

The assessment of losses consists of the analysis of interactions between phenomenon and goods, i.e. behaviour of structures and people that are exposed to landsliding. Hence, it is fundamental to determine the level of intensity of a potential phenomenon. An important characteristic of the movement comprised in the intensity analysis is the rate of landsliding (IUGS WG/L 1995).

3 METHODS OF LANDSLIDE HAZARD ZONATION

All the methods proposed are founded upon a single principle "the past and present are keys to the future" which implies that slope-failures in the future will be more likely to occur under those condition which led to past and present instability (Carrara et al. 1995). Application of the above principle requires mapping of the landslides and a set of geological-morphological causal factors, and a hazard model. There are three main approaches for the developing of hazard models: heuristic, statistical and deterministic approach. Each of them is based on different elements as shown in Table 2.

Table 2. Methods of landslide hazard zonation.

van Westen 1996	Leroi 1996
Heuristic approach	Expert evaluation
Statistical approach	Statistical return analysis
Deterministic approach	Mechanical models

3.1 *Heuristic approach*

In heuristic methods the expert opinion of the engineering geologist and/or geomorphologist is used to classify the hazard. Two types of heuristic analysis can be distinguished: geomorphic analysis and qualitative map combination.

The geomorphic method is also known as the direct mapping method (Hansen 1987). It consists of geomorphological and/or engineering geological mapping through which the surveyor identifies past and present landslides and makes assumptions on those sites where failures are likely to occur in the future. Direct hazard determination is based on individual experience. The decision rules vary from place to place and are difficult to formulate. In addition, the resulting documents generally are "paper" ones (Kienholz 1978).

To overcome the problem of the "hidden rules" in direct mapping, indirect mapping methods have been developed. Qualitative map combination is based on *a priori* knowledge of the causes of landsliding in the investigated area. Hence, instability factor are ranked and weighted according to their assumed or expected importance in causing a mass-movement. In this method expert's knowledge can be formalised into rules, but the result essentially depends on the experience of the surveyor. At present, maps obtained by this method cannot readily be evaluated in terms of reliability or certainty.

3.2 *Statistical approach*

In the statistical approach causality factor are defined *a posteriori*, through back analysis of

historical events. Therefore, the role of each factor (that led to landslides in the past) is determined on the basis of the observed relations with the past/present landslide distribution. The statistical approach can be applied following different techniques which essentially differ on the used statistical procedure: bivariate or multivariate.

In bivariate statistical analysis each instability factor map is combined with landslide distribution map, and weighting values based on landslide densities are calculated (Siddle 1991, van Westen 1993, Yin 1994).

Multivariate statistical analysis of important factors related to landslide occurrence give relative contribution of each of these factors to the total hazard within a defined land unit. For each sampling unit, the presence or absence of landslides is also determined. The model is conceptually fairly simple, but large data sets are needed to obtain enough cases to produce reliable results (Carrara 1995).

3.3 *Deterministic approach*

There are some examples of landslide hazard assessment by calculating safety factors over large areas (van Westen 1993, Leroi 1996). The resulting safety factors are only indicative and are used to test multiple scenarios based on variable triggering hypotheses. The most frequently considered are hydraulic and seismic triggers. The main problem with these methods lies in the choice of representative input parameters and the slope stability model.

For the rational consideration of the natural variability and uncertainty of each input variable in slope stability analyses a probabilistic approach is essential (Hammond et al. 1992, Terlien et al. 1995). The objective is to obtain probability distribution of the factor of safety and hence probability of failure. The most important limitation to the application of probabilistic methods in landslide hazard assessment may be the lack of statistical data on soil properties, pore water pressures and on loads (Chowdhury 1984).

4 LANDSLIDE RISK ASSESSMENT

Landslide risk assessment requires understanding, analysis and control of damages which are the consequences of the interaction between slope movements and exposed elements (property, people and various activities). However, due to the complexity of the phenomena and partly to an absence of conceptual knowledge of certain risk components, an unified approach to the problem hasn't emerged yet. As a result of technical and sociological advances several researchers and organisations started to develop methodology for landslide risk evaluation in the last decade (Fell 1994, Ragozin 1994, Leone et al. 1996, Leroueil et al. 1996).

Anderson's et al. (1996) proposal of a risk-based method for selecting alternatives for landslide risk mitigation is presented as follows. The proposal is interesting because it comprises the whole procedure: the identification of risk, the estimation of risk, and the evaluation of risk through either aversion or acceptance (Fig. 1). Risk identification involves development of the risk model for the evaluation of existing landslide risk. In order to achieve this it is necessary to recognise and list the various factors which could contribute to the landslide failure risk, and then to organise these into logical event sequences. The model is organised in the form of event tree which commences with events that can initiate failure and ends with the consequences of a failure (Fig. 2). In the later phase

the risk model serves for evaluation of the effectiveness of proposed rehabilitation alternatives. The second step involves risk estimation, i.e. assigning the probabilities and consequences to the occurrence of each failure mode. If these risks are unacceptable, the assessment proceeds to the third step - risk aversion. This involves formulation and evaluation of remedial action (rehabilitation) alternatives. The final step in the risk assessment process is taking the decision on what degree of safety is acceptable.

A crucial stage for a good understanding of slope movements and the risk associated with them is characterisation of the movement through factors having a mechanical significance. It requires establishing of relationship between the characteristics of a given movement, existence of definite predisposition factors, occurrence of triggering or aggravating factors, existence of definite revealing factors, and of the consequences

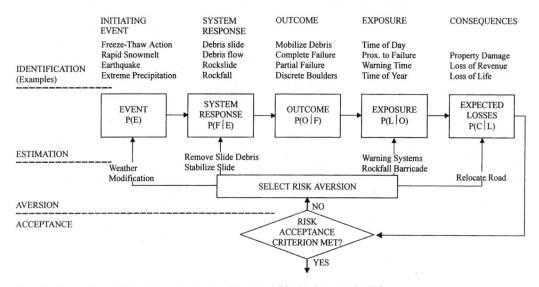

Figure 1. Framework model for risk-based method to mitigate landslides (Anderson et al. 1996).

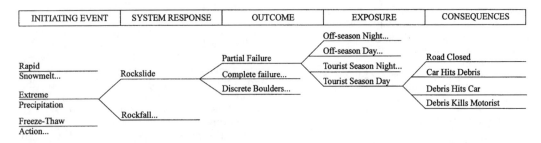

Figure 2. Hypothetical event tree branch for evaluating outcome probability for landslide risk assessment (Anderson et al. 1996).

of the movement. For this purpose, Vaunat et al. (1994) are developing a geotechnical characterisation of slope movements, taking into account slope movement type, involved material and movement stages. Such a characterisation constitutes an essential step for the development of expert systems on slope engineering, for the selection of numerical models for the simulation of specific aspects of slope behaviour, as well as for the design of remedial measures for stabilising a slope.

5 SCALE-RELATED RECOMMENDATIONS

Not all the methods of landslide hazard zonation are equally applicable at each scale of analysis, because of the difference in required input data and degree of precision of the obtained results. Table 3 provides an overview of the various methods of landslide hazard analysis and recommendations for their use at three most relevant scales. From the consideration of advantages and pitfalls of landslide hazard zonation techniques it follows that it is the best to draw hazard maps at the medium scale, i.e. 1:25,000. At this scale it is possible to obtain an overview of the hazard in its entirety, and also a reasonable cost of the work.

Regardless of the scale at which the hazard is evaluated, the risk maps should be drawn at scale above 1:5000. This is due to impossibility of acquisition of required input data at smaller scales.

The main field of application of landslide hazard and risk maps concerns land use planning, development and regulations. Hence, hazard and risk zonation have to comprise all levels of land use planning. To obtain an acceptable cost/benefit ratio and to ensure the practical applicability of the zonation, the development of a clear hierarchical methodology for structuring and analysing of the data is necessary.

5.1 Proposed future activities in Croatia

Based on the above paragraphs and in respect to the fact that the consideration of the instability in the planning processes in Croatia is only partial, and thus unsatisfactory (Stanić & Mihalić 1995), a proposal of a logical set of activities was constructed connected to the hazard and risk mapping which comprises all the levels of urban planning. The proposal is presented in the frame of recommendation for the application to urban planning documents currently valid in the Republic of Croatia with special emphasis to the territory of Zagreb City. These recommendations are divided in following levels (Fig. 3):

1) At the regional scale (1:100,000 or smaller) the national landslide inventory should be made. The objective of the inventory is to provide insight into spatial distribution of landslides in Croatia. It is also possible to analyse landslide density. Although the method does not enable the production of landslide hazard map, the quantitative presentation of landslide density would indicate areas in the Croatia where mass movements can be a constraint of development. The information obtained should serve for physical planning at the national and county level. Overlaying of landslide distribution map with the maps which display elements at risk (i.e. land use map) should also point at areas where making of landslide hazard maps is necessary. Additionally, the records of the most significant landslides in Croatia could also serve as an input for the World landslide inventory.

2) At the medium scale (1:25,000) statistical hazard analysis of geological-morphological causal factors is required. The detail of the hazard map should be such that adjacent slopes in the same lithology are evaluated separately, and may obtain different hazard scores, depending on other characteristics, such as slope angle and slope segments. This maps should present a base for a rational land-use planning in order to locate developments on stable ground. The field of application would be physical planning at the municipal and city level. Accordingly, landslide hazard map for the territory of Zagreb City should also become a component of the Physical plan of the Zagreb City, by replacing the existing Map of lithologic composition and the slope stability

Table 3. Hazard analysis techniques in relation to mapping scales (Soeters & van Westen 1996).

Type of analysis	Technique	Scale of use recommended		
		Regional (<1:100,000)	Medium (1:25,000-1:50,000)	Large (1:5000-1:10,000)
Heuristic analysis	Geomorphological analysis	Yes*	Yes*	Yes*
	Qualitative map combination	Yes**	Yes*	No
Statistical analysis	Bivariate statistical analysis	No	Yes	No
	Multivariate statistical analysis	No	Yes	No
Deterministic analysis	Safety factor analysis	No	No	Yes***

*But strongly supported by other more quantitative techniques to obtain an acceptable level of objectivity.
**But only if sufficient reliable data exist on the spatial distribution of the landslide controlling factors.
***But only under homogeneous terrain conditions, considering the variability of the geotechnical parameters.

851

categorisation of the Mt. Medvednica hillsides. On the basis of landslide hazard map, the legislation restricting development in the areas most susceptible to landslides could be enacted.

Enlargement of the landslide hazard map to the 1:10,000 scale could also serve as the basis for construction of the Physical development master plan of the Zagreb City. Overlaying of the hazard map with the map which displays elements at risk could indicate the level of risk. In the areas where risk is low, landslide hazard analysis will suffice.

3) At the large scale (>1:5.000) the information of the landslide risk is required. To achieve expected degree of precision it is necessary to undertake the complementary investigation, followed by geo-technical characterisation of slope movements, and thus the risk assessment. The application of risk maps lies in the construction of detailed physical development plans and of the regulation and implementation plans for the areas characterised by high risks.

6 CONCLUSIONS

Predictive models of landslide hazard and risk assessment constitute a major research field which may well take advantage of the potentials of the new technological advancements - GIS-driven data acquisition, manipulation and analysis. Consequently, the development of the methods of producing landslide hazard and risk maps is still in progress, and no uniform approach is accepted yet.

By evaluation of the methodological approaches on landslide hazard zonation practices the statistical analysis of geological-morphological causal factors is suggested, aimed to predict spatial probability of landslides (i.e. where failures are most likely to occur). This method makes possible to obtain landslide hazard maps at the scale 1:25,000 at an acceptable cost.

Due to the variety of geological situations, the diversity of materials, the complexity of acting mechanisms and the variability of controlling parameters, the clear indication of temporal probability of landsliding (i.e. when failures are likely to occur) can only be obtained by risk analysis at the detailed scale (>1:5000). Hence, of crucial importance for risk analysis is Vaunat's et al. (1994) geotechnical characterisation of slope movements. To be of value, in terms of evaluation and presentation of landslide mitigation alternative, risk analysis should encompass risk identification, estimation, aversion and acceptance, as proposed by Anderson's et al. (1996).

The priority areas for construction of risk maps are to be delimited on the basis of the hazard maps. The areas that are to be covered by hazard maps should be determined on the basis of data from national landslide inventory.

The development of general methodology for landslide hazard and risk mapping would require definition of the conceptual models and extraction of simplified operational models from the conceptual

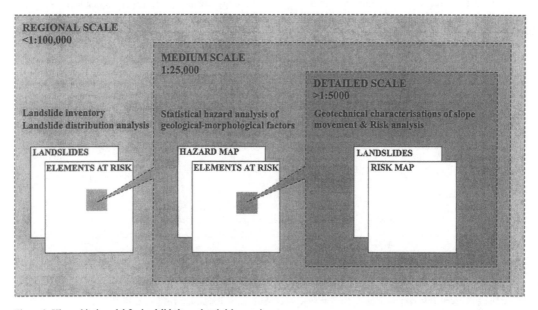

Figure 3. Hierarchical model for landslide hazard and risk mapping.

models. The choice of the models should also serve as a guide for development of appropriate data bases, having in mind that the availability of adequate data (both in quantity and quality) is crucial issue for the task to be accomplished.

REFERENCES

Anderson, L.R., Bowles, D.S., Pack, R.T. & J.R. Keaton 1996. A risk-based method for landslide mitigation. In K. Senneset (ed.), *Landslides - Proc. 7th Int. Symp. Landslides, Trondheim, 17-21 June 1996.* 1:135-140. Rotterdam: Balkema.

Brabb, E.E. 1984. Innovative Approaches to Landslide Hazard and Risk Mapping. *Proc. 4th Int. Symp. Landslides, Toronto, 1984.* 1:307-323.

Brown, W.M., III 1992. Information for disaster reduction: The National Landslide Information Center, US Geological Survey. In D.H. Bell (ed.), *Landslides - Proc. 6th Int. Symp. Landslides, Christchurch, 10-14 February.* 2:891-892. Rotterdam: Balkema.

Brown, W.M., III, Cruden, D.M. & J.S. Denison 1992. *The Directory of the World Landslide Inventory.* U.S. Geological Survey Open-File Report 92-427.

Carrara, A., Cardinali, M., Guzzetti, F. & P. Reichenbach 1995. GIS Technology in Mapping Landslide Hazard. In A. Carrara & F. Guzzetti (eds), *Geographical Information Systems in Assessing Natural Hazards*: 135-176. Dordrecht: Kluwer Academic Publishers.

Chowdhury, R.N. 1984. Recent Developments in Landslide Studies: Probabilistic Methods. State-of-the-Art-Report - Session VII (a). *Proc. 4th Int. Symp. Landslides, Toronto, 1984.* 1:209-228.

Cruden, D.M. & D.J. Varnes 1996. Landslide Types and Processes. In A.K. Turner & R.L.Schuster (eds), *Landslides: Investigation and Mitigation*: 36-75. Washington: National Academy Press.

Cruden, D.M. 1991. A Simple Definition of a Landslide. *Bulletin of the International Association of Engineering Geology.* 43:27-29.

Cruden, D.M. 1996. An inventory of landslides in Alberta, Canada. In K. Senneset (ed.), *Landslides - Proc. 7th Int. Symp. Landslides, Trondheim, 17-21 June 1996.* 3:1877-1882. Rotterdam: Balkema.

Fell, R. 1994. Landslide risk assessment and acceptable risk. *Can. Geotech. Journal.* 31:261-272.

Fernández, T., Irigaray, C. & J. Chacón 1996. Inventory and analysis of landslide determinant factors in Los Guajares Mountains, Granada (Southern Spain). In K. Senneset (ed.), *Landslides - Proc. 7th Int. Symp. Landslides, Trondheim, 17-21 June 1996.* 3:1891-1896. Rotterdam: Balkema.

Hammond, C.J., Prellwitz, R.W. & S.M. Miller 1992. Landslide hazard assessment using Monte Carlo simulation. In D.H. Bell (ed.), *Landslides - Proc. 6th Int. Symp. Landslides, Christchurch,10-14 February.* 2:959-964. Rotterdam: Balkema.

Hansen, A. 1987. Landslide Hazard Analysis. In D. Brunsden & D.B. Prior (eds), *Slope Instability*: 523-602. London: John Wiley and Sons.

IAEG Commission on Landslides 1990. Suggested Nomenclature for Landslides. *Bulletin of the International Association of Engineering Geology.* 41:13-16.

IUGS WG/L 1995. A Suggested Method for Describing the Rate of Movement of a Landslide. *Bulletin of the International Association of Engineering Geology.* 52:75-78.

Kienholz, H. 1978. Maps of Geomorfology and Natural Hazards of Grindewald, Switzerland, Scale 1:10.000. *Arctic and Alpine Research.* 10(2): 169-184.

Krauter, E., Lippomann, R., Moser, M., Müller, B. & H. Prinz 1996. Kinematical-geotechnical aspects of landslides in Germany. In K. Senneset (ed.), *Landslides - Proc. 7th Int. Symp. Landslides, Trondheim, 17-21 June 1996.* 1:251-256. Rotterdam: Balkema.

Leone, F., Asté, J.P. & E. Leroi 1996. Vulnerability assessment of elements exposed to mass-movemnet: Working toward a better risk perception. In K. Senneset (ed.), *Landslides - Proc. 7th Int. Symp. Landslides, Trondheim, 17-21 June 1996.* 1:263-270. Rotterdam: Balkema.

Leroi, E. 1996. Landslide hazard - Risk maps at different scales: Objectives, tools and developments. In K. Senneset (ed.), *Landslides - Proc. 7th Int. Symp. Landslides, Trondheim, 17-21 June 1996.* 1:35-52. Rotterdam: Balkema.

Leroueil, S., Locat, J., Vaunat, J., Picarelli, L., Lee, H. & R. Faure 1996. Geotechnical characterization of slope movements. In K. Senneset (ed.), *Landslides - Proc. 7th Int. Symp. Landslides, Trondheim, 17-21 June 1996.* 1:53-74. Rotterdam: Balkema.

McInnes, R.G. 1996. A Review of coastal landslide management on the Isle of Wight, UK. In K. Senneset (ed.), *Landslides - Proc. 7th Int. Symp. Landslides, Trondheim, 17-21 June 1996.* 1:301-307. Rotterdam: Balkema.

Mihalić, S. 1996. *Landslide hazard and risk zonation.* (In Croatian). Master thesis. University of Zagreb.

Popescu, M.E. 1994. A Suggested Method for Reporting Landslide Causes. *Bulletin of the International Association of Engineering Geology.* 50:71-74.

Ragozin, A.L. 1994. Basic principles of natural hazard risk assessment and management. *Proc. 7th Int. IAEG Congress, Lisbon, 5-9 September 1994.* 3:1277-1286. Rotterdam: Balkema.

Rezig, S., Favre, J.L. & E. Leroi 1996. The probabilistic evaluation of landslide risk. In K. Senneset (ed.), *Landslides - Proc. 7th Int. Symp. Landslides, Trondheim, 17-21 June 1996.* 1:351-356. Rotterdam: Balkema.

Rosenbaum, M.S. & M.E. Popescu 1996. Using a geographical information system to record and assess landslide-related risks in Romania. In K. Senneset (ed.), *Landslides - Proc. 7th Int. Symp. Landslides, Trondheim, 17-21 June 1996.* 1:363-370. Rotterdam: Balkema.

Schuster, R.L. 1991. Landslide hazard management - experience in the United States. *Slope stability engineering: developments and applications - Proc. Int. Conf. Slope Stability, Isle of Wight, 15-18 April 1991*: 253-263. London: Thomas Telford.

Siddle, H.J., Jones, D.B. & H.R. Payne 1991. Development of a methodology for landslip potential mapping in the Rhondda Valley. *Slope stability engineering: developments and applications - Proc. Int. Conf. Slope Stability, Isle of Wight, 15-18 April 1991*:253-263. London: Thomas Telford.

Soeters, R. & C.J. van Westen 1996. Slope Instability: Recognition, Analysis and Zonation. In A.K. Turner & R.L.Schuster (eds), *Landslides: Investigation and Mitigation*: 129-177. Washington: National Academy Press.

Stanić, B. & S. Mihalić 1995. Landslide hazard zoning. (In Croatian). In R. Mavar (ed.), *Geotechnical Engineering in Cities - Proc. 2nd Conf. Croatian Society for Soil Mechanics and Foundation Engineering, Varaždin, 4-6 October 1995*. 1:467-475.

Statham, I., Langer, M.F.B. & G. Bouckovalas 1995. The identification and monitoring of geohazards. *Interplay between Geotechnical Engineering and Engineering Geology - Proc. XI Europ. Conf. Soil Mechanics & Foundation Engineering, Copenhagen, 28 May - 1 June 1995*. 9:77-104.

Terlien, M.T.J., van Westen, C.J. & T.W.J. van Asch 1995. Deterministic modelling in GIS-based landslide hazard assessment. In A. Carrara & F. Guzzetti (eds), *Geographical Information Systems in Assessing Natural Hazards*: 57-77. Dordrecht: Kluwer Academic Publishers.

UNESCO Working Party on World Landslide Inventory 1993. A Suggested Method for Describing the Activity of a Landslide. *Bulletin of the International Association of Engineering Geology*. 47:53-57.

van Westen, C.J. 1993. *GISSIZ. Application of Geographic Information Systems to Landslide Hazard Zonation*. Enschede: ITC.

Varnes, D.J. 1984. Landslide hazard zonation: a review of principles and practice. *Natural Hazards*. 3. Paris: UNESCO.

Vaunat, J., Leroueil, S. & R.M. Faure 1994. Slope movements: A geotechnical perspective. *Proc. 7th Int. IAEG Congress, Lisbon, 5-9 September 1994*. 3:1637-1646. Rotterdam: Balkema.

Wieczorek, G.F. 1996. Landslide triggering mechanisms. In A.K. Turner & R.L.Schuster (eds), *Landslides: Investigation and Mitigation*: 76-90. Washington: National Academy Press.

WP/WLI 1993. A suggested method for a describing the activity of a landslide. *Bulletin of the International Association of Engineering Geology*. 47:53-57.

Yin, K.L. 1994. A computer-assisted mapping of landslide hazard evaluation. *Proc. 7th Int. IAEG Congress, Lisbon, 5-9 September 1994*. 6:4495-4499. Rotterdam: Balkema.

Geotechnical Hazards, Marić, Lisac & Szavits-Nossan (eds) © 1998 Taylor & Francis, ISBN 90 5410 957 2

Fuzzy logic concepts in limit states analysis of geotechnical structures

N.O. Nawari
Civil Engineering Department, University of Akron, Ohio, USA

R. Hartmann
IDAT GmbH, Darmstadt, Germany

ABSTRACT: In geotechnical engineering there are numerous sources of vague, imprecisely defined boundaries and subjective data. Conventionally, in the construction of the mathematical models in limit states design, these uncertain data are treated as random variables. Probability theory is viewed as the unique methodology to handle uncertainty in limit states analysis. But in geotechnical engineering, data and associated rules (geological maps, sampling, field tests, modes of failure, elasticity and plasticity theories, static and kinematics models...) are connected with fuzzy, dubious and to a great extent non-statistical data. In addition, there is impreciseness in the definition of system performance and failure events. That is, in the geotechnics, probability theory can not be considered as the only appropriate methodology in the limit states design.

The utilization of fuzzy logic provides a technique for the estimation of uncertainty in geotechnical structural analysis. Within the limit states design, this means that the risk of failure is not a crisp failure event but rather a fuzzy failure event. The environmental parameters are thus rendered into fuzzy variable, and the structural reliability is characterized by fuzziness.

In this paper, the application of fuzzy logic concepts within the framework of limit states analysis will be introduced. Further, disadvantages and difficulties of the conventional stochastic methods will be epitomised. Numerical examples illustrating this approach will be dealt with especially with reference to the European Code of practice EC 7.

1 GENERAL INTRODUCTION

Geotechnical Design is different and sometimes substantially distinct from other similar design problems in civil engineering. Geomaterials are so variable in nature that the chance for even specific locality is quite godforsaken. Geotechnical structures are extremely complex system, even with the use of Finite Element Method and modern computers, it is usually impractical and unfeasible to consider all the details in the mathematical model. Moreover, the damage path and failure behavior of most large structures remains unknown because of limited number of experimental results of full-scale structures.

In geotechnical analysis, the real problem must be reduced to a highly idealized mechanical model because of the heterogeneity of the physical real environment. There are often field measurements to examine and calibrate the computed model and sometimes the design is completely based on the continuos monitoring of the structure. The transfer of the laboratory test results and soils and rocks parameters to the real properties and behavior of geotechnical structures is not always certain and customarily correlative. The uncertainty space consists of commonly of the actual underground behavior, types and conditions of the soils and rocks and their geological spread, hydrogeological variability, environmental influences, and unexpected events. This incertitude can not be considered adequately with the probabilistic safety concept.

Probability theory is based upon the framework of randomness. The idea in this principle is that alternatives are to be treated equiprobable and there is no preference of one over the other. This means that there is a uniform, random process that generates the alternative such that all the alternatives are treated equitably, and therefore we assume they are equipossible. This statistical definition, although might be useful in some industrial applications, has difficulties from the geotechnical point of view. Thus, within this theoretical ambience, it should be appreciated and acclaimed that other types of uncertainties, which are not random in nature, do exist. These may include, for instance, judgmental opinion, vague terms and concepts, argumentation and causal relationship, ill-defined boundaries, ...etc.

2 LIMIT STATE DESIGN: BACKGROUND

At the beginning, the geotechnical engineers, rely primarily on the factor of safety at the design stage to reduce risk of potential adverse performance (collapse, excessive deformation, etc...). Factors of safety between 1.4 - 3 generally are considered to be adequate in most geotechnical design. However these values are recommended without reference to any other aspects of the design computational process or theoretical background. The assessment of the traditional factor of safety is essentially subjective, requiring only global appreciation of the uncertainty space against the backdrop of the previous experience. The sole reliance on the subjective definition of the global factor of safety had led to numerous inconsistencies. For instance, it suffers from major flaw in that it is not unique. Furthermore, ambiguity lies in the relationship between the factor of safety and the underlying level of risk.

In the 70's because of the rationalization of the probability of failure, the probability theory was introduced in the structural engineering. Herein, the statistical character of the basis variables of the Actions (loads, earthquake...) and Resistance (strength of concrete, steel...) are directly considered (s. Figure 1). In the 80's and 90's there are many studies about the application of the probability theory in geotechnical engineering with the objectives of improving the reliability of geotechnical structures and harmonizing with other disciplines in civil engineering. This approach can be seen in its practical form in the new European Code of Practice in the Geotechnics (EC7; Limit State Design in Geotechnical Engineering). The concept of limit state design refers to the design principles that entails the following basic requirements:

(a)- Identifying all potential collapse modes or limit states.

(b)- Scrutinize each limit state separately.
(c)- Assure that the likelihood of occurrence of each limit state event is sufficiently low.

Within this framework, the design risk is quantified by the probability of failure (P_f). The basis problem is then to evaluate P_f from some pertinent statistics of A and R (s. Figure 1). A simple closed form solution for P_f is available if A and R have a well defined probability density function, e.g. normal distribution. In such condition, the safety margin [(R − A) = M] will also be normally distributed with the following mean (μ_M) and standard deviation (σ_M):

$$\mu_M = \mu_R - \mu_A \qquad (1)$$
$$\sigma^2_M = \sigma^2_R - \sigma^2_A \qquad (2)$$

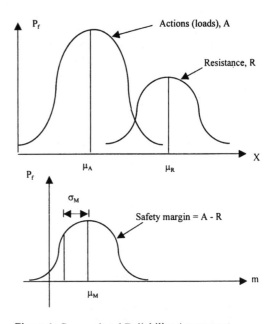

Figure 1. Conventional Reliability Assessment

Once the probability distribution of M is known, the probability of failure P_f can be evaluated:

$$P_f = Prob (A < R) = Prob.(M < 0) = \phi(- \mu_M/\sigma_M) \qquad (3)$$

Within this framework, a more convenient measure of design risk is the reliability index (β) which is defined as

$$\beta = - \phi^{-1}(P_f) \qquad (4)$$

The probabilistic concept as described above has some deficiency, when applied to geotechnical

856

structures. These drawbacks can be delineated in the following:

❖ It is necessary to collect lots of data to estimate the statistical parameters of the geotechnical properties. However, in general practice this will be expensive and difficult to achieve.

❖ The proper identification of potential failure mode is not always a trivial task. This effort generally requires an appreciation of the interaction between geologic environment, loading characteristics and foundation response. This identification depends intensely on experience and judgement.

❖ The specification of the safety level is also dependent on engineering surmise. If the failure probability of the system is estimated to be for instance 10^{-5}, would the system be assumed to be safe? If the system is assumed to be safe, does the assessment imply that the system has never a possibility to break down?

❖ Other types of uncertainty, which are not random in nature, can not be incorporated.

❖ System performance is considered to be either entirely non-failed or entirely failed which indicates a binary failure function (crisp failure). This exclude the middle stages between complete collapse and complete survival.

❖ There might be many possible sequences from cause events to the system failure. Such sequences can not be dealt with in the probability concept.

3 THE FUZZY-SAFETY CONCEPT

In the common practice, if the factor of safety is less than the nominal value, the structure is eventually considered as "dead". But in reality the structure may still perform its functions or resume its normal operation as soon as the unfavorable circumstances are repaired. But how to recognize the limits of the favorable operating conditions and the extent of their range? Answers to these questions require a combination of fuzzy variables like intuition, experiments, experience and theoretical tools.

Fuzzy logic theory approach is attractive in geotechnical engineering for the same reasons that the probabilistic concept has been found lacking. Subjective information, judgmental opinions, scant evidence, tentative facts, suspicious beliefs, ill

defined boundary, and disjoint data can be modified and manipulated using fuzzy logic theory in a manner which reflects the degree of imprecision in the original information and data. Fuzzy variables behave as a family of elastic constraints having a well defined semantic meaning, possess a natural ability to make all necessary perspective adjustments. Thus, they can be adopted to any variation of cognitive perspectives in geotechnical system.

3.1 Theoretical Setting

The essential theoretical backbone for the fuzzy-safety concept will be stated below:

3.1.1 The pattern space

Let Γ be an abstract space of generic element $k \in \Gamma$. The actual construction of F will depend upon the particular problem being modeled in much the same way as the sample space in probability theory. Suppose that $\Im$ is the discrete Topology on Γ. A Scale ξ will be defined as a special type of CHOQUET Capacity on $\Im$ which satisfies the following properties:

(i) $\xi(\phi) = 0$ and $\xi(\Gamma) = 1$

(ii)   For arbitrary collection of sets A_α on $\Im$:

$$\xi\left[\bigcup_\alpha A_\alpha\right] = \sup_\alpha \xi(A_\alpha) \qquad (5)$$

Then the triplet $(\Gamma, \Im, \xi)$ is referred to as the pattern space.

3.1.2 The fuzzy variable

A fuzzy variable, X, is a real valued function defined on a pattern space $(\Gamma, \Im, \xi)$. The preference Function of a fuzzy variable X, (λ_x) is a mapping from $\Re$ to the unit interval $[0, 1]$ and is given by:

$$\lambda_x(x) = \xi(\omega: X(\omega) = x) \qquad \forall\, x \in \Re \qquad (6)$$

where,

$$\sup_\alpha \lambda_x \xi\left(\bigcup_\alpha (\omega: X(\omega) = x)\right) = \xi(\Gamma) = 1.0$$

In general we use the briefer notation $(X = x)$ to denote the subset $(\omega: X(\omega) = x)$ on $\Im$.

857

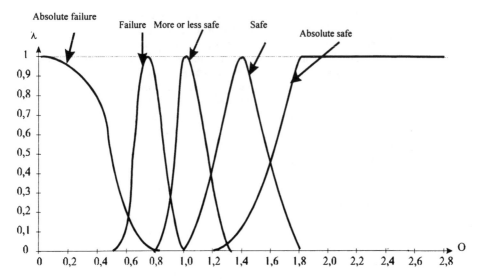

Figure 2. Failure and Safety limit states

3.1.3 Fuzzy Relations

A fuzzy relation R is a fuzzy set in a Cartesian product X x Y of universe of discourse X and Y. R(x, y) is the membership value of (x, y) in R. Fuzzy relations generalize ordinary relations. As such, they can be composed: let R and S be two fuzzy relations on X x Y and Y x Z respectively, the membership function of the fuzzy relation R o S, on X x Z is defined by:

$$\lambda_{RoS}(x,z) = \sup_{y \in Y} \min \left(\lambda_R(x,y), \lambda_S(x,z) \right) \qquad (7)$$

Note that in Eq. (7) min could be replaced by a product or other algebraic operations.
R can be interpreted as a fuzzy restriction on the value of a variable (u,v) ranging over (X x Y), i.e. R acts as an elastic constraint. When R is separable, u and v are said to be non-interactive in the sense that the choice of a value for u does not depend upon the choice of a value for v and conversely. Otherwise u and v are interactive.

3.1.4 The Extension Principle

Owing to this principle, any mathematical relationship between non-fuzzy elements can be fitted to deal with fuzzy entities. This principle will be stated below and its main applications will be seen later.

Let A_1, ..., A_n be fuzzy sets over X_i, ... X_n. respectively, their Cartesian product is defined by:

$$A_1 \times ... \times A_n = \int_{X_1 \times ... \times X_n} \min_{i=1,n} \lambda_{A_i}(x_i) / (x_1,...,x_n) \qquad (8)$$

Let f be a mapping f : X_1 x... x $X_n \to Y$. The fuzzy image B of A_1, ..., A_n through f has a membership function:

$$\lambda_B(y) = \sup_{X_1,...X_n \in X_1 \times ... \times X_n} \min_{i=1,n} \lambda_{A_i}(x_i) \qquad (9)$$

under the constraint y = f(x_1, ... x_n).

3.1.5 Fuzzy Entropy

Fuzzy entropy measures the level of imprecision of a fuzzy variable. It answers the question "How fuzzy is a fuzzy variable?" And it is a matter of degree. The fuzzy entropy of A, Ent(A) varies from 0 to 1. The mathematical description of fuzzy entropy can be obtained by different approaches. The classical definition uses the analog to SHAENON's random entropy:

$$Ent(A) = - \int [A \ln(A) - (1-A) \ln(1-A)]$$
$$= - \int [A \ln(A) - (A^C) \ln(A^C)] \qquad (10)$$

KOSKO used *the fuzzy Hamming Distance* to define fuzzy entropy:

$$Ent(A) = \frac{l^P(A \cap A^C)}{l^P(A \cup A^C)} \qquad (11)$$

where

858

$$l^P(A, A^C) = \sqrt[P]{\sum_1^n \left| \lambda_A(x_i) - \lambda_{A^C}(x_i) \right|^P} \qquad 1 \leq P \leq \infty$$

(12)

The simplest distance is the l^1 or the *fuzzy Hamming Distance*, the sum of the absolute-fit differences.

3.2 Fuzzy Logic in the Limit State Design

Multi-attribute engineering design starts with the specification of a set of functional requirements. These represent the functional space consisting of the minimum number of mutually independent and realizable elements, which describe completely the design properties of the structure to be designed. The elements of the set of solutions satisfying the functional requirements are called the fuzzy design variables. They represent the physical space in which the structure will be realized. Within the limit state context this can be described by the following:

Let x = a vector of algebraically independent design input variables in the mathematical model $\aleph$ (loads, geometric parameters, material properties, environmental conditions, ...)

$\aleph$ = a mathematical model of a quasi-static physical system which can be defined by a vector of mapping f of arbitrary large dimensions from the model input space X to a model performance space Z consisting of a vector output.

Then, formally the system model $\aleph$ can be defined as a trinomial set (**f**, X, Z), where X and Z are taken to be the universe of discourse of possible values of vector x and z.

The safety against a given collapse mode can be defined by specifying an inequality constraint:
$R_s \geq R_a$
where R_s = Resistance forces; R_a = External excitations (actions).

or $z = \dfrac{R_s}{R_a} \geq 1$ (13)

In the fuzzy failure event, there is no unique precise limit state surface to provide a crisp portioning of strict dilapidated and survival sets. Instead a family of limit state surfaces will be introduced to reflect the real structural environment:

(a)- Safety state (I): absolute safe
(b)- Safety state (II): safe
(c)- Safety state (III): more or less safe (slightly damaged)

(d)- failure state (I): partial collapse (require maintenance)
(e)- failure state (II): absolute collapse

To see the explanation of these definitions see figure 2.
This concept will be illustrated with the numerical example below.

4 CASE STUDY

This case study deals with the stability analysis of a disjointed rock slope. The case study is implemented to illustrate the application of the fuzzy safety concept outlined in the previous sections. As a comparison, the stability analysis is performed according to the European Code of practice in the Geotechnics (EC 7).

The geometry and the soil properties are shown in Figure 3. In this problem all dimensions, material properties and loading are fuzzy design variables.

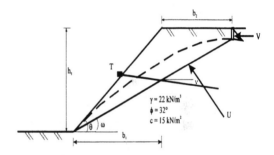

Figure 3. Stability of a Rock Slope

The stability of the slope is defined with the following relation:

$$O = \frac{cL + [W\cos\omega - U + V\sin\omega + T\cos v]\tan\varphi}{W\sin\omega + V\cos\omega - T\sin v} \geq 1.0$$

(14)

c = cohesion of the joint
L = length of the joint
W = total weight of the slope
φ = frictional angle of the joint
T = Anchor force
U, V = pore water pressure
θ = inclination of the slope from the horizontal
v = angle between the anchor and the sliding surface
ω = inclination of the sliding surface from horizontal

859

The computation algorithm of the Fussy-Safety will be now implemented.

The vector of fuzzy design variable is

$$X = \left\{ \tilde{h}_1, \tilde{b}_1, \tilde{b}_2, \tilde{\gamma}, \tilde{c}, \tilde{\varphi}, \tilde{T}, \tilde{\theta}, \tilde{\omega}, \tilde{\beta} \right\} \tag{15}$$

Table (1) summarizes the definition of the fuzzy design variables. The differences between constants and fuzzy variables will be shown with the sign ~ above the symbol. The constant vector is given by equation (16):

$$X = \begin{Bmatrix} h_1 \\ b_1 \\ b_2 \\ \gamma \\ c \\ \varphi \\ T \\ \theta \\ \omega \\ v \end{Bmatrix} = \begin{Bmatrix} 10.0 \text{ m} \\ 10.0 \text{ m} \\ 5.0 \text{ m} \\ 22.0 \text{ kN} / \text{m}^3 \\ 15.0 \text{ kN} / \text{m}^2 \\ 32.0° \\ 100.0 \text{ kN} \\ 45.0° \\ 30.0° \\ 20.0° \end{Bmatrix} \tag{16}$$

Table 1: Definition of the fuzzy Variables

Variable	Function parameters			
	a_1	a_2	v	u
$\tilde{h}_1$ [m]	9	1	1.5	9
$\tilde{b}_1$ [m]	10	10	1.5	0.5
$\tilde{b}_2$ [m]	5	5.5	0.5	0.5
$\tilde{\gamma}$ [kN/m³]	22	23	2	4
$\tilde{c}$ [kN/m²]	15	17	5	3
$\tilde{\varphi}$ [°]	30	32	3	3
$\tilde{T}$ [kN/m]	100	110	10	15
$\tilde{\theta}$ [°]	45	47	3	3
$\tilde{\omega}$ [°]	30	30	3	4
$\tilde{v}$ [°]	20	21	1	1

In Table 1 there are linear and non-linear functions, defined as follows

$$\lambda_x = \begin{cases} L(\xi) = L((a_1 - x) / u) & x \le a_1, \ u > 0 \\ R(\xi) = L((x - a_2) / v) & a_1 \le x_1 \le a_2 \\ 0 & x < a_1 - u, \ x > a_2 - v \end{cases} \tag{17}$$

In case of non-linear functions, the reference functions L(ξ) and R(ξ) are given by the following relations:

$$L(\xi) = \sqrt{1 + \frac{(x - a_1)}{u}} \tag{18}$$

$$R(\xi) = 1 - \left(\frac{x - a_2}{v} \right)^2 \tag{19}$$

For the design computations, the influence of the variance of the variables in the safety grade will be now investigated. The following six analysis cases will be considered:

(a) All design variables are fuzzy variables according to table 1.
(b) Gamma is constant (22.0 kN/m²); all other design variables are fuzzy variables.
(c) All design variables are constant except the cohesion ($\tilde{c}$).
(d) All design variables are constant except the friction angle ($\tilde{\varphi}$).
(e) All design variables are constant except anchor force ($\tilde{T}$).
(f) All design variables are constant except the inclination of the slope ($\tilde{\theta}$).

The results of the computations are depicted in Figure 4 and Figure 5. Figure 4 shows the analysis cases (a) and (b). This diagram illustrates that the influence of the fuzziness of the weight of the slope in the safety grade is very appreciable.

If we consider the cases (a) or (b), then we are dealing with fuzzy grad of safety (fuzzy-safety factor). Now, for the assessment of the safety state, we need to recognize the basic concept, namely the safety against Failure is not any more a crisp set. In Figure 4, it is clear that the domain less than 1 (< 1.0) (Failure domain) is smaller than the domain > 1.0 (safety domain). One can in this situation assess the slope stability as more or less safe (Safety state III) (see Figure 2).

The stability investigation according to the European Code of Practice EC 7 (Limit State Design in the Geotechnics). The partial safety

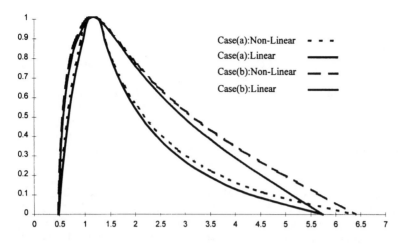

Figure 4. Fuzzy-Safety factor for the analysis cases (a) and (b)

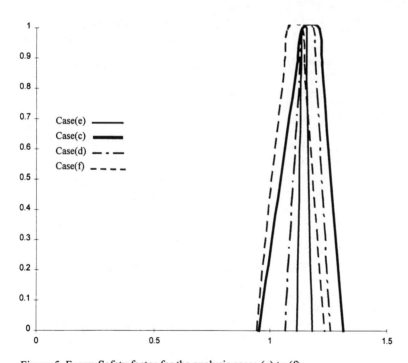

Figure 5. Fuzzy-Safety factor for the analysis cases (c) to (f)

factors are 1.6 for cohesion, 1.25 for friction angle and 1.3 for the anchor force. For determining the characteristic values, two assumptions were made:

(1) The constant values of the vector X in equation (22) are taken as characteristic values.

(2) The maximum values of the cohesion (c = 20 kN/m^2), frictional angle (φ =35°), and anchor force (T= 125 kN/m) are considered as characteristic values.

Assumption (l) resulted in a degree of utilization (l/f) of (1.0/0.786) and according to assumption (2) a value of (1/0,989). In other words, both cases represent unsafe structure! When the variance of all design variables is taken into account (not only c, φ and T, but also the geometry of the slope, joints and the rock unit weight), Figure 4 shows evidently that this slope is never in an absolute failure state.

5 CONCLUSIONS

A genuine essence of engineering design is to harmonize between a sound theory, a general methodological framework, efficient computation algorithm and detailed validation schemes. The theoretical development is commonly influenced by the practice. But when the theory gains appreciably from the application of other fields, so the theory will profit more utilities from this inclusion and in this case it is the theory which enhances and influences the practice. The Fuzzy-Safety Concept in the geotechnics is a typical example.

The probabilistic concept in geotechnical safety analysis can give only in special conditions reliable results. The obligation of the "mean value" as a representative value from partially defined data can adversely affect the model.

The fuzzy logic concept in the limit state design is patronized upon the inclusion and handling of the uncertainty in geotechnical structure through a combination of objective and mostly explicitly expressed data and from information that is eminently subjective. In this context, reflects the subjective preference in the correctness of the doubtful data the grade of its uncertainty. Thus, the designed structure will have less failure potential, more realistic functionality and aggrandized frugality.

Remarks

All the computations in this paper are performed using the Program WinFuzz2. This program can be ordered directly from the authors.

REFERENCES

CASAGRADE, A. 1965: Role of the Calculated Risk in Earthwork and Foundation Engineering. J. of Soil Mechanics Division, ASCE, July Vol. 9 1, No. SM4.

GUDEHUS, G. 1979: Statistische Sicherheitsnachweise im Grundbau, Geotechnik : 233-238.

HARTMANN, R und NAWARI, O. 1996: Implementing the Fuzzy Logic and Sets Theory in the Geotechnik: new approaches for safety and risk evaluation", (in German) German National Conference for the Geotechnik, Berlin, Sep., 1996: 505-516.

KOSKO, B. 1992: Neural Network and Fussy System – A Dynamic Approach to Machine Intelligence. Englewood Cliffs, NJ, Prentice-Hall.

NAWARI, O. and HARTMANN, R. 1997: Determination of the Characteristic values with respect to the new European Codes in Civil Engineering using Fuzzy Modeling ", (in German) Die Bautechnik 74, Heft 4, : 227-232, Berlin, April 1997.

NAWARI, O. and HARTMANLN, R. 1997: Risk Analysis of Landslides using Fuzzy System Theory", NAFEMS WORLD CONGRESS 97: Design Simulation & Optimization, Reliability & Application of Computational Methods, 9-11 April 1997 in Stuttgart: 912-920, Germany.

NOVAK, V. 1989: - Fuzzy Sets and Their Applications, Adam-Hilger, Bristol, 1989.

TERZAHGI, K. and PECK, R. B. 1961: Die Bodenmechanik in der Baupraxis. Springer-Veriag, 1961.

ZADEH, L. A. 1965: Fuzzy sets and systems, Proc. Symp. System Theory, Polytech. Inst. Brooklyn :20-37.

Geotechnical Hazards, Marić, Lisac & Szavits-Nossan (eds) © 1998 Taylor & Francis, ISBN 90 5410 957 2

Probabilistic risk assessment of landslide related geohazards

Mihail E. Popescu & Aurel Trandafir
University of Civil Engineering, Bucharest, Romania

Antonio Federico & Vincenzo Simeone
Politecnico di Bari, Faculty of Engineering, Taranto, Italy

ABSTRACT: The general mechanisms of landslide related geohazards are now fairly well understood but there remains the problem of establishing the risks to structural design and construction. This is being tackled by relating the local ground conditions to the regional geological surveys and integrate this with site-specific information using a Geographical Information System. For regional scales of assessment, influence based upon statistical analysis has been successfully applied to reflect their degree of influence, and then the factors combined to produce a hazard potential. The main aim of this contribution is to highlight the need for a probabilistic framework for slope stability analysis and emphasize that at present stage, the deterministic and probabilistic approaches must be regarded as complementary.

1 LANDSLIDES AS HAZARDS

At present society is greatly concerned with both natural and man-made hazards and the environmental impact of any engineering activity is considered to be extremely important all over the world. This decade (1990's) has been designated as the International Decade for Natural Disaster Reduction by the United Nations General Assembly.

To cope with hazards, whether natural or man-made, it is necessary to understand risk and try to quantify it. One site may be exposed to a spectrum of different hazardous events, such as storms, floods, earthquakes, rock avalanches and debris flows, which must be assessed separately, both in respect with their magnitude and probability of occurrence.

Frequently occurring cyclic events, such as storms, floods and earthquakes, can be observed over a period of time to obtain statistical data. Hazard probability can then be derived from an expected frequency. Some events, such as many landslides, are non-cyclic. Their probability of occurrence is not a frequency, but a measure of uncertainty whether they will ever take place or not (Popescu, 1996).

Risk engineering process includes (Boyd, 1994):
(1) identification of hazards;
(2) understanding the causes and sources of hazard;
(3) assessing the consequences which might arise as a result of the hazard occurring;

(4) assessing the probability of the hazard occurring;
(5) developing precautions to minimize the risk or mitigate the consequences, and
(6) assessing residual risk and its tolerability.

Within the framework of the United Nations International Decade for Natural Disaster Reduction, the International Union of Geological Sciences has established a Working Group on Landslides which is assisting the creation of a World Landslide Inventory. This has proposed a standard terminology for describing landslides; thus a working definition for a landslide is "the movement of a mass of rock, earth or debris down a slope" (Cruden, 1991).

In order to ascertain the potential hazard arising from the movement of a slope, what is needed is the likelihood of movement within the design life of the engineering works, and then be able to map its probable extent.

In order to achieve this, both geotechnical criteria (from field sampling and in-situ measurements, and laboratory investigations) and an appropriate theoretical model (using stress and stability calculations) will be required. The outcome will be a numerical "Factor of Safety" on which the assessment of stability can be made, but this will additionally require a value judgment concerning the appropriateness of the theoretical model and the quality of the available data.

The concept of likelihood requires a consideration of time, within which the geological environment exerts an influential role, as explored by Flageollet (1996) in his review of the temporal dimension of mass movement. For older landslides, climatically induced instability would be less likely to recur in the short term than would be seismically-induced instability since the threshold values of the former are currently lower (Palmquist, Bible, 1980).

The consideration of criteria can be refined by distinguishing between constraints, limiting the area in which slope instability is feasible, and factors, whereby the criteria can be measured on a relative scale indicating a variable degree of likelihood for slope instability occurring. A further consideration is the separation of the causal factors which are principally responsible for the general stability from those triggering factors which actually bring about failure (Popescu, 1996).

2 LANDSLIDE HAZARD ASSESSMENT

The keyword in hazard assessment is uncertainty and hence probabilistic methods are the most appropriate in both defining the areas at risk and analysing the mechanisms. The basis of a probabilistic framework is the recognition that the estimated factor of safety reflects imperfect knowledge and is, therefore, a variable.

The most important probabilistic method used in geotechnical risk assessment is the first order second moment method. One of the advantages of this method is that one can derive moments of the dependent variable from moments of the independent variables, without knowledge of probability density functions. The method is based on the mean value and coefficient of variation only.

The statistics characterising the soil variability and the variability of the spatial average values are necessary inputs into probabilistic methods for quantifying risk and reliability of soil structures. Soil properties should be modeled as spatially correlated variables or "random variables". The use of perfectly correlated soil properties gives rise to unrealistically large values of failure probabilities for geotechnical structures (White,1993).

The stability of slopes is undoubtedly the most popular application of probabilistic and reliability methods judging from the number of publications on this subject. Slope stability mechanics and analysis procedures are felt to be so well understood that variability of input parameters is considered to be the only significant unknown. Statistical distribution of input data such as soil strength and pore-pressure is analysed to estimate failure probability. However, 'ignorance factors' are more important than natural variability, yet they are almost never accounted for in probabilistic stability studies.

D'Appolonia (1977) proposed that an important consideration in a probabilistic analysis was what he referred to as « unknown unknowns ». These might include unanticipated geological conditions or some influence of the construction process. It is likely that provision for this component should be greater for new types of structures, new geological conditions and where either the structure or the geology is highly complex.

With the availability of friendly software, geotechnical reliability evaluations can be greatly facilitated. Recent changes in both computer hardware and software allow the responsible engineer to be now able to be more involved in personally running analyses and at the same time to be more readily able to construct a reasonably detailed model of the problem.

When mapping geological and geotechnical hazards one has to integrate regional geological surveys with site-specific information. Such integration suggests the use of a computer data management system and this would require the incorporation of a GIS (Geographical Information System) into the evaluation of ground related hazards. GIS is ideally suited to situations where there are many attributes at a specific location or across an area. An example of a GIS approach to landslides and structurally unstable soils related risk assessment in Romania was presented by Popescu and Rosenbaum (1993).

The ability to perform spatial operations is unique to GIS. The relational capability of GIS allows different databases to be linked together, spatial position being the common field. Data can be imported from external sources, principally as directly digitised files, scanned images or digital raster images derived from other systems, to which attributes can readily be assigned.

Unless considerable ground investigation data is already available, a didactic approach to map algebra can be difficult to apply and therefore in most situations, particularly for regional scale projects, an heuristic approach is usually more appropriate (Rosenbaum, Popescu, 1996).

The issue of scale has been discussed by Leroi (1996) whereby it is concluded that the GIS images and maps portraying the level of hazard need to be comparable to the regional scale of information likely to be available in support of the study, at 1:50,000 to 1:100,000. The final risk maps desired by engineers and planners are required at site specific scales of around 1:5,000. This order of magnitude difference requires careful judgment regarding the level of detail which can be realistically interpolated, and caution needs to be exercised where undue generalisation might be masking important local factors.

3 RELIABILITY ANALYSIS OF SLOPE STABILITY

Current procedures for evaluating the safety of slopes consist in determining a factor of safety which is compared with allowable values found to be satisfactory on the basis of previous experience. The factor of safety suffers from the following:

(1) Elements of uncertainty in analyses are not quantified when the factor of safety is used.

(2) The scale of the factor of safety is not known. For example, a slope with a factor of safety of 3.0 is not necessarily twice as safe as another with a factor of safety of 1.5.

(3) Allowable values to be selected for the factor of safety are the result of experience. In dealing with new or different problems for which there is no previous experience, there is no allowable factor of safety.

There is much to be gained from applying concepts of risk analysis to supplement conventional procedures for determining the factor of safety against shear failure of soil slopes.

The reliability index of a slope can be defined in several ways on the basis of the probability distribution of the random variable(s) governing the slope failure mechanism(s). Thus considering the factor of safety, one of the simplest definitions of the reliability index is:

$$\beta = (E[F] - 1.0) / \sigma[F] \qquad (1)$$

in which E[F] is the best estimate of the factor of safety and $\sigma[F]$ is the standard deviation of the factor of safety.

The reliability index provides a better indication of how close the slope is to failure than does the factor of safety alone. This is because it incorporates contributions due to randomness of soil parameters, geometry, environmental loading, and other physical effects as well as due to uncertainties in the computational method.

Slopes with large values of the reliability index are farther from failure than slopes with small values of the reliability index, regardless of the value of the best estimate of the factor of safety. This is illustrated in Fig.1 by the example of two slopes with different probability distribution functions for the factor of safety. The slope with the higher E[F] has a larger probability of failure than the slope with the lower E[F] though the conventional approach, as well as many regulations, would regard the former slope as significantly safer than the latter.

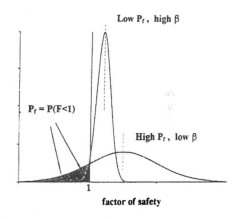

Figure 1 Example of two slopes with different pdf of factor of safety

A variety of techniques for evaluating the reliability index and the probability of failure is available. Two common methods are the Monte Carlo method and the Point Estimate method and these have been selected for use in this study.

The Monte Carlo simulation method determines the mean and standard deviation of a performance function of random variables by performing repeated computations with the randomly generated values for the component variables.

There are four stages in a Monte Carlo simulation:

(1) Generate random numbers (i.e. independent random variables uniformly distributed over the unit interval between zero and one), transform the random numbers from a uniform distribution to the distribution applicable to the component variable and calculate values of all component variables based on the appropriate random numbers.

(2) Using the randomly generated values of the component variables, compute the system performance function (i.e. factor of safety).

(3) Repeat steps (1) and (2) a large numbers of times. The number of times depends on the variability of the

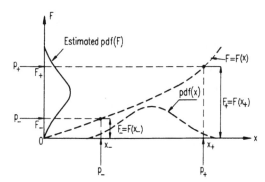

Figure 2 Transfer of information in the Point Estimate method

$$E[F] = \sum (P_{1\pm} \cdot P_{2\pm} \cdots P_{N\pm}) F(X_{1\pm}, X_{2\pm}, \ldots X_{N\pm}) \quad (3a)$$

$$E[F^2] = \sum (P_{1\pm} \cdot P_{2\pm} \cdots P_{N\pm}) F^2(X_{1\pm}, X_{2\pm}, \ldots X_{N\pm}) \quad (3b)$$

$$\sigma[F] = \sqrt{E[F^2] - (E[F])^2} \quad (3c)$$

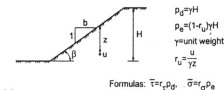

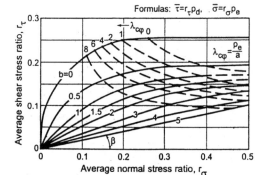

Figure 3 Dimensionless resistance envelopes developed by Janbu (1954)

input and output parameters and the desired accuracy of the output.

(4) Create a cumulative distribution of the system performance function (i.e. factor of safety) using the data obtained from the above simulations. Interpretation of the distribution of the system performance function provides the mean and standard deviation of the factor of safety.

The Monte Carlo simulation requires a high-speed computer so that a large number of trials can be conducted. Priest and Brown (1983) developed a probabilistic slope stability analysis method based on a Monte Carlo simulation routine with use of the Janbu method for which strength parameters were derived by the Hoek-Brown rock mass strength criterion.

An alternative approach for calculating the mean and standard deviation of the factor of safety is the Point Estimate method (Rosenblueth, 1975).

The basic principle of the method is illustrated in Fig.2 where the probability distribution function (pdf) of a random variable X is approximated by a two-point probability mass function. The mass function consists of concentrations P_+ and P_- at X_+ and X_-, respectively. If F(X) is a function of X, a two-point approximation of the pdf of F is obtained by evaluating the function F(X) at X_+ and X_- ($F_+ = F(X_+)$ and $F_- = F(X_-)$):

$$E[F] = P_+ F_+ + P_- F_- \quad (2a)$$

$$E[F]^2 = P_+ F_+^2 + P_- F_-^2 \quad (2b)$$

$$\sigma[F] = \sqrt{E[F^2] - (E[F])^2} \quad (2c)$$

In general if F is a function of N random independent variables, then 2^N points are needed to approximate the multivariate mass function and the entire procedure can be summarized as follows:

For correlated random variables, additional adjustment must be made to the probability concentrations, P_i. Where symmetrically distributed variables are assumed, the point estimates X_{i+} and X_{i-} are taken at one standard deviation above and below the expected value, respectively. The probability concentration or weighting factor P for the case of uncorrelated random variables with symmetrical distribution is equal to $(1/2)^N$.

4 QUICK EVALUATION OF THE FAILURE PROBABILITY WITH THE RESISTANCE ENVELOPE METHOD

In any slope stability evaluation, two basically different sets of data should be considered, namely (Janbu, 1954):

(a) the state of stress required to keep a given slope in equilibrium (obtainable analytically without any 'a priori' knowledge of the strength)

(b) the soil conditions, including the groundwater regime, and the shear strength parameters (obtainable only experimentally, without any 'a priori' knowledge of the level of safety).

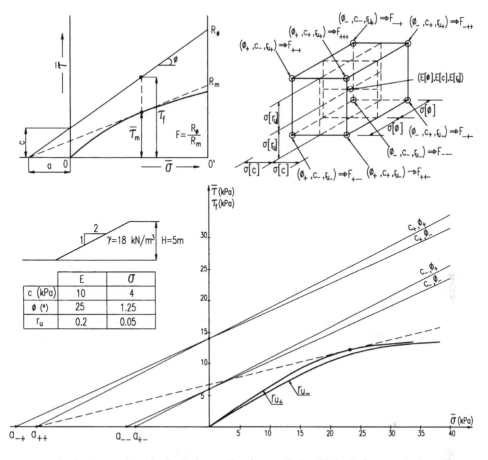

Figure 4 Simple slope analysed by the Resistance Envelope method and Point Estimate method

Casagrande (1950) developed a procedure to present the results of stability analyses, which was referred to as the Resistance Envelope method. It combines in a systematic way the two sets of data mentioned above. The resistance envelope represents the general relation between the average normal stress and the minimum value of the shear strength that should be developed by the material for the stability of the slope under consideration. Hence the resistance envelope is independent of the strength characteristics of the material.

A series of dimensionless resistance envelopes has been presented for use with simple uniform slopes where a failure model involving circular slip surfaces may be assumed (Fig.3). The strength envelope may be drawn on the same chart with the resistance envelope and the factor of safety visually assessed. Applications of the Resistance Envelope method have been discussed, among others, by Popescu (1985) and Federico et al. (1996).

There are now numerous deterministic software packages which have been coded for computing the factor of safety of an earth or rock slope based only on a fixed set of conditions which includes shear strength parameters, pore water pressure and slope geometry. Dai, Fredlund and Stolte (1993) illustrated how a deterministic slope stability package can be used in a probabilistic manner.

In the following a simple approach to the probability of failure of a slope based on Casagrande Resistance Envelope method with the Point Estimate method is introduced. The approach can be easily applied by hand computation and consequently can be used to check more sophisticated computer solutions.

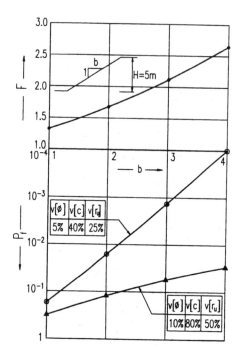

Figure 5 Factor of safety and probability of failure vs. slope ratio

The procedure is illustrated in Fig.4 for a simple slope. The soil shear strength parameters (i.e. cohesion and angle of internal friction) and the pore water pressure ratio have been considered as random variables that follow a normal distribution with known values of the mean and the standard deviation. The Point Estimate method was used to compute the magnitude of these variables at values of one standard deviation on either side of the mean values for all three variables (i.e. 2^3 terms). The Resistance Envelope method provided the point estimates of the factor of safety and the mean and the standard deviation of the factor of safety function have been obtained by the Point Estimate method. After assuming a normal statistical distribution for the factor of safety function, the following values of the probability of failure and the reliability index have been derived: $P_f = 1.82 \times 10^{-2}$ and $\beta = 2.09$.

The simple slope example illustrated in Fig.4 has been analysed with a computer program based on the procedure developed by Priest and Brown (1983). The program combines Janbu slope stability analysis method with Monte Carlo simulation method. The results of the computer program run are $P_f = 1.16 \times 10^{-2}$ and $\beta = 2.18$ which compare well with the results

obtained by the simple procedure based on the Resistance Envelope method with the Point Estimate method.

A major benefit of the probabilistic approach is that the sensitivity of the solution to various parameters can be determined. The parameters of most relevance for slope stability studies are the standard deviations of the cohesion, angle of internal friction and pore-water pressure ratio. This is illustrated in Fig.5 which gives the relationship between the factor of safety and the probability of failure as a function of the slope inclination, for a fixed height of the slope. The higher the standard deviation of the cohesion, angle of internal friction and pore-water pressure ratio, the higher the probability of failure. Using charts of the type presented in Fig.5 it becomes possible to design slopes for any desirable, or acceptable, risk of failure depending on the level of parameters scattering.

The simple slope presented in Fig.4 have been analysed by varying the standard deviation of each variable while maintaining the standard deviation for the other two variables at the original values given in Fig.4. The results plotted in Fig.6 show that the probability of failure is very sensitive to the standard deviation of the cohesion and is less sensitive in raport with the standard deviation of pore-water pressure ratio and angle of internal friction.

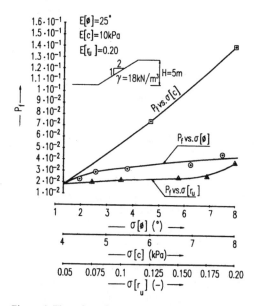

Figure 6 The effect of standard deviation of cohesion, angle of internal friction and pore-water pressure ratio on the probability of failure

Table 1 Probabilistic slope design criteria

CATEGORY OF SLOPE	CONSEQUENCES OF FAILURE	EXAMPLES	ACCEPTABLE VALUES		
			MINIMUM MEAN **F**	MAXIMA	
				P(F<1.0)	P(F<1.5)
1	Not serious	Individual benches; small* temporary slopes not adjacent to haulage roads	1.3	0.1	0.2
2	Moderately serious	Any slope of a permanent or semipermanent nature	1.6	0.01	0.1
3	Very serious	Medium-size and high slopes carrying major haulage roads or underlying permanent mine installations	2.0	0.003	0.05

* Small, height < 50m; medium, height 50-150m; high, height > 150 m

Guide to interpretation of slope performance

PERFORMANCE OF SLOPE	INTERPRETATION
Satisfies all three criteria	Stable slope
Exceeds minimum mean F, but violates one of both probabilistic criteria	Operation of slope presents risk that may or may not be acceptable; level of risk can be reduced by comprehensive monitoring programme
Falls below minimum mean F, but satisfies both probabilistic criteria	Marginal slope: minor modifications of slope geometry required to raise mean F to satisfactory level
Falls below minimum mean F and violates one or both probabilistic criteria	Unstable slope: major modifications of slope geometry required; rock improvement and slope monitoring may be necessary

5 CONCLUDING REMARKS

Slope stability engineering is concerned with decision-making based on information and analysis combined with observation. Concepts of statistics and probability have been used from time to time but, as a formal basis for analysis, the use of probabilistic framework has been advocated only in the last few decades.

Probabilistic slope stability analysis is a developing field with various models including different effects. A complete model for probabilistic analysis of a soil slope is still some way off. Nevertheless even a poor probabilistic analysis can add valuable information in guide the slope designer in decision making.

A guide of decision making and interpreting open-pit slope performance in the context of a probabilistic framework is given in Table1 according with Priest and Brown (1983).

ACKNOWLEDGEMENTS

This paper was drafted when the first author was with the Faculty of Engineering in Taranto as a CNR Visiting Professor. The hospitality of the Faculty is greatly appreciated. Gabriel Radulescu and Catalin Birtea, graduate students of the University of Civil Engineering, provided assistance with the computational work, Silviu Vasilica and Alina Rancea carefully prepared the figures and word processed the text.

REFERENCES

Boyd, R. D. 1994: Managing risk. *Ground Engineering*, vol.27, no.5, p. 30-33.

Casagrande, A. 1950: Notes on the design of earth dams, *Journ ASCE*, vol. 37, no. 4, Boston.

Christian, J.T. 1990: Reliability methods for slope stability of existing slopes, Proc. ASCE Geotechnical Conference on Slopes and Embankments, Berkeley, p. 409-418.

Cruden, D. M. 1991. A simple definition of a landslide. *Bulletin of the International Association of Engineering Geology.* 43:27-29.

Dai, Y., D.G. Fredlund, Stolte, W.J. 1993: A probabilistic slope stability analysis using deterministic computer software, *Proc. Conf. Probabilistic Methods in Geotechnical Engineering,* Camberra, Australia, p 267-274.

D'Appolonia, E.D. 1977: Relationship between design and construction in soil engineering. *Proc. Int. Conf. Soil Mech. Found. Eng.,* Tokyo, p 479-485.

Flageollet, J.C. 1996. The time dimension in the study of mass movements. *Geomorphology,* 15:185-190.

Federico, A., S. Adamo, V. Simeone 1996. Il metodi degli inviluppi di resistentza per l'analisi di stabilita dei pendii, *Geologia Applicata e Idrogeologia,* XXXI, p. 271-288.

Harr, M.E. 1977: Mechanics of particulate media-A probabilistic approach, McGraw-Hill, 543 p.

Janbu, N. 1954: Stability analysis of slopes with dimensionless parameters, *Harvard Soil Mech. Series,* no. 46, 81 p.

Leroi, E. 1996. Landslide hazard-risk maps at different scales: objectives, tools and developments. Special Lecture. *Proc. 7th Int. Symposium on Landslides,* Trondheim, vol. 1, p. 35-51.

Palmquist, R. C., Bible, G. 1980. Conceptual modelling of landslide distribution in time and space. *Bulletin of the International Association of Engineering Geology,* 21:178-186.

Popescu, M. 1985. Quelques considerations sur l'analyse de la stabilité au glissement des talus et des versants, *Rivista Italiana di Geotecnica,* XIX, no. 1, p. 7-22.

Popescu, M., Rosenbaum, M. 1993: Ground - related risks in Romania, *Geoscientist,* vol.3, no.3, p. 20-22.

Popescu, M. 1996. From landslide causes to landslide remediation, Special Lecture, *Proc. 7th Int. Symposium on Landslides,* Trondheim, vol. 1, p. 75-96.

Priest, S. D., E. T. Brown. 1983. Probabilistic staability analysis of variable rock slopes. *Trans Instn. Ming. Metall.* A92:1-12.

Rosenbaum, M., Popescu, M. 1996. Using a Geographical Information System to record and assess landslide-related risks in Romania. *Proc. 7th Int. Symposium on Landslides,* Trondheim, vol. 1,

p. 363-370.

Rosenblueth 1975: Point estimates for probability moments, *Proc. Natl. Academy of Science USA,* vol. 72, no. 10

White, W. 1993: Soil variability: Characterisation and modeling, *Proc. Conf. Probabilistic Methods in Geotechnical Engineering,* Canberra, p.111-120.

Geotechnical Hazards, Marić, Lisac & Szavits-Nossan (eds) © 1998 Taylor & Francis, ISBN 90 5410 957 2

Diminished shear strength surfaces within the southern Medvednica hill-sides in the city of Zagreb

Ž. Sokolić & N. Kralj
Geotehnički Studio d.o.o., Zagreb, Croatia

SUMMARY: In a part of the city of Zagreb built on southern hill-sides of the Medvednica mountain one is in the day-to-day geotechnical practice often confronted with the problem of soil instability (slide and creep). As usual, causes of the ground destabilization are various (geological and hydrogeological predispositions and influence of man). At certain of investigated sites existence of underground interfaces with specific characteristics was found. As it was assessed that their presence can have an essential influence on the terrain stability, a programme of investigating such phenomena was prepared. The paper presents results obtained so far by investigations performed at two sites.

1 GENERAL

The city of Zagreb has been built and is spreading over a low-lying part, as well as over southern hill-sides of the Medvednica mountain.

From the geotechnical point of view these two areas have their respective specific qualities.

The low-lying part is characterized by alluvial deposits of the Sava river. Upon the preconsolidated clay of variable depth, but most often at approximately 12 meters below the terrain surface, lie deposits of well-grained gravel upon which there is a surface cover of clayey-silty material. The ground water is to be found on clays and its table oscillates within the gravel layer and is in direct contact with the Sava river level.

Lower parts of the Medvednica mountain hill-sides are made of fresh-water clayey sediments. Sandy and gravelly material from erosion of higher parts has been transported by torrents and streams to the area of the clayey materials' sedimentation.

In such a way layers were formed, containing independent or interlinked lenses of more pervious material within clayey series. Clay sediments were consolidated due to their own weight. At a later time, tectonic movements caused folding, faulting and other deformations, accompanied by strong earthquakes.

Under such forces clays have consolidated into a preconsolidated cracked clay. More pervious sandy-gravelly layers within the clay are often saturated with water. Water penetrates into these layers in zones where they, in their upper horizons, appear at the terrain surface.

In such geological formation, considering also the morphology characteristics and activities of man (forest cutting and conversion of forested areas into cultivable ones, such as vineyards, building of traffic routes, building of water supply installations, but not always accompanied by sewage et sim.) has often as its consequence a soil destabilization in form of creeping and sliding.

These processes are activated within the soil along the surfaces of the least shear strength. At some of the investigated sites the presence of specific interfaces was recorded. Due to their situation in the underground and an obviously decreased possibility to take over shear stresses, a programme of testing the physical-mechanical properties was prepared.

2 DESCRIPTION OF INTERFACES

The interfaces being discussed here have been recorded at the contact of pervious and impervious soil layers.

The upper layer is more or less clay-mixed gravel or sand, while the lower layer is highly plastic clay. The interface is in fact bordered by a very thin layer of highly plastic clay, grey-blue in colour. Thickness of this layer does not surpass a couple of milimeters. In the course of exploratory drilling and excavation of trenches they are hard to register visually. The easiest way to find the interface is to make an excavation down to the depth of some 10 centimeters above the contact of pervious and impervious layers. Then the remaining part of the upper layer is sheared upon the lower one by dragging. For such treatment it is convenient to use a smaller pick, knife or a similar slab-like object. The cutting-edge is to be stuck into the upper layer to a depth somewhat lesser than the expected contact of interfaced layers and is then dragged parallel with the interface.

3 TESTING OF INTERFACES

The first site where some investigations of this phenomenon were performed is in the street called Pantovčak, at No. 72A. 1937 at this site a summer cottage was built. Since then until the present the building was expanded and storeys added to it a number of times. Presently it is used for a monastery of Franciscan sisters.

Due to the appearance of cracks on the building, the first geotechnical soil investigation in order to find the cause of appearance of cracks was for the first time performed in 1976. The geotechnical report, prepared 1976, comes to the conclusion that at the whole site exist signs of surface sliding and some suspicions about an increased activity of clay layers upon which the building is partly founded are voiced. Following such conclusions, a remedy was undertaken at the eastern part of the building, assuming form of reinforcing and deepening of foundation structure.

However, even after the remedy of the foundation structure below the eastern part of the building, cracks were still appearing on the building, becoming larger and larger in the course of time. Eighteen years after the first remedy, that is to say in 1994, cracks were of such intensity that the usability of building had become questionable because of them and a necessity to tackle the subject problem anew appeared. After performing new investigations, a solution was presented and the remedial works on a part of foundation structure performed in 1995, accompanied with installation of piezometers and inclinometers for the purpose of monitoring the ground water table and possible displacement of the soil.

In the course of excavating the drainage trench, which was performed after the commencement of monitoring developments along the slope, spreading of soil layers was monitored. It is significant that at the contact of pervious and impervious soil layers the previously described interfaces were observed.

In the manner described in Paragraph 2, a larger surface area of the interface was separated and the material from the contact scraped off with a thin knife. Samples of the material below and above the interface were also taken. The three thus obtained samples were subjected to an X-ray qualitative analysis. Mineral composition was determined by the X-ray diffraction method on silty samples. Determinations were reached on the Philips vertical X-ray goniometer, using a Cu tube, radiation of which was monochromatised by a graphite crystal monchromatiser. X-raying conditions were the same for all samples. Original samples were X-rayed, while, in order to enable a more precise determination of clays, samples being heated at the temperature of $400°C$, as well as samples being treated with ethylene glycol, or glycerine, were also X-rayed. Test results are as follows:

Sample 1 - clay above the interface
predominant ingredient: vermiculite (probably calcium-magnesium)
minor ingredients: quartz and plagioclase
accessory: mica, amphibole and epidote

Sample 2 - interface (clayey film)
predominant ingredient: vermiculite
minor ingredients: smectite, quartz and plagioclase
accessory: mica, amphibole and epidote

Sample 3 - clay beneath the interface
predominant ingredients: smectite and vermiculite
minor ingredients: quartz and plagioclase
accessory: mica, amphibole and epidote

On the basis of X-rays taken, one could say that the share of respective minerals is in all three samples more or less the same, excepting the vermiculite, share of which decreases and smectite, share of which increases from Sample 1 to Sample 3. Atterberg limits of plasticity of contact material were tested in the geomechanical lab, with the following results:

872

wl = 76.90%   wp = 23.18%   Ip = 53.72%

By monitoring the inclinometer shifts at the site Pantovčak 72A, which is in the course since 1995 until the present, displacements were recorded at the depths which correspond with depths at which the interface was recorded. Interface was recorded at the depth of 3.10 m beneath the ground surfaces (see Figure 1). Soil displacements were taking place at water tables as shown in Figure 1.

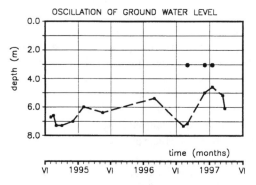

Time and depth of recorded displacement

Figure 1. Diagram of the ground water table oscillation and occurrence and depth of soil displacements.

It is obvious that the ground water table at the recorded displacements was deeper than the interface position. It can also be seen that displacements were taking place at extremely low or extremely high ground water table, which is connection with markedly dry (July 1996) or markedly rainy periods (December 1996 and January 1997). This fact can be linked with the clay swelling properties and activity of its minerals due to change in moisture. According to measured magnitudes of plasticity index and a passage percentage finer than 0.02 dia. sieve, using at the same time the experiences of Williams and Donaldson (1980) and van der Merve (1975), tested samples point to a high to very high susceptibility to swelling (see Figure 2). One could assume that such property of clay is influenced by minerals vermiculite and smectite.

In the place called Remete, which is also positioned on southern hill-sides of Medvednica mountain, in excavation of an exploratory cut, an interface of almost same visual marks as the one from the Pantovčak site was run into. After discovering this interface in the nature, it was decided to try and investigate physical-mechanical properties of the contact.

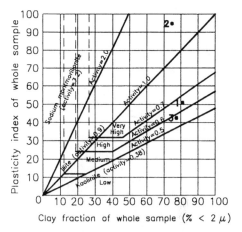

Figure 2 Expansiveness of soils and activity of clay minerals
R.E. Hunt Fig. 4.34 (1986)[166] (After Williams and Donaldson (1980)[51] ; van der Merwe (1975)[52])

Due to a very small thickness, it was impossible to take a standard undisturbed sample. A thin "film" of clay was therefore carefully scraped off the interface, using a thin knife. The sample obtained in this way was then in the lab inserted into the circular shear apparatus. Two tests were performed and residual parameters of shearing resistance obtained:

$c_{1r} = 0 \text{ kN/m}^2$ $c_{2r} = 0 \text{ kN/m}^2$
$\varphi_{1r} = 12.9°$ $\varphi_{1r} = 9.2°$

Shearing diagrams are shown in Figure 3. Second sample was taken in form of a cube with approx. 20 cm long sides. The interface was positioned at the approximate middle of cube's height. This sample was worked in the lab in such a way that lesser samples were prepared, which were then inserted into the circular shear apparatus. An effort was made to insert samples into the shear box in such manner that interfaces were laid down at the level of shearing surface. Results of these tests provided peak and residual shearing resistances (see Figure 3):

$c_v = 12.50 \text{ kN/m}^2$ $\varphi_v = 20.3°$
$c_r = 0 \text{ kN/m}^2$ $\varphi_r = 17.6°$

873

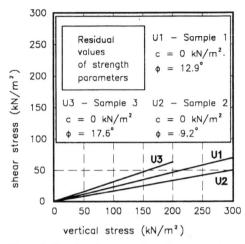

Figure 3. Shearing diagrams of tested samples (Samples U1 and U2 - circular shear, sample 3 - direct shear)

The grain-size distribution was also tested on three samples. The share of clay particles fluctuates from 76% to 80% (see Figure 4).

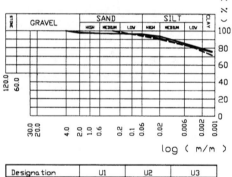

Figure 4. Grain-size diagrams of tested samples

On samples were also tested the Atterberg plasticity limits, with the following results:

sample 1: wl = 76.54% wp = 24.04% Ip = 52.50%
sample 2: wl = 122.94% wp = 28.43% Ip = 94.51%
sample 3 (cube): wl = 73.2 wp = 30.52%I p = 42.50%

4 DELIBERATION ON RESULTS

The question of destabilization of slopes along the southern hill-sides of Medvednica mountain, as well as the manner of their elimination engaged quite a number of soil mechanics engineers who had been confronted with them in the city of Zagreb. Involving themselves in resolving of these problems, authors were reaching very useful experiences. One of such, which we stress here, because of its link-up with topics of this paper, is investigation performed by prof. Nonveiller (1964), who was first one to call attention to presence of clays as materials with minimum residual parameters of shear resistance. His investigation has demonstrated that the sample of pure bluish clay, tested in a lab against shearing, (sample from a landslide at Prekriźje) has the following parameters:

$\varphi_r = 8.5°$
c = 27 kN/m² at moisture of 29% and
c = 10 kN/m² at moisture of 38%

Nonveiller quotes in his closing deliberation: "Conditions of origin of sliding in similar cases could be explained by the fact that in the slope very slow creeps occur already at shearing stresses which are considerably smaller than the finite strength, with a safety factor in order of magnitude of Fs = 2.0, which is correspondent to the lowest strength parameters determined on tri-axial samples. Due to such slow deformations, the shear strength of bluish clay interbedded lenses, which is lower than the average is gradually decreased, so that it is smeared over greater parts of the largest deformations surface. In such a way a sliding surface gradually comes into being, upon which only the blue clay opposes shearing by its resistance. In the process the bluish clay is being remoulded, its moisture and porosity are growing, while at the same time the share of cohesion in the total resistance against shearing decreases. In this manner the safety factor also gradually decreases, until it reaches the value of Fs = 1.0 and sliding becomes active. This result is in concordance with observations made by other investigators, i.e. that the permanent stability of a preconsolidated clay slope is dependent only on resistance, while the share of cohesion gradually decreases. Further on, it demonstrates that as the evidence of safety of a slope, strength of the material component with the least shear strength is competent..."

On the basis of lab sample tests in Bishop's rotary shear apparatus on samples from landslides Fazan and Prekriźje (Nonveiller 1964) and from the site Grmoščica (Ortolan 1995) diagrams were prepared, showing interlink between the plasticity index and residual values of the inner resistance angle.

If in this diagram values obtained at Pantovčak and Remete sites are plotted (see Figure 5), one can deduce that results decrease along the previously known line, translating the latter for the magnitude of inner resistance angle of 1.5° to 2.5° upwards.

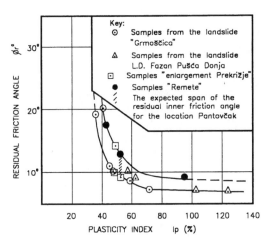

Figure 5. Diagram of interdependence of the plasticity index and the residual angle of inner resistance of highly plastic clays and interfaces in the sub-Medvednica area

One should mention that all samples were tested on Bishop's rotary shear apparatus under the same conditions. The tested materials were sedimented in similar environments and are correspondent by the time of origin.

We assess this fact to be valuable for two reasons. Firstly, it shows the concurrence of mechanical properties of described interfaces which in the package of Quaternary interchange of layers of the Medvednica southern slopes present critical surfaces in view of a possible origin of the sliding surface. The second reason is that the presented diagram (Figure 5) may in the day-to-day work be used in choosing residual shear resistance parameters. This is stressed by the fact that standard investigation methods are unable to explore real shear resistance parameters in a qualitative manner.

In doing so one should keep within limits of justifiableness of applying these directions only in such geotechnical environments where, on the basis of performed investigations, a possibility of existence of interfaces is assessed to be real.

5 CONCLUSION

In performing geotechnical investigative works for buildings within the sub-Medvednica areas of the city of Zagreb in Plio-Quaternary and Quaternary formation zones it is recommended that, beside usual operations, a possibility of existence of underground interfaces is assessed. Possibility of their existence is preponderantly tied to the existence of interbedding where pervious layers of gravel and sand, mixed with silts and clays are in contact with poorly pervious, highly plastic clays. Possibility of existence of interfaces becomes greater as the clay is of higher plasticity. Danger of slide or creep is increased if the inclination of interface is somewhat steeper than the inclination of terrain. In the past cases, colour of the clay layer was regularly a mix of grey and yellow-brown.

Reasons of originating of interface are also by all means interesting. It is estimated that they are primarily bound to different perviousness properties of contacting surfaces and a permanent or occasional streaming of water through the upper, better pervious layer. Streaming of water along the contact of layers may have influence upon chemical processes, may influence activity of the lower clay layer (minerals vermiculite and smectite), or their origin is tied to an old landslide active in the past.

The presented data are based upon a small number of performed tests, so that these will be amended in further work by new lab tests, while there also is an idea to test mechanical properties directly on the ground.

6 LITERATURE

Nonveiller 1964. *Landslide in preconsolidated cracked clay at Prekriźje in city of Zagreb*, Zagreb: Gradevinar.

Ortolan 1995 *Landslide Grmoščica in the line Ilica no. 342 till no 368*, Zagreb: Institut gradevinarstva Hrvatske - Zavod za geotehniku.

Hunt 1986. *Geotechnical engineering analysis and evaluation*, New York: McGraw-Hill Book Company

Geotechnical Hazards, Marić, Lisac & Szavits-Nossan (eds) © 1998 Taylor & Francis, ISBN 90 5410 957 2

Space stability of slopes: Kinematical approach

S.Škrabl & B.Macuh
University of Maribor, Faculty of Civil Engineering, Slovenia

ABSTRACT: The theoretical bases of the space slope stability analysis and the example of its implementation are presented. The plane slope stability analyses are usually used in practice and are suitable chiefly for analyses of slopes that are uniform in the direction perpendicular to the treated profile and whose the loads are uniformly distributed in that direction. When slopes have no uniformly layered soils, local loading or unloading, more complex procedures of the space stability analysis are needed to reach more real results. The energy or kinematical method (KEM) is used in the presented paper.
The considered problem is exacting, so some simplifications are considered: the peak shear strength of the soil, blocks are completely rigid, only associative soils and translator types of the failure are treated, no potential energy dissipation on the vertical surfaces, and the Mohr-Coulomb's yield criteria is assumed.

1. INTRODUCTION

In the history of construction, the soil objects were constructed on the basis of experiences acquired by builders that was being acquired by successful or unsuccessful object construction to retain water, watering canals and traffic ways.

In the nineteenth and at the beginning of the twentieth century the set of problems appeared especially during railway construction works and other traffic objects. The errors accumulated and the acquired experiences were not sufficient for an optimal and economical construction. The systematical research of the physical properties of soils began and first approximate procedures arose in that period. In that primary period the procedures (Fellenius 1927) were started that were the combination of experiences and numerical solutions. They were later essentially improved (Bishop 1955, Janbu 1957, etc.).

With the development of the computer science the unimaginable development of the area of stability analyses started. More exacting procedures were used that considered kinematical or energy conditions of the slopes and the methods of optimisation for evaluation of the critical failure surfaces (Chen 1969, Karal 1977, Sarma 1979, Michalowski 1995, etc.). Most of the mentioned authors considered only cases of the plane stability analysis. The general procedure of the space stability analysis was introduced in mathematical form by Leshchinsky and Huang 1992.

The procedure presented in the paper is based on the method of rigid blocks. On the basis of the relationship between dissipated or available mechanical energy of the system and the quantity of the absorbed energy, the safety coefficient of the slope is evaluated and/or stability conditions of analysed objects are established.

The given solutions of the space stability are useful for safety evaluation of the designed geotechnical objects during and after the construction works as well as for analysing and designing of sanitation of the space shaped landslides. When the results of the space stability analyses are compared to the comparable results of the plane stability analyses, less critical results are obtained.

2. THE PROPOSED SOLUTION

The problem of the space stability analysis could be done by the method of the ultimate limit state or in its complementary form by energy methods (KEM).

Because the space stability problem is an exacting one some simplifications should be considered. They have been already used in usual stability analyses:

1. The highest (peak) shear strength of the soil is considered in the analysis along the failure surface, that is reached at the transition from the process of hardening to softening area.

2. In the theory only associative soils are treated; whereas non associative soils are used only when

their yield rule could be substituted by energy equivalent parameters of associative soils.

3. In the paper only translator types of failure are treated, where prismatic soil blocks are moving translator in the direction of the chosen critical direction that is characteristic for the treated landslide.

4. The influences of the potential energy dissipation on vertical surfaces of individual blocks will not be considered. Further it will be assumed that individual blocks are completely rigid and that the change in the deformation energy has arisen only on the basis of the individual rigid space elements.

5. The yield criteria (for KEM) or failure condition of Mohr-Coulomb's shear law or its yield rule is considered.

The slope consists of several different layered soils that could be cohesive or non cohesive. The essential condition of the successfulness of the space stability analysis is chiefly that soils obey Mohr-Coulomb's criteria of the failure and plastification (yield rule). Figure 1 shows the elementary part or the rigid block of the failure body of ground-plan size $(dxdy)$ and the height that is equal to the depth of the failure surface at area.

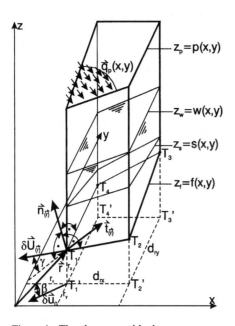

Figure 1: The elementary block

The potential failure surface is given by scalar function $z_f = f(x,y)$; individual layers are given as follows: upper layer $z_p = p(x,y)$; arbitrary layer (s) by function $z_s = s(x,y)$ and piezometric line by function $z_w = w(x,y)$, etc. The slope could be loaded with an arbitrary surface and/or inner loading that is given in the form of the distributed surface load $\bar{q} = \bar{q}(x, y)$; concentrated forces are not expected as they do not exist in the practice (wheel loads of the vehicle are substituted by uniformly distributed load as it would act on the plane of the lower structure considering the appropriate expansion of the concentrated loads).

The influences of the ground penetrated water and consolidation pore pressures are substituted with pore pressures on failure surface $u = u(x,y)$ that correspond to the sum of hydrostatic and consolidation pore pressures of the treated point of the treated slide surface. The soil own weight and possible accelerations or seismic forces could be considered as space distributed loads:

$$\bar{R} = (R_x, R_y, R_z)^T = (\rho \ddot{x}, \rho \ddot{y}, \rho(\ddot{z} - g))^T \quad (1)$$

where ρ, $\ddot{x}$ and g denote mass density, seismic accelerations and gravity acceleration of the treated landslide. The displacement vector in chosen direction β is evaluated in the space:

$$\delta \bar{U}(\bar{r}) = \{-1 \quad -tg\beta \quad tg\gamma\} \delta \|\bar{U}_{0x}\| \quad (2)$$

$$tg\gamma = \cfrac{\left[\begin{array}{c} tg\varphi(1+tg^2\beta)\sqrt{1+z_{f,x}^2+z_{f,y}^2} - \\ (z_{f,x}+tg\beta z_{f,y})\sqrt{1+tg^2\beta+(z_{f,x}+tg\beta z_{f,y})^2} \end{array} \right]}{\left[\begin{array}{c} \sqrt{1+tg^2\beta+(z_{f,x}+tg\beta z_{f,y})^2} + \\ tg\varphi(z_{f,x}+tg\beta z_{f,y})\sqrt{1+z_{f,x}^2+z_{f,y}^2} \end{array} \right]} \quad (3)$$

When using energy methods for space stability analyses the slope safety factor F could be expressed by quotient of highest absorbed energy of the system that is possible δW_s and the actually released (activated) energy of the system δW_a on the treated virtual displacements.

When the component of the virtual displacement in the direction x is taken as a unit ($\delta \|\bar{U}_{ox}\| = 1$) and the direction of the displacements is β, all displacements or kinematical relations between displacements of individual elements of the failure body are known for all translator modes of the relative displacements of the fictitious rigid blocks.

The free energy of the system (by treated virtual displacements) is created by volumetrically distributed forces of the failure body, by surface and other outer loads as well as by pore or hydrostatic and consolidation pressures in the landslide. The loss of the inner energy because of the soil softening and consequently the generation of the additional quantity of the free energy are not considered in that phase of the stability analysis. Because of the simplification only ideal elastoplastic materials are considered in the analysis.

The potential energy changes are considered as a part of the free energy only when it is positive; whereas the losses of the free energy or increase of potential energy of the system are added to the system capability to dissipate or absorb the free energy. Normal components of plastic deformations are negative (tensile deformations) on the plasticity surface (failure surface). Therefore the fluid that is in the landslide is giving its potential energy or increasing system's free energy. This could be an essential contribution to worsen slope stability conditions. The free energy of the system or space shaped landslide is evaluated:

$$\delta W_a = \iiint_V ((\langle -\rho \ddot{x} \rangle + \langle -tg\beta\rho \ddot{y} \rangle + \langle tg\gamma\rho \ddot{z} \rangle + \langle -tg\gamma g \rangle)dV +$$

$$\iint_{S_{xy}} ((\langle uz_f,_x \rangle + \langle -q_x \rangle + \langle tg\beta uz_f,_y \rangle + \langle -tg\beta q_y \rangle +$$

$$\langle tg\gamma q_z \rangle + \langle tg\gamma u \rangle)dS_{xy} \qquad (4)$$

where u denotes pore pressure or sum of hydrostatic and consolidation pore pressures that act on failure surface; symbol $\langle b \rangle$ denotes that only positive value of the scalar b is considered, otherwise it is zero; V and S_{xy} denote the domains of the volume and ground-plan area of the treated landslide. Similarly as the free energy of the system, the highest possible quantity of the system's absorbed energy on the treated virtual displacements is evaluated:

$$\delta W_s = \iiint_V ((\langle \rho \ddot{x} \rangle + \langle tg\beta\rho \ddot{y} \rangle + \langle -tg\gamma \ddot{z} \rangle + \langle tg\gamma g \rangle)dV +$$

$$\iint_{S_{xy}} ((\langle q_x \rangle + \langle tg\beta q_y \rangle + \langle -tg\gamma q_z \rangle)dS_{xy} + \iint_{S_{xy}} c(1 + tg^2\beta -$$

$$tg\gamma\sqrt{\frac{1 + z_f^2,_x + z_f^2,_y}{1 + tg^2\beta + (z_f,_x + tg\beta z_f,_y)^2}})dS_{xy} \qquad (5)$$

Sometimes it is due of unified concept of the safety coefficient more suitable to express the slope safety with the quotient of the failure and actual mobilised shear strength. In such a case the shear angle φ and cohesion c in Equations (5) and (7) are substituted by their mobilised values φ_m and c_m. It should be considered that for stabile systems the variation of the free energy must be equal to the change of the system's absorbed energy:

$$\delta W_a - \delta W_s = 0 \qquad (6)$$

In such a case the safety coefficient of the slope is evaluated iterative until Equation (8) is fulfilled.

3. NUMERICAL SOLUTION

The solutions of space stability could be evaluated in the analytical form only for very simple geometry and stratigraphy, chiefly for plane problems. The solutions should be given in a numerical form when solving general space landslides and non homogeneous slopes loaded by arbitrary surface and volume loads as well as arbitrary fields of hydrostatic and consolidation pore pressures. The integration that is needed is done numerically using standard Gauss's numerical integration.

To simplify the calculation of all integral expressions the discretisation of the space shaped failure surface into the finite number of triangular elements is done. The space course of the failure surface is approximated by flat planes as it is evident in Figure 3.

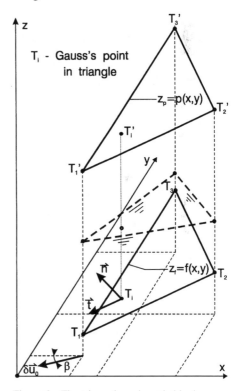

Figure 2: The triangular prismatic block

One of the more pointed Gauss's numerical integration is used for the integration of needed expressions. The co-ordinates of an arbitrary integration point T_i that is the part of the triangular segment k of the failure surface (Fig. 2) is evaluated:

$$x_i^k = \sum_{j=1}^{3} \eta_j^i x_j^k \qquad (7)$$

$$y_i^k = \sum_{j=1}^{3} \eta_j^i y_j^k \qquad (8)$$

$$z_i^k = \sum_{j=1}^{3} \eta_j^i z_j^k \qquad (9)$$

where x_j, y_j and z_j denote co-ordinates of the nodes of the triangular segments that illustrate space shaped failure surface. The value of the pore pressure in the integration point is evaluated:

$$u_i^k = \gamma_w \langle (z_w - z_i^k) \rangle + \int_{z_i^k}^{z_p(x_i^k, y_i^k)} r_u \rho g \, dz \qquad (10)$$

where r_u denotes the quotient of consolidation pore pressure according to the additional loading of the ground after the construction works. The free energy of the system in the numerical form is:

$$\delta W_a = \sum_{k=1}^{m} \sum_{i=1}^{n} \left[\int_{z_i^k}^{z_p(x_i^k, y_i^k)} \rho(\langle tg\gamma \, \ddot{z} \rangle + \langle -tg\gamma g \rangle) dz + \right.$$

$$\langle tg\gamma q_{z_i}^k \rangle + \langle tg\gamma u_i^k \rangle + \langle -\ddot{x} \rangle + \langle -\ddot{y} \rangle + \langle u_i^k z_{i'x}^k \rangle +$$

$$\left. \langle -q_{x_i}^k \rangle + \langle -q_{y_i}^k \rangle + \langle tg\beta u_i^k z_{i',y}^k \rangle \right] A_{xy}^k W_i \delta U_0 \qquad (11)$$

$$tg\gamma = \frac{\left[\begin{array}{l} tg\varphi(1+tg^2\beta)\sqrt{1+(z_{i',x}^k)^2+(z_{i',y}^k)^2} - \\ (z_{i',x}^k + tg\beta z_{i',y}^k)\sqrt{1+tg^2\beta+(z_{f',x}+tg\beta z_{f',y})^2} \end{array} \right]}{\left[\begin{array}{l} \sqrt{1+tg^2\beta+(z_{i',x}^k+tg\beta z_{i',y}^k)^2} + \\ tg\varphi(z_{i',x}^k+tg\beta z_{i',y}^k)\sqrt{1+(z_{i',x}^k)^2+(z_{i',y}^k)^2} \end{array} \right]} \qquad (12)$$

and the quantity of the energy, that belongs to the virtual displacement (equal one), that the system could absorb or take over before the failure, is:

$$\delta W_s = \sum_{k=1}^{m} \sum_{i=1}^{n} \left[\int_{z_i^k}^{z_p(x_i^k, y_i^k)} \rho(\langle tg\gamma \, g \rangle + \langle -tg\gamma \, \ddot{z} \rangle) dz + \right.$$

$$\langle -tg\gamma \, q_{z_i}^k \rangle + \langle tg\gamma u_i^k \rangle + \langle \ddot{x} \rangle + \langle \ddot{y} \rangle - \langle -u_i^k z_{i',x}^k \rangle +$$

$$\langle q_{x_i}^k \rangle + \langle q_{y_i}^k \rangle + \langle -tg\beta u_i^k z_{i',y}^k \rangle + c_i^k (1+tg^2\beta -$$

$$\left. tg\gamma \sqrt{\frac{1+(z_{i',x}^k)_x^2+(z_{i',y}^k)_y^2}{1+tg^2\beta+(z_{i',x}^k+tg\beta z_{i',y}^k)^2}} \right) \right] A_{xy}^k W_i \delta U_0 \qquad (13)$$

where m and n denote number of triangular segments that compose the curved failure surface, symbol $\langle b \rangle$ denotes positive value of the scalar b, when b is negative the value of the expression in parenthesis is zero. The slope safety is determined by the critical failure surface and the angle β that give the lowest ratio between free and dissipated quantity of the energy; their quotient is usually called the safety coefficient that is evaluated:

$$F = \frac{W_s}{W_a} \qquad (14)$$

4. PRACTICAL EXAMPLE

The applicability of the space stability analysis is shown on the practical example of analysing the instability conditions of the embankment between profiles P-98 and P-103 on highway Šentilj-Pesnica (Škrabl et al. 1997). Soon after construction works of the embankment started (after few months) on the area of about 8.0 m high embankment the cracks have appeared. They indicated the possibility of the stability decrease of the clay layers under the embankment. Very high water flow was appeared simultaneous close by existing surface of the slope under the embankment.

The geological and geotechnical conditions of the area were already more or less known. Enough in-situ and laboratory tests were done, and three inclinometers were built in. Many classical geological bores, penetration probes and other classical tests from the field of geology and geotechnics were carried out.

We have analysed the presented profiles of the embankment for two variants of the shear characteristics of the coherent soils:

(i) cohesion was equal to the undrained shear strength $c = S_u = 45.0$ kPa, and the angle of internal friction was equal $\varphi = 0.0$ °.

(ii) cohesion was equal $c = 21.0$ kPa, and the angle of internal friction was equal $\varphi = 17.0$ °. These usual shear characteristics of the clay soils were established with direct shear test. It was performed on the unsubmerged intact samples of the soil immediately after they were taken from the borehole and consolidated for 12 hours.

The central cross-section of the landslide, ground water level, and all soil characteristics are shown in Figure 3. The stratigraphy of the slope is given by three longitudinal sections P-99, P-100 and P-101 of the highway Šentilj-Pesnica. In the transverse direction of the embankment respectively in the longitudinal direction of the highway ten cross sections without intermediate point were given.

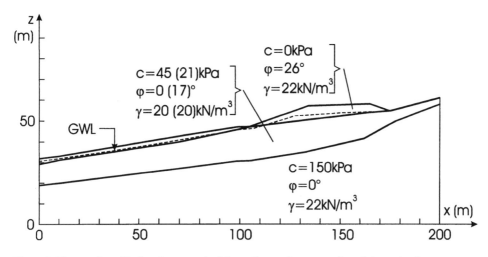

Figure 3: The stratigraphical and geometrical data of central cross-section of the embankment

The stratigraphy of the slope is determined on the bases of the primary and additional geological bores and geodetic copies that were done during the erection of the embankment.

The following analyses were performed:

(1) the first step of the three dimensional analysis of the sliding mass (3D-1st step) that was given by its length, width and depth and discretized in triangular prisms. During the optimising process the length, width and depth were optimised.

(2) the results of the analysis (1) were input data for next step of the three dimensional analysis (3D-2nd step) where during the optimising process the co-ordinates of the points that discretize the space slip surface were optimised.

(3) two dimensional analysis (2D) through the central longitudinal section. The initial slip surface was defined by the resulted points of the space slip surface that are on the central longitudinal section.

The initial point for determining the critical space slip surface was assumed to have length 150.0 m, width 100.0 m and depth 15.0 m. The numerical stability analysis was performed in more iteration cycles considering more detailed discretization of the sliding mass.

The appurtenant critical slip surface for the variant (i) after ten iterations of the 3D-1st step analysis and ten iterations of the 3D-2nd step analysis is shown in Figure 4. The factor of safety was 1.179. The convergence of the solution was distinctive slow because the selection of the kinematic admissible failure surfaces was partly restricted by rapid decrease of the low bearing capacity soils in the lateral direction of the treated embankment. At the same time we have to mention that the safety factors were already dropped under the stability limit.

The actual ground water level that was registered during the erection of the highway Šentilj-Pesnica was considered in the analyses.

The second variant of the stability analysis is performed assuming that the pore over pressure is activated in the magnitude of 100% of the additional vertical loading due to embankment ($r_u = 1$). The appurtenant critical slip surface for the variant (ii) after ten iterations of the 3D-1st step analysis and one iteration of the 3D-2nd step analysis is shown in Figure 5. The factor of safety was 1.055.

The convergence of the solution was distinctive slow because the selection of the kinematic admissible failure surfaces was partly restricted by rapid decrease of the low bearing capacity soils in the lateral direction of the treated embankment. At the same time we have to mention that the safety factors were already dropped under the stability limit. Such stability analysis gave us relatively good results proving us that the conditions to occur limit state of the construction were satisfied. The experiences found in the literature approved our stability analysis as real. There was found that the pore over pressures under high and wide embankments was decreasing very slowly. The failure of the highway embankment in Ontario (Canada) as an example could be given. The measurements in the silty soil showed that 20 years after the erection of the embankment the pore over pressure sunk only for 4% of its initial value. Therefore with the measurements the improvement of the soil shear characteristics was not possible proved by documents.

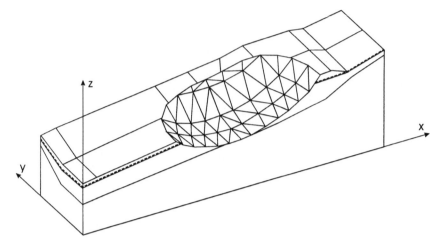

Figure 4: The shape of the critical slip surface for the variant (i), $F_s = 1.179$

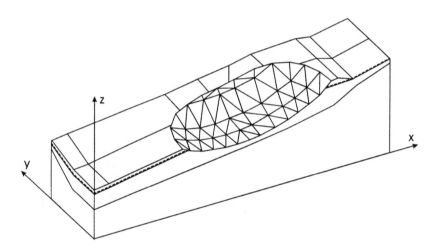

Figure 5: The shape of the critical slip surface for the variant (ii), $F_s = 1.055$

5. CONCLUSSION

The given results show that in the presented example the space stability analyses gave up to 15% better or more favourable results in comparison to comparable plane stability analysis. The first method allows us to consider even more exacting rheological soil properties (non associative models).

We estimate that a relatively exacting procedure of the stability analysis that is presented in the paper is mainly suitable for analysing very exacting soil objects as for instance: ground barriers, tunnel portals, high embankments and cut-offs, etc.

6. REFERENCES

Karal, K. 1977. Energy method for soil stability analyses. *ASCE Journal of Geotechnical Eng.* 103: 431-445.

Leshchinsky, D. & Huang, C.C. 1992. Generalised three-dimensional slope-stability analysis. *ASCE Journal of Geotechnical Eng.* 118: 1748-1764.

Michalowski, R.L. 1995. Slope stability analysis: a kinematical approach. *Geotechnique* 45: 283-293.

Škrabl, S. et al. 1997. Space slope stability analysis: final report of the research project, University of Maribor, Maribor.

Geotechnical Hazards, Marić, Lisac & Szavits-Nossan (eds) © 1998 Taylor & Francis, ISBN 90 5410 957 2

On the initiation of rainfall induced soil failure

Y. Tsukamoto & K. Ishihara
Department of Civil Engineering, Science University of Tokyo, Japan

Y. Nosaka
Obayashi Corporation, Japan

ABSTRACT: Heavy rainfall attacked Boso Peninsula, Chiba, Japan in September 1971, and caused slope failures at Omigawa area. In order to examine the cause and initiation of such shallow-depth soil failures due to rainfall, a series of tests are conducted on soil samples recovered from the Omigawa site. Isotropically consolidated and anisotropically consolidated undrained triaxial compression tests are performed. The characteristics of undrained stress paths of the samples, which are anisotropically consolidated to different values of a principal stress ratio K_c (= σ_1'/σ_3'), are examined in terms of the peak shear stress and the shear stress at steady states. Constant shear drained tests, which duplicate the hydrologic response of soils subject to rainfall, are also performed on anisotropically consolidated samples with different values of K_c. With these sets of experimental data, the influence of the hydrologic response of soils on the initiation of soil failures is addressed, with special reference to drained initiation and undrained mobilization of sloping soil masses.

1 INTRODUCTION

Rainfall-induced soil failures occur on slopes which are marginally stable and on various types of soils such as colluvial and residual soils. These failures typically occur at shallower depths of less than 3 metres. To examine the conditions leading to such shallow-depth soil failures, a number of studies addressed the causes of the failures to be twofold. They are the reduction in shear strength due to saturation and the hydrologic response of soils due to water infiltration, (Yoshida et al. 1991, Anderson & Sitar 1995). Natural soil slopes generally exist in an unsaturated condition, and the soil becomes saturated or nearly so during or after heavy rainfall. The soils within the slopes are likely subject to higher values of a principal stress ratio K_c (= σ_1'/σ_3'), which typically range from 0.2 to 0.5. The field stress path due to hydrologic response of soils to rainfall was examined by several researchers including Brand (1981), Brenner et al. (1985) and Anderson & Sitar (1995). The field stress path assumed by Anderson & Sitar (1995) is shown in Figure 1. The Mohr circle of the soil element at an unsaturated condition exists with a negative pore water pressure. Due to water infiltration caused by rainfall, the soil element becomes saturated and the

stress point moves towards the failure envelope, while the negative pore pressure becomes small. In particular, the soil element at the potential sliding surface becomes nearly saturated, and the positive pore pressure may develop primarily due to seepage flow and the soil element eventually reaches failure. Johnson and Sitar (1989) observed the positive pore pressure development during rainfall in the debris flow source area, and discussed the highly transient nature of the pore pressure response during storms. The magnitude of shear stress on the field stress path during the process of the hydrologic response can change depending on the direction of seepage flow, and the in-situ stress path may deviate from horizontal. However, because the deviation is relatively small and the exact nature of seepage flow is unknown, the horizontal stress path was considered to be a reasonable assumption for the hydrologic response due to rainfall, (Anderson & Sitar 1995). The influence of the hydrologic response of soils to rainfall can be illustrated as follows. The reduction of factor of safety F_s along the field stress path is demonstrated in Figure 2, which can be calculated by the following equation, assuming the infinite slope with a slope angle of θ and the soil having the internal friction angle ϕ and the unit weight γ for the unsaturated zone and γ_{sat}

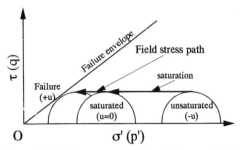

Figure 1. Field stress path (after Anderson & Sitar 1995)

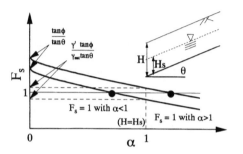

Figure 2. Schematic of reduction of safety factor

and γ' ($= \gamma_{sat} - 1$) for the saturated zone,

$$F_s = \frac{(1-\alpha)\gamma + \alpha\gamma'}{(1-\alpha)\gamma + \alpha\gamma_{sat}} \frac{\tan\phi}{\tan\theta}. \qquad (1)$$

In this diagram, the rate of saturation and the excess pore water pressure head is denoted by α, and is assumed as follows;

$$\alpha = \frac{H_s}{H} \quad \text{when} \quad \alpha \le 1, \qquad (2)$$

$$\alpha = 1 + \frac{\Delta u}{\gamma_w H} \quad \text{when} \quad \alpha > 1, \qquad (3)$$

where H_s is the height of the saturated zone, H is the height of the sloping soil mass and Δu is the excess pore water pressure. Although the pore water pressure rise is constrained when the water table reaches the surface of the sloping soil mass, the transient excess pore water pressure rise can occur due to seepage flow within the soil mass. The partially saturated soil element can exist at any values of α between 0 and 1, and the reduction of safety factor first occurs due to saturation of soils

along the stress path A in Figure 1, until the water table reaches the surface, in other words, α reaches unity, as shown in Figure 2. In some cases, the safety factor drops down to unity before α reaches unity. It is called a drained initiation. The factor of safety may further reduce along the stress path B in Figure 1, at which α exceeds unity due to the development of the excess pore water pressure. At any phase of a drained initiation, the soil elements within the sloping soil mass, particularly at the potential surface of the sliding soil mass, may become unstable when the safety factor becomes equal to unity and can cause undrained mobilization. The assumed scenario described above neglects the details of site conditions and may be oversimplified. However, it illustrates a likely scenario of how sloping soil masses come into the states of instability, and actually fail.

In this paper, the influence of the hydrologic response of soils due to rainfall on the collapse of sloping soil masses is examined with particular emphasis on drained initiation and undrained mobilization of soils.

2 EXPERIMENTAL DETAILS

Heavy rainfall attacked Boso Peninsula, Chiba, Japan in September 1971, during the passage of Typhoon No.25, and caused a numerous number of slope failures. The official report of the disasters caused by this stormy rainfall was published by Chiba Prefecture (1972). The areas where the slope failures occurred were located at the hillsides of the terrain on the eastern part of Chiba Prefecture, and about 50 metre above sea level. At Omigawa area, the precipitation during the evening of 6/September to the morning of 7/September amounted to 106 mm, and the maximum hourly precipitation was 22 mm/h. The rainfall ceased for a while, and the stormy rainfall started again. The precipitation during the evening of 7/September to the morning of 8/September was 207.5 mm, and the maximum hourly precipitation went up to 42 mm/h. The devastating slope failures finally occurred. The cross section of one of the slope failures at Omigawa area is shown in Figure 3, (Chiba Prefecture 1972). The soil material at the site mainly consists of cemented silty sand. The physical properties and grain size distribution of the soil material recovered from the site by the authors are summarized in Table 1 and Figure 4, respectively. The soil material is poorly-graded sand - silt mixtures, with non-plastic fines.

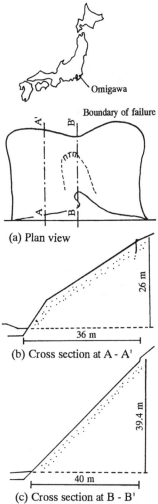

Omigawa

Boundary of failure

(a) Plan view

26 m

36 m

(b) Cross section at A - A'

39.4 m

40 m

(c) Cross section at B - B'

Figure 3. Cross section of slope failure at Omigawa site (after Chiba Prefecture 1972)

3 TEST RESULTS

Three types of tests are carried out on triaxial compression test apparatus, to evaluate the instability of the soil materials recovered from the site where the soil failure actually occurred at Omigawa site. They are isotropically consolidated undrained (ICU) tests, anisotropically consolidated undrained (ACU) tests, and constant shear drained (CSD) tests.

In all the tests, a moist placement (wet tamping) method is used for the preparation of soil specimens. The dry soil material is mixed with water to a

Table 1 Physical properties of tested soil material

Soil type	SM
Specific gravity	2.694
Plasticity index	NP
e_{max}	1.282
e_{min}	0.796

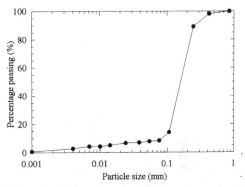

Figure 4. Grain size distribution of tested soil material

moisture content of 5 %. The moist soil material is poured into the mould and tamped equally by a tamping rod until a specified height is achieved. The same procedure is repeated to the full height of the specimen. This method is known to provide a fairly wide range of void ratios. Deaired water is then circulated in the specimen and saturated.

3.1 ICU tests

In the isotropically consolidated undrained tests, the soil specimens are first consolidated to isotropic states with $\sigma_c' = 100, 200$ and 300 kPa, and then subjected to undrained monotonic compressive loading in a strain - controlled manner. Figure 5(a) shows the p' - q diagram, in which $p' = (\sigma_v + \sigma_h)/2 - u$, $q = (\sigma_v - \sigma_h)/2$, σ_v is the axial stress, σ_h is the confining stress, and u is the excess pore water pressure. Some of the effective stress paths exhibit contractive behaviour when the soil is loose, and some are dilative when the soil is dense. However, all the effective stress paths converge to the steady states of $M = q/p' = 0.59$. Figure 5(b) shows the strain - strain behaviour of the isotropically consolidated specimens. The contractive specimens experience peaks at about 0.5 to 1 % axial strain ε_1, and strain-soften, while the dilative specimens pass through the states of phase transformation at similar axial strains and strain-harden, until the steady states are achieved.

3.2 ACU tests

In the anisotropically consolidated undrained tests, the soil specimens are first consolidated to isotropic states with $\sigma_c{}'$ = 50, 100 and 200 kPa, and then axial load is gradually increased with a drainage valve open until specified values of K_c ($= \sigma_h{}'/\sigma_v{}'$) are achieved. The K_c values of 0.3 to 0.7 are considered. The anisotropically consolidated specimens are then subjected to undrained monotonic compressive loading in a strain - controlled manner. Figure 5(a) shows the p' - q diagram for the specimens which are consolidated to K_c = 0.4. Even for the specimens anisotropically consolidated to the same K_c values, contractive and dilative behaviour exist, and they converge to the same steady states of M = 0.59. One of the characteristics of the effective stress paths for the anisotropically consolidated specimens is that the excess pore water pressure generates less than in the isotropically consolidated specimens, and they experience peaks and lose strength, and some of them gain strength again. Also noteworthy is that the axial strains necessary to achieve peaks are quite small, as shown in Figure 6(b). Another example is shown in Figure 7, in which two specimens

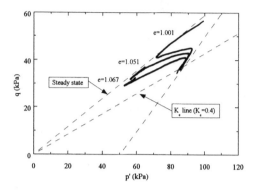

(a) p' - q diagram

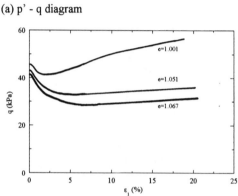

(b) ε_1 - q diagram
Figure 6. ACU tests

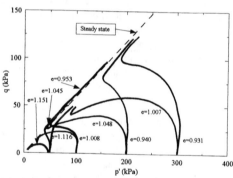

(a) p' - q diagram

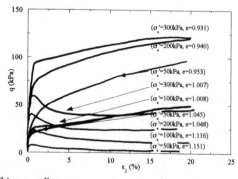

(b) e_1 - q diagram
Figure 5. ICU tests

consolidated to different values of K_c are plotted. Figures 7(a) and (b) clearly show peaks, quasi-steady states and steady states on the anisotropically consolidated specimens. The steady states and quasi-steady states are the concepts which have been often used for the characterization of flow behaviour for isotropically consolidated soil samples, (Ishihara 1993 and others). It is therefore of interest to examine them on anisotropically consolidated soil samples. Figure 8 summarizes the steady states achieved by ICU tests as well as ACU tests on the e - $p_s{}'$ diagram, where $p_s{}'$ is the effective mean stress at steady states.

For slope stability analysis, it is of importance to distinguish between gravity - induced shear stress and seepage - induced shear stress. The gravity - induced shear stress is represented by the initial shear stress in anisotropically consolidated samples. In ACU tests, the effective stress path experiences peak, quasi-steady state and steady state, as shown in Figure 9. The characterization of flow behaviour for anisotropically consolidated samples may be made with respect to the parameters shown in

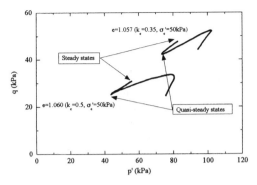

(a) p' - q diagram

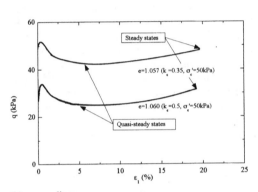

(b) ε_1 - q diagram

Figure 7. Steady states and quasi-steady states

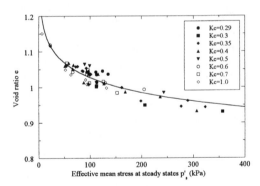

Figure 8. Steady state line on e - p_s' diagram

Figure 9, in which Δq_p is the difference between shear stresses at peak and at an initial shear stress, Δq_{qss} is the difference between shear stresses at peak and at a quasi-steady state, and $\Delta q_o/\Delta p_o$ is the inclination of the effective stress path at an initial stress state. Figure 10 shows the K_c - $\Delta q_o/\Delta p_o$ diagram. It is interesting to see that the $\Delta q_o/\Delta p_o$

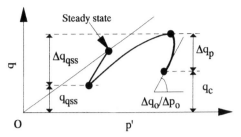

Figure 9. Definitions of $\triangle q_o/\triangle p_o$, $\triangle q_p$ and $\triangle q_{qss}$

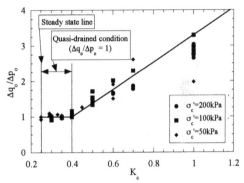

Figure 10. K_c - $\triangle q_o/\triangle p_o$ diagram

value stays at unity at 0.25 (K_c for steady state line) $< K_c < 0.4$, and then increases with increasing K_c values. It suggests that there is virtually no excess pore water pressure generation at $0.25 < K_c < 0.4$, although undrained straining is imposed on. Thus, it may be called a quasi-drained condition. Figures 11(a) and (b) show the K_c - Δq_p diagrams for the specimens consolidated to σ_c' = 50 kPa with e = 1.04 to 1.06, and to σ_c' = 100 kPa with e = 0.998 to 1.009, respectively. It is found that the Δq_p value reduces at a quasi-drained condition of $0.25 < K_c < 0.4$, and is lowest at K_c = 0.4, and then increases with increasing K_c values. Figure 12 shows the K_c - $\Delta q_{qss}/\Delta q_p$ diagram, for the specimens consolidated to σ_c' = 50 kPa with e = 1.04 to 1.06. It is found that the flow behaviour of anisotropically consolidated soil samples can be classified into four types at 0.25 (K_c for steady states) $< K_c < 1$ (K_c for isotropic states). At regions A and C, the soil exhibits contractive behaviour and the shear stress at a quasi-steady state is greater than the initial shear stress. At a region B, the soil exhibits contractive behaviour and the shear stress at a quasi-steady state becomes lower than the initial shear stress. Therefore, the soil loses strength down to the shear stress lower than the initial shear stress at an anisotropically

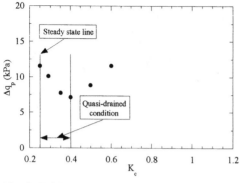

(a) σ_c'=50 kPa, e:1.04 - 1.06

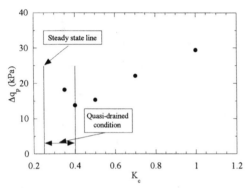

(b) σ_c'=100 kPa, e:0.998 - 1.009

Figure 11. K_c - $\triangle q_p$ diagram

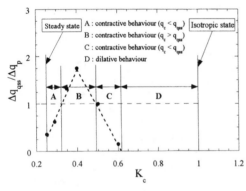

Figure 12. K_c - $\triangle q_{qss}/\triangle q_p$ diagram

consolidated state. At a region D, the soil exhibits dilative behaviour, and never loses strength. There ought be more study for soil samples with different void ratios, yet Figure 12 implies that there is a condition in which the soil is more susceptible to flow in terms of the K_c values.

3.3 CSD tests

The field stress path due to the hydrologic response of soils to rainfall can only be achieved by allowing the pore water within the soil specimen to drain, as noted by Anderson & Sitar (1995). There are practically two ways to perform the simulation of the field stress path in the triaxial soil specimens, by either increasing the back pressure within the specimens, or reducing the confining pressure. In this study, the constant shear drained (CSD) tests are conducted by the latter method. The soil specimens are first consolidated to σ_c' = 50 kPa, and then the axial load is increased until specified K_c values are attained. After anisotropically consolidated, the confining pressure is gradually reduced in a drained condition. Nine tests are performed in total, with different K_c values and void ratios. They are denoted as Tests A to I, as shown in Figure 13(a). In this diagram, almost upon touch to the steady state envelope, does every effective stress path bend down. Figure 13(c) shows the p' - ε_1 diagram. It can be seen that the axial strain begins to increase dramatically far before the field stress paths are touched upon the steady state envelope. It seems that it is a reasonable assumption that the rapid increase of the axial strain can actually trigger undrained straining in the field and cause flow. Anderson & Sitar also examined the mechanism of soil failures, based on the assumption that the stress transfer to neighboring soil elements occurs due to the reduction of shear stress along the field stress path observed in Figure 13(a). Figure 13(b) shows the p' - e diagram, in which the steady state line obtained from the ICU and ACU tests is also drawn. All the tests show dilative behaviour where the void ratios increase due to reducing effective mean stress p', except for Tests H and I which are extremely loose specimens. Also noteworthy is that they approach the steady state line. Figure 13(d) also shows the dilative behaviour for Tests A to G, where the negative volumetric strain occurs.

4 IMPLICATION TO FIELDS

For the analysis of surficial soil failures due to water infiltration, the assumption of the infinite slope has found its popularity. However, there are many factors that can affect the stability of sloping soil masses, aging, cementation, geologic history, permeability of soils, the characteristics of rainfall, geometry of slopes and others. All are site-specific, and should be carefully examined. This study

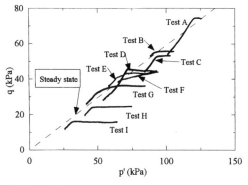

(a) p' - q diagram

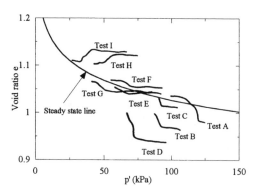

(b) p' - e diagram

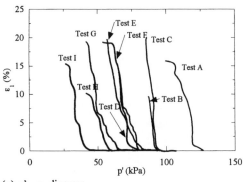

(c) p' - ε₁ diagram
Figure 13. CSD tests

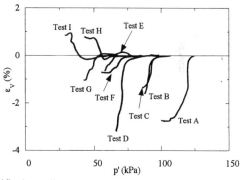

(d) p' - ε_v diagram

addressed and examined one of the likely scenarios of sloping soil failures due to rainfall. The hypothesis made is as follows. The marginally stable sloping soil masses, that are in an unsaturated condition, are subject to the field stress path of the hydrologic response of soils to rainfall. It is called a drained initiation. The soil elements, in particular at the potential sliding surface, become saturated and unable to sustain gravity - induced shear stress. The stress transfer to neighboring soil elements might then occur. The soil elements subject to stress transfer trigger undrained straining within themselves, and the instability might propagate from one to another. It is called an undrained mobilization. The triaxial tests conducted in this study seem to substantiate the hypothesis made above.

5 CONCLUSIONS

To examine the instability of sloping soil masses to rainfall, three types of triaxial tests were conducted.

They were isotropically consolidated undrained tests, anisotropically consolidated undrained tests, and constant shear drained tests. The soil material used for this study was recovered from the site where the soil actually failed due to heavy rainfall. One of the likely scenarios of sloping soil failures was addressed. It is drained initiation and undrained mobilization. The test results conducted in this study are found to substantiate the hypothesis made above.

ACKNOWLEDGEMENTS

The authors gratefully acknowledge Dr.Y.Yoshida for his help in recovering soil samples from Omigawa, Chiba Japan.

REFERENCES

Anderson, S.A. & N. Sitar 1995. Analysis of rainfall-induced debris flows. J. Geotech. Eng.,

ASCE, 121(7): 544-552.

Brand, E.W. 1981. Some thoughts on rain-induced slope failures. Proc. 10th Int. Conf. Soil Mech. Fdn. Eng., Stockholm, 3: 373-376.

Brenner, R.P., H.K. Tam & E.W. Brand 1985. Field stress path simulation of rain-induced slope failure. Proc. 11th Int. Conf. Soil Mech. Fdn. Eng., San Francisco, 2: 991-996.

Chiba Prefecture, Civil and River Division 1972. Report of disaster in Chiba due to autumn rain front on September 6 - 7, 1971, and Typhoon No.25. (in Japanese).

Ishihara, K. 1993. Liquefaction and flow failure during earthquakes. Geotechnique, 43(3): 351-415.

Johnson, K.A. & N. Sitar 1989. Significance of transient pore pressures and local slope conditions in debris flow initiation. Proc. 12th Int. Conf. Soil Mech. Fdn. Eng., Rio de Janeiro: 1619-1622.

Yoshida, Y., J. Kuwano & R. Kuwano 1991. Effects of saturation on shear strength of soils. Soils and Foundations, 31(1): 181-186.

Author index

Printed and bound by CPI Group (UK) Ltd, Croydon, CR0 4YY

28/10/2024

01780098-0001